Oxford University Press, Great Clarendon Street, Oxford OX2 6DP

Oxford New York
Athens Auckland Bangkok Bogota Bombay
Buenos Aires Calcutta Cape Town Dar es Salaam
Delhi Florence Hong Kong Istanbul Karachi
Kuala Lumpur Madras Madrid Melbourne
Mexico City Nairobi Paris Singapore
Taipei Tokyo Toronto Warsaw

and associated companies in
Berlin Ibadan

Oxford is a trade mark of Oxford University Press

First Edition published 1986
Second Edition first published 1997

British Library Cataloguing in Publication Data
available

Library of Congress Cataloging in Publication Data
available

860009 7
860123 9 (Italian cover edition only)

10 8 6 4 3 2 1

Printed in Great Britain by
(Scotland) Ltd.

The Oxford Italian Minidictionary

SECOND EDITION

ITALIAN–ENGLISH
ENGLISH–ITALIAN
ITALIANO–INGLE
INGLESE–ITAL

Joy

Britis

Data a

Library
Data avai
ISBN 0-19
ISBN 0-19

10 9 8 7 6 5

Printed in Gre
Charles Letts (S
Dalkeith, Scotla

OXFORD U

Contents/Indice

Editors/Redazione

Debora Mazza   Jane Goldie

Donatella Boi   Francesca Logi   Sonia Tinagli-Baxter

Peter Terrell   Carla Zipoli

Copy editors/Segreteria di redazione

Jacqueline Gregan   Daphne Trotter

Project management by/A cura di

LEXUS

Preface

This new edition of the Oxford Italian-English Minidictionary is an updated and expanded version of the dictionary edited by Joyce Andrews. Colloquial words and phrases figure largely, as do neologisms. Noteworthy additions include terms from special areas such as computers and business that have become a familiar feature of everyday language.

Prefazione

Questa nuova edizione del mini dizionario Oxford Italiano-Inglese è il risultato di un lavoro di ampliamento e aggiornamento della precedente edizione curata da Joyce Andrews. Un'attenzione particolare è stata rivolta a vocaboli ed espressioni colloquiali di coniazione recente e a termini relativi a settori specifici, quali l'informatica e il commercio, divenuti ricorrenti nella lingua di tutti i giorni.

Proprietary terms

This dictionary includes some words which are, or are asserted to be, proprietary names or trademarks. Their inclusion does not imply that they have acquired for legal purposes a non-proprietary or general significance, nor is any other judgment implied concerning their legal status. In cases where the editor has some evidence that a word is used as a proprietary name or trademark this is indicated by the symbol ®, but no judgment concerning the legal status of such words is made or implied thereby.

Marche depositate

Questo dizionario include alcune parole che sono o vengono considerate nomi di marche depositate. La loro presenza non implica che abbiano acquisito legalmente un significato generale, né si suggerisce alcun altro giudizio riguardo il loro stato giuridico. Qualora il redattore abbia trovato testimonianza dell'uso di una parola come marca depositata, quest'ultima è stata contrassegnata dal simbolo ®, ma nessun giudizio riguardo lo stato giuridico di tale parola viene espresso o suggerito in tal modo.

Introduction

In order to give the maximum information about English and Italian in the space available, this new dictionary uses certain space-saving conventions.

A swung dash ~ is used to replace the headword within the entry.

Where the headword contains a vertical bar | the swung dash replaces only the part of the headword that comes in front of the |. For example: **efficien|te** *a* efficient. **~za** *nf* efficiency (the second bold word reads **efficienza**).

Indicators are provided to guide the user to the best translation for a specific sense of a word. Types of indicator are:

field labels (see the list on pp ix–x), which indicate a general area of usage (commercial, computing, photography etc);

sense indicators, eg: **bore** *n* (*of gun*) calibro *m*; (*person*) seccatore, -trice *mf*;

typical subjects of verbs, eg: **bond** *vt* (*glue:*) attaccare;

typical objects of verbs, placed after the translation of the verb, eg: **boost** *vt* stimolare (*sales*); sollevare (*morale*);

nouns that typically go together with certain adjectives, eg: **rich** *a* ricco; (*food*) pesante.

A solid black circle means that the same word is being translated as a different part of speech, eg. **partition** *n* ... ● *vt* ...

English pronunciation is given for the Italian user in the International Phonetic Alphabet (see p viii).

Italian stress is shown by a ' placed in front of the stressed syllable in a word.

Square brackets are used around parts of an expression which can be omitted without altering the sense.

Introduzione

Allo scopo di fornire il maggior numero possibile di informazioni in inglese e in italiano, questo nuovo dizionario ricorre ad alcune convenzioni per sfruttare al massimo lo spazio disponibile.

Un trattino ondulato ~ è utilizzato al posto del lemma all'interno della voce.

Qualora il lemma contenga una barra verticale |, il trattino ondulato sostituisce solo la parte del lemma che precede |. Ad esempio: **dark|en** vt oscurare. **~ness** n buio m (la seconda parola in neretto va letta **darkness**).

Degli indicatori vengono forniti per indirizzare l'utente verso la traduzione corrispondente al senso voluto di una parola. I tipi di indicatori sono:

etichette semantiche (vedi la lista a pp ix-x), indicanti l'ambito specifico in cui la parola viene generalmente usata in quel senso (commercio, informatica, fotografia ecc);

indicatori di significato, es.: **redazione** nf (ufficio) editorial office; (di testi) editing;

soggetti tipici di verbi, es.: **trovarsi** vr (luogo:) be;

complementi oggetti tipici di verbi, collocati dopo la traduzione dello stesso verbo, es: **superare** vt overtake (veicolo); pass (esame);

sostantivi che ricorrono tipicamente con certi aggettivi, es.: **solare** a (energia, raggi) solar; (crema) sun.

Un pallino nero indica che la stessa parola viene tradotta come una diversa parte del discorso, es. **calcolatore** a ... ●nm ...

La pronuncia inglese è data usando l'Alfabeto Fonetico Internazionale (vedi p viii).

L'accento tonico nelle parole italiane è indicato dal segno ' collocato davanti alla sillaba accentata.

Delle parentesi quadre racchiudono parti di espressioni che possono essere omesse senza alterazioni di senso.

Pronunciation of Italian

Vowels:

a	is broad like *a* in *father*: **casa**.
e	has two sounds: closed like *ey* in *they*: **sera**; open like *e* in *egg*: **sette**.
i	is like *ee* in *feet*: **venire**.
o	has two sounds: closed like *o* in *show*: **bocca**; open like *o* in *dog*: **croma**.
u	is like *oo* in *moon*: **luna**.

When two or more vowels come together each vowel is pronounced separately: **buono; baia**.

Consonants:

b, d, f, l, m, n, p, t, v are pronounced as in English. When these are double they are sounded distinctly: **bello**.

c	before **a**, **o** or **u** and before consonants is like *k* in *king*: **cane**.
	before **e** or **i** is like *ch* in *church*: **cena**.
ch	is also like *k* in *king*: **chiesa**.
g	before **a**, **o**, or **u** is hard like *g* in *got*: **gufo**.
	before **e** or **i** is like *j* in *jelly*: **gentile**.
gh	is like *g* in *gun*: **ghiaccio**.
gl	when followed by **a**, **e**, **o**, or **u** is like *gl* in *glass*: **gloria**.
gli	is like *lli* in *million*: **figlio**.
gn	is like *ni* in *onion*: **bagno**.
h	is silent.
ng	is like *ng* in *finger* (not *singer*): **ringraziare**.
r	is pronounced distinctly.
s	between two vowels is like *s* in *rose*: **riso**.
	at the beginning of a word it is like *s* in *soap*: **sapone**.
sc	before **e** or **i** is like *sh* in *shell*: **scienza**.
z	sounds like *ts* within a word: **fazione**; like *dz* at the beginning: **zoo**.

The stress is shown by the sign ' printed before the stressed syllable.

Pronuncia inglese

SIMBOLI FONETICI

Vocali e dittonghi

æ bad	ʊ put	aʊ now
ɑ: ah	u: too	aʊə flour
e wet	ə ago	ɔɪ coin
ɪ sit	ɜ: work	ɪə here
i: see	eɪ made	eə hair
ɒ got	əʊ home	ʊə poor
ɔ: door	aɪ five	
ʌ cup	aɪə fire	

Consonanti

b boy	l leg	t ten
d day	m man	tʃ chip
dʒ page	n new	θ three
f foot	ŋ sing	ð this
g go	p pen	v verb
h he	r run	w wet
j yes	s speak	z his
k coat	ʃ ship	ʒ pleasure

Note: ' precede la sillaba accentata.
La vocale nasale in parole quali *nuance* è indicata nella trascrizione fonetica come ɒ̃: njuːɒ̃s.

Abbreviations/Abbreviazioni

adjective	*a*	aggettivo
abbreviation	*abbr*	abbreviazione
administration	*Admin*	amministrazione
adverb	*adv*	avverbio
aeronautics	*Aeron*	aeronautica
American	*Am*	americano
anatomy	*Anat*	anatomia
archaeology	*Archaeol*	archeologia
architecture	*Archit*	architettura
astrology	*Astr*	astrologia
attributive	*attrib*	attributo
automobiles	*Auto*	automobile
auxiliary	*aux*	ausiliario
biology	*Biol*	biologia
botany	*Bot*	botanica
British English	*Br*	inglese britannico
Chemistry	*Chem*	chimica
commerce	*Comm*	commercio
computers	*Comput*	informatica
conjunction	*conj*	congiunzione
cooking	*Culin*	cucina
definite article	*def art*	articolo determinativo
	ecc	eccetera
electricity	*Electr*	elettricità
et cetera	*etc*	
feminine	*f*	femminile
familiar	*fam*	familiare
figurative	*fig*	figurato
formal	*fml*	formale
geography	*Geog*	geografia
geology	*Geol*	geologia
grammar	*Gram*	grammatica
humorous	*hum*	umoristico
indefinite article	*indef art*	articolo indeterminativo
interjection	*int*	interiezione
interrogative	*inter*	interrogativo
invariable	*inv*	invariabile
(no plural form)		
law	*Jur*	legge/giuridico
literary	*liter*	letterario
masculine	*m*	maschile
mathematics	*Math*	matematica
mechanics	*Mech*	meccanica
medicine	*Med*	medicina

masculine or feminine	*mf*	maschile o femminile
military	*Mil*	militare
music	*Mus*	musica
noun	*n*	sostantivo
nautical	*Naut*	nautica
pejorative	*pej*	peggiorativo
personal	*pers*	personale
photography	*Phot*	fotografia
physics	*Phys*	fisica
plural	*pl*	plurale
politics	*Pol*	politica
possessive	*poss*	possessivo
past participle	*pp*	participio passato
prefix	*pref*	prefisso
preposition	*prep*	preposizione
present tense	*pres*	presente
pronoun	*pron*	pronome
psychology	*Psych*	psicologia
past tense	*pt*	tempo passato
	qcno	qualcuno
	qcsa	qualcosa
proprietary term	®	marca depositata
rail	*Rail*	ferrovia
reflexive	*refl*	riflessivo
religion	*Relig*	religione
relative pronoun	*rel pron*	pronome relativo
somebody	*sb*	
school	*Sch*	scuola
singular	*sg*	singolare
slang	*sl*	gergo
something	*sth*	
technical	*Techn*	tecnico
telephone	*Teleph*	telefono
theatrical	*Theat*	teatrale
television	*TV*	televisione
typography	*Typ*	tipografia
university	*Univ*	università
auxiliary verb	*v aux*	verbo ausiliare
intransitive verb	*vi*	verbo intransitivo
reflexive verb	*vr*	verbo riflessivo
transitive verb	*vt*	verbo transitivo
transitive and intransitive	*vt/i*	verbo transitivo e intransitivo
vulgar	*vulg*	volgare
cultural equivalent	≈	equivalenza culturale

Aa

a (**ad** *before vowel*) *prep* to; (*stato in luogo, tempo, età*) at; (*con mese, città*) in; (*mezzo, modo*) by; **dire qcsa a qcno** tell sb sth; **alle tre** at three o'clock; **a vent'anni** at the age of twenty; **a Natale** at Christmas; **a dicembre** in December; **ero al cinema** I was at the cinema; **vivo a Londra** I live in London; **a due a due** two by two; **a piedi** on *o* by foot; **maglia a maniche lunghe** long-sleeved sweater; **casa a tre piani** house with three floors; **giocare a tennis** play tennis; **50 km all'ora** 50 km an hour; **2 000 lire al chilo** 2,000 lire a kilo; **al mattino/alla sera** in the morning/evening; **a venti chilometri/due ore da qui** twenty kilometres/two hours away

a'bate *nm* abbot

abbacchi'ato *a* downhearted

ab'bacchio *nm* [young] lamb

abba'gli|ante *a* dazzling ●*nm* headlight, high beam

abba'gli|are *vt* dazzle. **ab'baglio** *nm* blunder; **prendere un ~** make a blunder

abbai'are *vi* bark

abba'ino *nm* dormer window

abbando'na|re *vt* abandon; leave (*luogo*); give up (*piani ecc*). **~rsi** *vr* let oneself go; **~rsi a** give oneself up to (*ricordi ecc*). **~to** *a* abandoned. **abban'dono** *nm* abandoning; *fig* abandon; (*stato*) neglect

abbassa'mento *nm* (*di temperatura, acqua, prezzi*) drop

abbas'sar|e *vt* lower; turn down (*radio, TV*); **~e i fari** dip the headlights. **~si** *vr* stoop; (*sole ecc*:) sink; *fig* demean oneself

ab'basso *adv* below ●*int* down with

abba'stanza *adv* enough; (*alquanto*) quite

ab'batter|e *vt* demolish; shoot down (*aereo*); put down (*animale*); topple (*regime*); (*fig: demoralizzare*) dishearten. **~si** *vr* (*cadere*) fall; *fig* be discouraged

abbatti'mento *nm* (*morale*) despondency

abbat'tuto *a* despondent, down-in-the-mouth

abba'zia *nf* abbey

abbel'lir|e *vt* embellish. **~si** *vr* adorn oneself

abbeve'ra|re *vt* water. **~'toio** *nm* drinking trough

abbi'ente *a* well-to-do

abbiglia'mento *nm* clothes *pl*; (*industria*) clothing industry, rag trade

abbigli'ar|e *vt* dress. **~si** *vr* dress up

abbina'mento *nm* combining

abbi'nare *vt* combine; match (*colori*)

abbindo'lare *vt* cheat

abbocca'mento *nm* interview; (*conversazione*) talk

abboc'care *vi* bite; (*tubi:*) join; *fig* swallow the bait

abboc'cato *a* (*vino*) fairly sweet

abbof'farsi *vr* stuff oneself

abbona'mento *nm* subscription; (*ferroviario ecc*) season-ticket; **fare l'~** take out a subscription

abbo'na|re *vt* make a subscription. **~rsi** *vr* subscribe (**a** to); take out a season-ticket (**a** for) (*teatro, stadio*). **~to, -a** *nmf* subscriber

abbon'dan|te *a* abundant; (*quantità*) copious; (*nevicata*) heavy; (*vestiario*) roomy. **~te di** abounding in. **~te'mente** *adv* ⟨*mangiare*⟩ copiously. **~za** *nf* abundance

abbon'dare *vi* abound

abbor'da|bile *a* ⟨*persona*⟩ approachable; ⟨*prezzo*⟩ reasonable. **~ggio** *nm* Mil boarding. **~re** *vt* board ⟨*nave*⟩; approach ⟨*persona*⟩; ⟨*fam: attaccar bottone a*⟩ chat up; tackle ⟨*compito ecc*⟩

abbotto'na|re *vt* button up. **~'tura** *nf* [row of] buttons. **~to** *a fig* tight-lipped

abboz'zare *vt* sketch [out]; **~ un sorriso** give a hint of a smile. **ab'bozzo** *nm* sketch

abbracci'are *vt* embrace; hug, embrace ⟨*persona*⟩; take up ⟨*professione*⟩; *fig* include. **ab'braccio** *nm* hug

abbrevi'a|re *vt* shorten; ⟨*ridurre*⟩ curtail; abbreviate ⟨*parola*⟩. **~zi'one** *nf* abbreviation

abbron'zante *nm* sun-tan lotion

abbron'za|re *vt* bronze; tan ⟨*pelle*⟩. **~rsi** *vr* get a tan. **~to** *a* tanned. **~'tura** *nf* [sun-]tan

abbrusto'lire *vt* toast; roast ⟨*caffè ecc*⟩

abbruti'mento *nm* brutalization. **abbru'tire** *vt* brutalize. **abbru'tirsi** *vr* become brutalized

abbuf'fa|rsi *vr fam* stuff oneself. **~ta** *nf* blowout

abbuo'nare *vt* reduce

abbu'ono *nm* allowance; *Sport* handicap

abdi'ca|re *vi* abdicate. **~zi'one** *nf* abdication

aber'rante *a* aberrant

aberrazi'one *nf* aberration

a'bete *nm fir*

abi'etto *a* despicable

'abi|le *a* able; ⟨*idoneo*⟩ fit; ⟨*astuto*⟩ clever. **~ità** *nf inv* ability; ⟨*idoneità*⟩ fitness; ⟨*astuzia*⟩ cleverness. **~'mente** *adv* ably; ⟨*con astuzia*⟩ cleverly

abili'ta|re *vt* qualify. **~to** *a* qualified. **~zi'one** *nf* qualification; ⟨*titolo*⟩ diploma

abis'sale *a* abysmal. **a'bisso** *nm* abyss

abi'tabile *a* inhabitable

abi'tacolo *nm* Auto passenger compartment

abi'tante *nmf* inhabitant

abi'ta|re *vi* live. **~to** *a* inhabited ● *nm* built-up area. **~zi'one** *nf* house

'abito *nm* ⟨*da donna*⟩ dress; ⟨*da uomo*⟩ suit. **~ da cerimonia/da sera** formal/evening dress

abitu'al|e *a* usual, habitual. **~'mente** *adv* usually

abitu'ar|e *vt* accustom. **~si a** *vr* get used to

abitu'dinario, -a *a* of fixed habits ● *nmf* person of fixed habits

abi'tudine *nf* habit; **d'~** usually; **per ~** out of habit; **avere l'~ di fare qcsa** be in the habit of doing sth

abnegazi'one *nf* self-sacrifice

ab'norme *a* abnormal

abo'li|re *vt* abolish; repeal ⟨*legge*⟩. **~zi'one** *nf* abolition; repeal

abomi'nevole *a* abominable

abo'rigeno, -a *a* & *nmf* aboriginal

abor'rire *vt* abhor

abor'ti|re *vi* miscarry; ⟨*volontariamente*⟩ have an abortion; *fig* fail. **~vo** *a* abortive. **a'borto** *nm* miscarriage; ⟨*volontario*⟩ abortion. **~sta** *a* pro-choice

abrasi'one *nf* abrasion. **abra'sivo** *a* & *nm* abrasive

abro'ga|re *vt* repeal. **~zi'one** *nf* repeal

'abside *nf* apse

abu'lia *nf* apathy. **a'bulico** *a* apathetic

abu's|are *vi* **~ di** abuse; over-indulge in ⟨*alcol*⟩; ⟨*approfittare di*⟩ take advantage of; ⟨*violentare*⟩ rape. **~ivo** *a* illegal

a'buso *nm* abuse. **~ di confidenza** breach of confidence

a.C. *abbr* **(avanti Cristo)** BC

'acca *nf fam* **non ho capito un'~** I understood damn all

acca'demia *nf* academy. **A~a di**

Belle Arti Academy of Fine Arts. **~co, -a** *a* academic ● *nmf* academician

acca'd|ere *vi* happen; **accada quel che accada** come what may. **~uto** *nm* event

accalappi'are *vt* catch; *fig* allure

accal'carsi *vr* crowd

accal'da|rsi *vr* get overheated; *fig* get excited. **~to** *a* overheated

accalo'rarsi *vr* get excited

accampa'mento *nm* camp. **accam'pare** *vt fig* put forth. **accam'parsi** *vr* camp

accani'mento *nm* tenacity; *(odio)* rage

acca'ni|rsi *vr* persist; *(infierire)* rage. **~to** *a* persistent; *(odio)* fierce; *fig* inveterate

ac'canto *adv* near; **~** *prep* next to

accanto'nare *vt* set aside; *Mil* billet

accaparra'mento *nm* hoarding; *Comm* cornering

accapar'ra|re *vt* hoard. **~rsi** *vr* grab; corner *(mercato)*. **~tore, ~trice** *nmf* hoarder

accapigli'arsi *vr* scuffle; *(litigare)* squabble

accappa'toio *nm* bathrobe; *(per spiaggia)* beachrobe

accappo'nare *vt* **fare ~ la pelle a** qcno make sb's flesh creep

accarez'zare *vt* caress, stroke; *fig* cherish

accartocci'ar|e *vt* scrunch up. **~si** *vr* curl up

acca'sarsi *vr* get married

accasci'arsi *vr* flop down; *fig* lose heart

accata'stare *vt* pile up

accatti'vante *a* beguiling

accatti'varsi *vr* **~ le simpatie/la stima/l'affetto di** qcno gain sb's sympathy/respect/affection

accatto'naggio *nm* begging. **accat'tone, -a** *nmf* beggar

accaval'lar|e *vt* cross *(gambe)*. **~si** *vr* pile up; *fig* overlap

acce'cante *a* *(luce)* blinding

acce'care *vt* blind ● *vi* go blind

ac'cedere *vi* **a** enter; *(acconsentire)* comply with

accele'ra|re *vi* accelerate ● *vt* speed up, accelerate; **~re il passo** quicken one's pace. **~to** *a* rapid. **~tore** *nm* accelerator. **~zi'one** *nf* acceleration

ac'cender|e *vt* light; turn on *(luce, TV ecc)*; *fig* inflame; **ha da ~?** have you got a light? **~si** *vr* catch fire; *(illuminarsi)* light up; *(TV ecc)* turn on; *fig* become inflamed

accendi'gas *nm inv* gas lighter; *(su cucina)* automatic ignition

accen'dino *nm* lighter

accendi'sigari *nm* cigar-lighter

accen'nare *vt* indicate; hum *(melodia)* ● *vi* **~** a beckon to; *fig* hint at; *(far l'atto di)* make as if to; **accenna a piovere** it looks like rain. **ac'cenno** *nm* gesture; *(con il capo)* nod; *fig* hint

accensi'one *nf* lighting; *(di motore)* ignition

accen'ta|re *vt* accent; *(con accento tonico)* stress. **~zi'one** *nf* accentuation. **ac'cento** *nm* accent; *(tonico)* stress

accentra'mento *nm* centralizing

accen'trare *vt* centralize

accentu'a|re *vt* accentuate. **~rsi** *vr* become more noticeable. **~to** *a* marked

accerchia'mento *nm* surrounding

accerchi'are *vt* surround

accerta'mento *nm* check

accer'tare *vt* ascertain; *(controllare)* check; assess *(reddito)*

ac'ceso *a* lighted; *(radio, TV ecc)* on; *(colore)* bright

acces'sibile *a* accessible; *(persona)* approachable; *(spesa)* reasonable

ac'cesso *nm* access; *(Med: di rabbia)* fit; **vietato l'~** no entry

acces'sorio *a* accessory; *(secondario)* of secondary importance ● *nm* accessory; **accessori** *pl* *(rifiniture)* fittings ac'cetta *nf* hatchet

accet'tabile *a* acceptable

accet'tare *vt* accept; ⟨*aderire a*⟩ agree to

accettazi'one *nf* acceptance; ⟨*luogo*⟩ reception. **~ [bagagli]** check-in. **[banco] ~** check-in [desk]

ac'cetto *a* agreeable; **essere bene ~** be very welcome

accezi'one *nf* meaning

acchiap'pare *vt* catch

acchito *nm* **di primo ~** at first

acciac'care *vt* crush; *fig* prostrate. **~to, -a** *a* **essere ~to** ache all over. **acci'acco** *nm* infirmity; **acciacchi** *pl* aches and pains

acciaie'ria *nf* steelworks

acci'aio *nm* steel; **~ inossidabile** stainless steel

acciden'tale *a* accidental. **~l'mente** *adv* accidentally. **~to** *a* ⟨*terreno*⟩ uneven

acci'dente *nm* accident; *Med* stroke; **non capisce/non vede un ~** *fam* he doesn't understand/can't see a damn thing. **acci'denti!** *int* damn!

accigli'arsi *vr* frown. **~to** *a* frowning

ac'cingersi *vr* **~ a** be about to

acci'picchia *int* good Lord!

acciuf'fare *vt* catch

acci'uga *nf* anchovy

accla'mare *vt* applaud; ⟨*eleggere*⟩ acclaim. **~zi'one** *nf* applause

acclima'tare *vt* acclimatize. **~si** *vr* get acclimatized

ac'cludere *vt* enclose. **~so** *a* enclosed

accocco'larsi *vr* squat

acco'gliente *a* welcoming; ⟨*confortevole*⟩ cosy. **~za** *nf* welcome

ac'cogliere *vt* receive; ⟨*conpiacere*⟩ welcome; ⟨*contenere*⟩ hold

accol'larsi *vr* take on ⟨*responsabilità, debiti, doveri*⟩. **accol'lato** *a* high-necked

accoltel'lare *vt* knife

accomia'tare *vt* dismiss. **~si** *vr* take one's leave ⟨*da*⟩

accomo'dante *a* accommodating

accomo'dar|e *vt* ⟨*riparare*⟩ mend; ⟨*disporre*⟩ arrange. **~si** *vr* make oneself at home; **si accomodi** come in!; ⟨*si sieda*⟩ take a seat!

accompagna'mento *nm* accompaniment; ⟨*seguito*⟩ retinue

accompa'gna|re *vt* accompany; **~re qcno a casa** see sb home; **~re qcno alla porta** show sb out. **~'tore, -'trice** *nmf* companion; ⟨*di comitiva*⟩ escort; *Mus* accompanist

accomu'nare *vt* pool

acconci'a|re *vt* arrange. **~'tura** *nf* hair-style; ⟨*ornamento*⟩ headdress

accondiscen'den|te *a* too obliging. **~za** *nf* excessive desire to please

accondi'scendere *vi* **~ a** condescend; comply with ⟨*desiderio*⟩; ⟨*acconsentire*⟩ consent to

acconsen'tire *vi* consent

acconten'tare *vt* satisfy. **~si** *vr* be content ⟨**di** with⟩

ac'conto *nm* deposit; **in ~** on account; **lasciare un ~** leave a deposit

accop'pare *vt fam* bump off

accoppia'mento *nm* coupling; ⟨*di animali*⟩ mating

accoppi'a|re *vt* couple; mate ⟨*animali*⟩. **~rsi** *vr* pair off; mate. **~ta** *nf* ⟨*scommessa*⟩ bet placed on two horses for first and second place

acco'rato *a* sorrowful

accorci'ar|e *vt* shorten. **~si** *vr* get shorter

accor'dar|e *vt* concede; match ⟨*colori ecc*⟩; *Mus* tune. **~si** *vr* agree

ac'cordo *nm* agreement; *Mus* chord; ⟨*armonia*⟩ harmony; **andare d'~** get on well; **d'~!** agreed!; **essere d'~** agree; **prendere accordi con qcno** make arrangements with sb

ac'corgersi *vr* **~ di** notice; ⟨*capire*⟩ realize

accorgi'mento *nm* shrewdness; ⟨*espediente*⟩ device

ac'correre *vi* hasten

accor'tezza nf (previdenza) fore-thought

ac'corto a shrewd; **mal ~** incautious

accosta'mento nm (di colori) combination

acco'star|e vt draw close to; approach ⟨persona⟩; set ajar ⟨porta ecc⟩. **~si** vr **~si a** come near to

accovacci'arsi vr crouch, squat down. **~to a** squatting

accoz'zaglia nf jumble; (di persone) mob

accoz'zare vt **~ colori** mix colours that clash

accredita'mento nm credit; **~ tramite bancogiro** Bank Giro Credit

accredi'tare vt confirm ⟨notizia⟩; Comm credit

ac'cresc|ere vt increase. **~ersi** vr grow larger. **~i'tivo** a augmentative

accucci'arsi vr ⟨cane:⟩ lie down; ⟨persona:⟩ crouch

accu'dire vi **~ a** attend to

accumu'la|re vt accumulate. **~rsi** vr pile up, accumulate. **~'tore** nm accumulator; Auto battery. **~zi'one** nf accumulation. **ac-'cumulo** nm (di merce) build-up

accura'tezza nf care

accu'rato a careful

ac'cusa nf accusation; Jur charge; **essere in stato di ~** Jur have been charged; **la Pubblica A~** Jur the public prosecutor

accu'sa|re vt accuse; Jur charge; complain of ⟨dolore⟩; **~re ricevuta di** Comm acknowledge receipt of. **~to, -a** nmf accused. **~'tore** nm Jur prosecutor

a'cerbo a sharp; (immaturo) unripe

'acero nm maple

a'cerrimo a implacable

ace'tone nm nail polish remover

a'ceto nm vinegar

A.C.I. abbr (**Automobile Club d'Italia**) Italian Automobile Association

acidità nf acidity. **~ di stomaco** acid stomach

'acido a acid; ⟨persona⟩ sour ● nm acid

a'cidulo a slightly sour

'acino nm berry; (chicco) grape

'acne nf acne

'acqua nf water; **fare ~** Naut leak; **~ in bocca!** fig mum's the word!. **~ di Colonia** eau de Cologne. **~ corrente** running water. **~ dolce** fresh water. **~ minerale** mineral water. **~ minerale gassata** fizzy mineral water. **~ naturale** still mineral water. **~ potabile** drinking water. **~ salata** salt water. **~ tonica** tonic water

acqua'forte nf etching

ac'quaio nm sink

acquama'rina a aquamarine

acqua'rello nm = **acquerello**

ac'quario nm aquarium; Astr Aquarius

acqua'santa nf holy water

acqua'scooter nm inv water-scooter

ac'quatico a aquatic

acquat'tarsi vr crouch

acqua'vite nf brandy

acquaz'zone nm downpour

acque'dotto nm aqueduct

'acqueo a vapore = water vapour

acque'rello nm water-colour

acqui'rente nmf purchaser

acqui'si|re vt acquire. **~to a** acquired. **~zi'one** nf attainment

acqui'st|are vt purchase; (ottenere) acquire. **ac'quisto** nm purchase; **uscire per ~i** go shopping; **fare ~i** shop

acqui'trino nm marsh

acquo'lina nf **far venire l'~ in bocca a qcno** make sb's mouth water

ac'quoso a watery

'acre a acrid; (al gusto) sour; fig harsh

a'crilico nm acrylic

a'croba|ta nmf acrobat. **~'zia** nf acrobatics pl

a'cronimo nm acronym

acu'ir|e vt sharpen. **~si** vr become more intense

a'culeo nm sting; Bot prickle

a'cume nm acumen

acumi'nato a pointed

a'custic|a nf acoustics pl. **~o** a acoustic

acu'tezza nf acuteness

acutiz'zarsi vr become worse

a'cuto a sharp; ⟨suono⟩ shrill; ⟨freddo, odore⟩ intense; Gram, Math, Med acute ● nm Mus high note

adagi'ar|e vt lay down. **~si** vr lie down

a'dagio adv slowly ● nm Mus adagio; ⟨proverbio⟩ adage

adattabi'lità nf adaptability

adatta'mento nm adaptation; **avere spirito di ~** be adaptable

adat'ta|re vt adapt; ⟨aggiustare⟩ fit. **~rsi** vr adapt. **~'tore** nm adaptor. **a'datto** a suitable (**a** for); ⟨giusto⟩ right

addebita'mento nm debit. **~ diretto** direct debit

addebi'tare vt debit; ascribe ⟨colpa⟩

ad'debito nm charge

addensa'mento nm thickening; ⟨di persone⟩ gathering

adden'sar|e vt thicken. **~si** vr thicken; ⟨affollarsi⟩ gather

adden'tare vt bite

adden'trarsi vr penetrate

ad'dentro adv deeply; **essere ~ in** be in on

addestra'mento nm training

adde'strar|e vt train. **~si** vr train

ad'det|to, -a a assigned ● nmf employee; ⟨diplomatico⟩ attaché; **addetti** pl **ai lavori** persons involved in the work. **~ stampa** information officer, press officer

addi'accio nm **dormire all'~** sleep in the open

addi'etro adv ⟨indietro⟩ back; ⟨nel passato⟩ before

ad'dio nm & int goodbye. **~ al celibato** stag night, stag party

addirit'tura adv ⟨perfino⟩ even;

⟨assolutamente⟩ absolutely; **~!** really!

ad'dirsi vr **~ a** suit

addi'tare vt point at; ⟨in mezzo a un gruppo⟩ point out; fig point to

addi'tivo a & nm additive

addizio'nal|e a additional. **~'mente** adv additionally

addizio'nare vt add [up]. **addizi'one** nf addition

addob'bare vt decorate. **ad'dobbo** nm decoration

addol'cir|e vt sweeten; tone down ⟨colore⟩; fig soften. **~si** vr fig mellow

addolo'rar|e vt grieve. **~rsi** vr be upset (**per** by). **~to** a pained, distressed

ad'dom|e nm abdomen. **~i'nale** a abdominal; [**muscoli**] **addominali** pl abdominals

addomesti'ca|re vt tame. **~'tore** nm tamer

addormen'ta|re vt put to sleep. **~rsi** vr go to sleep. **~to** a asleep; fig slow

addos'sar|e vt **~e a** ⟨appoggiare⟩ lean against; ⟨attribuire⟩ lay on. **~si** vr ⟨ammassarsi⟩ crowd; shoulder ⟨responsabilità ecc⟩

ad'dosso adv on; **~ a** prep on; ⟨molto vicino⟩ right next to; **mettere gli occhi ~ a qcno/qcsa** hanker after sb/sth; **non mettermi le mani ~!** keep your hands off me!; **stare ~ a qcno** fig be on sb's back

ad'durre vt produce ⟨prova, documento⟩; give ⟨pretesto, esempio⟩

adegua'mento nm adjustment

adegu'a|re vt adjust. **~rsi** vr conform. **~to** a adequate; ⟨conforme⟩ consistent

a'dempi|ere vt fulfil. **~'mento** nm fulfilment

ade'noidi nfpl adenoids

ade'ren|te a adhesive; ⟨vestito⟩ tight ● nmf follower. **~za** nf adhesion. **~ze** npl connections

ade'rire vi **~ a** stick to, adhere

to; support (*sciopero, petizione*); agree to (*richiesta*)

adesca'mento *nm Jur* soliciting

ade'scare *vt* bait; *fig* entice

adesi'one *nf* adhesion; *fig* agreement

ade'sivo *a* adhesive ● *nm* sticker; *Auto* bumper sticker

a'desso *adv* now; (*poco fa*) just now; (*tra poco*) any moment now; **da ~ in poi** from now on; **per ~** for the moment

adia'cente *a* adjacent; **~ a** next to

adi'bire *vt* **~ a** put to use as

'adipe *nm* adipose tissue

adi'ra|rsi *vr* get irate. **~to** *a* irate

a'dire *vt* resort to; **~ le vie legali** take legal proceedings

'adito *nm* **dare ~ a** give rise to

adocchi'are *vt* eye; (*con desiderio*) covet

adole'scen|te *a & nmf* adolescent. **~za** *nf* adolescence. **~zi'ale** *a* adolescent

adom'brar|e *vt* darken; *fig* veil. **~si** *vr* (*offendersi*) take offence

adope'rar|e *vt* use. **~si** *vr* take trouble

ado'rabile *a* adorable

ado'ra|re *vt* adore. **~zi'one** *nf* adoration

ador'nare *vt* adorn

adot't|are *vt* adopt. **~ivo** *a* adoptive. **adozi'one** *nf* adoption

ad *prep* = **a** (*davanti a vocale*)

adrena'lina *nf* adrenalin

adri'atico *a* Adriatic ● *nm* **l'A~** the Adriatic

adu'la|re *vt* flatter. **~'tore, ~'trice** *nmf* flatterer. **~zi'one** *nf* flattery

adulte'ra|re *vt* adulterate. **~to** *a* adulterated

adul'terio *nm* adultery. **a'dultero, -a** *a* adulterous ● *nm* adulterer ● *nf* adulteress

a'dulto, -a *a & nmf* adult; (*maturo*) mature

adu'nanza *nf* assembly

adu'na|re *vt* gather. **~ta** *nf Mil* parade

a'dunco *a* hooked

ae'rare *vt* air (*stanza*)

a'ereo *a* aerial; (*dell'aviazione*) air *attrib* ● *nm* aeroplane, plane

ae'robic|a *nf* aerobics. **~o** *a* aerobic

aerodi'namic|a *nf* aerodynamics *sg*. **~o** *a* aerodynamic

aero'nautic|a *nf* aeronautics *sg*; *Mil* Air Force. **~o** *a* aeronautical

aero'plano *nm* aeroplane

aero'porto *nm* airport

aero'scalo *nm* cargo and servicing area

aero'sol *nm inv* aerosol

'afa *nf* sultriness

affa'bil|e *a* affable. **~ità** *nf* affability

affaccen'da|rsi *vr* busy oneself (**a** with). **~to** *a* busy

affacci'arsi *vr* show oneself; **~ alla finestra** appear at the window

affa'ma|re *vt* starve [out]. **~to** *a* starving

affan'na|re *vt* leave breathless. **~rsi** *vr* busy oneself; (*agitarsi*) get worked up. **~to** *a* breathless; **dal respiro ~to** wheezy. **affanno** *nm* breathlessness; *fig* worry

af'fare *nm* matter; *Comm* transaction, deal; (*occasione*) bargain; **affari** *pl* business; **non sono affari tuoi** *fam* it's none of your business. **affa'rista** *nmf* wheeler-dealer

affasci'nante *a* fascinating; (*persona, sorriso*) bewitching

affasci'nare *vt* bewitch; *fig* charm

affati'camento *nm* fatigue

affati'car|e *vt* tire; (*sfinire*) exhaust. **~si** *vr* tire oneself out; (*affannarsi*) strive

af'fatto *adv* completely; **non...** not... at all; **niente ~!** not at all!

affer'ma|re *vt* affirm; (*sostenere*) assert. **~rsi** *vr* establish oneself

affermativa'mente *adv* in the affirmative

afferma'tivo *a* affirmative

affermazi'one *nf* assertion; (*successo*) achievement

affer'rar|e *vt* seize; catch ⟨oggetto⟩; ⟨capire⟩ grasp; **~e al volo** *fig* be quick on the uptake. **~si** *vr* **~si a** grasp at

affet'ta|re *vt* slice; ⟨ostentare⟩ affect. **~to a** sliced; ⟨sorriso, maniere⟩ affected ● *nm* cold meat, sliced meat. **~zi'one** *nf* affectation

affet'tivo *a* affective; **rapporto ~** emotional tie

af'fetto[1] *nm* affection; **con ~** affectionately

af'fetto[2] *a* **~ da** suffering from

affettuosità *nf inv* ⟨gesto⟩ affectionate gesture

affettu'oso *a* affectionate

affezio'na|rsi *vr* **~rsi a** grow fond of. **~to a** devoted ⟨a to⟩

affian'car|e *vt* put side by side; *Mil* flank; *fig* support. **~si** *vr* come side by side; *fig* stand together; **~si a qcno** *fig* help sb out

affiata'mento *nm* harmony

affia'ta|rsi *vr* get on well together. **~to a** close-knit; **una coppia ~ta** a very close couple

affibbi'are *vt* **~ qcsa a qcno** saddle sb with sth; **~ un pugno a qcno** let fly at sb

affi'dabile *a* dependable. **~ità** *nf* dependability

affida'mento *nm* ⟨Jur: dei minori⟩ custody; **fare ~ su qcno** rely on sb; **non dare ~** not inspire confidence

affi'dar|e *vt* entrust. **~si** *vr* **~si a** rely on

affievo'lirsi *vr* grow weak

af'figgere *vt* affix

affi'lare *vt* sharpen

affili'ar|e *vt* affiliate. **~si** *vr* become affiliated

affi'nare *vt* sharpen; ⟨perfezionare⟩ refine

affinché *conj* so that, in order that

af'fin|e *a* similar. **~ità** *nf* affinity

affiora'mento *nm* emergence; *Naut* surfacing

affio'rare *vi* emerge; *fig* come to light

af'fisso *nm* bill; *Gram* affix

affitta'camere *nm inv* landlord ● *nf inv* landlady

affit'tare *vt* ⟨dare in affitto⟩ let; ⟨prendere in affitto⟩ rent; **'af'fittasi'** 'to let', 'for rent'

af'fitto *nm* rent; **contratto d'~** lease; **dare in ~** to let; **prendere in ~** rent. **~u'ario, -a** *nmf Jur* lessee

af'fligger|e *vt* torment. **~si** *vr* distress oneself

af'flitto *a* distressed. **~zi'one** *nf* distress; *fig* affliction

afflosci'arsi *vr* become floppy; ⟨accasciarsi⟩ flop down; ⟨morale⟩ decline

afflu'en|te *a & nm* tributary. **~za** *nf* flow; ⟨di gente⟩ crowd

afflu'ire *vi* flow; *fig* pour in

af'flusso *nm* influx

affo'ga|re *vt/i* drown; *Culin* poach; **~re in** *fig* be swamped with. **~to a** ⟨persona⟩ drowned; ⟨uova⟩ poached. **~to al caffè** *nm* ice cream with hot espresso poured over it

affol'la|re *vt*, **~rsi** *vr* crowd. **~to a** crowded

affonda'mento *nm* sinking

affon'dare *vt/i* sink

affossa'mento *nm* pothole

affran'ca|re *vt* redeem ⟨bene⟩; stamp ⟨lettera⟩; free ⟨schiavo⟩. **~rsi** *vr* free oneself. **~'trice** *nf* franking machine. **~'tura** *nf* stamping; ⟨di spedizione⟩ postage

af'franto *a* prostrated; ⟨esausto⟩ worn out

af'fresco *nm* fresco

affret'ta|re *vt* speed up. **~rsi** *vr* hurry. **~ta'mente** *adv* hastily. **~to a** hasty

affron'tare *vt* face; confront ⟨il nemico⟩; meet ⟨le spese⟩. **~si** *vr* clash

af'fronto *nm* affront, insult; **fare un ~ a qcno** insult sb

affumi'ca|re *vt* fill with smoke; *Culin* smoke. **~to a** ⟨prosciutto, formaggio⟩ smoked

affuso'la|re *vt* taper [off]. **~to a** tapering

afo'risma *nm* aphorism

a'foso *a* sultry

'Africa *nf* Africa. **afri'cano, -a** *a* & *nmf* African

afrodi'siaco *a* & *nm* aphrodisiac

a'genda *nf* diary

agen'dina *nf* pocket-diary

a'gente *nm* agent; **agenti** *pl* **atmosferici** atmospheric agents. ~ **di cambio** stockbroker. ~ **di polizia** policeman

agen'zia *nf* agency; (*filiale*) branch office; (*di banca*) branch. ~ **di viaggi** travel agency. ~ **immobiliare** estate agency

agevo'la|re *vt* facilitate. **~zi'one** *nf* facilitation

a'gevol|e *a* easy; (*strada*) smooth. **~'mente** *adv* easily

agganci'ar|e *vt* hook up; *Rail* couple. **~si** *vr* (*vestito:*) hook up

ag'geggio *nm* gadget

agget'tivo *nm* adjective

agghiacci'ante *a* terrifying

agghiacci'ar|e *vt fig* — **qcno** make sb's blood run cold. **~si** *vr* freeze

agghin'da|re *vt fam* dress up. **~rsi** *vr fam* doll oneself up. **~to** *a* dressed up

aggiorna'mento *nm* up-date

aggior'na|re *vt* (*rinviare*) postpone; (*mettere a giorno*) bring up to date. **~rsi** *vr* get up to date. **~to** *a* up-to-date; (*versione*) updated

aggi'rar|e *vt* surround; (*fig: ingannare*) trick. **~si** *vr* hang about; **~si su** (*discorso ecc:*) be about; (*somma:*) be around

aggiudi'car|e *vt* award; (*all'asta*) knock down. **~si** *vr* win

aggi'un|gere *vt* add. **~ta** *nf* addition. **~tivo** *a* supplementary. **~to** *a* added *a* & *nm* (*assistente*) assistant

aggiu'star|e *vt* mend; (*sistemare*) settle; (*fam: mettere a posto*) fix. **~si** *vr* adapt; (*mettersi in ordine*) tidy oneself up; (*decidere*) sort things out; (*tempo:*) clear up

agglomera'mento *nm* conglomeration

agglome'rato *nm* built-up area

aggrap'par|e *vt* grasp. **~si** *vr* **~si a** cling to

aggra'vante *Jur nf* aggravation ● *a* aggravating

aggra'var|e *vt* (*peggiorare*) make worse; increase (*pena*); (*appesantire*) weigh down. **~si** *vr* worsen

aggrazi'ato *a* graceful

aggre'dire *vt* attack

aggre'ga|re *vt* add; (*associare a un gruppo ecc*) admit. **~rsi** *vr* **~rsi a** join. **~to a** associated ● *nm* aggregate; (*di case*) block

aggressi'one *nf* aggression; (*atto*) attack

aggres'sivo *a* aggressive. **~ività** *nf* aggressiveness. **~ore** *nm* aggressor

aggrin'zare, aggrin'zire *vt* wrinkle

aggrot'tare *vt* ~ **le ciglia/la fronte** frown

aggrovigli'a|re *vt* tangle. **~rsi** *vr* get entangled; *fig* get complicated. **~to** *a* entangled; *fig* confused

agguan'tare *vt* catch

aggu'ato *nm* ambush; (*tranello*) trap; **tendere un** ~ lie in wait

agguer'rito *a* fierce

agia'tezza *nf* comfort

agi'ato *a* (*persona*) well off; (*vita*) comfortable

a'gibil|e *a* (*palazzo*) fit for human habitation. **~ità** *nf* fitness for human habitation

'agil|e *a* agile. **~ità** *nf* agility

'agio *nf* ease; **mettersi a proprio** ~ make oneself at home

a'gire *vi* act; (*comportarsi*) behave; (*funzionare*) work; **~ su** affect

agi'ta|re *vt* shake; wave (*mano*); (*fig: turbare*) trouble. **~rsi** *vr* toss about; (*essere inquieto*) be restless; (*mare:*) get rough. **~to** *a* restless; (*mare*) rough. **~tore, ~trice** *nmf* (*persona*) agitator. **~zi'one** *nf* agitation; **mettere in ~zione** qcno make sb worried

'agli = a + gli

'aglio nm garlic

a'gnello nm lamb

agno'lotti nmpl ravioli sg

a'gnostico, -a a & nmf agnostic

'ago nm needle

ago'ni|a nf agony. ~z'zare vi be on one's deathbed

a'gnostic|a nf competition. ~o a competitive

agopun'tura nf acupuncture

a'gosto nm August

a'grari|a nf agriculture. ~o a agricultural ● nm landowner

a'gricol|o a agricultural. ~'tore nm farmer. ~'tura nf agriculture

agri'foglio nm holly

agritu'rismo nm farm holidays, agro-tourism

'agro a sour

agroalimen'tare a food attrib

agro'dolce a bitter-sweet; Culin sweet-and-sour; in ~ sweet and sour

agrono'mia nf agronomy

a'grume nm citrus fruit; (pianta) citrus tree

aguz'zare vt sharpen; ~ le orecchie prick up one's ears; ~ la vista look hard

aguz'zino nm slave-driver; (carceriere) jailer

ahimè int alas

'ai = a + i

'Aia nf L'~ The Hague

'aia nf threshing-floor

Aids nmf Aids

ai'rone nm heron

ai'tante a sturdy

aiu'ola nf flower-bed

aiu'tante nmf assistant ● nm Mil adjutant. ~ di campo aide-de-camp

aiu'tare vt help

ai'uto nm help, aid; (assistente) assistant

aiz'zare vt incite; ~ contro set on

al = a + il

'ala nf wing; fare ~ make way

ala'bastro nm alabaster

'alacre a brisk

a'lano nm Great Dane

'alba nf dawn

Alba'ni|a nf Albania. a~ese a & nmf Albanian

albeggi'are vi dawn

albe'ra|to a wooded; (viale) tree-lined. ~'tura nf Naut masts pl.

albe'rello nm sapling

al'bergo nm hotel. ~o diurno hotel where rooms are rented during the daytime. ~a'tore, ~a'trice nmf hotel-keeper. ~hi'ero a hotel attrib

'albero nm tree; Naut mast; Mech shaft. ~ genealogico family tree. ~ maestro Naut mainmast. ~ di Natale Christmas tree

albi'cocc|a nf apricot. ~o nm apricot-tree

al'bino nm albino

'albo nm register; (libro ecc) album; (per avvisi) notice board

'album nm album. ~ da disegno sketch-book

al'bume nm albumen

'alce nm elk

'alcol nm alcohol; Med spirit; (liquori forti) spirits pl; darsi all'~ take to drink. al'colici nmpl alcoholic drinks. al'colico a alcoholic. alco'lismo nm alcoholism. ~iz'zato, -a a & nmf alcoholic

alco'test® nm inv Breathalyser®

al'cova nf alcove

al'cun, al'cuno a & pron any; non ha ~ amico he hasn't any friends, he has no friends. alcuni pl some, a few; ~i suoi amici some of his friends

alea'torio a unpredictable

a'letta nf Mech fin

alfa'betico a alphabetical

alfabetizzazi'one nf ~ della popolazione teaching people to read and write

alfa'beto nm alphabet

alfi'ere nm (negli scacchi) bishop

al'fine adv eventually, in the end

'alga nf seaweed

'algebra nf algebra

Alge'ri|a *nf* Algeria. **a~no, -a** *a & nmf* Algerian

ali'ante *nm* glider

'alibi *nm inv* alibi

alie'na|re *vt* alienate. **~rsi** *vr* become estranged; **~rsi le simpatie di qcno** lose sb's good will. **~to, -a** *a* alienated ● *nmf* lunatic

a'lieno, -a *nmf* alien ● *a* **è ~ da invidia** envy is foreign to him

alimen'ta|re *vt* feed; *fig* foment ● *a* food *attrib*; *(abitudine)* dietary ● *nm* **~ri** *pl* food-stuffs. **~'tore** *nm* power unit. **~zi'one** *nf* feeding

ali'mento *nm* food; **alimenti** *pl* food; *Jur* alimony

a'liquota *nf* share; *(di imposta)* rate

ali'scafo *nm* hydrofoil

'alito *nm* breath

'alla = **a** + **la**

allaccia'mento *nm* connection

allacci'ar|e *vt* fasten *(cintura)*; lace up *(scarpe)*; do up *(vestito)*; *(collegare)* connect; form *(amicizia)*. **~si** *vr* do up, fasten *(vestito, cintura)*

allaga'mento *nm* flooding

alla'gar|e *vt* flood. **~si** *vr* become flooded

allam'pa'nato *a* lanky

allarga'mento *nm (di strada, ricerche)* widening

allar'gar|e *vt* widen; open *(braccia, gambe)*; let out *(vestito ecc)*; *fig* extend. **~si** *vr* widen

allar'mante *a* alarming

allar'ma|re *vt* alarm. **~to a** panicky

al'larme *nm* alarm; **dare l'~** raise the alarm; **falso ~** *fig* false alarm. **~ aereo** air raid warning

allar'mis|mo *nm* alarmism. **~ta** *nmf* alarmist

allatta'mento *nm (di animale)* suckling; *(di neonc|o)* feeding

allat'tare *vt* suckle *(animale)*; feed *(neonato)*

'alle = **a** + **le**

alle'a|nza *nf* alliance. **~to, -a a** allied ● *nmf* ally

alle'ar|e *vt* unite. **~si** *vr* form an alliance

alle'gare [1] *vt Jur* allege

alle'ga|re [2] *vt (accludere)* enclose; set on edge *(denti)*. **~to a** enclosed ● *nm* enclosure; **in ~to** attached, appended. **~zi'one** *nf Jur* allegation

allegge'rir|e *vt* lighten; *fig* alleviate. **~si** *vr* become lighter; *(vestirsi leggero)* put on lighter clothes

allego'ria *nf* allegory. **alle'gorico a** allegorical

allegra'mente *adv* breezily

alle'gria *nf* gaiety

al'legro *a* cheerful; *(colore)* bright; *(brillo)* tipsy ● *nm Mus* allegro

alle'luia *int* hallelujah!

allena'mento *nm* training

alle'na|re *vt*, **~rsi** *vr* train. **~'tore, ~'trice** *nmf* trainer, coach

allen'tar|e *vt* loosen; *fig* relax. **~si** *vr* become loose; *Mech* work loose

aller'gia *nf* allergy. **al'lergico a** allergic

al'lerta *nf* **stare ~** be alert, be on the alert

allesti'mento *nm* preparation. **~ scenico** *Theat* staging

alle'stire *vt* prepare; stage *(spettacolo)*; *Naut* fit out

allet'tante *a* alluring

allet'tare *vt* entice

alleva'mento *nm* breeding; *(processo)* bringing up; *(luogo)* farm; *(per piante)* nursery; **pollo di ~** battery chicken

alle'vare *vt* bring up *(bambini)*; breed *(animali)*; grow *(piante)*

allevi'are *vt* alleviate; *fig* lighten

alli'bito *a* astounded

allibra'tore *nm* bookmaker

allie'tar|e *vt* gladden. **~si** *vr* rejoice

alli'evo, -a *nmf* pupil ● *nm Mil* cadet

alliga'tore *nm* alligator

allinea'mento *nm* alignment

alline'ar|e *vt* line up; *Typ* align; *Fin* adjust. **~si** *vr* fall into line

'allo = a + lo

al'locco nm Zool tawny owl

al'lodola nf [skylark]

alloggi'are vt ⟨persona:⟩ put up; ⟨casa:⟩ provide accommodation for; Mil billet ● vi put up, stay; Mil be billeted. **al'loggio** nm ⟨appartamento⟩ flat; Mil billet

allontana'mento nm removal

allonta'nar|e vt move away; ⟨licenziare⟩ dismiss; avert ⟨pericolo⟩. **~si** vr go away

al'lora adv then; ⟨in quel tempo⟩ at that time; ⟨in tal caso⟩ in that case; **d'~ in poi** from then on; **e ~?** what now?; ⟨e con ciò?⟩ so what?; **fino ~** until then

al'loro nm laurel; Culin bay

'alluce nm big toe

alluci'na|nte a fam incredible; **sostanza ~nte** hallucinogen. **~to, -a** nmf fam space cadet. **~zi'one** nf hallucination

allucino'geno a ⟨sostanza⟩ hallucinatory

al'ludere vi **a ~** allude to

allu'minio nm aluminium

allun'gar|e vt lengthen; stretch [out] ⟨gamba⟩; extend ⟨tavolo⟩; ⟨diluire⟩ dilute; **~e il collo** crane one's neck. **~e le mani su qcno** touch sb up. **~e il passo** quicken one's step. **~si** vr grow longer; ⟨crescere⟩ grow taller; ⟨sdraiarsi⟩ lie down

allusi'one nf allusion

allu'sivo a allusive

alluvio'nale a alluvial

alluvi'one nf flood

al'meno adv at least; **[se] ~ venisse il sole!** if only the sun would come out!

a'logeno nm halogen ● a **lampada alogena** halogen lamp

a'lone nm halo

'Alpi nfpl **le ~** the Alps

alpi'nis|mo nm mountaineering. **~ta** nmf mountaineer

al'pino a Alpine ● nm Mil **gli alpini** the Alpine troops

al'quanto a a certain amount of ● adv rather

alt int stop

alta'lena nf swing; ⟨tavola in bilico⟩ see-saw

altale'nare vi fig vacillate

alta'mente adv highly

al'tare nm altar

alta'rino nm **scoprire gli altarini di qcno** reveal sb's guilty secrets

alte'ra|re vt alter; adulterate ⟨vino⟩; ⟨falsificare⟩ falsify. **~rsi** vr be altered; ⟨cibo:⟩ go bad; ⟨merci:⟩ deteriorate; ⟨arrabbiarsi⟩ get angry. **~to** a ⟨vino⟩ adulterated. **~zi'one** nf alteration; ⟨di vino⟩ adulteration

al'terco nm altercation

alter'nanza nf alternation

alter'na|re vt, **~rsi** vr alternate. **~'tiva** nf alternative. **~'tivo** a alternate. **~to** a alternating. **~'tore** nm Electr alternator

al'terno a alternate; **a giorni ~i** every other day

al'tero a haughty

al'tezza nf height; ⟨profondità⟩ depth; ⟨suono⟩ pitch; ⟨di tessuto⟩ width; ⟨titolo⟩ Highness; **essere all'~ di** be on a level with; fig be up to

altezzos|a'mente adv haughtily. **~ità** nf haughtiness

altez'zoso a haughty

al'ticcio a tipsy, merry

alti'piano nm plateau

alti'tudine nf altitude

'alto a high; ⟨di statura⟩ tall; ⟨profondo⟩ deep; ⟨suono⟩ high-pitched; ⟨tessuto⟩ wide; Geog northern; a **notte alta** in the middle of the night; **avere degli alti e bassi** have some ups and downs; **ad alta fedeltà** high-fidelity; **a voce alta, ad alta voce** in a loud voice; ⟨leggere⟩ aloud; **essere in ~ mare** be on the high seas. **alta finanza** nf high finance. **alta moda** nf high fashion. **alta tensione** nf high voltage ● adv high; **in ~** at the top; ⟨guardare:⟩ up; **mani in ~!** hands up!

alto'forno *nm* blast-furnace

altolà *int* halt there!

altolo'cato *a* highly placed

altopar'lante *nm* loudspeaker

altopi'ano *nm* plateau

altret'tanto *a & pron* as much; (*pl*) as many ●*adv* likewise; **buona fortuna! - grazie, ~ good luck! -** thank you, the same to you

altri'menti *adv* otherwise

'altro *a* other; **un ~, un'altra** another; **l'altr'anno** last year; **domani l'~** the day after tomorrow; **l'ho visto l'~ giorno** I saw him the other day ●*pron* other [one]; **un ~, un'altra** another [one]; **ne vuoi dell'~?** would you like some more?; **l'un l'~** one another; **nessun ~** nobody else; **gli altri** (*la gente*) other people ●*nm* something else; **non fa ~ che lavorare** he does nothing but work; **desidera ~?** (*in negozio*) anything else?; **più che ~,** **sono stanco** I'm tired more than anything; **se non ~** at least; **senz'~** certainly; **tra l'~** what's more; **~ che!** and how!

altri'eri *nm* **l'~** the day before yesterday

al'tronde *adv* **d'~** on the other hand

al'trove *adv* elsewhere

al'trui *a* other people's ●*nm* other people's belongings *pl*

altru'is|mo *nm* altruism. **~ta** *nmf* altruist

al'tura *nf* high ground; *Naut* deep sea

a'lunno, -a *nmf* pupil

alve'are *nm* hive

al'za|re *vt* lift, raise; (*costruire*) build; *Naut* hoist; **~re le spalle** shrug one's shoulders; **~re i tacchi** *fig* take to one's heels. **~rsi** *vr* rise; (*in piedi*) stand up; (*da letto*) get up; **~rsi in piedi** get to one's feet. **~ta** *nf* lifting; (*aumento*) rise; (*da letto*) getting up; *Archit* elevation. **~to** *a* up

a'mabile *a* lovable; (*vino*) sweet

a'maca *nf* hammock

amalga'mar|e *vt*, **~si** *vr* amalgamate

a'mante *a* **~ di** fond of ●*nm* lover ●*nf* mistress, lover

ama'rena *nf* sour black cherry

ama'retto *nm* macaroon

a'ma|re *vt* love; be fond of, like (*musica, sport ecc*). **~to, -a** *a* loved ●*nmf* beloved

ama'rezza *nf* bitterness; (*dolore*) sorrow

a'maro *a* bitter ●*nm* bitterness; (*liquore*) bitters *pl*

ama'rognolo *a* rather bitter

ama'tore, -'trice *nmf* lover

ambasci'a|ta *nf* embassy; (*messaggio*) message. **~'tore, -'trice** *nm* ambassador ●*nf* ambassadress

ambe'due *a & pron* both

ambien'ta|le *a* environmental. **~'lista** *a & nmf* environmentalist

ambien'tar|e *vt* acclimatize; set (*personaggio, film ecc*). **~si** *vr* get acclimatized

ambi'ente *nm* environment; (*stanza*) room; *fig* milieu

ambiguità *nf inv* ambiguity; (*di persona*) shadiness

am'biguo *a* ambiguous; (*persona*) shady

am'bire *vi* **~ a** aspire to

'ambito *nm* sphere

ambiva'len|te *a* ambivalent. **~za** *nf* ambivalence

ambizi'o|ne *nf* ambition. **~so** *a* ambitious

'ambra *nf* amber. **am'brato** *a* amber

ambu'lante *a* wandering; **vendi'tore ~** hawker

ambu'lanza *nf* ambulance

ambula'torio *nm* (*di medico*) surgery; (*di ospedale*) out-patients' [department]

a'meba *nf* amoeba

'amen *int* amen

a'meno *a* pleasant

A'merica *nf* America. **~ del Sud** South America. **ameri'cano, -a** *a & nmf* American

ame'tista *nf* amethyst

ami'anto *nm* asbestos

ami'chevole *a* friendly

ami'cizia *nf* friendship; **fare ~ con qcno** make friends with sb; **amicizie** *pl* ⟨*amici*⟩ friends

a'mico, -a *nm/f* friend; **~ del cuore** bosom friend

'amido *nm* starch

ammac'ca|re *vt* dent; bruise ⟨*frutto*⟩. **~rsi** *vr* ⟨*metallo:*⟩ get dented; ⟨*frutto:*⟩ bruise. **~to** *a* dented; ⟨*frutto*⟩ bruised. **~tura** *nf* dent; ⟨*livido*⟩ bruise

ammae'stra|re *vt* ⟨*istruire*⟩ teach; train ⟨*animale*⟩. **~to** *a* trained

ammai'nare *vt* lower ⟨*bandiera*⟩; furl ⟨*vele*⟩

amma'la|rsi *vr* fall ill. **~to, -a** *a* ill ● *nmf* sick person; ⟨*paziente*⟩ patient

ammali'are *vt* bewitch

am'manco *nm* deficit

ammanet'tare *vt* handcuff

ammani'cato *a* **essere ~** have connections

amma'raggio *nm* splashdown

amma'rare *vi* put down on the sea; ⟨*nave spaziale:*⟩ splash down

ammas'sar|e *vt* amass. **~si** *vr* crowd together. **am'masso** *nm* mass; ⟨*mucchio*⟩ pile

ammat'tire *vi* go mad

ammaz'zar|e *vt* kill. **~si** *vr* ⟨*suicidarsi*⟩ kill oneself; ⟨*rimanere ucciso*⟩ be killed

am'menda *nf* amends *pl*; ⟨*multa*⟩ fine; **fare ~ di qcsa** make amends for sth

am'messo *pp* *di* **ammettere** ● *conj* **~ che** supposing that

am'mettere *vt* admit; ⟨*riconoscere*⟩ acknowledge; ⟨*supporre*⟩ suppose

ammic'care *vi* wink

ammini'stra|re *vt* administer; ⟨*gestire*⟩ run. **~'tivo** *a* administrative. **~'tore, ~'trice** *nmf* administrator; ⟨*di azienda*⟩ manager; ⟨*di società*⟩ director. **~tore delegato** managing director. **~zi'one** *nf* ad-

ministration; **fatti di ordinaria ~zione** *fig* routine matters

ammi'raglio *nm* admiral. **~'ato** *nm* admiralty

ammi'ra|re *vt* admire. **~to** *a* **restare/essere ~to** be full of admiration. **~'tore, ~'trice** *nmf* admirer. **~zi'one** *nf* admiration.

ammi'revole *a* admirable

ammis'sibile *a* admissible

ammissi'one *nf* admission; ⟨*approvazione*⟩ acknowledgement

ammobili'a|re *vt* furnish. **~to** *a* furnished

am'modo *a* proper ● *adv* properly

am'mollo *nm* **in ~** soaking

ammo'niaca *nf* ammonia

ammoni'mento *nm* warning; ⟨*di rimprovero*⟩ admonishment

ammo'ni|re *vt* warn; ⟨*rimproverare*⟩ admonish. **~'tore** *a* admonishing. **~zi'one** *nf Sport* warning

ammon'tare *vi* **~ a** amount to ● *nm* amount

ammonticchi'are *vt* heap up

ammorbi'dente *nm* ⟨*per panni*⟩ softener

ammorbi'dir|e *vt*, **~si** *vr* soften

ammorta'mento *nm Comm* amortization

ammor'tare *vt* pay off ⟨*spesa*⟩; *Comm* amortize ⟨*debito*⟩

ammortiz'za|re *vt* *Comm* **~ ammortare**; *Mech* damp. **~'tore** *nm* shock-absorber

ammosci'ar|e *vt* make flabby. **~si** *vi* get flabby

ammucchi'a|re *vt*, **~rsi** *vr* pile up. **~ta** *nf* ⟨*sl: orgia*⟩ orgy

ammuf'fi|re *vi* go mouldy. **~to** *a* mouldy

ammutina'mento *nm* mutiny

ammuti'narsi *vr* mutiny

ammuto'lire *vi* be struck dumb

amne'sia *nf* amnesia

amni'stia *nf* amnesty

'amo *nm* hook; *fig* bait

amo'rale *a* amoral

a'more *nm* love; **fare l'~** make love; **per l'amor di Dio/del cielo!** for heaven's sake!; **andare d'~ e**

d'accordo get on like a house on fire; **~ proprio** self-respect; **è un ~** (*persona*) he/she is a darling; **per ~ di** for the sake of; **amori** *pl* love affairs. **~ggi'are** *vi* flirt.

amo'revole *a* loving

a'morfo *a* shapeless; (*persona*) colourless, grey

amo'roso *a* loving; (*sguardo ecc*) amorous; (*lettera, relazione*) love

ampi'ezza *nf* (*di esperienza*) breadth; (*di stanza*) spaciousness; (*di gonna*) fullness; (*importanza*) scale

'ampio *a* ample; (*esperienza*) wide; (*stanza*) spacious; (*vestito*) loose; (*gonna*) full; (*pantaloni*) baggy

am'plesso *nm* embrace

amplia'mento *nm* (*di casa, porto*) enlargement; (*di strada*) widening

ampli'are *vt* broaden (*conoscenze*)

amplifi'ca|re *vt* amplify; *fig* magnify. **~'tore** *nm* amplifier. **~zi'one** *nf* amplification

am'polla *nf* cruet

ampol'loso *a* pompous

ampu'ta|re *vt* amputate. **~zi'one** *nf* amputation

amu'leto *nm* amulet

anabba'gliante *a* *Auto* dipped ●*nmpl* **anabbaglianti** dipped headlights

anacro'nis|mo *nm* anachronism. **~tico** *a* **essere ~** be an anachronism

a'nagrafe *nf* (*ufficio*) registry office; (*registro*) register of births, marriages and deaths

ana'grafico *a* **dati** *nmpl* **anagrafici** personal data

ana'gramma *nm* anagram

anal'colico *a* non-alcoholic ●*nm* soft drink, non-alcoholic drink

a'nale *a* anal

analfa'be|ta *a* & *nmf* illiterate. **~tismo** *nm* illiteracy

anal'gesico *nm* painkiller

a'nalisi *nf inv* analysis; *Med* test. **~ grammaticale/del periodo/logica** parsing. **~ del sangue** blood test

ana'li|sta *nmf* analyst. **~tico** *a* analytical. **~z'zare** *vt* analyse; *Med* test

anal'lergico *a* hypoallergenic

analo'gia *nf* analogy. **a'nalogo** *a* analogous

'ananas *nm inv* pineapple

anar'chi|a *nf* anarchy. **a'narchico**, **-a** *a* anarchic ● *nmf* anarchist. **~smo** *nm* anarchism

A.N.A.S. *nf abbr* (**Azienda Nazionale Autonoma delle Strade**) *national road maintenance authority*

anato'mia *nf* anatomy. **ana'tomico** *a* anatomical; (*sedia*) contoured, ergonomic

'anatra *nf* duck

ana'troccolo *nm* duckling

'anca *nf* hip; (*di animale*) flank

ance'strale *a* ancestral

'anche *conj* also, too; (*persino*) even; **~ se** even if; **~ domani** tomorrow also *o* too, also tomorrow

anchilo'sato *a* *fig* stiff

an'cora *adv* still, yet; (*di nuovo*) again; (*di più*) some more; **~ una volta** once more

'ancora *nf* anchor; **gettare l'~ra** drop anchor. **~ raggio** *nm* anchorage. **~'rare** *vt* anchor

anda'mento *nm* (*del mercato, degli affari*) trend

an'dante *a* (*corrente*) current; (*di poco valore*) cheap ●*nm* *Mus* andante

an'da|re *vi* go; (*funzionare*) work; **~ via** (*partire*) leave; (*macchia:*) come out; **~ [bene]** (*confarsi*) suit; (*taglia:*) fit; **ti va bene alle tre?** does three o'clock suit you?; **non mi va di mangiare** I don't feel like eating; **~ di fretta** be in a hurry; **~ fiero di** be proud of; **~ di moda** be in fashion; **va per i 20 anni** he's nearly 20; **va va'** [là]! come on!; **come va?** how are things?; **~ a male** go off; **~ a fuoco** go up in flames; **va spedito [entro] stamattina** it must be sent this morning; **ne va del mio lavoro** my job is at stake; **come è andata**

a finire? how did it turn out?;
cosa vai dicendo? what are you
talking about?. **~rsene** go away;
⟨*morire*⟩ pass away ● *nm* going; **a
lungo ~re** eventually

'andito *nm* passage

an'drone *nm* entrance

a'neddoto *nm* anecdote

ane'lare *vt* ~ **a** long for. **a'nelito**
nm longing

a'nello *nm* ring; ⟨*di catena*⟩ link

ane'mia *nf* anaemia. **a'nemico** *a*
anaemic

a'nemone *nm* anemone

aneste'si|a *nf* anaesthesia; ⟨*so-
stanza*⟩ anaesthetic. **~'sta** *nmf*
anaesthetist. **ane'stetico** *a & nm*
anaesthetic

an'fibi *nmpl* ⟨*stivali*⟩ army boots

an'fibio *nm* ⟨*animale*⟩ amphibian
● *a* amphibious

anfite'atro *nm* amphitheatre

'anfora *nf* amphora

an'fratto *nm* ravine

an'gelico *a* angelic

an'gelo *nm* angel. **~ custode**
guardian angel

angli'c|ano *a* Anglican. **~ismo**
nm Anglicism

an'glofilo, -a *a & nmf* Anglophile

an'glofono, -a *nmf* English-
speaker

anglo'sassone *a & nmf* Anglo-
Saxon

ango'la|re *a* angular. **~zi'one** *nf*
angle shot

'angolo *nm* corner; *Math* angle. **~
[di] cottura** kitchenette

ango'loso *a* angular

an'gosci|a *nf* anguish. **~'are** *vt*
torment. **~'ato** *a* agonized. **~'oso**
a ⟨*disperato*⟩ anguished; ⟨*che dà
angoscia*⟩ distressing

angu'illa *nf* eel

an'guria *nf* water-melon

an'gusti|a *nf* ⟨*ansia*⟩ anxiety; ⟨*pe-
nuria*⟩ poverty. **~'are** *vt* distress.
~'arsi *vr* be very worried ⟨*per*
about⟩

an'gusto *a* narrow

'anice *nm* anise; *Culin* aniseed;
⟨*liquore*⟩ anisette

ani'dride *nf* ~ **carbonica** carbon
dioxide

'anima *nf* soul; **non c'era ~ viva**
there was not a soul about; **all'~!**
good grief!; **un'~ in pena** a soul in
torment. **~ gemella** soul mate

ani'ma|le *a & nm* animal; **~li
domestici** *pl* pets. **~'lesco** *a* animal

ani'ma|re *vt* give life to; ⟨*ravviva-
re*⟩ enliven; ⟨*incoraggiare*⟩ encourage. **~rsi** *vr* come to life; ⟨*acca-
lorarsi*⟩ become animated. **~to** *a*
⟨*animale*⟩ animated; ⟨*discussione*⟩ animated;
⟨*paese*⟩ lively. **~'tore**, **~'trice** *nmf*
leading spirit; *Cinema* animator.
~zi'one *nf* animation

'animo *nm* ⟨*mente*⟩ mind; ⟨*indole*⟩
disposition; ⟨*cuore*⟩ heart; **perdersi d'~** lose heart; **farsi ~** take
heart. **~sità** *nf* animosity

ani'moso *a* brave; ⟨*ostile*⟩ hostile

'anitra *nf* = **anatra**

annac'qua|re *vt anche fig* water
down. **~to** *a* watered down

annaffi'a|re *vt* water. **~'toio** *nm*
watering-can

an'nali *nmpl* annals

anna'spare *vi* flounder

an'nata *nf* year; ⟨*importo annuale*⟩
annual amount; ⟨*di vino*⟩ vintage

annebbia'mento *nm* fog build-up;
fig clouding

annebbi'ar|e *vt* cloud ⟨*vista, men-
te*⟩. **~si** *vr* become foggy; ⟨*vista,
mente*⟩ grow dim

annega'mento *nm* drowning

anne'ga|re *vt/i* drown

anne'rir|e *vt/i* blacken. **~si** *vr* be-
come black

annessi'one *nf* ⟨*di nazione*⟩ an-
nexation

an'nesso *pp di* **annettere** ● *a* at-
tached; ⟨*stato*⟩ annexed

an'nettere *vt* add; ⟨*accludere*⟩ enclose; annex ⟨*stato*⟩

annichi'lire *vt* annihilate

anni'darsi *vr* nest

annienta'mento *nm* annihilation

annien'tar|e *vt* annihilate. **~si** *vr* abase oneself

anniver'sario *a & nm* anniversary. **~ di matrimonio** wedding anniversary

'**anno** *nm* year; **Buon A~!** Happy New Year!; **quanti anni ha?** how old are you?; **Tommaso ha dieci anni** Thomas is ten [years old]. **~ bisestile** leap year

anno'dare *vt* knot; do up (*cintura*); *fig* form. **~si** *vr* become knotted

annoi'a|re *vt* bore; (*recare fastidio*) annoy. **~rsi** *vr* get bored; (*condizione*) be bored. **~to** *a* bored

anno'ta|re *vt* note down; annotate (*testo*). **~zi'one** *nf* note

annove'rare *vt* number

annu'a|le *a* annual, yearly. **~rio** *nm* year-book

annu'ire *vi* nod; (*acconsentire*) agree

annulla'mento *nm* annulment; (*di appuntamento*) cancellation

annul'lar|e *vt* annul; cancel (*appuntamento*); (*togliere efficacia a*) undo; disallow (*gol*); (*distruggere*) destroy. **~si** *vr* cancel each other out

annunci'a|re *vt* announce; (*preannunciare*) foretell. **~tore**, **~trice** *nmf* announcer. **~zi'one** *nf* Annunciation

an'nuncio *nm* announcement; (*pubblicitario*) advertisement; (*notizia*) news. **annunci** *pl* **economici** classified advertisements

'**annuo** *a* annual, yearly

annu'sare *vt* sniff

annuvo'lar|e *vt* cloud. **~si** *vr* cloud over

'**ano** *nm* anus

anoma'lia *nf* anomaly

a'nomalo *a* anomalous

anoni'mato *nm* **mantenere l'~** remain anonymous

a'nonimo, -a *a* anonymous ● *nm* (*pittore, scrittore*) anonymous painter/writer

anores'sia *nf* Med anorexia

ano'ressico, -a *a nmf* anorexic

anor'mal|e *a* abnormal ● *nmf* deviant, abnormal person. **~ità** *nf inv* abnormality

'**ansa** *nf* handle; (*di fiume*) bend

an'sare *vi* pant

'**ansia, ansi'età** *nf* anxiety; **stare/essere in ~ per** be anxious about

ansi'oso *a* anxious

antago'nis|mo *nm* antagonism. **~ta** *nmf* antagonist

an'tartico *a & nm* Antarctic

antece'dente *a* preceding ● *nm* precedent

ante'fatto *nm* prior event

ante'guerra *a* pre-war ● *nm* pre-war period

ante'nato, -a *nmf* ancestor

an'tenna *nf* Radio, TV aerial; (*di animale*) antenna; Naut yard. **~ parabolica** satellite dish

ante'porre *vt* put before

ante'prima *nf* preview; **vedere qcsa in ~** have a sneak preview of sth

anteri'ore *a* front *attrib*; (*nel tempo*) previous

antiade'rente *a* (*padella*) nonstick

antia'ereo *a* anti-aircraft *attrib*

antial'lergico *a* hypoallergenic

anti'atomico *a* **rifugio ~** fallout shelter

antibi'otico *a & nm* antibiotic

anti'caglia *nf* (*oggetto*) piece of old junk

antica'mente *adv* in ancient times, long ago

anti'camera *nf* ante-room; **far ~** be kept waiting

antichità *nf inv* antiquity; (*oggetto*) antique

antici'clone *nm* anticyclone

antici'pa|re *vt* advance; Comm pay in advance; (*prevedere*) anticipate; (*prevenire*) forestall ● *vi* be early. **~ta'mente** *adv* in advance. **~zi'one** *nf* anticipation; (*notizia*) advance news

an'ticipo *nm* advance; (*caparra*)

deposit; **in ~** early; *(nel lavoro)* ahead of schedule

an'tico *a* ancient; *(mobile ecc)* antique; *(vecchio)* old; **all'antica** old-fashioned ● *nmpl* **gli antichi** the ancients

anticoncezio'nale *a & nm* contraceptive

anticonfor'mis|mo *nm* unconventionality. **~ta** *nmf* nonconformist. **~tico** *a* unconventional, nonconformist

anticonge'lante *a & nm* antifreeze

anti'corpo *nm* antibody

anticostituzio'nale *a* unconstitutional

anti'crimine *a inv* *(squadra)* crime *attrib*

antidemo'cratico *a* undemocratic

antidolo'rifico *nm* painkiller

an'tidoto *nm* antidote

anti'droga *a inv* *(campagna)* antidrugs; *(squadra)* drug *attrib*

antie'stetico *a* ugly

antifa'scismo *nm* anti-fascism

antifa'scista *a & nmf* anti-fascist

anti'forfora *a inv* dandruff *attrib*

anti'furto *nm* anti-theft device; *(allarme)* alarm ● *a inv* *(sistema)* anti-theft

anti'gelo *nm* antifreeze; *(para-brezza)* defroster

antigi'enico *a* unhygienic

An'tille *nfpl* **le ~** the West Indies

an'tilope *nf* antelope

antin'cendio *a inv* **allarme ~** fire alarm; **porta ~** fire door

anti'nebbia *nm inv* Auto [faro] ~ foglamp, foglight

antinfiamma'torio *a & nm* anti-inflammatory

antinucle'are *a* anti-nuclear

antio'rario *a* anti-clockwise

anti'pasto *nm* hors d'oeuvre, starter

antipa'tia *nf* antipathy. **anti'patico** *a* unpleasant

an'tipodi *nmpl* antipodes; **essere agli ~** *fig* be poles apart

antiquari'ato *nm* antique trade

anti'quario, -a *nmf* antique dealer

anti'quato *a* antiquated

anti'ruggine *nm inv* rust-inhibitor

anti'rughe *a inv* anti-wrinkle *attrib*

anti'scippo *a inv* theft-proof

antise'mita *a* anti-Semitic

anti'settico *a & nm* antiseptic

antisoci'ale *a* anti-social

antista'minico *nm* antihistamine

anti'stante *a prep* in front of

anti'tarlo *nm inv* woodworm treatment

antiterro'ristico *a* antiterrorist *attrib*

an'titesi *nf inv* antithesis

antolo'gia *nf* anthology

'antro *nm* cavern

antropolo'gia *nf* anthropology. **antro'pologo, -a** *nmf* anthropologist

anu'lare *nm* ring-finger

'anzi *conj* in fact; *(o meglio)* or better still; *(al contrario)* on the contrary

anzianità *nf* old age; *(di servizio)* seniority

anzi'ano, -a *a* old, elderly; *(di grado ecc)* senior ● *nmf* elderly person

anziché *conj* rather than

anzi'tempo *adv* prematurely

anzi'tutto *adv* first of all

a'orta *nf* aorta

apa'tia *nf* apathy. **a'patico** *a* apathetic

'ape *nf* bee; **nido** *nm* **di api** honeycomb

aperi'tivo *nm* aperitif

aperta'mente *adv* openly

a'perto *a* open; **all'aria aperta** in the open air; **all'~** open-air

aper'tura *nf* opening; *(inizio)* beginning; *(ampiezza)* spread; *(di arco)* span; *Pol* overtures *pl*; *Phot* aperture; **~ mentale** openness

'apice *nm* apex

apicol'tura *nf* beekeeping

ap'nea *nf* immersione in ~ free
diving

a'polide *a* stateless ●*nmf* stateless person

a'postolo *nm* apostle

apostro'fare *vt* (mettere un apostrofo a) write with an apostrophe; reprimand 〈persona〉

a'postrofo *nm* apostrophe

appaga'mento *nm* fulfilment

appa'ga|re *vt* satisfy. ~**rsi** *vr* ~**rsi**
di be satisfied with

appai'are *vt* pair; mate 〈animali〉

appallot'tolare *vt* roll into a ball

appalta'tore *nm* contractor

ap'palto *nm* contract; **dare in ~**
contract out

appan'naggio *nm* (in denaro) annuity; *fig* prerogative

appan'nar|e *vt* mist 〈vetro〉; dim
〈vista〉. ~**si** *vr* mist over; 〈vista:〉
grow dim

appa'rato *nm* apparatus; (pompa)
display

apparecchi'a|re *vt* prepare ●*vi*
lay the table. ~**'tura** *nf* (impianti)
equipment

appa'recchio *nm* apparatus;
(congegno) device; (radio, tv ecc)
set; (aeroplano) aircraft. ~
acustico hearing aid

appa'ren|te *a* apparent. ~**te-
'mente** *adv* apparently. ~**za** *nf* appearance; **in ~za** apparently

appa'ri|re *vi* appear; (sembrare)
look. ~**'scente** *a* a striking; *pej*
gaudy. ~**zi'one** *nf* apparition

apparta'mento *nm* flat, apartment *Am*

appar'ta|rsi *vr* withdraw. ~**to** *a* se-
cluded

apparte'nenza *nf* membership

apparte'nere *vi* belong

appassio'nante *a* 〈storia, argo-
mento〉 exciting

appassio'na|re *vt* excite; (com-
muovere) move. ~**rsi** *vr* ~**rsi** a
become excited by. ~**to** *a* passionate; ~**to di** (entusiastico) fond
of

appas'sir|e *vi* wither. ~**si** *vr* fade

appel'larsi *vr* ~ a appeal to

ap'pello *nm* appeal; (chiamata per
nome) rollcall; (esami) exam session; **fare l'~** call the roll

ap'pena *adv* just; (a fatica) hardly
●*conj* [non] ~ as soon as, no
sooner... than

ap'pendere *vt* hang [up]

appendi'abiti *nm inv* hat-stand,
hallstand

appen'dice *nf* appendix. **appendi-
'cite** *nf* appendicitis

Appen'nini *nmpl* **gli ~** the Apen-
nines

appesan'tir|e *vt* weigh down. ~**si**
vr become heavy

ap'peso *pp di* **appendere** ●*a*
hanging; (impiccato) hanged

appe'ti|to *nm* appetite; **aver ~to**
be hungry; **buon ~to!** enjoy your
meal!. ~**'toso** *a* appetizing; *fig*
tempting

appezza'mento *nm* plot of land

appia'nar|e *vt* level; *fig* smooth
over. ~**si** *vr* improve

appiat'tir|e *vt* flatten. ~**si** *vr* flat-
ten oneself

appic'care *vt* ~ **il fuoco** a set fire
to

appicci'car|e *vt* stick; ~**e a** (fig:
appioppare) palm off on ●*vi* be
sticky. ~**si** *vr* stick; (cose:) stick together; ~**si a qcno** *fig* stick to sb
like glue

appicci'caticcio *a* sticky; *fig* clingy

appicci'coso *a* sticky; *fig* clingy

appie'dato *a* sono ~ I don't have
the car; **sono rimasto ~** I was
stranded

appi'eno *adv* fully

appigli'arsi *vr* ~ **a** get hold of; *fig*
stick to. **ap'piglio** *nm* fingerhold;
(per piedi) foothold; *fig* pretext

appiop'pare *vt* ~ **a** palm off on;
(fam: dare) give

appiso'larsi *vr* doze off

applau'dire *vt/i* applaud. **ap'plau-
so** *nm* applause

appli'cabile *a* applicable

appli'ca|re *vt* apply; enforce 〈legge
ecc〉. ~**rsi** *vr* apply oneself. ~**'tore**

nm applicator. **~zi'one** *nf* application; (*di legge*) enforcement

appoggi'ar|e *vt* lean (**a** against); (*mettere*) put; (*sostegno*) back. **~si** *vr* **~si a** lean against; *fig* rely on.

ap'poggio *nm* support

appollai'arsi *vr fig* perch

ap'porre *vt* affix

appor'tare *vt* bring; (*causare*) cause. **ap'porto** *nm* contribution

apposita'mente *adv* especially

ap'posito *a* proper

ap'posta *adv* on purpose; (*espressamente*) specially

apposta'mento *nm* ambush; (*caccia*) lying in wait

appo'star|e *vt* post (*soldati*). **~si** *vr* lie in wait

ap'prend|ere *vt* understand; (*imparare*) learn. **~i'mento** *nm* learning

appren'di|sta *nmf* apprentice. **~'stato** *nm* apprenticeship

apprensi'one *nf* apprehension; **essere in ~ per** be anxious about. **appren'sivo** *a* apprehensive

ap'presso *adv* & *prep* (*vicino*) near; (*dietro*) behind; **come ~** as follows

appre'star|e *vt* prepare. **~si** *vr* get ready

apprez'za|bile *a* appreciable. **~'mento** *nm* appreciation; (*giudizio*) opinion

apprez'za|re *vt* appreciate. **~to** *a* appreciated

ap'proccio *nm* approach

appro'dare *vi* land; **~ a** *fig* come to; **non ~ a nulla** come to nothing. **ap'prodo** *nm* landing; (*luogo*) landing-stage

approfit'ta|re *vi* take advantage (**di** of), profit (**di** by). **~'tore, ~'trice** *nmf* chancer

approfondi'mento *nm* deepening; **di ~** (*fig: esame*) further

approfon'di|re *vt* deepen. **~rsi** *vr* (*divario*) widen. **~to** *a* (*studio, ricerca*) in-depth

appropri'a|rsi *vr* (*essere adatto a*) suit; **~rsi di** take possession of.

~to *a* appropriate. **~zi'one** *nf Jur* appropriation. **~zione indebita** *Jur* embezzlement

approssi'ma|re *vt* **~re per eccesso/difetto** round up/down. **~rsi** *vr* draw near. **~tiva'mente** *adv* approximately. **~'tivo** *a* approximate. **~zi'one** *nf* approximation

appro'va|re *vt* approve of; approve (*legge*). **~zi'one** *nf* approval

approvvigiona'mento *nm* supplying; **approvvigionamenti** *pl* provisions

approvvigio'nar|e *vt* supply. **~si** *vr* stock up

appunta'mento *nm* appointment, date *fam*; **fissare un ~** make an appointment; **darsi ~** decide to meet

appun'tar|e *vt* (*annotare*) take notes; (*fissare*) fix; (*con spillo*) pin; (*appuntire*) sharpen. **~si** *vr* **~si su** (*teoria:*) to be based on

appun'ti|re *vt* sharpen. **~to** *a* (*mento*) pointed

ap'punto¹ *nm* note; (*piccola critica*) niggle

ap'punto² *adv* exactly; **per l'~!** exactly!; **stavo ~ dicendo...** I was just saying...

appu'rare *vt* verify

a'pribile *a* that can be opened

apribot'tiglie *nm inv* bottle-opener

a'prile *nm* April; **il primo d'~** April Fools' Day

a'prir|e *vt* open; turn on (*luce, acqua ecc*); (*con chiave*) unlock; open up (*ferita ecc*). **~si** *vr* open; (*spaccarsi*) split; (*confidarsi*) confide (**con** in)

apri'scatole *nf inv* tin-opener

aqua'planing *nm* **andare in ~** aquaplane

'aquil|a *nf* eagle; **non è un'~a!** he is no genius!. **~'lino** *a* aquiline

aqui'lone *nm* (*giocattolo*) kite

ara'besco *nm* arabesque; *hum* scribble

A'rabia Sau'dita *nf* l'**~** Saudi Arabia

'**arabo, -a** *a* Arab; ⟨*lingua*⟩ Arabic ● *nmf* Arab ● *nm* ⟨*lingua*⟩ Arabic

a'**rachide** *nf* peanut

ara'**gosta** *nf* lobster

a'**rancia** *nf* orange. ~'**ata** *nf* orangeade. ~**o** *nm* orange-tree; ⟨*colore*⟩ orange. ~'**one** *a & nm* orange

a'**ra**|**re** *vt* plough. ~'**tro** *nm* plough

ara'**tura** *nf* ploughing

a'**razzo** *nm* tapestry

arbi'**trar**|**e** *vt* arbitrate in; *Sport* referee. ~**ietà** *nf* arbitrariness. ~**io** *a* arbitrary

ar'**bitrio** *nm* will; **è un** ~ it's very high-handed

'**arbitro** *nm* arbiter; *Sport* referee; ⟨*nel baseball*⟩ umpire

ar'**busto** *nm* shrub

'**arca** *nf* ark; ⟨*cassa*⟩ chest

ar'**ca**|**ico** *a* archaic. ~'**ismo** *nm* archaism

ar'**cangelo** *nm* archangel

ar'**cata** *nf* arch; ⟨*serie di archi*⟩ arcade

arche|olo'**gia** *nf* archaeology. ~o'**logico** *a* archaeological. ~'**ologo, -a** *nmf* archaeologist

ar'**chetto** *nm* *Mus* bow

archi'**tet**'**tare** *vt* fig devise; **cosa state architettando?** *fig* what are you plotting?

archi'**tet**|**to** *nm* architect. ~'**tonico** *a* architectural. ~'**tura** *nf* architecture

archivi'**are** *vt* file; *Jur* close

ar'**chivio** *nm* archives *pl*; *Comput* file

archi'**vista** *nmf* filing clerk

arci'**cigno** *a* grim

arci'**pelago** *nm* archipelago

arci'**vescovo** *nm* archbishop

'**arco** *nm* arch; *Math* arc; ⟨*arma, Mus*⟩ bow; **nell'~ di una giornata/due mesi** in the space of a day/ two months

arcoba'**leno** *nm* rainbow

arcu'**a**|**re** *vt* bend. ~**rsi** *vr* bend. ~**to** *a* bent, curved; ⟨*schiena di gatto*⟩ arched

ar'**dente** *a* burning; *fig* ardent. ~'**mente** *adv* ardently

'**ardere** *vt/i* burn

ar'**desia** *nf* slate

ar'**di**|**re** *vi* dare. ~**to** *a* daring; ⟨*coraggioso*⟩ bold; ⟨*sfacciato*⟩ impudent

ar'**dore** *nm* ⟨*calore*⟩ heat; *fig* ardour

'**arduo** *a* arduous; ⟨*ripido*⟩ steep

'**area** *nf* area. ~ **di rigore** ⟨*in calcio*⟩ penalty area. ~ **di servizio** service area

a'**rena** *nf* arena

are'**narsi** *vr* run aground; ⟨*fig: trattative*⟩ reach deadlock; **mi sono arenato** I'm stuck

'**argano** *nm* winch

argen'**tato** *a* silver-plated

argente'**ria** *nf* silver[ware]

ar'**gento** *nm* silver

ar'**gil**|**la** *nf* clay. ~'**loso** *a* ⟨*terreno*⟩ clayey

argi'**nare** *vt* embank; *fig* hold in check, contain

'**argine** *nm* embankment; ⟨*diga*⟩ dike

argomen'**tare** *vi* argue

argo'**mento** *nm* argument; ⟨*motivo*⟩ reason; ⟨*soggetto*⟩ subject

argu'**ire** *vt* deduce

ar'**gu**|**to** *a* witty. ~**zia** *nf* wit; ⟨*battuta*⟩ witticism

'**aria** *nf* air; ⟨*aspetto*⟩ appearance; *Mus* tune; **andare all'**~ *fig* come to nothing; **avere l'**~... look...; **corrente d'**~ draught; **mandare all'**~ **qcsa** *fig* ruin sth

ari'**dità** *nf* aridity, dryness

'**arido** *a* arid

arieggi'**a**|**re** *vt* air. ~**to** *a* airy

a'**riete** *nm* ram. **A**~ *Astr* Aries

ari'**etta** *nf* ⟨*brezza*⟩ breeze

a'**ringa** *nf* herring

ari'**oso** *a* ⟨*locale*⟩ light and airy

aristo'**cra**|**tico, -a** *a* aristocratic ● *nmf* aristocrat. ~'**zia** *nf* aristocracy

arit'**metica** *nf* arithmetic

arlec'**chino** *nm* Harlequin; *fig* buffoon

'**arma** *nf* weapon; **armi** *pl* arms; ⟨*forze armate*⟩ [armed] forces;

chiamare alle armi call up; **sotto le armi** in the army; **alle prime armi** *fig* inexperienced, fledg[e]-ling. **~ da fuoco** firearm. **~ impropria** makeshift weapon. **~ a doppio taglio** *fig* double-edged sword

armadi'etto *nm* locker, cupboard

ar'madio *nm* cupboard; *(guardaroba)* wardrobe

armamen'tario *nm* tools *pl*; *fig* paraphernalia

arma'mento *nm* armament; *Naut* fitting out

ar'ma|re *vt* arm; *(equipaggiare)* fit out; *Archit* reinforce. **~rsi** *vr* arm oneself **(di** with). **~ta** *nf* army; *(flotta)* fleet. **~'tore** *nm* shipowner. **~'tura** *nf* framework; *(impalcatura)* scaffolding; *(di guerriero)* armour

armeggi'are *vi fig* manoeuvre

armi'stizio *nm* armistice

armo'ni|a *nf* harmony. **ar'monica** *nf* ~ **[a bocca]** mouth organ. **ar'monico** *a* harmonic. **~'oso** *a* harmonious

armoniz'zar|e *vt* harmonize ● *vi* match. **~si** *vr (colori:)* go together, match

ar'nese *nm* tool; *(oggetto)* thing; *(congegno)* gadget; **male in ~ in** bad condition

'arnia *nf* beehive

a'roma *nm* aroma; **aromi** *pl* herbs. **~tera'pia** *nf* aromatherapy

aro'matico *a* aromatic

aromatiz'zare *vt* flavour

'arpa *nf* harp

ar'peggio *nm* arpeggio

ar'pia *nf* harpy

arpi'one *nm* hook; *(pesca)* harpoon

arrabat'tarsi *vr* do all one can

arrabbi'ar|si *vr* get angry. **~to a** angry. **~tura** *nf* rage; **prendersi un'~tura** fly into a rage

arraf'fare *vt* grab

arrampi'ca|rsi *vr* climb [up]. **~ta** *nf* climb. **~'tore**, **~'trice** *nmf*

climber. **~'tore sociale** social climber

arran'care *vi* limp, hobble; *fig* struggle, limp along

arrangia'mento *nm* arrangement

arrangi'ar|e *vt* arrange. **~si** *vr* manage; **~si alla meglio** get by; **ar'rangiati!** get on with it!

arra'parsi *vr fam* get randy

arre'care *vt* bring; *(causare)* cause

arreda'mento *nm* interior decoration; *(l'arredare)* furnishing; *(mobili ecc)* furnishings *pl*

arre'da|re *vt* furnish. **~'tore**, **~'trice** *nmf* interior designer.

ar'redo *nm* furnishings *pl*

ar'rendersi *vr* surrender

arren'devo|le *a (persona)* yielding. **~'lezza** *nf* softness

arre'star|e *vt* arrest; *(fermare)* stop. **~si** *vr* halt. **ar'resto** *nm* stop; *Med, Jur* arrest; **la dichiaro in [stato d']arresto** you are under arrest; **mandato di arresto** warrant. **arresti** *pl* **domiciliari** *Jur* house arrest

arre'tra|re *vt/i* withdraw; pull back *(giocatore)*. **~to a** *(paese ecc)* backward; *(Mil: posizione)* rear; **numero ~to** *(di rivista)* back number; **del lavoro ~to** a backlog of work ● *nm (di stipendio)* back pay

arre'trati *nmpl* arrears

arricchi'mento *nm* enrichment

arric'chi|re *vt* enrich. **~rsi** *vr* get rich. **~to, -a** *nmf* nouveau riche

arricci'are *vt* curl; **~ il naso** turn up one's nose

ar'ringa *nf* harangue; *Jur* closing address

arrischi'a|rsi *vr* dare. **~to a** risky; *(imprudente)* rash

arri'va|re *vi* arrive; **~re a** *(raggiungere)* reach; *(ridursi)* be reduced to. **~to, -a** *a* successful; **ben ~to!** welcome! **~ nmf** successful person

arrive'derci *int* goodbye; **~ a domani** see you tomorrow

arri'vis|mo *nm* social climbing;

(nel lavoro) careerism. **~ta** *nmf* social climber; *(nel lavoro)* careerist

ar'rivo *nm* arrival; *Sport* finish

arro'gan|te *a* arrogant. **~za** *nf* arrogance

arro'garsi *vr* ~ **il diritto di fare** qcsa take it upon oneself to do sth

arrossa'mento *nm* reddening

arros'sar|e *vt* make red, redden *(occhi)*. **~si** *vr* go red

arros'sire *vi* blush, go red

arro'stire *vt* roast; toast *(pane)*; *(ai ferri)* grill. **ar'rosto** *a & nm* roast

arroto'lare *vt* roll up

arroton'dar|e *vt* make round; *Math ecc* round off. **~si** *vr* become round; *(persona:)* get plump

arrovel'larsi *vr* ~ **il cervello** rack one's brains

arroven'ta|re *vt* make red-hot. **~rsi** *vr* become red-hot. **~to** *a* red-hot

arruf'fa|re *vt* ruffle; *fig* confuse. **~to** *a (capelli)* ruffled

arruffianarsi *vr* ~ qcno *fig* butter sb up

arruggi'ni|re *vt* rust. **~rsi** *vr* go rusty; *fig (fisicamente)* stiffen up; *(conoscenze:)* go rusty. **~to** *a* rusty

arruola'mento *nm* enlistment

arruo'lar|e *vt/i*. **~si** *vr* enlist

arse'nale *nm* arsenal; *(cantiere)* [naval] dockyard

ar'senico *nm* arsenic

'arso *pp di* ardere ● *a* burnt; *(arido)* dry. **ar'sura** *nf* burning heat; *(sete)* parching thirst

'arte *nf* art; *(abilità)* craftsmanship; **le belle arti** the fine arts. **arti figurative** figurative arts

arte'fa|re *vt* adulterate *(vino)*; disguise *(voce)*. **~tto** *a* fake; *(vino)* adulterated

ar'tefice *nmf* craftsman; craftswoman; *fig* author

ar'teria *nf* artery. **~ [stradale]** arterial road

arterioscle'rosi *nf* arteriosclerosis, hardening of the arteries

'artico *a & nm* Arctic

artico'la|re *a* articular ● *vt* articulate; *(suddividere)* divide. **~rsi** *vr fig* **~rsi in** consist of. **~to** *a* Auto articulated; *fig* well-constructed. **~zi'one** *nf* Anat articulation

ar'ticolo *nm* article. **~ di fondo** leader

artifici'ale *a* artificial

arti'fic|io *nm* artifice; *(affettazione)* affectation. **~'oso** *a* artful; *(affettato)* affected

artigia'nale *a* made by hand; *hum* amateurish. **~'mente** *adv* with craftsmanship; *hum* amateurishly

artigia'nato *nm* craftsmanship; *(ceto)* craftsmen *pl.* **~'no, -a** *nm* craftsman ● *nm* craftswoman

artigli'ere *nm* artilleryman. **~e'ria** *nf* artillery

ar'tiglio *nm* claw; *fig* clutch

ar'tist|a *nmf* artist. **~ica'mente** *adv* artistically. **~ico** *a* artistic

'arto *nm* limb

ar'trite *nf* arthritis

ar'trosi *nf* rheumatism

arzigogo'lato *a* fantastic, bizarre

ar'zillo *a* sprightly

a'scella *nf* armpit

ascen'den|te *a* ascending ● *nm (antenato)* ancestor; *(influenza)* ascendancy; *Astr* ascendant

ascensi'one *nf* ascent; **l'A~** the Ascension

ascen'sore *nm* lift, elevator *Am*

a'scesa *nf* ascent; *(al trono)* accession; *(al potere)* rise

a'scesso *nm* abscess

a'sceta *nmf* ascetic

'ascia *nf* axe

asciugabianche'ria *nm inv (stenditoio)* clothes horse

asciuga'pelli *nm inv* hair dryer, hairdryer

asciuga'mano *nm* towel

asciu'gar|e *vt* dry. **~si** *vr* dry oneself; *(diventare asciutto)* dry up

asci'utto *a* dry; *(magro)* wiry; *(risposta)* curt; **essere all'~** *fig* be hard up

ascol'ta|re *vt* listen to ● *vi* listen. ~'tore, ~'trice *nmf* listener

a'scolto *nm* listening; **dare ~ a** listen to; **mettersi in ~ Radio** tune in

asfal'tare *vt* asphalt

a'sfalto *nm* asphalt

asfis'si|a *nf* asphyxia. ~'ante *a* ⟨caldo⟩ oppressive; ⟨fig: persona⟩ annoying. ~'are *vt* asphyxiate; *fig* annoy

'Asia *nf* Asia. **asi'atico, -a** *a* & *nmf* Asian

a'silo *nm* shelter; ⟨d'infanzia⟩ nursery school. **~ nido** day nursery. **~ politico** political asylum

asim'metrico *a* asymmetrical

'asino *nm* donkey; ⟨fig: persona stupida⟩ ass

'asma *nf* asthma. **a'smatico** *a* asthmatic

asoci'ale *a* asocial

'asola *nf* buttonhole

a'sparagi *nmpl* asparagus *sg*

a'sparago *nm* asparagus spear

asperità *nf inv* harshness; ⟨di terreno⟩ roughness

aspet'ta|re *vt* wait for; ⟨prevedere⟩ expect; **~re un bambino be** expecting [a baby]; **fare ~re qcno** keep sb waiting ● *vi* wait. **~rsi** *vr* expect. **~tiva** *nf* expectation

a'spetto¹ *nm* appearance; ⟨di problema⟩ aspect; **di bell'~** good-looking

a'spetto² *nm* **sala** *nf* **d'~** waiting room

aspi'rante *a* aspiring; ⟨pompa⟩ suction *attrib* ● *nmf* ⟨a un posto⟩ applicant; ⟨al trono⟩ aspirant; **gli aspiranti al titolo** the contenders for the title

aspira'polvere *nm inv* vacuum cleaner

aspi'ra|re *vt* inhale; *Mech* suck in ● *vi* **~re a** aspire to. **~tore** *nm* extractor fan. **~zi'one** *nf* inhalation; *Mech* suction; ⟨ambizione⟩ ambition

aspi'rina *nf* aspirin

aspor'tare *vt* take away

aspra'mente *adv* ⟨duramente⟩ severely

a'sprezza *nf* ⟨al gusto⟩ sourness; ⟨di clima⟩ severity; ⟨di suono⟩ harshness; ⟨di odore⟩ pungency

'aspro *a* ⟨al gusto⟩ sour; ⟨clima⟩ severe; ⟨suono, parole⟩ harsh; ⟨odore⟩ pungent; ⟨litigio⟩ bitter

assag|gi'are *vt* taste. **~'gini** *nmpl* *Culin* samples. **as'saggio** *nm* tasting; ⟨piccola quantità⟩ taste

as'sai *adv* very; ⟨moltissimo⟩ very much; ⟨abbastanza⟩ enough

assa'li|re *vt* attack. **~tore, ~'trice** *nmf* assailant

as'salto *nm* attack; **prendere d'~** storm ⟨città⟩; *fig* mob ⟨persona⟩; hold up ⟨banca⟩

assapo'rare *vt* savour

assassi'nare *vt* murder, assassinate; *fig* murder

assas'sin|io *nm* murder, assassination. **~o, -a** *a* murderous ● *nm* murderer ● *nf* murderess

'asse *nf* board ● *nm* *Techn* axle; *Math* axis. **~ da stiro** ironing board

asse con'dare *vt* satisfy; ⟨favorire⟩ support

assedi'are *vt* besiege. **as'sedio** *nm* siege

asse gna'mento *nm* allotment; **fare ~ su** rely on

asse'gna|re *vt* allot; award ⟨premio⟩. **~'tario** *nmf* recipient. **~zi'one** *nf* ⟨di alloggio, borsa di studio⟩ allocation; ⟨di premio⟩ award

as'segno *nm* allowance; ⟨bancario⟩ cheque; **contro ~** cash on delivery. **~ circolare** bank draft. **assegni** *pl* familiari family allowance. **~ non trasferibile** cheque made out to 'account payee only'

assem'blea *nf* assembly; ⟨adunanza⟩ gathering

assembra'mento *nm* gathering

assen'nato *a* sensible

as'senso *nm* assent

assen'tarsi *vr* go away; *(da stanza)* leave the room

as'sen|te *a* absent; *(distratto)* absent-minded ● *nmf* absentee. **~te'ismo** *nm* absenteeism. **~te'ista** *nmf* frequent absentee. **~za** *nf* absence; *(mancanza)* lack

asse'r|ire *vt* assert. **~'tivo** *a* assertive. **~zi'one** *nf* assertion

assesso'rato *nm* department

asses'sore *nm* councillor

assesta'mento *nm* settlement

asse'star|e *vt* arrange; **~e un colpo** deal a blow. **~si** *vr* settle oneself

asse'tato *a* parched

as'setto *nm* order; *Naut, Aeron* trim

assicu'ra|re *vt* assure; *Comm* insure; register *(posta)*; *(fissare)* secure; *(accertare)* ensure. **~rsi** *vr* *(con contratto)* insure oneself; *(legarsi)* fasten oneself; **~rsi che** make sure that. **~'tivo** *a* insurance *attrib.* **~'tore, ~'trice** *nmf* insurance agent ● *a* insurance *attrib.* **~zi'one** *nf* assurance; *(contratto)* insurance

assidera'mento *nm* exposure. **asside'rato** *a Med* suffering from exposure; *fam* frozen

assidu'a|mente *adv* assiduously. **~ità** *nf* assiduity

as'siduo *a* assiduous; *(cliente)* regular

assil'lante *a (persona, pensiero)* nagging

assil'lare *vt* pester

as'sillo *nm* worry

assimi'la|re *vt* assimilate. **~zi'one** *nf* assimilation

as'sise *nfpl* assizes; **Corte d'A~** Court of Assize[s]

assi'sten|te *nmf* assistant. **~te sociale** social worker. **~te di volo** flight attendant. **~za** *nf* assistance; *(presenza)* presence. **~za sociale** social work

assistenzi'a|le *a* welfare *attrib.* **~'lismo** *nm* welfare

as'sistere *vt* assist; *(curare)* nurse

● *vi* **~ a** *(essere presente)* be present at; watch *(spettacolo ecc)*

'asso *nm* ace; **piantare in ~** leave in the lurch

associ'a|re *vt* join; *(collegare)* associate. **~rsi** *vr* join forces; **~rsi a** join; subscribe to *(giornale ecc)*. **~zi'one** *nf* association

assogget'tar|e *vt* subject. **~si** *vr* submit

asso'lato *a* sunny

assol'dare *vt* recruit

as'solo *nm Mus* solo

as'solto *pp di* **assolvere**

assoluta'mente *adv* absolutely

assolu'tismo *nm* absolutism

asso'lu|to *a* absolute. **~zi'one** *nf* acquittal; *Relig* absolution

as'solvere *vt* perform *(compito)*; *Jur* acquit; *Relig* absolve

assomigli'ar|e *vi* **~e a** be like, resemble. **~si** *vr* resemble each other

assom'marsi *vr* combine; **~ a qcsa** add to sth

asso'nanza *nf* assonance

asson'nato *a* drowsy

asso'pirsi *vr* doze off

assor'bente *a & nm* absorbent. **~ igienico** sanitary towel

assor'bire *vt* absorb

assor'da|re *vt* deafen. **~nte** *a* deafening

assorti'mento *nm* assortment

assor'ti|re *vt* match *(colori)*. **~to** *a* assorted; *(colori, persone)* matched

as'sorto *a* engrossed

assottigli'ar|e *vt* make thin; *(aguzzare)* sharpen; *(ridurre)* reduce. **~si** *vr* grow thin; *(finanze:)* be whittled away

assue'fa|re *vt* accustom. **~rsi a** get used to. **~tto** *a* *(a caffè, aspirina)* immune to the effects; *(a droga)* addicted. **~zi'one** *nf* *(a caffè, aspirina)* immunity to the effects; *(a droga)* addiction

as'sumere *vt* assume; take on *(im-*

piegato); ~ **informazioni** make inquiries

as'sunto *pp di* **assumere** ●*nm* task. **assunzi'one** *nf* (*di impiegato*) employment

assurdità *nf inv* absurdity; ~ *pl* nonsense

as'surdo *a* absurd

'asta *nf* pole; *Mech* bar; *Comm* auction; **a mezz'~** at half-mast

a'stemio *a* abstemious

aste'nersi *vr* abstain (**da** from). **~si'one** *nf* abstention

aste'nuto, -a *nmf* abstainer

aste'risco *nm* asterisk

astig'ma|tico *a* astigmatic. **~'tismo** *nm* astigmatism

asti'nenza *nf* abstinence; **crisi di ~** cold turkey

'asti|o *nm* rancour; **avere ~o contro** qcno bear sb a grudge. **~'oso** *a* resentful

a'stratto *a* abstract

astrin'gente *a & nm* astringent

'astro *nm* star

astrolo'gia *nf* astrology. **a'strologo, -a** *nmf* astrologer

astro'nauta *nmf* astronaut

astro'nave *nf* spaceship

astrono'mia *nf* astronomy. **~o'nomico** *a* astronomical. **a'stronomo** *nm* astronomer

astrusità *nf* abstruseness

a'stuccio *nm* case

a'stu|to *a* shrewd; (*furbo*) cunning. **~zia** *nf* shrewdness; (*azione*) trick

ate'ismo *nm* atheism

A'tene *nf* Athens

'ateo, -a *a & nmf* atheist

a'tipico *a* atypical

at'lant|e *nm* atlas. **l'I[Oceano] A~ico** the Atlantic [Ocean]

at'let|a *nmf* athlete. **~ica** *nf* athletics *sg*. **~ica leggera** track and field events. **~ica pesante** weight-lifting, boxing, wrestling, *etc*. **~ico** *a* athletic

atmo'sfer|a *nf* atmosphere. **~ico** *a* atmospheric

a'tomic|a *nf* atom bomb. **~o** *a* atomic

'atomo *nm* atom

'atrio *nm* entrance hall

a'troce *a* atrocious; (*terrible*) dreadful. **~ità** *nf inv* atrocity

atrofiz'zarsi *vr* Med, fig atrophy

attaccabot'toni *nmf inv* (*crashing*) bore

attacca'brighe *nmf inv* troublemaker

attacca'mento *nm* attachment

attacca'panni *nm inv* (*coat-*) hanger; (*a muro*) clothes hook

attac'car|e *vt* attach; (*legare*) tie; (*appendere*) hang; (*cucire*) sew on; (*contagiare*) pass on; (*assalire*) attack; (*iniziare*) start ●*vi* stick; (*diffondersi*) catch on. **~si** *vr* cling; (*affezionarsi*) become attached; (*litigare*) quarrel

attacca'ticcio *a* sticky

at'tacco *nm* attack; (*punto d'unione*) junction

attar'darsi *vr* stay late; (*indugiare*) linger

attec'chire *vi* take; (*moda ecc*) catch on

atteggia'mento *nm* attitude

atteggi'ar|e *vt* assume. **~si** *vr* **~si a** pose as

attem'pato *a* elderly

at'tender|e *vt* wait for ●*vi* **~e a** attend to. **~si** *vr* expect

atten'dibil|e *a* reliable. **~ità** *nf* reliability

atte'nersi *vr* **~ a** stick to

attenta'mente *adv* attentively

atten'ta|re *vi* **~re a** make an attempt on. **~to** *nm* act of violence; (*contro politico ecc*) assassination attempt. **~'tore, ~'trice** *nmf* (*a scopo politico*) terrorist

at'tento *a* attentive; (*accurato*) careful; **~!** look out!; **stare ~** pay attention

attenu'ante *nf* extenuating circumstance

attenu'a|re *vt* attenuate; (*minimizzare*) minimize; subdue (*colori*

ecc); calm (*dolore*); soften (*colpo*).
~**rsi** *vr* diminish. ~**zi'one** *nf*
lessening

attenzi'one *nf* attention; ~**!** watch
out!

atter'ra|ggio *nm* landing. ~**re** *vt*
knock down ● *vi* land

atter'rir|e *vt* terrorize. ~**si** *vr* be
terrified

at'tes|a *nf* waiting; (*aspettativa*)
expectation; **in** ~**a di** waiting for.
~**o** *pp di* attendere

atte'sta|re *vt* state; (*certificare*)
certify. ~**to** *nm* certificate. ~**zi-
'one** *nf* certificate; (*dichiarazione*)
declaration

'attico *nm* attic

at'tiguo *a* adjacent

attil'lato *a* (*vestito*) close-fitting

'attimo *nm* moment

atti'nente *a* ~ **a** pertaining to

atti'rare *vt* attract

atti'tudine *nf* (*disposizione*) aptitude; (*atteggiamento*) attitude

atti'v|are *vt* activate. ~**ismo** *nm*
activism. ~**ista** *nmf* activist.
attività *nf inv* activity; *Comm* assets *pl*. ~**o** *a* active; *Comm* produc-
tive ● *nm* assets *pl*

attiz'za|re *vt* poke; *fig* stir up.
~**toio** *nm* poker

'atto *nm* act; (*azione*) action;
Comm, Jur deed; (*certificato*) certificate; **atti** *pl* (*di società ecc*) proceedings; **mettere in** ~ put into
effect

at'tonito *a* astonished

attorcigli'ar|e *vt* twist. ~**si** *vr* get
twisted

at'tore *nm* actor

attorni'ar|e *vt* surround. ~**si** *vr*
~**si di** surround oneself with

at'torno *adv* around, about ● *prep*
~ **a** around, about

attrac'care *vt/i* dock

attra'ente *a* attractive

at'tra|rre *vt* attract. ~**rsi** *vr* be at-
tracted to each other. ~**t'tiva** *nf*
charm. ~**zi'one** *nf* attraction.

~**zioni** *pl* **turistiche** tourist at-
tractions

attraversa'mento *nm* (*di strada*)
crossing. ~ **pedonale** pedestrian
crossing, crosswalk *Am*

attraver'sare *vt* cross; (*passare*)
go through

attra'verso *prep* through; (*obliquamente*) across

attrez'za|re *vt* equip; *Naut* rig.
~**rsi** *vr* kit oneself out; ~**tura** *nf*
equipment; *Naut* rigging

at'trezzo *nm* tool; **attrezzi** *pl*
equipment; *Sport* appliances *pl*;
Theat props *pl*

attribu'ir|e *vt* attribute. ~**si** *vr* as-
cribe to oneself; ~**si il merito di**
claim credit for

attri'buito *nm* attribute. ~**zi'one**
nf attribution

at'trice *nf* actress

at'trito *nm* friction

attu'abile *a* feasible

attu'al|e *a* present; (*di attualità*)
topical; (*effettivo*) actual. ~**ità** *nf*
topicality; (*avvenimento*) news;
programma di ~**ità** current affairs programme. ~**iz'zare** *vt* up-
date. ~**'mente** *adv* at present

attu'a|re *vt* carry out. ~**rsi** *vr* be
realized. ~**zi'one** *nf* carrying out

attu'tire *vt* deaden; ~ **il colpo** soften the blow

au'dac|e *a* daring, bold; (*insolente*)
audacious;. ~**ia** *nf* daring, bold-
ness; (*insolenza*) audacity

'audience *nf inv* (*telespettatori*)
audience

'audio *nm* audio

audiovi'sivo *a* audiovisual

audi'torio *nm* auditorium

audizi'one *nf* audition; *Jur* hearing

'auge *nm* height; **essere in** ~ be
popular

augu'rar|e *vt* wish. ~**si** *vr* hope.

au'gurio *nm* wish; (*presagio*)
omen; **auguri!** all the best!; (*a
Natale*) Happy Christmas!; **tanti
auguri** best wishes

'aula *nf* classroom; (*università*)

lecture-hall; (*sala*) hall. **~ magna** (*in università*) great hall. **~ del tribunale** courtroom

aumen'tare *vt/i* increase. **au'mento** *nm* increase; (*di stipendio*) [pay] rise

au'reola *nf* halo

au'rora *nf* dawn

auscul'tare *vt Med* auscultate

ausili'are *a & nmf* auxiliary

auspicabile *a* è **~ che...** it is to be hoped that...

auspi'care *vt* hope for

au'spicio *nm* omen; **auspici** (*pl: protezione*) auspices

austerità *nf* austerity

au'stero *a* austere.

Au'strali|a *nf* Australia. **a~'ano, -a** *a & nmf* Australian

'Austria *nf* Austria. **au'striaco, -a** *a & nmf* Austrian

autar'chia *nf* autarchy. **au'tarchico** *a* autarchic

autenti'c|are *vt* authenticate. **~ità** *nf* authenticity

au'tentico *a* authentic; (*vero*) true

au'tista *nm* driver

'auto *nf inv* car

'auto+ *pref* self+

autoabbron'zante *nm* self-tan ● *a* self-tanning

autoambu'lanza *nf* ambulance

autoartico'lato *nm* articulated lorry

autobio|gra'fia *nf* autobiography. **~'grafico** *a* autobiographical

auto'botte *nf* tanker

'autobus *nm inv* bus

auto'carro *nm* lorry

autocommiserazi'one *nf* self-pity

autoconcessio'nario *nm* car dealer

auto'critica *nf* self-criticism

autodi'datta *nmf* self-educated person, autodidact

autodi'fesa *nf* self-defence

auto'gol *nm inv* own goal

au'tografo *a & nm* autograph

autolesio'nis|mo *nm* fig self-destruction. **~tico** *a* self-destructive

auto'linea *nf* bus line

au'toma *nm* robot

automatica'mente *adv* automatically

auto'matico *a* automatic ● *nm* (*bottone*) press-stud; (*fucile*) automatic

automatiz'za|re *vt* automate. **~zi'one** *nf* automation

auto'mezzo *nm* motor vehicle

auto'mobil|e *nf* [motor] car. **~'lismo** *nm* motoring. **~'lista** *nmf* motorist. **~'listico** *a* (*industria*) automobile *attrib*

autonoma'mente *adv* autonomously

auto'no'mia *nf* autonomy; *Auto* range; (*di laptop, cellulare*) battery life. **au'tonomo** *a* autonomous

au'topsia *nf* autopsy

auto'radio *nf inv* car radio; (*veicolo*) radio car

au'tore, -'trice *nmf* author; (*di pittore*) painter; (*di furto ecc*) perpetrator; **quadro d'~** genuine master

auto'revo|le *a* authoritative; (*che ha influenza*) influential. **~'lezza** *nf* authority

autori'messa *nf* garage

autori'tà *nf inv* authority. **~'tario** *a* autocratic. **~ta'rismo** *nm* authoritarianism

autori'tratto *nm* self-portrait

autoriz'za|re *vt* authorize. **~zi'one** *nf* authorization

auto'scontro *nm inv* bumper car, dodgem

autoscu'ola *nf* driving school

auto'stop *nm* hitch-hiking; **fare l'~** hitch-hike. **~'pista** *nmf* hitch-hiker

auto'strada *nf* motorway

autostra'dale *a* motorway *attrib*

autosuffici'en|te *a* self-sufficient. **~za** *nf* self-sufficiency

autotrasporta'|tore, ~'trice *nmf* haulier, carrier

auto'treno *nm* articulated lorry, roadtrain

autovei'colo *nm* motor vehicle

auto'velox *nm inv* speed camera

autovet'tura *nf* motor vehicle

autun'nale *a* autumn[al]

au'tunno *nm* autumn

aval'lare *vt* endorse, back ⟨*cambiale*⟩; *fig* endorse

a'vallo *nm* endorsement

avam'braccio *nm* forearm

avan'guardia *nf* vanguard; *fig* avant-garde; **essere all'~** be in the forefront; *Techn* be at the leading edge

a'vanti *adv* ⟨*in avanti*⟩ forward; ⟨*davanti*⟩ in front; ⟨*prima*⟩ before; **~!** ⟨*entrate*⟩ come in!; ⟨*suvvia*⟩ come on!; ⟨*su semaforo*⟩ cross now, walk *Am*; **va' ~!** go ahead!; **andare ~** ⟨*precedere*⟩ go ahead; ⟨*orologio*:⟩ be fast; **~ e indietro** backwards and forwards ● *a* ⟨*precedente*⟩ before ● *prep* **~ a** before; ⟨*in presenza di*⟩ in the presence of

avan'tieri *adv* the day before yesterday

avanza'mento *nm* progress; ⟨*promozione*⟩ promotion

avan'zare *vi* advance; ⟨*progredire*⟩ progress; ⟨*essere d'avanzo*⟩ be left [over] ● *vt* advance; ⟨*superare*⟩ surpass; ⟨*promuovere*⟩ promote. **~rsi** *vr* advance; ⟨*avvicinarsi*⟩ approach. **~ta** *nf* advance. **~to** *a* advanced; ⟨*nella notte*⟩ late; **in età ~ta** elderly. **a'vanzo** *nm* remainder; *Comm* surplus; **avanzi** *pl* ⟨*rovine*⟩ remains; ⟨*di cibo*⟩ left-overs

ava'rìa *nf* ⟨*di motore*⟩ engine failure. **~'ato** *a* ⟨*frutta, verdura*⟩ rotten; ⟨*carne*⟩ tainted

ava'rizia *nf* avarice. **a'varo, -a** *a* stingy ● *nmf* miser

a'vena *nf* oats *pl*

a'vere *vt* have; ⟨*ottenere*⟩ get; ⟨*indossare*⟩ wear; ⟨*provare*⟩ feel; **ho trent'anni** I'm thirty; **ha avuto il posto** he got the job; **~ fame/freddo** be hungry/cold; **ho mal di denti** I've got toothache; **cos'ha a che fare con lui?** what has it got to do

with him?; **~ da fare** be busy; **che hai?** what's the matter with you?; **nei hai per molto?** will you be long?; **quanti ne abbiamo oggi?** what date is it today?; **avercela con qcno** have it in for sb ● *v aux* have; **non l'ho visto** I haven't seen him; **lo hai visto?** have you seen him?; **l'ho visto ieri** I saw him yesterday ● *nm* **averi** *pl* wealth *sg*

avia'tore *nm* flyer, aviator. **~zi'one** *nf* aviation; *Mil* Air Force

avidità *nf* avidness. **'avido** *a* avid

avio'getto *nm* jet

'avo, -a *nmf* ancestor

avo'cado *nm inv* avocado

a'vorio *nm* ivory

Avv. *abbr* avvocato

avva'lersi *vr* avail oneself ⟨**of** di⟩

avvalla'mento *nm* depression

avvalo'rare *vt* bear out ⟨*tesi*⟩; endorse ⟨*documento*⟩; ⟨*accrescere*⟩ enhance

avvam'pare *vi* flare up; ⟨*arrossire*⟩ blush

avvantaggi'ar|e *vt* favour. **~si** *vr* **~si di** benefit from; ⟨*approfittare*⟩ take advantage of

avve'd|ersi *vr* ⟨*accorgersi*⟩ notice; ⟨*capire*⟩ realize. **~uto** *a* shrewd

avvelena'mento *nm* poisoning

avvele'nare *vt* poison. **~rsi** *vr* poison oneself. **~to** *a* poisoned

avve'nente *a* attractive

avveni'mento *nm* event

avve'nire *vi* happen; ⟨*aver luogo*⟩ take place

avve'ni|re *nm* future. **~'ristico** *a* futuristic

avven'tarsi *vr* fling oneself. **~to** *a* ⟨*decisione*⟩ rash

av'vento *nm* advent; *Relig* Advent

avven'tore *nm* regular customer

avven'tu|ra *nf* adventure; ⟨*amorosa*⟩ affair; *d'*~ ⟨*film*⟩ adventure *attrib*. **~'rarsi** *vr* venture. **~ri'ero, -a** *nm* adventurer ● *nf* adventuress. **~'roso** *a* adventurous

avve'ra|bile *a* ⟨*previsione*⟩ that may come true. **~rsi** *vr* come true

av'verbio *nm* adverb

avver'sar|e *vt* oppose. **~io, -a** *a* opposing ● *nm/f* opponent

avversi|'one *nf* aversion. **~tà** *nf inv* adversity

av'verso *a* (*sfavorevole*) adverse; (*contrario*) averse

avver'tenza *nf* (*cura*) care; (*avvertimento*) warning; (*avviso*) notice; (*premessa*) foreword; **avvertenze** *pl* (*istruzioni*) instructions

avverti'mento *nm* warning

avver'tire *vt* warn; (*informare*) inform; (*sentire*) feel

avvez'zar|e *vt* accustom. **~si** *vr* accustom oneself. **avvezzo a** used to

avvia'mento *nm* starting; Comm goodwill

avvi'a|re *vt* start. **~rsi** *vr* set out. **~to** *a* under way; **bene ~to** thriving

avvicenda'mento *nm* (*in agricoltura*) rotation; (*nel lavoro*) replacement

avvicen'darsi *vr* take turns, alternate

avvici'namento *nm* approach

avvici'nar|e *vt* bring near; approach (*persona*). **~si** *vr* come nearer, approach; **~si a** come nearer to, approach

avvi'lente *a* demoralizing; (*umiliante*) humiliating

avvili'mento *nm* despondency; (*degradazione*) degradation

avvi'li|re *vt* dishearten; (*degradare*) degrade. **~rsi** *vr* lose heart; (*degradarsi*) degrade oneself. **~to** *a* disheartened; (*degradato*) degraded

avvilup'par|e *vt* envelop. **~si** *vr* wrap oneself up; (*aggrovigliarsi*) get entangled

avvinaz'zato *a* drunk

avvin'cente *a* ‹libro ecc› enthralling. **av'vincere** *vt* enthral

avvinghi'ar|e *vt* clutch. **~si** *vr* cling

av'vio *nm* start-up; **dare l'~** *a* qcsa get sth under way; **prendere l'~** get under way

avvi'sare *vt* inform; (*mettere in guardia*) warn

av'viso *nm* notice; (*annuncio*) announcement; (*avvertimento*) warning; (*pubblicitario*) advertisement; **a mio ~** in my opinion. **~ di garanzia** Jur notification that one is to be the subject of a legal enquiry

avvi'stare *vt* catch sight of

avvi'tare *vt* screw in; screw down ‹coperchio›

avviz'zire *vi* wither

avvo'ca|to *nm* lawyer; *fig* advocate. **~'tura** *nf* legal profession

av'volger|e *vt* wrap [up]. **~si** *vr* wrap oneself up

avvol'gibile *nm* roller blind

avvol'toio *nm* vulture

aza'lea *nf* azalea

azi'en|da *nf* business, firm. **~ agricola** farm. **~ di soggiorno** tourist bureau. **~'dale** *a* ‹politica, dirigente› company *attrib*; ‹giornale› in-house

aziona'mento *nm* operation

azio'nare *vt* operate

azio'nario *a* share *attrib*

azi'one *nf* action; Fin share; **d'~** ‹romanzo, film› action[-packed]. **azio'nista** *nm/f* shareholder

a'zoto *nm* nitrogen

azzan'nare *vt* seize with its teeth; sink its teeth into ‹gamba›

azzar'd|are *vt* risk. **~arsi** *vr* dare. **~ato** *a* risky; (*precipitoso*) rash. **az'zardo** *nm* hazard; **gioco d'azzardo** game of chance

azzec'care *vt* hit; (*fig: indovinare*) guess

azzuf'farsi *vr* come to blows

az'zur|ro *a* & *nm* blue; **il principe ~** Prince Charming. **~'rognolo** *a* bluish

Bb

bab'beo a foolish ● nm idiot
'babbo nm fam dad, daddy. **B~ Natale** Father Christmas
bab'buccia nf slipper
babbu'ino nm baboon
ba'bordo nm Naut port side
baby'sitter nmf inv baby-sitter; **fare la ~** babysit
ba'cato a wormeaten
'bacca nf berry
baccalà nm inv dried salted cod
bac'cano nm din
bac'cello nm pod
bac'chetta nf rod; (magica) wand; (di direttore d'orchestra) baton; (di tamburo) drumstick
ba'checa nf showcase; (in ufficio) notice board. **~ elettronica** Comput bulletin board
bacia'mano nm kiss on the hand; **fare il ~ a qcno** kiss sb's hand
baci'ar|e vt kiss. **~si** vr kiss [each other]
ba'cillo nm bacillus
baci'nella nf basin
ba'cino nm basin; Anat pelvis; (di porto) dock; (di minerali) field
'bacio nm kiss
'baco nm worm. **~ da seta** silkworm
ba'cucco a **un vecchio ~** a senile old man
'bada nf **tenere qcno a ~** keep sb at bay
ba'dare vi take care (a of); (fare attenzione) look out; **bada ai fatti tuoi!** mind your own business!
ba'dia nf abbey
ba'dile nm shovel
'badminton nm badminton
'baffi nmpl moustache sg; (di animale) whiskers; **mi fa un baffo** I don't give a damn; **ridere sotto i ~** laugh up one's sleeve
baf'futo a moustached

ba'gagli nmpl luggage, baggage. **~'aio** nm Rail luggage van; Auto boot
ba'gaglio nm luggage; **un ~** a piece of luggage. **~ a mano** hand luggage, hand baggage
baggia'nata nf **non dire baggianate** don't talk nonsense
bagli'ore nm glare; (improvviso) flash; (fig: di speranza) glimmer
ba'gnante nmf bather
ba'gna|re vt wet; (inzuppare) soak; (immergere) dip; (innaffiare) water; (mare, lago:) wash; (fiume:) flow through. **~rsi** vr get wet; (al mare ecc) swim, bathe.
bagnasci'uga nm inv edge of the water, waterline
ba'gnato a wet
ba'gnino, -a nmf life guard
'bagno nm bath; (stanza) bathroom; (gabinetto) toilet; (in casa) toilet, bathroom; (al mare) swim, bathe; **bagni** pl (stabilimento) lido; **fare il ~** have a bath; (nel mare ecc) [have a] swim or bathe; **andare in ~** go to the bathroom or toilet; **mettere a ~** soak. **~ turco** Turkish bath
bagnoma'ria nm **cuocere a ~** cook in a double saucepan
bagnoschi'uma nm inv bubble bath
'baia nf bay
baio'netta nf bayonet
'baita nf mountain chalet
bala'ustra, balaus'trata nf balustrade
balbet'tare vt/i stammer; (bambino:) babble. **~io** nm stammering; babble
bal'buzi|e nf stutter. **~'ente** a stuttering ● nmf stutterer
Bal'can|i nmpl Balkans. **b~ico** a Balkan
balco'nata nf Theat balcony, dress circle
balcon'cino nm **reggiseno a ~** underwired bra
bal'cone nm balcony
baldac'chino nm canopy; **letto a ~** four-poster bed

bal'dan|za nf boldness. **~'zoso** a bold

bal'doria nf revelry; **far ~** have a riotous time

Bale'ari nfpl le [isole] **~** the Balearics, the Balearic Islands

ba'lena nf whale

bale'nare vi lighten; fig flash; **mi è balenata un'idea** I've just had an idea

bale'niera nf whaler

ba'leno nm **in un ~** in a flash

ba'lera nf dance hall

ba'lia nf wetnurse

ba'lia nf **in ~ di** at the mercy of

ba'listico a ballistic; **perito ~** ballistics expert

balla nf bale; (fam: frottola) tall story

bal'labile a good for dancing to

bal'la|re vi dance. **~ta** nf ballad

balla'toio nm (nelle scale) landing

balle'rino, -a nf dancer; (classico) ballet dancer; **ballerina** (classica) ballet dancer, ballerina

bal'letto nm ballet

bal'lista nmf fam bull-shitter

'ballo nm dance; (il ballare) dancing; **sala da ~** ballroom; **essere in ~** (lavoro, vita:) be at stake; (persona:) be committed; **tirare qcno in ~** involve sb

ballonzo'lare vi skip about

ballot'taggio nm second count (of votes)

balne'a|re a bathing attrib. **stagione ~** swimming season. **stazione ~** seaside resort. **~zi'one** nf è vietata la **~zione** no swimming

ba'lordo a foolish; (stordito) stunned; **tempo ~** nasty weather

'balsamo nm balsam; (per capelli) conditioner; (lenimento) balm

'baltico a Baltic. **il** [mar] **B~** the Baltic [Sea]

balu'ardo nm bulwark

'balza nf crag; (di abito) flounce

bal'zano a (idea) weird

bal'zare vi bounce; (saltare) jump; **~ in piedi** leap to one's feet.

'balzo nm bounce; (salto) jump;

prendere la palla al balzo seize an opportunity

bam'bagia nf cotton wool; **vivere nella ~** fig be in clover

bambi'nata nf childish thing to do/say

bam'bi|no, -a nmf child; (appena nato) baby; **avere un ~no** have a baby. **~'none, -a** nmf pej big or overgrown child

bam'boccio nm chubby child; (sciocco) simpleton; (fantoccio) rag doll

'bambo|la nf doll. **~'lotto** nm male doll

bambù nm bamboo

ba'nal|e a banal; **~ità** nf inv banality; **~iz'zare** vt trivialize

ba'nan|a nf banana. **~o** nm banana-tree

'banca nf bank. **~ [di] dati** databank

banca'rella nf stall

ban'cario, -a a a banking attrib; **trasferimento ~** bank transfer ●nmf bank employee

banca'rotta nf bankruptcy; **fare ~** go bankrupt

banchet'tare vi banquet. **ban'chetto** nm banquet

banchi'ere nm banker

ban'china nf Naut quay; (in stazione) platform; (di strada) path; **non transitabile** soft verge

ban'chisa nf floe

'banco nm (di scuola) desk; (di negozio) counter; (di officina) bench; (di gioco, banca) bank; (di mercato) stall; (degli imputati) dock; **sotto ~** under the counter; **medicinale da ~** over the counter medicines. **~ informazioni** information desk. **~ di nebbia** fog bank

'bancomat® nm inv autobank, cashpoint; (carta) bank card, cash card

ban'cone nm counter; (in bar) bar

banco'nota nf banknote, bill Am; **banco'note** pl paper currency '**banda** *nf* band; (*di delinquenti*) gang. **~ d'atterraggio** landing strip. **~ rumorosa** rumble strip

banderu'ola *nf* weathercock; *Naut* pennant

bandi'e|ra *nf* flag; **cambiare ~ra** change sides, switch allegiances. **~'rina** *nf* (*nel calcio*) corner flag. **~'rine** *pl* bunting *sg*

ban'di|re *vt* banish; (*pubblicare*) publish; *fig* dispense with (*formalità, complimenti*). **~to** *nm* bandit. **~'tore** *nm* (*di aste*) auctioneer

'**bando** *nm* proclamation; **~ di concorso** job advertisement (*published in an official gazette for a job for which a competitive examination has to be taken*)

bar *nm inv* bar

'**bara** *nf* coffin

ba'rac|ca *nf* hut; (*catapecchia*) hovel; **mandare avanti la ~ca** keep the ship afloat. **~'cato** *nm* person living in a makeshift shelter. **~'chino** *nm* (*di gelati, giornali*) kiosk; *Radio* CB radio. **~'cone** *nm* (*roulotte*) circus caravan; (*in luna park*) booth. **~'copoli** *nf inv* shanty town

bara'onda *nf* chaos; **non fare ~** don't make a mess

ba'rare *vi* cheat

'**baratro** *nm* chasm

barat'tare *vt* barter. **ba'ratto** *nm* barter

ba'rattolo *nm* jar; (*di latta*) tin

'**barba** *nf* beard; (*fam: noia*) bore; **farsi la ~** shave; **è una ~** (*noia*) it's boring

barbabi'etola *nf* beetroot. **~ da zucchero** sugar-beet

bar'barico *a* barbaric. **bar'barie** *nf* barbarity. '**barbaro** *a* barbarous ● *nm* barbarian

'**barbecue** *nm inv* barbecue

barbi'ere *nm* barber; (*negozio*) barber's

barbi'turico *nm* barbiturate

bar'bone *nm* (*vagabondo*) vagrant; (*cane*) poodle

bar'boso *a fam* boring

barbu'gliare *vi* mumble

bar'buto *a* bearded

'**barca** *nf* boat; **una ~ di** *fig* a lot of. **~ a motore** motorboat. **~ da pesca** fishing boat. **~ a remi** rowing boat, rowboat *Am*. **~ di salvataggio** lifeboat. **~ a vela** sailing boat, sailboat *Am*. **~'iolo** *nm* boatman

barcame'narsi *vr* manage

barcol'lare *vi* stagger

bar'cone *nm* barge; (*di ponte*) pontoon

bar'dar|e *vt* harness. **~si** *vr hum* dress up

ba'rel|la *nf* stretcher. **~li'ere** *nm* stretcher-bearer

'**Barents: il mare di ~** the Barents Sea

bari'centro *nm* centre of gravity

ba'ri|le *nm* barrel. **~'lotto** *nm fig* tub of lard

ba'rista *nm* barman ● *nf* barmaid

ba'ritono *nm* baritone

bar'lume *nm* glimmer; **un ~ di speranza** a glimmer of hope

'**barman** *nm inv* barman

'**baro** *nm* cardsharper

ba'rocco *a & nm* baroque

ba'rometro *nm* barometer

ba'rone *nm* baron; **i baroni** *fig* the top brass. **baro'nessa** *nf* baroness

'**barra** *nf* bar; (*lineetta*) oblique; *Naut* tiller. **~ spazio** *Comput* space bar. **~ strumenti** *Comput* tool bar

bar'rare *vt* block off (*strada*)

barri'ca|re *vt* barricade. **~ta** *nf* barricade

barri'era *nf* barrier; (*stradale*) road-block; *Geol* reef. **~ razziale** colour bar

bar'ri|re *vi* trumpet. **~to** *nm* trumpeting

barzel'letta *nf* joke; **~ sporca** *o* **spinta** dirty joke

basa'mento *nm* base

ba'sar|e *vt* base. **~si** *vr* **~si su** be based on; **mi baso su ciò che ho visto** I'm going on [the basis of] what I saw

'basco, -a nmf & a Basque ●nm (copricapo) beret

'base nf basis; (fondamento) foundation; Mil base; Pol rank and file; **a ~ di** containing; **in ~ a** on the basis of ● **dati** database

'baseball nm baseball

ba'setta nf sideburn

basi'lare a basic

ba'silica nf basilica

ba'silico nm basil

ba'sista nm grass roots politician; (di un crimine) mastermind

'basket nm basketball

bas'sezza nf lowness; (di statura) shortness; (viltà) vileness

bas'sista nmf bassist

'basso, -a a low; (di statura) short; (acqua) shallow; (televisione) quiet; (vile) despicable; **parlare a bassa voce** speak quietly, speak in a low voice; **la bassa Italia** southern Italy ●nm lower part; Mus bass. **guardare in ~** look down

basso'fondo nm (pl bassifondi) shallows pl; **bassifondi** pl (quartieri poveri) slums

bassorili'evo nm bas-relief

bas'sotto nm dachshund

ba'stardo, -a a bastard; (di animale) mongrel ●nmf bastard; (animale) mongrel

ba'stare vi be enough; (durare) last; **basta!** that's enough!, that'll do!; **basta che** (purché) provided that; **basta così** that's enough; **basta così?** is that enough?, will that do?; (in negozio) else?; **basta andare alla posta** you only have to go to the post office

Basti'an con'trario nm contrary old so-and-so

basti'one nm bastion

basto'nare vt beat

baston'cino nm (da sci) ski pole. **~ di pesce** fish finger, fish stick Am

ba'stone nm stick; (da golf) club; (da passeggio) walking stick

ba'tosta nf blow

bat'taglia nf battle; (lotta) fight. **~'are** vi battle; fig fight

bat'taglio nm (di campana) clapper; (di porta) knocker

battagli'one nm battalion

bat'tello nm boat; (motonave) steamer

bat'tente nm (di porta) wing; (di finestra) shutter; (battaglio) knocker

'batter|e vt beat; (percorrere) scour; thresh (grano); break (record) ● vi (bussare, urtare) knock; (cuore:) beat; (ali ecc:) flap; Tennis serve; **~e a macchina** type; **~e gli occhi** blink; **~e le mani** clap [one's hands]; **~e le ore** strike the hours. **~si** vr fight

bat'teri nmpl bacteria

batte'ria nf battery; Mus drums pl

bat'terio nm bacterium. **~'logico** a bacteriological

batte'rista nmf drummer

bat'tesimo nm baptism, christening

battez'zare vt baptize, christen

battiba'leno nm **in un ~** in a flash

batti'becco nm squabble

batticu'ore nm palpitation; **mi venne il ~** I was scared

bat'tigia nf water's edge

batti'mano nm applause

batti'panni nm inv carpetbeater

batti'stero nm baptistery

batti'strada nm inv outrider; (di pneumatico) tread; Sport pacesetter

battitap'peto nm inv carpet sweeper

'battito nm (del cuore) [heart]beat; (alle tempie) throbbing; (di orologio) ticking; (della pioggia) beating

bat'tuta nf beat; (colpo) knock; (spiritosaggine) wisecrack; (osservazione) remark; Mus bar; Tennis service; Theat cue; (dattilografia) stroke

ba'tuffolo nm flock

ba'ule nm trunk

'**bava** nf dribble; (di cane ecc) slobber; **aver la ~ alla bocca** foam at the mouth

bava'glino nm bib

ba'vaglio nm gag

'**bavero** nm collar

ba'zar nm inv bazaar

baz'zecola nf trifle

bazzi'care vt/i haunt

be'arsi vr delight (**di** in)

beati'tudine nf bliss. **be'ato** a blissful; Relig blessed; **beato te!** lucky you!

beauty-'case nm inv toilet bag

bebè nm inv baby

bec'caccia nf woodcock

bec'ca|re vt peck; fig catch. **~rsi** vr (litigare) quarrel. **~ta** nf peck

beccheggi'are vi pitch

bec'chino nm grave-digger

'**bec|co** nm beak; (di caffettiera ecc) spout. **~'cuccio** nm spout

be'fana nf Epiphany; (donna brutta) old witch

'**beffa** nf hoax; **farsi beffe di qcno** mock sb. **beffardo** a derisory; (persona) mocking

beffar|e vt mock. **~si** vr **~si di** make fun of

'**bega** nf quarrel; **è una bella ~** it's really annoying

be'gonia nf begonia

beige a & nm beige

be'la|re vi bleat. **~to** nm bleating

'**belga** a & nmf Belgian

'**Belgio** nm Belgium

'**bella** nf (in carte, Sport) decider

bel'lezza nf beauty; **che ~!** how lovely!; **chiudere/finire in ~** end on a high note

'**belli|co** a war attrib. **~'coso** a warlike. **~ge'rante** a & nmf belligerent

'**bello** a nice; (di aspetto) beautiful; ⟨uomo⟩ handsome; (moralmente) good; **cosa fai di ~ stasera?** what are you up to tonight?; **oggi fa ~** it's a nice day; **una bella cifra** a lot; **un bel piatto di pasta** a big plate of pasta; **nel bel mezzo** right in the middle; **un bel niente**

absolutely nothing; **bell'e fatto** over and done with; **bell'amico/[a] fine friend he is/you are!**; **questa è bella!** that's a good one!; **scamparla bella** have a narrow escape ●nm (bellezza) beauty; (innamorato) sweetheart; **sul più ~** at the crucial moment; **il ~ è che...** the funny thing is that...

'**belva** nf wild beast

be'molle nm Mus flat

ben vedi **bene**

benché conj though, although

'**benda** nf bandage; (per occhi) blindfold. **ben'dare** vt bandage; blindfold ⟨occhi⟩

'**bene** adv well; **ben ~** thoroughly; **~! good!**; **star ~** (di salute) be well; (vestito, stile:) suit; (finanziariamente) be well off; **non sta ~** (non è educato) it's not nice; **sta/va ~!** all right!; **ti sta ~!** [it] serves you right!; **ti auguro ~** I wish you well; **di ~ in meglio** better and better; **fare ~** (aver ragione) do the right thing; **fare ~** a ⟨cibo:⟩ be good for; **una persona per ~** a good person; **per ~** ⟨fare⟩ properly; **è ben difficile** it's very difficult; **come tu ben sai** as you well know; **lo credo ~!** I can well believe it! ●nm good; **per il tuo ~** for your own good. **beni** nmpl (averi) property sg; **un ~ di famiglia** a family heirloom

bene'detto a blessed

bene'di|re vt bless. **~zi'one** nf blessing

benedu'cato a well-mannered

benefat'tore, -'trice nm benefactor ●nf benefactress

benefi'care vt help

benefi'cenza nf charity

benefi'ci|are vi **~e di** profit by. **~io, -a** a & nmf beneficiary. **bene'ficio** nm benefit. **be'nefico** a beneficial; (di beneficenza) charitable

bene'placito nm consent, approval

be'nessere nm well-being

bene'stante a well-off ● nmf well-off person

bene'stare nm consent

benevo'lenza nf benevolence. **be'nevolo** a benevolent

ben'fatto a well-made

'**beni** nmpl property sg; Fin assets; ~ **di consumo** consumer goods

benia'mino nm favourite

be'nigno a kindly; Med benign

beninfor'mato a well-informed

beninten zio'nato, -a a well-meaning ● nmf well-meaning person

benin'teso adv needless to say, of course

benpen'sante a & nmf self-righteous

benser'vito nm **dare il ~ a qcno** give sb the sack

bensì conj but rather

benve'nuto a & nm welcome

ben'visto a **essere ~** go down well (**da** with)

benvo'lere vt **farsi ~ da qcno** win sb's affection; **prendere qcno in ~** take a liking to sb; **essere benvoluto da tutti** to be well-liked by everyone

ben'zina nf petrol, gas Am; **far ~** get petrol. **~ verde** unleaded petrol. **benzi'naio, -a** nmf petrol station attendant

'**bere** vt drink; (assorbire) absorb; fig swallow ● nm drinking; (bevande) drinks pl

berga'motto nm bergamot

ber'lina nf Auto saloon

Ber'lino nm Berlin

ber'muda nfpl (pantaloni) Bermuda shorts

ber'noccolo nm bump; (disposizione) flair

ber'retto nm beret, cap

bersagli'are vt fig bombard. **ber'saglio** nm target

be'stemmi|a nf swear-word; (maledizione) oath; (sproposito) blasphemy. **~'are** vi swear

'**besti|a** nf animal; (persona brutale) beast; (persona sciocca) fool;

andare in ~a fam blow one's top. **~'ale** a bestial; (espressione, violenza) brutal; (fam: freddo, fame) terrible. **~alità** nf inv bestiality; fig nonsense. **~'ame** nm livestock

'**bettola** nf fig dive

be'tulla nf birch

be'vanda nf drink

bevi'tore, -'trice nmf drinker

be'vut|a nf drink. **-o** pp di **bere**

bi'ada nf fodder

bianche'ria nf linen. **~ intima** underwear

bi'anco a white; (foglio, pagina ecc) blank ● nm white; **mangiare in ~** not eat any fried or heavy foods; **andare in ~** fam not score; **in ~ e nero** (film, fotografia) black and white, monochrome; **passare una notte in ~** have a sleepless night

bian'core nm (bianchezza) whiteness

bianco'spino nm hawthorn

biasci'care vt (mangiare) eat noisily; (parlare) mumble

biasi'mare vt blame. **bi'asimo** nm blame

'**Bibbia** nf Bible

bibe'ron nm inv (baby's) bottle

bi'bita nf (soft) drink

'**biblico** a biblical

bibliogra'fia nf bibliography

biblio'te|ca nf library; (mobile) bookcase. **~'cario, -a** nmf librarian

bicarbo'nato nm bicarbonate. **~ di sodio** bicarbonate of soda

bicchi'ere nm glass

bicchie'rino nm fam tipple

bici'cletta nf bicycle; **andare in ~** ride a bicycle

bico'lore a two-coloured

bidè nm inv bidet

bi'dello, -a nmf janitor, [school] caretaker

bido'nata nf fam swindle

bi'done nm bin; (fam: truffa) swindle; **fare un ~ a qcno** fam stand sb up

bien'nale a biennial

bi'ennio nm two-year period

bi'etola nf beet

bifo'cale a bifocal

bi'folco, -a nmf fig boor

bifor'c|arsi vr fork. **~azi'one** nf fork. **~uto** a forked

biga'mia nf bigamy. **'bigamo, -a** a bigamous ● nmf bigamist

bighello'nare vi loaf around. **bighel'lone** nm loafer

bigiotte'ria nf costume jewellery; (negozio) jeweller's

bigliet't|aio nm booking clerk; (sui treni) ticket-collector. **~e'ria** nf ticket-office; Theat box-office

bigli'et|to nm ticket; (lettera breve) note; (cartoncino) card; (di banca) banknote. **~to da visita** business card. **~'tone** nm (fam: soldi) big one

bignè nm inv cream puff

bigo'dino nm roller

bi'gotto nm bigot

bi'kini nm inv bikini

bi'lanc|ia nf scales pl; (di orologio, Comm) balance; **B~a** Astr Libra. **~'are** vt balance; fig weigh. **~o** nm budget; Comm balance sheet; **fare il ~o** to balance the books; fig take stock

'bile nf bile; fig rage

bili'ardo nm billiards sg

'bilico nm equilibrium; **in ~** in the balance

bi'lingue a bilingual

bili'one nm billion

bilo'cale a two-room

'bimbo, -a nmf child

bimen'sile a fortnightly

bime'strale a bimonthly

bi'nario nm track; (piattaforma) platform

bi'nocolo nm binoculars pl

bio'chimica nf biochemistry

biodegra'dabile a biodegradable

bio'etica nf bioethics

bio'fisica nf biophysics

biogra'fia nf biography. **bio'grafi-co** a biographical. **bi'ografo, -a** nmf biographer

biolo'gia nf biology. **bio'logico** a biological. **bi'ologo, -a** nmf biologist

bi'ond|a nf blonde. **~o** a blond ● nm fair colour; (uomo) fair-haired man

bio'sfera nf biosphere

bi'ossido nm **~ di carbonio** carbon dioxide

biparti'tismo nm two-party system

'birba nf, **bir'bante** nm rascal, rogue. **bir'bone** a wicked

biri'chino, -a a naughty ● nmf little devil

bi'rillo nm skittle

'birr|a nf beer; **a tutta ~a** fig flat out. **~a chiara** lager. **~a scura** brown ale. **~e'ria** nf beer-house; (fabbrica) brewery

bis nm inv encore

bi'saccia nf haversack

bi'sbetic|a nf shrew. **~o** a bad-tempered

bisbigli'are vt/i whisper. **bi'sbiglio** nm whisper

'bisca nf gambling-house

'biscia nf snake

bi'scotto nm biscuit

bisessu'ale a & nmf bisexual

bise'stile a **anno ~** leap year

bisetti'na'nale a fortnightly

bi'slacco a peculiar

bis'nonno, -a nmf great-grand-father; great-grandmother

biso'gn|are vi **~a agire subito** we must act at once; **~a farlo** it is necessary to do it; **non ~a venire** you don't have to come. **~o** nm need; (povertà) poverty; **aver ~o di** need. **~oso** a needy; (povero) poor; **~oso di** in need of

bi'sonte nm bison

bi'stecca nf steak

bisticci'are vi quarrel. **bi'sticcio** nm quarrel; (gioco di parole) pun

bistrat'tare vt mistreat

bi'sturi nm inv scalpel

bi'torzolo nm lump

'**bitter** *nm inv* (bitter) apéritif

bi'vacco *nm* bivouac

'**bivio** *nm* crossroads; (*di strada*) fork

bizan'tino *a* Byzantine

'**bizza** *nf* tantrum; **fare le bizze** (*bambini:*) play up

biz'zarro *a* bizarre

biz'zeffe *adv* **a ~** galore

blan'dire *vt* soothe; (*allettare*) flatter. '**blando** *a* mild

bla'sone *nm* coat of arms

blate'rare *vi* blether, blather

'**blatta** *nf* cockroach

blin'da|re *vt* armour-plate. **~to** *a* armoured

blitz *nm inv* blitz

bloc'car|e *vt* block; (*isolare*) cut off; *Mil* blockade; *Comm* freeze. **~si** *vr* *Mech* jam

blocca'sterzo *nm* steering lock

'**blocco** *nm* block; *Mil* blockade; (*dei fitti*) restriction; (*di carta*) pad; (*unione*) coalition; **in ~** *Comm* in bulk. **~ stradale** road-block

bloc-'notes *nm inv* writing pad

blu *a & nm* blue

blue-'jeans *nmpl* jeans

bluff *nm inv* (*carte, fig*) bluff. **bluf'fare** *vi* (*carte, fig*) bluff

'**blusa** *nf* blouse

'**boa** *nm* boa [constrictor]; (*sciarpa*) [feather] boa ● *nf Naut* buoy

bo'ato *nm* rumbling

bo'bina *nf* spool; (*di film*) reel; *Electr* coil

'**bocca** *nf* mouth; **a ~ aperta** *fig* dumbfounded; **in ~ al lupo!** *fam* break a leg!; **fare la respirazione ~ a ~** a qcno give sb mouth to mouth resuscitation *or* the kiss of life

boc'caccia *nf* grimace; **far boc-cacce** make faces

boc'caglio *nm* nozzle

boc'cale *nm* jug; (*da birra*) tankard

bocca'porto *nm Naut* hatch

boc'cata *nf* (*di fumo*) puff; **prendere una ~ d'aria** get a breath of fresh air

boc'cetta *nf* small bottle

boccheggi'are *vi* gasp

boc'chino *nm* cigarette holder; (*di pipa, Mus*) mouthpiece

'**bocc|ia** *nf* (*palla*) bowl; **~e** *pl* (*gioco*) bowls *sg*

bocci'a|re *vt* (*agli esami*) fail; (*respingere*) reject; (*alle bocce*) hit; **essere ~to** *fig* fail; (*ripetere*) repeat a year. **~tura** *nf* failure

bocci'olo *nm* bud

boc'cone *nm* mouthful; (*piccolo pasto*) snack

boc'coni *adv* face downwards

'**boia** *nm* executioner

boi'ata *nf fam* rubbish

boicot'tare *vt* boycott

bo'lero *nm* bolero

'**bolgia** *nf* (*caos*) bedlam

bo'lide *nm* meteor; **passare come un ~** shoot past [like a rocket]

Bo'livi|a *nf* Bolivia. **b~ano, -a** *a & nmf* Bolivian

'**bolla** *nf* bubble; (*pustola*) blister

bol'la|re *vt* stamp; *fig* brand. **~to a** *fig* branded; **carta ~ta** paper with stamp showing payment of duty

bol'lente *a* boiling [hot]

bol'let|ta *nf* bill; **essere in ~ta** be hard up. **~tino** *nm* bulletin; *Comm* list

bol'lino *nm* coupon

bol'li|re *vt/i* boil. **~to** *nm* boiled meat. **~tore** *nm* boiler; (*per l'acqua*) kettle. **~tura** *nf* boiling

'**bollo** *nm* stamp

bol'lore *nm* boil; (*caldo*) intense heat; *fig* ardour

'**bomba** *nf* bomb; **a prova di ~** bomb-proof

bombarda'mento *nm* shelling; (*con aerei*) bombing; *fig* bombardment. **~ aereo** air raid

bombar'da|re *vt* shell; (*con aerei*) bomb; *fig* bombard. **~i'ere** *nm* bomber

bom'betta *nf* bowler [hat]

'**bombola** *nf* cylinder. **~ di gas** gas bottle, gas cylinder

bombo'lone nm doughnut

bomboni'era nf wedding keepsake

bo'naccia nf Naut calm

bonacci'one, -a nmf good-natured person ● a good-natured

bo'nario a kindly

bo'nifica nf land reclamation. **bonifi'care** vt reclaim

bo'nifico nm Comm discount; (bancario) [credit] transfer

bontà nf goodness; (gentilezza) kindness

'bora nf bora (cold north-east wind in the upper Adriatic)

borbot't|are vi mumble; (stomaco:) rumble. **~io** nm mumbling; (di stomaco) rumbling

'borchia nf stud. **~'ato** a studded

bor'da|re vt border. **~'tura** nf border

bor'deaux a inv (colore) claret

bor'dello nm brothel; fig bedlam; (disordine) mess

'bordo nm border; (estremità) edge; a ~ Naut, Aeron on board

bor'gata nf hamlet

bor'ghese a bourgeois; (abito) civilian; **in** ~ in civilian dress; (poliziotto) in plain clothes

borghe'sia nf middle classes pl

'borgo nm village; (quartiere) district

'bori|a nf conceit. **~'oso** a conceited

bor'lotto nm [fagiolo] ~ borlotto bean

boro'talco nm talcum powder

bor'raccia nf flask

'bors|a nf bag; (borsetta) handbag; (valori) Stock Exchange. **~a dell'acqua calda** hot-water bottle. **~a frigo** cool-box. **~a della spesa** shopping bag. **~a di studio** scholarship. **~ai'olo** nm pickpocket. **~el'lino** nm purse. **bor'sista** nmf Fin speculator; Sch scholarship holder

bor'sello nm (portamonete) purse; (borsetto) man's handbag.

~tta nf handbag. **~tto** nm man's handbag

bo'scaglia nf woodlands pl

bosca'iolo nm woodman; (guardaboschi) forester

'bosco nm wood. **bo'scoso** a wooded

'bossolo nm cartridge case

bo'tanic|a nf botany. **~o** a botanical ● nm botanist

'botola nf trapdoor

'botta nf blow; (rumore) bang; **fare a botte** come to blows. **~ e risposta** fig thrust and counter-thrust

'botte nf barrel

bot'te|ga nf shop; (di artigiano) workshop. **~'gaio, -a** nmf shopkeeper. **~'ghino** nm Theatr box-office; (del lotto) lottery-shop

bot'tigli|a nf bottle; **in ~a** bottled. **~e'ria** nf wine shop

bot'tino nm loot; Mil booty

'botto nm bang; **di ~** all of a sudden

bot'tone nm button; Bot bud

bo'vino a bovine; **bovini** pl cattle

box nm inv (per cavalli) loosebox; (recinto per bambini) play-pen

'boxe nf boxing

'bozza nf draft; Typ proof; (bernoccolo) bump. **boz'zetto** nm sketch

'bozzolo nm cocoon

brac'care vt hunt

brac'cetto nm **a ~** arm in arm

bracci'a|le nm bracelet; (fascia) armband. **~'letto** nm bracelet; (di orologio) watch-strap

bracci'ante nm day labourer

bracci'ata nf (nel nuoto) stroke

'bracci|o nm (pl nf **braccia**) arm; (di fiume, pl **bracci**) arm. **~'olo** nm (di sedia) arm[rest]; (da nuoto) armband

'bracco nm hound

bracconi'ere nm poacher

'brac|e nf embers pl; **alla ~e** chargrilled. **~i'ere** nm brazier. **~'ola** nf chop

'brado a **allo stato ~** in the wild

'brama *nf* longing. bra'mare *vt* long for. bramo'sia *nf* yearning

'branca *nf* branch

'branchia *nf* gill

'branco *nm* (*di cani*) pack; (*pej: di persone*) gang

branco'lare *vi* grope

'branda *nf* camp-bed

bran'dello *nm* scrap; **a brandelli** in tatters

bran'dire *vt* brandish

'brano *nm* piece; (*di libro*) passage

Bra'sile *nm* Brazil. **b~i'ano, -a** *a* & *nmf* Brazilian

bra'vata *nf* bragging

'bravo *a* good; (*abile*) clever; (*coraggioso*) brave; **~!** well done!. bra'vura *nf* skill

'breccia *nf* breach; **sulla ~** *fig* very successful, at the top

bre'saola *nf* dried, salted beef sliced thinly and eaten cold

bre'tella *nf* shoulder-strap; **bretelle** *pl* (*di calzoni*) braces

'breve *a* brief, short; **in ~** briefly; **tra ~** shortly

brevet'tare *vt* patent. bre'vetto *nm* patent; (*attestato*) licence

brevità *nf* shortness

'brezza *nf* breeze

'bricco *nm* jug

bric'cone *nm* blackguard; *hum* rascal

'briciola *nf* crumb; *fig* grain. **~o** *nm* fragment

'briga *nf* (*fastidio*) trouble; (*lite*) quarrel; **attaccar** ~ pick a quarrel; **prendersi la ~ di fare qcsa** go to the trouble of doing sth

brigadi'ere *nm* (*dei carabinieri*) sergeant

bri'gante *nm* bandit; *hum* rogue

bri'gare *vi* intrigue

bri'gata *nf* brigade; (*gruppo*) group

briga'tista *nmf* *Pol* member of the Red Brigades

'briglia *nf* rein; **a ~ sciolta** at breakneck speed

bril'lante *a* brilliant; (*scintillante*) sparkling ● *nm* diamond

bril'lare *vi* shine; (*metallo:*) glitter; (*scintillare*) sparkle

'brillo *a* tipsy

'brina *nf* hoar-frost

brin'dare *vi* toast; **~ a qcno** drink a toast to sb

'brindisi *nm inv* toast

bri'tannico *a* British

'brivido *nm* shiver; (*di paura ecc*) shudder; (*di emozione*) thrill

brizzo'lato *a* greying

'brocca *nf* jug

broc'cato *nm* brocade

'broccoli *nmpl* broccoli *sg*

bro'daglia *nf pej* dishwater

'brodo *nm* broth; (*per cucinare*) stock. **~ ristretto** consommé

'broglio *nm* **~ elettorale** gerrymandering

bron'chite *nf* bronchitis

'broncio *nm* sulk; **fare il ~** sulk

bronto'lare *vi* grumble; (*tuono ecc:*) rumble. **~io** *nm* grumbling; (*di tuono*) rumbling. **~one, -a** *nmf* grumbler

'bronzo *nm* bronze

bros'sura *nf* **edizione in ~** paperback

bru'care *vt* (*pecora:*) graze

bruciacchi'are *vt* scorch

brucia'pelo *adv* **a ~** point-blank

bruci'are *vt* burn; (*scottare*) scald; (*incendiare*) set fire to ● *vi* burn; (*scottare*) scald. **~rsi** *vr* burn oneself. **~to** *a* burnt; *fig* burnt-out. **~'tore** *nm* burner. **~'tura** *nf* burn. bruci'ore *nm* burning sensation

'bruco *nm* grub

'brufolo *nm* spot

brughi'era *nf* heath

bruli'care *vi* swarm. **~hio** *nm* swarming

'brullo *a* bare

'bruma *nf* mist

'bruno *a* brown; (*occhi, capelli*) dark

brusca'mente *adv* (*di colpo*) suddenly

bru'schetta *nf* toasted bread rubbed with garlic and sprinkled with olive oil

'**brusco** *a* sharp; (*persona*) brusque, abrupt; (*improvviso*) sudden

bru'sio *nm* buzzing

bru'tal|e *a* brutal. **~ità** *nf inv* brutality. **~iz'zare** *vt* brutalize. '**bruto** *a & nm* brute

brut'tezza *nf* ugliness

'**brut|to** *a* ugly; (*tempo, tipo, situazione, affare*) nasty; (*cattivo*) bad; **~ta copia** rough copy; **~to tiro** dirty trick. **~'tura** *nf* ugly thing

'**buca** *nf* hole; (*avvallamento*) hollow. **~ delle lettere** (*a casa*) letterbox

buca'neve *nm inv* snowdrop

bu'car|e *vt* make a hole in; (*pungere*) prick; punch (*biglietti*) ● *vi* have a puncture. **~si** *vr* prick oneself; (*con droga*) shoot up

bu'cato *nm* washing

'**buccia** *nf* peel, skin

bucherel'lare *vt* riddle

'**buco** *nm* hole

bu'dello *nm* (*pl f* **budella**) bowel

bu'dino *nm* pudding

'**bue** *nm* (*pl* **buoi**) ox; **carne di ~** beef

'**bufalo** *nm* buffalo

bu'fera *nf* storm; (*di neve*) blizzard

buf'fetto *nm* cuff

'**buffo** *a* funny; *Theat* comic ● *nm* funny thing. **~'nata** *nf* (*scherzo*) joke. **buf'fone** *nm* buffoon; **fare il buffone** play the fool

bu'gi|a *nf* lie; **~a pietosa** white lie. **~'ardo, -a** *a* lying ● *nmf* liar

bugi'gattolo *nm* cubby-hole

'**buio** *a* dark ● *nm* darkness; **al ~** in the dark; **~ pesto** pitch dark

'**bulbo** *nm* bulb; (*dell'occhio*) eyeball

Bulga'ria *nf* Bulgaria. '**bulgaro, -a** *a & nmf* Bulgarian

buli'mia *nf* bulimia. **bu'limico** *a* bulimic

'**bullo** *nm* bully

bul'lone *nm* bolt

'**bunker** *nm inv* bunker

buona'fede *nf* good faith

buona'notte *int* good night

buona'sera *int* good evening

buon'giorno *int* good morning; (*di pomeriggio*) good afternoon

buon'grado: di ~ *adv* willingly

buongu'staio, -a *nmf* gourmet. **buon'gusto** *nm* good taste

bu'ono *a* good; (*momento*) right; **dar ~** (*convalidare*) accept; **alla buona** easy-going; (*cena*) informal; **buona notte/sera** good night/evening; **buon compleanno/Natale!** happy birthday/merry Christmas!; **~** *senso* common sense; **di buon'ora** early; **una buona volta** once and for all; **buona parte di** the best part of; **tre ore buone** three good hours ● *nm* good; (*in film*) goody; (*tagliando*) voucher; (*titolo*) bond; **con le buone** gently; **~ sconto** money-off coupon ● *nmf* **buono, -a a nulla** dead loss

buontem'pone, -a *nmf* happy-go-lucky person

buonu'more *nm* good temper

buonu'scita *nf* retirement bonus; (*di dirigente*) golden handshake

burat'tino *nm* puppet

'**burbero** *a* surly; (*nei modi*) rough

buro'crat|e *nm* bureaucrat. **buro'cratico** *a* bureaucratic. **~'zia** *nf* bureaucracy

bur'rasca *nf* storm. **~'scoso** *a* stormy

'**burro** *nm* butter

bur'rone *nm* ravine

bu'scar|e *vt*, **~si** *vr* catch; **~le** *fam* get a hiding

bus'sare *vt* knock

'**bussola** *nf* compass; **perdere la ~** lose one's bearings

'**busta** *nf* envelope; (*astuccio*) case. **~ paga** pay packet. **~'rella** *nf* bribe. **bu'stina** *nf* (*di tè*) tea bag; (*per medicine*) sachet

'**busto** *nm* bust; (*indumento*) girdle

but'tar|e *vt* throw; **~e giù** (*demolire*) knock down; (*inghiottire*) gulp down; scribble down (*scritto*); *fam* put on (*pasta*); (*scoraggiare*) dishearten; **~e via** throw away.

~**si** *vr* throw oneself; (*saltare*) jump

butte'rato *a* pock-marked

buz'zurro *nm fam* yokel

....

Cc

....

caba'ret *nm inv* cabaret

ca'bina *nf Naut, Aeron* cabin; (*balneare*) beach hut. ~ **elettorale** polling booth. ~ **di pilotaggio** cockpit. ~ **telefonica** telephone box. **cabi'nato** *nm* cabin cruiser

ca'cao *nm* cocoa

'**cacca** *nf fam* pooh

'**caccia** *nf* hunt; (*con fucile*) shooting; (*inseguimento*) chase; (*selvaggina*) game ● *nm inv Aeron* fighter; *Naut* destroyer

cacciabombardi'ere *nm* fighter-bomber

cacciagi'one *nf* game

cacci'a|re *vt* hunt; (*mandar via*) chase away; (*scacciare*) drive out; (*ficcare*) shove ● *vi* go hunting. ~**rsi** *vr* (*nascondersi*) hide; (*andare a finire*) get to; ~**rsi nei guai** get into trouble; **alla ~tora** *a Culin* chasseur. ~**tore** *nm* hunter. ~**tore di frodo** poacher

caccia'vite *nm inv* screwdriver

ca'chet *nm inv Med* capsule; (*colorante*) colour rinse; (*stile*) cachet

'**cachi** *nm inv* (*albero, frutta*) persimmon

'**cacio** *nm* (*formaggio*) cheese

'**caco** *nm fam* (*frutto*) persimmon

'**cactus** *nm inv* cactus

ca'da|vere *nm* corpse. ~**verico** *a fig* deathly pale

ca'dente *a* falling; (*casa*) crumbling

ca'denza *nf* cadence; (*ritmo*) rhythm; *Mus* cadence

ca'dere *vi* fall; (*capelli ecc:*) fall out; (*capitombolare*) tumble; (*vestito ecc:*) hang; **far** ~ (*di mano*)

drop; ~ **dal sonno** feel very sleepy; **lasciar** ~ drop; ~ **dalle nuvole** *fig* be taken aback

ca'detto *nm* cadet

ca'duta *nf* fall; (*di capelli*) loss; *fig* downfall

caffè *nm inv* coffee; (*locale*) café. ~ **corretto** espresso coffee with a dash of liqueur. ~ **lungo** weak black coffee. ~ **macchiato** coffee with a dash of milk. ~ **ristretto** extra-strong espresso coffee. ~ **solubile** instant coffee. ~'**ina** *nf* caffeine. ~'**latte** *nm inv* white coffee.

caffetti'era *nf* coffee-pot

cafo'naggine *nf* boorishness

cafo'nata *nf* boorishness

ca'fone, -a *nmf* boor

ca'gare *vi fam* crap

cagio'nare *vt* cause

cagio'nevole *a* delicate

cagli'ar|e *vi*, ~**si** *vr* curdle

'**cagna** *nf* bitch

ca'gnara *nf fam* din

ca'gnesco *a* **guardare qcno in** ~ scowl at sb

'**cala** *nf* creek

cala'brone *nm* hornet

cala'maio *nm* inkpot

cala'mari *nmpl* squid *sg*

cala'mita *nf* magnet

calamità *nf inv* calamity

ca'lar|e *vi* come down; (*vento:*) drop; (*diminuire*) fall; (*tramontare*) set ● *vt* (*abbassare*) lower; (*nei lavori a maglia*) decrease ● *nm* (*di luna*) waning. ~**si** *vr* lower oneself

'**calca** *nf* throng

cal'cagno *nm* heel

cal'care[1] *nm* limestone

cal'care[2] *vt* tread; (*premere*) press [down]; ~ **la mano** *fig* exaggerate; ~ **le orme di qcno** *fig* follow in sb's footsteps

'**calce**[1] *nf* lime

'**calce**[2] *nm* **in** ~ at the foot of the page

calce'struzzo *nm* concrete

cal'cetto nm Sport five-a-side [football]

calci'a|re vt kick. **~tore** nm footballer

cal'cina nf mortar

calci'naccio nm (pezzo di intonaco) flake of plaster

'calcio¹ nm kick; Sport football; (di arma da fuoco) butt; **dare un ~** a kick. **~ d'angolo** corner [kick]

'calcio² nm (chimica) calcium

'calco nm (con carta) tracing; (arte) cast

calco'la|re vt calculate; (considerare) consider. **~'tore** a calculating ● nm calculator; (macchina elettronica) computer

'calcolo nm calculation; Med stone

cal'daia nf boiler

caldar'rosta nf roast chestnut

caldeggi'are vt support

'caldo a warm; (molto caldo) hot ● nm heat; **avere ~** be warm/hot; **fa ~** it is warm/hot

calen'dario nm calendar

'calibro nm calibre; (strumento) callipers pl; **di grosso ~** (persona) top attrib

'calice nm goblet; Relig chalice

ca'ligine nf fog; (industriale) smog

calli'grafia nf handwriting; (cinese) calligraphy

cal'lista nmf chiropodist. **'callo** nm corn; **fare il callo a** become hardened to. **cal'loso** a callous

'calma nf calm. **cal'mante** a calming ● nm sedative. **cal'mare** vt calm [down]; (lenire) soothe. **cal'marsi** vr calm down; (vento:) drop; (dolore:) die down. **calmo** a calm

'calo nm Comm fall; (di volume) shrinkage; (di peso) loss

calorosa'mente adv (cordialmente) warmly

ca'lore nm heat; (moderato) warmth; **in ~** (animale) on heat. **calo'roso** a warm

calo'ria nf calorie

ca'lorico a calorific

calo'rifero nm radiator

calpe'stare vt trample [down]; fig trample on (diritti, sentimenti); **vietato ~ l'erba** keep off the grass

calpe'stio nm (passi) footsteps

ca'lunnia nf slander. **~'are** vt slander. **~'oso** a slanderous

ca'lura nf heat

cal'vario nm Calvary; fig trial

cal'vizie nf baldness. **'calvo** a bald

'calza nf (da donna) stocking; (da uomo) sock. **~a'maglia** nf tights pl; (per danza) leotard

cal'zante a fig fitting

cal'za|re vt (indossare) wear; (mettersi) put on ● vi fit

calza'scarpe nm inv shoehorn

calza'tura nf footwear

calzatu'rificio nm shoe factory

cal'zetta nf è **una mezza ~** fig he's no use

calzet'tone nm knee-length woollen sock. **cal'zino** nm sock

calzo'la|io nm shoemaker. **~e'ria** nf (negozio) shoe shop

calzon'cini nmpl shorts. **~ da bagno** swimming trunks

cal'zone nm Culin folded pizza with tomato and mozzarella or ricotta inside

cal'zoni nmpl trousers, pants Am

camale'onte nm chameleon

cambi'ale nf bill of exchange

cambia'mento nm change

cambi'ar|e vt/i change; move (casa); (fare cambio di) exchange; **~e rotta** Naut alter course. **~si** vr change. **'cambio** nm change; (Comm, scambio) exchange; Mech gear; **dare il cambio a** qcno relieve sb; **in cambio di** in exchange for

'camera nf room; (mobili) [bedroom] suite; Phot camera; **C~** Pol, Comm Chamber. **~ ardente** funeral parlour. **~ d'aria** inner tube. **C~ di Commercio** Chamber of Commerce. **C~ dei Deputati** Pol ≈ House of Commons. **~ doppia**

double room. ~ **da letto** bedroom. ~ **matrimoniale** double room. ~ **oscura** darkroom. ~ **singola** single room

came'rata¹ *nf (dormitorio)* dormitory; *Mil* barrack room

came'ra|ta² *nmf (amico)* mate; *Pol* comrade. **~'tismo** *nm* comradeship

cameri'era *nf* maid; *(di ristorante)* waitress; *(in albergo)* chambermaid; *(di bordo)* stewardess

cameri'ere *nm* manservant; *(di ristorante)* waiter; *(di bordo)* steward

came'rino *nm* dressing-room

'camice *nm* overall. **cami'cetta** *nf* blouse. **ca'micia** *nf* shirt; **uovo in camicia** poached egg. **camicia da notte** nightdress

cami'netto *nm* fireplace

ca'mino *nm* chimney; *(focolare)* fireplace

'camion *nm inv* lorry *Br*, truck

camion'cino *nm* van

camio'netta *nf* jeep

camio'nista *nmf* lorry driver *Br*, truck driver

cam'mello *nm* camel; *(tessuto)* camel-hair ● *a inv (colore)* camel

cam'meo *nm* cameo

cammi'na|re *vi* walk; *(auto, orologio:)* go. **cam'mino** *nm* way; **essere in cammino** be on the way; **mettersi in cammino** set out. **~ta** *nf* walk; **fare una ~ta** go for a walk

camo'milla *nf* camomile; *(bevanda)* camomile tea

ca'morra *nf* local mafia

ca'moscio *nm* chamois; *(pelle)* suede

cam'pagna *nf* country; *(paesaggio)* countryside; *Comm, Mil* campaign; **in ~** in the country. ~ **elettorale** election campaign. ~ **pubblicitaria** marketing campaign. **campa'gnolo, -a** *a* rustic ● *nm* countryman ● *nf* countrywoman

cam'pale *a* field *attrib*; **giornata** ~ *fig* strenuous day

campa'na *nf* bell; *(di vetro)* belljar. **~'nella** *nf (di tenda)* curtain ring. **~'nello** *nm* door-bell; *(cicalino)* buzzer

campa'nile *nm* belfry

campani'lismo *nm* parochialism

campani'lista *nmf* person with a parochial outlook

cam'panula *nf Bot* campanula

cam'pare *vi* live; *(a stento)* get by

cam'pato *a* ~ **in aria** unfounded

campeggi'a|re *vi* camp; *(spiccare)* stand out. **cam'peggio** *nm* camping; *(terreno)* campsite. **~'tore, ~'trice** *nmf* camper

cam'pestre *a* rural

'camping *nm inv* campsite

campio'nari|o *nm* [set of] samples ● *a* samples; **fiera ~a** a trade fair

campio'nato *nm* championship

campiona'tura *nf (di merce)* range of samples

campi'on|e *nm* champion; *Comm* sample; *(esemplare)* specimen. **~'essa** *nf* ladies' champion

'campo *nm* field; *(accampamento)* camp. ~ **da calcio** football pitch. ~ **di concentramento** concentration camp. ~ **da golf** golf course. ~ **da tennis** tennis court

campo'santo *nm* cemetery

camuf'far|e *vt* disguise. **~si** *vr* disguise oneself

'Cana|da *nm* Canada. **~'dese** *a* & *nmf* Canadian

ca'naglia *nf* scoundrel; *(plebaglia)* rabble

ca'nal|e *nm* channel; *(artificiale)* canal. **~iz'zare** *vt* channel *(acque)*. **~izzazi'one** *nf* channelling; *(rete)* pipes *pl*

'canapa *nf* hemp

cana'rino *nm* canary

cancel'la|re *vt* cross out; *(con la gomma)* rub out; *fig* wipe out; *(annullare)* cancel; *Comput* delete, erase. **~'tura** *nf* erasure. **~zi'one** *nf* cancellation; *Comput* deletion

cancelle'ria *nf* chancellery; *(articoli per scrivere)* stationery

cancelli'ere *nm* chancellor; *(di tribunale)* clerk

can'cello *nm* gate

cance'ro|geno *nm* carcinogen
● *a* carcinogenic. **~'roso** *a* cancerous

can'crena *nf* gangrene

'cancro *nm* cancer. **C~** *Astr* Cancer

cande'gina *nf* bleach. **~'are** *vt* bleach. **can'deggio** *nm* bleaching

can'de|la *nf* candle; *Auto* spark plug; **~'labro** *nm* candelabra. **~li'ere** *nm* candlestick

cande'lotto *nm (di dinamite)* stick

candida'mente *adv* candidly

candi'da|rsi *vr* stand as a candidate. **~to, -a** *nmf* candidate. **~'tura** *nf Pol* candidacy; *(per lavoro)* application

'candido *a* snow-white; *(sincero)* candid; *(puro)* pure

can'dito *a* candied

can'dore *nm* whiteness; *fig* innocence

'cane *nm* dog; *(di arma da fuoco)* cock; **un tempo da cani** foul weather. **~ da caccia** hunting dog

ca'nestro *nm* basket

cangi'ante *a* iridescent; **seta ~** shot silk

can'guro *nm* kangaroo

ca'nile *nm* kennel; *(di allevamento)* kennels *pl*. **~ municipale** dog pound

ca'nino *a & nm* canine

'canna *nf* reed; *(da zucchero)* cane; *(di fucile)* barrel; *(bastone)* stick; *(di bicicletta)* crossbar; *(asta)* rod; *(fam: hascish)* joint; **povero in ~** destitute. **~ da pesca** fishing-rod

can'nella *nf* cinnamon

can'neto *nm* bed of reeds

canni'ba|le *nm* cannibal. **~'lismo** *nm* cannibalism

cannocchi'ale *nm* telescope

canno'nata *nf* cannon shot; **è una ~** *fig* it's brilliant

cannon'cino *nm (dolce)* cream horn

can'none *nm* cannon; *fig* ace

can'nuccia *nf* [drinking] straw; *(di pipa)* stem

ca'noa *nf* canoe

ca'none *nm* canon; *(affitto)* rent; **equo ~** fair rents act

ca'noni|co *nm* canon. **~z'zare** *vt* canonize. **~zzazi'one** *nf* canonization

ca'noro *a* melodious

ca'notta *nf (estiva)* vest top

canot'taggio *nm* canoeing; *(voga)* rowing

canotti'era *nf* singlet

canotti'ere *nm* oarsman

ca'notto *nm* [rubber] dinghy

cano'vaccio *nm (trama)* plot; *(straccio)* duster

can'tante *nmf* singer

can't|are *vt/i* sing. **~au'tore, ~a'trice** *nmf* singer-songwriter. **~icchi'are** *vt* sing softly; *(a bocca chiusa)* hum

canti'ere *nm* yard; *Naut* shipyard; *(di edificio)* construction site. **~ navale** naval dockyard

canti'lena *nf* singsong; *(ninna-nanna)* lullaby

can'tina *nf* cellar; *(osteria)* wine shop

'canto¹ *nm* singing; *(canzone)* song; *Relig* chant; *(poesia)* poem

'canto² *nm (angolo)* corner; *(lato)* side; **dal ~ mio** for my part; **d'altro ~** on the other hand

canto'nata *nf* **prendere una ~** *fig* be sadly mistaken

can'tone *nm* canton; *(angolo)* corner

can'tuccio *nm* nook

canzo'na|re *vt* tease. **~'torio** *a* teasing. **~'tura** *nf* teasing

can'zo|ne *nf* song. **~'netta** *nf fam* pop song. **~ni'ere** *nm* songbook

'caos *nm* chaos. **ca'otico** *a* chaotic

C.A.P. *nm abbr (Codice di Avviamento Postale)* post code, zip code *Am*

ca'pace *a* able; *(esperto)* skilled;

⟨stadio, contenitore⟩ big; **~e di** ⟨disposto a⟩ capable of. **~ità** nf inv ability; ⟨attitudine⟩ skill; ⟨capienza⟩ capacity

capaci'tarsi vr **~ di** ⟨rendersi conto⟩ understand; ⟨accorgersi⟩ realize

ca'panna nf hut

capan'nello nm **fare ~ intorno a** qcno/qcsa gather round sb/sth

capan'none nm shed; Aeron hangar

ca'parbio a obstinate

ca'parra nf deposit

capa'tina nf short visit; **fare una ~ in città/da** qcno pop into town/ in on sb

ca'pello nm hair; **~li** pl ⟨capigliatura⟩ hair sg. **~'lone** nm hippie. **~'luto** a hairy

capez'zale nm bolster; fig bedside

ca'pezzolo nm nipple

capi'en|te a capacious. **~za** nf capacity

capiglia'tura nf hair

ca'pire vt understand; **~ male** misunderstand; **si capisce!** naturally!; **sì, ho capito** yes, I see

capi'ta|le a Jur capital; ⟨principale⟩ main ● nf ⟨città⟩ capital ● nm Comm capital. **~'lismo** nm capitalism. **~'lista** nmf capitalist. **~'listico** a capitalist

capitane'ria nf **~ di porto** port authorities pl

capi'tano nm captain

capi'tare vi ⟨giungere per caso⟩ come; ⟨accadere⟩ happen

capi'tello nm Archit capital

capito'la|re vi capitulate. **~zi'one** nf capitulation

ca'pitolo nm chapter

capi'tombolo nm headlong fall; **fare un ~** tumble down

'capo nm head; ⟨chi comanda⟩ boss fam; ⟨di vestiario⟩ item; Geog cape; ⟨in tribù⟩ chief; ⟨parte estrema⟩ top; **a ~** ⟨in dettato⟩ new paragraph; **da ~** over again; **in ~ a un mese** within a month; **giramento di ~** dizziness; **mal di ~** head-

ache; **~ d'abbigliamento** item of clothing. **~ d'accusa** Jur charge, count. **~ di bestiame** head of cattle

capo'banda nm Mus bandmaster; ⟨di delinquenti⟩ ringleader

ca'poccia nm ⟨fam: testa⟩ nut

capocci'one, -a nmf fam brainbox

capo'danno nm New Year's Day

capofa'miglia nm head of the family

capo'fitto nm **a ~** headlong

capo'giro nm giddiness

capola'voro nm masterpiece

capo'linea nm terminus

capo'lino nm **fare ~** peep in

capolu'ogo nm main town

capo'rale nm lance-corporal

capo'squadra nmf Sport team captain

capo'stipite nm ⟨di famiglia⟩ progenitor

capo'tavola nmf head of the table

capo'treno nm guard

capouf'ficio nmf head clerk

capo'verso nm first line

capo'vol|gere vt overturn; fig reverse. **~gersi** vr overturn; ⟨barca:⟩ capsize; fig be reversed. **~to** pp di **capovolgere** ● a upside-down

'cappa nf cloak; ⟨di camino⟩ cowl; ⟨di cucina⟩ hood

cap'pel|la nf chapel. **~'lano** nm chaplain

cap'pello nm hat. **~ a cilindro** top hat

'cappero nm caper

'cappio nm noose

cap'pone nm capon

cap'potto nm [over]coat

cappuc'cino nm ⟨frate⟩ Capuchin; ⟨bevanda⟩ white coffee

cap'puccio nm hood; ⟨di penna stilografica⟩ cap

'capra nf goat. **ca'pretto** nm kid

ca'priccio nm whim; ⟨bizzarria⟩ freak; **fare i capricci** have tantrums. **~'oso** a capricious; ⟨bambino⟩ naughty

Capri'corno nm Astr Capricorn

capri'ola nf somersault

capri'olo nm roe-deer

'capro nm [billy-]goat. ● **espiatorio** scapegoat. **ca'prone** nm [billy] goat

'capsula nf capsule; ⟨di proiettile⟩ cap; ⟨di dente⟩ crown

cap'tare vt Radio, TV pick up; catch ⟨attenzione⟩

cara'bina nf carbine

carabini'ere nm carabiniere; **carabini'eri** pl Italian police force (which is a branch of the army)

ca'raffa nf carafe

Ca'raibi nmpl ⟨zona⟩ Caribbean sg; ⟨isole⟩ Caribbean Islands; **il mar dei ~** the Caribbean [Sea]

cara'mella nf sweet

cara'mello nm caramel

ca'rato nm carat

ca'ratte|re nm character; ⟨caratteristica⟩ characteristic; Typ type; **di buon ~re** good-natured. **~'ristico, -a** a characteristic; ⟨pittoresco⟩ quaint ● nf characteristic. **~riz'zare** vt characterize

carbon'cino nm ⟨per disegno⟩ charcoal

car'bone nm coal

carboniz'zare vt burn to a cinder

carbu'rante nm fuel

carbura'tore nm carburettor

car'cassa nf carcass; fig old wreck

carce'ra|rio a prison attrib. **~to, -a** nmf prisoner. **~zi'one** nf imprisonment. **~zione preventiva** preventive detention

'carcer|e nm prison; ⟨punizione⟩ imprisonment. **~i'ere, -a** nmf gaoler

carci'ofo nm artichoke

car'diaco a cardiac

'cardine nm hinge

cardio|chi'rurgo nm heart surgeon. **~lo'gia** nf cardiology. **cardi'ologo** nm heart specialist. **~'tonico** nm heart stimulant

'cardo nm thistle

ca'rena nf Naut bottom

ca'ren|te a **~te di** lacking in. **~za** nf lack; ⟨scarsità⟩ scarcity

care'stia nf famine; ⟨mancanza⟩ dearth

ca'rezza nf caress; **fare una ~** a caress

cari'a|rsi vi decay. **~to** a decayed

'carica nf office; Mil, Electr charge; fig drive. **cari'care** vt load; Mil, Electr charge; wind up ⟨orologio⟩. **~'tore** nm ⟨per proiettile⟩ magazine

carica'tura nf caricature. **~'rale** a grotesque. **~'rista** nmf caricaturist

'carico a loaded ⟨di with⟩; ⟨colore⟩ strong; ⟨orologio⟩ wound [up]; ⟨batteria⟩ charged ● nm load; ⟨di nave⟩ cargo; ⟨il caricare⟩ loading; **a ~ di** Comm to be charged to; ⟨persona⟩ dependent on

'carie nf [tooth] decay

ca'rino a pretty; ⟨piacevole⟩ agreeable

ca'risma nm charisma. **cari'smatico** a charismatic

cari'tà nf charity; **per ~a!** ⟨come rifiuto⟩ God forbid!. **~a'tevole** a charitable

carnagi'one nf complexion

car'naio nm fig shambles

car'nale a carnal; **cugino ~** first cousin

'carne nf flesh; ⟨alimento⟩ meat; **~ di manzo/maiale/vitello** beef/pork/veal

car'nefi|ce nm executioner. **~'cina** nf slaughter

carne'va|le nm carnival. **~'lesco** a carnival

car'nivoro nm carnivore ● a carnivorous

car'noso a fleshy

'caro, -a a dear; **cari saluti** kind regards ● nmf fam darling, dear; **i miei cari** my nearest and dearest

ca'rogna nf carcass; fig bastard

caro'sello nm merry-go-round

ca'rota nf carrot

caro'vana nf caravan; ⟨di veicoli⟩ convoy

caro'vita nm high cost of living

'carpa nf carp

carpenti'ere nm carpenter

car'pire vt seize; (con difficoltà) extort

car'pone, car'poni adv on all fours

car'rabile a suitable for vehicles; **passo ~** vedi **carraio**

car'raio a **passo ~** nm ~ entrance to driveway, garage etc where parking is forbidden

carreggi'ata nf roadway; **doppia ~** dual carriageway, divided highway Am

carrel'lata nf TV pan

car'rello nm trolley; (di macchina da scrivere) carriage; Aeron undercarriage; Cinema, TV dolly. **~ d'atterraggio** Aeron landing gear

car'retto nm cart

carri'e|ra nf career; **di gran ~ra** at full speed; **fare ~ra** get on. **~'rismo** nm careerism

carri'ola nf wheelbarrow

'carro nm cart. **~ armato** tank. **~ attrezzi** breakdown vehicle, wrecker Am. **~ funebre** hearse. **~ merci** truck

car'rozza nf carriage; Rail car, coach. **~ cuccette** sleeping car. **~ ristorante** restaurant car

carroz'zella nf (per bambini) pram; (per invalidi) wheelchair

carrozze'ria nf bodywork; (officina) bodyshop

carroz'zina nf pram; (pieghevole) push-chair, stroller Am

carroz'zone nm (di circo) caravan

'carta nf paper; (da gioco) card; (statuto) charter; Geog map. **~ d'argento** ≈ senior citizens' railcard. **~ assorbente** blotting-paper. **~ di credito** credit card. **~ geografica** map. **~ d'identità** identity card. **~ igienica** toilet-paper. **~ di imbarco** boarding card or pass. **~ da lettere** writing-paper. **~ da parati** wallpaper. **~ stagnola** silver paper; Culin aluminium foil. **~ straccia** waste

paper. **~ stradale** road map. **~ velina** tissue-paper. **~ verde** Auto green card. **~ vetrata** sandpaper

carta'car'bone nf carbon paper

car'taccia nf waste paper

carta'modello nm pattern

carta'mo'neta nf paper money

carta'pesta nf papier mâché

carta'straccia nf waste paper

carte'trare vt sand [down]

car'tel|la nf (per documenti ecc) briefcase; (di cartone) folder; (di scolaro) satchel. **~la clinica** medical record. **~'lina** nf document wallet, folder

cartel'lino nm (etichetta) label; (dei prezzi) price-tag; (di presenza) time-card; **timbrare il ~** clock in; (all'uscita) clock out

car'tel|lo nm sign; (pubblicitario) poster; (stradale) road sign; (di protesta) placard; Comm cartel. **~'lone** nm poster; Theat bill

carti'era nf paper-mill

carti'lagine nf cartilage

car'tina nf map

car'toccio nm paper bag; **al ~** Culin baked in foil

carto'|laio, -a nmf stationer. **~le'ria** nf stationer's. **~libre'ria** nf stationer's and book shop

carto'lina nf postcard. **~ postale** postcard

carto'mante nmf fortune-teller

carton'cino nm (materiale) card

car'tone nm cardboard; (arte) cartoon. **~ animato** [animated] cartoon

car'tuccia nf cartridge

'casa nf house; (abitazione propria) home; (ditta) firm; **amico di ~** family friend; **andare a ~** go home; **essere di ~** be like one of the family; **fatto in ~** homemade; **padrone di ~** (di pensione ecc) landlord; (proprietario) house owner. **~ di cura** nursing home. **~ popolare** council house. **~ dello studente** hall of residence

ca'sacca nf military coat; (giacca) jacket

ca'saccio *adv* **a** ~ at random
casa'ling|a *nf* housewife. **~o** *a* domestic; *(fatto in casa)* home-made; *(amante della casa)* home-loving; *(semplice)* homely
ca'scante *a* falling; *(floscio)* flabby
ca'sca|re *vi* fall [down]. **~ta** *nf (di acqua)* waterfall
ca'schetto *nm* **[capelli a]** ~ bob
ca'scina *nf* farm building
'casco *nm* crash-helmet; *(asciugacapelli)* [hair-]drier; ~ **di banane** bunch of bananas
caseggi'ato *nm* block of flats *Br*, apartment block
case'ificio *nm* dairy
ca'sella *nf* pigeon-hole. ~ **postale** post office box; *Comput* mailbox
casel'lante *nmf (per treni)* signalman
casel'lario *nm* ~ **giudiziario** record of convictions; **avere il ~ giudiziario vergine** have no criminal record
ca'sello **[autostra'dale]** *nm* [motorway] toll booth
case'reccio *a* home-made
ca'serma *nf* barracks *pl*; *(dei carabinieri)* [police] station
casi'nista *nmf fam* muddler.
ca'sino *nm fam (bordello)* brothel; *(fig: confusione)* racket; *(disordine)* mess; **un casino di** loads of
casinò *nm inv* casino
ca'sistica *nf (classificazione)* case records *pl*
'caso *nm* chance; *(fatto, circostanza, Med, Gram)* case; **a** ~ at random; ~ **mai** if need be; **far** ~ **a** pay attention to; **non far** ~ a take no account of; **per** ~ by chance. ~ **[giudiziario]** [legal] case
caso'lare *nm* farmhouse
'caspita *int* good gracious!
'cassa *nf* till; *Comm* cash; *(luogo di pagamento)* cash desk; *(mobile)* chest; *(istituto bancario)* bank. ~ **automatica prelievi** cash dispenser, automatic teller. ~ **da**

morto coffin. ~ **toracica** *Anat* ribcage
cassa'forte *nf* safe
cassa'panca *nf* linen chest
casseru'ola *nf* saucepan
cas'setta *nf* case; *(per registratore)* cassette. ~ **delle lettere** postbox, letterbox. ~ **di sicurezza** strong-box
cas'set|to *nm* drawer. **~'tone** *nm* chest of drawers
cassi'ere, -a *nmf* cashier; *(di supermercato)* checkout assistant, checkout operator; *(di banca)* teller
'casta *nf* caste
ca'stagn|a *nf* chestnut. **casta'gneto** *nm* chestnut grove. **~o** *nm* chestnut[-tree]
ca'stano *a* chestnut
ca'stello *nm* castle; *(impalcatura)* scaffold
casti'gare *vt* punish
casti'gato *a (casto)* chaste
ca'stigo *nm* punishment
castità *nf* chastity. **'casto** *a* chaste
'castoro *nm* beaver
ca'strare *vt* castrate
casu'al|e *a* chance *attrib*. **~'mente** *adv* by chance
ca'supola *nf* little house
cata'clisma *nm fig* upheaval
cata'comba *nf* catacomb
cata'fascio *nm* **andare a** ~ go to rack and ruin
cata'litico *a* **marmitta catalitica** *Auto* catalytic converter
cataliz'za|re *vt fig* heighten. **~'tore** *nm Auto* catalytic converter
catalo'gare *vt* catalogue. **ca'talogo** *nm* catalogue
catama'rano *nm (da diporto)* catamaran
cata'pecchia *nf* hovel; *fam* dump
catapul'tar|e *vt (scaraventare fuori)* eject. **~si** *vr (precipitarsi)* dive
catarifran'gente *nm* reflector
ca'tarro *nm* catarrh
ca'tasta *nf* pile
ca'tasto *nm* land register

ca'tastrofe nf catastrophe. **cata'strofico** a catastrophic

cate'chismo nm catechism

cate|go'ria nf category. **~'gorico** a categorical

ca'tena nf chain. **~ montuosa** mountain range. **catene** pl **da neve** tyre-chains. **catene'naccio** nm bolt

cate|'nella nf (collana) chain. **~'nina** nf chain

cate'ratta nf cataract

ca'terva nf **una ~ di** heaps of

cati'nel|la nf basin; **piovere a ~e** bucket down

ca'tino nm basin

ca'torcio nm fam old wreck

ca'trame nm tar

'cattedra nf (tavolo di insegnante) desk; (di università) chair

catte'drale nf cathedral

catti'veria nf wickedness; (azione) wicked action

cattività nf captivity

cat'tivo a bad; (bambino) naughty

catto'licesimo nm Catholicism

cat'tolico, -a a & nmf [Roman] Catholic

cat'tu|ra nf capture. **~'rare** vt capture

caucciù nm rubber

'causa nf cause; Jur lawsuit; **far ~ a** qcno sue sb. **cau'sare** vt cause

'caustico a caustic

cauta'mente adv cautiously

cau'tela nf caution

caute'lar|e vt protect. **~si** vr take precautions

cauteriz'z|are vt cauterize. **~i'one** nf cauterization

'cauto a cautious

cauzi'one nf security; (per libertà provvisoria) bail

'cava nf quarry; fig mine

caval'ca|re vt ride; (stare a cavalcioni) sit astride. **~ta** nf ride; (corteo) cavalcade. **~'via** nm flyover

cavalci'oni: a **~** adv astride

cavali'ere nm rider; (titolo) knight; (accompagnatore) escort; (al ballo) partner

cavalle'resco a chivalrous. **~'ria** nf chivalry; Mil cavalry. **~'rizzo, -a** nm horseman ● nf horsewoman

caval'letta nf grasshopper

caval'letto nm trestle; (di macchina fotografica) tripod; (di pittore) easel

caval'lina nf (ginnastica) horse

ca'vallo nm horse; (misura di potenza) horsepower; (scacchi) knight; (dei pantaloni) crotch; **a ~** on horseback; **andare a ~** go horse-riding. **~ a dondolo** rocking-horse

caval'lone nm (ondata) roller

caval'luccio ma'rino nm sea horse

ca'var|e vt take out; (di dosso) take off; **~sela** get away with it; **se la cava bene** he's doing all right

cava'tappi nm inv corkscrew

ca'ver|na nf cave. **~'noso** a (voce) deep

'cavia nf guinea-pig

cavi'ale nm caviar

ca'viglia nf ankle

cavil'lare vi quibble. **ca'villo** nm quibble

cavità nf inv cavity

'cavo a hollow ● nm cavity; (di metallo) cable; Naut rope

cavo'lata nf fam rubbish

cavol'fiore nm cauliflower

'cavolo nm cabbage; **~!** fam sugar!

caz'zo int vulg fuck!

caz'zot|to nm punch; **prendere qcno a ~i** beat sb up

cazzu'ola nf trowel

c/c abbr (conto corrente) c/a

CD-Rom nm inv CD-Rom

ce pers pron (a noi) (to) us ● adv there; **~ ne sono molti** there are many

'cece nm chick-pea

cecità nf blindness

ceco, -a a & nmf Czech; **la Repubblica Ceca** the Czech Republic

Cecoslo'vacc|hia nf Czechoslovakia. **c~o, -a** a & nmf Czechoslovak 'cedere *vi* (*arrendersi*) surrender; (*concedere*) yield; (*sprofondare*) subside ● *vt* give up; make over (*proprietà ecc*). ce'devole *a* (*terreno ecc*) soft; *fig* yielding. cedi'mento *nm* (*di terreno*) subsidence

'cedola *nf* coupon

'cedro *nm* (*albero*) cedar; (*frutto*) citron

C.E.E. *nf abbr* (**Comunità Economica Europea**) E|E|C

'ceffo *nm* (*muso*) snout; (*pej: persona*) mug

cef'fone *nm* slap

ce'la|re *vt* conceal. ~si *vr* hide

cele'bra|re *vt* celebrate. ~zi'one *nf* celebration

'celebre *a* famous. ~ità *nf inv* celebrity

'celere *a* swift

ce'leste *a* (*divino*) heavenly ● *a & nm* (*colore*) sky-blue

celi'bato *nm* celibacy

'celibe *a* single ● *nm* bachelor

'cella *nf* cell

'cellofan *nm inv* cellophane; *Culin* cling film

'cellula *nf* cell. ~ fotoelettrica electronic eye

cellu'lare *nm* (*telefono*) cellular phone ● *a* [**furgone**] ~ *nm* police van. [telefono] ~ *nm* cellular phone

cellu'lite *nf* cellulite

cellu'loide *a* celluloid

cellu'losa *nf* cellulose

'celt|a *nm* Celt. ~ico *a* Celtic

cemen'tare *vt* cement. ce'mento *nm* cement. cemento armato reinforced concrete

'cena *nf* dinner; (*leggera*) supper

ce'nacolo *nm* circle

ce'nare *vi* have dinner

'cenci|o *nm* rag; (*per spolverare*) duster. ~'oso *a* in rags

'cenere *nf* ash; (*di carbone ecc*) cinders

ce'netta *nf* (*cena semplice*) informal dinner

'cenno *nm* sign; (*col capo*) nod;

(*con la mano*) wave; (*allusione*) hint; (*breve resoconto*) mention

ce'none *nm* il ~ di Capodanno/Natale special New Year's Eve/Christmas Eve dinner

censi'mento *nm* census

cen'sore *nm* censor. ~ura *nf* censorship. ~u'rare *vt* censor

centelli'nare *vt* sip

cente'nari|o, -a *a & nmf* centenarian ● *nm* (*commemorazione*) centenary. ~'nale *a* centennial

cen'tesimo *a* hundredth ● *nm* (*di dollaro*) cent; non avere un ~ be penniless

centi'grado *a* centigrade. ~metro *nm* centimetre

centi'naio *nm* hundred

'cento *a & nm* a *or* one hundred; per ~ per cent

cento'metrista *nmf Sport* one hundred metres runner

cento'mila *nm* a *or* one hundred thousand

cen'trale *a* central ● *nf* (*di società ecc*) head office. ~ atomica atomic power station. ~ elettrica power station. ~ nucleare nuclear power station. ~ telefonica [telephone] exchange

centra'li|na *nf Teleph* switchboard. ~'nista *nmf* operator

centra'lino *nm Teleph* exchange; (*di albergo ecc*) switchboard

centra'li|smo *nm* centralism. ~z'zare *vt* centralize

cen'trare *vt* ~ qcsa hit sth in the centre; (*fissare nel centro*) centre; *fig* hit on the head (*idea*)

centri'fu|ga *nf* spin-drier. ~ [asciugaverdure] shaker. ~'gare *vt Techn* centrifuge; (*lavatrice:*) spin

cen'trino *nm* doily

'centro *nm* centre. ~ [città] city centre. ~ commerciale shopping centre, mall. ~ sociale community centre

'ceppo *nm* (*di albero*) stump; (*da ardere*) log; (*fig: gruppo*) stock

cera | che

52

un po' ~ non lo vedo it's been a while since I saw him; **mi piace più Roma ~ Milano** I like Rome better than Milan; ~ **ti piaccia o no** whether you like it or not; ~ **io sappia** as far as I know

checché *indef pron* whatever

chemiotera'pia *nf* chemotherapy

chero'sene *nm* paraffin

cheru'bino *nm* cherub

cheti'chella: alla ~ *adv* silently

'**cheto** *a* quiet

chi *rel pron* whoever; *(coloro che)* people who; **ho trovato ~ ti può aiutare** I found somebody who can help you; **c'è ~ dice che...** some people say that...; **senti ~ parla!** listen to who's talking! ● *pron inter (soggetto)* who; *(oggetto, con preposizione)* who, whom *fml*; *(possessivo)* **di ~** whose; ~ **sei?** who are you?; ~ **hai incontrato?** who did you meet?; **di ~ sono questi libri?** whose books are these?; **con ~ parli?** who are you talking to?; **a ~ lo dici!** tell me about it!

chi'acchie|ra *nf* chat; *(pettegolezzo)* gossip. **~'rare** *vi* chat; *(far pettegolezzi)* gossip. **~'rato** *a* essere **~rato** *(persona:)* be the subject of gossip; **~re** *pl* chitchat; **far quattro ~re** have a chat. **~'rone, -a** *a* talkative ● *nmf* chatterer

chia'ma|re *vt* call; *(far venire)* send for; **come ti chiami?** what's your name?; **mi chiamo Roberto** my name is Robert; **~re alle armi** call up. **~rsi** *vr* be called. **~ta** *nf* call; *Mil* call-up

chi'appa *nf* *fam* cheek

chiara'mente *adv* clearly

chia'rezza *nf* clarity; *(limpidezza)* clearness

chiarifi'ca|re *vt* clarify. **~'tore a** clarificatory. **~zi'one** *nf* clarification

chiari'mento *nm* clarification

chia'rire *vt* make clear; *(spiegare)* explain. **~si** *vr* become clear

chi'aro *a* clear; *(luminoso)* bright;

(colore) light. **chia'rore** *nm* glimmer

chiaroveg'gente *a* clear-sighted ● *nmf* clairvoyant

chi'as|so *nm* din. **~'soso** *a* rowdy

chi'av|e *nf* key; **chiudere a ~e** lock. **~e inglese** monkey-wrench. **~i'stello** *nm* latch

chiaz'za *nf* stain. **~'zare** *vt* stain

chic *a inv* chic

chicches'sia *pron* anybody

'chicco *nm* grain; *(di caffè)* bean; *(d'uva)* grape

chi'eder|e *vt* ask; *(per avere)* ask for; *(esigere)* demand. **~si** *vr* wonder

chi'esa *nf* church

chi'esto *pp di* chiedere

'chiglia *nf* keel

'chilo *nm* kilo

chilo'grammo *nm* kilogram[me]

chilome'traggio *nm* *Auto* mileage

chilo'metrico *a* in kilometres

chi'lometro *nm* kilometre

chi'mera *nf* fig illusion

'chimic|a *nf* chemistry. **~o, -a** *a* chemical ● *nm* chemist

'china *nf* *(declivio)* slope; **inchiostro di ~** Indian ink

chi'nar|e *vt* lower. **~si** *vr* stoop

chincaglie'rie *nfpl* knick-knacks

chinesitera'pia *nf* physiotherapy

chi'nino *nm* quinine

chi'no *a* bent

chi'notto *nm* sparkling soft drink

chi'occia *nf* sitting hen

chi'occiola *nf* snail; **scala a ~** spiral staircase

chi'odo *nm* nail; *(idea fissa)* obsession. **~ di garofano** clove

chi'oma *nf* head of hair; *(fogliame)* foliage

chi'osco *nm* kiosk; *(per giornali)* news-stand

chi'ostro *nm* cloister

chiro'man|te *nmf* palmist. **~'zia** *nf* palmistry

chirur'gia *nf* surgery. **chi'rurgico** *a* surgical. **chi'rurgo** *nm* surgeon

chissà *adv* who knows; ~ **quando**

arriverà I wonder when he will arrive

chi'tarra nf guitar. **~rista** nmf guitarist

chi'udere vt shut, close; (con la chiave) lock; turn off ‹luce, acqua ecc›; (per sempre) close down ‹negozio, fabbrica ecc›; (recingere) enclose ● vi shut, close. **~si** vr shut; ‹tempo:› cloud over; ‹ferita:› heal over; fig withdraw into oneself

chi'unque pron anyone, anybody ● rel pron whoever

chi'usa nf enclosure; (di canale) lock; (conclusione) close

chi'u|so pp di **chiudere** ● a shut; ‹tempo› overcast; ‹persona› reserved. **~sura** nf closing; (sistema) lock; (allacciatura) fastener. **~sura lampo** zip, zipper Am

ci pron (personale) us; (riflessivo) ourselves; (reciproco) each other; (a ciò, di ciò ecc) about it; **non ci disturbare** don't disturb us; **aspettateci** wait for us; **ci ha detto tutto** he told us everything; **ce lo manderanno** they'll send it to us; **ci consideriamo...** we consider ourselves...; **ci laviamo le mani** we wash our hands; **ci odiamo** we hate each other; **non ci penso mai** I never think about it; **pensaci** think about it! ● adv ‹qui› here; ‹lì› there; (moto per luogo) through it; **ci siamo** we are here; **ci siete?** are you there?; **ci siamo passati tutti** we all went through it; **c'è** there is; **ce ne sono molti** there are many; **ci vuole pazienza** it takes patience; **non ci vedo/sento** I can't see/hear

cia'bat|ta nf slipper. **~'tare** vi shuffle

ciabat'tino nm cobbler

ci'alda nf wafer

cial'trone nm (mascalzone) scoundrel

ciam'bella nf Culin ring-shaped cake; (salvagente) lifebelt; (gonfiabile) rubber ring

cianci'are vi gossip

cianfru'saglie nfpl knick-knacks

cia'notico a (colorito) puce

ci'ao int fam (all' arrivo) hello!, hi!; (alla partenza) bye-bye!, cheerio!

ciar'la|re vi chat. **~'tano** nm charlatan

cias'cuno a each ● pron everyone, everybody; (distributivo) each [one]; **per ~** each

ci'bar|e vt feed. **~ie** nfpl provisions. **~si** vr eat; **~si di** live on

ciber'netico a cybernetic

'cibo nm food

ci'cala nf cicada

cica'lino nm buzzer

cica'tri|ce nf scar. **~z'zante** nm ointment

cicatriz'zarsi vr heal [up]. **cicatrizzazi'one** nf healing

'cicca nf cigarette end; (fam: sigaretta) fag; (fam: gomma) [chewing] gum

cic'chetto nm fam (bicchierino) nip; (rimprovero) telling-off

'cicci|a nf fam fat, flab. **~'one, -a** nmf fam fatty, fatso

cice'rone nm guide

cicla'mino nm cyclamen

ci'clis|mo nm cycling. **~ta** nmf cyclist

'ciclo nm cycle; (di malattia) course

ciclomo'tore nm moped

ci'clone nm cyclone

ci'cogna nf stork

ci'coria nf chicory

ci'eco, -a a blind ● nm blind man / nf blind woman

ci'elo nm sky; Relig heaven; **santo ~!** good heavens!

'cifra nf figure; (somma) sum; (monogramma) monogram; (codice) code

ci'fra|re vt embroider with a monogram; (codificare) code. **~to** a monogrammed; coded

'ciglio nm (bordo) edge; (pl nf ciglia: delle palpebre) eyelash

'cigno nm swan

cigo'l|are *vt* squeak. **~io** *nm* squeak

'Cile *nm* Chile

ci'lecca *nf* far ~ miss

ci'leno, -a *a & nmf* Chilean

cili'egi|a *nf* cherry. **~o** *nm* cherry [tree]

cilin'drata *nf* cubic capacity, c.c.; **macchina di alta ~** highpowered car

ci'lindro *nm* cylinder; (*cappello*) top hat

'cima *nf* top; (*fig: persona*) genius; **da ~ a fondo** from top to bottom

ci'melio *nm* relic

cimen'tar|e *vt* put to the test. **~si** *vr* (*provare*) try one's hand

'cimice *nf* bug; (*puntina*) drawing pin, thumbtack *Am*

cimini'era *nf* chimney; *Naut* funnel

cimi'tero *nm* cemetery

ci'murro *nm* distemper

'Cina *nf* China

cin cin! *int* cheers!

cincischi'are *vi* fiddle

'cine *nm fam* cinema

'cinema *nm inv* cinema. cine'presa *nf* cine-camera

ci'nese *a & nmf* Chinese

cine'teca *nf* (*raccolta*) film collection

ci'netico *a* kinetic

'cingere *vt* (*circondare*) surround

'cinghia *nf* strap; (*cintura*) belt

cinghi'ale *nm* wild boar; **pelle di ~** pigskin

cinguet'tare *vi* twitter. **~io** *nm* twittering

'cinico *a* cynical

ci'niglia *nf* (*tessuto*) chenille

ci'nismo *nm* cynicism

ci'nofilo *a* (*unità*) dog-loving

cin'quanta *a & nm* fifty. cinquan'tenne *a & nmf* fifty-year-old. cinquan'tesimo *a* fiftieth. cinquan'tina *nf* una cinquantina di about fifty

'cinque *a & nm* five

cinquecen'tesco *a* sixteenth-century

cinque'cento *a* five hundred ●*nm* il C~ the sixteenth century

cinque'mila *a & nm* five thousand

'cinta *nf* (*di pantaloni*) belt; **muro di ~** [boundary] wall. cin'tare *vt* enclose

'cintola *nf* (*di pantaloni*) belt

cin'tura *nf* belt. **~ di salvataggio** lifebelt. **~ di sicurezza** *Aeron, Auto* seat-belt

cintu'rino *nm* **~ dell'orologio** watch-strap

ciò *pron* this; that; **~ che** what; **~ nondimeno** nevertheless

ci'occa *nf* lock

ciocco'la|ta *nf* chocolate; (*bevanda*) [hot] chocolate. **~'tino** *nm* chocolate. **~to** *nm* chocolate. **~to al latte/fondente** milk/plain chocolate

cioè *adv* that is

ciondo'l|are *vi* dangle. ci'ondolo *nm* pendant. **~oni** *adv fig* hanging about

ciono'nstante *adv* nonetheless

ci'otola *nf* bowl

ci'ottolo *nm* pebble

ci'polla *nf* onion; (*bulbo*) bulb

ci'presso *nm* cypress

'cipria *nf* [face] powder

'Cipro *nm* Cyprus. cipri'ota *a & nmf* Cypriot

'circa *adv & prep* about

'circo *nm* circus

circo'la|re *a* circular ●*nf* circular; (*di metropolitana*) circle line ●*vi* circulate. **~'torio** *a Med* circulatory. **~zi'one** *nf* circulation; (*traffico*) traffic

'circolo *nm* circle; (*società*) club

circon'ci|dere *vt* circumcise. **~si'one** *nf* circumcision

circon'dar|e *vt* surround. **~io** *nm* (*amministrativo*) administrative district. **~si di** *vr* surround oneself with

circonfe'renza *nf* circumference. **~ dei fianchi** hip measurement

circonvallazi'one *nf* ring road

circo'scritto *a* limited

circoscrizi'one *nf* area. **~ eletto-rale** constituency

circo'spetto *a* wary

circospezi'one *nf* con **~** warily

circo'stante *a* surrounding

circo'stanza *nf* circumstance; *(occasione)* occasion

circu'ire *vt (ingannare)* trick

cir'cuito *nm* circuit

circumnavi'ga|re *vt* circumnavigate. **~zi'one** *nf* circumnavigation

'ciste *nf inv* cyst

ci'sterna *nf* cistern; *(serbatoio)* tank

'cisti *nf inv* cyst

ci'ta|re *vt (riportare brani ecc)* quote; *(come esempio)* cite; *Jur* summons. **~zi'one** *nf* quotation; *Jur* summons *sg*

citofo'nare *vt* buzz. **ci'tofono** *nm* entry phone; *(in ufficio, su aereo ecc)* intercom

ciucci'are *vt fam* suck. **ci'uccio** *nm fam* dummy

ci'uco *nm* ass

ci'uffo *nm* tuft

ci'urma *nf Naut* crew

ci'vet|ta *nf* owl; *(fig: donna)* flirt; **[auto] ~ta** unmarked police car. **~'tare** *vi* flirt. **~te'ria** *nf* coquettishness

'civico *a* civic

ci'vil|e *a* civil. **~iz'zare** *vt* civilize. **~iz'zato** *a (paese)* civilized. **~izzazi'one** *nf* civilization. **~'mente** *adv* civilly

civiltà *nf inv* civilization; *(cortesia)* civility

'clacson *nm inv* horn. **clacso-'nare** *vi* beep the horn, hoot

cla'mo|re *nm* clamour; **fare ~re** cause a sensation. **~rosa'mente** *adv (sbagliare)* sensationally.

~'roso *a* noisy; *(sbaglio)* sensational

clan *nm inv* clan; *fig* clique

clandestina|'mente *adv* secretly. **~ità** *nf* secrecy

clande'stino *a* clandestine; **movimento ~** underground movement; **passeggero ~** stowaway

clari'netto *nm* clarinet

'classe *nf* class. **~ turistica** tourist class

classi'cis|mo *nm* classicism. **~ta** *nmf* classicist

'classico *a* classical; *(tipico)* classic ● *nm* classic

clas'sifi|ca *nf* classification; *Sport* results *pl*. **~'care** *vt* classify. **~'carsi** *vr* be placed. **~ca'tore** *nm (cartella)* folder. **~cazi'one** *nf* classification

clas'sista *nmf* class-conscious person

'clausola *nf* clause

claustro|fo'bia *nf* claustrophobia. **~'fobico** *a* claustrophobic

clau'sura *nf Relig* enclosed order

clavi'cembalo *nm* harpsichord

cla'vicola *nf* collar-bone

cle'men|te *a* merciful; *(tempo)* mild. **~za** *nf* mercy

cleri'cale *a* clerical. **'clero** *nm* clergy

clic *nm Comput* click; **fare ~ su** click on

cli'en|te *nmf* client; *(di negozio)* customer. **~'tela** *nf* customers *pl*

'clima *nm* climate. **cli'matico** *a* climatic; **stazione climatica** health resort

'clinica *nf* clinic. **clinico** *a* clinical ● *nm* clinician

clo'aca *nf* sewer

'cloro *nm* chlorine. **~'formio** *nm* chloroform

clou *a inv* **i momenti ~** the highlights

coabi'ta|re *vi* live together. **~zi'one** *nf* cohabitation

coagu'la|re *vt, ~rsi vr* coagulate. **~zi'one** *nf* coagulation

coaliz|i'one nf coalition. **~'zarsi** vr unite

co'atto a Jur compulsory

'**cobra** nm inv cobra

coca'ina nf cocaine. **cocai'noma- ne** nmf cocaine addict

cocci'nella nf ladybird

'**coccio** nm earthenware; (fram- mento) fragment

cocciu'taggine nf stubbornness. **~'uto** a stubborn

'**cocco** nm coconut palm; fam love; **noce di ~** coconut

cocco'drillo nm crocodile

cocco'lare vt cuddle

co'cente a ⟨sole⟩ burning

'**cocktail** nm inv (ricevimento) cocktail party

co'comero nm watermelon

co'cuzzolo nm top; (di testa, cap- pello) crown

'**coda** nf tail; (di abito) train; (fila) queue; **fare la ~** queue [up], stand in line Am. **~ di cavallo** (acconcia- tura) ponytail. **~ dell'occhio** corner of one's eye **~ di paglia** guilty conscience

co'dardo, -a a cowardly ● nmf coward

'**codice** nm code. **~ di avviamento postale** postal code, zip code Am. **~ a barre** bar-code. **~ fiscale** tax code. **~ della strada** highway code.

codifi'care vt codify

coe'ren|te a consistent. **~za** nf consistency

coesi'one nf cohesion

coe'sistere vi coexist

coe'taneo, -a a & nmf contempo- rary

cofa'netto nm casket. '**cofano** nm (forziere) chest; Auto bonnet, hood Am

'**cogliere** vt pick; (sorprendere) catch; (afferrare) seize; (colpire) hit

co'gnato, -a nmf brother-in-law; sister-in-law

cogni'zione nf knowledge

co'gnome nm surname

'**coi** = con + i

coinci'denza nf coincidence; (di treno ecc) connection

coin'cidere vi coincide

coinqui'lino nm flatmate

coin'vol|gere vt involve. **~gi- 'mento** nm involvement. **~to a** in- volved

co'ito nm coitus

col = con + il

colà adv there

cola'brodo nm inv strainer; **ri- dotto a un ~brodo** fam full of holes. **~'pasta** nm inv colander

co'la|re vt strain; (versare lenta- mente) drip ● vi (gocciolare) drip; (perdere) leak; **~re a picco** Naut sink. **~ta** nf (di metallo) casting; (di lava) flow

colazi'one nf (del mattino) break- fast; (di mezzogiorno) lunch; **prima ~** breakfast; **far ~** have breakfast/lunch. **~ al sacco** packed lunch

co'lei pron f the one

co'lera nm cholera

coleste'rolo nm cholesterol

colf nf abbr (**collaboratrice fami- liare**) home help

'colica nf colic

co'lino nm ⟨tea⟩ strainer

'**colla** nf glue; (di farina) paste. **~ di pesce** gelatine

collabo'ra|re vi collaborate. **~'tore, ~'trice** nmf collaborator. **~zi'one** nf collaboration

col'lana nf necklace; (serie) series

col'lant nm tights pl

col'lare nm collar

col'lasso nm collapse

collau'dare vt test. **col'laudo** nm test

'**colle** nm hill

col'lega nmf colleague

collega'mento nm connection; Mil liaison; Radio ecc link. **colle'ga|re** vt connect. **~si** vr TV, Radio link up

collegi'ale nmf boarder ● a ⟨re- sponsabilità, decisione⟩ collective

col'legio nm (convitto) boarding-school. ~ elettorale constituency

'collera nf anger; andare in ~ get angry. col'lerico a irascible

col'letta nf collection

collet|tività nf inv community. ~'tivo a collective; (interesse) general; biglietto ~tivo group ticket

col'letto nm collar

collezio|'nare vt collect. ~'one nf collection. ~o'nista nmf collector

colli'mare vi coincide

col'li|na nf hill. ~'noso a (terreno) hilly

col'lirio nm eyewash

collisi'one nf collision

'collo nm neck; (pacco) package; a ~ alto high-necked. ~ del piede instep

colloca'mento nm placing; (impiego) employment

collo'ca|re vt place. ~rsi vr take one's place. ~zi'one nf placing

colloqui'ale a (termine) colloquial. col'loquio nm conversation; (udienza ecc) interview; (esame) oral [exam]

collusi'one nf collusion

colluttazi'one nf scuffle

col'mare vt fill [to the brim]; bridge (divario) ● qcno di gentilezze overwhelm sb with kindness. 'colmo a full ● nm top; fig height; al colmo della disperazione in the depths of despair; questo è il colmo! (con indignazione) this is the last straw!; (con stupore) I don't believe it!

co'lomb|a nf dove. ~o nm pigeon

co'loni|a¹ nf colony; ~a (estiva) (per bambini) holiday camp. ~'ale a colonial

co'lonia² nf [acqua di] ~ [eau de] Cologne

co'lonico a (terreno, casa) farm

coloniz'za|re vt colonize. ~'tore, ~'trice nmf colonizer

co'lon|na nf column. ~ sonora sound-track. ~ vertebrale spine. ~'nato nm colonnade

colon'nello nm colonel

co'lono nm tenant farmer

colo'rante nm colouring

colo'rare vt colour; colour in (disegno)

co'lore nm colour; a colori in colour; di ~ coloured. colo'rito a coloured; (viso) rosy; (racconto) colourful ● nm complexion

co'loro pron pl the ones

colos'sale a colossal. co'losso nm colossus

'colpa nf fault; (biasimo) blame; (colpevolezza) guilt; (peccato) sin; dare la ~ a blame; essere in ~ be at fault; per ~ di because of. col'pevole a guilty ● nmf culprit

col'pire vt hit, strike; ~ nel segno hit the nail on the head

'colpo nm blow; (di arma da fuoco) shot; (urto) knock; (emozione) shock; Med, Sport stroke; (furto) raid; di ~ suddenly; far ~ make a strong impression; far venire un ~ a qcno fig give sb a fright; perdere colpi (motore:) keep missing; a ~ d'occhio at a glance; a ~ sicuro for certain. ~ d'aria chill. ~ basso blow below the belt. ~ di scena coup de théâtre. ~ di sole sunstroke; colpi pl di sole (su capelli) highlights. ~ di stato coup [d'état]. ~ di telefono ring; dare un ~ di telefono a qn give sb a ring. ~ di testa [sudden] impulse. ~ di vento gust of wind

col'poso a omicidio ~ manslaughter

coltel'lata nf stab. col'tello nm knife

colti'va|re vt cultivate. ~'tore, ~'trice nmf farmer. ~zi'one nf farming; (di piante) growing

'colto pp di cogliere ● a cultured

'coltre nf blanket

col'tura nf cultivation

co'lui pron inv m the one

'coma nm inv coma; in ~ in a coma

comanda'mento nm commandment

coman'dante nm commander; Naut, Aeron captain

coman'dare vt command; Mech control ●vi be in charge. **co'mando** nm command; (di macchina) control

co'mare nf (madrina) godmother

combaci'are vi fit together; ⟨testimonianze:⟩ concur

combat'tente a fighting ●nm combatant. **ex ~** ex-serviceman **com'bat|tere** vt/i fight. **~ti'mento** nm fight; Mil battle; **fuori ~timento** ⟨pugilato⟩ knocked out. **~'tuto** a ⟨gara⟩ hard fought

combi'na|re vt/i arrange; (mettere insieme) combine; ⟨fam: fare⟩ do; **cosa stai ~ndo?** what are you doing?. **~rsi** vr combine; (mettersi d'accordo) come to an agreement. **~zi'one** nf combination; (caso) coincidence; **per ~zione** by chance

com'briccola nf gang

combu'sti|bile a combustible ●nm fuel. **~'one** nf combustion

com'butta nf gang; **in ~** in league

'come adv like; (in qualità di) as; (interrogativo, esclamativo) how; **questo vestito è ~ il tuo** this dress is like yours; **~ stai?** how are you?; **~ va?** how are things?; **~ mai?** how come?; **~?** what?; **non sa ~ fare** he doesn't know what to do; **~ sta bene!** how well he looks!; **~ no!** that will be right!; **~ tu sai** as you know; **fa ~ vuoi** do as you like; **~ se** as if ●conj (non appena) as soon as

co'meta nf comet

'comico, -a a comic[al]; ⟨teatro⟩ comic ●nm funny side ●nm (attore) comedian, comic actor ●nf (a torte in faccia) slapstick sketch

co'mignolo nm chimney-pot

cominci'are vt/i begin, start; **a ~ da oggi** from today; **per ~** to begin with

comi'tato nm committee

comi'tiva nf party, group

co'mizio nm meeting

com'mando nm inv commando

com'medi|a nf comedy; ⟨opera teatrale⟩ play; fig sham. **~a musicale** musical. **~'ante** nmf comedian; fig pej phoney. **~'ografo, -a** nmf playwright

commemo'ra|re vt commemorate. **~zi'one** nf commemoration

commen'sale nmf fellow diner

commen't|are vt comment on; (annotare) annotate. **~'ario** nm commentary. **~a'tore, -a'trice** nmf commentator. **com'mento** nm comment

commerci'a|le a commercial; ⟨relazioni, trattative⟩ trade; ⟨attività⟩ business. **centro ~le** shopping centre. **~'lista** nmf business consultant; (contabile) accountant. **~'lizzare** vt market. **~lizzazi'one** nf marketing

commerci'ante nmf trader, merchant; (negoziante) shopkeeper. **~ all'ingrosso** wholesaler

commerci'are vi **~ in** deal in

com'mercio nm commerce; (internazionale) trade; (affari) business; **in ~** ⟨prodotto⟩ on sale. **~ all'ingrosso** wholesale trade. **~ al minuto** retail trade

com'messo, -a pp di **commettere** ●nmf shop assistant. **~ viaggiatore** commercial traveller ●nf (ordine) order

comme'stibile a edible. **commestibili** nmpl groceries

com'mettere vt commit; make ⟨sbaglio⟩

commi'ato nm leave; **prendere ~ da** take leave of

commise'rar|e vt commiserate. **~si** vr feel sorry for oneself

commissari'ato nm (di polizia) police station

commis's|ario nm [police] superintendent; (membro di commissione) commissioner; Sport steward; Comm commission agent. **~'ario d'esame** examiner. **~i'one** nf (incarico) errand; (comitato ecc) commission; (Comm: di merce) order; (Comm: di acquisti) **~'ioni** pl (acquisti) trade **~ione** go shopping. **~ione d'esa-**

me board of examiners. **C~ione Europea** European Commission

commit'tente *nf* purchaser

com'mo|sso *pp di* **commuovere** ● *a* moved. ~**vente** *a* moving

commozi'one *nf* emotion. ~ **cerebrale** concussion

commu'over|e *vt* touch, move. ~**si** *vr* be touched

commu'tare *vt* change; *Jur* commute

comò *nm inv* chest of drawers

comoda'mente *adv* comfortably

como'dino *nm* bedside table

comodità *nf inv* comfort; (*convenienza*) convenience

'comodo *a* comfortable; (*conveniente*) convenient; (*spazioso*) roomy; (*facile*) easy; **stia ~!** don't get up!; **far ~** be useful ● *nm* comfort; **fare il proprio ~** do as one pleases

compae'sano, -a *nmf* fellow countryman

com'pagine *nf* (*squadra*) team

compa'gnia *nf* company; (*gruppo*) party; **fare ~ a qcno** keep sb company; **essere di ~** be sociable. ~ **aerea** airline

com'pagno, -a *nmf* companion; (*Comm, Sport, in coppia*) partner; *Pol* comrade. ~ **di scuola** schoolmate

compa'rabile *a* comparable

compa'ra|re *vt* compare. ~**'tivo** *a & nm* comparative. ~**zi'one** *nf* comparison

com'pare *nm* (*padrino*) godfather; (*testimone di matrimonio*) witness

compa'rire *vi* appear; (*spiccare*) stand out; ~ **in giudizio** appear in court

com'parso, -a *pp di* **comparire** ● *nf* appearance; *Cinema* extra; *Theat* walk-on

compartecipazi'one *nf* sharing; (*quota*) share

comparti'mento *nm* compartment; (*amministrativo*) department

compas'sato *a* calm and collected

compassi'o|ne *nf* compassion;

aver ~ **per** feel pity for; **far ~** arouse pity. ~**'nevole** *a* compassionate

com'passo *nm* [pair of] compasses *pl*

compa'tibil|e *a* (*conciliabile*) compatible; (*scusabile*) excusable. ~**ità** *nf* compatibility. ~**'mente** *adv* ~**mente con i miei impegni** if my commitments allow

compa'tire *vt* pity; (*scusare*) make allowances for

compatri'ota *nmf* compatriot

compat'tezza *nf* (*di materia*) compactness. **com'patto** *a* compact; (*denso*) dense; (*solido*) solid; *fig* united

compene'trare *vt* pervade

compen'sar|e *vt* compensate; (*supplire*) make up for. ~**si** *vr* balance each other out

compen'sato *nm* (*legno*) plywood

compensazi'one *nf* compensation

com'penso *nm* compensation; (*retribuzione*) remuneration; **in ~** (*in cambio*) in return; (*d'altra parte*) on the other hand; (*invece*) instead

'comper|a *nf* purchase; **far ~e** do some shopping

compe'rare *vt* buy

compe'ten|te *a* competent. ~**za** *nf* competence; (*responsabilità*) responsibility

compe'tere *vi* compete; ~ **a** (*compito*) be the responsibility of

competi'tiv|ità *nf* competitiveness. ~**'tivo** *a* (*prezzo, carattere*) competitive. ~**'tore, ~'trice** *nmf* competitor. ~**zi'one** *nf* competition

compia'cen|te *a* obliging. ~**za** *nf* obligingness

compia'c|ere *vt/i* please. ~**ersi** *vr* (*congratularsi*) congratulate. ~**ersi di** (*degnarsi*) condescend. ~**i'mento** *nm* satisfaction; *pej* smugness. ~**i'uto** *a* satisfied; (*aria, sorriso*) smug

compi'an|gere *vt* pity; (*per lutto*

ecc) sympathize with. **~to** *a* lamented ●*nm* grief

'compier|e *vt* (*concludere*) complete; commit (*delitto*); **~e gli anni** have one's birthday. **~si** *vr* end; (*avverarsi*) come true

compi'la|re *vt* compile; fill in (*modulo*). **~zi'one** *nf* compilation

compi'mento *nm* **portare a ~** **qcsa** conclude sth

com'pire *vt* = **compiere**

compi'tare *vt* spell

com'pito¹ *a* polite

'compito² *nm* task; *Sch* homework

compi'uto *a* **a avere 30 anni ~i** be over 30

comple'anno *nm* birthday

complemen'tare *a* complementary; (*secondario*) subsidiary

comple'mento *nm* complement; *Mil* draft. **~ oggetto** direct object

complessi|tà *nf* complexity. **~siva'mente** *adv* on the whole. **~'sivo** *a* comprehensive; (*totale*) total. **com'plesso** *a* complex; (*difficile*) complicated ●*nm* complex; (*di cantanti ecc*) group; (*di circostanze, fattori*) combination; **in ~so** on the whole

completa'mente *adv* completely

comple'tare *vt* complete

com'pleto *a* complete; (*pieno*) full [up]; **essere al ~** (*teatro:*) be sold out; **la famiglia al ~** the whole family ●*nm* (*vestito*) suit; (*insieme di cose*) set

compli'ca|re *vt* complicate. **~rsi** *vr* become complicated. **~to** complicated. **~zi'one** *nf* complication; **salvo ~zioni** all being well

'complic|e *nmf* accomplice ●*a* (*sguardo*) knowing. **~ità** *nf* complicity

complimen'tar|e *vt* compliment. **~si** *vr* **~si con** congratulate

compli'menti *nmpl* (*ossequi*) regards; (*congratulazioni*) congratulations; **far ~** stand on ceremony

compli'mento *nm* compliment

complot'tare *vi* plot. **com'plotto** *nm* plot

compo'nente *a & nm* component ●*nmf* member

compo'nibile *a* (*cucina*) fitted; (*mobili*) modular

componi'mento *nm* composition; (*letterario*) work

com'por|re *vt* compose; (*ordinare*) put in order; *Typ* set. **~si** *vr* **~si di** be made up of

comporta'mento *nm* behaviour

compor'tar|e *vt* involve; (*consentire*) allow. **~si** *vr* behave

composi'tore, -'trice *nmf* composer; *Typ* compositor. **~zi'one** *nf* composition

com'posta *nf* stewed fruit; (*concime*) compost

compo'stezza *nf* composure

com'posto *pp di* **comporre** ●*a* composed; (*costituito*) comprising; **stai ~!** sit properly! ●*nm* *Chem* compound

com'pra|re *vt* buy. **~ tore**, **~ 'trice** *nmf* buyer

compra'vendita *nf* buying and selling

com'pren|dere *vt* understand; (*includere*) comprise. **~'sibile** *a* understandable. **~sibil'mente** *adv* understandably. **~si'one** *nf* understanding. **~'sivo** *a* understanding; (*che include*) inclusive. **com'preso** *pp di* **comprendere** ●*a* included; **tutto compreso** (*prezzo*) all-in

com'pressa *nf* compress; (*pastiglia*) tablet

compressi'one *nf* compression. **com'presso** *pp di* **comprimere** ●*a* compressed

com'primere *vt* press; (*reprimere*) repress

compro'me|sso *pp di* **compromettere** ●*nm* compromise. **~t'tente** *a* compromising. **~t'tere** *vt* compromise

comproprietà *nf* multiple ownership

compro'vare *vt* prove

com'punto a contrite

compu'tare vt calculate

com'puter nm computer. **~iz'zare** vt computerize. **~iz'zato** a computerized

computiste'ria nf book-keeping. **'computo** nm calculation

comu'nale a municipal

co'mune a common; (condiviso) mutual; (ordinario) ordinary ●nm borough, council; (amministrativo) commune; **fuori del ~** out of the ordinary. **~'mente** adv commonly

comuni'ca|re vt communicate; pass on (malattia); Relig administer Communion to. **~rsi** vr receive Communion. **~'tiva** nf communicativeness. **~'tivo** a communicative. **~to** nm communiqué. **~to stampa** press release. **~zi'one** nf communication; Teleph [phone] call; **avere la ~zione** get through; **dare la ~zione a qcno** put sb through

comuni'one nf communion; Relig [Holy] Communion

comu'nis|mo nm communism. **~ta** a & nmf communist

comunità nf inv community. **C~ [Economica] Europea** European [Economic] Community

co'munque conj however ●adv anyhow

con prep with; (mezzo) by; **~ facilità** easily; **~ mia grande gioia** to my great delight; **è gentile ~ tutti** he is kind to everyone; **col treno** by train; **~ questo tempo** in this weather

co'nato nm **~ di vomito** retching

'conca nf basin; (valle) dell

conca'te|na|re vt link together. **~zi'one** nf connection

'concavo a concave

con'ceder|e vt grant; award (premio); (ammettere) admit. **~si** vr allow oneself (pausa)

concentra'mento nm concentration

concen'tra|re vt, **~rsi** vr concen-

trate. **~to** a concentrated ●nm **~to di pomodoro** tomato purée. **~zi'one** nf concentration

concepi'mento nm conception

conce'pire vt conceive (bambino); (capire) understand; (figurarsi) conceive of; devise (piano ecc)

con'cernere vt concern

concer'tar|e vt Mus harmonize; (organizzare) arrange. **~si** vr agree

concer'tista nmf concert performer. **con'certo** nm concert; (composizione) concerto

concessio'nario nm agent

concessi'one nf concession

con'cesso pp di concedere

con'cetto nm concept; (opinione) opinion

concezi'one nf conception; (idea) concept

con'chiglia nf [sea] shell

'concia nf tanning; (di tabacco) curing

conci'a|re vt tan; cure (tabacco); **~re qcno per le feste** give sb a good hiding. **~rsi** vr (sporcarsi) get dirty; (vestirsi male) dress badly. **~to** a (pelle, cuoio) tanned

concili'abile a compatible

concili'ante a conciliatory

concili'a|re vt reconcile; settle (contravvenzione); (favorire) induce. **~rsi** vr go together; (mettersi d'accordo) become reconciled. **~zi'one** nf reconciliation; Jur settlement

con'cilio nm Relig council; (riunione) assembly

conci'mare vt feed (pianta). **con'cime** nm manure; (chimico) fertilizer

concisi'one nf conciseness. **con'ciso** a concise

conci'tato a excited

concit'tadino, -a nmf fellow citizen

con'clu|dere vt conclude; (finire con successo) achieve. **~dersi** vr come to an end. **~si'one** nf conclusion; **in ~sione** (insomma) in short. ~'sivo a conclusive. ~so pp di concludere

concomi'tanza nf (di circostanze, fatti) combination

concor'da|nza nf agreement. ~re vt agree; Gram make agree. ~to nm agreement; Jur, Comm arrangement

con'cord|e a in agreement; (unanime) unanimous

concor'ren|te a concurrent; (rivale) ●nmf Comm, Sport competitor; (candidato) candidate. ~za nf competition. ~zi'ale a competitive

con'cor|rere vi (contribuire) concur; (andare insieme) go together; (competere) compete. ~so pp di concorrere ●nm competition. fuori ~so not in the official competition. ~so di bellezza beauty contest

concreta'mente adv concretely

concre|'tare vt (concludere) achieve. ~tiz'zare vt put into concrete form (idea, progetto)

con'creto a concrete; in ~ in concrete terms

concussi'one nf extortion

con'danna nf sentence; pronunziare una ~ pass a sentence. condan'nare vt condemn; Jur sentence. condan'nato, -a nmf convict

conden'sare vt, ~rsi vr condense. ~zi'one nf condensation

condi'mento nm seasoning; (salsa) dressing. con'dire vt flavour; dress (insalata)

condiscen'den|te a indulgent; pej condescending. ~za nf indulgence; pej condescension

condi'videre vt share

condizio'na|le a & nm conditional ●nf Jur suspended sentence. ~'mento nm Psych conditioning condizio'na|re vt condition. ~to a conditional. ~tore nm air conditioner

condizi'one nf condition; a ~ che on condition that

condogli'anze nfpl condolences; fare le ~ a offer condolences to

condomini'ale a (spese) common. condo'minio nm joint ownership; (edificio) condominium

condo'nare vt remit. con'dono nm remission

con'dotta nf conduct, (circoscrizione di medico) district; (di gara ecc) management; (tubazione) piping

con'dotto pp di condurre ● a medico ~ district doctor ● nm pipe; Anat duct

condu'cente nm driver

con'du|rre vt lead; drive (veicoli); (accompagnare) take; conduct (gas, elettricità ecc); (gestire) run. ~rsi vr behave. ~'tore, ~'trice nmf TV presenter; (di veicolo) driver ●nm Electr conductor. ~t'tura nf duct

confabu'lare vi have a confab

confa'cente a suitable. con'farsi vr confarsi a suit

confederazi'one nf confederation

confe'renz|a nf (discorso) lecture; (congresso) conference. ~a stampa news conference. ~i'ere, -a nmf lecturer

confe'rire vt (donare) give ●vi confer

con'ferma nf confirmation. confer'mare vt confirm

confes'sa|re vt, ~arsi vr confess. ~io'nale a & nm confessional. ~i'one nf confession. ~ore nm confessor

con'fetto nm sugared almond

confet'tura nf jam

confezio'na|re vt manufacture; make (abiti); package (merci). ~to a (vestiti) off-the-peg; (gelato) wrapped

confezi'one nf manufacture; (di abiti) tailoring; (di pacchi) packaging; confezioni pl clothes. ~ regalo gift pack

confic'car|e vt thrust. ~si vr run into

confi'd|are vi ~are in trust ●vt

confide. **~arsi** vr **~arsi con** confide in. **~ente** a confident ● nmf confidant

confi'denz|a nf confidence; (familiarità) familiarity; **prendersi delle ~e** take liberties. **~l'ale** a confidential; (rapporto, tono) familiar

configu'ra|re vt Comput configure. **~zi'one** nf configuration

confi'nante a neighbouring

confi'na|re vi (relegare) confine ● vi **~re con** border on. **~rsi** vr withdraw. **~to** a confined

con'fin|e nm border; (tra terreni) boundary. **~o** nm political exile

con'fisca nf (di proprietà) forfeiture. **~'scare** vt confiscate

con'flitt|o nm conflict. **~u'ale** a adversarial

conflu'enza nf confluence; (di strade) junction

conflu'ire vi (fiumi:) flow together; (strade:) meet

con'fonder|e vt confuse; (turbare) confound; (imbarazzare) embarrass. **~si** a (mescolarsi) mingle; (turbarsi) become confused; vr (sbagliarsi) be mistaken

confor'ma|re vt, **~rsi** vr conform. **~zi'one** nf conformity (a with); (del terreno) composition

con'forme a according. **~'mente** adv accordingly

confor'mi|smo nm conformity. **~sta** nmf conformist. **~tà** nf (a norma) conformity

confor'tante a comforting

confor'ta|re vt comfort. **~evole** a (comodo) comfortable. **con'forto** nm comfort

confron'tare vt compare

con'fronto nm comparison; **in ~ a** by comparison with; **nei tuoi confronti** towards you; **senza ~** far and away

confusi|o'nario a (persona) muddle-headed. **~'one** nf confusion; (baccano) racket; (disordine) mess; (imbarazzo) embarrassment. **con'fuso** pp di **confondere**

● a confused; (indistinto) indistinct; (imbarazzato) embarrassed

confu'tare vt confute

conge'dar|e vt dismiss; Mil discharge. **~si** vr take one's leave

con'gedo nm leave; **essere in ~** be on leave. **~ malattia** sick leave. **~ maternità** maternity leave

conge'gnare vt devise; (mettere insieme) assemble. **con'gegno** nm device

congela'mento nm freezing; Med frost-bite

conge'la|re vt freeze. **~to** a (cibo) deep-frozen. **~'tore** nm freezer

congeni'ale a congenial

con'genito a congenital

congesti'ona|re vt congest. **~to** a (traffico) congested; (viso) flushed. **congesti'one** nf congestion

conget'tura nf conjecture

congi'unger|e vt join; combine (sforzi). **~si** vr join

congiunti'vite nf conjunctivitis

congiun'tivo nm subjunctive

congi'unto pp di **congiungere** ● a joined ● nm relative

congiun'tu|ra nf joint; (circostanza) juncture; (situazione) situation. **~'rale** a economic

congiunzi'one nf Gram conjunction

congi'u|ra nf conspiracy. **~'rare** vi conspire

conglome'rato nm conglomerate; fig conglomeration; (da costruzione) concrete

congratu'la|rsi vr **~rsi con qcno per** congratulate sb on. **~zi'oni** nfpl congratulations

con'grega nf band

congre'ga|re vt, **~rsi** vr congregate. **~zi'one** nf congregation

con'gresso nm congress

'congruo a proper; (giusto) fair

conguagli'are vt balance. **con-gu'aglio** nm balance

coni'are vt coin

'conico a conical

co'nifera nf conifer

co'niglio *nm* rabbit

coniu'gale *a* marital; *(vita)* married

coniu'ga|re *vt* conjugate. **~rsi** *vr* get married. **~zi'one** *nf* conjugation

'coniuge *nmf* spouse

connazio'nale *nmf* compatriot

connessi'one *nf* connection. **con'nesso** *pp di* **connettere**

con'nettere *vt* connect ● *vi* think rationally

conni'vente *a* conniving

conno'ta|re *vt* connote. **~to** *nm* distinguishing feature; **~ti** *pl* description

con'nubio *nm fig* union

'cono *nm* cone

cono'scen|te *nmf* acquaintance. **~za** *nf* knowledge; *(persona)* acquaintance; *(sensi)* consciousness; **perdere ~za** lose consciousness; **riprendere ~za** regain consciousness, come to

co'nosc|ere *vt* know; *(essere a conoscenza di)* be acquainted with; *(fare la conoscenza di)* meet. **~i'tore**, **~i'trice** *nmf* connoisseur. **~i'uto** *pp di* **conoscere** ● *a* well-known

con'quist|a *nf* conquest. **conqui-'stare** *vt* conquer; *fig* win

consa'cra|re *vt* consecrate; *(or-dain* ⟨*sacerdote*⟩; *(dedicare)* dedicate. **~rsi** *vr* devote oneself. **~zi'one** *nf* consecration

consangu'ineo, **-a** *nmf* blood-relation

consa'pevol|e *a* conscious. **~'lez-za** *nf* consciousness. **~l'mente** *adv* consciously

'con•cio *a* conscious

consecu'tivo *a* consecutive; *(se-guente)* next

con'segna *nf* delivery; *(merce)* consignment; *(custodia)* care; *(di prigioniero)* handover; *(Mil: ordi-ne)* orders *pl*; *(Mil: punizione)* confinement; **pagamento alla ~** cash on delivery

conse'gnare *vt* deliver; *(affidare)*

give in charge; *Mil* confine to barracks

consegu'en|te *a* consequent. **~za** *nf* consequence; **di ~za** *(perciò)* consequently

consegui'mento *nm* achievement

consegu'ire *vt* achieve ● *vi* follow

con'senso *nm* consent

consensu'ale *a* consensus-based

consen'tire *vi* consent ● *vt* allow

con'serto *a* **a braccia conserte** with one's arms folded

con'serva *nf* preserve; *(di frutta)* jam; *(di agrumi)* marmalade. **~ di pomodoro** tomato sauce

conser'var|e *vt* preserve; *(mante-nere)* keep. **~si** *vr* keep; **~si in sa-lute** keep well

conserva'tore, **-'trice** *nmf Pol* conservative

conserva'torio *nm* conservatory

conservazi'one *nf* preservation; **a lunga ~** long-life

conside'ra|re *vt* consider; *(stima-re)* regard. **~to** *a* *(stimato)* esteemed. **~zi'one** *nf* consideration; *(osservazione, riflessione)* remark

conside'revole *a* considerable

consigli'abile *a* advisable

consigli'|are *vt* advise; *(raccoman-dare)* recommend. **~'arsi** *vr* **~arsi con qcno** ask sb's advice. **~'ere**, **-a** *nmf* adviser; *(membro di con-siglio)* councillor

con'siglio *nm* advice; *(ente)* coun-cil. **~ d'amministrazione** board of directors. **C~ dei Ministri** Cabinet

consi'sten|te *a* substantial; *(spes-so)* thick; *(fig: argomento)* valid. **~za** *nf* consistency; *(spessore)* thickness

con'sistere *vi* **in** consist of

consoci'ata *nf (azienda)* associate company

conso'lar|e¹ *vt* console; *(rallegra-re)* cheer. **~si** *vr* console oneself

conso'lare² *a* consular. **~to** *nm* consulate

consolazi'one *nf* consolation; *(gioia)* joy

con'sole *nm inv (tastiera)* console

'console *nm* consul

consoli'dar|e *vt*, ~si *vr* consolidate

conso'nante *nf* consonant

'consono *a* consistent

con'sorte *nmf* consort

con'sorzio *nm* consortium

con'stare *vi* ~ di consist of; *(risultare)* appear; a quanto mi consta as far as I know; mi consta che it appears that

consta'ta|re *vt* ascertain. ~zi'one *nf* observation

consu'e|to *a* & *nm* usual. ~tudi-'nario *a (diritto)* common; *(persona)* set in one's ways. ~'tudine *nf* habit; *(usanza)* custom

consu'len|te *nmf* consultant. ~za *nf* consultancy

consul'ta|re *vt* consult. ~rsi con consult with. ~zi'one *nf* consultation

consul'tivo *a* consultative. ~orio *nm* clinic

consu'ma|re *vt (usare)* consume; wear out *(abito, scarpe)*; consummate *(matrimonio)*; commit *(delitto)*. ~rsi *vr* consume; *(abito, scarpe)* wear out; *(struggersi)* pine

consu'mato *a (politico)* seasoned; *(scarpe, tappeto)* worn

consuma'tore, -'trice *nmf* consumer. ~zi'one *nf (bibita)* drink; *(spuntino)* snack

consu'mismo *nm* consumerism. ~ta *nmf* consumerist

con'sumo *nm* consumption; *(di abito, scarpe)* wear; *(uso)* use; generi di ~ consumer goods or items. ~ [di carburante] [fuel] consumption

consun'tivo *nm* [bilancio] ~ final statement

conta'balle *nmf fam* storyteller

con'tabil|e *a* book-keeping ●*nmf* accountant. ~ità *nf* accounting; tenere la ~ità keep the accounts

contachi'lometri *nm inv* mileometer, odometer *Am*

conta'dino, -a *nmf* farm-worker; *(medievale)* peasant

contagi'are *vt* infect. con'tagio *nm* infection. ~'oso *a* infectious

conta'gocce *nm inv* dropper

contami'na|re *vt* contaminate. ~zi'one *nf* contamination

con'tante *nm* cash; pagare in contanti pay cash

con'tare *vt/i* count; *(tenere conto di)* take into account; *(proporsi)* intend

conta'scatti *nm inv Teleph* time-unit counter

conta'tore *nm* meter

contat'tare *vt* contact. con'tatto *nm* contact

'conte *nm* count

conteggi'are *vt* put on the bill ●*vi* calculate. con'teggio *nm* calculation. conteggio alla rovescia countdown

con'te|gno *nm* behaviour; *(atteggiamento)* attitude. ~'gnoso *a* dignified

contem'pla|re *vt* contemplate; *(fissare)* gaze at. ~zi'one *nf* contemplation

con'tempo *nm* nel ~ in the meantime

contempo'ranea'mente *adv* at once. ~'raneo, -a *a* & *nmf* contemporary

conten'dente *nmf* competitor. con'tendere *vi* compete; *(litigare)* quarrel ●*vt* contend

con'ten|ere *vt* contain; *(reprimere)* repress. ~ersi *vr* contain oneself. ~i'tore *nm* container

conten'tarsi *vr* ~ di be content with

conten'tezza *nf* joy

conten'tino *nm* placebo

con'tento *a* glad; *(soddisfatto)* contented

conte'nuto *nm* contents *pl*; *(soggetto)* content

contenzi'oso *nm* legal department

con'tes|a *nf* disagreement; *Sport*

contest. **~o** pp di **contendere ●** a contested

con'tessa nf countess

conte'sta|re vt contest; Jur notify. **~'tario** a anti-establishment. **~'tore**, **~'trice** nmf protester. **~zi'one** nf (disputa) dispute

con'testo nm context

con'tiguo a adjacent

continen'tale a continental. **conti'nente** nm continent

conti'nenza nf continence

contin'gen|te nm contingent; (quota) quota. **~za** nf contingency

continua'mente adv (senza interruzione) continuously; (frequentemente) continually

continu'are vt/i continue; (riprendere) resume. **~ativo** a permanent. **~azi'one** nf continuation. **~ità** nf continuity

con'tinuo a continuous; (molto frequente) continual. **corrente ~** a direct current; **di ~** continually

'conto nm calculation; (in banca, negozio) account; (di ristorante ecc) bill; (stima) consideration; **a conti fatti** all things considered; **far ~ di** (supporre) suppose; (proporsi) intend; **far ~ su** rely on; **in fin dei conti** when all is said and done; **per ~ di** on behalf of; **per ~ mio** (a mio parere) in my opinion; (da solo) on my own; **starsene per ~ proprio** be on one's own; **rendersi ~ di** qcsa realize sth; **sul ~ di** qcno (voci, informazioni) about sb; **tener ~ di** qcsa take sth into account; **tenere da ~** keep look after sth; **fare i conti con** qcno fig sort sb out. **~ corrente** current account, checking account Am. **~ alla rovescia** countdown

con'torcere vt twist. **~si** vr twist about

contor'nare vt surround

con'torno nm contour; Culin vegetables pl

contorsi'one nf contortion. **con'torto** pp di contorcere **●** a twisted

contrabban'dare vt smuggle. **~di'ere**, **-a** nmf smuggler. **con'trab'bando** nm contraband

contrab'basso nm double bass

contraccambi'are vt return. con'trac'cambio nm return

contracce|t'tivo nm contraceptive. **~zi'one** nf contraception

contrac'col|po nm rebound; (di arma da fuoco) recoil; fig repercussion

con'trada nf (rione) district

contrad'detto pp di contraddire

contrad'di|re vt contradict. **~t'torio** a contradictory. **~zi'one** nf contradiction

contraddi'stin|guere vt differentiate. **~to** a distinct

contra'ente nmf contracting party

contra'ereo a anti-aircraft

contraf'fa|re vt disguise; (imitare) imitate; (falsificare) forge. **~tto** a forged. **~zi'one** nf disguising; (imitazione) imitation; (falsificazione) forgery

con'tralto nm countertenor **●** nf contralto

contrap'peso nm counterbalance

contrap'por|re vt counter; (confrontare) compare. **~si** vr contrast; **~si a** be opposed to

contraria'mente adv contrary (a to)

contrari'are vt oppose; (infastidire) annoy. **~'arsi** vr get annoyed. **~età** nf inv adversity; (ostacolo) set-back

con'trario a contrary, opposite; (direzione) opposite; (sfavorevole) unfavourable **●** nm contrary, opposite; **al ~** on the contrary

con'trarre vt contract

contras'se|gnare vt mark. **~'segno** nm mark; [in] **~segno** (spedizione) cash on delivery, COD

contra'stante a contrasting

contra'stare vt oppose; (contestare) contest **●** vi clash. **con'trasto** nm contrast; (litigio) dispute

contrattac'care vt counter-attack. **contrat'tacco** *nm* counter-attack

contrat'ta|re *vt/i* negotiate; (*mercanteggiare*) bargain. **~zi'one** *nf* (*salariale*) bargaining

contrat'tempo *nm* hitch

con'tratt|o *pp di* **contrarre ●** *nm* contract. **~o a termine** fixed-term contract. **~u'ale** *a* contractual

contravve'n|ire *vi* contravene. **~zi'one** *nf* contravention; (*multa*) fine

contrazi'one *nf* contraction; (*di prezzi*) reduction

contribu'ente *nmf* contributor; (*del fisco*) taxpayer

contribu'ire *vi* contribute. **contri'buto** *nm* contribution

'contro *prep* against; **~ di me** against me **●** *nm* **il pro e il ~** the pros and cons *pl*

contro'battere *vt* counter

controbilanci'are *vt* counterbalance

controcor'rente *a* (*idee, persona*) non-conformist **●** *adv* upriver; *fig* upstream

controffen'siva *nf* counter-offensive

controfi'gura *nf* stand-in

controfir'mare *vt* countersign

controindicazi'one *nf* *Med* contraindication

control'la|re *vt* control; (*verificare*) check; (*collaudare*) test. **~rsi** *vr* have self-control. **~to** *a* controlled

con'troll|o *nm* control; (*verifica*) check; *Med* check-up. **~lo delle nascite** birth control. **~'lore** *nm* controller; (*sui treni ecc*) [ticket] inspector. **~lore di volo** air-traffic controller

contro'luce *nf* in **~** against the light

contro'mano *adv* in the wrong direction

contromi'sura *nf* countermeasure

contropi'ede *nm* **prendere in ~** catch off guard

controprodu'cente *a* self-defeating

con'trordine *nm* counter order; **salvo ~i** unless I/you hear to the contrary

contro'senso *nm* contradiction in terms

controspio'naggio *nm* counter-espionage

contro'vento *adv* against the wind

contro'vers|ia *nf* controversy; *Jur* dispute. **~o** *a* controversial

contro'voglia *adv* unwillingly

contu'macia *nf* default; **in ~** in one's absence

contun'dente *a* (*corpo, arma*) blunt

contur'ba|nte *a* perturbing

contusi'one *nf* bruise

convale'scen|te *a* convalescent. **~za** *nf* convalescence; **essere in ~za** be convalescing

con'vali|da *nf* validation. **~'dare** *vt* confirm; validate (*atto, biglietto*)

con'vegno *nm* meeting; (*congresso*) congress

conve'nevol|e *a* suitable; **~i** *pl* pleasantries

conveni'en|te *a* convenient; (*prezzo*) attractive; (*vantaggioso*) advantageous. **~za** *nf* convenience; (*interesse*) advantage; (*di prezzo*) attractiveness

conve'nire *vi* (*riunirsi*) gather; (*concordare*) agree; (*ammettere*) admit; (*essere opportuno*) be convenient **●** *vt* agree on; **ci conviene andare** it is better to go; **non mi conviene stancarmi** I'd better not tire myself out

con'vento *nm* (*di suore*) convent; (*di frati*) monastery

conve'nuto *a* fixed

convenzio'nale *a* conventional. **~'one** *nf* convention

conver'gen|te *a* converging. **~za** *nf fig* confluence

con'vergere *vi* converge

conver'sa|re *vi* converse. **~zi'one** *nf* conversation

conversi'one *nf* conversion

con'verso *pp di* **convergere**

conver'tibile *nf Auto* convertible

conver'ti|re *vt* convert. **~rsi** *vr* be converted. **~to, -a** *nmf* convert

con'vesso *a* convex

convin'cente *a* convincing

con'vin|cere *vt* convince. **~to a** convinced. **~zi'one** *nf* conviction

con'vitto *nm* boarding school

convi'ven|te *nm* common-law husband ● *nf* common-law wife. **~za** *nf* cohabitation. **con'vivere** *vi* live together

convivi'ale *a* convivial

convo'ca|re *vt* convene. **~zi'one** *nf* convening

convogli'are *vt* convey; ⟨navi:⟩ convoy. **con'voglio** *nm* convoy; ⟨ferroviario⟩ train

convulsi'one *nf* convulsion. **con'vulso** *a* convulsive; ⟨febbrile⟩ feverish

coope'ra|re *vi* co-operate. **~'tiva** *nf* co-operative. **~zi'one** *nf* co-operation

coordina'mento *nm* co-ordination

coordi'na|re *vt* co-ordinate. **~ta** *nf Math* coordinate. **~zi'one** *nf* co-ordination

co'perchio *nm* lid; ⟨copertura⟩ cover

co'perta *nf* blanket; ⟨copertura⟩ cover; *Naut* deck

coper'tina *nf* cover; ⟨di libro⟩ dust-jacket

co'perto *pp di* **coprire** ● *a* covered; ⟨cielo⟩ overcast ● *nm* ⟨a tavola⟩ place; ⟨prezzo del coperto⟩ cover charge; **al ~** under cover

coper'tone *nm* tarpaulin; ⟨gomma⟩ tyre

coper'tura *nf* covering; *Comm, Fin* cover

'copia *nf* copy; **bella/brutta ~** fair/rough copy. **~ su carta** hardcopy. **copi'are** *vt* copy

copi'one *nm* script

copi'oso *a* plentiful

'coppa *nf* ⟨calice⟩ goblet; ⟨per gelato ecc⟩ dish; *Sport* cup. **~ [di] gelato** ice-cream ⟨served in a dish⟩

cop'petta *nf* ⟨di ceramica, vetro⟩ bowl; ⟨di gelato⟩ small tub

'coppia *nf* couple; ⟨in carte⟩ pair

co'prente *a* ⟨cipria, vernice⟩ covering

copri'capo *nm* headgear

coprifu'oco *nm* curfew

copri'letto *nm* bedspread

copripiu'mino *nm* duvet cover

co'pri|re *vt* cover; drown ⟨suono⟩; hold ⟨carica⟩. **~si** *vr* ⟨vestirsi⟩ cover up; *fig* cover oneself; ⟨cielo:⟩ become overcast

coque *sf* **alla ~** ⟨uovo⟩ soft-boiled

co'raggi|o *nm* courage; ⟨sfaccia-taggine⟩ nerve; **~o!** come on. **~'oso** *a* courageous

co'rale *a* choral

co'rallo *nm* coral

co'rano *nm* Koran

co'raz|za *nf* armour; ⟨di animali⟩ shell. **~'zata** *nf* battleship. **~'zato** *a* ⟨nave⟩ armour-clad

corbelle'ria *nf* nonsense; ⟨sproposito⟩ blunder

'corda *nf* cord; ⟨spago, Mus⟩ string; ⟨fune⟩ rope; ⟨cavo⟩ cable; **essere giù di ~** be depressed; **dare ~ a** *qcno* encourage sb. **corde** *pl vocali* vocal cords

cor'data *nf* roped party

cordi'a|le *a* cordial ● *nm* ⟨bevanda⟩ cordial; **saluti ~i** best wishes. **~ità** *nf* cordiality

cor'doglio *nm* grief; ⟨lutto⟩ mourning

cor'done *nm* cord; ⟨schieramento⟩ cordon. **~ ombelicale** umbilical cord

core'ogra'fia *nf* choreography. **~'ografo, -a** *nmf* choreographer

cori'andoli *nmpl* confetti *sg*

cori'andolo *nm* ⟨spezia⟩ coriander

cori'car|e *vt* put to bed. **~si** *vr* go to bed

co'rista *nmf* choir member

cor'nacchia nf crow

corna vedi **corno**

corna'musa nf bagpipes pl

cornea nf cornea

cor'nett|a nf Mus cornet; (del telefono) receiver. **~o** nm (brioche) croissant

cor'ni|ce nf frame. **~ci'one** nm cornice

'corno nm (pl nf **corna**) horn; **fare le corna** a qcno be unfaithful to sb; **fare le corna** (per scongiuro) touch wood. **cor'nuto** a horned ● nm (fam: marito tradito) cuckold; (insulto) bastard

'coro nm chorus; Relig choir

co'rolla nf corolla

co'rona nf crown; (di fiori) wreath; (rosario) rosary. **~'mento** nm (di impresa) crowning. **coro'nare** vt crown; (sogno) fulfil

cor'petto nm bodice

'corpo nm body; (Mil, diplomatico) corps inv; **a ~ a ~** man to man; **andare di ~** move one's bowels. **~ di ballo** corps de ballet. **~ insegnante** teaching staff. **~ del reato** incriminating item

corpo'rale a corporal

corporati'vismo nm corporatism

corpora'tura nf build

corporazi'one nf corporation

cor'poreo a bodily

cor'poso a full-bodied

corpu'lento a stout

cor'puscolo nm corpuscle

corre'dare vt equip

corre'dino nm (per neonato) layette

cor'redo nm (nuziale) trousseau

cor'reggere vt correct; lace (bevanda)

corre'lare vt correlate

cor'rente a running; (in vigore) current; (frequente) everyday; (inglese ecc) fluent ● nf current; (d'aria) draught; **essere al ~** be up to date. **~'mente** adv (parlare) fluently

'correre vi run; (affrettarsi) hurry; Sport race; (notizie:) circulate; **~**

dietro a run after ● vt run; **~ un pericolo** run a risk; **lascia ~!** don't bother!

corre|tta'mente adv correctly.

cor'retto pp di **correggere** ● a correct; (caffè) with a drop of alcohol. **~zi'one** nf correction. **~zione di bozze** proof-reading

cor'rida nf bullfight

corri'doio nm corridor; Aeron aisle

corri|'dore, -'trice nmf racer; (a piedi) runner

corri'era nf coach, bus

corri'ere nm courier; (posta) mail; (spedizioniere) carrier

corri'mano nm bannister

corrispet'tivo nm amount due

corrispon'den|te a corresponding ● nmf correspondent. **~za** nf correspondence; **scuola/corsi per ~za** correspondence course; **vendite per ~za** mail-order [shopping]. **corri'spondere** vi correspond; (stanza:) communicate; **corrispondere a** (contraccambiare) return

corri'sposto a (amore) reciprocated

corrobo'rare vt strengthen; fig corroborate

cor'roder|e vt, **~si** vr corrode

cor'rompere vt corrupt; (con denaro) bribe

corrosi'one nf corrosion. **corro'sivo** a corrosive

cor'roso pp di **corrodere**

cor'rotto pp di **corrompere** ● a corrupt

corrucci'a|rsi vr be vexed. **~to** a upset

corru'gare vt wrinkle; **~ la fronte** knit one's brows

corruzi'one nf corruption; (con denaro) bribery

'corsa nf running; (rapida) dash; Sport race; (di treno ecc) journey; **di ~** at a run; **fare una ~** run

cor'sia nf gangway; (di ospedale) ward; Auto lane; (di supermercato) aisle

cor'sivo *nm* italics *pl*

'corso *pp di* **correre** ● *nm* course; (*strada*) main street; *Comm* circulation; **lavori in ~** work in progress; **nel ~ di** during. **~ d'acqua** watercourse

'corte *nf* [court]yard; (*Jur, regale*) court; **fare la ~ a** qcno court sb. **~ d'appello** court of appeal

cor'teccia *nf* bark

corteggia'mento *nm* courtship

coreggi'a|re *vt* court. **~'tore** *nm* admirer

cor'teo *nm* procession

cor'te|se *a* courteous. **~'sia** *nf* courtesy; **per ~sia** please

cortigi'ano, -a *nmf* courtier ● *nf* courtesan

cor'tile *nm* courtyard

cor'tina *nf* curtain; (*schermo*) screen

'corto *a* short; **per farla corta** in short; **essere a ~ di** be short of. **~ circuito** *nm* short [circuit]

cortome'traggio *nm Cinema* short

cor'vino *a* jet-black

'corvo *nm* raven

'cosa *nf* thing; (*faccenda*) matter; *inter, rel* what; [**che**] **~** what; **nessuna ~** nothing; **ogni ~** everything; **per prima ~** first of all; **tante cose** so many things; (*augurio*) all the best

'cosca *nf* clan

'coscia *nf* thigh; *Culin* leg

cosci'en|te *a* conscious. **~za** *nf* conscience; (*consapevolezza*) consciousness

co'scri|tto *nm* conscript. **~zi'one** *nf* conscription

così *adv* so; (*in questo modo*) like this, like that; (*perciò*) therefore; **le cose stanno ~** that's how things stand; **fermo ~!** hold it; **proprio ~!** exactly!; **basta ~!** that will do!; **ah, è ~?** it's like that, is it?; **~ ~** so-so; **e ~ via** and so on; **per ~ dire** so to speak; **più di ~** any more; **una ~ cara ragazza!** such a nice girl!; **è stato ~**

generoso da aiutarti he was kind enough to help you ● *conj* (*allora*) so ● *a inv* (*tale*) like that, such; **una ragazza ~** a girl like that, such a girl

cosicché *conj* and so

cosid'detto *a* so-called

co'smesi *nf* cosmetics

co'smetico *a* & *nm* cosmetic

'cosmico *a* cosmic

'cosmo *nm* cosmos

cosmopo'lita *a* cosmopolitan

co'spargere *vt* sprinkle; (*disseminare*) scatter

co'spetto *nm* **al ~ di** in the presence of

co'spicuo *a* conspicuous; (*somma ecc*) considerable

cospi'ra|re *vi* conspire. **~'tore**, **~'trice** *nmf* conspirator. **~zi'one** *nf* conspiracy

'costa *nf* coast, coastline; *Anat* rib

costà *adv* there

co'stan|te *a* & *nf* constant. **~za** *nf* constancy

co'stare *vi* cost; **quanto costa?** how much is it?

co'stata *nf* chop

costeggi'are *vt* (*per mare*) coast; (*per terra*) skirt

co'stei *pers pron vedi* **costui**

costellazi'one *nf* constellation

coster'na|to *a* dismayed. **~zi'one** *nf* consternation

costi'er|a *nf* stretch of coast. **~o** *a* coastal

costi'pa|to *a* constipated. **~zi'one** *nf* constipation; (*raffreddore*) bad cold

costitu'ire *vt* constitute; (*formare*) form; (*nominare*) appoint. **~si** *vr Jur* give oneself up

costituzio'nale *a* constitutional.

costituzi'one *nf* constitution; (*fondazione*) setting up

'costo *nm* cost; **ad ogni ~** at all costs; **a nessun ~** on no account

'costola *nf* rib; (*di libro*) spine

costo'letta *nf* cutlet

co'storo *pron vedi* **costui**

co'stoso *a* costly

co'stretto pp di **costringere**

co'stringere vt compel; (stringere) constrict. **~'tivo** a coercive. **~zi'one** nf constraint

costru'ire vt build, construct. **~'tivo** a constructive. **~zi'one** nf building, construction

co'stui, co'stei, pl **co'storo** prons (soggetto) he, she, pl they; (complemento) him, her, pl them

co'stume nm (usanza) custom; (condotta) morals pl; (indumento) costume. **~ da bagno** swim-suit; (da uomo) swimming trunks

co'tenna nf pigskin; (della pancetta) rind

coto'letta nf cutlet

co'tone nm cotton. **~ idrofilo** cotton wool, absorbent cotton Am

'cotta nf (fam: innamoramento) crush

'cottimo nm **lavorare a** ~ do piece-work

'cotto pp di **cuocere ●** a done; (fam: infatuato) in love; (fam: sbronzo) drunk; **ben ~** (carne) well done

'cotton fi'oc® nm inv cotton bud

cot'tura nf cooking

co'vare vt hatch; sicken for (malattia); harbour (odio) **●** vi smoulder

'covo nm den

co'vone nm sheaf

'cozza nf mussel

coz'zare vi **~ contro** bump into.

'cozzo nm fig clash

C.P. abbr (**Casella Postale**) PO Box

'crampo nm cramp

'cranio nm skull

cra'tere nm crater

cra'vatta nf tie; (a farfalla) bow-tie

cre'anza nf politeness; **mala ~** bad manners

cre'are vt create; (causare) cause. **~tività** nf creativity; **~'tivo** a creative. **~to** nm creation. **~'tore,** **~'trice** nmf creator. **~zi'one** nf creation

crea'tura nf creature; (bambino) baby; **povera ~!** poor thing!

cre'den|te nmf believer. **~za** nf belief; Comm credit; (mobile) sideboard. **~zi'ali** nfpl credentials

'creder|e vt believe; (pensare) think **●** vi **~e in** believe in; **credo di sì** I think so; **non ti credo** I don't believe you. **cre'dibile** a credible. **~si** vr think oneself to be; **si crede uno scrittore** he flatters himself he is a writer. **cre'dibile** a credible.

credibilità nf credibility

'credi|to nm credit; (stima) esteem; **comprare a ~to** buy on credit. **~'tore,** **~'trice** nmf creditor

'credo nm inv credo

credulità nf credulity

'credu|lo a credulous. **~'lone, -a** nmf simpleton

'crema nf cream; (di uova e latte) custard. **~ idratante** moisturizer. **~ pasticciera** egg custard. **~ solare** suntan lotion

cre'ma|re vt cremate. **~'torio** nm crematorium. **~zi'one** nf cremation

crème cara'mel nf crème caramel

cre'meria nf dairy (also selling ice cream and cakes)

Crem'lino nm Kremlin

'crepa nf crack

cre'paccio nm cleft; (di ghiaccio) crevasse

crepacu'ore nm heart-break

crepa'pelle: a ~ adv fit to burst; **ridere a ~** split one's sides with laughter

cre'pare vi crack; (fam: morire) kick the bucket; **~ dal ridere** laugh fit to burst

crepa'tura nf crevice

crêpe nf inv pancake

crepi'tare vi crackle

cre'puscolo nm twilight

cre'scendo nm crescendo

'cresc|ere vi grow; (aumentare) increase **●** vt (allevare) bring up; (aumentare) increase. **~ita** nf growth; (aumento) increase. **~i'uto** pp di **crescere**

'cresi|ma *nf* confirmation. ~'mare *vt* confirm

'crespo *a* ⟨capelli⟩ frizzy ●*nm* crêpe

'cresta *nf* crest; ⟨cima⟩ peak

'creta *nf* clay

'Creta *nf* Crete

cre'tino, -a *a* stupid ●*nmf* idiot

cric *nm* jack

'cricca *nf* gang

cri'ceto *nm* hamster

crimi'nal|e *a* & *nmf* criminal. ~ità *nf* crime. 'crimine *nm* crime

crimi'noso *a* criminal

'crin|e *nm* horsehair. ~'iera *nf* mane

'cripta *nf* crypt

crisan'temo *nm* chrysanthemum

'crisi *nf inv* crisis; *Med* fit

cristal'lino *nm* crystalline

cristalliz'zar|e *vt*, ~si *vr* crystallize; ⟨fig: parola, espressione⟩ become part of the language

cri'stallo *nm* crystal

Cristia'nesimo *nm* Christianity

cristi'ano, -a *a* & *nmf* Christian

'Cristo *nm* Christ; un povero c~ *a* poor beggar

cri'terio *nm* criterion; ⟨buon senso⟩ [common] sense

'criti|ca *nf* criticism; ⟨recensione⟩ review. criti'care *vt* criticize. ~co *a* critical ●*nm* critic. ~cone, -a *nmf* faultfinder

crivel'lare *vt* riddle (di with)

cri'vello *nm* sieve

croc'cante *a* crisp ●*nm* type of crunchy nut biscuit

croc'chetta *nf* croquette

'croce *nf* cross; a occhio e ~ roughly; fare testa e ~ spin a coin. C~ Rossa Red Cross

croce'via *nm inv* crossroads *sg*

croci'ata *nf* crusade

cro'cicchio *nm* crossroads *sg*

croci'era *nf* cruise; *Archit* crossing

croci'fi|ggere *vt* crucify. ~ssi'one *nf* crucifixion. ~sso *pp di* crocifiggere ●*a* crucified ●*nm* crucifix

crogio'larsi *vr* bask

crogi[u]'olo *nm* crucible; *fig* melting pot

crol'lare *vi* collapse; ⟨prezzi:⟩ slump. 'crollo *nm* collapse; ⟨dei prezzi⟩ slump

cro'mato *a* chromium-plated.

'cromo *nm* chrome. cromo'soma *nm* chromosome

'cronaca *nf* chronicle; ⟨di giornale⟩ news; *TV, Radio* commentary; fatto di ~ news item. ~ nera crime news

'cronico *a* chronic

cro'nista *nmf* reporter

crono'logico *a* chronological

crono'traggio *nm* timing

crono'trare *vt* time

cro'nometro *nm* chronometer

'crosta *nf* crust; ⟨di formaggio⟩ rind; ⟨di ferita⟩ scab; ⟨quadro⟩ daub

cro'staceo *nm* shellfish

cro'stata *nf* tart

cro'stino *nm* croûton

crucci'arsi *vr* worry. 'cruccio *nm* worry

cruci'ale *a* crucial

cruci'verba *nm inv* crossword [puzzle]

cru'del|e *a* cruel. ~tà *nf inv* cruelty

'crudo *a* raw; ⟨rigido⟩ harsh

cru'ento *a* bloody

cru'miro *nm* blackleg, scab

'crusca *nf* bran

cru'scotto *nm* dashboard

'Cuba *nf* Cuba

cu'betto *nm* ~ di ghiaccio ice cube

'cubico *a* cubic

cubi'tal|e *a* a caratteri ~i in enormous letters

'cubo *nm* cube

cuc'cagna *nf* abundance; ⟨baldoria⟩ merry-making; paese della ~ land of plenty

cuc'cetta *nf* ⟨su un treno⟩ couchette; *Naut* berth

cucchia'ino *nm* teaspoon

cucchi'a|io *nm* spoon; **al ~io** *(dolce)* creamy. **~i'ata** *nf* spoonful

'cuccia *nf* dog's bed; **fa la ~! lie down!**

cuccio'lata *nf* litter

cucciolo *nm* puppy

cu'cina *nf* kitchen; *(il cucinare)* cooking; *(cibo)* food; *(apparecchio)* cooker; **far da ~** cook; **libro di ~** cook[ery] book. **~ a gas** gas cooker

cuci'n|are *vt* cook. **~ino** *nm* kitchenette

cu'ci|re *vt* sew; **macchina per ~re** sewing-machine. **~to** *nm* sewing. **~'tura** *nf* seam

cucù *nm inv* cuckoo

'cuculo *nm* cuckoo

'cuffia *nf* bonnet; *(da bagno)* bathing-cap; *(ricevitore)* headphones *pl*

cu'gino, -a *nmf* cousin

'cui *pron rel (persona: con prep)* who, whom *fml*; *(cose, animali: con prep)* which; *(tra articolo e nome)* whose; **la persona con ~ ho parlato** the person [who] I spoke to; **la ditta per ~ lavoro** the company I work for, the company for which I work; **l'amico il ~ libro è stato pubblicato** the friend whose book was published; **in ~** *(dove)* where; *(quando)* that; **per ~** *(perciò)* so; **la città in ~ vivo** the city I live in, the city where I live; **il giorno in ~ l'ho visto** the day [that] I saw him

culi'nari|a *nf* cookery. **~o a** culinary

'culla *nf* cradle. **cul'lare** *vt* rock

culmi'na|nte *a* culminating. **~re** *vi* culminate. **'culmine** *nm* peak

'culo *nm vulg* arse; *(fortuna)* luck

'culto *nm* cult; *Relig* religion; *(adorazione)* worship

cul'tu|ra *nf* culture. **~ra generale** general knowledge. **~'rale** *a* cultural

cultu'ris|mo *nm* body-building. **~ta** *nmf* body-builder

cumula'tivo *a* cumulative; **biglietto ~** group ticket

'cumulo *nm* pile; *(mucchio)* heap; *(nuvola)* cumulus

'cuneo *nm* wedge

cu'netta *nf* gutter

cu'ocere *vt/i* cook; fire *(ceramica)*

cu'oco, -a *nmf* cook

cu'oio *nm* leather. **~ capelluto** scalp

cu'ore *nm* heart; **cuori** *pl (carte)* hearts; **nel profondo del ~** in one's heart of hearts; **di [buon] ~** *(persona)* kind-hearted; **nel ~ della notte** in the middle of the night; **stare a ~ a qcno** be very important to sb

cupi'digia *nf* greed

'cupo *a* gloomy; *(suono)* deep

'cupola *nf* dome

'cura *nf* care; *(amministrazione)* management; *Med* treatment; **a ~ di** edited by; **in ~** under treatment. **~ dimagrante** *(slimming)* diet. **cu'rante** *a* **medico curante** GP, doctor

cu'rar|e *vt* take care of; *Med* treat; *(guarire)* cure; edit *(testo)*. **~si** *vr* take care of oneself; *Med* follow a treatment; **~si di** *(badare a)* mind

cu'rato *nm* parish priest

cura'tore, -'trice *nmf* trustee; *(di testo)* editor

'curia *nf* curia

curio's|are *vi* be curious; *(mettere il naso)* pry *(in* into); *(nei negozi)* look around. **~ità** *nf inv* curiosity. **curi'oso** *a* curious; *(strano)* odd

cur'sore *nm* Comput cursor

'curva *nf* curve; *(stradale)* bend. **~ a gomito** U-bend. **cur'vare** *vt* curve; *(strada:)* bend. **cur'varsi** *vr* bend. **'curvo** *a* curved; *(piegato)* bent

cusci'netto *nm* pad; *Mech* bearing

cu'scino *nm* cushion; *(guanciale)* pillow. **~ d'aria** air cushion

'cuspide *nf* spire

cu'stod|e *nm* caretaker. **~e giudiziario** official receiver. **~ia**

nf care; *Jur* custody; *(astuccio)* case. **~ia cautelare** remand.

custo'dire *vt* keep; *(badare)* look after

cu'taneo *a* skin *attrib*

'cute *nf* skin

cu'ticola *nf* cuticle

Dd

da *prep* from; *(con verbo passivo)* by; *(moto a luogo)* to; *(moto per luogo)* through; *(stato in luogo)* at; *(temporale)* since; *(continuativo)* for; *(causale)* with; *(in qualità di)* as; *(con caratteristica)* with; *(come)* like; **da Roma a Milano** from Rome to Milan; **staccare un quadro dalla parete** take a picture off the wall; **i bambini dai 5 ai 10 anni** children between 5 and 10; **vedere qcsa da vicino/lontano** see sth from up close/from a distance; **scritto da** written by; **andare dal panettiere** go to the baker's; **passo da te più tardi** I'll come over to your place later; **passiamo da qui** let's go this way; **un appuntamento dal dentista** an appointment at the dentist's; **il treno passa da Venezia** the train goes through Venice; **dall'anno scorso** since last year; **vivo qui da due anni** I've been living here for two years; **da domani** from tomorrow; **piangere dal dolore** cry with pain; **ho molto da fare** I have a lot to do; **occhiali da sole** sunglasses; **qualcosa da mangiare** something to eat; **un uomo dai capelli scuri** a man with dark hair; **un oggetto da poco** it's not worth much; **l'ho fatto da solo** I did it by myself; **si è fatto da sé** he is a self-made man; **non è da lui** it's not like him

dac'capo *adv* again; *(dall'inizio)* from the beginning

dacché *conj* since

'dado *nm* dice; *Culin* stock cube; *Techn* nut

daf'fare *nm* work

dagli = da + gli. 'dai = da + i

dai *int* come on!

'daino *nm* deer; *(pelle)* buckskin

dal = da + il. 'dalla = da + la. 'dalle = da + le. 'dallo = da + lo

'dalia *nf* dahlia

dal'tonico *a* colour-blind

'dama *nf* lady; *(nei balli)* partner; *(gioco)* draughts *sg*

dami'gella *nf (di sposa)* bridesmaid

damigi'ana *nf* demijohn

dam'meno *adv* **non essere ~ (di qcno)** be no less good (than sb)

da'naro *nm* = **denaro**

dana'roso *a (fam: ricco)* loaded

da'nese *a* Danish ● *nmf* Dane ● *nm (lingua)* Danish

Dani'marca *nf* Denmark

dan'na|re *vt* damn; **far ~re qcno** drive sb mad. **~to** *a* damned. **~zi'one** *nf* damnation

danneggia'mento *nm* damage. **~'are** *vt* damage; *(nuocere)* harm

'danno *nm* damage; *(a persona)* harm. **dan'noso** *a* harmful

Da'nubio *nm* Danube

'danza *nf* dance; *(il danzare)* dancing. **dan'zare** *vi* dance

dapper'tutto *adv* everywhere

dap'poco *a* worthless

dap'prima *adv* at first

'dardo *nm* dart

'dar|e *vt* give; sit *(esame)*; have *(festa)*; **~ qcsa a qcno** give sb sth; **~ da mangiare a qcno** give sb something to eat; **~ la buonanotte a qcno** say good night to sb; **~ del tu/del lei a qcno** address sb as "tu"/"lei"; **~ del cretino a qcno** call sb an idiot; **~ qcsa per scontato** take sth for granted; **cosa danno alla TV stasera?** what's on TV tonight? ● *vi* **~ nel-**

l'occhio be conspicuous; ~ **alla testa** go to one's head; ~ **su** ⟨finestra, casa⟩ look on to; ~ **sui** o **ai nervi** a qcno get on sb's nerves ●nm Comm debit. ~**si** vr ⟨scambiarsi⟩ give each other; ~**si da fare** get down to it; **si è dato tanto da fare!** he went to so much trouble!; ~**si a** ⟨cominciare⟩ take up; ~**si al bere** take to drink; ~**si per** ⟨malato, assente⟩ pretend to be; ~**si per vinto** give up; **può** ~**si** maybe

'**darsena** nf dock

'**data** nf date; ~ **di emissione** date of issue. ~ **di nascita** date of birth. ~ **di scadenza** cut-off date

da'ta|re vt date; a ~**re da** as from. ~**to** a dated

dato a given; ⟨dedito⟩ addicted; ~ **che** seeing that, given that ●nm datum. ~ **di fatto** well-established fact; **dati** pl data. **da'tore** nm giver. **datore, datrice** nmf **di lavoro** employer

'**dattero** nm date

dattilogra'f|are vt type. ~**ia** nf typing. **datti'lografo, -a** nmf typist

dattilo'scritto a ⟨copia⟩ typewritten

dat'torno adv **togliersi** ~ clear off

da'vanti adv before; ⟨dirimpetto⟩ opposite; ⟨di fronte⟩ in front ●a inv front ●nm front; ~ **a** prep before, in front of

davan'zale nm window sill

da'vanzo adv more than enough

dav'vero adv really; **per** ~ in earnest; **dici** ~? honestly?

'**dazio** nm duty; ⟨ufficio⟩ customs pl

d.C. abbr ⟨dopo Cristo⟩ AD

'**dea** nf goddess

debel'lare vt defeat

debili'ta|nte a weakening. ~**rsi** vr weaken. ~**zi'one** nf debilitation

debita'mente adv duly

'**debi|to** a due; **a tempo** ~ in due course ●nm debt. ~**'tore, -'trice** nmf debtor

'**debo|le** a weak; ⟨luce⟩ dim; ⟨suono⟩ faint ●nm weak point; ⟨preferenza⟩ weakness. ~**'lezza** nf weakness

debor'dare vi overflow

debosci'ato a debauched

debut'ta|nte nm ⟨attore⟩ actor making his début ●nf actress making her début. ~**re** vi make one's début. **de'butto** nm début

deca'den|te a decadent. ~**'tismo** nm decadence. ~**za** nf decline; Jur loss. **deca'dere** vi lapse. **decadi'mento** nm ⟨delle arti⟩ decline

decaffei'nato a decaffeinated ●nm decaffeinated coffee, decaf fam

decan'tare vt ⟨lodare⟩ praise

decapi'ta|re vt decapitate; behead ⟨condannato⟩. ~**zi'one** nf decapitation; beheading

decappot'tabile a convertible

de'ce|dere vi ⟨morire⟩ die. ~'**duto** a deceased

dece'lerare vt decelerate, slow down

decen'nale a ten-yearly. **de'cennio** nm decade

de'cen|te a decent. ~**te'mente** adv decently. ~**za** nf decency

decentra'mento nm decentralization

de'cesso nm death; **atto di** ~ death certificate

de'ci|dere vt decide; settle ⟨questione⟩. ~**si** vr make up one's mind

deci'frare vt decipher; ⟨documenti cifrati⟩ decode

deci'male a decimal

deci'mare vt decimate

'**decimo** a tenth

de'cina nf Math ten; **una** ~ **di** ⟨circa dieci⟩ about ten

decisa'mente adv definitely, decidedly

decisio'nale a decision-making

deci'si'one nf decision. ~'**sivo** a decisive. **de'ciso** pp di **decidere** ●a decided

decla'ma|re *vt/i* declaim. **~'torio** *a (stile)* declamatory

declas'sare *vt* downgrade

decli'na|re *vt* decline; **~re ogni responsabilità** disclaim all responsibility ● *vi* go down; *(tramontare)* set. **~zi'one** *nf* Gram declension. **de'clino** *nm* decline; **in declino** *(popolarità:)* on the decline

decodificazi'one *nf* decoding

decol'lare *vi* take off

décolle'té *nm inv* décolleté, low neckline

de'collo *nm* take-off

decolo'ra|nte *nm* bleach. **~re** *vt* bleach

decolorazi'one *nf* bleaching

decom'po|rre *vt*, **~rsi** *vr* decompose. **~sizi'one** *nf* decomposition

deconcen'trarsi *vr* become distracted

deconge'lare *vt* defrost

decongestio'nare *vt* Med, fig relieve congestion in

deco'ra|re *vt* decorate. **~'tivo** *a* decorative. **~to** *a (ornato)* decorated. **~'tore**, **~'trice** *nmf* decorator. **~zi'one** *nf* decoration

de'coro *nm* decorum

decorosa'mente *adv* decorously. **decoroso** *a* dignified

decor'renza *nf* **~ dal...** starting from...

de'correre *vi* pass; **a ~ da** with effect from. **de'corso** *pp di* **decorrere** ● *nm* passing; Med course

de'crepito *a* decrepit

decre'scente *a* decreasing. **decre'scere** *vi* decrease; *(prezzi:)* go down; *(acque:)* subside

decre'tare *vt* decree. **de'creto** *nm* decree. **decreto legge** *decree which has the force of law*

'dedalo *nm* maze

'dedica *nf* dedication

dedi'ca|re *vt* dedicate. **~si** *vr* dedicate oneself

'dedi|to *a* **~** given to; *(assorto)* engrossed in; addicted to *(vizi)*. **~zi'one** *nf* dedication

de'dotto *pp di* **dedurre**

dedu'cibile *a (tassa)* allowable

de'du|rre *vt* deduce; *(sottrarre)* deduct. **~'ttivo** *a* deductive. **~zi'one** *nf* deduction

defal'care *vt* deduct

defe'rire *vt* Jur remit

defezi|o'nare *vi (abbandonare)* defect. **~'one** *nf* defection

defici'en|te *a (mancante)* deficient; Med mentally deficient ● *nmf* mental defective; pej half-wit. **~za** *nf* deficiency; *(lacuna)* gap; Med mental deficiency

'deficit *nm inv* deficit. **~'tario** *a (bilancio)* deficit *attrib*

defi'larsi *vr (scomparire)* slip away

défilé *nm inv* fashion show

defi'ni|re *vt* define; *(risolvere)* settle. **~tiva'mente** *adv* for good. **~'tivo** *a* definitive. **~to** *a* definite. **~zi'one** *nf* definition; *(soluzione)* settlement

deflazi'one *nf* deflation

deflet'tore *nm* Auto quarterlight

deflu'ire *vi (liquidi:)* flow away; *(persone:)* stream out

de'flusso *nm (di marea)* ebb

defor'ma|re *vt* deform *(arto)*; fig distort. **~si** *vr* lose its shape. **de'forme** *a* deformed. **~ità** *nf* deformity

defor'mato *a* warped. **~zi'one** *nf (di fatti)* distortion; **è una ~zione professionale** put it down to the job

defrau'dare *vt* defraud

de'funto, **-a** *a & nmf* deceased

degene'ra|re *vi* degenerate. **~to** *a* degenerate. **~zi'one** *nf* degeneration. **de'genere** *a* degenerate

de'gen|te *a* bedridden ● *nmf* patient. **~za** *nf* confinement

'degli = **di** + **gli**

deglu'tire *vt* swallow

de'gna|re *vt* **~ qcno di uno sguardo** deign to look at sb. **~si** *vr* deign, condescend

'degno *a* worthy; *(meritevole)* deserving

degrada'mento *nm* degradation

degra'dante *a* demeaning

degra'da|re *vt* degrade. **~rsi** *vr* lower oneself; ⟨*città:*⟩ fall into a state of disrepair. **~zi'one** *nf* degradation

de'grado *nm* damage; **~ ambien'tale** *nm* environmental damage

degu'sta|re *vt* taste. **~zi'one** *nf* tasting

'dei = **di** + **i**. **'del** = **di** + **il**

dela'tore, -'trice *nmf* [police] informer. **~zi'one** *nf* informing

'delega *nf* proxy

dele'ga|re *vt* delegate. **~to** *nm* delegate. **~zi'one** *nf* delegation

dele'terio *a* harmful

del'fino *nm* dolphin; ⟨*stile di nuoto*⟩ butterfly [stroke]

de'libera *nf* bylaw

delibe'ra|re *vt/i* deliberate; **~ su/ in** rule on/in. **~to** *a* deliberate

delicata'mente *adv* delicately

delica'tezza *nf* delicacy; ⟨*fragilità*⟩ frailty; ⟨*tatto*⟩ tact

deli'cato *a* delicate; ⟨*salute*⟩ frail; ⟨*suono, colore*⟩ soft

delimi'tare *vt* delimit

deline'a|re *vt* outline. **~rsi** *vr* be outlined; *fig* take shape. **~to** *a* defined

delin'quen|te *nmf* delinquent. **~za** *nf* delinquency

deli'rante *a* *Med* delirious; ⟨*assurdo*⟩ insane

deli'rare *vi* be delirious. **de'lirio** *nm* delirium; *fig* frenzy

de'litt|o *nm* crime. **~u'oso** *a* criminal

de'lizia *nf* delight. **~'are** *vt* delight. **~'oso** *a* delightful; ⟨*cibo*⟩ delicious

'della = **di** + **la**. **'delle** = **di** + **le**. **'dello** = **di** + **lo**

'delta *nm inv* delta

delta'plano *nm* hang-glider; **fare ~** go hang-gliding

delucidazi'one *nf* clarification

delu'dente *a* disappointing

de'lu|dere *vt* disappoint. **~si'one** *nf* disappointment. **de'luso** *a* disappointed

dema'gogico *a* popularity-seeking, demagogic

demar'ca|re *vt* demarcate. **~zi'one** *nf* demarcation

de'men|te *a* demented. **~za** *nf* dementia. **~zi'ale** *a* ⟨*assurdo*⟩ zany

demilitariz'za|re *vt* demilitarize. **~zi'one** *nf* demilitarization

demistificazi'one *nf* debunking

demo'cratico *a* democratic. **~'zia** *nf* democracy

democristi'ano, -a *a* & *nmf* Christian Democrat

demogra'fia *nf* demography. **demo'grafico** *a* demographic

demo'li|re *vt* demolish. **~zi'one** *nf* demolition

'demone *nm* demon. **de'monio** *nm* demon

demoraliz'zar|e *vt* demoralize. **~si** *vr* become demoralized

de'mordere *vi* give up

demoti'vato *a* demotivated

de'nari *nmpl* ⟨*nelle carte*⟩ diamonds

de'naro *nm* money

deni'gra|re *vt* denigrate. **~'torio** *a* denigratory

denomi'na|re *vt* name. **~'tore** *nm* denominator. **~zi'one** *nf* denomination; **~zione di origine controllata** mark guaranteeing the quality of a wine

deno'tare *vt* denote

densità *nf inv* density. **'denso** *a* thick, dense

den'ta|le *a* dental. **~rio** *a* dental. **~ta** *nf* bite. **~tura** *nf* teeth *pl*

'dente *nm* tooth; ⟨*di forchetta*⟩ prong; **al ~** *Culin* just slightly firm. **~ del giudizio** wisdom tooth. **~ di latte** milk tooth. **denti'era** *nf* dentures *pl*, false teeth *pl*

denti'fricio *nm* toothpaste

den'tista *nmf* dentist

'dentro *adv* in, inside; ⟨*in casa*⟩ indoors; **da ~** from within; **qui ~** in here ●*prep* in, inside; ⟨*di tempo*⟩ within, by ●*nm* inside

denuclearizzazi'one *nf* denuclearization

denu'dar|e *vt* bare. **~si** *vr* strip

de'nuncia, de'nunzia *nf* denunciation; (*alla polizia*) reporting; (*dei redditi*) [income] tax return. **~'are** *vt* denounce; (*accusare*) report

denu'trito *a* underfed. **~zi'one** *nf* malnutrition

deodo'rante *a* & *nm* deodorant

dépendance *nf inv* outbuilding

depe'ri|bile *a* perishable. **~'mento** *nm* wasting away; (*dei merci*) deterioration. **~re** *vi* waste away

depi'la|re *vt* depilate. **~rsi** *vr* shave (*gambe*); pluck (*sopracciglia*). **~'torio** *nm* depilatory

deplo'rabile *a* deplorable

deplo'r|are *vt* deplore; (*dolersi di*) grieve over. **~evole** *a* deplorable

de'porre *vt* put down; lay down (*armi*); lay (*uova*); (*togliere da una carica*) depose; (*testimoniare*) testify

depor'ta|re *vt* deport. **~to, -a** *nmf* deportee. **~zi'one** *nf* deportation

deposi'ta|re *vt* deposit; (*lasciare in custodia*) leave; (*in magazzino*) store. **~io, -a** *nmf* (*di segreto*) repository. **~si** *vr* settle

de'posi|to *nm* deposit; (*luogo*) warehouse; *Mil* depot. **~to bagagli** left-luggage office. **~zi'one** *nf* deposition; (*da una carica*) removal

depra'va|re *vt* deprave. **~to** *a* depraved. **~zi'one** *nf* depravity

depre'cabile *a* appalling. **~re** *vt* deprecate

depre'dare *vt* plunder

depressi'one *nf* depression. **de'presso** *pp di* **deprimere** ●*a* depressed

deprez'zar|e *vt* depreciate. **~si** *vr* depreciate

depri'mente *a* depressing

de'primer|e *vt* depress. **~si** *vr* become depressed

depu'ra|re *vt* purify. **~'tore** *nm* purifier

depu'ta|re *vt* delegate. **~to, -a** *nmf* Member of Parliament, MP

deraglia'mento *nm* derailment

deragli'are *vi* go off the lines; **far ~** derail

'derby *nm inv* *Sport* local Derby

deregolamentazi'one *nf* deregulation

dere'litto *a* derelict

de'ri|dere *vt* deride. **~si'one** *nf* derision. **~'sorio** *a* derisory

de'riva *nf* drift; **andare alla ~** drift

deri'va|re *vi* **~re da** (*provenire*) derive from ●*vt* derive; (*sviare*) divert. **~zi'one** *nf* derivation; (*di fiume*) diversion

dermato|lo'gia *nf* dermatology. **~'logico** *a* dermatological. **der'ma'tologo, -a** *nmf* dermatologist

'deroga *nf* dispensation. **dero'ga|re** *vi* **derogare a** depart from

der'ra|ta *nf* merchandise. **~e alimentari** foodstuffs

deru'bare *vt* rob

descrit'tivo *a* descriptive. **des'critto** *pp di* **descrivere**

des'cri|vere *vt* describe. **~'vibile** *a* describable. **~zi'one** *nf* description

de'serto *a* uninhabited ●*nm* desert

deside'rabile *a* desirable

deside'rare *vt* wish; (*volere*) want; (*intensamente*) long for; (*bramare*) desire; **desidera?** what would you like?, can I help you?; **lasciare a ~** leave a lot to be desired

desi'de|rio *nm* wish; (*brama*) desire; (*intenso*) longing. **~'roso** *a* desirous; (*bramoso*) longing

desi'gnare *vt* designate; (*fissare*) fix

desi'nenza *nf* ending

de'sistere *vi* **~ da** desist from

'desktop 'publishing *nm inv* desktop publishing

deso'lante *a* distressing

deso'la|re *vt* distress. **~to** *a* desolate; (*spiacente*) sorry. **~zi'one** *nf* desolation 'despota *nm* despot

de'star|e *vt* waken; *fig* awaken. ~si *vr* waken; *fig* awaken

desti'na|re *vt* destine; (*nominare*) appoint; (*assegnare*) assign; (*indirizzare*) address. ~'tario *nm* (*di lettera, pacco*) addressee. ~zi'one *nf* destination; *fig* purpose

de'stino *nm* destiny; (*fato*) fate

destitu'ire *vt* dismiss. ~zi'one *nf* dismissal

'desto *a liter* awake

'destra *nf* (*parte*) right; (*mano*) right hand; **prendere a** ~ turn right

destreggi'ar|e *vi*, ~si *vr* manoeuvre

de'strezza *nf* dexterity; (*abilità*) skill

'destro *a* right; (*abile*) skilful

detei'nato *a* tannin-free

dete'n|ere *vt* hold; (*polizia:*) detain. ~uto, -a *nmf* prisoner. ~zi'one *nf* detention

deter'gente *a* cleaning; (*latte, crema*) cleansing ● *nm* detergent; (*per la pelle*) cleanser

deteriora'mento *nm* deterioration

deterio'rar|e *vt* cause to deteriorate. ~si *vr* deteriorate

determi'nante *a* decisive

determi'na|re *vt* determine. ~rsi *vr* ~rsi a resolve to. ~'tezza *nf* determination. ~'tivo *a* Gram definite. ~to *a* (*risoluto*) determined; (*particolare*) specific. ~zi'one *nf* determination; (*decisione*) decision

deter'rente *a* & *nm* deterrent

deter'sivo *nm* detergent. ~ per i piatti washing-up liquid

dete'stare *vt* detest, hate

deto'nare *vi* detonate

de'tra|rre *vt* deduct (da from). ~zi'one *nf* deduction

detri'mento *nm* detriment; a ~ di to the detriment of

de'trito *nm* debris

'detta *nf* a ~ di according to

dettagli'ante *nmf* Comm retailer

dettagli'a|re *vt* detail. ~ta'mente *adv* in detail

det'taglio *nm* detail; al ~ Comm retail

det'ta|re *vt* dictate; ~re legge *fig* lay down the law. ~to *nm*, ~'tura *nf* dictation

'detto *a* said; (*chiamato*) called; (*soprannominato*) nicknamed; ~ fatto no sooner said than done ● *nm* saying

detur'pare *vt* disfigure

deva'sta|re *vt* devastate. ~to *a* devastated. ~zi'one *nf* devastation; *fig* ravages *pl*

devi'a|re *vi* deviate ● *vt* divert. ~zi'one *nf* deviation; (*stradale*) diversion

devitaliz'zare *vt* deaden (*dente*)

devo'lu|to *pp di* devolvere ● *a* devolved. ~zi'one *nf* devolution

de'volvere *vt* devolve

de'vo|to *a* devout; (*affezionato*) devoted. ~zi'one *nf* devotion

di *prep* of, (*partitivo*) some; (*scritto da*) by; (*parlare, pensare ecc*) about; (*con causa, mezzo*) with; (*con provenienza*) from; (*in comparazioni*) than; (*con infinito*) to; la casa di mio padre/dei miei genitori my father's house/my parents' house; compra del pane buy some bread; hai del pane? do you have any bread?; un film di guerra a war film; piangere di dolore cry with pain; coperto di neve covered with snow; sono di Genova I'm from Genoa; uscire di casa leave one's house; più alto di te taller than you; è ora di partire it's time to go; crede di aver ragione he thinks he's right; dire di sì say yes; di domenica on Sundays; di sera in the evening; una pausa di un'ora an hour's break; un corso di due mesi a two-month course

dia'be|te *nm* diabetes. ~ico, -a *a* & *nmf* diabetic

dia'bolico *a* diabolical

dia'dema *nm* diadem; (*di donna*) tiara

di'afano *a* diaphanous

dia'framma *nm* diaphragm; (*divisione*) screen

di'agnos|i *nf inv* diagnosis. **~ti'care** *vt* diagnose

diago'nale *a & nf* diagonal

dia'gramma *nm* diagram

dialet'tale *a* dialect. **dia'letto** *nm* dialect

dialo'gante *a* **unità ~** Comput interactive terminal

di'alogo *nm* dialogue

dia'mante *nm* diamond

di'ametro *nm* diameter

di'amine *int* **che ~...** what on earth...

diaposi'tiva *nf* slide

di'ario *nm* diary

diar'rea *nf* diarrhoea

di'avolo *nm* devil; **va al ~** go to hell!; **che ~ fai?** what the hell are you doing?

di'batt|ere *vt* debate. **~ersi** *vr* struggle. **~ito** *nm* debate; (*meno formale*) discussion

dica'stero *nm* office

di'cembre *nm* December

dice'ria *nf* rumour

dichia'ra|re *vt* state; (*ufficialmente*) declare. **~rsi** *vr* **si dichiara innocente** he says he's innocent. **~zi'one** *nf* statement; (*documento, di guerra*) declaration

dician'nove *a & nm* nineteen

dicias'sette *a & nm* seventeen

dici'otto *a & nm* eighteen

dici'tura *nf* wording

didasca'lia *nf* (*di film*) subtitle; (*di illustrazione*) caption

di'dattic|a *nf* didactics. **~o** *a* didactic; (*televisione*) educational

di'dentro *adv* inside

didi'etro *adv* behind ● *nm hum* hindquarters *pl*

di'eci *a & nm* ten

die'cina = **decina**

'diesel *a & nf inv* diesel

di'esis *nm inv* sharp

di'eta *nf* diet; **essere a ~** be on

a diet. **die'tetico** *a* diet. **die'tista** *nmf* dietician. **die'tologo** *nmf* dietician

di'etro *adv* behind ● *prep* behind; (*dopo*) after ● *a* back; (*di zampe*) hind ● *nm* back; **le stanze di ~** the back rooms; **le zampe di ~** the hind legs

dietro'front *nm inv* about-turn; *fig* U-turn

di'fatti *adv* in fact

di'fen|dere *vt* defend. **~dersi** *vr* defend oneself. **~siva** *nf* **stare sulla ~siva** be on the defensive. **~sivo** *a* defensive. **~sore** *nm* defender; **avvocato ~sore** defence counsel

di'fesa *nf* defence; **prendere le ~e di qcno** come to sb's defence. ● *pp di* **difendere**

difet'tare *vi* be defective; **~are di** lack. **~ivo** *a* defective

di'fet|to *nm* defect; (*morale*) fault, flaw; (*mancanza*) lack; (*di zampe, abito*) flaw; **essere in ~to** be at fault; **far ~to** to be lacking. **~'toso** *a* defective; (*abito*) flawed

diffa'ma|re *vt* (*con parole*) slander; (*per iscritto*) libel. **~'torio** *a* slanderous; (*per iscritto*) libellous. **~zi'one** *nf* slander; (*scritta*) libel

diffe'ren|te *a* different. **~za** *nf* difference; **a ~za di** unlike; **non fare ~za** make no distinction (**fra** between). **~zi'ale** *a & nm* differential

differenzi'ar|e *vt* differentiate. **~si** *vr* **~si da** differ from

diffe'ri|re *vt* postpone ● *vi* be different. **~ta** *nf* **in ~ta** TV prerecorded

diffi'cil|e *a* difficult; (*duro*) hard; (*improbabile*) unlikely ● *nm* difficulty. **~'mente** *adv* with difficulty

difficoltà *nf inv* difficulty

dif'fida *nf* warning

diffi'd|are *vi* **~are di** distrust ● *vt* warn. **~ente** *a* mistrustful. **~enza** *nf* mistrust

dif'fondere *vt* spread; diffuse (*calore, luce ecc*). **~si** *vr* spread.

diffusi'one *nf* diffusion; *(di giornale)* circulation

dif'fu|so *pp di* **diffondere** ● *a* common; *(malattia)* widespread; *(luce)* diffuse. **~'sore** *nm (per asciugacapelli)* diffuser

difi'lato *adv* straight; *(subito)* straightaway

'diga *nf* dam; *(argine)* dike

dige'ribile *a* digestible

dige'rire *vt* digest; *fam* stomach. **~sti'one** *nf* digestion. **~'stivo** *a* digestive ● *nm* digestive; *(dopo cena)* liqueur

digi'tale *a* digital; *(delle dita)* finger *attrib* ● *nf (fiore)* foxglove

digi'tare *vt* key in

digiu'nare *vi* fast

digi'uno *a* **essere ~** have an empty stomach ● *nm* fast; **a ~** *(bere ecc)* on an empty stomach

digni'tà *nf* dignity. **~'tario** *nm* dignitary. **~'toso** *a* dignified

digressi'one *nf* digression

digri'gnare *vi* **~ i denti** grind one's teeth

dila'gare *vi* flood; *fig* spread

dilani'are *vt* tear to pieces

dilapi'dare *vt* squander

dila'ta|re *vt*, **~rsi** *vr* dilate; *(metallo, gas:)* expand. **~zi'one** *nf* dilation

dilazio'nabile *a* postponable

dilazio|'nare *vt* delay. **~'one** *nf* delay

dilegu'ar|e *vt* disperse. **~si** *vr* disappear

di'lemma *nm* dilemma

dilet'tan|te *nmf* amateur. **~'tistico** *a* amateurish

dilet'tare *vt* delight

di'letto, -a *a* beloved ● *nm (piacere)* delight ● *nmf (persona)* beloved

dili'gen|te *a* diligent; *(lavoro)* accurate. **~za** *nf* diligence

dilu'ire *vt* dilute

dilun'gar|e *vt* prolong. **~si** *vr* **~si su** dwell on *(argomento)*

diluvi'are *vi* pour [down]. **di'luvio** *nm* downpour; *fig* flood

---

dima'gr|ante *a* slimming, diet. **~i'mento** *nm* loss of weight. **~ire** *vi* slim

dime'nar|e *vt* wave; wag *(coda)*. **~si** *vr* be agitated

dimensi'one *nf* dimension; *(misura)* size

dimenti'canza *nf* forgetfulness; *(svista)* oversight

dimenti'car|e *vt*, **~si** *vr* ~ **[di]** forget. **dimentico** *a* **dimentico di** *(che non ricorda)* forgetful of

di'messo *pp di* **dimettere** ● *a* humble; *(trasandato)* shabby; *(voce)* low

dimesti'chezza *nf* familiarity

di'metter|e *vt* dismiss; *(da ospedale ecc)* discharge. **~si** *vr* resign

dimez'zare *vt* halve

dimi'nu|ire *vt/i* diminish; *(in maglia)* decrease. **~'tivo** *a & nm* diminutive. **~zi'one** *nf* decrease; *(riduzione)* reduction

dimissi'oni *nfpl* resignation *sg*; **dare le ~** resign

di'mo|ra *nf* residence. **~'rare** *vi* reside

dimo'strante *nmf* demonstrator

dimo'stra|re *vt* demonstrate; *(provare)* prove; *(mostrare)* show. **~rsi** *vr* prove [to be]. **~'tivo** *a* demonstrative. **~zi'one** *nf* demonstration; *Math* proof

di'namico, -a *a* dynamic ● *nf* dynamics *sg*. **dina'mismo** *nm* dynamism

dinami'tardo *a* **attentato ~** bomb attack

dina'mite *nf* dynamite

'dinamo *nf inv* dynamo

di'nanzi *adv* in front ● *prep* **~ a** in front of

dina'stia *nf* dynasty

dini'ego *nm* denial

dinocco'lato *a* lanky

dino'sauro *nm* dinosaur

din'torn|i *nmpl* outskirts; **nei ~i di** in the vicinity of. **~o** *adv* around

'dio *nm (pl 'dei)* god; **D~** God

di'ocesi *nf inv* diocese

dipa'nare vt wind into a ball; fig unravel

diparti'mento nm department

dipen'den|te a depending ● nmf employee. **~za** nf dependence; (edificio) annexe

di'pendere vi ~ **da** depend on; (provenire) derive from; **dipende** it depends

di'pingere vt paint; (descrivere) describe. **~si** vr (truccarsi) make up. **di'pinto** pp di **dipingere** ● a painted ● nm painting

di'plo|ma nm diploma. **~'marsi** vr graduate

diplo'matico a diplomatic ● nm diplomat; (pasticcino) millefeuille (with alcohol)

diplo'mato nmf person with school qualification ● a qualified

diplo'mazia nf diplomacy

di'porto nm **imbarcazione da ~** pleasure craft

dira'dare vt thin out; make less frequent (visite). **~si** vr thin out; (nebbia:) clear

dira'ma|re vt issue ● vi, **~rsi** vr branch out; (diffondersi) spread. **~zi'one** nf (di strada) fork

'dire vt say; (raccontare, riferire) tell; **~ quello che si pensa** speak one's mind; **voler ~** mean; **volevo ben ~!** I wondered!; **~ di sì/no** say yes/no; **si dice che...** rumour has it that...; **come si dice "casa" in inglese?** what's the English for "casa"?; **questo nome mi dice qualcosa** the name rings a bell; **che ne dici di...?** how about...?; **non c'è che ~** there's no disputing that; **e ~ che...** to think that...; **a dir poco/tanto** at least/most ● vi ~ **bene/male** di speak highly/ill of sb; **dica pure** (in negozio) how can I help you?; **dici sul serio?** are you serious?; **per modo di ~** in a manner of speaking

diretta'mente adv directly

diret'tissima nf **per ~** Jur without

going through the normal procedures

diret'tissimo nm fast train

diret'tiva nf directive

di'retto pp di **dirigere** ● a direct. **~ a** (inteso) meant for. **essere ~** a be heading for. **in diretta** (trasmissione) live ● nm (treno) through train

diret'|tore, -'trice nmf manager; manageress; (di scuola) headmaster; headmistress. **~tore d'orchestra** conductor

direzi'one nf direction; (di società) management; Sch headmaster's/ headmistress's office (primary school)

diri'gen|te a ruling ● nmf executive; Pol leader. **~za** nf management. **~zi'ale** a management attrib; managerial

di'rigere vt direct; conduct (orchestra); run (impresa). **~si** vr **~si verso** head for

dirim'petto adv opposite ● prep **~ a** facing

di'ritto¹, dritto a straight; (destro) right ● adv straight; **andare ~** go straight on ● nm right side; Tennis forehand; **fare un ~** (a maglia) knit one

di'ritto² nm right; Jur law. **~i** pl **d'autore** royalties

dirit'tura nf straight line; fig honesty. **~ d'arrivo** Sport home straight

diroc'cato a tumbledown

dirom'pente a fig explosive

dirot'ta|re vt reroute (treno, aereo); (illegalmente) hijack; divert (traffico) ● vi alter course. **~'tore, ~'trice** nmf hijacker

di'rotto a (pioggia) pouring; (pianto) uncontrollable; **piovere a ~** rain heavily

di'rupo nm precipice

dis'abile nmf disabled person

disabi'tato a uninhabited

disabitu'arsi vr **~ a** get out of the habit of

disac'cordo nm disagreement

disadat'tato, -a *a* maladjusted ● *nmf* misfit

disa'dorno *a* unadorned

disa'gevole *a* (*scomodo*) uncomfortable

disagi'ato *a* poor; (*vita*) hard

di'sagio *nm* discomfort; (*difficoltà*) inconvenience; (*imbarazzo*) embarrassment; **sentirsi a ~** feel uncomfortable; **disagi** *pl* (*privazioni*) hardships

disappro'vare *vt* disapprove of. **~zi'one** *nf* disapproval

disap'punto *nm* disappointment

disar'mante *a fig* disarming

disar'mare *vt/i* disarm. **di'sarmo** *nm* disarmament

disa'strato, -a *a* devastated ● *nmf* disaster victim

di'sastro *nm* disaster; (*fam: grande confusione*) mess; (*fam: persona*) disaster area. **disa'stroso** *a* disastrous

disat'tento *a* inattentive. **~zi'one** *nf* inattention; (*svista*) oversight

disatti'vare *vt* de-activate

disa'vanzo *nm* deficit

disavven'tura *nf* misadventure

dis'brigo *nm* dispatch

dis'capito **nm a ~ di** to the detriment of

dis'carica *nf* scrap-yard

discen'dente *a* descending ● *nmf* descendant. **~za** *nf* descent; (*discendenti*) descendants *pl*

di'scendere *vt/i* descend; (*dal treno*) get off; (*da cavallo*) dismount; (*sbarcare*) land. **~ da** (*trarre origine da*) be a descendant of

di'scepolo, -a *nmf* disciple

di'scernere *vt* discern

di'scesa *nf* descent; (*pendio*) slope; **~a in picchiata** (*di aereo*) nosedive; **essere in ~a** (*strada:*) go downhill. **~ a libera** (*in sci*) downhill race. **disce'sista** *nmf* (*sciatore*) downhill skier. **~o** *pp di* **discendere**

dis'chetto *nm Comput* diskette

dischi'udere *vt* open; (*svelare*) disclose. **~si** *vr* open up

disci'ogliere *vt*, **~si** *vr* dissolve; (*neve:*) thaw; (*fondersi*) melt. **disci'olto** *pp di* **disciogliere**

disci'plina *nf* discipline. **~'nare** *a* disciplinary ● *vt* discipline. **~'nato** *a* disciplined

'disco *nm Disc; Comput* disk; *Sport* discus; *Mus* record; **ernia del ~** slipped disc. **~ fisso** *Comput* hard disk. **~ volante** flying saucer

discogra'fia *nf* (*insieme di incisioni*) discography. **disco'grafico** *a* (*industria*) record *attrib*, recording; **casa discografica** record company, recording company

'discolo *nmf* rascal ● *a* unruly

discol'pare *vt* clear. **~si** *vr* clear oneself

disco'noscere *vt* disown (*figlio*)

discontinuità *nf* (*nel lavoro*) irregularity. **discon'tinuo** *a* intermittent; (*fig: impegno, rendimento*) uneven

discor'dante *a* discordant. **~za** *nf* mismatch

discor'dare *vi* (*opinioni:*) conflict. **dis'corde** *a* clashing. **dis'cordia** *nf* discord; (*dissenso*) dissension

dis'correre *vi* talk (**di** about). **~'sivo** *a* colloquial. **dis'corso** *pp di* **discorrere** ● *nm* speech; (*conversazione*) talk

dis'costo *a* distant ● *adv* far away; **stare ~** stand apart

disco'teca *nf* disco; (*raccolta*) record library. **~'caro** *nmf pej* disco freak

discre'pante *a* contradictory. **~za** *nf* discrepancy

dis'creto *a* discreet; (*moderato*) moderate; (*abbastanza buono*) fairly good. **~zi'one** *nf* discretion; (*giudizio*) judgement; **a ~zione di** at the discretion of

discrimi'nante *a* extenuating

discrimi'nare *vt* discriminate. **~'torio** *a* (*atteggiamento*) discriminatory. **~zi'one** *nf* discrimination

discussi'one *nf* discussion; (*alter-*

co) argument. **dis'cusso** *pp di*
**discutere ● ** *a* controversial

dis'cutere *vt* discuss; (*formale*)
debate; (*litigare*) argue; **~ sul**
prezzo bargain. **discu'tibile** *a* debatable; (*gusto*) questionable

disde'gnare *vt* disdain. **dis'degno**
nm disdain

dis'dett|a *nf* retraction; (*sfortuna*)
bad luck; *Comm* cancellation. **~o**
pp di **disdire**

disdi'cevole *a* unbecoming

dis'dire *vt* retract; (*annullare*) cancel

diseduca'tivo *a* boorish, uncouth

dise'gna|re *vt* draw; (*progettare*)
design. **~tore**, **~trice** *nmf* designer. **di'segno** *nm* drawing;
(*progetto, linea*) design

diser'bante *nm* herbicide, weed-
killer ● *a* herbicidal, weed-killing

disere'da|re *vt* disinherit. **~to** *a*
dispossessed ● *nmf* **i ~ti** the dispossessed

diser|'tare *vt/i* desert; **~tare la**
scuola stay away from school.
~'tore *nm* deserter. **~zi'one** *nf* desertion

disfaci'mento *nm* decay

dis'fa|re *vt* undo; strip (*letto*);
(*smantellare*) take down; (*annientare*) defeat; **~re le valigie** unpack
[one's bags]. **~rsi** *vr* fall to pieces;
(*sciogliersi*) melt; **~rsi di** (*liberarsi di*) get rid of; **~rsi in lacrime**
dissolve into tears. **~tta** *nf* defeat.
~tto *a fig* worn out

disfat'tis|mo *nm* defeatism. **~ta** *a*
& *nmf* defeatist

disfunzi'one *nf* disorder

dis'gelo *nm* thaw

dis'grazi|a *nf* misfortune; (*incidente*) accident; (*sfavore*) disgrace.
~ata'mente *adv* unfortunately.
~'ato, -a *a* unfortunate ● *nmf*
wretch

disgre'gare *vt* break up. **~si** *vr*
disintegrate

disgu'ido *nm* **~ postale** mistake
in delivery

disgu'st|are *vt* disgust. **~arsi** *vr*

~arsi di be disgusted by.
dis'gusto *nm* disgust. **~oso** *a* disgusting

disidra'ta|re *vt* dehydrate. **~to** *a*
dehydrated

disil'lu|dere *vt* disenchant. **~si'o-**
ne *nf* disenchantment. **~so** *a* disillusioned

disimbal'lare *vt* unpack

disimpa'rare *vt* forget

disimpe'gnar|e *vt* release; (*compiere*) fulfil; redeem (*oggetto dato*
in pegno). **~si** *vr* disengage
oneself; (*cavarsela*) manage.
disim'pegno *nm* (*locale*) vestibule

disincan'tato *a* (*disilluso*) disillusioned

disinfe'sta|re *vt* disinfest.
~zi'one *nf* disinfestation

disinfet'tante *a* & *nm* disinfectant

disinfe|t'tare *vt* disinfect. **~zi'one**
nf disinfection

disinfor'mato *a* uninformed

disini'bito *a* uninhibited

disinne'scare *vt* defuse (*mina*).
disin'nesco *nm* (*di bomba*) bomb
disposal

disinse'rire *vt* disconnect

disinte'gra|re *vt*, **~rsi** *vr* disintegrate. **~zi'one** *nf* disintegration

disinteres'sarsi *vr* **~ di** take no
interest in. **disinte'resse** *nm* indifference; (*oggettività*) disinterestedness

disintossi'ca|re *vt* detoxify. **~rsi**
vr come off drugs. **~zi'one** *nf* giving up alcohol/drugs

disin'volto *a* natural. **disinvol-**
'tura *nf* confidence

disles'sia *nf* dyslexia. **disl'essico** *a*
a dyslexic

disli'vello *nm* difference in height;
fig inequality

dislo'care *vt Mil* post

dismenor'rea *nf* dysmenorrhoea

dismi'sura *nf* excess; **a ~** excessively

disobbedi'ente *a* disobedient

disobbe'dire *vt* disobey

disoccu'pa|to, -a *a* unemployed

● *nmf* unemployed person. **~zi'o ne** *nf* unemployment

disonestà *nf* dishonesty. **diso 'nesto** *a* dishonest

disono'rare *vt* dishonour. **diso 'nore** *nm* dishonour

di'sopra *adv* above ● *a* upper ● *nm* top

disordi'na|re *vt* disarrange. **~ta'mente** *adv* untidily. **~to** *a* untidy; (*sregolato*) immoderate. **di'sordine** *nm* disorder, untidi ness; (*sregolatezza*) debauchery

disorganiz'za|re *vt* disorganize. **~to** *a* disorganized. **~zi'one** *nf* disorganization

disorienta'mento *nm* disorienta tion

disorien'ta|re *vt* disorientate. **~rsi** *vr* lose one's bearings. **~to** *a fig* bewildered

di'sotto *adv* below ● *a* lower ● *nm* bottom

dis'paccio *nm* dispatch

dispa'rato *a* disparate

'dispari *a* odd, uneven. **~tà** *nf inv* disparity

dis'parte *adv* in ~ apart; **stare in** ~ stand aside

dis'pendi|o *nm* (*spreco*) waste. **~'oso** *a* expensive

dis'pen|sa *nf* pantry; (*distribuzio ne*) distribution; (*mobile*) cup board; *Jur* exemption; *Relig* dispensation; (*pubblicazione pe riodica*) number. **~'sare** *vt* distribute; (*esentare*) exonerate

dispe'ra|re *vi* despair (**di** of). **~rsi** *vr* despair. **~ta'mente** (*piangere*) desperately. **~to** *a* desperate. **~zi'one** *nf* despair

dis'per|dere *vt*, **~dersi** *vr* scatter, disperse. **~si'one** *nf* dispersion; (*di truppe*) dispersal. **~'sivo** *a* disorganized. **~so** *pp* **di disperdere** ● *a* scattered; (*smar rito*) lost ● *nm* missing soldier

dis'pet|to *nm* spite; **a ~to di** in spite of; **fare un ~to a qcno** spite sb. **~'toso** *a* spiteful

dispia'c|ere *nm* upset; (*rammari*

co) regret; (*dolore*) sorrow; (*preoc cupazione*) worry ● *vi* **mi dispia ce** I'm sorry; **non mi dispiace** I don't dislike it; **se non ti dispiace** if you don't mind. **~i'uto** *a* upset; (*dolente*) sorry

dispo'nibil|e *a* available; (*gentile*) helpful. **~ità** *nf* availability; (*gen tilezza*) helpfulness

dis'por|re *vt* arrange ● *vi* dispose; (*stabilire*) order; **~re di** have at one's disposal. **~rsi** *vr* (*in fila*) line up

disposi'tivo *nm* device

disposizi'one *nf* disposition; (*or dine*) order; (*libera disponibilità*) disposal. **dis'posto** *pp* **di disporre** ● *a* ready; (*incline*) disposed; **es sere ben disposto verso** be fa vourably disposed towards

dis'potico *a* despotic. **dispo'tismo** *nm* despotism

dispregia'tivo *a* disparaging

disprez'zare *vt* despise. **dis'prez zo** *nm* contempt

'disputa *nf* dispute

dispu'tar|e *vi* dispute; (*gareggia re*) compete. **~si** *vr* **~si qcsa** contend for sth

dissacra'torio *a* debunking

dissangua'mento *nm* loss of blood

dissangu'a|re *vt*, **~rsi** *vr* bleed. **~rsi** *vr fig* become impoverished. **~to** *a* bloodless; *fig* impover ished

dissa'pore *nm* disagreement

dissec'car|e *vt*, **~si** *vr* dry up

dissemi'nare *vt* disseminate; (*no tizie*) spread

dis'senso *nm* dissent; (*disaccordo*) disagreement

dissen'teria *nf* dysentery

dissen'tire *vi* disagree (**da** with)

dissertazi'one *nf* dissertation

disser'vizio *nm* poor service

disse'sta|re *vt* upset; *Comm* dam age. **~to** *a* (*strada*) uneven. **dis'sesto** *nm* ruin

disse'tante *a* thirst-quenching

disse'ta|re vt ~ **re** qcno quench sb's thirst

dissi'dente a & nmf dissident

dis'sidio nm disagreement

dis'simile a unlike, dissimilar

dissimu'lare vt conceal; (*fingere*) dissimulate

dissi'pa|re vt dissipate; (*sperperare*) squander. **~rsi** vr (*nebbia:*) clear; (*dubbio:*) disappear. **~to** a dissipated. **~zi'one** nf squandering

dissoci'ar|e vt, **~si** vr dissociate

disso'dare vt till

dis'solto pp di **dissolvere**

disso'luto a dissolute

dis'solver|e vt, **~si** vr dissolve; (*disperdere*) dispel

disso'nanza nf dissonance

dissua|'dere vt dissuade. **~si'one** nf dissuasion. **~'sivo** a dissuasive

distac'car|e vt detach; Sport leave behind. **~si** vr be detached.

di'stacco nm detachment; (*separazione*) separation; (Sport) lead

di'stan|te a far away; (fig: *persona*) detached ● adv far away. **~za** nf distance. **~zi'are** vt space out; Sport outdistance

di'stare vi be distant; **quanto dista?** how far is it?

di'sten|dere vt stretch out (*p* r*te del corpo*); (*spiegare*) spr1ad; (*deporre*) lay. **~dersi** vr stretch; (*sdraiarsi*) lie down; (*rilassarsi*) relax. **~si'one** nf stretching; (*rilassamento*) relaxation; Pol détente. **~'sivo** a relaxing

di'steso, -a pp di **distendere** ● nf expanse

distil'l|are vt/i distil. **~azi'one** nf distillation. **~e'ria** nf distillery

di'stinguer|e vt distinguish. **~si** vr distinguish oneself. **distin'guibile** a distinguishable

di'stinta nf Comm list. ~ **di pagamento** receipt. ~ **di versamento** paying-in slip

distinta'mente adv (*separatamente*) individually, separately; (*chiaramente*) clearly

distin'tivo a distinctive ● nm badge

di'stin|to, -a pp di **distinguere** ● a distinct; (*signorile*) distinguished; **~ti saluti** Yours faithfully. **~zi'one** nf distinction

di'stogliere vt ~ **da** (*allontanare*) remove from; (*dissuadere*) dissuade from. **di'stolto** pp di **distogliere**

di'storcere vt twist

distorsi'one nf Med sprain; (*alterazione*) distortion

di'stra|rre vt distract; (*divertire*) amuse. **~rsi** vr get distracted; (*svagarsi*) amuse oneself; **non ti distrarre!** pay attention!. **~rsi** vr (*deconcentrarsi*) be distracted. **~tta'mente** adv absently. **~tto** pp di **distrarre** ● a absent-minded; (*disattento*) inattentive. **~zi'one** nf absent-mindedness; (*errore*) inattention; (*svago*) amusement

di'stretto nm district

distribu|'ire vt distribute; (*disporre*) arrange; deal (*carte*). **~'tore** nm distributor; (*di benzina*) petrol pump; (*automatico*) slot-machine. **~zi'one** nf distribution

distri'car|e vt disentangle; **~si** vr fig get out of it

di'strugger|e vt destroy. **~t'tivo** a destructive; (*critica*) negative. **~tto** pp di **distruggere** ● a destroyed; **un uomo ~tto** a broken man. **~zi'one** nf destruction

distur'bar|e vt disturb; (*sconvolgere*) upset. **~si** vr trouble oneself.

di'sturbo nm bother; (*indisposizione*) trouble; Med problem; Radio, TV interference; **disturbi** pl Radio, TV static. **disturbi di stomaco** stomach trouble

disubbidi'en|te a disobedient. **~za** nf disobedience

disubbi'dire vi ~ **a** disobey

disugu|agli'anza nf disparity. **~'ale** a unequal; (*irregolare*) irregular

disu'mano a inhuman

di'suso *nm* cadere in ~ fall into disuse

di'tale *nm* thimble

di'tata *nf* poke; (*impronta*) fingermark

'dito *nm* (*pl inf* **dita**) finger; (*di vino, acqua*) finger. ~ **del piede** toe

'ditta *nf* firm

dit'tafono *nm* dictaphone

ditta'tor|e *nm* dictator. ~**i'ale** *a* dictatorial. **ditta'tura** *nf* dictatorship

dit'tongo *nm* diphthong

di'urno *a* daytime; **spettacolo** ~ matinée

'diva *nf* diva

diva'ga|re *vi* digress. ~**zi'one** *nf* digression

divam'pare *vi* burst into flames; *fig* spread like wildfire

di'vano *nm* settee, sofa. ~ **letto** sofa bed

divari'care *vt* open

di'vario *nm* discrepancy; **un** ~ **di opinioni** a difference of opinion

dive'n|ire *vi* = **diventare**. ~**uto** *pp di* **divenire**

diven'tare *vi* become; (*lentamente*) grow; (*rapidamente*) turn

di'verbio *nm* squabble

diver'gen|te *a* divergent. ~**za** *nf* divergence; ~**za di opinioni** difference of opinion. **di'vergere** *vi* diverge

diversa'mente *adv* (*altrimenti*) otherwise; (*in modo diverso*) differently

diversifi'ca|re *vt* diversify. ~**rsi** *vr* differ, be different. ~**zi'one** *nf* diversification

diver|si'one *nf* diversion. ~**sità** *nf inv* difference. ~**'sivo** *nm* diversion. **di'verso** *a* different; **diversi** *pl* (*parecchi*) several ● *pron* several [people]

diver'tente *a* amusing. **diverti'mento** *nm* amusement

diver'tir|e *vt* amuse. ~**si** *vr* enjoy oneself

divi'dendo *nm* dividend

di'vider|e *vt* divide; (*condividere*) share. ~**si** *vr* (*separarsi*) separate

divi'eto *nm* prohibition; ~ **di sosta** no parking

divinco'larsi *vr* wriggle

divinità *nf inv* divinity. **di'vino** *a* divine

di'visa *nf* uniform; *Comm* currency

divisi'one *nf* division

di'vismo *nm* worship; (*atteggiamento*) superstar mentality

di'vi|so *pp di* **dividere**. ~'**sore** *nm* divisor. ~'**sorio** *a* dividing; **muro** ~**sorio** partition wall

'divo, -a *nmf* star

divo'rar|e *vt* devour. ~**si** *vr* ~**si da** be consumed with

divorzi'a|re *vi* divorce. ~**to, -a** *nmf* divorcee. **di'vorzio** *nm* divorce

divul'ga|re *vt* divulge; (*rendere popolare*) popularize. ~**rsi** *vr* spread. ~'**tivo** *a* popular. ~**zi'one** *nf* popularization

dizio'nario *nm* dictionary

dizi'one *nf* diction

do *nm* *Mus* (*chiave, nota*) C

'doccia *nf* shower; (*grondaia*) gutter; **fare la** ~ have a shower

do'cen|te *a* teaching ● *nmf* teacher; (*di università*) lecturer. ~**za** *nf* university teacher's qualification

'docile *a* docile

documen'tar|e *vt* document. ~**si** *vr* gather information (**su** about)

documen'tario *a* & *nm* documentary

documen'ta|to *a* well-documented; (*persona*) well-informed. ~**zi'one** *nf* documentation

docu'mento *nm* document

dodi'cesimo *a* & *nm* twelfth. **'dodici** *a* & *nm* twelve

do'gan|a *nf* customs *pl*; (*dazio*) duty. **doga'nale** *a* customs. ~'**iere** *nm* customs officer

'doglie *nfpl* labour pains

'dogma *nm* dogma. **dog'matico** *a* dogmatic. ~'**tismo** *nm* dogmatism 'dolce *a* sweet; ⟨*clima*⟩ mild; ⟨*voce, consonante*⟩ soft; ⟨*acqua*⟩ fresh ● *nm* ⟨*portata*⟩ dessert; ⟨*torta*⟩ cake; **non mangio dolci** I don't eat sweet things. ~'mente *adv* sweetly. dol'cezza *nf* sweetness; ⟨*di clima*⟩ mildness

dolce'vita *a inv* ⟨*maglione*⟩ roll-neck

dolci'ario *a* confectionery

dolci'astro *a* sweetish

dolcifi'cante *nm* sweetener ● *a* sweetening

dolci'umi *nmpl* sweets

do'lente *a* painful; ⟨*spiacente*⟩ sorry

do'le|re *vi* ache, hurt; ⟨*dispiacere*⟩ regret. ~rsi *vr* ⟨*protestare*⟩ complain; ~rsi di be sorry for

'dollaro *nm* dollar

'dolo *nm Jur* malice; ⟨*truffa*⟩ fraud

Dolo'miti *nfpl* le ~ the Dolomites

do'lore *nm* pain; ⟨*morale*⟩ sorrow. dolo'roso *a* painful

do'loso *a* malicious

do'manda *nf* question; ⟨*richiesta*⟩ request; ⟨*scritta*⟩ application; *Comm* demand; **fare una ~ (a qcno)** ask (sb) a question. **~ di impiego** job application

doman'dar|e *vt* ask; ⟨*esigere*⟩ demand; **~e qcsa a qcno** ask sb for sth. **~si** *vr* wonder

do'mani *adv* tomorrow; **~ sera** tomorrow evening ● *nm* il **~** the future; **a ~** see you tomorrow

do'ma|re *vt* tame; *fig* control ⟨*emozioni*⟩. **~'tore** *nm* tamer

domat'tina *adv* tomorrow morning

do'meni|ca *nf* Sunday. **~'cale** *a* Sunday *attrib*

do'mestico, -a *a* domestic ● *nm* servant ● *nf* maid

domicili'are *a* arresti domiciliari *Jur* house arrest

domicili'arsi *vr* settle

domi'cilio *nm* domicile; ⟨*abitazione*⟩ home; **recapitiamo a ~** we do home deliveries

domi'na|re *vt* dominate; ⟨*controllare*⟩ control ● *vi* rule over; ⟨*prevalere*⟩ be dominant. **~rsi** *vr* control oneself. **~'tore**, **~'trice** *nmf* ruler **~zi'one** *nf* domination

do'minio *nm* control; *Pol* dominion; ⟨*ambito*⟩ field; **di ~ pubblico** common knowledge

don *nm inv* ⟨*ecclesiastico*⟩ Father

do'na|re *vt* give; donate ⟨*sangue, organo*⟩ ● *vi* **~re a** ⟨*giovane esteticamente*⟩ suit. **~'tore**, **~'trice** *nmf* donor. **~zi'one** *nf* donation

dondo'l|are *vt* swing; ⟨*cullare*⟩ rock ● *vi* sway. **~arsi** *vr* swing. **~io** *nm* rocking. 'dondolo *nm* swing; **cavallo/sedia a dondolo** rocking-horse/chair

dongio'vanni *nm inv* Romeo

'donna *nf* woman. **~ di servizio** domestic help

don'naccia *nf pej* whore

donnai'olo *a* philanderer

'donnola *nf* weasel

'dono *nm* gift

'dopo *prep* after; ⟨*a partire da*⟩ since ● *adv* after, afterwards; ⟨*più tardi*⟩ later; ⟨*in seguito*⟩ later on; **~ di me** after me

dopo'barba *nm inv* aftershave

dopo'cena *nm inv* evening

dopodiché *adv* after which

dopodo'mani *adv* the day after tomorrow

dopogu'erra *nm inv* post-war period

dopo'pranzo *nm inv* afternoon

dopo'sci *a & nm inv* après-ski

doposcu'ola *nm inv* after-school activities *pl*

dopo-'shampoo *nm inv* conditioner ● *a inv* conditioning

dopo'sole *nm inv* aftersun cream ● *a inv* aftersun

dopo'tutto *adv* after all

doppi'aggio *nm* dubbing

doppia'mente *adv* ⟨*in misura doppia*⟩ doubly

doppi'a|re *vt Naut* double; *Sport* lap; *Cinema* dub. **~'tore**, **~'trice** *nmf* dubber

'doppio a & adv double. **~ clic** nm Comput double click. **~ fallo** nm Tennis double fault. **~ gioco** nm double-dealing. **~ mento** nm double chin. **~ senso** nm double entendre. **doppi vetri** nmpl double glazing ●nm double, twice the quantity; Tennis doubles pl. **~ misto** Tennis mixed doubles ●adv double

doppi'one nm duplicate

doppio'petto a double-breasted

dop'pista nmf Tennis doubles player

do'ra|re vt gild; Culin brown. **~to** a gilt; (color oro) golden. **~'tura** nf gilding

dormicchi'are vi doze

dormigli'one, -a nmf sleepyhead; fig lazy-bones

dor'mi|re vi sleep; (essere addormentato) be asleep; fig be asleep. **~ta** nf good sleep. **~'tina** nf nap. **~'torio** nm dormitory

dormi'veglia essere in ~ be half asleep

dor'sale a dorsal ●nf (di monte) ridge

'dorso nm back; (di libro) spine; (di monte) crest; (nel nuoto) backstroke

do'saggio nm dosage

do'sare vt dose; fig measure; **~ le parole** weigh one's words

dosa'tore nm measuring jug

'dose nf dose; **in buona ~** fig in good measure. **~ eccessiva** overdose

dossi'er nm inv (raccolta di dati, fascicolo) file

'dosso nm (dorso) back; **levarsi di ~ gli abiti** take off one's clothes

do'ta|re vt endow; (di accessori) equip. **~to** a (persona) gifted; (fornito) equipped. **~zi'one** nf (attrezzatura) equipment; **in ~zione** at one's disposal

'dote nf dowry; (qualità) gift

'dotto a learned ●nm scholar; Anat duct

dotto'|rato nm doctorate. **dot'tore, ~'ressa** nmf doctor

dot'trina nf doctrine

'dove adv where; **di ~ sei?** where do you come from; **fin ~?** how far?; **per ~?** which way?

do'vere vi (obbligo) have to, must; **devo andare** I have to go, I must go; **devo venire anch'io?** do I have to come too?; **avresti dovuto dirmelo** you should have told me, you ought to have told me; **devo sedermi un attimo** I must sit down for a minute, I need to sit down for a minute; **dev'essere successo qualcosa** something must have happened; **come si deve** properly ●vt (essere debitore di, derivare) owe; **essere dovuto a** be due to ●nm duty; **per ~** out of duty. **do'veroso** a only right and proper

do'vunque adv (dappertutto) everywhere; (in qualsiasi luogo) anywhere ●conj wherever

do'vuto a due; (debito) proper

doz'zi|na nf dozen. **~'nale** a cheap

dra'gare vt dredge

'drago nm dragon

dramm|a nm drama. **dram'matico** a dramatic. **~atiz'zare** vt dramatize. **~a'turgo** nm playwright. **dram'mone** nm (film) tear-jerker

drappeggi'are vt drape. **drap'peggio** nm drapery

drap'pello nm Mil squad; (gruppo) band

'drastico a drastic

dre'na|ggio nm drainage. **~re** vt drain

drib'blare vt (in calcio) dribble. **'dribbling** nm inv (in calcio) dribble

'dritta nf (mano destra) right hand; Naut starboard; (informazione) pointer, tip; **a ~ e a manca** (dappertutto) left, right and centre

dritto a = **diritto'** ●nmf fam crafty so-and-so

driz'zar|e *vt* straighten; *(rizzare)* prick up. **~si** *vr* straighten [up]; *(alzarsi)* raise

'dro|ga *nf* drug. **~'gare** *vt* drug. **~'garsi** *vr* take drugs. **~'gato, -a** *nmf* drug addict

drogh|e'ria *nf* grocery. **~'i'ere, -a** *nmf* grocer

drome'dario *nm* dromedary

'dubbi|o *a* doubtful; *(ambiguo)* dubious ● *nm* doubt; *(sospetto)* suspicion; **mettere in ~o** doubt; **essere fuori ~o** be beyond doubt; **essere in ~o** be doubtful. **~'oso** *a* doubtful

dubi'ta|re *vi* doubt; **~re di** doubt; *(diffidare)* mistrust; **dubito che venga** I doubt whether he'll come. **~'tivo** *a* *(ambiguo)* ambiguous

'duca, du'chessa *nmf* duke; duchess

'due *a & nm* two

due'cento *a & nm* two hundred

due'mila *a & nm* two thousand

due'pezzi *nm inv* (bikini) bikini

'duna *nf* dune

'dunque *conj* therefore; *(allora)* well [then]

'duo *nm inv* duo; *Mus* duet

du'omo *nm* cathedral

'duplex *nm* *Teleph* party line

dupli'ca|re *vt* duplicate. **~to** *nm* duplicate. **'duplice** *a* double; **in duplice** in duplicate

dura'mente *adv* *(lavorare)* hard; *(rimproverare)* harshly

du'rante *prep* during

du'r|are *vi* last; *(cibo:)* keep; *(resistere)* hold out. **~ata** *nf* duration. **~a'turo, ~evole** *a* lasting, enduring

du'rezza *nf* hardness; *(di carne)* toughness; *(di voce, padre)* harshness

'duro, -a *a* hard; *(persona, carne* tough; *(voce)* harsh; *(pane)* stale; **tieni ~!** *(resistere)* hang in there!

● *nmf* *(persona)* tough person, toughie *fam*

du'rone *nm* hardened skin

'duttile *a* *(materiale)* ductile; *(carattere)* malleable

- -

Ee

- -

e, ed *conj* and

'ebano *nm* ebony

eb'bene *conj* well [then]

eb'brezza *nf* inebriation; *(euforia)* elation; **guida in stato di ~** drink-driving, drunken driving. **'ebbro** *a* inebriated; **ebbro di gioia** delirious with joy

'ebete *a* stupid

ebolliz'one *nf* boiling

e'braico *a* Hebrew ● *nm* *(lingua)* Hebrew. **e'br|eo, -a** *a* Jewish ● *nmf* Jew; Jewess

'Ebridi *nfpl* **le ~** the Hebrides

eca'tombe *nf* **fare un'~** wreak havoc

ecc *abbr* **(eccetera)** etc

ecce'den|te *a* *(peso, bagaglio)* excess. **~za** *nf* excess; *(d'avanzo)* surplus; **avere qcsa in ~za** have an excess of sth; **bagagli in ~za** excess baggage. **~za di cassa** surplus. **ec'cedere** *vt* exceed ● *vi* go too far; **eccedere nel mangiare** overeat; **eccedere nel bere** drink to excess

eccel'len|te *a* excellent. **~za** *nf* excellence; *(titolo)* Excellency; **per ~za** par excellence. **ec'cellere** *vi* excel (**in** at)

eccentricità *nf* eccentricity. **ec'centrico, -a** *a & nmf* eccentric

eccessiva'mente *adv* excessively. **ecces'sivo** *a* excessive

ec'cesso *nm* excess; **andare agli eccessi** go to extremes; **all'~** to excess. **~ di velocità** speeding

ec'cetera *adv* et cetera

ec'cetto *prep* except; **~ che** *(a*

eccezionale *meno che*) unless. **eccettu'are** *vt* except

eccezio'nal|e *a* exceptional. **~'mente** *adv* exceptionally; (*contrariamente alla regola*) as an exception

eccezi'one *nf* exception; *Jur* objection; **a ~ di** with the exception of

eccita'mento *nm* excitement. **ecci'tante** *a* exciting; ⟨*sostanza*⟩ stimulant ● *nm* stimulant

ecci'ta|re *vt* excite. **~rsi** *vr* get excited. **~to** *a* excited

eccitazi'one *nf* excitement

ecclesi'astico *a* ecclesiastical ● *nm* priest

'ecco *adv* ⟨*qui*⟩ here; ⟨*là*⟩ there; **~!** exactly!; **~ fatto** there we are; **~ la tua borsa** here is your bag; **~** **[li] mio figlio** there is my son; **~mi** here I am; **~ tutto** that is all

ec'come *adv* & *int* and how!

echeggi'are *vi* echo

e'clissi *nf inv* eclipse

'eco *nmf* (*pl m* **echi**) echo

ecogra'fia *nf* scan

ecolo'gia *nf* ecology. **eco'logico** *a* ecological; ⟨*prodotto*⟩ environmentally friendly

e commerci'ale *nf* ampersand

econo'm|ia *nf* economy; ⟨*scienza*⟩ economics; **fare ~ia** economize ⟨*di* on⟩. **eco'nomico** *a* economic; ⟨*a buon prezzo*⟩ cheap. **~ista** *nmf* economist. **~iz'zare** *vt/i* economize; save ⟨*tempo, denaro*⟩. **e'conomo, -a** *a* thrifty ● *nmf* ⟨*di collegio*⟩ bursar

é'cru *a inv* raw

'Ecu *nm inv* ECU, ecu

ec'zema *nm* eczema

ed *conj vedi* **e**

'edera *nf* ivy

e'dicola *nf* [newspaper] kiosk

edifi'cabile *a* ⟨*area, terreno*⟩ classified as suitable for development

edifi'cante *a* edifying

edifi'care *vt* build; ⟨*indurre al bene*⟩ edify

edi'ficio *nm* building; *fig* structure

e'dile *a* building *attrib*

edi'lizi|a *nf* building trade. **~o** *a* building *attrib*

edi'tore, -'trice *a* publishing ● *nmf* publisher; ⟨*curatore*⟩ editor. **~to'ria** *nf* publishing. **~tori'ale** *a* publishing ● *nm* ⟨*articolo*⟩ editorial, leader

edizi'one *nf* edition; ⟨*di manifestazione*⟩ performance. **~ ridotta** abridg[e]ment. **~ della sera** ⟨*di telegiornale*⟩ evening news

edu'ca|re *vt* educate; ⟨*allevare*⟩ bring up. **~'tivo** *a* educational. **~to** *a* polite. **~'tore, -'trice** *nmf* educator. **~zi'one** *nf* education; ⟨*di bambini*⟩ upbringing; ⟨*buone maniere*⟩ [good] manners *pl.* **~zione fisica** physical education

e'felide *nf* freckle

effemi'nato *a* effeminate

efferve'scente *a* effervescent; ⟨*frizzante*⟩ fizzy; ⟨*aspirina*⟩ soluble

effettiva'mente *adv* **è troppo tardi - ~** it's too late – so it is

effet'tivo *a* actual; ⟨*efficace*⟩ effective; ⟨*personale*⟩ permanent; *Mil* regular ● *nm* ⟨*somma totale*⟩ sum total

ef'fetto *nm* effect; ⟨*impressione*⟩ impression; **in ~i** in fact; **a tutti gli ~i** to all intents and purposes; **~i personali** personal belongings. **~u'are** *vt* effect; carry out ⟨*controllo, sondaggio*⟩. **~u'arsi** *vr* take place

effi'cac|e *a* effective. **~ia** *nf* effectiveness

effici'en|te *a* efficient. **~za** *nf* efficiency

ef'fimero *a* ephemeral

effusi'one *nf* effusion

E'geo *nm* l'~ the Aegean [Sea]

E'gitto *nm* Egypt. **egizi'ano, -a** *a* & *nmf* Egyptian

'egli *pers pron* he; **~ stesso** he himself

ego'centrico, -a *a* egocentric ● *nmf* egocentric person

ego'is|mo nm selfishness. **~ta** a selfish ● nmf selfish person. **~tico** a selfish

e'gregio a distinguished; **E~ Signore** Dear Sir

eguali'tario a & nm egalitarian

eiaculazi'one nf ejaculation

elabo'ra|re vt elaborate; process ⟨dati⟩. **~to** a elaborate. **~zi'one** nf elaboration; ⟨di dati⟩ processing. **~zione [di] testi** word processing

elar'gire vt lavish

elastici'tà nf elasticity. **~z'zato** a ⟨stoffa⟩ elasticated. **e'lastico** a elastic; ⟨tessuto⟩ stretch; ⟨orario, mente⟩ flexible; ⟨persona⟩ easygoing ● nm elastic; ⟨fascia⟩ rubber band

ele'fante nm elephant

ele'gan|te a elegant. **~za** nf elegance

e'leggere vt elect. **eleg'gibile** a eligible

elemen'tare a elementary; **scuola ~** primary school

ele'mento nm element; **elementi** pl ⟨fatti⟩ data; ⟨rudimenti⟩ elements

ele'mosina nf charity; **chiedere l'~** beg. **elemosi'nare** vt/i beg

elen'care vt list

e'lenco nm list. **~ abbonati** telephone directory. **~ telefonico** telephone directory

elet'tivo a ⟨carica⟩ elective. **e'letto, -a** pp di **eleggere** ● a chosen ⟨nominato⟩ elected member; **per pochi eletti** for the chosen few

eletto'ra|le a electoral. **~to** nm electorate

elet'tore, -'trice nmf voter

elet'trauto nm garage for electrical repairs

elettri'cista nm electrician

elettri'cità nf electricity. **e'lettrico** a electric. **~z'zante** a ⟨notizia, gara⟩ electrifying. **~z'zare** vt fig electrify. **~z'zato** a fig electrified

elettrocardio'gramma nm electrocardiogram

e'lettrodo nm electrode

elettrodo'mestico nm [electrical] household appliance

elet'trone nm electron

elet'tronico, -a a electronic ● nf electronics

ele'va|re vt raise; ⟨promuovere⟩ promote; ⟨erigere⟩ erect; ⟨fig: migliorare⟩ better; **~ al quadrato/ cubo** square/cube. **~rsi** vr rise; ⟨edificio⟩ stand. **~to** a high. **~zi'one** nf elevation

elezi'one nf election

'elica nf Naut screw, propeller; Aeron propeller; ⟨del ventilatore⟩ blade

eli'cottero nm helicopter

elimi'na|re vt eliminate. **~'toria** nf Sport preliminary heat. **~zi'one** nf elimination

é'li|te nf inv élite. **~'tista** a élitist

'ella pers pron she

el'lepi nm inv LP

el'metto nm helmet

elogi'are vt praise. **e'logio** nm praise; ⟨discorso, scritto⟩ eulogy

elo'quen|te a eloquent; fig telltale. **~za** nf eloquence

e'lu|dere vt elude; evade ⟨sorveglianza, controllo⟩. **~'sivo** a elusive

el'vetico a Swiss

emaci'ato a emaciated

'E-mail nf e-mail

ema'na|re vt give off; pass ⟨legge⟩ ● vi emanate. **~zi'one** nf giving off; ⟨di legge⟩ enactment

emanci'pa|re vt emancipate. **~rsi** vr become emancipated. **~to** a emancipated. **~zi'one** nf emancipation

emargi'na|to nm marginalized person. **~zi'one** nf marginalization

ema'toma nm haematoma

em'bargo nm embargo

em'ble|ma nm emblem. **~'matico** a emblematic

embo'lia nf embolism

embri'o'nale a Biol, fig embryonic. **embri'one** nm embryo

emen│da'mento *nm* amendment.
~'dare *vt* amend

emer'gen│te *a* emergent. **~za** *nf*
emergency; **in caso di ~za** in an
emergency

e'mergere *vi* emerge; ⟨sottomari-
no:⟩ surface; ⟨distinguersi⟩ stand
out

e'merito *a* ⟨professore⟩ emeritus;
un ~ imbecille a prize idiot

e'merso *pp di* **emergere**

e'messo *pp di* **emettere**

e'mettere *vt* emit; give out ⟨luce,
suono⟩; let out ⟨grido⟩; ⟨mettere in
circolazione⟩ issue

emi'crania *nf* migraine

emi'gra│re *vi* emigrate. **~to, -a**
nmf immigrant. **~zi'one** *nf* emi-
gration

emi'nen│te *a* eminent. **~za** *nf* emi-
nence

e'miro *nm* emir

emis'fero *nm* hemisphere

emis'sario *nm* emissary

emissi'one *nf* emission; ⟨di dena-
ro⟩ issue; ⟨trasmissione⟩ broadcast

emit'tente *a* issuing; ⟨trasmitten-
te⟩ broadcasting ● *nf Radio* transmitter

emorra'gia *nf* haemorrhage

emor'roidi *nfpl* piles

emotività *nf* emotional make-up.
emo'tivo *a* emotional

emozio'na│nte *a* exciting; ⟨com-
movente⟩ moving. **~re** *vt* excite;
⟨commuovere⟩ move. **~rsi** *vr*
become excited; ⟨commuoversi⟩ be
moved. **~to** *a* excited; ⟨commosso⟩
moved. **emozi'one** *nf* emotion;
⟨agitazione⟩ excitement

'empio *a* impious; ⟨spietato⟩ pitiless; ⟨malvagio⟩ wicked

em'pirico *a* empirical

em'porio *nm* emporium; ⟨negozio⟩
general store

emu'la│re *vt* emulate. **~zi'one** *nf*
emulation

emulsi'one *nf* emulsion

en'ciclica *nf* encyclical

enciclope'dia *nf* encyclopaedia

encomi'are *vt* commend. **en'comio** *nm* commendation

en'demico *a* endemic

endo've│na *nf* intravenous injection. **~'noso** *a* intravenous; **per
via ~nosa** intravenously

E.N.I.T. *nm abbr* (Ente Nazionale
Italiano per il Turismo) Italian
State Tourist Office

ener'getico *a* ⟨risorse, crisi⟩ energy
attrib; ⟨alimento⟩ energy-giving

ener'gia *nf* energy. **e'nergico** *a* energetic; ⟨efficace⟩ strong

ener'gumeno *nm* Neanderthal

'enfasi *nf* emphasis

en'fati│co *a* emphatic. **~z'zare** *vt*
emphasize

e'nigma *nm* enigma. **enig'matico**
a enigmatic. **enig'mistica** *nf* puz-
zles

en'nesimo *a Math* nth; *fam* ump-
teenth

e'norm│e *a* enormous. **~e'mente**
adv massively. **~ità** *nf inv* enormity; ⟨assurdità⟩ absurdity

eno'teca *nf* wine-tasting shop

'ente *nm* board; ⟨società⟩ company;
⟨filosofia⟩ being

entità *nf inv* ⟨filosofia⟩ entity;
⟨gravità⟩ seriousness; ⟨dimensio-
ne⟩ extent

entou'rage *nm inv* entourage

en'trambi *a & pron* both

en'tra│re *vi* go in, enter; **~re in** go
into; ⟨stare in, trovar posto in⟩ fit
into; ⟨arruolarsi⟩ join; **~rci** ⟨avere
a che fare⟩ have to do with; **tu che
c'entri?** what has it got to do with
you? **~ta** *nf* entry, entrance; **~te**
pl Comm takings; ⟨reddito⟩ income
sg

'entro *prep* ⟨tempo⟩ within

entro'terra *nm inv* hinterland

entusias'mante *a* fascinating, exciting

entusias'mar│e *vt* arouse enthusiasm in. **~si** *vr* be enthusiastic
⟨per about⟩

entusi'as│mo *nm* enthusiasm.
~ta *a* enthusiastic ● *nmf* enthusi-
ast. **~tico** *a* enthusiastic

enume'ra|re *vt* enumerate. **~zi-'one** *nf* enumeration

enunci'a|re *vt* enunciate. **~zi'one** *nf* enunciation

epa'tite *nf* hepatitis

'epico *a* epic

epide'mia *nf* epidemic

epi'dermide *nf* epidermis

Epifa'nia *nf* Epiphany

epi'gramma *nm* epigram

epiles'sia *nf* epilepsy. **epi'lettico, -a** *à* & *nmf* epileptic

e'pilogo *nm* epilogue

epi'sodi|co *a* episodic; **caso ~co** one-off case. **~o** *nm* episode

e'piteto *nm* epithet

'epoca *nf* age; (*periodo*) period; **a quell'~** in those days; **auto d'~** vintage car

ep'pure *conj* [and] yet

epu'rare *vt* purge

equa'tore *nm* equator. **equatori-'ale** *a* equatorial

equazi'one *nf* equation

e'questre *a* equestrian; **circo ~** circus

equi'latero *a* equilateral

equili'bra|re *vt* balance. **~to** *a* (*persona*) well-balanced. **equi-'librio** *nm* balance; (*buon senso*) common sense; (*di bilancia*) equilibrium

equili'brismo *nm* **fare ~** do a balancing act

e'quino *a* horse *attrib*

equi'nozio *nm* equinox

equipaggia'mento *nm* equipment

equipaggi'are *vt* equip; (*di persone*) man

equi'paggio *nm* crew; *Aeron* cabin crew

equipa'rare *vt* make equal

é'quipe *nf inv* team

equità *nf* equity

equitazi'one *nf* riding

equiva'lente *a* & *nm* equivalent. **~za** *nf* equivalence

equiva'lere *vi* **~ a** be equivalent to

equivo'care *vi* misunderstand

e'quivoco *a* equivocal; (*sospetto*) suspicious; **un tipo ~** a shady character ●*nm* misunderstanding

'equo *a* fair, just

'era *nf* era

'erba *nf* grass; (*aromatica, medicinale*) herb. **~ cipollina** chives *pl*. **er'baccia** *nf* weed. **er'baceo** *a* herbaceous

erbi'cida *nm* weed-killer

erbo'rist|a *nmf* herbalist. **~e'ria** *nf* herbalist's shop

er'boso *a* grassy

er'culeo *a* (*forza*) herculean

e'rede *nmf* heir; heiress. **~ità** *nf inv* inheritance; *Biol* heredity. **~i'tare** *vt* inherit. **~itarietà** *nf* heredity. **~i'tario** *a* hereditary

ere'mita *nm* hermit

ere'sia *nf* heresy. **e'retico, -a** *a* heretical ●*nmf* heretic

e're|tto *pp di* **erigere** *a* erect. **~zi'one** *nf* erection; (*costruzione*) building

er'gastolo *nm* life sentence; (*luogo*) prison

'erica *nf* heather

e'rigere *vt* erect; (*fig: fondare*) found

eri'tema *nm* (*cutaneo*) inflammation; (*solare*) sunburn

ermel'lino *nm* ermine

ermetica'mente *adv* hermetically. **er'metico** *a* hermetic; (*a tenuta d'aria*) airtight

'ernia *nf* hernia

e'rodere *vi* erode

e'ro|e *nm* hero. **~'ico** *a* heroic. **~'ismo** *nm* heroism

ero'ga|re *vt* distribute; (*fornire*) supply. **~zi'one** *nf* supply

ero'ina *nf* heroine; (*droga*) heroin

ero'sione *nf* erosion

e'rotico *a* erotic. **ero'tismo** *nm* eroticism

er'rante *a* wandering. **er'rare** *vi* wander; (*sbagliare*) be mistaken

er'rato *a* (*sbagliato*) mistaken

'erre *nf* **~ moscia** burr

erronea'mente *adv* mistakenly

er'rore *nm* error, mistake; (*di stampa*) misprint; **essere in ~** be wrong

'erta *nf* **stare all'~** be on the alert

eru'di|rsi *vr* get educated. **~to** *a* learned

erut'tare *vt* (*vulcano:*) erupt ● *vi* (*ruttare*) belch. **eruzi'one** *nf* eruption; *Med* rash

esacer'bare *vt* exacerbate

esage'ra|re *vt* exaggerate ● *vi* exaggerate; (*nel comportamento*) go over the top; **~re nel mangiare** eat too much. **~ta'mente** *adv* excessively. **~to** *a* exaggerated; (*prezzo*) exorbitant ● *nm* **è un ~to** he exaggerates. **~zi'one** *nf* exaggeration; **è costato un'~zione** it cost the earth

esa'lare *vt/i* exhale

esal'ta|re *vt* exalt; (*entusiasmare*) elate. **~to** *a* (*fanatico*) fanatical ● *nm* fanatic. **~zi'one** *nf* exaltation; (*in discorso*) fervour

e'same *nm* examination, exam; **dare un ~** take an exam; **prendere in ~** examine. **~ del sangue** blood test. **esami** *pl* **di maturità** ≈ A-levels

esami'na|re *vt* examine. **~'tore,** **~'trice** *nmf* examiner

e'sangue *a* bloodless

e'sanime *a* lifeless

esaspe'rante *a* exasperating

esaspe'ra|re *vt* exasperate. **~rsi** *vr* get exasperated. **~zi'one** *nf* exasperation

esat|ta'mente *adv* exactly. **~'tezza** *nf* exactness; (*precisione*) precision; (*di risposta, risultato*) accuracy

e'satto *pp di* **esigere** ● *a* exact; (*risposta, risultato*) correct; (*orologio*) right; **hai l'ora esatta?** do you have the right time?; **sono le due esatte** it's two o'clock exactly

esat'tore *nm* collector

esau'dire *vt* grant; fulfil (*speranze*)

esauri'ente *a* exhaustive

esau'ri|re *vt* exhaust. **~rsi** *vr* ex-

haust oneself; (*merci ecc:*) run out. **~to** *a* exhausted; (*merci*) sold out; (*libro*) out of print; **fare il tutto ~to** (*spettacolo:*) play to a full house

'esca *nf* bait

escande'scenz|a *nf* outburst; **dare in ~e** lose one's temper

escla'ma|re *vi* exclaim. **~'tivo** *a* exclamatory. **~zi'one** *nf* exclamation

es'clu|dere *vt* exclude; rule out (*possibilità, ipotesi*). **~si'one** *nf* exclusion. **~'siva** *nf* exclusive right, sole right; **in ~siva** exclusive. **~siva'mente** *adv* exclusively. **~'sivo** *a* exclusive. **~so** *pp di* **escludere** **a non è ~so che ci sia** it's not out of the question that he'll be there

escogi'tare *vt* contrive

escre'mento *nm* excrement

escursi'one *nf* excursion; (*scorreria*) raid; (*di temperatura*) range

ese'cra|bile *a* abominable, abhor

esecu|'tivo *a* & *nm* executive. **~'tore, ~'trice** *nmf* executor; *Mus* performer. **~zi'one** *nf* execution; *Mus* performance

esegu'ire *vt* carry out; *Jur* execute; *Mus* perform

e'sempio *nm* example; **ad o per ~** for example; **dare l'~** set sb an example; **fare un ~** give an example. **esem'plare** *a* exemplary ● *nm* specimen; (*di libro*) copy. **esemplifi'care** *vt* exemplify

esen'tar|e *vt* exempt. **~si** *vr* free oneself. **e'sente** *a* exempt. **esente da imposta** duty-free. **esente da IVA** VAT-exempt

esen'tasse *a* duty-free

e'sequie *nfpl* funeral rites

eser'cente *nmf* shopkeeper

eserci'ta|re *vt* exercise; (*addestrare*) train; (*fare uso di*) exert; (*professione*) practise. **~rsi** *vr* practise. **~zi'one** *nf* exercise; *Mil* drill

e'sercito *nm* army

eser'cizio *nm* exercise; (*pratica*) practice; *Comm* financial year; (*azienda*) business; **essere fuori ~** be out of practice

esi'bi|re *vt* show off; produce ⟨*documenti*⟩. **~rsi** *vr Theat* perform; *fig* show off. **~zi'one** *nf Theat* performance; (*di documenti*) production

esibizio'nis|mo *nm* showing off. **~ta** *nmf* exhibitionist

esi'gen|te *a* exacting; (*pignolo*) fastidious. **~za** *nf* demand; (*bisogno*) need. **e'sigere** *vt* demand; (*riscuotere*) collect

e'siguo *a* meagre

esila'ran|te *a* exhilarating

'esile *a* slender; ⟨*voce*⟩ thin

esili'a|re *vt* exile. **~rsi** *vr* go into exile. **~to, -a** *a* exiled ● *nmf* exile. **e'silio** *nm* exile

e'simer|e *vt* release. **~si** *vr* **~si da** get out of

esi'sten|te *a* existing. **~za** *nf* existence. **~zi'ale** *a* existential. **~zia'lismo** *nm* existentialism

e'sistere *vi* exist

esi'tante *a* hesitating; ⟨*voce*⟩ faltering

esi'ta|re *vi* hesitate. **~zi'one** *nf* hesitation

'esito *nm* result; **avere buon ~** be a success

'esodo *nm* exodus

e'sofago *nm* oesophagus

esone'rare *vt* exempt. **e'sonero** *nm* exemption

esorbi'tante *a* exorbitant

esorciz'zare *vt* exorcize

esordi'ente *nmf* person making his/her début. **e'sordio** *nm* opening; (*di attore*) début. **esor'dire** *vi* début

esor'tare *vt* (*pregare*) beg; (*incitare*) urge

eso'terico *a* esoteric

e'sotico *a* exotic

espa'drillas *nfpl* espadrilles

es'pan|dere *vt* expand. **~dersi** *vr* expand; (*diffondersi*) extend. **~si'one** *nf* expansion. **~'sivo** *a* expansive; ⟨*persona*⟩ friendly

espatri'are *vi* leave one's country. **es'patrio** *nm* expatriation

espedi'ente *nm* expedient; **vivere di ~** live by one's wits

es'pellere *vt* expel

esperi'enza *nf* experience; **parlare per ~enza** speak from experience. **~'mento** *nm* experiment

es'perto, -a *a* & *nmf* expert

espi'a|re *vt* atone for. **~'torio** *a* expiatory

espi'rare *vt/i* breathe out

espli'care *vt* carry on

esplicita'mente *adv* explicitly. **es'plicito** *a* explicit

es'plodere *vi* explode ● *vt* fire

esplo'ra|re *vt* explore. **~'tore, ~'trice** *nmf* explorer; **giovane ~tore** boy scout. **~zi'one** *nf* exploration

esplo|si'one *nf* explosion. **~'sivo** *a* & *nm* explosive

espo'nente *nm* exponent

es'por|re *vt* expose; display ⟨*merci*⟩; (*spiegare*) expound; exhibit ⟨*quadri ecc*⟩. **~si** *vr* (*compromettersi*) compromise oneself; (*al sole*) expose oneself; (*alle critiche*) lay oneself open

espor'ta|re *vt* export. **~'tore, ~'trice** *nmf* exporter. **~zi'one** *nf* export

esposi'zione *nf* (*mostra*) exhibition; (*in vetrina*) display; (*spiegazione ecc*) exposition; (*posizione, fotografia*) exposure. **es'posto** *pp di* **esporre** ● *a* exposed; **esposto a** ⟨*rivolto*⟩ facing ● *nm Jur ecc* statement

espressa'mente *adv* expressly; **non l'ha detto ~** he didn't put it in so many words

espres|si'one *nf* expression. **~'sivo** *a* expressive

es'presso *pp di* **esprimere** ● *a* express ● *nm* (*lettera*) express letter; (*treno*) express train; (*caffè*) espresso; **per ~** ⟨*spedire*⟩ [by] express [post]

es'primer|e *vt* express. **~si** *vr* express oneself

espropri'a|**re** vt dispossess. **~zi'one** nf Jur expropriation. **es'proprio** nm expropriation

espulsi'one nf expulsion. **es'pulso** pp di **espellere**

es'senz|**a** nf essence. **~i'ale** a essential ●nm important thing. **~ial'mente** a essentially

'essere vi be; **c'è** there is; **ci sono** there are; **che ora è?** – **sono le dieci** what time is it? – it's ten o'clock; **chi è?** – **sono io** who is it? – it's me; **ci sono!** (ho capito) I've got it!; **ci siamo!** (siamo arrivati) here we are at last!; **è stato detto che** it has been said that; **siamo in due** there are two of us; **questa camicia è da lavare** this shirt is to be washed; **non è da te** it's not like you; **~ di** (provenire da) be from; **~ per** (favorevole) be in favour of; **se fossi in te,...** if I were you,...; **sarà!** if you say so!; **come sarebbe a dire?** what are you getting at? ●v aux have; (in passivi) be; **siamo arrivati** we have arrived; **ci sono stati ieri** I was there yesterday; **sono nato a Torino** I was born in Turin; **è riconosciuto come...** he is recognized as... ●nm being. ~ **umano** human being. ~ **vivente** living creature

essic'cato a dried

'esso, -a pers pron he, she; (cosa, animale) it

est nm east

'estasi nf ecstasy; **andare in ~ per** go into raptures over. **~'are** vt enrapture

e'state nf summer

e'sten|**dere** vt extend. **~dersi** vr spread; (allungarsi) stretch. **~si'one** nf extension; (ampiezza) expanse; Mus range. **~'sivo** a extensive

estenu'ante a exhausting

estenu'a|**re** vt wear out; deplete (risorse, casse). **~rsi** vr wear oneself out

esteri'or|**e** a & nm exterior.

~'mente adv externally; (di persone) outwardly

esterna'mente adv on the outside

ester'nare vt express, show

e'sterno a external; **per uso ~** for external use only ●nm (allievo) day-boy; Archit exterior; (scala) outside; (in film) location shot

'estero a foreign ●nm foreign countries pl; **all'~** abroad

esterre'fatto a horrified

e'steso pp di **estendere** ●a extensive; (diffuso) widespread; **per ~** (scrivere) in full

e'stetic|**a** nf aesthetics sg. **~a'mente** adv aesthetically. **~o, -a** a aesthetic; (chirurgia, chirurgo) plastic. **este'tista** nf beautician

'estimo nm estimate

e'stin|**guere** vt extinguish. **~guersi** vr die out. **~to, -a** pp di **estinguere** ●nmf deceased. **~'tore** nm [fire] extinguisher. **~zi'one** nf extinction; (di incendio) putting out

estir'pa|**re** vt uproot; extract (dente); fig eradicate (crimine, malattia). **~zi'one** nf eradication; (di dente) extraction

e'stivo a summer

e'stor|**cere** vt extort. **~si'one** nf extortion. **~to** pp di **estorcere**

estradizi'one nf extradition

e'straneo, -a a extraneous; (straniero) foreign ●nmf stranger

estrani'ar|**e** vt estrange. **~si** vr become estranged

e'stra|**rre** vt extract; (sorteggiare) draw. **~tto** pp di **estrarre** ●nm extract; (brano) excerpt; (documento) abstract. **~tto conto** statement [of account], bank statement. **~zi'one** nf extraction; (a sorte) draw

estrema'mente adv extremely

estre'mis|**mo** nm extremism. **~ta** nmf extremist

estremità nf inv extremity; (di una corda) end ●nfpl Anat extremities

e'stremo *a* extreme; *(ultimo)* last; **misure estreme** drastic measures; **l'E~ Oriente** the Far East ● *nm (limite)* extreme. **estremi** *pl (di documento)* main points; *(di reato)* essential elements; **essere agli estremi** be at the end of one's tether

'estro *nm (disposizione artistica)* talent; *(ispirazione)* inspiration; *(capriccio)* whim. **e'stroso** *a* talented; *(capriccioso)* unpredictable

estro'mettere *vt* expel

estro'verso *a* extroverted ● *nm* extrovert

estu'ario *nm* estuary

esube'ran|te *a* exuberant. **~za** *nf* exuberance

'esule *nmf* exile

esul'tante *a* exultant

esul'tare *vi* rejoice

esu'mare *vt* exhume

età *nf inv* age; **raggiungere la maggiore ~** come of age; **un uomo di mezz'~** a middle-aged man

'etere *nm* ether. **e'tereo** *a* ethereal

eterna'mente *adv* eternally

eter'nità *nf* eternity; **è un'~ che non la vedo** I haven't seen her for ages

e'terno *a* eternal; *(questione, problema)* age-old; **in ~** *fam* for ever

etero'geneo *a* diverse, heterogeneous

eterosessu'ale *nmf* heterosexual

'etic|a *nf* ethics. **~o** *a* ethical

eti'chetta¹ *nf* label; *(con il prezzo)* price-tag

eti'chetta² *nf (cerimoniale)* etiquette

etichet'tare *vt* label

eti'lometro *nm* Breathalyzer®

etimolo'gia *nf* etymology

Eti'opia *nf* Ethiopia

'etnico *a* ethnic. **etnolo'gia** *nf* ethnology

e'trusco *a & nmf* Etruscan

'ettaro *nm* hectare

'etto, etto'grammo *nm* hundred grams, ≈ quarter pound

euca'lipto *nm* eucalyptus

eucari'stia *nf* Eucharist

eufe'mismo *nm* euphemism

eufo'ria *nf* elation; *Med* euphoria. **eu'forico** *a* elated; *Med* euphoric

Euro'city *nm* international Intercity

eurodepu'tato *nm* Euro MP, MEP

Eu'ropa *nf* Europe. **euro'peo, -a** *a & nmf* European

eutana'sia *nf* euthanasia

evacu|a|re *vt* evacuate. **~zi'one** *nf* evacuation

e'vadere *vi* evade; *(sbrigare)* deal with ● *vi ~* **da** escape from

evane'scente *a* vanishing

evan'gel|ico *a* evangelical. **evange'lista** *nm* evangelist. **~o** *nm* = **vangelo**

evapo'ra|re *vi* evaporate. **~zi'one** *nf* evaporation

evasi'one *nf* escape; *(fiscale)* evasion; *fig* escapism. **eva'sivo** *a* evasive

e'vaso *pp di* **evadere** ● *nm* fugitive. **eva'sore** *nm ~* **fiscale** tax evader

e'vento *nm* event

eventu'al|e *a* possible. **~ità** *nf inv* eventuality

evi'dente *a* evident; **è ~te che** it is obvious that. **~te'mente** *adv* evidently. **~za** *nf* evidence; **mettere in ~za** emphasize; **mettersi in ~za** make oneself conspicuous

evidenzi'a|re *vt* highlight. **~'tore** *nm (penna)* highlighter

evi'tare *vt* avoid; *(risparmiare)* spare

evo'care *vt* evoke

evo'lu|to *pp di* **evolvere** ● *a* evolved; *(progredito)* progressive; *(civiltà, nazione)* advanced; **una donna evoluta** a modern woman. **~zi'one** *nf* evolution; *(di ginnasta, aereo)* circle

e'volver|e *vt* develop. **~si** *vr* evolve

ev'viva *int* hurray; **~ il Papa!** long live the Pope!; **gridare ~** cheer

ex+ *pref* ex+, former

'**extra** *a inv* extra; (*qualità*) first-class ● *nm inv* extra
extracomuni'tario *a* non-EC
extraconiu'gale *a* extramarital
extrater'restre *nmf* extra-terrestrial

Ff

fa[1] *nm inv Mus* (*chiave, nota*) F
fa[2] *adv* ago; **due mesi ~** two months ago
fabbi'sogno *nm* requirements *pl*, needs *pl*
'**fabbrica** *nf* factory
fabbri'cabile *a* (*area, terreno*) that can be built on
fabbri'cante *nm* manufacturer
fabbri'ca|re *vt* build; (*produrre*) manufacture; (*fig: inventare*) fabricate. ~**to** *nm* building. ~**zi'one** *nf* manufacturing; (*costruzione*) building
'**fabbro** *nm* blacksmith
fac'cend|a *nf* matter; ~**e** *pl* (*lavori domestici*) housework *sg*. ~**i'ere** *nm* wheeler-dealer
fac'chino *nm* porter
'**facci|a** *nf* face; (*di foglio*) side; ~**a** a ~**a** face to face; ~**a tosta** cheek; **voltar** ~**a** change sides; **di** ~**a** (*palazzo*) opposite; **alla** ~**a di** (*fam: a dispetto di*) in spite of. ~'**ata** *nf* façade; (*di foglio*) side; (*fig: esteriorità*) outward appearance
fa'ceto *a* facetious; **tra il serio e il** ~ half joking
fa'chiro *nm* fakir
'**facil|e** *a* easy; (*affabile*) easygoing; **essere** ~**e alle critiche** be quick to criticize; **essere** ~ **e al riso** laugh a lot; ~**e a farsi** easy to do; **è** ~**e che piova** it's likely to rain. ~**ità** *nf inv* ease; (*disposizione*) aptitude; **avere** ~**ità di parola** express oneself well

facili'ta|re *vt* facilitate. ~**zi'one** *nf* facility; ~**zioni** *pl* special terms
facil'mente *adv* (*con facilità*) easily; (*probabilmente*) probably
faci'lone *a* slapdash. ~'**ria** *nf* slapdash attitude
facino'roso *a* violent
facoltà *nf inv* faculty; (*potere*) power. ~'**tivo** *a* optional; **fermata** ~**tiva** request stop
faco'ltoso *a* wealthy
fac'simile *nm* facsimile
fac'totum *nmf* man/girl Friday, factotum
'**faggio** *nm* beech
fagi'ano *nm* pheasant
fagio'lino *nm* French bean
fagi'olo *nm* bean; **a** ~ (*arrivare, capitare*) at the right time
fagoci'tare *vt* gobble up (*società*)
fa'gotto *nm* bundle; *Mus* bassoon
'**faida** *nf* feud
fai da te *nm* do-it-yourself, DIY
fal'cata *nf* stride
'**falc|e** *nf* scythe. **fal'cetto** *nm* sickle. ~**i'are** *vt* cut; *fig* mow down. ~**ia'trice** *nf* [lawn-]mower
'**falco** *nm* hawk
fal'cone *nm* falcon
'**falda** *nf* stratum; (*di neve*) flake; (*di cappello*) brim; (*pendio*) slope
fale'gname *nm* carpenter. ~'**ria** *nf* carpentry
'**falla** *nf* leak
fal'lace *a* deceptive
'**fallico** *a* phallic
falli'mentare *a* disastrous; *Jur* bankruptcy. **falli'mento** *nm* *Fin* bankruptcy; *fig* failure
fal'li|re *vi* *Fin* go bankrupt; *fig* fail ● *vt* miss (*colpo*). ~**to, -a** *a* unsuccessful; *Fin* bankrupt ● *nmf* failure; *Fin* bankrupt
'**fallo** *nm* fault; (*errore*) mistake; *Sport* foul; (*imperfezione*) flaw; **senza** ~ without fail
falò *nm inv* bonfire
fal'sar|e *vt* alter; (*falsificare*) falsify. ~**io, -a** *nmf* forger; (*di documenti*) counterfeiter
falsifi'care *vt* fake; (*contraffare*)

forge. **~zi'one** *nf* (*di documento*) falsification

falsità *nf* falseness

'falso *a* false; (*sbagliato*) wrong; (*opera d'arte ecc*) fake; (*gioielli, oro*) imitation ● *nm* forgery; **giurare il ~** commit perjury

'fama *nf* fame; (*reputazione*) reputation

'fame *nf* hunger; **aver ~** be hungry; **fare la ~** barely scrape a living. **fa'melico** *a* ravenous

famige'rato *a* infamous

fa'miglia *nf* family

famili'ar|e *a* family *attrib*; (*ben noto*) familiar; (*senza cerimonie*) informal ● *nmf* relative, relation **~ità** *nf* familiarity; (*informalità*) informality. **~iz'zarsi** *vr* familiarize oneself

fa'moso *a* famous

fa'nale *nm* lamp; *Auto ecc* light. **fanali** *pl* **posteriori** *Auto* rear lights

fa'natico, -a *a* fanatical; **essere ~ di calcio/cinema** be a football/cinema fanatic ● *nmf* fanatic. **fana'tismo** *nm* fanaticism

fanci'ul|la *nf* young girl. **~'lezza** *nf* childhood. **~lo** *nm* young boy

fan'donia *nf* lie; **fandonie!** nonsense!

fan'fara *nf* fanfare; (*complesso*) brass band

fanfaro'nata *nf* brag. **fanfa'rone, -a** *nmf* braggart

fan'ghiglia *nf* mud. **'fango** *nm* mud. **fan'goso** *a* muddy

fannul'lone, -a *nmf* idler

fantasci'enza *nf* science fiction

fanta'si|a *nf* fantasy; (*immaginazione*) imagination; (*capriccio*) fancy; (*di tessuto*) pattern. **~'oso** *a* (*stilista, ragazzo*) imaginative; (*resoconto*) improbable

fan'tasma *nm* ghost

fantasti'c|are *vi* day-dream. **~he'ria** *nf* day-dream. **fan'tastico** *a* fantastic; (*racconto*) fantasy

'fante *nm* infantryman; (*nelle carte*) jack. **~'ria** *nf* infantry

fan'tino *nm* jockey

fan'toccio *nm* puppet

fanto'matico *a* (*inafferrabile*) phantom *attrib*

fara'butto *nm* trickster

fara'ona *nf* (*uccello*) guinea-fowl

far'ci|re *vt* stuff; fill (*torta*). **~to** *a* stuffed; (*dolce*) filled

far'dello *nm* bundle; *fig* burden

'fare *vt* do; make (*dolce, letto ecc*); (*recitare la parte di*) play; (*trascorrere*) spend; **~ una pausa/un sogno** have a break/a dream; **~ colpo su** impress; **~ paura a** frighten; **~ piacere a** please; **farla finita** put an end to it; **~ l'insegnante** be a teacher; **~ lo scemo** play the idiot; **~ una settimana al mare** spend a week at the seaside; **3 più 3 fa 6** 3 and 3 makes 6; **quanto fa? - fanno 10 000 lira** how much is it? - it's 10,000 lire; **far ~ qcsa a qcno** get sb to do sth; (*costringere*) make sb do sth; **~ vedere** show; **fammi parlare** let me speak; **niente a che ~ con** nothing to do with; **non c'è niente da ~** (*per problema*) there is nothing we/you/etc. can do; **fa caldo/buio** it's warm/dark; **non fa niente** it doesn't matter; **strada facendo** on the way. **farcela** (*riuscire*) manage ● *vi* **fai in modo di venire** try and come; **~ da** act as; **~ per** make as if to; **~ presto** be quick; **non fa per me** it's not for me ● *nm* way; **sul far dei giorno** at daybreak. **farsi** *vr* (*diventare*) get; (*sl: drogarsi*) shoot up; **farsi avanti** come forward; **farsi i fatti propri** mind one's own business; **farsi la barba** shave; **farsi la villa** *fam* buy a villa; **farsi il ragazzo** *fam* find a boyfriend; **farsi due risate** have a laugh; **farsi male** hurt oneself; **farsi strada** (*aver successo*) make one's way in the world

fa'retto *nm* spot[light]

far'falla *nf* butterfly

farfal'lino *nm* (*cravatta*) bow tie

farfugli'are *vt* mutter

fa'rina *nf* flour. **fari'nacei** *nmpl* starchy food *sg*

fa'ringe *nf* pharynx

fari'noso *a* ⟨neve⟩ powdery; ⟨mela⟩ soft; ⟨patata⟩ floury

farma|'ceutico *a* pharmaceutical. **~'cia** *nf* pharmacy; ⟨negozio⟩ chemist's [shop]. **~cia di turno** duty chemist. **~'cista** *nmf* chemist. **'farmaco** *nm* drug

'faro *nm* Auto headlight; Aeron beacon; ⟨costruzione⟩ lighthouse

'farsa *nf* farce

'fasci|a *nf* band; ⟨zona⟩ area; ⟨ufficiale⟩ sash; ⟨benda⟩ bandage. **~'are** *vt* bandage; cling to ⟨fianchi⟩. **~a'tura** *nf* dressing; ⟨azione⟩ bandaging

fa'scicolo *nm* file; ⟨di rivista⟩ issue; ⟨libretto⟩ booklet

'fascino *nm* fascination

'fascio *nm* bundle; ⟨di fiori⟩ bunch

fa'sci|smo *nm* fascism. **~ta** *nmf* fascist

'fase *nf* phase

fa'stidi|o *nm* nuisance; ⟨scomodo⟩ inconvenience; **dar ~o a qcno** bother sb; **~i** *pl* ⟨preoccupazioni⟩ worries; ⟨disturbi⟩ troubles. **~'oso** *a* tiresome

'fasto *nm* pomp. **fa'stoso** *a* sumptuous

fa'sullo *a* bogus

'fata *nf* fairy

fa'ta|le *a* fatal; ⟨inevitabile⟩ fated

fata'l|ismo *nm* fatalism. **~ista** *nmf* fatalist. **~ità** *nf inv* fate; ⟨caso sfortunato⟩ misfortune. **~'mente** *adv* inevitably

fa'tica *nf* effort; ⟨lavoro faticoso⟩ hard work; ⟨stanchezza⟩ fatigue; **a ~** with great difficulty; **è ~ sprecata** it's a waste of time; **fare ~ a fare qcsa** find it difficult to do sth; **fare ~ a finire qcsa** struggle to finish sth. **fati'caccia** *nf* pain

fati'ca|re *vi* toil; **~re a** ⟨stentare⟩ find it difficult to. **~ta** *nf* effort;

⟨sfacchinata⟩ grind. **fati'coso** *a* tiring; ⟨difficile⟩ difficult

'fato *nm* fate

fat'taccio *nm hum* foul deed

fat'tezze *nfpl* features

fat'tibile *a* feasible

'fatto *pp di* **fare** ● *a* done, made; **~ a mano/in casa** handmade/home-made ● *nm* fact; ⟨azione⟩ action; ⟨avvenimento⟩ event; **bada ai fatti tuoi!** mind your own business; **sa il ~ suo** he knows his business; **di ~** in fact; **in ~ di** as regards

fat'to|re *nm* ⟨causa, Math⟩ factor; ⟨di fattoria⟩ farm manager. **~'ria** *nf* farm; ⟨casa⟩ farmhouse

fatto'rino *nm* messenger [boy]

fattuc'chi'era *nf* witch

fat'tura *nf* ⟨stile⟩ cut; ⟨lavorazione⟩ workmanship; Comm invoice

fattu'ra|re *vt* invoice; ⟨adulterare⟩ adulterate. **~to** *nm* turnover, sales *pl*. **~zi'one** *nf* invoicing, billing

'fatuo *a* fatuous

'fauna *nf* fauna

fau'tore *nm* supporter

'fava *nf* broad bean

fa'vella *nf* speech

fa'villa *nf* spark

'favo|la *nf* fable; ⟨fiaba⟩ story; ⟨oggetto di pettegolezzi⟩ laughing-stock; ⟨meraviglia⟩ dream. **~'loso** *a* fabulous

fa'vo|ri|re *vt* favour; ⟨promuovere⟩ promote; **vuol ~re?** ⟨a cena, pranzo⟩ will you have some?; ⟨entrare⟩ will you come in?. **~to, -a** *a a* favoured

favo're *nm* favour; **essere a ~ di** be in favour of; **per ~** please; **di ~** ⟨condizioni, trattamento⟩ preferential. **~ggia'mento** *nm Jur* aiding and abetting. **favo'revole** *a* favourable. **~vol'mente** *adv* favourably

favo'ri|re *vt* favour; ⟨promuovere⟩ promote; **vuol ~re?** ⟨a cena, pranzo⟩ will you have some?; ⟨entrare⟩ will you come in?. **~to, -a a** *a* favoured

fax *nm inv* fax. **fa'xare** *vt* fax

fazi'one *nf* faction

faziosità *nf* bias. **fazi'oso** *nm* sectarian

fazzolet'tino nm ~ [**di carta**] [paper] tissue

fazzo'letto nm handkerchief; ⟨da testa⟩ headscarf

feb'braio nm February

'febbre nf fever; **avere la ~** have o run a temperature. **~ da fieno** hay fever. **febbrici'tante** a fevered. **feb'brile** a feverish

'feccia nf dregs pl

'fecola nf potato flour

fecon'da|re vt fertilize. **~'tore** nm fertilizer. **~zi'one** nf fertilization. **~zione artificiale** artificial insemination. **fe'condo** a fertile

'fede nf faith; ⟨fiducia⟩ trust; ⟨anello⟩ wedding-ring; **in buona/mala ~** in good/bad faith; **prestar ~ a** believe; **tener ~ alla parola** keep one's word. **fe'dele** a faithful ● nmf believer; ⟨seguace⟩ follower. **~l'mente** adv faithfully. **~ltà** nf faithfulness; **alta ~ltà** high fidelity

federa nf pillowcase

fede'ra|le a federal. **~'lismo** nm federalism. **~zi'one** nf federation

fe'dina nf **avere la ~ penale sporca/pulita** have a/no criminal record

'fegato nm liver; fig guts pl

'felce nf fern

fe'lic|e a happy; ⟨fortunato⟩ lucky. **~ità** nf happiness

felici'ta|rsi vr **~rsi con** congratulate. **~zi'oni** nfpl congratulations

fe'lino a feline

'felpa nf ⟨indumento⟩ sweatshirt

fel'pato a brushed; ⟨passo⟩ stealthy

'feltro nm felt; ⟨cappello⟩ felt hat

'femmin|a nf female. **femm'nile** a feminine; ⟨rivista, abbigliamento⟩ women's; ⟨sesso⟩ female ● nm feminine. **~ilità** nf femininity. **femmi'nismo** nm feminism

'femore nm femur

'fend|ere vt split. **~i'tura** nf split; ⟨in roccia⟩ crack

feni'cottero nm flamingo

fenome'nale a phenomenal. **fe'nomeno** nm phenomenon

'feretro nm coffin

feri'ale a weekday; **giorno ~** weekday

'ferie nfpl holidays; ⟨di università, tribunale ecc⟩ vacation sg; **andare in ~** go on holiday

feri'mento nm wounding

fe'ri|re vt wound; ⟨in incidente⟩ injure; fig hurt. **~rsi** vr injure oneself. **~ta** nf wound. **~to** a wounded ● nm wounded person; Mil casualty

'ferma nf Mil period of service

ferma'pelli nm inv hairslide

ferma'carte nm inv paperweight

ferma'vatta nm inv tiepin

fer'maglio nm clasp; ⟨spilla⟩ brooch; ⟨per capelli⟩ hair slide

ferma'mente adv firmly

fer'ma|re vt stop; ⟨fissare⟩ fix; Jur detain ● vi stop. **~rsi** vr stop. **~ta** nf stop. **~ta dell'autobus** bus-stop. **~ta a richiesta** request stop

fermen'ta|re vi ferme. **~zi'one** nf fermentation. **fer'mento** nm ferment; ⟨lievito⟩ yeast

fer'mezza nf firmness

'fermo a still; ⟨veicolo⟩ stationary; ⟨stabile⟩ steady; ⟨orologio⟩ not working ● nm Jur detention; Mech catch; **in stato di ~** in custody

fe'roc|e a ferocious; ⟨bestia⟩ wild; ⟨freddo, dolore⟩ unbearable. **~emente** adv fiercely, ferociously. **~ia** nf ferocity

fer'raglia nf scrap iron

ferra'gosto nm 15 August ⟨bank holiday in Italy⟩; ⟨periodo⟩ August holidays pl

ferra'menta nfpl ironmongery sg; **negozio di ~** ironmonger's

fer'ra|re vt shoe ⟨cavallo⟩. **~to in** ⟨preparato in⟩ well up in

'ferreo a iron

'ferro nm iron; ⟨attrezzo⟩ tool; ⟨di chirurgo⟩ instrument; **bistecca ai ferri** grilled steak; **di ~** ⟨memoria⟩ excellent; ⟨alibi⟩ cast-iron; **salute**

di ~ iron constitution. ~ **battuto**
wrought iron. ~ **da calza** knitting
needle. ~ **di cavallo** horseshoe. ~
da stiro iron

ferro'vecchio nm scrap merchant
ferro'vi|a nf railway. ~'ario a railway. ~'ere nm railwayman

fertil|e a fertile. ~ità nf fertility.
~iz'zante nm fertilizer

fer'vente a blazing; fig fervent

'fervere vi ⟨preparativi:⟩ be well
under way

'fervido a fervent; ~i **auguri** best
wishes

fer'vore nm fervour

fesse'ria nf nonsense

'fesso pp di **fendere** ● a cracked;
(fam: sciocco) foolish ● nm fam
(idiota) fool; **far** ~ **qcno** con sb

fes'sura nf crack; (per gettone ecc)
slot

'festa nf feast; (giorno festivo) holiday; (compleanno) birthday; (ricevimento) party; fig joy; **fare** ~ **a**
qcno welcome sb; **essere in** ~ be
on holiday; **far** ~ celebrate. ~i'olo
a festive

festeggia'mento nm celebration;
(manifestazione) festivity

festeggi'are vt celebrate; (accogliere festosamente) give a hearty
welcome to

fe'stino nm party

festività nfpl festivities. **fe'stivo**
a holiday; (lieto) festive. **festivi**
nmpl public holidays

fe'stone nm (nel cucito) scallop,
scollop

fe'stoso a merry

fe'tente a evil smelling; fig revolting ● nmf fam bastard

fe'ticcio nm fetish

'feto nm foetus

fe'tore nm stench

'fetta nf slice; **a fette** sliced. ~
biscottata slices of crispy toast-
like bread

fet'tuccia nf tape; (con nome)
name tape

feu'dale a feudal. **'feudo** nm feud

FFSS abbr (**Ferrovie dello Stato**)
Italian state railways

fi'aba nf fairy-tale. **fia'besco** a
fairy-tale

fi'acc|a nf weariness; (indolenza)
laziness; **battere la** ~**a** be sluggish. **fiac'care** vt weaken. ~**o** a
weak; (indolente) slack; (stanco)
weary; ⟨partita⟩ dull

fi'acco|la nf torch. ~**'lata** nf torch-
light procession

fi'ala nf phial

fi'amma nf flame; Naut pennant;
in fiamme aflame. **andare in**
fiamme go up in flames. ~
ossidrica blowtorch

fiam'ma|nte a flaming; **nuovo**
~**nte** brand new. ~**ta** nf blaze

fiammeggi'are vi blaze

fiam'mifero nm match

fiam'mingo, -a a Flemish ● nmf
Fleming ● nm (lingua) Flemish

fiancheggi'are vt border; fig support

fi'anco nm side; (di persona) hip;
(di animale) flank; Mil wing; **al**
mio ~ by my side; ~ **a** ~
⟨lavorare⟩ side by side

fi'asco nm flask; fig fiasco; **fare** ~
be a fiasco

fia'tare vi breathe; (parlare)
breathe a word

fi'ato nm breath; (vigore) stamina;
strumenti a ~ wind instruments;
senza ~ breathlessly; **tutto d'un**
~ ⟨bere, leggere⟩ all in one go

'fibbia nf buckle

'fibra nf fibre; **fibre** pl (alimentari)
roughage. ~ **ottica** optical fibre

ficca'naso nmf nosey parker

fic'car|e vt thrust; drive ⟨chiodo
ecc⟩; (fam: mettere) shove. ~**si** vr
thrust oneself; (nascondersi) hide;
~**si nei guai** get oneself into trouble

fiche nf ⟨gettone⟩ chip

'fico nm (albero) fig-tree; (frutto)
fig. ~ **d'India** prickly pear

'fico, -a fam nmf cool sort ● a cool

fidanza'mento nm engagement

fidan'za|rsi *vr* get engaged. **~to, -a** *nmf (ufficiale)* fiancé; fiancée

fi'da|rsi *vr* **~rsi di** trust. **~to** *a* trustworthy

'fido *nm* devoted follower; *Comm* credit

fi'duci|a *nf* confidence; **degno di ~a** trustworthy; **persona di ~a** reliable person; **di ~a** *(fornitore, banca)* regular, usual. **~'oso** *a* trusting

fi'ele *nm* bile; *fig* bitterness

fie'nile *nm* barn. **fi'eno** *nm* hay

fi'era *nf* fair

fie'rezza *nf (dignità)* pride. **fi'ero** *a* proud

fi'evole *a* faint; *‹luce›* dim

'fifa *nf fam* jitters; **aver ~** have the jitters. **fi'fone, -a** *nmf fam* chicken

'figli|a *nf* daughter; **~a unica** only child. **~'astra** *nf* stepdaughter. **~'astro** *nm* stepson. **~o** *nm* son; *(generico)* child. **~o di papà** spoilt brat. **~o unico** only child

figli'occi|a *nf* goddaughter. **~o** *nm* godson

figli'o|la *nf* girl. **~'lanza** *nf* offspring. **~lo** *nm* boy

'figo, -a *vedi* **fico, -a**

fi'gura *nf* figure; *(aspetto esteriore)* shape; *(illustrazione)* illustration; **far bella/brutta ~** make a good/ bad impression; **mi hai fatto fare una brutta ~** you made me look a fool; **che ~!** how embarrassing!

figu'raccia *nf* bad impression

figu'ra|re *vt* represent; *(simboleggiare)* symbolize; *(immaginare)* imagine ▸ *vi (far figura)* cut a fine figure; *(in lista)* appear, figure. **~rsi** *vr (immaginarsi)* imagine; **~ti!** imagine that!; **posso? – [ma] ~ti!** may I? – of course!. **~'tivo** *a* figurative

figu'rina *nf (da raccolta)* ≈ cigarette card

figu|ri'nista *nmf* dress designer. **~'rino** *nm* fashion sketch. **~'rone** *nm* **fare un ~rone** make an excellent impression

'fila *nf* line; *(di soldati ecc)* file; *(di* oggetti) row; *(coda)* queue; **di ~** in succession; **fare la ~** queue [up], stand in line *Am*; **in ~ indiana** single file

fila'mento *nm* filament

filan'tro'pia *nf* philanthropy

fi'lare *vt* spin; *Naut* pay out ▸ *vi (andarsene)* run away; *‹liquido:›* trickle; **fila!** *fam* scram!; **~ con** *(fam: amoreggiare)* go out with; **~ dritto** toe the line

filar'monica *nf (orchestra)* orchestra

fila'strocca *nf* rigmarole; *(per bambini)* nursery rhyme

filate'lia *nf* philately

fi'la|to *a* spun; *(ininterrotto)* running; *(continuato)* uninterrupted; **di ~to** *(subito)* immediately ▸ *nm* yarn. **~'tura** *nf* spinning; *(filanda)* spinning mill

fil di 'ferro *nm* wire

fi'letto *nm (bordo)* border; *(di vite)* thread; *Culin* fillet

fili'ale *a* filial ▸ *nf Comm* branch

fili'grana *nf* filigree; *(su carta)* watermark

film *nm inv* film. **~ giallo** thriller. **~ a lungo metraggio** feature film

fil'ma|re *vt* film. **~to** *nm* short film. **fil'mino** *nm* cine film

'filo *nm* thread; *(tessile)* yarn; *(metallico)* wire; *(di lama)* edge; *(venatura)* grain; *(di perle)* string; *(d'erba)* blade; *(di luce)* ray; **con un ~ di voce** in a whisper; **per ~ e per segno** in detail; **fare il ~ a** *qcno* fancy sb; **perdere il ~** lose the thread. **~ spinato** barbed wire

'filobus *nm inv* trolleybus

filodiffusi'one *nf* rediffusion

fi'lone *nm* vein; *(di pane)* long loaf

filoso'fia *nf* philosophy. **fi'losofo, -a** *nmf* philosopher

fil'trare *vt* filter. **'filtro** *nm* filter

'filza *nf* string

fin *vedi* **fine, fino**[1]

fi'nal|e *a* final ▸ *nm* end ▸ *nf Sport* final. **fina'lista** *nmf* finalist. **~ità** *nf inv* finality; *(scopo)* aim.

~'**mente** adv at last; (in ultimo) finally

fi'nanz|a nf finance; ~'i**ario** a financial. ~'**ere** nm financier; (guardia di finanza) customs officer. ~ia'**mento** nm funding

finanzi'a|re vt fund, finance. ~'tore, ~'trice nmf backer

finché conj until; (per tutto il tempo che) as long as

'fine a fine; (sottile) thin; (udito, vista) keen; (raffinato) refined ● nf end; alla ~ in the end; alla fin ~ after all; in fin dei conti when all's said and done; te lo dico a fin di bene I'm telling you for your own good; senza ~ endless ● nm aim. ~ settimana weekend

fi'nestr|a nf window. fine'strella nf di aiuto Comput help window, help box. fine'strino nm Rail, Auto window

fi'nezza nf fineness; (sottigliezza) thinness; (raffinatezza) refinement

'finger|e vt pretend; feign (affetto ecc). ~si vr pretend to be

fini'menti nmpl finishing touches; (per cavallo) harness sg

fini'mondo nm end of the world; fig pandemonium

fi'ni|re vt/i finish, end; (smettere) stop; (diventare, andare a finire) end up; ~**scila!** stop it!. ~to a finished; (abile) accomplished. ~'tura nf finish

finlan'dese a Finnish ● nmf Finn ● nm (lingua) Finnish

Fin'landia nf Finland

'fino[1] prep ~ a till, until; (spazio) as far as; ~ all'ultimo to the last; fin da (tempo) since; (spazio) from; fin qui as far as here; fin troppo too much; ~ a che punto how far

'fino[2] a fine; (acuto) subtle; (puro) pure

fi'nocchio nm fennel; (fam: omosessuale) poof

fi'nora adv so far, up till now

'finta nf pretence, sham; Sport feint; far ~ di pretend to; far ~ di

niente act as if nothing had happened; per ~ (per scherzo) for a laugh

'fint|o, -a pp di fingere ● a false; (artificiale) artificial; fare il ~o tonto act dumb

finzi'one nf pretence

fi'occo nm bow; (di neve) flake; (nappa) tassel; coi fiocchi fig excellent. ~ di neve snowflake

fi'ocina nf harpoon

fi'oco a weak; (luce) dim

fio'raio, -a nmf florist

fior'daliso nm cornflower

fi'ordo nm fiord

fi'ore nm flower; (parte scelta) cream; fiori pl (nelle carte) clubs; a fior d'acqua on the surface of the water; fior di (abbondanza) a lot of; ha i nervi a fior di pelle his nerves are on edge; a fiori flowery

fioren'tino a Florentine

fio'retto nm (scherma) foil; Relig act of mortification

fio'rire vi flower; (albero:) blossom; fig flourish

fio'rista nmf florist

fiori'tura nf (di albero) blossoming

fi'otto nm scorrere a fiotti pour out; piove a fiotti the rain is pouring down

Fi'renze nf Florence

'firma nf signature; (nome) name

fir'ma|re vt sign. ~'tario, -a nmf signatory. ~to a (abito, borsa) designer attrib

fisar'monica nf accordion

fi'scale a fiscal

fischi'are vi whistle ● vt whistle; (in segno di disapprovazione) boo

fischiet't|are vt whistle. ~io nm whistling

fischi'etto nm whistle. 'fischio nm whistle

'fisco nm treasury; (tasse) taxation; il ~ the taxman

'fisica nf physics

'fisica'mente adv physically

'fisico, -a a physical ● nmf physicist ● nm physique

fisima *nf* whim

fisio|lo'gia *nf* physiology. **~'logico** *a* physiological

fisiono'mia *nf* features, face; *(di paesaggio)* appearance

fisiotera'pi|a *nf* physiotherapy. **~sta** *nmf* physiotherapist

fis'sa|re *vt* fix, fasten; *(guardare fissamente)* stare at; arrange *(appuntamento, ora)*. **~rsi** *vr (stabilirsi)* settle; *(fissare lo sguardo)* stare; **~rsi su** *(ostinarsi)* set one's mind on; **~rsi di fare qcsa** become obsessed with doing sth. **~to** *nm (persona)* person with an obsession. **~zi'one** *nf* fixation; *(osessione)* obsession

'fisso *a* fixed; **un lavoro ~** a regular job; **senza fissa dimora** of no fixed abode

'fitta *nf* sharp pain

fit'tizio *a* fictitious

'fitto¹ *a* thick; **~ di** full of ● *nm* depth

fitto² *nm (affitto)* rent; **dare a ~** let; **prendere a ~** rent; *(noleggiare)* hire

fiu'mana *nf* swollen river; *fig* stream

fi'ume *nm* river; *fig* stream

fiu'tare *vt* smell. **fi'uto** *nm* [sense of] smell; *fig* nose

'flaccido *a* flabby

fla'cone *nm* bottle

fla'gello *nm* scourge

fla'grante *a* flagrant; **in ~** in the act

fla'nella *nf* flannel

'flash *nm inv* Journ newsflash

'flauto *nm* flute

'flebile *a* feeble

'flemma *nf* calm; *Med* phlegm. **flem'matico** *a* phlegmatic

fles'sibil|e *a* flexible. **~ità** *nf* flexibility

flessi'one *nf (del busto in avanti)* forward bend

'flesso *pp di* flettere

flessu'oso *a* supple

'flettere *vt* bend

flir'tare *vi* flirt

F.lli *abbr* **(fratelli)** Bros

'floppy disk *nm inv* floppy disk

'flora *nf* flora

'florido *a* flourishing

'floscio *a* limp; *(flaccido)* flabby

'flotta *nf* fleet. **flot'tiglia** *nf* flotilla

flu'ente *a* fluent

flu'ido *nm* fluid

flu'ire *vi* flow

fluore'scente *a* fluorescent

flu'oro *nm* fluorine

'flusso *nm* flow; *Med* flux; *(del mare)* flood[-tide]; **~ e riflusso** ebb and flow

fluttu'ante *a* fluctuating

fluttu'a|re *vi (prezzi, moneta:)* fluctuate. **~zi'one** *nf* fluctuation

fluvi'ale *a* river

fo'bia *nf* phobia

'foca *nf* seal

fo'caccia *nf (pane)* flat bread; *(dolce)* ≈ raisin bread

fo'cale *a (distanza, punto)* focal. **focaliz'zare** *vt* get into focus *(fotografia)*; focus *(attenzione)*; define *(problema)*

'foce *nf* mouth

foco'laio *nm* *Med* focus; *fig* centre

foco'lare *nm* hearth; *(caminetto)* fireplace; *Techn* furnace

fo'coso *a* fiery

fode'ra *nf* lining; *(di libro)* dustjacket; *(di poltrona ecc)* loose cover. **fode'rare** *vt* line; cover *(libro)*. **~o** *nm* sheath

'foga *nf* impetuosity

'foggi|a *nf* fashion; *(maniera)* manner; *(forma)* shape. **~'are** *vt* mould

'fogli|a *nf* leaf; *(di metallo)* foil. **~'ame** *nm* foliage

fogli'etto *nm (pezzetto di carta)* piece of paper

'foglio *nm* sheet; *(pagina)* leaf. **~ elettronico** *Comput* spreadsheet. **~ rosa** provisional driving licence

'fogna *nf* sewer. **~'tura** *nf* sewerage

fo'lata *nf* gust

fol'clo|re nm folklore. **~'ristico** a folk; (bizzarro) weird

folgo'ra|re vi (splendere) shine ● vt (con un fulmine) strike. **~zi'one** nf (da fulmine, elettrica) electrocution; (idea) brainwave

'folgore nf thunderbolt

'folia nf crowd

'folle a mad; in ~ Auto in neutral; andare in ~ Auto coast

folle'mente adv madly

fol'lia nf madness; alla ~ (amare) to distraction

'folto a thick

fomen'tare vt stir up

fond'ale nm Theat backcloth

fonda'men|ta nfpl foundations. **~'tale** a fundamental. **~to** nm (di principio, teoria) foundation

fon'da|re vt establish; base (ragionamento, accusa). **~to** a (ragionamento) well-founded. **~zi'one** nf establishment; **~zioni** pl (di edificio) foundations

fon'delli nmpl prendere qcno per i ~ fam pull sb's leg

fon'dente a (cioccolato) dark

'fonder|e vt/i melt; (colori:) blend. **~si** vr melt; Comm merge. **fonde'ria** nf foundry

'fondi nmpl (denaro) funds; (di caffè) grounds

'fondo a deep; è notte fonda it's the middle of the night ● nm bottom; (fine) end; (sfondo) background; (indole) nature; (somma di denaro) fund; (feccia) dregs pl; andare a ~ (nave:) sink; da cima a ~ from beginning to end; in ~ after all; in ~ in ~ deep down; fino in ~ right to the end; (capire) thoroughly. ~ d'investimento investment trust

fondo'tinta nm foundation cream

fon'duta nf fondue made with cheese, milk and eggs

fo'netic|a nf phonetics. **~o** a phonetic

fon'tana nf fountain

'fonte nf spring; fig source ● nm font

fo'raggio nm forage

fo'rar|e vt pierce; punch (biglietto) ● vi puncture. **~si** vr (gomma, pallone:) go soft

'forbici nfpl scissors

forbi'cine nfpl (per le unghie) nail scissors

for'bito a erudite

'forca nf fork; (patibolo) gallows pl

for'cella nf fork; (per capelli) hairpin

for'chet|ta nf fork. **~'tata** nf (quantità) forkful

for'cina nf hairpin

'forcipe nm forceps pl

for'cone nm pitchfork

fo'resta nf forest. **fore'stale** a forest attrib

foresti'ero, -a a foreign ● nmf foreigner

for'fait nm inv fixed price; dare ~ (abbandonare) give up

'forfora nf dandruff

'forgi|a nf forge. **~'are** vt forge

'forma nf form; (sagoma) shape; Culin mould; (da calzolaio) last; essere in ~ be in good form; a ~ di in the shape of; forme pl (del corpo) figure sg; (convenzioni) appearances

formag'gino nm processed cheese. **for'maggio** nm cheese

for'mal|e a formal. **~ità** nf inv formality. **~iz'zarsi** vr stand on ceremony. **~'mente** adv formally

for'ma|re vt form. **~rsi** vr form; (svilupparsi) develop. **~to** nm size; (di libro) format; **~to tessera** (fotografia) passport-size

format'tare vt format

formazi'one nf formation; Sport line-up. **~ professionale** vocational training

for'mi|ca nf ant. **~'caio** nm anthill

'formica® nf (laminato plastico) Formica®

formico'l|are vi (braccio ecc:) tingle; **~are di** be swarming with; mi **~a la mano** I have pins and needles in my hand. **~io** nm swarming; (di braccio ecc) pins and needles pl

formi'dabile a (tremendo) formidable; (eccezionale) tremendous

for'mina nf mould

for'moso, -a a shapely

'formula nf formula. **formu'lare** vt formulate; (esprimere) express

for'nace nf furnace; (per laterizi) kiln

for'naio nm baker; (negozio) bakery

for'nello nm stove; (di pipa) bowl

for'ni∥re vt supply (**di** with). **~'tore** nm supplier. **~'tura** nf supply

'forno nm oven; (panetteria) bakery; **al ~** roast. **~ a microonde** microwave [oven]

'foro nm hole; (romano) forum; (tribunale) [law] court

'forse adv perhaps, maybe; **essere in ~** be in doubt

forsen'nato, -a a mad ● nmf madman; madwoman

'forte a strong; (colore) bright; (suono) loud; (resistente) tough; (spesa) considerable; (dolore) severe; (pioggia) heavy; (a tennis, calcio) good; (fam: simpatico) great; (taglia) large ● adv strongly; (parlare) loudly; (velocemente) fast; (piovere) heavily ● nm (fortezza) fort; (specialità) strong point

for'tezza nf fortress; (forza morale) fortitude

fortifi'care vt fortify

for'tino nm Mil blockhouse

for'tuito a fortuitous; **incontro ~** chance encounter

for'tuna nf fortune; (successo) success; (buona sorte) luck. **atterraggio di ~** forced landing; **per ~** be lucky; **buona ~!** good luck!; **di ~** makeshift; **per ~** luckily. **fortu'nato** a lucky, fortunate; (impresa) successful. **~ta'mente** adv fortunately

fo'runcolo nm pimple; (grosso) boil

'forza nf strength; (potenza)

power; (fisica) force; **di ~** by force; **a ~ di** by dint of; **con ~** hard; **~! come on!; ~ di volontà** will-power; **~ maggiore** circumstances beyond one's control; **la ~ pubblica** the police; **per ~** against one's will; (naturalmente) of course; **farsi ~** bear up; **mare ~ 8** force 8 gale; **bella ~! fam** big deal!. **le forze armate** the armed forces. **~ di gravità** [force of] gravity

for'za∥re vt force; (scassare) break open; (sforzare) strain. **~to a** forced; (sorriso) strained ● nm convict

forzi'ere nm coffer

for'zuto a strong

fo'schia nf haze

'fosco a dark

fo'sfato nm phosphate

'fosforo nm phosphorus

'fossa nf pit; (tomba) grave. **~ biologica** cesspool. **fos'sato** nm (di fortificazione) moat

fos'setta nf dimple

'fossile nm fossil

'fosso nm ditch; Mil trench

'foto nf inv fam photo; **fare delle ~** take some photos

foto'cellula nf photocell

fotocomposizi'one nf filmsetting, photocomposition

foto'copi∥a nf photocopy. **~'are** vt photocopy. **~a'trice** nf photocopier

foto'finish nm inv photo finish

foto'genico a photogenic

fotogra'fare vt photograph. **~'fia** nf (arte) photography; (immagine) photograph; **fare ~fie** take photographs. **foto'grafico** a photographic; **macchina fotografica** camera. **fo'tografo, -a** nmf photographer

foto'gramma nm frame

fotomo'dello, -a nmf [photographer's] model

fotomon'taggio nm photomontage

fotoro'manzo nm photo story

'fotter|e vt (fam: rubare) nick; vulg fuck, screw. ~sene vr vulg not give a fuck

fot'tuto a (fam: maledetto) bloody

fou'lard nm inv scarf

fra prep (in mezzo a due) between; (in un insieme) among; (tempo, distanza) in; detto ~ noi between you and me; ~ sé e sé to oneself; ~ l'altro what's more; ~ breve soon; ~ quindici giorni in two weeks' time; ~ tutti, siamo in venti there are twenty of us altogether

fracas'sar|e vt smash. ~si vr shatter

fra'casso nm din; (di cose che cadono) crash

'fradicio a (bagnato) soaked; (guasto) rotten; ubriaco ~ blind drunk

'fragil|e a fragile; fig frail. ~ità nf fragility; fig frailty

'fragola nf strawberry

fra'go|re nm uproar; (di cose rotte) clatter; (di tuono) rumble. ~roso a uproarious; (tuono) rumbling; (suono) clanging

fra'gran|te a fragrant. ~za nf fragrance

frain'ten|dere vt misunderstand. ~ndersi vr be at cross-purposes. ~so pp di fraintendere

frammen'tario a fragmentary. fram'mento nm fragment

'frana nf landslide; (fam: persona) walking disaster area. fra'nare vi slide down

franca'mente adv frankly

fran'cese a French ● nmf Frenchman; Frenchwoman ● nm (lingua) French

fran'chezza nf frankness

'Francia nf France

'franco¹ a frank; Comm free; farla franca get away with sth

'franco² nm (moneta) franc

franco'bollo nm stamp

fran'gente nm (onda) breaker; (scoglio) reef; (fig: momento diffici-

le) crisis; in quel ~ given the situation

'frangia nf fringe

fra'noso a subject to landslides

fran'toio nm olive-press

frantu'mar|e vt, ~si vr shatter. fran'tumi nmpl splinters; andare in frantumi be smashed to smithereens

frappé nm inv milkshake

frap'por|re vt interpose. ~si vr intervene

fra'sario nm vocabulary; (libro) phrase book

'frase nf sentence; (espressione) phrase. ~ fatta cliché

'frassino nm ash[-tree]

frastagli'a|re vt make jagged. ~to a jagged

frastor'na|re vt daze. ~to a dazed

frastu'ono nm racket

'frate nm friar; (monaco) monk

fratel'la|nza nf brotherhood. ~stro nm half-brother

fra'tell|i nmpl (fratello e sorella) brother and sister. ~o nm brother

fraterniz'zare vi fraternize.

frat'taglie nfpl (di pollo ecc) giblets

frat'tanto adv in the meantime

frat'tempo nm nel ~ meanwhile, in the meantime

frat'tu|ra nf fracture. ~'rare vt, ~'rarsi vr break

fraudo'lento a fraudulent

frazi'one nf fraction; (borgata) hamlet

'freccia nf arrow; Auto indicator. ~'ata nf (osservazione pungente) cutting remark

fredda'mente adv coldly

fred'dare vt cool; (fig: con sguardo, battuta) cut down; (uccidere) kill

fred'dezza nf coldness

'freddo a & nm cold; aver ~ be cold; fa ~ it's cold

freddo'loso a sensitive to cold, chilly

fred'dura nf pun

fre'ga|re vt rub; (fam: truffare)

cheat; (fam: rubare) swipe.
~**rsene** fam not give a damn; **chi
se ne frega!** what the heck!. ~**si**
vr rub (occhi). ~**ta** nf rub. ~**'tura**
nf fam (truffa) swindle; (delusione)
letdown

'**fregio** nm Archit frieze; (ornamento) decoration

fre'**mente** a quivering

'**frem|ere** vi quiver. ~**ito** nm
quiver

fre'**na|re** vt brake; fig restrain;
hold back (lacrime, impazienza)
● vi brake. ~**rsi** vr check oneself.
~**ta** nf **fare una ~ta brusca** hit
the brakes

fre'**nesia** nf frenzy; (desiderio
smodato) craze. fre'**netico** a frenzied

'**freno** nm brake; fig check;
togliere il ~ release the brake;
usare il ~ apply the brake; **tenere
a ~** restrain. ~ **a mano** handbrake

frequen'**tare** vt frequent (scuola
ecc); mix with (persone)

fre'**quen|te** a frequent; **di ~te** frequently. ~**za** nf frequency;
(assiduità) attendance

fre'**schezza** nf freshness; (di temperatura) coolness

'**fresco** a fresh; (temperatura)
cool; **stai ~!** you're for it! ● nm
coolness; **far ~** be cool; **mettere/
tenere in ~** put/keep in a cool
place

'**fretta** nf hurry, haste; **aver ~** be in
a hurry; **far ~ a qcno** hurry sb; **in
~ e furia** in a great hurry.
fretto·losa'**mente** adv hurriedly.
fretto'**loso** a (persona) in a hurry;
(lavoro) rushed, hurried

fri'**abile** a crumbly

'**friggere** vt fry; **vai a farti ~!** I get
lost! ● vi sizzle

friggi'**trice** nf chip pan

frigi'**dità** nf frigidity. '**frigido** a
frigid

fri'**gnare** vi whine

'**frigo** nm fridge

frigo'**bar** nm inv minibar

frigo'**rifero** a refrigerating ● nm
refrigerator

fringu'**ello** nm chaffinch

frit'**tata** nf omelette

frit'**tella** nf fritter; (fam: macchia
d'unto) grease stain

'**fritto** pp di friggere ● a fried;
essere ~ be done for ● nm fried
food. ~ **misto** mixed fried fish/
vegetables. frit'**tura** nf (pietanza)
fried dish

frivo'**lezza** nf frivolity. '**frivolo** a
frivolous

frizio'**nare** vt rub. frizi'**one** nf friction; Mech clutch; (di pelle) rub

friz'**zante** a fizzy; (vino) sparkling;
(aria) bracing

'**frizzo** nm gibe

fro'**dare** vt defraud

'**frode** nf fraud. ~ **fiscale** tax evasion

'**frollo** a tender; (selvaggina) high;
(persona) spineless; **pasta frolla**
short[crust] pastry

'**fronda** nf [leafy] branch; fig rebellion. fron'**doso** a leafy

fron'**tale** a frontal; (scontro) head-on

'**fronte** nf forehead; (di edificio)
front; **di ~** opposite; **di ~ a** opposite, facing; (a paragone) compared with; **far ~ a** face ● nm Mil,
Pol front. ~**ggi'are** vt face

fronte'**spizio** nm title page

fronti'**era** nf frontier, border

fron'**tone** nm pediment

fronzolo nm frill

'**frotta** nf swarm; (di animali) flock

frot'**tola** nf fib; **frottole** pl nonsense sg

fru'**gale** a frugal

fru'**gare** vi rummage ● vt search

frul'**la|re** vt Culin whisk ● vi (ali:)
whirr. ~**to** nm ~**to di frutta** fruit
drink with milk and crushed ice.
~**tore** nm [electric] mixer.
frul'**lino** nm whisk

fru'**mento** nm wheat

frusci'**are** vi rustle

fru'scio nm rustle; (radio, giradischi) background noise; (di acque) murmur

'frusta nf whip; (frullino) whisk

fru'sta|re vt whip. ~ta nf lash.

fru'stino nm riding crop

fru'stra|re vt frustrate. ~to a frustrated. ~zi'one nf frustration

'frutta nf fruit; (portata) dessert. frut'tare vi bear fruit ● vt yield. frut'teto nm orchard. ~i'vendolo, -a nmf greengrocer. ~o nm anche fig fruit; Fin yield; ~i di bosco fruits of the forest. ~i di mare seafood sg. ~u'oso a profitable

f.to abbr (firmato) signed

fu a (defunto) late; il ~ signor Rossi the late Mr Rossi

fuci'la|re vt shoot. ~ta nf shot

fu'cile nm rifle

fu'cina nf forge

'fucsia nf fuchsia

'fuga nf escape; (perdita) leak; Mus fugue; darsi alla ~ take to flight

fu'gace a fleeting

fug'gevole a short-lived

fuggi'asco, -a nmf fugitive

fuggi'fuggi nm stampede

fug'gi|re vi flee; (innamorati:) elope; fig fly. ~'tivo, -a nmf fugitive

'fulcro nm fulcrum

ful'gore nm splendour

fu'liggine nf soot

fulmi'nar|e vt strike by lightning; (con sguardo) look daggers at; (con scarica elettrica) electrocute. ~si vr burn out. 'fulmine nm lightning. ful'mineo a rapid

'fulvo a tawny

fumai'olo nm funnel; (di casa) chimney

fu'ma|re vt/i smoke; (in ebollizione) steam. ~'tore, ~'trice nmf smoker; non fumatori non-smoker, non-smoking

fu'metto nm comic strip; fumetti pl comics

'fumo nm smoke; (vapore) steam;

fig hot air; andare in ~ vanish.

fu'moso a (ambiente) smoky; (discorso) vague

fu'nambolo, -a nmf tightrope walker

'fune nf rope; (cavo) cable

'funebre a funeral; (cupa) gloomy

fune'rale nm funeral

fu'nereo a (aria) funereal

fu'nesto a sad

'fungere vi ~ da act as

'fungo nm mushroom; Bot, Med fungus

funico'lare nf funicular [railway]

funi'via nf cableway

funzio'nal|e a functional. ~ità nf functionality

funziona'mento nm functioning

funzio'nare vi work, function; ~ da (fungere da) act as

funzio'nario nm official

funzi'one nf function; (carica) office; Relig service; entrare in ~ take up office

fu'oco nm fire; (fisica, fotografia) focus; far ~ fire; dar ~ a set fire to; prendere ~ catch fire. fuochi pl d'artificio fireworks. ~ di paglia nine-days' wonder

fuorché prep except

fu'ori adv out; (all'esterno) outside; (all'aperto) outdoors; andare di ~ (traboccare) spill over; essere di sé be beside oneself; essere in ~ (sporgere) stick out; far ~ fam do in; ~ luogo (inopportuno) out of place; ~ mano out of the way; ~ moda old-fashioned; ~ pasto between meals; ~ pericolo out of danger; ~ questione out of the question; ~ uso out of use ● nm outside

fuori'bordo nm speedboat (with outboard motor)

fuori'classe nmf inv champion

fuori'gioco nm & adv offside

fuori'legge nmf outlaw

fuori'serie a custom-made ● nf Auto custom-built model

fuori'strada nm off-road vehicle

fuorvi'are *vt* lead astray ● *vi* go astray

furbacchi'one *nm* crafty old devil

furbe'ria *nf* cunning. **fur'bizia** *nf* cunning

'furbo *a* cunning; (*intelligente*) clever; (*astuto*) shrewd; **bravo ~!** nice one!; **fare il ~** try to be clever

fu'rente *a* furious

fur'fante *nm* scoundrel

furgon'cino *nm* delivery van. **fur'gone** *nm* van

'furia *nf* fury; (*fretta*) haste; **a ~ di** by dint of. **~'bondo, ~'oso** *a* furious

fu'rore *nm* fury; (*veemenza*) frenzy; **far ~** be all the rage. **~ggi'are** *vi* be a great success

furtiva'mente *adv* covertly. **fur'tivo** *a* furtive

'furto *nm* theft; (*con scasso*) burglary; **commettere un ~** steal

'fusa *nfpl* **fare le ~** purr

fu'scello *nm* (*di legno*) twig; (*di paglia*) straw; **sei un ~** you're as light as a feather

fu'seaux *mpl* leggings

fu'sibile *nm* fuse

fusi'one *nf* fusion; *Comm* merger

'fuso *pp di* **fondere** ● *a* melted ● *nm* spindle; (*di fuso*) ~ **orario** time zone. **~ orario** time zone

fusoli'era *nf* fuselage

fu'stagno *nm* corduroy

fu'stino *nm* (*di detersivo*) box

'fusto *nm* stem; (*tronco*) trunk; (*recipiente di metallo*) drum; (*di legno*) barrel

'futile *a* futile

fu'turo *a & nm* future

Gg

gab'bar|e *vt* cheat. **~si** *vr* **~si di** make fun of

'gabbia *nf* cage; (*da imballaggio*) crate. **~ degli imputati** dock. **~ toracica** rib cage

gabbi'ano *nm* [sea]gull

gabi'netto *nm* (*di medico*) consulting room; *Pol* cabinet; (*toletta*) lavatory; (*laboratorio*) laboratory

'gaffe *nf* blunder

gagli'ardo *a* vigorous

gai'ezza *nf* gaiety. **'gaio** *a* cheerful

'gala *nf* gala

ga'lante *a* gallant. **~'ria** *nf* gallantry. **galantu'omo** *nm* (*pl* **galantuomini**) gentleman

ga'lassia *nf* galaxy

gala'teo *nm* [good] manners *pl*; (*trattato*) book of etiquette

gale'otto *nm* (*rematore*) galley-slave; (*condannato*) convict

ga'lera *nf* (*nave*) galley; *fam* prison

'galla *nf Bot* gall; **a ~** *adv* afloat; **venire a ~** surface

galleggi'ante *a* floating ● *nm* craft; (*boa*) float

galleggi'are *vi* float

galle'ria *nf* (*traforo*) tunnel; (*d'arte*) gallery; *Theat* circle; (*arcata*) arcade. **~ d'arte** art gallery

'Galles *nm* Wales. **gal'lese** *a* welsh ● *nm* Welshman; (*lingua*) Welsh ● *nf* Welshwoman

gal'letto *nm* cockerel; **fare il ~** show off

gal'lina *nf* hen

gal'lismo *nm* machismo

'gallo *nm* cock

gal'lone *nm* stripe; (*misura*) gallon

galop'pare *vi* gallop. **ga'loppo** *nm* gallop; **al galoppo** at a gallop

galvaniz'zare *vt* galvanize

'gamba *nf* leg; (*di lettera*) stem; **a quattro gambe** on all fours;

darsela a gamba take to one's heels; **essere in ~** (*essere forte*) be strong; (*capace*) be smart

gamba'letto *nm* pop sock

gambe'retto *nm* shrimp. '**gambero** *nm* prawn; (*di fiume*) crayfish

'**gambo** *nm* stem; (*di pianta*) stalk

'**gamma** *nf Mus* scale; *fig* range

ga'**nascia** *nf* jaw; **ganasce** *pl* **del freno** brake shoes

'**gancio** *nm* hook

'**ganghero** *nm* **uscire dai gangheri** *fig* get into a temper

'**gara** *nf* competition; (*di velocità*) race; **fare a ~** compete. **~ d'appalto** call for tenders

ga'**rage** *nm inv* garage

ga'**ran|te** *nmf* guarantor. **~'tire** *vt* guarantee; (*rendersi garante*) vouch for; (*assicurare*) assure. **~'zia** *nf* guarantee; **in ~zia** under guarantee

gar'**ba|re** *vi* like; **non mi garba** I don't like it. **~to** *a* courteous

'**garbo** *nm* courtesy; (*grazia*) grace; **con ~** graciously

gareggi'**are** *vi* compete

garga'**nella** *nf* **a ~** from the bottle

garga'**rismo** *nm* gargle; **fare i gargarismi** gargle

ga'**rofano** *nm* carnation

gar'**rire** *vi* chirp

'**garza** *nf* gauze

gar'**zone** *nm* boy. **~ di stalla** stable-boy

gas *nm inv* gas; **dare ~** *Auto* accelerate; **a tutto ~** flat out. **~ lacrimogeno** tear gas. **~ di scarico** *pl* exhaust fumes

gas'**dotto** *nm* natural gas pipeline

ga'**solio** *nm* diesel oil

ga'**someter** *nm* gasometer

gas'**sa|re** *vt* aerate; (*uccidere col gas*) gas. **~ato** *a* gassy. **~oso, -a** *a* gassy; (*bevanda*) fizzy ● *nf* lemonade

'**gastrico** *a* gastric. **ga'strite** *nf* gastritis

gastro|no'**mia** *nf* gastronomy. **~'nomico** *a* gastronomic. **ga'stronomo, -a** *nmf* gourmet

'**gatta** *nf* **una ~ da pelare** a headache

gatta'buia *nf hum* clink

gat'**tino, -a** *nmf* kitten

'**gatto, -a** *nmf* cat. **~ delle nevi** snowmobile

gat'**toni** *adv* on all fours

ga'**vetta** *nf* mess tin; **fare la ~** rise through the ranks

gay *a inv* gay

'**gazza** *nf* magpie

gaz'**zarra** *nf* racket

gaz'**zella** *nf* gazelle; *Auto* police car

gaz'**zetta** *nf* gazette

gaz'**zosa** *nf* clear lemonade

'**geco** *nm* gecko

ge'**la|re** *vt/i* freeze. **~ta** *nf* frost

gela'**t|aio, -a** *nmf* ice-cream seller; (*negozio*) ice-cream shop. **~e'ria** *nf* ice-cream parlour. **~i'era** *nf* ice-cream maker

gela'**tina** *nf* gelatine; (*dolce*) jelly. **~na di frutta** fruit jelly. **~'noso** *a* gelatinous

ge'**lato** *a* frozen ● *nm* ice-cream

'**gelido** *a* freezing

'**gelo** *nm* (*freddo intenso*) freezing cold; (*brina*) frost; *fig* chill

ge'**lone** *nm* chilblain

gelosa'**mente** *adv* jealously

gelo'**sia** *nf* jealousy. **ge'loso, -a** *a* jealous

'**gelso** *nm* mulberry[-tree]

gel'**somino** *nm* jasmine

geme'**llaggio** *nm* twinning

ge'**mello, -a** *a* & *nmf* twin; (*di polsino*) cuff-link; **Gemelli** *pl Astr* Gemini *sg*

'**gem|ere** *vi* groan; (*tubare*) coo. **~ito** *nm* groan

'**gemma** *nf* gem; *Bot* bud

'**gene** *nm* gene

genea|lo'**gia** *nf* genealogy. **gene'ral¹** *a* general; **spese ~i** overheads

gene'**rale²** *nm Mil* general

generali'**tà** *nf* (*qualità*) generality, general nature; **~ pl** (*dati personali*) particulars

generaliz'**za|re** *vt* generalize.

~zi'one *nf* generalization. **general'mente** *adv* generally

gene'ra|re *vt* give birth to; *(causare)* breed; *Techn* generate. **~'tore** *nm Techn* generator. **~zi'one** *nf* generation

'genere *nm* kind; *Biol* genus; *Gram* gender; *(letterario, artistico)* genre; *(prodotto)* product; **il ~ umano** mankind; **in ~** generally. **generi** *pl* **alimentari** provisions

generica'mente *adv* generically. **ge'nerico** *a* generic; **medico ~ generico** general practitioner

'genero *nm* son-in-law

generosità *nf* generosity. **gene'roso** *a* generous

'genesi *nf* genesis

ge'netico, -a *a* genetic ● *nf* genetics

gen'giva *nf* gum

geni'ale *a* ingenious; *(congeniale)* congenial

'genio *nm* genius; **andare a ~** be to one's taste. **~ civile** civil engineering. **~ [militare]** Engineers

geni'tale *a* genital. **genitali** *nmpl* genitals

geni'tore *nm* parent

gen'naio *nm* January

'Genova *nf* Genoa

gen'taglia *nf* rabble

'gente *nf* people *pl*

gen'til|e *a* kind; **G~e Signore** *(in lettere)* Dear Sir. **genti'lezza** *(in lettere)* Dear Sir. **genti'lezza** *nf* kindness; **per gentilezza** *(per favore)* please. **~'mente** *adv* kindly. **~'uomo** *(pl* **~u'omini)** *nm* gentleman

genu'ino *a* genuine; *⟨cibo, prodotto⟩* natural

geogra'fia *nf* geography. **geo'grafico** *a* geographical. **ge'ografo, -a** *nmf* geographer

geolo'gia *nf* geology. **geo'logico** *a* geological. **ge'ologo, -a** *nmf* geologist

ge'ometra *nmf* surveyor

geome'tria *nf* geometry. **geo'metrico** *a* geometric[al]

ge'ranio *nm* geranium

gerar'chia *nf* hierarchy. **ge'rarchico** *a* hierarchic[al]

ge'rente *nm* manager ● *nf* manageress

'gergo *nm* slang; *(di professione ecc)* jargon

geria'tria *nf* geriatrics *sg*

Ger'mania *nf* Germany

'germe *nm* germ; *(fig: principio)* seed

germogli'are *vi* sprout. **ger'moglio** *nm* sprout

gero'glifico *nm* hieroglyph

'gesso *nm* chalk; *(Med, scultura)* plaster

gestazi'one *nf* gestation

gestico'lare *vi* gesticulate

gesti'one *nf* management

ge'stir|e *vi* manage. **~si** *vr* budget one's time and money

'gesto *nm* gesture; *(azione pl nf* **gesta)** deed

ge'store *nm* manager

Gesù *nm* Jesus. **~ bambino** baby Jesus

gesu'ita *nm* Jesuit

get'ta|re *vt* throw; *(scagliare)* fling; *(emettere)* spout; *Techn, fig* cast; **~re via** throw away. **~rsi** *vr* throw oneself; **~rsi in** *(fiume:)* flow into. **~ta** *nf* throw; *Techn* casting

'getto *nm* throw; *(di liquidi, gas)* jet; **a ~ continuo** in a continuous stream; **di ~** straight off

getto'nato *a* ⟨canzone⟩ popular. **get'tone** *nm* token; *(per giochi)* counter

ghe'pardo *nm* cheetah

ghettiz'zare *vt* ghettoize. **'ghetto** *nm* ghetto

ghiacci'aio *nm* glacier

ghiacci'a|re *vt/i* freeze. **~to** *a* frozen; *(freddissimo)* ice-cold

ghi'acci|o *nm* ice; *Auto* black ice. **~'olo** *nm* icicle; *(gelato)* ice lolly

ghi'aia *nf* gravel

ghi'anda *nf* acorn

ghi'andola *nf* gland

ghigliot'tina *nf* guillotine

ghi'gnare *vi* sneer. **'ghigno** *nm* sneer

ghi'ot|to *a* greedy, gluttonous; (*appetitoso*) appetizing. **~'tone, -a** *nmf* glutton. **~tone'ria** *nf* (*qualità*) gluttony; (*cibo*) tasty morsel

ghir'landa *nf* (*corona*) wreath; (*di fiori*) garland

'ghiro *nm* dormouse; **dormire come un ~** sleep like a log

'ghisa *nf* cast iron

già *adv* already; (*un tempo*) formerly; **~!** indeed!; **~ da ieri** since yesterday

gi'acca *nf* jacket. **~ a vento** windcheater

giacché *conj* since

giac'cone *nm* jacket

gia'cere *vi* lie

giaci'mento *nm* deposit. **~ di petrolio** oil deposit

gia'cinto *nm* hyacinth

gi'ada *nf* jade

giaggi'olo *nm* iris

giagu'aro *nm* jaguar

gial'lastro *a* yellowish

gi'allo *a & nm* yellow; **[libro] ~** thriller

Giap'pone *nm* Japan. **giappo'nese** *a & nmf* Japanese

giardi'n|aggio *nm* gardening. **~i'ere, -a** *nmf* gardener ● *nf* Auto estate car; (*sottaceti*) pickles *pl*

giar'dino *nm* garden. **~ d'infanzia** kindergarten. **~ pensile** roofgarden. **~ zoologico** zoo

giarretti'era *nf* garter

giavel'lotto *nm* javelin

gi'gan|te *a* gigantic ● *nm* giant. **~'tesco** *a* gigantic

gigantogra'fia *nf* blow-up

'giglio *nm* lily

gilè *nm inv* waistcoat

gin *nm inv* gin

gineco|lo'gia *nf* gynaecology. **~'logico** *a* gynaecological. **gine-'cologo, -a** *nmf* gynaecologist

gi'nepro *nm* juniper

gi'nestra *nf* broom

gingil'larsi *vr* fiddle; (*perder tem-*

po) potter. **gin'gillo** *nm* plaything; (*ninnolo*) knick-knack

gin'nasio *nm* (*scuola*) ≈ grammar school

gin'nast|a *nmf* gymnast. **~ica** *nf* gymnastics; (*esercizi*) exercises *pl*

ginocchi'ata *nf* **prendere una ~** bang one's knee

gi'nocchio *nm* (*pl m* **ginocchi** *o f* **ginocchia**) knee; **in ~o** on one's knees; **mettersi in ~o** kneel down; (*per supplicare*) go down on one's knees; **al ~o** (*gonna*) kneelength. **~oni** *adv* kneeling

gio'ca|re *vt/i* play; (*giocherellare*) toy; (*d'azzardo*) gamble; (*puntare*) stake; (*ingannare*) trick. **~rsi la carriera** throw one's career away. **~'tore, ~'trice** *nmf* player; (*d'azzardo*) gambler

gio'cattolo *nm* toy

giocherel'l|are *vi* toy; (*nervosamente*) fiddle. **~one** *a* skittish

gi'oco *nm* game; (*di bambini, Techn*) play; (*d'azzardo*) gambling; (*scherzo*) joke; (*insieme di pezzi ecc*) set; **essere in ~** be at stake; **fare il doppio ~ con qcno** double-cross sb

giocoli'ere *nm* juggler

gio'coso *a* playful

gi'ogo *nm* yoke

gi'oia *nf* joy; (*gioiello*) jewel; (*appellativo*) sweetie

gioiel|le'ria *nf* jeweller's [shop]. **~i'ere, -a** *nmf* jeweller; (*negozio*) jeweller's. **gioi'ello** *nm* jewel; **gioielli** *pl* jewellery

gioi'oso *a* joyous

gio'ire *vi* **~ per** rejoice at

Gior'dania *nf* Jordan

giorna'laio, -a *nmf* newsagent, newsdealer

gior'nale *nm* [news]paper; (*diario*) journal. **~ di bordo** logbook. **~ radio** news bulletin

giornali'ero *a* daily ● *nm* (*per sciare*) day pass

giorna'lino *nm* comic

giorna'lis|mo *nm* journalism. **~ta** *nmf* journalist

gior'nal'mente *adv* daily

gior'nata *nf* day; **in ~** today; **vivere alla ~** live from day to day

gi'orno *nm* day; **al ~** per day; **al ~ d'oggi** nowadays; **di ~** by day; **in pieno ~** in broad daylight; **un ~ sì, un ~** no every other day

gi'ostra *nf* merry-go-round

giova'mento *nm* **trarre ~ da** derive benefit from

gi'ova|ne *a* young; (*giovanile*) youthful ●*nm* youth, young man ●*nf* girl, young woman. **~'nile** *a* youthful. **~'notto** *nm* young man

gio'var|e *vi* **~e a** be useful to; (*far bene a*) be good for. **~si** *vr* **~si di** avail oneself of

giovedì *nm inv* Thursday. **~ grasso** *last Thursday before Lent*

gioventù *nf* youth; (*i giovani*) young people *pl*

giovi'ale *a* jovial

giovi'nezza *nf* youth

gira'dischi *nm inv* record-player

gi'raffa *nf* giraffe; *Cinema* boom

giran'dola *nf* (*fuoco d'artificio*) Catherine wheel; (*giocattolo*) windmill; (*banderuola*) weathercock

gi'ra|re *vt* turn; (*andare intorno, visitare*) go round; *Comm* endorse; *Cinema* shoot ●*vi* turn; (*aerei, uccelli:*) circle; (*andare in giro*) wander; **far ~re** le scatole a qcno *fam* drive sb round the twist; **~re al largo** steer clear. **~rsi** *vr* turn [round]; **mi gira la testa** I feel dizzy. **~ta** *nf* turn; *Comm* endorsement; (*in macchina ecc*) ride; **fare una ~ta** (*a piedi*) go for a walk; (*in macchina*) go for a ride

girar'rosto *nm* spit

gira'sole *nm* sunflower

gira'volta *nf* spin; *fig* U-turn

gi'rello *nm* (*per bambini*) babywalker; *Culin* topside

gi'revole *a* revolving

gi'rino *nm* tadpole

'giro *nm* turn; (*circolo*) circle; (*percorso*) round; (*viaggio*) tour; (*passeggiata*) short walk; (*in*

macchina) drive; (*in bicicletta*) ride; (*circolazione di denaro*) circulation; **nel ~ di un mese** within a month; **prendere in ~ qcno** pull sb's leg; **senza giri di parole** without beating about the bush; **a ~ di posta** by return mail. **~ d'affari** *Comm* turnover. **~ [della] manica** armhole. **giri** *pl* al minuto rpm. **~ turistico** sightseeing tour. **~ vita** waist measurement

giro'collo *nm* choker; **a ~** crewneck

gi'rone *nm* round

gironzo'lare *vi* wander about

giro'tondo *nm* ring-a-ring-o'-roses

girova'gare *vi* wander about.

gi'rovago *nm* wanderer

'gita *nf* trip; **andare in ~** go on a trip. **~ scolastica** school trip.

gi'tante *nmf* tripper

giù *adv* down; (*sotto*) below; (*dabbasso*) downstairs; **a testa in ~** (*a capofitto*) headlong; **essere ~** be down; (*di salute*) be run down; **~ di corda** down; **~ di lì, su per ~** more or less; **non andare ~ a qcno** stick in sb's craw

gi'ub|ba *nf* jacket; *Mil* tunic. **~'botto** *nm* bomber jacket, jerkin

giudi'care *vt* judge; (*ritenere*) consider

gi'udice *nm* judge. **~ conciliatore** justice of the peace. **~ di gara** umpire. **~ di linea** linesman

giu'dizi|o *nm* judg[e]ment; (*opinione*) opinion; (*senno*) wisdom; (*processo*) trial; (*sentenza*) sentence; **mettere ~o** become wise. **~'oso** *a* sensible

gi'ugno *nm* June

giu'menta *nf* mare

gi'unco *nm* reed

gi'ungere *vi* arrive; **~ a** (*riuscire*) succeed in ●*vt* (*unire*) join

gi'ungla *nf* jungle

gi'unta *nf* addition; *Mil* junta; **per ~** in addition. **~ comunale** district council

gi'unto *pp di* **giungere** ●*nm* *Mech* joint

giun'tura nf joint

gluo'care, giu'oco = **giocare, gioco**

giura'mento nm oath; **prestare ~** take the oath

giu'ra|re vt/i swear. **~to, -a** a sworn ● nmf juror

giu'ria nf jury

giu'ridico a legal

giurisdizi'one nf jurisdiction

giurispru'denza nf jurisprudence

giu'rista nmf jurist

giustifi'ca|re vt justify. **~zi'one** nf justification

giu'stizia nf justice. **~'are** vt execute. **~'ere** nm executioner

gi'usto a just, fair; (adatto) right; (esatto) exact ● nm (uomo retto) just man; (cosa giusta) right ● adv exactly; **~ ora** just now

glaci'ale a glacial

gla'diolo nm gladiolus

'glassa nf Culin icing

gli def art impl (before vowel and s + consonant, gn, ps, z) the; vedi **il** ● pron (a lui) [to] him; (a esso) [to] it; (a loro) [to] them

glice'rina nf glycerine

gli'cine nm wisteria

gli'e|lo, -a pron [to] him/her/them; (forma di cortesia) [to] you; **~ chiedo** I'll ask him/her/them/ you; **glie'l'ho prestato** I've lent it to him/her/them/you. **~ne** (di ciò) [of] it; **~ne ho dato un po'** I gave him/her/them/you some

glo'bal|e a global; fig overall. **~'mente** adv globally

'globo nm globe. **~ oculare** eyeball. **~ terrestre** globe

'globulo nm globule; Med corpuscle. **~ bianco** white cell, white corpuscle. **~ rosso** red cell, red corpuscle

'glori|a nf glory. **~'arsi** vr ~**arsi di** be proud of. **~'oso** a glorious

glos'sario nm glossary

glu'cosio nm glucose

'gluteo nm buttock

'gnomo nm gnome

'gnorri nm **fare lo ~** play dumb

'gobba nf hump. **~o, -a** a hunchbacked ● nmf hunchback

'gocci|a nf drop; (di sudore) bead; **è stata l'ultima ~a** it was the last straw. **~o'lare** vi drip. **~o'lio** nm dripping

go'der|e vi (sessualmente) come; **~e di** enjoy. **~sela** have a good time. **~si** vr **~si qcsa** enjoy sth

godi'mento nm enjoyment

goffa'mente adv awkwardly. **'goffo** a awkward

'gola nf throat; (ingordigia) gluttony; Geog gorge; (di camino) flue; **avere mal di ~** have a sore throat; **far ~ a qcno** tempt sb

golf nm inv jersey; Sport golf

'golfo nm gulf

golosi'tà nf inv greediness; (cibo) tasty morsel. **go'loso** a greedy

'golpe nm inv coup

gomi'tata nf nudge

'gomito nm elbow; **alzare il ~** raise one's elbow

go'mitolo nm ball

'gomma nf rubber; (colla, da masticare) gum; (pneumatico) tyre. **~ da masticare** chewing gum

gommapi'uma nf foam rubber

gom'mista nm tyre specialist

gom'mone nm [rubber] dinghy

gom'moso a chewy

'gondola nf gondola. **~l'ere** nm gondolier

gonfa'lone nm banner

gonfi'abile a inflatable

gonfi'ar|e vi swell ● vt blow up; pump up (pneumatico); (esagerare) exaggerate. **~si** vr swell; (acque:) rise. **'gonfio** a swollen; (pneumatico) inflated; **a gonfie vele** splendidly. **gonfi'ore** nm swelling

gongo'la|nte a overjoyed. **~re** vi be overjoyed

'gonna nf skirt. **~ pantalone** culottes pl

'gonzo nm simpleton

gorgheggi'are vi warble.

gor'gheggio nm warble

'gorgo nm whirlpool

gorgogli'are *vi* gurgle

go'rilla *nm inv* gorilla; (*guardia del corpo*) bodyguard, minder

'gotico *a* & *nm* Gothic

gover'nante *nf* housekeeper

gover'na|re *vt* govern; (*dominare*) rule; (*dirigere*) manage; (*curare*) look after. **~tivo** *a* government. **~tore** *nm* governor

go'verno *nm* government; (*dominio*) rule; **al ~** in power

gracchi'are *vi* caw; (*fig: persona:*) screech

graci'dare *vi* croak

'gracile *a* delicate

gra'dasso *nm* braggart

grada'tamente *adv* gradually

gradazi'one *nf* gradation. **~ alcoolica** alcohol[ic] content

gra'de'vol|e *a* agreeable. **~'mente** *adv* pleasantly, agreeably

gradi'men|to *nm* liking; **indice di ~** *Radio, TV* popularity rating; **non è di mio ~** it's not to my liking

gradi'nata *nf* flight of steps; (*di stadio*) stand; (*di teatro*) tiers *pl*

gra'dino *nm* step

gra'di|re *vt* like; (*desiderare*) wish. **~to** *a* pleasant; (*bene accetto*) welcome

'grado *nm* degree; (*rango*) rank; **di buon ~** willingly; **essere in ~ di fare qcsa** be in a position to do sth; (*essere capace a*) be able to do sth

gradu'ale *a* gradual

gradu'a|re *vt* graduate. **~to a** graded; (*provvisto di scala graduata*) graduated ● *nm* *Mil* noncommissioned officer. **~'toria** *nf* list. **~zi'one** *nf* graduation

'graffa *nf* clip; (*segno grafico*) brace

graf'fetta *nf* staple

graffi'a|re *vt* scratch. **~'tura** *nf* scratch

'graffio *nm* scratch

gra'fia *nf* [hand]writing; (*ortografia*) spelling

'grafic|a *nf* graphics; **~a pubblici-**

taria commercial art. **~a'mente** *adv* in graphics, graphically. **~o** *a* graphic ● *nm* graph; (*persona*) graphic designer

gra'migna *nf* weed

gram'mati|ca *nf* grammar. **~'cale** *a* grammatical

'grammo *nm* gram[me]

gran *a vedi* **grande**

'grana *nf* grain; (*formaggio*) parmesan; (*fam: seccatura*) trouble; (*fam: soldi*) readies *pl*

gra'naio *nm* barn

gra'nat|a *nf* *Mil* grenade; (*frutto*) pomegranate. **~i'ere** *nm* *Mil* grenadier

Gran Bre'tagna *nf* Great Britain

'granchio *nm* crab; (*fig: errore*) blunder; **prendere un ~** make a blunder

grandango'lare *nm* wide-angle lens

'grande (*a volte* **gran**) *a* (*ampio*) large; (*grosso*) big; (*alto*) tall; (*largo*) wide; (*fig: senso morale*) great; (*grandioso*) grand; (*adulto*) grown-up; **ho una gran fame** I'm very hungry; **fa un gran caldo** it is very hot; **in ~** on a large scale; **in gran parte** to a great extent; **non è un gran che** it is nothing much; **un gran ballo** a grand ball ● *nmf* (*persona adulta*) grown-up; (*persona eminente*) great man/woman. **~ggi'are** *vi* **~ggiare su** tower over; (*darsi arie*) show off

gran'dezza *nf* greatness; (*ampiezza*) largeness; (*larghezza*) width, breadth; (*dimensione*) size; (*fasto*) grandeur; (*prodigalità*) lavishness; **a ~ naturale** life-size

grandi'nare *vi* hail; **grandina** it's hailing. **'grandine** *nf* hail

grandiosità *nf* grandeur. **grandi'oso** *a* grand

gran'duca *nm* grand duke

gra'nello *nm* grain; (*di frutta*) pip

gra'nita *nf* crushed ice drink

gra'nito *nm* granite

'grano *nm* grain; (*frumento*) wheat

gran'turco *nm* maize

'granulo nm granule

'grappa nf grappa; (morsa) cramp

'grappolo nm bunch. **~ d'uva** bunch of grapes

gras'setto nm bold [type]

gras'sezza nf fatness; (untuosità) greasiness

'gras|so a fat; (cibo) fatty; (unto) greasy; (terreno) rich; (grossolano) coarse ●nm fat; (sostanza) grease. **~'soccio** a plump

'grata nf grating. **gra'tella**, **gra'ticola** nf Culin grill

gra'tifica nf bonus. **~zi'one** nf satisfaction

grati'na|re vt cook au gratin. **~to** a au gratin

'gratis adv free

grati'tudine nf gratitude. **'grato** a grateful; (gradito) pleasant

gratta'capo nm trouble

grattaci'elo nm skyscraper

grat'tar|e vt scratch; (raschiare) scrape; (grattugiare) grate; (fam: rubare) pinch ●vi grate. **~si** vr scratch oneself

grat'tugi|a nf grater. **~'are** vt grate

gratuita'mente adv free [of charge]. **gra'tuito** a free [of charge]; (ingiustificato) gratuitous

gra'vare vt burden ●vi **~ su** weigh on

'grave a (pesante) heavy; (serio) serious; (difficile) hard; (voce, suono) low; (fonetica) grave; **essere ~** (gravemente ammalato) be seriously ill. **~'mente** adv seriously, gravely

gravi'danza nf pregnancy. **'gravido** a pregnant

gravità nf seriousness; Phys gravity

gravi'tare vi gravitate

gra'voso a onerous

'grazi|a nf grace; (favore) favour; Jur pardon; **entrare nelle ~e di qcno** get into sb's good books. **~'are** vt pardon

'grazie int thank you!, thanks!; **~ mille!** many thanks!, thanks a lot!

grazi'oso a charming; (carino) pretty

'Grec|ia nf Greece. **g~o**, **-a** a & nmf Greek

'gregge nm flock

'greggio a raw ●nm (petrolio) crude [oil]

grembi'ale, **grembi'ule** nm apron

'grembo nm lap; (utero) womb; fig bosom

gre'mi|re vt pack. **~rsi** vr become crowded (**di** with). **~to** a packed

'gretto a stingy; (di vedute ristrette) narrow-minded

'grezzo a = greggio

gri'dare vi shout; (di dolore) scream; (animale:) cry ●vt shout

'grido nm (pl m gridi o f grida) shout, cry; (di animale) cry; **l'ultimo ~** the latest fashion; **scrittore di ~** celebrated writer

'grigio a & nm grey

'griglia nf grill; **alla ~** grilled

gril'letto nm trigger

'grillo nm cricket; (fig: capriccio) whim

grimal'dello nm picklock

'grinfia nf fig clutch

'grin|ta nf grit. **~'toso** a determined

'grinza nf wrinkle; (di stoffa) crease

grip'pare vi Mech seize

gris'sino nm bread-stick

'gronda nf eaves pl

gron'daia nf gutter

gron'dare vi pour; (essere bagnato fradicio) be dripping

'groppa nf back

'groppo nm knot; **avere un ~ alla gola** have a lump in one's throat

gros'sezza nf size; (spessore) thickness

gros'sista nmf wholesaler

'grosso a big, large; (spesso) thick; (grossolano) coarse; (grave) serious ●nm big part; (massa) bulk; **farla grossa** do a stupid thing

grosso|**lanità** *nf inv* (*qualità*) coarseness; (*di errore*) grossness; (*azione, parola*) coarse thing. **~'lano** *a* coarse; (*errore*) gross

grosso'modo *adv* roughly

'grotta *nf* cave, grotto

grot'tesco *a & nm* grotesque

grovi'era *nmf* Gruyère

gro'viglio *nm* tangle; *fig* muddle

gru *nf inv* (*uccello, edilizia*) crane

'gruccia *nf* (*stampella*) crutch; (*per vestito*) hanger

gru'gni|**re** *vi* grunt. **~to** *nm* grunt

'grugno *nm* snout

'grullo *a* silly

'grumo *nm* clot; (*di farina ecc*) lump. **gru'moso** *a* lumpy

'gruppo *nm* group; (*comitiva*) party. **~ sanguigno** blood group

gruvi'era *nmf* Gruyère

'gruzzolo *nm* nest-egg

guada'gnare *vt* earn; gain (*tempo, forza ecc*). **gua'dagno** *nm* gain; (*profitto*) profit; (*entrate*) earnings *pl*

gu'ado *nm* ford; passare a **~** ford

gua'ina *nf* sheath; (*busto*) girdle

gu'aio *nm* trouble; **che ~!** that's just brilliant!; essere nei guai be in a fix; **guai a te se lo tocchi!** don't you dare touch it!

gua'i|**re** *vi* yelp. **~to** *nm* yelp

gu'anci|**a** *nf* cheek. **~'ale** *nm* pillow

gu'anto *nm* glove. **guantoni** *pl* [da boxe] boxing gloves

guarda'coste *nm inv* coastguard

guarda'linee *nm inv* *Sport* linesman

guar'dar|**e** *vt* look at; (*osservare*) watch; (*badare a*) look after; (*dare su*) look out on ●*vi* look; (*essere orientato verso*) face. **~si** *vr* look at oneself; **~si da** beware of; (*astenersi*) refrain from

guarda'rob|**a** *nm inv* wardrobe; (*di locale pubblico*) cloakroom. **~i'ere, -a** *nmf* cloakroom attendant

gu'ardia *nf* guard; (*poliziotto*) policeman; (*vigilanza*) watch; es-

sere di **~** be on guard; (*medico:*) be on duty; fare la **~** a keep guard over; mettere in **~** qcno warn sb; stare in **~** be on one's guard. **~ carceraria** prison warder. **~ del corpo** bodyguard, minder. **~ di finanza** ≈ Fraud Squad. **~ forestale** forest ranger. **~ medica** duty doctor

guardi'ano, -a *nmf* caretaker. **~ notturno** night watchman

guar'dingo *a* cautious

guardi'ola *nf* gatekeeper's lodge

guarigi'one *nf* recovery

gua'rire *vt* cure ●*vi* recover; (*ferita:*) heal [up]

guarnigi'one *nf* garrison

guar'ni|**re** *vt* trim; *Culin* garnish. **~zi'one** *nf* trimming; *Culin* garnish; *Mech* gasket

guasta'feste *nmf inv* spoilsport

gua'star|**e** *vt* spoil; (*rovinare*) ruin; break (*meccanismo*). **~si** *vr* spoil; (*andare a male*) go bad; (*tempo:*) change for the worse; (*meccanismo:*) break down. **gua'sto** *a* broken; (*ascensore, telefono*) out of order; (*cibo, dente*) bad ●*nm* breakdown; (*danno*) damage

guazza'buglio *nm* muddle

guaz'zare *vi* wallow

gu'ercio *a* cross-eyed

guer'r|**a** *nf* war; (*tecnica bellica*) warfare. **~ fredda** Cold War. **~ mondiale** world war. **~afon'daio** *nm* warmonger. **~eggi'are** *vi* wage war. **guer'resco** *a* (*di guerra*) war; (*bellicoso*) warlike. **~i'ero** *nm* warrior

guer'rigli|**a** *nf* guerrilla warfare. **~'ero, -a** *nmf* guerrilla

'gufo *nm* owl

'guglia *nf* spire

gu'id|**a** *nf* guide; (*direzione*) guidance; (*comando*) leadership; *Auto* driving; (*tappeto*) runner; **~ a destra/sinistra** right-/left-hand drive. **~a telefonica** telephone directory. **~a turistica** tourist guide. **gui'dare** *vt* guide; *Auto*

drive; steer ⟨nave⟩. **~a'tore,
~a'trice** nmf driver
guin'zaglio nm leash
guiz'zare vi dart; ⟨luce:⟩ flash.
gu'izzo nm dart; ⟨di luce⟩ flash
'guscio nm shell
gu'stare vt taste ● vi like. **'gusto**
nm taste; ⟨piacere⟩ liking. **mangiare di gusto** eat heartily; **prenderci gusto** come to enjoy it,
develop a taste for it. **gu'stoso** a
tasty; fig delightful
guttu'rale a guttural

....

Hh

....

habitué nmf inv regular [customer]
ham'burger nm inv hamburger
'handicap nm inv Sport handicap
handicap'pa|re vt handicap. **~to,
-a** nmf disabled person ● a disabled
'harem nm inv harem
'hascisc nm hashish
hennè nm henna
hi-fi nm inv hi-fi
'hippy a hippy
hockey nm hockey. **~ su ghiaccio**
ice hockey. **~ su prato** hockey
hollywoodi'ano a Hollywood
attrib
ho'tel nm inv hotel

....

Ii

....

i def art mpl the; vedi **il**
i'ato nm hiatus
iber'na|re vi hibernate. **~zi'one** nf
hibernation
i'bisco nm hibiscus
'ibrido a & nm hybrid
'iceberg nm inv iceberg
i'cona nf icon

ld'dio nm God
i'dea nf idea; ⟨opinione⟩ opinion;
⟨ideale⟩ ideal; ⟨indizio⟩ inkling;
⟨piccola quantità⟩ hint; ⟨intenzione⟩ intention; **cambiare ~** change
one's mind; **neanche per ~!** not
on your life!; **chiarirsi le idee** get
one's ideas straight. **~ fissa**
obsession
ide'a|le a & nm ideal. **~'lista** nmf
idealist. **~liz'zare** vt idealize
ide'a|re vt conceive. **~'tore,
~ trice** nmf originator
'idem adv the same
i'dentico a identical
identifi'cabile a identifiable
identifi'ca|re vt identify. **~zi'one**
nf identification
identi'kit nm inv identikit®
identità nf inv identity
ideolo'gia nf ideology. **ideo'logico**
a ideological
i'dilli|co a idyllic. **~o** nm idyll
idi'oma nm idiom. **idio'matico** a
idiomatic
idi'ota a idiotic ● nmf idiot.
idio'zia nf ⟨cosa stupida⟩ idiocy
idola'trare vt worship
idoleggi'are vt idolize. **'idolo** nm
idol
idoneità nf suitability; Mil fitness;
esame di ~ qualifying examination. **i'doneo a idoneo a** suitable
for; Mil fit for
i'drante nm hydrant
idra'ta|re vt hydrate; ⟨cosmetico:⟩
moisturize. **~nte** a ⟨crema, gel⟩
moisturizing. **~zi'one** nf moisturizing
i'draulico a hydraulic ● nm
plumber
'idrico a water attrib
idrocar'buro nm hydrocarbon
idroe'lettrico a hydroelectric
i'drofilo a vedi **cotone**
i'drogeno nm hydrogen
idromas'saggio nm ⟨sistema⟩
whirlpool bath
idrovo'lante nm seaplane
i'ella nf fam bad luck; **portare ~** be

bad luck. **iel'lato** *a fam* jinxed, plagued by bad luck

i'ena *nf* hyena

i'eri *adv* yesterday; **~ l'altro, l'altro ~** the day before yesterday; **~ pomeriggio** yesterday afternoon; **il giornale di ~** yesterday's paper

ietta'tore, -'trice *nmf* jinx. **~'tura** *nf (sfortuna)* bad luck

igi'en|e *nf* hygiene. **~ico** *a* hygienic. **igie'nista** *nmf* hygienist

i'gnaro *a* unaware

i'gnobile *a* base; *(non onorevole)* dishonourable

igno'ran|te *a* ignorant ● *nmf* ignoramus. **~za** *nf* ignorance

igno'rare *vt (non sapere)* be unaware of; *(trascurare)* ignore

i'gnoto *a* unknown

il *def art m* the; **il latte fa bene** milk is good for you; **il signor Magnetti** Mr Magnetti; **il dottor Piazza** Dr Piazza; **ha il naso storto** he has a bent nose; **mettiti il cappello** put your hat on; **il lunedì** on Mondays; **il 1986** 1986; **5 000 lire il chilo** 5,000 lire the *o* a kilo

'ilar|e *a* merry. **~ità** *nf* hilarity

illazi'one *nf* inference

illecita'mente *adv* illicitly. **il'lecito** *a* illicit

ille'gal|e *a* illegal. **~ità** *nf* illegality. **~'mente** *adv* illegally

illeg'gibile *a* illegible; *(libro)* unreadable

illegittimità *nf* illegitimacy. **ille'gittimo** *a* illegitimate

il'leso *a* unhurt

illette'rato, -a *a & nmf* illiterate

illi'bato *a* chaste

illimi'tato *a* unlimited

illivi'dire *vt* bruise ● *vi (per rabbia)* turn livid

il'logico *a* illogical

il'luder|e *vt* deceive. **~si** *vr* deceive oneself

illumi'na|re *vt* light [up]; *fig* enlighten; **~ a giorno** floodlight. **~rsi** *vr* light up. **~zi'one** *nf* lighting; *fig* enlightenment

illumi'nismo *nm* Enlightenment

illusi'one *nf* illusion; **farsi illusioni** delude oneself

illusio'nis|mo *nm* conjuring. **~ta** *nmf* conjurer

il'lu|so, -a *pp di* **illudere** ● *a* deluded ● *nmf* day-dreamer. **~'sorio** *a* illusory

illu'stra|re *vt* illustrate. **~'tivo** *a* illustrative. **~'tore, ~'trice** *nmf* illustrator. **~zi'one** *nf* illustration

il'lustre *a* distinguished

imbacuc'ca|re *vt*, **~rsi** *vr* wrap up. **~to** *a* wrapped up

imbal'la|ggio *nm* packing. **~re** *vt* pack; *Auto* race

imbalsa'ma|re *vt* embalm; stuff *(animale)*. **~to** *a* embalmed; *(animale)* stuffed

imbam'bolato *a* vacant

imbaraz'zante *a* embarrassing

imbaraz'za|re *vt* embarrass; *(ostacolare)* encumber. **~to** *a* embarrassed

imba'razzo *nm* embarrassment; *(ostacolo)* hindrance; **trarre qcno d'~** help sb out of a difficulty; **avere l'~ della scelta** be spoilt for choice. **~ di stomaco** indigestion

imbarca'dero *nm* landing-stage

imbar'ca|re *vt* embark; *(fam: rimorchiare)* score. **~rsi** *vr* embark, go on board. **~zi'one** *nf* boat. **~zione di salvataggio** lifeboat. **im'barco** *nm* embarkation, boarding; *(banchina)* landing-stage

imba'sti|re *vt* tack; *fig* sketch. **~'tura** *nf* tacking, basting

im'batter|si *vr* **~** in run into

imbat'tibile *a* unbeatable. **~uto** *a* unbeaten

imbavagli'are *vt* gag

imbec'cata *nf* *Theat* prompt

imbe'cille *a* stupid ● *nmf* *Med* imbecile

imbel'lire *vt* embellish

im'berbe *a* beardless; *fig* inexperienced

imbestia'li|re *vi*, **~rsi** *vr* fly into a rage. **~to** *a* enraged

im'bever|e *vt* imbue (**di** with). **~si** *vr* absorb

imbe'v|ibile *a* undrinkable. **~uto** *a* **~uto di** (*acqua*) soaked in; (*nozioni*) imbued with

imbian'c|are *vt* whiten ● *vi* turn white. **~hino** *nm* house painter

imbizzar'rir|e *vi*, **~si** *vr* become restless; (*arrabbiarsi*) become angry

imboc'ca|re *vt* feed; (*entrare*) enter; *fig* prompt. **~'tura** *nf* opening; (*ingresso*) entrance; (*Mus: di strumento*) mouthpiece. **im'bocco** *nm* entrance

imbo'scar|e *vt* hide. **~si** *vr Mil* shirk military service

imbo'scata *nf* ambush

imbottigli'a|re *vt* bottle. **~rsi** *vr* get snarled up in a traffic jam. **~to** *a* (*vino, acqua*) bottled

imbot'ti|re *vt* stuff; pad (*giacca*); *Culin* fill. **~rsi** *vr* **~rsi di** (*fig: di pasticche*) stuff oneself with. **~ta** *nf* quilt. **~to** *a* (*spalle*) padded; (*cuscino*) stuffed; (*panino*) filled. **~'tura** *nf* stuffing; (*di giacca*) padding; *Culin* filling

imbracci'are *vt* shoulder (*fucile*)

imbra'nato *a* clumsy

imbrat'tar|e *vt* mark. **~si** *vr* dirty oneself

imbroc'car|e *vt* hit; **~la giusta** hit the nail on the head

imbrogli'|are *vt* muddle; (*raggirare*) cheat. **~arsi** *vr* get tangled; (*confondersi*) get confused. **im'broglio** *nm* tangle; (*pasticcio*) mess; (*inganno*) trick. **~'one, -a** *nmf* cheat

imbronci'a|re *vi*, **~rsi** *vr* sulk. **~to** *a* sulky

imbru'nire *vi* get dark; **all'~ at** dusk

imbrut'tire *vt* make ugly ● *vi* become ugly

imbu'care *vt* post, mail; (*nel biliardo*) pot

imbur'rare *vt* butter

im'buto *nm* funnel

imi'ta|re *vt* imitate. **~'tore, ~'trice** *nmf* imitator, impersonator. **~zi'one** *nf* imitation

immaco'lato *a* immaculate

immagazzi'nare *vt* store

immagi'na|re *vt* imagine; (*supporre*) suppose; **s'immagini!** imagine that!. **~rio** *a* imaginary. **~zi'one** *nf* imagination

im'magine *nf* image; (*rappresentazione, idea*) picture

imman'cabil|e *a* unfailing. **~'mente** *adv* without fail

im'mane *a* huge; (*orribile*) terrible

imma'nente *a* immanent

immangi'abile *a* inedible

immatrico'la|re *vt* register. **~rsi** *vr* (*studente*) matriculate. **~zi'one** *nf* registration; (*di studente*) matriculation

immaturità *nf* immaturity. **imma'turo** *a* unripe; (*persona*) immature; (*precoce*) premature

immedesi'ma|rsi *vr* **~rsi in** identify oneself with. **~zi'one** *nf* identification

immediata'mente *adv* immediately. **~'tezza** *nf* immediacy. **imme'diato** *a* immediate

immemo'rabile *a* immemorial

immensa'mente *adv* enormously. **~ità** *nf* immensity. **im'menso** *a* immense

immensu'rabile *a* immeasurable

im'merger|e *vt* immerse. **~si** *vr* plunge; (*sommergibile:*) dive; **~si in** immerse oneself in

immeri'tato *a* undeserved. **~evole** *a* undeserving

immersi'one *nf* immersion; (*di sommergibile*) dive. **im'merso** *pp di* immergere

immi'gra|nte *a & nmf* immigrant. **~re** *vi* immigrate. **~to, -a** *nmf* immigrant. **~zi'one** *nf* immigration

immi'nen|te *a* imminent. **~za** *nf* imminence

immischi'ar|e *vt* involve. **~si** *vr* **~si in** meddle in

immis'sario *nm* tributary

immissi'one *nf* insertion
im'mobile *a* motionless
im'mobili *nmpl* real estate. **~'are**
a **società ~are** building society,
savings and loan *Am*
immobili|tà *nf* immobility. **~z'za-**
re *vt* immobilize; *Comm* tie up
immo'desto *a* immodest
immo'lare *vt* sacrifice
immondez'zaio *nm* rubbish tip.
immon'dizia *nf* filth; (*spazzatura*)
rubbish. **im'mondo** *a* filthy
immo'ral|e *a* immoral. **~ità** *nf* im-
morality
immorta'lare *vt* immortalize. **im-**
mor'tale *a* immortal
immoti'vato *a* (*gesto*) unjustified
im'mun|e *a* exempt; *Med* immune.
~ità *nf* immunity. **~iz'zare** *vt* im-
munize. **~izzazi'one** *nf* immuni-
zation
immunodefici'enza *nf* immuno-
deficiency
immuso'nirsi *vr* sulk. **~to** *a*
sulky
immu'ta|bile *a* unchangeable. **~to**
a unchanging
impacchet'tare *vt* wrap up
impacci'a|re *vt* hamper; (*distur-*
bare) inconvenience; (*imbarazza-*
re) embarrass. **~to** *a* embarrassed; (*goffo*) awkward. **im'pac-**
cio *nm* embarrassment; (*ostacolo*)
hindrance; (*situazione difficile*)
awkward situation
im'pacco *nm* compress
impadro'nirsi *vr* **~ di** take posses-
sion of; (*fig: imparare*) master
impa'gabile *a* priceless
impagi'na|re *vt* paginate. **~zi'one**
nf pagination
impagli'are *vt* stuff ⟨*animale*⟩
impa'lato *a* fig stiff
impalca'tura *nf* scaffolding; *fig*
structure
impalli'dire *vi* turn pale; (*fig:*
perdere d'importanza) pale into
insignificance
impa'nare *vt* *Culin* roll in
breadcrumbs
impanta'narsi *vr* get bogged down

impape'rarsi *vr,* **impappi'narsi**
vr falter, stammer
impa'rare *vt* learn
impareggi'abile *a* incomparable
imparen'ta|rsi *vr* **~ con** become
related to. **~to** *a* related
impari *a* unequal; (*dispari*) odd
impar'tire *vt* impart
imparzi'al|e *a* impartial. **~ità** *nf*
impartiality
impas'sibile *a* impassive
impas'ta|re *vt* *Culin* knead; blend
⟨*colori*⟩. **~'tura** *nf* kneading.
im'pasto *nm* *Culin* dough; (*miscu-*
glio) mixture
impastic'carsi *vr* pop pills
im'patto *nm* impact
impau'rir|e *vt* frighten. **~si** *vr* be-
come frightened
im'pavido *a* fearless
impazi'en|te *a* impatient; **~te di**
fare qcsa eager to do sth. **~'tirsi**
vr lose patience. **~za** *nf* impa-
tience
impaz'zata *nf* **all'~** at breakneck
speed
impaz'zire *vi* go mad; ⟨*maionese:*⟩
separate; **far ~ qcno** drive sb
mad; **~ per** be crazy about; **da ~**
⟨*mal di testa*⟩ blinding
impec'cabile *a* impeccable
impedi'mento *nm* hindrance;
(*ostacolo*) obstacle
impe'dire *vt* **~ di** prevent from;
(*impacciare*) hinder; (*ostruire*)
obstruct; **~ a qcno di fare qcsa**
prevent sb [from] doing sth
impe'gna|re *vt* (*dare in pegno*)
pawn; (*vincolare*) bind; (*prenota-*
re) reserve; (*assorbire*) take up.
~rsi *vr* apply oneself; **~rsi a fare**
qcsa commit oneself to doing sth.
~'tiva *nf* referral. **~'tivo** *a* binding;
⟨*lavoro*⟩ demanding. **~ato** *a* en-
gaged; *Pol* committed. **im'pegno**
nm engagement; *Comm* commit-
ment; (*zelo*) zeal
impel'lente *a* pressing
impene'trabile *a* impenetrable
impen'na|rsi *vr* ⟨*cavallo:*⟩ rear; *fig*
bristle. **~ta** *nf* (*di prezzi*) sharp

rise; *(di cavallo)* rearing; *(di moto)* wheelie

impen'sa|bile *a* unthinkable. **~to** *a* unexpected

impensie'rir|e *vt*, **~si** *vr* worry

impe'ra|nte *a* prevailing. **~re** *vi* reign; ‹tendenza:› prevail, hold sway

impera'tivo *a & nm* imperative

impera'tore, -'trice *nm* emperor ● *nf* empress

impercet'tibile *a* imperceptible

imperdo'nabile *a* unforgivable

imper'fe|tto *a & nm* imperfect. **~zi'one** *nf* imperfection

imperi'a|le *a* imperial. **~'lismo** *nm* imperialism. **~'lista** *a* imperialist. **~'listico** *a* imperialistic

imperi'oso *a* imperious; *(impellente)* urgent

impe'rizia *nf* lack of skill

imperme'abile *a* waterproof ● *nm* raincoat

imperni'ar|e *vt* pivot; *(fondare)* base. **~si** *vr* **~si su** be based on

im'pero *nm* empire; *(potere)* rule

imperscru'tabile *a* inscrutable

imperso'nale *a* impersonal

imperso'nare *vt* personify; *(interpretare)* act [the part of]

imper'territo *a* undaunted

imperti'nen|te *a* impertinent. **~za** *nf* impertinence

impertur'ba|bile *a* imperturbable. **~to** *a* unperturbed

imperver'sare *vi* rage

im'pervio *a* inaccessible

'impet|o *nm* impetus; *(impulso)* impulse; *(slancio)* transport. **~u'oso** *a* impetuous; *(vento)* blustering

impet'tito *a* stiff

impian'tare *vt* install; set up ‹azienda›

impi'anto *nm* plant; *(sistema)* system; *(operazione)* installation. **~ radio** *Auto* car stereo system

impia'strare *vt* plaster; *(sporcare)* dirty. **impi'astro** *nm* poultice; *(persona noiosa)* bore; *(pasticcione)* cack-handed person

impic'car|e *vt* hang. **~si** *vr* hang oneself

impicci'arsi *vr* meddle. **im'piccio** *nm* hindrance; *(seccatura)* bother. **~'one, -a** *nmf* nosey parker

impie'ga|re *vt* employ; *(usare)* use; spend *(tempo, denaro)*; *Fin* invest; **l'autobus ha ~to un'ora** it took the bus an hour. **~rsi** *vr* get [oneself] a job

impiega'tizio *a* clerical

impie'gato, -a *nmf* employee. **~ di banca** bank clerk. **impi'ego** *nm* employment; *(posto)* job; *Fin* investment

impieto'sir|e *vt* move to pity. **~si** *vr* be moved to pity

impie'trito *a* petrified

impigli'ar|e *vt* entangle. **~si** *vr* get entangled

impi'grir|e *vt* make lazy. **~si** *vr* get lazy

impla'cabile *a* implacable

impli'ca|re *vt* implicate; *(sottintendere)* imply. **~rsi** *vr* become involved. **~zi'one** *nf* implication

implicita'mente *adv* implicitly. **im'plicito** *a* implicit

implo'ra|re *vt* implore. **~zi'one** *nf* entreaty

impolve'ra|re *vt* cover with dust. **~rsi** *vr* get covered with dust. **~to** *a* dusty

impon'derabile *a* imponderable; *(causa, evento)* unpredictable

impo'nen|te *a* imposing. **~za** *nf* impressiveness

impo'nibile *a* taxable ● *nm* taxable income

impopo'lar|e *a* unpopular. **~ità** *nf* unpopularity

im'por|re *vt* impose; *(ordinare)* order. **~si** *vr* assert oneself; *(aver successo)* be successful; **~si di** *(prefiggersi di)* set oneself the task of

impor'tan|te *a* important ● *nm* important thing. **~za** *nf* importance

impor'ta|re *vt* *Comm, Comput* import; *(comportare)* cause ● *vi* matter; (*essere necessario*) be necessary. **non** ~**!** it doesn't matter!; **non me ne** ~ **niente!** I couldn't care less!. ~'**tore**, ~'**trice** *nmf* importer. ~**zi'one** *nf* importation; (*merce importata*) import

im'porto *nm* amount

importu'nare *vt* pester. **impor'tuno** *a* troublesome; (*inopportuno*) untimely

imposizi'one *nf* imposition; (*imposta*) tax

imposses'sarsi *vr* ~ **di** seize

impos'sibile *a* impossible ● **non fare l'**~**e** do absolutely all one can. ~**ità** *nf* impossibility

im'posta[1] *nf* tax; ~ **sul reddito** income tax; ~ **sul valore aggiunto** value added tax

im'posta[2] *nf* (*di finestra*) shutter

impo'sta|**re** *vt* (*progettare*) plan; (*basare*) base; *Mus* pitch; (*imbucare*) post, mail; set out (*domanda, problema*). ~**zi'one** *nf* planning; (*di voce*) pitching

im'posto *pp di* **imporre**

impo'store, -a *nmf* impostor

impo'ten|**te** *a* powerless; *Med* impotent. ~**za** *nf* powerlessness; *Med* impotence

impove'rir|**e** *vt* impoverish. ~**si** *vr* become poor

imprati'cabile *a* impracticable; (*strada*) impassable

imprati'chir|**e** *vt* train. ~**si** *vr* ~**si in** *o a* get practice in

impre'ca|**re** *vi* curse. ~**zi'one** *nf* curse

impreci's|**abile** *a* indeterminable. ~**ato** *a* indeterminate. ~**i'one** *nf* inaccuracy. **impre'ciso** *a* inaccurate

impre'gnar|**e** *vt* impregnate; (*imbevere*) soak; *fig* imbue. ~**si** *vr* become impregnated with

imprendi'tor|**e, -'trice** *nmf* entrepreneur. ~**i'ale** *a* entrepreneurial

imprepa'rato *a* unprepared

im'presa *nf* undertaking; (*gesta*) exploit; (*azienda*) firm

impre'sario *nm* impresario; (*appaltatore*) contractor

imprescin'dibile *a* inescapable

impressio'na|**bile** *a* impressionable. ~**nte** *a* impressive; (*spaventoso*) frightening

impressio'nare *vt* impress; (*spaventare*) frighten; expose (*foto*). ~**o'narsi** *vr* be affected; (*spaventarsi*) be frightened. ~**'one** *nf* impression; (*sensazione*) sensation; (*impronta*) mark; **far** ~**one a** **qcno** upset sb

impressio'nis|**mo** *nm* impressionism. ~**ta** *nmf* impressionist

im'presso *pp di* **imprimere** ● *a* printed

impre'stare *vt* lend

impreve'dibile *a* unforeseeable; (*persona*) unpredictable

imprevi'dente *a* improvident

impre'visto *a* unforeseen ● *nm* unforeseen event; **salvo imprevisti** all being well

imprigio'na'mento *nm* imprisonment. ~**'nare** *vt* imprison

im'primere *vt* impress; (*stampare*) print; (*comunicare*) impart

impro'babil|**e** *a* unlikely, improbable. ~**ità** *nf* improbability

improdut'tivo *a* unproductive

im'pronta *nf* impression; *fig* mark. ~ **digitale** fingerprint. ~ **del piede** footprint

impro'perio *nm* insult; **improperi** *pl* abuse *sg*

im'proprio *a* improper

improvvisa'mente *adv* suddenly

improvvi'sa|**re** *vt/i* improvise. ~**rsi** *vr* turn oneself into a. ~**ta** *nf* surprise. ~**to a** (*discorso*) unrehearsed. ~**zi'one** *nf* improvisation

improv'viso *a* sudden; **all'**~ unexpectedly

impru'den|**te** *a* imprudent. ~**za** *nf* imprudence

impu'gna|**re** *vt* grasp; *Jur* contest. ~**tura** *nf* grip; (*manico*) handle

impulsività *nf* impulsiveness. **impul'sivo** *a* impulsive

im'pulso *nm* impulse; **agire d'~** act on impulse

impune'mente *adv* with impunity. **impu'nito** *a* unpunished

impun'tarsi *vr fig* dig one's heels in

impun'tura *nf* stitching

impurità *nf inv* impurity. **im'puro** *a* impure

impu'tabile *a* attributable (**a** to)

impu'tare *vt* attribute; (*accusare*) charge. **~to, -a** *nmf* accused. **~zi'one** *nf* charge

imputri'dire *vi* rot

in *prep* in; (*moto a luogo*) to; (*su*) on; (*entro*) within; (*mezzo*) by; (*con materiale*) made of; **essere in casa/ufficio** be at home/at the office; **in mano/tasca** in one's hand/pocket; **andare in Francia/campagna** go to France/the country; **salire in treno** get on the train; **versa la birra nel bicchiere** pour the beer into the glass; **in alto** up there; **in giornata** within the day; **nel 1997** in 1997; **una borsa in pelle** a bag made of leather, a leather bag; **in macchina** (*viaggiare, venire*) by car; **in contanti** [in] cash; **in vacanza** on holiday; **di giorno in giorno** from day to day; **se fossi in te** if I were you; **siamo in sette** there are seven of us

inabbor'dabile *a* unapproachable

i'nabile *a* incapable; (*fisicamente*) unfit. **~ità** *nf* incapacity

inabi'tabile *a* uninhabitable

inacces'sibile *a* inaccessible; (*persona*) unapproachable

inaccet'tabile *a* unacceptable. **~ità** *nf* unacceptability

inacer'bire *vt* embitter; exacerbate (*rapporto*). **~si** *vr* grow bitter

inaci'dire *vt* turn sour. **~si** *vr* go sour; (*persona:*) become embittered

ina'datto *a* unsuitable

inadegu'ato *a* inadequate

inadempi'ente *nmf* defaulter. **~'mento** *nm* nonfulfilment

inaffer'rabile *a* elusive

ina'la|re *vt* inhale. **~'tore** *nm* inhaler. **~zi'one** *nf* inhalation

inalbe'rare *vt* hoist. **~si** *vr* (*cavallo:*) rear [up]; (*adirarsi*) lose one's temper

inalte'rabile *a* unchangeable; (*colore*) fast. **~to** *a* unchanged

inami'da|re *vt* starch. **~to** *a* starched

inammis'sibile *a* inadmissible

inamovi'bile *a* irremovable

inani'mato *a* inanimate; (*senza vita*) lifeless

inappa'gabile *a* unsatisfiable. **~to** *a* unfulfilled

inappel'labile *a* final

inappe'tenza *nf* lack of appetite

inappli'cabile *a* inapplicable

inappun'tabile *a* faultless

inar'ca|re *vt* arch; raise (*sopracciglia*). **~si** *vr* (*legno:*) warp; (*ripiano:*) sag; (*linea:*) curve

inari'dire *vt* parch; empty of feelings (*persona*). **~si** *vr* dry up; (*persona:*) become empty of feelings

inarti'colato *a* inarticulate

inaspettata'mente *adv* unexpectedly. **inaspet'tato** *a* unexpected

inaspri'mento *nm* (*di carattere*) embitterment; (*di conflitto*) worsening

inasprir|e *vt* embitter. **~si** *vr* become embittered

inattac'cabile *a* unassailable; (*irreprensibile*) irreproachable

inatten'dibile *a* unreliable. **inat'teso** *a* unexpected

inattività *nf* inactivity. **inat'tivo** *a* inactive

inattu'abile *a* impracticable

inau'dito *a* unheard of

inaugu'rale *a* inaugural; **viaggio ~** maiden voyage

inaugu'ra|re *vt* inaugurate; open (*mostra*); unveil (*statua*); christen (*lavastoviglie ecc*). **~zi'one** *nf* inauguration; (*di mostra*) opening; (*di statua*) unveiling

inavver'tenza *nf* inadvertence. **~ita'mente** *adv* inadvertently

incagli'ar|e vi ground ● vt hinder. **~si** vr run aground

incalco'labile a incalculable

incal'li|rsi vr grow callous; (abituarsi) become hardened. **~to** a callous; (abituato) hardened

incal'za|nte a (ritmo) driving; (richiesta) urgent. **~re** vt pursue; fig press

incame'rare vt appropriate

incammi'nar|e vt get going; (fig: guidare) set off. **~si** vr set out

incana'lar|e vt canalize; fig channel. **~si** vr converge on

incande'scen|te a incandescent; (discussione) burning. **~za** nf incandescence

incan'ta|re vt enchant. **~rsi** vr stand spellbound; (inceppparsi) jam. **~tore**, **~'trice** nm enchanter ● nf enchantress

incan'tesimo nm spell

incan'tevole a enchanting

in'canto nm spell; fig delight; (asta) auction; **come per ~** as if by magic

incanu'ti|re vt turn white. **~to** a white

inca'pac|e a incapable. **~ità** nf incapability

incapo'nirsi vr be set (**a fare** on doing)

incap'pare vi **~ in** run into

incappucci'arsi vr wrap up

incapricci'arsi vr **~ di** take a fancy to

incapsu'lare vt seal; crown (dente)

incarce'ra|re vt imprison. **~zi'one** nf imprisonment

incari'ca|re vt charge. **~rsi** vr take upon oneself; **me ne incarico io I** will see to it. **~to, -a** a in charge ● nmf representative. **in'carico** nm charge; **per incarico di** on behalf of

incar'na|re vt embody. **~rsi** vr become incarnate. **~zi'one** nf incarnation

incarta'mento nm documents pl. **incar'tare** vt wrap [in paper]

incasi'nato a fam (vita) screwed up; (stanza) messed up

incas'sa|re vt pack; Mech embed; box in (mobile, frigo); (riscuotere) cash; take (colpo). **~to** a set; (fiume) deeply embanked. **in'casso** nm collection; (introito) takings pl

incasto'na|re vt set. **~tura** nf setting. **~to** a embedded; (anello) inset (**di** with)

inca'strare vt fit in; (fam: in situazione) corner. **~si** vr fit. **in'castro** nm joint; **a incastro** (pezzi) interlocking

incate'nare vt chain

incatra'mare vt tar

incat'tivire vt turn nasty

in'cauto a imprudent

inca'va|re vt hollow out. **~to** a hollow. **~tura** nf hollow. **in'cavo** nm hollow; (scanalatura) groove

incavo'la|rsi vr fam get shirty. **~to** a fam shirty

incendi'ar|e vt set fire to; fig inflame. **~si** vr catch fire. **~io, -a** a incendiary; (fig: discorso) inflammatory; (fig: bellezza) sultry ● nmf arsonist. **in'cendio** nm fire. **incendio doloso** arson

incene'ri|re vt burn to ashes; (cremare) cremate. **~rsi** vr be burnt to ashes. **~tore** nm incinerator

in'censo nm incense

incensu'rato a blameless; **essere ~** Jur have a clean record

incenti'vare vt motivate. **incen'tivo** nm incentive

incen'trarsi vr **~ su** centre on

incep'par|e vt block; fig hamper. **~si** vr jam

ince'rata nf oilcloth

incerot'tato a with a plaster on

incer'tezza nf uncertainty. **in'cer-to** a uncertain ● nm uncertainty

inces'sante a unceasing. **~'mente** adv incessantly

in'cest|o nm incest. **~u'oso** a incestuous

in'cetta nf buying up; **fare ~ di** stockpile

inchi'esta nf investigation
inchi'nar|e vt, ~**si** vr bow. **in'chino** nm bow; (di donna) curtsy
inchio'dare vt nail; nail down ⟨coperchio⟩; ~ **a letto** ⟨malattia:⟩ confine to bed
inchi'ostro nm ink
inciam'pare vi stumble; ~ **in** (imbattersi) run into. **inci'ampo** nm hindrance
inci'dente a incidental
inci'den|te nm (episodio) incident; (infortunio) accident. ~**za** nf incidence
in'cidere vt cut; ⟨arte⟩ engrave; (registrare) record ● vi ~ **su** ⟨gravare⟩ weigh upon
in'cinta a pregnant
incipi'ente a incipient
incipri'ar|e vt powder; ~**si** vr powder one's face
in'circa adv **all'~** more or less
incisi'one nf incision; ⟨arte⟩ engraving; ⟨acquaforte⟩ etching; (registrazione) recording
inci'sivo a incisive ● nm ⟨dente⟩ incisor
in'ciso nm **per ~** incidentally
incita'mento nm incitement. **inci'tare** vt incite
inci'vil|e a uncivilized; (maleducato) impolite. ~**tà** nf barbarism; (maleducazione) rudeness
incle'men|te a harsh. ~**za** nf harshness
incli'nabile a reclining
incli'nar|e vt tilt ● vi ~**re a** be inclined to. ~**rsi** vr list. ~**to** a tilted; ⟨terreno⟩ sloping. ~**zi'one** nf slope, inclination. **in'cline** a inclined
in'clu|dere vt include; (allegare) enclose. ~**si'one** nf inclusion. ~**sivo** a inclusive. ~**so** pp di **includere** ● a included; (compreso) inclusive; (allegato) enclosed
incoe'ren|te a (contraddittorio) inconsistent. ~**za** nf inconsistency
in'cognit|a nf unknown quantity. ~**o** a unknown ● nm **in** ~**o** incognito

incol'lar|e vt stick; (con colla liquida) glue. ~**si** vr stick to; ~**si a** qcno stick close to sb
incolle'ri|rsi vr lose one's temper. ~**to** a enraged
incol'mabile a ⟨differenza⟩ unbridgeable; (vuoto) unfillable
incolon'nare vt line up
inco'lore a colourless
incol'pare vt blame
in'colto a uncultivated; ⟨persona⟩ uneducated
in'colume a unhurt
incom'ben|te a impending. ~**za** nf task
in'combere vi ~ **su** hang over; ~ **a** ⟨spettare⟩ be incumbent on
incomo'dar|e vt inconvenience. ~**si** vr trouble. **in'comodo** a uncomfortable; (inopportuno) inconvenient ● nm inconvenience
incompa'rabile a incomparable
incompa'tibil|e a incompatible. ~**ità** nf incompatibility
incompe'ten|te a incompetent. ~**za** nf incompetence
incompi'uto a unfinished
incom'pleto a incomplete
incompren'si|bile a incomprehensible. ~**'one** nf lack of understanding; (malinteso) misunderstanding. **incom'preso** a misunderstood
inconce'pibile a inconceivable
inconcili'abile a irreconcilable
inconclu'dente a inconclusive; ⟨persona⟩ ineffectual
incondizio'nata|mente adv unconditionally. ~**'nato** a unconditional
inconfes'sabile a unmentionable
inconfon'dibile a unmistakable
inconfu'tabile a irrefutable
incongru'ente a inconsistent
in'congruo a inadequate
inconsa'pevol|e a unaware; (inconscio) unconscious. ~**'mente** adv unwittingly
in'conscio a unconscious ● nm unconscious
inconscia'mente adv uncon-

sciously. **in'conscio** a & nm Psych unconscious

inconsi'sten|te a insubstantial; ⟨notizia ecc⟩ unfounded. **~za** nf (di ragionamento, prove) flimsiness

inconso'labile a inconsolable

inconsu'eto a unusual

incon'sulto a rash

incontami'nato a uncontaminated

inconte'nibile a irrepressible

inconten'tabile a insatiable; ⟨esigente⟩ hard to please

inconte'stabile a indisputable

inconti'nen|te a incontinent. **~za** nf incontinence

incon'trar|e vt meet; encounter, meet with ⟨difficoltà⟩. **~si** vr meet ⟨con qcno sb⟩

incon'trario: all'~ adv the other way around; (in modo sbagliato) the wrong way around

incontra'stabile a incontrovertible. **~to** a undisputed

in'contro nm meeting; Sport match. **~ al vertice** summit meeting ● prep **~ a** towards; **andare ~ a qn** go to meet sb; fig meet sb half way

inconveni'ente nm drawback

incoraggi|a'mento nm encouragement. **~'ante** a encouraging. **~'are** vt encourage

incornici'a|re vt frame. **~'tura** nf framing

incoro'na|re vt crown. **~zi'one** nf coronation

incorpo'rar|e vt incorporate; (mescolare) blend. **~si** vr blend; ⟨territori:⟩ merge

incorreg'gibile a incorrigible

in'correre vi incur; **~ nel pericolo di...** run the risk of...

incorrut'tibile a incorruptible

incosci'en|te a unconscious; (irresponsabile) reckless ● nmf irresponsible person. **~za** nf unconsciousness; recklessness

inco'stan|te a changeable; ⟨per-

sona⟩ fickle. **~za** nf changeableness; (di persona) fickleness

incostituzio'nale a unconstitutional

incre'dibile a unbelievable, incredible

incredulità nf incredulity. **in'credulo** a incredulous

incremen'tare vt increase; (intensificare) step up. **incre'mento** nm increase. **incremento demografico** population growth

incresci'oso a regrettable

incre'spar|e vt ruffle; wrinkle ⟨tessuto⟩; make frizzy ⟨capelli⟩; **~e la fronte** frown. **~si** vr ⟨acqua:⟩ ripple; ⟨tessuto:⟩ wrinkle; ⟨capelli:⟩ go frizzy

incrimi'na|re vt indict; fig incriminate. **~zi'one** nf indictment

incri'na|re vt crack; fig affect ⟨amicizia⟩. **~rsi** vr crack; ⟨amicizia:⟩ be affected. **~'tura** nf crack

incroci'a|re vt cross ● vi Naut, Aeron cruise. **~rsi** vr cross. **~'tore** nm cruiser

in'crocio nm crossing; (di strade) crossroads sg

incrol'labile a indestructible

incro'sta|re vt encrust. **~zi'one** nf encrustation

incuba'|trice nf incubator. **~zi'one** nf incubation

'incubo nm nightmare

in'cudine nf anvil

incu'rabile a incurable

incu'rante a careless

incurio'sir|e vt make curious. **~si** vr become curious

incursi'one nf raid. **~ aerea** air raid

incurva'mento nm bending

incur'va|re vt, **~rsi** vr bend. **~'tura** nf bending

in'cusso pp di incutere

incu'stodito a unguarded

in'cutere vt arouse; **~ spavento a qcno** strike fear into sb

'indaco nm indigo

indaffa'rato a busy

inda'gare vt/i investigate

in'dagine *nf* research; (*giudizia-ria*) investigation. **~ di mercato** market survey

indebi'tar|e *vt*, **~si** *vr* get into debt

in'debito *a* undue

indeboli'mento *nm* weakening

indebo'lir|e *vt*, **~si** *vr* weaken

inde'cen|te *a* indecent. **~za** *nf* indecency; (*vergogna*) disgrace

indeci'frabile *a* indecipherable

indecisi'one *nf* indecision. **inde-'ciso** *a* undecided

inde'fesso *a* tireless

indefi'ni|bile *a* indefinable. **~to** *a* indefinite

indefor'mabile *a* crushproof

in'degno *a* unworthy

inde'lebile *a* indelible

indelica'tezza *nf* indelicacy; (*azione*) tactless act. **indeli'cato** *a* indiscreet; (*grossolano*) indelicate

indemoni'ato *a* possessed

in'denn|e *a* uninjured; (*da malat-tia*) unaffected. **~ità** *nf inv* allow-ance; (*per danni*) compensation. **~ità di trasferta** travel allow-ance. **~iz'zare** *vt* compensate. **inden'nizzo** *nm* compensation

indero'gabile *a* binding

indescri'vibile *a* indescribable

indeside'ra|bile *a* undesirable. **~to** *a* (*figlio, ospite*) unwanted

indetermi'na|bile *a* indetermin-able. **~'tezza** *nf* vagueness. **~to** *a* indeterminate

'Indi|a *nf* India. **i~'ano, -a** *a & nmf* Indian; **in fila i~ana** in single file

indiavo'lato *a* possessed; (*vivace*) wild

indi'ca|re *vt* show, indicate; (*col dito*) point at; (*far notare*) point out; (*consigliare*) advise. **~'tivo** *a* indicative ● *nm* Gram indicative. **~'tore** *nm* indicator; (*prontuario*) directory. **~zi'one** *nf* indication; (*istruzione*) direction

'indice *nm* (*dito*) forefinger; (*lan-cetta*) pointer; (*di libro, statistica*) index; (*fig: segno*) sign

indi'cibile *a* inexpressible

indietreggi'are *vi* draw back; *Mil* retreat

indi'etro *adv* back, behind; **all'~** backwards; **avanti e ~** back and forth; **essere ~** be behind; (*mentalmente*) be backward; (*con pagamenti*) be in arrears; (*di orologio*) be slow; **fare marcia ~** reverse; **rimandare ~** send back; **rimanere ~** be left behind; **torna ~!** come back!

indi'feso *a* undefended; (*inerme*) helpless

indiffe'ren|te *a* indifferent; **mi è ~te** it is all the same to me. **~za** *nf* indifference

in'digeno, -a *a* indigenous ● *nmf* native

indi'gen|te *a* needy. **~za** *nf* pov-erty

indigesti'one *nf* indigestion. **indi'gesto** *a* indigestible

indi'gna|re *vt* make indignant. **~rsi** *vr* be indignant. **~to** *a* indig-nant. **~zi'one** *nf* indignation

indimenti'cabile *a* unforgettable

indipen'den|te *a* independent. **~te'mente** *adv* independently; **~temente dal tempo** regardless of the weather, whatever the weather. **~za** *nf* independence

in'dire *vt* announce

indiretta'mente *adv* indirectly. **indi'retto** *a* indirect

indiriz'zar|e *vt* address; (*mandare*) send; (*dirigere*) direct. **~si** *vr* di-rect one's steps. **indi'rizzo** *nm* ad-dress; (*direzione*) direction

indisci'pli|na *nf* lack of discipline. **~'nato** *a* undisciplined

indi'scre|to *a* indiscreet. **~zi'one** *nf* indiscretion

indiscrimi'nata'mente *adv* indis-criminately. **~'nato** *a* indiscrimi-nate

indi'scusso *a* unquestioned

indiscu'tibil|e *a* unquestionable. **~'mente** *adv* unquestioningly

indispen'sabile *a* essential, indis-pensable

indispet'tir|e vt irritate. **~si** vr get irritated

indi'spo|rre vt antagonize. **~sto** pp di **indisporre ● a** indisposed. **~sizi'one** nf indisposition

indisso'lubile a indissoluble

indissolubil'mente adv indissolubly

indistin'guibile a indiscernible

indistinta'mente adv without exception. **indi'stinto** a indistinct

indistrut'tibile a indestructible

indistur'bato a undisturbed

in'divia nf endive

individu'a|le a individual. **~'lista** nmf individualist. **~lità** nf individuality. **~re** vt individualize; (localizzare) locate; (riconoscere) single out

indi'viduo nm individual

indivi'sibile a indivisible. **indi'viso** a undivided

indizi'a|re vt throw suspicion on. **~to, -a** a suspected **●** nmf suspect. **in'dizio** nm sign; Jur circumstantial evidence

'indole nf nature

indo'len|te a indolent. **~za** nf indolence

indolenzi'mento nm stiffness

indolen'zi|rsi vr go stiff. **~to** a stiff

indo'lore a painless

indo'mani nm l'**~** the following day

Indo'nesia nf Indonesia

indo'rare vt gild

indos'sa|re vt wear; (mettere addosso) put on. **~'tore, ~'trice** nmf model

in'dotto pp di **indurre**

indottri'nare vt indoctrinate

indovi'n|are vt guess; (predire) foretell. **~ato** a successful; (scelta) well-chosen. **~ello** nm riddle. **indo'vino, -a** nmf fortune-teller

indubbia'mente adv undoubtedly. **in'dubbio** a undoubted

indugi'ar|e vi, **~si** vr linger. **in'dugio** nm delay

indul'gen|te a indulgent. **~za** nf indulgence

in'dul|gere vi **~gere a** indulge in. **~to** pp di **indulgere ●** nm Jur pardon

indu'mento nm garment; **indumenti** pl clothes

induri'mento nm hardening

indu'rir|e vt, **~si** vr harden

in'durre vt induce

in'dustri|a nf industry. **~'ale** a industrial **●** nmf industrialist

industrializ'za|re vt industrialize. **~to** a industrialized. **~zi'one** nf industrialization

industrial'mente adv industrially

industri|'arsi vr try one's hardest. **~'oso** a industrious

induzi'one nf induction

inebe'tito a stunned

inebri'ante a intoxicating, exciting

inecce'pibile a unexceptionable

i'nedia nf starvation

i'nedito a unpublished

ineffi'cace a ineffective

ineffici'en|te a inefficient. **~za** nf inefficiency

ineguagli'abile a incomparable

inegu'ale a unequal; (superficie) uneven

inelut'tabile a inescapable

ine'rente a ~ a concerning

i'nerme a unarmed; fig defenceless

inerpi'carsi vr **~ su** clamber up; (pianta:) climb up

i'ner|te a inactive; Phys inert. **~zia** nf inactivity; Phys inertia

inesat'tezza nf inaccuracy. **ine'satto** a inaccurate; (erroneo) incorrect; (non riscosso) uncollected

inesau'ribile a inexhaustible

inesi'sten|te a non-existent. **~za** nf non-existence

ineso'rabile a inexorable

inesperi'enza nf inexperience. **ine'sperto** a inexperienced

inespli'cabile a inexplicable

ine'sploso *a* unexploded

inespri'mibile *a* inexpressible

inesti'mabile *a* inestimable

inetti'tudine *nf* ineptitude. **i'netto** *a* inept; **inetto a** unsuited to

ine'vaso *a* ⟨*pratiche*⟩ pending; ⟨*corrispondenza*⟩ unanswered

inevi'tabile *a* inevitable. **~'mente** *adv* inevitably

i'nezia *nf* trifle

infagot'tare *vt* wrap up. **~si** *vr* wrap [oneself] up

infal'libile *a* infallible

infa'ma|re *vt* defame. **~'torio** *a* defamatory

in'fam|e *a* infamous; ⟨*fam: orrendo*⟩ awful, shocking. **~ia** *nf* infamy

infan'garsi *vr* get muddy

infan'tile *a* ⟨*letteratura, abbigliamento*⟩ children's; ⟨*ingenuità*⟩ childlike; *pej* childish

in'fanzia *nf* childhood; ⟨*bambini*⟩ children *pl*; **prima ~** infancy

infar'cire *vi* pepper ⟨*discorso*⟩ ⟨*di* with⟩

infari'na|re *vt* flour; **~re di** sprinkle with. **~tura** *nf* fig smattering

in'farto *nm* coronary

infasti'dir|e *vt* irritate. **~si** *vr* get irritated

infati'cabile *a* untiring

in'fatti *conj* as a matter of fact; ⟨*veramente*⟩ indeed

infatu'a|rsi *vr* become infatuated ⟨*di* with⟩. **~to a** infatuated. **~zi'one** *nf* infatuation

in'fausto *a* ill-omened

infe'condo *a* infertile

infe'del|e *a* unfaithful. **~tà** *nf* unfaithfulness; **~** *pl* affairs

infe'lic|e *a* unhappy; ⟨*inappropriato*⟩ unfortunate; ⟨*cattivo*⟩ bad. **~ità** *nf* unhappiness

infel'tri|rsi *vr* get matted. **~to a** matted

inferi'or|e *a* ⟨*più basso*⟩ lower; ⟨*qualità*⟩ inferior ● *nmf* inferior. **~ità** *nf* inferiority

inferme'ria *nf* infirmary; ⟨*di nave*⟩ sick-bay

infermi'er|a *nf* nurse. **~e** *nm* [male] nurse

infermità *nf* sickness. **~ mentale** mental illness. **in'fermo, -a** *a* sick ● *nmf* invalid

infer'nale *a* infernal; ⟨*spaventoso*⟩ hellish

in'ferno *nm* hell; **va all'~!** go to hell!

infero'cirsi *vr* become fierce

inferri'ata *nf* grating

infervo'rar|e *vt* arouse enthusiasm in. **~si** *vr* get excited

infe'stare *vt* infest

infet'tare *vt* infect. **~arsi** *vr* become infected. **~ivo** *a* infectious. **in'fetto** *a* infected. **infezi'one** *nf* infection

infiac'chir|e *vt/i*, **~si** *vr* weaken

infiam'mabile *a* [in]flammable

infiam'ma|re *vt* set on fire; *Med, fig* inflame. **~rsi** *vr* catch fire; *Med* become inflamed. **~zi'one** *nf Med* inflammation

in'fido *a* treacherous

infie'rire *vi* ⟨*imperversare*⟩ rage; **~ su** attack furiously

in'figger|e *vt* drive. **~si** *vr* **~si in** penetrate

infi'lare *vt* thread; ⟨*mettere*⟩ insert; ⟨*indossare*⟩ put on. **~si** *vr* slip on ⟨*vestito*⟩; **~si in** ⟨*introdursi in*⟩ slip into

infil'tra|rsi *vr* infiltrate. **~zi'one** *nf* infiltration; ⟨*d'acqua*⟩ seepage; ⟨*Med: iniezione*⟩ injection

infil'zare *vt* pierce; ⟨*infilare*⟩ string; ⟨*conficcare*⟩ stick

'infimo *a* lowest

in'fine *adv* finally; ⟨*insomma*⟩ in short

infinità *nf* infinity; **un'~ di** masses of. **~'mente** *adv* infinitely.

infi'nito *a* infinite; *Gram* infinitive ● *nm* infinite; *Gram* infinitive; *Math* infinity; **all'infinito** *adv* indefinitely

infinocchi'are *vt fam* hoodwink

infischi'arsi *vr* **~ di** not care about; **me ne infischio** *fam* I couldn't care less in'fisso *pp di* **infiggere** ● *nm* fixture; (*di porta, finestra*) frame

infit'tir|e *vt/i*, ~si *vr* thicken

inflazi'one *nf* inflation

infles'sibil|e *a* inflexible. ~ità *nf* inflexibility

inflessi'one *nf* inflexion

in'fli|ggere *vt* inflict. ~tto *pp di* **infliggere**

influ'en|te *a* influential. ~za *nf* influence; *Med* influenza

influen'za|bile *a* ⟨mente, opinione⟩ impressionable. ~re *vt* influence. ~to *a* ⟨malato⟩ with the flu

influ'ire *vi* ~ su influence

in'flusso *nm* influence

info'carsi *vr* catch fire; ⟨viso:⟩ go red; ⟨discussione:⟩ become heated

info'gnarsi *vr fam* get into a mess

infol'tire *vt/i* thicken

infon'dato *a* unfounded

in'fondere *vt* instil

infor'care *vt* fork up; get on ⟨bici⟩; put on ⟨occhiali⟩

infor'male *a* informal

infor'ma|re *vt* inform. ~rsi *vr* inquire (**di** about). ~'tivo *a* informative.

infor'matic|a *nf* computing, IT. ~o *a* computer *attrib*

infor'ma|tivo *a* informative. infor'mato *a* informed; **male informato** ill-informed. ~'tore, ~'trice *nmf* (*di polizia*) informer. ~zi'one *nf* information (*solo sg*); un'~zione a piece of information

in'forme *a* shapeless

infor'nare *vt* put into the oven

infortu'narsi *vr* have an accident.

infor'tu|nio *nm* accident. ~nio sul lavoro industrial accident. ~'nistica *nf* study of industrial accidents

infos'sa|rsi *vr* sink; ⟨guance, occhi:⟩ become hollow. ~to *a* sunken, hollow

infradici'ar|e *vt* drench. ~si *vr* get drenched; (*diventare marcio*) rot

infra'dito *nm inv* ⟨scarpe⟩ flip-flop

in'frang|ere *vt* break; (*in mille*

pezzi*) shatter. ~ersi *vr* break. ~'gibile *a* unbreakable

in'franto *pp di* **infrangere** ● *a* shattered; (*fig: cuore*) broken

infra'rosso *a* infra-red

infrastrut'tura *nf* infrastructure

infrazi'one *nf* offence

infredda'tura *nf* cold

infreddo'li|rsi *vr* feel cold. ~to *a* cold

infruttu'oso *a* fruitless

infuo'ca|re *vt* make red-hot. ~to *a* burning

infu'ori *adv* all'~ outwards; all'~ di except

infuri'a|re *vi* rage. ~rsi *vr* fly into a rage. ~to *a* blustering

infusi'one *nf* infusion. in'fuso *pp di* **infondere** ● *nm* infusion

Ing. *abbr* **ingegnere**

ingabbi'are *vt* cage; (*fig: mettere in prigione*) jail

ingaggi'are *vt* engage; sign up ⟨calciatori ecc⟩; begin ⟨lotta, battaglia⟩. in'gaggio *nm* engagement; (*di calciatore*) signing [up]

ingan'nar|e *vt* deceive; (*essere infedele a*) be unfaithful to. ~si *vr* deceive oneself; **se non m'inganno** if I am not mistaken

ingan'nevole *a* deceptive. in'ganno *nm* deceit; (*frode*) fraud

ingarbugli'a|re *vt* entangle; (*confondere*) confuse. ~rsi *vr* get entangled; (*confondersi*) become confused. ~to *a* confused

in'gegn|e *nm* engineer. ingegne-'ria *nf* engineering

in'gegno *nm* brains *pl*; (*genio*) genius; (*abilità*) ingenuity. ~sa-'mente *adv* ingeniously

ingegnosità *nf* ingenuity. inge-'gnoso *a* ingenious

ingelo'sir|e *vt* make jealous. ~si *vr* become jealous

in'gente *a* huge

ingenu'a'mente *adv* artlessly. ~ità *nf* ingenuousness. in'genuo *a* ingenuous; (*credulone*) naïve

inge'renza *nf* interference

inge'rire vt swallow

inges'sa|re vt put in plaster. **~'tura** nf plaster

Inghil'terra nf England

inghiot'tire vt swallow

in'ghippo nm trick

ingial'li|re vi, **~rsi** vr turn yellow. **~to** a yellowed

ingigan'tir|e vt magnify ● vi, **~si** vr grow to enormous proportions

inginocchi'a|rsi vr kneel [down]. **~to** a kneeling. **~'toio** nm prie-dieu

ingioiel'larsi vr put on one's jewels

ingiù adv down; **all'~** downwards; **a testa ~** head downwards

ingi'un|gere vt order. **~zi'one** nf injunction. **~zione di pagamento** final demand

ingi'uri|a nf insult; (torto) wrong; (danno) damage. **~'are** vt insult; (fare un torto a) wrong. **~'oso** a insulting

ingiusta'mente adv unjustly, unfairly. **ingiu'stizia** nf injustice. **ingi'usto** a unjust, unfair

in'glese a English ● nm Englishman; (lingua) English ● nf Englishwoman

ingoi'are vt swallow

ingol'far|e vt flood (motore). **~si** vr fig get involved; (motore:) flood

ingom'bra|nte a cumbersome. **~re** vt clutter up; fig cram (mente)

in'gombro nm encumbrance; **essere d'~** be in the way

ingor'digia nf greed. **in'gordo** a greedy

ingor'gar|e vt block. **~si** vr be blocked [up]. **in'gorgo** nm blockage; (del traffico) jam

ingoz'zar|e vt gobble up; (nutrire eccessivamente) stuff; fatten (animali). **~si** vr stuff oneself (di with)

ingra'na|ggio nm gear; fig mechanism. **~re** vt engage ● vi be in gear

ingrandi'mento nm enlargement

ingran'di|re vt enlarge; (esagera-

re) magnify. **~rsi** vr become larger; (aumentare) increase

ingras'sar|e vt fatten up; Mech grease ● vi, **~si** vr put on weight

ingrati'tudine nf ingratitude.

in'grato a ungrateful; (sgradevole) thankless

ingrazi'arsi vr ingratiate oneself with

ingredi'ente nm ingredient

in'gresso nm entrance; (accesso) admittance; (sala) hall; **~ gratuito/libero** admission free; **vietato l'~** no entry; no admittance

ingros'sar|e vt make big; (gonfiare) swell ● vi, **~si** vr grow big; (gonfiare) swell

in'grosso: all'~ adv wholesale; (pressappoco) roughly

ingua'ribile a incurable

'inguine nm groin

ingurgi'tare vt gulp down

ini'bi|re vt inhibit; (vietare) forbid. **~to** a inhibited. **~zi'one** nf inhibition; (divieto) prohibition

iniet'tar|e vt inject. **~si** vr **~si di sangue** (occhi:) become bloodshot. **iniezi'one** nf injection

inimi'carsi vr make an enemy of. **inimi'cizia** nf enmity

inimi'tabile a inimitable

ininter'rotta'mente adv continuously. **~'rotto** a continuous

iniquità nf iniquity. **i'niquo** a iniquitous

inizi'al|e a & nf initial. **~'mente** adv initially

inizi'are vt begin; (avviare) open; **~ qcno a qcsa** initiate sb in sth ● vi begin

inizia'tiva nf initiative; **prendere l'~** take the initiative

inizi'a|to, -a a initiated ● nmf initiate; **gli ~ti** the initiated. **~'tore, ~'trice** nmf initiator. **~zi'one** nf initiation

i'nizio nm beginning, start; **dare ~** a start; **avere ~** get under way

innaffi'a|re vt water. **~'toio** nm watering-can

innal'zar|e vt raise; (*erigere*) erect. **~si** vr rise

innamo'ra|rsi vr fall in love (**di** with). **~ta** nf girl-friend. **~to** a in love ● nm boy-friend

in'nanzi adv (*stato in luogo*) in front; (*di tempo*) ahead; (*avanti*) forward; (*prima*) before; **d'ora** ~ from now on ● *prep* (*prima*) before; ~ **a** in front of. **~'tutto** adv first of all; (*soprattutto*) above all

in'nato a innate

innatu'rale a unnatural

inne'gabile a undeniable

innervo'sir|e vt make nervous. **~si** vr get irritated

inne'scare vt prime. **in'nesco** nm primer

inne'stare vt graft; *Mech* engage; (*inserire*) insert. **in'nesto** nm graft; *Mech* clutch; *Electr* connection

inne'vato a covered in snow

'inno nm hymn. **~ nazionale** national anthem

inno'cen|te a innocent **~te'mente** adv innocently. **~za** nf innocence.

in'nocuo a innocuous

inno'va|re vt make changes in. **~'tivo** a innovative. **~'tore** a trailblazing. **~zi'one** nf innovation

innume'revole a innumerable

ino'doro a odourless

inoffen'sivo a harmless

inol'trar|e vt forward. **~si** vr advance

inol'trato a late

i'noltre adv besides

inon'da|re vt flood. **~zi'one** nf flood

inope'roso a idle

inoppor'tuno a untimely

inorgo'glir|e vt make proud. **~si** vr become proud

inorri'dire vt horrify ● vi be horrified

inospi'tale a inhospitable

inosser'vato a unobserved; (*non rispettato*) disregarded; **passare** ~ go unnoticed

inossi'dabile a stainless

'inox a inv (*acciaio*) stainless

inqua'dra|re vt frame; *fig* put in context (*scrittore, problema*). **~rsi** vr fit into. **~'tura** nf framing

inqualifi'cabile a unspeakable

inquie'tar|e vt worry. **~si** vr get worried; (*impazientirsi*) get cross. **inqui'eto** a restless; (*preoccupato*) worried. **inquie'tudine** nf anxiety

inqui'lino, -a nmf tenant

inquina'mento nm pollution

inqui'na|re vt pollute. **~to** a polluted

inqui'rente a *Jur* (*magistrato*) examining; **commissione** ~ commission of enquiry

inqui'si|re vt/i investigate. **~to** a under investigation. **~tore, ~'trice** a inquiring ● nmf inquisitor. **~zi'one** nf inquisition

insab'biare vt shelve

insa'la|ta nf salad. **~a belga** endive. **~i'era** nf salad bowl

insa'lubre a unhealthy

insa'nabile a incurable

insangui'na|re vt cover with blood. **~to** a bloody

insa'pone vt soap

insa'po|re a tasteless. **~'rire** vt flavour

insa'puta nf all'~ di unknown to

insazi'abile a insatiable

insce'nare vt stage

inscin'dibile a inseparable

inse'dia|mento nm installation

inse'dia|rsi vr install. **~si** vr install oneself

in'segna nf sign; (*bandiera*) flag; (*decorazione*) decoration; (*emblema*) insignia pl; (*stemma*) symbol. **~ luminosa** neon sign

insegna'mento nm teaching.

inse'gnante a teaching ● nmf teacher

inse'gnare vt/i teach; ~ **qcsa a qcno** teach sb sth

insegui'mento nmf pursuit

insegui're vt pursue. **~'tore, ~'trice** nmf pursuer

inselvati'chir|e vt make wild ● vi. **~si** vr grow wild

insemi'na|re vt inseminate. **~zi'o-ne** nf insemination. **~zione artifi-ciale** artificial insemination

insena'tura nf inlet

insen'sato a senseless; (folle) crazy

insen'sibil|e a insensitive; (braccio ecc) numb. **~ità** nf insensitivity

insepa'rabile a inseparable

inseri'mento nm insertion

inse'rir|e vt insert; place (annuncio); Electr connect. **~si** vr **~si in** get into. **in'serto** nm file; (in un film ecc) insert

inservi'ente nmf attendant

inserzi'o|ne nf insertion; (avviso) advertisement. **~'nista** nmf advertiser

insetti'cida nm insecticide

in'setto nm insect

insicu'rezza nf insecurity. **insi'cu-ro** a insecure

in'sid|ia nf trick; (tranello) snare. **~'are** vt/i lay a trap for. **~'oso** a insidious

insi'eme adv together; (contemporaneamente) at the same time ● prep **~ a** [together] with ● nm whole; (completo) outfit; Theat ensemble; Math set; **nell'~** as a whole; **tutto ~** all together; (bere) at one go

in'signe a renowned

insignifi'cante a insignificant

insi'gnire vt decorate

insinda'cabile a final

insinu'ante a insinuating

insinu'a|re vt insinuate. **~rsi** vr penetrate; **~rsi in** fig creep into. **~zi'one** nf insinuation

in'sipido a insipid

insi'sten|te a insistent. **~te'men-te** adv repeatedly. **~za** nf insistence. **in'sistere** vi insist; (perseverare) persevere

insoddisfa'cente a unsatisfactory

insoddi'sfa|tto a unsatisfied; (scontento) dissatisfied. **~zi'one** nf dissatisfaction

insoffe'ren|te a intolerant. **~za** nf intolerance

insolazi'one nf sunstroke

inso'len|te a rude, insolent. **~za** nf rudeness, insolence; (commento) insolent remark

in'solito a unusual

inso'lubile a insoluble

inso'luto a unsolved; (non pagato) unpaid

insol'ven|za nf insolvency

in'somma adv in short; **~!** well really!; (così così) so so

in'sonne a sleepless. **~ia** nf insomnia

insonno'lito a sleepy

insonoriz'zato a soundproofed

insoppor'tabile a unbearable

insor'genza nf onset

in'sorgere vi revolt, rise up; (sorgere) arise; (difficoltà) crop up

insormon'tabile a (ostacolo, difficoltà) insurmountable

in'sorto pp di **insorgere** ● a rebellious ● nm rebel

insospet'tabile a unsuspected

insospet'tir|e vt make suspicious ● vi, **~si** vr become suspicious

insoste'nibile a untenable; (insopportabile) unbearable

insostitu'ibile a irreplaceable

inspe'ra|bile a una sua vittoria è **~bile** there is no hope of him winning. **~to** a unhoped-for

inspie'gabile a inexplicable

inspi'rare vt breathe in

in'stabil|e a unstable; (tempo) changeable. **~ità** nf instability; (di tempo) changeability

instal'la|re vt install. **~rsi** vr settle in. **~zi'one** nf installation

instan'cabile a untiring

instau'ra|re vt found. **~rsi** vr become established. **~zi'one** nf foundation

instra'dare vt direct

insù adv all'**~** upwards

insubordinazi'one nf insubordination

insuc'cesso nm failure

insudici'ar|e *vt* dirty. **~si** *vr* get dirty

insuffici'en|te *a* insufficient; (*inadeguato*) inadequate ● *nf Sch* fail. **~za** *nf* insufficiency; (*inadeguatezza*) inadequacy; *Sch* fail. **~za cardiaca** heart failure. **~za di prove** lack of evidence

insu'lare *a* insular

insu'lina *nf* insulin

in'sulso *a* insipid; (*sciocco*) silly

insul'tare *vt* insult. **in'sulto** *nm* insult

insupe'rabile *a* insuperable; (*eccezionale*) incomparable

insurrezi'one *nf* insurrection

insussi'stente *a* groundless

intac'care *vt* nick; (*corrodere*) corrode; draw on (*capitale*); (*danneggiare*) damage

intagli'are *vt* carve. **in'taglio** *nm* carving

intan'gibile *a* untouchable

in'tanto *adv* meanwhile; (*per ora*) for the moment; (*avversativo*) but; **~ che** while

intarsi'a|re *vt* inlay. **~to a ~to di** inset with. **in'tarsio** *nm* inlay

inta'sa|re *vt* clog; block (*traffico*). **~rsi** *vr* get blocked. **~to a** blocked

inta'scare *vt* pocket

in'tatto *a* intact

intavo'lare *vt* start

inte'gra|le *a* whole; **edizione ~le** unabridged edition; **pane ~le** wholemeal bread. **~l'mente** *adv* fully. **~nte** *a* integral. **In'tegro** *a* complete; (*retto*) upright

inte'gra|re *vt* integrate; (*aggiungere*) supplement. **~rsi** *vr* integrate. **~'tivo** *a* (*corso*) supplementary. **~zi'one** *nf* integration

integrità *nf* integrity

intelaia'tura *nf* framework

intel'letto *nm* intellect

intellettu'al|e *a & nmf* intellectual. **~'mente** *adv* intellectually

intelli'gen|te *a* intelligent. **~'mente** *adv* intelligently. **~za** *nf* intelligence

intelli'gibile *a* intelligible. **~'mente** *adv* intelligibly

intempe'ranza *nf* intemperance

intem'perie *nfpl* bad weather

inten'den|te *nm* superintendent. **~za** *nf* **~za di finanza** inland revenue office

in'tender|e *vt* (*comprendere*) understand; (*udire*) hear; (*avere intenzione*) intend; (*significare*) mean. **~sela con** have an understanding with; **~si** *vr* (*capirsi*) understand each other; **~si di** (*essere esperto*) have a good knowledge of

inten|di'mento *nm* understanding; (*intenzione*) intention. **~'tore**, **~'trice** *nmf* connoisseur

intene'ri|re *vt* soften; (*commuovere*) touch. **~si** *vr* be touched

intensa'mente *adv* intensely

intensifi'car|e *vt*, **~si** *vr* intensify

intensità *nf inv* intensity. **inten'sivo** *a* intensive. **in'tenso** *a* intense

inten'tare *vt* start up; **~ causa contro qcno** bring o institute proceedings against sb

in'tento *a* engrossed (**a in**) ● *nm* purpose

intenzio'nato *a* **essere ~ a fare qcsa** have the intention of doing sth

intenzio'nale *a* intentional. **inten'zi'one** *nf* intention; **senza ~ne** unintentionally; **avere ~ne di fare qcsa** intend to do sth, have the intention of doing sth.

intera'gire *vi* interact

intera'mente *adv* completely, entirely

interat'tivo *a* interactive. **~zi'one** *nf* interaction

interca'lare[1] *nm* stock phrase

interca'lare[2] *vt* insert

intercambi'abile *a* interchangeable

interca'pedine *nf* cavity

inter'ce|dere *vi* intercede. **~ssi'one** *nf* intercession

intercet'ta|re *vt* intercept; tap (*telefono*). **~zi'one** *nf* interception. **~zione telefonica** telephone tapping

inter'city nm inv inter-city

intercontinen'tale a intercontinental

inter'correre vi (tempo:) elapse; (esistere) exist

interco'stale a intercostal

inter'detto pp di interdire ● a astonished; (proibito) forbidden; **rimanere ~** be taken aback

inter'dire vt forbid; Jur deprive of civil rights. **~zi'one** nf prohibition

interessa'mento nm interest

interes'sante a interesting; **essere in stato ~** be pregnant

interes'sare vt interest; (riguardare) concern ● vi **~re a matter** to. **~rsi** vr **~rsi a** take an interest in. **~rsi di** take care of. **~to, -a** nmf interested party ● a (interessato); **essere ~to** pej have an interest

inte'resse nm interest; **fare qcsa per ~** do sth out of self-interest

inter'faccia nf Comput interface

interfe'renza nf interference

interfe'rire vi interfere

interiezi'one nf interjection

interi'ora nfpl entrails

interi'ore a interior

inter'ludio nm interlude

intermedi'ario, -a a & nmf intermediary

inter'medio a in-between

inter'mezzo nm Theat, Mus intermezzo

intermi'nabile a interminable

intermit'ten|te a intermittent; (luce) flashing. **~za** nf luce a **~za** flashing light

interna'mento nm internment; (in manicomio) committal

inter'nare vt intern; (in manicomio) commit [to a mental institution]

in'terno a internal; Geog inland; (interiore) inner; (politica) national; **alunno ~** boarder ● nm interior; (di condominio) flat; Teleph

extension; Cinema interior shot; **all'~** inside

internazio'nale a international

in'tero a whole, entire; (intatto) intact; (completo) complete; **per ~ in** full

interpel'lare vt consult

inter'por|re vt place (ostacolo). **~si** vr come between

interpre'ta|re vt interpret; Mus perform. **~zi'one** nf interpretation; Mus performance. **in'terprete** nmf interpreter; Mus performer

inter'ra|re vt (seppellire) bury; plant (pianta, seme). **~to** nm basement

interro'ga|re vt question; Sch test; examine (studenti). **~'tivo** a interrogative; (sguardo) questioning; **punto ~tivo** question mark ● nm question. **~'torio** a & nm questioning. **~zi'one** nf question; Sch oral [test]

inter'romper|e vt interrupt; (sospendere) stop; cut off (collegamento). **~si** vr break off

interrut'tore nm switch

interruzi'one nf interruption; **senza ~** non-stop. **~ di gravidanza** termination of pregnancy

interse'|care vt, **~carsi** vr intersect. **~zi'one** nf intersection

inter'stizio nm interstice

interur'ban|a nf long-distance call. **~o** a inter-city; **telefonata ~a** long-distance call

inter'vallo nm space out. (spazio) space; Sch break. **intervallo pubblicitario** commercial break

interve'nire vi intervene; (Med: operare) operate; **~ a** take part in. **inter'vento** nm intervention; (presenza) presence; (chirurgico) operation; **pronto intervento** emergency services

inter'vista nf interview

intervi'sta|re vt interview. **~'tore, -'trice** nmf interviewer

in'tes|a nf understanding; **cenno**

d'~a acknowledgement. ~o *pp di*
intendere ● *a* resta ~o che...
needless to say,...; ~il agreed!; ~o
a meant to; **non darsi per ~o**
refuse to understand

inte'sta|re *vt* head; write one's
name and address at the top of
⟨*lettera*⟩; Comm register. ~rsi *vr*
~rsi a fare qcosa take it into one's
head to do sth. ~'tario, -a *nmf*
holder. ~zi'one *nf* heading; (*su
carta da lettere*) letterhead

intesti'nale *a* intestinal

inte'stino *a* ⟨*lotte*⟩ internal ●*nm*
intestine

intima'mente *adv* ⟨*conoscere*⟩ inti-
mately

inti'ma|re *vt* order; ~**re l'alt a**
qcno order sb to stop. ~**zi'one** *nf*
order

intimida'torio *a* threatening.
~**zi'one** *nf* intimidation

intimi'dire *vt* intimidate

intimità *nf* cosiness. **'intimo** *a* intimate; (*interno*) innermost; ⟨*ami-
co*⟩ close ●*nm* ⟨*amico*⟩ close
friend; (*dell'animo*) heart

intimo'ri|re *vt* frighten. ~**rsi** *vr* get
frightened. ~**to** *a* frightened

in'tingere *vt* dip

in'tingolo *nm* sauce; (*pietanza*)
stew

intiriz'zi|re *vt* numb. ~**rsi** *vr* grow
numb. ~**to a essere** ~**to** (*dal
freddo*) be perished

intito'lar|e *vt* entitle; (*dedicare*)
dedicate. ~**si** *vr* be called

intolle'rabile *a* intolerable

intona'care *vt* plaster. **in'tonaco**
nm plaster

into'na|re *vt* start to sing; tune
⟨*strumento*⟩; (*accordare*) match.
~**rsi** *vr* match. ~**to a** ⟨*persona*⟩
able to sing in tune; ⟨*colore*⟩
matching

intonazi'one *nf* (*inflessione*) intonation; (*ironico*) tone

inton'ti|re *vt* daze; ⟨*gas:*⟩ make
dizzy ●*vi* be dazed. ~**to** *a* dazed

intop'pare *vi* ~ **in** run into

in'toppo *nm* obstacle

---

in'torno *adv* around ●*prep* ~ **a**
around; (*circa*) about

intorpi'di|re *vt* numb. ~**rsi** *vr* be-
come numb. ~**to** *a* torpid

intossi'ca|re *vt* poison. ~**rsi** *vr* be
poisoned. ~**zi'one** *nf* poisoning

intral'ciare *vt* hamper

in'tralcio *nm* hitch; **essere d'~** be
a hindrance (*a*)

intrallaz'zare *vi* intrigue. **intral-
'lazzo** *nm* racket

intramon'tabile *a* timeless

intramusco'lare *a* intramuscular

intransi'gen|te *a* intransigent,
uncompromising. ~**za** *nf* intran-
sigence

intransi'tivo *a* intransitive

intrappo'lato a rimanere ~ be
trapped

intrapren'den|te *a* enterprising.
~**za** *nf* initiative

intra'prendere *vt* undertake

intrat'tabile *a* very difficult

intratte'n|ere *vt* entertain. ~**ersi**
vr linger. ~**i'mento** *nm* entertainment

intrave'dere *vt* catch a glimpse of;
(*presagire*) foresee

intrecci'ar|e *vt* interweave; plait
⟨*capelli, corda*⟩. ~**si** *vr* intertwine;
(*aggrovigliarsi*) become tangled;
~**e le mani** clasp one's hands

in'treccio *nm* (*trama*) plot

in'trepido *a* intrepid

intri'cato *a* tangled

intri'gante *a* scheming; (*affascinante*) intriguing

intri'gar|e *vt* entangle; (*incuriosi-
re*) intrigue ●*vi* intrigue, scheme.
~**rsi** *vr* meddle. **in'trigo** *nm* plot;
intrighi *pl* intrigues

in'trinseco *a* intrinsic

in'triso *a* ~ **di** soaked in

intri'stirsi *vr* grow sad

intro'du|rre *vt* introduce; (*inseri-
re*) insert; ~**rre a** (*iniziare a*)
introduce to. ~**rsi** *vr* get in (**in**
to). ~**t'tivo** *a* (*pagine, discorso*) introductory. ~**zi'one** *nf* introduction

in'troito *nm* income, revenue; (*in-casso*) takings *pl*

intro'metter|e *vt* introduce. **~si** *vr* interfere; (*interporsi*) intervene. **intromissi'one** *nf* intervention

intro'vabile *a* that can't be found; (*prodotto*) unobtainable

intro'verso, -a *a* introverted ● *nmf* introvert

intrufo'larsi *vr* sneak in

in'truglio *nm* concoction

intrusi'one *nf* intrusion. **in'truso, -a** *nmf* intruder

intu'i|re *vt* perceive

intui'tiva|mente *adv* intuitively. **~'tivo** *a* intuitive. **in'tuito** *nm* intuition. **~zi'one** *nf* intuition

inuguagli'anza *nf* inequality

inu'mano *a* inhuman

inu'mare *vt* inter

inumi'dir|e *vt* dampen; moisten (*labbra*). **~si** *vr* become damp

i'nutil|e *a* useless; (*superfluo*) unnecessary. **~ità** *nf* uselessness

inutiliz'za|bile *a* unusable. **~to** *a* unused

inutil'mente *adv* fruitlessly

inva'dente *a* intrusive

in'vadere *vt* invade; (*affollare*) overrun

invali'd|are *vt* invalidate. **~ità** *nf* disability; *Jur* invalidity. **in'vali-do, -a** *a* invalid; (*handicappato*) disabled ● *nmf* disabled person

in'vano *adv* in vain

invari'abil|e *a* invariable

invari'ato *a* unchanged

invasi'one *nf* invasion. **in'vaso** *pp* *di* invadere. **inva'sore** *a* invading ● *nm* invader

invecchia'mento *nm* (*di vino*) maturation

invecchi'are *vt/i* age

in'vece *adv* instead; (*anzi*) but; **~ di** instead of

inve'ire *vi* **~ contro** inveigh against

inven'd|ibile *a* unsaleable. **~uto** *a* unsold

inven'tare *vt* invent

inventari'are *vt* make an inventory of. **inven'tario** *nm* inventory

inven'tivo, -a *a* inventive ● *nf* inventiveness. **~'tore, ~'trice** *nmf* inventor. **~zi'one** *nf* invention

inver'nale *a* wintry. **in'verno** *nm* winter

invero'simile *a* improbable

inversa'mente *adv* inversely; **~ proporzionale** in inverse proportion

inversi'one *nf* inversion; *Mech* reversal. **in'verso** *a* inverse; (*opposto*) opposite ● *nm* opposite

inverte'brato *a & nm* invertebrate

inver'ti|re *vt* reverse; (*capovolgere*) turn upside down. **~to, -a** *nmf* homosexual

investi'ga|re *vt* investigate. **~'tore** *nm* investigator. **~zi'one** *nf* investigation

investi'mento *nm* investment; (*incidente*) crash

inve'sti|re *vt* invest; (*urtare*) collide with; (*travolgere*) run over; **~ qcno di** invest sb with. **~'tura** *nf* investiture

invet'tiva *nf* invective

invi'a|re *vt* send. **~to, -a** *nmf* envoy; (*di giornale*) correspondent

invidi|a *nf* envy. **~'are** *vt* envy. **~'oso** *a* envious

invigo'rir|e *vt* invigorate. **~si** *vr* become strong

invin'cibile *a* invincible

in'vio *nm* dispatch; *Comput* enter

invio'labile *a* inviolable

invipe'ri|rsi *vr* get nasty. **~to** *a* furious

invi'sibil|e *a* invisible. **~ità** *nf* invisibility

invi'tante *a* (*piatto, profumo*) enticing

invi'ta|re *vt* invite. **~to, -a** *nmf* guest. **in'vito** *nm* invitation

invo'ca|re *vt* invoke; (*implorare*) beg. **~zi'one** *nf* invocation

invogli'ar|e *vt* tempt; (*indurre*) induce. **~si** *vr* **~si di** take a fancy to

involon'taria|mente *adv* involuntarily. **~'tario** *a* involuntary

invol'tino nm Culin beef olive

in'volto nm parcel; (fagotto) bundle

in'volucro nm wrapping

invulne'rabile a invulnerable

inzacche'rare vt splash with mud

inzup'par|e vt soak; (intingere) dip. **~si** vr get soaked

'io pers pron I; **chi è?** – **[sono] io** who is it? – [it's] me; **l'ho fatto io [stesso]** I did it myself ● nm **l'~** the ego

i'odio nm iodine

l'onio nm **lo** ~ the Ionian [Sea]

i'osa: a ~ adv in abundance

iperat'tivo a hyperactive

ipermer'cato nm hypermarket

iper'metrope a long-sighted

ipersen'sibile a hypersensitive

ipertensi'one nf high blood pressure

ip'no|si nf hypnosis. **~tico** a hypnotic. **~'tismo** nm hypnotism. **~tiz'zare** vt hypnotize

ipoca'lorico a low-calorie

ipocon'driaco, -a a & nmf hypochondriac

ipocri'sia nf hypocrisy. **i'pocrita** a hypocritical ● nmf hypocrite

ipo'te|ca nf mortgage. **~'care** vt mortgage

i'potesi nf inv hypothesis; (caso, eventualità) eventuality. **ipo'teti-co a** hypothetical. **ipotiz'zare** vt hypothesize

'ippico, -a a horse attrib ● nf riding

ippoca'stano nm horse-chestnut

ip'podromo nm racecourse

ippo'potamo nm hippopotamus

'ira nf anger. **~'scibile** a irascible

i'rato a irate

'iride nf Anat iris; (arcobaleno) rainbow

Ir'lan|da nf Ireland. **~da del Nord** Northern Ireland. **i~'dese** a Irish ● nm Irishman; (lingua) Irish ● nf Irishwoman

iro'nia nf irony. **i'ronico** a ironic[al]

irradia|re vt/i radiate. **~zi'one** nf radiation

irraggiun'gibile a unattainable

irragio'nevole a unreasonable; (speranza, timore) irrational; (assurdo) absurd

irrazio'nal|e a irrational. **~ità** a irrationality. **~'mente** adv irrationally

irre'a|le a unreal. **~'listico** a unrealistic. **~liz'zabile** a unattainable. **~ità** nf unreality

irrecupe'rabile a irrecoverable

irrego'lar|e a irregular. **~ità** nf inv irregularity

irremo'vibile a fig adamant

irrepa'rabile a irreparable

irrepe'ribile a not to be found; **sarò** ~ I won't be contactable

irrepren'sibile a irreproachable

irrepri'mibile a irrepressible

irrequi'eto a restless

irresi'stibile a irresistible

irrespon'sabil|e a irresponsible. **~ità** nf irresponsibility

irrever'sibile a irreversible

irrevo'cabile a irrevocable

irricono'scibile a unrecognizable

irri'ga|re vt irrigate; (fiume:) flow through. **~zi'one** nf irrigation

irrigi'dimento nm stiffening

irrigi'dir|e vt, **~si** vr stiffen

irrile'vante a unimportant

irrime'diabile a irreparable

irripe'tibile a unrepeatable

irri'sorio a derisive; (differenza, particolare, somma) insignificant

irri'ta|bile a irritable. **~nte** a aggravating

irri'ta|re vt irritate. **~rsi** vr get annoyed. **~to** a irritated; (gola) sore. **~zi'one** nf irritation

irrobu'stir|e vt fortify. **~si** vr get stronger

ir'rompere vi burst (**in** into)

irro'rare vt sprinkle

irru'ente a impetuous

irruzi'one nf **fare** ~ **in** burst into

i'scritto, -a pp di **iscrivere** ● a registered ● nmf member; **per** ~ in writing

i'scriver|e vt register. **~si** vr ~**si a** register at, enrol at (scuola); join *(circolo ecc)*. **iscrizi'one** *nf* registration; *(epigrafe)* inscription

i'sla|mico *a* Islamic. **~'mismo** *nm* Islam

i'slan|da *nf* Iceland. **i~'dese** *a* Icelandic ● *nmf* Icelander

'isola *nf* island. **le isole britanniche** the British Isles. **~ pedonale** traffic island. **~ spartitraffico** traffic island. **iso'lano, -a** *a* insular ● *nmf* islander

iso'lante *a* insulating ● *nm* insulator

iso'la|re *vt* isolate; *Mech, Electr* insulate; *(acusticamente)* soundproof. **~to** *a* isolated ● *nm (di appartamenti)* block

ispes'sir|e *vt, ~si* *vr* thicken

ispet'to|rato *nm* inspectorate. **ispet'tore** *nm* inspector. **ispezio'nare** *vt* inspect. **ispezi'one** *nf* inspection

'ispido *a* bristly

ispi'ra|re *vt* inspire; suggest *(idea, soluzione)*. **~rsi** *vr* **~rsi a** be based on. **~to** *a* inspired. **~zi'one** *nf* inspiration; *(idea)* idea

Isra'el|e *nm* Israel. **i~i'ano, -a** *a* & *nmf* Israeli

is'sare *vt* hoist

istan'taneo, -a *a* instantaneous ● *nf* snapshot

i'stante *nm* instant; **all'~** instantly

i'stanza *nf* petition

i'sterico *a* hysterical. **iste'rismo** *nm* hysteria

isti'ga|re *vt* instigate; **~re qcno al male** incite sb to evil. **~'tore, ~'trice** *nmf* instigator. **~zi'one** *nf* instigation

istin'tiva'mente *adv* instinctively. **~'tivo** *a* instinctive. **i'stinto** *nm* instinct; **d'istinto** instinctively

istitu'ire *vt* institute; *(fondare)* found; initiate *(manifestazione)*

isti'tu|to *nm* institute; *(universitario)* department; *Sch* secondary school. **~to di bellezza** beauty salon. **~'tore, ~'trice** *nmf (insegnante)* tutor; *(fondatore)* founder

istituzio'nale *a* institutional. **istituzi'one** *nf* institution

'istmo *nm* isthmus

'istrice *nm* porcupine

istru'i|re *vt* instruct; *(addestrare)* train; *(informare)* inform; *Jur* prepare. **~to** *a* educated

istrut'tivo *a* instructive. **~ore, ~rice** *nmf* instructor; **giudice ~ore** examining magistrate. **~oria** *nf Jur* investigation. **istruzi'one** *nf* education; *(indicazione)* instruction

I'tali|a *nf* Italy. **i~'ano, -a** *a* & *nmf* Italian

itine'rario *nm* route, itinerary

itte'rizia *nf* jaundice

'ittico *a* fishing *attrib*

I.V.A. *nf abbr (imposta sul valore aggiunto)* VAT

Jj

jack *nm inv* jack

jazz *nm* jazz. **jaz'zista** *nmf* jazz player

jeep *nf inv* jeep

'jolly *nm inv (carta da gioco)* joker

Jugo'slav|ia *nf* Yugoslavia. **j~o, -a** *a* & *nmf* Yugoslav[ian]

ju'niores *nmfpl Sport* juniors

Kk

ka'jal *nm inv* kohl

kara'oke *nm inv* karaoke

kara'te *nm* karate

kg *abbr (chilogrammo)* kg

km *abbr (chilometro)* km

LI

l' *def art mf* (*before vowel*) the; *vedi* **il**

la *def art f* the; *vedi* **il** ● *pron* (*oggetto, riferito a persona*) her; (*riferito a cosa, animale*) it; (*forma di cortesia*) you ● *nm inv Mus* (*chiave, nota*) A

là *adv* there; **di là** (*in quel luogo*) in there; **eccolo là!** there he is!; **farsi più in là** (*far largo*) make way; **là dentro** in there; **là fuori** out there; **[ma] va là!** come off it!; **più in là** (*nel tempo*) later on; (*nello spazio*) further on

'labbro *nm* (*pl mf Anat* **labbra**) lip

labi'rinto *nm* labyrinth; (*di sentieri ecc*) maze

labora'torio *nm* laboratory; (*di negozio, officina ecc*) workshop

labori'oso *a* (*operoso*) industrious; (*faticoso*) laborious

labu'rista *a* Labour ● *nmf* member of the Labour Party

'lacca *nf* lacquer; (*per capelli*) hairspray, lacquer. **lac'care** *vt* lacquer

'laccio *nm* noose; (*lazo*) lasso; (*trappola*) snare; (*stringa*) lace

lace'rante *a* (*grido*) earsplitting

lace'rare *vt* tear; lacerate (*carne*). **~rsi** *vr* tear. **~zi'one** *nf* laceration. **'lacero** *a* torn; (*cencioso*) ragged

la'conico *a* laconic

'lacri|ma *nf* tear; (*goccia*) drop. **~'mare** *vi* weep. **~'mevole** *a* tearjerking

lacri'mogeno *a* **gas** **~** tear gas

lacri'moso *a* tearful

la'cuna *nf* gap. **lacu'noso** *a* (*preparazione, resoconto*) incomplete

la'custre *a* lake *attrib*

'ladro, -a *nmf* thief; **al ~!** stop

thief! **~'cinio** *nm* theft. **la'druncolo** *nm* petty thief

'lager *nm inv* concentration camp

laggiù *adv* down there; (*lontano*) over there

'lagna *nf* (*fam: persona*) moaning Minnie; (*film*) bore

la'gna|nza *nf* complaint. **~rsi** *vr* moan; (*protestare*) complain (**di** about). **la'gnoso** *a* (*persona*) moaning

'lago *nm* lake

la'guna *nf* lagoon

'laico, -a *a* lay; (*vita*) secular ● *nm* layman ● *nf* laywoman

'lama *nf* blade ● *nm inv* (*animale*) llama

lambic'carsi *vr* **~ il cervello** rack one's brains

lam'bire *vt* lap

lamé *nm inv* lamé

lamen'tar|e *vt* lament. **~si** *vr* moan. **~si di** (*lagnarsi*) complain about

lamen'te|la *nf* complaint. **~vole** *a* mournful; (*pietoso*) pitiful. **la'mento** *nm* moan

la'metta *nf* **~** [**da barba**] razor blade

lami'era *nf* sheet metal

'lamina *nf* foil. **~ d'oro** gold leaf

lami'na|re *vt* laminate. **~to** *a* laminated ● *nm* laminate; (*tessuto*) lamé

'lampa|da *nf* lamp. **~da abbronzante** sunlamp. **~da a pila** torch. **~'dario** *nm* chandelier. **~'dina** *nf* light bulb

lam'pante *a* clear

lampeggi'a|re *vi* flash. **~'tore** *nm* Auto indicator

lampi'one *nm* street lamp

'lampo *nm* flash of lightning; (*luce*) flash; **lampi** *pl* lightning *sg*. **~ di genio** stroke of genius. (*cerniera*) **~** zip [*fastener*], zipper *Am*

lam'pone *nm* raspberry

'lana *nf* wool; **di ~** woollen. **~ d'acciaio** steel wool. **~ vergine** new wool. **~ di vetro** glass wool

lan'cetta nf pointer; (di orologio) hand

'lancia nf (arma) spear, lance; Naut launch

lanci'ar|e vt throw; (da un aereo) drop; launch (missile, prodotto); give (grido); **~e uno sguardo a** glance at. **~si** vr fling oneself; (intraprendere) launch out

lanci'nante a piercing

'lancio nm throwing; (da aereo) drop; (di missile, prodotto) launch. **~ del disco** discus [throwing]. **~ del giavellotto** javelin [throwing]. **~ del peso** putting the shot

'landa nf heath

languido a languid

langu'ore nm languor

lani'ero a wool

lani'ficio nm woollen mill

lan'terna nf lantern; (faro) lighthouse

la'nugine nf down

lapi'dare vt stone; fig demolish

lapi'dario a (conciso) terse

'lapide nf tombstone; (commemorativa) memorial tablet

'lapis nm inv pencil

'lapsus nm inv lapse, error

'lardo nm lard

larga'mente adv (ampiamente) widely

lar'ghezza nf width, breadth; fig liberality. **~ di vedute** broadmindedness

'largo a wide; (ampio) broad; (abito) loose; (liberale) liberal; (abbondante) generous; **stare alla larga** keep away; **~ di manica** fig generous; **~ di spalle/vedute** broad-shouldered/-minded ● nm width; **andare al ~** Naut go out to sea; **fare ~** make room; **farsi ~** make one's way; **al ~ di** off the coast of

'larice nm larch

la'ringe nf larynx. **larin'gite** nf laryngitis

'larva nf larva; (persona emaciata) shadow

la'sagne nfpl lasagna sg

lasciapas'sare nm inv pass

lasci'ar|e vt leave; (rinunciare) give up; (rimetterci) lose; (smettere di tenere) let go [of]; (concedere) let; **~e di fare qcsa** (smettere) stop doing sth; **lascia perdere!** forget it!; **lasci**alo venire, lascia che venga let him come. **~si** vr (reciproco) leave each other, split up; **~si andare** let oneself go

'lascito nm legacy

'laser a & nm inv (raggio) **~** laser [beam]

lassa'tivo a & nm laxative

'lasso nm. **~ di tempo** period of time

lassù adv up there

'lastra nf slab; (di ghiaccio) sheet; (di metallo, Phot) plate; (radiografia) X-ray [plate]

lastri'car|e vt pave. **~to, 'lastrico** nm pavement; **sul lastrico** on one's beam-ends

la'tente a latent

late'rale a side attrib; Med, Techn ecc lateral; **via ~** side street

late'rizi nmpl bricks

lati'fondo nm large estate

la'tino a & nm Latin

lati'tan|te a in hiding ● nmf fugitive [from justice]

lati'tudine nf latitude

'lato a (ampio) broad; **in senso ~** broadly speaking ● nm side; (aspetto) aspect; **a ~ di** beside; **dal ~ mio** (punto di vista) for my part; **d'altro ~** fig on the other hand

la'tra|re vi bark. **~to** nm barking

la'trina nf latrine

'latta nf tin, can

lat'taio, -a nm milkman ● nf milkwoman

lat'tante a breast-fed ● nmf suckling

'latt|e nm milk. **~e acido** sour milk. **~e condensato** condensed milk. **~e detergente** cleansing milk. **~e in polvere** powdered milk. **~e scremato** skimmed milk. **~eo** a milky. **~e'ria** nf dairy. **~i'cini** nmpl dairy products. **~i'era** nf milk jug

lat'tina nf can
lat'tuga nf lettuce
lau'rea nf degree; **prendere la ~a** graduate. **~'ando, -a** nmf final-year student
laure'a|rsi vr graduate. **~to, -a** a & nmf graduate
'lauro nm laurel
'lauto a lavish; **~ guadagno** handsome profit
'lava nf lava
la'vabile a washable
la'vabo nm wash-basin
la'vaggio nm washing. **~ automatico** (per auto) carwash. **~ del cervello** brainwashing. **~ a secco** dry-cleaning
la'vagna nf slate; Sch blackboard
la'van|da nf wash; Bot lavender; **fare una ~da gastrica** have one's stomach pumped. **~'daia** nf washerwoman. **~de'ria** nf laundry. **~deria automatica** launderette
lavan'dino nm sink; (hum: persona) bottomless pit
lavapi'atti nmf inv dishwasher
la'var|e vt wash; **~e i piatti** wash up. **~si** vr wash, have a wash; **~si i denti** brush one's teeth; **~si le mani** wash one's hands
lava'secco nmf inv dry-cleaner's
lavasto'viglie nf inv dishwasher
la'vata nf wash; **darsi una ~** have a wash; **~ di capo** fig scolding
lava'tivo, -a nmf idler
lava'trice nf washing-machine
lavo'rante nmf worker
lavo'ra|re vi work ● vt work; knead (pasta ecc); till (la terra); **~re a maglia** knit. **~to** a working. **~to** a (pietra, legno) carved; (cuoio) tooled; (metallo) wrought. **~tore, ~trice** nmf worker ● a working. **~zi'one** nf manufacture; (di terra) working; (artigianale) workmanship; (del terreno) cultivation. **lavo'rio** nm intense activity
la'voro nm work; (faticoso, sociale) labour; (impiego) job; Theat play; **mettersi al ~** set to work (su

on). **~ a maglia** knitting. **~ nero** moonlighting. **~ straordinario** overtime. **~ a tempo pieno** full-time job. **lavori** pl di casa housework. **lavori** pl in corso roadworks. **lavori** pl forzati hard labour. **lavori** pl stradali roadworks
le def art fpl the; vedi **il** ● pers pron (oggetto) them; (a lei) her; (forma di cortesia) you
le'al|e a loyal. **~'mente** adv loyally. **~tà** nf loyalty
'lebbra nf leprosy
'lecca 'lecca nm inv lollipop
leccapi'edi nmf inv pej bootlicker
lec'ca|re vt lick; fig suck up to. **~rsi** vr lick; (fig: agghindarsi) doll oneself up; **da ~rsi i baffi** mouth-watering. **~ta** nf lick
leccor'nia nf delicacy
'lecito a lawful; (permesso) permissible
'ledere vt damage; Med injure
'lega nf league; (di metalli) alloy; **far ~ con** qcno take up with sb
le'gaccio nm string; (delle scarpe) shoelace
le'gal|e a legal ● nm lawyer. **~ità** nf legality. **~iz'zare** vt authenticate; (rendere legale) legalize. **~'mente** adv legally
le'game nm tie; (amoroso) liaison; (connessione) link
lega'mento nm Med ligament
le'gar|e vt tie up (persona); tie together (due cose); (unire, rilegare) bind; alloy (metalli); (connect; **~sela al dito** fig bear a grudge ● vi (far lega) get on well. **~si** vr bind oneself; **~si a qcno** become attached to sb
le'gato nm legacy; Relig legate
lega'tura nf tying; (di libro) binding
le'genda nf legend
'legge nf law; (parlamentare) act; **a norma di ~** by law
leg'genda nf legend; (didascalia) caption. **leggen'dario** a legendary
'leggere vt/i read

legge'r|ezza *nf* lightness; (*frivolezza*) frivolity; (*incostanza*) fickleness. **~'mente** *adv* slightly

leg'gero *a* light; (*bevanda*) weak; (*lieve*) slight; (*frivolo*) frivolous; (*incostante*) fickle; **alla leggera** frivolously

leg'gibile *a* (*scrittura*) legible; (*stile*) readable

leg'gio *nm* lectern; *Mus* music stand

legife'rare *vi* legislate

legio'nario *nm* legionary. **legi'one** *nf* legion

legisla'tivo *a* legislative. **~'tore** *nm* legislator. **~'tura** *nf* legislature. **~zi'one** *nf* legislation

legittimità *nf* legitimacy. **le'gittimo** *a* legitimate; (*giusto*) proper; **legittima difesa** self-defence

'legna *nf* firewood

le'gname *nm* timber

le'gnata *nf* blow with a stick

'legno *nm* wood; **di ~** wooden. **~ compensato** plywood. **le'gnoso** *a* woody

le'gume *nm* pod

'lei *pers pron* (*soggetto*) she; (*oggetto, con prep*) her; (*forma di cortesia*) you; **lo ha fatto ~ stessa** she did it herself

'lembo *nm* edge; (*di terra*) strip

'lemma *nm* headword

'lena *nf* vigour

le'nire *vt* soothe

lenta'mente *adv* slowly

'lente *nf* lens. **~ a contatto** contact lens. **~ d'ingrandimento** magnifying glass

len'tezza *nf* slowness

len'ticchia *nf* lentil

len'tiggine *nf* freckle

'lento *a* slow; (*allentato*) slack; (*abito*) loose

'lenza *nf* fishing-line

len'zuolo *nm* (*pl f* **lenzuola**) *nm* sheet

le'one *nm* lion; *Astr* Leo

leo'pardo *nm* leopard

'lepre *nf* hare

'lercio *a* filthy

'lesbica *nf* lesbian

lesi'nare *vt* grudge ● *vi* be stingy

lesio'nare *vt* damage. **lesi'one** *nf* lesion

'leso *pp* **di ledere** ● *a* injured

'lessare *vt* boil

'lessico *nm* vocabulary

'lesso *a* boiled ● *nm* boiled meat

'lesto *a* quick; (*mente*) sharp

le'tale *a* lethal

leta'maio *nm* dunghill; *fig* pigsty.

le'tame *nm* dung

le'targico *a* lethargic. **~o** *nm* lethargy; (*di animali*) hibernation

le'tizia *nf* joy

'lettera *nf* letter; **alla ~** literally; **~ maiuscola** capital letter; **~ minuscola** small letter; **lettere** *pl* (*letteratura*) literature *sg*; *Univ* Arts; **dottore in lettere** BA, Bachelor of Arts

lette'rale *a* literal

lette'rario *a* literary

lette'rato *a* well-read

lettera'tura *nf* literature

let'tiga *nf* stretcher

let'tino *nm* cot; *Med* couch

'letto *nm* bed. **~ a castello** bunkbed. **~ a una piazza** single bed. **~ a due piazze** double bed. **~ matrimoniale** double bed

letto'rato *nm* (*corso*) ≈ tutorial

let'tore, **-'trice** *nmf* reader; *Univ* language assistant ● *nm Comput* disk drive. **~ di CD-ROM** CD-Rom drive

let'tura *nf* reading

leuce'mia *nf* leukaemia

'leva *nf* lever; *Mil* call-up; **far ~** lever. **~ del cambio** gear lever

le'vante *nm* East; (*vento*) east wind

le'va|re *vt* (*alzare*) raise; (*togliere*) take away; (*rimuovere*) take off; (*estrarre*) pull out; **~rsi di mezzo** get sth out of the way. **~rsi** *vr* rise; (*da letto*) get up; **~rsi di mezzo**, **~rsi dai piedi** get out of the way. **~ta** *nf* rising; (*di posta*) collection

leva'taccia *nf* fare una ~ get up at the crack of dawn

leva'toio *a* ponte ~ drawbridge

levi'ga|re *vt* smooth; (con carta vetro) rub down. **~to** *a* (superficie) polished

levri'ero *nm* greyhound

lezi'one *nf* lesson; *Univ* lecture; (rimprovero) rebuke

lezi'oso *a* (stile, modi) affected

li *pers pron mpl* them

lì *adv* there; **fin lì** as far as there; **giù di lì** thereabouts; **lì per lì** there and then

Li'bano *nm* Lebanon

'libbra *nf* (peso) pound

li'beccio *nm* south-west wind

li'bellula *nf* dragon-fly

libe'rale *a* liberal; (generoso) generous ●*nmf* liberal

libe'ra|re *vt* free; release (prigioniero); vacate (stanza); (salvare) rescue. **~rsi** *vr* (stanza:) become vacant; *Teleph* become free; (da impegno) get out of it; **~rsi di** get rid of. **~'tore, ~'trice** *a* liberating ●*nmf* liberator. **~'torio** *a* liberating. **~zi'one** *nf* liberation; **la L~zione** (ricorrenza) Liberation Day

'liber|o *a* free; (strada) clear. **~o docente** qualified university lecturer. **~o professionista** self-employed person. **~tà** *nf inv* freedom; (di prigioniero) release. **~tà provvisoria** *Jur* bail; **~tà** *pl* (confidenze) liberties

'liberty *nm & a inv* Art Nouveau

'Libi|a *nf* Libya. **l~co, -a** *a & nmf* Libyan

li'bidi|ne *nf* lust. **~'noso** *a* lustful.

li'bido *nf* libido

li'braio *nm* bookseller

libre'ria *nf* (negozio) bookshop; (mobile) bookcase; (biblioteca) library

li'bretto *nm* booklet; *Mus* libretto. **~ degli assegni** cheque book. **~ di circolazione** logbook. **~ d'istruzioni** instruction booklet. **~ di risparmio** bankbook. **~ uni-**

versitario book held by students which records details of their exam performances

'libro *nm* book. **~ giallo** thriller. **~ paga** payroll

lice'ale *nmf* secondary-school student ●*a* secondary-school *attrib*

li'cenza *nf* licence; (permesso) permission; *Mil* leave; *Sch* school-leaving certificate; **essere in ~** be on leave

licenzia'mento *nm* dismissal

licenzi'a|re *vt* dismiss, sack *fam*. **~rsi** *vr* (da un impiego) resign; (accomiatarsi) take one's leave

li'ceo *nm* secondary school, high school. **~ classico** secondary school with an emphasis on humanities. **~ scientifico** secondary school with an emphasis on sciences

'lichene *nm* lichen

'lido *nm* beach

li'eto *a* glad; (evento) happy; **molto ~!** pleased to meet you!

li'eve *a* light; (debole) faint; (trascurabile) slight

lievi'tare *vi* rise ●*vt* leaven. **li'evito** *nm* yeast. **lievito in polvere** baking powder

'lifting *nm inv* face-lift

'ligio *a* essere ~ al dovere have a sense of duty

'lilla *nf Bot* lilac ●*nm* (colore) lilac

'lima *nf* file

limacci'oso *a* slimy

li'mare *vt* file

'limbo *nm* limbo

li'metta *nf* nail-file

limi'ta|re *nm* threshold ●*vt* limit. **~rsi** *vr* **~rsi a fare qcsa** restrict oneself to doing sth; **~rsi in qcsa** cut down on sth. **~to** *a* limited. **~zi'one** *nf* limitation

'limite *nm* limit; (confine) boundary. **~ di velocità** speed limit

li'mitrofo *a* neighbouring

limo'nata *nf* (bibita) lemonade; (succo) lemon juice

li'mone *nm* lemon; (*albero*) lemon tree

'limpido *a* clear; (*occhi*) limpid

'lince *nf* lynx

linci'are *vt* lynch

'lindo *a* neat; (*pulito*) clean

'linea *nf* line; (*di autobus, aereo*) route; (*di metro*) line; (*di abito*) cut; (*di auto, mobile*) design; (*fisico*) figure; **in ~ d'aria** as the crow flies; **è caduta la ~** I've been cut off; **in ~ di massima** as a rule; **a grandi linee** in outline; **mantenere la ~** keep one's figure; **in prima ~** in the front line; **mettersi in ~** line up; **nave di ~** liner; **volo di ~** scheduled flight. **~ d'arrivo** finishing line. **~ continua** unbroken line

linea'menti *nmpl* features

line'are *a* linear; (*discorso*) to the point; (*ragionamento*) consistent

line'etta *nf* (*tratto lungo*) dash; (*d'unione*) hyphen

lin'gotto *nm* ingot

'lingu|a *nf* tongue; (*linguaggio*) language. **~'accia** *nf* (*persona*) backbiter. **~'aggio** *nm* language. **~'etta** *nf* (*di scarpa*) tongue; (*di strumento*) reed; (*di busta*) flap

lingu'ist|a *nmf* linguist. **~ica** *nf* linguistics *sg*. **~ico** *a* linguistic

'lino *nm* Bot flax; (*tessuto*) linen

li'noleum *nm* linoleum

liofiliz'za|re *vt* freeze-dry. **~to** *a* freeze-dried

liposuzi'one *nf* liposuction

lique'far|e *vt*, **~si** *vr* liquefy; (*sciogliersi*) melt

liqui'da|re *vt* liquidate; settle (*conto*); pay off (*debiti*); clear (*merce*); (*fam: uccidere*) get rid of. **~zi'one** *nf* liquidation; (*di conti*) settling; (*di merce*) clearance sale

'liquido *a & nm* liquid

liqui'rizia *nf* liquorice

li'quore *nm* liqueur; **liquori** *pl* (*bevande alcooliche*) liquors

'lira *nf* lira; *Mus* lyre

'lirico, -a *a* lyrical; (*poesia*) lyric;

(cantante, musica) opera *attrib* ● *nf* lyric poetry; *Mus* opera

'lisca *nf* fishbone; **avere la ~** (*fam: nel parlare*) have a lisp

lisci'are *vt* smooth; (*accarezzare*) stroke. **'liscio** *a* smooth; (*capelli*) straight; (*liquore*) neat; (*acqua minerale*) still; **passarla liscia** get away with it

'liso *a* worn [out]

'list|a *nf* list; (*striscia*) strip. **~ di attesa** waiting list; **in ~ di attesa** *Aeron* stand-by. **~ elettorale** electoral register. **~ nera** blacklist. **~ di nozze** wedding list. **li'stare** *vt* edge; *Comput* list

li'stino *nm* list. **~ prezzi** price list

Lit. *abbr* (**lire italiane**) Italian lire

'lite *nf* quarrel; (*baruffa*) row; *Jur* lawsuit

liti'gare *vi* quarrel. **li'tigio** *nm* quarrel. **litigi'oso** *a* quarrelsome

lito'rale *a* coastal ● *nm* coast

'litro *nm* litre

li'turgico *a* liturgical

li'vella *nf* level. **~ a bolla d'aria** spirit level

livel'lar|e *vt* level. **~si** *vr* level out

li'vello *nm* level; **passaggio a ~** level crossing; **sotto/sul ~ del mare** below/above sea level

'livido *a* livid; (*per il freddo*) blue; (*per una botta*) black and blue ● *nm* bruise

Li'vorno *nf* Leghorn

'lizza *nf* lists *pl*; **essere in ~ per qcsa** be in the running for sth

lo *def art m* (*before s* + *consonant, gn, ps, z*) the; *vedi* **il** ● *pron* (*riferito a persona*) him; (*riferito a cosa*) it; **non lo so** I don't know

'lobo *nm* lobe

lo'cal|e *a* local ● *nm* (*stanza*) room; (*treno*) local train; **~i** *pl* (*edifici*) premises. **~e notturno** night-club. **~ità** *nf inv* locality

localiz'zare *vt* localize; (*trovare*) locate

lo'canda *nf* inn

locan'dina *nf* bill, poster

loca'tario, -a *nmf* tenant. **~'tore,**

~'trice *nm* landlord ● *nf* landlady.
~zi'one *nf* tenancy
locomo'tiva *nf* locomotive.
~zi'one *nf* locomotion; **mezzi di
~zione** means of transport
'loculo *nm* burial niche
lo'custa *nf* locust
locuzi'one *nf* expression
lo'dare *vt* praise. **'lode** *nf* praise;
laurea con lode first-class degree
'loden *m inv* (*cappotto*) loden coat
lo'devole *a* praiseworthy
'lodola *nf* lark
'loggia *nf* loggia; (*massonica*)
lodge
loggi'one *nm* gallery, the gods
'logica *nf* logic
logica'mente *adv* (*in modo logico*)
logically; (*ovviamente*) of course
'logico *a* logical
lo'gistica *nf* logistics *sg*
logo'rante *a* (*esperienza*) wearing
logo'ra|re *vt* wear out; (*sciupare*)
waste. **~rsi** *vr* wear out; (*persona:*) wear oneself out. **logo'rio**
nm wear and tear. **'logoro** *a* worn-
out
lom'baggine *nf* lumbago
Lombar'dia *nf* Lombardy
lom'ba|ta *nf* loin. **'lombo** *nm Anat*
loin
lom'brico *nm* earthworm
'Londra *nf* London
lon'gevo *a* long-lived
longi'lineo *a* tall and slim
longi'tudine *nf* longitude
lontana'mente *adv* distantly; (*vagamente*) vaguely; **neanche ~** not
for a moment
lonta'nanza *nf* distance; (*separazione*) separation; **in ~** in the
distance
lon'tano *a* far; (*distante*) distant;
(*nel tempo*) far-off, distant; (*parente*) distant; (*vago*) vague; (*assente*) absent; **più ~** further ● *adv*
far [away]; **da ~** from a distance;
tenersi ~ da keep away from
'lontra *nf* otter
lo'quace *a* talkative
'lordo *a* dirty; (*somma, peso*) gross

'loro[1] *pron pl* (*soggetto*) they;
(*oggetto*) them; (*forma di cortesia*)
you; **sta a ~** it is up to them
'loro[2] (**il ~** *m*, **la ~** *f*, **i ~** *mpl*, **le ~**
fpl) *a* their; (*forma di cortesia*)
your; **un ~** **amico** a friend of
theirs; (*forma di cortesia*) a friend
of yours ● *pron* theirs; (*forma di
cortesia*) yours; **i ~** their folk
lo'sanga *nf* lozenge; **a losanghe**
diamond-shaped
'losco *a* suspicious
'loto *nm* lotus
'lott|a *nf* fight, struggle; (*contrasto*)
conflict; *Sport* wrestling. **lot'tare**
vi fight, struggle; *Sport, fig* wrestle. **~a'tore** *nm* wrestler
lotte'ria *nf* lottery
'lotto *nm* [national] lottery; (*porzione*) lot; (*di terreno*) plot
lozi'one *nf* lotion
lubrifi'ca|nte *a* lubricating ● *nm*
lubricant. **~re** *vt* lubricate
luc'chetto *nm* padlock
lucci'ca|nte *a* sparkling. **~re** *vi*
sparkle. **lucci'chio** *nm* sparkle
'luccio *nm* pike
'lucciola *nf* glow-worm
'luce *nf* light; **far ~ su** shed light
on; **dare alla ~** give birth to. **~
della luna** moonlight. **luci** *pl* di
posizione sidelights. **~ del sole**
sunlight
lu'cen|te *a* shining. **~'tezza** *nf*
shine
lucer'nario *nm* skylight
lu'certola *nf* lizard
'lucida'labbra *nm inv* lip gloss
luci'da|re *vt* polish. **~'trice** *nf*
[floor-]polisher. **'lucido** *a* shiny;
(*pavimento, scarpe*) polished;
(*chiaro*) clear; (*persona, mente*)
lucid; (*occhi*) watery ● *nm* shine.
lucido [**da scarpe**] [shoe] polish
lucra'tivo *a* lucrative. **'lucro** *nm*
lucre
'luglio *nm* July
'lugubre *a* gloomy
'lui *pers pron* (*soggetto*) he; (*oggetto,
con prep*) him; **lo ha fatto ~
stesso** he did it himself

lu'maca nf (mollusco) snail; fig slowcoach

'lume nm lamp; (luce) light; **a ~ di candela** by candlelight

luminosità nf brightness. **lumi-'noso** a luminous; (stanza, cielo ecc) bright

'luna nf moon; **chiaro di ~** moonlight; **avere la ~ storta** be in a bad mood. **~ di miele** honeymoon

luna park nm inv fairground

lu'nare a lunar

lu'nario nm almanac; **sbarcare il ~** make both ends meet

lu'natico a moody

lunedì nm inv Monday

lu'netta nf half-moon [shape]

lun'gaggine nf slowness

lun'ghezza nf length. **~ d'onda** wavelength

'lungi adv **ero [ben] ~ dall'imma-ginare che...** I never dreamt for a moment that...

lungimi'rante a far-sighted, far-seeing

'lungo a long; (diluito) weak; (lento) slow; **saperla lunga** be shrewd ● nm length; **di gran lunga** by far; **andare per le lunghe** drag on ● prep (durante) throughout; (per la lunghezza di) along

lungofi'ume nm riverside

lungo'lago nm lakeside

lungo'mare nm sea front

lungome'traggio nm feature film

lu'notto nm rear window

lu'ogo nm place; (punto preciso) spot; (passo d'autore) passage; **aver ~** take place; **dar ~ a** give rise to; **del ~** (usanze) local. **~ comune** platitude. **~ pubblico** public place

luogote'nente nm Mil lieutenant

lu'petto nm Cub [Scout]

'lupo nm wolf

'luppolo nm hop

'lurido a filthy. **luri'dume** nm filth

lu'singa nf flattery

lusin'g|are vt flatter. **~arsi** vr flatter oneself; (illudersi) fool oneself. **~hi'ero** a flattering

lus'sa|re vt, **~rsi** vr dislocate. **~zi'one** nf dislocation

Lussem'burgo nm Luxembourg

'lusso nm luxury; **di ~** luxury attrib

lussu'oso a luxurious

lussureggi'ante a luxuriant

lus'suria nf lust

lu'strare vt polish

lu'strino nm sequin

'lustro a lustrous ● nm sheen; fig prestige; (quinquennio) five-year period

'lutt|o nm mourning; **~o stretto** deep mourning. **~u'oso** a mournful

••••••••••••••••••••••••••••••••••••

Mm

m abbr (metro) m

ma conj but; (eppure) yet; **ma!** (dubbio) I don't know; (indignazione) really!; **ma davvero?** really?; **ma sì!** why not!; (certo che sì) of course!

'macabro a macabre

macché int of course not!

macche'roni nmpl macaroni sg

macche'ronico a (italiano) broken

'macchia¹ nf stain; (di diverso colore) spot; (piccola) speck; **senza ~** spotless

'macchia² nf (boscaglia) scrub; **darsi alla ~** take to the woods

macchi'a|re vt, **~rsi** vr stain. **~to a** (caffè) with a dash of milk; **~to di** (sporco) stained with

'macchina nf machine; (motore) engine; (automobile) car. **~ da cucire** sewing machine. **~ da presa** cine camera. **~ da scrivere** typewriter

macchinal'mente adv mechanically

macchi'nare vt plot

macchi'nario nm machinery

macchi'netta nf (per i denti) brace

macchi'nista *nm* Rail engine-driver; *Naut* engineer; *Theat* stagehand

macchi'noso *a* complicated

mace'donia *nf* fruit salad

macel'la|io *nm* butcher. **~re** *vt* slaughter, butcher. **macelle'ria** *nf* butcher's [shop]. **ma'cello** *nm* (*mattatoio*) slaughterhouse; *fig* shambles *sg*; **andare al macello** *fig* go to the slaughter; **mandare al macello** *fig* send to his/her death

mace'rar|e *vt* macerate; *fig* distress. **~si** *vr* be consumed

ma'cerie *nfpl* rubble *sg*; (*rottami*) debris *sg*

ma'cigno *nm* boulder

maci'lento *a* emaciated

'macina *nf* millstone

macinacaffè *nm* inv coffee mill

macina'pepe *nm* inv pepper mill

maci'na|re *vt* mill. **~to** *a* ground ● *nm* (*carne*) mince. **maci'nino** *nm* mill; (*hum: macchina*) old banger

maciul'lare *vt* (*stritolare*) crush

macrobiotic|a *nf* **negozio di ~a** health-food shop. **~o** *a* macrobiotic

macro'scopico *a* macroscopic

macu'lato *a* spotted

'madido *a* ~ **di** moist with

Ma'donna *nf* Our Lady

mador'nale *a* gross

'madre *nf* mother. **~'lingua** *a* inv inglese **~'lingua** English native speaker. **~'patria** *nf* native land. **~'perla** *nf* mother-of-pearl

ma'drina *nf* godmother

maestà *nf* majesty

maestosità *nf* majesty. **mae'stoso** *a* majestic

mae'strale *nm* northwest wind

mae'stranza *nf* workers *pl*

mae'stria *nf* mastery

ma'estro, -a *nmf* teacher ● *nm* master; *Mus* maestro. **~ di cerimonie** master of ceremonies ● *a* (*principale*) chief; (*di grande abilità*) skilful

'mafi|a *nf* Mafia. **~'oso** *a* of the Mafia ● *nm* member of the Mafia, Mafioso

'maga *nf* sorceress

ma'gagna *nf* fault

ma'gari *adv* (*forse*) maybe ● *int* I wish! ● *conj* (*per esprimere desiderio*) if only; (*anche se*) even if

magazzini'ere *nm* storesman, warehouseman. **magaz'zino** *nm* warehouse; (*emporio*) shop; **grande magazzino** department store

'maggio *nm* May

maggio'lino *nm* May bug

maggio'rana *nf* marjoram

maggio'ranza *nf* majority

maggio'rare *vt* increase

maggior'domo *nm* butler

maggi'ore *a* (*di dimensioni, numero*) bigger, larger; (*superlativo*) biggest, largest; (*di età*) older; (*superlativo*) oldest; (*di importanza, Mus*) major; (*superlativo*) greatest; **la maggior parte di** most; **la maggior parte del tempo** most of the time ● *pron* (*di dimensioni*) the bigger, the larger; (*superlativo*) the biggest, the largest; (*di età*) the older; (*superlativo*) the oldest; (*di importanza*) the major; (*superlativo*) the greatest ● *nm* Mil major; *Aeron* squadron leader.

maggio'renne *a* of age ● *nmf* adult

maggiori'tario *a* (*sistema*) first-past-the-post attrib. **~'mente** *adv* [all] the more; (*più di tutto*) most

'Magi *nmpl* **i re ~** the Magi

ma'gia *nf* magic; (*trucco*) magic trick **magica'mente** *adv* magically. **'magico** *a* magic

magi'stero *nm* (*insegnamento*) teaching; (*maestria*) skill; **facoltà di ~** arts faculty

magi'strale *a* masterly; **istituto ~e** teachers' training college

magi'stra|to *nm* magistrate. **~'tura** *nf* magistrature. **la ~tura** the Bench

'magli|a *nf* stitch; (*lavoro ai ferri*) knitting; (*tessuto*) jersey; (*di rete*) mesh; (di catena) link; (indumento) vest; **fare la ~ a** knit. **~a diritta** knit. **~a rosa** (ciclismo) ≈ yellow jersey. **~a rovescia** purl. **~e'ria** nf knitwear. **~'etta** nf **~etta [a maniche corte]** tee-shirt. **~'ficio** nm knitwear factory. **ma'glina** nf (tessuto) jersey

magli'one nm sweater

'**magma** nm magma

ma'gnanimo a magnanimous

ma'gnate nm magnate

ma'gnesi|a nf magnesia. **~o** nm magnesium

ma'gne|te nm magnet. **~tico** a a magnetic. **~'tismo** nm magnetism

magne'tofono nm tape recorder

magnifi|ca'mente adv magnificently. **~'cenza** nf magnificence; (generosità) munificence. **ma'gnifico** a magnificent; (generoso) munificent

ma'gnolia nf magnolia

'**mago** nm magician

ma'gone nm **avere il ~ be down; mi è venuto il ~** I've got a lump in my throat

'**magr|a** nf low water. **ma'grezza** nf thinness. **~o** a thin; (carne) lean; (scarso) meagre

'**mai** adv never; (inter, talvolta) ever; **caso ~** if anything; **caso ~ tornasse** in case he comes back; **come ~?** why?; **cosa ~?** what on earth?; **~ più** never again; **più che ~** more than ever; **quando ~?** whenever?; **quasi ~** hardly ever

mai'ale nm pig; (carne) pork

mai'olica nf majolica

maio'nese nf mayonnaise

'**mais** nm maize

mai'uscol|a nf capital [letter]. **~o** a capital

mal vedi **male**

'**mala** nf la ~ sl the underworld

mala'fede nf bad faith

malaf'fare nm **gente di ~** shady characters pl

mala'lingua nf backbiter

mala'mente adv (ridotto) badly

malan'dato a in bad shape; (di salute) in poor health

ma'lanimo nm ill will

ma'lanno nm misfortune; (malattia) illness; **prendersi un ~** catch something

mala'pena: a ~ adv hardly

ma'laria nf malaria

mala'ticcio a sickly

ma'lato, -a a ill, sick; (pianta) diseased ● nmf sick person. **~ di mente** mentally ill person.

malat'tia nf disease, illness; **ho preso due giorni di malattia** I had two days off sick. **malattia venerea** venereal disease

malaugu'rato a ill-omened. **malau'gurio** nm bad o ill omen

mala'vita nf underworld

mala'voglia nf unwillingness; **di ~** unwillingly

malcapi'tato a wretched

malce'lato a ill-concealed

mal'concio a battered

malcon'tento nm discontent

malco'stume nm immorality

mal'destro a awkward; (inesperto) inexperienced

maldi'cen|te a slanderous. **~za** nf slander

maldi'sposto a ill-disposed

'**male** adv badly; **funzionare ~** not work properly; **star ~** be ill; **star ~ a qcno** (vestito ecc.) not suit sb; **rimanerci ~** be hurt; **non c'è ~!** not bad at all! ● nm evil; (dolore) pain; (malattia) illness; (danno) harm. **distinguere il bene dal ~** know right from wrong; **andare a ~** go off; **aver ~ a** have a pain in; **dove hai ~?** where does it hurt?; **far ~ a qcno** (provocare dolore) hurt sb; (cibo:) be bad for sb; **le cipolle mi fanno ~** onions don't agree with me; **mi fa ~ la schiena** my back is hurting; **mal d'auto** car-sickness. **mal di denti** toothache. **mal di gola** sore throat. **mal di mare** sea-sickness; **avere il mal di mare** be sea-sick. **mal di pan-**

cia stomach ache. **mal di testa** headache

male'detto a cursed; ⟨*orribile*⟩ awful

male'di|re vt curse. **~zi'one** nf curse; **~zione!** damn!

maledu|cata'mente adv rudely. **~'cato** a ill-mannered. **~cazi'one** nf rudeness

male'fatta nf misdeed

male'ficio nm witchcraft. **ma- 'lefico** a ⟨*azione*⟩ evil; ⟨*nocivo*⟩ harmful

maleodo'rante a foul-smelling

ma'lessere nm indisposition; *fig* uneasiness

ma'levolo a malevolent

malfa'mato a of ill repute

mal'fat|to a badly done; ⟨*malformato*⟩ ill-shaped. **~'tore** nm wrongdoer

mal'fermo a unsteady; ⟨*salute*⟩ poor

malfor'ma|to a misshapen. **~zi'one** nf malformation

malgo'verno nm misgovernment

mal'grado prep in spite of ● conj although

ma'lia nf spell

mali'gn|are vi malign. **~ità** nf malice; *Med* malignancy. **ma'ligno** a malicious; ⟨*perfido*⟩ evil; *Med* malignant

malinco'ni|a nf melancholy. **~ca- 'mente** adv melancholically. **mal- in'conico** a melancholy

malincu'ore: a ~ adv unwillingly, reluctantly

malinfor'mato a misinformed

malintenzio'nato, -a nmf miscreant

malin'teso a mistaken ● nm misunderstanding

ma'lizi|a nf malice; ⟨*astuzia*⟩ cunning; ⟨*espediente*⟩ trick. **~'oso** a malicious; ⟨*birichino*⟩ mischievous

malle'abile a malleable

mal'loppo nm *fam* loot

malme'nare vt ill-treat

mal'messo a ⟨*vestito male*⟩ shab-

bily dressed; ⟨*casa*⟩ poorly furnished; ⟨*fig: senza soldi*⟩ hard up

malnu'tri|to a undernourished. **~zi'one** nf malnutrition

'malo a in **~ modo** badly

ma'locchio nm evil eye

ma'lora nf ruin; **della ~** awful; **andare in ~** go to ruin

ma'lore nm illness; **essere colto da ~** be suddenly taken ill

mairi'dotto a ⟨*persona*⟩ in a sorry state

mal'sano a unhealthy

'malta nf mortar

mal'tempo nm bad weather

'malto nm malt

maltrat|ta'mento nm ill-treatment. **~'tare** vt ill-treat

malu'more nm bad mood; **di ~** in a bad mood

mal'vagi|o a wicked. **~tà** nf wickedness

malversazi'one nf embezzlement

mal'visto a unpopular ⟨*da* with⟩

malvi'vente nm criminal

malvolenti'eri adv unwillingly

malvo'lere vt **farsi ~** make oneself unpopular

'mamma nf mummy, mum; **~ mia!** good gracious!

mam'mella nf breast

mam'mifero nm mammal

'mammola nf violet

ma'nata nf ⟨*handful*⟩; ⟨*colpo*⟩ slap

'manca nf vedi **manco**

manca'mento nm **avere un ~** faint

man'can|te a missing. **~za** nf lack; ⟨*assenza*⟩ absence; ⟨*insufficienza*⟩ shortage; ⟨*fallo*⟩ fault; ⟨*imperfezione*⟩ defect; **in ~za d'altro** failing all else; **sento la sua ~za** I miss him

man'care vi be lacking; ⟨*essere assente*⟩ be missing; ⟨*venir meno*⟩ fail; ⟨*morire*⟩ pass away; **~ di** be lacking in; **~ a** fail to keep ⟨*promessa*⟩; **mi manca casa** I miss home; **mi manchi** I miss you; **mi è mancato il tempo** I didn't have [the] time; **mi mancano**

1000 lire I'm 1,000 lire short; **quanto manca alla partenza?** how long before we leave?; **è mancata la corrente** there was a power failure; **sentirsi** ~ feel faint; **sentirsi** ~ **il respiro** be unable to breathe [properly] ● *vt* miss ⟨*bersaglio*⟩; **è mancato poco che cadesse** he nearly fell

'manche *nf inv* heat

man'chevole *a* defective

'mancia *nf* tip

manci'ata *nf* handful

man'cino *a* left-handed

'manco, -a *a* left ● *nf* left hand ● *adv* (*nemmeno*) not even

man'dante *nmf* (*di delitto*) instigator

manda'rancio *nm* clementine

man'dare *vt* send; (*emettere*) give off; utter ⟨*suono*⟩; ~ **a chiamare** send for; ~ **avanti la casa** run the house; ~ **giù** (*ingoiare*) swallow

manda'rino *nm* Bot mandarin

man'data *nf* (*di consignment*); (*di serratura*) turn; **chiudere a doppia** ~ double lock

man'dato *nm* (*incarico*) mandate; Jur warrant; (*di pagamento*) money order. ~ **di comparizione** [**in giudizio**] subpoena. ~ **di perquisizione** search warrant

man'dibola *nf* jaw

mando'lino *nm* mandolin

'mandorla *nf* almond; **a** ~**la** (*occhi*) almond-shaped. ~**lato** *nm* nut brittle (*type of nougat*). ~**lo** *nm* almond[-tree]

'mandria *nf* herd

maneg'gevole *a* easy to handle.

maneggi'are *vt* handle

ma'neggio *nm* handling; (*intrigo*) plot; (*scuola di equitazione*) riding school

ma'nesco *a* quick to hit out

ma'netta *nf* hand lever; **manette** *pl* handcuffs

man'forte *nm* **dare** ~ **a qcno** support sb

manga'nello *nm* truncheon

manga'nese *nm* manganese

mange'reccio *a* edible

mangia'dischi® *nm inv* type of portable record player

mangia'fumo *a inv* **candela** ~ air-purifier in the form of a candle

mangia'nastri *nm inv* cassette player

mangi'a|re *vt/i* eat; (*consumare*) eat up; (*corrodere*) eat away; take ⟨*scacchi, carte ecc*⟩ ● *nm* eating; (*cibo*) food; (*pasto*) meal. ~**rsi** *vr* ~**rsi le parole** mumble; ~**rsi le unghie** bite one's nails

mangi'ata *nf* big meal; **farsi una bella** ~ **di... feast on...**

mangia'toia *nf* manger

man'gime *nm* fodder

mangi'one, -a *nmf fam* glutton

mangiucchi'are *vt* nibble

'mango *nm* mango

ma'nia *nf* mania. ~ **di grandezza** delusions of grandeur. ~**co, -a** *a* maniacal ● *nmf* maniac

'manica *nf* sleeve; (*fam: gruppo*) band; **a maniche lunghe** long-sleeved; **essere in maniche di camicia** be in shirt sleeves; **essere di** ~ **larga** be free with one's money. ~ **a vento** wind sock

'Manica *nf* **la** ~ the [English] Channel

manica'retto *nm* tasty dish

mani'chetta *nf* hose

mani'chino *nm* (*da sarto, vetrina*) dummy

'manico *nm* handle; Mus neck

mani'comio *nm* mental home; (*fam: confusione*) tip

mani'cotto *nm* muff; Mech sleeve

mani'cure *nf* manicure ● *nmf inv* (*persona*) manicurist

mani'e|ra *nf* manner; **in** ~**ra che** so that. ~**rato** *a* affected; (*stile*) mannered. ~**rismo** *nm* mannerism

manifat'tura *nf* manufacture; (*fabbrica*) factory

manife'stante *nmf* demonstrator

manife'sta|re *vt* show; (*esprimere*) express ● *vi* demonstrate. ~**rsi** *vr* show oneself. ~**zi'one** *nf* show;

(espressione) expression; *(sintomo)* manifestation; *(dimostrazione pubblica)* demonstration

mani'festo *a* evident ● *nm* poster; *(dichiarazione pubblica)* manifesto

ma'niglia *nf* handle; *(sostegno, in autobus ecc)* strap

manipo'la|re *vt* handle; *(massaggiare)* massage; *(alterare)* adulterate; *fig* manipulate. **~'tore, ~'trice** *nmf* manipulator. **~zi'one** *nf* handling; *(massaggio)* massage; *(alterazione)* adulteration; *fig* manipulation

mani'scalco *nm* smith

man'naia *nf (scure)* axe; *(da macellaio)* cleaver

man'naro *a* lupo **~** werewolf

'mano *nf* hand; *(strato di vernice ecc)* coat; **alla ~** informal; **fuori ~** out of the way; **man ~** little by little; **man ~ che** as; **sotto ~** to hand

mano'dopera *nf* labour

ma'nometro *nm* gauge

mano'mettere *vt* tamper with; *(violare)* violate

ma'nopola *nf (di apparecchio)* knob; *(guanto)* mitten; *(su pullman)* handle

mano'scritto *a* handwritten ● *nm* manuscript

mano'vale *nf* labourer

mano'vella *nf* handle; *Techn* crank

ma'no|vra *nf* manoeuvre; *Rail* shunting; **fare le ~vre** *Auto* manoeuvre. **~'vrabile** *a fig* easy to manipulate. **~'vrare** *vt (azionare)* operate; *fig* manipulate *(persona)* ● *vi* manoeuvre

manro'vescio *nm* slap

man'sarda *nf* attic

mansi'one *nf* task; *(dovere)* duty

mansu'eto *a* meek; *(animale)* docile

man'tel|la *nf* cape. **~o** *nm* cloak; *(soprabito, di animale)* coat; *(di neve)* mantle

mante'ner|e *vt (conservare)* keep;

(in buono stato, sostentare) maintain. **~si** *vr* **~si in forma** keep fit.

manteni'mento *nm* maintenance

'mantice *nm* bellows *pl*; *(di automobile)* hood

'manto *nm* cloak; *(coltre)* mantle

manto'vana *nf (di tende)* pelmet

manu'al|e *a & nm* manual. **~e d'uso** user manual. **~'mente** *adv* manually

ma'nubrio *nm* handle; *(di bicicletta)* handlebars *pl*; *(per ginnastica)* dumb-bell

manu'fatto *a* manufactured

manutenzi'one *nf* maintenance

'manzo *nm* steer; *(carne)* beef

'mappa *nf* map

mappa'mondo *nm* globe

mar *vedi* **mare**

ma'rasma *nm fig* decline

mara'to|na *nf* marathon. **~'neta** *nmf* marathon runner

'marca *nf* mark; *Comm* brand; *(fabbricazione)* make; *(scontrino)* ticket. **~ da bollo** revenue stamp

mar'ca|re *vt* mark; *Sport* score. **~ta'mente** *adv* markedly. **~to** *a* *(tratto, accento)* strong, marked. **~'tore** *nm (nel calcio)* scorer

mar'chese, ~a *nm* marquis ● *nf* marchioness

marchi'are *vt* brand

'marchio *nm* brand; *(caratteristica)* mark. **~ di fabbrica** trademark. **~ registrato** registered trademark

'marcia *nf* march; *Auto* gear; *Sport* walk; **mettere in ~** put into gear; **mettersi in ~** start off. **~ funebre** funeral march. **~ indietro** reverse gear; **fare ~ indietro** reverse; *fig* back-pedal. **~ nuziale** wedding march

marciapi'ede *nm* pavement; *(di stazione)* platform

marci'a|re *vi* march; *(funzionare)* go, work. **~'tore, ~'trice** *nmf* walker

'marcio *a* rotten ● *nm* rotten part; *fig* corruption. **mar'cire** *vi* go bad, rot

'marco nm (moneta) mark

'mare nm sea; (luogo di mare) seaside; **sul** ~ (casa) at the seaside; (città) on the sea; **in alto** ~ on the high seas; **essere in alto** ~ fig not know which way to turn. ~ **Adriatico** Adriatic Sea. **mar Ionio** Ionian Sea. **mar Mediterraneo** Mediterranean. **mar Tirreno** Tyrrhenian Sea

ma'rea nf tide; **una** ~ **di** hundreds of; **alta** ~ high tide; **bassa** ~ low tide

mareggi'ata nf [sea] storm

mare'moto nm tidal wave, seaquake

maresci'allo nm (ufficiale) marshal; (sottufficiale) warrant-officer

marga'rina nf margarine

marghe'rita nf marguerite. **margheri'tina** nf daisy

margi'nale a marginal. ~'**mente** adv marginally

'margine nm margin; (orlo) brink; (bordo) border. ~ **di errore** margin of error. ~ **di sicurezza** safety margin

ma'rina nf navy; (costa) seashore; (quadro) seascape. ~ **mercantile** merchant navy. ~ **militare** navy

mari'naio nm sailor

mari'na|re vt marinate; ~**re la scuola** play truant. ~**ta** nf marinade. ~**to** a Culin marinated

ma'rino a sea attrib, marine

mario'netta nf puppet

ma'rito nm husband

ma'rittimo a maritime

mar'maglia nf rabble

marmel'lata nf jam; (di agrumi) marmalade

mar'mitta nf pot; Auto silencer. ~ **catalitica** catalytic converter

'marmo nm marble

mar'mocchio nm fam brat

mar'mor|eo a marble. ~**iz'zato** a marbled

mar'motta nf marmot

Ma'rocco nm Morocco

ma'roso nm breaker

mar'rone a brown ●nm brown; (castagna) chestnut; **marroni** pl **canditi** marrons glacés

mar'sina nf tails pl

mar'supio nm (borsa) bumbag

martedi nm inv Tuesday. ~ **grasso** Shrove Tuesday

martel'lante a (mal di testa) pounding

martel'la|re vt hammer ●vi throb. ~**ta** nf hammer blow

martel'letto nm (di giudice) gavel

mar'tello nm hammer; (di battente) knocker. ~ **pneumatico** pneumatic drill

marti'netto nm Mech jack

'martire nmf martyr. **mar'tirio** nm martyrdom

'martora nf marten

martori'are vt torment

mar'xis|mo nm Marxism. ~**ta** a & nmf Marxist

marza'pane nm marzipan

marzi'ale a martial

marzi'ano, -a nmf Martian

'marzo nm March

mascal'zone nm rascal

ma'scara nm inv mascara

mascar'pone nm full-fat cream cheese often used for desserts

ma'scella nf jaw

'mascher|a nf mask; (costume) fancy dress; Cinema, Theat usher m, usherette f; (nella commedia dell'arte) stock character. ~**a antigas** gas mask. ~**a di bellezza** face pack. ~**a ad ossigeno** oxygen mask. ~**a'mento** nm masking; Mil camouflage. **masche'rare** vt mask; fig camouflage. ~**arsi** vr put on a mask; ~**arsi da** dress up as. ~**ata** nf masquerade

maschi'accio nm (ragazza) tomboy

ma'schi|le a masculine; (sesso) male ●nm masculine [gender]. ~**lista** a sexist. **'maschio** a male; (virile) manly ●nm male; (figlio) son. **masco'lino** a masculine

ma'scotte nf inv mascot

maso'chis|mo *nm* masochism. **~ta** *a* & *nmf* masochist

'massa *nf* mass; *Electr* earth, ground *Am*; **comunicazioni di ~** mass media

massa'cra|nte *a* gruelling. **~re** *vt* massacre. **mas'sacro** *nm* massacre; *fig* mess

massaggi'a|re *vt* massage. **massaggio** *nm* massage. **~tore**, **~'trice** *nm* masseur ●*nf* masseuse

mas'saia *nf* housewife

masse'rizie *nfpl* household effects

mas'siccio *a* massive; *(oro ecc)* solid; *(corporatura)* heavy ●*nm* massif

'massim|a *nf* maxim; *(temperatura)* maximum. **~o** *a* greatest; *(quantità)* maximum, greatest ●*nm* il **~o** the maximum; **al ~o** at [the] most, as a maximum

'masso *nm* rock

mas'sone *nm* [Free]mason. **~'ria** Freemasonry

ma'stello *nm* wooden box for the grape or olive harvest

masti'care *vt* chew; *(borbottare)* mumble

'mastice *nm* mastic; *(per vetri)* putty

ma'stino *nm* mastiff

masto'dontico *a* gigantic

'mastro *nm* master; **libro ~** ledger

mastur'ba|rsi *vr* masturbate. **~zi'one** *nf* masturbation

ma'tassa *nf* skein

mate'matic|a *nf* mathematics, maths. **~o**, **-a** *a* mathematical ●*nmf* mathematician

materas'sino *nm* **~ gonfiabile** air bed

mate'rasso *nm* mattress. **~ a molle** spring mattress

ma'teria *nf* matter; *(materiale)* material; *(di studio)* subject. **~ prima** raw material

materi'a|le *a* material; *(grossolano)* coarse ●*nm* material. **~'lismo** *nm* materialism. **~'lista** *a* materialistic ●*nmf* materialist.

~liz'zarsi *vr* materialize. **~l'mente** *adv* physically

mater'nità *nf* motherhood; **ospedale di ~** maternity hospital

ma'terno *a* maternal; **lingua materna** mother tongue

ma'tita *nf* pencil

ma'trice *nf* matrix; *(origini)* roots *pl*; *Comm* counterfoil

ma'tricola *nf (registro)* register; *Univ* fresher

ma'trigna *nf* stepmother

matrimoni'ale *a* matrimonial; **vita ~** married life. **matri'monio** *nm* marriage; *(cerimonia)* wedding

ma'trona *nf* matron

'matta *nf (nelle carte)* joker

mattacchi'one, **-a** *nmf* rascal

matta'toio *nm* slaughterhouse

matte'rello *nm* rolling-pin

mat'ti|na *nf* morning; **la ~na** in the morning. **~'nata** *nf* morning; *Theat* matinée. **~ni'ero** *a* essere **~niero** be an early riser. **~no** *nm* morning

'matto, **-a** *a* mad, crazy; *Med* insane; *(falso)* false; *(opaco)* matt; **~ da legare** barking mad; **avere una voglia matta di** be dying for ●*nmf* madman; madwoman

mat'tone *nm* brick; *(libro)* bore

matto'nella *nf* tile

mattu'tino *a* morning *attrib*

matu'rare *vt* ripen. **maturità** *nf* maturity; *Sch* school-leaving certificate. **ma'turo** *a* mature; *(frutto)* ripe

ma'tusa *nf* old fogey

mauso'leo *nm* mausoleum

maxi+ *pref* maxi+

'mazza *nf* club; *(martello)* hammer; *(da baseball, cricket)* bat. **~ da golf** golf-club. **maz'zata** *nf* blow

maz'zetta *nf (di banconote)* bundle

'mazzo *nm* bunch; *(carte da gioco)* pack

me *pers pron* me; **me lo ha dato** he gave it to me; **fai come me** do as I

meandro | menare

do; **è più veloce di me** he is faster than me o faster than I am

me'andro nm meander

M.E.C. nm abbr (**Mercato Comune Europeo**) EEC

mec'canica nf mechanics sg

meccanica'mente adv mechanically

mec'canico a mechanical ● nm mechanic. **mecca'nismo** nm mechanism

mèche nfpl [farsi] **fare le ~** have one's hair streaked

me'daglia nf medal. **~'one** nm medallion; (gioiello) locket

me'desimo a same

'media nf average; Sch average mark; Math mean; **essere nella ~a** be in the mid-range. **~'ano** a middle ● nm (calcio) half-back

medi'ante prep by

media're vt act as intermediary in. **~'tore**, **~'trice** nmf mediator; Comm middleman. **~zi'one** nf mediation

medica'mento nm medicine

medi'care vt treat; dress (ferita). **~zi'one** nf medication; (di ferita) dressing

medi'cina nf medicine. **~ina legale** forensic medicine. **~i'nale** a medicinal ● nm medicine

'medico a medical ● nm doctor. **~ generico** general practitioner. **~ legale** forensic scientist. **~ di turno** duty doctor

medie'vale a medieval

'medio a average; (punto) middle; (statura) medium ● nm (dito) middle finger

medi'ocre a mediocre; (scadente) poor

medio'evo nm Middle Ages pl

medi'tare vt meditate; (progettare) plan; (considerare attentamente) think over ● vi meditate. **~zi'one** nf meditation

mediter'raneo a Mediterranean; **il [mar] M~** the Mediterranean [Sea]

me'dusa nf jellyfish

me'gafono nm megaphone

mega'lattico a fam gigantic

mega'lomane nmf megalomaniac

me'gera nf hag

'meglio adv better; **tanto ~**, **così** so much the better ● a better; (superlativo) best ● nmf best ● nf **avere la ~ su** have the better of; **fare qcsa alla [bell'e] ~** do sth as best one can ● nm **fare del proprio ~** do one's best; **fare qcsa il ~ possibile** make an excellent job of sth; **al ~** to the best of one's ability; **per il ~** for the best

'mela nf apple. **~ cotogna** quince

mela'grana nf pomegranate

mela'nina nf melanin

melan'zana nf aubergine, eggplant Am

me'lassa nf molasses sg

me'lenso a (persona, film) dull

mel'lifluo a (parole) honeyed; (voce) sugary

'melma nf slime. **mel'moso** a slimy

melo nm apple[-tree]

melo'dia nf melody. **me'lodico** a melodic. **~'oso** a melodious

melo'dram|ma nm melodrama. **~'matico** a melodramatic

melo'grano nm pomegranate tree

me'lone nm melon

mem'brana nf membrane

'membro nm member; (pl nf **membra** Anat) limb

memo'rabile a memorable

'memore a mindful; (riconoscente) grateful

me'moria nf memory; (oggetto ricordo) souvenir. **imparare a ~a** learn by heart. **~a permanente** Comput non-volatile memory. **~a tampone** Comput buffer. **~a volatile** Comput volatile memory; **memorie** pl (biografiche) memories. **~'ale** nm memorial. **~z'zare** vt memorize; Comput save, store

mena'dito: a ~ adv perfectly

me'nare vt lead; (fam: picchiare) hit

mendi'ca|nte *nmf* beggar. **~re** *vt/i* beg

menefre'ghista *a* devil-may-care

me'ningi *nfpl* spremersi le **~** rack one's brains

menin'gite *nf* meningitis

me'nisco *nm* meniscus

'meno *adv* less; *(superlativo)* least; *(in operazioni, con temperatura)* minus; **far** qcsa alla **~ peggio** do sth as best one can; **fare a ~ di** qcsa do without sth; **non posso fare a ~ di ridere** I can't help laughing; **~ male!** thank goodness!; **sempre ~** less and less; **venir ~** *(svenire)* faint; **venir ~ a** qcno *(coraggio:)* fail sb; **sono le tre ~ un quarto** it's a quarter to three; **che tu venga o ~** whether you're coming or not; **quanto ~** at least ● *a inv* less; *(con nomi plurali)* fewer ● *nm* least; *Math* minus sign; **il ~ possibile** as little as possible; **per lo ~** at least ● *prep* except [for] ● *conj* a **~ che** unless

meno'ma|re *vt (incidente:)* maim. **~to** *a* disabled

meno'pausa *nf* menopause

'mensa *nf* table; *Mil* mess; *Sch, Univ* refectory

men'sil|e *a* monthly ● *nm (stipendio)* [monthly] salary; *(rivista)* monthly. **~ità** *nf inv* monthly salary. **~'mente** *adv* monthly

'mensola *nf* bracket; *(scaffale)* shelf

'menta *nf* mint. **~ peperita** peppermint

men'tal|e *a* mental. **~ità** *nf inv* mentality

'mente *nf* mind; **a ~ fredda** in cold blood; **venire in ~ a** qcno occur to sb; **mi è uscito di ~** it slipped my mind

men'tina *nf* mint

men'tire *vi* lie

'mento *nm* chin

'mentre *conj (temporale)* while; *(invece)* whereas

menù *nm inv* menu. **~ fisso** set

menu. **~ a tendina** *Comput* pulldown menu

menzio'nare *vt* mention. **menzi'one** *nf* mention

men'zogna *nf* lie

mera'viglia *nf* wonder; **a ~** marvellously; **che ~!** how wonderful!; **con mia grande ~** much to my amazement; **mi fa ~ che...** I am surprised that...

meravigli'ar|e *vt* surprise. **~si** *vr* **~si di** be surprised at

meravigli'osa|mente *adv* marvellously. **~'oso** *a* marvellous

mer'can|te *nm* merchant. **~teggi'are** *vi* trade; *(sul prezzo)* bargain. **~tile** *a* mercantile. **~'zia** *nf* merchandise, goods *pl* ● *nm* merchant ship

mer'cato *nm* market; *Fin* market [-place]. **a buon ~** *(comprare)* cheap[ly]; *(articolo)* cheap. **~ dei cambi** foreign exchange market. **M~ Comune [Europeo]** [European] Common Market. **~ coperto** covered market. **~ libero** free market. **~ nero** black market

'merce *nf* goods *pl*

mercé *nf* alla **~ di** at the mercy of

merce'nario *a & nm* mercenary

merce'ria *nf* haberdashery; *(negozio)* haberdasher's

mercoledì *nm inv* Wednesday. **~ delle Ceneri** Ash Wednesday

mer'curio *nm* mercury

me'renda *nf* afternoon snack; **far ~** have an afternoon snack

meridi'ana *nf* sundial

meridi'ano *a* midday ● *nm* meridian

meridio'nale *a* southern ● *nmf* southerner. **meridi'one** *nm* south

me'ringa *nf* meringue. **~'gata** *nf* meringue pie

meri'tare *vt* deserve. **~'tevole** *a* deserving

'meri|to *nm* merit; *(valore)* worth; **in ~to a** as to; **per ~to di** thanks to. **~'torio** *a* meritorious

mer'letto *nm* lace

'**merlo** nm blackbird
mer'luzzo nm cod
'**mero** a mere
meschine'ria nf meanness. **me-**
'**schino** a wretched; (gretto) mean
● nm wretch
mesco|la'mento nm mixing.
~ **lanza** nf mixture
mesco|la're vt mix; shuffle
⟨carte⟩; (confondere) mix up; blend
⟨tè, tabacco ecc⟩. ~**rsi** vr mix;
(immischiarsi) meddle. ~**ta** nf (a
carte) shuffle; Culin stir
'**mese** nm month
me'setto nm un ~ about a month
'**messa¹** nf Mass
'**messa²** nf (il mettere) putting. ~
in moto Auto starting. ~ **in piega**
(di capelli) set. ~ **a punto** adjustment. ~ **in scena** production. ~ **a**
terra earthing, grounding Am
messag'gero nm messenger. **mes-**
'**saggio** nm message
mes'sale nm missal
'**messe** nf harvest
Mes'sia nm Messiah
messi'cano, -a a & nmf Mexican
'**Messico** nm Mexico
messin'scena nf staging; fig act
'**messo** pp di **mettere** ● nm messenger
mesti'ere nm trade; (lavoro) job;
essere del ~ be an expert, know
one's trade
'**mesto** a sad
'**mestola** nf (di cuoco) ladle
mestru'a|le a menstrual. ~**zi'one**
nf menstruation. ~**zi'oni** pl period
'**meta** nf destination; fig aim
metà nf inv half; (centro) middle; a
~ **strada** half-way; **fare a** ~ **con**
qcno go halves with sb
metabo'lismo nm metabolism
meta'done nm methadone
meta'fisico a metaphysical
me'tafora nf metaphor. **meta-**
'**forico** a metaphorical
me'talli|co a metallic. ~**z'zato** a
⟨grigio⟩ metallic

me'tall|o nm metal. ~**ur'gia** nf
metallurgy
metalmec'canico a engineering
● nm engineering worker
meta'morfosi nf metamorphosis
me'tano nm methane. ~'**dotto** nm
methane pipeline
meta'nolo nm methanol
me'teora nf meteor. **meteo'rite** nm
meteorite
meteoro|lo'gia nf meteorology.
~'**logico** a meteorological
me'ticcio, -a nmf half-caste
metico'loso a meticulous
me'tod|ico a methodical. '**metodo**
nm method. ~**olo'gia** nf method-
ology
me'traggio nm length (in metres)
'**metrico, -a** a metric; (in poesia)
metrical ● nf metrics sg
'**metro** nm metre; (nastro) tape
measure ● nf inv (fam: metropolitana) tube Br, subway
me'tronomo nm metronome
metro'notte nmf inv night security guard
me'tropoli nf inv metropolis.
~'**tana** nf subway, underground
Br. ~'**tano** a metropolitan
'**metter|e** vt put; (indossare) put
on; (fam: installare) put in; ~**e al**
mondo bring into the world; ~**e**
da parte set aside; ~**e fiducia** inspire trust; ~**e qcsa in chiaro**
make sth clear; ~**e in mostra** display; ~**e a posto** tidy up; ~**e in**
vendita put up for sale; ~**e su** set
up ⟨casa, azienda⟩; **metter su**
famiglia start a family; **ci ho**
messo un'ora it took me an hour;
mettiamo che... let's suppose
that... ~**si** vr (indossare) put on;
(diventare) turn out; ~**si a** start
to; ~**si con qcno** (fam: formare
una coppia) start to go out with sb;
~**si a letto** go to bed; ~**si a**
sedere sit down; ~**si in viaggio**
set out
'**mezza** nf **è la** ~ it's half past
twelve; **sono le quattro e** ~ it's
half past four

mezza'luna *nf* half moon; (*simbolo islamico*) crescent; (*coltello*) two-handled chopping knife; **a ~** half-moon shaped

mezza'manica *nf* **a ~** (*maglia*) short-sleeved

mez'zano *a* middle

mezza'notte *nf* midnight

mezz'asta *a* **~** *adv* at half mast

'**mezzo** *a* half; **di mezza età** middle-aged; **~ bicchiere** half a glass; **una mezza idea** a vague idea; **siamo mezzi morti** we're half dead; **sono le quattro e ~** it's half past four. **mezz'ora** *nf* half an hour. **mezza pensione** *nf* half board. **mezza stagione** *nf* **una giacca di mezza stagione** a spring/autumn jacket ● *adv* (*a metà*) half ● *nm* (*metà*) half; (*centro*) middle; (*per raggiungere un fine*) means (*sg*; **uno e ~** one and a half; **tre anni e ~** three and a half years; **in ~ a** in the middle of; **il giusto ~** the happy medium; **levare di ~** clear away; **per ~ di** by means of; **a ~ posta** by mail; **via di ~** *fig* halfway house; (*soluzione*) middle way. **mezzi** *pl* (*denaro*) means *pl*. **mezzi pubblici** public transport. **mezzi di trasporto** [means of] transport

mezzo'busto *a* **a ~** (*foto, ritratto*) half-length

mezzo'fondo *nm* middle-distance running

mezzogi'orno *nm* midday; (*sud*) South. **il M~** Southern Italy. **~ in punto** high noon

mi *pers pron* me; (*refl*) myself; **mi ha dato un libro** he gave me a book; **mi lavo le mani** I wash my hands; **eccomi** here I am ● *nm Mus* (*chiave, nota*) E

miago'l|are *vi* miaow. **~io** *nm* miaowing

'**mica**[1] *nf* mica

'**mica**[2] *adv fam* (*per caso*) by any chance; **hai ~ visto Paolo?** have you seen Paolo, by any chance?;

non è ~ bello it is not at all nice; **~ male** not bad

'**miccia** *nf* fuse

micidi'ale *a* deadly

'**micio** *nm* pussy-cat

'**microbo** *nm* microbe

micro'cosmo *nm* microcosm

micro'fiche *nf inv* microfiche

micro'film *nm inv* microfilm

mi'crofono *nm* microphone

microorga'nismo *nm* microorganism

microproces'sore *nm* microprocessor

micro'scopi|o *nm* microscope. **~co** *a* microscopic

micro'solco *nm* (*disco*) long-playing record

mi'dollo *nm* (*pl f* **midolla**, *Anat*) marrow; **fino al ~** through and through. **~ osseo** bone marrow. **~ spinale** spinal cord

'**mie**, **mi'ei** *vedi* **mio**

mi'eie *nm* honey

mi'et|ere *vt* reap. **~i'trice** *nf Mech* harvester. **~i'tura** *nf* harvest

migli'aio *nm* (*pl f* **migliaia**) thousand. **a migliaia** in thousands

'**miglio** *nm Bot* millet; (*misura*: *pl f* **miglia**) mile

migliora'mento *nm* improvement

miglio'rare *vt/i* improve

migli'ore *a* better; (*superlativo*) the best ● *nmf* **il/la ~** the best

'**mignolo** *nm* little finger; (*del piede*) little toe

mi'gra|re *vi* migrate. **~zi'one** *nf* migration

'**mila** *vedi* **mille**

Mi'lano *nf* Milan

miliar'dario, **-a** *nm* millionaire; (*plurimiliardario*) billionaire ● *nf* millionairess; billionairess

mili'ardo *nm* billion

mili'are *a* **pietra ~** milestone

milio'nario, **-a** *nm* millionaire ● *nf* millionairess

mili'one *nm* million

milio'nesimo *a* millionth

mili'tante *a* & *nmf* militant

mili'tare *vi* **~ in** be a member of

〈*partito ecc*〉● *a* military ● *nm* soldier; **fare il** ~ do one's military service. ~ **di leva** National Serviceviceman

'**milite** *nm* soldier. **mil'izia** *nf* militia

'**mille** *a* & *nm* 〈*pl* **mila**〉 *a* o one thousand; **due/tre mila** two/three thousand; ~ **grazie!** thanks a lot!

mille'foglie *nm inv* Culin vanilla slice

mil'lennio *nm* millennium

millepi'edi *nm inv* centipede

mil'lesimo *a* & *nm* thousandth

milli'grammo *nm* milligram

mil'limetro *nm* millimetre

'**milza** *nf* spleen

mi'mare *vt* mimic 〈*persona*〉 ● *vi* mime

mi'metico *a* camouflage *attrib*

mimetiz'zare *vt* camouflage. ~**si** *vr* camouflage oneself

'**mim|ica** *nf* mime. ~**ico** *a* mimic. ~**o** *nm* mime

mi'mosa *nf* mimosa

'**mina** *nf* mine; 〈*di matita*〉 lead

mi'naccia *nf* threat

minacci'are *vt* threaten. ~'**oso** *a* threatening

mi'nare *vt* mine; *fig* undermine

mina'tor|e *nm* miner. ~**io** *a* threatening

mine'ra|le *a* & *nm* mineral. ~**rio** *a* mining *attrib*

mi'nestra *nf* soup. **mine'strone** *nm* vegetable soup; 〈*fam: insieme confuso*〉 hotchpotch

mingher'lino *a* skinny

mini+ *pref* mini+

minia'tura *nf* miniature. **minia-turiz'zato** *a* miniaturized

mini'era *nf* mine

mini'golf *nm* miniature golf

mini'gonna *nf* miniskirt

minima'mente *adv* minimally

mini'market *nm inv* minimarket

minimiz'zare *vt* minimize

'**minimo** *a* least, slightest; 〈*il più basso*〉 lowest; 〈*salario, quantità ecc*〉 minimum ● *nm* minimum; **girare al** ~ *Auto* idle

mini'stero *nm* ministry; 〈*governo*〉 government

mi'nistro *nm* minister. **M~ del Tesoro** Finance Minister, Chancellor of the Exchequer *Br*

mino'ranza *nf* minority *attrib*

mino'rato, -a *a* disabled ● *nmf* disabled person

mi'nore *a* 〈*gruppo, numero*〉 smaller; 〈*superlativo*〉 smallest; 〈*distanza*〉 shorter; 〈*superlativo*〉 shortest; 〈*prezzo*〉 lower; 〈*superlativo*〉 lowest; 〈*di età*〉 younger; 〈*superlativo*〉 youngest; 〈*di importanza*〉 minor; 〈*superlativo*〉 least important ● *nmf* younger; 〈*superlativo*〉 youngest; *Jur* minor; **il** ~ **dei mali** the lesser of two evils; **i minori di 14 anni** children under 14. **mino'renne** *a* under age ● *nmf* minor

minori'tario *a* minority *attrib*

minu'etto *nm* minuet

mi'nuscolo, -a *a* tiny ● *nf* small letter

mi'nuta *nf* rough copy

mi'nuto¹ *a* minute; 〈*persona*〉 delicate; 〈*ricerca*〉 detailed; 〈*pioggia, neve*〉 fine; **al** ~ *Comm* retail

mi'nuto² *nm* 〈*di tempo*〉 minute; **spaccare il** ~ be dead on time

mi'nuzia *nf* trifle. ~'**oso** *a* detailed; 〈*persona*〉 meticulous

'**mio** 〈**il mio** *m*, **la mia** *f*, **i miei** *mpl*, **le mie** *fpl*〉 *a poss* my; **questa macchina è mia** this car is mine; ~ **padre** my father; **un** ~ **amico** a friend of mine ● *poss pron* mine; **i miei** 〈*genitori ecc*〉 my folks

'miope *a* short-sighted. **mio'pia** *nf* short-sightedness

'**mira** *nf* aim; 〈*bersaglio*〉 target; **prendere la** ~ take aim; **prendere di** ~ qcno *fig* have it in for sb

mi'racolo *nm* miracle. ~**sa'mente** *adv* miraculously. **miraco'loso** *a* miraculous

mi'raggio *nm* mirage

mi'rar|e *vi* [take] aim. ~**si** *vr* 〈*guardarsi*〉 look at oneself

mi'riade *nf* myriad

mi'rino nm sight; Phot view-finder

mir'tillo nm blueberry

mi'santropo, -a nmf misanthropist

mi'scela nf mixture; (di caffè, tabacco ecc) blend. **~'tore** nm (di acqua) mixer tap

miscel'lanea nf miscellany

'mischia nf scuffle; (nel rugby) scrum

mischi'ar|e vt mix; shuffle (carte da gioco). **~si** vr mix; (immischiarsi) interfere

misco'noscere vt not appreciate

mi'scuglio nm mixture; fig medley

mise'rabile a wretched

misera'mente adv (finire) miserably; (vivere) in abject poverty

mi'seria nf poverty; (infelicità) misery; **guadagnare una ~** earn a pittance; **porca ~**! hell!; **miserie** pl (disgrazie) misfortunes

miseri'cordi|a nf mercy. **~'oso** a merciful

'misero a (miserabile) wretched; (povero) poor; (scarso) paltry

mi'sfatto nm misdeed

mi'sogino nm misogynist

mis'saggio nm vision mixer

'missile nm missile

missio'nario, -a nmf missionary. **missi'one** nf mission

misteri|osa'mente adv mysteriously. **~'oso** a mysterious. **mi'stero** nm mystery

'misti|ca nf mysticism. **~'cismo** nm mysticism. **~co** a mystic[al] ● nm mystic

mistifi'ca|re vt distort (verità). **~zi'one** nf (della verità) distortion

'misto a mixed; **scuola mista** mixed or co-educational school ● nm mixture; **~ lana/cotone** wool/cotton mix

mi'sura nf measure; (dimensione) measurement; (taglia) size; (limite) limit; **su ~** (abiti) made to measure; (mobile) custom-made; **a ~** (andare, calzare) perfectly; a

~ che as. **~ di sicurezza** safety measure. **misu'rare** vt measure; try on (indumenti); (limitare) limit. **misu'rarsi** vr **misurarsi con** (gareggiare) compete with. **misu'rato** a measured. **misu'rino** nm measuring spoon

'mite a mild; (prezzo) moderate

'mitico a mythical

miti'gar|e vt mitigate. **~si** vr calm down; (clima:) become mild

mitiz'zare vt mythicize

'mito nm myth. **~lo'gia** nf mythology. **~lo'logico** a mythological

mi'tomane nmf compulsive liar

'mitra nf Relig mitre ● nm inv Mil machine-gun

mitragli'a|re vt machine-gun; **~re di domande** fire questions at. **~'trice** nf machine-gun

mit'tente nmf sender

mne'monico a mnemonic

mo' nm **a ~ di** by way of (esempio, consolazione)

'mobile[1] a mobile; (volubile) fickle; (che si può muovere) movable; **beni mobili** personal estate; **squadra ~** flying squad

'mobile[2] nm piece of furniture; **mobili** pl furniture sg. **mo'bilia** nf furniture. **~l'ficio** nf furniture factory

mo'bilio nm furniture

mobilità nf mobility

mobili'ta|re vt mobilize. **~zi'one** nf mobilization

mocas'sino nm moccasin

mocci'oso, -a nmf brat

'moccolo nm (di candela) candle-end; (moccio) snot

'moda nf fashion; **di ~** in fashion; **alla ~** (musica, vestiti) up-to-date; **fuori ~** unfashionable

modalità nf inv formality; **~ d'uso** instruction

mo'della nf model. **model'lare** vt model

model'li|no nm model. **~sta** nmf designer

mo'dello nm model; (stampo)

mould; (*di carta*) pattern; (*modulo*) form

'**modem** *nm inv* modem; **mandare per ~** modem, send by modem

mode'ra|re *vt* moderate; (*diminuire*) reduce. **~rsi** *vr* control oneself. **~ta'mente** *adv* moderately **~to** *a* moderate. **~'tore**, **~'trice** *nmf* (*in tavola rotonda*) moderator. **~zi'one** *nf* moderation

modern|a'mente *adv* (*in modo moderno*) in a modern style. **~iz'zare** *vt* modernize. **mo-'derno** *a* modern

mo'dest|ia *nf* modesty. **~o** *a* modest

'**modico** *a* reasonable

mo'difica *nf* modification

modifi'ca|re *vt* modify. **~zi'one** *nf* modification

mo'dista *nf* milliner

'**modo** *nm* way; (*garbo*) manners *pl*; (*occasione*) chance; *Gram* mood; **ad ogni ~** anyhow; **di ~ che** so that; **fare in ~ di** try to; **in che ~** (*inter*) how; **in qualche ~** somehow; **in questo ~** like this; **di dire** idiom; **per ~ di dire** so to speak

modu'la|re *vt* modulate. **~zi'one** *nf* modulation. **~zione di frequenza** frequency modulation. **~'tore** *nm* **~tore di frequenza** frequency modulator

'**modulo** *nm* form; (*lunare, di comando*) module. **~ continuo** continuous paper

'mogano *nm* mahogany

'**mogio** *a* dejected

'**moglie** *nf* wife

'**mola** *nf* millstone; *Mech* grindstone

mo'lare *nm* molar

'**mole** *nf* mass; (*dimensione*) size

mo'lecola *nf* molecule

mole'stare *vt* bother; (*più forte*) molest. **mo'lestia** *nf* nuisance. **mo'lesto** *a* bothersome

'**molla** *nf* spring; **molle** *pl* tongs

mol'lare *vt* let go; (*fam: lasciare*) leave; *fam* give (*ceffone*); *Naut* cast

off ● *vi* cease; **mollala!** *fam* stop that!

'**molle** *a* soft; (*bagnato*) wet

mol'letta *nf* (*per capelli*) hair-grip; (*per bucato*) clothes-peg; **mollette** *pl* (*per ghiaccio ecc*) tongs

mol'lezz|a *nf* softness; **~e** *pl fig* luxury

mol'lica *nf* crumb

mol'lusco *nm* mollusc

'**molo** *nm* pier; (*banchina*) dock

mol'teplic|e *a* manifold; (*numeroso*) numerous. **~ità** *nf* multiplicity

moltipli'ca|re *vt*, **~rsi** *vr* multiply. **~'tore** *nm* multiplier. **~'trice** *nf* calculating machine. **~zi'one** *nf* multiplication

molti'tudine *nf* multitude

'**molto** *a* a lot of; (*con negazione e interrogazione*) much, a lot of; (*con nomi plurali*) many, a lot of; **non ~ tempo** not much time, not a lot of time ● *adv* very; (*con verbi*) a lot; (*con avverbi*) much; **~ stupido** very stupid; **mangiare ~** eat a lot; **~ più veloce** much faster; **non mangiare ~** not eat a lot, not eat much ● *pron* a lot; (*molto tempo*) a lot of time; (*con negazione e interrogazione*) much, a lot; (*plurale*) many; **non ne ho ~** I don't have much, I don't have a lot; **non ne ho molti** I don't have many, I don't have a lot; **non lo metterò ~** I won't be long; **fra non ~** before long; **molti** (*persone*) a lot of people; **eravamo in molti** there were a lot of us

momentanea'mente *adv* momentarily; **è ~ assente** he's not here at the moment. **momen-'taneo** *a* momentary

mo'mento *nm* moment; **a momenti** (*a volte*) sometimes; (*fra un momento*) in a moment; **dal ~ che** since; **per il ~** for the time being; **da un ~ all'altro** (*cambiare idea ecc*) from one moment to the next; (*aspettare qcno ecc*) at any moment

'**monac|a** *nf* nun. **~o** *nm* monk

'**Monaco** *nm* Monaco ●*nf* (*di Baviera*) Munich

mo'**narca** *nm* monarch. **monar**-'**chia** *nf* monarchy. ~**hico, -a** *a* monarchic ●*nmf* monarchist

mona'**stero** *nm* (*di monaci*) monastery; (*di monache*) convent. mo'**nastico** *a* monastic

monche'**rino** *nm* stump

'**monco** *a* maimed; (*fig: troncato*) truncated; ~ **di un braccio** one-armed

mon'**dano** *a* worldly; **vita mondana** social life

mondi'**ale** *a* world *attrib*; **di fama** ~ world-famous

'**mondo** *nm* world; **il bel** ~ fashionable society; **un** ~ (*molto*) a lot

mondovisi'**one** *nf* **in** ~ transmitted worldwide

mo'**nello, -a** *nmf* urchin

mo'**neta** *nf* coin; (*denaro*) money; (*denaro spicciolo*) [small] change. ~ **estera** foreign currency. ~ **legale** legal tender. ~ **unica** single currency. **mone'tario** *a* monetary

mongolfi'**era** *nf* hot air balloon

mo'**nile** *nm* jewel

'**monito** *nm* warning

moni'**tore** *nm* monitor

mo'**nocolo** *nm* monocle

monoco'**lore** *a* *Pol* one-party

mono'**dose** *a* *inv* individually packaged

monogra'**fia** *nf* monograph

mono'**gramma** *nm* monogram

mono'**kini** *nm* *inv* monokini

mono'**lingue** *a* monolingual

monolo'**cale** *nm* studio flat *Br*, studio apartment

mo'**nologo** *nm* monologue

mono'**pattino** *nm* [child's] scooter

mono'**polio** *nm* monopoly. ~**o di Stato** state monopoly. ~'**z'zare** *vt* monopolize

mono'**sci** *nm* *inv* monoski

monosil'**labico** *a* monosyllabic. mono'**sillabo** *nm* monosyllable

monoto'**nia** *nf* monotony. mo-'**notono** *a* monotonous

mono'**uso** *a* disposable

monsi'**gnore** *nm* monsignor

mon'**sone** *nm* monsoon

monta'**carichi** *nm* *inv* hoist

mon'**taggio** *nm* *Mech* assembly; *Cinema* editing; **catena di** ~ production line

mon'**tagna** *nf* mountain; (*zona*) mountains *pl.* **montagne** *pl* **russe** big dipper. ~'**gnoso** *a* mountainous. ~'**naro, -a** *nmf* highlander. ~**no** *a* mountain *attrib*

mon'**tante** *nm* (*di finestra, porta*) upright

mon'**tare** *vt/i* mount; get on (*veicolo*); (*aumentare*) rise; *Mech* assemble; frame (*quadro*); *Culin* whip; edit (*film*); (*a cavallo*) ride; *fig* blow up; ~**rsi la testa** *nf* get big-headed. ~**to, -a** *nmf* poser. ~'**tura** *nf Mech* assembling; (*di occhiali*) frame; (*di gioiello*) mounting; *fig* exaggeration

'**monte** *nm* **anche** *fig* mountain; **a** ~ up-stream; **andare a** ~ be ruined; **mandare a** ~ **qcsa** ruin sth. ~ **di pietà** pawnshop

monte'**premi** *nm* *inv* jackpot

mont'**gomery** *nm* *inv* duffle coat

mon'**tone** *nm* ram; **carne di** ~ mutton

montu'**oso** *a* mountainous

monumen'**tale** *a* monumental. monu'**mento** *nm* monument

mo'**quette** *nf* (*tappeto*) fitted carpet

'**mora** *nf* (*del gelso*) mulberry; (*del rovo*) blackberry

mo'**rale** *a* moral ●*nf* morals *pl*; (*di storia*) moral ●*nm* morale. mora'**lista** *nm* moralist. ~**ità** *nf* morality; (*condotta*) morals *pl.* ~**iz'zare** *vt/i* moralize. ~'**mente** *adv* morally

morbi'**dezza** *nf* softness

'**morbido** *a* soft

mor'**billo** *nm* measles *sg*

'**morbo** *nm* disease. ~**sità** *nf* (*qualità*) morbidity

mor'**boso** *a* morbid

mor'**dace** *a* cutting

mor'dente a biting. **'mordere** vt bite; (corrodere) bite into. **mordic-chi'are** vt gnaw

mor'fina nf morphine. **morfi-'nomane** nmf morphine addict

mori'bondo a dying; (istituzione) moribund

morige'rato a moderate

mo'rire vi die; (fig die out; **fa un freddo da ~** it's freezing cold, it's perishing; **~ di noia** be bored to death; **c'era da ~ dal ridere** it was killingly or hilariously funny

mor'mone nmf Mormon

mormo'r|are vt/i murmur; (brontolare) mutter. **~io** nm murmuring; (lamentela) grumbling

'moro a dark ● nm Moor

mo'roso a in arrears

'morsa nf vice; fig grip

'morse a alfabeto ~ Morse code

mor'setto nm clamp

morsi'care vt bite. **'morso** nm bite; (di cibo, briglia) bit; **i morsi della fame** hunger pangs

morta'della nf mortadella (type of salted pork)

mor'taio nm mortar

mor'tal|e a mortal; (simile a morte) deadly; **di una noia ~e** deadly. **~ità** nf mortality. **~'mente** adv (ferito) fatally; (offeso) mortally

morta'retto nm firecracker

'morte nf death

mortifi'cante a mortifying

mortifi'ca|re vt mortify. **~rsi** vr be mortified. **~to** a mortified. **~zi'one** nf mortification

'morto, -a pp di morire ● a dead; **~ di freddo** frozen to death; **stanco ~** dead tired ● nm dead man ● nf dead woman

mor'torio nm funeral

mo'saico nm mosaic

'Mosca nf Moscow

'mosca nf fly; (barba) goatee. **~ cieca** blindman's buff

mo'scato a muscat; **noce mosca-ta** nutmeg ● nm muscatel

mosce'rino nm midge; (fam: persona) midget

mo'schea nf mosque

moschi'cida a fly attrib

'moscio a limp; **avere l'erre moscia** not be able to say one's r's properly

mo'scone nm bluebottle; (barca) pedalo

'moss|a nf movement; (passo) move. **~o** pp di muovere ● a (mare) rough; (capelli) wavy; (fotografia) blurred

mo'starda nf mustard

'mostra nf show; (d'arte) exhibition; **far ~ di** pretend; **in ~** on show; **mettersi in ~** make oneself conspicuous

mo'stra|re vt show; (indicare) point out; (spiegare) explain. **~rsi** vr show oneself; (apparire) appear

'mostro nm monster; (fig: persona) genius; **~ sacro** fig sacred cow

mostru|osa'mente adv tremendously. **~'oso** a monstrous; (incredibile) enormous

mo'tel nm inv motel

moti'va|re vt cause; Jur justify. **~to** a (persona) motivated. **~zi'one** nf motivation; (giustificazione) justification

mo'tivo nm reason; (movente) motive; (in musica, letteratura) theme; (disegno) motif

'moto nm motion; (esercizio) exercise; (gesto) movement; (sommossa) rising ● nf inv (motocicletta) motor bike; **mettere in ~** start (motore)

moto'carro nm three-wheeler

motoci'cl|etta nf motor cycle. **~ismo** nm motorcycling. **~ista** nmf motor-cyclist

moto'cros|s nm motocross. **~'sista** nmf scrambler

moto'lancia nf motor launch

moto'nave nf motor vessel

mo'tore a motor ● nm motor, engine. **moto'retta** nf motor scooter.

moto'rino nm moped. **motorino d'avviamento** starter

motoriz'za|to a Mil motorized. **~zi'one** nf (ufficio) vehicle licensing office

moto'scafo nm motorboat

motove'detta nf patrol vessel

'motto nm motto; (facezia) witticism; (massima) saying

mountain bike nf inv mountain bike

mouse nm inv Comput mouse

mo'vente nm motive

movimen'ta|re vt enliven. **~to** a lively. **movi'mento** nm movement; **essere sempre in movimento** be always on the go

mozi'one nf motion

mozzafi'ato a inv nail-biting

moz'zare vt cut off; dock (coda); **~ il fiato a qcno** take sb's breath away

mozza'rella nf mozzarella, mild, white cheese

mozzi'cone nm (di sigaretta) stub

'mozzo nm Mech hub; Naut ship's boy ●a (coda) truncated; (testa) severed

'mucca nf cow. **morbo della ~ pazza** mad cow disease

'mucchio nm heap, pile; **un ~ di** fig lots of

'muco nm mucus

'muffa nf mould; **fare la ~** go mouldy. **muf'fire** vi go mouldy

muf'fole nfpl mittens

mug'gi|re vi (mucca:) moo, low; (toro:) bellow. **~to** nm moo; bellow; (azione) mooing; bellowing

mu'ghetto nm lily of the valley

mugo'lare vi whine; (persona:) moan. **mugo'lio** nm whining

mugu'gnare vi fam mumble

mulat'ti'era nf mule track

mu'latto, -a nmf mulatto

muli'nello nm (d'acqua) whirlpool; (di vento) eddy; (giocattolo) windmill

mu'lino nm mill. **~ a vento** windmill

'mulo nm mule

'multa nf fine. **mul'tare** vt fine

multico'lore a multicoloured

multi'lingue a multilingual

multi'media nmf multimedia

multimedi'ale a multimedia attrib

multimiliar'dario, -a nmf multimillionaire

multinazio'nale nf multinational

'multiplo a & nm multiple

multiproprietà nf inv time-share

multi'uso a (utensile) all-purpose

'mummia nf mummy

'mungere vt milk

mungi'tura nf milking

munici'pa|le a municipal. **~ità** nf inv town council. **muni'cipio** nm town hall

mu'nifico a munificent

mu'nire vt fortify; **~ di** (provvedere) supply with

muni'zioni nfpl ammunition sg

'munto pp di mungere

mu'over|e vt move; (suscitare) arouse. **~si** vr move; **muoviti!** hurry up!, come on!

mura nfpl (cinta di città) walls

mu'raglia nf wall

mu'rale a mural; (pittura) wall attrib

mu'ra|re vt wall up. **~'tore** nm bricklayer; (con pietre) mason; (operaio edile) builder. **~'tura** nf (di pietra) masonry, stonework; (di mattoni) brickwork

mu'rena nf moray eel

'muro nm wall; (di nebbia) bank; a **~** (armadio) built-in. **~ portante** load-bearing wall. **~ del suono** sound barrier

'muschio nm Bot moss

musco'la|re a muscular. **~'tura** nf muscles pl. **'muscolo** nm muscle

mu'seo nm museum

museru'ola nf muzzle

'musi|ca nf music. **~cal** nm inv musical. **~'cale** a musical. **~'cista** nmf musician

'muso nm muzzle; (pej: di persona) mug; (di aeroplano) nose; **fare il ~** sulk. **mu'sone, -a** nmf sulker

'mussola *nf* muslin

musul'mano, -a *nmf* Moslem

'muta *nf* (*cambio*) change; (*di penne*) moult; (*di cani*) pack; (*per immersione subacquea*) wetsuit

muta'mento *nm* change

mu'tande *nfpl* pants; (*da donna*) knickers. ~'doni *nmpl* (*da uomo*) long johns; (*da donna*) bloomers

mu'tare *vt* change

'mutevole *a* changeable

muti'la|re *vt* mutilate. ~'to, -a *nmf* disabled person. ~to di guerra disabled ex-serviceman. ~zi'one *nf* mutilation

mu'tismo *nm* dumbness; *fig* obstinate silence

'muto *a* dumb; (*silenzioso*) silent; (*fonetica*) mute

'mutu|a *nf* (*cassa nf*) ~ sickness benefit fund. ~'ato, -a *nmf* ≈ NHS patient

'mutuo[1] *a* mutual

'mutuo[2] *nm* loan; (*per la casa*) mortgage; fare un ~ take out a mortgage. ~ ipotecario mortgage

••••••••••••••••

Nn

••••••••••••••••

'nacchera *nf* castanet

'nafta *nf* naphtha; (*per motori*) diesel oil

'naia *nf* cobra; (*sl: servizio militare*) national service

'nailon *nm* nylon

'nanna *nf* (*sl: infantile*) byebyes; andare a ~ go byebyes; fare la ~ sleep

'nano, -a *a & nmf* dwarf

napole'tano, -a *a & nmf* Neapolitan

'Napoli *nf* Naples

'nappa *nf* tassel; (*pelle*) soft leather

narci'sismo *nm* narcissism. ~ta *a & nmf* narcissist

nar'ciso *nm* narcissus

nar'cotico *a & nm* narcotic

na'rice *nf* nostril

nar'ra|re *vt* tell. ~'tivo, -a *a* narrative ● *nf* fiction. ~'tore, ~'trice *nmf* narrator. ~zi'one *nf* narration; (*racconto*) story

na'sale *a* nasal

'nasc|ere *vi* (*venire al mondo*) be born; (*germogliare*) sprout; (*sorgere*) rise; ~ere da *fig* arise from. ~ita *nf* birth. ~'ituro *nm* unborn child

na'scond|ere *vt* hide. ~si *vr* hide

nascon'diglio *nm* hiding-place. ~no *nm* hide-and-seek. na'scosto *pp di* nascondere ● *a* hidden; di ~ secretly

na'sello *nm* (*pesce*) hake

'naso *nm* nose

'nastro *nm* ribbon; (*di registratore ecc*) tape. ~ adesivo adhesive tape. ~ isolante insulating tape. ~ trasportatore conveyor belt

na'tal|e *a* (*paese*) of one's birth. N~e *nm* Christmas; ~i *pl* parentage. ~ità *nf* [number of] births.

nata'lizio, -a (*del Natale*) Christmas *attrib*; (*di nascita*) of one's birth

na'tante *a* floating ● *nm* craft

'natica *nf* buttock

na'tio *a* native

Nativi'tà *nf* Nativity. na'tivo, -a *a & nmf* native

'nato *pp di* nascere ● *a* born; uno scrittore ~ a born writer; nata Rossi née Rossi

NATO *nf* Nato, NATO

na'tura *nf* nature; pagare in ~ pay in kind. ~ morta still life

natu'ra|le *a* natural; al ~ (*alimento*) plain, natural; ~le! naturally, of course. ~'lezza *nf* naturalness. ~liz'zare *vt* naturalize. ~l'mente *adv* (*ovviamente*) naturally, of course

natu'rista *nmf* naturalist

naufra'gare *vi* be wrecked; (*persona:*) be shipwrecked. nau'fragio *nm* shipwreck; *fig* wreck. 'naufrago, -a *nmf* survivor

'nause|a *nf* nausea; avere la ~a

feel sick. **~a'bondo** a nauseating. **~'ante** a nauseating. **~'are** vt nauseate

'nautic|a nf navigation. **~o** a nautical

na'vale a naval

na'vata nf (centrale) nave; (laterale) aisle

'nave nf ship. **~ cisterna** tanker. **~ da guerra** warship. **~ spaziale** spaceship

na'vetta nf shuttle

navicella nf **~ spaziale** nose cone

navi'gabile a navigable

navi'ga|re vi sail; **~re in Internet** surf the Net. **~tore, ~'trice** nf navigator. **~zi'one** nf navigation

na'viglio nm fleet; (canale) canal

nazio'na|le a national ● nf Sport national team. **~'lismo** nm nationalism. **~'lista** nmf nationalist **~lità** nf inv nationality. **~liz'zare** vt nationalize. **~zi'one** nf nation

na'zista a nmf Nazi

N.B. abbr (nota bene) N.B.

ne pers pron (di lui) about him; (di lei) about her; (di loro) about them; (di ciò) about it; (da ciò) from that; (di un insieme) of it; (di un gruppo) of them; **non ne conosco nessuno** I don't know any of them; **ne ho I have some; non ne ho più** I don't have any left ● adv from there; **ne vengo ora** I've just come from there; **me ne vado** I'm off

né conj né... né... neither... nor...; **non ne ho il tempo né la voglia** I don't have either the time or the inclination; **né tu né io vogliamo andare** neither you nor I want to go; **né l'uno né l'altro** neither (of them/us)

ne'anche adv (neppure) not even; (senza neppure) without even ● conj (e neppure) neither... nor; **non parlo inglese, e lui ~** I don't speak English, neither does he o and he doesn't either

'nebbi|a nf mist; (in città, su strada) fog. **~'oso** a misty; foggy

necessaria'mente adv necessarily. **neces'sario** a necessary

necessità nf inv necessity; (bisogno) need

necessi'tare vi **~ di** need; (essere necessario) be necessary

necro'logio nm obituary

ne'cropoli nf inv necropolis

ne'fando a wicked

ne'fasto a ill-omened

ne'ga|re vt deny; (rifiutare) refuse; **essere ~to per qcsa** be no good at sth. **~'tivo, -a** a negative ● nf negative. **~zi'one** nf negation; (diniego) denial; Gram negative

ne'gletto a neglected

'negli = in + gli

negli'gen|te a negligent. **~za** nf negligence

negozi'abile a negotiable

negozi'ante nmf dealer; (bottegaio) shopkeeper

negozi'a|re vt negotiate ● vi **~re in trade.** in. **~ti** nmpl negotiations

ne'gozio nm shop

'negro, -a a Negro, black ● nmf Negro, black; (scrittore) ghost writer

'nei = in + i. **'nel** = in + il. **'nella** = in + la. **'nelle** = in + le. **'nello** = in + lo

'nembo nm nimbus

ne'mico, -a a hostile ● nmf enemy

nem'meno conj not even

'nenia nf dirge; (per bambini) lullaby; (piagnucolio) wail

'neo nm mole; (applicato) beauty spot

'neo+ pref neo+

neofa'scismo nm neofascism

neo'litico a Neolithic

neolo'gismo nm neologism

'neon nm neon

neo'nato, -a a newborn ● nmf newborn baby

neozelan'dese a New Zealand ● nmf New Zealander

nep'pure conj not even

'nerb|o nm (forza) strength; fig backbone. **~o'ruto** a brawny

ne'retto nm Typ bold [type]

'nero a black; (fam: arrabbiato) fuming **in** ~ **nm** black; **mettere** ~ **su bianco** put in writing

nerva'tura nf nerves pl; Bot veining; (di libro) band

'nervo nm nerve; Bot vein; **avere i nervi** be bad-tempered; **dare ai nervi a qcno** get on sb's nerves. ~**'sismo** nm nervousness

ner'voso a nervous; (irritabile) bad-tempered; **avere il** ~ be irritable; **esaurimento** nm ~ nervous breakdown

'nespo|la nf medlar. ~**o** nm medlar[-tree]

'nesso nm link

nes'suno a no, not... any; (qualche) any; **non ho nessun problema** I don't have any problems, I have no problems; **non lo trovo da nessuna parte** I can't find it anywhere; **in nessun modo** on no account; **nessuna notizia?** any news? ● pron nobody, no one, not... anybody, not... anyone; (qualcuno) anybody, anyone; **hai delle domande?** – **nessuna** do you have any questions? – none; ~ **di voi** none of you; ~ **dei due** (di voi due) neither of you; **non ho visto** ~ **dei tuoi amici** I haven't seen any of your friends; **c'è** ~? is anybody there?

'nettare nm nectar

net'tare vt clean

net'tezza nf cleanliness. ~ **urbana** cleansing department

'netto a clean; (chiaro) clear; Comm net; **di** ~ just like that

nettur'bino nm dustman

neu'tra|le a & nm neutral. ~**ità** nf neutrality. ~**iz'zare** vt neutralize.

'neutro a neutral; Gram neuter ● nm Gram neuter

neu'trone nm neutron

'neve nf snow

nevi'care vi snow; ~**ca** it is snowing. ~**'cata** nf snowfall. **ne'vischio** nm sleet. **ne'voso** a snowy

nevral'gia nf neuralgia. **ne'vralgi-co** a neuralgic

ne'vro|si nf inv neurosis. ~**tico** a neurotic

'nibbio nm kite

'nicchia nf niche

nicchi'are vi shilly-shally

'nichel nm nickel

nichi'lista a & nmf nihilist

nico'tina nf nicotine

nidi'ata nf brood. **'nido** nm nest; (giardino d'infanzia) crèche

ni'ente pron nothing, not... anything; (qualcosa) anything; **non ho fatto** ~ **di male** I didn't do anything wrong, I did nothing wrong; **grazie! – di** ~! thank you! – don't mention it!; **non serve a** ~ it is no use; **vuoi** ~? do you want anything?; **da** ~ (poco importante) minor; (di poco valore) worthless ● a inv **per non ho** ~ **fame** I'm not the slightest bit hungry ● adv non **fa** ~ (non importa) it doesn't matter; **per** ~ at all; (litigare) over nothing; ~ **affatto!** no way! ● nm **un bel** ~ absolutely nothing

niente'meno ... **niente'meno** adv ~ **che** no less than ● int fancy that!

'ninfa nf nymph

nin'fea nf water-lily

ninna'nanna nf lullaby

'ninnolo nm plaything; (fronzolo) knick-knack

ni'pote nm (di zii) nephew; (di nonni) grandson, grandchild ● nf (di zii) niece; (di nonni) granddaughter, grandchild

'nisba pron (sl: niente) zilch

'nitido a neat; (chiaro) clear

ni'trato nm nitrate

ni'tri|re vi neigh. ~**to** nm (di cavallo) neigh

n° abbr (numero) No

no adv no; (con congiunzione) not; **dire di no** say no; **credo di no** I don't think so; **perché no?** why not?; **io no** not me; **ha detto così, no?** he said so, didn't he?; **fa freddo, no?** it's cold, isn't it?

'**nobil**|**e** *a* noble ● *nm* noble, nobleman ● *nf* noble, noblewoman. ~**l'are** *a* noble. ~**tà** *nf* nobility

'**nocca** *nf* knuckle

'**noccio'la** *nf* hazelnut. ~**o** *nm* (*albero*) hazel

'**nocciolo** *nm* stone; *fig* heart

'**noce** *nf* walnut ● *nm* (*albero*, *legno*) walnut. ~ **moscata** nutmeg. ~'**pesca** *nf* nectarine

no'**civo** *a* harmful

'**nodo** *nm* knot; *fig* lump; *Comput* node; **fare il ~ della cravatta** do up one's tie. ~ **alla gola** lump in the throat. **no'doso** *a* knotty.

'**nodulo** *nm* nodule

'**noi** *pers pron* (*soggetto*) we; (*oggetto*, *con prep*) us; **chi è? ~ siamo** ~ who is it? – it's us

'**noia** *nf* boredom; (*fastidio*) bother; (*persona*) bore; **dar ~** annoy

noi'**altri** *pers pron* we

noi'**oso** *a* boring; (*fastidioso*) tiresome

noleggi'**are** *vt* hire; (*dare a noleggio*) hire out; charter (*nave*, *aereo*). **no'leggio** *nm* hire; (*di nave*, *aereo*) charter. '**nolo** *nm* hire; *Naut* freight; **a nolo** for hire

'**nomade** *a* nomadic ● *nmf* nomad

'**nome** *nm* name; *Gram* noun; **a ~ di** in the name of; **di ~** by name; **farsi un ~** ~ make a name for oneself. ~ **di famiglia** surname. ~ **da ragazza** maiden name. no'**mea** *nf* reputation

nomencla'**tura** *nf* nomenclature

no'**mignolo** *nm* nickname

'**nomina** *nf* appointment. **nomi'nale** *a* nominal; *Gram* noun *attrib*

nomi'**na**|**re** *vt* name; (*menzionare*) mention; (*eleggere*) appoint. ~'**tivo** *a* nominative; *Comm* registered ● *nm* nominative; (*nome*) name

non *adv* not; ~ **ti amo** I do not don't love you; ~ **c'è di che** not at all

non**ché** *conj* (*tanto meno*) let alone; (*e anche*) as well as

noncu'**ran**|**te** *a* nonchalant; (*negligente*) indifferent. ~**za** *nf* nonchalance; (*negligenza*) indifference

nondi'**meno** *conj* nevertheless

'**nonna** *nf* grandmother, grandma *fam*

'**nonno** *nm* grandfather, grandpa *fam*; **nonni** *pl* grandparents

non'**nulla** *a* & *nm* inv trifle

'**nono** *a* & *nm* ninth

nono'**stante** *prep* in spite of ● *conj* although

nontiscordardi'**mé** *nm* inv forget-me-not

nonvio'**lento** *a* nonviolent

nord *nm* north; **del ~** northern

nord-'**est** *nm* northeast; **a ~** northeasterly

'**nordico** *a* northern

nordocciden'**tale** *a* northwestern

nordorien'**tale** *a* northeastern

nord-'**ovest** *nm* northwest; **a ~** northwesterly

'**norma** *nf* rule; (*istruzione*) instruction; **a ~ di legge** according to law; **è buona ~** it's advisable

nor'**mal**|**e** *a* normal. ~**ità** *nf* normality. ~**iz'zare** *vt* normalize. ~'**mente** *adv* normally

norve'**gese** *a* & *nmf* Norwegian. **Nor'vegia** *nf* Norway

nossi'**gnore** *adv* no way

nostal'**gia** *nf* (*di casa*, *patria*) homesickness; (*del passato*) nostalgia; **aver ~** be homesick; **aver ~ di** qcno miss sb. **no'stalgico**, -**a** *a* nostalgic ● *a* reactionary

no'**strano** *a* local; (*fatto in casa*) home-made

'**nostro** (**il nostro** *m*, **la nostra** *f*, **i nostri** *mpl*, **le nostre** *fpl*) *poss a* our; **quella macchina è nostra** that car is ours; ~ **padre** our father; **un ~ amico** a friend of ours ● *poss pron* ours

'**nota** *nf* (*segno*) sign; (*comunicazione*, *commento*, *Mus*) note; (*conto*) bill; (*lista*) list; **degno di ~** noteworthy; **prendere ~** take note. **note** *pl* **caratteristiche** distinguishing marks

no'tabile a & nm notable

no'taio nm notary

no'ta|re vt (segnare) mark; (annotare) note down; (osservare) notice; **far ~re qcsa** point sth out; **farsi ~re** get oneself noticed. **~zi'one** nf marking; (annotazione) notation

notes nm inv notepad

no'tevole a (degno di nota) remarkable; (grande) considerable

no'tifica nf notification. **notifi'care** vt notify; Comm advise. **~zi'one** nf notification

no'tizi|a nf una ~a a piece of news, some news; (informazione) a piece of information, some information; **le ~e** the news sg. **~'ario** nm news sg

'noto a [well-]known; **rendere ~** (far sapere) announce

notorietà nf fame; **raggiungere la ~** become famous. **no'torio** a well-known; pej notorious

not'tambulo nm night-bird

not'tata nf night; **far ~** stay up all night

'notte nf night; **di ~** at night; **~ bianca** sleepless night; **peggio che andar di ~** worse than ever. **~'tempo** adv at night

not'turno a nocturnal; (servizio ecc) night

no'vanta a & nm ninety

novan't|enne a & nmf ninety-year-old. **~esimo** a ninetieth. **~ina** nf about ninety. **'nove** a & nm nine. **nove'cento** a & nm nine hundred. **il N~cento** the twentieth century

no'vella nf short story

novel'lino, -a a inexperienced ● nmf novice, beginner. **no'vello** a new

no'vembre nm November

novità nf inv novelty; (notizie) news sg; **l'ultima ~** (moda) the latest fashion

novizi'ato nm Relig novitiate; (tirocinio) apprenticeship

nozi'one nf notion; **nozioni** pl rudiments

'nozze nfpl marriage sg; (cerimonia) wedding sg. **~ d'argento** silver wedding [anniversary]. **~ d'oro** golden wedding [anniversary]

'nub|e nf cloud. **~e tossica** toxic cloud. **~'ifragio** nm cloudburst

'nubile a unmarried ● nf unmarried woman

'nuca nf nape

nucle'are a nuclear

'nucleo nm nucleus; (unità) unit

nu'di|smo nm nudism. **~sta** nmf nudist. **~tà** nf inv nudity, nakedness

'nudo a naked; (spoglio) bare; **a occhio ~** to the naked eye

'nugolo nm large number

'nulla pron = niente; **da ~** worthless

nulla'osta nm inv permit

nulla'nente nm & nm; **i nullatenenti** the have-nots

nullità nf inv (persona) nonentity

'nullo a Jur null and void

nume'rabile a countable. **~le a &** nm numeral

nume'ra|re vt number. **~zi'one** nf numbering. **nu'merico** a numerical

'numero nm number; (romano, arabo) numeral; (di scarpe ecc) size; **dare i numeri** be off one's head. **~ cardinale** cardinal [number]. **~ decimale** decimal. **~ ordinale** ordinal [number]. **~ di telefono** phone number. **nume'roso** a numerous

'nunzio nm nuncio

nu'ocere vi ~ a harm

nu'ora nf daughter-in-law

nuo'ta|re vi swim; fig wallow; **~re nell'oro** be stinking rich, be rolling in it. **nu'oto** nm swimming. **~'tore, ~'trice** nmf swimmer

nu'ov|a nf (notizia) news sg. **~a'mente** adv again; **di ~o** again; **rimettere a ~o** give a new lease of life to

nutri'ente *a* nourishing. **~'mento** *nm* nourishment

nu'trire *vt* nourish; harbour (*sentimenti*). **~rsi** eat; **~rsi di** *fig* live on. **~'tivo** *a* nourishing. **~zi'one** *nf* nutrition

'nuvola *nf* cloud. **nuvo'loso** *a* cloudy

nuzi'ale *a* nuptial; (*vestito, anello ecc*) wedding *attrib*

Oo

O *abbr* (**ovest**) W

o *conj* or; **l'uno ~ l'altro** one or the other, either

'oasi *nf inv* oasis

obbedi'ente *ecc* = **ubbidiente** *ecc*

obbli'gare *vt* force, oblige; **~rsi** *vr* **~rsi a** undertake to. **~to a** obliged. **~'torio** *a* compulsory. **~zi'one** *nf* obligation; *Comm* bond. **'obbligo** *nm* obligation; (*dovere*) duty; **avere obblighi verso** be under an obligation to; **d'obbligo** obligatory

obbligatoria'mente *adv* **fare qcsa ~** be obliged to do sth; **bisogna ~ farlo** you absolutely have to do it

ob'brobrio *nm* disgrace. **~'brioso** *a* disgraceful

obe'lisco *nm* obelisk

obe'rare *vt* overburden

obesità *nf* obesity. **o'beso** *a* obese

obiet'tare *vt/i* object; **~ su** object to

obietti'va'mente *adv* objectively. **~vità** *nf* objectivity. **obiet'tivo** *a* objective ● *nm* objective; (*scopo*) object

obiet'tore *nm* objector. **~ttore di coscienza** conscientious objector. **~zi'one** *nf* objection

obi'torio *nm* mortuary

o'blio *nm* oblivion

o'bliquo *a* oblique; *fig* underhand

oblite'rare *vt* obliterate

oblò *nm inv* porthole

'oboe *nm* oboe

obso'leto *a* obsolete

'oca *nf* (*pl* **oche**) goose; (*donna*) silly girl

occasio'nale *a* occasional. **~'mente** *adv* occasionally

occasi'one *nf* occasion; (*buon affare*) bargain; (*motivo*) cause; (*opportunità*) chance; **d'~** secondhand

occhi'aia *nf* eye socket; **occhiaie** *pl* shadows under the eyes

occhi'ali *nmpl* glasses, spectacles. **~ da sole** sunglasses. **~ da vista** glasses, spectacles

occhi'ata *nf* look; **dare un'~ a** have a look at

occhieggi'are *vt* ogle ● *vi* (*far capolino*) peep

occhi'ello *nm* buttonhole; (*asola*) eyelet

'occhio *nm* eye; **~!** watch out!; **a quattr'occhi** in private; **tenere d'~ qcno** keep an eye on sb; **a ~ [e croce]** roughly; **chiudere un'~** turn a blind eye; **dare nell'~** attract attention; **pagare** *o* **spendere un ~ [della testa]** pay an arm and a leg; **saltare agli occhi** be blindingly obvious. **~ nero** (*pesto*) black eye. **~ di pernice** (*callo*) corn. **~'lino** *nm* **fare l'~ino a qcno** wink at sb

occiden'tale *a* western ● *nmf* westerner. **occi'dente** *nm* west

oc'cludere *vt* obstruct. **occlusi'one** *nf* occlusion

occor'rente *a* necessary ● *nm* the necessary. **~za** *nf* need; **all'~za** if need be

oc'correre *vi* be necessary

occulta'mento *nm* **~ di prove** concealment of evidence

occul'tare *vt* hide. **~ismo** *nm* occult. **oc'culto** *a* hidden; (*magico*) occult

occu'pante *nmf* occupier; (*abusivo*) squatter

occu'pa|re *vt* occupy; spend ⟨*tempo*⟩; take up ⟨*spazio*⟩; ⟨*dar lavoro a*⟩ employ. **~rsi** *vr* occupy oneself; ⟨*trovare lavoro*⟩ find a job; **~rsi di** ⟨*badare*⟩ look after. **~to** *a* engaged; ⟨*persona*⟩ busy; ⟨*posto*⟩ taken. **~zi'one** *nf* occupation; **trovarsi un'~zione** ⟨*interesse*⟩ find oneself something to do

o'ceano *nm* ocean. **~ Atlantico** Atlantic [Ocean]. **~ Pacifico** Pacific [Ocean]

'ocra *nf* ochre

ocu'lare *a* ocular; ⟨*testimone, bagno*⟩ eye *attrib*

ocula'tezza *nf* care. **ocu'lato** *a* ⟨*scelta*⟩ wise

ocu'lista *nmf* optician; ⟨*per malattie*⟩ ophthalmologist

od *conj* or

'ode *nf* ode

odi'are *vt* hate

odi'erno *a* of today; ⟨*attuale*⟩ present

'odi|o *nm* hatred; **avere in ~o** hate. **~'oso** *a* hateful

odo'ra|re *vt* smell; ⟨*profumare*⟩ perfume ● *vi* **~re di** smell of. **~to** *nm* sense of smell. **o'dore** *nm* smell; ⟨*profumo*⟩ scent; **c'è odore di...** there's a smell of...; **sentire odore di** smell; *odori pl* Culin herbs. **odo'roso** *a* fragrant

of'fender|e *vt* offend; ⟨*ferire*⟩ injure. **~si** *vr* take offence

offen'siv|a *nf* Mil offensive. **~o a** offensive

offe'rente *nmf* offerer; ⟨*in aste*⟩ bidder

of'fert|a *nf* offer; ⟨*donazione a*⟩ donation; Comm supply; ⟨*nelle aste*⟩ bid; **in ~a speciale** on special offer. **~o** *pp di* **offrire**

of'fesa *nf* offence. **~o** *pp di* **offendere** ● *a* offended

offi'ciare *vt* officiate

offi'cina *nf* workshop; **~ [meccanica]** garage

of'frir|e *vt* offer. **~si** *vr* offer oneself; ⟨*occasione:*⟩ present itself; **~si di fare qcsa** offer to do sth

offu'scar|e *vt* darken; *fig* dull ⟨*memoria, bellezza*⟩; blur ⟨*vista*⟩. **~si** *vr* darken; ⟨*fig: memoria, bellezza:*⟩ fade away; ⟨*vista:*⟩ become blurred

of'talmico *a* ophthalmic

oggettivi'tà *nf* objectivity. **ogget-'tivo** *a* objective

og'getto *nm* object; ⟨*argomento*⟩ subject; **oggetti** *pl* **smarriti** lost property, lost and found Am

'oggi *adv & nm* today; ⟨*al giorno d'oggi*⟩ nowadays; **da ~ in poi** from today on; **~ a otto** a week today; **dall'~ al domani** overnight; **il giornale di ~** today's paper; **al giorno d'~** these days, nowadays. **~'gl'orno** *adv* nowadays

'ogni *a inv* every; ⟨*qualsiasi*⟩ any; **~ tre giorni** every three days; **ad ~ costo** at any cost; **ad ~ modo** anyway; **~ cosa** everything; **~ tanto** now and then; **~ volta che** every time, whenever

o'gnuno *pron* everyone, everybody; **~ di voi** each of you

ohimè *int* oh dear!

'ola *nf inv* Mexican wave

O'lan|da *nf* Holland. **o~'dese** *a* Dutch ● *nm* Dutchman; ⟨*lingua*⟩ Dutch ● *nf* Dutchwoman

ole'andro *nm* oleander

ole'at|o *a* oiled; **carta ~a** greaseproof paper

oleo'dotto *nm* oil pipeline. **ole'oso** *a* oily

ol'fatto *nm* sense of smell

oli'are *vt* oil

oli'era *nf* cruet

olim'piadi *nfpl* Olympic Games. **o'limpico** *a* Olympic. **olim'pionico** *a* ⟨*primato, squadra*⟩ Olympic

'olio *nm* oil; **sott'~** in oil; **colori a ~** oils; **quadro a ~** oil painting. **~ di mais** corn oil. **~ d'oliva** olive oil. **~ di semi** vegetable oil. **~ solare** sun-tan oil

o'liv|a *nf* olive. **ol'vastro** *a* olive. **oli'veto** *nm* olive grove. **~o** *nm* olive tree '**olmo** nm elm

oltraggi'are vt offend. **ol'traggio** nm offence

ol'tranza nf **ad ~** to the bitter end

'**oltre** adv (di luogo) further; (di tempo) longer ● prep (di luogo) over; (di tempo) later than; (più di) more than; (in aggiunta) besides; **~ a** (eccetto) except, apart from; **per ~ due settimane** for more than two weeks; **una settimana e ~ a** week and more. **~'mare** adv overseas. **~'modo** adv extremely

oltrepas'sare vt go beyond; (eccedere) exceed

o'maggio nm homage; (dono) gift; **in ~ con** free with; **omaggi** pl (saluti) respects

ombeli'cale a umbilical; **cordone ~** umbilical cord. **ombe'lico** nm navel

'**ombr|a** nf (zona) shade; (immagine oscura) shadow; **all'~a** in the shade. **~eggi'are** vt shade

om'brello nm umbrella. **ombrel-'lone** nm beach umbrella

om'bretto nm eye-shadow

om'broso a shady; (cavallo) skittish

ome'lette nf inv omelette

ome'lia nf Relig sermon

omeopa'tia nf homoeopathy. **omeo'patico** a homoeopathic ● nm homoeopath

omertà nf conspiracy of silence

o'messo pp di **omettere**

o'mettere vt omit

omi'cid|a a murderous ● nmf murderer. **~io** nm murder. **~io colposo** manslaughter

omissi'one nf omission

omogeneiz'zato a homogenized. **omo'geneo** a homogeneous

omolo'gare vt approve

o'monimo, -a nmf namesake ● nm (parola) homonym

omosessu'al|e a & nmf homosexual. **~ità** nf homosexuality

On. abbr (onorevole) MP

'**oncia** nf ounce

'**onda** nf wave; **andare in ~** Radio

go on the air. **a onde** in waves. **onde** pl **corte** short wave. **onde** pl **lunghe** long wave. **onde** pl **medie** medium wave. **on'data** nf wave

'**onde** conj so that ● pron whereby

ondeggi'are vi (onde) (barca) roll

ondula'torio a undulating. **~'zi-one** nf undulation; (di capelli) wave

'oner|e nm burden. **~'oso** a onerous

onestà nf honesty; (rettitudine) integrity. **o'nesto** a honest; (giusto) just

'**onice** nf onyx

onnipo'tente a omnipotent

onnipre'sente a ubiquitous; Rel omnipresent

ono'mastico nm name-day

ono'ra|bile a honourable. **~re** vt (fare onore a) be a credit to; honour (promessa). **~rio** a honorary ● nm fee. **~rsi** vr **~rsi di** be proud of

o'nore nm honour; **in ~ di** (festa, ricevimento) in honour of; **fare ~ a** do justice to (pranzo); **farsi ~ in** excel in time; **fare gli onori di casa** do the honours

ono'revole a honourable ● nmf Member of Parliament

onorifi'cenza nf honour; (decorazione) decoration. **ono'rifico** a honorary

'**onta** nf shame

O.N.U. nf abbr (**Organizzazione delle Nazioni Unite**) UN

o'paco a opaque; (colori ecc) dull; (fotografia, rossetto) matt

o'pale nf opal

'**opera** nf (lavoro) work; (azione) deed; Mus opera; (teatro) opera house; (ente) institution; **mettere in ~** put into effect; **mettersi all'~** get to work; **opere** pl **pubbliche** public works. **~ d'arte** work of art. **~ lirica** opera

ope'raio, -a a working ● nmf worker; **~ specializzato** skilled worker

ope'ra|re vt Med operate on; **farsi ~re** have an operation ● vi operate; (*agire*) work. ~'tivo, ~'torio *a* operating *attrib.* ~'tore, ~'trice *nmf* operator; TV cameraman. ~tore turistico tour operator. ~zi'one *nf* operation; *Comm* transaction

ope'retta *nf* operetta

ope'roso *a* industrious

opini'one *nf* opinion; rimanere della propria ~ still feel the same way. ~ pubblica public opinion, vox pop

'oppio *nm* opium

oppo'nente *a* opposing ● *nmf* opponent

op'por|re *vt* oppose; (*obiettare*) object; (*~re resistenza*) offer resistance. ~si *vr* ~si a oppose

opportu'ni|smo *nm* expediency. ~sta *nmf* opportunist. ~tà *nf inv* opportunity; (*l'essere opportuno*) timeliness. **oppor'tuno** *a* opportune; (*adeguato*) appropriate; ritenere opportuno fare qcsa think it appropriate to do sth; il momento opportuno the right moment

opposi'tore *nm* opposer. ~zi'one *nf* opposition; d'~zione (*giornale, partito*) opposition

op'posto *pp di* opporre ● *a* opposite; (*opinioni*) opposing ● *nm* opposite; all'~ on the contrary

oppres|si'one *nf* oppression. ~'siva *a* oppressive. op'presso *pp di* opprimere ● *a* oppressed. ~'sore *nm* oppressor

oppri'me|nte *a* oppressive. op'primere *vt* oppress; (*gravare*) weigh down

op'pure *conj* otherwise, or [else]; lunedì ~ martedì Monday or Tuesday

op'tare *vi* ~ per opt for

opu'lento *a* opulent

o'puscolo *nm* booklet; (*pubblicitario*) brochure

opzio'nale *a* optional. opzi'one *nf* option

'ora¹ *nf* time; (*unità*) hour; di buon'~ early; che ~ è?, che ore

sono? what time is it?; mezz'~ half an hour; a ore (*lavorare, pagare*) by the hour; 50 km all'~ 50 km an hour; a un'~ di macchina one hour by car; non vedo l'~ di vederti I can't wait to see you; fare le ore piccole stay up until the small hours. l'~ d'arrivo arrival time. l'~ esatta *Teleph* speaking clock. ~ legale daylight saving time. ~ di punta, ore *pl* di punta peak time; (*per il traffico*) rush hour

'ora² *adv* now; (*tra poco*) presently; ~ come ~ just now, at the moment; d'~ in poi from now on; per ~ for the time being, for now; è ~ di finirla! that's enough now! ● *conj* (*dunque*) now [then]; ~ che ci penso,... now that I come to think about it,...

o'racolo *nm* oracle

'orafo *nm* goldsmith

o'rale *a & nm* oral; per via ~ by mouth

ora'mai *adv* = ormai

o'rario *a* (*tariffa*) hourly; (*segnale*) time *attrib;* (*velocità*) per hour ● *nm* time; (*tabella dell'orario*) timetable, schedule *Am;* essere in ~ be on time; in senso ~ clockwise. ~ di chiusura closing time. ~ flessibile flexitime. ~ di sportello banking hours. ~ d'ufficio business hours. ~ di visita *Med* consulting hours

o'rata *nf* gilthead

ora'tore, -'trice *nmf* speaker

ora'torio, -a *a* oratorical ● *nm Mus* oratorio ● *nmf* oratory.

orazi'one *nf Relig* prayer

'orbita *nf* orbit; *Anat* [eye-]socket

or'chestra *nf* orchestra; (*parte del teatro*) pit

orche'stra|le *a* orchestral ● *nmf* member of an/the orchestra. ~re *vt* orchestrate

orchi'dea *nf* orchid

'orco *nm* ogre

'orda *nf* horde

or'digno *nm* device; (*arnese*) tool. ~ esplosivo explosive device

ordi'nale *a & attr* ordinal

ordina'mento *nm* order; (*leggi*) rules *pl*

ordi'nanza *nf* (*del sindaco*) bylaw; d'~ (*soldato*) on duty

ordi'na|re *vt* (*sistemare*) arrange; (*comandare*) order; (*prescrivere*) prescribe; *Relig* ordain

ordi'nario *a* ordinary; (*grossolano*) common; (*professore*) with a permanent position; di ordinaria amministrazione routine ● *nm* ordinary; *Univ* professor

ordi'nato *a* (*in ordine*) tidy

ordinazi'one *nf* order; fare un'~ place an order

'ordine *nm* order; (*di avvocati, medici*) association; mettere in ~ put in order; tidy up (*appartamento ecc*); di prim'~ first-class; di terz'~ (*film, albergo*) third-rate; di ~ pratico/economico (*problema*) of a practical/economic nature; fino a nuovo ~ until further notice; parola d'~ password; l'~ del giorno agenda. ordini sacri *pl* Holy Orders

or'dire *vt* (*tramare*) plot

orec'chino *nm* ear-ring

o'recchio *nm* (*pl inf* orecchie) ear; avere ~o have a good ear; mi è giunto all'~o che... I've heard that...; parlare all'~o a qcno whisper in sb's ear; suonare a ~o play by ear; '~oni *pl Med* mumps *sg*

o'refice *nm* jeweller. ~'ria *nf* (*arte*) goldsmith's art; (*negozio*) goldsmith's [shop]

'orfano, -a *a* orphan ● *nmf* orphan. ~'trofio *nm* orphanage

orga'netto *nm* barrel-organ; (*a bocca*) mouth-organ; (*fisarmonica*) accordion

or'ganico *a* organic ● *nm* personnel

orga'nismo *nm* organism; (*corpo umano*) body

orga'nista *nmf* organist

organiz'za|re *vt* organize. ~rsi *vr* get organized. ~'tore, ~'trice *nmf* organizer. ~zi'one *nf* organization

'organo *nm* organ

or'gasmo *nm* orgasm; *fig* agitation

'orgia *nf* orgy

or'goglio *nm* pride. ~'oso *a* proud

orien'tale *a* eastern; (*cinese ecc*) oriental

orienta'mento *nm* orientation; perdere l'~ lose one's bearings; senso dell'~ sense of direction

orien'ta|re *vt* orientate. ~rsi *vr* find one's bearings; (*tendere*) tend

ori'ente *nm* east. l'Estremo O~ the Far East. il Medio O~ the Middle East

o'rigano *nm* oregano

origi'na|le *a* original; (*eccentrico*) odd ● *nm* original. ~lità *nf* originality. ~re *vt/i* originate. ~rio *a* (*nativo*) native

o'rigine *nf* origin; in ~ originally; aver ~ da originate from; dare ~ a give rise to

origli'are *vi* eavesdrop

o'rina *nf* urine. ori'nale *nm* chamber-pot. ori'nare *vi* urinate

ori'undo *a* native

orizzon'tale *a* horizontal

orizzon'tare *vt* = orientare. oriz'zonte *nm* horizon

or'la|re *vt* hem. ~'tura *nf* hem. 'orlo *nm* edge; (*di vestito ecc*) hem

'orma *nf* track; (*di piede*) footprint; (*impronta*) mark

or'mai *adv* by now; (*passato*) by then; (*quasi*) almost

ormeg'gi|are *vt* moor. or'meggio *nm* mooring

ormo'nale *a* hormonal. or'mone *nm* hormone

orna'mentale *a* ornamental. orna'mento *nm* ornament

or'na|re *vt* decorate. ~rsi *vr* deck oneself. ~to *a* (*stile*) ornate

ornitolo'gia *nf* ornithology

'**oro** *nm* gold; **d'~** gold; *fig* golden; **una persona d'~** a wonderful person

orologi'aio, -a *nmf* clockmaker, watchmaker

oro'logio *nm* (*portatile*) watch; (*da tavolo, muro ecc*) clock. ~ **a pendolo** grandfather clock. ~ **da polso** wrist-watch. ~ **a sveglia** alarm clock

o'roscopo *nm* horoscope

or'rendo *a* awful, dreadful

or'ribile *a* horrible

orripi'lante *a* horrifying

or'rore *nm* horror; **avere qcsa in** ~ hate sth

orsacchi'otto *nm* teddy bear

'**orso** *nm* bear; (*persona scontrosa*) hermit. ~ **bianco** polar bear

or'taggio *nm* vegetable

or'tensia *nf* hydrangea

or'tica *nf* nettle. **orti'caria** *nf* nettle-rash

orticol'tura *nf* horticulture. '**orto** *nm* vegetable plot

or'todosso *a* orthodox

ortogo'nale *a* perpendicular

orto'gra·fia *nf* spelling. ~'**grafico** *a* spelling *attrib*

orto'lano *nm* market gardener; (*negozio*) greengrocer's

ortope'dia *nf* orthopaedics *sg*. ~'**pedico** *a* orthopaedic ● *nm* orthopaedist

orzai'olo *nm* sty

or'zata *nf* barley-water

osan'nato *a* praised to the skies

o'sare *vt/i* dare; (*avere audacia*) be daring

oscenità *nf inv* obscenity. o'sceno *a* obscene

oscil'la·re *vi* swing; (*prezzi ecc:*) fluctuate; *Tech* oscillate; (*fig: essere indeciso*) vacillate. ~'**zi'one** *nf* swinging; (*di prezzi*) fluctuation; *Tech* oscillation

oscura'mento *nm* darkening; (*fig: di vista, mente*) dimming; (*totale*) black-out

oscu'r·are *vt* darken; *fig* obscure.

~'**arsi** *vr* get dark. ~**ità** *nf* darkness. o'**scuro** *a* dark; (*triste*) gloomy; (*incomprensibile*) obscure

ospe'da·le *nm* hospital. ~'**iero** *a* hospital *attrib*

ospi'ta·le *a* hospitable. ~**lità** *nf* hospitality. ~**re** *vt* give hospitality to. '**ospite** *nm* (*chi ospita*) host; (*chi viene ospitato*) guest ● *nf* hostess; guest

o'spizio *nm* (*per vecchi*) [old people's] home

ossa'tura *nf* bone structure; (*di romanzo*) structure, framework. '**osseo** *a* bone *attrib*

ossequi'are *vt* pay one's respects to. os'**sequio** *nm* homage; **ossequi** *pl* respects. ~'**oso** *a* obsequious

osser'van·te *a* (*cattolico*) practising. ~**za** *nf* observance

osser'va·re *vt* observe; (*notare*) notice; keep (*ordine, silenzio*). ~'**tore**, ~'**trice** *nmf* observer. ~'**torio** *nm* *Astr* observatory; *Mil* observation post. ~**zi'one** *nf* observation; (*rimprovero*) reproach

ossessio'na·nte *a* haunting; (*persona*) nagging. ~**re** *vt* obsess; (*infastidire*) nag. **ossessi'one** *nf* obsession; (*assillo*) pain in the neck. **osses'sivo** *a* obsessive. os'**sesso** *a* obsessed

os'sia *conj* that is

ossi'dabile *a* liable to tarnish

ossi'dar·e *vt*, ~**si** *vr* oxidize

os'sido *nm* oxide. ~ **di carbonio** carbon monoxide

ossi'drica *a* fiamma ossidrica blowlamp

ossige'na·re *vt* oxygenate; (*decolorare*) bleach. ~**si** *vr fig* put back on its feet (*azienda*); ~**si i capelli** dye one's hair blonde os'**sigeno** *nm* oxygen

'**osso** *nm* (*Anat: pl nf* **ossa**) bone; (*di frutto*) stone

osso'buco *nm* marrowbone

os'suto *a* bony

ostaco'lare *vt* hinder, obstruct.

o'stacolo nm obstacle; *Sport* hurdle

o'staggio nm hostage; **prendere in ~** take hostage

o'stello nm ~ **della gioventù** youth hostel

osten'ta|re vt show off; **~re indifferenza** pretend to be indifferent. **~zi'one** nf ostentation

oste'ria nf inn

o'stetrico, -a a obstetric ● nmf obstetrician

'ostia nf host; (*cialda*) wafer

'ostico a tough

o'stil|e a hostile. **~ità** nf inv hostility

osti'na|rsi vr persist (**a** in). **~to a** obstinate. **~zi'one** nf obstinacy

ostra'cismo nm ostracism

'ostrica nf oyster

ostro'goto nm parlare ~ talk double Dutch

ostru'ire vt obstruct. **~zi'one** nf obstruction

otorinolaringoi'atra nmf ear, nose and throat specialist

ottago'nale a octagonal. **ot'tagono** nm octagon

ot'tan|ta a & nm eighty. **~'tenne** a & nmf eighty-year-old. **~'tesimo** a eightieth. **~'tina** nf about eighty

ot'tav|a nf octave. **~o** a eighth

otte'nere vt obtain; (*più comune*) get; (*conseguire*) achieve

'ottico, -a a optic[al] ● nmf optician ● nf (*scienza*) optics sg; (*di lenti ecc*) optics pl

otti'ma|le a optimum. **~'mente** adv very well

otti'mis|mo nm optimism. **~ta** nmf optimist. **~tico** a optimistic

'ottimo a very good ● nm optimum

'otto a & nm eight

ot'tobre nm October

otto'cento a & nm eight hundred; **l'O~** the nineteenth century

ot'tone nm brass

ottuage'nario, -a a & nmf octogenarian

ottu'ra|re vt block; fill (*dente*). **~rsi** vr clog. **~'tore** nm *Phot* shutter. **~zi'one** nf stopping; (*di dente*) filling

ot'tuso pp di ottundere ● a obtuse

o'vaia nf ovary

o'vale a & nm oval

o'vat|ta nf cotton wool. **~'tato** a (*suono, passi*) muffled

ovazi'one nf ovation

over'dose nf inv overdose

'ovest nm west

o'vile nm sheep-fold. **~no** a sheep attrib

ovo'via nf two-seater cable car

ovulazi'one nf ovulation

o'vunque adv = **dovunque**

ov'vero conj or; (*cioè*) that is

ovvia'mente adv obviously

ovvi'are vi ~ **a** qcsa counter sth. **'ovvio** a obvious

ozi'are vi laze around. **'ozio** nm idleness; **stare in ozio** idle about. **ozi'oso** a idle; (*questione*) pointless

o'zono nm ozone; **buco nell'~** hole in the ozone layer

••••••••••••••••••••

Pp

pa'ca|re vt quieten. **~to** a quiet

pac'chetto nm packet; (*postale*) parcel, package; (*di sigarette*) pack, packet. **~ software** software package

'pacchia nf (*fam: situazione*) bed of roses

pacchi'ana|ta nf **è una ~** it's so garish. **pacchi'ano** a garish

'pacco nm parcel; (*involto*) bundle. **~ regalo** gift-wrapped package

paccot'tiglia nf (*roba scadente*) junk, rubbish

'pace nf peace; **darsi ~** forget it; **fare ~ con qcno** make it up with sb; **lasciare in ~ qcno** leave sb in peace

pachi'derma nm (animale) pachyderm

pachi'stano, -a nmf & a Pakistani

pacifi'ca|re vt reconcile; (mettere pace) pacify. **~zi'one** nf reconciliation

pa'cifico a pacific; (calmo) peaceful; **il P~** the Pacific

paci'fis|mo nm pacifism. **~ta** nmf pacifist

pacioc'cone, -a nmf fam chubbychops

pa'dano a pianura padana Po Valley

pa'del|la nf frying-pan; (per malati) bedpan. **~lata** nf una ~lata di a frying-panful of

padigli'one nm pavilion

'padr|e nm father; **~i** pl (antenati) forefathers. **pa'drino** nm godfather. **~e'nostro** nm **il ~e-nostro** the Lord's Prayer. **~e-'terno** nm God Almighty

padro'nanza nf mastery. **~ di sé** self-control

pa'drone, -a nmf master; mistress; (datore di lavoro) boss; (proprietario) owner. **~ggi'are** vt master

pae'saggio nm scenery; (pittura) landscape. **~gista** nmf landscape architect

pae'sano, -a a country ● nmf villager

pa'ese nm (nazione) country; (territorio) land; (villaggio) village; **il Bel P~** Italy; **va' a quel ~!** get lost!; **Paesi** pl **Bassi** Netherlands

paf'futo a plump

'paga nf pay, wages pl

pa'gabile a payable

pa'gaia nf paddle

paga'mento nm payment; **a ~** (parcheggio) which you have to pay to use. **~ anticipato** Comm advance payment. **~ alla consegna** cash on delivery, COD

paga'nesimo nm paganism

pa'gano, -a a & nmf pagan

pa'gare vt/i pay; **~ da bere a qcno**

buy sb a drink; **te la faccio ~** you'll pay for this

pa'gella nf [school] report

'pagina nf page. **Pagine** pl **Gialle** Yellow Pages. **~ web** Comput web page

'paglia nf straw

pagliac'cetto nm (per bambini) rompers pl

pagliac'ciata nf farce

pagli'accio nm clown

pagli'aio nm haystack

paglie'riccio nm straw mattress

pagli'etta nf (cappello) boater; (per pentole) steel wool

pagli'uzza nf wisp of straw; (di metallo) particle

pa'gnotta nf [round] loaf

pail'lette nf inv sequin

'paio nm (pl nf **paia**) pair; **un ~** (circa due) a couple; **un ~ di scarpe, forbici** a pair of

'Pakistan nm Pakistan

'pala nf shovel; (di remo, elica) blade; (di ruota) paddle

pala'fitta nf pile-dwelling

pa'lasport nm inv indoor sports arena

pa'late nfpl **a ~** (fare soldi) hand over fist

pa'lato nm palate

palaz'zetto nm **~ dello sport** indoor sports arena

palaz'zina nf villa

pa'lazzo nm palace; (edificio) building. **~ delle esposizioni** exhibition centre. **~ di giustizia** law courts pl, courthouse. **~ dello sport** indoor sports arena

'palco nm (pedana) platform; Theat box. **~['scenico]** stage

pale'sare vt disclose. **~si** vr reveal oneself. **pa'lese** a evident

Pale'sti|na nf Palestine. **~'nese** nmf Palestinian

pa'lestra nf gymnasium, gym; (ginnastica) gymnastics pl

pa'letta nf spade; (per focolare) shovel. **~ [della spazzatura]** dustpan

pa'letto nm peg

'palio *nm* (*premio*) prize. **il P~** horse-race held at Siena

paliz'zata *nf* fence

'palla *nf* ball; (*proiettile*) bullet; (*fam: bugia*) porkie; **che palle!** *vulg* this is a pain in the arse!. **~ di neve** snowball. **~ al piede** *fig* millstone round one's neck

pallaca'nestro *nf* basketball

palla'mano *nf* handball

pallanu'oto *nf* water polo

palla'volo *nf* volley-ball

palleggi'are *vi* (*calcio*) practise ball control; *Tennis* knock up

pallia'tivo *nm* palliative

'pallido *a* pale; **non ne ho la più pallida idea** I don't have the faintest idea

pal'lina *nf* (*di vetro*) marble

pal'lino *nm* avere il **~ del calcio** be crazy about football

pallon'cino *nm* balloon; (*lanterna*) Chinese lantern; (*fam: etilometro*) Breathalyzer®

pal'lone *nm* ball; (*calcio*) football; (*aerostato*) balloon

pal'lore *nm* pallor

pal'loso *a* sl boring

pal'lottola *nf* pellet; (*proiettile*) bullet

'palm|a *nf* Bot palm. **~o** *nm* Anat palm; (*misura*) hand's-breadth; **restare con un ~o di naso** feel disappointed

'palo *nm* pole; (*di sostegno*) stake; (*in calcio*) goalpost; **fare il ~** (*ladro:*) keep a lookout. **~ della luce** lamppost

palom'baro *nm* diver

pal'pare *vt* feel

'palpebra *nf* eyelid

palpi'ta|re *vi* throb; (*fremere*) quiver. **~zi'one** *nf* palpitation.

'palpito *nm* throb; (*del cuore*) beat

pa'lude *nf* marsh, swamp

palu'doso *a* marshy

pa'lustre *a* marshy; (*piante, uccelli*) marsh *attrib*

pam'pino *nm* vine leaf

pana'cea *nf* panacea

'panca *nf* bench; (*in chiesa*) pew

pancar'ré *nm* sliced bread

pan'cetta *nf* Culin bacon; (*di una certa età*) paunch

pan'chetto *nm* [foot]stool

pan'china *nf* garden seat; (*in calcio*) bench

'pancia *nf* belly, tummy *fam*; **mal di ~** stomach-ache; **metter su ~** develop a paunch; **a ~ in giù** lying face down. **panci'era** *nf* corset

panci'olle *nf* stare in **~** lounge about

panci'one *nm* (*persona*) pot belly

panci'otto *nm* waistcoat

pande'monio *nm* pandemonium

pan'doro *nm* kind of sponge cake traditionally eaten at Christmas

'pane *nm* bread; (*pagnotta*) loaf; (*di burro*) block. **~ a cassetta** sliced bread. **pan grattato** breadcrumbs *pl*. **~ di segale** rye bread. **pan di Spagna** sponge cake. **~ tostato** toast

panet'te'ria *nf* bakery; (*negozio*) baker's [shop]. **~i'ere, -a** *nmf* baker

panet'tone *nf* dome-shaped cake with sultanas and candied fruit eaten at Christmas

pan'filo *nm* yacht

pan'forte *nm* nougat-like spicy delicacy from Siena

'panico *nm* panic; **lasciarsi prendere dal ~** panic

pani'ere *nm* basket; (*cesta*) hamper

pani'ficio *nm* bakery; (*negozio*) baker's [shop]

pani'naro *nm* sl preppie

pa'nino *nm* [bread] roll. **~ imbottito** filled roll. **~ al prosciutto** ham roll. **~'teca** *nf* sandwich bar

'panna *nf* cream. **~ da cucina** [single] cream. **~ montata** whipped cream

'panne *nf* Mech in **~** broken down; **restare in ~** break down

pan'nello *nm* panel. **~ solare** solar panel

'panno *nm* cloth; **panni** *pl* (*abiti*)

clothes; **mettersi nei panni di qcno** *fig* put oneself in sb's shoes

pan'nocchia *nf* (*di granoturco*) cob

panno'lino *nm* (*per bambini*) nappy; (*da donna*) sanitary towel

pano'rama *nm* panorama; *fig* overview. **~ico** *a* panoramic

pantacol'lant *nmpl* leggings

pantalon'cini *nmpl* ~ [**corti**] shorts

panta'loni *nmpl* trousers, pants *Am*

pan'tano *nm* bog

pan'tera *nf* panther; (*auto della polizia*) high-speed police car

pan'tofola *nf* slipper. **~'laio, -a** *nmf fig* stay-at-home

pan'zana *nf* fib

pao'nazzo *a* purple

'papa *nm* Pope

papà *nm inv* dad[dy]

pa'pale *a* papal

papa'lina *nf* skull-cap

papa'razzo *nm* paparazzo

pa'pato *nm* papacy

pa'pavero *nm* poppy

'paper|a *nf* (*errore*) slip of the tongue. **~o** *nm* gosling

papil'lon *nm inv* bow tie

pa'piro *nm* papyrus

'pappa *nf* (*per bambini*) pap

pappa'gallo *nm* parrot

pappa'molle *nmf* wimp

'para *nf* suole *nfpl* di ~ crêpe soles

pa'rabola *nf* parable; (*curva*) parabola

para'bolico *a* parabolic

para'brezza *nm inv* windscreen, windshield *Am*

paracadu'tar|e *vt* parachute. **~si** *vr* parachute

paraca'du|te *nm inv* parachute. **~'tismo** *nm* parachuting. **~'tista** *nmf* parachutist

para'carro *nm* roadside post

paradi'siaco *a* heavenly

para'diso *nm* paradise. **~ terrestre** Eden, earthly paradise

parados'sale *a* paradoxical. **para'dosso** *nm* paradox

para'fango *nm* mudguard

paraf'fina *nf* paraffin

parafra'sare *vt* paraphrase

para'fulmine *nm* lightning-conductor

pa'raggi *nmpl* neighbourhood *sg*

parago'na|bile *a* comparable (**a** to). **~re** *vt* compare. **para'gone** *nm* comparison; **a paragone di** in comparison with

pa'ragrafo *nm* paragraph

pa'ra|lisi *nf inv* paralysis. **~'litico, -a** *a & nmf* paralytic. **~liz'zare** *vt* paralyse. **~liz'zato** *a* (*dalla paura*) transfixed

paral'lel|a *nf* parallel line. **~a'mente** *adv* in parallel. **~o** *a &* *nm* parallel; **~e** *pl* parallel bars. **~o'gramma** *nm* parallelogram

para'lume *nm* lampshade

para'medico *nm* paramedic

pa'rametro *nm* parameter

para'noi|a *nf* paranoia. **~co, -a** *a & nmf* paranoid

paranor'male *a* (*fenomeno, facoltà*) paranormal

para'occhi *nmpl* blinkers. **parao'recchie** *nm inv* earmuffs

para'petto *nm* parapet

para'piglia *nm* turmoil

para'plegico, -a *a & nmf* paraplegic

pa'rar|e *vt* (*addobbare*) adorn; (*riparare*) shield; save (*tiro, pallone*); ward off, parry (*schiaffo, pugno*) ●*vi* (*mirare*) lead up to. **~si** *vr* (*abbigliarsi*) dress up; (*da pioggia, pugni*) protect oneself; **~si dinanzi a qcno** appear in front of sb

para'sole *nm inv* parasol

paras'sita *a* parasitic ●*nm* parasite

parasta'tale *a* a government-controlled

pa'rata *nf* parade; (*in calcio*) save; (*in scherma, pugilato*) parry

para'urti *nm inv* *Auto* bumper, fender *Am*

para'vento *nm* screen

par'cella *nf* bill

parcheggi'a|re *vt* park. **par-**

'**cheggio** nm parking; (posteggio) carpark, parking lot Am. ~'**tore**, ~'**trice** nmf parking attendant. ~**tore** abusivo person who illegally earns money by looking after parked cars

par'**chimetro** nm parking-meter

'**parco**[1] a sparing; (moderato) moderate

'**parco**[2] nm park. ~ **di divertimen-ti** fun-fair. ~ **giochi** playground. ~ **naturale** wildlife park. ~ **na-zionale** national park. ~ **regiona-le** [regional] wildlife park

pa'**recchi** a a good many ● pron several

pa'**recchio** a quite a lot of ● pron quite a lot ● adv rather; (parec-chio tempo) quite a time

pareggi'**are** vt level; (eguagliare) equal; Comm balance ● vi draw

pa'**reggio** nm Comm balance; Sport draw

paren'**tado** nm relatives pl; (vinco-lo di sangue) relationship

pa'**rente** nmf relative. ~ **stretto** close relation

paren'**tela** nf relatives pl; (vincolo di sangue) relationship

paren'**tesi** nf inv parenthesis; (se-gno grafico) bracket; (fig: pausa) break. ~ **pl graffe** curly brackets. ~ **quadre** square brackets. ~ **ton-de** round brackets

pa'**reo** nm (copricostume) sarong; **a ~** (gonna) wrap-around

pa'**rere**[1] nm opinion; **a mio ~** in my opinion

pa'**rere**[2] vi seem; (pensare) think; **che te ne pare?** what do you think of it?; **pare di sì** it seems so

pa'**rete** nf wall; (in alpinismo) face. ~ **divisoria** partition wall

'**pari** a inv equal; (numero) even; **andare di ~ passo** keep pace; **essere ~** be even o quits; **arrivare ~** draw; ~ (copiare, ripetere) word for word; **fare ~ o dispari** toss a coin ● nmf inv equal, peer; **ragazza alla ~** au pair [girl]; **mettersi in ~ con qcsa** catch up

with sth ● nm (titolo nobiliare) peer

Pa'**rigi** nf Paris

pa'**riglia** nf pair

pari'**tà** nf equality; Tennis deuce. ~'**tario** a parity attrib

parlamen'**tare** a parliamentary ● nmf Member of Parliament ● vi discuss. parla'**mento** nm Parlia-ment. **il Parlamento europeo** the European Parliament

parlan'**tina** nf **avere la ~** be a chatterbox

par'**la**|**re** vt/i speak, talk; (confessa-re) talk; ~ **bene/male di qcno** speak well/ill of somebody; **non parliamone più** let's forget about it; **non se ne parla nemmeno!** don't even mention it!. ~**to** a (lingua) spoken. ~'**torio** nm par-lour; (in prigione) visiting room

parlot'**tare** vi mutter. parlot'**tio** nm muttering

parmigi'**ano** nm Parmesan

pa'**rodia** nf parody

pa'**rola** nf word; (facoltà) speech; **è una ~!** it is easier said than done!; **parole** pl (di canzone) words, lyr-ics; **rivolgere la ~ a** address; **dare a qcno la propria ~** give sb one's word; **in parole povere** crudely speaking. **parole** pl **incrociate** crossword [puzzle] sg. ~ **d'onore** word of honour. ~ **d'ordine** pass-word. paro'**laccia** nf swear-word

par'**quet** nm inv (pavimento) par-quet flooring

par'**rocchi**|**a** nf parish. ~'**ale** a parish attrib. ~'**ano, -a** nmf pa-rishioner. '**parroco** nm parish priest

par'**rucca** nf wig

parrucchi'**ere, -a** nmf hairdresser

parruc'**chino** nm toupée, hair-piece

parsi'**moni**|**a** nf thrift. ~'**oso** a thrifty

'**parso** pp di parere

'**parte** nf part; (lato) side; (partito) party; (porzione) share; **a ~** apart from; **in ~** in part; **la maggior**

di the majority of; **d'altra ~** on the other hand; **da ~** aside; (*in disparte*) to one side; **farsi da ~** stand aside; **da ~ di** from; (*per conto di*) on behalf of; **è gentile da ~ tua** it is kind of you; **fare una brutta ~** behave badly towards sb; **da che ~ è...?** whereabouts is...?; **da una ~...**, **dall'altra...** on the one hand..., on the other hand...; **dall'altra ~ di** on the other side of; **da nessuna ~** nowhere; **da tutte le parti** (*essere*) everywhere; **da questa ~** (*in questa direzione*) this way; **da un anno a questa ~** for about a year now; **essere dalla ~ di** be on sb's side; **prendere le parti di qcno** take sb's side; **essere ~ in causa** be involved; **fare ~** (*appartenere a*) be a member of; **rendere ~ a** take part in. **~ civile** plaintiff

parteci'pante *nmf* participant

parteci'pa|re *vi* **~re a** participate in, take part in; (*condividere*) share in. **~zi'one** *nf* participation; (*annuncio*) announcement; *Fin* shareholding; (*presenza*) presence. **~'tecipe** *a* participating

parteggi'are *vi* **~ per** side with

par'tenza *nf* departure; *Sport* start; **in ~ per** leaving for

parti'cella *nf* particle

parti'cipio *nm* participle

partico'la|re *a* particular; (*privato*) private **●** *nm* detail, particular; **fin nei minimi ~i** down to the smallest detail. **~eggi'ato** *a* detailed. **~ità** *nf inv* particularity; (*dettaglio*) detail

partigi'ano, -a *a & nmf* partisan

par'tire *vi* leave; (*aver inizio*) start; **a ~ da** [beginning] from

par'tita *nf* game; (*incontro*) match; *Comm* lot; (*contabilità*) entry. **~ di calcio** football match. **~ a carte** game of cards

par'tito *nm* party; (*scelta*) choice; (*occasione di matrimonio*) match;

per ~ preso out of sheer pigheadedness

'parto *nm* childbirth; **un ~ facile** an easy birth o labour; **dolori** *pl* **del ~** labour pains. **~ cesareo** Caesarean section. **~rire** *vt* give birth to

par'venza *nf* appearance

parzi'al|e *a* partial. **~ità** *nf* partiality. **~'mente** *adv* (*non completamente*) partially; **~mente scremato** semi-skimmed

pasco'lare *vt* graze. **'pascolo** *nm* pasture

'Pasqua *nf* Easter. **pa'squale** *a* Easter *attrib*

'passa: e ~ *adv* (*e oltre*) plus

pas'sabile *a* passable

pas'saggio *nm* passage; (*traversata*) crossing; *Sport* pass; (*su veicolo*) lift; **essere di ~** be passing through. **~ a livello** level crossing, grade crossing *Am*. **~ pedonale** pedestrian crossing

passamon'tagna *nm inv* balaclava

pas'sante *nmf* passer-by **●** *nm* (*di cintura*) loop **●** *a Tennis* passing

passa'porto *nm* passport

pas'sa|re *vi* pass; (*attraversare*) pass through; (*far visita*) call; (*andare*) go; (*essere approvato*) be passed; **~re alla storia** go down in history; **mi è ~to di mente** it slipped my mind; **~re per un genio/idiota** be taken for a genius/an idiot; **farsi ~re per qcno** pass oneself off as sb **●** *vt* (*far scorrere*) pass over; (*sopportare*) go through; (*al telefono*) put through; (*Culin*) strain; **~re di moda** go out of fashion; **le passo il signor Rossi** I'll put you through to Mr Rossi; **~rsela bene** be well off; **come te la passi?** how are you doing?. **~ta** *nf* (*di vernice*) coat; (*spolverata*) dusting; (*occhiata*) look

passa'tempo *nm* pastime

pas'sato *a* past; **l'anno ~** last year; **sono le tre passate** it's past

o after three o'clock ● *nm* past;
Culin purée; *Gram* past tense. ~
prossimo *Gram* present perfect.
~ **remoto** *Gram* [simple] past. ~
di verdure cream of vegetable
soup

passaver'dure *nm inv* food mill

passeg'gero, -a *a* passing ● *nmf*
passenger

passeggi'a|re *vi* walk, stroll. **~ta**
nf walk, stroll; (*luogo*) public
walk; (*in bicicletta*) ride; **fare una
~ta** go for a walk

passeg'gino *nm* pushchair,
stroller *Am*

pas'seggio *nm* walk; (*luogo*)
promenade; **andare a ~** go for a
walk; **scarpe da ~** walking
shoes

passe-partout *nm inv* master-key

passe'rella *nf* gangway; *Aeron*
boarding bridge; (*per sfilate*) catwalk

'passero *nm* sparrow. **passe'rotto**
nm (*passero*) sparrow

pas'sibile *a* ~ **di** liable to

passio'nale *a* passionate. **passi'o-**
ne *nf* passion

pas'sivo *a* passive ● *nm* passive;
Comm liabilities *pl*; **in ~** (*bilancio*) loss-making

'passo *nm* step; (*orma*) footprint;
(*andatura*) pace; (*brano*) passage;
(*valico*) pass; **a due passi da qui** a
stone's throw away; **a ~ d'uomo**
at walking pace; **di buon ~** at
a spanking pace; **fare due passi**
go for a stroll; **di pari ~** *fig* hand
in hand. **~ carrabile, ~ carraio**
driveway

'past|a *nf* (*impasto per pane ecc*)
dough; (*per dolci, pasticcino*) pastry; (*pastasciutta*) pasta; (*massa
molle*) paste; *fig* nature. **~a frolla**
shortcrust pastry. **pa'stella** *nf* batter

pastasci'utta *nf* pasta

pa'stello *nm* pastel

pa'sticca *nf* pastille; (*fam: pastiglia*) pill

pasticce'ria *nf* cake shop, patisse-

rie; (*pasticcini*) pastries *pl*; (*arte*)
confectionery

pasticci'are *vi* make a mess ● *vt*
make a mess of

pasticci'ere, -a *nmf* confectioner

pastic'cino *nm* little cake

pa'sticci|o *nm Culin* pie; (*lavoro
disordinato*) mess; **mettersi nei
~** get into trouble. **~'one,
-a** *nmf* bungler ● *a* a bungling

pasti'ficio *nm* pasta factory

pa'stiglia *nf Med* pill, tablet; (*di
menta*) sweet. **~ dei freni** brake
pad

'pasto *nm* meal

pasto'rale *a* pastoral. **pa'store** *nm*
shepherd; *Relig* pastor. **pastore
tedesco** German shepherd, Alsatian

pastoriz'za|re *vt* pasteurize. **~to** *a*
pasteurized. **~zi'one** *nf* pasteuri-
zation

pa'stoso *a* doughy; *fig* mellow

pa'stura *nf* pasture; (*per pesci*)
bait

pa'tacca *nf* (*macchia*) stain; (*fig:
oggetto senza valore*) piece of junk

pa'tata *nf* potato. **patate** *pl* **fritte**
chips *Br*, French fries. **pata'tine**
nfpl [potato] crisps, chips *Am*

pata'trac *nm inv* (*crollo*) crash

pâté *nm inv* pâté

pa'tella *nf* limpet

pa'tema *nm* anxiety

pa'tente *nf* licence. **~ di guida**
driving licence, driver's license
Am

pater'na|le *nf* scolding. **~'lista** *nm*
paternalist

paternità *nf* paternity. **pa'terno** *a*
paternal; (*affetto ecc*) fatherly

pa'tetico *a* pathetic. **'pathos** *nm*
pathos

pa'tibolo *nm* gallows *sg*

'patina *nf* patina; (*sulla lingua*)
coating

pa'ti|re *vt/i* suffer. **~to, -a** *a* suffer-
ing ● *nmf* fanatic. **~to della
musica** music lover

patolo'gia *nf* pathology. **pato-
'logico** *a* pathological

'**patria** *nf* native land
patri'arca *nm* patriarch
pa'trigno *nm* stepfather
patrimoni'ale *a* property *attrib.*
patri'monio *nm* estate
patri'o|ta *nmf* patriot. **~tico** *a* patriotic. **~'tismo** *nm* patriotism
pa'trizio, -a *a & nmf* patrician
patro|ci'nare *vt* support. **~'cinio** *nm* support
patro'nato *nm* patronage. **pa'trono** *nm* Relig patron saint; Jur counsel
'**patta** *nf* (*di tasca*) flap
'**patta²** *nf* (*pareggio*) draw
patteggi|a'mento *nm* bargaining. **~'are** *vt/i* negotiate
patti'naggio *nm* skating. **~ su ghiaccio** ice skating. **~ a rotelle** roller skating
patti'na|re *vi* skate; (*auto:*) skid. **~'tore, ~'trice** *nmf* skater. '**pattino** *nm* skate; Aeron skid. **pattino da ghiaccio** iceskate. **pattino a rotelle** roller-skate
'**patto** *nm* deal; Pol pact; **a ~ che** on condition that
pat'tuglia *nf* patrol. **~ stradale** patrol car; police motorbike, highway patrol Am
pattu'ire *vt* negotiate
pattumi'era *nf* dustbin, trashcan Am
pa'ura *nf* fear; (*spavento*) fright; **aver ~** be afraid; **mettere ~ a** frighten. **pau'roso** *a* (*che fa paura*) frightening; (*che ha paura*) fearful; (*fam: enorme*) awesome
'**pausa** *nf* pause; (*nel lavoro*) break; **fare una ~** pause; (*nel lavoro*) have a break
pavimen'ta|re *vt* pave (*strada*). **~zi'one** *nf* (*operazione*) paving. **pavi'mento** *nm* floor
pa'vone *nm* peacock. **~ggi'arsi** *vr* strut
pazien'tare *vi* be patient
pazi'ente *a & nmf* patient. **~'mente** *adv* patiently. **pazi'enza** *nf* patience; **pazienza!** never mind!

'**pazza** *nf* madwoman. **~'mente** *adv* madly
paz'z|esco *a* foolish; (*esagerato*) crazy. **~ia** *nf* madness; (*azione*) [act of] folly. '**pazzo** *a* mad; fig crazy ● *nm* madman; **essere pazzo di/per** be crazy about; **pazzo di gioia** mad with joy; **da pazzi** fam crackpot; **darsi alla pazza gioia** live it up. **paz'zoide** *a* whacky
'**pecca** *nf* fault; **senza ~** flawless. **peccami'noso** *a* sinful
pec'ca|re *vi* sin; **~re di** be guilty of (*ingratitudine*). **~to** *nm* sin; **~to che...** it's a pity that...; [**che**] **~to!** [what a] pity!. **~'tore, ~'trice** *nmf* sinner
'**pece** *nf* pitch
'**peco|ra** *nf* sheep. **~ra nera** black sheep. **~'raio** *nm* shepherd. **~'rella** *nf* cielo a **~relle** sky full of fluffy white clouds. **~'rino** *nm* (*formaggio*) sheep's milk cheese
peculi'ar|e *a* **~ di** peculiar to. **~ità** *nf inv* peculiarity
pe'daggio *nm* toll
peda'go'gia *nf* pedagogy. **peda'gogico** *a* pedagogical
peda'lare *vi* pedal. **pe'dale** *nm* pedal. **pedalò** *nm inv* pedalo
pe'dana *nf* footrest; Sport springboard
pe'dante *a* pedantic. **~'ria** *nf* pedantry. **pedan'tesco** *a* pedantic
pe'data *nf* (*in calcio*) kick; (*impronta*) footprint
pede'rasta *nm* pederast
pe'destre *a* pedestrian
pedi'atra *nmf* paediatrician. **pedia'tria** *nf* paediatrics *sg*
pedi'cure *nmf inv* chiropodist, podiatrist Am ● *nm* (*cura dei piedi*) pedicure
pedi'gree *nm inv* pedigree
pe'dina *nf* (*alla dama*) piece; fig pawn. **~'mento** *nm* shadowing. **pedi'nare** *vt* shadow
pedo'filo, -a *nmf* paedophile
pedo'nale *a* pedestrian. **pe'done, -a** *nmf* pedestrian

peeling *nm inv* exfoliation treatment

'**peggio** *adv* worse; **~ per te!** too bad!; **~ di cosi** any worse; **la persona ~ vestita** the worst dressed person ● *a* worse; **niente di ~** nothing worse ● *nm* il **~ è che...** the worst of it is that...; **pensare al ~** think the worst ● *nf* **alla ~** at worst; **avere la ~** get the worst of it; **alla meno ~** as best I can

peggiora'mento *nm* worsening

peggio'ra|re *vt* make worse, worsen ● *vi* get worse, worsen. **~tivo** *a* pejorative

peggi'ore *a* worse; (*superlativo*) worst; **nella ~ delle ipotesi** if the worst comes to the worst ● *nmf* il/la **~** the worst

'**pegno** *nm* pledge; (*nei giochi di società*) forfeit; *fig* token

pelan'drone *nm* slob

pe'la|re *vt* (*spennare*) pluck; (*spellare*) skin; (*sbucciare*) peel; (*fam: spillare denaro*) fleece. **~rsi** *vr fam* lose one's hair. **~to** *a* bald. **~ti** *nmpl* (*pomodori*) peeled tomatoes

pel'lame *nm* skins *pl*

pelle *nf* skin; (*cuoio*) leather; (*buccia*) peel; **avere la ~ d'oca** have goose-flesh

pellegri'naggio *nm* pilgrimage. **pelle'grino, -a** *nmf* pilgrim

pelle'rossa *nmf* Red Indian, Redskin

pellette'ria *nf* leather goods *pl*

pelli'cano *nm* pelican

pellicce'ria *nf* furrier's [shop]. **pel'licc|ia** *nf* fur; (*indumento*) fur coat. **~i'aio, -a** *nmf* furrier

pel'licola *nf* Phot, Cinema film. **~ [trasparente]** cling film

'**pelo** *nm* hair; (*di animale*) coat; (*di lana*) pile; **per un ~** by the skin of one's teeth; **cavarsela per un ~** have a narrow escape. **pe'loso** *a* hairy

'**peltro** *nm* pewter

pe'luche *nm inv* **giocattolo di ~** soft toy

pe'luria *nf* down

pelvico *a* pelvic

'**pena** *nf* (*punizione*) punishment; (*sofferenza*) pain; (*dispiacere*) sorrow; (*disturbo*) trouble; **a mala ~** hardly; **mi fa ~** I pity him; **vale la ~ andare** it is worth [while] going. **~ di morte** death sentence

pe'na|le *a* criminal; **diritto** *nm* **~e** criminal law. **~ità** *nf inv* penalty

penaliz'za|re *vt* penalize. **~zi'one** *nf* (*penalità*) penalty

pe'nare *vi* suffer; (*faticare*) find it difficult

pen'daglio *nm* pendant

pen'dant *nm inv* **fare ~ [con]** match

pen'den|te *a* hanging; *Comm* outstanding ● *nm* (*ciondolo*) pendant; **~ti** *pl* drop earrings. **~za** *nf* slope; *Comm* outstanding account

'**pendere** *vi* hang; (*superficie:*) slope; (*essere inclinato*) lean

pen'dio *nm* slope; **in ~** sloping

pendo'l|are *a* pendulum ● *nmf* commuter. **~ino** *nm* (*treno*) special, first class only, fast train

'pendolo *nm* pendulum

pene *nm* penis

pene'trante *a* penetrating; (*freddo*) biting

pene'tra|re *vt/i* penetrate; (*trafiggere*) pierce ● *vt* (*odore:*) get into ● *vi* (*entrare furtivamente*) steal in. **~zi'one** *nf* penetration

penicil'lina *nf* penicillin

pe'nisola *nf* peninsula

peni'ten|te *a & nmf* penitent. **~za** *nf* penitence; (*punizione*) penance; (*in gioco*) forfeit. **~zi'ario** *nm* penitentiary

'**penna** *nf* (*da scrivere*) pen; (*di uccello*) feather. **~ a feltro** felt-tip[ped pen]. **~ a sfera** ball-point [pen]. **~ stilografica** fountain-pen

pen'nacchio *nm* plume

penna'rello *nm* felt-tip[ped pen]

pennel'la|re *vt* paint. **~ta** *nf* brushstroke. **pen'nello** *nm* brush; **a pennello** (*a perfezione*) perfectly

pen'nino *nm* nib

pen'none nm (di bandiera) flag-pole

pen'nuto a feathered

pe'nombra nf half-light

pe'noso a (fam: pessimo) painful

pen'sa|re vi think; **penso di sì** I think so; **~re a** think of; remember to ⟨chiudere il gas ecc⟩; **pensa ai fatti tuoi!** mind your own business!; **ci penso io** I'll take care of it; **~re di fare qcsa** think of doing sth; **~re tra sé e sé** think to oneself ● vt think. **~re** nf idea

pensi'e|ro nm thought; (mente) mind; (preoccupazione) worry; **stare in ~ro per** be anxious about. **~roso** a pensive

'pensi|le a hanging; **giardino ~le** roof-garden ● nm (mobile) wall unit. **~'lina** nf (di fermata d'autobus) bus shelter

pensio'nante nmf boarder; (ospite pagante) lodger

pensio'nato, -a nmf pensioner ● nm (per anziani) [old folks'] home; (per studenti) hostel. **pensi'one** nf pension; (albergo) boarding-house; (vitto e alloggio) board and lodging; **andare in pensione** retire; **mezza pensione** half board. **pensione completa** full board

pen'soso a pensive

pen'tagono nm pentagon

Pente'coste nf Whitsun

penti'mento nm repentance

pen'ti|rsi vr **~rsi di** repent of; (rammaricarsi) regret. **~'tismo** nm turning informant. **~to** nm Mafioso turned informant

'pentola nf saucepan; (contenuto) potful. **~ a pressione** pressure cooker

pe'nultimo a last but one, penultimate

pe'nuria nf shortage

penzo'l|are vi dangle. **~o** a peppery

pe'pa|re vt pepper. **~to** a peppery

'pepe nm pepper; **grano di ~** peppercorn. **~ in grani** whole pepper-corns. **~ macinato** ground pepper

pepero'n|ata nf peppers cooked in olive oil with onion, tomato and garlic. **~'cino** nm chilli pepper. **pepe'rone** nm pepper. **peperone verde** green pepper

pe'pita nf nugget

per prep for; (attraverso) through; (stato in luogo) in, on; (distributivo) per; (mezzo, entro) by; (causa) with; (in qualità di) as; **~ strada** on the street; **~ la fine del mese** by the end of the month; **in fila ~ due** in double file; **l'ho sentito ~ telefono** I spoke to him on the phone; **~ iscritto** in writing; **~ caso** by chance; **ho aspettato ~ ore** I've been waiting for hours; **~ tempo** in time; **~ sempre** forever; **~ scherzo** as a joke; **gridare ~ il dolore** scream with pain; **vendere ~ 10 milioni** sell for 10 million; **uno ~ volta** one at a time; **uno ~ uno** one by one; **venti ~ cento** twenty per cent; **~ fare qcsa** [in order to] do sth; **stare ~** be about to; **è troppo bello ~ essere vero** it's too good to be true

'pera nf pear; **farsi una ~** (sl: di eroina) shoot up

perbe'nismo nm prissiness. **~ta** a inv prissy

per'cento adv per cent. **percen-tu'ale** nf percentage

perce'pibile a perceivable; (somma) payable

perce'pi|re vt perceive; (riscuotere) cash

perce'tti|bile a perceptible. **~zi'one** nf perception

perché conj (in interrogazioni) why; (per il fatto che) because; (affinché) so that; **~ non vieni?** why don't you come?; **dimmi ~** tell me why; **~ no/sì** because!; **la ragione ~ l'ho fatto** the reason [that] I did it, the reason why I did it; **è troppo difficile ~ lo possa capire** it's too difficult for me to understand ● nm inv reason

[why]; **senza un** ~ without any
reason

perciò *conj* so

per'correre *vt* cover *(distanza)*;
(viaggiare) travel. **per'corso** *pp
di* **percorrere** ●*nm (tragitto)*
course, route; *(distanza)* distance;
(viaggio) journey

per'cossa *nf* blow. ~o *pp di*
percuotere. **percu'otere** *vt* strike

percussi'one *nf* percussion;
strumenti *pl a* ~**e** percussion
instruments. ~'**nista** *nmf* percus-
sionist

per'dente *nmf* loser

'perder|e *vt* lose; *(sprecare)* waste;
(non prendere) miss; *(fig: vizio)*
ruin; ~**e tempo** waste time ● *vi*
lose; *(recipiente)* leak; **lascia** ~**e**
forget it!. ~**si** *vr* get lost; *(reciproco)* lose touch

perdifi'ato: **a** ~ *adv (gridare)* at
the top of one's voice

perdigi'orno *nmf inv* idler

'perdita *nf* loss; *(spreco)* waste;
(falla) leak; **a** ~ **d'occhio** as far
as the eye can see. ~ **di tempo**
waste of time. **perditempo** *nm*
waste of time

perdo'nare *vt* forgive; *(scusare)*
excuse. **per'dono** *nm* forgiveness;
Jur pardon

perdu'rare *vi* last; *(perseverare)*
persist

perduta'mente *adv* hopelessly.
per'duto *pp di* **perdere** ●*a* lost;
(rovinato) ruined

pe'renne *a* everlasting; *Bot* perennial; **nevi perenni** perpetual
snow. ~'**mente** *adv* perpetually

peren'torio *a* peremptory

per'fetto *a* perfect ● *nm Gram* perfect [tense]

perfezio'nar|e *vt* perfect; *(miglio-
rare)* improve. ~**si** *vr* improve
oneself; *(specializzarsi)* specialize

perfezi'o|ne *nf* perfection; **alla**
~**ne** to perfection. ~'**nismo** *nm*
perfectionism. ~'**nista** *nmf* per-
fectionist

per'fidia *nf* wickedness; *(atto)*

wicked act. **'perfido** *a* treacherous; *(malvagio)* perverse

per'fino *adv* even

perfo'ra|re *vt* pierce; punch *(sche-
de)*; *Mech* drill. ~'**tore**, ~'**trice**
nmf punch-card operator ●*nm*
perforator. ~**zi'one** *nf* perfora-
tion; *(di schede)* punching

per'formance *nf inv* performance

perga'mena *nf* parchment

perico'lante *a* precarious; *(azienda)* shaky

pe'rico|lo *nm* danger; *(rischio)*
risk; **mettere in** ~**lo** endanger.
~**lo pubblico** danger to society.
~'**loso** *a* dangerous

perife'ria *nf* periphery; *(di città)*
outskirts *pl*; *fig* fringes *pl*

peri'feric|a *nf* periphery; *(strada)*
ring road. ~**o** *a (quartiere)* outlying

pe'rifrasi *nf inv* circumlocution

pe'rimetro *nm* perimeter

peri'odico *nm* periodical ● *a* periodical; *(vento, mal di testa, Math)*
recurring. **pe'riodo** *nm* period;
Gram sentence. **periodo di prova**
trial period

peripe'zie *nfpl* misadventures

pe'rire *vi* perish

peri'scopio *nm* periscope

pe'ri|to, -a *a* skilled ● *nmf* expert

perito'nite *nf* peritonitis

pe'rizia *nf* skill; *(valutazione)* survey

'perla *nf* pearl. **per'lina** *nf* bead

perlo'meno *adv* at least

perlu'stra|re *vt* patrol. ~**zi'one** *nf*
patrol; **andare in** ~**zione** go on
patrol

perma'loso *a* touchy

perma'ne|nte *a* permanent ● *nf*
perm; **farsi [fare] la** ~**nte** have a
perm. ~**nza** *nf* permanence; *(soggiorno)* stay; **in** ~**nza** permanently. ~**re** *vi* remain

perme'are *vt* permeate

per'messo *pp di* **permettere** ● *nm*
permission; *(autorizzazione)* permit; *Mil* leave; [**è**] ~**?** *(posso
entrare?)* may I come in?; *(posso*

passare?) excuse me. ~ **di lavoro** work permit

per'mettere *vt* allow, permit; **potersi** ~ **qcsa** *(finanziariamente)* be able to afford sth; **come si permette?** how dare you?. **permis'sivo** *a* permissive

permutazi'one *nf* exchange; *Math* permutation

per'nacchia *nf (sl: con la bocca)* raspberry *sl*

per'nice *nf* partridge. ~**i'oso** *a* pernicious

'perno *nm* pivot

pernot'tare *vi* stay overnight

'pero *nm* pear-tree

però *conj* but; *(tuttavia)* however

pero'rare *vt* plead

perpendico'lare *a & nf* perpendicular

perpe'trare *vt* perpetrate

perpetu'are *vt* perpetuate. **per'petuo** *a* perpetual

perplessità *nf inv* perplexity; *(dubbio)* doubt. **per'plesso** *a* perplexed

perqui'si|re *vt* search. ~**zi'one** *nf* search. ~**zione domiciliare** search of the premises

persecu'tore, -'**trice** *nmf* persecutor. ~**zi'one** *nf* persecution

persegu'ire *vt* pursue

persegui'tare *vt* persecute

perseve'ra|nte *a* persevering. ~**nza** *nf* perseverance. ~**re** *vi* persevere

persi'ano, -a *a* Persian ● *nf (di finestra)* shutter. **'persico** *a* Persian

per'sino *adv* = **perfino**

persi'sten|te *a* persistent. ~**za** *nf* persistence. **per'sistere** *vi* persist

'perso *pp di* **perdere** ● *a* lost; **a tempo** ~ in one's spare time

per'sona *nf* person; *(un tale)* somebody; **di** ~, **in** ~ in person, personally; **per** ~ per person, a head; **per interposta** ~ through an intermediary; **persone** *pl* people

perso'naggio *nm (persona di*

riguardo) personality; *Theat* ecc character

perso'na|le *a* personal ● *nm* staff. ~**e di terra** ground crew. ~**ità** *nf inv* personality. ~**iz'zare** *vt* customize *(auto ecc)*; personalize *(penna ecc)*

personifi'ca|re *vt* personify. ~**zi'one** *nf* personification

perspi'cac|e *a* shrewd. ~**ia** *nf* shrewdness

persua'de|re *vt* convince; impress *(critici)*; ~**dere qcno a fare qcsa** persuade sb to do sth. ~**si'one** *nf* persuasion. ~**'sivo** *a* persuasive. **persu'aso** *pp di* **persuadere**

per'tanto *conj* therefore

'pertica *nf* pole

perti'nente *a* relevant

per'tosse *nf* whooping cough

pertur'ba|re *vt* perturb. ~**rsi** *vr* be perturbed. ~**zi'one** *nf* disturbance. ~**zione atmosferica** atmospheric disturbance

per'va|dere *vt* pervade. ~**so** *pp di* **pervadere**

perve'nire *vi* reach; **far** ~ **qcsa a qcno** send sth to sb

perversi|'one *nf* perversion. ~**ità** *nf* perversity. **per'verso** *a* perverse

perver'ti|re *vt* pervert. ~**to** *a* perverted ● *nm* pervert

per'vinca *nm (colore)* blue with a touch of purple

p. es. *abbr (per esempio)* e.g.

pesa *nf* weighing; *(bilancia)* weighing machine; *(per veicoli)* weighbridge

pe'sante *a* heavy; *(stomaco)* overfull ● *adv (vestirsi)* warmly. ~**'mente** *adv (cadere)* heavily. **pesan'tezza** *nf* heaviness

pe'sar|e *vt/i* weigh; ~**e su** *fig* lie heavy on; ~**e le parole** weigh one's words. ~**si** *vr* weigh oneself

'pesca[^1] *nf (frutto)* peach

'pesca[^2] *nf* fishing; **andare a** ~ go fishing. ~ **subacquea** underwater fishing. **pe'scare** *vt (andare a pesca di)* fish for; *(prendere)* catch;

(*fig: trovare*) fish out. **~'tore** *nm* fisherman

'pesce *nm* fish. **~ d'aprile!** April Fool!. **~ grosso** *fig* big fish. **~ piccolo** *fig* small fry. **~ rosso** goldfish. **~ spada** swordfish. **Pesci** *Astr* Pisces

pesce'cane *nm* shark

pesche'reccio *nm* fishing boat

pesc|he'ria *nf* fishmonger's [shop]. **~hi'era** *nf* fish-pond. **~i-'vendolo** *nm* fishmonger

'pesco *nm* peach-tree

'peso *nm* weight; **essere di ~ per qcno** be a burden to sb; **di poco ~** (*senza importanza*) not very important; **non dare ~ a qcsa** not attach any importance to sth

pessi'mis|mo *nm* pessimism. **~ta** *nmf* pessimist ● *a* pessimistic. **'pessimo** *a* very bad

pe'staggio *nm* beating-up. **pe'stare** *vt* tread on; (*schiacciare*) crush; (*picchiare*) beat; crush (*aglio, prezzemolo*)

'peste *nf* plague; (*persona*) pest

pe'stello *nm* pestle

pesti'cida *nm* pesticide. **pe'stifero** *a* (*fastidioso*) pestilential

pesti'len|za *nf* pestilence; (*fetore*) stench. **~zi'ale** *a* (*odore aria*) noxious

'pesto *a* ground; **occhio ~** black eye ● *nm* basil and garlic sauce

'petalo *nm* petal

pe'tardo *nm* banger

petizi'one *nf* petition; **fare una ~** draw up a petition

petro|li'era *nf* [oil] tanker. **~lifero** *a* oil-bearing. **pe'trolio** *nm* oil

pette'gare *vi* gossip. **~lezzo** *nm* piece of gossip; **far ~lezzi** gossip

pet'tegolo, -a *a* gossipy ● *nmf* gossip

petti'na|re *vt* comb. **~rsi** *vr* comb one's hair. **~'tura** *nf* combing; (*acconciatura*) hair-style. **'pettine** *nm* comb

'petting *nm* petting

petti'nino *nm* (*fermaglio*) comb

petti'rosso *nm* robin [redbreast]

'petto *nm* chest; (*seno*) breast; **a doppio ~** double-breasted

petto'rale *nm* (*in gare sportive*) number.. **~'rina** *nf* (*di salopette*) bib. **~'ruto** *a* (*donna*) full-breasted; (*uomo*) broad-chested

petu'lante *a* impertinent

'pezza *nf* cloth; (*toppa*) patch; (*rotolo di tessuto*) roll

pez'zente *nmf* tramp; (*avaro*) miser

'pezzo *nm* piece; (*parte*) part; **un bel ~ d'uomo** a fine figure of a man; **un ~** (*di tempo*) some time; (*di spazio*) a long way; **al ~** (*costare*) each; **essere a pezzi** (*stanco*) be shattered; **fare a pezzi** tear to shreds. **~ grosso** bigwig

pia'cente *a* attractive

pia'ce|re *nm* pleasure; (*favore*) favour; **a ~re** as much as one likes; **per ~re!** please!; **~re** [**di conoscerla**]! pleased to meet you!; **con ~re** with pleasure ● *vi* **la Scozia mi piace** I like Scotland; **mi piacciono i dolci** I like sweets; **faccio come mi pare e piace** I do as I please; **ti piace?** do you like it?; **lo spettacolo è piaciuto** the show was a success. **~vole** *a* pleasant

piaci'mento *nm* **a ~** as much as you like

pia'dina *nf* unleavened focaccia bread

pi'aga *nf* sore; *fig* scourge; (*fig: persona noiosa*) pain; (*fig: ricordo doloroso*) wound

piagni'steo *nm* whining

piagnuco'lare *vi* whimper

pi'alla *nf* plane. **pial'lare** *vt* plane

pi'ana *nf* (*pianura*) plane. **pianeg-gi'ante** *a* level

piane'rottolo *nm* landing

pia'neta *nm* planet

pi'angere *vi* cry; (*disperatamente*) weep ● *vt* (*lamentare*) lament; (*per un lutto*) mourn

pianifi'ca|re *vt* plan. **~zi'one** *nf* planning

pia'nista *nmf* *Mus* pianist

pi'ano a flat; (a livello) flush; (regolare) smooth; (facile) easy ● adv slowly; (con cautela) gently; **andarci ~** go carefully ● nm plain; (di edificio) floor; (livello) plane; (progetto) plan; Mus piano; **di primo ~** first-rate; **primo ~** Phot close-up; **in primo ~** in the foreground. **~ regolatore** town plan. **~ di studi** syllabus

piano'forte nm piano. **~ a coda** grand piano

piano'terra nm inv ground floor, first floor Am

pi'anta nf plant; (del piede) sole; (disegno) plan; (di sana ~ (totalmente) entirely; **in ~ stabile** permanently. **~ stradale** road map. **~gi'one** nf plantation

piantagrane nmf fam è un/una ~ he's/she's bolshie

pian'tar|e vt plant; (conficcare) drive; (fam: abbandonare) dump; **piantala!** fam stop it!. **~si** vr plant oneself; (fam: lasciarsi) leave each other

pianter'reno nm ground floor, first floor Am

pi'anto pp di piangere ● nm crying; (disperato) weeping; (lacrime) tears pl

pianto'nare vt guard. **~'tone** nm guard

pia'nura nf plain

pi'astra nf plate; (lastra) slab; Culin griddle. **~ elettronica** circuit board. **~ madre** Comput motherboard

pia'strella nf tile

pia'strina nf Mil identity disc; Med platelet; Comput chip

piatta'forma nf platform. **~ di lancio** launch pad

piat'tino nm saucer

pi'atto a flat ● nm plate; (da portata, vivanda) dish; (portata) course; (parte piatta) flat; (di giradischi) turntable; **piatti** pl Mus cymbals; **lavare i piatti** do the dishes, do the washing-up. **~**

fondo soup plate. **~ piano** [ordinary] plate

pi'azza nf square; Comm market; **letto a una ~** single bed; **letto a due piazze** double bed; **far ~ pulita** make a clean sweep. **~'forte** nf stronghold. **piaz'zale** nm large square. **~'mento** nm (in classifica) placing

piaz'za|re vt place. **~rsi** vr Sport be placed; **~rsi secondo** come second. **~to** a (cavallo) placed; **ben ~to** (robusto) well built

piaz'zista nm salesman

piaz'zuola nf **~ di sosta** pull-in

pic'cante a hot; (pungente) sharp; (salace) spicy

pic'carsi vr (risentirsi) take offence; **~ di** (vantarsi di) claim to

'picche nfpl (in carte) spades

picchet'tare vt stake; (scioperanti) picket. **pic'chetto** nm picket

picchi'a|re vt beat, hit ● vi (bussare) knock; Aeron nosedive; **~re in testa** (motore) knock. **~ta** nf beating; Aeron nosedive; **scendere in ~ta** nosedive

picchiet'tare vt tap; (punteggiare) spot

picchiet'tio nm tapping

'picchio nm woodpecker

pic'cino a tiny; (gretto) mean; (di poca importanza) petty ● nm little one, child

picci'one nm pigeon

'picco nm peak; **a ~** vertically; **colare a ~** sink

'piccolo, -a a small, little; (di età) young; (di statura) short; (gretto) petty ● nmf child, little one; **da ~** as a child

pic'co|ne nm pickaxe. **~zza** nf ice axe

pic'nic nm inv picnic

pi'docchio nm louse

piè nm inv **a ~ di pagina** at the foot of the page; **saltare a ~ pari** skip

pi'ede nm foot; **a piedi** on foot; **andare a piedi** walk; **a piedi nudi** barefoot; **a ~ libero** free; **in piedi**

standing; **alzarsi in piedi** stand up; **in punta di piedi** on tiptoe; **ai piedi di** (*montagna*) at the foot of; **prendere ~** *fig* gain ground; (*moda:*) catch on; **mettere in piedi** (*allestire*) set up; **togliti dai piedi** get out of the way!. **~ di porco** (*strumento*) jemmy

pie'dino *nm* **fare ~ a** qcno *fam* play footsie with sb

piedi'stallo *nm* pedestal

pi'ega *nf* (*piegatura*) fold; (*di gonna*) pleat; (*di pantaloni*) crease; (*grinza*) wrinkle; (*andamento*) turn; **non fare una ~** (*ragionamento:*) be flawless

pie'ga|re *vt* fold; (*flettere*) bend ● *vi* bend. **~rsi** *vr* bend. **~rsi a** *fig* yield to. **~tura** *nf* folding

pieghet'ta|re *vt* pleat. **~to** *a* pleated. **pie'ghevole** *a* pliable; (*tavolo*) folding ● *nm* leaflet

piemon'tese *a* Piedmontese

pi'en|a *nf* (*di fiume*) flood; (*folla*) 'crowd. **~o** *a* full; (*massiccio*) solid; **in ~a estate** in the middle of summer; **a ~i voti** (*diplomarsi*) ≈ with A-grades, with first class honours ● *nm* (*colmo*) height; (*carico*) full load; **in ~o** (*completamente*) fully; **fare il ~o** (*di benzina*) fill up

pie'none *nm* **c'era il ~** the place was packed

pietà *nf* pity; (*misericordia*) mercy; **senza ~** (*persona*) pitiless; (*spietatamente*) pitilessly; **avere ~ di** qcno take pity on sb; **far ~** (*far pena*) be pitiful

pie'tanza *nf* dish

pie'toso *a* pitiful, merciful; (*fam: pessimo*) terrible

pi'etr|a *nf* stone. **~a dura** semi-precious stone. **~a preziosa** precious stone. **~a dello scandalo** cause of the scandal. **pie'trame** *nm* stones *pl.* **~ifi'care** *vt* petrify. **pie'trina** *nf* (*di accendino*) flint. **pie'troso** *a* stony

'piffero *nm* fife

pigi'ama *nm* pyjamas *pl*

'pigia 'pigia *nm inv* crowd, crush. **pigi'are** *vt* press

pigi'one *nf* rent; **dare a ~** let, rent out; **prendere a ~** rent

pigli'are *vt* (*fam: afferrare*) catch. **'piglio** *nm* air

pig'mento *nm* pigment

pig'meo, -a *a & nmf* pygmy

'pigna *nf* cone

pi'gnolo *a* pedantic

pigo'lare *vi* chirp. **pigo'lio** *nm* chirping

pi'grizia *nf* laziness. **'pigro** *a* lazy; (*intelletto*) slow

'pila *nf* pile; *Electr* battery; (*fam: lampadina tascabile*) torch; (*vasca*) basin; **a pile** battery operated, battery powered

pi'lastro *nm* pillar

'pillola *nf* pill; **prendere la ~** be on the pill

pi'lone *nm* pylon; (*di ponte*) pier

pi'lota *nmf* pilot ● *nm* *Auto* driver. **pilo'tare** *vt* pilot; drive (*auto*)

pinaco'teca *nf* art gallery

'Pinco Pallino *nm* so-and-so

pi'neta *nf* pine-wood

ping'pong *nm* table tennis, ping-pong *fam*

'pingue *a* fat. **~'edine** *nf* fatness

pingu'ino *nm* penguin; (*gelato*) choc ice on a stick

'pinna *nf* fin; (*per nuotare*) flipper

'pino *nm* pine[-tree]. **pi'nolo** *nm* pine kernel. **~ marittimo** cluster or maritime pine

'pinta *nf* pint

'pinza *nf* pliers *pl; Med* forceps *pl*

pin'za|re *vt* (*con pinzatrice*) staple. **~'trice** *nf* stapler

pin'zette *nfpl* tweezers *pl*

pinzi'monio *nm* sauce for crudités

'pio *a* pious; (*benefico*) charitable

pi'oggia *nf* rain; (*fig: di pietre, insulti*) hail, shower; **sotto la ~** in the rain. **~ acida** acid rain

pi'olo *nm* (*di scala*) rung

piom'ba|re *vi* fall heavily; **~re su** fall upon ● *vt* fill (*dente*). **~'tura** *nf* (*di dente*) filling. **piom'bino**

(sigillo) [lead] seal; *(da pesca)* sinker; *(in gonne)* weight
pi'ombo *nm* lead; *(sigillo)* [lead] seal; **a ~** plumb; **senza ~** *(benzina)* lead-free
pioni'ere, -a *nmf* pioneer
pi'oppo *nm* poplar
pio'vano *a* **acqua piovana** rain-water
pi'ov|ere *vi* rain; **~e** it's raining; **~iggi'nare** *vi* drizzle. **pio'voso** *a* rainy
'pipa *nf* pipe
pipi *nf* **fare [la] ~** pee, piddle; **andare a fare [la] ~** go for a pee
pipi'strello *nm* bat
pi'ramide *nf* pyramid
pi'ranha *inv* piranha
pi'rat|a *nm* pirate. **~a della strada** road-hog ● *a inv* pirate. **~e'ria** *nf* piracy
piro'etta *nf* pirouette
pi'rofil|a *nf* *(tegame)* oven-proof dish. **~o** *a* heat-resistant
pi'romane *nm* pyromaniac
pi'roscafo *nm* steamer. **~ di linea** liner
pisci'are *vi* *vulg* piss
pi'scina *nf* swimming pool. **~ coperta** indoor swimming pool. **~ scoperta** outdoor swimming pool
pi'sello *nm* pea; *(fam: pene)* willie
piso'lino *nm* nap; **fare un ~** have a nap
'pista *nf* track; *Aeron* runway; *(orma)* footprint; *(sci)* slope, piste. **~ d'atterraggio** airstrip. **~ da ballo** dance floor. **~ ciclabile** cycle track
pi'stacchio *nm* pistachio
pi'stola *nf* pistol; *(per spruzzare)* spray-gun. **~ a spruzzo** paint spray
pi'stone *nm* piston
pi'tone *nm* python
pit'to|re, -'trice *nmf* painter. **~'resco** *a* picturesque. **pit'torico** *a* pictorial
pit'tu|ra *nf* painting. **~'rare** *vt* paint
più *adv* more; *(superlativo)* most;

Math plus; **~ importante** more important; **il ~ importante** the most important; **~ caro** dearer; **il ~ caro** the dearest; **di ~** more; **una coperta in ~** an extra blanket; **non ho ~ soldi** I don't have any more money; **non vive ~ a Milano** he no longer lives in Milan, he doesn't live in Milan any longer; **~ o meno** more or less; **il ~ lentamente possibile** as slow as possible; **per di ~** what's more; **mai ~!** I never again!; **~ di** more than; **sempre ~** more and more ● *a* more; *(superlativo)* most; **~ tempo** more time; **la classe con ~ alunni** the class with most pupils; **~ volte** several times ● *nm* most; *Math* plus sign; **il ~ è fatto** the worst is over; **parlare del ~ e del meno** make small talk; **i ~** the majority
piuccheper'fetto *nm* pluperfect
pi'uma *nf* feather. **piu'maggio** *nm* plumage. **piu'mino** *nm* *(di cigni)* down; *(copriletto)* eiderdown; *(per cipria)* powder-puff; *(per spolverare)* feather duster; *(giacca)* down jacket. **piu'mone®** *nm* duvet, continental quilt
piut'tosto *adv* rather; *(invece)* instead
pi'vello *nm* *fam* greenhorn
'pizza *nf* pizza; *Cinema* reel.
pizzai'ola *nf* slices of beef in tomato sauce, oregano and anchovies
pizze'ria *nf* pizza restaurant, pizzeria
pizzi'c|are *vt* pinch; *(pungere)* sting; *(di sapore)* taste sharp; *(fam: sorprendere)* catch; *Mus* pluck ● *vi* scratch; *(cibo:)* be spicy. **'pizzico** *nm*, **~otto** *nm* pinch
'pizzo *nm* lace; *(di montagna)* peak
pla'car|e *vt* placate; assuage *(fame, dolore)*. **~si** *vr* calm down
'placca *nf* plate; *(commemorativa, dentale)* plaque; *Med* patch
plac'ca|re *vt* plate. **~to a ~to d'argento** silver-plated. **~to d'oro** gold-plated. **~'tura** *nf* plating

pla'centa *nf* placenta

'placido *a* placid

plagi'are *vt* plagiarize; pressure *(persona)*. 'plagio *nm* plagiarism

plaid *nm inv* tartan rug

pla'nare *vi* glide

'plancia *nf Naut* bridge; *(passerella)* gangplank

plane'tario *a* planetary ●*nm* planetarium

pla'smare *vt* mould

'plastic|a *nf (arte)* plastic art; *Med* plastic surgery; *(materia)* plastic. ~o *a* plastic ●*nm* plastic model

'platano *nm* plane[-tree]

pla'tea *nf* stalls *pl*; *(pubblico)* audience

'platino *nm* platinum

pla'tonico *a* platonic

plau'sibil|e *a* plausible. ~ità *nf* plausibility

ple'baglia *nf pej* mob

pleni'lunio *nm* full moon

'plettro *nm* plectrum

pleu'rite *nf* pleurisy

'plico *nm* packet; in ~ a parte under separate cover

plissé *a inv* plissé; *(gonna)* accordeon-pleated

plo'tone *nm* platoon; *(di ciclisti)* group. ~ d'esecuzione firing-squad

'plumbeo *a* leaden

plu'ral|e *a & nm* plural; al ~e in the plural. ~ità *nf (maggioranza)* majority

pluridiscipli'nare *a* multidisciplinary

plurien'nale *a* ~ esperienza many years' experience

pluripar'titico *a Pol* multi-party

plu'tonio *nm* plutonium

pluvi'ale *a* rain *attrib*

pneu'matico *a* pneumatic ●*nm* tyre

pneu'monia *nf* pneumonia

po' *vedi* poco

po'chette *nf inv* clutch bag

po'chino *nm* un ~ a little bit

'poco *a* little; *(tempo)* short; *(con nomi plurali)* few ●*pron* little;

(poco tempo) a short time; *(plurale)* few ●*nm* little; un po' a little [bit]; un po' di a little, some; *(con nomi plurali)* a few; a ~ a ~ little by little; fra ~ soon; per ~ *(a poco prezzo)* cheap; *(quasi)* nearly; ~ fa a little while ago; *(quasi)* nearly; ~ fa a little while ago; sono arrivato da ~ I have just arrived; un bel po' quite a lot; un ~ di buono a shady character ●*adv (con verbi)* not much; *(con avverbi)* not very; parla ~ he doesn't speak much; lo conosco ~ I don't know him very well; ~ spesso not very often

po'dere *nm* farm

pode'roso *a* powerful

'podio *nm* dais; *Mus* podium

po'dismo *nm* walking. ~ta *nmf* walker

po'e|ma *nm* poem. ~'sia *nf* poetry; *(componimento)* poem. ~ta *nm* poet. ~'tessa *nf* poetess. ~tico *a* poetic

poggiapi'edi *nm inv* footrest

poggi'a|re *vt* lean; *(posare)* place ●*vi* ~re su to be based on. ~'testa *nm inv* head-rest

'poggio *nm* hillock

poggi'olo *nm* balcony

poi *adv (dopo)* then; *(più tardi)* later [on]; *(finalmente)* finally. d'ora in ~ from now on; questa ~! well!

poiché *conj* since

pois *nm inv* a ~ polka-dot

'poker *nm* poker

po'lacco, -a *a* Polish ●*nmf* Pole ●*nm (lingua)* Polish

po'lar|e *a* polar. ~iz'zare *vt* polarize

'polca *nf* polka

po'lemi|ca *nf* controversy. ~ca-'mente *adv* controversially. ~co *a* controversial. ~z'zare *vi* engage in controversy

po'lenta *nf* cornmeal porridge

poli'clinico *nm* general hospital

poli'estere *nm* polyester

poliga'mia *nf* polygamy. po'ligamo *a* polygamous

polio[mie'lite] *nf* polio[myelitis]

'poli**po** nm polyp

poli**sti'rolo** nm polystyrene

poli**tecnico** nm polytechnic

po'litic|a nf politics sg; (linea di condotta) policy; **fare ~a** be in politics. **~iz'zare** vt politicize. **~o, -a** a political ● nmf politician

poliva'lente a catch-all

poli'zi|a nf police. **~a giudiziaria** ≈ Criminal Investigation Department, CID. **~a stradale** traffic police. **~'esco** a police attrib; ⟨romanzo, film⟩ detective attrib. **~'otto** nm policeman

'polizza nf policy

pol'la|io nm chicken run; (fam: luogo chiassoso) mad house. **~me** nm poultry. **~'strello** nm spring chicken. **~stro** nm cockerel

'pollice nm thumb; (unità di misura) inch

'polline nm pollen; **allergia al ~** hay fever

polli'vendolo, -a nmf poulterer

'pollo nm chicken; (fam: sempliciotte) simpleton. **~ arrosto** roast chicken

polmo'nare a pulmonary. pol'mone nm lung. **polmone d'acciaio** iron lung. **~'nite** nf pneumonia

'polo nm pole; Sport polo; (maglietta) polo top. **~ nord** North Pole. **~ sud** South Pole

Po'lonia nf Poland

'polpa nf pulp

pol'paccio nm calf

polpa'strello nm fingertip

pol'pet|ta nf meatball. **~'tone** nm meat loaf

'polpo nm octopus

pol'poso a fleshy

pol'sino nm cuff

'polso nm pulse; Anat wrist; fig authority; **avere ~** be strict

pol'tiglia nf mush

pol'trire vi lie around

pol'tron|a nf armchair; Theat seat in the stalls. **~e a** lazy

'polve|re nf dust; (sostanza polverizzata) powder; **in ~re** powdered; **sapone in ~re** soap powder. **~re**

da sparo gun powder. **~'rina** nf (medicina) powder. **~riz'zare** vt pulverize; (nebulizzare) atomize. **~'rone** nm cloud of dust. **~'roso** a dusty

po'mata nf ointment, cream

po'mello nm knob; (guancia) cheek

pomeridi'ano a afternoon attrib; **alle tre pomeridiane** at three in the afternoon, at three p.m. **pome'riggio** nm afternoon

'pomice nf pumice

'pomo nm (oggetto) knob. **~ d'Adamo** Adam's apple

pomo'doro nm tomato

'pompa nf pump; (sfarzo) pomp. **pompe pl funebri** (funzione) funeral. **pom'pare** vt pump; (gonfiare d'aria) pump up; (fig: esagerare) exaggerate; **pompare fuori** pump out

pom'pelmo nm grapefruit

pompi'ere nm fireman; **i pompieri** the fire brigade

pom'pon nm inv pompom

pom'poso a pompous

ponde'rare vt ponder

po'nente nm west

'ponte nm bridge; Naut deck; (impalcatura) scaffolding; **fare il ~ fig** make a long weekend of it

pon'tefice nm pontiff

pontifi'ca|re vi pontificate. **~to** nm pontificate

ponti'ficio a papal

pon'tile nm jetty

popò nf inv fam pooh

popo'lano a of the [common] people

popo'la|re a popular; (comune) common ● vt populate. **~rsi** vr get crowded. **~rità** nf popularity. **~zi'one** nf population. **'popolo** nm people. **popo'loso** a populous

'poppa nf Naut stern; (mammella) breast; **a ~** astern

pop'pa|re vt suck. **~ta** nf (pasto) feed. **~'toio** nm [feeding-]bottle

popu'lista nmf populist

199

porcata | portiera

por'cata nf load of rubbish; **porcate** pl (fam: cibo) junk food

porcel'lana nf porcelain, china

porcel'lino nm piglet. ~ **d'India** guinea-pig

porche'ria nf dirt; (fig: cosa orrenda) piece of filth; (fam: robaccia) rubbish

por'cile nm pigsty. ~**no** a pig attrib ● **nm** (fungo) edible mushroom. '**porco** nm pig; (carne) pork

porco'spino nm porcupine

'**porgere** vt give; (offrire) offer; **porgo distinti saluti** (in lettera) I remain, yours sincerely

porno|gra'fia nf pornography. **~'grafico** a pornographic

'**poro** nm pore. **po'roso** a porous

'**porpora** nf purple

'**porre** vt put; (collocare) place; (supporre) suppose; ask (domanda); present (candidatura); **poniamo il caso che...** let us suppose that...; **~re fine** o **termine** a put an end to. **~si** vr put oneself; **~si a sedere** sit down; **~si in cammino** set out

'**porro** nm Bot leek; (verruca) wart

'**porta** nf door; Sport goal; (di città) gate; Comput port. **~ a ~** door-to-door; **mettere alla ~** show sb the door. **~ di servizio** tradesmen's entrance

portaba'gagli nm inv (facchino) porter; (di treno ecc) luggage rack; Auto boot, trunk Am; (sul tetto di un'auto) roof rack

portabot'tiglie nm inv bottle rack, wine rack

porta'cenere nm inv ashtray

portachi'avi nm inv keyring

porta'cipria nm inv compact

portadocu'menti nm inv document wallet

porta'erei nf inv aircraft carrier

portafi'nestra nf French window

porta'foglio nm wallet; (per documenti) portfolio; (ministero) ministry

portafor'tuna nm inv lucky charm ● a inv lucky

portagi'oie nm inv jewellery box

por'tale nm door

portama'tite nm inv pencil case

porta'mento nm carriage; (condotta) behaviour

porta'mina nm inv propelling pencil

portamo'nete nm inv purse

por'tante a bearing attrib

portaom'brelli nm inv umbrella stand

porta'pacchi nm inv roof rack; (su bicicletta) luggage rack

porta'penne nm inv pencil case

por'tare vt (verso chi parla) bring; (lontano da chi parla) take; (sorreggere, Math) carry; (condurre) lead; (indossare) wear; (avere) bear. **~rsi** vr (trasferirsi) move; (comportarsi) behave; **~rsi bene/male gli anni** look young/old for one's age

portari'viste nm inv magazine rack

porta'sci nm inv ski rack

portasiga'rette nm inv cigarette-case

porta'spilli nm inv pin-cushion

por'tata nf (di pranzo) course; Auto carrying capacity; (di arma) range; (fig: abilità) capability; **a ~ta di mano** within reach; **alla ~ta di tutti** accessible to all; (finanziariamente) within everybody's reach. **por'tatile** a & nm portable. **~to** a (indumento) worn; (dotato) gifted; **essere ~to per qcsa** have a gift for sth; **essere ~to a** (tendere a) be inclined to. **~'tore**, **~'trice** nmf bearer; **al ~tore** to the bearer: **~tore di handicap** disabled person

portatovagli'olo nm napkin ring

portau'ovo nm inv egg-cup

porta'voce nm inv spokesman ● nf inv spokeswoman

por'tento nm marvel; (persona dotata) prodigy

'**portico** nm portico

porti'era nf door; (tendaggio)

door curtain. **~e** nm porter, doorman; *Sport* goalkeeper. **~e di notte** night porter

porti'n|aio, -a nmf caretaker, concierge. **~e'ria** nf concierge's room; (*di ospedale*) porter's lodge

'porto pp di **porgere** ●nm harbour; (*complesso*) port; (*vino*) port [wine]; (*spesa di trasporto*) carriage; **andare in ~** succeed. **~ d'armi** gun licence

Porto'g|allo nm Portugal. **p~hese** a & nmf Portuguese

por'tone nm main door

portu'ale nm dockworker, docker

porzi'one nf portion

'posa nf laying; (*riposo*) rest; *Phot* exposure; (*atteggiamento*) pose; **mettersi in ~** pose

po'sa|re vt put; (*giù*) put [down] ●vi (*poggiare*) rest; (*per un ritratto*) pose. **~rsi** vr alight; (*sostare*) rest; *Aeron* land. **~ta** nf piece of cutlery; **~te** pl cutlery sg. **~to** a sedate

po'scritto nm postscript

posi'tivo a positive

posizio'nare vt position

posizi'one nf position; **farsi una ~** get ahead

posolo'gia nf dosage

po'spo|rre vt place after; (*posticipare*) postpone. **~sto** pp di **posporre**

posse'd|ere vt possess, own. **~i'mento** nm possession

posses|'sivo a possessive. **pos-'sesso** nm ownership; (*bene*) possession. **~'sore** nm owner

pos'sibil|e a possible; **il più presto ~e** as soon as possible ●nm **fare [tutto] il ~e** do one's best. **~ità** nf inv possibility; (*occasione*) chance ●nfpl (*mezzi*) means

possi'dente nmf land-owner

'posta nf post, mail; (*ufficio postale*) post office; (*al gioco*) stake; **spese di ~** postage; **per ~** by post, by mail; **la ~ in gioco è...** fig what's at stake is...; **a bella ~** on purpose; **Poste e Telecomunica-**

zioni pl [Italian] Post Office. **~ elettronica** electronic mail, e-mail. **~ elettronica vocale** voice-mail

posta'giro nm postal giro

po'stale a postal

postazi'one nf position

postda'tare vt postdate ⟨assegno⟩

posteggi'a|re vt/i park. **~'tore, ~'trice** nmf parking attendant. **po'steggio** nm car-park, parking lot Am; (*di taxi*) taxi-rank

po'steri nmpl descendants. **~'ore** a rear; (*nel tempo*) later ●nm fam posterior, behind. **~tà** nf posterity

po'sticcio a artificial; ⟨baffi, barba⟩ false ●nm hair-piece

postic'ipare vt postpone

po'stilla nf note; *Jur* rider

po'stino nm postman, mailman Am

'posto pp di **porre** ●nm place; (*spazio*) room; (*impiego*) job; *Mil* post; (*sedile*) seat; **a/fuori ~** in/out of place; **prendere ~** take up room; **sul ~** on-site; **essere a ~** ⟨casa, libri⟩ be tidy; **mettere a ~** tidy ⟨stanza⟩; **fare ~ a** make room for; **al ~ di** (*invece di*) in place of, instead of. **~ di blocco** check-point. **~ di guida** driving seat. **~ di lavoro** workstation. **posti** pl **in piedi** standing room. **~ di polizia** police station. **posti** pl **a sedere** seating

post-'partum a post-natal

'postumo a posthumous ●nm after-effect

po'tabile a drinkable; **acqua ~** drinking water

po'tare vt prune

po'tassio nm potassium

po'ten|te a powerful; (*efficace*) potent. **~za** nf power; (*efficacia*) potency. **~zi'ale** a & nm potential

po'tere nm power; **al ~** in power ●vi can, be able to; **posso entra-re?** can I come in?; (*formale*) may I come in?; **posso fare qualche cosa?** can I do something?; **che tu possa essere felice!** may you be happy!; **non ne posso più** (*sono stanco*) I can't go on; (*sono stufo*) I can't take any more; **può darsi** perhaps; **può darsi che sia vero** perhaps it's true; **potrebbe aver ragione** he could be right, he might be right; **avresti potuto telefonare** you could have phoned, you might have phoned; **spero di poter venire** I hope to be able to come; **senza poter telefonare** without being able to phone

potestà *nf inv* power

'pover|o, -a *a* poor; (*semplice*) plain ●*nm* poor man ●*nf* poor woman; **i ~i** the poor. **~tà** *nf* poverty

'pozza *nf* pool. **poz'zanghera** *nf* puddle

'pozzo *nm* well; (*minerario*) pit. **~ petrolifero** oil-well

PP.TT. *abbr* (Poste e Telegrafi) [Italian] Post Office

prag'matico *a* pragmatic

prali'nato *a* ⟨*mandorla, gelato*⟩ praline-coated

pram'matica **essere di ~** be customary

pran'zare *vi* dine; (*a mezzogiorno*) lunch. **'pranzo** *nm* dinner; (*a mezzogiorno*) lunch. **pranzo di nozze** wedding breakfast

'prassi *nf* standard procedure

prate'ria *nf* grassland

'prati|ca *nf* practice; (*esperienza*) experience; (*documentazione*) file; **avere ~ca di qcsa** be familiar with sth; **far ~ca** gain experience; **fare le pratiche per** gather the necessary papers for. **~'cabile** *a* practicable; ⟨*strada*⟩ passable. **~ca'mente** *adv* practically. **~'cante** *nmf* apprentice; *Relig* [regular] church-goer

prati'ca|re *vt* practise; (*frequentare*) associate with; (*fare*) make

praticità *nf* practicality. **'pratico** *a* practical; (*esperto*) experienced; **essere pratico di qcsa** know about sth

'prato *nm* meadow; (*di giardino*) lawn

pre'ambolo *nm* preamble

preannunci'are *vt* give advance notice of

preavvi'sare *vt* forewarn. **preav-'viso** *nm* warning

pre'cario *a* precarious

precauzi'one *nf* precaution; (*cautela*) care

prece'den|te *a* previous ●*nm* precedent. **~te'mente** *adv* previously. **~za** *nf* precedence; (*di veicoli*) right of way; **dare la ~za** give way. **pre'cedere** *vt* precede

pre'cetto *nm* precept

precipi'ta|re *vt* **~re le cose** precipitate events; **~re qcno nella disperazione** cast sb into a state of despair ●*vi* fall headlong; ⟨*situazione, eventi:*⟩ come to a head. **~rsi** *vr* (*gettarsi*) throw oneself; (*affrettarsi*) rush; **~rsi a fare qcsa** rush to do sth. **~zi'one** *nf* (*fretta*) haste; (*atmosferica*) precipitation. **precipi'toso** *a* hasty; (*avventato*) reckless; ⟨*caduta*⟩ headlong

preci'pizio *nm* precipice; **a ~** headlong

precisa'mente *adv* precisely

preci'sa|re *vt* specify; (*spiegare*) clarify. **~zi'one** *nf* clarification

precisi'one *nf* precision. **pre'ciso** *a* precise; ⟨*ore*⟩ sharp; (*identico*) identical

pre'clu|dere *vt* preclude. **~so** *pp di* **precludere**

pre'coc|e *a* precocious; (*prematuro*) premature. **~ità** *nf* precociousness

precon'cetto *a* preconceived ●*nm* prejudice

pre'corre|re *vt* **~ere i tempi** be ahead of one's time

precur'sore *nm* forerunner, precursor

'preda *nf* prey; (*bottino*) booty; **essere in ~ al panico** be panic-stricken; **in ~ alle fiamme** engulfed in flames. **pre'dare** vt plunder. ~'**tore** nm predator

predeces'sore nm predecessor

pre'del|la nf platform. ~'**lino** nm step

predesti'na|re vt predestine. ~**to** a Relig predestined, preordained

predetermi'nato a predetermined, preordained

pre'detto pp di predire

'predica nf sermon; fig lecture

predi'ca|re vt preach. ~**to** nm predicate

predi'let|to, -a pp di **prediligere** ● a favourite ● nmf pet. ~**zi'one** nf predilection. **predi'ligere** vt prefer

pre'di|re vt foretell

predi'spo|rre vt arrange. ~**rsi** vr prepare oneself for. ~**sizi'one** nf predisposition; (al disegno ecc) bent (a for). ~**sto** pp di predisporre

predizi'one nf prediction

predomi'na|nte a predominant. ~**re** vi predominate. **predo'minio** nm predominance

pre'done nm robber

prefabbri'cato a prefabricated ● nm prefabricated building

prefazi'one nf preface

prefe'renz|a nf preference; **di** ~**a** preferably. ~**i'ale** a preferential; **corsia ~iale** bus and taxi lane

prefe'ribile a preferable. ~'**mente** adv preferably

prefe'ri|re vt prefer. ~**to, -a** a & nmf favourite

pre'fet|to nm prefect. ~'**tura** nf prefecture

pre'figgersi vr be determined

pre'fisso pp di **prefiggere** ● nm prefix; Teleph (dialling) code

pre'ga|re vt/i pray; (supplicare) beg; **farsi** ~ need persuading

preghi'era nf prayer; (richiesta) request

pregi'ato a esteemed; (prezioso) valuable. **pregio** nm esteem; (va-

lore) value; (di persona) good point; **di pregio** valuable

pregiudi'ca|re vt prejudice; (danneggiare) harm. ~**to** a prejudiced ● nm Jur previous offender

pregiu'dizio nm prejudice; (danno) detriment

'**prego** int (non c'è di che) don't mention it!; (per favore) please; ~? I beg your pardon?

pregu'stare vt look forward to

prei'storia nf prehistory. **prei-'storico** a prehistoric

pre'lato nm prelate

prela'vaggio nm prewash

preleva'mento nm withdrawal. **prele'vare** vt withdraw (denaro); collect (merci); Med take. **preli'evo** nm (di soldi) withdrawal. **prelievo di sangue** blood sample

prelimi'nare a preliminary ● nm **preliminari** pl preliminaries

pre'ludio nm prelude

prema'man nm inv maternity dress ● a maternity attrib

prematrimoni'ale a premarital

prema'turo, -a a premature ● nmf premature baby

premedi'ta|re vt premeditate. ~**zi'one** nf premeditation

'**premere** vt press; Comput hit (tasto) ● vi a (importare) matter to; **mi preme sapere** I need to know; ~ **su** press on; push (pulsante)

pre'messa nf introduction

pre'me|sso pp di **premettere**. ~**sso che** bearing in mind that. ~**ttere** vt put forward; (mettere prima) put before.

premi'a|re vt give a prize to; (ricompensare) reward. ~**zi'one** nf prize giving

premi'nente a pre-eminent

'**premio** nm prize; (ricompensa) reward; Comm premium. ~ **di consolazione** booby prize

premoni'tore a (sogno, segno) premonitory. ~**zi'one** nf premonition

premu'nir|e vt fortify. ~**si** vr take

protective measures; **~si di pro-**
vide oneself with; **~si contro** protect oneself against

pre'mu|ra nf (fretta) hurry; (cura)
care. **~'roso** a thoughtful

prena'tale a antenatal

'prender|e vt take; (afferrare)
seize; catch (treno, malattia, ladro, pesce) have (cibo, bevanda)
(far pagare) charge; (assumere)
take on; **~e informazioni** make
inquiries; **~e a calci/pugni** kick/
punch; **che ti prende?** what (got
into you)? **quanto prende?** what
do you charge? **~e una persona
per un'altra** mistake a person for
someone else ●vi (voltare) turn;
(attecchire) take root; (rapprendersi) set; **~e a destra/sinistra** turn
right/left; **~e a fare qcsa** start
doing sth. **~si** vr **~si a pugni**
come to blows; **~si cura di** take
care of (ammalato) **~sela** take it
to heart

prendi'sole nm sundress

preno'ta|re vt book, reserve. **~to** a
booked, reserved **~zi'one** nf booking, reservation

'prensile a prehensile

preoccu'pante a alarming

preoccu'pa|re vt worry. **~rsi** vr
~rsi worry (di about); **~rsi di fare
qcsa** take the trouble to do sth. **~to**
a (ansioso) worried. **~zi'one** nf
worry; (apprensione) concern

prepa'ra|re vt prepare. **~rsi** vr get
ready. **~tivi** nmpl preparations.
~to nm (prodotto) preparation.
~torio a preparatory. **~zi'one** nf
preparation

prepensiona'mento nm early retirement

prepon'de|ran|te a predominant.
~za nf prevalence

pre'porre vt place before

preposizi'one nf preposition

pre'posto pp di **preporre** ●a ~ a
(addetto a) in charge of

prepo'tente a overbearing

●nmf bully. **~za** nf high-
handedness

preroga'tiva nf prerogative

'presa nf taking; (conquista) capture; (stretta) hold; (di cemento
ecc) setting; Electr socket; (pizzi-
co) pinch; **essere alle prese con**
be struggling ∘ grappling with; **~
rapida** (cemento, colla) quick-
setting; **fare ~ su qcno** influence
sb. **~ d'aria** air vent. **~ in giro** leg-
pull. **~ multipla** adaptor

pre'sagio nm omen. **presa'gire** vt
foretell

'presbite a long-sighted

presbiteri'ano, **-a** a & nmf Pres-
byterian. **presbi'terio** nm pres-
bytery

pre'scelto a selected

pre'scindere vi **~ da** leave aside;
a ~ da apart from

presco'lare a in età ~ pre-school

pre'scritto pp di **prescrivere**

pre'scri|vere vt prescribe. **~zi-
'one** nf prescription; (norma) rule

preselezi'one nf **chiamare qcno
in ~** call sb via the operator

presen'ta|re vt present; (far cono-
scere) introduce; show (documento) (inoltrare) submit. **~rsi** vr
present oneself; (farsi conoscere)
introduce oneself; (a ufficio)
attend; (alla polizia ecc) report;
(come candidato) stand; run;
(occasione:) occur; **~rsi bene/
male** (persona:) make a good/bad
impression; (apparizione:) look
good/bad. **~'tore**, **~'trice** nmf
presenter; (di notizie) announcer.
~zi'one nf presentation; (per co-
noscersi) introduction

pre'sente a present; (attuale) current; (questo) this; **aver ~** remember ●nm present; **i presenti** those
present ●nf **allegato alla ~** (in
lettera) enclosed

presenti'mento nm foreboding

pre'senza nf presence; (aspetto)
appearance; **in ~ di**, **alla ~ di**
in the presence of; **di bella ~** per-

sonable. ~ **di spirito** presence of mind

presenzi'are vi ~ **a** attend

pre'sepe nm, **pre'sepio** nm crib

preser'va|re vt preserve; (proteggere) protect (da from). ~'tivo nm condom. ~zi'one nf preservation

'**preside** nm headmaster; Univ dean ●nf headmistress; Univ dean

presi'den|te nm chairman; Pol president ●nf chairwoman; Pol president. ~ **dei consiglio [dei ministri]** Prime Minister. ~ **della repubblica** President of the Republic. ~**za** nf presidency; (di assemblea) chairmanship. ~**zi'ale** a presidential

presidi'are vt garrison. **pre'sidio** nm garrison

presi'edere vt preside over

'**preso** pp di **prendere**

'**pressa** nf Mech press

pres'sante a urgent

pressap'poco adv about

pres'sare vt press

pressi'one nf pressure; **far ~ su** put pressure on. ~ **del sangue** blood pressure

'**presso** prep near; (a casa di) with; (negli indirizzi) care of, c/o; ⟨lavorare⟩ for ●pressi nmpl: **nei pressi di...** in the neighbourhood o vicinity of...

pressoché adv almost

pressuriz'za|re vt pressurize. ~**to** a pressurized

prestabi'li|re vt arrange in advance. ~**to** a agreed

prestam'pato ●nm (modulo) form

pre'stante a good-looking

pre'star|e vt lend; ~**e attenzione** pay attention; ~**e aiuto** lend a hand; **farsi** ~**e** borrow (da from). ~**si** vi ⟨frase:⟩ lend itself; ⟨persona:⟩ offer

prestazi'one nf performance; **prestazioni** pl ⟨servizi⟩ services

prestigia'tore, -'trice nmf conjurer

pre'stigi|o nm prestige; **gioco di** ~**o** conjuring trick. ~'**oso** nm prestigious

'**prestito** nm loan; **dare in** ~ lend; **prendere in** ~ borrow

'**presto** adv soon; (di buon'ora) early; (in fretta) quickly; **a** ~ see you soon; **al più** ~ as soon as possible; ~ **o tardi** sooner or later; **far** ~ be quick

pre'sumere vt presume; (credere) think

presu'mibile a ~ **che...** presumably,...

pre'sunto a ⟨colpevole⟩ presumed

presun|tu'oso a presumptuous ●nmf presumptuous person. ~**zi'one** nf presumption

presup'po|rre vt suppose; (richiedere) presuppose. ~**sizi'one** nf presupposition. ~**sto** nm essential requirement

'**prete** nm priest

preten'den|te nmf pretender ●nm (corteggiatore) suitor

pre'ten|dere vt (sostenere) claim; (esigere) demand ●vi ~**dere a** claim to; (esigere) demand to. ~**si'one** nf pretension. ~**zi'oso** a pretentious

pre'tes|a nf pretension; (esigenza) claim; **senza** ~ unpretentious. ~**o** pp di **pretendere**

pre'testo nm pretext

pre'tore nm magistrate

pretta'mente adv decidedly

pre'tura nf magistrate's court

preva'le|nte a prevalent. ~**nte'mente** adv primarily. ~**nza** nf prevalence. ~**re** vi prevail

pre'valso pp di **prevalere**

preve'dere vt foresee; forecast (tempo); (legge ecc) provide for

preve'nire vt precede; (evitare) prevent; (avvertire) forewarn

preven|ti'vare vt estimate; (aspettarsi) budget for. ~'**tivo** a preventive ●nm Comm estimate

preve'nuto a forewarned; (mal disposto) prejudiced. ~**zi'one** nf prevention; (*preconcetto*) prejudice

previ'den|te *a* provident. **~za** *nf* foresight. **~za sociale** social security, welfare *Am.* **~zi'ale** *a* provident

'previo *a* **~ pagamento** on payment

previsi'one *nf* forecast; **in ~ di** in anticipation of

pre'visto *pp di* **prevedere** ● *a* foreseen ● *nm* **più/meno/prima del ~** more/less/earlier than expected

prezi'oso *a* precious

prez'zemolo *nm* parsley

'prezzo *nm* price. **~ di fabbrica** factory price. **~ all'ingrosso** wholesale price. **[a] metà ~** half price

prigi'on|e *nf* prison; (*pena*) imprisonment. **prigio'nia** *nf* imprisonment. **~i'ero, -a** *a* imprisoned ● *nmf* prisoner

'prima *adv* before; (*più presto*) earlier; (*in primo luogo*) first; **~, finiamo questo** let's finish this first; **puoi venire ~?** can't you come any sooner?; (*di ore*) can't you come any earlier?; **~ o poi** sooner or later; **quanto ~** as soon as possible ● *prep* **~ di** before; **~ d'ora** before now ● *conj* **~ che** before **~ di** first class; *Theat* first night; *Auto* first [gear]

pri'mario *a* primary; (*principale*) principal

pri'mat|e *nm* primate. **~o** *nm* supremacy; *Sport* record

prima've|ra *nf* spring. **~'rile** *a* spring *attrib*

primeggi'are *vi* excel

primi'tivo *a* primitive; (*originario*) original

pri'mizie *nfpl* early produce *sg*

'primo *a* first; (*fondamentale*) principal; (*precedente di due*) former; (*iniziale*) early; (*migliore*) best ● *nm* first; **primi** *pl* (*i primi giorni*) the beginning; **in un ~ tempo** at first. **prima copia** master copy

primo'genito, -a *a* & *nmf* firstborn

primordi'ale *a* primordial

'primula *nf* primrose

princi'pale *a* main ● *nm* head, boss *fam*

princi'pato *nm* principality. **'principe** *nm* prince. **principe ereditario** crown prince. **~'pesco** *a* princely. **~'pessa** *nf* princess

principi'ante *nmf* beginner

prin'cipio *nm* beginning; (*concetto*) principle; (*causa*) cause; **per ~** on principle

pri'ore *nm* prior

priori'tà *nf inv* priority. **~'tario** *a* having priority

'prisma *nm* prism

pri'va|re *vt* deprive. **~rsi** *vr* deprive oneself

privatizzazi'one *nf* privatization. **pri'vato, -a** *a* private ● *nmf* private citizen

privazi'one *nf* deprivation

privilegi'are *vt* privilege; (*considerare più importante*) favour. **privi'legio** *nm* privilege

'privo *a* **~ di** devoid of; (*mancante*) lacking in

pro *prep* for ● *nm* advantage; **a che ~?** what's the point?; **il ~ e il contro** the pros and cons

pro'babil|e *a* probable. **~ità** *nf inv* probability. **~'mente** *adv* probably

pro'ble|ma *nm* problem. **~'matico** *a* problematic

pro'boscide *nf* trunk

procacci'ar|e *vt*, **~si** *vr* obtain

pro'cace *a* (*ragazza*) provocative

pro'ced|ere *vi* proceed; (*iniziare*) start; **~ere contro** *Jur* start legal proceedings against. **~i'mento** *nm* process; *Jur* proceedings *pl*. **proce'dura** *nf* procedure

proces'sare *vt Jur* try

processi'one *nf* procession

pro'cesso *nm* process; *Jur* trial

proces'sore *nm Comput* processor

processu'ale *a* trial

pro'cinto nm essere in ~ di about to

pro'clama nm proclamation

procla'ma|re vt proclaim. ~zi'one nf proclamation

procrasti'na|re vt liter postpone

procreazi'one nf procreation

pro'cura nf power of attorney; per ~ by proxy

procu'ra|re vt/i procure; (causare) cause; (cercare) try. ~'tore nm attorney. P~tore Generale Attorney General. ~tore legale lawyer. ~tore della repubblica public prosecutor

'prode a brave. **pro'dezza** nf bravery

prodi'gare vt lavish. ~si vr do one's best

pro'digi|o nm prodigy. ~'oso a prodigious

pro'dotto pp di **produrre** ● nm product. **prodotti agricoli** farm produce sg. ~ **derivato** by-product. ~ **interno lordo** gross domestic product. ~ **nazionale lordo** gross national product

pro'du|rre vt produce. ~rsi vr ⟨attore:⟩ play; ⟨accadere⟩ happen. ~ttività nf productivity. ~'tivo a productive. ~'tore, ~t'trice nmf producer. ~zi'one nf production

profa'na|re vt desecrate. ~zi'one nf desecration. **pro'fano** a profane

profe'rire vt utter

Prof.essa abbr (**Professoressa**) Prof.

profes'sare vt profess; practise ⟨professione⟩

professio'nale a professional

professi'o|ne nf profession; **libera ~ne** profession. ~'nismo nm professionalism. ~'nista nmf professional

profes'sor|e, -'essa nmf Sch teacher; Univ lecturer; (titolare di cattedra) professor

pro'fe|ta nm prophet. ~tico a prophetic. ~tiz'zare vt prophesy. ~'zia nf prophecy

pro'ficuo a profitable

profi'lar|e vt outline; ⟨ornare⟩ border; Aeron streamline. ~si vr stand out

profi'lattico a prophylactic ● nm condom

pro'filo nm profile; (breve studio) outline; **di** ~ in profile

profit'tare vi ~ **di** ⟨avvantaggiarsi⟩ profit by; ⟨approfittare⟩ take advantage of. **pro'fitto** nm profit; (vantaggio) advantage

profonda'mente adv deeply, profoundly. ~**ità** nf inv depth

pro'fondo a deep; fig profound; ⟨cultura⟩ great

'profugo, -a nmf refugee

profu'mar|e vt perfume. ~si vr put on perfume

profumata'mente adv pagare ~ pay through the nose

profu'mato a ⟨fiore⟩ fragrant; ⟨fazzoletto ecc⟩ scented

profume'ria nf perfumery. **pro'fumo** nm perfume, scent

profusi'one nf profusion; **a** ~ in profusion. **pro'fuso** pp di **profondere** ● a profuse

proget'tare vt plan. ~'tista nmf designer. **pro'getto** nm plan; ⟨di lavoro importante⟩ project. **progetto di legge** bill

prog'nosi nf inv prognosis; **in** ~ **riservata** on the danger list

pro'gramma nm programme; Comput program. ~ **scolastico** syllabus

program'ma|re vt programme; Comput program. ~'tore, ~'trice nmf [computer] programmer. ~zi'one nf programming

progre'dire vi [make] progress

progres'si|one nf progression. ~'sivo a progressive. **pro'gresso** nm progress

proi'bi|re vt forbid. ~'tivo a prohibitive. ~to a forbidden. ~zi'one nf prohibition

proiet'tare vt project; show ⟨film⟩. ~t'tore nm projector; Auto headlight

proi'ettile nm bullet

proiezi'one *nf* projection

'prole *nf* offspring. **proletari'ato** *nm* proletariat. **prole'tario** *a & nm* proletarian

prolife'rare *vi* proliferate. **pro'lifico** *a* prolific

pro'lisso *a* verbose, prolix

'prologo *nm* prologue

pro'lunga *nf Electr* extension

prolun'gar|e *vt* prolong; (*allungare*) lengthen; extend ⟨*contratto, scadenza*⟩. **~si** *vr* continue; **~si su** (*dilungarsi*) dwell upon

prome'moria *nm* memo; (*per se stessi*) reminder; note; (*formale*) memorandum

pro'me|ssa *nf* promise. **~sso** *pp di* **promettere. ~ttere** *vt/i* promise

promet'tente *a* promising

promi'nente *a* prominent

promiscuità *nf* promiscuity. **pro'miscuo** *a* promiscuous

promon'torio *nm* promontory

pro'mo|sso *pp di* **promuovere ● a** *Sch* who has gone up a year; *Univ* who has passed an exam. **~'tore**, **~'trice** *nmf* promoter

promozio'nale *a* promotional. **promozi'one** *nf* promotion

promul'gare *vt* promulgate

promu'overe *vt* promote; *Sch* move up a class

proni'pote *nm* (*di bisnonno*) great-grandson; (*di prozio*) great-nephew ● *nf* (*di bisnonno*) great-granddaughter; (*di prozio*) great-niece

pro'nome *nm* pronoun

pronosti'care *vt* forecast, predict. **pro'nostico** *nm* forecast

pron'tezza *nf* readiness; (*rapidità*) quickness

'pronto *a* ready; (*rapido*) quick; *Teleph* hallo!; **tenersi ~** be ready (**per** for); **pronti, via!** (*in gare*) ready! steady! go!. **~ soccorso** first aid; (*in ospedale*) accident and emergency

prontu'ario *nm* handbook

pro'nuncia *nf* pronunciation

pronunci'a|re *vt* pronounce; (*dire*)

utter; deliver ⟨*discorso*⟩. **~rsi** *vr* (*su un argomento*) give one's opinion. **~to** *a* pronounced; (*prominente*) prominent

pro'nunzia ecc = **pronuncia** ecc

propa'ganda *nf* propaganda

propa'ga|re *vt* propagate. **~rsi** *vr* spread. **~zi'one** *nf* propagation

prope'deutico *a* introductory

pro'pen|dere *vi* **~dere per** be in favour of. **~si'one** *nf* inclination, propensity. **~so** *pp di* **propendere ● a essere ~so a fare** qcsa be inclined to do sth

propi'nare *vt* administer

pro'pizio *a* favourable

proponi'mento *nm* resolution

pro'por|re *vt* propose; (*suggerire*) suggest. **~si** *vr* set oneself ⟨*obiettivo, meta*⟩; **~si di** intend to

proporzio'na|le *a* proportional. **~re** *vt* proportion. **~to** *a* proportioned. **proporzi'one** *nf* proportion

pro'posito *nm* purpose; **a ~** by the way; **a ~ di** with regard to; **di ~** (*apposta*) on purpose; **capitare a ~**, **giungere a ~** come at just the right time

proposizi'one *nf* clause; (*frase*) sentence

pro'post|a *nf* proposal. **~o** *pp di* **proporre**

proprietà *nf inv* property; (*diritto*) ownership; (*correttezza*) propriety. **~ immobiliare** property. **~ privata** private property. **proprie'taria** *nf* owner; (*di casa affittata*) landlady. **proprie'tario** *nm* owner; (*di casa affittata*) landlord

'proprio *a* one's [own]; (*caratteristico*) typical; (*appropriato*) proper ● *adv* just; (*veramente*) really; **non ~** not really, not exactly; (*affatto*) not... at all ● *pron* one's own ● *nm* one's [own]. **lavorare in ~** be one's own boss; **mettersi in ~** set up on one's own

propul|si'one *nf* propulsion. **~'sore** *nm* propeller

'proroga *nf* extension

proro'ga|bile *a* extendable. **~re** *vt* extend

pro'rompere *vi* burst out

'**prosa** *nf* prose. **pro'saico** *a* prosaic

pro'scio|gliere *vt* release; *Jur* acquit. **~lto** *pp di* **prosciogliere**

prosciu'gar|e *vt* dry up; (*bonificare*) reclaim. **~si** *vr* dry up

prosci'utto *nm* ham. **~ cotto** cooked ham. **~ crudo** type of dry-cured ham, Parma ham

pro'scri|tto, -a *pp di* **proscrivere** ● *nmf* exile

prosecuzi'one *nf* continuation

prosegui'mento *nm* continuation; **buon ~!** (*viaggio*) have a good journey!; (*festa*) enjoy the rest of the party!

prosegu'ire *vt* continue ● *vi* go on, continue

prospe'r|are *vi* prosper. **~ità** *nf* prosperity. '**prospero** *a* prosperous; (*favorevole*) favourable. **~oso** *a* flourishing; (*ragazza*) buxom

prospet'tar|e *vt* show. **~si** *vr* seem

prospet'tiva *nf* perspective; (*panorama*) view; *fig* prospect. **pro'spetto** *nm* (*vista*) view; (*facciata*) façade; (*tabella*) table

prospici'ente *a* facing

prossima'mente *adv* soon

prossimità *nf* proximity

'**prossimo, -a** *a* near; (*seguente*) next; (*molto vicino*) close; **l'anno ~** next year ● *nmf* neighbour

prosti'tu|ta *nf* prostitute. **~zi'one** *nf* prostitution

pro'stra|re *vt* prostrate. **~rsi** *vr* prostrate oneself. **~to** *a* prostrate

protago'nista *nmf* protagonist

pro'te|ggere *vt* protect; (*favorire*) favour

prote'ina *nf* protein

pro'tender|e *vt* stretch out. **~si** *vr* (*in avanti*) lean out. **pro'teso** *pp di* **protendere**

pro'te|sta *nf* protest; (*dichiarazione*) protestation. **~'stante** *a & nmf* Protestant. **~'stare** *vt/i* protest

protet'tivo *a* protective. **~tto** *pp*

di **proteggere**. **~t'tore, ~t'trice** *nmf* protector; (*sostenitore*) patron ● *nm* (*di prostituta*) pimp. **~zi'one** *nf* protection

protocol'lare *a* (*visita*) protocol ● *vt* register

proto'collo *nm* protocol; (*registro*) register; **carta ~** official stamped paper

pro'totipo *nm* prototype

pro'tra|rre *vt* protract; (*differire*) postpone. **~rsi** *vr* go on, continue. **~tto** *pp di* **protrarre**

protube'ran|te *a* protuberant. **~za** *nf* protuberance

'**prova** *nf* test; (*dimostrazione*) proof; (*tentativo*) try; (*di abito*) fitting; *Sport* heat; *Theat* rehearsal; (*bozza*) proof; **fino a ~ contraria** until I'm told otherwise; **in ~** (*assumere*) for a trial period; **mettere alla ~** put to the test. **~ generale** dress rehearsal

pro'var|e *vt* test; (*dimostrare*) prove; (*tentare*) try; try on (*abiti ecc*); (*sentire*) feel; *Theat* rehearse. **~si** *vr* try

proveni'enza *nf* origin. **prove'nire** *vi* **provenire da** come from

pro'vento *nm* proceeds *pl*

prove'nuto *pp di* **provenire**

pro'verbio *nm* proverb

pro'vetta *nf* test-tube; **bambino in ~** test-tube baby

pro'vetto *a* skilled

pro'vinc|ia *nf* province; (*strada*) B road, secondary road. **~'ale** *a* provincial; **strada ~ale** B road, secondary road

pro'vino *nm* specimen; *Cinema* screen test

provo'ca|nte *a* provocative. **~re** *vt* provoke; (*causare*) cause. **~'tore, ~'trice** *nmf* trouble-maker. **~'torio** *a* provocative. **~zi'one** *nf* provocation

provve'der|e *vi* **~ere a** provide for. **~i'mento** *nm* measure; (*previdenza*) precaution

provvi'den|za *nf* providence. **~i'ale** *a* providential

provvigi'one *nf Comm* commission

provvi'sorio *a* provisional

prov'vista *nf* supply

pro'zio, -a *nm* great-uncle ● *nf* great-aunt

'prua *nf* prow

pru'den|te *a* prudent. **~za** *nf* prudence; **per ~za** as a precaution

'prudere *vi* itch

'prugn|a *nf* plum. **~a secca** prune. **~o** *nm* plum[-tree]

prurigi'noso *a* itchy. **pru'rito** *nm* itch

pseu'donimo *nm* pseudonym

psica'na|lisi *nf* psychoanalysis. **~'lista** *nmf* psychoanalyst. **~liz'zare** *vt* psychoanalyse

'psiche *nf* psyche

psichi'a|tra *nmf* psychiatrist. **~'tria** *nf* psychiatry. **~trico** *a* psychiatric

'psichico *a* mental

psico|lo'gia *nf* psychology. **~'logico** *a* psychological. **psi'cologo, -a** *nmf* psychologist

psico'patico, -a *a* psychopathic ● *nmf* psychopath

PT *abbr* (**Posta e Telecomunicazioni**) PO

pubbli'ca|re *vt* publish. **~zi'one** *nf* publication. **~zioni** *pl* (*di matrimonio*) banns

pubbli'cista *nmf Journ* correspondent

pubblicità *nf inv* publicity, advertising; (*annuncio*) advertisement, advert; **fare ~ a qcsa** advertise sth; **piccola ~** small advertisements. **pubbli'tario** *a* advertising

'pubblico *a* public; **scuola pubblica** state school ● *nm* public; (*spettatori*) audience; **grande ~** general public. **Pubblica Sicurezza** Police. **~ ufficiale** civil servant

'pube *nm* pubis

pubertà *nf* puberty

pu'dico *a* modest. **pu'dore** *nm* modesty

pue'rile *a* children's; *pej* childish

pugi'lato *nm* boxing. **'pugile** *nm* boxer

pugna'la|re *vt* stab. **~ta** *nf* stab. **pu'gnale** *nm* dagger

'pugno *nm* fist; (*colpo*) punch; (*manciata*) fistful; (*fig: numero limitato*) handful; **dare un ~ a** punch

'pulce *nf* flea; (*microfono*) bug

pul'cino *nm* chick; (*nel calcio*) junior

pu'ledra *nf* filly

pu'ledro *nm* colt

pu'li|re *vt* clean. **~re a secco** dryclean. **~to** *a* clean. **~'tura** *nf* cleaning. **~'zia** *nf* (*il pulire*) cleaning; (*l'essere pulito*) cleanliness; **~zie** *pl* housework; **fare le ~zie** do the cleaning

'pullman *nm inv* bus, coach; (*urbano*) bus

pul'mino *nm* minibus

'pulpito *nm* pulpit

pul'sante *nm* button; *Electr* [push-]button. **~ di accensione** on/off switch

pul'sa|re *vi* pulsate. **~zi'one** *nf* pulsation

pul'viscolo *nm* dust

'puma *nm inv* puma

pun'gente *a* (*insetto*) stinging; (*odore ecc*) sharp

'punger|e *vt* prick; (*insetto:*) sting. **~si** *vr* **~si un dito** prick one's finger

pungigli'one *nm* sting

pu'ni|re *vt* punish. **~'tivo** *a* punitive. **~zi'one** *nf* punishment; *Sport* free kick

'punta *nf* point; (*estremità*) tip; (*di monte*) peak; (*un po'*) pinch; *Sport* forward; **doppie punte** (*di capelli*) split ends

pun'tare *vt* point; (*spingere con forza*) push; (*scommettere*) bet; (*fam: appuntare*) fasten ● *vi* **~ su** fig rely on; **~ verso** (*dirigersi*) head for; **~ a** aspire to

punta'spilli *nm inv* pincushion

pun'tat|a *nf* (*di una storia*) instalment; (*televisiva*) episode; (*al gioco*) stake, bet; (*breve visita*) flying visit; **a ~e** serialized, in instalments; **fare una ~a a/in** pop over to (*luogo*)

punteggia'tura *nf* punctuation

pun'teggio *nm* score

puntel'lare *vt* prop. **pun'tello** *nm* prop

pun'tiglio *nm* spite; (*ostinazione*) obstinacy. **~'oso** *a* punctilious, pernickety *pej*

pun'tina *nf* (*da disegno*) drawing pin, thumb tack *Am*; (*di giradischi*) stylus. **~o** *nm* dot; **a ~o** perfectly; (*cotto*) to a T

'punto *nm* point; (*in cucito, Med*) stitch; (*in punteggiatura*) full stop; **in che ~?** where, exactly?; **di ~ in bianco** all of a sudden; **due punti** colon; **in ~** sharp; **mettere a ~** put right; *fig* fine tune; **tune up** (*motore*); **essere sul ~ di fare qcsa** be about to do sth, be on the point of doing sth. **punti pl cardinali** points of the compass. **~ debole** blind spot. **~ esclamativo** exclamation mark. **~ interrogativo** question mark. **~ nero** *Med* blackhead. **~ di riferimento** landmark; (*per la qualità*) benchmark. **~ di vendita** point of sale. **~ e virgola** semicolon. **~ di vista** point of view

puntu'al|e *a* punctual. **~ità** *f* punctuality. **~'mente** *adv* punctually, on time

pun'tura *nf* (*di insetto*) sting; (*di ago ecc*) prick; *Med* puncture; (*iniezione*) injection; (*fitta*) stabbing pain

punzecchi'are *vt* prick; *fig* tease

'pupa *nf* doll. **pu'pazzo** *nm* puppet. **pupazzo di neve** snowman

pu'pilla *nf Anat* pupil

pu'pillo, -a *nmf* (*di professore*) favourite

purché *conj* provided

'pure *adv* too, also; (*concessivo*) **fate ~!** please do! ● *conj* (*tuttavia*)

yet; (*anche se*) even if; **pur di** just to

purè *nm inv* purée. **~ di patate** mashed potatoes, creamed potatoes

pu'rezza *nf* purity

'purga *nf* purge. **pur'gante** *nm* laxative. **pur'gare** *vt* purge

purga'torio *nm* purgatory

purifi'care *vt* purify

puri'tano, -a *a & nmf* Puritan

'puro *a* pure; (*vino ecc*) undiluted; **per ~ caso** by sheer chance, purely by chance

puro'sangue *a & nm* thoroughbred

pur'troppo *adv* unfortunately

'pus *nm* pus. **'pustola** *nf* pimple

puti'ferio *nm* uproar

putre'far|e *vi*, **~si** *vr* putrefy

'putrido *a* putrid

put'tana *nf vulg* whore

'puzza *nf* = **puzzo**

puz'zare *vi* stink; **~ di bruciato** *fig* smell fishy

'puzzo *nm* stink, bad smell. **~la** *nf* polecat. **~'lente** *a* stinking

p.zza *abbr* (**piazza**) Sq.

Qq

qua *adv* here; **da un anno in ~** for the last year; **da quando in ~?** since when?; **di ~** this way; **di ~ di** on this side of; **~ dentro** in here; **~ sotto** under here; **~ vicino** near here; **~ e là** here and there

qua'derno *nm* exercise book; (*per appunti*) notebook

quadrango'lare *a* (*forma*) quadrangular. **qua'drangolo** *nm* quadrangle

qua'drante *nm* quadrant; (*di orologio*) dial

qua'dra|re *vt* square; (*contabilità*) balance ● *vi* fit in. **~to** *a* square; (*equilibrato*) level-headed ● *nm* square; (*pugilato*) ring; **al ~to** squared

quadret'tato *a* squared; (*carta*) graph *attrib.* **qua'dretto** *nm* square; (*piccolo quadro*) small picture; **a quadretti** (*tessuto*) check

quadricro'mia *nf* four-colour printing

quadrien'nale *a* (*che dura quattro anni*) four-year

quadri'foglio *nm* four-leaf clover

quadri'latero *nm* quadrilateral

quadri'mestre *nm* four-month period

'**quadro** *nm* picture; painting; (*quadrato*) square; (*fig: scena*) sight; (*tabella*) table; *Theat* scene; *Comm* executive **quadri** *pl* (*carte*) diamonds; **a quadri** (*tessuto, giacca, motivo*) check. **quadri** *pl* **direttivi** senior management

qua'drupede *nm* quadruped

quaggiù *adv* down here

'**quaglia** *nf* quail

'**qualche** *a* (*alcuni*) a few, some; (*un certo*) some; (*in interrogazioni*) any; **ho ~ problema** I have a few problems, I have some problems; **~ tempo fa** some time ago; **hai ~ libro italiano?** have you any Italian books?; **posso prendere ~ libro?** can I take some books?; **in ~ modo** somehow; **in ~ posto** somewhere; **~ volta** sometimes; **~ cosa = qualcosa**

qual'cosa *pron* something; (*in interrogazioni*) anything; **~'altro** something else; **vuoi ~'altro?** would you like anything else?; **~ di strano** something strange; **vuoi ~a da mangiare?** would you like something to eat?

qual'cuno *pron* someone, somebody; (*in interrogazioni*) anyone, anybody; (*alcuni*) some; (*in interrogazioni*) any; **c'è ~?** is anybody in?; **qualcun altro** someone else, somebody else; **c'è qualcun altro che aspetta?** is anybody else waiting?; **ho letto ~ dei suoi libri** I've read some of his books; **cono-**

sci ~ dei suoi amici? do you know any of his friends?

'**quale** *a* which; (*indeterminato*) what; (*come*) as, like; **~ macchina è la tua?** which car is yours?; **~ motivo avrà di parlare così?** what reason would he have to speak like that?; **~ onore!** what an honour!; **città quali Venezia** towns like Venice; **~ che sia la tua opinione** whatever you may think ● *pron inter* which [one]; **~ preferisci?** which [one] do you prefer? ● *pron rel* il/la ~ (*persona*) who; (*animale, cosa*) that, which; (*oggetto: con prep*) whom; (*animale, cosa*) which; **ho incontrato tua madre, la ~ mi ha detto...** I met your mother, who told me...; **l'ufficio nel ~ lavoro** the office in which I work; **l'uomo con il ~ parlavo** the man to whom I was speaking ● *adv* (*come*) as

qua'lifica *nf* (*titolo*) title

qualifi'care *vt* qualify; (*definire*) define. **~rsi** *vr* be placed. **~'tivo** *a* qualifying. **~ato** *a* (*operaio*) semi-skilled. **~zi'one** *nf* qualification

qualità *nf inv* quality; (*specie*) kind; **in ~ di** in one's capacity as. **~tiva'mente** *adv* qualitatively. **~'tivo** *a* qualitative

qua'lora *conj* in case

qual'siasi, qua'lunque *a* any; (*non importa quale*) whatever; (*ordinario*) ordinary; **dammi una penna** ~ give me any pen [whatsoever]; **farei ~ cosa** I would do anything; **~ cosa io faccia** whatever I do; **~ persona** anyone; **in ~ caso** in any case; **uno ~** any one, whichever; **l'uomo qualunque** the man in the street; **vivo in una casa** ~ I live in an ordinary house

qualunqu'ismo *nm* lack of political views

'**quando** *conj & adv* when; **da ~ ti ho visto** since I saw you; **da ~ esci con lui?** how long have you been going out with him?; **da ~ in qua?** since when?; **~... ~...** sometimes..., sometimes...

quantifi'care vt quantify

quantità nf inv quantity; **una ~ di** (gran numero) a great deal of. **~tiva'mente** adv quantitatively. **~'tivo** nm amount ●a quantitative

'quanto a inter how much; (con nomi plurali) how many; (in esclamazione) what a lot of; **~ tempo?** how long?; **quanti anni hai?** how old are you? ●a rel as much... as; (con nomi plurali) as many... as; **prendi ~ denaro ti serve** take as much money as you need; **prendi quanti libri vuoi** take as many books as you like ●pron inter how much; (quanto tempo) how long; (plurale) how many; **quanti ne abbiamo oggi?** what date is it today?, what's the date today? ●pron rel so much as; (quanto tempo) as long as; (plurale) as many as; **prendine ~/quanti ne vuoi** take as much/as many as you like; **stai ~ vuoi** stay as long as you like; **questo è ~** that's it ●adv inter how much; (quanto tempo) how long; **~ sei alto?** how tall are you?; **~ hai aspettato?** how long did you wait for?; **~ costa?** how much is it?; **~ mi dispiace!** I'm so sorry!; **~ è bello!** how nice! ●adv rel as much as; **lavoro ~ posso** I work as much as I can; **è tanto intelligente ~ bello** he's as intelligent as he's good-looking; **in ~** (in qualità di) as; (poiché) since; **in ~ a me** as far as I'm concerned; **per ~** however; **per ~ ne sappia** as far as I know; **per ~ mi riguarda** as far as I'm concerned; **per ~ mi sia simpatico** much as I like him; **~ a se;** **per ~ prima** (al più presto) as soon as possible

quan'tunque conj although

qua'ranta a & nm forty

quaran'tena nf quarantine

quaran'tenn|e a forty-year-old. **~io** nm period of forty years

quaran'tesimo a fortieth. **~ina** nf una **~ina** about forty

qua'resima nf Lent

quar'tetto nm quartet

quarti'ere nm district; Mil quarters pl. **~ generale** headquarters

'quarto a fourth ●nm fourth; (quarta parte) quarter; **le sette e un ~** a quarter past seven. **quarti pl di finale** quarterfinals. **~ d'ora** a quarter of an hour. **quar'tultimo, -a** nmf fourth from the end, fourth last

'quarzo nm quartz

'quasi adv almost, nearly; **~ mai** hardly ever ●conj (come se) as if; **~ ~ sto a casa** I'm tempted to stay home

quas'sù adv up here

'quatto a crouching; (silenzioso) silent; **starsene ~ ~** keep very quiet

quat'tordici a & nm fourteen

quat'trini nmpl money sg, dosh sg fam

'quattro a & nm four; **dirne ~ a** qcno give sb a piece of one's mind; **farsi in ~** (per qcno/per fare qcsa) go to a lot of trouble (for sb/ to do sth); **in ~ e quat-tr'otto** in a flash. **~ per ~** nm inv Auto four-wheel drive [vehicle]

quat'trocchi: a ~ adv in private

quattro'cento a & nm four hundred; **il ~cento** the fifteenth century

quattro'mila a & nm four thousand

'quell|o a that (pl those); **quell'albero** that tree; **quegli alberi** those trees; **quel cane** that dog; **quei cani** those dogs ●pron that [one] (pl those [ones]); **~o li** that one over there; **~o che** the one that; (ciò che) what; **quelli che** the ones that, those that; **~o a destra** the one on the right

'quercia nf oak

que'rela nf [legal] action

quere'lare vt bring an action against

que'sito nm question

questio'nario nm questionnaire

quest'ione nf question; (faccenda) matter; (litigio) quarrel; in ~ in doubt; è fuori ~ it's out of the question; è di vita o di morte it's a matter of life and death

'quest|o a this (pl these) ● pron this [one] (pl these [ones]); ~o qui, ~o qua this one here; ~o è quello che a detto that's what he said; per ~o for this or that reason. **quest'oggi** today

que'store nm chief of police

que'stura nf police headquarters pl

qui adv here; da ~ in poi from now on; fin ~ (di tempo) up till now, until now; ~ dentro in here; ~ sotto under here; ~ vicino adv near here ● nm ~ pro quo misunderstanding

quie'scienza nf trattamento di ~ retirement package

quie'tanza nf receipt

quie'tar|e vt calm. ~**si** vr quieten down

qui'et|e nf quiet; disturbo della ~e pubblica breach of the peace. ~**o** a quiet

'quindi adv then ● conj therefore

'quindi|ci a & nm fifteen. ~**cina** nf una ~cina about fifteen; una ~cina di giorni a fortnight Br, two weeks pl

quinquen'nale a (che dura cinque anni) five-year. **quin'quennio** nm [period of] five years

quin'tale nm a hundred kilograms

'quinte nfpl Theat wings

quin'tetto nm quintet

'quinto a fifth

'quintuplo a quintuple

qui'squiglia nf perdersi in quisquiglie get bogged down in details

'quota nf quota; (rata) instalment; (altitudine) height; Aeron altitude, height; (ippica) odds pl; per-

dere ~ lose altitude; prendere ~ gain altitude. ~ **di iscrizione** entry fee

quo'ta|re vt Comm quote. ~**to** a quoted; essere ~**to in Borsa** be quoted on the Stock Exchange. ~**zi'one** nf quotation

quotidi|ana'mente adv daily. ~**'ano** a daily; (ordinario) everyday ● nm daily [paper]

quozi'ente nm quotient. ~ **d'intelligenza** intelligence quotient, IQ

Rr

ra'barbaro nm rhubarb

'rabbia nf rage; (ira) anger; Med rabies sg; che ~! what a nuisance!; mi fa ~ it makes me angry

rab'bino nm rabbi

rabbiosa'mente adv furiously. **rabbi'oso** a hot-tempered; Med rabid; (violento) violent

rabbo'nir|e vt pacify. ~**si** vr calm down

rabbrivi'dire vi shudder; (di freddo) shiver

rabbui'ars|i vr become dark

raccapez'zar|e vt put together. ~**si** vr see one's way ahead

raccapricci'ante a horrifying

raccatta'palle nm inv ball boy ● nf inv ball girl

raccat'tare vt pick up

rac'chetta nf racket. ~ **da ping pong** table-tennis bat. ~ **da sci** ski stick, ski pole. ~ **da tennis** tennis racket

'racchio a fam ugly

racchi'udere vt contain

rac'cogli|ere vt pick; (da terra) pick up; (mietere) harvest; (collezionare) collect; (radunare) gather; win (voti ecc); (dare asilo a) take in. ~**ersi** vr gather; (concentrarsi) collect one's thoughts. ~**'mento** nm concentration.

~'tore, ~'trice *nmf* collector
● *nm* (*cartella*) ring-binder

rac'colto, -a *pp di* raccogliere
● *a* (*rannicchiato*) hunched; (*intimo*) cosy; (*concentrato*) engrossed
● *nm* (*mietitura*) harvest ~ *nf* collection; (*di scritti*) compilation; (*del grano ecc*) harvesting; (*adunata*) gathering

raccoman'dabile *a* recommendable; **poco** ~ (*persona*) shady

raccoman'da|re *vt* recommend; (*affidare*) entrust. ~rsi *vr* (*implorare*) beg. ~ta *nf* registered letter; ~ta con ricevuta di ritorno recorded delivery. ~espresso *nf* guaranteed next-day delivery of recorded items. ~zi'one *nf* recommendation

raccon'tare *vt* tell. rac'conto *nm* story

raccorci'are *vt* shorten

raccor'dare *vt* join. rac'cordo *nm* connection; (*stradale*) feeder. raccordo anulare ring road. raccordo ferroviario siding

ra'chitico *a* rickety; (*poco sviluppato*) stunted

racimo'lare *vt* scrape together

'racket *nm inv* racket

'radar *nm* radar

raddol'cir|e *vt* sweeten; *fig* soften. ~si *vr* become milder; (*carattere:*) mellow

raddoppi'are *vt* double. rad'doppio *nm* doubling

raddriz'zare *vt* straighten

'rader|e *vt* shave; graze (*muro*); ~e al suolo raze [to the ground]. ~si *vr* shave

radi'are *vt* strike off; ~ dall'albo strike off

radia|'tore *nm* radiator. ~zi'one *nf* radiation

'radica *nf* briar

radi'cale *a* radical ● *nm* *Gram* root; *Pol* radical

ra'dicchio *nm* chicory

ra'dice *nf* root; mettere [le] radici *fig* put down roots. ~ quadrata square root

'radio *nf inv* radio; via [~ by radio.
~ a transistor transistor radio
● *nm* *Chem* radium.

radioama'tore, -'trice *nmf* [radio] ham

radioascolta'tore, -'trice *nmf* listener

radioat|tività *nf* radioactivity.
~'tivo *a* radioactive

radio'cro|naca *nf* radio commentary; fare ~ (**naca di** commentate on. ~'nista *nf* radio reporter

radiodiffusi'one *nf* broadcasting

radiogra|'fare *vt* X-ray. ~'fia *nf* X-ray [photograph]; (*radiologia*) radiography; fare una ~fia (*paziente:*) have an X-ray; (*dottore:*) take an X-ray

radio'fonico *a* radio *attrib*

radio'lina *nf* transistor

radi'ologo, -a *nmf* radiologist

radi'oso *a* radiant

radio'sveglia *nf* radio alarm

radio'taxi *nm inv* radio taxi

radiote'lefono *nm* radiotelephone; (*privato*) cordless [phone]

radiotelevi'sivo *a* broadcasting *attrib*

'rado *a* sparse; (*non frequente*) rare; di ~ seldom

radu'nar|e *vt*, ~si *vr* gather [together]. ra'duno *nm* meeting; *Sport* rally

ra'dura *nf* clearing

'rafano *nm* horseradish

raffazzo'nato *a* (*discorso, lavoro*) botched

raf'fermo *a* stale

'raffica *nf* gust; (*di armi da fuoco*) burst; (*di domande*) barrage

raffigu'ra|re *vt* represent. ~zi'one *nf* representation

raffi'na|re *vt* refine. ~'ta'mente *adv* elegantly. ~'tezza *nf* refinement. ~'to *a* refined. raffine'ria *nf* refinery

rafforza|'mento *nm* reinforcement; (*di muscolatura*) strengthening. ~re *vt* reinforce. ~'tivo *nm* *Gram* intensifier

raffredda'mento nm (processo) cooling

raffred'd|are vi cool. **~arsi** vr get cold; (prendere un raffreddore) catch a cold. **~ore** nm cold. **~ore da fieno** hay fever

raf'fronto nm comparison

'rafia nf raffia

Rag. abbr **ragioniere**

ra'gaz|za nf girl; (fidanzata) girlfriend. **~za alla pari** au pair [girl]. **~zata** nf prank. **~zo** nm boy; (fidanzato) boyfriend; **da ~zo** (da giovane) as a boy

ragge'lar|e vt fig freeze. **~si** vr fig turn to ice

raggi'ante a radiant; **~ di successo** flushed with success

raggi'era nf a **~** with a pattern like spokes radiating from a centre

'raggio nm ray; Math radius; (di ruota) spoke; **~ d'azione** range. **~ laser** laser beam

raggi'rare vt trick. **rag'giro** nm trick

raggi'un|gere vt reach; (conseguire) achieve. **~'gibile** a (luogo) within reach

raggomito'lar|e vt wind. **~si** vr curl up

raggrane'llare vt scrape together

raggrin'zir|e vt, **~si** vr wrinkle

raggrup|pa'mento nm (gruppo) group; (azione) grouping. **~'pare** vt group together

ragguagli'are vt compare; (informare) inform. **raggu'aglio** nm comparison; (informazione) information

ragguar'devole a considerable

'ragia nf resin; **acqua ~** turpentine

ragiona'mento nm reasoning; (discussione) discussion. **ragio'nare** vi reason; (discutere) discuss

ragi'one nf reason; (ciò che è giusto) right; **a ~ o a torto** rightly or wrongly; **aver ~** be right; **perdere la ~** go out of one's mind;

a ragion veduta after due consideration

ragione'ria nf accountancy

ragio'nevol|e a reasonable. **~'mente** adv reasonably

ragioni'ere, -a nmf accountant

ragli'are vi bray

'ragno nm cobweb. **'ragno** nm spider

ragù nm inv meat sauce

RAI nf abbr (**Radio Audizioni Italiane**) Italian public broadcasting company

ralle'gra|re vt gladden. **~rsi** vr rejoice; **~rsi con qcno** congratulate sb. **~'menti** nmpl congratulations

rallenta'mento nm slowing down

rallen'ta|re vt/i slow down; (allentare) slacken. **~rsi** vr slow down. **~tore** nm (su strada) speed bump; **al ~tore** in slow motion

raman'zina nf reprimand

ra'marro nm type of lizard

ra'mato a (capelli) copper[-coloured]

'rame nm copper

ramifi'ca|re vi, **~rsi** vr branch out; (strada:) branch. **~zi'one** nf ramification

rammari'carsi vr **~ di** regret; (lamentarsi) complain (di about). **ram'marico** nm regret

rammen'dare vt darn. **ram'mendo** nm darning

rammen'tar|e vt remember; **~e qcsa a qcno** (richiamare alla memoria) remind sb of sth. **~si** vr remember

rammol'li|re vt soften. **~rsi** vr go soft. **~to, -a** nmf wimp

'ramo nm branch. **~'scello** nm twig

'rampa nf (di scale) flight. **~ d'accesso** slip road. **~ di lancio** launch[ing] pad

ram'pante a **giovane ~** yuppie

rampi'cante a climbing ● nm Bot creeper

ram'pollo nm hum brat; (discendente) descendant

ram'pone nm harpoon; (per scarpe) crampon

'rana nf frog; (nel nuoto) breaststroke; **uomo ~** frogman

'rancido a rancid

ran'core nm resentment

ran'dagio a stray

'rango nm rank

rannicchi'arsi vr huddle up

rannuvola'mento nm clouding over. **rannuvo'larsi** vr cloud over

ra'nocchio nm frog

ranto'lare vi wheeze. **'rantolo** nm wheeze; (di moribondo) death-rattle

'rapa nf turnip

ra'pace a rapacious; (uccello) predatory

ra'pare vt crop

'rapida nf rapids pl. **~'mente** adv rapidly

rapidità nf speed

'rapido a swift ● nm (treno) express [train]

rapi'mento nm (crimine) kidnapping

ra'pina nf robbery; **~ a mano armata** armed robbery. **~ in banca** bank robbery. **rapi'nare** vt rob. **~'tore** nm robber

ra'pi|re vt abduct; (a scopo di riscatto) kidnap; (estasiare) ravish. **~tore**, **~trice** nmf kidnapper

rappacifi'ca|re vt pacify. **~rsi** vr be reconciled, make it up. **~zi'one** nf reconciliation

rappor'tare vt reproduce (disegno); (confrontare) compare

rap'porto nm report; (connessione) relation; (legame) relationship; Math, Techn ratio; **rapporti** pl relationship; **essere in buoni rapporti** be on good terms. **~ di amicizia** friendship. **~ di lavoro** working relationship. **rapporti** pl **sessuali** sexual intercourse

rap'prendersi vr set; (latte:) curdle

rappre'saglia nf reprisal

rappresen'tan|te nmf representa-

tive. **~te di classe** class representative. **~te di commercio** sales representative, [sales] rep fam. **~za** nf delegation; Comm agency; **spese** pl **di ~za** entertainment expenses; **di ~za** (appartamento ecc) company

rappresen'ta|re vt represent; Theat perform. **~'tivo** a representative. **~zi'one** nf representation; (spettacolo) performance

rap'preso pp di **rapprendersi**

rapso'dia nf rhapsody

'raptus nm inv fit of madness

rara'mente adv rarely, seldom

rare'fa|re vt, **~rsi** vr rarefy. **~tto** a rarefied

rarità nf inv rarity. **'raro** a rare

ra'sar|e vt shave; trim (siepe ecc). **~si** vr shave

raschia'mento nm Med curettage

raschi'are vt scrape; (togliere) scrape off

rasen'tare vt go close to. **ra'sente** prep very close to

'raso pp di **radere** ● a smooth; (colmo) full to the brim; (barba) close-cropped; **~ terra** close to the ground; **un cucchiaio ~** a level spoonful ● nm satin

ra'soio nm razor

ras'segna nf review; (mostra) exhibition; (musicale, cinematografica) festival; **passare in ~** review; Mil inspect

rasse'gna|re vt present. **~rsi** vr resign oneself. **~to** a (persona, aria, tono) resigned. **~zi'one** nf resignation

rasse're'nar|e vt clear; fig cheer up. **~si** vr become clear; fig cheer up

rasset'tare vt tidy up; (riparare) mend

rassicu'ra|nte a (persona, parole, presenza) reassuring. **~re** vt reassure. **~zi'one** nf reassurance

rasso'dare vt harden; fig strengthen

rassomigli'a|nza *nf* resemblance. **~re** *vi* **~re a** resemble

rastrella'mento *nm* (*di fieno*) raking; (*perlustrazione*) combing. **rastrel'iare** *vt* rake; (*perlustrare*) comb

rastrelli'era *nf* rack; (*per biciclette*) bicycle rack; (*scolapiatti*) [plate] rack. **ra'strello** *nm* rake

'rata *nf* instalment; **pagare a rate** pay by instalments; **comprare qcsa a rate** buy sth on hire purchase, buy sth on the installment plan *Am.* **rate'ale** *a* (*pagamento*) instalments; **pagamento rateale** payment by instalments

rate'are, rateiz'zare *vt* divide into instalments

ra'tifica *nf Jur* ratification

ratifi'care *vt Jur* ratify

'ratto *nm* abduction; (*roditore*) rat

rattop'pare *vt* patch. **rat'toppo** *nm* patch

rattrap'pir|e *vt* make stiff. **~si** *vr* become stiff

rattri'star|e *vt* sadden. **~si** *vr* become sad

rau'cedine *nf* hoarseness. **'rauco** *a* hoarse

rava'nello *nm* radish

ravi'oli *nmpl* ravioli *sg*

ravve'dersi *vr* mend one's ways

ravvicina'mento *nm* (*tra persone*) reconciliation; *Pol* rapprochement

ravvici'nar|e *vt* bring closer; (*riconciliare*) reconcile. **~si** *vr* be reconciled

ravvi'sare *vt* recognize

ravvi'var|e *vt* revive; *fig* brighten up. **~si** *vr* revive

'rayon *nm* rayon

razio'cinio *nm* rational thought; (*buon senso*) common sense

razio'nal|e *a* rational. **~ità** *nf* (*raziocinio*) rationality; (*di ambiente*) functional nature. **~iz'zare** *vt* rationalize (*programmi, metodi, spazio*). **~'mente** *adv* (*con raziocinio*) rationally

razio'nare *vt* ration. **razi'one** *nf* ration

'razza *nf* race; (*di cani ecc*) breed; (*genere*) kind; **che ~ di idiota!** *fam* what an idiot!

raz'zia *nf* raid

razzi'ale *a* racial

raz'zis|mo *nm* racism. **~ta** *a* & *nmf* racist

'razzo *nm* rocket. **~ da segnalazione** flare

razzo'lare *vi* (*polli:*) scratch about

re *nm inv* king; *Mus* (*chiave, nota*) D

rea'gire *vi* react

re'ale *a* real; (*di re*) royal

rea'lismo *nm* realism. **~ta** *nmf* realist; (*fautore del re*) royalist

realistica'mente *adv* realistically. **rea'listico** *a* realistic

realiz'zabile *a* (*programma*) feasible

realiz'za|re *vt* (*attuare*) carry out, realize; *Comm* make; score (*gol, canestro*); (*rendere conto di*) realize. **~rsi** *vr* come true; (*nel lavoro ecc*) fulfil oneself. **~zi'one** *nf* realization; (*di sogno, persona*) fulfilment. **~zione scenica** production

rea'lizzo *nm* (*vendita*) proceeds *pl*; (*riscossione*) yield

real'mente *adv* really

realtà *nf inv* reality. **~ virtuale** virtual reality

re'ato *nm* crime, criminal offence

reat'tivo *a* reactive

reat'tore *nm* reactor; *Aeron* jet [aircraft]

reazio'nario, -a *a* & *nmf* reactionary

reazi'one *nf* reaction. **~ a catena** chain reaction

'rebus *nm inv* rebus; (*enigma*) puzzle

recapi'tare *vt* deliver. **re'capito** *nm* address; (*consegna*) delivery. **recapito a domicilio** home delivery. **recapito telefonico** contact telephone number

re'car|e vt bear; (*produrre*) cause. **~si** vr go

re'cedere vi recede; *fig* give up

recensi'one nf review

recen's|ire vt review. **~ore** nm reviewer

re'cente a recent; **di ~** recently. **~'mente** adv recently

recessi'one nf recession

reces'sivo a Biol recessive.

re'cesso nm recess

re'cidere vt cut off

reci'divo, -a a Med recurrent ● nmf repeat offender

recin'|tare vt close off. **re'cinto** nm enclosure; (*per animali*) pen; (*per bambini*) play-pen. **~zi'one** nf (*muro*) wall; (*rete*) wire fence; (*cancellata*) railings pl

recipi'ente nm container

re'ciproco a reciprocal

re'ciso pp di recidere

'recita nf performance. **reci'tare** vt recite; Theat act; play (*ruolo*). **~zi'one** nf recitation; Theat acting

recla'mare vi protest ● vt claim

ré'clame nf inv advertising; (*avviso pubblicitario*) advertisement

re'clamo nm complaint; **ufficio reclami** complaints department or office

recli'na|bile a reclining; **sedile ~bile** reclining seat. **~re** vt tilt (*sedile*); lean (*capo*)

reclusi'one nf imprisonment. **re'cluso, -a** a secluded ● nmf prisoner

'recluta nf recruit

reclu'ta|mento nm recruitment. **~tare** vt recruit

'record nm inv record ● a inv (*cifra*) record attrib

recrimi'na|re vi recriminate. **~zi'one** nf recrimination

recupe'rare vt recover. **re'cupero** nm recovery; **corso di recupero** additional classes; **minuti di recupero** Sport injury time

redargu'ire vt rebuke

re'datto pp di redigere

redat'tore, -'trice nmf editor; (*di testo*) writer

redazi'one nf (*ufficio*) editorial office; (*di testi*) editing

reddi'tizio a profitable

'reddito nm income. **~ imponibile** taxable income

re'den|to pp di redimere. **~'tore** nm redeemer. **~zi'one** nf redemption

re'digere vt write; draw up (*documento*)

re'dimer|e vt redeem. **~si** vr redeem oneself

'redini nfpl reins

re'duce a **~ da** back from ● nmf survivor

refe'rendum nm inv referendum

refe'renza nf reference

refet'torio nm refectory

refrat'tario a refractory; **essere ~ a** have no aptitude for

refrige'ra|re vt refrigerate. **~zi'one** nf refrigeration

refur'tiva nf stolen goods pl

rega'lare vt give

re'gale a regal

re'galo nm present, gift

re'gata nf regatta

reg'gen|te nmf regent. **~za** nf regency

'regger|e vt (*sorreggere*) bear; (*tenere in mano*) hold; (*dirigere*) run; (*governare*) govern; Gram take ● vi (*resistere*) hold out; (*durare*) last; fig stand. **~si** vr stand

'reggia nf royal palace

reggi'calze nm inv suspender belt

reggi'mento nm regiment; (*fig: molte persone*) army

reggi'petto, reggi'seno nm bra

re'gia nf Cinema direction; Theat production

re'gime nm regime; (*dieta*) diet; Mech speed. **~ militare** military regime

re'gina nf queen

'regio a royal

regio'na|le a regional. **~'lismo** nm (*parola*) regionalism

regi'one nf region

re'gista *nmf* Cinema director; Theat, TV producer

regi'stra|re *vt* register; Comm enter; (*incidere su nastro*) tape, record; (*su disco*) record. **~'tore** *nm* recorder; (*magnetofono*) tape-recorder. **~tore di cassa** cash register. **~zi'one** *nf* registration; Comm entry; (*di programma*) recording

re'gistro *nm* register; (*ufficio*) registry. **~ di cassa** ledger

re'gnare *vi* reign

'regno *nm* kingdom; (*sovranità*) reign. **R~ Unito** United Kingdom

'regola *nf* rule; **essere in ~** be in order; (*persona:*) have one's papers in order. **rego'labile** *a* (*meccanismo*) adjustable. **~'mento** *nm* regulation; Comm settlement. **~mento di conti** settling of scores

rego'lar|e *a* regular ● *vt* regulate; (*ridurre, moderare*) limit; (*sistemare*) settle. **~si** *vr* (*agire*) act; (*moderarsi*) control oneself. **~ità** *nf inv* regularity. **~iz'zare** *vt* settle (*debito*)

rego'la|ta *nf* darsi una **~ta** pull oneself together. **~'tore, ~'trice** *a* piano **~tore** urban development plan

'regolo *nm* ruler

regre'dire *vi* Biol, Psych regress

regres|si'one *nf* regression. **~'sivo** *a* regressive. **re'gresso** *nm* decline

reinseri'mento *nm* (*di persona*) reintegration

reinser'irsi *vr* (*in ambiente*) reintegrate

reinte'grare *vt* restore

relativa'mente *adv* relatively; **~** as regards. **relativ'ità** *nf* relativity. **rela'tivo** *a* relative

rela'tore, -'trice *nmf* (*in una conferenza*) speaker

re'lax *nm* relaxation

relazi'one *nf* relation[ship]; (*rapporto amoroso*) [love] affair; (*resoconto*) report; **pubbliche relazioni** *pl* public relations

rele'gare *vt* relegate

religi'o|ne *nf* religion. **~so, -a** *a* religious ● *nm* monk ● *nf* nun

re'liqui|a *nf* relic. **~'ario** *nm* reliquary

re'litto *nm* wreck

re'ma|re *vi* row. **~'tore, ~'trice** *nmf* rower

remini'scenza *nf* reminiscence

remissi'one *nf* remission; (*sottomissione*) submissiveness. **remis'sivo** *a* submissive

'remo *nm* oar

re'mora *nf* **senza remore** without hesitation

re'moto *a* remote

remune'ra|re *vt* remunerate. **~'tivo** *a* remunerative. **~zi'one** *nf* remuneration

'render|e *vt* (*restituire*) return; (*esprimere*) render; (*fruttare*) yield; (*far diventare*) make. **~si** *vr* become; **~si conto di qcsa** realize sth; **~si utile** make oneself useful

rendi'conto *nm* report

rendi'mento *nm* rendering; (*produzione*) yield

'rendita *nf* income; (*dello Stato*) revenue; **vivere di ~** *fig* rest on one's laurels

'rene *nm* kidney. **~ artificiale** kidney machine

'reni *nfpl* (*schiena*) back

reni'tente *a* **essere ~ a** (*consigli di qcno*) be unwilling to accept

'renna *nf* (*animale*) reindeer (*pl inv*); (*pelle*) buckskin

'Reno *nm* Rhine

'reo, -a *a* guilty ● *nmf* offender

re'parto *nm* department; Mil unit

repel'lente *a* repulsive

repen'taglio *nm* **mettere a ~** risk

repen'tino *a* sudden

repe'ribile *a* available; **non è ~** (*perduto*) it's not to be found

repe'rire *vt* trace (*fondi*)

re'perto *nm* **~ archeologico** find

reper'torio *nm* repertory; (*elenco*)

index; **immagini** pl di ~ archive footage

'replica nf reply; (obiezione) objection; (copia) replica; Theat repeat performance. **repli'care** vt reply; Theat repeat

repor'tage nm inv report

repres|si'one nf repression. **~si'vo** a repressive. **re'presso** pp di **reprimere**. **re'primere** vt repress

re'pubbli|ca nf republic. **~'cano, -a** a & nmf republican

repu'tare vt consider

reputazi'one nf reputation

requi'si|re vt requisition. **~to** nm requirement

requisi'toria nf (arringa) closing speech

requisizi'one nf requisition

'resa nf surrender; Comm rendering. **~ dei conti** rendering of accounts

'residence nm inv residential hotel

resi'den|te a & nmf resident. **~za** nf residence; (soggiorno) stay. **~zi'ale** a residential; **zona ~ziale** residential district

re'siduo a residual ● nm remainder

'resina nf resin

resi'sten|te a resistant; **~te all'acqua** water-resistant. **~za** nf resistance; (fisica) stamina; Electr resistor; **la R~za** the Resistance

re'sistere vi ~ [a] resist; (a colpi, scosse) stand up to; **~ alla pioggia/al vento** be rain-/wind-resistant

'reso pp di **rendere**

reso'conto nm report

respin'gente nm Rail buffer

re'spin|gere vt repel; (rifiutare) reject; (bocciare) fail. **~to** pp di **respingere**

respi'ra|re vt/i breathe. **~tore** nm respirator. **~tore [a tubo]** snorkel **~'torio** a respiratory. **~zi'one** nf breathing; Med respiration. **~zione bocca a bocca** mouth-to-mouth rescuscitation, kiss of life.

re'spiro nm breath; (il respirare) breathing; fig respite

respon'sabi|le a responsible (**di** for); Jur liable ● nmf person responsible. **~e della produzione** production manager. **~ità** nf inv responsibility; Jur liability. **~iz'zare** vt give responsibility to

re'sponso nm response

'ressa nf crowd

re'stante a remaining ● nm remainder

re'stare vi = rimanere

restau'ra|re vt restore. **~'tore, ~'trice** nmf restorer. **~zi'one** nf restoration. **re'stauro** nm (riparazione) repair

re'stio a restive; ~ a reluctant to

restitu|'ire vt return; (reintegrare) restore. **~zi'one** nf return; Jur restitution

'resto nm remainder; (saldo) balance; (denaro) change; **resti** pl (avanzi) remains; **del ~** besides

re'strin|gere vt contract; take in (vestiti); (limitare) restrict; shrink (stoffa). **~si** vr contract; (farsi più vicini) close up; (stoffa:) shrink. **restringi'mento** nm (di tessuto) shrinkage

restrit'tivo a (legge, clausola) restrictive. **~zi'one** nf restriction

resurrezi'one nf resurrection

resusci'tare vt/i revive

re'tata nf round-up

'rete nf net; (sistema) network; (televisiva) channel; (in calcio, hockey) goal; fig trap; (per la spesa) string bag. **~ locale** Comput local [area] network, LAN. **~ stradale** road network. **~ televisiva** television channel

reti'cen|te a reticent. **~za** nf reticence

retico'lato nm grid; (rete metallica) wire netting. **re'ticolo** nm network

'retina nf retina

re'tina nf (per capelli) hair net

re'torico, -a a rhetorical; **domanda retorica** rhetorical question ● nf rhetoric

retribu'ire *vt* remunerate. **~zi'o-ne** *nf* remuneration

'**retro** *adv* behind; **vedi** ~ see over ● *nm inv* back. **~ di copertina** outside back cover

retroat'tivo *a* retroactive

retro'ce|dere *vi* retreat ● *vt* Mil demote; *Sport* relegate. **~ssi'one** *nf* Sport relegation

retroda'tare *vt* backdate

re'trogrado *a* retrograde; *fig* old-fashioned; *Pol* reactionary

retrogu'ardia *nf* Mil rearguard

retro'marcia *nf* reverse [gear]

retro'scena *nm inv* Theat back-stage; *fig* background details *pl*

retrospet'tivo *a* retrospective

retro'stante *a* **il palazzo** ~ the building behind

retrovi'sore *nm* rear-view mirror

'**retta**[1] *nf* Math straight line; **(di collegio, pensionato)** fee

'**retta**[2] *nf* **dar** ~ **a qcno** take sb's advice

rettango'lare *a* rectangular. **ret-'tangolo** *a* right-angled ● *nm* rec-tangle

ret'tifi|ca *nf* rectification. **~'care** *vt* rectify

'**rettile** *nm* reptile

retti'lineo *a* rectilinear; **(retto)** up-right ● *nm* Sport back straight

retti'tudine *nf* rectitude

'**retto** *pp di* **reggere** ● *a* straight; *fig* upright; **(giusto)** correct; **angolo** ~ right angle

ret'tore *nm* Relig rector; *Univ* chancellor

reu'matico *a* rheumatic

reuma'tismi *nmpl* rheumatism

reve'rendo *a* reverend

rever'sibile *a* reversible

revisio'nare *vt* revise; *Comm* au-dit; *Auto* overhaul. **revisi'one** *nf* revision; *Comm* audit; *Auto* over-haul. **revi'sore** *nm* **(di conti)** audi-tor; **(di bozze)** proof-reader; **(di traduzioni)** revisor

re'vival *nm inv* revival

'**revoca** *nf* repeal. **revo'care** *vt* re-peal

riabili'tare *vt* rehabilitate. **~zi'o-ne** *nf* rehabilitation

riabitu'ar|e *vt* reaccustom. **~si** *vr* reaccustom oneself

riac'cender|e *vt* rekindle **(fuoco)**. **~si** *vr* **(luce)** come back on

riacqui'stare *vt* buy back; regain **(libertà, prestigio)**; recover **(vista, udito)**

riagganci'are *vt* replace **(ricevito-re)**; ~ **la cornetta** hang up ● *vi* hang up

riallac'ciare *vt* refasten; recon-nect **(corrente)**; renew **(amicizia)**

rial'zare *vt* raise ● *vi* rise. **ri'alzo** *nm* rise

riani'mar|e *vt* Med resuscitate; **(ri-dare forza a)** revive; **(dare corag-gio a)** cheer up. **~si** *vr* re-gain consciousness; **(riprendere forza)** revive; **(riprendere corag-gio)** cheer up

riaper'tura *nf* reopening

ria'prir|e *vt*, **~si** *vr* reopen

ri'armo *nm* rearmament

rias'sumere *vt* **(ricapitolare)** re-sume

riassun'tivo *a* summarizing. **rias-'sunto** *pp di* **riassumere** ● *nm* summary

ria'ver|e *vt* get back; regain **(sa-lute, vista)**. **~si** *vr* recover

riavvia'mento *nm* **(tra perso-ne)** reconciliation

riavvici'nar|e *vt* reconcile **(paesi, persone)**. **~si** *vr* **(riconciliarsi)** be reconciled, make it up

riba'dire *vt* **(confermare)** reaffirm

ri'balta *nf* flap; *Theat* footlights *pl*; *fig* limelight

ribal'tabile *a* tip-up

ribal'tar|e *vt/i*, **~si** *vr* tip over; *Naut* capsize

ribas'sare *vt* lower ● *vi* fall. **ri'basso** *nm* fall; **(sconto)** discount

ri'battere *vt* **(a macchina)** retype; **(controbattere)** deny ● *vi* answer back

ribel'l|arsi *vr* rebel. **ri'belle** *a* re-bellious ● *nmf* rebel. **~'lone** *nf* re-bellion '**ribes** *nm inv* (*rosso*) redcurrant; (*nero*) blackcurrant

ribol'lire *vi* (*fermentare*) ferment; *fig* seethe

ri'brezzo *nm* disgust; **far ~ a** disgust

rica'dere *vi* fall back; (*nel peccato ecc*) lapse; (*pendere*) hang [down]; **~ su** (*riversarsi*) fall on. **rica'duta** *nf* relapse

rical'care *vt* trace

ricalci'trante *a* recalcitrant

rica'mare *vt* embroider. **~to a** embroidered

ri'cambi *nmpl* spare parts

ricambi'are *vt* return; reciprocate (*sentimento*); **~ qcsa a qcno** repay sb for sth. **ri'cambio** *nm* replacement; *Biol* metabolism; **pezzo di ricambio** *Mech* spare [part]

ri'camo *nm* embroidery

ricapito'la|re *vt* sum up. **~zi'one** *nf* summary, recap *fam*

ri'carica *nf* (*di sveglia*) rewinding

ricari'care *vt* reload (*macchina fotografica, fucile, camion*); recharge (*batteria*); *Comput* reboot

ricat'ta|re *vt* blackmail. **~tore**, **~'trice** *nmf* blackmailer. **ri'catto** *nm* blackmail

rica'va|re *vt* get; (*ottenere*) obtain; (*dedurre*) draw. **~to** *nm* proceeds *pl*. **ri'cavo** *nm* proceeds *pl*

'ricca *nf* rich woman. **~'mente** *adv* lavishly

ric'chezza *nf* wealth; *fig* richness; **ricchezze** *pl* riches

'riccio *a* curly ● *nm* curl; (*animale*) hedgehog. **~ di mare** sea-urchin. **~lo** *nm* curl. **~'luto a** curly. **ricci'uto** *a* (*barba*) curly

'ricco *a* rich ● *nm* rich man

ri'cerca *nf* search; (*indagine*) investigation; (*scientifica*) research; *Sch* project

ricer'ca|re *vt* search for; (*fare ricerche su*) research. **~ta** *nf* wanted woman. **~'tezza** *nf* refinement. **~to** *a* sought-after; (*raffina-*

to) refined; (*affettato*) affected ● *nm* (*polizia*) wanted man

ricetrasmit'tente *nf* transceiver

ri'cetta *nf* *Med* prescription; *Culin* recipe

ricet'tacolo *nm* receptacle

ricet'tario *nm* (*di cucina*) recipe book

ricetta'|tore, -'trice *nmf* fence, receiver of stolen goods. **~zi'one** *nf* receiving [stolen goods]

rice'vente *a* (*apparecchio, stazione*) receiving ● *nmf* receiver

ri'cev|ere *vt* receive; (*dare il benvenuto*) welcome; (*di albergo*) accommodate. **~i'mento** *nm* receiving; (*accoglienza*) welcome; (*trattenimento*) reception

ricevi'tor|e *nm* receiver. **~'la** *nf* **~la del lotto** agency authorized to sell lottery tickets

rice'vuta *nf* receipt. **~ fiscale** tax receipt

ricezi'one *nf* *Radio, TV* reception

richia'mare *vt* (*al telefono*) call back; (*far tornare*) recall; (*rimproverare*) rebuke; (*attirare*) draw; **~ alla mente** call to mind. **richi'amo** *nm* recall; (*attrazione*) call

richi'edere *vt* ask for; (*di nuovo*) ask again for; **~ a qcno di fare qcsa** ask *o* request sb to do sth. **richi'esta** *nf* request; *Comm* demand

ri'chiuder|e *vt* shut again, close again. **~si** *vr* (*ferita:*) heal

rici'claggio *nm* recycling

rici'clar|e *vt* recycle. ● **~si** *vr* retrain; (*cambiare lavoro*) change one's line of work

'ricino *nm* **olio di ~** castor oil

ricognizi'one *nf* *Mil* reconnaissance

ri'colmo *a* full

ricomin'care *vt/i* start again

ricompa'rire *vi* reappear

ricom'pen|sa *nf* reward. **~'sare** *vt* reward

ricom'por|re *vt* (*riscrivere*) rewrite; (*ricostruire*) reform; *Typ* reset. **~si** vr regain one's composure

riconcili·a|re vt reconcile. **~rsi** vr be reconciled. **~zi'one** nf reconciliation

ricono'scen|te a grateful. **~za** nf gratitude

rico'nosc|ere vt recognize; (ammettere) acknowledge. **~i'mento** nm recognition; (ammissione) acknowledgement; (per la polizia) identification. **~i'uto** a recognized

riconqui'stare vt Mil retake, reconquer

riconside'rare vt rethink

rico'prire vt re-cover; (rivestire) coat; (di insulti) shower (**di** with); hold (carica)

ricor'da|re vt remember; (richiamare alla memoria) recall; (far ricordare) remind; (rassomigliare) look like. **~si** vr **~si** [**di**] remember. **ri'cordo** nm memory; (oggetto) memento; (di viaggio) souvenir; **ricordi** pl (memorie) memoirs

ricor'ren|te a recurrent. **~za** nf recurrence; (anniversario) anniversary

ri'correre vi recur; (accadere) occur; (data:) fall; **~ a** have recourse to; (rivolgersi a) turn to.

ri'corso pp di **ricorrere** ● nm recourse; Jur appeal

ricostitu'ente nm tonic

ricostitu'ire vt re-establish

ricostru'i|re vt reconstruct. **~zi'one** nf reconstruction

ricove'ra|re vt give shelter to; **~re in ospedale** admit to hospital, hospitalize. **~to, -a** nmf hospital patient. **ri'covero** nm shelter; (ospizio) home

ricre'a|re vt re-create; (ristorare) restore. **~rsi** vr amuse oneself. **~tivo** a recreational. **~zi'one** nf recreation; Sch break

ri'credersi vr change one's mind

ricupe'rare vt recover; rehabilitate (tossicodipendente); **~ il tem-**

po perduto make up for lost time.

ri'cupero nm recovery; (di tossicodipendente) rehabilitation; (salvataggio) rescue; [**minuti** nmpl **di**] **ricupero** injury time

ri'curvo a bent

ridacchi'are vi giggle

ri'dare vt give back, return

ri'dente a (piacevole) pleasant

'ridere vi laugh; **~ di** (deridere) laugh at

ri'detto pp di **ridire**

ridicoliz'zare vt ridicule. **ri'dicolo** a ridiculous

ridimensio'nare vt reshape; fig see in the right perspective

ri'dire vt repeat; (criticare) find fault with; **trova sempre da ~** he's always finding fault

ridon'dante a redundant

ri'dotto pp di **ridurre** ● nm Theat foyer ● a reduced

ri'du|rre vt reduce. **~rsi** vr diminish. **~rsi a** be reduced to. **~t'tivo** a reductive. **~zi'one** nf reduction; (per cinema, teatro) adaptation

rieducazi'one nf (di malato) rehabilitation

riem'pi|re vt fill [up]; fill in (moduli ecc). **~rsi** vr fill [up]. **~'tivo** a filling ● nm filler

rien'tranza nf recess

rien'trare vi go/come back in; (tornare) return; (piegare indentro) recede; **~ in** (far parte) fall within. **ri'entro** nm return; (di astronave) re-entry

riepilo'gare vt recapitulate. **rie'pilogo** nm roundup

riesami'nare vt reappraise

ri'essere vi ci risiamo! here we go again!

riesu'mare vt exhume

rievo'ca|re vt (commemorare) commemorate. **~zi'one** nf (commemorazione) commemoration

rifaci'mento nm remake

ri'fa|re vt do again; (creare) make again; (riparare) repair; (imitare) imitate; make (letto). **~rsi** vr (rimettersi) recover; (vendicarsi) get even; **~rsi una vita/carriera** make a new life/career for oneself; **~rsi il trucco** touch up one's makeup; **~rsi** make up for. **~tto** pp di **rifare**

riferi'mento nm reference

rife'rir|e vt report; (descrivere) to **● vi** make a report. **~si** vr **~si a** refer to

rifi'lare vt (tagliare a filo) trim; (fam: affibbiare) saddle

rifi'ni|re vt finish off. **~tura** nf finish

rifio'rire vi blossom again; fig flourish again

rifiu'tare vt refuse. **rifi'uto** nm refusal; **rifiuti** pl (immondizie) rubbish sg. **rifiuti** pl **urbani** urban waste sg

riflessi'one nf reflection; (osservazione) remark. **rifles'sivo** a thoughtful; Gram reflexive

ri'flesso pp di **riflettere ● nm** (luce) reflection; Med reflex; **per ~** indirectly

riflett'er|e vt reflect **● vi** think. **~si** vr be reflected

riflet'tore nm reflector; (proiettore) searchlight

ri'flusso nm ebb

rifocil'lar|e vt restore. **~si** vr liter, hum take some refreshment

ri'fondere vt (rimborsare) refund

ri'forma nf reform; Relig reformation; Mil exemption on medical grounds

rifor'ma|re vt re-form; (migliorare) reform; Mil declare unfit for military service. **~to** a (chiesa) Reformed. **~tore**, **~trice** nmf reformer. **~torio** nm reformatory. **rifor'mista** a reformist

riforni'mento nm supply; (scorta) stock; (di combustibile) refuelling. **stazione** nf di **~** petrol station

rifor'nir|e vt **~e di** provide with. **~si** vr restock, stock up (di with)

ri'fra|ngere vt refract. **~tto** pp di **rifrangere**. **~zi'one** nf refraction

rifug'gire vi **~ da** fig shun

rifugi'a|rsi vr take refuge. **~to, -a** nmf refugee

ri'fugio nm shelter; (nascondiglio) hideaway

'riga nf line; (fila) row; (striscia) stripe; (scriminatura) parting; (regolo) rule; **a righe** (stoffa) striped; (quaderno) ruled; **mettersi in ~** line up

ri'gagnolo nm rivulet

ri'gare vt rule (foglio) **● vi ~ dritto** behave well

rigatti'ere nm junk dealer

rigene'rare vt regenerate

riget'tare vt (gettare indietro) throw back; (respingere) reject; (vomitare) throw up. **ri'getto** nm rejection

ri'ghello nm ruler

rigida'mente adv rigidly. **~ità** nf rigidity; (di clima) severity; (severità) strictness. **'rigido** a rigid; (freddo) severe; (severo) strict

rigi'rar|e vt turn again; (ripercorrere) go round; fig twist (argomentazione) **● vi** walk about. **~si** vr turn round; (nel letto) turn over. **ri'giro** nm (imbroglio) trick

'rigo nm line; Mus staff

ri'goglio nm bloom. **~'oso** a luxuriant

ri'gonfio a swollen

ri'gore nm rigours pl; **a ~** strictly speaking; **calcio di ~** penalty [kick]; **area di ~** penalty area; **essere di ~** be compulsory

rigorosa'mente adv (giudicare) severely. **~'roso** a (severo) strict; (scrupoloso) rigorous

riguada'gnare vt regain (quota, velocità)

riguar'dar|e vt look at again; (considerare) regard; (concernere) concern; **per quanto riguarda** with regard to. **~si** vr take care of oneself. **rigu'ardo** nm care; (considerazione) consideration; **nei riguardi di** towards; **riguardo a** with regard to

ri'gurgito nm regurgitation

rilanci'are vt throw back (palla);

⟨*di nuovo*⟩ throw again; increase ⟨*offerta*⟩; revive ⟨*moda*⟩; relaunch ⟨*prodotto*⟩ ● *vi* (*a carte*) raise the stakes

rilasci'ar|e *vt* (*concedere*) grant; (*liberare*) release; issue ⟨*documento*⟩. **~si** *vr* relax. **ri'lascio** *nm* release; (*di documento*) issue

rilassa'mento *nm* (*relax*) relaxation

rilas'sa|re *vt*, **~rsi** *vr* relax. **~to a** ⟨*ambiente*⟩ relaxed

rile'ga|re *vt* bind ⟨*libro*⟩. **~to a** bound. **~tura** *nf* binding

ri'leggere *vt* reread

ri'lento: a ~ *adv* slowly

rileva'mento *nm* survey; Comm buyout

rile'van|te *a* considerable

rile'va|re *vt* (*trarre*) get; (*mettere in evidenza*) point out; (*notare*) notice; (*topografia*) survey; Comm take over; Mil relieve. **~zi'one** *nf* (*statistica*) survey

rili'evo *nm* relief; Geog elevation; (*topografia*) survey; (*importanza*) importance; (*osservazione*) remark; **mettere in ~** qcsa point sth out

rilut'tan|te *a* reluctant. **~za** *nf* reluctance

'rima *nf* rhyme; **far ~ con** qcsa rhyme with sth

riman'dare *vt* (*posporre*) postpone; (*mandare indietro*) send back; (*mandare di nuovo*) send again; (*far ridare un esame*) make resit an examination. **ri'mando** *nm* return; (*in un libro*) cross-reference

rima'nen|te *a* remaining ● *nm* remainder. **~za** *nf* remainder; **~ze** *pl* remnants

rima'ne|re *vi* stay, remain; (*essere d'avanzo*) be left; (*venirsi a trovare*) be left; (*restare stupito*) be astonished; (*restare d'accordo*) agree

rimar'chevole *a* remarkable

ri'mare *vt/i* rhyme

rimargi'nar|e *vt*, **~si** *vr* heal

ri'masto *pp di* **rimanere**

rima'sugli *nmpl* (*di cibo*) leftovers

rimbal'zare *vi* rebound ⟨*proiettile:*⟩ ricochet; **far ~** bounce. **rim'balzo** *nm* rebound; (*di proiettile*) ricochet

rimbam'bi|re *vi* be in one's dotage ● *vt* stun. **~to a** in one's dotage

rimboc'care *vt* turn up; roll up ⟨*maniche*⟩; tuck in ⟨*coperte*⟩

rimbom'bare *vi* resound

rimbor'sare *vt* reimburse, repay. **rim'borso** *nm* reimbursement, repayment. **rimborso spese** reimbursement of expenses

rimedi'are *vi* **~ a** a remedy; make up for ⟨*errore*⟩; (*procurare*) scrape up. **ri'medio** *nm* remedy

rimesco'la|re *vt* mix [up]; shuffle ⟨*carte*⟩; (*rivangare*) rake up

ri'messa *nf* (*locale per veicoli*) garage; (*per aerei*) hangar; (*per autobus*) depot; (*di denaro*) remittance; (*di merci*) consignment

ri'messo *pp di* **rimettere**

ri'metter|e *vt* (*a posto*) put back; (*restituire*) return; (*affidare*) entrust; (*perdonare*) remit; (*rimandare*) put off; (*vomitare*) bring up; **~ci** (*fam: perdere*) lose [out]. **~si** *vr* (*ristabilirsi*) recover; *tempo:*⟩ clear up; **~si a** a start again

'rimmel® *nm inv* mascara

rimoder'nare *vt* modernize

rimon'tare *vt* (*risalire*) go up; Mech reassemble ● *vi* remount; **~ a** (*risalire*) go back to

rimorchi'a|re *vt* tow; fam pick up ⟨*ragazza*⟩. **~'tore** *nm* tug⟨boat⟩. **ri'morchio** *nm* tow; (*veicolo*) trailer

ri'morso *nm* remorse

rimo'stranza *nf* complaint

rimozi'one *nf* removal; (*da un incarico*) dismissal. **~ forzata** (*da un gally parked vehicles removed illeowner's expense*)

rim'pasto *nm* Pol reshuffle

rimpatri'are *vt/i* repatriate. **rim'patrio** *nm* repatriation. **rim-**

rim'pian|gere *vt* regret. **~to** *pp di* **rimpiangere** ● *nm* regret

rimpiat'tino *nm* hide-and-seek

rimpiaz'zare *vt* replace

rimpiccio'lire *vi* become smaller

rimpinz'ar|e *vt* ~e **di** stuff with. ~**si** *vr* stuff oneself

rimprove'rare *vt* reproach; ~ **qcsa a qcno** reproach sb for sth. **rim'provero** *nm* reproach

rimugi'nare *vt* rummage; *fig* ~ **su** brood over

rimune'ra|re *vt* remunerate. ~**tivo** *a* remunerative. ~**zi'one** *nf* remuneration

ri'muovere *vt* remove

ri'nascere *vi* be reborn, born again

rinascimen'tale *a* Renaissance. **Rinasci'mento** *nm* Renaissance

ri'nascita *nf* rebirth

rincal'zare *vt* (*sostenere*) support; (*rimboccare*) tuck in. **rin'calzo** *nm* support; **rincalzi** *pl Mil* reserves

rincan'tucci'arsi *vr* hide oneself away in a corner

rinca'rare *vt* increase the price of ● *vi* become more expensive. **rin'caro** *nm* price increase

rinca'sare *vi* return home

rinchi'uder|e *vt* shut up. ~**si** *vr* shut oneself up

rin'correre *vt* run after

rin'corsa *nf* run-up. ~**o** *pp* **di** rincorrere

rin'cresc|ere *vi* **mi rincresce che non...** I'm sorry that I non...; **se non ti** ~**e** if you don't can't...; **se non ti** ~**e** if you don't mind. ~**i'mento** *nm* regret. ~**i'uto** *pp* **di** rincrescere

rincreti'nire *vi* be stupid

rincu'lare *vi* (*arma:*) recoil; (*cavallo:*) shy. **rin'culo** *nm* recoil

rincuo'rar|e *vt* encourage. ~**si** *vr* take heart

rinfacci'are *vt* ~ **qcsa a qcno** throw sth in sb's face

rinfor'zar|e *vt* strengthen; (*rendere più saldo*) reinforce. ~**si** *vr* become stronger. **rin'forzo** *nm* reinforcement; *fig* support

rinfran'care *vt* reassure

rinfre'scante *a* cooling

rinfre'scar|e *vt* cool; (*rinnovare*) freshen up ● *vi* get cooler. ~**si** *vr* freshen [oneself] up. **rin'fresco** *nm* light refreshment; (*ricevimento*) party

rin'fusa *nf* **alla** ~ at random

ringhi'are *vi* snarl

ringhi'era *nf* railing; (*di scala*) banisters *pl*

ringiova'nire *vt* rejuvenate (*pelle, persona*); (*vestito:*) make look younger ● *vi* become young again; (*sembrare*) look young again

ringrazi'a|mento *nm* thanks *pl*. ~'**are** *vt* thank

rinne'ga|re *vt* disown. ~**to, -a** *nmf* renegade

rinnova'mento *nm* renewal; (*di edifici*) renovation

rinno'var|e *vt* renew; renovate (*edifici*). ~**si** *vr* be renewed; (*ripetersi*) recur, happen again. **rin'novo** *nm* renewal

rinoce'ronte *nm* rhinoceros

rino'mato *a* renowned

rinsal'dare *vt* consolidate

rinsa'vire *vi* come to one's senses

rinsec'chi|re *vi* shrivel up. ~**to** *a* shrivelled up

rinta'narsi *vr* hide oneself away; (*animale:*) retreat into its den

rintoc'care *vi* (*campana:*) toll; (*orologio:*) strike. **rin'tocco** *nm* toll; (*di orologio*) stroke

rinton'ti|re *vt* anche *fig* stun. ~**to** *a* (*stordito*) dazed

rintracci'are *vt* trace

rintro'nare *vt* stun ● *vi* boom

ri'nuncia *nf* renunciation

rinunci'a|re *vi* ~**re a** renounce, give up. ~'**tario** *a* defeatist

ri'nunzia, rinunzi'are = rinuncia, rinunciare

rinveni'mento *nm* (*di reperti*) discovery; (*di refurtiva*) recovery.

rinve'nire *vt* find ● *vi* (*riprendere i sensi*) come round; (*ridiventare fresco*) revive

rinvi'are *vt* put off; (*mandare indietro*) return; (*in libro*) refer; ~ **a giudizio** indict

rin'vio nm Sport goal kick; (in libro) cross-reference; (di appuntamento) postponement; (di merce) return

rio'nale a local. **ri'one** nm district

riordi'nare vt tidy [up]; (ordinare di nuovo) reorder; (riorganizzare) reorganize

riorganiz'zare vt reorganize

ripa'gare vt repay

ripa'ra|re vt (proteggere) shelter, protect; (aggiustare) repair; (porre rimedio) remedy ●vi **~re a** make up for. **~rsi** vr take shelter. **~to a** ⟨luogo⟩ sheltered. **~zi'one** nf repair; fig reparation. **ri'paro** nm shelter; (rimedio) remedy

ripar'ti|re vt (dividere) divide ●vi leave again. **~zi'one** nf division

ripas'sare vt recross; (rivedere) revise ●vi pass again. **~ta** nf (di vernice) second coat. **ri'passo** nm (di lezione) revision

ripensa'mento nm second thoughts pl

ripen'sare vi (cambiare idea) change one's mind; **~ a** think of; **ripensaci!** think again!

riper'correre vt (con la memoria) go back over

riper'corso pp di **ripercuotere**

ripercu'oter|e vt strike again. **~si** vr ⟨suono:⟩ reverberate; **~si su** (fig: avere conseguenze) impact on. **ripercussi'one** nf repercussion

ripe'scare vt fish out ⟨oggetti⟩

ripe'tente nmf student repeating a year

ri'pet|ere vt repeat. **~ersi** vr ⟨evento:⟩ recur. **~izi'one** nf repetition; (di lezione) revision; (lezione privata) private lesson. **~uta'mente** adv repeatedly

ri'piano nm (di scaffale) shelf; (terreno pianeggiante) terrace

ri'picc|a nf **fare qcsa per ~a** do sth out of spite. **~o** nm spite

'ripido a steep

ripie'gar|e vt refold; (abbassare) lower ●vi (indietreggiare) retreat.

~si vr bend; ⟨sedile:⟩ fold. **ripi'ego** nm expedient; (via d'uscita) way out

ripi'eno a full; Culin stuffed ●nm filling; Culin stuffing

ripo'pol|are vt repopulate. **~si** vr be repopulated

ri'porre vt put back; (mettere da parte) put away; (collocare) place; repeat ⟨domanda⟩

ripor'ta|re vt (restituire) bring/take back; (riferire) report; (subire) suffer; Math carry; win ⟨vittoria⟩; transfer ⟨disegno⟩. **~si** vr go back; (riferirsi) refer. **ri'porto** nm **cane da riporto** gun dog

ripo'sante a ⟨colore⟩ restful, soothing

ripo'sa|re vi rest. **~rsi** vr rest. **~to a** ⟨mente⟩ fresh. **ri'poso** nm rest; **andare a riposo** retire; **riposo!** Mil at ease!; **giorno di riposo** day off

ripo'stiglio nm cupboard

ri'posto pp di **riporre**

ri'prend|ere vt take again; (prendere indietro) take back; (riconquistare) recapture; (ricuperare) recover; (ricominciare) resume; (rimproverare) reprimand; take in ⟨cucitura⟩; Cinema shoot. **~si** vr recover; (correggersi) correct oneself

ri'presa nf resumption; (ricupero) recovery; Theat revival; Cinema shot; Auto acceleration; Mus repeat. **~ aerea** bird's-eye view

ripresen'tare vt resubmit ⟨domanda, certificato⟩. **~si** vr (a ufficio) go/come back again; (come candidato) stand o run again; (occasione:) arise again

ri'preso pp di **riprendere**

ripristi'nare vt restore

ripro'dotto pp di **riprodurre**

ripro'du|rre vt, **~rsi** vr reproduce. **~t'tivo** a reproductive. **~zi'one** nf reproduction

ripro'mettersi vr (intendere) intend

ri'prova nf confirmation

ripudi'are *vt* repudiate

ripu'gnan|te *a* repugnant. **~za** *nf* disgust. **ripu'gnare** *vi* **ripugnare a** a disgust

ripu'li|re *vt* clean [up]; *fig* polish. **~ta** *nf* **darsi una ~ta** have a wash and brushup

ripuls|i'one *nf* repulsion. **~'ivo** *a* a repulsive

ri'quadro *nm* square; (*pannello*) panel

ri'sacca *nf* undertow

ri'saia *nf* rice field, paddy field

risa'lire *vt* go back up ● *vi* ~ **a** (*nel tempo*) go back to; (*essere datato a*) date back to, go back to

risal'tare *vi* (*emergere*) stand out. **ri'salto** *nm* prominence; (*rilievo*) relief

risa'nare *vt* heal; (*bonificare*) reclaim

risa'puto *a* a well-known

risarci'mento *nm* compensation. **risar'cire** *vt* indemnify

ri'sata *nf* laugh

riscalda'mento *nm* heating. **~ autonomo** central heating (*for one flat*)

riscal'dar|e *vt* heat; warm (*persona*). **~si** *vr* warm up

riscat'tar|e *vt* ransom. **~si** *vr* redeem oneself. **ri'scatto** *nm* ransom; (*morale*) redemption

rischia'rar|e *vt* light up; brighten (*colore*). **~si** *vr* light up; (*cielo*) clear up

rischi|'are *vt* risk ● *vi* run the risk. **'rischio** *nm* risk. **~'oso** *a* a risky

risciac'quare *vt* rinse. **risci'acquo** *nm* rinse

riscon'trare *vt* (*confrontare*) compare; (*verificare*) check; (*rilevare*) find. **ri'scontro** *nm* comparison; check; (*Comm: risposta*) reply

ri'scossa *nf* revolt; (*riconquista*) recovery

riscossi'one *nf* collection

ri'scosso *pp di* **riscuotere**

riscu'oter|e *vt* shake; (*percepire*)

draw; (*ottenere*) gain; cash (*assegno*). **~si** *vr* rouse oneself

risen'ti|re *vt* hear again; (*provare*) feel ● *vi* ~ **re di** feel the effect of. **~rsi** *vr* (*offendersi*) take offence. **~to** *a* resentful

ri'serbo *nm* reserve; **mantenere il ~** remain tight-lipped

ri'serva *nf* reserve; (*di caccia, pesca*) preserve; *Sport* substitute, reserve. **~ di caccia** game reserve. **~ indiana** Indian reservation. **~ naturale** wildlife reserve

riser'va|re *vt* reserve; (*prenotare*) book; (*per occasione*) keep. **~rsi** *vr* (*ripromettersi*) plan for oneself (*cambiamento*). **~'tezza** *nf* reserve. **~to** *a* a reserved

ri'siedere *vi* ~ **a** reside in

'riso[1] *pp di* **ridere** ● *nm* (*pl nf* **risa**) laughter; (*singolo*) laugh. **~'lino** *nm* giggle

'riso[2] *nm* (*cereale*) rice

ri'solto *pp di* **risolvere**

risolu'tezza *nf* determination. **riso'luto** *a* resolute, determined. **~zi'one** *nf* resolution

ri'solver|e *vt* resolve; *Math* solve. **~si** *vr* (*decidersi*) decide; **~si in** turn into

riso'na|nza *nf* resonance; **aver ~nza** *fig* arouse great interest. **~re** *vi* resound; (*rimbombare*) echo

ri'sorgere *vi* rise again

risorgi'mento *nm* revival; (*storico*) Risorgimento

ri'sorsa *nf* resource; (*espediente*) resort

ri'sorto *pp di* **risorgere**

ri'sotto *nm* risotto

ri'sparmi *nmpl* (*soldi*) savings

risparmi'a|re *vt* save; (*salvare*) spare. **~'tore**, **~'trice** *nmf* saver

ri'sparmio *nm* saving

rispecchi'are *vt* reflect

rispet'tabile *a* respectable. **~ità** *nf* respectability

rispet'tare *vt* respect; **farsi ~** command respect

rispet'tivo *a* respective

ri'spetto *nm* respect; ~ **a** as regards; *(in paragone a)* compared to

rispet|tosa'mente *adv* respectfully. ~**toso** *a* respectful

risplen'dente *a* shining. **ri'splendere** *vi* shine

rispon'den|te ~**te a** in keeping with. ~**za** *nf* correspondence

ri'spondere *vi* answer; *(rimbeccare)* answer back; *(obbedire)* respond; ~ **a** a reply to; ~ **di** *(rendersi responsabile)* answer for

ri'spost|a *nf* answer, reply; *(reazione)* response. ~**o pp di rispondere**

'**rissa** *nf* brawl. **ris'soso** *a* pugnacious

ristabi'lire *vt* re-establish. ~**si** *vr* *(in salute)* recover

ristagnare *vi* stagnate; *(sangue:)* coagulate. **ri'stagno** *nm* stagnation

ri'stampa *nf* reprint; *(azione)* reprinting. **ristam'pare** *vt* reprint

risto'rante *nm* restaurant

risto'ra|re *vt* refresh. ~**rsi** *vr* liter take some refreshment; *(riposarsi)* take a rest. ~**tore**, ~**trice** *nmf* *(proprietario di ristorante)* restaurateur; *(fornitore)* caterer ●*a* refreshing. **ri'storo** *nm* refreshment; *(sollievo)* relief

ristret'tezza *nf* narrowness; *(povertà)* poverty; **vivere in ristrettezze** live in straitened circumstances

ri'stretto *pp di restringere* ●*a* narrow; *(condensato)* condensed; *(limitato)* restricted; **di idee ristrette** narrow-minded

ristruttu'rare *vt* restructure, reorganize *(ditta)*; refurbish *(casa)*

risucchi'are *vt* suck in. **ri'succhio** *nm* whirlpool; *(di corrente)* undertow

risul'ta|re *vi* result; *(riuscire)* turn out. ~**to** *nm* result

risuo'nare *vi* *(grida, parola:)* echo; *Phys* resonate

risurrezi'one *nf* resurrection

risusci'tare *vt* resuscitate; *fig* revive ●*vi* return to life

risvegli'ar|e *vt* reawaken *(interesse)*. ~**si** *vr* wake up; *(natura:)* awake; *(desiderio:)* be aroused. **ri'sveglio** *nm* waking up; *(dell'interesse)* revival; *(del desiderio)* arousal

ri'svolto *nm* *(di giacca)* lapel; *(di pantaloni)* turn-up, cuff *Am*; *(di manica)* cuff; *(di tasca)* flap; *(di libro)* inside flap

ritagli'are *vt* cut out. **ri'taglio** *nm* cutting; *(di stoffa)* scrap

ritar'da|re *vi* be late; *(orologio:)* be slow ●*vt* delay; slow down *(progresso)*; *(differire)* postpone. ~**tario, -a** *nmf* late-comer. ~**to** *a* *Psych* retarded

ri'tardo *nm* delay; **essere in** ~ be late; *(volo:)* be delayed

ri'tegno *nm* reserve

rite'ne|re *vt* retain; deduct *(somma)*; *(credere)* believe. ~**uta** *nf* *(sul salario)* deduction

riti'ra|re *vt* throw back *(palla)*; *(prelevare)* withdraw; *(riscuotere)* draw; collect *(pacco)*. ~**rsi** *vr* withdraw; *(stoffa:)* shrink; *(da attività)* retire; *(marea:)* recede. ~**ta** *nf* retreat; *(WC)* toilet. **ri'tiro** *nm* withdrawal; *Relig* retreat; *(da attività)* retirement. **ritiro bagagli** baggage reclaim

'ritmo *nm* rhythm

'rito *nm* rite; **di** ~ customary

ritoc'care *vt* *(correggere)* touch up. **ri'tocco** *nm* retouch

ritor'na|re *vi* return; *(andare/venire indietro)* go/come back; *(ricorrere)* recur; *(ridiventare)* become again

ritor'nello *nm* refrain

ri'torno *nm* return

ritorsi'one *nf* retaliation

ri'trarre *vt* *(ritirare)* withdraw; *(distogliere)* turn away; *(rappresentare)* portray

ritrat'ta|re *vt* deal with again; retract *(dichiarazione)*. ~**zi'one** *nf* withdrawal, retraction

ritrat'tista *nmf* portrait painter.
　ri'tratto *pp di* **ritrarre** ● *nm* portrait

ritro'sia *nf* shyness. **ri'troso** *a* backward; (*timido*) shy; **a ritroso** backwards; **ritroso a** a reluctant to

ritrova'mento *nm* (*azione*) finding

ritro'va|re *vt* find [again]; regain (*salute*). **~rsi** *vr* meet; (*di nuovo*) meet again; (*capitare*) find oneself; (*raccapezzarsi*) see one's way. **~to** *nm* discovery. **ri'trovo** *nm* meeting-place; (*notturno*) night-club

'ritto *a* upright; (*diritto*) straight

ritu'ale *a & nm* ritual

riunifi'ca|re *vt* reunify. **~rsi** *vr* be reunited. **~zi'one** *nf* reunification

riuni'one *nf* meeting; (*fra amici*) reunion

riu'nir|e *vt* (*unire*) join together; (*radunare*) gather. **~si** *vr* be reunited; (*adunarsi*) meet

riusc'i|re *vi* (*aver successo*) succeed; (*in matematica ecc*) be good (**in** at); (*aver esito*) turn out; **le è riuscito simpatico** she found him likeable. **~ta** *nf* (*esito*) result; (*successo*) success

'riva *nf* (*di mare, lago*) shore; (*di fiume*) bank

ri'val|e *nmf* rival. **~ità** *nf inv* rivalry

rivalutazi'one *nf* revaluation

rivan'gare *vt* dig up again

rive'dere *vt* see again; revise (*lezione*); (*verificare*) check

rive'la|re *vt* reveal. **~rsi** *vr* (*dimostrarsi*) turn out. **~tore** *a* revealing ● *nm Techn* detector. **~zi'one** *nf* revelation

ri'vendere *vt* resell

rivendi'ca|re *vt* claim. **~zi'one** *nf* claim

ri'vendi|ta *nf* (*negozio*) shop. **~tore**, **~'trice** *nmf* retailer. **~tore autorizzato** authorized dealer

ri'verbero *nm* reverberation; (*bagliore*) glare

rive'renza *nf* reverence; (*inchino*) curtsy; (*di uomo*) bow

rive'rire *vt* respect; (*ossequiare*) pay one's respects to

river'sar|e *vt* pour. **~si** *vr* (*fiume:*) flow

river'sibile *a* reversible

rivesti'mento *nm* covering

rive'sti|re *vt* (*rifornire di abiti*) clothe; (*ricoprire*) cover; (*internamente*) line; hold (*carica*). **~rsi** *vr* get dressed again; (*per una festa*) dress up

rivi'era *nf* coast; **la ~ ligure** the Italian Riviera

ri'vincita *nf Sport* return match; (*vendetta*) revenge

rivis'suto *pp di* **rivivere**

ri'vista *nf* review; (*pubblicazione*) magazine; *Theat* revue; **passare in ~** review

ri'vivere *vi* come to life again; (*riprendere le forze*) revive ● *vt* relive

ri'volger|e *vt* turn; (*indirizzare*) address; **~a da** (*distogliere*) turn away from. **~si** *vr* turn round; **~si a** (*indirizzarsi*) turn to

ri'volta *nf* revolt

rivol'tante *a* disgusting

rivol'tar|e *vt* turn [over]; (*mettendo l'interno verso l'esterno*) turn inside out; (*sconvolgere*) upset. **~si** *vr* (*ribellarsi*) revolt

rivol'tella *nf* revolver

ri'volto *pp di* **rivolgere**

rivoluzio'nare *vt* revolutionize. **~io**, **-a** *a & nmf* revolutionary. **rivoluzi'one** *nf* revolution; (*fig: disordine*) chaos

riz'zare *vt* raise; (*innalzare*) erect; prick up (*orecchie*). **~si** *vr* stand up; (*capelli:*) stand on end; (*orecchie:*) prick up

'roba *nf* stuff; (*personale*) belongings *pl*, stuff; (*faccenda*) thing; (*sl: droga*) drugs *pl*; **~ da mattir** absolute madness!; **~ da mangiare** food, things to eat

ro'baccia *nf* rubbish

ro'bot *nm inv* robot. **~iz'zato** *a* robotic

robu'stezza *nf* sturdiness, robustness; (*forza*) strength. **ro'busto** *a* sturdy, robust; (*forte*) strong

'rocca *nf* fortress. **~'forte** *nf* stronghold

roc'chetto *nm* reel

'roccia *nf* rock

ro'da|ggio *nm* running in. **~re** *vt* run in

'roder|e *vt* gnaw; (*corrodere*) corrode. **~si** *vr* **~si da** (*logorarsi*) be consumed with. **rodi'tore** *nm* rodent

rodo'dendro *nm* rhododendron

'rogna *nf* scabies *sg*; *fig* nuisance

ro'gnone *nm* Culin kidney

'rogo *nm* (*supplizio*) stake; (*per cadaveri*) pyre

'Roma *nf* Rome

Roma'nia *nf* Romania

ro'manico *a* Romanesque

ro'mano, -a *a & nmf* Roman

romanti'cismo *nm* romanticism.

ro'mantico *a* romantic

ro'man|za *nf* romance. **~'zato** *a* romanticized. **~'zesco** *a* fictional; (*stravagante*) wild, unrealistic. **~zi'ere** *nm* novelist

ro'manzo *a* Romance ● *nm* novel. **~ d'appendice** serial story. **~ giallo** thriller

'rombo *nm* rumble; *Math* rhombus; (*pesce*) turbot

'romper|e *vt* break; break off (*relazione*); **non ~ [le scatole]!** (*fam: seccare*) don't be a pain [in the neck]!. **~si** *vr* break; **~si una gamba** break one's leg

rompi'capo *nm* nuisance; (*indovinello*) puzzle

rompi'collo *nm* daredevil; **a ~** at breakneck speed

rompighi'accio *nm* ice-breaker

rompi'scatole *nmf inv fam* pain

'ronda *nf* rounds *pl*

ron'della *nf Mech* washer

'rondine *nf* swallow

ron'done *nm* swift

ron'fare *vi* (*russare*) snore

ron'zare *vi* buzz; **~ attorno a qcno** *fig* hang about sb

ron'zino *nm* jade

ron'zio *nm* buzz

'rosa *nf* rose. **~ dei venti** wind rose ● *a & nm* (*colore*) pink. **ro'saio** *nm* rose-bush

ro'sario *nm* rosary

ro'sato *a* rosy ● *nm* (*vino*) rosé

'roseo *a* pink

ro'seto *nm* rose garden

rosicchi'are *vt* nibble; (*rodere*) gnaw

rosma'rino *nm* rosemary

'roso *pp di* **rodere**

roso'lare *vt* brown

roso'lia *nf* German measles

ro'sone *nm*. rosette; (*apertura*) rose-window

'rospo *nm* toad

ros'setto *nm* (*per labbra*) lipstick

'rosso *a & nm* red; **passare con il ~** jump a red light. **~ d'uovo** [egg] yolk. **ros'sore** *nm* redness; (*della pelle*) flush

rosticce'ria *nf* shop selling cooked meat and other prepared food

ro'tabile *a* **strada ~** carriageway

ro'taia *nf* rail; (*solco*) rut

ro'ta|re *vt/i* rotate. **~zi'one** *nf* rotation

rote'are *vt/i* roll

ro'tella *nf* small wheel; (*di mobile*) castor

roto'la|re *vt/i* roll. **~si** *vr* roll [about]. **'rotolo** *nm* roll; **andare a rotoli** go to rack and ruin

rotondità *nf* (*qualità*) roundness; ● *pl* (*curve femminili*) curves.

ro'tondo, -a *a* round ● *nf* (*spiazzo*) terrace

ro'tore *nm* rotor

'rotta *nf Naut, Aeron* course; **far ~ per** make course for; **fuori ~** off course

'rotta *nf a* **~ di collo** at breakneck speed; **essere in ~ con** be on bad terms with

rot'tame *nm* scrap; *fig* wreck

'rotto *pp di* **rompere** ● *a* broken; (*stracciato*) torn

rot'tura *nf* break; **che ~ di scato-
lei** *fam* what a pain!

'rotula *nf* kneecap

rou'lette *nf inv* roulette

rou'lotte *nf inv* caravan, trailer
Am

rou'tine *nf* routine; **di ~
⟨operazioni, controlli⟩** routine

ro'vente *a* scorching

'rovere *nm ⟨legno⟩* oak

rovesci'ar|e *vt ⟨buttare a terra⟩*
knock over; *⟨sottosopra⟩* turn upside down; *⟨rivoltare⟩* turn inside
out; spill *⟨liquido⟩*; overthrow
⟨governo⟩; reverse *⟨situazione⟩*.
~si *vr ⟨capovolgersi⟩* overturn;
⟨riversarsi⟩ pour. **ro'vescio** *a ⟨con-
trario⟩* reverse; **alla rovescia**
⟨capovolto⟩ upside down; *⟨con
l'interno all'esterno⟩* inside out
● *nm* reverse; *⟨nella maglia⟩* purl;
⟨di pioggia⟩ downpour; *Tennis*
backhand

ro'vina *nf* ruin; *⟨crollo⟩* collapse

rovi'na|re *vt* ruin; *⟨guastare⟩* spoil
● *vi* crash. **~rsi** *vr* be ruined. **~to
a ⟨oggetto⟩** ruined. **rovi'noso** *a* ru-
inous

rovi'stare *vt* ransack

'rovo *nm* bramble

'rozzo *a* rough

R.R. *abbr ⟨ricevuta di ritorno⟩* return receipt for registered mail

'ruba *nf* **andare a ~** sell like hot
cakes

ru'bare *vt* steal

rubi'netto *nm* tap, faucet *Am*

ru'bino *nm* ruby

ru'brica *nf ⟨in giornale⟩* column;
⟨in programma televisivo⟩ TV report; *⟨quaderno con indice⟩* address book. **~ telefonica** telephone and address book

'rude *a* rough

'rudere *nm* ruin

rudimen'tale *a* rudimentary. **rudi-
'menti** *nmpl* rudiments

ruffi'an|a *nf* procuress. **~o** *nm*
pimp; *⟨adulatore⟩* bootlicker

'ruga *nf* wrinkle

'ruggine *nf* rust; **fare la ~** go rusty

rug'gi|re *vi* roar. **~to** *nm* roar

rugi'ada *nf* dew

ru'goso *a* wrinkled

rul'lare *vi* roll; *Aeron* taxi

rul'lino *nm* film

rul'lio *nm* rolling; *Aeron* taxiing

'rullo *nm* roll; *Techn* roller

rum *nm inv* rum

ru'meno, -a *a & nmf* Romanian

rumi'nare *vt* ruminate

ru'mor|e *nm* noise; *fig* rumour.
~eggi'are *vi* rumble. **rumo'roso** *a*
noisy; *⟨sonoro⟩* loud

ru'olo *nm* roll; *Theat* role; **di ~** on
the staff

ru'ota *nf* wheel; **andare a ~ libera**
free-wheel. **~ di scorta** spare
wheel

'rupe *nf* cliff

ru'rale *a* rural

ru'scello *nm* stream

'ruspa *nf* bulldozer

rus'sare *vi* snore

'Russ|ia *nf* Russia. **r~o, -a** *a & nmf*
Russian; *⟨lingua⟩* Russian

'rustico *a* rural; *⟨carattere⟩* rough

rut'tare *vi* belch. **'rutto** *nm* belch

'ruvido *a* coarse

ruzzo'l|are *vi* tumble down. **~one**
nm tumble; **cadere ruzzoloni**
tumble down

●●

Ss

●●

'sabato *nm* Saturday

'sabbi|a *nf* sand. **~e** *pl* **mobili**
quicksand. **~'oso** *a* sandy

sabo'ta|ggio *nm* sabotage. **~re** *vt*
sabotage. **~'tore, ~'trice** *nmf*
saboteur

'sacca *nf* bag. **~ da viaggio** travel-
ling-bag

sacca'rina *nf* saccharin

sac'cente *a* pretentious ● *nmf*
know-all

saccheggi'ar|e *vt* sack; *hum* raid

⟨*frigo*⟩. **~'tore, ~'trice** *nmf* plunderer. **sac'cheggio** *nm* sack

sac'chetto *nm* bag

'**sacco** *nm* sack; *Anat* sac; **mettere nel ~** *fig* swindle; **un ~** (*moltissimo*) a lot; **un ~ di** (*gran quantità*) lots of. **~ a pelo** sleeping-bag

sacer'do|te *nm* priest. **~zio** *nm* priesthood

sacra'mento *nm* sacrament

sacrifi'ca|re *vt* sacrifice. **~rsi** *vr* sacrifice oneself. **~to a** (*non valorizzato*) wasted. **sacri'ficio** *nm* sacrifice

sacri'legio *nm* sacrilege. **sa'crilego** *a* sacrilegious

'**sacro** *a* sacred ● *nm Anat* sacrum

sacro'santo *a* sacrosanct

'**sadico, -a** *a* sadistic ● *nmf* sadist. **sa'dismo** *nm* sadism

sa'etta *nf* arrow

sa'fari *nm inv* safari

'**saga** *nf* saga

sa'gace *a* shrewd

sag'gezza *nf* wisdom

saggi'are *vt* test

'**saggio**[1] *nm* (*scritto*) essay; (*prova*) proof; (*di metallo*) assay; (*campione*) sample; (*esempio*) example

'**saggio**[2] *a* wise ● *nm* (*persona*) sage

sag'gistica *nf* non-fiction

Sagit'tario *nm Astr* Sagittarius

'sagoma *nf* shape; (*profilo*) outline; **che ~!** *fam* what a character!. **sago'mato** *a* shaped

'**sagra** *nf* festival

sagre'|stano *nm* sacristan. **~'stia** *nf* sacristy

'**sala** *nf* hall; (*stanza*) room; (*salotto*) living room. **~ d'attesa** waiting room. **~ da ballo** ballroom. **~ d'imbarco** departure lounge. **~ macchine** engine room. **~ operatoria** operating theatre *Br*, operating room *Am*. **~ parto** delivery room. **~ da pranzo** dining room

sa'lame *nm* salami

sala'moia *nf* brine

sa'lare *vt* salt

sa'lario *nm* wages *pl*

sa'lasso *nm* **essere un ~** *fig* cost a fortune

sala'tini *nmpl* savouries (*eaten with aperitifs*)

sa'lato *a* salty; (*costoso*) dear

sal'ciccia *nf* = **salsiccia**

sal'dar|e *vt* weld; **set** (*osso*) pay off (*debito*), settle (*conto*); **~e a stagno** solder. **~si** *vr* (*Med: osso:*) knit

salda'trice *nf* welder; (*a stagno*) soldering iron

salda'tura *nf* weld; (*azione*) welding; (*di osso*) knitting

'**saldo** *a* firm; (*resistente*) strong ● *nm* (*di conto*) settlement; (*svendita*) sale; *Comm* balance

'**sale** *nm* salt; **restare di ~** be struck dumb [with astonishment]. **~ fine** table salt. **~ grosso** cooking salt. **sali** *pl* **e tabacchi** tobacconist's shop

'**salice** *nm* willow. **~ piangente** weeping willow

sali'ente *a* outstanding; **i punti salienti di un discorso** the main points of a speech

sali'era *nf* salt-cellar

sa'lina *nf* salt-works *sg*

sa'li|re *vi* go/come up; (*levarsi*) rise; (*su treno ecc*) get on; (*in macchina*) get in ● *vt* go/come up (*scale*). **~ta** *nf* climb; (*aumento*) rise; **in ~ta** uphill

sa'liva *nf* saliva

'**salma** *nf* corpse

'**salmo** *nm* psalm

sal'mone *nm & a inv* salmon

sa'lone *nm* hall; (*salotto*) living room; (*di parrucchiere*) salon. **~ di bellezza** beauty parlour

salo'pette *nf inv* dungarees *pl*

salot'tino *nm* bower

sa'lotto *nm* drawing room; (*soggiorno*) sitting room; (*mobili*) [three-piece] suite; **fare ~** chat

sal'pare *vt/i* sail; **~ l'ancora** weigh anchor

'**salsa** *nf* sauce. **~ di pomodoro** tomato sauce

sal'sedine *nf* saltiness

sal'siccia *nf* sausage

salsi'era *nf* sauce-boat

sal'ta|re *vi* jump; (*venir via*) come off; (*balzare*) leap; (*esplodere*) blow up; **~r fuori** spring from nowhere; (*oggetto cercato:*) turn up; **è ~to fuori che...** it emerged that...; **~re fuori con...** come out with...; **~re in aria** blow up; **~re in mente** spring to mind ● *vt* jump [over]; skip (*pasti, lezioni*); *Culin* sauté. **~to** *a Culin* sautéed

saltel'lare *vi* hop; (*di gioia*) skip

saltim'banco *nm* acrobat

'salto *nm* jump; (*balzo*) leap; (*dislivello*) drop; (*fig: omissione, lacuna*) gap; **fare un ~ da** (*visitare*) drop in on; **in un ~** *fig* in a jiffy. **~ in alto** high jump. **~ con l'asta** pole-vault. **~ in lungo** long jump. **~ pagina** *Comput* page down

saltuaria'mente *adv* occasionally. **saltu'ario** *a* desultory; **lavoro saltuario** casual work

sa'lubre *a* healthy

salume'ria *nf* delicatessen. **sa'lumi** *nmpl* cold cuts

salu'tare *vt* greet; (*congedandosi*) say goodbye to; (*portare i saluti a*) give one's regards to; *Mil* salute ● *a* healthy

sa'lute *nf* health; **~!** (*dopo uno starnuto*) bless you!; (*a un brindisi*) cheers!

sa'luto *nm* greeting; (*di addio*) goodbye; *Mil* salute; **saluti** *pl* (*ossequi*) regards

'salva *nf* salvo; **sparare a salve** fire blanks

salvada'naio *nm* money box

salva'gente *nm* lifebelt; (*a giubbotto*) life-jacket; (*ciambella*) rubber ring; (*spartitraffico*) traffic island

salvaguar'dare *vt* safeguard. **salvagu'ardia** *nf* safeguard

sal'var|e *vt* save; (*proteggere*) protect. **~si** *vr* save oneself

salva'slip *nm* inv panty-liner

salva'taggio *nm* rescue; *Naut* salvage; *Comput* saving; **battello di**

~taggio lifeboat. **~'tore**, **~'trice** *nmf* saviour

sal'vezza *nf* safety; *Relig* salvation

'salvia *nf* sage

salvi'etta *nf* serviette

'salvo *a* safe ● *prep* except [for] ● *conj* ~ **che** (*a meno che*) unless; (*eccetto che*) except that

samari'tano, **-a** *a & nmf* Samaritan

sam'buco *nm* elder

san *nm* S~ **Francesco** Saint Francis

sa'nare *vt* heal

sana'torio *nm* sanatorium

san'cire *vt* sanction

'sandalo *nm* sandal; *Bot* sandalwood

'sangu|e *nm* blood; **al ~e** (*carne*) rare; **farsi cattivo ~e** per worry about; **occhi iniettati di ~e** bloodshot eyes. **~e freddo** composure; **a ~e freddo** in cold blood. **~'igno** *a* blood

sangui'naccio *nm Culin* black pudding

sangui'nante *a* bleeding

sangui'nar|e *vi* bleed. **~io** *a* bloodthirsty

sangui'noso *a* bloody

sangui'suga *nf* leech

sanità *nf* soundness; (*salute*) health. **~ mentale** sanity, mental health

sani'tario *a* sanitary; **Servizio S~** National Health Service

'sano *a* sound; (*salutare*) healthy; **~ di mente** sane; **~ come un pesce** as fit as a fiddle

San Sil'vestro *nm* New Year's Eve

santifi'care *vt* sanctify

'santo *a* holy; (*con nome proprio*) saint ● *nm* saint. **san'tone** *nm* guru. **santu'ario** *nm* sanctuary

sanzi'one *nf* sanction

sa'pere *vt* know; (*essere capace di*) be able to; (*venire a sapere*) hear; **saperla lunga** know a thing or two ● *vi* ~ **di** know about; (*aver sapore di*) taste of; (*aver odore di*) smell of; **saperci fare** have the know-how ● *nm* knowledge

sapi'en|te *a* wise; (*esperto*) expert ● *nm* (*uomo colto*) sage. **~za** *nf* wisdom

sa'pone *nm* soap. **~ da bucato** washing soap. **sapo'netta** *nf* bar of soap

sa'pore *nm* taste. **saporita'mente** *adv* (*dormire*) soundly. **sapo'rito** *a* tasty

sapu'tello, -a *a* & *nm* *sl* know-all, know-it-all *Am*

saraci'nesca *nf* roller shutter

sar'cas|mo *nm* sarcasm. **~tico** *a* sarcastic

Sar'degna *nf* Sardinia

sar'dina *nf* sardine

'sardo, -a *a* & *nmf* Sardinian

sar'donico *a* sardonic

'sarto, -a *nm* tailor ● *nf* dressmaker. **~'ria** *nf* tailor's; dressmaker's; (*arte*) couture

sas'sata *nf* blow with a stone; **prendere a sassate** stone. **'sasso** *nm* stone; (*ciottolo*) pebble

sassofo'nista *nmf* saxophonist. **sas'sofono** *nm* saxophone

sas'soso *a* stony

'Satana *nm* Satan. **sa'tanico** *a* satanic

sa'tellite *a inv* & *nm* satellite

sati'nato *a* glossy

'satira *nf* satire. **sa'tirico** *a* satirical

satu'ra|re *vt* saturate. **~zi'one** *f* saturation. **'saturo** *a* saturated; (*pieno*) full

'sauna *nf* sauna

savo'iardo *nm* (*biscotto*) sponge finger

sazi'ar|e *vt* satiate. **~si** *vr* **~si di** *fig* grow tired of

sazietà *nf* **mangiare a ~** eat one's fill. **'sazio** *a* satiated

sbaciucchi'ar|e *vt* smother with kisses. **~si** *vr* kiss and cuddle

sbada'ta|ggine *nf* carelessness; **è stata una ~ggine** it was careless. **~'mente** *adv* carelessly. **sba'dato** *a* careless

sbadigli'are *vi* yawn. **sba'diglio** *nm* yawn

sba'fa|re *vt* sponge. **~ta** *nf sl* nosh

'sbafo *nm* sponging; **a ~** (*gratis*) without paying

sbagli'ar|e *vi* make a mistake; (*aver torto*) be wrong ● *vt* make a mistake in; **~e strada** go the wrong way; **~e numero** get the number wrong; *Teleph* dial a wrong number. **~si** *vr* make a mistake. **'sbaglio** *nm* mistake; **per sbaglio** by mistake

sbal'la|re *vt* unpack; *fam* screw up (*conti*) ● *vi fam* go crazy. **~ato** *a* (*squilibrato*) unbalanced. **'sballo** *nm fam* scream; (*per droga*) trip; **da sballo** *sl* terrific

sballot'tare *vt* toss about

sbalor'di|re *vt* stun ● *vi* be stunned. **~tivo** *a* amazing. **~to** *a* stunned

sbal'zare *vt* throw; (*da una carica*) dismiss ● *vi* bounce; (*saltare*) leap. **'sbalzo** *nm* bounce; (*sussulto*) jolt; (*di temperatura*) sudden change; **a sbalzi** in spurts; **a sbalzo** (*lavoro a rilievo*) embossed

sban'care *vt* bankrupt; **~ il banco** break the bank

sbanda'mento *nm* *Auto* skid; *Naut* list; *fig* going off the rails

sban'da|re *vi* *Auto* skid; *Naut* list. **~rsi** *vr* (*disperdersi*) disperse. **'sbanda'ta** *nf* skid; *Naut* list; **prendere una ~ta per** get a crush on. **~to, -a** *a* mixed-up ● *nmf* mixed-up person

sbandie'rare *vt* wave; *fig* display

sbaracca're *vt/i* clear up

sbaragli'are *vt* rout. **sba'raglio** *nm* rout; **mettere allo sbaraglio** rout

sbaraz'zar|e *vt* clear. **~si** *vr* **~si di** get rid of

sbaraz'zino, -a *a* mischievous ● *nmf* scamp

sbar'bar|e *vt*, **~si** *vr* shave

sbar'care *vt/i* disembark; **~ il lunario** make ends meet. **'sbarco** *nm* landing; (*di merci*) unloading

'sbarra *nf* bar; (*di passaggio a li-*

vello) barrier. **~'mento** nm barricade. **sbar'rare** vt bar; (*ostruire*) block; cross (*assegno*); (*spalancare*) open wide

sbatacchi'are vt/i sl bang, slam

'sbatter|e vt bang, slam, bang (*porta*); (*urtare*) knock; Culin beat; flap (*ali*); shake (*tappeto* ● sl bang; (*porta:*) slam, bang. **~si** vr sl rush around; **~sene** di qcsa not give a damn about sth. **sbat'tuto** a tossed; Culin beaten; fig run down

sba'va|re vi dribble; (*colore:*) smear. **~'tura** nf smear; **senza** **~ture** fig faultless

sbelli'carsi vr **~ dalle risa** split one's sides [with laughter]

'sberla nf slap

sbia'di|re vt/i, **~rsi** vr fade. **~to** a faded; fig colourless

sbian'car|e vt/i, **~si** vr whiten

sbi'eco a slanting; **di ~** on the slant; (*guardare*) sidelong; **guardare qcno di ~** look askance at sb; **tagliare di ~** cut on the bias

sbigot'ti|re vt dismay ● vi, **~rsi** vr be dismayed. **~to** a dismayed

sbilanci'ar|e vt unbalance ● vi (*perdere l'equilibrio*) overbalance. **~si** vr lose one's balance

sbirci'a|re vt cast sidelong glances at. **~ta** nf furtive glance. **~'tina** nf **dare una ~tina** a sneak a glance at

sbizzar'rirsi vr satisfy one's whims

sbloc'care vt unblock; Mech release; decontrol (*prezzi*)

sboc'care vi **~ in** (*fiume:*) flow into; (*strada:*) lead to; (*folla:*) pour into

sboc'cato a foul-mouthed

sbocci'are vi blossom

'sbocco nm flowing; (*foce*) mouth; Comm outlet

sbolo'gnare vt fam get rid of

'sbornia nf **prendere una ~** get drunk

sbor'sare vt pay out

sbot'tare vi burst out

sbotto'nar|e vt unbutton. **~si** vr (*fam: confidarsi*) open up; **~si la** **camicia** unbutton one's shirt

sbra'carsi vr put on something more comfortable; **~ dalle risate** fam kill oneself laughing

sbracci'a|rsi vr wave one's arms. **~to** a bare-armed; (*abito*) sleeveless

sbrai'tare vi bawl

sbra'nare vt tear to pieces

sbricio'lar|e vt, **~si** vr crumble

sbri'ga|re vt expedite; (*occuparsi di*) attend to. **~rsi** vr be quick. **~'tivo** a quick

sbrindel'la|re vt tear to shreds. **~to** a in rags

sbro'do|lare vt stain. **~one** nm messy eater, dribbler

'sbronz|a nf prendersi una ~ get tight. **sbron'zarsi** vr get tight. **~o** a (*ubriaco*) tight

sbruffo'nata nf boast. **sbruf'fone, -a** nmf boaster

sbu'care vi come out

sbucci'ar|e vt peel; shell (*piselli*). **~si** vr graze oneself

sbuf'fare vi snort; (*per impazienza*) fume. **'sbuffo** nm puff

'scabbia nf scabies sg

sca'broso a rough; fig difficult; (*scena*) indecent

scacci'are vt chase away

'scacc|o nm check; **~hi** pl (*gioco*) chess; (*pezzi*) chessmen; **dare ~** **matto** a checkmate; **a ~hi** (*tessuto*) checked. **~hi'era** nf chess-board

sca'dente a shoddy

sca'de|nza nf (*di contratto*) expiry; Comm maturity; (*di progetto*) deadline; **a breve/lunga ~nza** short-/long-term. **~re** vi expire; (*valore:*) decline; (*debito:*) be due. **sca'duto** a (*biglietto*) out-of-date

sca'fandro nm diving suit; (*di astronauta*) spacesuit

scaf'fale nm shelf; (*libreria*) bookshelf

'scafo nm hull

scagion'are vt exonerate

'**scaglia** *nf* scale; (*di sapone*) flake; (*scheggia*) chip

scagli'ar|e *vt* fling. **~si** *vr* fling oneself; **~si contro** *fig* rail against

scagli'o'nare *vt* space out. **~'one** *nm* group; **a ~oni** in groups. **~one di reddito** tax bracket

'**scala** *nf* staircase; (*portatile*) ladder; (*Mus, misura, fig*) scale; (*dei salari*) cost of living index

sca'la|re *vt* climb; (*capelli*); (*detrarre*) deduct. **~ta** *nf* climb; (*dell'Everest ecc*) ascent; **fare delle ~te** go climbing. **~'tore, ~'trice** *nmf* climber

scalca'gnato *a* down at heel

scalci'are *vi* kick

scalci'nato *a* shabby

scalda'bagno *nm* water heater

scalda'muscoli *nm inv* leg-warmer

scal'dar|e *vt* heat. **~si** *vr* warm up; (*eccitarsi*) get excited

scal'fi|re *vt* scratch. **~t'tura** *nf* scratch

scali'nata *nf* flight of steps. **sca'lino** *nm* step; (*di scala a pioli*) rung

scalma'narsi *vr* get worked up

'**scalo** *nm* slipway; *Aeron, Naut* port of call; **fare ~** a call at; *Aeron* land at

sca'lo|gna *nf* bad luck. **~'gnato** *a* unlucky

scalop'pina *nf* escalope

scal'pello *nm* chisel

scalpi'tare *vi* paw the ground; *fig* champ at the bit

'**scalpo** *nm* scalp

scal'trezza *nf* shrewdness. **scal'trirsi** *vr* get shrewder. '**scaltro** *a* shrewd

scal'zare *vt* bare the roots of (*albero*); *fig* undermine; (*da una carica*) oust

'**scalzo** *a* & *adv* barefoot

scambi'are *vt* exchange; **~are**

qcno per qualcun altro mistake sb for somebody else. **~'evole** *a* reciprocal

'**scambio** *nm* exchange; *Comm* trade; **libero ~** free trade

scamosci'ato *a* suede

scampa'gnata *nf* trip to the country

scampa'nato *a* (*gonna*) flared

scampanel'lata *nf* (*loud*) ring

scam'pare *vt* save; (*evitare*) escape; **scamparla bella** have a lucky escape. '**scampo** *nm* escape

'**scampolo** *nm* remnant

scanala'tura *nf* groove

scandagli'are *vt* sound

scanda'listico *a* sensational

scandaliz'zare *vt* scandalize. **~iz'zarsi** *vr* be scandalized

'**scanda|lo** *nm* scandal. **~'loso** *a* (*somma ecc*) scandalous; (*fortuna*) outrageous

Scandi'navia *nf* Scandinavia. **scandi'navo, -a** *a* & *nmf* Scandinavian

scan'dire *vt* scan (*verso*); pronounce clearly (*parole*)

scan'nare *vt* slaughter

scanneriz'zare *vi* Comput scan

scansafa'tiche *nmf inv* lazybones *sg*

scan'sar|e *vt* shift; (*evitare*) avoid. **~si** *vr* get out of the way

scansi'one *nf* Comput scanning

'**scanso** *nm* a **~ di** in order to avoid; a **~ di equivoci** to avoid any misunderstanding

scanti'nato *nm* basement

scanto'nare *vi* turn the corner; (*svignarsela*) sneak off

scanzo'nato *a* easy-going

scapacci'one *nm* smack

scape'strato *a* dissolute

sca'pito *nm* loss; a **~ di** to the detriment of

'**scapola** *nf* shoulder-blade

'**scapolo** *nm* bachelor

scappa'mento *nm* Auto exhaust

scap'pa|re *vi* escape; (*andarsene*) dash [off]; (*sfuggire*) slip; **mi ~ da ridere!** I want to burst out laughing; **mi ~ la pipi** I'm bursting, I need a pee. **~ta** nf short visit. **~'tella** nf escapade; (infedeltà) fling. **~'toia** nf way out

scappel'lotto nm cuff

scara'bocchio nm scribble

scara'faggio nm cockroach

scara'mantico a (gesto) to ward off the evil eye

scara'muccia nf skirmish

scarabocchi'are vt scribble

scaraven'tare vt hurl

scarce'rare vt release [from prison]

scardi'nare vt unhinge

'scarica nf discharge; (di arma da fuoco) volley; fig shower

scari'ca|re vt discharge; (di arma, merci) unload ⟨arma, merci⟩; fig unburden. **~rsi** vr ⟨fiume:⟩ flow; ⟨orologio, batteria:⟩ run down; fig unwind. **~'tore** nm loader; (di porto) docker. **'scarico** a unloaded; (vuoto) empty; ⟨orologio⟩ run-down; ⟨batteria⟩ flat; fig untroubled ●nm unloading; (di rifiuti) dumping; (di acqua) draining; (di sostanze inquinanti) discharge; (luogo) [rubbish] dump; Auto exhaust; ⟨idraulico⟩ drain; (tubo) waste pipe

scarlat'tina nf scarlet fever

scar'latto a scarlet

'scarno a thin; (fig: stile) bare

sca'ro|gna nf fam bad luck. **~'gnato** a fam unlucky

'scarpa nf shoe; (fam: persona) dead loss. **scarpe** pl **da ginnastica** trainers, gym shoes

scar'pata nf slope; (burrone) escarpment

scarpi'nare vi hike

scar'pone nm boot. **scarponi** pl **da sci** ski boot. **scarponi** pl **da trekking** walking boots

scarroz'zare vt/i drive around

scarseggi'are vi be scarce; ~ **di** (mancare) be short of

scar'sezza nf scarcity, shortage. **scarsità** nf shortage. **'scarso** a scarce; (manchevole) short

scarta'mento nm Rail gauge. **~ ridotto** narrow gauge

scar'tare vt discard; unwrap ⟨pacco⟩; (respingere) reject ●vi (deviare) swerve. **'scarto** nm scrap; (in carte) discard; (deviazione) swerve; (distacco) gap

scar'toffie nfpl bumf, bumph

scas'sa|re vt break. **~to** a fam clapped out

scassi'nare vt force open

scassina'tore, -'trice nmf burglar. **'scasso** nm (furto) housebreaking

scate'na|re vt fig stir up. **~rsi** vr break out; (fig: temporale:) break; (fam: infiammarsi) get excited. **~to** a crazy

'scatola nf box; (di latta) can, tin Br; **in ~** ⟨cibo⟩ canned, tinned Br; **rompere le scatole a qcno** fam get on sb's nerves

scat'tare vi go off; (balzare) spring up; (adirarsi) lose one's temper; take ⟨foto⟩. **'scatto** nm (balzo) spring; (d'ira) outburst; (di telefono) unit; (dispositivo) release; **a scatti** jerkily; **di scatto** suddenly

scatu'rire vi spring

scaval'care vt jump over ⟨muretto⟩; climb over ⟨muro⟩; (fig: superare) overtake

sca'vare vt dig ⟨buca⟩; dig up ⟨tesoro⟩; excavate ⟨città sepolta⟩. **'scavo** nm excavation

scazzot'tata nf fam punch-up

'scegliere vt choose, select

scelle'rato a wicked

'scelt|a nf choice; (di articoli) range; **...a ~a** (in menù) choice of...; **prendine uno a ~a** take your choice o pick; **di prima ~a** top-grade, choice. **~o** pp di **scegliere** ● select; (merce ecc) choice

sce'mare vt/i diminish

sce'menza nf silliness; (azione) silly thing to do/say. **'scemo** a silly

'scempio nm havoc; (fig: di pae-

saggio) ruination; **fare ~ di** play havoc with

'scena *nf* scene; *(palcoscenico)* stage; **entrare in ~** go/come on; *fig* enter the scene; **fare ~** put on an act; **fare una ~** make a scene; **andare in ~** *Theat* be staged, be put on. **sce'nario** *nm* scenery

sce'nata *nf* row, scene

'scendere *vi* go/come down; *(da treno, autobus)* get off; *(da macchina)* get out; *(strada:)* slope; *(notte, prezzi:)* fall ● *vt* go/come down *(scale)*

sceneggi'are *vt* dramatize. **~to** *nm* television serial. **~'tura** *nf* screenplay

'scenico *a* scenic

scervel'larsi *vr* rack one's brains. **~to** *a* brainless

'sceso *pp di* **scendere**

scetti'cismo *nm* scepticism. **'scettico, -a** *a* sceptical ● *nmf* sceptic

'scettro *nm* sceptre

'scheda *nf* card. **~ elettorale** ballot-paper. **~ di espansione** *Comput* expansion card. **~ perforata** punch card. **~ telefonica** phonecard. **sche'dare** *vt* file. **sche'dario** *nm* file; *(mobile)* filing cabinet

sche'dina *nf* ≈ pools coupon; **giocare la ~** do the pools

'scheggia|a *nf* fragment; *(di legno)* splinter. **~'arsi** *vr* chip; *(legno:)* splinter

'scheletro *nm* skeleton

'schema *nm* diagram; *(abbozzo)* outline. **sche'matico** *a* schematic. **~tiz'zare** *vt* schematize

'scherma *nf* fencing

scher'mirsi *vr* protect oneself

'schermo *nm* screen; **grande ~** big screen

scher'nire *vt* mock. **'scherno** *nm* mockery

scher'zare *vi* joke; *(giocare)* play

'scherzo *nm* joke; *(trucco)* trick; *(effetto)* play; *Mus* scherzo; **fare uno ~ a** qcno play a joke on sb;

per ~ for fun; **stare allo ~** take a joke. **scher'zoso** *a* playful

schiaccia'noci *nm inv* nutcrackers *pl*

schiacci'ante *a* damning

schiacci'are *vt* crush; *Sport* smash; press *(pulsante)*; crack *(noce)*; **~ un pisolino** grab forty winks

schiaffeggi'are *vt* slap. **schi'affo** *nm* slap; **dare uno schiaffo a** slap

schiamaz'zare *vi* make a racket; *(galline:)* cackle

schian'tar|e *vt* break. **~si** *vr* crash ● *vi* **schianto dalla fatica** I'm wiped out. **'schianto** *nm* crash; *fam* knock-out; *(divertente)* scream

schia'rir|e *vt* clear; *(sbiadire)* fade ● *vi*, **~si** *vr* brighten up; **~si la gola** clear one's throat

schiavitù *nf* slavery. **schi'avo, -a** *nmf* slave

schi'ena *nf* back; **mal di ~** backache. **schie'nale** *nm (di sedia)* back

schi'er|a *nf Mil* rank; *(moltitudine)* crowd. **~'amento** *nm* lining up

schie'rar|e *vt* draw up. **'~si** *vr* draw up; **~si con** *(parteggiare)* side with

schiet'tezza *nf* frankness. **schi'etto** *a* frank; *(puro)* pure

schi'fezza *nf* una **~** rubbish. **schifil'toso** *a* fussy. **'schifo** *nm* disgust; **mi fa schifo** it makes me sick. **schi'foso** *a* disgusting; *(di cattiva qualità)* rubbishy

schioc'car|e *vt* crack; snap *(dita)*. **schi'occo** *nm (di frusta)* crack; *(di bacio)* smack; *(di dita, lingua)* click

schi'oppo *nm* **ad un tiro di ~** a stone's throw away

schi'uder|e *vt*, **~si** *vr* open

schi'u|ma *nf* foam; *(di sapone)* lather; *(feccia)* scum. **~ma da barba** shaving foam. **~'mare** *vt* skim ● *vi* foam

schi'uso *pp di* **schiudere**

schi'vare vt avoid. **'schivo** a bashful

schizo'frenico a schizophrenic

schiz'zare vt squirt; ‹inzaccherare› splash; ‹abbozzare› sketch ● vi spurt; ~ **via** scurry away

schiz'zato, -**a** a & nmf sl loony

schizzi'noso a squeamish

'schizzo nm squirt; ‹di fango› splash; ‹abbozzo› sketch

sci nm inv ski; ‹sport› skiing. ~ **d'acqua** water-skiing

'scia nf wake; ‹di fumo ecc› trail

sci'abola nf sabre

sciabor'dare vt/i lap

scia'callo nm jackal; fig profiteer

sciac'quare vt rinse. ~**si** vr rinse oneself. **sci'acquo** nm mouthwash

scia'gura nf disaster. ~'**rato** a unfortunate; ‹scellerato› wicked

scialac'quare vt squander

scia'lare vi spend money like water

sci'albo a pale; fig dull

sci'alle nm shawl

scia'luppa nf dinghy. ~ **di salvataggio** lifeboat

sci'ame nm swarm

sci'ampo nm shampoo

scian'cato a lame

sci'are vi ski

sci'arpa nf scarf

sci'atica nf Med sciatica

scia'tore, -'**trice** nmf skier

sci'atto a slovenly; ‹stile› careless. **sciat'tone**, -**a** nmf slovenly person

scienti'fico a scientific

sci'enza nf science; ‹sapere› knowledge. ~'**iato**, -**a** nmf scientist

'scimmi|a nf monkey. ~**ot'tare** vt ape

scimpanzé nm inv chimpanzee, chimp

scimu'nito a idiotic

'scindere vt, ~**si** vr split

scin'tilla nf spark. **scintil'lante** a sparkling. **scintil'lare** vi sparkle

scioc'ca|nte a shocking. ~**re** vt shock

scioc'chezza nf foolishness; ‹assurdità› nonsense. **sci'occo** a foolish

sci'ogliere vt untie; undo, untie ‹nodo›; ‹liberare› release; ‹liquefare› melt; dissolve ‹contratto, qcsa nell'acqua›; loosen up ‹muscoli›. ~**si** vr release oneself; ‹liquefarsi› melt; ‹contratto:› be dissolved; ‹pastiglia:› dissolve

sciogli'lingua nm inv tonguetwister

scio'lina nf wax

sciol'tezza nf agility; ‹disinvoltura› ease

sci'olto pp di **sciogliere** ● a loose; ‹agile› agile; ‹disinvolto› easy; **versi sciolti** blank verse sg

sciope'ra|nte nmf striker. ~**re** vi go on strike, strike. **sci'opero** nm strike. **sciopero a singhiozzo** on-off strike

sciori'nare vt fig show off

sci'pito a insipid

scip'pa|re vt fam snatch. ~'**tore**, ~'**trice** nmf bag snatcher. '**scippo** nm bag-snatching

sci'rocco nm sirocco

sci'roppo a ‹frutta› in syrup. **sci'roppo** nm syrup

'scisma nm schism

scis'sione nf division

'scisso pp di **scindere**

sciu'par|e vt spoil; ‹sperperare› waste. ~**si** vr get spoiled; ‹deperire› wear oneself out. **sciu'pio** nm waste

scivo'l|are vi slide; ‹involontariamente› slip. '**scivolo** nm slide; Techn chute. ~**oso** a slippery

scle'rosi nf sclerosis

scoc'care vt shoot ● vi ‹scintilla:› shoot out; ‹ora:› strike

scocci'a|re vt ‹dare noia a› bother. ~**rsi** vr be bothered. ~**to** a fam narked. ~'**tore**, ~'**trice** nmf bore. ~'**tura** nf nuisance

sco'della nf bowl

scodinzo'lare vi wag its tail

scogli'era nf cliff; ‹a flor d'acqua›

reef. **'scoglio** nm rock; (fig: ostacolo) stumbling block

scoi'attolo nm squirrel

scola'pasta nm inv colander. **~pi'atti** nm inv dish drainer

sco'lara nf schoolgirl

sco'lare vt drain; strain (pasta, verdura) ● vi drip

sco'la|ro nm schoolboy. **~'resca** nf pupils pl. **~stico** a school attrib

scoli'osi nf curvature of the spine

scol'la|re vt cut away the neck of (abito); (staccare) unstick. **~to** a (abito) low-necked. **~'tura** nf neckline

'scolo nm drainage

scolo'ri|re vt, **~rsi** vr fade. **~to** a faded

scol'pire vt carve; (imprimere) engrave

scombi'nare vt upset

scombusso'lare vt muddle up

scom'messa nf bet. **~o** pp di **scommettere**. **scom'mettere** vt bet

scomo'dar|e vt, **~si** vr trouble. **scomodità** nf discomfort. **'scomodo** a uncomfortable ● nm essere di scomodo a qcno be a trouble to sb

scompa'rire vi disappear; (morire) pass on. **scom'parsa** nf disappearance; (morte) passing, death. **scom'parso, -a** pp di **scomparire** ● nmf departed

scomparti'mento nm compartment. **scom'parto** nm compartment

scom'penso nm imbalance

scompigli'are vt disarrange. **scom'piglio** nm confusion

scom'po|rre vt take to pieces; (fig: turbare) upset. **~rsi** vr get flustered, lose one's composure. **~sto** pp di **scomporre** ● a (sguaiato) unseemly; (disordinato) untidy

sco'muni|ca nf excommunication. **~'care** vt excommunicate

sconcer'ta|re vt disconcert; (rendere perplesso) bewilder. **~to** a disconcerted; bewildered

scon'cezza nf obscenity. **'sconcio** a (osceno) dirty ● nm è uno sconcio che... it's a disgrace that...

sconclusio'nato a incoherent

scon'dito a unseasoned; (insalata) with no dressing

scon'fes'sare vt disown

scon'figgere vt defeat

sconfi'na|re vi cross the border; (in proprietà privata) trespass. **~to** a unlimited

scon'fitt|a nf defeat. **~o** pp di **sconfiggere**

scon'forto nm dejection

scon'gelare vt thaw out (cibo), defrost

scongi'u|rare vt beseech; (evitare) avert. **~'uro** nm fare gli scongiuri touch wood, knock on wood Am

scon'nesso pp di **sconnettere** ● a fig incoherent. **scon'nettere** vt disconnect

sconosci'uto, -a a unknown ● nmf stranger

sconquas'sare vt smash; (sconvolgere) upset

sconside'rato a inconsiderate

sconsigli'a|bile a not advisable. **~re** vt advise against

sconso'lato a disconsolate

scon'ta|re vt discount; (dedurre) deduct; (pagare) pay off; serve (pena). **~to** a discount; (ovvio) expected; **~to del 10%** with 10% discount; **dare qcsa per ~to** take sth for granted

scon'tento a displeased ● nm discontent

'sconto nm discount; **fare uno ~** give a discount

scon'trarsi vr clash; (urtare) collide

scon'trino nm ticket; (di cassa) receipt

'scontro nm clash; (urto) collision

scon'troso a unsociable

conveni'ente a unprofitable; (scorretto) unseemly

sconvol'gente a mind-blowing

scon'vol|gere vt upset; (mettere in disordine) disarrange. **~gi'mento**

nm upheaval. **~to** *pp di* **sconvolgere ● a** distraught

'scopa *nf* broom. **sco'pare** *vt* sweep; *vulg* shag, screw

scoperchi'are *vt* take the lid off (*pentola*); take the roof off (*casa*)

sco'pert|a *nf* discovery. **~o** *pp di* **scoprire ● a** uncovered; (*senza riparo*) exposed; (*conto*) overdrawn; (*spoglio*) bare

'scopo *nm* aim; **allo ~ di** in order to

scoppi'are *vi* burst; *fig* break out. **scoppiet'tare** *vi* crackle. **'scoppio** *nm* burst; (*di guerra*) outbreak; (*esplosione*) explosion

sco'prire *vt* discover; (*togliere la copertura*) uncover

scoraggi'ante *a* discouraging. **scoraggi'a|re** *vt* discourage. **~rsi** *vr* lose heart

scor'butico *a* peevish

scorcia'toia *nf* short cut

'scorcio *nm* (*di epoca*) end; (*di cielo*) patch; (*in arte*) foreshortening; **di ~** (*vedere*) from an angle. **~ panoramico** panoramic view

scor'da|re *vt*, **~rsi** *vr* forget. **~to** *a* *Mus* out of tune

sco'reggi|a *nf fam* fart. **~are** *vi fam* fart

'scorgere *vt* make out; (*notare*) notice

'scoria *nf* waste; (*di metallo, carbone*) slag; **scorie** *pl* **radioattive** radioactive waste

scor'nato *a fig* hangdog. **'scorno** *nm* humiliation

scorpacci'ata *nf* bellyful; **fare una ~ di** stuff oneself with

scorpi'one *nm* scorpion; *Astr* Scorpio

scorraz'zare *vi* run about

'scorrere *vt* (*dare un'occhiata*) glance through **● vi** run; (*scivolare*) slide; (*fluire*) flow; *Comput* scroll. **scor'revole ● a porta scorrevole** sliding door

scorre'ria *nf* raid

scorret'tezza *nf* (*mancanza di educazione*) bad manners *pl*. **scor-**

'retto *a* incorrect; (*sconveniente*) improper

scorri'banda *nf* raid; *fig* excursion

scor'revole *nf* glance. **~o** *pp di* **scorrere ● a** last

scor'soio *a* **nodo ~** noose

scor'ta *nf* escort; (*provvista*) supply. **~'tare** *vt* escort

scor'te|se *a* discourteous. **~'sia** *nf* discourtesy

scorti'ca|re *vt* skin. **~tura** *nf* graze

'scorto *pp di* **scorgere**

'scorza *nf* peel; (*crosta*) crust; (*corteccia*) bark

sco'sceso *a* steep

'scossa *nf* shake; *Electr, fig* shock; **prendere la ~** get an electric shock. **~ elettrica** electric shock. **~ sismica** earth tremor

'scosso *pp di* **scuotere ● a** shaken; (*sconvolto*) upset

sco'stante *a* off-putting

sco'sta|re *vt* push away. **~rsi** *vr* stand aside

scostu'mato *a* dissolute; (*maleducato*) ill-mannered

scot'tante *a* (*argomento*) dangerous

scot'ta|re *vt* scald **● vi** burn; (*bevanda:*) be too hot; (*sole, pentola:*) be very hot. **~rsi** *vr* burn oneself; (*al sole*) get sunburnt; *fig* get one's fingers burnt. **~tura** *nf* burn; (*da liquido*) scald; **~tura solare** sunburn; *fig* painful experience

'scotto *a* overcooked

sco'vare *vt* (*scoprire*) discover

'Scozia *nf* Scotland. **~'zese** *a* Scottish **● nmf** Scot

scredi'tare *vt* discredit

scre'mare *vt* skim

screpo'la|re *vt*, **~rsi** *vr* crack. **~to** *a* (*labbra*) chapped. **~'tura** *nf* crack

screzi'ato *a* speckled

'screzio *nm* disagreement

scribacchi|are *vt* scribble. **~'chino, -a** *nmf* scribbler; (*impiegato*) penpusher

scricchio'l|are *vi* creak. **~io** *nm* creaking

'scricciolo *nm* wren

'scrigno *nm* casket

scrimina'tura *nf* parting

'scrit|ta *nf* writing; (*su muro*) graffiti. **~to** *pp di* **scrivere** ● *a* written ● *nm* writing; (*lettera*) letter. **~toio** *nm* writing-desk. **~tore**, **~trice** *nmf* writer. **~tura** *nf* writing; *Relig* scripture

scrittu'rare *vt* engage

scriva'nia *nf* desk

'scrivere *vt* write; (*descrivere*) write about; **a macchina** type

scroc'c|are *vt* **~are a** scrounge off. **'scrocco** *nm fam* a scrocco fam without paying; **vivere a scrocco** sponge off other people. **~one**, -a *nmf* sponger

'scrofa *nf* sow

scrol'lare *vt* shake; **~e le spalle** shrug one's shoulders. **~si** *vr* shake oneself; **~si qcsa di dosso** shake sth off

scrosci'are *vi* roar; (*pioggia:*) pelt down. **'scroscio** *nm* roar; (*di pioggia*) pelting; **uno scroscio di applausi** thunderous applause

scro'stare *vt* scrape. **~si** *vr* peel off

'scrupo|lo *nm* scruple; (*diligenza*) care; **senza scrupoli** unscrupulous, without scruples. **~loso** *a* scrupulous

scru'ta|re *vt* scan; (*indagare*) search. **~tore** *nm* (*alle elezioni*) returning officer

scruti'nare *vt* scrutinize. **scru-'tinio** *nm* (*di voti alle elezioni*) poll; *Sch* assessment of progress

scu'cire *vt* unstitch; **scuci i soldi** *fam* cough up (the money)!

scude'ria *nf* stable

scu'detto *nm Sport* championship shield

'scudo *nm* shield

sculacci'|are *vt* spank. **~ata** *nf* spanking. **~one** *nm* spanking

scule'tare *vi* wiggle one's hips

scul'tore, -'trice *nm* sculptor ● *nf* sculptress. **~tura** *nf* sculpture

scu'ola *nf* school. **~ elementare** primary school. **~ guida** driving school. **~ materna** day nursery. **~ media** secondary school. **~ media [inferiore]** secondary school (*10-13*). **~ [media] superiore** secondary school (*13-18*). **~ dell'obbligo** compulsory education

scu'oter|e *vt* shake. **~si** *vr* (*destarsi*) rouse oneself; **~si di dosso** shake off

'scure *nf* axe

scu'reggia *nf fam* fart. **scureggi'are** *vi fam* fart

scu'rire *vt/i* darken

'scuro *a* dark ● *nm* darkness; (*imposta*) shutter

scur'rile *a* scurrilous

'scusa *nf* excuse; (*giustificazione*) apology; **chiedere ~** apologize; **chiedo ~!** I'm sorry!

scu'sar|e *vt* excuse. **~si** *vr* **~si** apologize (**di** for); **[mi] scusi!** excuse me!; (*chiedendo perdono*) [I'm] sorry!

sdebi'tarsi *vr* (*disobbligarsi*) repay a kindness

sde'gna|re *vt* despise. **~rsi** *vr* get angry. **~to** *a* indignant. **'sdegno** *nm* disdain. **sde'gnoso** *a* disdainful

sden'tato *a* toothless

sdolci'nato *a* sentimental, schmaltzy

sdoppi'are *vt* halve

sdrai'arsi *vr* lie down. **'sdraio** *nm* [**sedia a**] **sdraio** deckchair

sdrammatiz'zare *vi* provide some comic relief

sdruccio'l|are *vi* slither. **~evole** *a* slippery

se *conj* if; (*interrogativo*) whether, if; **se mai** (*caso mai*) if need be; **se mai telefonasse,...** should he call,...; if he calls,...; **se no** otherwise, or else; **se non altro** at least, if nothing else; **se pure** (*sebbene*) even though; (*anche se*) even if; **non so se sia vero** I don't know whether it's true, I don't know if it's true; **come se** as if; **se lo avessi saputo prima!** if only I had known before!; **e se andassimo fuori a cena?** how about going out for dinner? ● *nm inv* if

sé *pers pron* oneself; (*lui*) himself; (*lei*) herself; (*esso, essa*) itself; (*loro*) themselves; **l'ha fatto da sé** he did it himself; **ha preso i soldi con sé** he took the money with him; **si sono tenuti le notizie per sé** they kept the news to themselves

seb'bene *conj* although

'secca *nf* shallows *pl*; **in ~** (*nave*) aground

sec'cante *a* annoying

sec'ca|re *vt* dry; (*importunare*) annoy ● *vi* dry up. **~rsi** *vr* dry up; (*irritarsi*) get annoyed; (*annoiarsi*) get bored. **~'tore, ~'trice** *nmf* nuisance. **~'tura** *nf* bother

secchi'ello *nm* pail

'secchio *nm* bucket. **~ della spazzatura** rubbish bin, trash can *Am*

'secco, -a *a* dry; (*disseccato*) dried; (*magro*) thin; (*brusco*) curt; (*preciso*) sharp; **restare a ~** be left penniless; **restarci ~** (*fam: morire di colpo*) be killed on the spot ● *nm* (*siccità*) drought; **lavare a ~** dry-clean

secessi'one *nf* secession

seco'lare *a* age-old; (*laico*) secular.

'secolo *nm* century; (*epoca*) age; **è un secolo che non lo vedo** *fam* I haven't seen him for ages *o* yonks

se'cond|a *nf Sch*, *Rail* second class; *Auto* second [gear]. **~o** *a* second ● *nm* second; (*secondo piatto*) main course ● *prep* according to; **~o me** in my opinion

secondo'genito, -a *a & nm* second-born

secrezi'one *nf* secretion

'sedano *nm* celery

seda'tivo *a & nm* sedative

'sede *nf* seat; (*centro*) centre; *Relig* see; *Comm* head office. **~ sociale** registered office

seden'tario *a* sedentary

se'der|e *vi* sit. **~si** *vr* sit down ● *nm* (*deretano*) bottom

'sedia *nf* chair. **~ a dondolo** rocking chair. **~ a rotelle** wheelchair

sedi'cente *a* self-styled

'sedici *a & nm* sixteen

se'dile *nm* seat

sedizi'o|ne *nf* sedition. **~so** *a* seditious

se'dotto *pp di* **sedurre**

sedu'cente *a* seductive; (*allettante*) enticing

se'durre *vt* seduce

se'duta *nf* session; (*di posa*) sitting. **~ stante** *adv* here and now

seduzi'one *nf* seduction

'sega *nf* saw; *vulg* wank

'segala *nf* rye

se'gare *vt* saw

sega'tura *nf* sawdust

'seggio *nm* seat. **~ elettorale** polling station

seg'gio|la *nf* chair. **~lino** *nm* seat; (*da bambino*) child's seat. **~lone** *nm* (*per bambini*) high chair

seggio'via *nf* chair lift

seghe'ria *nf* sawmill

se'ghetto *nm* hacksaw

seg'mento *nm* segment

segna'la|re *vt* signal; (*annunciare*) announce; (*indicare*) point out. **~rsi** *vr* distinguish oneself

se'gna|le *nm* signal; (*stradale*) sign. **~le acustico** beep. **~le orario** time signal. **~letica** *nf* signals *pl*. **~letica stradale** road signs *pl*

segna'libro *nm* bookmark

se'gnar|e *vt* mark; (*prendere nota*) note; (*indicare*) indicate; *Sport* score. **~si** *vr* cross oneself. **'segno** *nm* sign; (*traccia, limite*) mark; (*bersaglio*) target; **far segno** (*col capo*) nod; (*con la mano*) beckon.

segno zodiacale birth sign

segre'ga|re *vt* segregate. **~zi'one** *nf* segregation

segretari'ato *nm* secretariat

segre'tario, -a *nmf* secretary. **~ comunale** town clerk

segrete'ria *nf* (*ufficio*) [administrative] office; (*segretariato*) secretariat. **~ telefonica** answering machine, answerphone

segre'tezza *nf* secrecy

se'greto *a & nm* secret; **in ~** in secret

segu'ace *nmf* follower

segu'ente *a* following, next

se'gugio *nm* bloodhound

segu'ire *vt/i* follow; (*continuare*) continue

segui'tare *vt/i* continue

'seguito *nm* retinue; (*sequela*) series; (*continuazione*) continuation; **di ~** in succession; **in ~** later on; **in ~ a** following; **al ~** in his/her wake; (*a causa di*) owing to; **fare ~ a** *Comm* follow up

'sei *a & nm* six. **sei'cento** *a & nm* six hundred; **il Seicento** the seventeenth century. **sei'mila** *a & nm* six thousand

sel'ciato *nm* paving

selet'tivo *a* selective. **selezio'nare** *vt* select. **selezi'one** *nf* selection

'sella *nf* saddle. **sel'lare** *vt* saddle

seltz *nm* soda water

'selva *nf* forest

selvag'gina *nf* game

sel'vaggio, -a *a* wild; (*primitivo*) savage ● *nmf* savage

sel'vatico *a* wild

se'maforo *nm* traffic lights *pl*

se'mantica *nf* semantics *sg*

sem'brare *vi* seem; (*assomigliare*) look like; **che te ne sembra?** what do you think?; **mi sembra che...** I think...

'seme *nm* seed; (*di mela*) pip; (*di carte*) suit; (*sperma*) semen

se'mestre *nm* half-year

semi'cerchio *nm* semicircle

semifi'nale *nf* semifinal

semi'freddo *nm* ice cream and sponge dessert

'semina *nf* sowing

semi'nare *vt* sow; *fam* shake off (*inseguitori*)

semi'nario *nm* seminar; *Relig* seminary

seminter'rato *nm* basement

se'mitico *a* Semitic

sem'mai *conj* in case ● *adv* **è lui, ~, che...** if anyone, it's him who...

'semola *nf* bran. **semo'lino** *nm* semolina

'semplice *a* simple; **in parole semplici** in plain words. **~'cemente** *adv* simply. **~'ci'otto, -a** *nmf* simpleton. **~'cistico** *a* simplistic. **~'cità** *nf* simplicity. **~'fi'care** *vt* simplify

'sempre *adv* always; (*ancora*) still; **per ~** for ever

sempre'verde *a & nm* evergreen

'senape *nf* mustard

se'nato *nm* senate. **sena'tore** *nm* senator

se'nile *a* senile. **~'ità** *nf* senility

'senno *nm* sense

'seno *nm* (*petto*) breast; *Math* sine; **in ~ a** in the bosom of

sen'sato *a* sensible

sensazi'o'nale *a* sensational. **~'o'ne** *nf* sensation

sen'sibile *a* sensitive; (*percepibile*) perceptible; (*notevole*) considerable. **~'ità** *nf* sensitivity. **~'iz'zare** *vt* make more aware (**a** of)

sensi'tivo, -a *a* sensory ● *nmf* sensitive person; (*medium*) medium

'senso *nm* sense; (*significato*) meaning; (*direzione*) direction; **far ~ a** qcno make sb shudder; **non ha ~** it doesn't make sense; **senza ~** meaningless; **perdere i sensi** lose consciousness; **~ dell'umorismo** sense of humour; **~ unico** (*strada*) one-way; **~ vietato** no entry

sensu'ale *a* sensual. **~'ità** *nf* sensuality

sen'tenza *nf* sentence; (*massima*) saying. **~'iare** *vi* *Jur* pass judgment

senti'ero *nm* path

sentimen'tale *a* sentimental

senti'mento *nm* feeling

senti'nella *nf* sentry

sen'tire *vt* feel; (*udire*) hear; (*ascoltare*) listen to; (*gustare*) taste;

(*odorare*) smell ● *vi* feel; (*udire*) hear; **~re caldo/freddo** feel hot/cold. **~rsi** *vr* feel; **~rsi di fare qcsa** feel like doing sth; **~rsi bene** feel well; **~rsi poco bene** feel unwell; **~rsela di fare qcsa** feel up to doing sth. **~to** *a* (*sincero*) sincere; **per ~to dire** by hearsay

sen'tore *nm* inkling

'**senza** *prep* without; **~ correre** without running; **senz'altro** certainly; **~ ombrello** without an umbrella

senza'tetto *nm inv* **i ~** the homeless

sepa'ra|re *vt* separate. **~rsi** *vr* separate; (*amici*:) part; **~rsi da** be separated from. **~ta'mente** *adv* separately. **~zi'one** *nf* separation

se'pol|cro *nm* sepulchre. **~to** *pp di* **seppellire. ~'tura** *nf* burial

seppel'lire *vt* bury

'**seppia** *nf* cuttle fish; **nero di ~** sepia

sep'pure *conj* even if

se'quenza *nf* sequence

seque'strare *vt* (*rapire*) kidnap; *Jur* impound; (*confiscare*) confiscate. **se'questro** *nm Jur* impounding; (*di persona*) kidnap[ping]

'**sera** *nf* evening; **di ~** in the evening. **se'rale** *a* evening. **se'rata** *nf* evening; (*ricevimento*) party

ser'bare *vt* keep; harbour (*odio*); cherish (*speranza*)

serba'toio *nm* tank. **~ d'acqua** water tank; (*per una città*) reservoir

'**serbo, -a** *a & nmf* Serbian ● *nm* (*lingua*) Serbian; **mettere in ~** put aside

sere'nata *nf* serenade

serenità *nf* serenity. **se'reno** *a* serene; (*cielo*) clear

ser'gente *nm* sergeant

seria'mente *adv* seriously

'**serie** *nf inv* series; (*complesso*) set; *Sport* division; **fuori ~** custom-built; **produzione in ~** mass production; **di ~ B** second-rate

serietà *nf* seriousness. '**serio** *a* serious; (*degno di fiducia*) reliable; **sul serio** seriously; (*davvero*) really

ser'mone *nm* sermon

'**serpe** *nf liter* viper. **~ggi'are** *vi* meander; (*diffondersi*) spread

ser'pente *nm* snake. **~ a sonagli** rattlesnake

'**serra** *nf* greenhouse; **effetto ~** greenhouse effect

ser'randa *nf* shutter

ser'ra|re *vt* shut; (*stringere*) tighten; (*incalzare*) press on. **~'tura** *nf* lock

ser'vi|re *vt* serve; (*al ristorante*) wait on ● *vi* serve; (*essere utile*) be of use; **non serve** it's no good. **~rsi** *vr* (*di cibo*) help oneself; **~si da** buy from; **~si di** use

servitù *nf inv* servitude; (*personale di servizio*) servants *pl*

servizi'evole *a* obliging

ser'vizio *nm* service; (*da caffè ecc*) set; (*di cronaca, sportivo*) report; **servizi** *pl* bathroom; **essere di ~** be on duty; **fare ~** (*autobus ecc*:) run; **fuori ~** (*bus*) not in service; (*ascensore*) out of order; **~ compreso** service charge included. **~ in camera** room service. **~ civile** civilian duties done instead of national service. **~ militare** military service. **~ pubblico** utility company. **~ al tavolo** waiter service

'**servo, -a** *nm* servant

servo'sterzo *nm* power steering

ses'san|ta *a & nm* sixty. **~'tina** *nf* **una ~tina** about sixty

sessi'one *nf* session

'**sesso** *nm* sex

sessu'al|e *a* sexual. **~ità** *nf* sexuality

'**sesto**[1] *a* sixth

'**sesto**[2] *nm* (*ordine*) order

'**seta** *nf* silk

setacci'are *vt* sieve. **se'taccio** *nm* sieve

'**sete** *nf* thirst; **avere ~** be thirsty

'**setola** *nf* bristle

'**setta** *nf* sect

set'tan|ta *a & nm* seventy. **~'tina** *nf* una **~tina** about seventy

'**sette** *a & nm* seven. **~'cento** *a & nm* seven hundred; **il S~cento** the eighteenth century

set'tembre *nm* September

settentri|o'nale *a* northern ● *nmf* northerner. **~'one** *nm* north

setti'ma|na *nf* week. **~'nale** *a & nm* weekly

'**settimo** *a* seventh

set'tore *nm* sector

severi'tà *nf* severity. **se'vero** *a* severe; (*rigoroso*) strict

se'vizi|a *nf* torture; **se'vizie** *pl* torture *sg*. **~'are** *vt* torture

sezio'nare *vt* divide; *Med* dissect. **sezi'one** *nf* section; (*reparto*) department; *Med* dissection

sfaccen'dato *a* idle

sfacchi'na|re *vi* toil. **~ta** *nf* drudgery

sfacci|a'taggine *nf* cheek, insolence. **~'ato** *a* cheeky, fresh *Am*

sfa'celo *nm* ruin; **in ~** in ruins

sfal'darsi *vr* flake off

sfa'mar|e *vt* feed. **~si** *vr* satisfy one's hunger, eat one's fill

'**sfar|zo** *nm* pomp. **~'zoso** *a* sumptuous

sfa'sato *a fam* confused; (*motore*) which needs tuning

sfasci'a|re *vt* unbandage; (*fracassare*) smash. **~rsi** *vr* fall to pieces. **~to** *a* beat-up

sfa'tare *vt* explode

sfati'cato *a* lazy

sfavil'la|nte *a* sparkling. **~re** *vi* sparkle

sfavo'revole *a* unfavourable

sfavo'rire *vt* disadvantage, put at a disadvantage

'**sfer|a** *nf* sphere. **~ico** *a* spherical

sfer'rare *vt* unshoe (*cavallo*); (*scagliare*) land

sfer'zare *vt* whip

sfian'carsi *vr* wear oneself out

sfi'bra|re *vt* exhaust. **~to** *a* exhausted

'**sfida** *nf* challenge. **sfi'dare** *vt* challenge

sfi'duci|a *nf* mistrust. **~'ato** *a* discouraged

'**sfiga** *nf vulg* bloody bad luck

sfigu'rare *vt* disfigure ● *vi* (*far cattiva figura*) look out of place

sfilacci'ar|e *vt*, **~si** *vr* fray

sfi'la|re *vt* unthread; (*togliere di dosso*) take off ● *vi* (*truppe:*) march past; (*in parata*) parade. **~rsi** *vr* come unthreaded; (*collant:*) ladder; take off (*pantaloni*). **~ta** *nf* parade; (*sfilza*) series. **~ta di moda** fashion show

sfilza *nf* (*di errori, domande*) string

'**sfinge** *nf* sphinx

sfi'nito *a* worn out

sfio'rare *vt* skim; touch on (*argomento*)

sfio'rire *vi* wither; (*bellezza:*) fade

sfitto *a* vacant

'**sfizio** *nm* whim, fancy; **togliersi uno ~** satisfy a whim

sfo'cato *a* out of focus

sfoci'are *vi* ~ **in** flow into

sfode'ra|re *vt* draw (*pistola, spada*). **~to** *a* unlined

sfo'gar|e *vt* vent. **~si** *vr* give vent to one's feelings

sfoggi'are *vt/i* show off. '**sfoggio** *nm* show, display; **fare sfoggio di** show off

'**sfoglia** *nf* sheet of pastry; **pasta ~** puff pastry

sfogli'are *vt* leaf through

'**sfogo** *nm* outlet; *fig* outburst; *Med* rash; **dare ~ a** give vent to

sfolgo'ra|nte *a* blazing. **~re** *vi* blaze

sfol'lare *vt* clear ● *vi* *Mil* be evacuated

sfol'tire *vt* thin [out]

sfon'dare *vt* break down ● *vi* (*aver successo*) make a name for oneself

'**sfondo** *nm* background

sfor'ma|re *vt* pull out of shape (*tasche*). **~rsi** *vr* lose its shape; (*persona:*) lose one's figure. **~to** *nm* Culin flan

sfor'nito *a* ~ **di** ⟨*negozio*⟩ out of

sfor'tuna *nf* bad luck. ~**ta'mente** *adv* unfortunately. ~**to'nato** *a* unlucky

sfor'zare *vt* force. ~**si** *vr* try hard. **'sforzo** *nm* effort; ⟨*tensione*⟩ strain

'sfottere *vt sl* tease

sfracel'larsi *vr* smash

sfrat'tare *vt* evict. **'sfratto** *nm* eviction

sfrecci'are *vi* flash past

sfregi'a|re *vt* slash. ~**to** *a* scarred **'sfregio** *nm* slash

sfre'na|rsi *vr* run wild. ~**to** *a* wild

sfron'tato *a* shameless

sfrutta'mento *nm* exploitation. **sfrut'tare** *vt* exploit

sfug'gente *a* elusive; ⟨*mento*⟩ receding

sfug'gi|re *vi* escape; ~**re a** escape [from]; **mi sfugge** it escapes me; **mi è sfuggito di mano** I lost hold of it ● *vt* avoid. ~**ta** *nf* **di** ~**ta** in passing

sfu'ma|re *vi* ⟨*svanire*⟩ vanish; ⟨*colore:*⟩ shade off ● *vt* soften ⟨*colore*⟩. ~**'tura** *nf* shade

sfuri'ata *nf* outburst [of anger]

sga'bello *nm* stool

sgabuz'zino *nm* cupboard

sgam'bato *a* ⟨*costume da bagno*⟩ high-cut

sgambet'tare *vi* kick one's legs; ⟨*camminare*⟩ trot. **sgam'betto** *nm* **fare lo sgambetto a qcno** trip sb up

sganasci'arsi *vr* ~ **dalle risa** roar with laughter

sganci'a|re *vt* unhook; *Rail* uncouple; drop ⟨*bombe*⟩; *fam* cough up ⟨*denaro*⟩. ~**si** *vr* become unhooked; *fig* get away

sganghe'rato *a* ramshackle

sgar'bato *a* rude. **'sgarbo** *nm* discourtesy; **fare uno sgarbo a** be rude

sgargi'ante *a* garish

sgar'rare *vi* be wrong; ⟨*da regola*⟩ stray from the straight and narrow. **'sgarro** *nm* mistake, slip

sgattaio'lare *vi* sneak away; ~ **via** decamp

sghignaz'zare *vi* laugh scornfully, sneer

sgob'b|are *vi* slog; ⟨*fam: studente:*⟩ swot. ~**one, -a** *nmf* slogger; ⟨*fam: studente*⟩ swot

sgoccio'lare *vi* drip

sgo'larsi *vr* shout oneself hoarse

sgomb'e|rare *vt* clear [out]. **'sgombro** *a* clear ● *nm* ⟨*trasloco*⟩ removal; ⟨*pesce*⟩ mackerel

sgomen'tar|e *vt* dismay. ~**si** *vr* be dismayed. **sgo'mento** *nm* dismay

sgomi'nare *vt* defeat

sgom'mata *nf* screech of tyres

sgonfi'ar|e *vt* deflate. ~**si** *vr* go down. **'sgonfio** *a* flat

'sgorbio *nm* scrawl; ⟨*fig: vista sgradevole*⟩ sight

sgor'gare *vi* gush [out] ● *vt* flush out, unblock ⟨*lavandino*⟩

sgoz'zare *vt* ~ **qcno** cut sb's throat

sgra'd|evole *a* disagreeable. ~**ito** *a* unwelcome

sgrammati'cato *a* ungrammatical

sgra'nare *vt* shell ⟨*piselli*⟩; open wide ⟨*occhi*⟩

sgran'chir|e *vt*, ~**si** *vr* stretch

sgranocchi'are *vt* munch

sgras'sare *vt* remove the grease from

sgrazi'ato *a* ungainly

sgreto'lar|e *vt*, ~**si** *vr* crumble

sgri'da|re *vt* scold. ~**ta** *nf* scolding

sgros'sare *vt* rough-hew ⟨*marmo*⟩; *fig* polish

sguai'ato *a* coarse

sgual'cire *vt* crumple

sgual'drina *nf* slut

sgu'ardo *nm* look; ⟨*breve*⟩ glance

'sguattero, -a *nmf* skivvy

sguaz'zare *vi* splash; ⟨*nel fango*⟩ wallow

sguinzagli'are *vt* unleash

sgusci'are *vt* shell ● *vi* ⟨*sfuggire*⟩ slip away; ~ **fuori** slip out

shake'rare *vt* shake

si *pers pron* ⟨*riflessivo*⟩ oneself; ⟨*lui*⟩

himself; (*lei*) herself; (*esso, essa*) itself; (*loro*) themselves; (*reciproco*) each other; (*tra più di due*) one another; (*impersonale*) you, one; **lavarsi** wash [oneself]; **si è lavata** she washed [herself]; **lavarsi le mani** wash one's hands; **si è lavata le mani** she washed her hands; **si è mangiato un pollo intero** he ate an entire chicken by himself; **incontrarsi** meet each other; **la gente si aiuta a vicenda** people help one another; **non si sa mai** you never know, one never knows *fml*; **queste cose si dimenticano facilmente** these things are easily forgotten ● *nm* (*chiave, nota*) B

si *adv* yes

'sia[1] *vedi* **essere**

'sia[2] *conj* **~...~...** (*entrambi*) both...and...; (*o l'uno o l'altro*) either...or...**~ che venga, ~ che non venga** whether he comes or not; **scegli ~ questo ~ quello** choose either this one or that one; **voglio ~ questo che quello** I want both this one and that one

sia'mese *a* Siamese

sibi'lare *vi* hiss. **'sibilo** *nm* hiss

sic'cario *nm* hired killer

sicché *conj* (*perciò*) so [that]; (*allora*) then

siccità *nf* drought

sic'come *conj* as

Si'cilia *nf* Sicily. **s~'ano, -a** *a & nmf* Sicilian

si'cura *nf* safety catch; (*di portiera*) child-proof lock. **~'mente** *adv* definitely

sicu'rezza *nf* (*certezza*) certainty; (*salvezza*) safety; **uscita di ~** emergency exit

si'curo *a* (*non pericoloso*) safe; (*certo*) sure; (*saldo*) steady; *Comm* sound ● *adv* certainly ● *nm* safety; **al ~** safe; **andare sul ~** play [it] safe; **di ~** definitely; **di ~, sarà arrivato** he must have arrived

siderur'gia *nf* iron and steel in-

dustry. **side'rurgico** *a* iron and steel *attrib*

'sidro *nm* cider

si'epe *nf* hedge

si'ero *nm* serum

sieroposi'tivo *a* HIV positive

si'esta *nf* afternoon nap, siesta

si'fone *nm* siphon

Sig. *abbr* (**signore**) Mr

Sig.a *abbr* (**signora**) Mrs, Ms

siga'retta *nf* cigarette; **pantaloni a ~** drainpipes

'sigaro *nm* cigar

Sigg. *abbr* (**signori**) Messrs

sigil'lare *vt* seal. **si'gillo** *nm* seal

'sigla *nf* initials *pl*. **~ musicale** signature tune. **si'glare** *vt* initial

Sig.na *abbr* (**signorina**) Miss, Ms

signifi'ca|re *vt* mean. **~'tivo** *a* significant. **~to** *nm* meaning

si'gnora *nf* lady; (*davanti a nome proprio*) Mrs; (*non sposata*) Miss; (*in lettere ufficiali*) Dear Madam; **il signor Venè e ~** Mr and Mrs Venè

si'gnore *nm* gentleman; *Relig* lord; (*davanti a nome proprio*) Mr; (*in lettere ufficiali*) Dear Sir. **signo'rile** *a* gentlemanly; (*di lusso*) luxury

signo'rina *nf* young lady; (*seguito da nome proprio*) Miss

silenzia'tore *nm* silencer

si'lenzi|o *nm* silence. **~'oso** *a* silent

silhou'ette *nf* silhouette, outline

si'licio *nm* **piastrina di ~** silicon chip

sili'cone *nm* silicone

'sillaba *nf* syllable

silu'rare *vt* torpedo. **si'luro** *nm* torpedo

simboleggi'are *vt* symbolize

sim'bolico *a* symbolic[al]

'simbolo *nm* symbol

similarità *nf* *inv* similarity

'simil|e *a* similar; (*tale*) such; **~e a** like ● *nm* (*il prossimo*) fellow man. **~'mente** *adv* similarly. **~'pelle** *nf* Leatherette®

simme'tria *nf* symmetry. **sim'metrico** *a* symmetric[al]

simpa'ti|a nf liking; (compenetrazione) sympathy; **prendere qcno in ~a** take a liking to sb. **sim'patico** a nice. **~iz'zante** nmf well-wisher. **~iz'zare** vt **~izzare con** take a liking to; **~izzare per qcsa/qcno** lean towards sth/sb

sim'posio nm symposium

simu'la|re vt simulate; feign (amicizia, interesse). **~zi'one** nf simulation

simul'tane|a nf **in ~a** simultaneously. **~o** a simultaneous

sina'goga nf synagogue

sincerità nf sincerity. **sin'cero** a sincere

'sincope nf syncopation; Med fainting fit

sincron'ia nf synchronization; **in ~** with synchronized timing

sincroniz'za|re vt synchronize. **~zi'one** nf synchronization

sinda'ca|le a [trade] union, [labor] union Am. **~lista** nmf trade unionist, labor union member Am. **~re** nt inspect. **~to** nm [trade] union, [labor] union Am; (associazione) syndicate

'sindaco nm mayor

'sindrome nf syndrome

sinfo'nia nf symphony. **sin'fonico** a symphonic

singhioz'zare vi (di pianto) sob. **~'ozzo** nm hiccup; (di pianto) sob; **avere il ~ozzo** have the hiccups

singo'lar|e a singular ● nm singular. **~'mente** adv individually; (stranamente) peculiarly

'singolo a single ● nm individual; Tennis singles pl

si'nistra nf left; **a ~** on the left; **girare a ~** turn to the left; **con la guida a ~** (auto) with left-hand drive

sini'strato a injured

si'nistr|o, -a a left[-hand]; (avverso) sinister ● nm accident ● nf left [hand]; Pol left [wing]

'sino prep = **fino'**

si'nonimo a synonymous ● nm synonym

sin'ta|ssi nf syntax. **~ttico** a syntactic[al]

'sintesi nf synthesis; (riassunto) summary

sin'teti|co a synthetic; (conciso) summary. **~z'zare** vt summarize

sintetizza'tore nm synthesizer

sinto'matico a symptomatic. **'sintomo** nm symptom

sinto'nia nf tuning; **in ~** on the same wavelength

sinu'oso a (strada) winding

sinu'site nf sinusitis

si'pario nm curtain

si'rena nf siren

'Siri|a nf Syria. **s~'ano, -a** a & nmf Syrian

si'ringa nf syringe

'sismico a seismic

si'stema nm system. **S~a Monetario Europeo** European Monetary System. **~a operativo** Comput operating system

siste'ma|re vt (mettere) put; tidy up (casa, camera); (risolvere) sort out; (procurare lavoro a) fix up with a job; (trovare alloggio a) find accommodation for; (sposare) marry off; (fam: punire) sort out. **~rsi** vr settle down; (trovare un lavoro) find a job; (trovare alloggio) find accommodation; (sposarsi) marry. **~tico** a systematic. **~zi'one** nf arrangement; (di questione) settlement; (lavoro) job; (alloggio) accommodation; (matrimonio) marriage

'sito nm site. **~ web** Comput web site

situ'are vt place

situazi'one nf situation

ski-'lift nm ski tow

slacci'are vt unfasten

slanci'a|rsi vr hurl oneself. **~to** a slender; (slancio) nm impetus; (impulso) impulse

sla'vato a (carnagione, capelli) fair

'slavo a Slav[onic]

sle'al|e a disloyal. **~tà** nf disloyalty

sle'gare vt untie

'slitta nf sledge, sleigh. **~'mento**

nm (*di macchina*) skid; (*fig: di riunione*) postponement

slit'ta|re *vt* (*riunione:*) be put off. **~ta** *nf* skid

slit'tino *nm* toboggan

'slogan *nm inv* slogan

slo'ga|re *vt* dislocate. **~rsi** *vr* **~rsi una caviglia** sprain one's ankle. **~'tura** *nf* dislocation

sloggi'are *vt* dislodge ● *vi* move out

Slo'vacchia *nf* Slovakia

Slo'venia *nf* Slovenia

smacchi'a|re *vt* clean. **~'tore** *nm* stain remover

'smacco *nm* humiliating defeat

smagli'ante *a* dazzling

smagli'a|rsi *vr* (*calza:*) ladder *Br*, run. **~'tura** *nf* ladder *Br*, run

smalizi'ato *a* cunning

smal'ta|re *vt* enamel; glaze (*ceramica*); varnish (*unghie*). **~to** *a* enamelled

smalti'mento *nm* disposal; (*di merce*) selling off. **~ rifiuti** waste disposal; (*di grassi*) burning off

smal'tire *vt* burn off; (*merce*) sell off; *fig* get through (*corrispondenza*); **~ la sbornia** sober up

'smalto *nm* enamel; (*di ceramica*) glaze; (*per le unghie*) nail varnish

'smania *nf* fidgets *pl*; (*desiderio*) longing. **~'are** *vi* have the fidgets; **~are per** long for. **~'oso** *a* restless

smantella'mento *nm* dismantling. **~'lare** *vt* dismantle

smarri'mento *nm* loss; (*psicologico*) bewilderment

smar'ri|re *vt* lose; (*temporaneamente*) mislay. **~rsi** *vr* get lost; (*turbarsi*) be bewildered

smasche'rare *vt* unmask. **~si** *vr* (*tradirsi*) give oneself away

SME *nm abbr* (**Sistema Monetario Europeo**) EMS

smemo'rato, -a *a* forgetful ● *nmf* scatterbrain

smen'ti|re *vt* deny. **~ta** *nf* denial

sme'raldo *nm* & *a* emerald

smerci'are *vt* sell off

smerigli'ato *a* emery; **vetro ~** frosted glass. **sme'riglio** *nm* emery

'smesso *pp di* **smettere** ● *a* (*abiti*) cast-off

smett|ere *vt* stop; stop wearing (*abiti*); **~ila!** stop it!

smidol'lato *a* spineless

sminu'ir|e *vt* diminish. **~si** *vr fig* belittle oneself

sminuz'zare *vt* crumble; (*fig: analizzare*) analyse in detail

smista'mento *nm* clearing; (*postale*) sorting. **smi'stare** *vt* sort; *Mil* post

smisu'rato *a* boundless; (*esorbitante*) excessive

smobili'ta|re *vt* demobilize. **~zi'one** *nf* demobilization

smo'dato *a* immoderate

smog *nm* smog

'smoking *nm inv* dinner jacket, tuxedo *Am*

smon'tabile *a* jointed

smon'tar|e *vt* take to pieces; (*scoraggiare*) dishearten ● *vi* (*da veicolo*) get off; (*da cavallo*) dismount; (*dal servizio*) go off duty. **~si** *vr* lose heart

'smorfi|a *nf* grimace; (*moina*) simper; **fare ~e** make faces. **~'oso** *a* affected

'smorto *a* pale; (*colore*) dull

smor'zare *vt* dim (*luce*); tone down (*colori*); deaden (*suoni*); quench (*sete*)

'smosso *pp di* **smuovere**

smotta'mento *nm* landslide

'smunto *a* emaciated

smu'over|e *vt* shift; (*commuovere*) move. **~si** *vr* move; (*commuoversi*) be moved

smus'sar|e *vt* round off; (*fig: attenuare*) tone down. **~si** *vr* go blunt

snatu'rato *a* inhuman

snel'lir|e *vt* slim down. **~si** *vr* slim [down]. **'snello** *a* slim

sner'vante *a* enervating

sner'va|re *vt* enervate. **~rsi** *vr* get exhausted

sni'dare vt drive out

snif'fare vt snort

snob'bare vt snub. **sno'bismo** nm snobbery

snoccio'lare vt stone; fig blurt out

sno'da|re vt untie; (sciogliere) loosen. **~rsi** vr come untied; (strada:) wind. **~to** a (persona) double-jointed; (dita) flexible

so'ave a gentle

sobbal'zare vi jerk; (trasalire) start. **sob'balzo** nm jerk; (trasalimento) start

sobbar'carsi vr **~ a** undertake

sob'borgo nm suburb

sobil'la|re vt stir up

'sobrio a sober

soc'chiu|dere vt half-close. **~so** pp di **socchiudere** ● a (occhi) half-closed; (porta) ajar

soc'combere vi succumb

soc'cor|rere vt assist. **~so** pp di **soccorrere** ● nm assistance; **soccorsi** pl rescuers; (dopo disastro) relief workers. **~so stradale** breakdown service

socialdemo'cra|tico, -a a Social Democratic ● nmf Social Democrat. **~'zia** nf Social Democracy

soci'ale a social

socia'li|smo nm Socialism. **~sta** a & nmf Socialist. **~z'zare** vt socialize

società nf inv society; Comm company. **~ per azioni** plc. **~ a responsabilità limitata** limited liability company

soci'evole a sociable

'socio, -a nmf member; Comm partner

sociolo'gia nf sociology. **socio'logico** a a sociological

'soda nf soda

soddisfa'cente a satisfactory

soddi'sfa|re vt/i satisfy; meet (richiesta); make amends for (offesa). **~tto** pp di **soddisfare** ● a satisfied. **~zi'one** nf satisfaction

'sodo a hard; fig firm; (uovo) hard-boiled ● adv hard; **dormire ~**
sleep soundly ● nm **venire al ~** get to the point

sofà nm inv sofa

soffe'ren|te a (malato) ill. **~za** nf suffering

soffer'marsi vr pause; **~ su** dwell on

sof'ferto pp di **soffrire**

soffi'a|re vt blow; reveal (segreto); (rubare) pinch fam ● vi blow. **~ta** nf fig sl tip-off

'soffice a soft

'soffio nm puff; Med murmur

sof'fitt|a nf attic. **~o** nm ceiling

soffo|ca'mento nm suffocation

soffo'ca|nte a suffocating. **~re** vt/i suffocate; (con cibo) choke; fig stifle

sof'friggere vt fry lightly

sof'frire vt/i suffer; (sopportare) bear; **~ di** suffer from

sof'fritto pp di **soffriggere**

sof'fuso a (luce) soft

sofisti'ca|re vt (adulterare) adulterate ● vi (sottilizzare) quibble. **~to** a sophisticated

sogget|tiva'mente adv subjectively. **~'tivo** a subjective

sog'getto nm subject ● a subject; **essere ~ a** be subject to

sogge|zi'one nf subjection; (rispetto) awe

sogghi'gnare vi sneer. **sog'ghigno** nm sneer

soggio'gare vt subdue

soggior'nare vi stay. **soggi'orno** nm stay; (stanza) living room

soggi'ungere vt add

'soglia nf threshold

'sogliola nf sole

so'gna|re vt/i dream; **~re a occhi aperti** daydream. **~'tore**, **~'trice** nmf dreamer. **'sogno** nm dream; **fare un sogno** have a dream; **neanche per sogno!** not at all!

'soia nf soya

sol nm Mus (chiave, nota) G

so'laio nm attic

sola'mente adv only

so'lar|e a (energia, raggi) solar;

(crema) sun attrib. **~ium** nm inv solarium

sol'care vt plough. **'solco** nm furrow; (di ruota) track; (di nave) wake; (di disco) groove

sol'dato nm soldier

'soldo nm **non ha un ~** he hasn't got a penny to his name; **senza un ~** penniless; **soldi** pl (denaro) money sg

'sole nm sun; (luce del sole) sun[light]; **al ~** in the sun; **prendere il ~** sunbathe

soleggi'ato a sunny

so'lenn|e a solemn. **~ità** nf solemnity

so'lere vi be in the habit of; **come si suol dire** as they say

sol'fato nm sulphate

soli'da|le a in agreement. **~rietà** nf solidarity

solidifi'car|e vt/i, **~si** vr solidify

solidità nf solidity; (di colori) fastness. **'solido** a solid; (robusto) sturdy; (colore) fast ● nm solid

soli'loquio nm soliloquy

so'lista a solo ● nmf soloist

solita'mente adv usually

soli'tario a solitary; (isolato) lonely ● nm (brillante) solitaire; (gioco di carte) patience, solitaire

'solito a usual; **essere ~ fare qcsa** be in the habit of doing sth ● nm usual; **di ~** usually

soli'tudine nf solitude

solleci'ta|re vt speed up; urge (persona). **~zi'one** nf (richiesta) request; (preghiera) entreaty

sol'leci|to a prompt ● nm reminder. **~'tudine** nf promptness; (interessamento) concern

solle'one nm noonday sun; (periodo) dog days of summer

solleti'care vt tickle. **sol'letico** nm tickling; **fare il solletico a qcno** tickle sb; **soffrire il solletico** be ticklish

solleva'mento nm **~ pesi** weight-lifting

solle'var|e vt lift; (elevare) raise;

(confortare) comfort. **~si** vr rise; (riaversi) recover

solli'evo nm relief

'solo, -a a alone; (isolato) lonely; (unico) only; Mus solo; **da ~** by myself/yourself/himself etc ● nmf **il ~, la sola** the only one ● nm Mus solo ● adv only

sol'stizio nm solstice

sol'tanto adv only

so'lubile a soluble; (caffè) instant

soluzi'one nf solution; Comm payment

sol'vente a & nm solvent; **~ per unghie** nail polish remover

'soma nf **bestia da ~** beast of burden

so'maro nm ass; Sch dunce

so'matico a somatic

somigli'an|te a similar. **~za** nf resemblance

somigli'ar|e vi **~e a** resemble. **~si** vr be alike

'somma nf sum; Math addition

som'mare vt add; (totalizzare) add up

som'mario a & nm summary

som'mato a **tutto ~** all things considered

sommeli'er nm inv wine waiter

som'mer|gere vt submerge. **~'gibile** nm submarine. **~so** pp di **sommergere**

som'messo a soft

sommini'stra|re vt administer. **~zi'one** nf administration

sommità nf inv summit

'sommo a highest; fig supreme ● nm summit

som'mossa nf rising

sommozza'tore nm frogman

so'naglio nm bell

so'nata nf sonata; fig fam beating

'sonda nf Mech drill; (spaziale, Med) probe. **son'daggio** nm drilling; (spaziale, Med) probe; (indagine) survey. **sondaggio d'opinioni** opinion poll. **son'dare** vt sound; (investigare) probe

so'netto nm sonnet

sonnambu'lismo nm sleepwalking. **son'nambulo, -a** nmf sleep-walker

sonnecchi'are vi doze

son'nifero nm sleeping-pill

'sonno nm sleep; **aver ~** be sleepy. **~'lenza** nf sleepiness

so'noro a resonant; (rumoroso) loud; (onde, scheda) sound attrib

sontu'oso a sumptuous

sopo'rifero a soporific

sop'palco nm platform

soppe'rire vi **~ a qcsa** provide for sth

soppe'sare vt weigh up (situazione)

soppi'atto: di ~ adv furtively

soppor'tare vt support; (tollerare) stand; bear (dolore)

soppressi'one nf removal; (di legge) abolition; (di diritti, pubblicazione) suppression; (annullamento) cancellation. **sop'presso** pp di **sopprimere**

sop'primere vt get rid of; abolish (legge); suppress (diritti, pubblicazione); (annullare) cancel

'sopra adv on top; (più in alto) higher [up]; (al piano superiore) upstairs; (in testo) above; **mettilo lì ~** put it up there; **di ~** upstairs; **dormirci ~** fig sleep on it; **pensarci ~** think about it; **vedi ~** see above ● prep **~ [a]** on; (senza contatto, oltre) over; (riguardo a) about; **è ~ al tavolo, è ~ il tavolo** it's on the table; **il quadro è appeso ~ al camino** the picture is hanging over the fireplace; **il ponte passa ~ all'autostrada** the bridge crosses over the motorway; **è caduto ~ il tetto** it fell on the roof; **l'uno ~ l'altro** one on top of the other; (senza contatto) one above the other; **abita ~ di me** he lives upstairs from me; **i bambini ~ i dieci anni** children over ten; **20° ~ lo zero** 20° above zero; **~ il livello del mare** above sea level; **rifletti ~ quello che è successo** think about what happened; **non ha nessuno ~ di sé** he has no-

body above him; **al di ~ di** over ● nm **il [di] ~** the top

so'prabito nm overcoat

soprac'ciglio nm (pl nf **sopracciglia**) eyebrow

sopracco'perta nf (di letto) bedspread; (di libro) [dust-]jacket. **~'tina** nf book jacket

soprad'detto a above-mentioned

sopraele'vata nf elevated railway

sopraf'fare vt overwhelm. **~'tto** pp di **sopraffare**. **~zi'one** nf abuse of power

sopraf'fino a excellent; (gusto, udito) highly refined

sopraggi'ungere vi (persona:) turn up; (accadere) happen

soprallu'ogo nm inspection

sopram'mobile nm ornament

soprannatu'rale a & nm supernatural

sopran'nome nm nickname. **~i'nare** vt nickname

so'prano nmf soprano

soprappensi'ero adv lost in thought

sopras'salto nm **di ~** with a start

soprasse'dere vi **~ a** postpone

soprat'tutto adv above all

sopravvalu'tare vt overvalue

sopravve'nire vi turn up; (accadere) happen. **~'vento** nm fig upper hand

sopravvis'suto pp di **sopravvivere**. **~'venza** nf survival. **sopravvivere** vi survive; **sopravvivere a** outlive (persona)

soprinten'den|te nmf supervisor; (di museo ecc) keeper. **~za** nf supervision; (ente) board

so'pruso nm abuse of power

soq'quadro nm **mettere a ~** turn upside down

sor'betto nm sorbet

sor'bire vt sip; fig put up with

'sordido a sordid; (avaro) stingy

sor'dina nf mute; **in ~** fig on the quiet

sordità nf deafness. **'sordo, -a** a deaf; (rumore, dolore) dull ● nmf deaf person. **sordo'muto, -a** *a* deaf-and-dumb ● *nmf* deaf mute

so'rel|la *nf* sister. **~lastra** *nf* stepsister

sor'gente *nf* spring; *(fonte)* source

'sorgere *vi* rise; *fig* arise

sormon'tare *vt* surmount

sorni'one *a* sly

sorpas'sa|re *vt* surpass; *(eccedere)* exceed; overtake, pass *Am (veicolo)*. **~to** *a* old-fashioned. **sor'passo** *nm* overtaking, passing *Am*

sorpren'dente *a* surprising; *(straordinario)* remarkable

sor'prendere *vt* surprise; *(cogliere in flagrante)* catch

sor'pres|a *nf* surprise; **di ~** by surprise. **~o** *pp di* **sorprendere**

sor're|ggere *vt* support; *(tenere)* hold up. **~ggersi** *vr* support oneself. **~tto** *pp di* **sorreggere**

sorri'dente *a* smiling

sor'ri|dere *vi* smile. **~so** *pp di* **sorridere** ● *nm* smile

sorseggi'are *vt* sip. **'sorso** *nm* sip; *(piccola quantità)* drop

'sorta *nf* sort; **di ~** whatever; **ogni ~ di** all sorts of

'sorte *nf* fate; *(caso imprevisto)* chance; **tirare a ~** draw lots. **~ggi'are** *vt* draw lots for. **sor'teggio** *nm* draw

sorti'legio *nm* witchcraft

sor'ti|re *vi* come out. **~ta** *nf Mil* sortie; *(battuta)* witticism

'sorto *pp di* **sorgere**

sorvegli'an|te *nmf* keeper; *(controllore)* overseer. **~za** *nf* watch; *Mil ecc* surveillance

sorvegli'are *vt* watch over; *(controllare)* oversee; *(polizia)* watch, keep under surveillance

sorvo'lare *vt* fly over; *fig* skip

'sosia *nm inv* double

so'spen|dere *vt* hang; *(interrompere)* stop; *(privare di una carica)* suspend. **~si'one** *nf* suspension

so'speso, -a *pp di* **sospendere** ● *a (impiegato, alunno)* suspended; **~ a** hanging from; **~ a un filo** *fig*

hanging by a thread ● *nm* **in ~** pending; *(emozionato)* in suspense

sospet'tare *vt* suspect. **so'spetto** *a* suspicious; **persona sospetta** suspicious person ● *nm* suspicion; *(persona)* suspect. **~'toso** *a* suspicious

so'spin|gere *vt* drive. **~to** *pp di* **sospingere**

sospi'rare *vi* sigh ● *vt* long for. **so'spiro** *nm* sigh

'sosta *nf* stop; *(pausa)* pause; **senza ~** non-stop; **"divieto di ~"** "no parking"

sostan'tivo *nm* noun

so'stan|za *nf* substance; **~e** *pl (patrimonio)* property *sg*; **in ~a** to sum up. **~i'oso** *a* substantial; *(cibo)* nourishing

so'stare *vi* stop; *(fare una pausa)* pause

so'stegno *nm* support

soste'ne|re *vt* support; *(sopportare)* bear; *(resistere)* withstand; *(affermare)* maintain; *(nutrire)* sustain; *sit (esame)*; take; **~re le spese** meet the costs. **~si** *vr* support oneself

sosteni'tore, -'trice *nmf* supporter

sostenta'mento *nm* maintenance

soste'nuto *a (stile)* formal; *(prezzi, velocità)* high

sostitu'ir|e *vt* substitute (**a** for), replace (**con** with). **~si** *vr* **~si a** replace

sosti'tu|to, -a *nmf* replacement, stand-in ● *nm (surrogato)* substitute. **~zi'one** *nf* substitution

sotta'ceto *a* pickled; **sottaceti** *pl* pickles

sot'tana *nf* petticoat; *(di prete)* cassock

sotter'fugio *nm* subterfuge; **di ~** secretly

sotter'raneo *a* underground ● *nm* cellar

sotter'rare *vt* bury

sottigli'ezza *nf* slimness; *fig* subtlety

sot'til|e *a* thin; *(udito, odorato)*

keen; ⟨*osservazione, distinzione*⟩ subtle. **~iz'zare** *vi* split hairs

sottin'te|ndere *vt* imply. **~so** *pp di* **sottintendere** ● *nm* allusion; **senza ~si** openly ● *a* implied

'sotto *adv* below; ⟨*più in basso*⟩ lower [down]; ⟨*al di sotto*⟩ underneath; ⟨*al piano di sotto*⟩ downstairs; **è li ~** it's underneath; **~ ~** deep down; ⟨*di nascosto*⟩ on the quiet; **di ~** downstairs; **mettersi ~ fig** get down to it; **mettere ~** ⟨*fam: investire*⟩ knock down; **fatti ~!** *fam* get stuck in! ● *prep* ~ **[a]** under; ⟨*al di sotto di*⟩ under[neath]; **abita ~ di me** he lives downstairs from me; **i bambini ~ i dieci anni** children under ten; **20° ~ zero** 20° below zero; **~ il livello del mare** below sea level; **~ la pioggia** in the rain; **~ Elisabetta I** under Elizabeth I; **~ calmante** under sedation; **~ condizione che...** on condition that...; **~ giuramento** under oath; **~ sorveglianza** under surveillance; **~ Natale/gli esami** around Christmas/exam time; **al di ~ di** under; **andare ~ i 50 all'ora** do less than 50km an hour ● *nm* **il [di] ~** the bottom

sotto'banco *adv* under the counter

sottobicchi'ere *nm* coaster
sotto'bosco *nm* undergrowth
sotto'braccio *adv* arm in arm
sotto'fondo *nm* background
sottoline'are *vt* underline; *fig* stress
sott'olio *adv* in oil
sotto'mano *adv* within reach
sottoma'rino *a & nm* submarine
sotto'messo *pp di* **sottomettere** ● *a* ⟨*remissivo*⟩ submissive
sotto'mette|re *vt* submit; subdue ⟨*popolo*⟩. **~si** *vr* submit. **sottomissi'one** *nf* submission
sottopa'gare *vt* underpay
sottopas'saggio *nm* underpass; ⟨*pedonale*⟩ subway
sotto'por|re *vt* submit; ⟨*costringe-*

re⟩ subject. **~si** *vr* submit oneself; **~si a** undergo. **sotto'posto** *pp di* **sottoporre**
sotto'scala *nm* cupboard under the stairs
sotto'scritto *pp di* **sottoscrivere** ● *nm* undersigned
sotto'scri|vere *vt* sign; ⟨*approva-re*⟩ sanction, subscribe to. **~zi'one** *nf* ⟨*petizione*⟩ petition; ⟨*approvazione*⟩ sanction; ⟨*raccolta di dena-ro*⟩ appeal
sottosegre'tario *nm* undersecretary
sotto'sopra *adv* upside down
sotto'stante *a* **la strada ~** the road below
sottosu'olo *nm* subsoil
sottosvi'lup'pato *a* underdeveloped. **~'luppo** *nm* underdevelopment
sotto'terra *adv* underground
sotto'titolo *nm* subtitle
sottovalu'tare *vt* underestimate
sotto'veste *nf* slip
sotto'voce *adv* in a low voice
sottovu'oto *a* vacuum-packed
sot'tra|rre *vt* remove; embezzle ⟨*fondi*⟩; *Math* subtract. **~rsi** *vr* **~rsi a** escape from; avoid ⟨*re-sponsabilità*⟩. **~tto** *pp di* **sottrarre**. **~zi'one** *nf* removal; ⟨*di fondi*⟩ embezzlement; *Math* subtraction
sottuffici'ale *nm* non-commissioned officer; *Naut* petty officer
sou'brette *nf inv* showgirl
so'vietico, -a *a & nmf* Soviet
sovraccari'care *vt* overload. **sovrac'carico** *a* overloaded ⟨**di** with) ● *nm* overload
sovraffati'carsi *vr* overexert oneself
sovrannatu'rale *a & nm* = **soprannaturale**
so'vrano, -a *a* sovereign; *fig* supreme ● *nmf* sovereign
sovrap'por|re *vt* superimpose. **~si** *vr* overlap. **sovrapposizi'one** *nf* superimposition
sovra'stare *vt* dominate; ⟨*fig: perico-lo:*⟩ hang over

sovrinten'den|te, **~za = soprin-
tendente, soprintendenza**
sovru'mano *a* superhuman
sovvenzi'one *nf* subsidy
sover'sivo *a* subversive
'sozzo *a* filthy
S.p.A. *abbr* (**società per azioni**)
plc
spac'ca|re *vt* split; chop (*legna*).
~rsi *vr* split. **~'tura** *nf* split
spacci'a|re *vt* deal in, push
(*droga*); **~re qcsa per qcsa** pass
sth off as sth; **essere ~to** be done
for, be a goner. **~rsi** *vr* **~rsi per**
pass oneself off as. **~'tore, ~'trice**
nmf (*di droga*) pusher; (*di denaro
falso*) distributor of forged bank
notes. **'spaccio** *nm* (*di droga*)
dealer, pusher; (*negozio*) shop
'spacco *nm* split
spac'cone, -a *nmf* boaster
'spada *nf* sword. **~c'cino** *nm*
swordsman
spadroneggi'are *vi* act the boss
spae'sato *a* disorientated
spa'ghetti *nmpl* spaghetti *sg*
spa'ghetto *nm* (*fam:* spavento)
fright
'Spagna *nf* Spain
spa'gnolo, -a *a* Spanish ● *nmf*
Spaniard ● *nm* (*lingua*) Spanish
'spago *nm* string; **dare ~ a** qcno
encourage sb
spai'ato *a* odd
spalan'ca|re *vt*, **~rsi** *vr* open
wide. **~to** *a* wide open
spa'lare *vt* shovel
spall|a *nf* shoulder; (*di comico*)
straight man; **~e** *pl* (*schiena*)
back; **alle ~e di** qcno (*ridere*) behind sb's back. **~eggi'are** *vt* back
up
spal'letta *nf* parapet
spalli'era *nf* back; (*di letto*) head-
board; (*ginnastica*) wall bars *pl*
spal'lina *nf* strap; (*imbottitura*)
shoulder pad
spal'mare *vt* spread
spande|re *vt* spread; (*versare*)
spill. **~rsi** *vr* spread
spappo'lare *vt* crush

spa'ra|re *vt/i* shoot; **~rle grosse**
talk big. **~ta** *nf fam* tall story.
~'toria *nf* shooting
sparecchi'are *vt* clear
spa'reggio *nm Comm* deficit;
Sport play-off
'sparg|ere *vt* scatter; (*diffondere*)
spread; shed (*lacrime, sangue*).
~ersi *vr* spread. **~i'mento** *nm*
scattering; (*di lacrime, sangue*)
shedding; **~imento di sangue**
bloodshed
spa'ri|re *vi* disappear; **~scil** get
lost!. **~zi'one** *nf* disappearance
spa'rlare *vi* **~ di** run down
'sparo *nm* shot
sparpagli'a|re *vt*, **~si** *vr* scatter
'sparso *pp di* spargere ● *a* scat-
tered; (*sciolto*) loose
spar'tire *vt* share out; (*separare*)
separate
sparti'traffico *nm inv* traffic island; (*di autostrada*) central reservation, median strip *Am*
spartizi'one *nf* division
spa'ruto *a* gaunt; (*gruppo*) small;
(*peli, capelli*) sparse
sparvi'ero *nm* sparrow-hawk
spasi'ma|nte *nm* admirer. **~re**
vi suffer agonies
'spasimo *nm* spasm
spa'smodico *a* spasmodic
spas'sar|si *vr* amuse oneself;
~sela have a good time
spassio'nato *a* (*osservatore*) dis-
passionate, impartial
'spasso *nm* fun; **essere uno ~** be
hilarious; **andare a un ~** go for a
walk. **spas'soso** *a* hilarious
'spatola *nf* spatula
spau'racchio *nm* scarecrow; *fig*
bugbear. **spau'rire** *vt* frighten
spa'valdo *a* defiant
spaventa'passeri *nm inv* scarecrow
spaven'tar|e *vt* frighten, scare.
~si *vr* be frightened, be scared.
spa'vento *nm* fright. **spaven'toso**
a frightening; (*fam: enorme*) incredible

spazi|ale a spatial; (cosmico) space attrib

spazi|are vt space out ● vi range

spazien|tirsi vr lose [one's] patience

'**spazi|o** nm space. ~'**oso** a spacious

spazzaca|mino nm chimney sweep

spaz'z|are vt sweep; ~**are via** sweep away; (fam: mangiare) devour. ~**a'tura** nf (immondizia) rubbish. ~**ino** nm road sweeper; (netturbino) dustman

'**spazzo|la** nf brush; (di tergicristallo) blade. ~'**lare** vt brush. ~'**lino** nm small brush. ~**lino da denti** toothbrush. ~'**lone** nm scrubbing brush

specchi'arsi vr look at oneself in a/the mirror; (riflettersi) be mirrored; ~ **in qcno** model oneself on sb

specchi'etto nm ~ **retrovisore** driving mirror

'**specchio** nm mirror

speci'a|le a special ● nm TV special [programme]. ~'**lista** nmf specialist. ~**lità** nf inv speciality, specialty Am

specializ'za|re vt, ~**rsi** vr specialize. ~**to** a (operaio) skilled

special'mente adv especially

'**specie** nf inv (scientifico) species; (tipo) kind; **fare** ~ **a** surprise

specifi'care vt specify. **spe'cifico** a specific

specu'lare¹ vi speculate; ~ **su** (indagare) speculate on; Fin speculate in

specu'lare² a mirror attrib

specula|'tore, -'trice nmf speculator. ~**zi'one** nf speculation

spe'di|re vt send. ~**to** pp di **spedire** ● a quick; (parlata) fluent. ~**zi'one** nf (di lettere ecc) dispatch; Comm consignment; (scientifica) expedition

'**spegnere** vt put out; turn off (gas, luce); switch off (motore);

slake (sete). ~**si** vr go out; (morire) pass away

spelacchi'ato a (tappeto) threadbare; (cane) mangy

spe'lar|e vt skin (coniglio). ~**si** vr (cane:) moult

speleolo'gia nf potholing, speleology

spel'lar|e vt skin; fig fleece. ~**si** vr peel off

spe'lonca nf cave; fig dingy hole

spendacci'one, -a nmf spendthrift

'**spendere** vt spend; ~ **fiato** waste one's breath

spen'nare vt pluck; fam fleece (cliente)

spennel'lare vt brush

spensie|ra'tezza nf lightheartedness. ~'**rato** a carefree

'**spento** pp di **spegnere** ● a off; (gas) out; (smorto) dull

spe'ranza nf hope; **pieno di** ~ hopeful; **senza** ~ hopeless

spe'rare vt hope for; (aspettarsi) expect ● vi ~ **in** trust in; **spero di sì** I hope so

'**sperder|si** vr get lost. ~'**duto** a lost; (isolato) secluded

sper'giuro, -a nmf perjurer ● nm perjury

spericolato a swashbuckling

sperimen'ta|le a experimental. ~**re** vt experiment with; test (resistenza, capacità, teoria). ~**zi'one** nf experimentation

'**sperma** nm sperm

spe'rone nm spur

sperpe'rare vt squander. '**sperpero** nm waste

'**spesa** nf expense; (acquisto) purchase; **andare a far** ~ go shopping; **fare la** ~ do the shopping; **fare le** ~**e di** pay for. ~**e** pl **bancarie** bank charges. ~**e a carico del destinatario** carriage forward. ~**e di spedizione** shipping costs. **spe'sato** a all-expenses-paid. ~**o** pp di **spendere**

'**spesso¹** a thick

'**spesso²** adv often

spes'sore nm thickness; (fig: consistenza) substance

spet'tabile a (Comm abbr **Spett.**) **S~ ditta Rossi** Messrs Rossi

spettaco'lare a spectacular. **spet'tacolo** nm spectacle; (rappresentazione) show. **~'loso** a spectacular

spet'tare vi ~ **a** be up to; (diritto:) be due to

spetta'tore, -'trice nmf spectator; **spettatori** pl (di cinema ecc) audience sg

spettego'lare vi gossip

spetti'nar|e vt ~**e** qcno ruffle sb's hair. **~si** vr ruffle one's hair

spet'trale a ghostly. **'spettro** nm ghost; Phys spectrum

'spezie nfpl spices

spez'zar|e vt, **~si** vr break

spezza'tino nm stew

spez'zato nm coordinated jacket and trousers

spezzet'tare vt break into small pieces

'spia nf spy; (della polizia) informer; (di porta) peep-hole; **fare la ~** sneak. **~ [luminosa** light. **~ dell'olio** oil [warning] light

spiacci'care vt squash

spia'ce|nte a sorry. **~vole** a unpleasant

spi'aggia nf beach

spia'nare vt level; (rendere liscio) smooth; roll out (pasta); raze to the ground (edificio)

spi'ano nm **a tutto ~** flat out

spian'tato a **~re un soldo** penniless

spi'are vt spy on; wait for (occasione ecc)

spiattel'lare vt blurt out; shove (oggetto)

spiaz'zare vt wrong-foot

spi'azzo nm (radura) clearing

spic'ca|re vt **~re un salto** jump; **~re il volo** take flight ● vi stand out. **~to** a marked

'spicchio nm (di agrumi) segment; (di aglio) clove

spicci'a|rsi vr hurry up. **~'tivo** a speedy

'spicciolo a (comune) banal; (denaro, 10 000 lire) in change. **spiccioli** pl change sg

'spicco nm relief; **fare ~** stand out

'spider nmf inv open-top sports car

spie'dino nm kebab. **spi'edo** nm spit; **allo spiedo** on a spit, spit-roasted

spie'ga|re vt explain; open out (cartina); unfurl (vele). **~rsi** vr explain oneself; (vele, bandiere:) unfurl. **~zi'one** nf explanation

spiegaz'zato a crumpled

spie'tato a ruthless

spiffe'rare vt blurt out ● vi (vento:) whistle. **'spiffero** nm (corrente d'aria) draught

'spiga nf spike; Bot ear

spigli'ato a self-possessed

'spigolo nm edge; (angolo) corner

'spilla nf (gioiello) brooch. **~ da balia** safety pin. **~ di sicurezza** safety pin

spil'lare vt tap

'spillo nm pin. **~ di sicurezza** safety pin; (in arma) safety catch

spi'lorcio a stingy

spilun'gone, -a nmf beanpole

'spina nf thorn; (di pesce) bone; Electr plug. **~ dorsale** spine

spi'naci nmpl spinach

spi'nale a spinal

spi'nato a (filo) barbed; (pianta) thorny

spi'nello nm fam joint

'spinger|e vt push; fig drive. **~si** vr (andare) proceed

spi'noso a thorny

'spint|a nf push; (violenta) thrust; fig spur. **~o** pp di spingere

spio'naggio nm espionage, spying

spio'vente a (tetto) sloping

spi'overe vi liter stop raining; (ricadere) fall; (scorrere) flow down

'spira nf coil

spi'raglio nm small opening; (soffio d'aria) breath of air; (raggio di luce) gleam of light

spi'rale a spiral ● nf spiral; (negli

orologi) hairspring; *(anticoncezionale)* coil

spi'rare *vi (soffiare)* blow; *(morire)* pass away

spiri'tato *a* possessed; *(espressione)* wild. **~ismo** *nm* spiritualism.

'spirito *nm* spirit; *(arguzia)* wit; *(intelletto)* mind; **fare dello spirito** be witty; **sotto spirito** in brandy. **~o'saggine** *nf* witticism. **spiri'toso** *a* witty

spiritu'ale *a* spiritual

splen'dente *a* shining

'splen|dere *vi* shine. **~dido** *a* splendid. **~'dore** *nm* splendour

spode'stare *vt* dispossess; depose *(re)*

'spoglia *nf (di animale)* skin; **spoglie** *pl (salma)* mortal remains; *(bottino)* spoils

spogli'a|re *vt* strip; *(svestire)* undress; *(fare lo spoglio di)* go through. **~'rello** *nm* strip-tease. **~rsi** *vr* strip, undress. **~'toio** *nm* dressing room; *Sport* changing room; *(guardaroba)* cloakroom, checkroom *Am.* **'spoglio** *a* undressed; *(albero, muro)* bare ● *nm (scrutinio)* perusal

'spola *nf* shuttle; **fare la ~** shuttle

spol'pare *vt* take the flesh off; *fig* fleece

spolve'rare *vt* dust; *fam* devour *(cibo)*

'sponda *nf (di mare, lago)* shore; *(di fiume)* bank; *(bordo)* edge

sponsoriz'zare *vt* sponsor

spon'taneo *a* spontaneous

spopo'lar|e *vt* depopulate ● *vi (avere successo)* draw the crowds. **~si** *vr* become depopulated

sporadica'mente *adv* sporadically. **spo'radico** *a* sporadic

sporcacci'one, -a *nmf* dirty pig

spor'c|are *vt* dirty; *(macchiare)* soil. **~arsi** *vr* get dirty. **~izia** *nf* dirt. **'sporco** *a* dirty; **avere la coscienza sporca** have a guilty conscience ● *nm* dirt

spor'gen|te *a* jutting. **~za** *nf* projection

'sporger|e *vt* stretch out; **~e querela contro** take legal action against ● *vi* jut out. **~si** *vr* lean out

sport *nm inv* sport

'sporta *nf* shopping basket

spor'tello *nm* door; *(di banca ecc)* window. **~ automatico** cash dispenser

spor'tivo, -a *a* sports *attrib; (persona)* sporty ● *nm* sportsman ● *nf* sportswoman

'sporto *pp di* **sporgere**

'sposa *nf* bride. **~'lizio** *nm* wedding

spo'sa|re *vt* marry; *fig* espouse. **~rsi** *vr* get married; *(vino:)* go (con with). **~to** *a* married. **'sposo** *nm* bridegroom; **sposi** *pl* [**novelli**] newlyweds

spossa'tezza *nf* exhaustion. **spos'sato** *a* exhausted, worn out

spo'sta|re *vt* move; *(differire)* postpone; *(cambiare)* change. **~rsi** *vr* move. **~to, -a** *a* ill-adjusted ● *nmf (disadattato)* misfit

'spranga *nf* bar. **spran'gare** *vt* bar

'sprazzo *nm (di colore)* splash; *(di luce)* flash; *fig* glimmer

spre'care *vt* waste. **'spreco** *nm* waste

spre'g|evole *a* despicable. **~ia'tivo** *a* pejorative. **'spregio** *nm* contempt

spregiudi'cato *a* unscrupulous

'spremer|e *vt* squeeze. **~si** **~si le meningi** rack one's brains

spremi'agrumi *nm* lemon squeezer

spre'muta *nf* juice. **~ d'arancia** fresh orange [juice]

sprez'zante *a* contemptuous

sprigio'nar|e *vt* emit. **~si** *vr* burst out

spriz'zare *vt/i* spurt; **be bursting with** *(salute, gioia)*

sprofon'dar|e *vi* sink; *(crollare)* collapse. **~si** *vr* **~si in** sink into; *fig* be engrossed in

spro'nare *vt* spur on. **'sprone** *nm* spur; *(sartoria)* yoke

sproporzio'nato *a* disproportion-
ate. **~'one** *nf* disproportion

sproposi'tato *a* full of blunders;
⟨*enorme*⟩ huge. **spro'posito** *nm*
blunder; ⟨*eccesso*⟩ excessive
amount; **a sproposito** inopportunely

sprovve'duto *a* unprepared; **~ di**
lacking in

sprov'visto *a* **~ di** out of; lacking
in ⟨*fantasia, pazienza*⟩; **alla
sprovvista** unexpectedly

spruz'zare *vt* sprinkle; ⟨*vaporizzare*⟩ spray; ⟨*inzaccherare*⟩ spatter.
~'tore *nm* spray; **'spruzzo** *nm*
spray; ⟨*di fango*⟩ splash

spudora'tezza *nf* shamelessness.
~'rato *a* shameless

'spugna *nf* sponge; ⟨*tessuto*⟩ towelling. **spu'gnoso** *a* spongy

'spuma *nf* foam; ⟨*schiuma*⟩ froth;
Culin mousse. **spu'mante** *nm*
sparkling wine, spumante.
spumeggi'are *vi* foam

spun'tare *vt* ⟨*rompere la punta di*⟩
break the point of; trim ⟨*capelli*⟩;
~rla *fig* win ● *vi* ⟨*pianta*⟩ sprout;
⟨*capelli*⟩ begin to grow; ⟨*sorgere*⟩
rise; ⟨*apparire*⟩ appear. **~rsi** *vr* get
blunt. **~ta** *nf* trim

spun'tino *nm* snack

'spunto *nm* cue; *fig* starting point;
dare ~ a give rise to

spur'gare *vt* purge. **~si** *vr Med* expectorate

spu'tare *vt/i* spit; **~ sentenze** pass
judgment; **'sputo** *nm* spit

'squadra *nf* ⟨*gruppo*⟩ team, squad;
⟨*di polizia ecc*⟩ squad; ⟨*da disegno*⟩
square. **squa'drare** *vt* square;
⟨*guardare*⟩ look up and down

squa'driglia *nf*, **~one** *nf* squadron

squagli'are *vt*, **~si** *vr* melt; **~sela**
⟨*fam: svignarsela*⟩ steal out

squalifi'care *vt* disqualification.
~'care *vt* disqualify

'squallido *a* squalid. **squal'lore**
nm squalor

'squalo *nm* shark

'squama *nf* scale; ⟨*di pelle*⟩ flake

squa'mare *vt* scale. **~arsi** *vr*
⟨*pelle*⟩ flake off. **~'moso** *a* scaly;
⟨*pelle*⟩ flaky

squarcia'gola: a ~ *adv* at the top
of one's voice

squarci'are *vt* rip. **'squarcio** *nm*
rip; ⟨*di ferita, in nave*⟩ gash; ⟨*di
cielo*⟩ patch

squar'tare *vt* quarter; dismember
⟨*animale*⟩

squat'trinato *a* penniless

squilib'rare *vt* unbalance. **~to, -a**
a unbalanced ● *nmf* lunatic.
squi'librio *nm* imbalance

squil'lante *a* shrill. **~re** *vi*
⟨*campana*⟩ peal; ⟨*tromba:*⟩ blare;
⟨*telefono:*⟩ ring. **'squillo** *nm* blare;
Teleph ring; ⟨*ragazza*⟩ call girl

squi'sito *a* exquisite

squit'tire *vi* ⟨*pappagallo, fig:*⟩
squawk; ⟨*topo:*⟩ squeak

sradi'care *vt* uproot; eradicate
⟨*vizio, male*⟩

srego'latezza *nf* dissipation.
~'lato *a* inordinate; ⟨*dissoluto*⟩
dissolute

s.r.l. *abbr* ⟨*società a responsabilità limitata*⟩ Ltd

sroto'lare *vt* uncoil

SS *abbr* ⟨*strada statale*⟩ national
road

'stabile *a* stable; ⟨*permanente*⟩
lasting; ⟨*saldo*⟩ steady; **compagnia ~** *Theat* repertory company
● *nm* ⟨*edificio*⟩ building

stabili'mento *nm* factory; ⟨*industriale*⟩ plant; ⟨*edificio*⟩ establishment. **~ balneare** lido

stabi'lire *vt* establish; ⟨*decidere*⟩
decide. **~rsi** *vr* settle. **~tà** *nf* stability

stabiliz'zare *vt* stabilize. **~rsi** *vr*
stabilize. **~'tore** *nm* stabilizer

stac'care *vt* detach; pronounce
clearly ⟨*parole*⟩; ⟨*separare*⟩ separate; turn off ⟨*corrente*⟩; **~ gli
occhi da** take one's eyes off ● *vi*
⟨*fam: finire di lavorare*⟩ knock off.
~si *vr* come off; **~si da** break
away from ⟨*partito, famiglia*⟩

staccio'nata nf fence

'**stacco** nm gap

'**stadio** nm stadium

'**staffa** nf stirrup

staf'fetta nf dispatch rider

stagio'nale a seasonal

stagio'na|re vt season ‹legno›; mature ‹formaggio›. **~to** a ‹legno› seasoned; ‹formaggio› matured

stagi'one nf season; **alta/bassa ~** high/low season

stagli'arsi vr stand out

sta'gna|nte a stagnant. **~re** vt ‹saldare› solder; ‹chiudere ermeticamente› seal ● vi ‹acqua:› stagnate. '**stagno** a ‹a tenuta d'acqua› watertight ● nm ‹acqua ferma› pond; ‹metallo› tin

sta'gnola nf tinfoil

stalag'mite nf stalagmite

stalat'tite nf stalactite

'**stall|a** nf stable; ‹per buoi› cowshed. **~l'ere** nm groom

stal'lone nm stallion

sta'mani, stamat'tina adv this morning

stam'becco nm ibex

stam'berga nf hovel

'**stampa** nf Typ printing; ‹giornali, giornalisti› press; ‹riproduzione› print

stam'pa|nte nf printer. **~nte ad aghi** dot matrix printer. **~nte laser** laser printer. **~re** vt print. **~'tello** nm block letters pl

stam'pella nf crutch

'**stampo** nm mould; **di vecchio ~** ‹persona› of the old school

sta'nare vt drive out

stan'car|e vt tire; ‹annoiare› bore. **~si** vr get tired

stan'chezza nf tiredness. '**stanco** a tired; **stanco di** ‹stufo› fed up with. **stanco morto** dead tired, knackered fam

'**standard** a & nm inv standard. **~iz'zare** vt standardize

'**stan|ga** nf bar; ‹persona› beanpole. **~'gata** nf fig blow; ‹fam: nel calcio› big kick; **prendere una ~gata** ‹fam: agli esami, economi-

ca› come a cropper. **stan'ghetta** nf ‹di occhiali› leg

sta'notte nf tonight; ‹la notte scorsa› last night

'**stante** prep on account of; **a sé** separate

stan'tio a stale

stan'tuffo nm piston

'**stanza** nf room; ‹metrica› stanza

stanzi'are vt allocate

stap'pare vt uncork

'**stare** vi ‹rimanere› stay; ‹abitare› live; ‹con gerundio› be; **sto solo cinque minuti** I'll stay only five minutes; **sto in piazza Peyron** I live in Peyron Square; **sta dormendo** he's sleeping; **~ a** ‹attenersi› keep to; ‹spettare› be up to; **~ bene** ‹economicamente› be well off; ‹di salute› be well; ‹addirsi› suit; **~ dietro a** ‹seguire› follow; ‹sorvegliare› keep an eye on; ‹corteggiare› run after; **~ in piedi** stand; **~ per** be about to; **ben ti sta!** it serves you right!; **come stai/sta?** how are you?; **lasciar ~** leave alone; **starci** ‹essere contenuto› go into; ‹essere d'accordo› agree; **il 3 nel 12 ci sta 4 volte** 3 into 12 goes 4; **non sa ~ agli scherzi** he can't take a joke; **~ su** ‹con la schiena› sit up straight; **~ sulle proprie** keep oneself to oneself. **starsene** vr ‹rimanere› stay

starnu'tire vi sneeze. **star'nuto** nm sneeze

sta'sera adv this evening, tonight

sta'tale a state attrib ● nmf state employee ● nf ‹strada› main road, trunk road

'**statico** a static

sta'tista nm statesman

sta'tistic|a nf statistics sg. **~o** a statistical

'**stato** pp di **essere**, **stare** ● nm state; ‹posizione sociale› position; Jur status. **~ d'animo** frame of mind. **~ civile** marital status. **S~ Maggiore** Mil General Staff. **Stati pl Uniti [d'America]** United States [of America] 'statua nf statue

statuni'tense a United States
attrib, US attrib ● nmf citizen of
the United States, US citizen

sta'tura nf height; di alta ~ tall; di
bassa ~ short

sta'tuto nm statute

stazio'nario a stationary

stazi'one nf station; (città) resort.
~ balneare seaside resort. ~ ferroviaria railway station Br, train
station. ~ di servizio petrol
station Br, service station. ~ termale spa

'stecca nf stick; (di ombrello) rib;
(da biliardo) cue; Med splint; (di
sigarette) carton; (di reggiseno)
stiffener

stec'cato nm fence

stec'chito a skinny; (rigido) stiff;
(morto) stone cold dead

'stella nf star; salire alle stelle
〈prezzi:〉 rise sky-high. ~ alpina
edelweiss. ~ cadente shooting
star. ~ filante streamer. ~ di
mare starfish

stel'lare a star attrib; 〈grandezza〉
stellar. ~to a starry

'stelo nm stem; lampada nf a ~
standard lamp

'stemma nm coat of arms

stempi'ato a bald at the temples

sten'dardo nm standard

'stender|e vt spread out; (appende-
re) hang out; (distendere) stretch
[out]; (scrivere) write down. ~si
vr stretch out

stendibianche'ria nm inv, sten-
di'toio nm clothes horse

stenodattilogra'fia nf shorthand
typing. ~'lografo, -a nmf shorthand typist

stenogra'f|are vt take down in
shorthand. ~ia nf shorthand

sten'ta|re vi ~re a find it hard to.
~to a laboured. 'stento nm (fatica) effort; a stento with difficulty;
stenti pl hardships, privations

'sterco nm dung

'stereo['fonico] a stereo[phonic]

stereoti'pato a stereotyped; 〈sor-
riso〉 insincere. stere'otipo nm
stereotype

'steril|e a sterile; 〈terreno〉 barren.
~ità nf sterility. ~iz'zare vt steri-
lize. ~izzazi'one nf sterilization

stermi'nare vt exterminate

stermi'nato a immense

ster'minio nm extermination

'sterno nm breastbone

ste'roide nm steroid

ster'zare vi steer. 'sterzo nm steering

'steso pp di stendere

'stesso a same; io ~ myself; tu ~
yourself; me ~ myself; se ~ himself; in quel momento ~ at that
very moment; dalla stessa regina (in persona) by the Queen
herself; tuo fratello ~ dice che
hai torto even your brother says
you're wrong; coi miei stessi occhi with my own eyes ● pron lo ~
the same one; (la stessa cosa) the
same; fa lo ~ it's all the same; ci
vado lo ~ I'll go just the same

ste'sura nf drawing up; (documento) draft

stick nm colla a ~ glue stick;
deodorante a ~ stick deodorant

'stigma nm stigma. ~te nfpl stig-
mata

sti'lare vt draw up

'stil|e nm style. ~e libero (nel nuo-
to) freestyle, crawl. sti'lista nmf
stylist. ~iz'zato a stylized

stil'lare vi ooze

stilo'grafic|a nf fountain pen. ~o
a penna a fountain pen

'stima nf esteem; (valutazione) estimate. sti'mare vt esteem; (valutare) estimate; (ritenere) consider

stimo'la|nte a stimulating ● nm
stimulant. ~re vt stimulate; (incitare) incite

'stimolo nm stimulus; (fitta) pang

'stinco nm shin bone

'stinger|e vt/i fade. ~si vr fade.
'stinto pp di stingere

sti'par|e vt cram. **~si** vr crowd together

stipendi'ato a salaried ● nm salaried worker. **sti'pendio** nm salary

'stipite nm doorpost

stipu'la|re vt stipulate. **~zi'one** nf stipulation; (accordo) agreement

stira'mento nm sprain

sti'ra|re vt iron; (distendere) stretch. **~rsi** vr (distendersi) stretch; pull (muscolo). **~tura** nf ironing. **'stiro** nm **ferro da stiro** iron

'stirpe nf stock

stiti'chezza nf constipation. **'stitico** a constipated

'stiva nf Naut hold

sti'vale nm boot. **stivali** pl **di gomma** Wellington boots, Wellingtons

'stizza nf anger

stiz'zi|re vt irritate. **~rsi** vr become irritated. **~to** a irritated. **stiz'zoso** a peevish

stocca'fisso nm stockfish

stoc'cata nf stab; (battuta pungente) gibe

'stoffa nf material; fig stuff

'stola nf stole

'stolto a foolish

stoma'chevole a revolting

'stomaco nm stomach; **mal di ~** stomach-ache

sto'na|re vt/i sing/play out of tune ● vi (non intonarsi) clash. **~to** a out of tune; (discordante) clashing; (confuso) bewildered. **~tura** nf false note; (discordanza) clash

'stoppia nf stubble

stop'pino nm wick

stop'poso a tough

'storcer|e vt, **~si** vr twist

stor'di|re vt stun; (intontire) daze. **~rsi** vr dull one's senses. **~to** a stunned; (intontito) dazed; (sventato) heedless

'storia nf history; (racconto, bugia) story; (pretesto) excuse; **senza storie** no fuss!; **fare [delle] storie** make a fuss

'storico, -a a historical; (di impor-

tanza storica) historic ● nmf historian

stori'one nm sturgeon

'stormo nm flock

'storno nm starling

storpi'a|re vt cripple; mangle (parole). **~tura** nf deformation. **'storpio, -a** a crippled ● nmf cripple

'stort|a nf (distorsione) sprain; **prendere una ~ alla caviglia** sprain one's ankle. **~o** pp di **storcere** ● a crooked; (ritorto) twisted; (gambe) bandy; fig wrong

sto'viglie nfpl crockery sg

'strabico a cross-eyed; **essere ~** be cross-eyed, have a squint

strabili'ante a astonishing

stra'bismo nm squint

straboc'care vi overflow

stra'carico a overloaded

strac'cia|re vt tear; (fam: vincere) thrash. **~'ato** a torn; (persona) in rags; (prezzi) slashed; **a un prezzo ~ato** dirt cheap. **'straccio** a torn ● nm rag; (strofinaccio) cloth. **~'one** nm tramp

stra'cotto a overdone; (fam: innamorato) head over heels ● nm stew

'strada nf road; (di città) street; (fig: cammino) way; **essere fuori ~** be on the wrong track; **fare ~** lead the way; **farsi ~** make one's way. **~ maestra** main road. **~ a senso unico** one-way street. **~ senza uscita** blind alley. **stra'dale** a road attrib

strafalci'one nm blunder

stra'fare vi overdo it, overdo things

stra'foro: di ~ adv on the sly

strafot'ten|te a arrogant. **~za** nf arrogance

'strage nf slaughter

stra'icio nm (parte) extract

stralu'na|re ~re gli occhi open one's eyes wide. **~to** a (occhi) staring; (persona) distraught

stramaz'zare vi fall heavily

strambe'ria nf oddity. **'strambo** a strange

strampa'lato a odd

stra'nezza nf strangeness

strango'lare vt strangle

strani'ero, -a a foreign ● nmf foreigner

'strano a strange

straordi|naria'mente adv extraordinarily. **~'nario** a extraordinary; (notevole) remarkable; (edizione) special; **lavoro ~nario** overtime; **treno ~nario** special train

strapaz'zar|e vt ill-treat; scramble (uova). **~si** vr tire oneself out. **stra'pazzo** nm strain; **da strapazzo** fig worthless

strapi'eno a overflowing

strapi'ombo nm projection; **a ~** sheer

strap'par|e vt tear; (per distruggere) tear up; pull out (dente, capelli); (sradicare) pull up; (estorcere) wring. **~si** vr get torn; (allontanarsi) tear oneself away. **'strappo** nm tear; (strattone) jerk; (fam: passaggio) lift; **fare uno strappo alla regola** make an exception to the rule. **~ muscolare** muscle strain

strapun'tino nm folding seat

strari'pare vi flood

strasci'c|are vt tear; shuffle (piedi); drawl (parole). **'strascico** nm train; fig after-effect

strass nm inv rhinestone

strata'gemma nm stratagem

strate'gia nf strategy. **stra'tegico** a strategic

'strato nm layer; (di vernice ecc) coat, layer; (roccioso, sociale) stratum. **~'sfera** nf stratosphere. **~'sferico** a stratospheric; fig sky-high

stravac'ca|rsi vr fam slouch. **~to** a fam slouching

strava'gan|te a extravagant; (eccentrico) eccentric. **~za** nf extravagance; (eccentricità) eccentricity

stra'vecchio a ancient

strave'dere vt **~ per** worship

stravizi'are vi indulge oneself. **stra'vizio** nm excess

stra'volg|ere vt twist; (turbare) upset. **~i'mento** nm twisting. **stra'volto** a distraught; (fam: stanco) done in

strazi'a|nte a heartrending; (dolore) agonizing. **~re** vt grate on (orecchie); break (cuore). **'strazio** nm agony; **essere uno strazio** be agony; **che strazio!** fam it's awful!

'strega nf witch. **stre'gare** vt bewitch. **stre'gone** nm wizard

'stregua nf **alla ~ di** like

stre'ma|re vt exhaust. **~to** a exhausted

'stremo nm **ridotto allo ~** at the end of one's tether

'strenuo a strenuous

strepi'tare vi make a din. **'strepito** nm noise. **~'toso** a noisy; fig resounding

stres'sa|nte a (lavoro, situazione) stressful. **~to** a stressed [out]

'stretta nf grasp; (dolore) pang; **essere alle strette** be in dire straits; **mettere alle strette** qcno have sb's back up against the wall. **~ di mano** handshake

stret'tezza nf narrowness; **strettezze** pl (difficoltà finanziarie) financial difficulties

'stret|to pp di **stringere** ● a narrow; (serrato) tight; (vicino) close; (dialetto) broad; (rigoroso) strict; **lo ~to necessario** the bare minimum ● nm Geog strait. **~'toia** nf bottleneck; (fam: difficoltà) tight spot

stri'a|to a striped. **~'tura** nf streak

stri'dente a strident

'stridere vi squeak; fig clash. **stri'dore** nm screech

stri'dulo a shrill

strigli'a|re vt groom. **~ta** nf grooming; fig dressing down

stril'l|are vi/t scream. **'strillo** nm scream

strimin'zito *a* skimpy; (*magro*) skinny

strimpel'lare *vt* strum

'strin|ga *nf* lace; *Comput* string. **~'gato** *a fig* terse

'stringer|e *vt* press; (*serrare*) squeeze; (*tenere stretto*) hold tight; take in (*abito*); (*comprimere*) be tight; (*restringere*) tighten; **~e la mano a** shake hands with ● *vi* (*premere*) press. **~si** *vr* (*accostarsi*) draw close a (*a* to); (*avvicinarsi*) squeeze up

'striscia *nf* strip; (*riga*) stripe. **strisce** *pl* [**pedonali**] zebra crossing *sg*

strisci'ar|e *vi* crawl; (*sfiorare*) graze ● *vt* drag (*piedi*). **~si** *vr* **~si a** a rub against. **'striscio** *nm* graze; *Med* smear; **colpire di striscio** graze

strisci'one *nm* banner

strito'lare *vt* grind

striz'zare *vt* squeeze; (*torcere*) wring [out]; **~ l'occhio** wink

'strofa *nf* strophe

strofi'naccio *nm* cloth; (*per spolverare*) duster. **~ da cucina** tea towel

strofi'nare *vt* rub

strombaz'zare *vt* boast about ● *vi* hoot

strombaz'zata *nf* (*di clacson*) hoot

stron'care *vt* cut off; (*reprimere*) crush; (*criticare*) tear to shreds

'stronzo *nm vulg* shit

stropicci'are *vt* rub; crumple (*vestito*)

stroz'za|re *vt* strangle. **~'tura** *nf* strangling; (*di strada*) narrowing

strozzi'naggio *nm* loan-sharking

stroz'zino *nm pej* usurer; (*truffatore*) shark

strug'gente *a* all-consuming

'struggersi *vr liter* pine [away]

strumen'tale *a* instrumental

strumentaliz'zare *vt* make use of

strumentazi'one *nf* instrumentation

stru'mento *nm* instrument; (*arnese*) tool. **~ a corda** string instru-

ment. **~ musicale** musical instrument

strusci'are *vt* rub

'strutto *nm* lard

strut'tura *nf* structure. **struttu'rale** *a* structural

struttu'rare *vt* structure

strutturazi'one *nf* structuring

'struzzo *nm* ostrich

stuc'care *vt* stucco

stuc'chevole *a* nauseating

'stucco *nm* stucco

stu'den|te, -'essa *nmf* student; (*di scuola*) schoolboy; schoolgirl. **~'tesco** *a* student; (*di scolaro*) school *attrib*

studi'ar|e *vt* study. **~si** *vr* **~si di** try to

'studi|o *nm* studying; (*stanza, ricerca*) study; (*di artista, TV ecc*) studio; (*di professionista*) office. **~'oso, -a** *a* studious ● *nmf* scholar

'stufa *nf* stove. **~ elettrica** electric fire

stu'fa|re *vt* *Culin* stew; (*dare fastidio*) bore. **~rsi** *vr* get bored. **~to** *nm* stew

'stufo *a* bored; **essere ~ di** be fed up with

stu'oia *nf* mat

stupefa'cente *a* amazing ● *nm* drug

stu'pendo *a* stupendous

stupi'd|aggine *nf* (*azione*) stupid thing; (*cosa da poco*) nothing. **~ata** *nf* stupid thing. **~ità** *nf* stupidity. **'stupido** *a* stupid

stu'pir|e *vt* astonish ● *vi*, **~si** *vr* be astonished. **stu'pore** *nm* amazement

stu'pra|re *vt* rape. **~'tore** *nm* rapist. **'stupro** *nm* rape

sturalavan'dini *nm inv* plunger

stu'rare *vt* uncork; unblock (*lavandino*)

stuzzica'denti *nm inv* toothpick

stuzzi'care *vt* prod [at]; pick (*denti*); poke (*fuoco*); (*molestare*) tease; whet (*appetito*)

stuzzi'chino *nm Culin* appetizer

su *prep* on; (*senza contatto*) over; (*riguardo a*) about; (*circa, intorno a*) about; around; **le chiavi sono sul tavolo** the keys are on the table; **il quadro è appeso sul camino** the picture is hanging over the fireplace; **un libro sull'antico Egitto** a book on *o* about Ancient Egypt; **costa sulle 50 000 lire** it costs about 50,000 lire; **decidere sul momento** decide at the time; **su commissione** on commission; **su due piedi** on the spot; **uno su dieci** one out of ten ● *adv* (*sopra*) up; (*al piano di sopra*) upstairs; (*addosso*) on; **ho su il cappotto** I've got my coat on; **in su** (*guardare*) up; **dalla vita in su** from the waist up; **su!** come on!

su'bacqueo *a* underwater

subaffit'tare *vt* sublet. **subaf'fitto** *nm* sublet

subal'terno *a* & *nm* subordinate

sub'buglio *nm* turmoil

sub'conscio *a nm* subconscious

subdola'mente *adv* deviously. **'subdolo** *a* devious, underhand

suben'trare *vi* ⟨*circostanze:*⟩ come up; **~ a** take the place of

su'bire *vt* undergo; (*patire*) suffer

subis'sare *vt fig* **~ di** overwhelm with

'subito *adv* at once; **~ dopo** straight after

su'blime *a* sublime

subo'dorare *vt* suspect

subordi'nato, -a *a* & *nmf* subordinate

subur'bano *a* suburban

suc'cedere *vi* (*accadere*) happen; **~e a** succeed; (*venire dopo*) follow; **~e al trono** succeed to the throne. **~si** *vr* happen one after the other

succes'sione *nf* succession; **in ~** in succession

succes|siva'mente *adv* subsequently. **~'sivo** *a* successive

suc'ces|so *pp di* **succedere** ● *nm* success; (*esito*) outcome; (*disco ecc*) hit. **~'sone** *nm* huge success

succes'sore *nm* successor

succhi'are *vt* suck [up]

suc'cinto *a* (*conciso*) concise; ⟨*abito*⟩ scanty

'succo *nm* juice; *fig* essence; **~ di frutta** fruit juice. **suc'coso** *a* juicy

'succube *nm* **essere ~ di qcno** be totally dominated by sb

succu'lento *a* succulent

succur'sale *nf* branch [office]

sud *nm* south; **del ~** southern

su'da|re *vi* sweat, perspire; (*faticare*) sweat blood; **~re freddo** be in a cold sweat. **~ta** *nf* sweat *anche fig* sweat. **~'ticcio** *a* sweaty. **~to** *a* sweaty

sud'detto *a* above-mentioned

'suddito, -a *nmf* subject

suddi'vi|dere *vt* subdivide. **~si'o-ne** *nf* subdivision

su'd-est *nm* southeast

'sudici|o *a* dirty, filthy. **~'ume** *nm* dirt, filth

sudorazi'one *nf* perspiring. **su-'dore** *nm* sweat, perspiration; *fig* sweat

su'd-ovest *nm* southwest

suffici'en|te *a* sufficient; (*presuntuoso*) conceited ● *nm* bare essentials *pl*; *Sch* pass mark. **~za** *nf* sufficiency; (*presunzione*) conceit; *Sch* pass; **a ~za** enough

suf'fisso *nm* suffix

suf'fragio (*voto*) vote. **~ universale** universal suffrage

sugge'rimento *nm* suggestion

sugge'ri|re *vt* suggest; *Theat* prompt. **~'tore, ~'trice** *nmf Theat* prompter

suggestio'nabile *a* suggestible

suggestio'na|re *vt* influence. **~to** *a* influenced. **suggesti'one** *nf* influence

sugge'stivo *a* suggestive; ⟨*musica ecc*⟩ evocative

'sughero *nm* cork

'sugli = **su + gli**

'sugo *nm* (*di frutta*) juice; (*di carne*) gravy; (*salsa*) sauce; (*sostanza*) substance

'sui = **su + i**

sui'cid|a a suicidal ● nmf suicide. **suici'darsi** vr commit suicide. **~io** nm suicide

su'ino a carne suina pork ● nm swine

sul = su + il. **'sullo** = su + lo. **'sulla** = su + la. **'sulle** = su + le

sul'ta|na nf sultana. **~'nina** a uva **~nina** sultana. **~no** nm sultan

'sunto nm summary

'suo, -a poss a **il ~, i suoi** his; (di cosa, animale) its; (forma di cortesia) your; **la sua, le sue** her; (di cosa, animale) its; (forma di cortesia) your; **questa macchina è sua** this car is his/hers; **~ padre** his/her/your father; **un ~ amico** a friend of his/hers/yours ● poss pron **il ~, i suoi** his; (di cosa, animale) its; (forma di cortesia) yours; **la sua, le sue** hers; (di cosa, animale) its; (forma di cortesia) yours; **i suoi** his/her folk

su'ocera nf mother-in-law

su'ocero nm father-in-law

su'ola nf sole

su'olo nm ground; (terreno) soil

suo'na|re vt/i Mus play; ring (campanello); sound (allarme, clacson); (orologio:) strike. **~tore, ~'trice** nmf player. **suone'ria** nf alarm. **su'ono** nm sound

su'ora nf nun; **Suor Maria** Sister Maria

superal'colico nm spirit ● a **bevande** pl **superalcoliche** spirits

supera'mento nm (di timidezza) overcoming; (di esame) success (di in)

supe'rare vt surpass; (eccedere) exceed; (vincere) overcome; over-take, pass Am (veicolo); pass (esame)

su'perb|ia nf haughtiness. **~o** a haughty; (magnifico) superb

super'donna nf superwoman

superdo'tato a highly gifted

superfici'al|e a superficial ● nmf superficial person. **~ità** nf superficiality. **super'ficie** nf surface; (area) area

su'perfluo a superfluous

superi'or|e a superior; (di grado) senior; (più elevato) higher; (sovrastante) upper; (al di sopra) above ● nmf superior. **~ità** nf superiority

superla'tivo a & nm superlative

supermer'cato nm supermarket

super'sonico a supersonic

su'perstite a surviving ● nmf survivor

superstizi'o|ne nf superstition. **~so** a superstitious

super'strada nf toll-free motor-way

supervi'si|one nf supervision. **~'sore** nm supervisor

su'pino a supine

suppel'lettili nfpl furnishings

suppergiù adv about

supplemen'tare a additional, supplementary

supple'mento nm supplement; **~ rapido** express train supplement

sup'plen|te a temporary ● nmf Sch supply teacher. **~za** nf temporary post

'suppli|ca nf plea; (domanda) petition. **~'care** vt beg. **~'chevole** a imploring

sup'plire vt replace ● vi **~ a** (compensare) make up for

sup'plizio nm torture

sup'porre vt suppose

sup'porto nm support

supposizi'one nf supposition

sup'posta nf suppository

sup'posto pp di **supporre**

supre'mazia nf supremacy. **su-'premo** a supreme

sur'fare vi **~ in Internet** surf the Net

surge'la|re vt deep-freeze. **~ti** nmpl frozen food sg. **~to** a frozen

surrea'lis|mo nm surrealism. **~ta** nmf surrealist

surris'cal'dare vt overheat

surro'gato nm substitute

suscet'tibile a touchy. **~ità** nf touchiness

susci'tare *vt* stir up; arouse ⟨*ammirazione ecc*⟩

su'sina *nf* plum. **~o** *nm* plumtree

su'spense *nf* suspense

sussegu'ente *a* subsequent. **~'irsi** *vr* follow one after the other

sussidi'ar|e *vt* subsidize. **~io** *a* subsidiary. **sus'sidio** *nm* subsidy; ⟨*aiuto*⟩ aid. **sussidio di disoccupazione** unemployment benefit

sussi'ego *nm* haughtiness

sussi'stenza *nf* subsistence. **sus'sistere** *vi* subsist; ⟨*essere valido*⟩ hold good

sussul'tare *vi* start. **sus'sulto** *nm* start

sussur'rare *vt* whisper. **sus'surro** *nm* whisper

su'tu|ra *nf* suture. **~'rare** *vt* suture

sva'gar|e *vt* amuse. **~si** *vr* amuse oneself. **'svago** *nm* relaxation; ⟨*divertimento*⟩ amusement

svaligi'are *vt* rob; burgle ⟨*casa*⟩

svalu'ta|re *vt* devalue; *fig* underestimate. **~rsi** *vr* lose value. **~zi'one** *nf* devaluation

svam'pito *a* absent-minded

sva'nire *vi* vanish

svantaggi'ato *a* at a disadvantage; ⟨*bambino, paese*⟩ disadvantaged. **svan'taggio** *nm* disadvantage; **essere in svantaggio** *Sport* be losing; **in svantaggio di tre punti** three points down; **~'oso** *a* disadvantageous

svapo'rare *vi* evaporate

svari'ato *a* varied

sva'sato *a* flared

'svastica *nf* swastika

sve'dese *a & nm* ⟨*lingua*⟩ Swedish ● *nmf* Swede

'sveglia *nf* ⟨*orologio*⟩ alarm [clock]; **~! I get up!; mettere la ~** set the alarm [clock]

svegli'ar|e *vt* wake up; *fig* awaken. **~si** *vr* wake up. **'sveglio** *a* awake; ⟨*di mente*⟩ quick-witted

sve'lare *vt* reveal

svel'tezza *nf* speed; *fig* quick-wittedness

svel'tir|e *vt* quicken. **~si** *vr* ⟨*per-*

sona:⟩ liven up. **'svelto** *a* quick; ⟨*slanciato*⟩ svelte; **alla ~svelta** quickly

'svend|ere *vt* undersell. **~ita** *nf* [clearance] sale

sveni'mento *nm* fainting fit. **sve'nire** *vi* faint

sven'ta|re *vt* foil. **~to** *a* thoughtless ● *nmf* thoughtless person

'sventola *nf* slap; **orecchie** *nfpl* a **~** protruding ears

svento'lare *vt/i* wave

sven'trare *vt* disembowel; *fig* demolish ⟨*edificio*⟩

sven'tura *nf* misfortune. **sventu'rato** *a* unfortunate

sve'nuto *pp di* **svenire**

svergo'gnato *a* shameless

sver'nare *vi* winter

sve'stir|e *vt* undress. **~si** *vr* undress, get undressed

'Svezia *nf* Sweden

svezza'mento *nm* weaning. **svez'zare** *vt* wean

svi'ar|e *vt* divert; ⟨*corrompere*⟩ lead astray. **~si** *vr* *fig* go astray

svico'lare *vi* turn down a side street; ⟨*fig: dalla questione ecc*⟩ evade the issue; ⟨*fig: da una persona*⟩ dodge out of the way

svi'gnarsela *vr* slip away

svi'lire *vt* debase

svilup'par|e *vt*, **~si** *vr* develop. **svi'luppo** *nm* development; **paese in via di sviluppo** developing country

svinco'lar|e *vt* release; clear ⟨*merce*⟩. **~si** *vr* free oneself. **'svincolo** *nm* clearance; ⟨*di autostrada*⟩ exit

svi'sce|rare *vt* gut; *fig* dissect. **~to** *a* ⟨*amore*⟩ passionate; ⟨*ossequioso*⟩ obsequious

'svista *nf* oversight

svi'ta|re *vt* unscrew. **~to** *a* ⟨*fam: matto*⟩ cracked, nutty

'Svizzera *nf* Switzerland. **s~o, -a** *a & nmf* Swiss

svogli|a'tezza *nf* half-heartedness. **~'ato** *a* lazy

svolaz'za|nte *a* ⟨*capelli*⟩ windswept. **~re** *vi* flutter

'svolg|ere vt unwind; unwrap ⟨pacco⟩; (risolvere) solve; (portare a termine) develop. **~si** vr (accadere) take place. **svolgi'mento** nm course; (sviluppo) development

'svolta nf turning; fig turning-point. **svol'tare** vi turn

'svolto pp di svolgere

svuo'tare vt empty [out]

Tt

tabac'c|aio, -a nmf tobacconist. **~he'ria** nf tobacconist's (which also sells stamps, postcards etc). **ta'bacco** nm tobacco

ta'bel|la nf table; (lista) list. **~la dei prezzi** price list. **~'lina** nf Math multiplication table. **~'lone** nm wall chart. **~lone del cane-stro** backboard

taber'nacolo nm tabernacle

tabù a & nm inv taboo

tabu'lato nm Comput [data] print-out

'tacca nf notch; **di mezza ~** (attore, giornalista) second-rate

tac'cagno a fam stingy

tac'cheggio nm shoplifting

tac'chetto nm Sport stud

tac'chino nm turkey

tacci'are vt ~ qcno di qcsa accuse sb of sth

'tacco nm heel; **alzare i tacchi** take to one's heels; **scarpe senza ~** flat shoes. **tacchi** pl a spillo stiletto heels

taccu'ino nm notebook

ta'cere vi be silent ● vt say nothing about; **mettere a ~ qcsa** (scandalo) hush sth up; **mettere a ~ qcno** silence sb

ta'chimetro nm speedometer

'tacito a silent; (inespresso) tacit. **taci'turno** a taciturn

ta'fano nm horsefly

taffe'ruglio nm scuffle

'taglia nf (riscatto) ransom; (ricompensa) reward; (statura) height; (misura) size. **~ unica** one size

taglia'carte nm inv paperknife

taglia'erba nm inv lawn-mower

tagliafu'oco a inv **porta ~** fire door; **striscia ~** fire break

tagli'ando nm coupon; **fare il ~** ≈ put one's car in for its MOT

tagli'ar|e vt cut; (attraversare) cut across; (interrompere) cut off; (togliere) cut out; carve ⟨carne⟩; mow ⟨erba⟩; **farsi ~e i capelli** have a haircut ● vi cut. **~si i capelli** have a hair-cut oneself; **~si i capelli** have a hair-cut

taglia'telle nfpl tagliatelle sg, thin, flat strips of egg pasta

taglieggi'are vt extort money from

tagli'ente a sharp ● nm cutting edge. **~re** nm chopping board

'taglio nm cut; (il tagliare) cutting; (di stoffa) length; (parte tagliente) edge; **a doppio ~** double-edged. **~ cesareo** Caesarean section

tagli'ola nf trap

tagli'one nm **legge del ~** an eye for an eye and a tooth for a tooth

tagliuz'zare vt cut into small pieces

tail'leur nm inv [lady's] suit

talassotera'pia nf thalasso-therapy

'talco nm talcum powder

'tale a such a; (con nomi plurali) such; **c'è un ~ disordine** there is such a mess; **non accetto tali scuse** I won't accept such excuses; **il rumore era ~ che non si sentiva nulla** there was so much noise you couldn't hear yourself think; **il ~ giorno** on such and such a day; **quel tal signore** that gentleman; **~ quale** just like ● pron un ~ someone; **quel ~** that man; **il tal dei tali** such and such a person

ta'lento nm talent

tali'smano nm talisman

tallo'nare *vt* be hot on the heels of

tallon'cino *nm* coupon

tal'lone *nm* heel

tal'mente *adv* so

ta'lora *adv* = **talvolta**

'talpa *nf* mole

tal'volta *adv* sometimes

tamburel'lare *vi* (con le dita) drum; (pioggia:) beat, drum. **tambu'rello** *nm* tambourine. **tambu'rino** *nm* drummer. **tam'buro** *nm* drum

Ta'migi *nm* Thames

tampona'mento *nm* Auto collision; (di ferita) dressing; (di falla) plugging. **~ a catena** pile-up. **tampo'nare** *vt* (urtare) crash into; (otturare) plug. **tam'pone** *nm* swab; (per timbri) pad; (per mestruazioni) tampon; (per treni, Comput) buffer

'tana *nf* den

'tanfo *nm* stench

'tanga *nm inv* tanga

tan'gente *a* tangent ● *nf* tangent; (somma) bribe. **~'topoli** *nf* widespread corruption in Italy in the early 90s. **~zi'ale** *nf* orbital road

tan'gibile *a* tangible

'tango *nm* tango

tan'tino: un ~ *adv* a little [bit]

'tanto *a* [so] much; (con nomi plurali [so] many, [such] a lot of; **~ tempo** [such] a long time; **non ha tanta pazienza** he doesn't have much patience; **~ tempo quanto ti serve** as much time as you need; **non è ~ intelligente quanto suo padre** he's not as intelligent as his father; **tanti amici quanti parenti** as many friends as relatives ● *pron* much; (plurale) many; (tanto tempo) a long time; **è un uomo come tanti** he's just an ordinary man; **tanti** (molte persone) many people; **non ci vuole così ~** it doesn't take that long; **~ quanto** as much as; **tanti quanti** as many as ● (comunque) anyway, in any case ● *adv* (così) so; (con verbi) so much; **~ debole**

so weak; **è ~ ingenuo da crederle** he's naïve enough to believe her; **di ~ in ~** every now and then; **~ l'uno come l'altro** both; **~ quanto** as much as; **tre volte ~** three times as much; **una volta ~** once in a while; **~ meglio così** so much the better!; **tant'è** so much so; **~ per cambiare** for a change

'tappa *nf* stop; (parte di viaggio) stage

tappa'buchi *nm inv* stopgap

tap'pare *vt* plug; (cork (bottiglia); **~e la bocca a qcno** fam shut sb up. **~si vr** ~si gli occhi cover one's eyes; **~si il naso** hold one's nose; **~si le orecchie** put one's fingers in one's ears

tappa'rella *nf fam* roller blind

tappe'tino *nm* mat; Comput mouse mat. **~ antiscivolo** safety bathmat

tap'peto *nm* carpet; (piccolo) rug; **andare al ~** (pugilato:) hit the canvas; **mandare qcno al ~** knock sb down

tappez'z|are *vt* paper (pareti); (rivestire) cover. **~e'ria** *nf* tapestry; (di carta) wallpaper; (arte) upholstery. **~i'ere** *nm* upholsterer; (imbianchino) decorator

'tappo *nm* plug; (di sughero) cork; (di metallo, per penna) top; (fam: persona piccola) dwarf. **~ di sughero** cork

'tara *nf* (difetto) flaw; (ereditaria) hereditary defect; (peso) tare

ta'rantola *nf* tarantula

ta'ra|re *vt* calibrate (strumento). **~to** *a* Comm discounted; Techn calibrated; Med with a hereditary defect; fam crazy

tarchi'ato *a* stocky

tar'dare *vi* be late ● *vt* delay

'tard|i *adv* late; **al più ~i** at the latest; **più ~i** later [on]; **sul ~i** late in the day; **far ~i** (essere in ritardo) be late; (con gli amici) stay up late; **a più ~i** see you later ● *a inv* late. **tar'divo** *a* late; (bambino) retarded. **~o a** *a* slow; (tempo) late

'targ|a *nf* plate; *Auto* numberplate. **~ di circolazione** numberplate.

tar'gato *a* **un'auto targata...** a car with the registration number... **~'hetta** *nf* (*su porta*) nameplate; (*sulla valigia*) name tag

ta'rif|fa *nf* rate, tariff. **~'fario** *nm* price list

tar'larsi *vr* get wormeaten. **'tarlo** *nm* woodworm

'tarma *nf* moth. **tar'marsi** *vr* get moth-eaten

ta'rocco *nm* tarot; **ta'rocchi** *nm* tarot

tartagli'are *vi* stutter

'tartaro *a & nm* tartar

tarta'ruga *nf* tortoise; (*di mare*) turtle; (*per pettine ecc*) tortoiseshell

tartas'sare *vt* (*angariare*) harass

tar'tina *nf* canapé

tar'tufo *nm* truffle

'tasca *nf* pocket; (*in borsa*) compartment; (*da ~* pocket *attrib*; **avere le tasche piene di qcsa** *fam* have had a bellyful of sth. **~ da pasticciere** icing bag

ta'scabile *a* pocket *attrib* ● *nm* paperback

tasca'pane *nm inv* haversack

ta'schino *nm* breast pocket

'tassa *nf* tax; (*discrizione ecc*) fee; (*doganale*) duty. **~ di circolazione** road tax. **~ d'iscrizione** registration fee

tas'sametro *nm* taximeter

tas'sare *vt* tax

tassa|tiva'mente *adv* without question. **~'tivo** *a* peremptory

tassazi'one *nf* taxation

tas'sello *nm* wedge; (*di stoffa*) gusset

tassì *nm inv* taxi. **tas'sista** *nmf* taxi driver

'tasso¹ *nm* *Bot* yew; (*animale*) badger

'tasso² *nm* *Comm* rate. **~ di cambio** exchange rate. **~ di interesse** interest rate

ta'stare *vt* feel; (*sondare*) sound; **~**

il terreno *fig* test the water or the ground

tasti'e|ra *nf* keyboard. **~'rista** *nmf* keyboarder

'tasto *nm* key; (*tatto*) touch. **~ delicato** *fig* touchy subject. **~ funzione** *Comput* function key. **~ tabulatore** tab key

ta'stoni: a ~ *adv* gropingly

'tattica *nf* tactics *pl*

'tattico *a* tactical

'tatto *nm* (*senso*) touch; (*accortezza*) tact; **aver ~** be tactful

tatu'a|ggio *nm* tattoo. **~re** *vt* tattoo

'tavola *nf* table; (*illustrazione*) plate; (*asse*) plank. **~ calda** snackbar

tavo'lato *nm* boarding; (*pavimento*) wood floor

tavo'letta *nf* bar; (*medicinale*) tablet; **andare a ~** *Auto* drive flat out

tavo'lino *nm* small table

'tavolo *nm* table. **~ operatorio** *Med* operating table

tavo'lozza *nf* palette

'tazza *nf* cup; (*del water*) bowl. **~ da caffè/tè** coffee-cup/teacup

taz'zina *nf* **~ da caffè** espresso coffee cup

T.C.I. *abbr* (**Touring Club Italiano**) Italian Touring Club

te *pers pron* you; **te l'ho dato** I gave it to you

tè *nm inv* tea

tea'trale *a* theatrical

te'atro *nm* theatre. **~ all'aperto** open-air theatre. **~ di posa** *Cinema* set. **~ tenda** marquee for *theatre performances*

'tecnico, -a *a* technical ● *nmf* technician ● *nf* technique

tec'nigrafo *nm* drawing board

tecno|lo'gia *nf* technology. **~'logico** *a* technological

te'desco, -a *a & nmf* German

tedi|o *nm* tedium. **~'oso** *a* tedious

te'game *nm* saucepan

'teglia *nf* baking tin

'tegola *nf* tile; *fig* blow

tei'era *nf* teapot

tek nm teak

'tela nf cloth; (per quadri, vele) canvas; Theat curtain. **~ cerata** oilcloth. **~ di lino** linen

te'laio nm (di bicicletta, finestra) frame; Auto chassis; (per tessere) loom

tele'camera nf television camera

teleco|man'dato a remote-controlled, remote control attrib. **~'mando** nm remote control

Telecom Italia nf Italian State telephone company

telecomunicazi'oni nfpl telecommunications

tele'cro|naca nf [television] commentary. **~naca diretta** live [television] coverage. **~naca regi-strata** recording. **~'nista** nmf television commentator

tele'ferica nf cableway

telefo'na|re vt/i [tele]phone, ring. **~ta** nf call. **~ta interurbana** long-distance call

telefonica'mente adv by [tele]phone

tele'fo|nico a [tele]phone attrib. **~'nino** nm mobile [phone]. **~'nista** nmf operator

te'lefono nm [tele]phone. **~ senza filo** cordless [phone]. **~ a gettoni** pay phone, coin-box. **~ interno** internal telephone. **~ a schede** cardphone

telegior'nale nm television news sg

telegra'fare vt telegraph. **tele-'grafico** a telegraphic; (risposta) monosyllabic; **sii telegrafico** keep it brief

tele'gramma nm telegram

tele'matica nf data communications, telematics

teleno'vela nf soap opera

teleobiet'tivo nm telephoto lens

telepa'tia nf telepathy

telero'manzo nm television serial

tele'schermo nm television screen

tele'scopio nm telescope

teleselezi'one nf subscriber trunk dialling, STD; **chiamare in ~** dial direct

telespetta'tore, **-'trice** nmf viewer

tele'text® nm Teletext®

tele'video nm videophone

televisi'one nf television; **guarda-re la ~** watch television

televi'sivo a television attrib; **operatore ~** television camera-man; **apparecchio ~** television set

televi'sore nm television [set]

'tema nm theme; Sch essay. **te'ma-tica** nf main theme

teme'rario a reckless

te'me|re vt be afraid of, fear ● vi be afraid, fear

tem'paccio nm filthy weather

temperama'tite nm inv pencil-sharpener

tempera'mento nm temperament

tempe'ra|re vt temper; sharpen (matita). **~to** a temperate. **~'tura** nf temperature. **~'tura ambiente** room temperature

tempe'rino nm penknife

tem'pe|sta nf storm. **~sta di neve** snowstorm. **~sta di sabbia** sand-storm

tempe|stiva'mente adv quickly. **~'stivo** a timely. **~'stoso** a stormy

'tempia nf Anat temple

'tempio nm Relig temple

tem'pismo nm timing

'tempo nm time; (atmosferico) weather; Mus tempo; Gram tense; (di film) part; (di partita) half; **a suo ~** in due course; **~ fa** some time ago; **un ~** once; **ha fatto il suo ~** it's superannuated. **~ supplementare** Sport extra time, overtime Am. **~'rale** a temporal ● nm [thunder]storm. **~ranea-'mente** adv temporarily. **~'raneo** a temporary. **~'reggi'are** vi play for time

tem'prare vt temper

te'nac|e a tenacious. **~ia** nf tenac-ity

te'naglia *nf* pincers *pl*

'tenda *nf* curtain; (*per campeggio*) tent; (*tendone*) awning. ~ a ossigeno oxygen tent

ten'denz|a *nf* tendency. ~ial'mente *adv* by nature. ~l'oso *a* tendentious

'tendere *vt* (*allargare*) stretch [out]; (*tirare*) tighten; (*porgere*) hold out; fig lay (*trappola*) ● *vi* ~ a aim at; (*essere portato a*) tend to

'tendine *nm* tendon

ten'do|ne *nm* awning; (*di circo*) tent. ~poli *nf inv* tent city

'tenebre *nfpl* darkness. tene'broso *a* gloomy

te'nente *nm* lieutenant

tenera'mente *adv* tenderly

te'ner|e *vt* hold; (*mantenere*) keep; (*gestire*) run; (*prendere*) take; (*seguire*) follow; (*considerare*) consider ● *vi* hold; ~ci a, ~e a be keen on; ~e per support (*squadra*). ~si *vr* hold on (a to); (*in una condizione*) keep oneself; (*seguire*) stick to; ~si indietro stand back

tene'rezza *nf* tenderness. 'tenero *a* tender

'tenia *nf* tapeworm

'tennis *nm* tennis. ~ da tavolo table tennis. ten'nista *nmf* tennis player

te'nore *nm* standard; *Mus* tenor; a ~ di legge by law. ~ di vita standard of living

tensi'one *nf* tension; *Electr* voltage; alta ~ high voltage

ten'tacolo *nm* tentacle

ten'ta|re *vt* attempt; (*sperimentare*) try; (*indurre in tentazione*) tempt. ~'tivo *nm* attempt. ~zi'one *nf* temptation

tenten|na'mento *nm* wavering. ~'nare *vi* waver

te'nue *a* fine; (*debole*) weak; (*esiguo*) small; (*leggero*) slight

te'nuta *nf* (*capacità*) capacity; (*Sport: resistenza*) stamina; (*possedimento*) estate; (*divisa*) uniform; (*abbigliamento*) clothes *pl*; a

~ d'aria airtight. ~ di strada road holding

teolo'gia *nf* theology. teo'logico *a* theological. te'ologo *nm* theologian

teo'rema *nm* theorem

teo'ria *nf* theory

teorica'mente *adv* theoretically. te'orico *a* theoretical

te'pore *nm* warmth

'teppa *nf* mob. tep'pismo *nm* hooliganism. tep'pista *nmf* hooligan

tera'peutico *a* therapeutic. tera'pia *nf* therapy

tergicri'stallo *nm* windscreen wiper, windshield wiper *Am*

tergilu'notto *nm* rear windscreen wiper

tergiver'sare *vi* hesitate

'tergo *nm* a ~ behind; segue a ~ please turn over, PTO

ter'male *a* thermal; stazione ~ spa. 'terme *nfpl* thermal baths

'termico *a* thermal

termi'na|le *a & nm* terminal; malato ~e terminally ill person. ~re *vt/i* finish, end. 'termine *nm* (*limite*) limit; (*fine*) end; (*condizione, espressione*) term

terminolo'gia *nf* terminology

ter'mite *nf* termite

termoco'perta *nf* electric blanket

ter'mometro *nm* thermometer

'termos *nm inv* thermos®

termosi'fone *nm* radiator; (*sistema*) central heating

ter'mostato *nm* thermostat

'terra *nf* earth; (*regione*) land; (*terreno*) ground; (*argilla*) clay; (*cosmetico*) dark face powder (*which gives the impression of a tan*); a ~ (*sulla costa*) ashore; (*installazioni*) onshore; per ~ on the ground; sotto ~ underground. ~'cotta *nf* terracotta; vasellame di ~cotta earthenware. ~'ferma *nf* dry land. ~pi'eno *nm* embankment

ter'razza *nf*, ~o *nm* balcony

terre'moto, -a *a* (*zona*) affected by an earthquake ● *nmf* earth-

quake victim. **terre'moto** *nm*
earthquake

ter'reno *a* earthly ●*nm* ground;
(*suolo*) soil; (*proprietà terriera*)
land; **perdere/guadagnare ~**
lose/gain ground. **~ di gioco**
playing field

ter'restre *a* terrestrial; **esercito
~** land forces *pl*

ter'ribil|e *a* terrible. **~'mente** *adv*
terribly

terrifi'cante *a* terrifying

ter'riccio *nm* potting compost

territori'ale *a* territorial. **terri'torio** *nm* territory

ter'rore *nm* terror

terro'ris|mo *nm* terrorism. **~ta**
nmf terrorist

terroriz'zare *vt* terrorize

'terso *a* clear

ter'zetto *nm* trio

terzi'ario *a* tertiary

'terzo *a* third; **di terz'ordine** (*locale, servizio*) third-rate; **fare il ~
grado a qn** give sb the third degree; **la terza età** the third age
●*nm* third; **terzi** *pl* Jur third
party *sg*. **terz'ultimo, -a** *a* & *nmf*
third from last

'tesa *nf* brim

'teschio *nm* skull

'tesi *nf inv* thesis

'teso *pp* **di tendere** ●*a* taut; *fig*
tense

tesor|e'ria *nf* treasury. **~i'ere** *nm*
treasurer

te'soro *nm* treasure; (*tesoreria*)
treasury

'tessera *nf* card; (*abbonamento
all'autobus*) season ticket

'tessere *vt* weave; hatch (*complotto*)

tesse'rino *nm* travel card

'tessile *a* textile. **tessili** *nmpl* tex-
tiles; (*operai*) textile workers

tessi|'tore, -'trice *nmf* weaver.
~'tura *nf* weaving

tes'suto *nm* fabric; *Anat* tissue

'testa *nf* head; (*cervello*) brain;
essere in ~ a be ahead of; **in ~**
Sport in the lead; **~ o croce?**

heads or tails?; **fare ~ o croce**
have a toss-up to decide

'testa-'coda *nm inv* **fare un ~**
spin right round

testa'mento *nm* will; **T~** *Relig*
Testament

testar'daggine *nf* stubbornness.
te'stardo *a* stubborn

te'stata *nf* head; (*intestazione*)
heading; (*colpo*) butt

'teste *nmf* witness

te'sticolo *nm* testicle

testi'mon|e *nmf* witness. **~e ocu-
lare** eye witness

testi'monial *nmf inv* celebrity who
promotes a brand of cosmetics

testimoni'|anza *nf* testimony;
falsa **~anza** *Jur* perjury. **~'are** *vt*
testify to ● *vi* testify, give evidence

'testo *nm* text; **far ~** be an authority

te'stone, -a *nmf* blockhead

testu'ale *a* textual

'tetano *nm* tetanus

'tetro *a* gloomy

tetta'rella *nf* teat

'tetto *nm* roof. **~ apribile** (*di auto*)
sunshine roof. **tet'toia** *nf* roofing.
tet'tuccio *nm* **tettuccio apribile**
sun-roof

'Tevere *nm* Tiber

ti *pers pron* you; (*riflessivo*) yourself; **ti ha dato un libro** he gave
you a book; **lavati le mani** wash
your hands; **eccoti!** here you are!;
sbrigati! hurry up!

ti'ara *nf* tiara

tic *nm inv* tic

ticchet't|are *vi* tick. **~io** *nm* tick-
ing

'ticchio *nm* tic; (*ghiribizzo*) whim

'ticket *nm inv* (*per farmaco, esame*) amount paid by National
Health patients

tiepida'mente *adv* half-heartedly.
ti'epido *a anche fig* lukewarm

ti'fare *vi* **~ per** shout for. **'tifo** *nm*
Med typhus; **fare il tifo per** *fig* be a
fan of

tifo'idea *nf* typhoid

ti'fone *nm* typhoon

ti'foso, -a nmf fan

'tiglio nm lime

ti'grato a gatto ~ tabby [cat]

'tigre nf tiger

'tilde nmf tilde

tim'ballo nm Culin pie

tim'brare vt stamp; **~ il cartellino** clock in/out

'timbro nm stamp; (di voce) tone

timida'mente adv timidly, shyly. **timi'dezza** nf timidity, shyness. **'timido** a timid, shy

'timo nm thyme

ti'mone nm rudder. **~i'ere** nm helmsman

ti'more nm fear; (soggezione) awe. **timo'roso** a timorous

'timpano nm eardrum; Mus kettle-drum

ti'nello nm dining-room

'tingere vt dye; (macchiare) stain. **~si** ⟨viso, cielo:⟩ be tinged (di with); **~si i capelli** have one's hair dyed; (da solo) dye one's hair

'tino nm, **ti'nozza** nf tub

'tint|a nf dye; (colore) colour; **in ~a unita** plain. **~a'rella** nf fam suntan

tintin'nare vi tinkle

'tinto pp di **tingere**. **~'ria** nf (negozio) cleaner's. **tin'tura** nf dyeing; (colorante) dye.

'tipico a typical

'tipo nm type; (fam: individuo) chap, guy

tipogra'fia nf printery; (arte) typography. **tipo'grafico** a typographic[al]. **ti'pografo** nm printer

tip tap nm tap dancing

ti'raggio nm draught

tiramisù nm inv dessert made of coffee-soaked sponge, eggs, Marsala, cream and cocoa powder

tiran|neggi'are vt tyrannize. **~'nia** nf tyranny. **ti'ranno, -a** a tyrannical ● nmf tyrant

tirapi'edi nm inv pej hanger-on

ti'rar|e vt pull; (gettare) throw; kick ⟨palla⟩; (sparare) fire; (trac-ciare) draw; (stampare) print ● vi pull; ⟨vento:⟩ blow; ⟨abito:⟩ be

tight; (sparare) fire; **~e avanti** get by; **~e su** (crescere) bring up; (da terra) pick up; **tirar su col naso** sniffle. **~si** vr **~si indietro** fig back out, pull out

tiras'segno nm target shooting; (alla fiera) rifle range

ti'rata nf (strattone) pull, tug; **in una ~** in one go

tira'tore nm shot. **~ scelto** marksman

tira'tura nf printing; (di giornali) circulation; (di libri) [print] run

tirchie'ria nf meanness. **'tirchio** a mean

tiri'tera nf spiel

'tiro nm (traino) draught; (lancio) throw; (sparo) shot; (scherzo) trick. **~ con l'arco** archery. **~ alla fune** tug-of-war. **~ a segno** rifle-range

tiro'cinio nm apprenticeship

ti'roide nf thyroid

Tir'reno nm **Il [mar] ~** the Tyrrhenian Sea

ti'sana nf herb[al] tea

tito'lare a regular ● nmf principal; (proprietario) owner; (calcio) regular player

'titolo nm title; (accademico) qualification; Comm security; **a ~ di** as; **a ~ di favore** as a favour. **titoli** pl **di studio** qualifications

titu'ba|nte a hesitant. **~nza** nf hesitation. **~re** vi hesitate

tivù nf inv fam TV, telly

'tizio nm fellow

tiz'zone nm brand

toc'cante a touching

toc'ca|re vt touch; touch on (argomento); (tastare) feel; (riguardare) concern ● vi **~re a** (capitare) happen to; **mi tocca aspettare** I'll have to wait; **tocca a te** it's your turn; (da pagare da bere) it's your round

tocca'sana nm inv cure-all

'tocco nm touch; (di pennello, orologio) stroke; (di pane ecc) chunk ● a fam crazy, touched '**toga** *nf* toga; *(accademica, di magistrato)* gown

'**toglier|e** *vt* take off *(coperta)*; take away *(bambino da scuola, sete, Math)*; take out, remove *(dente)*; **~e qcsa di mano a qcno** take sth away from sb; **~e qcno dei guai** get sb out of trouble; **ciò non toglie che...** nevertheless... **~si** *vr* take off *(abito)*; **~si la vita** take one's [own] life; **togliti dai piedi!** get out of here!

toilette *nf inv*, **to'letta** *nf* toilet; *(mobile)* dressing table

tolle'ra|nte *a* tolerant. **~nza** *nf* tolerance. **~re** *vt* tolerate

'**tolto** *pp di* **togliere**

to'**maia** *nf* upper

'**tomba** *nf* grave, tomb

tom'**bino** *nm* manhole cover

'**tombola** *nf* bingo; *(caduta)* tumble

'**tomo** *nm* tome

'**tonaca** *nf* habit

tonalità *nf inv Mus* tonality

'**tondo** *a* ● *nm* circle

'**tonfo** *nm* thud; *(in acqua)* splash

'**tonico** *a* & *nm* tonic

tonifi'**care** *vt* brace

tonnel'la|ggio *nm* tonnage. **~ta** *nf* ton

'**tonno** *nm* tuna [fish]

'**tono** *nm* tone

ton'**sil|la** *nf* tonsil. **~'lite** *nf* tonsillitis

'**tonto** *a fam* thick

top *nm inv (indumento)* sun-top

to'**pazio** *nm* topaz

'**topless** *nm inv* **in ~** topless

'**topo** *nm* mouse. **~ di biblioteca** *fig* bookworm

topogra'**fia** *nf* topography. **topo-'grafico** *a* topographic[al]

to'**ponimo** *nm* place name

'**toppa** *nf (rattoppo)* patch; *(serratura)* keyhole

to'**race** *nm* chest. **to'racico** *a* thoracic; **gabbia toracica** rib cage

'**torba** *nf* peat

'**torbido** *a* cloudy; *fig* troubled

'**torcer|e** *vt* twist; wring [out] *(biancheria)*. **~si** *vr* twist

'**torchio** *nm* press

'**torcia** *nf* torch

torci'**collo** *nm* stiff neck

'**tordo** *nm* thrush

to'**rero** *nm* bullfighter

To'**rino** *nf* Turin

tor'**menta** *nf* snowstorm

tormen'**tare** *vt* torment. **tor'mento** *nm* torment

torna'**conto** *nm* benefit

tor'**nado** *nm* tornado

tor'**nante** *nm* hairpin bend

tor'**nare** *vi* return, go/come back; *(ridiventare)* become again; *(conto:)* add up; **~ a sorridere** become happy again

tor'**neo** *nm* tournament

'**tornio** *nm* lathe

'**torno** *nm* **togliersi di ~** get out of the way

'**toro** *nm* bull; *Astr* Taurus

tor'**pedin|e** *nf* torpedo. **~'i'era** *nf* torpedo boat

tor'**pore** *nm* torpor

'**torre** *nf* tower; *(scacchi)* castle. **~ di controllo** control tower

torrefazi'**one** *nf* roasting

tor'**ren|te** *nm* torrent, mountain stream; *(fig: di lacrime)* flood. **~zi'ale** *a* torrential

tor'**retta** *nf* turret

'**torrido** *a* torrid

torri'**one** *nm* keep

tor'**rone** *nm* nougat

'**torso** *nm* torso; *(di mela, pera)* core; **a ~ nudo** bare-chested

'**torsolo** *nm* core

'**torta** *nf* cake; *(crostata)* tart

tortel'**lini** *nmpl* tortellini, *small packets of pasta stuffed with pork, ham, Parmesan and nutmeg*

torti'**era** *nf* baking tin

tor'**tino** *nm* pie

'**torto** *pp di* **torcere** ● *a* twisted ● *nm* wrong; *(colpa)* fault; **aver ~** be wrong; **a ~** wrongly

tor'**tora** *nf* turtle-dove

tortu'**oso** *a* winding; *(ambiguo)* tortuous

tor'tu|ra *nf* torture. **~'rare** *vt* torture

'torvo *a* grim

to'sare *vt* shear

tosa'tura *nf* shearing

To'scana *nf* Tuscany

'tosse *nf* cough

'tossico *a* toxic ●*nm* poison. **tossi'comane** *nmf* drug addict, drug user

tos'sire *vi* cough

tosta'pane *nm inv* toaster

to'stare *vt* toast ⟨*pane*⟩; roast ⟨*caffè*⟩

'tosto *adv* ⟨*subito*⟩ soon ●*a fam* cool

tot *a inv* una cifra ~ such and such a figure ●*nm un* ~ so much

to'tal|e *a & nm* total. **~ità** *nf* entirety; **la ~ità dei presenti** all those present

totali'tario *a* totalitarian

totaliz'zare *vt* total; score ⟨*punti*⟩

total'mente *adv* totally

'totano *nm* squid

toto'calcio *nm* ≈ [football] pools *pl*

tournée *nf inv* tour

to'vagli|a *nf* tablecloth. **~'etta** *nf* ~**etta [all'americana]** place mat. **~'olo** *nm* napkin

'tozzo *a* squat ●*nm* ~ **di pane** stale piece of bread

tra = fra

trabal'la|nte *a* staggering; ⟨*sedia*⟩ rickety, wonky. **~re** *vi* stagger; ⟨*veicolo:*⟩ jolt

tra'biccolo *nm fam* contraption; ⟨*auto*⟩ jalopy

traboc'care *vi* overflow

traboc'chetto *nm* trap

tracan'nare *vt* gulp down

'tracc|ia *nf* track; ⟨*orma*⟩ footstep; ⟨*striscia*⟩ trail; ⟨*residuo*⟩ trace; *fig* sign. **~'are** *vt* trace; sketch out ⟨*schema*⟩; draw ⟨*linea*⟩. **~'ato** *nm* ⟨*schema*⟩ layout

tra'chea *nf* windpipe

tra'colla *nf* shoulder-strap; **borsa a ~** shoulder-bag

tra'collo *nm* collapse

tradi'mento *nm* betrayal; *Pol* treason

tra'di|re *vt* betray; be unfaithful to ⟨*moglie, marito*⟩. **~'tore**, **~'trice** *nmf* traitor

tradizio'na|le *a* traditional. **~'lista** *nmf* traditionalist. **~l'mente** *adv* traditionally. **tradizi'one** *nf* tradition

tra'dotto *pp di* tradurre

tra'du|rre *vt* translate. **~t'tore**, **~t'trice** *nmf* translator. **~ttore elettronico** electronic phrasebook. **~zi'one** *nf* translation

tra'ente *nmf Comm* drawer

trafe'lato *a* breathless

traffi'ca|nte *nmf* dealer. **~nte di droga** [drug] pusher. **~re** *vi ⟨affaccendarsi⟩* busy oneself; **~re in** *pej* traffic in. **'traffico** *nm* traffic; *Comm* trade

tra'figgere *vt* stab; ⟨*straziare*⟩ pierce

tra'fila *nf fig* rigmarole

trafo'rare *vt* bore, drill. **tra'foro** *nm* boring; ⟨*galleria*⟩ tunnel

trafu'gare *vt* steal

tra'gedia *nf* tragedy

traghet'tare *vt* ferry. **tra'ghetto** *nm* ferrying; ⟨*nave*⟩ ferry

tragica'mente *adv* tragically. **'tragico** *a* tragic ●*nm* ⟨*autore*⟩ tragedian

tra'gitto *nm* journey; ⟨*per mare*⟩ crossing

tragu'ardo *nm* finishing post; ⟨*meta*⟩ goal

traiet'toria *nf* trajectory

trai'nare *vt* drag; ⟨*rimorchiare*⟩ tow

tralasci'are *vt* interrupt; ⟨*omettere*⟩ leave out

'tralcio *nm Bot* shoot

tra'liccio *nm* ⟨*graticcio*⟩ trellis

tram *nm inv* tram, streetcar *Am*

'trama *nf* weft; ⟨*di film ecc*⟩ plot

traman'dare *vt* hand down

tra'mare *vt* weave; ⟨*macchinare*⟩ plot

tram'busto *nm* turmoil, hullaballoo

trame'stio *nm* bustle

tramez'zino *nm* sandwich

tra'mezzo *nm* partition

'**tramite** *prep* through ● *nm* link; **fare da** ~ act as go-between

tramon'tana *nf* north wind

tramon'tare *vi* set; (*declinare*) decline. **tra'monto** *nm* sunset; (*declino*) decline

tramor'tire *vt* stun ● *vi* faint

trampo'lino *nm* springboard; (*per lo sci*) ski-jump

'**trampolo** *nm* stilt

tramu'tare *vt* transform

'**trancia** *nf* shears *pl*; (*fetta*) slice

tra'nello *nm* trap

trangugi'are *vt* gulp down, gobble up

'**tranne** *prep* except

tranquilla'mente *adv* peacefully

tranquil'lante *nm* tranquillizer

tranquil'lità *nf* calm; (*di spirito*) tranquillity. ~**z'zare** *vt* reassure. **tran'quillo** *a* quiet; (*pacifico*) peaceful; (*coscienza*) easy

transat'lantico *a* transatlantic ● *nm* ocean liner

tran'sa|tto *pp* di **transigere**. ~**zi'one** *nf* Comm transaction

tran'senna *nf* (*barriera*) barrier

tran'sigere *vi* reach an agreement; (*cedere*) yield

transi'tabile *a* passable. ~**re** *vi* pass

transi'tivo *a* transitive

'**transi|to** *nm* transit; **diritto di** ~**to** right of way; "**divieto di** ~**to**" "no thoroughfare". ~'**torio** *a* transitory. ~**zi'one** *nf* transition

tran'tran *nm* fam routine

tranvi'ere *nm* tram driver, streetcar driver Am

'**trapano** *nm* drill

trapas'sare *vt* go [right] through ● *vi* (*morire*) pass away

tra'passo *nm* passage

trape'lare *vi* (*liquido, fig:*) leak out

tra'pezio *nm* trapeze; Math trapezium

trapian'tare *vt* transplant. ~'**anto** *nm* transplant

'**trappola** *nf* trap

tra'punta *nf* quilt

trarre *vt* draw; (*ricavare*) obtain; ~ **in inganno** deceive

trasa'lire *vi* start

trasan'dato *a* shabby

trasbor'dare *vt* transfer; Naut tran[s]ship ● *vi* change. **tra'sbordo** *nm* trans[s]hipment

tra'scendere *vt* transcend ● *vi* (*eccedere*) go too far

trasci'nar|e *vt* drag; (*fig: entusiasmo:*) carry away. ~**si** *vr* drag oneself

tra'scorrere *vt* spend ● *vi* pass

tra'scri|tto *pp* di **trascrivere**. ~**vere** *vt* transcribe. ~**zi'one** *nf* transcription

trascu'ra|bile *a* negligible. ~**re** *vt* neglect; (*non tenere conto di*) disregard. ~'**tezza** *nf* negligence. ~**to** *a* negligent; (*curato male*) neglected; (*nel vestire*) slovenly

trasecolato *a* amazed

trasferi'mento *nm* transfer; (*trasloco*) move

trasfe'ri|re *vt* transfer. ~**rsi** *vr* move

tra'sferta *nf* transfer; (*indennità*) subsistence allowance; Sport away match; **in** ~ (*impiegato*) on secondment; **giocare in** ~ play away

trasfigu'rare *vt* transfigure

trasfor'ma|re *vt* transform; (*in rugby*) convert. ~'**tore** *nm* transformer. ~**zi'one** *nf* transformation; (*in rugby*) conversion

trasfor'mista *nmf* (*artista*) quick-change artist

trasfusi'one *nf* transfusion

trasgre'dire *vt* disobey; Jur infringe

trasgre'ditrice *nf* transgressor

trasgres|si'one *nf* infringement. ~'**sivo** *a* intended to shock. ~'**sore** *nm* transgressor

tra'slato *a* metaphorical

traslo'car|e *vt* move ● *vi*, ~**si** *vr* move house. **tra'sloco** *nm* removal

tra'smesso pp di trasmettere

tra'smett|ere vt pass on; TV, Radio broadcast; Techn, Med transmit. **~i'tore** nm transmitter

trasmis'si|bile a transmissible. **~'one** nf transmission; TV, Radio programme

trasmit'tente transmitter ● nf broadcasting station

traso'gna|re vi day-dream. **~to** a dreamy

traspa'ren|te a transparent. **~za** nf transparency; **in** ~**za** against the light. **traspa'rire** vi show [through]

traspi'ra|re vi perspire; fig transpire. **~zi'one** nf perspiration

tra'sporre vt transpose

traspor'tare vt transport; **lasciarsi** ~ **da** get carried away by. **tra'sporto** nm transport; (passione) passion

trastul'lar|e vt amuse. **~si** vr amuse oneself

trasu'dare vt ooze with ● vi sweat

trasver'sale a transverse

trasvo'la|re vt fly over ● vi **~re su** fig skim over. **~ta** nf crossing [by air]

'tratta nf (traffico illegale) trade; Comm draft

trat'tabile a or nearest offer, o.n.o.

tratta'mento nm treatment. **~ di riguardo** special treatment

trat'ta|re vt treat; (commerciare in) deal in; (negoziare) negotiate ● vi **~re di** deal with. **~rsi** vr **di che si tratta?** what is it about?; **si tratta di...** it's about... **~to** nm treaty; (opera scritta) treatise

tratteggi'are vt outline; (descrivere) sketch

tratte'ner|e vt (far restare) keep; hold (respiro, in questura); hold back (lacrime, riso); (frenare) restrain; (da paga) withhold; **sono stato trattenuto** (ritardato) I was o got held up. **~si** vr restrain oneself; (fermarsi) stay; **~si su** (indugiare) dwell on. **tratteni'mento**

nm entertainment; (ricevimento) party

tratte'nuta nf deduction

trat'tino nm dash; (in parole composte) hyphen

'tratto pp di trarre ● nm (di spazio, tempo) stretch; (di penna) stroke; (linea) line; (brano) passage; **tratti** pl (lineamenti) features; **a tratti** at intervals; **ad un** ~ suddenly

trat'tore nm tractor

tratto'ria nf restaurant

'trauma nm trauma. **trau'matico** a traumatic. **~tiz'zare** vt traumatize

tra'vaglio nm labour; (angoscia) anguish

trava'sare vt decant

'trave nf beam

tra'veggole nfpl **avere le** ~ be seeing things

tra'versa nf crossbar; **è una** ~ **di Via Roma** it's off Via Roma, it crosses via Roma

traver'sa|re vt cross. **~ta** nf crossing

traver'sie nfpl misfortunes

traver'sina nf Rail sleeper

tra'vers|o a crosswise ● adv **di** ~**o** crossways; **andare di** ~**o** (cibo) go down the wrong way; **camminare di** ~**o** not walk in a straight line; **guardare qcno di** ~**o** look askance at sb. **~one** nm (in calcio) cross

travesti'mento nm disguise

trave'sti|re vt disguise. **~rsi** vr disguise oneself. **~to** a disguised ● nm transvestite

travi'are vt lead astray

travi'sare vt distort

travol'gente a overwhelming

tra'vol|gere vt sweep away; (sopraffare) overwhelm. **~to** pp di travolgere

trazi'one nf traction. ~ **anteriore/posteriore** front-/rear-wheel drive

tre a & nm three

trebbi'a|re vt thresh

'treccia nf plait, braid

tre'cento *a & nm* three hundred; **il T~** the fourteenth century

tredi'cesima *nf* extra month's salary paid as a Christmas bonus

'tredici *a & nm* thirteen

'tregua *nf* truce; *fig* respite

tre'mare *vi* tremble; *(di freddo)* shiver. **trema'rella** *nf fam* jitters *pl*

tremenda'mente *adv* terribly. **tre'mendo** *a* terrible; **ho una fame tremenda** I'm terribly hungry

tremen'tina *nf* turpentine

tre'mila *a & nm* three thousand

'tremito *nm* tremble

tremo'lare *vi* shake; *(luce:)* flicker. **tre'more** *nm* trembling

tre'nino *nm* miniature railway

'treno *nm* train

'tren|ta *a & nm* thirty; **~ta e lode** top marks. **~tatré giri** *nm inv* LP. **~'tenne** *a & nmf* thirty-year-old. **~'tesimo** *a & nm* thirtieth. **~'tina** *nf* **una ~tina** of about thirty

trepi'dare *vi* be anxious. **'trepido** *a* anxious

treppi'ede *nm* tripod

'tresca *nf* intrigue; *(amorosa)* affair

'trespolo *nm* perch

triango'lare *a* triangular. **tri-'angolo** *nm* triangle

tri'bale *a* tribal

tribo'la|re *vi (soffrire)* suffer; *(fare fatica)* go through all kinds of trials and tribulations. **~zi'one** *nf* tribulation

tribù *nf inv* tribe

tri'buna *nf* tribune; *(per uditori)* gallery; *Sport* stand. **~ coperta** stand

tribu'nale *nm* court

tribu'tare *vt* bestow

tribu'tario *a tax attrib.* **tri'buto** *nm* tribute; *(tassa)* tax

tri'checo *nm* walrus

tri'ciclo *nm* tricycle

trico'lore *a* three-coloured ● *nm (bandiera)* tricolour

tri'dente *nm* trident

trien'nale *a (ogni tre anni)* three-yearly; *(lungo tre anni)* three-year.

tri'ennio *nm* three-year period

tri'foglio *nm* clover

trifo'lato *a* sliced thinly and cooked with olive oil, parsley and garlic

'triglia *nf* mullet

trigonome'tria *nf* trigonometry

tril'lare *vi* trill

'trillo *nm* trill

trilo'gia *nf* trilogy

tri'mestre *nm* quarter; *Sch* term

'trina *nf* lace

trin'cea *nf* trench. **~'rare** *vt* entrench

trincia'pollo *nm inv* poultry shears *pl*

trinci'are *vt* cut up

Trinità *nf* Trinity

'trio *nm* trio

trion'fa|le *a* triumphal. **~nte** *a* triumphant. **~re** *vi* triumph; **~re su** triumph over. **tri'onfo** *nm* triumph

tripli'care *vt* triple. **'triplice** *a* triple; **in triplice [copia]** in triplicate. **'triplo** *a* treble ● *nm* **il triplo (di)** three times as much (as)

'trippa *nf* tripe; *(fam: pancia)* belly

'trist|e *a* sad; *(luogo)* gloomy. **tri-'stezza** *nf* sadness. **~o** *a* wicked; *(meschino)* miserable

trita'carne *nm inv* mincer. **~ghi'accio** *nm inv* ice-crusher

tri'tare *vt* mince. **'trito** *a* **trito e ritrito** well-worn, trite

'trittico *nm* triptych

tritu'rare *vt* chop finely

triumvi'rato *nm* triumvirate

tri'vella *nf* drill. **trivel'lare** *vt* drill

trivi'ale *a* vulgar

tro'feo *nm* trophy

'trogolo *nm (per maiali)* trough

'troia *nf* sow; *vulg* bitch; *(sessuale)* whore

'tromba *nf* trumpet; *Auto* horn; *(delle scale)* well. **~ d'aria** whirlwind

trom'bare *vt vulg* screw; *(fam: in esame)* fail

trom'b|etta *nm* toy trumpet. **~one** *nm* trombone

trom'bosi *nf* thrombosis

tron'care *vt* sever; truncate ⟨*parola*⟩

'tronco *a* truncated; **licenziare in ~** fire on the spot ● *nm* trunk; ⟨*di strada*⟩ section. **tron'cone** *nm* stump

troneggi'are *vi* ~ **su** tower over

'trono *nm* throne

tropi'cale *a* tropical. **'tropico** *nm* tropic

'troppo *a* too much; ⟨*con nomi plurali*⟩ too many ● *pron* too much; ⟨*plurale*⟩ too many; ⟨*troppo tempo*⟩ too long; **troppi** ⟨*troppa gente*⟩ too many people ● *adv* too; ⟨*con verbi*⟩ too much; **~ stanco** too tired; **ho mangiato ~** I ate too much; **hai fame? - non ~** are you hungry? - not very; **sentirsi di ~** feel unwanted

'trota *nf* trout

trot'tare *vi* trot. **trotterel'lare** *vi* trot along; ⟨*bimbo:*⟩ toddle

'trotto *nm* trot; **andare al ~** trot

'trottola *nf* [spinning] top; ⟨*movimento*⟩ spin

troupe *nf inv* ~ **televisiva** camera crew

tro'vare *vt* find; ⟨*scoprire*⟩ find out; ⟨*incontrare*⟩ meet; ⟨*ritenere*⟩ think; **andare a ~re** go and see. **~rsi** *vr* find oneself; ⟨*luogo:*⟩ be; ⟨*sentirsi*⟩ feel. **~ta** *nf* bright idea. **~ta pubblicitaria** advertising gimmick

truc'care *vt* make up; ⟨*falsificare*⟩ fix *sl*. **~rsi** *vr* make up. **~tore**, **~'trice** *nmf* make-up artist

'trucco *nm* ⟨*cosmetico*⟩ make-up; ⟨*imbroglio*⟩ trick

'truce *a* fierce; ⟨*delitto*⟩ appalling

truci'dare *vt* slay

truci'olo *nm* shaving

trucu'lento *a* truculent

'truffa *nf* fraud. **truf'fare** *vt* swindle. **~'tore**, **~'trice** *nmf* swindler

'truppa *nf* troops *pl*; ⟨*gruppo*⟩ group

tu *pers pron* you; **sei tu?** is that you?; **l'hai fatto tu?** did you do it

yourself?; **a tu per tu** in private; **darsi del tu** *use the familiar tu*

'tuba *nf Mus* tuba; ⟨*cappello*⟩ top hat

tu'bare *vi* coo

tuba'tura, tubazi'one *nf* piping

tubazi'oni *nfpl* piping *sg*, pipes

tuberco'losi *nf* tuberculosis

tu'betto *nm* tube

tu'bino *nm* ⟨*vestito*⟩ shift

'tubo *nm* pipe; *Anat* canal; **non ho capito un ~** *fam* I understood zilch. **~ di scappamento** exhaust [pipe]

tubo'lare *a* tubular

tuf'fare *vt* plunge. **~rsi** *vr* dive. **~'tore**, **~'trice** *nmf* diver

'tuffo *nm* dive; ⟨*bagno*⟩ dip; **ho avuto un ~ al cuore** my heart missed a beat. **~ di testa** dive

'tufo *nm* tufa

tu'gurio *nm* hovel

tuli'pano *nm* tulip

'tulle *nm* tulle

tume'fatto *a* swollen. **~zi'one** *nf* swelling. **'tumido** *a* swollen

tu'more *nm* tumour

tumulazi'one *nf* burial

tu'multo *nm* turmoil; ⟨*sommossa*⟩ riot. **~u'oso** *a* uproarious

'tunica *nf* tunic

Tuni'sia *nf* Tunisia

'tunnel *nm inv* tunnel

'tuo (il ~ *m*, la tua *f*, i ~i *mpl*, le tue *fpl*) *poss a* your; **è tua questa macchina?** is this car yours?; **un ~ amico** a friend of yours; **~ padre** your father ● *poss pron* yours; **i tuoi** your folks

tuo'nare *vi* thunder. **tu'ono** *nm* thunder

tu'orlo *nm* yolk

tu'racciolo *nm* stopper; ⟨*di sughero*⟩ cork

tu'rare *vt* stop; cork ⟨*bottiglia*⟩. **~si** *vr* become blocked; **~si le orecchie** stick one's fingers in one's ears; **~si il naso** hold one's nose

turba'mento *nm* disturbance; ⟨*sconvolgimento*⟩ upsetting. **~ del-**

la quiete pubblica breach of the peace

tur'bante *nm* turban

tur'ba|re *vt* upset. ~rsi *vr* get upset. ~to *a* upset

tur'bina *nf* turbine

turbi'nare *vi* whirl. 'turbine *nm* whirl. turbine di vento whirlwind

turbo'len|to *a* turbulent. ~za *nf* turbulence

turboreat'tore *nm* turbo-jet

tur'chese *a & nmf* turquoise

Tur'chia *nf* Turkey

tur'chino *a & nm* deep blue

'turco, -a *a* Turkish ● nmf Turk ● nm (lingua) Turkish; fig double Dutch; fumare come un ~ smoke like a chimney; bestemmiare come un ~ swear like a trooper

tu'ris|mo *nm* tourism. ~ta *nmf* tourist. ~tico *a* tourist

'turno *nm* turn; a ~ in turn; di ~ on duty; fare a ~ take turns. ~ di notte night shift

'turpe *a* base. ~i'loquio *nm* foul language

'tuta *nf* overalls *pl*; *Sport* tracksuit. ~ da ginnastica tracksuit. ~ da lavoro overalls *pl*. ~ mimetica camouflage. ~ spaziale spacesuit. ~ subacquea wetsuit

tu'tela *nf* *Jur* guardianship; (protezione) protection. tute'lare *vt* protect

tu'tina *nf* sleepsuit; (da danza) leotard

tu'tore, -'trice *nmf* guardian

'tutta *nf* mettercela ~ per fare qcsa go flat out for sth

tutta'via *conj* nevertheless, still

'tutto *a* whole; (con nomi plurali) all; (ogni) every; tutta la classe the whole class, all the class; tutti gli alunni all the pupils; a tutta velocità at full speed; ho aspettato ~ il giorno I waited all day [long]; in ~ il mondo all over the world; noi tutti all of us; era tutta contenta she was delighted; tutti e due both; tutti e tre all three ● pron all; (tutta la gente)

everybody; (tutte le cose) everything; (qualunque cosa) anything; l'ho mangiato ~ I ate it all; le ho lavate tutte I washed them all; raccontami ~ tell me everything; lo sanno tutti everybody knows; è capace di ~ he's capable of anything; ~ compreso all in; del ~ quite; in ~ altogether completely; tutt'a un tratto all at once; tutt'altro not at all; tutt'altro che anything but ● nm whole; tentare il ~ per ~ go for broke. ~'fare *a inv & nmf* [impiegato] ~ general handyman; donna ~ general maid

tut'tora *adv* still

tutù *nm inv* tutu, ballet dress

tv *nf inv* TV

Uu

ubbidi'en|te *a* obedient. ~za *nf* obedience. ubbi'dire *vi* ~ (a) obey

ubi'ca|to *a* located. ~zi'one *nf* location

ubria'car|e *vt* get drunk. ~si *vr* get drunk; ~si *fig* become intoxicated with

ubria'chezza *nf* drunkenness; in stato di ~ inebriated

ubri'aco, -a *a* drunk; ~ fradicio dead *o* blind drunk ● nmf drunk

ubria'cone *nm* drunkard

uccelli'era *nf* aviary. uc'cello *nm* bird; (vulg: pene) cock

uc'cider|e *vt* kill. ~si *vr* kill oneself

ucci'si'one *nf* killing. uc'ciso *pp di* uccidere. ~'sore *nm* killer

u'dente *a* i non udenti the hearing-impaired

u'dibile *a* audible

udi'enza *nf* audience; (colloquio) interview; *Jur* hearing

u'di|re *vt* hear. ~'tivo *a* auditory. ~to *nm* hearing. ~'tore, ~'trice *nmf* listener; **Sch** unregistered student (*allowed to sit in on lectures*). ~**torio** *nm* audience

'uffa *int* (*con impazienza*) come on!; (*con tono seccato*) damn!

uffici'al|e *a* official ● *nm* officer; (*funzionario*) official; **pubblico** ~**e** public official; ~**e giudiziario** clerk of the court. ~**iz'zare** *vt* make official, officialize

uf'ficio *nm* office; (*dovere*) duty. ~ **di collocamento** employment office. ~ **informazioni** information office. ~ **del personale** personnel department. ~**sa'mente** *adv* unofficially

uffici'oso *a* unofficial

'ufo[1] *nm inv* UFO

'ufo[2]: **a** ~ *adv* without paying

uggi'oso *a* boring

uguagli'a|nza *nf* equality. ~**re** *vt* make equal; (*essere uguale*) equal; (*livellare*) level. ~**rsi** *vr* ~**rsi a** compare oneself to

ugu'al|e *a* equal; (*lo stesso*) the same; (*simile*) like. ~**'mente** *adv* equally; (*malgrado tutto*) all the same

'ulcera *nf* ulcer

uli'veto *nm* olive grove

ulteri'or|e *a* further. ~**'mente** *adv* further

ultima'mente *adv* lately

ulti'ma|re *vt* complete. ~**tum** *nm inv* ultimatum

ulti'missime *nfpl* **Journ** stop press, latest news *sg*

'ultimo *a* last; (*notizie ecc*) latest; (*più lontano*) farthest; *fig* ultimate ● *nm* last; **fino all'**~ to the last; **per** ~ at the end; **l'**~ **piano** the top floor

ultrà *nmf inv* **Sport** fanatical supporter

ultramo'derno *a* ultramodern

ultra'rapido *a* extra-fast

ultrasen'sibile *a* ultrasensitive

ultra's|onico *a* ultrasonic. ~**u'o-no** *nm* ultrasound

ultrater'reno *a* (*vita*) after death

ultravio'letto *a* ultraviolet

ulu'la|re *vi* howl. ~**to** *nm* howling; **gli** ~**ti** the howls, the howling

umana'mente *adv* (*trattare*) humanely; ~ **impossibile** not humanly possible

uma'nesimo *nm* humanism

umani|tà *nf* humanity. ~**'tario** *a* humanitarian. **u'mano** *a* human; (*benevolo*) humane

umidifica'tore *nm* humidifier

umidità *nf* dampness; (*di clima*) humidity. **'umido** *a* damp; (*clima*) humid; (*mani, occhi*) moist ● *nm* dampness; **in umido** Culin stewed

'umile *a* humble

umili'a|nte *a* humiliating. ~**re** *vt* humiliate. ~**rsi** *vr* humble oneself. ~**zi'one** *nf* humiliation.

umiltà *nf* humility. **umil'mente** *adv* humbly

u'more *nm* humour; (*stato d'animo*) mood; **di cattivo/buon** ~ in a bad/good mood

umo'ris|mo *nm* humour. ~**ta** *nmf* humorist. ~**tico** *a* humorous

un *indef art* **a**; (*davanti a vocale o h muta*) an; *vedi* **uno**

una *indef art* **f** a; *vedi* **un**

u'nanim|e *a* unanimous. ~**e'men-te** *adv* unanimously. ~**ità** *nf* unanimity; **all'**~**ità** unanimously

unci'nato *a* hooked; (*parentesi*) angle

unci'netto *nm* crochet hook

un'cino *nm* hook

'undici *a & nm* eleven

'unger|e *vt* grease; (*sporcare*) get greasy; **Relig** anoint; (*blandire*) flatter. ~**si** *vr* (*con olio solare*) oil oneself; ~**si le mani** get one's hands greasy

unghe'rese *a & nmf* Hungarian. **Unghe'ria** *nf* Hungary; (*lingua*) Hungarian

'unghi|a *nf* nail; (*di animale*) claw. ~**'ata** *nf* (*graffio*) scratch

ungu'ento *nm* ointment

unica'mente *adv* only. **'unico** *a* only; (*singolo*) single; (*incomparabile*) unique

unifi'ca|re *vt* unify. **~zi'one** *nf* unification

unifor'mar|e *vt* level. **~si** *vr* conform (**a** to)

uni'forme *a* & *nf* uniform. **~ità** *nf* uniformity

unilate'rale *a* unilateral

uni'one *nf* union; (*armonia*) unity. **U~ Europea** European Union. **U~ Monetaria Europea** European Monetary Union. **~ sindacale** trade union, labor union *Am.* **U~ Sovietica** Soviet Union

u'ni|re *vt* unite; (*collegare*) join; blend (*colori ecc*). **~rsi** *vr* unite; (*collegarsi*) join

'unisex *a inv* unisex

unità *nf inv* unity; *Math, Mil* unit; *Comput* drive. **~ di misura** unit of measurement. **~rio** *a* unitary

u'nito *a* united; (*tinta*) plain

univer'sale *a* universal. **~iz'zare** *vt* universalize. **~'mente** *adv* universally

università *nf inv* university. **~rio, -a** *a* university *attrib* ● *nmf* (*insegnante*) university lecturer; (*studente*) undergraduate

uni'verso *nm* universe

uno, -a *indef art* (*before s + consonant, gn, ps, z*) a ● *pron* one; **a ~ a ~** one by one; **l'~ e l'altro** both [of them]; **né l'~ né l'altro** neither [of them]; **~ di noi** one of us; **~ fa quello che può** you do what you can ● *a* a, one ● *nm* (*numerale*) one; (*un tale*) some man ● *nf* some woman

'unt|o *pp di* **ungere** ● *a* greasy ● *nm* grease. **~u'oso** *a* greasy.

unzi'one *nf* l'Estrema Unzione Extreme Unction

u'omo *nm* (*pl* uomini) man. **~ d'affari** business man. **~ di fiducia** right-hand man. **~ di Stato** statesman

u'ovo *nm* (*pl* uova) egg. **~ in camicia** poached egg. **~ alla coque** boiled egg. **~ di Pasqua** Easter egg. **~ sodo** hard-boiled egg. **~ strapazzato** scrambled egg

ura'gano *nm* hurricane

u'ranio *nm* uranium

urba'nesimo *nm* urbanization. **~ista** *nmf* town planner. **~istica** *nf* town planning. **~istico** *a* urban. **urbanizzazi'one** *nf* urbanization. **ur'bano** *a* urban; (*cortese*) urbane

ur'gen|te *a* urgent. **~te'mente** *adv* urgently. **~za** *nf* urgency; **in caso d'~za** in an emergency; **d'~za** (*misura, chiamata*) emergency

'urgere *vi* be urgent

u'rina *nf* urine. **uri'nare** *vi* urinate

ur'lare *vi* shout, yell; (*cane, vento:*) howl. **'urlo** *nm* (*pl nm* urli, *pl* urla) shout; (*di cane, vento*) howling

'urna *nf* urn; (*elettorale*) ballot box; **andare alle urne** to go to the polls

urrà *int* hurrah!

U.R.S.S. *nf abbr* (**Unione delle Repubbliche Socialiste Sovietiche**) USSR

ur'tar|e *vt* knock against; (*scontrarsi*) bump into; *fig* irritate. **~si** *vr* collide; *fig* clash

'urto *nm* knock; (*scontro*) crash; (*contrasto*) conflict; *fig* clash; **d'~** (*misura, terapia*) shock

usa e getta *a inv* (*rasoio, siringa*) throw-away, disposable

u'sanza *nf* custom; (*moda*) fashion

u'sa|re *vt* use; (*impiegare*) employ; (*esercitare*) exercise; **~re l'abitudine** *qcsa* be in the habit of doing sth ● *vi* (*essere di moda*) be fashionable; **non si usa più** it is out of fashion; (*attrezzatura, espressione:*) it's not used any more. **~to** *a* used; (*non nuovo*) second-hand

U.S.A. *nmpl abbr* US[A] *sg*

u'scente *a* (*presidente*) outgoing

usci'ere *nm* usher. **'uscio** *nm* door

u'sci|re *vi* come out; (*andare fuori*) go out; (*sfuggire*) get out; (*essere sorteggiato*) come up; (*giornale:*) come out; **~re da** *Comput* exit from, quit; **~re di strada** leave the road. **~ta** *nf* exit, way out; (*spesa*) outlay; (*di autostrada*) junction; (*battuta*) witty remark; **essere in**

libera ~ta be off duty. ~ta di servizio back door. ~ta di sicurezza emergency exit

usi'gnolo nm nightingale

'uso nm use; (abitudine) custom; (usanza) usage; fuori ~ out of use; per ~ esterno (medicina) for external use only

U.S.S.L. nf abbr (Unità Socio-Sanitaria Locale) local health centre

ustio'na|rsi vr burn oneself. ~to, -a nmf burns case ●a burnt. usti'one nf burn

usu'ale a usual

usufru'ire vi ~ di take advantage of

u'sura nf usury. usu'raio nm usurer

usur'pare vt usurp

u'tensile nm tool; Culin utensil; cassetta degli utensili tool box

u'tente nmf user. ~ finale end user

u'tenza nf use; (utenti) users pl. ~ finale end users

ute'rino a uterine. 'utero nm womb

'util|e a useful ●nm Comm profit. ~ità nf usefulness, utility; Comput utility. ~i'taria nf Auto small car. ~i'tario a utilitarian

utiliz'za|re vt utilize. ~zi'one nf utilization. uti'lizzo nm (utilizzazione) use

uto'pistico a Utopian

'uva nf grapes pl; chicco d'~ grape. ~ passa raisins pl. ~ sultanina currants pl

---

Vv

---

va'cante a vacant

va'canza nf holiday; (posto vacante) vacancy. essere in ~ be on holiday

'vacca nf cow. ~ da latte dairy cow

vacci'nare vt vaccinate. ~inazi'one nf vaccination. vac'cino nm vaccine

vacil'la|nte a tottering; (oggetto) wobbly; (luce) flickering; fig wavering. ~re vi totter; (oggetto:) wobble; (luce:) flicker; fig waver

'vacuo a (vano) vain; fig empty ●nm vacuum

vagabon'dare vi wander. vagabondo, -a a (cane) stray; gente vagabonda tramps pl ●nmf tramp

va'gare vi wander

vagheggi'are vt long for

va'gina nf vagina. ~'nale a vaginal

va'gire vi whimper. ~to nm whimper

'vaglia nm inv money order. ~ bancario bank draft. ~ postale postal order

vagli'are vt sift; fig weigh

'vago a vague

vagon'cino nm (di funivia) car

va'gone nm (per passeggeri) carriage; (per merci) wagon. ~ letto sleeper. ~ ristorante restaurant car

vai'olo nm smallpox

va'langa nf avalanche

va'lente a skilful

va'ler|e vi be worth; (contare) count; (regola:) be valid (per to); (essere valido) be valid; far ~ i propri diritti assert one's rights; farsi ~e assert oneself; non vale! that's not fair!; tanto vale che me ne vada I might as well go ●vt ~re qcsa a qcno (procurare) earn sb sth; ~ne la pena be worth it; vale la pena di vederlo it's worth seeing; ~si di avail oneself of

valeri'ana nf valerian

va'levole a valid

vali'care vt cross. 'valico nm pass

validità nf validity; con ~ illimitata valid indefinitely

'valido a valid; (efficace) efficient; (contributo) valuable

valige'ria *nf (fabbrica)* leather factory; *(negozio)* leather goods shop

va'ligia *nf* suitcase; **fare le valigie** pack; *fig* pack one's bags. ~ **diplomatica** diplomatic bag

val'lata *nf* valley. **'valle** *nf* valley; **a valle** downstream

val'lett|a *nf* TV assistant. ~**o** *nm* valet; *TV* assistant

val'lone *nm (valle)* deep valley

va'lor|e *nm* value, worth; *(merito)* merit; *(coraggio)* valour; ~**i** *pl Comm* securities; **di** ~**e** *(oggetto)* valuable; **oggetti** *pl* **di** ~**e** valuables; **senza** ~**e** worthless. ~**iz'zare** *vt (mettere in valore)* use to advantage; *(aumentare di valore)* increase the value of; *(migliorare l'aspetto di)* enhance

valo'roso *a* courageous

'valso *pp di* **valere**

va'luta *nf* currency. ~ **estera** foreign currency

valu'ta|re *vt* value; weigh up *(situazione)*. ~**rio** *a (mercato, norme)* currency. ~**zi'one** *nf* valuation

'valva *nf* valve. **'valvola** *nf* valve; *Electr* fuse

'valzer *nm inv* waltz

vam'pata *nf* blaze; *(di calore)* blast; *(al viso)* flush

vam'piro *nm* vampire; *fig* bloodsucker

vana'mente *adv (inutilmente)* in vain

van'da|lico *a* **atto** ~**lico** act of vandalism. ~**lismo** *nm* vandalism. **'vandalo** *nm* vandal

vaneggi'are *vi* rave

'vanga *nf* spade. **van'gare** *vt* dig

van'gelo *nm* Gospel; *(fam: verità)* gospel [truth]

vanifi'care *vt* nullify

va'nigli|a *nf* vanilla. ~'ato *a (zucchero)* vanilla *attrib*

vanil'lina *nf* vanillin

vanità *nf* vanity. **van'toso** *a* vain

'vano *a* vain ● *nm (stanza)* room; *(spazio vuoto)* hollow

van'taggi|o *nm* advantage; *Sport* lead; *Tennis* advantage; **trarre** ~**o**

da qcsa derive benefit from sth. ~'oso *a* advantageous

van't|are *vt* praise; *(possedere)* boast. ~**arsi** *vr* boast. ~**e'ria** *nf* boasting. **'vanto** *nm* boast

'vanvera *nf* **a** ~ at random; **parlare a** ~ talk nonsense

va'por|e *nm* steam; *(di benzina, cascata)* vapour; **a** ~**e** steam *attrib*; **al** ~**e** *Culin* steamed. ~**e acqueo** steam, water vapour; **battello a** ~ **e** a steamboat. **vapo'retto** *nm* ferry. ~**i'era** *nf* steam engine

vaporiz'za|re *vt* vaporize. ~'tore *nm* spray

vapo'roso *a (vestito)* filmy; **capelli vaporosi** big hair *sg*

va'rare *vt* launch

var'care *vt* cross. **'varco** *nm* passage; **aspettare al varco** lie in wait

vari'abil|e *a* changeable, variable ● *nf* variable. ~**ità** *nf* changeableness, variability

vari'a|nte *nf* variant. ~**re** *vt/i* vary; ~**re di umore** change one's mood. ~**zi'one** *nf* variation

va'rice *nf* varicose vein

vari'cella *nf* chickenpox

vari'coso *a* varicose

varie'gato *a* variegated

varietà *nf inv* variety ● *nm inv* variety show

'vario *a* varied; *(al pl, parecchi)* various; **vari** *pl (molti)* several; **varie ed eventuali** any other business

vario'pinto *a* multicoloured

'varo *nm* launch

va'saio *nm* potter

'vasca *nf* tub; *(piscina)* pool; *(lunghezza)* length. ~ **da bagno** bath

va'scello *nm* vessel

va'schetta *nf* tub

vase'lina *nf* Vaseline®

vasel'lame *nm* china. ~ **d'oro/d'argento** gold/silver plate

'vaso *nm* pot; *(da fiori)* vase; *Anat* vessel; *(per cibi)* jar. ~ **da notte** chamber pot

vas'soio *nm* tray

vastità *nf* vastness. **'vasto** *a* vast; **di vaste vedute** broad-minded

Vati'cano *nm* Vatican

vattela'pesca *adv fam* God knows!

ve *pers pron* you; **ve l'ho dato** I gave it to you

vecchia *nf* old woman. **vecchi'aia** *nf* old age. **'vecchio** *a* old ● *nmf* old man; **i vecchi** old people

'vece *nf* **in ~ di** in place of; **fare le veci di** qcno take sb's place

ve'dente *a* **i non vedenti** the visually handicapped

ve'der|e *vt/i* see; **far ~e** show; **farsi ~e** show one's face; **non vedo l'ora di...** I can't wait to.... **~si** *vr* see oneself; (*reciproco*) see each other

ve'detta *nf* (*luogo*) lookout; *Naut* patrol vessel

'vedovo, -a *nm* widower ● *nf* widow

ve'duta *nf* view

vee'mente *a* vehement

vege'ta|le *a & nm* vegetable. **~li'ano** *a & nmf* vegan. **~re** *vi* vegetate. **~ri'ano, -a** *a & nmf* vegetarian. **~zi'one** *nf* vegetation

'vegeto *a vedi* **vivo**

veg'gente *nmf* clairvoyant

'veglia *nf* watch; **fare la ~** keep watch. **~ funebre** vigil

vegli'are *vi* be awake; **~are su** watch over. **~'one** *nm* **~one di capodanno** New Year's Eve celebration

ve'icolo *nm* vehicle

'vela *nf* sail; *Sport* sailing; **far ~** set sail

ve'la|re *vt* veil; (*fig: nascondere*) hide. **~rsi** *vr* (*vista:*) mist over; (*voce:*) go husky. **~ta'mente** *adv* indirectly. **~to** *a* veiled; (*occhi*) misty; (*collant*) sheer

'velcro® *nm* velcro®

veleggi'are *vi* sail

ve'leno *nm* poison. **vele'noso** *a* poisonous

veli'ero *nm* sailing ship

ve'lina *nf* (**carta**) ~ tissue paper; (*copia*) carbon copy

ve'lista *nm* yachtsman ● *nf* yachtswoman

ve'livolo *nm* aircraft

velle'ità *nf inv* foolish ambition. **~'tario** *a* unrealistic

'vello *nm* fleece

vellu'tato *a* velvety. **vel'luto** *nm* velvet. **velluto a coste** corduroy

'velo *nm* veil; (*di zucchero, cipria*) dusting; (*tessuto*) voile

ve'loc|e *a* fast. **~e'mente** *adv* quickly. **velo'cista** *nmf* Sport sprinter. **~ità** *nf inv* speed; (*Auto: marcia*) gear. **~ità di crociera** cruising speed. **~iz'zare** *vt* speed up

ve'lodromo *nm* cycle track

'vena *nf* vein; **essere in ~ di** be in the mood for

ve'nale *a* venal; (*persona*) mercenary, venal

ve'nato *a* grainy

vena'torio *a* a hunting *attrib*

vena'tura *nf* (*di legno*) grain; (*di foglia, marmo*) vein

ven'demmia *nf* grape harvest. **~'are** *vt* harvest

'vender|e *vt* sell. **~si** *vr* sell oneself; **vendesi** for sale

ven'detta *nf* revenge

vendi'ca|re *vt* avenge. **~rsi** *vr* get one's revenge. **~'tivo** *a* vindictive

'vendita *nf* sale; **in ~ta** on sale. **~ta all'asta** sale by auction. **~ta al dettaglio** retailing. **~ta all'ingrosso** wholesaling. **~ta al minuto** retailing. **~ta porta a porta** door-to-door selling. **~'tore, ~'trice** *nmf* seller. **~tore ambulante** hawker, pedlar

vene'ra|bile, ~ndo *a* venerable **vene'rare** *vt* revere

venerdì *nm inv* Friday. **V~ Santo** Good Friday

'Venere *nf* Venus. **ve'nereo** *a* venereal

Ve'nezi|a *nf* Venice. **v~'ano, -a** *a & nmf* Venetian ● *nf* (*persiana*) Venetian blind; *Culin* sweet bun

veni·ale *a* venial
ve'nire *vi* come; (*riuscire*) turn out; (*costare*) cost; (*in passivi*) be **∼ a sapere** learn; **∼ in mente** occur; **∼ meno** (*svenire*) faint; **∼ meno a un contratto** go back on a contract; **∼ via** come away; (*staccarsi*) come off; **mi viene da piangere** I feel like crying; **vieni a prendermi** come and pick me up
ven'taglio *nm* fan
ven'tata *nf* gust [of wind]; *fig* breath
ven'tenne *a & nmf* twenty-year-old. **∼simo** *a & nm* twentieth.
'venti *a & nm* twenty
venti'la|re *vt* air. **∼tore** *nm* fan. **∼zi'one** *nf* ventilation
ven'tina *nf* **una ∼** (*circa venti*) about twenty
ventiquat'trore *nf inv* (*valigia*) overnight case
'vento *nm* wind; **farsi ∼** fan oneself
ven'tosa *nf* sucker
ven'toso *a* windy
'ventre *nm* stomach. **ven'triloquo** *nm* ventriloquist
ven'tura *nf* fortune; **andare alla ∼** trust to luck
ven'turo *a* next
ve'nuta *nf* coming
ve'randa *nf* veranda
vera'mente *adv* really
ver'ba|le *a* verbal ●*nm* (*di riunione*) minutes *pl.* **∼'mente** *adv* verbally
'verbo *nm* verb. **∼ ausiliare** auxiliary [verb]
'verde *a* green ●*nm* green; (*vegetazione*) greenery; (*semaforo*) green light; **essere al ∼** be broke. **∼ oliva** olive green. **∼ pisello** pea green. **∼rame** *nm* verdigris
ver'detto *nm* verdict
ver'dura *nf* vegetables *pl*; **una ∼** a vegetable
'verga *nf* rod
vergi'n|ale *a* virginal. **'vergine** *nf* virgin; *Astr* Virgo ●*a* virgin; (*cassetta*) blank. **∼ità** *nf* virginity

ver'gogna *nf* shame; (*timidezza*) shyness
vergo'gn|arsi *vr* feel ashamed; (*essere timido*) feel shy. **∼oso** *a* ashamed; (*timido*) shy; (*disonorevole*) shameful
ve'rifica *nf* check. **verifi'cabile** *a* verifiable
verifi'car|e *vt* check. **∼si** *vr* come true
ve'rismo *nm* realism
veri'tà *nf* truth. **∼i'ero** *a* truthful
'verme *nm* worm. **∼ solitario** tapeworm
ver'miglio *a & nm* vermilion
ver'mut *nm inv* vermouth
ver'nacolo *nm* vernacular
ver'nic|e *nf* paint; (*trasparente*) varnish; (*pelle*) patent leather; *fig* veneer; **"∼e fresca"** "wet paint". **∼i'are** *vt* paint; (*con vernice trasparente*) varnish. **∼la'tura** *nf* painting; (*strato*) paintwork; *fig* veneer
'vero *a* true; (*autentico*) real; (*perfetto*) perfect; **è ∼?** is that so?; **∼ e proprio** full-blown; **sei stanca, ∼** you're tired, aren't you ●*nm* truth; (*realtà*) life
verosimigli'anza *nf* probability. **vero'simile** *a* probable
ver'ruca *nf* wart; (*sotto la pianta del piede*) verruca
versa'mento *nm* (*pagamento*) payment; (*in banca*) deposit
ver'sante *nm* slope
ver'sa|re *vt* pour; (*spargere*) shed; (*rovesciare*) spill; pay (*denaro*). **∼rsi** *vr* spill; (*sfociare*) flow
ver'satil|e *a* versatile. **∼ità** *nf* versatility
ver'setto *nm* verse
versi'one *nf* version; (*traduzione*) translation; **"∼ integrale"** "unabridged version"; **"∼ ridotta"** "abridged version"
'verso¹ *nm* verse; (*grido*) cry; (*gesto*) gesture; (*senso*) direction; (*modo*) manner; **fare il ∼ a qcno** ape sb; **non c'è ∼ di** there is no way of

'**verso²** prep towards; (nei pressi di) round about; ~ **dove?** which way?

'**vertebra** nf vertebra

'**vertere** vi ~ **su** focus on

verti'cal|e a vertical; (in parole crociate) down ●nm vertical ●nf handstand. ~'mente adv vertically

'**vertice** nm summit; Math vertex; **conferenza al** ~ summit conference

ver'tigine nf dizziness; Med vertigo. vertigini pl giddy spells; **aver le vertigini** feel dizzy

vertigi|nosa'mente adv dizzily. ~'noso a dizzy; (velocità) breakneck; (prezzi) sky-high; (scollatura) plunging

ve'scica nf bladder; (sulla pelle) blister

'**vescovo** nm bishop

'**vespa** nf wasp

vespasi'ano nm urinal

'**vespro** nm vespers pl

ves'sillo nm standard

ve'staglia nf dressing gown

'**vest|e** nf dress; (rivestimento) covering; **in ~e di** in the capacity of; **in ~e ufficiale** in an official capacity. ~i'ario nm clothing

ve'stibolo nm hall

ve'stigio nm (pl nm **vestigi**, pl nf **vestigia**) trace

ve'sti|re vt dress. ~**rsi** vr get dressed. ~**ti** pl clothes. ~**to** a dressed ● nm (da uomo) suit; (da donna) dress

vete'rano, -a a & nmf veteran

veteri'naria nf veterinary science

veteri'nario a veterinary ●nm veterinary surgeon

'**veto** nm inv veto

ve'tra|io nm glazier. ~**ta** nf big window; (in chiesa) stained-glass window; (porta) glass door. ~**to** a glazed. **vetre'ria** nf glass works

ve'tri|na nf [shop-]window; (mobile) display cabinet. ~'**nista** nmf window dresser

vetri'olo nm vitriol

'**vetro** nm glass; (di finestra, porta) pane. ~'**resina** nf fibreglass

'**vetta** nf peak

vet'tore nm vector

vetto'vaglie nfpl provisions

vet'tura nf coach; (ferroviaria) carriage; Auto car. vettu'rino nm coachman

vezzeggi'a|re vt fondle. ~'**tivo** nm pet name. '**vezzo** nm habit; (attrattiva) charm; **vezzi** pl (moine) affectation sg. **vez'zoso** a charming; pej affected

vi pers pron you; (riflessivo) yourselves; (reciproco) each other; (tra più persone) one another; **vi ho dato un libro** I gave you a book; **lavatevi le mani** wash your hands; **eccovi** here you are! ●adv = **ci**

'**via¹** nf street, road; fig way; Anat tract; **in** ~ **di** in the course of; **per** ~ **di** on account of; ~ ~ **che** as; **per** ~ **aerea** by airmail

'**via²** adv away; (fuori) out; **andar** ~ go away; **e così** ~ and so on; **e** ~ **dicendo** and whatnot ●int ~! go away!; Sport go!; (andiamo) come on! ●nm starting signal

viabilità nf road conditions pl; (rete) road network; (norme) road and traffic laws pl

via'card nf inv motorway card

via'dotto nm viaduct

viaggi'a|re vi travel. ~'**tore**, ~'**trice** nmf traveller

vi'aggio nm journey; (breve) trip; **buon** ~! safe journey!, have a good trip!; **fare un** ~ go on a journey. ~ **di nozze** honeymoon

vi'ale nm avenue; (privato) drive

via'vai nm coming and going

vi'bra|nte a vibrant. ~**re** vi vibrate; (fremere) quiver. ~**zi'one** nf vibration

vi'cario nm vicar

'**vice+** pref vice+

'**vice** nm deputy. ~**diret'tore** nm assistant manager

vi'cenda nf event; **a** ~ (fra due) each other; (a turno) in turn[s]

vice'versa *adv* vice versa
vici'na|nza *nf* nearness; **~nze** *pl* (*paraggi*) neighbourhood. **~to** *nm* neighbourhood; (*vicini*) neighbours *pl*
vi'cino, -a *a* near; (*accanto*) next ● *adv* near, close. **~ a** *prep* near [to] ● *nmf* neighbour. **~ di casa** nextdoor neighbour
vicissi'tudine *nf* vicissitude
'vicolo *nm* alley
'video *nm* video. **~'camera** *nf* camcorder. **~cas'setta** *nf* video cassette
videoci'tofono *nm* video entry phone
video'clip *nm inv* video clip
videogi'oco *nm* video game
videoregistra'tore *nm* video-recorder
video'teca *nf* video library
video'tel® *nm* ≈ Videotex®
videotermi'nale *nm* visual display unit, VDU
vidi'mare *vt* authenticate
vie'ta|re *vt* forbid; **sosta ~ta** 'no parking; **~to fumare** 'no smoking; **~to ai minori di 18 anni** prohibited to children under the age of 18
vi'gente *a* in force. **'vigere** *vi* be in force
vigi'la|nte *a* vigilant. **~nza** *nf* vigilance. **~re** *vt* keep an eye on ● *vi* keep watch
'vigile *a* watchful ● *nm* ~ [urbano] policeman. **~ del fuoco** fireman
vi'gilia *nf* eve
vigliacche'ria *nf* cowardice. **vi-gli'acco, -a** *a* cowardly ● *nmf* coward
'vigna *nf*, **vi'gneto** *nm* vineyard
vi'gnetta *nf* cartoon
vi'gore *nm* vigour; **entrare in ~** come into force. **vigo'roso** *a* vigorous
'vile *a* cowardly; (*abietto*) vile
'villa *nf* villa
vil'laggio *nm* village. **~ turistico** holiday village
vil'lano *a* rude ● *nm* boor; (*contadino*) peasant

villeggi'a|nte *nmf* holiday-maker. **~re** *vi* spend one's holidays. **~'tura** *nf* holiday[s] [*pl*], vacation *Am*
vil'l|etta *nf* small detached house. **~ino** *nm* detached house
viltà *nf* cowardice
'vimine *nm* wicker
'vinc|ere *vt* win; (*sconfiggere*) beat; (*superare*) overcome. **~ita** *nf* win; (*somma vinta*) winnings *pl*. **~i'tore, ~i'trice** *nmf* winner
vinco'la|nte *a* binding. **~re** *vt* bind; *Comm* tie up. **'vincolo** *nm* bond
vi'nicolo *a* wine *attrib*
vinil'pelle® *nm* Leatherette®
'vino *nm* wine. **~ spumante** sparkling wine. **~ da taglio** blending wine. **~ da tavola** table wine
'vinto *pp di* vincere
vi'ola *nf Bot* violet; *Mus* viola. **vio'laceo** *a* purplish; (*labbra*) blue
vio'la|re *vt* violate. **~zi'one** *nf* violation. **~zione di domicilio** breaking and entering
violen'tare *vt* rape
violente'mente *adv* violently
vio'len|to *a* violent. **~za** *nf* violence. **~za carnale** rape
vio'letta *nf* violet
vio'letto *a & nm* (*colore*) violet
violi'nista *nmf* violinist. **vio'lino** *nm* violin. **violon'cello** *nm* cello
vi'ottolo *nm* path
'vipera *nf* viper
vi'ra|ggio *nm Phot* toning; *Naut*, *Aeron* turn. **~re** *vi* turn; **~re di bordo** veer
'virgola *nf* comma. **~ette** *nfpl* inverted commas
vi'ril|e *a* virile; (*da uomo*) manly. **~ità** *nf* virility; manliness
virtù *nf inv* virtue; **in ~ di** (*legge*) under. **~'ale** *a* virtual. **~'oso** *a* virtuous ● *nm* virtuoso
viru'lento *a* virulent
'virus *nm inv* virus
visa'gista *nmf* beautician

visce'rale a visceral; (odio) deep-seated; (reazione) gut

'viscere nm internal organ ● nfpl guts

'vischi|o nm mistletoe. **~'oso** a viscous; (appiccicoso) sticky

'viscido a slimy

vi'scont|e nm viscount. **~'essa** nf viscountess

vi'scoso a viscous

vi'sibile a visible

visi'bilio nm profusion; **andare in ~** go into ecstasies

visibilità nf visibility

visi'era nf (di elmo) visor; (di berretto) peak

visio'nare vt examine; Cinema screen. **visi'one** nf vision; **prima visione** Cinema first showing

'visit|a nf visit; (breve) call; Med examination; **fare ~a a** qcno pay sb a visit. **~a di controllo** Med checkup. **visi'tare** vt visit; (brevemente) call on; Med examine; **~a'tore, ~a'trice** nmf visitor

vi'sivo a visual

'viso nm face

vi'sone nm mink

'vispo a lively

vis'suto pp di **vivere** ● a experienced

'vist|a nf sight; (veduta) view; **a ~a d'occhio** (crescere) visibly; (estendersi) as far as the eye can see; **in ~a di** in view of; **perdere di ~a** qcno lose sight of sb; fig lose touch with sb. **~o** pp di **vedere** ● nm visa. **vi'stoso** a showy; (notevole) considerable

visu'al|e a visual. **~izza'tore** nm Comput display, VDU. **~izzazi'one** nf Comput display

'vita nf life; (durata della vita) lifetime; Anat waist; **a ~** for life; **essere in fin di ~** to be at death's door; **essere in ~** be alive

vi'tal|e a vital. **~ità** nf vitality

vita'lizio a life attrib ● nm [life] annuity

vita'min|a nf vitamin. **~iz'zato** a vitamin-enriched

'vite nf Mech screw; Bot vine

vi'tello nm calf; Culin veal; (pelle) calfskin

vi'ticcio nm tendril

viticol't|ore nm wine grower. **~ura** nf wine growing

'vitreo a vitreous; (sguardo) glassy

'vittima nf victim

'vitto nm food; (pasti) board. **~ e alloggio** board and lodging

vit'toria nf victory

vittori'ano a Victorian

vittori'oso a victorious

vi'uzza nf narrow lane

'viva int hurrah!; **~ la Regina!** long live the Queen!

vi'vac|e a vivacious; (mente) lively; (colore) bright. **~ità** nf vivacity; (di mente) liveliness; (di colore) brightness. **~iz'zare** vt liven up

vi'vaio nm nursery; (per pesci) pond; fig breeding ground

viva'mente adv (ringraziare) warmly

vi'vanda nf food; (piatto) dish

vi'vente a living ● nmpl **i viventi** the living

'vivere vi live; **~ di** live on ● vt (passare) go through ● nm life

vi'veri nmpl provisions

'vivido a vivid

vivisezi'one nf vivisection

'vivo a alive; (vivente) living; (vivace) lively; (colore) bright; **~ e vegeto** alive and kicking; **farsi ~** keep in touch; (arrivare) turn up ● nm colpire qcno sul **~** cut sb to the quick; **dal ~** (trasmissione) live; (disegnare) from life; **i vivi** the living

vizi'are vt spoil (bambino ecc); (guastare) vitiate. **~'ato** a spoilt; (aria) stale. **'vizio** nm vice; (cattiva abitudine) bad habit; (difetto) flaw. **~'oso** a dissolute; (difettoso) faulty; **circolo ~oso** vicious circle

vocabo'lario nm dictionary;

(*lessico*) vocabulary. **vo'cabolo** *nm* word

vo'cale *a* vocal ●*nf* vowel. **vo-'calico** *a* ⟨*corde*⟩ vocal; ⟨*suono*⟩ vowel *attrib*

vocazi'one *nf* vocation

'voce *nf* voice; (*diceria*) rumour; (*di bilancio, dizionario*) entry

voci'are *vi* (*spettegolare*) gossip ●*nm* buzz of conversation

vocife'rare *vi* shout; **si vocifera che...** it is rumoured that...

'vogl|a *nf* rowing; (*lena*) enthusiasm; (*moda*) vogue; **essere in ~a** be in fashion. **vo'gare** *vi* row. **~a'tore** *nm* oarsman; (*attrezzo*) rowing machine

'vogli|a *nf* desire; (*volontà*) will; (*della pelle*) birthmark; **aver ~a di fare qcsa** feel like doing sth. **~'oso** *a* ⟨*occhi, persona*⟩ covetous

'voi *pers pron* you; **siete ~?** is that you?; **l'avete fatto ~?** did you do it yourself?. **~a'ltri** *pers pron* you

vo'lano *nm* shuttlecock; *Mech* flywheel

vo'lante *a* flying; (*foglio*) loose ●*nm* steering-wheel

volan'tino *nm* leaflet

vo'la|re *vi* fly. **~'ta** *nf Sport* final sprint; **di ~ta** in a rush

vo'latile *a* (*liquido*) volatile ●*nm* bird

volée *nf inv Tennis* volley

vo'lente *a ~ o nolente* whether you like it or not

volente'roso *a* willing

volenti'eri *adv* willingly; **~!** with pleasure!

vo'lere *vt* want; (*chiedere di*) ask for; (*aver bisogno di*) need; **vuole che lo faccia io** he wants me to do it; **fai come vuoi** do as you like; **se tuo padre vuole, ti porto al cinema** if your father agrees, I'll take you to the cinema; **vorrei un caffè** I'd like a coffee; **la leggenda vuole che...** legend has it that...; **la vuoi smettere?** will you stop that!; **senza ~** without meaning to; **voler bene/male a qcno** love/

have something against sb; **voler dire** mean; **ci vuole il latte** we need milk; **ci vuole tempo/pazienza** it takes time/patience; **volerne a** have a grudge against; **vuoi...vuoi...** either...or... ●*nm* will; volenti *pl* wishes

vol'gar|e *a* vulgar; (*popolare*) common. **~ità** *nf inv* vulgarity. **~iz'zare** *vt* popularize. **~'mente** *adv* (*grossolanamente*) vulgarly, coarsely; (*comunemente*) commonly

'volg|ere *vt/i* turn. **~si** *vr* turn (*round*); **~si a** (*dedicarsi*) take up

voli'era *nf* aviary

voli'tivo *a* strong-minded

'volo *nm* flight; **al ~** (*fare qcsa*) quickly; (*prendere qcsa*) in midair; **alzarsi in ~** (*uccello*) take off; **in ~** airborne. **~ di linea** scheduled flight. **~ nazionale** domestic flight. **~ a vela** gliding.

volontà *nf inv* will; (*desiderio*) wish; **a ~** (*mangiare*) as much as you like. **~ria'mente** *adv* voluntarily. **volon'tario** *a* voluntary ●*nm* volunteer

volonte'roso *a* willing

'volpe *nf* fox

volt *nm inv* volt

'volta *nf* time; (*turno*) turn; (*curva*) bend; *Archit* vault; **4 volte** 4 times 4; **a volte** sometimes; **c'era una ~...** once upon a time, there was...; **una ~** once; **due volte** twice; **tre/quattro volte** three/ four times; **una ~ per tutte** once and for all; **uno ~ per** one at a time; **uno alla ~** one at a time; **alla ~ di** in the direction of

volta'faccia *nm inv* volte-face

vol'taggio *nm* voltage

vol'ta|re *vt/i* turn; (*rigirare*) turn round; (*rivoltare*) turn over; **~re pagina** *fig* forget the past. **~rsi** *vr* turn (*round*)

volta'stomaco *nm* nausea; *fig* disgust

volteggi'are *vi* circle; (*ginnastica*) vault

'volto pp di **volgere ●** nm face; **mi ha mostrato il suo vero ~** he revealed his true colours
vo'lubile a fickle
vo'lum|e nm volume. **~i'noso** a voluminous
voluta'mente adv deliberately
voluttu|osità nf voluptuousness. **~'oso** a voluptuous
vomi'tare vt vomit, be sick. **vomi'tevole** a nauseating. **'vomito** nm vomit.
'vongola nf clam
vo'race a voracious. **~'mente** adv voraciously
vo'ragine nf abyss
'vortice nm whirl; (gorgo) whirlpool; (di vento) whirlwind
'vostro (il ~ m, la vostra f, i vostri mpl, le vostre fpl) poss a your; **è vostra questa macchina?** is this car yours?; **un ~ amico** a friend of yours; **~ padre** your father **●** poss pron yours; **i vostri** your folks
vo'ta|nte nmf voter. **~re** vi vote. **~zi'one** nf voting; Sch marks pl. **'voto** nm vote; Sch mark; Relig vow
vs. abbr Comm (vostro) yours
vul'canico a volcanic. **vul'cano** nm volcano
vulne'rabil|e a vulnerable. **~ità** nf vulnerability
vuo'tare vt, **vuo'tarsi** vr empty
vu'oto a empty; (non occupato) vacant; **~** (sprovvisto) devoid of **●** nm empty space; Phys vacuum; fig void; **assegno a ~** dud cheque; **sotto ~** (prodotto) vacuum-packed; **~ a perdere** no deposit. **~ d'aria** air pocket

Ww

W abbr (viva) long live
'wafer nm inv (biscotto) wafer
walkie-'talkie nm inv walkie-talkie
water nm inv toilet, loo fam
watt nm inv watt
wat'tora nm inv Phys watt-hour
WC nm WC
'western a inv cowboy attrib **●** nm Cinema western

Xx

X, x a raggi pl X X-rays; **il giorno X** D-day
xenofo'bia nf xenophobia. **xe'nofobo, -a** a xenophobic **●** nmf xenophobe
xe'res nm inv sherry
xi'lofono nm xylophone

Yy

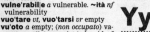

yacht nm inv yacht
yen nm inv Fin yen
'yeti nm yeti
'yoga nm yoga; (praticante) yogi
'yogurt nm inv yoghurt. **~i'era** nf yoghurt-maker
'yorkshire nm inv (cane) Yorkshire terrier
yo-yo nm inv yoyo®

Zz

zaba[gl]i'one *nm* zabaglione (*dessert made from eggs, wine or marsala and sugar*)

'zacchera *nf* (*schizzo*) splash of mud

zaf'fata *nf* whiff; (*di fumo*) cloud

zaffe'rano *nm* saffron

zaf'firo *nm* sapphire

'zaino *nm* rucksack

'zampa *nf* leg; **a quattro zampe** ⟨*animale*⟩ four-legged; (*carponi*) on all fours. **zampe** *pl* **di gallina** crow's feet

zampil'la|nte *a* spurting. **~re** *vi* spurt. **zam'pillo** *nm* spurt

zam'pogna *nf* bagpipe. **zampo-'gnaro** *nm* piper

zam'pone *nfpl* stuffed pig's trotter with lentils

'zanna *nf* fang; (*di elefante*) tusk

zan'zar|a *nf* mosquito. **~i'era** *nf* (*velo*) mosquito net; (*su finestra*) insect screen

'zappa *nf* hoe. **zap'pare** *vt* hoe

zat'tera *nf* raft

zatte'roni *nmpl* (*scarpe*) wedge shoes

za'vorra *nf* ballast; (*fig* dead wood

'zazzera *nf* mop of hair

'zebra *nf* zebra; **zebre** *pl* (*passaggio pedonale*) zebra crossing

'zecca¹ *nf* mint; **nuovo di ~** brand-new

'zecca² *nf* (*parassita*) tick

zec'chino *nm* sequin; **oro ~ pure gold**

ze'lante *a* zealous. **'zelo** *nm* zeal

'zenit *nm* zenith

'zenzero *nm* ginger

'zeppa *nf* wedge

'zeppo *a* packed full; **pieno ~ di** crammed *o* packed with

zer'bino *nm* doormat

'zero *nm* zero, nought; (*in calcio*) nil; *Tennis* love; **due a ~** (*in*

partite) two nil; **ricominciare da ~** *fig* start again from scratch

'zeta *nf* zed, zee *Am*

'zia *nf* aunt

zibel'lino *nm* sable

'zigomo *nm* cheek-bone

zigri'nato *a* ⟨*pelle*⟩ grained; (*metallo*) milled

zig'zag *nm inv* zigzag; **andare a ~** zigzag

zim'bello *nm* decoy; (*oggetto di scherno*) laughing-stock

'zinco *nm* zinc

'zingaro, -a *nmf* gypsy

'zio *nm* uncle

zi'tel|la *nf* spinster; *pej* old maid. **~'lona** *nf pej* old maid

zit'tire *vi* fall silent ● *vt* silence. **'zitto** *a* silent; **sta' zitto!** keep quiet!

ziz'zania *nf* (*discordia*) discord; **seminare ~** cause trouble

'zoccolo *nm* clog; (*di cavallo*) hoof; (*di terra*) clump; (*di parete*) skirting board, baseboard *Am*; (*di colonna*) base

zodia'cale *a* of the zodiac. **zo'diaco** *nm* zodiac

'zolfo *nm* sulphur

'zolla *nf* clod; (*di zucchero*) lump

zol'letta *nf* sugar cube, sugar lump

'zombi *nmf inv fig* zombie

'zona *nf* zone; (*area*) area. **~ di depressione** area of low pressure. **~ disco** area for parking discs only. **~ pedonale** pedestrian precinct. **~ verde** green belt

'zonzo *adv* **andare a ~** stroll about

zoo *nm inv* zoo

zoolo'gia *nf* zoology. **zoo'logico** *a* zoological. **zo'ologo, -a** *nmf* zoologist

zoo sa'fari *nm inv* safari park

zoppi'ca|nte *a* limping; *fig* shaky. **~re** *vi* limp; (*essere debole*) be shaky. **'zoppo, -a** *a* lame ● *nmf* cripple

'zoti'cone *nm* boor

zu'ava *nf* **calzoni** *pl* **alla ~** plusfours

'**zucca** *nf* marrow; (*fam: testa*) head; (*fam: persona*) thickie

zucche'r|are *vt* sugar. **~i'era** *nf* sugar bowl. **~i'ficio** *nm* sugar refinery. **zucche'rino** *a* sugary ● *nm* sugar cube, sugar lump

'**zucchero** *nm* sugar. **~ di canna** cane sugar. **~ vanigliato** vanilla sugar. **~ a velo** icing sugar. **zucche'roso** *a fig* honeyed

zuc'chin|a *nf*, **~o** *nm* courgette, zucchini *Am*

zuc'cone *nm* blockhead

'**zuffa** *nf* scuffle

zufo'lare *vt/i* whistle

zu'mare *vi* zoom

'**zuppa** *nf* soup. **~ inglese** trifle

zup'petta *nf* fare **~ [con]** dunk

zuppi'era *nf* soup tureen

'**zuppo** *a* soaked

Aa

A /eɪ/ n Mus la m inv

a /ə/, accentato /eɪ/ (davanti a una vocale **an**) indef art un m, una f; (before s + consonant, gn, ps and z) uno; (before feminine noun starting with a vowel) un'; (each) a; **I am a lawyer** sono avvocato; **a tiger is a feline** la tigre è un felino; **a knife and fork** un coltello e una forchetta; **a Mr Smith is looking for you** un certo signor Smith ti sta cercando; **£2 a kilo/a head** due sterline al chilo/a testa

aback /ə'bæk/ adv **be taken ~** essere preso in contropiede

abandon /ə'bændən/ vt abbandonare; (give up) rinunciare a ● n abbandono m. **~ed** a abbandonato

abashed /ə'bæʃt/ a imbarazzato

abate /ə'beɪt/ vi calmarsi

abattoir /'æbətwɑ:(r)/ n mattatoio m

abbey /'æbɪ/ n abbazia f

abbreviat|e /ə'bri:vɪeɪt/ vt abbreviare. **~ion** /-'eɪʃn/ n abbreviazione f

abdicat|e /'æbdɪkeɪt/ vi abdicare ● vt rinunciare a. **~ion** /-'keɪʃn/ n abdicazione f

abdom|en /'æbdəmən/ n addome m. **~inal** /-'dɒmɪnl/ a addominale

abduct /əb'dʌkt/ vt rapire. **~ion** /-'ʌkʃn/ n rapimento m

aberration /æbə'reɪʃn/ n aberrazione f

abet /ə'bet/ vt (pt/pp **abetted**) **aid and ~** Jur essere complice di

abeyance /ə'beɪəns/ n **in ~** in sospeso; **fall into ~** cadere in disuso

abhor /əb'hɔ:(r)/ vt (pt/pp **abhorred**) aborrire. **~rence** /-'hɒrəns/ n orrore m

abid|e /ə'baɪd/ vt (pt/pp **abided**) (tolerate) sopportare ● **abide by** vi rispettare. **~ing** a perpetuo

ability /ə'bɪlətɪ/ n capacità f inv

abject /'æbdʒekt/ a (poverty) degradante; (apology) umile; (coward) abietto

ablaze /ə'bleɪz/ a in fiamme; **be ~ with light** risplendere di luci

able /'eɪbl/ a capace, abile; **be ~ to do sth** poter fare qcsa; **were you ~ to...?** sei riuscito a...? **~-bodied** a robusto; Mil abile

ably /'eɪblɪ/ adv abilmente

abnormal /æb'nɔ:ml/ a anormale. **~ity** /-'mælətɪ/ n anormalità f inv. **~ly** adv in modo anormale

aboard /ə'bɔ:d/ adv & prep a bordo

abolish /ə'bɒlɪʃ/ vt abolire. **~ition** /æbə'lɪʃn/ n abolizione f

abominable /ə'bɒmɪnəbl/ a abominevole

Aborigine /æbə'rɪdʒənɪ/ n aborigeno, -a mf d'Australia

abort /ə'bɔ:t/ vt fare abortire; fig annullare. **~ion** /-ɔ:ʃn/ n aborto m; **have an ~ion** abortire. **~ive** /-tɪv/ a (attempt) infruttuoso

abound /ə'baʊnd/ vi abbondare; **~ in** abbondare di

about /ə'baʊt/ adv (here and there) [di] qua e [di] là; (approximately) circa; **be ~** (illness, tourists:) essere in giro; **be up and ~** essere alzato; **leave sth lying ~** lasciare in giro qcsa ● prep (concerning) su; (in the region of) intorno a; (here and there in) per; **what is the book/the film ~?** di cosa parla il libro/il film?; **talk/know ~** parlare/sapere di; **I know nothing ~ it** non ne so niente; **~ 5 o'clock** intorno alle 5; **travel ~ the world**

viaggiare per il mondo; **be ~ to do sth** stare per fare qcsa; **how going to the cinema?** e se andassimo al cinema?

about: **~-'face** n, **~-'turn** n dietro front m inv

above /ə'bʌv/ adv & prep sopra; **~ all** soprattutto

above: **~-'board** a onesto. **~-'mentioned** a suddetto

abrasive /ə'breɪsɪv/ a abrasivo; (remark) caustico ● n abrasivo m

abreast /ə'brest/ adv fianco a fianco; **come ~ of** allinearsi con; **keep ~ of** tenersi al corrente di

abridged /ə'brɪdʒd/ a ridotto

abroad /ə'brɔːd/ adv all'estero

abrupt /ə'brʌpt/ a brusco

abscess /'æbsɪs/ n ascesso m

abscond /əb'skɒnd/ vi fuggire

absence /'æbsəns/ n assenza f; (lack) mancanza f

absent¹ /'æbsənt/ a assente

absent² /æb'sent/ vt **~ oneself** essere assente

absentee /æbsən'tiː/ n assente mf

absent-minded /æbsənt'maɪndɪd/ a distratto

absolute /'æbsəluːt/ a assoluto; **an ~ idiot** un perfetto idiota. **~ly** adv assolutamente; (fam: indicating agreement) esattamente

absolution /æbsə'luːʃn/ n assoluzione f

absolve /əb'zɒlv/ vt assolvere

absorb /əb'sɔːb/ vt assorbire; **~ed in** assorto in. **~ent** /-ənt/ a assorbente

absorption /əb'sɔːpʃn/ n assorbimento m; (in activity) concentrazione f

abstain /əb'steɪn/ vi astenersi (**from** da)

abstemious /əb'stiːmɪəs/ a moderato

abstention /əb'stenʃn/ n Pol astensione f

abstinence /'æbstɪnəns/ n astinenza f

abstract /'æbstrækt/ a astratto

● n astratto m; (summary) estratto m

absurd /əb'sɜːd/ a assurdo. **~ity** n assurdità f inv

abundan|ce /ə'bʌndəns/ n abbondanza f. **~t** a abbondante

abuse¹ /ə'bjuːz/ vt (misuse) abusare di; (insult) insultare; (ill-treat) maltrattare

abus|e² /ə'bjuːs/ n abuso m; (verbal) insulti mpl; (ill-treatment) maltrattamento m. **~ive** /-ɪv/ a offensivo

abut /ə'bʌt/ vi (pt/pp abutted) confinare (**onto** con)

abysmal /ə'bɪzml/ a fam pessimo; (ignorance) abissale

abyss /ə'bɪs/ n abisso m

academic /ækə'demɪk/ a teorico; (qualifications, system) scolastico; **be ~** ⟨person:⟩ avere predisposizione allo studio ● n docente mf universitario, -a

academy /ə'kædəmɪ/ n accademia f; (of music) conservatorio m

accede /ək'siːd/ vi **~ to** accedere a (request); salire a (throne)

accelerat|e /ək'seləreɪt/ vt/i accelerare. **~ion** /-'reɪʃn/ n accelerazione f. **~or** n Auto acceleratore m

accent /'æksənt/ n accento m

accentuate /ək'sentjʊeɪt/ vt accentuare

accept /ək'sept/ vt accettare. **~able** /-əbl/ a accettabile. **~ance** n accettazione f

access /'ækses/ n accesso m. **~ible** /ək'sesɪbl/ a accessibile

accession /ək'seʃn/ n (to throne) ascesa f al trono

accessory /ək'sesərɪ/ n accessorio m; Jur complice mf

accident /'æksɪdənt/ n incidente m; (chance) caso m; **by ~** per caso; (unintentionally) senza volere; **I'm sorry, it was an ~** mi dispiace, non l'ho fatto apposta. **~al** /-'dentl/ a (meeting) casuale; (death) incidentale; (unintentional) involontario. **~ally** adv per caso; (*unintentionally*) inavvertitamente

acclaim /ə'kleɪm/ *n* acclamazione *f* • *vt* acclamare (**as come**)

acclimatize /ə'klaɪmətaɪz/ *vt* be- **come** ~**d** acclimatarsi

accolade /'ækəleɪd/ *n* riconoscimento *m*

accommodat|e /ə'kɒmədeɪt/ *vt* ospitare; (*oblige*) favorire. ~**ing** *a* accomodante. ~**ion** /-'deɪʃn/ *n* (*place to stay*) sistemazione *f*

accompan|iment /ə'kʌmpənɪmənt/ *n* accompagnamento *m*. ~**ist** *n* Mus accompagnatore, -trice *mf*

accompany /ə'kʌmpənɪ/ *vt* (*pt/pp* **-ied**) accompagnare

accomplice /ə'kʌmplɪs/ *n* complice *mf*

accomplish /ə'kʌmplɪʃ/ *vt* (*achieve*) concludere; realizzare (*aim*). ~**ed** *a* dotato; (*fact*) compiuto. ~**ment** *n* realizzazione *f*; (*achievement*) risultato *m*; (*talent*) talento *m*

accord /ə'kɔːd/ *n* (*treaty*) accordo *m*; **with one** ~ tutti d'accordo; **of his own** ~ di sua spontanea volontà. ~**ance** *n* **in** ~**ance with** in conformità *f* o *a*

according /ə'kɔːdɪŋ/ *adv* ~ **to** secondo. ~**ly** *adv* di conseguenza

accordion /ə'kɔːdɪən/ *n* fisarmonica *f*

accost /ə'kɒst/ *vt* abbordare

account /ə'kaʊnt/ *n* conto *m*; (*report*) descrizione *f*; (*of eyewitness*) resoconto *m*; ~**s** *pl* Comm conti *mpl*; **on** ~ a causa di; **on no** ~ per nessun motivo; **on this** ~ per questo motivo; **on my** ~ per causa mia; **on no** ~ di nessuna importanza; **take into** ~ tener conto di • **account for** *vi* (*explain*) spiegare; (*person*) render conto di; (*constitute*) costituire. ~**ability** *n* responsabilità *f inv*. ~**able** *a* responsabile (**for** di)

accountant /ə'kaʊntənt/ *n* (book-

keeper) contabile *mf*; (consultant) commercialista *mf*

accredited /ə'kredɪtɪd/ *a* accreditato

accrue /ə'kruː/ *vi* (*interest:*) maturare

accumulat|e /ə'kjuːmjʊleɪt/ *vt* accumulare • *vi* accumularsi. ~**ion** /-'leɪʃn/ *n* accumulazione *f*

accura|cy /'ækjʊrəsɪ/ *n* precisione *f*. ~**te** /-rət/ *a* preciso. ~**tely** *adv* con precisione

accusation /ækjʊ'zeɪʃn/ *n* accusa *f*

accusative /ə'kjuːzətɪv/ *a & n* ~ **[case]** Gram accusativo *m*

accuse /ə'kjuːz/ *vt* accusare; ~ **sb of doing sth** accusare qcno di fare qcsa. ~**d** *n* **the** ~**d** /ə'kjuːzd/ *m*, l'accusata *f*

accustom /ə'kʌstəm/ *vt* abituare (**to** a); **grow** *or* **get** ~**ed to** abituarsi a. ~**ed** *a* abituato

ace /eɪs/ *n* Cards asso *m*; (*tennis*) ace *m inv*

ache /eɪk/ *n* dolore *m* • *vi* dolere, far male; ~ **all over** essere tutto indolenzito

achieve /ə'tʃiːv/ *vt* ottenere (*success*); realizzare (*goal, ambition*). ~**ment** *n* (*feat*) successo *m*

acid /'æsɪd/ *a* acido • *n* acido *m*. ~**ity** /ə'sɪdətɪ/ *n* acidità *f*. ~ '**rain** *n* pioggia *f* acida

acknowledge /ək'nɒlɪdʒ/ *vt* riconoscere; rispondere a (*greeting*); far cenno di aver notato (*sb's presence*); ~ **receipt of** accusare ricevuta di. ~**ment** *n* riconoscimento *m*; **send an** ~**ment of a letter** confermare il ricevimento di una lettera

acne /'æknɪ/ *n* acne *f*

acorn /'eɪkɔːn/ *n* ghianda *f*

acoustic /ə'kuːstɪk/ *a* acustico. ~**s** *npl* acustica *fsg*

acquaint /ə'kweɪnt/ *vt* ~ **sb with** metter qcno al corrente di; **be** ~**ed with** conoscere (*person*); essere a conoscenza di (*fact*). ~**ance** *n* (*person*) conoscente *mf*; **make**

sb's ~ance fare la conoscenza di qcno

acquiesce /ækwɪ'es/ vi acconsentire (to, in a). ~nce n acquiescenza f

acquire /ə'kwaɪə(r)/ vt acquisire

acquisition /ækwɪ'zɪʃn/ n acquisizione f. ~ive /ə'kwɪzɪtɪv/ a avido

acquit /ə'kwɪt/ vt (pt/pp acquitted) assolvere; ~ oneself well cavarsela bene. ~tal n assoluzione f

acre /'eɪkə(r)/ n acro m (= 4 047 m²)

acrid /'ækrɪd/ a acre

acrimoni|ous /ækrɪ'məʊnɪəs/ a aspro. ~y /'ækrɪmənɪ/ n asprezza f

acrobat /'ækrəbæt/ n acrobata mf. ~ic /-'bætɪk/ a acrobatico

across /ə'krɒs/ adv dall'altra parte; (wide) in larghezza; not (lengthwise) attraverso; (in crossword) orizzontale; come ~ sb imbattersi in qcsa; go ~ attraversare ● prep (crosswise) di traverso su; (on the other side of) dall'altra parte di

act /ækt/ n atto m; (in variety show) numero m; put on an ~ fam fare scena ● vi agire; (behave) comportarsi; Theat recitare; (pretend) fingere; ~ as fare da ● vt recitare ⟨role⟩. ~ing a ⟨deputy⟩ provvisorio ● n Theat recitazione f; (profession) teatro m. ~ing profession n professione f dell'attore

action /'ækʃn/ n azione f; Mil combattimento m; Jur azione f legale; out of ~ ⟨machine:⟩ fuori uso; take ~ agire. ~ 'replay n replay m inv

activ|e /'æktɪv/ a attivo. ~ely adv attivamente. ~ity /-'tɪvətɪ/ n attività f inv

act|or /'æktə(r)/ n attore m. ~ress n attrice f

actual· /'æktʃʊəl/ a ⟨real⟩ reale. ~ly adv in realtà

acumen /'ækjʊmən/ n acume m

acupuncture /'ækjʊ-/ n agopuntura f

acute /ə'kju:t/ a acuto; ⟨shortage, hardship⟩ estremo

AD abbr (**Anno Domini**) d.C.

ad /æd/ n fam pubblicità f inv

adamant /'ædəmənt/ a categorico (that sul fatto che)

adapt /ə'dæpt/ vt adattare ⟨play⟩ ● vi adattarsi. ~ability /-ə'bɪlətɪ/ n adattabilità f. ~able /-əbl/ a adattabile

adaptation /ædæp'teɪʃn/ n Theat adattamento m

adapter, adaptor /ə'dæptə(r)/ n adattatore m; (two-way) presa f multipla

add /æd/ vt aggiungere; Math addizionare ● vi addizionare; ~ to (fig: increase) aggravare. **add up** vt addizionare ⟨figures⟩ ● vi addizionare; ~ up to ammontare a; it doesn't ~ up fig non quadra

adder /'ædə(r)/ n vipera f

addict /'ædɪkt/ n tossicodipendente mf; fig fanatico, -a f

addict|ed /ə'dɪktɪd/ a assuefatto (to a); ~ed to drugs tossicodipendente; he's ~ed to television è videodipendente. ~ion /-kʃn/ n dipendenza f; (to drugs) tossicodipendenza f. ~ive /-ɪv/ a be ~ive dare assuefazione

addition /ə'dɪʃn/ n Math addizione f; (thing added) aggiunta f; in ~ in aggiunta. ~al a supplementare. ~ally adv in più

additive /'ædɪtɪv/ n additivo m

address /ə'dres/ n indirizzo m; (speech) discorso m; form of ~ formula f di cortesia ● vt indirizzare; (speak to) rivolgersi a ⟨person⟩; tenere un discorso a ⟨meeting⟩. ~ee /ædre'si:/ n destinatario, -a mf

adenoids /'ædənɔɪdz/ npl adenoidi fpl

adept /'ædept/ a & n esperto, -a mf (at in)

adequate /'ædɪkwət/ a adeguato. ~ly adv adeguatamente

adhere /əd'hɪə(r)/ vi aderire; ~ to attenersi a ⟨principles, rules⟩

adhesive /əd'hi:sɪv/ *a* adesivo ● *n* adesivo *m*

adjacent /ə'dʒeɪsənt/ *a* adiacente

adjective /'ædʒɪktɪv/ *n* aggettivo *m*

adjoin /ə'dʒɔɪn/ *vt* essere adiacente a. ~**ing** *a* adiacente

adjourn /ə'dʒɜːn/ *vt/i* aggiornare (*until* a). ~**ment** *n* aggiornamento *m*

adjudicate /ə'dʒuːdɪkeɪt/ *vi* decidere; (*in competition*) giudicare

adjust /ə'dʒʌst/ *vt* modificare; regolare (*focus, sound etc*) ● *vi* adattarsi. ~**able** /-əbl/ *a* regolabile. ~**ment** *n* adattamento *m*, Techn regolamento *m*

ad lib /æd'lɪb/ *a* improvvisato ● *adv* a piacere ● *vi* (*pt/pp* **ad libbed**) *fam* improvvisare

administer /əd'mɪnɪstə(r)/ *vt* amministrare; somministrare (*medicine*)

administrat|ion /ədmɪnɪ'streɪʃn/ *n* amministrazione *f*; Pol governo *m*. ~**or** /əd'mɪnɪstreɪtə(r)/ *n* amministratore, -trice *mf*

admirable /'ædmərəbl/ *a* ammirevole

admiral /'ædmərəl/ *n* ammiraglio *m*

admiration /ædmə'reɪʃn/ *n* ammirazione *f*

admire /əd'maɪə(r)/ *vt* ammirare. ~**r** *n* ammiratore, -trice *mf*

admissible /əd'mɪsəbl/ *a* ammissibile

admission /əd'mɪʃn/ *n* ammissione *f*; (*to hospital*) ricovero *m*; (*entry*) ingresso *m*

admit /əd'mɪt/ *vt* (*pt/pp* **admitted**) (*let in*) far entrare; (*to hospital*) ricoverare; (*acknowledge*) ammettere ● *vi* ~ **to sth** ammettere qcsa. ~**tance** *n* ammissione *f*; **'no ~tance'** 'vietato l'ingresso'. ~**tedly** *adv* bisogna riconoscersi

admonish /əd'mɒnɪʃ/ *vt* ammonire

ado /ə'duː/ *n* **without more ~** senza ulteriori indugi

adolescen|ce /ædə'lesns/ *n* adolescenza *f*. ~**t** *a & n* adolescente *mf*

adopt /ə'dɒpt/ *vt* adottare; Pol scegliere (*candidate*). ~**ion** /-ɒpʃn/ *n* adozione *f*. ~**ive** /-ɪv/ *a* adottivo *m*

ador|able /ə'dɔːrəbl/ *a* adorabile. ~**ation** /ædə'reɪʃn/ *n* adorazione *f*

adore /ə'dɔː(r)/ *vt* adorare

adrenalin /ə'drenəlɪn/ *n* adrenalina *f*

Adriatic /eɪdrɪ'ætɪk/ *a & n* **the ~ [Sea]** il mare Adriatico, l'Adriatico *m*

adrift /ə'drɪft/ *a* alla deriva; **be ~** andare alla deriva; **come ~** staccarsi

adroit /ə'drɔɪt/ *a* abile

adulation /ædjʊ'leɪʃn/ *n* adulazione *f*

adult /'ædʌlt/ *n* adulto, -a *mf*

adulterate /ə'dʌltəreɪt/ *vt* adulterare (*wine*)

adultery /ə'dʌltərɪ/ *n* adulterio *m*

advance /əd'vɑːns/ *n* avanzamento *m*; Mil avanzata *f*; (*payment*) anticipo *m*; **in ~** in anticipo ● *vi* avanzare; (*make progress*) fare progressi ● *vt* avanzare (*theory*); promuovere (*cause*); anticipare (*money*). ~ **booking** *n* prenotazione *f* [in anticipo]. ~**d** *a* avanzato. ~**ment** *n* promozione *f*

advantage /əd'vɑːntɪdʒ/ *n* vantaggio *m*; **take ~ of** approfittare di. ~**ous** /ædvən'teɪdʒəs/ *a* vantaggioso

advent /'ædvent/ *n* avvento *m*

adventur|e /əd'ventʃə(r)/ *n* avventura *f*. ~**ous** /-rəs/ *a* avventuroso

adverb /'ædvɜːb/ *n* avverbio *m*

adversary /'ædvəsərɪ/ *n* avversario, -a *mf*

advers|e /'ædvɜːs/ *a* avverso. ~**ity** /əd'vɜːsətɪ/ *n* avversità *f*

advert /'ædvɜːt/ *n* *fam* = **advertisement**

advertise /'ædvətaɪz/ *vt* reclamizzare; mettere un annuncio per (*job, flat*) ● *vi* fare pubblicità (*for job, flat*) mettere un annuncio

advertisement /əd'vɜːtɪsmənt/ n
pubblicità f inv; (in paper) inserzione f, annuncio m

advertis|er /'ædvətaɪzə(r)/ n (in
newspaper) inserzionista mf.
~ing n pubblicità f ● attrib pubblicitario

advice /əd'vaɪs/ n consigli mpl;
piece of ~ consiglio m

advisable /əd'vaɪzəbl/ a consigliabile

advis|e /əd'vaɪz/ vt consigliare;
(inform) avvisare; **~e sb to do
sth** consigliare a qcno di fare
qcsa; **~e sth against sth** sconsigliare qcsa a qcno. **~er** n consulente mf. **~ory** a consultivo

advocate¹ /'ædvəkət/ n (supporter) fautore, -trice mf

advocate² /'ædvəkeɪt/ vt propugnare

aerial /'eərɪəl/ a aereo ● n antenna f

aerobics /eə'rəubɪks/ n aerobica
fsg

aero|drome /'eərədrəum/ n aerodromo m. **~plane** n aeroplano m

aerosol /'eərəsɒl/ n bomboletta f
spray

aesthetic /iːs'θetɪk/ a estetico

afar /ə'fɑː(r)/ adv **from ~** da lontano

affable /'æfəbl/ a affabile

affair /ə'feə(r)/ n affare m; (scandal) caso m; (sexual) relazione f

affect /ə'fekt/ vt influire su;
(emotionally) colpire; (concern) riguardare. **~ation** /æfek'teɪʃn/ n
affettazione f. **~ed** a affettato

affection /ə'fekʃn/ n affetto m.
~ate /-ət/ a affettuoso

affiliated /ə'fɪlɪeɪtɪd/ a affiliato

affinity /ə'fɪnətɪ/ n affinità f inv

affirm /ə'fɜːm/ vt affermare; Jur
dichiarare solennemente

affirmative /ə'fɜːmətɪv/ a affermativo ● n **in the ~** affermativamente

afflict /ə'flɪkt/ vt affliggere. **~ion**
/-ɪkʃn/ n afflizione f

affluen|ce /'æfluəns/ n agiatezza f.
~t a agiato

afford /ə'fɔːd/ vt **be able to ~ sth**
potersi permettere qcsa. **~able**
/-əbl/ a abbordabile

affray /ə'freɪ/ n rissa f

affront /ə'frʌnt/ n affronto m

afield /ə'fiːld/ adv **further ~** più
lontano

afloat /ə'fləut/ a a galla

afoot /ə'fut/ a **there's something
~** si sta preparando qualcosa

aforesaid /ə'fɔːsed/ a Jur suddetto

afraid /ə'freɪd/ a **be ~** aver paura;
I'm ~ not purtroppo no; **I'm ~ so**
temo di sì; **I'm ~ I can't help you**
mi dispiace, ma non posso esserle
d'aiuto

afresh /ə'freʃ/ adv da capo

Africa /'æfrɪkə/ n Africa f. **~n** a &
n africano, -a mf

after /'ɑːftə(r)/ adv dopo; **the day
~** il giorno dopo; **be ~** cercare
● prep dopo; **~ all** dopotutto; **~ the
day ~ tomorrow** dopodomani
● conj dopo che

after: **~-effect** n conseguenza f.
~math /-mɑːθ/ n conseguenze fpl;
the ~math of war il dopoguerra;
in the ~math of nel periodo successivo a. **~'noon** n pomeriggio
m; **good ~noon!** buon giorno!
~-sales service n servizio m assistenza clienti. **~shave** n (lozione
f) dopobarba m inv. **~thought** n
added as an ~thought aggiunto
in un secondo momento; **~wards**
adv in seguito

again /ə'gein/ adv di nuovo; [then]
~ (besides) inoltre; (on the other
hand) d'altra parte; **~ and ~** continuamente

against /ə'geinst/ prep contro

age /eidʒ/ n età f inv; (era) era f. **~s**
fam secoli; **what ~ are you?**
quanti anni hai?; **be under ~** non
avere l'età richiesta; **he's two
years of ~** ha due anni ● vt/i
(pres p ageing) invecchiare

aged¹ /eidʒd/ a **~ two** di due anni

aged[2] /ˈeɪdʒɪd/ a anziano ● n the ~ pl gli anziani

ageless /ˈeɪdʒlɪs/ a senza età

agency /ˈeɪdʒənsɪ/ n agenzia f; **have the ~ for** essere in concessione di

agenda /əˈdʒendə/ n ordine m del giorno; **on the ~** all'ordine del giorno; fig in programma

agent /ˈeɪdʒənt/ n agente mf

aggravat|e /ˈægrəveɪt/ vt aggravare; (annoy) esasperare. **~ion** /-ˈveɪʃn/ n aggravamento m; (annoyance) esasperazione f

aggregate /ˈægrɪgət/ a totale ● n totale m; **on ~** nel complesso

aggres.|ion /əˈgreʃn/ n aggressione f. **~ive** /-sɪv/ a aggressivo. **~iveness** n aggressività f. **~or** n aggressore m

aggro /ˈægrəʊ/ n fam aggressività f; (problems) grane fpl

aghast /əˈgɑːst/ a inorridito

agil|e /ˈædʒaɪl/ a agile. **~ity** /əˈdʒɪlətɪ/ n agilità f

agitat|e /ˈædʒɪteɪt/ vt mettere in agitazione; (shake) agitare ● vi fig **~e for** creare delle agitazioni per. **~ed** a agitato. **~ion** /-ˈteɪʃn/ n agitazione f. **~or** n agitatore m, -trice mf

agnostic /ægˈnɒstɪk/ n agnostico, -a mf

ago /əˈgəʊ/ adv fa; **a long time/a month ~** molto tempo/un mese fa

agog /əˈgɒg/ a eccitato

agoniz|e /ˈægənaɪz/ vi angosciarsi (over per). **~ing** a angosciante

agony /ˈægənɪ/ n agonia f; (mental) angoscia f; **be in ~** avere dei dolori atroci

agree /əˈgriː/ vt accordarsi su; **~ to do sth** accettare di fare qcsa; **~ that** essere d'accordo [sul fatto] che ● vi essere d'accordo; (figures:) concordare; (reach agreement) mettersi d'accordo; (get on) andare d'accordo; (consent) acconsentire (to a); **it doesn't ~ with me** mi fa male; **~ with sth** (approve of) approvare qcsa

agreeable /əˈgriːəbl/ a gradevole; (willing) d'accordo

agreed /əˈgriːd/ a convenuto

agreement /əˈgriːmənt/ n accordo m; **in ~** d'accordo

agricultur|al /ægrɪˈkʌltʃərəl/ a agricolo. **~e** /ˈægrɪkʌltʃə(r)/ n agricoltura f

aground /əˈgraʊnd/ adv **run ~** (ship:) arenarsi

ahead /əˈhed/ adv avanti; **be ~ of** essere davanti a; fig essere avanti rispetto a; **draw ~** passare davanti (of a); **get ~** (in life) riuscire; **go ~!** fai pure!; **look ~** pensare all'avvenire; **plan ~** fare progetti per l'avvenire

aid /eɪd/ n aiuto m; **in ~ of** a favore di ● vt aiutare

aide /eɪd/ n assistente mf

Aids /eɪdz/ n AIDS m

ail|ing /ˈeɪlɪŋ/ a malato. **~ment** n disturbo m

aim /eɪm/ n mira f; fig scopo m; **take ~** prendere la mira ● vt puntare (gun) (at contro) ● vi mirare; **~ to do sth** aspirare a fare qcsa. **~less·a, ~lessly** adv senza scopo

air /eə(r)/ n aria f; **be on the ~** (programme:) essere in onda; **put on ~s** darsi delle arie; **by ~** in aereo; (airmail) per via aerea ● vt arieggiare; far conoscere (views)

air: **~-bed** n materassino m [gonfiabile]. **~-conditioned** a con aria condizionata. **~-conditioning** n aria f condizionata. **~craft** n aereo m. **~craft carrier** n portaerei f inv. **~fare** n tariffa f aerea. **~field** n campo m d'aviazione. **~ force** n aviazione f. **~ freshener** n deodorante m per l'ambiente. **~gun** n fucile m pneumatico. **~ hostess** n hostess f inv. **~ letter** n aerogramma m. **~line** n compagnia f aerea. **~lock** n bolla f d'aria. **~mail** n posta f aerea. **~plane** n Am aereo m. **~ pocket** n vuoto m d'aria. **~port** n aeroporto m. **~-raid** n incursione f aerea. **~-raid shelter** n rifugio m an-

tiaereo. **~ship** *n* dirigibile *m*.
~tight *a* ermetico. **~ traffic** *n*
traffico *m* aereo. **~-traffic con-**
troller *n* controllore *m* di volo.
~worthy *a* idoneo al volo

airy /'eərɪ/ *a* (**-ier, -iest**) arieggiato;
(*manner*) noncurante

aisle /aɪl/ *n* corridoio *m*; (*in*
supermarket) corsia *f*; (*in church*)
navata *f*

ajar /ə'dʒɑː(r)/ *a* socchiuso

akin /ə'kɪn/ *a* **~ to** simile a

alacrity /ə'lækrətɪ/ *n* alacrità *f* inv

alarm /ə'lɑːm/ *n* allarme *m*; **set**
the ~ (*of alarm clock*) mettere la
sveglia ● *vt* allarmare. **~ clock** *n*
sveglia *f*

alas /ə'læs/ *int* ahimè

album /'ælbəm/ *n* album *m* inv

alcohol /'ælkəhɒl/ *n* alcool *m*. **~ic**
/-'hɒlɪk/ *a* alcolico ● *n* alcolizzato.
-a *mf*. **~ism** *n* alcolismo *m*

alcove /'ælkəʊv/ *n* alcova *f*

alert /ə'lɜːt/ *a* sveglio; (*watchful*)
vigile ● *n* segnale *m* d'allarme;
be on the ~ stare allerta ● *vt*
allertare

algae /'ældʒiː/ *npl* alghe *fpl*

algebra /'ældʒɪbrə/ *n* algebra *f*

Algeria /æl'dʒɪərɪə/ *n* Algeria *f*.
~n *a & n* algerino, -a *mf*

alias /'eɪlɪəs/ *n* pseudonimo *m*
● *adv* alias

alibi /'ælɪbaɪ/ *n* alibi *m* inv

alien /'eɪlɪən/ *a* straniero; *fig* estraneo ● *n* straniero, -a *mf*; (*from*
space) alieno, -a *mf*

alienat|e /'eɪlɪəneɪt/ *vt* alienare.
~ion /-'neɪʃn/ *n* alienazione *f*

alight[1] /ə'laɪt/ *vi* scendere; (*bird*:)
posarsi

alight[2] *a* **be ~** essere in fiamme;
set ~ dar fuoco a

align /ə'laɪn/ *vt* allineare. **~ment** *n*
allineamento *m*; **out of ~ment**
non allineato

alike /ə'laɪk/ *a* simile; **be ~** rassomigliarsi ● *adv* in modo simile;
look ~ rassomigliarsi; **summer**
and winter ~ sia d'estate che d'inverno

alimony /'ælɪmənɪ/ *n* alimenti *mpl*

alive /ə'laɪv/ *a* vivo; **~ with** bruli-
cante di; **~ to** sensibile a; **~ and**
kicking vivo e vegeto

alkali /'ælkəlaɪ/ *n* alcali *m*

all /ɔːl/ *a* tutto; **~ the children, ~**
children tutti i bambini; **~ day**
tutto il giorno; **he refused ~ help**
ha rifiutato qualsiasi aiuto; **for ~**
that (*nevertheless*) ciononostante;
in ~ sincerity in tutta sincerità;
be ~ for essere favorevole a
● *pron* tutto; **~ of you/them** tutti
voi/loro; **~ of it** tutto; **~ of the**
town tutta la città; **in ~** in tutto;
in ~ tutto sommato; **most of ~**
più di ogni altra cosa; **once and for**
~ una volta per tutte ● *adv*
completamente; **~ but** quasi; **~ at**
once (*at the same time*) tutto in
una volta; **~ at once, ~ of a**
sudden all'improvviso; **~ too**
soon troppo presto; **~ the same**
(*nevertheless*) ciononostante;
~ the better meglio ancora; **she's**
not ~ that così brava come attrice non è
● *n* in
tutto; *fam* esausto; **thirty/three**
~ (*in sport*) trenta/tre pari; **~**
over (*finished*) tutto finito; (*every-*
where) dappertutto; **it's ~ right** (*I*
don't mind) non fa niente; **I'm ~**
right (*not hurt*) non ho niente; **~**
right! va bene!

allay /ə'leɪ/ *vt* placare (*suspicions*,
anger)

allegation /ælɪ'geɪʃn/ *n* accusa *f*

allege /ə'ledʒ/ *vt* dichiarare. **~d a**
presunto. **~dly** /-ɪdlɪ/ *adv* a quanto si dice

allegiance /ə'liːdʒəns/ *n* fedeltà *f*

allegor|ical /ælɪ'gɒrɪkl/ *a* allegori-
co. **~y** /'ælɪgərɪ/ *n* allegoria *f*

allerg|ic /ə'lɜːdʒɪk/ *a* allergico. **~y**
/'ælədʒɪ/ *n* allergia *f*

alleviate /ə'liːvɪeɪt/ *vt* alleviare

alley /'ælɪ/ *n* vicolo *m*; (*for*
bowling) corsia *f*

alliance /ə'laɪəns/ *n* alleanza *f*

allied /'ælaɪd/ *a* alleato; (*fig:*
related) connesso (**to** a)

alligator /ˈælɪgeɪtə(r)/ *n* alligatore *m*

allocat|e /ˈæləkeɪt/ *vt* assegnare; distribuire (*resources*). **~ion** /-ˈkeɪʃn/ *n* assegnazione *f*; (*of resources*) distribuzione *f*

allot /əˈlɒt/ *vt* (*pt/pp* **allotted**) distribuire. **~ment** *n* distribuzione *f*; (*share*) parte *f*; (*land*) piccolo lotto *m* di terreno

allow /əˈlaʊ/ *vt* permettere; (*grant*) accordare; (*reckon on*) contare; (*agree*) ammettere; **~ for** tener conto di; **~ sb to do sth** permettere a qcno di fare qcsa; **you are not ~ed to...** è vietato...

allowance /əˈlaʊəns/ *n* sussidio *m*; (*Am: pocket money*) paghetta *f*; (*for petrol etc*) indennità *f inv*; (*of luggage, duty free*) limite *m*; **make ~s for** essere indulgente verso (*sb*); tener conto di (*sth*)

alloy /ˈælɔɪ/ *n* lega *f*

allude /əˈluːd/ *vi* alludere

allusion /əˈluːʒn/ *n* allusione *f*

ally[1] /ˈælaɪ/ *n* alleato, -a *mf*

ally[2] /əˈlaɪ/ *vt* (*pt/pp* **-ied**) alleare; **~ oneself with** allearsi con

almighty /ɔːlˈmaɪti/ *a* (*fam: big*) mega *inv* ● *n* **the A~** l'Onnipotente *m*

almond /ˈɑːmənd/ *n* mandorla *f*; (*tree*) mandorlo *m*

almost /ˈɔːlməʊst/ *adv* quasi

alone /əˈləʊn/ *a* solo; **leave me ~!** lasciami in pace!; **let ~** (*not to mention*) figurarsi ● *adv* da solo

along /əˈlɒŋ/ *prep* lungo ● *adv* **~ with** assieme a; **all ~** tutto il tempo; **come ~!** (*hurry up*) vieni qui!; **I'll be ~ in a minute** arrivo tra un attimo; **move ~** spostarsi; **move ~!** circolare!

along|side *adv* lungo bordo ● *prep* lungo; **work ~ sb** lavorare fianco a fianco con qcno

aloof /əˈluːf/ *a* distante

aloud /əˈlaʊd/ *adv* ad alta voce

alphabet /ˈælfəbet/ *n* alfabeto *m*. **~ical** /-ˈbetɪkl/ *a* alfabetico

alpine /ˈælpaɪn/ *a* alpino

Alps /ælps/ *npl* Alpi *fpl*

already /ɔːlˈredɪ/ *adv* già

Alsatian /ælˈseɪʃn/ *n* (*dog*) pastore *m* tedesco

also /ˈɔːlsəʊ/ *adv* anche; **~, I need...** [e] inoltre, ho bisogno di...

altar /ˈɔːltə(r)/ *n* altare *m*

alter /ˈɔːltə(r)/ *vt* cambiare; aggiustare (*clothes*) ● *vi* cambiare. **~ation** /-ˈreɪʃn/ *n* modifica *f*

alternate[1] /ˈɔːltəneɪt/ *vi* alternarsi ● *vt* alternare

alternate[2] /ɔːlˈtɜːnət/ *a* alterno; **on ~ days** a giorni alterni

alternating current *n* corrente *f* alternata

alternative /ɔːlˈtɜːnətɪv/ *a* alternativo ● *n* alternativa *f*. **~ly** *adv* alternativamente

although /ɔːlˈðəʊ/ *conj* benché, sebbene

altitude /ˈæltɪtjuːd/ *n* altitudine *f*

altogether /ɔːltəˈgeðə(r)/ *adv* (*in all*) in tutto; (*completely*) completamente; **I'm not ~ sure** non sono del tutto sicuro

altruistic /æltrʊˈɪstɪk/ *a* altruistico

aluminium /æljʊˈmɪnɪəm/ *n*, *Am* **aluminum** /əˈluːmɪnəm/ *n* alluminio *m*

always /ˈɔːlweɪz/ *adv* sempre

am /æm/ *see* **be**

a.m. *abbr* (*ante meridiem*) del mattino

amalgamate /əˈmælgəmeɪt/ *vt* fondere ● *vi* fondersi

amass /əˈmæs/ *vt* accumulare

amateur /ˈæmətə(r)/ *n* non professionista *mf*, *pej* dilettante *mf* ● *attrib* dilettante; **~ dramatics** filodrammatica *f*. **~ish** *a* dilettantesco

amaze /əˈmeɪz/ *vt* stupire. **~d** *a* stupito. **~ment** *n* stupore *m*

amazing /əˈmeɪzɪŋ/ *a* incredibile

ambassador /æmˈbæsədə(r)/ *n* ambasciatore, -trice *mf*

amber /ˈæmbə(r)/ *n* ambra *f* ● *a* (*colour*) ambra *inv*

ambidextrous /ˈæmbɪˈdekstrəs/ a ambidestro

ambience /ˈæmbɪəns/ n atmosfera f

ambiguity /ˈæmbɪˈgjuːəti/ n ambiguità f inv. **~ous** /-ˈbɪgjʊəs/ a ambiguo

ambition /æmˈbɪʃn/ n ambizione f; (aim) aspirazione f. **~ous** /-ʃəs/ a ambizioso

ambivalent /æmˈbɪvələnt/ a ambivalente

amble /ˈæmbl/ vi camminare senza fretta

ambulance /ˈæmbjʊləns/ n ambulanza f

ambush /ˈæmbʊʃ/ n imboscata f ● vt tendere un'imboscata a

amenable /əˈmiːnəbl/ a conciliante; **~ to** sensibile a

amend /əˈmend/ vt modificare. **~ment** n modifica f. **~s** npl make **~s** fare ammenda (for di, per)

amenities /əˈmiːnətiz/ npl comodità fpl

America /əˈmerɪkə/ n America f. **~n** a & n americano, -a mf

amiable /ˈeɪmɪəbl/ a amabile

amicable /ˈæmɪkəbl/ a amichevole

amiss /əˈmɪs/ a **there's something ~** c'è qualcosa che non va ● adv **take sth ~** prendersela [a male]; **it won't come ~** non sarebbe sgradito

ammonia /əˈməʊnɪə/ n ammoniaca f

ammunition /æmjʊˈnɪʃn/ n munizioni fpl

amnesia /æmˈniːzɪə/ n amnesia f

amnesty /ˈæmnəstɪ/ n amnistia f

among[st] /əˈmʌŋ[st]/ prep tra, fra

amoral /eɪˈmɒrəl/ a amorale

amorous /ˈæmərəs/ a amoroso

amount /əˈmaʊnt/ n quantità f inv; (sum of money) importo m ● vi **to ~ to** ammontare a; fig equivale a

amp /æmp/ n ampère m inv

amphibian /æmˈfɪbɪən/ n anfibio m. **~ous** /-ɪəs/ a anfibio

amphitheatre /ˈæmfɪ-/ n anfiteatro m

ample /ˈæmpl/ a (large) grande; (proportions) ampio; (enough) largamente sufficiente

amplifier /ˈæmplɪfaɪə(r)/ n amplificatore m. **~y** /-faɪ/ vt (pt/pp **-ied**) amplificare (sound)

amputate /ˈæmpjʊteɪt/ vt amputare. **~ion** /-ˈteɪʃn/ n amputazione f

amuse /əˈmjuːz/ vt divertire. **~ment** n divertimento m. **~ment arcade** n sala f giochi

amusing /əˈmjuːzɪŋ/ a divertente

an /ən/, accentato /æn/ see **a**

anaemia /əˈniːmɪə/ n anemia f. **~ic** a anemico

anaesthetic /ænəsˈθetɪk/ n anestesia f

anaesthetist /əˈniːsθətɪst/ n anestesista mf

analogue /ˈænəlɒg/ a analogico

analogy /əˈnælədʒɪ/ n analogia f

analyse /ˈænəlaɪz/ vt analizzare

analysis /əˈnæləsɪs/ n analisi f inv

analyst /ˈænəlɪst/ n analista mf

analytical /ænəˈlɪtɪkl/ a analitico

anarchist /ˈænəkɪst/ n anarchico, -a mf. **~y** n anarchia f

anatomical /ænəˈtɒmɪkl/ a anatomico. **~ically** adv anatomicamente. **~y** /əˈnætəmɪ/ n anatomia f

ancestor /ˈænsestə(r)/ n antenato, -a mf. **~ry** n antenati mpl

anchor /ˈæŋkə(r)/ n ancora f ● vi gettar l'ancora ● vt ancorare

anchovy /ˈæntʃəvɪ/ n acciuga f

ancient /ˈeɪnʃənt/ a antico; fam vecchio

ancillary /ænˈsɪlərɪ/ a ausiliario

and /ənd/, accentato /ænd/ conj e; **two ~ two** due più due; **six hundred ~ two** seicentodue; **more ~ more** sempre più; **nice ~ warm** bello caldo; **try ~ come** cerca di venire; **go ~ get** vai a prendere

anecdote /ˈænɪkdəʊt/ n aneddoto m

anew /ə'nju:/ *adv* di nuovo

angel /'emdʒl/ *n* angelo *m*. **~ic** /æn'dʒelɪk/ *a* angelico

anger /'æŋgə(r)/ *n* rabbia *f* ● *vt* far arrabbiare

angle[1] /'æŋgl/ *n* angolo *m*; *fig* angolazione *f*; **at an ~** storto

angle[2] *vi* pescare con la lenza; **~ for** *fig* cercare di ottenere. **~r** *n* pescatore, -trice *mf*

Anglican /'æŋglɪkən/ *a* & *n* anglicano, -a *mf*

Anglo-Saxon /æŋgləʊ'sæksn/ *a* & *n* anglo-sassone *m*

angr|y /'æŋgrɪ/ *a* (**-ier, -iest**) arrabbiato; **get ~y** arrabbiarsi; **~y with** or **at** *sb* arrabbiato con qcno; **~y at** or **about** *sth* arrabbiato per qcsa. **~ily** *adv* rabbiosamente

anguish /'æŋgwɪʃ/ *n* angoscia *f*

angular /'æŋgjʊlə(r)/ *a* angolare

animal /'ænɪml/ *a* & *n* animale *m*

animate[1] /'ænɪmeɪt/ *vt* animare

animat|e[2] /'ænɪmət/ *a* animato; (*person*) vivace. **~ed** *a* animato; (*person*) vivace. **~ion** /-'meɪʃn/ *n* animazione *f*

animosity /ænɪ'mɒsətɪ/ *n* animosità *f inv*

ankle /'æŋkl/ *n* caviglia *f*

annex /ə'neks/ *vt* annettere

annex[e] /'æneks/ *n* annesso *m*

annihilat|e /ə'naɪəleɪt/ *vt* annientare. **~ion** /-'leɪʃn/ *n* annientamento *m*

anniversary /ænɪ'vɜ:sərɪ/ *n* anniversario *m*

announce /ə'naʊns/ *vt* annunciare. **~ment** *n* annuncio *m*. **~r** *n* annunciatore, -trice *mf*

annoy /ə'nɔɪ/ *vt* dare fastidio a; **get ~ed** essere infastidito. **~ance** *n* seccatura *f*; (*anger*) irritazione *f*. **~ing** *a* fastidioso

annual /'ænjʊəl/ *a* annuale; (*income*) annuo ● *n* *Bot* pianta *f* annua; (*children's book*) almanacco *m*

annuity /ə'nju:ətɪ/ *n* annualità *f inv*

annul /ə'nʌl/ *vt* (*pt/pp* **annulled**) annullare

anomaly /ə'nɒmǝlɪ/ *n* anomalia *f*

anonymous /ə'nɒnɪməs/ *a* anonimo

anorak /'ænəræk/ *n* giacca *f* a vento

anorex|ia /ænə'reksɪə/ *n* anoressia *f*. **~ic** *a* anoressico

another /ə'nʌðə(r)/ *a* & *pron*; **~ [one]** un altro, un'altra; **in ~ way** diversamente; **one ~** l'un l'altro

answer /'ɑ:nsə(r)/ *n* risposta *f*; (*solution*) soluzione *f* ● *vt* rispondere a (*person, question, letter*); esaudire (*prayer*); **~ the door** aprire la porta; **~ the telephone** rispondere al telefono ● *vi* rispondere; **~ back** ribattere; **~ for** rispondere di. **~able** /-əbl/ *a* responsabile; **be ~able to sb** rispondere a qcno. **~ing machine** *n* *Teleph* segreteria *f* telefonica

ant /ænt/ *n* formica *f*

antagonis|m /æn'tægənɪzm/ *n* antagonismo *m*. **~tic** /-'nɪstɪk/ *a* antagonistico

antagonize /æn'tægənaɪz/ *vt* provocare l'ostilità di

Antarctic /æn'tɑ:ktɪk/ *n* Antartico *m* ● *a* antartico

antenatal /æntɪ'neɪtl/ *a* prenatale

antenna /æn'tenə/ *n* antenna *f*

anthem /'ænθəm/ *n* inno *m*

anthology /æn'θɒlədʒɪ/ *n* antologia *f*

anthropology /ænθrə'pɒlədʒɪ/ *n* antropologia *f*

anti-'aircraft /æntɪ-/ *a* antiaereo

antibiotic /æntɪbaɪ'ɒtɪk/ *n* antibiotico *m*

'antibody /æntɪ-/ *n* anticorpo *m*

anticipat|e /æn'tɪsɪpeɪt/ *vt* prevedere; (*forestall*) anticipare. **~ion** /-'peɪʃn/ *n* anticipo *m*; (*excitement*) attesa *f*

anti'climax *n* delusione *f*

anti'clockwise *a* & *adv* in senso antiorario

antics /'æntɪks/ *npl* gesti *mpl* buffi

anti'cyclone *n* anticiclone *m*

antidote /'æntɪdəʊt/ *n* antidoto *m*

'antifreeze *n* antigelo *m*

antipathy /ænˈtɪpəθɪ/ n antipatia f
antiquated /ˈæntɪkweɪtɪd/ a antiquato
antique /ænˈtiːk/ a antico ● n antichità f inv. ~ **dealer** n antiquario, -a mf
antiquity /ænˈtɪkwətɪ/ n antichità f
anti-Semitic /æntɪsɪˈmɪtɪk/ a antisemita
anti'septic a & n antisettico m
anti'social a ⟨behaviour⟩ antisociale; ⟨person⟩ asociale
anti'virus program n Comput programma m di antivirus
antlers /ˈæntləz/ npl corna fpl
anus /ˈeɪnəs/ n ano m
anxiety /ænˈzaɪətɪ/ n ansia f
anxious /ˈæŋkʃəs/ a ansioso. ~**ly** adv con ansia
any /ˈenɪ/ a ⟨no matter which⟩ qualsiasi, qualunque; **have we ~ wine/biscuits?** abbiamo del vino/dei biscotti?; **have we ~ jam/apples?** abbiamo della marmellata/delle mele?; ~ **colour/number you like** qualsiasi colore/numero ti piaccia; **we don't have ~ wine/biscuits** non abbiamo vino/biscotti; **I don't have ~ reason to lie** non ho nessun motivo per mentire; **for ~ reason** per qualsiasi ragione ● pron ⟨some⟩ ne; ⟨no matter which⟩ uno qualsiasi; **I don't want ~ [of it]** non ne voglio [nessuno]; **there aren't ~** non ce ne sono; **have we ~?** ne abbiamo?; **have you read ~ of her books?** hai letto qualcuno dei suoi libri? ● adv **I can't go ~ quicker** non posso andare più in fretta; **is it ~ better?** va un po' meglio?; **would you like ~ more?** ne vuoi ancora?; **I can't eat ~ more** non posso mangiare più niente
'anybody pron chiunque; ⟨after negative⟩ nessuno; **I haven't seen ~** non ho visto nessuno
'anyhow adv ad ogni modo, co-

munque; ⟨badly⟩ non importa come
'anyone pron = anybody
'anything pron qualche cosa, qualcosa; ⟨no matter what⟩ qualsiasi cosa; ⟨after negative⟩ niente; **take/buy ~ you like** prendi/compra quello che vuoi; **I don't remember ~** non mi ricordo niente; **he's ~ but stupid** è tutto, ma non stupido; **I'll do ~ but that** farò qualsiasi cosa, tranne quello
'anyway adv ad ogni modo, comunque
'anywhere adv dovunque; ⟨after negative⟩ da nessuna parte; **put it ~** mettilo dove vuoi; **I can't find it ~** non lo trovo da nessuna parte; ~ **else** da qualch'altra parte; ⟨after negative⟩ da nessun'altra parte; **I don't want to go ~ else** non voglio andare da nessun'altra parte
apart /əˈpɑːt/ adv lontano; **live ~** vivere separati; **100 miles ~** lontani 100 miglia; ~ **from** a parte; **you can't tell them ~** non si possono distinguere; **joking ~** scherzi a parte
apartment /əˈpɑːtmənt/ n ⟨Am: flat⟩ appartamento m; **in my ~** a casa mia
apathy /ˈæpəθɪ/ n apatia f
ape /eɪp/ n scimmia f ● vt scimmiottare
aperitif /əˈperɪtiːf/ n aperitivo m
aperture /ˈæpətʃə(r)/ n apertura f
apex /ˈeɪpeks/ n vertice m
apiece /əˈpiːs/ adv ciascuno
apologetic /əpɒləˈdʒetɪk/ a ⟨air, remark⟩ di scusa; **be ~** essere spiacente
apologize /əˈpɒlədʒaɪz/ vi scusarsi (**for** per)
apology /əˈpɒlədʒɪ/ n scusa f; fig **an ~ for a dinner** una sottospecie di cena
apostle /əˈpɒsl/ n apostolo m
apostrophe /əˈpɒstrəfɪ/ n apostrofo m

appal /əˈpɔːl/ vt (pt/pp appalled) sconvolgere. **~ing** a sconvolgente

apparatus /æpəˈreɪtəs/ n apparato m

apparent /əˈpærənt/ a evidente; (seeming) apparente. **~ly** adv apparentemente

apparition /æpəˈrɪʃn/ n apparizione f

appeal /əˈpiːl/ n appello m; (attraction) attrattiva f ● vi fare appello; **~ to** (be attractive to) attrarre. **~ing** a attraente

appear /əˈpɪə(r)/ vi apparire; (seem) sembrare; (publication:) uscire; Theat esibirsi. **~ance** n apparizione f; (look) aspetto m; **to all ~ances** a giudicare dalle apparenze; **keep up ~ances** salvare le apparenze

appease /əˈpiːz/ vt placare

appendicitis /əpendɪˈsaɪtɪs/ n appendicite f

appendix /əˈpendɪks/ n (pl -ices /-ɪsiːz/) (of book) appendice f; (pl -es) Anat appendice f

appetite /ˈæpɪtaɪt/ n appetito m

appetiz|er /ˈæpɪtaɪzə(r)/ n stuzzichino m. **~ing** a appetitoso

applaud /əˈplɔːd/ vt/i applaudire. **~se** n applauso m

apple /ˈæpl/ n mela f. **~-tree** n melo m

appliance /əˈplaɪəns/ n attrezzo m; [electrical] ~ elettrodomestico m

applicable /ˈæplɪkəbl/ a **be ~ to** essere valido per; **not ~** (on form) non applicabile

applicant /ˈæplɪkənt/ n candidato, -a mf

application /æplɪˈkeɪʃn/ n applicazione f; (request) domanda f; (for job) candidatura f; **~ form** n modulo m di domanda

applied /əˈplaɪd/ a applicato

apply /əˈplaɪ/ vt (pt/pp -ied) applicare; **~ oneself** applicarsi ● vi applicarsi; (law:) essere applicabile; **~ to** (ask) rivolgersi a; **~ for** domanda per (job etc)

appoint /əˈpɔɪnt/ vt nominare; fissare (time). **~ment** n appuntamento m; (to job) nomina f; (job) posto m

appraisal /əˈpreɪz(ə)l/ n valutazione f

appreciable /əˈpriːʃəbl/ a sensibile

appreciat|e /əˈpriːʃeɪt/ vt apprezzare; (understand) comprendere ● vi (increase in value) aumentare di valore. **~ion** /-ˈeɪʃn/ n (gratitude) riconoscenza f; (enjoyment) apprezzamento m; (understanding) comprensione f; (in value) aumento m. **~ive** /-ətɪv/ a riconoscente

apprehend /æprɪˈhend/ vt arrestare

apprehens|ion /æprɪˈhenʃn/ n arresto m; (fear) apprensione f. **~ive** /-sɪv/ a apprensivo

apprentice /əˈprentɪs/ n apprendista mf. **~ship** n apprendistato m

approach /əˈprəʊtʃ/ n avvicinamento m; (to problem) approccio m; (access) accesso m; **make ~es to** fare degli approcci con ● vi avvicinarsi a vt avvicinarsi a; (with request) rivolgersi a; affrontare (problem). **~able** /-əbl/ a accessibile

appropriate[1] /əˈprəʊprɪət/ a appropriato

appropriate[2] /əˈprəʊprɪeɪt/ vt appropriarsi di

approval /əˈpruːvl/ n approvazione f; **on ~** in prova

approv|e /əˈpruːv/ vt approvare ● vi **~e of** approvare (sth); avere una buona opinione di (sb). **~ing** a (smile, nod) d'approvazione

approximate /əˈprɒksɪmət/ a approssimativo. **~ly** adv approssimativamente

approximation /əprɒksɪˈmeɪʃn/ n approssimazione f

apricot /ˈeɪprɪkɒt/ n albicocca f

April /ˈeɪprəl/ n aprile m; **~ Fool's Day** il primo d'aprile

apron /ˈeɪprən/ n grembiule m

apt /æpt/ *a* appropriato; **be ~ to do sth** avere tendenza a fare qcsa

aptitude /ˈæptɪtjuːd/ *n* disposizione *f*. **~ test** *n* test *m* *inv* attitudinale

aqualung /ˈækwəlʌŋ/ *n* autorespiratore *m*

aquarium /əˈkweərɪəm/ *n* acquario *m*

Aquarius /əˈkweərɪəs/ *n Astr* Acquario *m*

aquatic /əˈkwætɪk/ *a* acquatico

Arab /ˈærəb/ *a* & *n* arabo, -a *mf*. **~ian** /əˈreɪbɪən/ *a* arabo

Arabic /ˈærəbɪk/ *a* arabo; **~ numerals** numeri *mpl* arabici ● *n* arabo *m*

arable /ˈærəbl/ *a* coltivabile

arbitrary /ˈɑːbɪtrərɪ/ *a* arbitrario

arbitrat|e /ˈɑːbɪtreɪt/ *vi* arbitrare. **~ion** /-ˈtreɪʃn/ *n* arbitraggio *m*

arc /ɑːk/ *n* arco *m*

arcade /ɑːˈkeɪd/ *n* portico *m*; (*shops*) galleria *f*

arch /ɑːtʃ/ *n* arco *m*; (*of foot*) dorso *m* del piede

archaeological /ɑːkɪəˈlɒdʒɪkl/ *a* archeologico

archaeolog|ist /ɑːkɪˈɒlədʒɪst/ *n* archeologo, -a *mf*. **~y** *n* archeologia *f*

archaic /ɑːˈkeɪɪk/ *a* arcaico

arch'bishop /ɑːtʃ-/ *n* arcivescovo *m*

arch-'enemy *n* acerrimo nemico *m*

architect /ˈɑːkɪtekt/ *n* architetto *m*. **~ural** /ɑːkɪˈtektʃərəl/ *a* architettonico

architecture /ˈɑːkɪtektʃə(r)/ *n* architettura *f*

archives /ˈɑːkaɪvz/ *npl* archivi *mpl*

archiving /ˈɑːkaɪvɪŋ/ *n Comput* archiviazione *f*

archway /ˈɑːtʃweɪ/ *n* arco *m*

Arctic /ˈɑːktɪk/ *a* artico ● *n* **the ~** l'Artico

ardent /ˈɑːdənt/ *a* ardente

arduous /ˈɑːdjʊəs/ *a* arduo

are /ɑː(r)/ *see* **be**

area /ˈeərɪə/ *n* area *f*; (*region*) zona

f; (*fig: field*) campo *m*. **~ code** *n* prefisso *m* [telefonico]

arena /əˈriːnə/ *n* arena *f*

aren't /ɑːnt/ = **are not** *see* **be**

Argentina /ɑːdʒənˈtiːnə/ *n* Argentina *f*

Argentinian /ɑːdʒənˈtɪnɪən/ *a* & *n* argentino, -a *mf*

argue /ˈɑːgjuː/ *vi* litigare (**about** su); (*debate*) dibattere; **don't ~!** non discutere! ● *vt* (*debate*) dibattere; (*reason*) ➤ **that** sostenere che

argument /ˈɑːgjʊmənt/ *n* argomento *m*; (*reasoning*) ragionamento *m*; **have an ~** litigare. **~ative** /-ˈmentətɪv/ *a* polemico

aria /ˈɑːrɪə/ *n* aria *f*

arid /ˈærɪd/ *a* arido

Aries /ˈeəriːz/ *n Astr* Ariete *m*

arise /əˈraɪz/ *vi* (*pt* **arose**, *pp* **arisen**) (*opportunity, need, problem:*) presentarsi; (*result*) derivare

aristocracy /ærɪˈstɒkrəsɪ/ *n* aristocrazia *f*

aristocrat /ˈærɪstəkræt/ *n* aristocratico, -a *mf*. **~ic** /-ˈkrætɪk/ *a* aristocratico

arithmetic /əˈrɪθmətɪk/ *n* aritmetica *f*

arm /ɑːm/ *n* braccio *m*; (*of chair*) bracciolo *m*; **~s** *pl* (*weapons*) armi *fpl*; **~ in ~** a braccetto; **up in ~s** *fam* furioso (**about** per) ● *vt* armare

armaments /ˈɑːməmənts/ *npl* armamenti *mpl*

'armchair *n* poltrona *f*

armed /ɑːmd/ *a* armato; **~ forces** forze *fpl* armate; **~ robbery** rapina *f* a mano armata

armistice /ˈɑːmɪstɪs/ *n* armistizio *m*

armour /ˈɑːmə(r)/ *n* armatura *f*. **~ed** *a* (*vehicle*) blindato

'armpit *n* ascella *f*

army /ˈɑːmɪ/ *n* esercito *m*; **join the ~** arruolarsi

aroma /əˈrəʊmə/ *n* aroma *f*. **~tic** /ærəˈmætɪk/ *a* aromatico

arose /əˈrəʊz/ *see* **arise**

around /ə'raʊnd/ *adv* intorno; **all ~** tutt'intorno; **I'm not from ~ here** non sono di qui; **he's not ~** non c'è ● *prep* intorno a; in giro per ⟨*room, shops, world*⟩

arouse /ə'raʊz/ *vt* svegliare; ⟨*sexually*⟩ eccitare

arrange /ə'reɪndʒ/ *vt* sistemare ⟨*furniture, books*⟩; organizzare ⟨*meeting*⟩; fissare ⟨*date, time*⟩; **~ to do sth** combinare di fare qcsa. **~ment** *n* ⟨*of furniture*⟩ sistemazione *f*; *Mus* arrangiamento *m*; ⟨*agreement*⟩ accordo *m*; ⟨*of flowers*⟩ composizione *f*. **make ~ments** prendere disposizioni

arrears /ə'rɪəz/ *npl* arretrati *mpl*. **be in ~** essere in arretrato; **paid in ~** pagato a lavoro eseguito

arrest /ə'rest/ *n* arresto *m*; **under ~** in stato d'arresto ● *vt* arrestare

arrival /ə'raɪvl/ *n* arrivo *m*; **new ~s** *pl* nuovi arrivati *mpl*

arrive /ə'raɪv/ *vi* arrivare; **~ at** *fig* raggiungere

arrogan|ce /'ærəgəns/ *n* arroganza *f*. **~t** *a* arrogante

arrow /'ærəʊ/ *n* freccia *f*

arse /ɑːs/ *n* *vulg* culo *m*

arsenic /'ɑːsənɪk/ *n* arsenico *m*

arson /'ɑːsn/ *n* incendio *m* doloso. **~ist** /-sənɪst/ *n* incendiario, -a *mf*

art /ɑːt/ *n* arte *f*; **~s and crafts** *pl* artigianato *m*; **the A~s** *pl* l'arte *f*; **A~s degree** *Univ* laurea *f* in Lettere

artery /'ɑːtərɪ/ *n* arteria *f*

artful /'ɑːtfl/ *a* scaltro

'art gallery *n* galleria *f* d'arte

arthritis /ɑː'θraɪtɪs/ *n* artrite *f*

artichoke /'ɑːtɪtʃəʊk/ *n* carciofo *m*

article /'ɑːtɪkl/ *n* articolo *m*; **~ of clothing** capo *m* d'abbigliamento

articulate[1] /ɑː'tɪkjʊlət/ *a* ⟨*speech*⟩ chiaro; **be ~** esprimersi bene

articulate[2] /ɑː'tɪkjʊleɪt/ *vt* scandire ⟨*words*⟩. **~d lorry** *n* autotreno *m*

artifice /'ɑːtɪfɪs/ *n* artificio *m*

artificial /ɑːtɪ'fɪʃl/ *a* artificiale.

~ly *adv* artificialmente; ⟨*smile*⟩ artificiosamente

artillery /ɑː'tɪlərɪ/ *n* artiglieria *f*

artist /'ɑːtɪst/ *n* artista *mf*

artiste /ɑː'tiːst/ *n* *Theat* artista *mf*

artistic /ɑː'tɪstɪk/ *a* artistico

as /æz/ *conj* come; ⟨*since*⟩ siccome; ⟨*while*⟩ mentre; **as he grew older** diventando vecchio; **as you get to know her** conoscendola meglio; **young as she is** per quanto sia giovane ● *prep* come; **as a friend** come amico; **as a child** da bambino; **as a foreigner** in quanto straniero; **disguised as** travestito da ● *adv* as well ⟨*also*⟩ anche; **as soon as I get home** [non] appena arrivo a casa; **as quick as you** veloce quanto te; **as quick as you can** più veloce che puoi; **as far as** ⟨*distance*⟩ fino a; **as far as I'm concerned** per quanto mi riguarda; **as long as** finché; ⟨*provided that*⟩ purché

asbestos /æz'bestɒs/ *n* amianto *m*

ascend /ə'send/ *vi* salire ● *vt* salire a ⟨*throne*⟩

Ascension /ə'senʃn/ *n* *Relig* Ascensione *f*

ascent /ə'sent/ *n* ascesa *f*

ascertain /æsə'teɪn/ *vt* accertare

ascribe /ə'skraɪb/ *vt* attribuire

ash[1] /æʃ/ *n* ⟨*tree*⟩ frassino *m*

ash[2] *n* cenere *f*

ashamed /ə'ʃeɪmd/ *a* be/feel **~** vergognarsi

ashore /ə'ʃɔː(r)/ *adv* a terra; **go ~** sbarcare

ash: **~tray** *n* portacenere *m*. **A~ Wednesday** *n* mercoledì *m inv* delle Ceneri

Asia /'eɪʒə/ *n* Asia *f*. **~n** *a & n* asiatico, -a *mf*. **~tic** /eɪʒɪ'ætɪk/ *a* asiatico

aside /ə'saɪd/ *adv* take **sb ~** prendere qcno a parte; **put sth ~** mettere qcsa da parte; **~ from you** *Am* a parte te

ask /ɑːsk/ *vt* fare ⟨*question*⟩; ⟨*invite*⟩ invitare; **~ sb sth** domandare *or* chiedere qcsa a qcno; **~ sb**

to do sth domandare or chiedere a qcno di fare qcsa ● vi ~ about sth informarsi su qcsa; ~ after chiedere [notizie] di; ~ for chiedere ‹sth›; chiedere di ‹sb›; ~ for trouble *fam* andare in cerca di guai. ask in *vt* ~ sb in invitare qcno ad entrare. ask out *vt* ~ sb out chiedere a qcno di uscire

askance /əˈskɑːns/ *adv* look ~ at sb/sth guardare qcno/qcsa di traverso

askew /əˈskjuː/ *a & adv* a traverso

asleep /əˈsliːp/ *a* be ~ dormire; fall ~ addormentarsi

asparagus /əˈspærəgəs/ *n* asparagi *mpl*

aspect /ˈæspekt/ *n* aspetto *m*

aspersions /əˈspɜːʃnz/ *npl* **cast ~ on** diffamare

asphalt /ˈæsfælt/ *n* asfalto *m*

asphyxia /əsˈfɪksɪə/ *n* asfissia *f*. **~te** /əsˈfɪksɪeɪt/ *vt* asfissiare. **~tion** /-ˈeɪʃn/ *n* asfissia *f*

aspirations /æspəˈreɪʃnz/ *npl* aspirazioni *fpl*

aspire /əˈspaɪə(r)/ *vi* ~ to aspirare a

ass /æs/ *n* asino *m*

assailant /əˈseɪlənt/ *n* assalitore, -trice *mf*

assassin /əˈsæsɪn/ *n* assassino, -a *mf*. **~ate** *vt* assassinare. **~ation** /-ˈneɪʃn/ *n* assassinio *m*

assault /əˈsɔːlt/ *n* *Mil* assalto *m*; *Jur* aggressione *f* ● *vt* aggredire

assemble /əˈsembl/ *vi* radunarsi ● *vt* radunare; *Techn* montare

assembly /əˈsemblɪ/ *n* assemblea *f*; *Sch* assemblea *f* giornaliera di alunni e professori di una scuola; *Techn* montaggio *m*. ~ **line** catena *f* di montaggio

assent /əˈsent/ *n* assenso *m* ● *vi* acconsentire

assert /əˈsɜːt/ *vt* asserire; far valere ‹one's rights›; **~ oneself** farsi valere. **~ion** /-ɜːʃn/ *n* asserzione *f*. **~ive** /-tɪv/ *a* **be ~ive** farsi valere

assess /əˈses/ *vt* valutare; (*for tax purposes*) stabilire l'imponibile

di. **~ment** *n* valutazione *f*; (*of tax*) accertamento *m*

asset /ˈæset/ *n* (*advantage*) vantaggio *m*; (*person*) elemento *m* prezioso. **~s** *pl* beni *mpl*; (*on balance sheet*) attivo *msg*

assign /əˈsaɪn/ *vt* assegnare. **~ment** *n* (*task*) incarico *m*

assimilate /əˈsɪmɪleɪt/ *vt* assimilare; integrare *‹person›*

assist /əˈsɪst/ *vt/i* assistere; ~ **sb to do sth** assistere qcno nel fare qcsa. **~ance** *n* assistenza *f*. **~ant** *a* ~ant **manager** vicedirettore, -trice *mf* ● *n* assistente *mf*; (*in shop*) commesso, -a *mf*

associate¹ /əˈsəʊʃɪeɪt/ *vt* associare (**with** *a*); **be ~ed with sth** (*involved in*) essere coinvolto in qcsa ● *vi* ~e **with** frequentare. **~ion** /-ˈeɪʃn/ *n* associazione *f*. **A~ion** ˈ**Football** *n* [gioco *m* del] calcio *m*

associate² /əˈsəʊʃɪət/ *a* associato ● *n* collega *mf*; (*member*) socio, -a *mf*

assorted /əˈsɔːtɪd/ *a* assortito. **~ment** *n* assortimento *m*

assume /əˈsjuːm/ *vt* presumere; assumere *‹control›*. **~e office** entrare in carica; **~ing that you're right,...** ammettendo che tu abbia ragione,...

assumption /əˈsʌmpʃn/ *n* supposizione *f*; **on the ~ that** partendo dal presupposto che; **the A~** *Relig* l'Assunzione *f*

assurance /əˈʃʊərəns/ *n* assicurazione *f*; (*confidence*) sicurezza *f*

assure /əˈʃʊə(r)/ *vt* assicurare. **~d** *a* sicuro

asterisk /ˈæstərɪsk/ *n* asterisco *m*

astern /əˈstɜːn/ *adv* a poppa

asthma /ˈæsmə/ *n* asma *f*. **~tic** /-ˈmætɪk/ *a* asmatico

astonish /əˈstɒnɪʃ/ *vt* stupire. **~ing** *a* stupefacente. **~ment** *n* stupore *m*

astound /əˈstaʊnd/ *vt* stupire

astray /əˈstreɪ/ *adv* **go ~** smarrirsi; (*morally*) uscire dalla retta via; **lead ~** traviare

astride /əˈstraɪd/ *adv* [a] cavalcioni ● *prep* a cavalcioni di

astrolog|er /əˈstrɒlədʒə(r)/ *n* astrologo, -a *mf*. **~y** *n* astrologia *f*

astronaut /ˈæstrənɔːt/ *n* astronauta *mf*

astronom|er /əˈstrɒnəmə(r)/ *n* astronomo, -a *mf*. **~ical** /æstrəˈnɒmɪkl/ *a* astronomico. **~y** *n* astronomia *f*

astute /əˈstjuːt/ *a* astuto

asylum /əˈsaɪləm/ *n* [**political**] **~** asilo *m* politico; [**lunatic**] **~** manicomio *m*

at /ət/, *accentato* /æt/ *prep* a; **at the station/the market** alla stazione/al mercato; **at the office/the bank** in ufficio/banca; **at the beginning** all'inizio; **at John's** da John; **at the hairdresser's** dal parrucchiere; **at home** a casa; **at work** al lavoro; **at school** a scuola; **at a party/wedding** a una festa/un matrimonio; **at 1 o'clock** all'una; **at 50 km an hour** ai 50 all'ora; **at Christmas/Easter** a Natale/Pasqua; **at times** talvolta; **two at a time** due alla volta; **good at languages** bravo nelle lingue; **at sb's request** su richiesta di qcno; **are you at all worried?** sei preoccupato?

ate /et/ *see* **eat**

atheist /ˈeɪθɪɪst/ *n* ateo, -a *mf*

athlet|e /ˈæθliːt/ *n* atleta *mf*. **~ic** /-ˈletɪk/ *a* atletico. **~ics** /-ˈletɪks/ *n* atletica *fsg*

Atlantic /ətˈlæntɪk/ *a* & *n* **the ~** [**Ocean**] l'[Oceano *m*] Atlantico *m*

atlas /ˈætləs/ *n* atlante *m*

atmospher|e /ˈætməsfɪə(r)/ *n* atmosfera *f*. **~ic** /-ˈferɪk/ *a* atmosferico

atom /ˈætəm/ *n* atomo *m*. **~ bomb** *n* bomba *f* atomica

atomic /əˈtɒmɪk/ *a* atomico

atone /əˈtəʊn/ *vi* **~ for** pagare per. **~ment** *n* espiazione *f*

atrocious /əˈtrəʊʃəs/ *a* atroce; (*fam*: *meal, weather*) abominevole

atrocity /əˈtrɒsəti/ *n* atrocità *f inv*

attach /əˈtætʃ/ *vt* attaccare; attribuire (*importance*); **be ~ed to** *fig* essere attaccato a

attaché /əˈtæʃeɪ/ *n* addetto *m*. **~ case** *n* ventiquattrore *f inv*

attachment /əˈtætʃmənt/ *n* (*affection*) attaccamento *m*; (*accessory*) accessorio *m*

attack /əˈtæk/ *n* attacco *m*; (*physical*) aggressione *f* ● *vt* attaccare; (*physically*) aggredire. **~er** *n* assalitore, -trice *mf*; (*critic*) detrattore, -trice *mf*

attain /əˈteɪn/ *vt* realizzare (*ambition*); raggiungere (*success, age, goal*)

attempt /əˈtempt/ *n* tentativo *m* ● *vt* tentare

attend /əˈtend/ *vt* essere presente a; (*go regularly to*) frequentare; (*doctor*:) avere in cura ● *vi* essere presente; (*pay attention*) prestare attenzione. **attend to** *vt* occuparsi di; (*in shop*) servire. **~ance** *n* presenza *f*. **~ant** *n* guardiano, -a *mf*

attention /əˈtenʃn/ *n* attenzione *f*; **~! Mil** attenti!; **pay ~** prestare attenzione; **need ~** aver bisogno di attenzioni; (*skin, hair, plant*:) dover essere curato; (*car, tyres*:) dover essere riparato; **for the ~ of** all'attenzione di

attentive /əˈtentɪv/ *a* (*pupil, audience*) attento

attest /əˈtest/ *vt/i* attestare

attic /ˈætɪk/ *n* soffitta *f*

attitude /ˈætɪtjuːd/ *n* atteggiamento *m*

attorney /əˈtɜːnɪ/ *n* (*Am*: *lawyer*) avvocato *m*; **power of ~** delega *f*

attract /əˈtrækt/ *vt* attirare. **~ion** /-ækʃn/ *n* attrazione *f*; (*feature*) attrattiva *f*. **~ive** /-tɪv/ *a* (*person*) attraente; (*proposal, price*) allettante

attribute[1] /ˈætrɪbjuːt/ *n* attributo *m*

attribute[2] /əˈtrɪbjuːt/ *vt* attribuire

attrition /ə'trɪʃn/ n war of ~ guerra f di logoramento

aubergine /'əʊbəʒiːn/ n melanzana f

auburn /'ɔːbən/ a castano ramato

auction /'ɔːkʃn/ n asta f ● vt vendere all'asta. **~eer** /-ʃə'nɪə(r)/ n banditore m

audaci|ous /ɔː'deɪʃəs/ a sfacciato; (daring) audace. **~ty** /-'dæsɪtɪ/ n sfacciataggine f; (daring) audacia f

audible /'ɔːdəbl/ a udibile

audience /'ɔːdɪəns/ n Theat pubblico m; TV telespettatori mpl; Radio ascoltatori mpl; (meeting) udienza f

audio /'ɔːdɪəʊ/: **~tape** n audiocassetta f. ~ **typist** n dattilografo, -a mf (che trascrive registrazioni). **~'visual** a audiovisivo

audit /'ɔːdɪt/ n verifica f del bilancio ● vt verificare

audition /ɔː'dɪʃn/ n audizione f ● vi fare un'audizione

auditor /'ɔːdɪtə(r)/ n revisore m di conti

auditorium /ɔːdɪ'tɔːrɪəm/ n sala f

augment /ɔːg'ment/ vt aumentare

augur /'ɔːgə(r)/ vi ~ **well/ill** essere di buon/cattivo augurio

August /'ɔːgəst/ n agosto m

aunt /ɑːnt/ n zia f

au pair /əʊ'peə(r)/ n ~ **[girl]** ragazza f alla pari

aura /'ɔːrə/ n aura f

auspices /'ɔːspɪsɪz/ npl under the ~ of sotto l'egida di

auspicious /ɔː'spɪʃəs/ a di buon augurio

auster|e /ɒ'stɪə(r)/ a austero. **~ity** /-terɪtɪ/ n austerità f

Australia /ɒ'streɪlɪə/ n Australia f. **~n** a & n australiano, -a mf

Austria /'ɒstrɪə/ n Austria f. **~n** a & n austriaco, -a mf

authentic /ɔː'θentɪk/ a autentico. **~ate** vt autenticare. **~ity** /-'tɪsɪtɪ/ n autenticità f

author /'ɔːθə(r)/ n autore m

authoritarian /ɔːθɒrɪ'teərɪən/ a autoritario

authoritative /ɔː'θɒrɪtətɪv/ a autorevole; (manner) autoritario

authority /ɔː'θɒrətɪ/ n autorità f; (permission) autorizzazione f; be in ~ over avere autorità su

authorization /ɔːθəraɪ'zeɪʃn/ n autorizzazione f

authorize /'ɔːθəraɪz/ vt autorizzare

autobi'ography /ɔːtə-/ n autobiografia f

autocratic /ɔːtə'krætɪk/ a autocratico

autograph /'ɔːtə-/ n autografo m

automate /'ɔːtəmeɪt/ vt automatizzare

automatic /ɔːtə'mætɪk/ a automatico ● n (car) macchina f col cambio automatico; (washing machine) lavatrice f automatica. **~ally** adv automaticamente

automation /ɔːtə'meɪʃn/ n automazione f

automobile /'ɔːtəməbiːl/ n automobile f

autonom|ous /ɔː'tɒnəməs/ a autonomo. **~y** n autonomia f

autopsy /'ɒtɒpsɪ/ n autopsia f

autumn /'ɔːtəm/ n autunno m. **~al** /-'tʌmnl/ a autunnale

auxiliary /ɔːg'zɪlɪərɪ/ a ausiliario ● n ausiliare m

avail /ə'veɪl/ n to no ~ invano ● vi ~ **oneself of** approfittare di

available /ə'veɪləbl/ a disponibile; (book, record etc) in vendita

avalanche /'ævəlɑːnʃ/ n valanga f

avarice /'ævərɪs/ n avidità f

avenge /ə'vendʒ/ vt vendicare

avenue /'ævənjuː/ n viale m; fig strada f

average /'ævərɪdʒ/ a medio; (mediocre) mediocre ● n media f; **on** ~ in media ● vt (sales, attendance etc:) raggiungere una media di. **average out** vi risultare in media

averse /ə'vɜːs/ a not be ~e to sth

non essere contro qcsa. **~ion** /-ˈɜːʃn/ n avversione f (to per)

avert /əˈvɜːt/ vt evitare ⟨crisis⟩; distogliere ⟨eyes⟩

aviary /ˈeɪvɪərɪ/ n uccelliera f

aviation /eɪvɪˈeɪʃn/ n aviazione f

avid /ˈævɪd/ a avido ⟨for di⟩; ⟨reader⟩ appassionato

avocado /ævəˈkɑːdəʊ/ n avocado m

avoid /əˈvɔɪd/ vt evitare. **~able** /-əbl/ a evitabile

await /əˈweɪt/ vt attendere

awake /əˈweɪk/ a sveglio; **wide ~** completamente sveglio ● vi (pt awoke, pp awoken) svegliarsi

awaken /əˈweɪkn/ vt svegliare. **~ing** n risveglio m

award /əˈwɔːd/ n premio m; ⟨medal⟩ riconoscimento m; (of prize) assegnazione f ● vt assegnare; ⟨hand over⟩ consegnare

aware /əˈweə(r)/ a be ~ of ⟨sense⟩ percepire; ⟨know⟩ essere conscio di; **become ~ of** accorgersi di; ⟨learn⟩ venire a sapere di; **be ~ that** rendersi conto che. **~ness** n percezione f; ⟨knowledge⟩ consapevolezza f

awash /əˈwɒʃ/ a inondato ⟨with di⟩

away /əˈweɪ/ adv via; **go/stay ~** andare/stare via; **he's ~ from his desk/the office** non è alla sua scrivania/in ufficio; **far ~** lontano; **four kilometres ~** a quattro chilometri; **play ~** Sport giocare fuori casa. **~ game** n partita f fuori casa

awe /ɔː/ n soggezione f

awful /ˈɔːfl/ a terribile. **~ly** adv /ˈɔːf(ʊ)lɪ/ terribilmente; ⟨pretty⟩ estremamente

awhile /əˈwaɪl/ adv per un po'

awkward /ˈɔːkwəd/ a ⟨movement⟩ goffo; ⟨moment, situation⟩ imbarazzante; ⟨time⟩ scomodo. **~ly** adv ⟨move⟩ goffamente; ⟨say⟩ con imbarazzo

awning /ˈɔːnɪŋ/ n tendone m

awoke(n) /əˈwəʊk(ən)/ see awake

awry /əˈraɪ/ adv storto

axe /æks/ n scure f ● vt ⟨pres p axing⟩ fare dei tagli a ⟨budget⟩; sopprimere ⟨jobs⟩; annullare ⟨project⟩

axis /ˈæksɪs/ n (pl axes /-siːz/) asse m

axle /ˈæksl/ n Techn asse m

ay[e] /aɪ/ adv sì ● n sì m invar

Bb

B /biː/ n Mus si m inv

BA n abbr Bachelor of Arts

babble /ˈbæbl/ vi farfugliare; ⟨stream:⟩ gorgogliare

baby /ˈbeɪbɪ/ n bambino, -a mf; ⟨fam: darling⟩ tesoro m

baby: **~ carriage** n Am carrozzina f. **~ish** a bambinesco. **~-sit** vi fare da baby-sitter. **~-sitter** n baby-sitter mf

bachelor /ˈbætʃələ(r)/ n scapolo m; B~ **of Arts/Science** laureato, -a mf in lettere/in scienze

back /bæk/ n schiena f; (of horse, hand) dorso m; (of chair) schienale m; (of house, cheque, page) retro m; (in football) difesa f; **at the ~** in fondo; **in the ~** Auto dietro; **~ to front** ⟨sweater⟩ il davanti di dietro; **at the ~ of beyond** in un posto sperduto ● a posteriore; ⟨taxes, payments⟩ arretrato ● adv indietro; ⟨returned⟩ di ritorno; **turn/ move ~** tornare/spostarsi indietro; **put it ~ here/there** rimettilo qui/là; **~ at home** di ritorno a casa; **I'll be ~ in five minutes** torno fra cinque minuti; **I'm just ~** sono appena tornato; **when do you want the book ~?** quando rivuoi il libro?; **pay ~** ripagare ⟨sb⟩; restituire ⟨money⟩; **~ in power** di nuovo al potere ● vt ⟨support⟩ sostenere; ⟨with money⟩ finanziare; puntare su ⟨horse⟩; ⟨cover the back of⟩ rivestire il retro

di ● *vi Auto* fare retromarcia.
back down *vi* battere in ritirata.
back in *vi Auto* entrare in retromarcia; ⟨*person:*⟩ entrare camminando all'indietro. **back out** *vi Auto* uscire in retromarcia; ⟨*person:*⟩ uscire camminando all'indietro; *fig* tirarsi indietro (**of** da). **back up** *vt* sostenere; confermare ⟨*person's alibi*⟩; *Comput* fare una copia di salvataggio di; **be ~ed up** ⟨*traffic:*⟩ essere congestionato ● *vi Auto* fare retromarcia

back: **~ache** *n* mal *m* di schiena. **~bencher** *n* parlamentare *mf* ordinario, -a. **~biting** *n* maldicenza *f*. **~bone** *n* spina *f* dorsale. **~chat** *n* risposta *f* impertinente. **~date** *vt* retrodatare ⟨*cheque*⟩; **~dated to** valido a partire da. **~'door** *n* porta *f* di servizio

backer /ˈbækə(r)/ *n* sostenitore, -trice *mf*; (*with money*) finanziatore, -trice *mf*

back: **~'fire** *vi Auto* avere un ritorno di fiamma; ⟨*fig: plan*⟩ fallire. **~ground** *n* sfondo *m*; (*environment*) ambiente *m*. **~hand** *n* (*tennis*) rovescio *m*. **~'handed** *a* ⟨*compliment*⟩ implicito. **~'hander** *n* (*fam: bribe*) bustarella *f*

backing /ˈbækɪŋ/ *n* (*support*) supporto *m*; (*material*) riserva *f*; *Mus* accompagnamento *m*; **~ group** gruppo *m* d'accompagnamento

back: **~lash** *n fig* reazione *f* opposta. **~log** *n* **~log of work** lavoro *m* arretrato. **~ 'seat** *n* sedile *m* posteriore. **~side** *n fam* fondoschiena *m inv.* **~slash** *n Typ* barra *f* retroversa. **~stage** *a & adv* dietro le quinte. **~stroke** *n* dorso *m*. **~up** *n* rinforzi *mpl*; *Comput* riserva *f*. **~up copy** *n Comput* copia *f* di riserva

backward /ˈbækwəd/ *a* (*step*) indietro; ⟨*child*⟩ lento nell'apprendimento; ⟨*country*⟩ arretrato ● *adv* **~s** (*also Am:* **~**) indietro; ⟨*fall, walk*⟩ all'indietro; **~s and forwards** avanti e indietro

back: **~water** *n fig* luogo *m* allo scarto. **~ 'yard** *n* cortile *m*

bacon /ˈbeɪkn/ *n* pancetta *f*

bacteria /bækˈtɪərɪə/ *npl* batteri *mpl*

bad /bæd/ *a* (**worse, worst**) cattivo; ⟨*weather, habit, news, accident*⟩ brutto; ⟨*apple etc*⟩ marcio; **the light is ~** non c'è una buona luce; **use ~ language** dire delle parolacce; **feel ~** sentirsi male; (*feel guilty*) sentirsi in colpa; **have a ~ back** avere dei problemi alla schiena; **smoking is ~ for you** fumare fa male; **go ~** andare a male; **that's just too ~!** pazienza!; **not ~** niente male

bade /bæd/ *see* bid

badge /bædʒ/ *n* distintivo *m*

badger /ˈbædʒə(r)/ *n* tasso *m* ● *vt* tormentare

badly /ˈbædlɪ/ *adv* male; ⟨*hurt*⟩ gravemente; **~ off** povero; **~ behaved** maleducato; **need ~** aver estremamente bisogno di

bad-mannered *a* maleducato

badminton /ˈbædmɪntən/ *n* badminton *m*

bad-'tempered *a* irascibile

baffle /ˈbæfl/ *vt* confondere

bag /bæg/ *n* borsa *f*; (*of paper*) sacchetto *m*; **old ~** *sl* megera *f*; **~s under the eyes** occhiaie *fpl*; **~s of** *fam* un sacco di

baggage /ˈbægɪdʒ/ *n* bagagli *mpl*

baggy /ˈbægɪ/ *a* ⟨*clothes*⟩ ampio

bagpipes *npl* cornamusa *fsg*

Bahamas /bəˈhɑːməz/ *npl* **the ~** le Bahamas

bail /beɪl/ *n* cauzione *f*; **on ~** su cauzione ● **bail out** *vt Naut* aggottare; **~ sb out** *Jur* pagare la cauzione per qcno ● *vi Aeron* paracadutarsi

bait /beɪt/ *n* esca *f* ● *vt* innescare; (*fig: torment*) tormentare

bake /beɪk/ *vt* cuocere al forno; (*make*) fare ● *vi* cuocersi al forno

baker /ˈbeɪkə(r)/ *n* fornaio, -a *mf*, panettiere, -a *mf*; **~'s [shop]** panetteria f. **~y** n panificio m, forno m

baking /ˈbeɪkɪŋ/ n cottura f al forno. **~-powder** n lievito m in polvere. **~-tin** n teglia f

balance /ˈbæləns/ n equilibrio m; Comm bilancio m; (outstanding sum) saldo m; [bank] ~ saldo m; **be** or **hang in the** ~ fig essere in sospeso ● vt bilanciare; equilibrare (budget); Comm fare il bilancio di (books) ● vi bilanciarsi; Comm essere in pareggio. **~d** a equilibrato. ~ **sheet** n bilancio m [d'esercizio]

balcony /ˈbælkənɪ/ n balcone m

bald /bɔːld/ a (person) calvo; (tyre) liscio; (statement) nudo e crudo; **go** ~ perdere i capelli

bald|ing /ˈbɔːldɪŋ/ a **be** ~**ing** stare perdendo i capelli. **~ness** n calvizie f

bale /beɪl/ n balla f

baleful /ˈbeɪlfl/ a malvagio; (sad) triste

balk /bɔːlk/ vt ostacolare ● vi ~ **at** (horse:) impennarsi davanti a; fig tirarsi indietro davanti a

Balkans /ˈbɔːlknz/ npl Balcani mpl

ball[1] /bɔːl/ n palla f; (football) pallone m; (of yarn) gomitolo m; **on the** ~ fam sveglio

ball[2] n (dance) ballo m

ballad /ˈbæləd/ n ballata f

ballast /ˈbæləst/ n zavorra f

ball-'bearing n cuscinetto m a sfera

ballerina /bæləˈriːnə/ n ballerina f [classica]

ballet /ˈbæleɪ/ n balletto m; (art form) danza f. ~ **dancer** n ballerino, -a mf [classico, -a]

ballistic /bəˈlɪstɪk/ a balistico. ~**s** n balistica fsg

balloon /bəˈluːn/ n pallone m; Aeron mongolfiera f

ballot /ˈbælət/ n votazione f. ~**-box** n urna f. ~**-paper** n scheda f di votazione

ball: ~**-point** [ˈpen] n penna f a sfera. ~**room** n sala f da ballo

balm /bɑːm/ n balsamo m

balmy /ˈbɑːmɪ/ a (-ier, -iest) mite; (fam: crazy) strampalato

Baltic /ˈbɔːltɪk/ a & n **the** ~ [Sea] il [mar] Baltico

bamboo /bæmˈbuː/ n bambù m inv

bamboozle /bæmˈbuːzl/ vt (fam: mystify) confondere

ban /bæn/ n proibizione f ● vt (pt/pp banned) proibire; ~ **from** espellere da (club); **she was** ~**ed from driving** le hanno ritirato la patente

banal /bəˈnɑːl/ a banale. ~**ity** /-ˈnælətɪ/ n banalità f inv

banana /bəˈnɑːnə/ n banana f

band /bænd/ n banda f; (stripe) nastro m; (Mus: pop group) complesso m; (Mus: brass ~) banda f; Mil fanfara f ● **band together** vi riunirsi

bandage /ˈbændɪdʒ/ n benda f ● vt fasciare (limb)

b. & b. abbr **bed and breakfast**

bandit /ˈbændɪt/ n bandito m

band: ~**stand** n palco m coperto [dell'orchestra]. ~**wagon** n **jump on the** ~**wagon** fig seguire la corrente

bandy[1] /ˈbændɪ/ vt (pt/pp -ied) scambiarsi (words). **bandy about** vt far circolare

bandy[2] a (-ier, -iest) **be** ~ avere le gambe storte

bang /bæŋ/ n (noise) fragore m; (of gun, firework) scoppio m; (blow) colpo m ● adv ~ **in the middle of** fam proprio nel mezzo di; **go** ~ (gun:) sparare; (balloon:) esplodere ● int bum! ● vt battere (fist); battere su (table); sbattere (door, head) ● vi scoppiare; (door:) sbattere

banger /ˈbæŋə(r)/ n (firework) petardo m; (fam: sausage) salsiccia f; **old** ~ (fam: car) macinino m

bangle /ˈbæŋgl/ n braccialetto m

banish /ˈbænɪʃ/ vt bandire

banisters /ˈbænɪstəz/ npl ringhiera fsg

bank[1] /bæŋk/ n (of river) sponda f;

(slope) scarpata f ● vi Aeron inclinarsi in virata

bank² n banca f ● vt depositare in banca ● vi ~ **with** avere un conto [bancario] presso. **bank on** vt contare su

'**bank account** n conto m in banca

'**bank card** n carta f assegno.

banker /'bæŋkə(r)/ n banchiere m

bank: ~ '**holiday** n giorno m festivo. ~**ing** n bancario m. ~ '**manager** n direttore, -trice mf di banca. ~**note** n banconota f

bankrupt /'bæŋkrʌpt/ a fallito; **go** ~ fallire ● n persona f che ha fatto fallimento ● vt far fallire. ~**cy** n bancarotta f

banner /'bænə(r)/ n stendardo m; (of demonstrators) striscione m

banns /bænz/ npl Relig pubblicazioni fpl [di matrimonio]

banquet /'bæŋkwɪt/ n banchetto m

banter /'bæntə(r)/ n battute fpl di spirito

baptism /'bæptɪzm/ n battesimo m

Baptist /'bæptɪst/ a & n battista mf

baptize /bæp'taɪz/ vt battezzare

bar /bɑː(r)/ n sbarra f; Jur ordine m degli avvocati; (of chocolate) tavoletta f; (café) bar m inv; (counter) banco m; Mus battuta f; (fig: obstacle) ostacolo m; ~ **of soap/gold** saponetta f/lingotto m; **behind ~s** fam dietro le sbarre ● vt (pt/pp **barred**) sbarrare (way); sprangare (door); escludere (person) ● prep tranne; ~ **none** in assoluto

barbarian /bɑː'beərɪən/ n barbaro, -a mf

barbar|ic /bɑː'bærɪk/ a barbarico. ~**ity** n barbarie f inv. ~**ous** /'bɑːbərəs/ a barbaro

barbecue /'bɑːbɪkjuː/ n barbecue m inv; (party) grigliata f, barbecue m inv ● vt arrostire sul barbecue

barbed /bɑːbd/ a ~ **wire** filo m spinato

barber /'bɑːbə(r)/ n barbiere m

barbiturate /bɑː'bɪtjʊrət/ n barbiturico m

'**bar code** n codice m a barre

bare /beə(r)/ a nudo; (tree, room) spoglio; (floor) senza moquette ● vt scoprire; mostrare (teeth)

bare: ~**back** adv senza sella. ~**faced** a sfacciato. ~**foot** adv scalzo. ~'**headed** a a capo scoperto

barely /'beəlɪ/ adv appena

bargain /'bɑːgɪn/ n (agreement) patto m; (good buy) affare m; **into the** ~ per di più ● vi contrattare; (haggle) trattare. **bargain for** vt (expect) aspettarsi

barge /bɑːdʒ/ n barcone m ● **barge in** vi fam (to room) piombare dentro; (into conversation) interrompere bruscamente. ~ **into** vt piombare dentro a (room); venire addosso a (person)

baritone /'bærɪtəʊn/ n baritono m

bark¹ /bɑːk/ n (of tree) corteccia f

bark² n abbaiamento m ● vi abbaiare

barley /'bɑːlɪ/ n orzo m

bar: ~**maid** n barista f. ~**man** n barista m

barmy /'bɑːmɪ/ a fam strampalato

barn /bɑːn/ n granaio m

barometer /bə'rɒmɪtə(r)/ n barometro m

baron /'bærn/ n barone m. ~**ess** n baronessa f

baroque /bə'rɒk/ a & n barocco m

barracks /'bærəks/ npl caserma fsg

barrage /'bærɑːʒ/ n Mil sbarramento m; (fig: of criticism) sfilza f

barrel /'bærl/ n barile m, botte f; (of gun) canna f. ~**-organ** n organetto m [a cilindro]

barren /'bærən/ a sterile; (landscape) brullo

barricade /'bærɪkeɪd/ n barricata f ● vt barricare

barrier /'bærɪə(r)/ n barriera f; Rail cancello m; fig ostacolo m

barring /'bɑːrɪŋ/ prep ~ **accidents** tranne imprevisti

barrister /'bærɪstə(r)/ *n* avvocato *m*

barrow /'bærəʊ/ *n* carretto *m*; (*wheel*~) carriola *f*

barter /'bɑːtə(r)/ *vi* barattare (**for** con)

base /beɪs/ *n* base *f* ● *a* vile ● *vt* basare; **be ~d on** basarsi su

base: **~ball** *n* baseball *m*. **~less** *a* infondato. **~ment** *n* seminterrato *m*. **~ment flat** *n* appartamento *m* nel seminterrato

bash /bæʃ/ *n* colpo *m* [violento] ● *vt* colpire [violentemente]; (*dent*) ammaccare; **~ed in** *a* ammaccato

bashful /'bæʃfl/ *a* timido

basic /'beɪsɪk/ *a* di base; (*condition, requirement*) basilare; (*living conditions*) povero; **my Italian is pretty ~** il mio italiano è abbastanza rudimentale; **the ~s** (*of language, science*) i rudimenti; (*essentials*) l'essenziale *m*. **~ally** *adv* fondamentalmente

basil /'bæzɪl/ *n* basilico *m*

basilica /bə'zɪlɪkə/ *n* basilica *f*

basin /'beɪsn/ *n* bacinella *f*; (*wash-hand* ~) lavabo *m*; (*for food*) recipiente *m*; *Geog* bacino *m*

basis /'beɪsɪs/ *n* (*pl* **-ses** /-siːz/) base *f*

bask /bɑːsk/ *vi* crogiolarsi

basket /'bɑːskɪt/ *n* cestino *m*. **~ball** *n* pallacanestro *f*

Basle /bɑːl/ *n* Basilea *f*

bass /beɪs/ *a* basso; **~ voice** voce *f* di basso ● *n* basso *m*

bastard /'bɑːstəd/ *n* (*illegitimate child*) bastardo, -a *mf*; *sl* figlio *m* di puttana

bastion /'bæstɪən/ *n* bastione *m*

bat¹ /bæt/ *n* mazza *f*; (*for table tennis*) racchetta *f*; **off one's own ~** *fam* tutto da solo ● *vt* (*pt/pp* **batted**) battere; **she didn't ~ an eyelid** *fig* non ha battuto ciglio

bat² /bæt/ *n Zool* pipistrello *m*

batch /bætʃ/ *n* gruppo *m*; (*of goods*) partita *f*; (*of bread*) infornata *f*

bated /'beɪtɪd/ *a* **with ~ breath** col fiato sospeso

bath /bɑːθ/ *n* (*pl* **~s** /bɑːðz/) bagno *m*; (*tub*) vasca *f* da bagno; **~s** *pl* piscina *f*; **have a ~** fare un bagno ● *vt* fare il bagno a

bathe /beɪð/ *n* bagno *m* ● *vi* fare il bagno ● *vt* lavare (*wound*). **~r** *n* bagnante *mf*

bathing /'beɪðɪŋ/ *n* bagni *mpl*. **~cap** *n* cuffia *f*. **~costume** *n* costume *m* da bagno

bath: **~mat** *n* tappetino *m* da bagno. **~robe** *n* accappatoio *m*. **~room** *n* bagno *m*. **~towel** *n* asciugamano *m* da bagno

baton /'bætn/ *n Mus* bacchetta *f*

battalion /bə'tælɪən/ *n* battaglione *m*

batter /'bætə(r)/ *n Culin* pastella *f*. **~ed** *a* (*car*) malandato; (*wife, baby*) maltrattato

battery /'bætərɪ/ *n* batteria *f*; (*of torch, radio*) pila *f*

battle /'bætl/ *n* battaglia *f*; *fig* lotta *f* ● *vi fig* lottare

battle: **~field** *n* campo *m* di battaglia. **~ship** *n* corazzata *f*

bawdy /'bɔːdɪ/ *a* (**-ier, -iest**) piccante

bawl /bɔːl/ *vt/i* urlare

bay¹ /beɪ/ *n Geog* baia *f*

bay² *n* **keep at ~** tenere a bada

bay³ *n Bot* alloro *m*. **~-leaf** *n* foglia *f* d'alloro

bayonet /'beɪənɪt/ *n* baionetta *f*

bay window *n* bay window *f inv* (*grande finestra sporgente*)

bazaar /bə'zɑː(r)/ *n* bazar *m inv*

BC *abbr* (**before Christ**) a.C.

be /biː/ *vi* (*pres* **am, are, is, are**; *pt* **was, were**; *pp* **been**) essere; **he is a teacher** è insegnante, fa l'insegnante; **what do you want to be?** cosa vuoi fare? **be quiet!** sta' zitto!; **I am cold/hot** ho freddo/caldo; **it's cold/hot, isn't it?** fa freddo/caldo, vero?; **how are you?** come stai?; **I am well** sto bene; **there is** c'è; **there are** ci sono; **I have been to Venice** sono stato a

Venezia; **has the postman been?** è passato il postino?; **you're coming too, aren't you?** vieni anche tu, no?; **it's yours, is it?** è tuo, vero?; **was John there? - yes, he was** c'era John? - sì; **John wasn't there - yes he was!** John non c'era - sì che c'era!; **three and three are six** tre più tre fanno sei; **he is five** ha cinque anni; **that will be £10, please** fanno 10 sterline, per favore; **how much is it?** quanto costa?; **that's £5 you owe me** mi devi 5 sterline ● *v aux* **I am coming/reading** sto venendo/leggendo; **I'm staying** (*not leaving*) resto; **I am being lazy** sono pigro; **I was thinking of you** stavo pensando a te; **you are not to tell him** non devi dirglielo; **you are to do that immediately** devi farlo subito ● *passive* essere; **I have been robbed** sono stato derubato

beach /biːtʃ/ *n* spiaggia *f*. ~**wear** *n* abbigliamento *m* da spiaggia

bead /biːd/ *n* perlina *f*

beak /biːk/ *n* becco *m*

beaker /'biːkə(r)/ *n* coppa *f*

beam /biːm/ *n* trave *f*; (*of light*) raggio *m* ● *vi* irradiare; (*person:*) essere raggiante. ~**ing** *a* raggiante

bean /biːn/ *n* fagiolo *m*; (*of coffee*) chicco *m*

bear[1] /beə(r)/ *n* orso *m*

bear[2] *v* (*pt* bore, *pp* borne) ● *vt* (*endure*) sopportare; mettere al mondo (*child*); (*carry*) portare; ~ **in mind** tener presente ● *vi* ~ **left/right** andare a sinistra/a destra. **bear with** *vt* aver pazienza con. ~**able** /-əbl/ *a* sopportabile

beard /biəd/ *n* barba *f*. ~**ed** *a* barbuto

bearer /'beərə(r)/ *n* portatore, -trice *mf*; (*of passport*) titolare *mf*

bearing /'beərɪŋ/ *n* portamento *m*; Techn cuscinetto *m* (a sfera); **have a ~ on** avere attinenza con; **get one's ~s** orientarsi

beast /biːst/ *n* bestia *f*; (*fam: person*) animale *m*

beat /biːt/ *n* battito *m*; (*rhythm*) battuta *f*; (*of policeman*) giro *m* d'ispezione ● *v* (*pt* beat, *pp* beaten) ● *vt* battere; picchiare (*person*); ~ **it!** *fam* darsela a gambe!; **it ~s me why...** *fam* non capisco proprio perché... **beat up** *vt* picchiare

beat|en /'biːtn/ *a* **off the ~en track** fuori mano. ~**ing** *n* bastonata *f*; **get a ~ing** (*with fists*) essere preso a pugni; (*team, player:*) prendere una batosta

beautician /bjuː'tɪʃn/ *n* estetista *mf*

beauti|ful /'bjuːtɪfl/ *a* bello. ~**fully** *adv* splendidamente

beauty /'bjuːtɪ/ *n* bellezza *f*. ~ **parlour** *n* istituto *m* di bellezza. ~ **spot** *n* neo *m*; (*place*) luogo *m* pittoresco

beaver /'biːvə(r)/ *n* castoro *m*

became /bɪ'keɪm/ *see* become

because /bɪ'kɒz/ *conj* perché; ~ **you didn't tell me, I...** poiché non me lo hai detto,... ● *adv* ~ **of** a causa di

beck /bek/ *n* **at the ~ and call of** a completa disposizione di

beckon /'bekn/ *vt/i* ~ **[to]** chiamare con un cenno

become /bɪ'kʌm/ *v* (*pt* became, *pp* become) ● *vi* diventare ● *vi* diventare; **what has ~e of her?** che ne è di lei? ~**ing** *a* (*clothes*) bello

bed /bed/ *n* letto *m*; (*of sea, lake*) fondo *m*; (*layer*) strato *m*; (*of flowers*) aiuola *f*; **in ~** a letto; **go to ~** andare a letto; ~ **and breakfast** pensione *f* familiare in cui il prezzo della camera comprende la prima colazione. ~**clothes** *npl* lenzuola *fpl* e coperte *fpl*. ~**ding** *n* biancheria *f* per il letto, materasso e guanciali

bedlam /'bedləm/ *n* baraonda *f*

bedraggled /bɪ'dræɡld/ *a* inzaccherato

bed: ~**ridden** *a* costretto a letto. ~**room** *n* camera *f* da letto

'**bedside** *n* at his ~ al suo capezzale. ~ '**lamp** *n* abat-jour *m inv*. ~ '**table** *n* comodino *m*

bed: ~'**sit** *n*, ~'**sitter** *n*, ~-'**sitting-room** *n* = camera *f* ammobiliata fornita di cucina. ~**spread** *n* copriletto *m*. ~**time** *n* l'ora *f* di andare a letto

bee /biː/ *n* ape *f*

beech /biːtʃ/ *n* faggio *m*

beef /biːf/ *n* manzo *m*. ~**burger** *n* hamburger *m inv*

bee: ~**hive** *n* alveare *m*. ~**line** *n* make a ~**line** for *fam* precipitarsi verso

been /biːn/ *see* **be**

beer /bɪə(r)/ *n* birra *f*

beetle /'biːtl/ *n* scarafaggio *m*

beetroot /'biːtruːt/ *n* barbabietola *f*

before /bɪ'fɔː(r)/ *prep* prima di; the day ~ yesterday ieri l'altro; ~ long fra poco ● *adv* prima; never ~ have I seen... non ho mai visto prima...; ~ that prima; ~ going prima di andare ● *conj* (time) prima che; ~ you go prima che tu vada. ~**hand** *adv* in anticipo

befriend /bɪ'frend/ *vt* trattare da amico

beg /beg/ *v* (*pt/pp* **begged**) ● *vi* mendicare ● *vt* pregare; chiedere (*favour, forgiveness*)

began /bɪ'gæn/ *see* **begin**

beggar /'begə(r)/ *n* mendicante *mf*; poor ~! povero cristo!

begin /bɪ'gɪn/ *vt/i* (*pt* **began**, *pp* **begun**, *pres p* **beginning**) cominciare. ~**ner** *n* principiante *mf*. ~**ning** *n* principio *m*

begonia /bɪ'gəʊnɪə/ *n* begonia *f*

begrudge /bɪ'grʌdʒ/ *vt* (*envy*) essere invidioso di; dare malvolentieri (*money*)

begun /bɪ'gʌn/ *see* **begin**

behalf /bɪ'hɑːf/ *n* on ~ of a nome di; on my ~ a nome mio

behave /bɪ'heɪv/ *vi* comportarsi; ~ [oneself] comportarsi bene

behaviour /bɪ'heɪvjə(r)/ *n* comportamento *m*; (*of prisoner, soldier*) condotta *f*

behead /bɪ'hed/ *vt* decapitare

behind /bɪ'haɪnd/ *prep* dietro; be ~ sth *fig* stare dietro qcsa ● *adv* dietro, indietro; (*late*) in ritardo; a long way ~ molto indietro ● *n* *fam* dietro *m*. ~**hand** *adv* indietro

beholden /bɪ'həʊldn/ *a* obbligato (to verso)

beige /beɪʒ/ *a* & *n* beige *m inv*

being /'biːɪŋ/ *n* essere *m*; come into ~ nascere

belated /bɪ'leɪtɪd/ *a* tardivo

belch /beltʃ/ *vi* ruttare ● *vt* [out] eruttare (*smoke*)

belfry /'belfrɪ/ *n* campanile *m*

Belgian /'beldʒən/ *a* & *n* belga *mf*

Belgium /'beldʒəm/ *n* Belgio *m*

belief /bɪ'liːf/ *n* fede *f*; (*opinion*) convinzione *f*

believable /bɪ'liːvəbl/ *a* credibile

believe /bɪ'liːv/ *vt/i* credere. ~ in *Relig* credente *mf*; be a great ~r in credere fermamente in

belittle /bɪ'lɪtl/ *vt* sminuire (*person, achievements*)

bell /bel/ *n* campana *f*; (*on door*) campanello *m*

belligerent /bɪ'lɪdʒərənt/ *a* belligerante; (*aggressive*) bellicoso

bellow /'beləʊ/ *vi* gridare a squarciagola; (*animal:*) muggire

bellows /'beləʊz/ *npl* (*for fire*) soffietto *msg*

belly /'belɪ/ *n* pancia *f*

belong /bɪ'lɒŋ/ *vi* appartenere (to a); (*be member*) essere socio (to di). ~**ings** *npl* cose *fpl*

beloved /bɪ'lʌvɪd/ *a* & *n* amato, -a *mf*

below /bɪ'ləʊ/ *prep* sotto; (*with numbers*) al di sotto di ● *adv* sotto, di sotto; *Naut* sotto coperta; see ~ guardare qui di seguito

belt /belt/ *n* cintura *f*; (*area*) zona *f*; *Techn* cinghia *f* ● *vi* ~ along (*fam: rush*) filare velocemente ● *vt* (*fam: hit*) picchiare

bemused /br'mju:zd/ a confuso

bench /bentʃ/ n panchina f; (work~) piano m da lavoro; the B~ Jur la magistratura

bend /bend/ n curva f; (of river) ansa f ● v (pt/pp bent) ● vt piegare ● vi piegarsi; (road:) curvare; ~ [down] chinarsi. bend over vi inchinarsi

beneath /br'ni:θ/ prep sotto, al di sotto di; he thinks it's ~ him fig pensa che sia sotto al suo livello ● adv giù

benediction /benr'dɪkʃn/ n Relig benedizione f

benefactor /'benɪfæktə(r)/ n benefattore, -trice mf

beneficial /benɪ'fɪʃl/ a benefico

beneficiary /benɪ'fɪʃərɪ/ n beneficiario, -a mf

benefit /'benɪfɪt/ n vantaggio m; (allowance) indennità f inv ● v (pt/pp -fited, pres p -fiting) ● vt giovare a ● vi trarre vantaggio (from da)

benevolence /br'nevələns/ n benevolenza f. ~t a benevolo

benign /br'nam/ a benevolo; Med benigno

bent /bent/ see bend ● a (person) ricurvo; (distorted) curvato; (fam: dishonest) corrotto; be ~ on doing sth essere ben deciso a fare qcsa ● n predisposizione f

bequeath /br'kwi:ð/ vt lasciare in eredità. ~quest /-'kwest/ n lascito m

bereaved /br'ri:vd/ n the ~d pl i familiari del defunto. ~ment n lutto m

bereft /br'reft/ a ~ of privo di

beret /'bereɪ/ n berretto m

berry /'berɪ/ n bacca f

berserk /bə'sɜ:k/ a go ~ diventare una belva

berth /bɜ:θ/ n (bed) cuccetta f; (anchorage) ormeggio m ● vi ormeggiare

beseech /br'si:tʃ/ vt (pt/pp beseeched or besought) supplicare

beside /br'saɪd/ prep accanto a; ~ oneself fuori di sé

besides /br'saɪdz/ prep oltre a ● adv inoltre

besiege /br'si:dʒ/ vt assediare

besought /br'sɔ:t/ see beseech

best /best/ a migliore; the ~ part of a year la maggior parte dell'anno; ~ before Comm preferibilmente prima di ● n the ~ il meglio; (person) il/la migliore; at ~ tutt'al più; all the ~! tanti auguri!; do one's ~ fare del proprio meglio; to the ~ of my knowledge per quel che ne so; make the ~ of it cogliere il lato buono della cosa ● adv meglio, nel modo migliore; as ~ I could meglio che potevo. ~ 'man n testimone m

bestow /br'stəʊ/ vt conferire (on a)

best'seller n bestseller m inv

bet /bet/ n scommessa f ● vt/i (pt/pp bet or betted) scommettere

betray /br'treɪ/ vt tradire. ~al n tradimento m

better /'betə(r)/ a migliore, meglio; (after illness) rimettersi ● adv meglio; ~ off meglio; (wealthier) più ricco; all the ~ tanto meglio; the sooner the ~ prima è, meglio è; I've thought ~ of it ci ho ripensato; you'd ~ stay faresti meglio a restare; I'd ~ not è meglio che non lo faccia ● vt migliorare; ~ oneself migliorare le proprie condizioni

'betting shop n ricevitoria f (dell'allibratore)

between /br'twi:n/ prep fra, tra; ~ you and me detto fra di noi; ~ us (together) tra me e te ● adv [in] in mezzo; (time) frattempo

beverage /'bevərɪdʒ/ n bevanda f

beware /br'weə(r)/ vi guardarsi (of da); ~ of the dog! attenti al cane!

bewilder /br'wɪldə(r)/ vt disorientare; ~ed perplesso. ~ment n perplessità f

beyond /br'jɒnd/ prep oltre; ~

reach irraggiungibile; ~ **doubt** senza alcun dubbio; ~ **belief** da non credere; **it's ~ me** *fam* non riesco proprio a capire ● *adv* più in là

bias /'baɪəs/ *n* (preference) preferenza *f*; *pej* pregiudizio *m* ● *vt* (*pt/pp* **biased**) (influence) influenzare. **~ed** *a* parziale

bib /bɪb/ *n* bavaglino *m*

Bible /'baɪbl/ *n* Bibbia *f*

biblical /'bɪblɪkl/ *a* biblico

bicarbonate /baɪ'kɑːbənɪt/ *n* ~ **of soda** bicarbonato *m* di sodio

biceps /'baɪseps/ *n* bicipite *m*

bicker /'bɪkə(r)/ *vi* litigare

bicycle /'baɪsɪkl/ *n* bicicletta *f* ● *vi* andare in bicicletta

bid[1] /bɪd/ *n* (offerta *f*; (attempt) tentativo *m* ● *vt/i* (*pt/pp* **bid**, *pres p* **bidding**) offrire; (in cards) dichiarare

bid[2] *vt* (*pt* **bade** or **bid**, *pp* **bidden** or **bid**, *pres p* **bidding**) liter (command) comandare; ~ **sb welcome** dare il benvenuto a qcno

bidder /'bɪdə(r)/ *n* offerente *mf*

bide /baɪd/ *vt* ~ **one's time** aspettare il momento buono

biennial /baɪ'enɪəl/ *a* biennale

bifocals /baɪ'fəʊklz/ *npl* occhiali *mpl* bifocali

big /bɪg/ *a* (**bigger, biggest**) grande; (brother, sister) più grande; (fam: generous) generoso ● *adv* **talk ~** *fam* sparare grosse

bigam|ist /'bɪgəmɪst/ *n* bigamo, -a *mf*. **~y** *n* bigamia *f*

'big-head *n* *fam* gasato, -a *mf*

big-'headed *a* *fam* gasato

bigot /'bɪgət/ *n* fanatico, -a *mf*. **~ed** *a* di mentalità ristretta

'bigwig *n* *fam* pezzo *m* grosso

bike /baɪk/ *n* *fam* bici *f* *inv*

bikini /bɪ'kiːnɪ/ *n* bikini *m* *inv*

bile /baɪl/ *n* bile *f*

bilingual /baɪ'lɪŋgwəl/ *a* bilingue

bill[1] /bɪl/ *n* fattura *f*; (in restaurant etc) conto *m*; (poster) manifesto *m*; *Pol* progetto *m* di legge; (Am: note) biglietto *m* di banca ● *vt* fatturare

bill[2] *n* (beak) becco *m*

billfold *n* *Am* portafoglio *m*

billiards /'bɪljədz/ *n* biliardo *m*

billion /'bɪljən/ *n* (thousand million) miliardo *m*; (old-fashioned Br: million million) mille miliardi *mpl*

billy-goat /'bɪlɪ-/ *n* caprone *m*

bin /bɪn/ *n* bidone *m*

bind /baɪnd/ *vt* (*pt/pp* **bound**) legare (**to** a); (bandage) fasciare; *Jur* obbligare. **~ing** *a* (promise, contract) vincolante ● *n* (of book) rilegatura *f*; (on ski) attacco *m* [di sicurezza]

binge /bɪndʒ/ *n* *fam* **have a ~** fare baldoria; (eat a lot) abbuffarsi ● *vi* abbuffarsi (**on** di)

binoculars /bɪ'nɒkjʊləz/ *npl* [**pair of**] ~ binocolo *msg*

bio'chemist /baɪəʊ-/ *n* biochimico, -a *mf*. **~ry** *n* biochimica *f*

biodegradable /-dɪ'greɪdəbl/ *a* biodegradabile

biograph|er /baɪ'ɒgrəfə(r)/ *n* biografo, -a *mf*. **~y** *n* biografia *f*

biological /baɪə'lɒdʒɪkl/ *a* biologico

biolog|ist /baɪ'ɒlədʒɪst/ *n* biologo, -a *mf*. **~y** *n* biologia *f*

birch /bɜːtʃ/ *n* (tree) betulla *f*

bird /bɜːd/ *n* uccello *m*; (fam: girl) ragazza *f*

Biro® /'baɪrəʊ/ *n* biro *f* *inv*

birth /bɜːθ/ *n* nascita *f*

birth: ~ **certificate** *n* certificato *m* di nascita. **~-control** *n* controllo *m* delle nascite. **~day** *n* compleanno *m*. **~mark** *n* voglia *f*. **~-rate** *n* natalità *f*

biscuit /'bɪskɪt/ *n* biscotto *m*

bisect /baɪ'sekt/ *vt* dividere in due [parti]

bishop /'bɪʃəp/ *n* vescovo *m*; (in chess) alfiere *m*

bit[1] /bɪt/ *n* pezzo *m*; (smaller) pezzetto *m*; (for horse) morso *m*; *Comput* bit *m* *inv*; **a ~** *n* un pezzo di (cheese, paper); un po' di (time, rain, silence); **~ by ~** poco a poco; **do one's ~** fare la propria parte

bit² *see* bite

bitch /bɪtʃ/ *n* cagna *f*; *sl* stronza *f*. **~y** *a* velenoso

bit|e /baɪt/ *n* morso *m*; *(insect* ~*)* puntura *f*; *(mouthful)* boccone *m* ● *vt* (*pt* **bit**, *pp* **bitten**) mordere; *(insect:)* pungere; **~e one's nails** mangiarsi le unghie ● *vi* mordere; *(insect:)* pungere. **~ing** *a* *(wind, criticism)* pungente; *(remark)* mordace

bitter /'bɪtə(r)/ *a* amaro ● *n* *Br* birra *f* amara. **~ly** *adv* amaramente; **it's ~ly cold** c'è un freddo pungente. **~ness** *n* amarezza *f*

bitty /'bɪtɪ/ *a* *Br fam* frammentario

bizarre /bɪ'zɑ:(r)/ *a* bizzarro

blab /blæb/ *vi* (*pt/pp* **blabbed**) spifferare

black /blæk/ *a* nero; **be ~ and blue** essere pieno di lividi ● *n* negro, -a *mf* ● *vt* boicottare *(goods)*. **black out** *vt* cancellare ● *vi* *(lose consciousness)* perdere coscienza

black: ~berry *n* mora *f*. **~bird** *n* merlo *m*. **~board** *n* *Sch* lavagna *f*. **~'currant** *n* ribes *m inv* nero; **~'eye** *n* occhio *m* nero. **~ 'ice** *n* ghiaccio *m* *(sulla strada)*. **~leg** *n* *Br* crumiro *m*. **~list** *vt* mettere sulla lista nera. **~mail** *n* ricatto *m* ● *vt* ricattare. **~mailer** *n* ricattatore, -trice *mf*. **~ 'market** *n* mercato *m* nero. **~-out** *n* blackout *m inv*; **have a ~-out** *Med* perdere coscienza. **~smith** *n* fabbro *m*

bladder /'blædə(r)/ *n* *Anat* vescica *f*

blade /bleɪd/ *n* lama *f*; *(of grass)* filo *m*

blame /bleɪm/ *n* colpa *f* ● *vt* dare la colpa a; **~ sb for doing sth** dare la colpa a qcno per aver fatto qcsa; **no one is to ~** non è colpa di nessuno. **~less** *a* innocente

blanch /blɑ:ntʃ/ *vi* sbiancare ● *vt* *Culin* sbollentare

blancmange /blə'mɒnʒ/ *n* biancomangiare *m inv*

bland /blænd/ *a* *(food)* insipido; *(person)* insulso

blank /blæŋk/ *a* bianco; *(look)* vuoto ● *n* spazio *m* vuoto; *(cartridge)* a salve. **~ 'cheque** *n* assegno *m* in bianco

blanket /'blæŋkɪt/ *n* coperta *f*

blank 'verse *n* versi *mpl* sciolti

blare /bleə(r)/ *vi* suonare a tutto volume. **blare out** *vt* far risuonare ● *vi* *(music, radio:)* strillare

blasé /'blɑ:zeɪ/ *a* vissuto, blasé *inv*

blaspheme /blæs'fi:m/ *vi* bestemmiare

blasphem|ous /'blæsfəməs/ *a* blasfemo. **~y** *n* bestemmia *f*

blast /blɑ:st/ *n* *(gust)* raffica *f*; *(sound)* scoppio *m* ● *vt* *(with explosive)* far saltare ● *int sl* maledizione!. **~ed** *a* *sl* maledetto

blast: ~-furnace *n* altoforno *m*. **~-off** *n* *(of missile)* lancio *m*

blatant /'bleɪtənt/ *a* sfacciato

blaze /bleɪz/ *n* incendio *m*; **a ~ of colour** un'esplosione *f* di colori ● *vi* ardere

blazer /'bleɪzə(r)/ *n* blazer *m inv*

bleach /bli:tʃ/ *n* decolorante *m*; *(for cleaning)* candeggina *f* ● *vt* sbiancare; ossigenare *(hair)*

bleak /bli:k/ *a* desolato; *(fig: prospects, future)* tetro

bleary-eyed /blɪərɪ'aɪd/ *a* **look ~** avere gli occhi assonnati

bleat /bli:t/ *vi* belare ● *n* belato *m*

bleed /bli:d/ *v* (*pt/pp* **bled**) ● *vi* sanguinare ● *vt* spurgare *(brakes, radiator)*

bleep /bli:p/ *n* bip *m* ● *vi* suonare ● *vt* chiamare *(col cercapersone) (doctor)*. **~er** *n* cercapersone *m inv*

blemish /'blemɪʃ/ *n* macchia *f*

blend /blend/ *n* *(of tea, coffee, whisky)* miscela *f*; *(of colours)* insieme *m* ● *vt* mescolare ● *vi* *(colours, sounds:)* fondersi (**with** con). **~er** *n* *Culin* frullatore *m*

bless /bles/ *vt* benedire. **~ed** /'blesɪd/ *a also sl* benedetto. **~ing** *n* benedizione *f*

blew /blu:/ *see* blow²

blight /blaɪt/ n Bot ruggine f ● vt far avvizzire ⟨plants⟩

blind[1] /blaɪnd/ a cieco; **the ~** npl i ciechi mpl; **~ man/woman** cieco/cieca ● vt accecare

blind[2] n ⟨roller⟩ ~ avvolgibile m; [Venetian] ~ veneziana f

blind: ~ 'alley n vicolo m cieco. **~fold** a be **~fold** avere gli occhi bendati ● n benda f ● vt bendare gli occhi a. **~ly** adv ciecamente. **~ness** n cecità f

blink /blɪŋk/ vi sbattere le palpebre; ⟨light:⟩ tremolare

blinkered /'blɪŋkəd/ adj fig be **~** avere i paraocchi

blinkers /'blɪŋkəz/ npl paraocchi mpl

bliss /blɪs/ n Rel beatitudine f; ⟨happiness⟩ felicità f. **~ful** a beato; ⟨happy⟩ meraviglioso

blister /'blɪstə(r)/ n Med vescica f; ⟨in paint⟩ bolla f ● vi ⟨paint:⟩ formare una bolla/delle bolle

blitz /blɪts/ n bombardamento m aereo; **have a ~ on** sth fig darci sotto con qcsa

blizzard /'blɪzəd/ n tormenta f

bloated /'bləʊtɪd/ a gonfio

blob /blɒb/ n goccia f

bloc /blɒk/ n Pol blocco m

block /blɒk/ n blocco m; ⟨building⟩ isolato m; ⟨building ~⟩ cubo m ⟨per giochi di costruzione⟩; **~ of flats** palazzo m ● vt bloccare. **block up** vt bloccare

blockade /blɒ'keɪd/ n blocco m ● vt bloccare

blockage /'blɒkɪdʒ/ n ostruzione f

block: **~head** n fam testone, -a mf. **~ 'letters** npl stampatello m

bloke /bləʊk/ n fam tizio m

blonde /blɒnd/ a biondo ● n bionda f

blood /blʌd/ n sangue m

blood: **~ bath** n bagno m di sangue. **~ count** n esame m emocromocitometrico. **~ donor** n donatore m di sangue. **~ group** n gruppo m sanguigno. **~hound** n segugio m. **~-poisoning** n setti-cemia f. **~ pressure** n pressione f del sangue. **~shed** n spargimento m di sangue. **~shot** a iniettato di sangue. **~ sports** npl sport mpl cruenti. **~-stained** a macchiato di sangue. **~stream** n sangue m. **~ test** n analisi f del sangue. **~thirsty** a assetato di sangue. **~ transfusion** n trasfusione f del sangue

bloody /'blʌdɪ/ a (-ier, -iest) insan-guinato; sl maledetto ● adv sl **~ easy/difficult** facile/difficile da matti. **~-'minded** a scorbutico

bloom /bluːm/ n fiore m; in ~ ⟨flower:⟩ sbocciato; ⟨tree:⟩ in fiore ● vi fiorire; fig essere in forma smagliante

bloomer /'bluːmə(r)/ n fam papera f. **~ing** a fam maledetto. **~ers** npl mutandoni mpl ⟨da donna⟩

blossom /'blɒsəm/ n fiori mpl ⟨d'albero⟩; ⟨single one⟩ fiore m ● vi sbocciare

blot /blɒt/ n also fig macchia f ● **blot out** vt (pt/pp blotted) fig cancellare

blotch /blɒtʃ/ n macchia f. **~y** a chiazzato

'blotting-paper n carta f assorbente

blouse /blaʊz/ n camicetta f

blow[1] /bləʊ/ n colpo m

blow[2] v (pt blew, pp blown) ● vi ⟨wind:⟩ soffiare; ⟨fuse:⟩ saltare ● vt ⟨fam: squander⟩ sperperare; **~ one's nose** soffiarsi il naso.

blow away vt far volar via ⟨papers⟩ ● vi ⟨papers:⟩ volare via. **blow down** vt abbattere ● vi abbattersi al suolo. **blow out** vt ⟨extinguish⟩ spegnere. **blow over** vi ⟨storm:⟩ passare; ⟨fuss, trouble:⟩ dissiparsi. **blow up** vt ⟨inflate⟩ gonfiare; ⟨enlarge⟩ ingrandire ⟨photograph⟩; ⟨by explosion⟩ far esplodere ● vi esplodere

blow: **~-dry** vt asciugare col fon. **~lamp** n fiamma f ossidrica

blown /bləʊn/ see blow[2]

'blowtorch n fiamma f ossidrica

blowy /'bləʊi/ a ventoso

blue /blu:/ a (pale) celeste; (navy) blu inv; (royal) azzurro; ~ **with cold** livido per il freddo ●n blu m inv; **have the ~s** essere giù [di tono]; **out of the ~** inaspettatamente

blue: ~**bell** n giacinto m di bosco. ~**berry** n mirtillo m. ~**bottle** n moscone m. ~ **film** n film m inv a luci rosse. ~**print** n fig riferimento m

bluff /blʌf/ n bluff m inv ●vi bluffare

blunder /'blʌndə(r)/ n gaffe f inv ●vi fare una/delle gaffe

blunt /blʌnt/ a spuntato; (person) reciso. ~**ly** adv schiettamente

blur /blɜ:(r)/ n **it's all a** ~ è tutto un insieme confuso ●vt (pt/pp **blurred**) rendere confuso. ~**red** a (vision, photo) sfocato

blurb /blɜ:b/ n soffietto m editoriale

blurt /blɜ:t/ vt ~ **out** spifferare

blush /blʌʃ/ n rossore m ●vi arrossire

blusher /'blʌʃə(r)/ n fard m

bluster /'blʌstə(r)/ n sbruffonata f. ~**y** a (wind) furioso; (day, weather) molto ventoso

boar /bɔ:(r)/ n cinghiale m

board /bɔ:d/ n tavola f; (for notices) tabellone m; (committee) assemblea f; (of directors) consiglio m; **full** ~ Br pensione f completa; **half** ~ Br mezza pensione f; ~ **and lodging** vitto e alloggio m; **go by the** ~ fam andare a monte ●vt Naut, Aviat salire a bordo di ●vi (passengers:) salire a bordo. **board up** vt sbarrare con delle assi. **board with** vt stare a pensione da.

boarder /'bɔ:də(r)/ n pensionante mf; Sch convittore, -trice mf

board: ~-**game** n gioco m da tavolo. ~-**ing-house** n pensione f. ~**ing-school** n collegio m

boast /bəʊst/ vi vantarsi (about di). ~**ful** a vanaglorioso

boat /bəʊt/ n barca f; (ship) nave f. ~**er** n (hat) paglietta f

bob /bɒb/ n (hairstyle) caschetto m ●vi (pt/pp **bobbed**) (also ~ **up and down**) andare su e giù

bob-sleigh n bob m inv

bode /bəʊd/ vi **~ well/ill** essere di buono/cattivo augurio

bodily /'bɒdɪlɪ/ a fisico ●adv (forcibly) fisicamente

body /'bɒdɪ/ n corpo m; (organization) ente m; (amount: of poems etc) quantità f. ~**guard** n guardia f del corpo. ~**work** n Auto carrozzeria f

bog /bɒg/ n palude f ●vt (pt/pp **bogged**) **get** ~**ged down** impantanarsi

boggle /'bɒgl/ vi **the mind** ~**s** non posso neanche immaginarlo

bogus /'bəʊgəs/ a falso

boil¹ /bɔɪl/ n Med foruncolo m

boil² n **bring/come to the** ~ portare/arrivare ad ebollizione ●vt [far] bollire ●vi bollire; (fig: with anger) ribollire; **the water or kettle's** ~**ing** l'acqua bolle. **boil down to** vt fig ridursi a. **boil over** vi straboccare (bollendo). **boil up** vt far bollire

boiler /'bɔɪlə(r)/ n caldaia f. ~**suit** n tuta f

boiling point n punto m di ebollizione

boisterous /'bɔɪstərəs/ a chiassoso

bold /bəʊld/ a audace ●n Typ neretto m. ~**ness** n audacia f

bollard /'bɒlɑ:d/ n colonnina m di sbarramento al traffico

bolster /'bəʊlstə(r)/ n cuscino m (lungo e rotondo) ●vt ~ [**up**] sostenere

bolt /bəʊlt/ n (for door) catenaccio m; (for fixing) bullone m ●vt fissare (con i bulloni) (**to** a); chiudere col chiavistello (door); ingurgitare (food) ●vi svignarsela; (horse:) scappar via ●adv ~ **upright** diritto come un fuso

bomb /bɒm/ n bomba f ● vt bombardare

bombard /bɒmˈbɑːd/ vt also fig bombardare

bombastic /bɒmˈbæstɪk/ a ampolloso

bomb|er /ˈbɒmə(r)/ n Aviat bombardiere m; (person) dinamitardo m. **~er jacket** giubbotto m, bomber m inv. **~shell** n (fig: news) bomba f

bond /bɒnd/ n fig legame m; Comm obbligazione f ● vt (glue:) attaccare

bondage /ˈbɒndɪdʒ/ n schiavitù f

bone /bəʊn/ n osso m; (of fish) spina f ● vt disossare (meat); togliere le spine da (fish). **~-dry** a secco

bonfire /ˈbɒn-/ n falò m inv. **~ night** festa celebrata la notte del 5 novembre con fuochi d'artificio e falò

bonnet /ˈbɒnɪt/ n cuffia f; (of car) cofano m

bonus /ˈbəʊnəs/ n (individual) gratifica f; (production ~) premio m; (life insurance) dividendo m; **a ~** fig qualcosa in più

bony /ˈbəʊnɪ/ a (-ier, -iest) ossuto; (fish) pieno di spine

boo /buː/ interj (to surprise or frighten) bu! ● vt/i fischiare

boob /buːb/ n (fam: mistake) gaffe f inv; (breast) tetta f ● vi fam fare una gaffe

book /bʊk/ n libro m; (of tickets) blocchetto m; **keep the ~s** Comm tenere la contabilità; **be in sb's bad/good ~s** essere nel libro nero/nelle grazie di qcno ● vt (reserve) prenotare; (for offence) multare ● vi (reserve) prenotare

book: **~case** n libreria f. **~ends** npl reggilibri mpl. **~ing-office** n biglietteria f. **~keeping** n contabilità f. **~let** n opuscolo m. **~maker** n allibratore m. **~mark** n segnalibro m. **~seller** n libraio, -a mf. **~shop** n libreria f. **~worm** n topo m di biblioteca

boom /buːm/ n Comm boom m inv;

(upturn) impennata f; (of thunder, gun) rimbombo m ● vi (thunder, gun:) rimbombare; fig prosperare

boon /buːn/ n benedizione f

boor /bʊə(r)/ n zoticone m. **~ish** a maleducato

boost /buːst/ n spinta f ● vt stimolare (sales); sollevare (morale); far crescere (hopes). **~er** n Med dose f supplementare

boot /buːt/ n stivale m; (up to ankle) stivaletto m; (football) scarpetta f; (climbing) scarpone m; Auto portabagagli m inv ● vt Comput inizializzare

booth /buːð/ n (Teleph, voting) cabina f; (at market) bancarella f

'boot-up n Comput boot m inv

booty /ˈbuːtɪ/ n bottino m

booze /buːz/ fam n alcolici mpl. **~-up** n bella bevuta f

border /ˈbɔːdə(r)/ n bordo m; (frontier) frontiera f; (in garden) bordura f ● vi **~ on** confinare con; fig essere ai confini di (madness). **~line** n linea f di demarcazione; **~line case** caso m di dubbio

bore[1] /bɔː(r)/ see **bear**[2]

bore[2] vt Techn forare

bor|e[3] n (of gun) calibro m; (person) seccatore, -trice mf; (thing) seccatura f ● vt annoiare. **~edom** n noia f. **be ~ed** (to tears or to death) annoiarsi (da morire). **~ing** a noioso

born /bɔːn/ pp **be ~** nascere; **I was ~ in 1966** sono nato nel 1966 ● a nato; **a ~ liar/actor** un bugiardo/attore nato

borne /bɔːn/ see **bear**[2]

borough /ˈbʌrə/ n municipalità f inv

borrow /ˈbɒrəʊ/ vt prendere a prestito (from da); **can I ~ your pen?** mi presti la tua penna?

bosom /ˈbʊzm/ n seno m

boss /bɒs/ n direttore, -trice mf ● vt (also **~ about**) comandare a bacchetta. **~y** a autoritario

botanical /bəˈtænɪkl/ a botanico

botan|ist /'bɒtənɪst/ n botanico, -a mf. **~y** n botanica f

botch /bɒtʃ/ vt fare un pasticcio con

both /bəʊθ/ adj & pron tutti e due, entrambi ● adv **~ men and women** entrambi uomini e donne; **~ [of] the children** tutti e due i bambini; **they are ~ dead** sono morti entrambi; **~ of them** tutti e due

bother /'bɒðə(r)/ n preoccupazione f, (minor trouble) fastidio m; **it's no ~** non c'è problema ● int fam che seccatura! ● vt (annoy) dare fastidio a; (disturb) disturbare ● vi preoccuparsi (about di); **don't ~** lascia perdere

bottle /'bɒtl/ n bottiglia f; (baby's) biberon m inv ● vt imbottigliare. **bottle up** vt fig reprimere

bottle: ~ bank n contenitore m per la raccolta del vetro. **~-neck** n fig ingorgo m. **~-opener** n apribottiglie m inv

bottom /'bɒtm/ a ultimo; **the ~ shelf** l'ultimo scaffale in basso ● n (of container) fondo m; (of river) fondale m; (of hill) piedi mpl; (buttocks) sedere m; **at the ~ of the page** in fondo alla pagina; **get to the ~ of** fig vedere cosa c'è sotto. **~less** a senza fondo

bough /baʊ/ n ramoscello m

bought /bɔːt/ see **buy**

boulder /'bəʊldə(r)/ n masso m

bounce /baʊns/ vi rimbalzare; (fam: cheque:) essere respinto ● vt far rimbalzare (ball)

bouncer /'baʊnsə(r)/ n fam buttafuori m inv

bound[1] /baʊnd/ n balzo m ● vi balzare

bound[2] see **bind** ● a **~ for** (ship) diretto a; **be ~ to do** (likely) dovere fare per forza; (obliged) essere costretto a fare

boundary /'baʊndərɪ/ n limite m

boundless a illimitato

bounds /baʊndz/ npl fig limiti mpl; **out of ~** fuori dai limiti

bouquet /bʊ'keɪ/ n mazzo m di fiori; (of wine) bouquet m

bourgeois /'bʊəʒwɑ:/ a pej borghese

bout /baʊt/ n Med attacco m; Sport incontro m

bow[1] /bəʊ/ n (weapon) arco m; Mus archetto m; (knot) nodo m

bow[2] /baʊ/ n inchino m ● vi inchinarsi ● vt piegare (head)

bow[3] /baʊ/ n Naut prua f

bowel /'baʊəl/ n intestino m; **~s** pl intestini mpl

bowl[1] /bəʊl/ n (for soup, cereal) scodella f; (of pipe) fornello m

bowl[2] n (ball) boccia f ● vt lanciare ● vi Cricket servire; (in bowls) lanciare. **bowl over** vt buttar giù; (fig: leave speechless) lasciar senza parole

bow-legged /bəʊ'legd/ a dalle gambe storte

bowler[1] /'bəʊlə(r)/ n Cricket lanciatore m; Bowls giocatore m di bocce

bowler[2] n **~ [hat]** bombetta f

bowling /'bəʊlɪŋ/ n gioco m delle bocce. **~-alley** n pista f da bowling

bowls /bəʊlz/ n gioco m delle bocce

bow-tie /bəʊ-/ n cravatta f a farfalla

box[1] /bɒks/ n scatola f; Theat palco m

box[2] vi Sport fare il pugile ● vt **~ sb's ears** dare uno scappaccione a qcno

box|er /'bɒksə(r)/ n pugile m. **~ing** n pugilato m. **B~ing Day** n (giorno m di) Santo Stefano m

box: ~-office n Theat botteghino m. **~-room** n Br sgabuzzino m

boy /bɔɪ/ n ragazzo m; (younger) bambino m

boycott /'bɔɪkɒt/ n boicottaggio m ● vt boicottare

boy: ~friend n ragazzo m. **~ish** a da ragazzino

bra /brɑ:/ n reggiseno m

brace /breɪs/ n sostegno m; (dental) apparecchio m; **~s** pl

bretelle *fpl* ● *vt* ~ **oneself** *fig* farsi forza (**for** per affrontare)

bracelet /'breɪslɪt/ *n* braccialetto *m*

bracing /'breɪsɪŋ/ *a* tonificante

bracken /'brækn/ *n* felce *f*

bracket /'brækɪt/ *n* mensola *f*; (*group*) categoria *f*; *Typ* parentesi *f inv* ● *vt* mettere fra parentesi

brag /bræg/ *vi* (*pt/pp* **bragged**) vantarsi (**about** di)

braid /breɪd/ *n* (*edging*) passamano *m*

braille /breɪl/ *n* braille *m*

brain /breɪn/ *n* cervello *m*; ~s *pl fig* testa *fsg*

brain: ~**child** *n* invenzione *f* personale. ~ **dead** *a Med* celebralmente morto; *fig* fam senza cervello. ~**less** *a* senza cervello. ~**wash** *vt* fare il lavaggio del cervello a. ~**wave** *n* lampo *m* di genio

brainy /'breɪnɪ/ *a* (**-ier, -iest**) intelligente

braise /breɪz/ *vt* brasare

brake /breɪk/ *n* freno *m* ● *vi* frenare. ~**-light** *n* stop *m inv*

bramble /'bræmbl/ *n* rovo *m*; (*fruit*) mora *f*

bran /bræn/ *n* crusca *f*

branch /brɑːntʃ/ *n also fig* ramo *m*; *Comm* succursale *f* ● *vi* (*road:*) biforcarsi. **branch off** *vi* biforcarsi. **branch out** *vi* ~ **out** into allargare le proprie attività nel ramo di

brand /brænd/ *n* marca *f*; (*on animal*) marchio *m* ● *vt* marcare (*animal*); *fig* tacciare (**as** di)

brandish /'brændɪʃ/ *vt* brandire

brand-'new *a* nuovo fiammante

brandy /'brændɪ/ *n* brandy *m inv*

brash /bræʃ/ *a* sfrontato

brass /brɑːs/ *n* ottone *m*; the ~ *Mus* gli ottoni *mpl*; **top** ~ *fam* pezzi *mpl* grossi. ~ **band** *n* banda *f* (di soli ottoni)

brassiere /'bræzɪə(r)/ *n* fml, Am reggipetto *m*

brat /bræt/ *n pej* marmocchio, -a *mf*

bravado /brə'vɑːdəʊ/ *n* bravata *f*

brave /breɪv/ *a* coraggioso ● *vt* affrontare. ~**ry** /-ərɪ/ *n* coraggio *m*

brawl /brɔːl/ *n* rissa *f* ● *vi* azzuffarsi

brawn /brɔːn/ *n Culin* ≈ soppressata *f*

brawny /'brɔːnɪ/ *a* muscoloso

brazen /'breɪzn/ *a* sfrontato

brazier /'breɪzɪə(r)/ *n* braciere *m*

Brazil /brə'zɪl/ *n* Brasile *m*. ~**ian** *a & n* brasiliano, -a *mf*. ~ **nut** *n* noce *f* del Brasile

breach /briːtʃ/ *n* (*of law*) violazione *f*; (*gap*) breccia *f*; (*fig*: *in party*) frattura *f*; ~ **of contract** inadempienza *f* di contratto; ~ **of the peace** violazione *f* della quiete pubblica ● *vt* recedere (*contract*)

bread /bred/ *n* pane *m*; **a slice of** ~ **and butter** una fetta di pane imburrato

bread: ~ **bin** *n* cassetta *f* portapane *inv*. ~**crumbs** *npl* briciole *fpl*; *Culin* pangrattato *m*. ~**line** *n* **be on the** ~**line** essere povero in canna

breadth /bredθ/ *n* larghezza *f*

'bread-winner *n* quello, -a *mf* che porta i soldi a casa

break /breɪk/ *n* rottura *f*; (*interval*) intervallo *m*; (*interruption*) interruzione *f*; (*fam: chance*) opportunità *f inv* ● *vt* (*pt* **broke**, *pp* **broken**) ● *vt* rompere; (*interrupt*) interrompere; ~ **one's arm** rompersi un braccio ● *vi* rompersi; (*day:*) spuntare; (*storm:*) scoppiare; (*news:*) diffondersi; (*boy's voice:*) cambiare. **break away** *vi* scappare; *fig* chiudere (**from** con). **break down** *vi* (*machine, car:*) guastarsi; (*emotionally*) cedere (*psicologicamente*) ● *vt* sfondare (*door*); ripartire (*figures*). **break into** *vt* introdursi (con la forza) in; forzare (*car*). **break off** *vt* rompere (*engagement*) ● *vi* (*part of whole:*) rompersi. **break out** *vi* (*fight, war:*) scoppiare. **break up** *vt* far cessare (*fight*); disperdere 〈crowd:〉 ● vi 〈crowd:〉 disperdersi; 〈couple:〉 separarsi; **Sch** iniziare le vacanze

'break|able /'breɪkəbl/ a fragile. **~age** /-ɪdʒ/ n rottura f. **~down** n (of car, machine) guasto m; **Med** esaurimento n nervoso; (of figures) analisi f inv. **~er** n 〈wave〉 frangente m

breakfast /'brekfəst/ n [prima] colazione f

break: **~through** n scoperta f. **~water** n frangiflutti m inv

breast /brest/ n seno m. **~-feed** vt allattare [al seno]. **~-stroke** n nuoto m a rana

breath /breθ/ n respiro m, fiato m; **out of ~** senza fiato

breathalyse /'breθəlaɪz/ vt sottoporre alla prova [etilica] del palloncino. **~r®** n Br alcoltest m inv

breathe /briːð/ vt/i respirare. **breathe in** vi inspirare ● vt respirare 〈scent, air〉. **breathe out** vt/i espirare

breath|er /'briːðə(r)/ n pausa f. **~ing** n respirazione f

breath /breθ/: **~less** a senza fiato. **~-taking** a mozzafiato. **~ test** n prova [etilica] f del palloncino

bred /bred/ see **breed**

breed /briːd/ n razza f ● v (pt/pp bred) ● vt allevare; 〈give rise to〉 generare ● vi riprodursi. **~er** n allevatore, -trice mf. **~ing** n allevamento m; fig educazione f

breez|e /briːz/ n brezza f. **~y** a ventoso

brew /bruː/ n infuso m ● vt mettere in infusione 〈tea〉; produrre 〈beer〉 ● vi fig 〈trouble:〉 essere nell'aria. **~er** n birraio m. **~ery** n fabbrica f di birra

bribe /braɪb/ n 〈of money〉 bustarella f; 〈large sum of money〉 tangente f ● vt corrompere. **~ry** /-ərɪ/ n corruzione f

brick /brɪk/ n mattone m. **'~layer** n muratore m ● **brick up** vt murare

bridal /'braɪdl/ a nuziale

bride /braɪd/ n sposa f. **~groom** n sposo m. **~smaid** n damigella f d'onore

bridge¹ /brɪdʒ/ n ponte m; (of nose) setto m nasale; (of spectacles) ponticello m ● vt fig colmare 〈gap〉

bridge² n Cards bridge m

bridle /'braɪdl/ n briglia f

brief¹ /briːf/ a breve

brief² n istruzioni fpl; 〈Jur: case〉 causa f ● vt dare istruzioni a; **Jur** affidare la causa a. **~case** n cartella f

brief|ing /'briːfɪŋ/ n briefing m inv. **~ly** adv brevemente. **~ly,...** in breve,... **~ness** n brevità f

briefs /briːfs/ npl slip m inv

brigad|e /brɪ'geɪd/ n brigata f. **~ier** /-ə'dɪə(r)/ n generale m di brigata

bright /braɪt/ a 〈metal, idea〉 brillante; 〈day, room, future〉 luminoso; 〈clever〉 intelligente; **~ red** rosso m acceso

bright|en /'braɪtn/ v **~en [up]** ● vt ravvivare; rallegrare 〈person〉 ● vi 〈weather:〉 schiarirsi; 〈face:〉 illuminarsi; 〈person:〉 rallegrarsi. **~ly** adv 〈shine〉 intensamente; 〈smile〉 allegramente. **~ness** n luminosità f; 〈intelligence〉 intelligenza f

brilliance /'brɪljəns/ n luminosità f; (of person) genialità f

brilliant /'brɪljənt/ a 〈very good〉 eccezionale; 〈very intelligent〉 brillante; 〈sunshine〉 splendente

brim /brɪm/ n bordo m; (of hat) tesa f ● **brim over** vi (pt/pp brimmed) traboccare

brine /braɪn/ n salamoia f

bring /brɪŋ/ vt (pt/pp brought) portare 〈person, object〉. **bring about** vt causare. **bring along** vt portare [con sé]. **bring back** vt restituire 〈sth borrowed〉; reintrodurre 〈hanging〉; fare ritornare in mente 〈memories〉. **bring down** vt portare giù; fare cadere 〈government〉; fare abbassare 〈price〉.

bring off *vt* ~ sth off riuscire a fare qcsa. **bring on** *vt* (*cause*) provocare. **bring out** *vt* (*emphasize*) mettere in evidenza; pubblicare (*book*). **bring round** *vt* portare; (*persuade*) convincere; far rinvenire (*unconscious person*). **bring up** *vt* (*vomit*) rimettere; allevare (*children*); tirare fuori (*question, subject*)

brink /brɪŋk/ *n* orlo *m*

brisk /brɪsk/ *a* svelto; (*person*) sbrigativo; (*trade, business*) redditizio; (*walk*) a passo spedito

bristle /'brɪsl/ *n* setola *f* • *vi* ~**ling with** pieno di. ~**ly** *a* (*chin*) ispido

Britain /'brɪtn/ *n* Gran Bretagna *f*. ~**ish** *a* britannico; (*ambassador*) della Gran Bretagna • *npl* **the** ~**ish** il popolo britannico. ~**on** *n* cittadino, -a britannico, -a *mf*

brittle /'brɪtl/ *a* fragile

broach /brəʊtʃ/ *vt* toccare (*subject*)

broad /brɔːd/ *a* ampio; (*hint* chiaro; (*accent*) marcato. **two metres** ~ largo due metri; **in** ~ **daylight** in pieno giorno. ~ **beans** *npl* fave *fpl*

broadcast *n* trasmissione *f* • *vt/i* (*pt/pp* -**cast**) trasmettere. ~**er** *n* giornalista *mf* radiotelevisivo, -a. ~**ing** *n* diffusione *f* radiotelevisiva; **be in** ~**ing** lavorare per la televisione/radio

broaden /'brɔːdn/ *vt* allargare • *vi* allargarsi

broadly /'brɔːdlɪ/ *adv* largamente; ~ [**speaking**] generalmente

broad|minded *a* di larghe vedute

broccoli /'brɒkəlɪ/ *n* inv broccoli *mpl*

brochure /'brəʊʃə(r)/ *n* opuscolo *m*; (*travel*) dépliant *m* inv

broke /brəʊk/ *see* **break** • *a fam* al verde

broken /'brəʊkn/ *see* **break** • *a* rotto; (*fig: marriage*) fallito. ~ **English** inglese *m* stentato. ~**-hearted** *a* affranto

broker /'brəʊkə(r)/ *n* broker *m inv*

brolly /'brɒlɪ/ *n fam* ombrello *m*

bronchitis /brɒŋ'kaɪtɪs/ *n* bronchite *f*

bronze /brɒnz/ *n* bronzo *m* • *attrib* di bronzo

brooch /brəʊtʃ/ *n* spilla *f*

brood /bruːd/ *n* covata *f*; (*hum: children*) prole *f* • *vi fig* rimuginare

brook /brʊk/ *n* ruscello *m*

broom /bruːm/ *n* scopa *f*. ~**stick** *n* manico *m* di scopa

broth /brɒθ/ *n* brodo *m*

brothel /'brɒθl/ *n* bordello *m*

brother /'brʌðə(r)/ *n* fratello *m*

brother: ~**-in-law** *n* (*pl* ~**s-in-law**) cognato *m*. ~**ly** *a* fraterno

brought /brɔːt/ *see* **bring**

brow /braʊ/ *n* fronte *f*; (*of hill*) cima *f*

browbeat *vt* (*pt* -**beat**, *pp* -**beaten**) intimidire

brown /braʊn/ *a* marrone; castano (*hair*) • *n* marrone *m* • *vt* rosolare (*meat*) • *vi* (*meat*) rosolarsi. ~ **paper** *n* carta *f* da pacchi

Brownie /'braʊnɪ/ *n* coccinella *f* (*negli scout*)

browse /braʊz/ *vi* (*read*) leggicchiare; (*in shop*) curiosare

bruise /bruːz/ *n* livido *m*; (*on fruit*) ammaccatura *f* • *vt* ammaccare (*fruit*); ~ **one's arm** farsi un livido sul braccio. ~**d** *a* contuso

brunette /bruː'net/ *n* bruna *f*

brunt /brʌnt/ *n* **bear the** ~ **of** sth subire maggiormente qcsa

brush /brʌʃ/ *n* spazzola *f*; (*with long handle*) spazzolone *m*; (*for paint*) pennello *m*; (*bushes*) boscaglia *f*; (*fig: conflict*) breve scontro *m* • *vt* spazzolare (*hair*); lavarsi (*teeth*); scopare (*stairs, floor*). **brush against** *vt* sfiorare. **brush aside** *vt fig* ignorare. **brush off** *vt* spazzolare; (*with hands*) togliere; ignorare (*criticism*). **brush up** *vt/i fig* ~ **up** [**on**] rinfrescare

brusque /brʊsk/ *a* brusco

Brussels /'brʌslz/ *n* Bruxelles *f*. ~

sprouts *npl* cavoletti *mpl* di Bruxelles

brutal /ˈbruːtl/ *a* brutale. **~ity** /-ˈtælətɪ/ *n* brutalità *f inv*

brute /bruːt/ *n* bruto *m*. **~ force** forza *f* bruta

BSc *n abbr* **Bachelor of Science**

BSE *n abbr* (**bovine spongiform encephalitis**) encefalite *f* bovina spongiforme

bubble /ˈbʌbl/ *n* bolla *f*; (*in drink*) bollicina *f*

buck[1] /bʌk/ *n* maschio *m* del cervo; (*rabbit*) maschio *m* del coniglio ● *vi* (*horse:*) saltare a quattro zampe. **buck up** *vi fam* tirarsi su; (*hurry*) sbrigarsi

buck[2] *n Am fam* dollaro *m*

buck[3] *n* **pass the ~** scaricare la responsabilità

bucket /ˈbʌkɪt/ *n* secchio *m*

buckle /ˈbʌkl/ *n* fibbia *f* ● *vt* allacciare ● *vi* (*shelf:*) piegarsi; (*wheel:*) storcersi

bud /bʌd/ *n* bocciolo *m*

Buddhism /ˈbʊdɪzm/ *n* buddismo *m*. **~t** *a & n* buddista *mf*

buddy /ˈbʌdɪ/ *n fam* amico, -a *mf*

budge /bʌdʒ/ *vt* spostare ● *vi* spostarsi

budgerigar /ˈbʌdʒərɪɡɑː(r)/ *n* cocorita *f*

budget /ˈbʌdʒɪt/ *n* bilancio *m*; (*allotted to specific activity*) budget *m inv* ● *vi* (*pt/pp* **budgeted**) prevedere le spese; **~ for sth** includere qcsa nelle spese previste

buff /bʌf/ *a* (*colour*) [color] camoscio ● *n fam* fanatico, -a *mf*

buffalo /ˈbʌfələʊ/ *n* (*inv or pl* **-es**) bufalo *m*

buffer /ˈbʌfə(r)/ *n Rail* respingente *m*; **old** ~ *fam* vecchio bacucco *m*. ~ **zone** *n* zona *f* cuscinetto

buffet[1] /ˈbʊfeɪ/ *n* buffet *m inv*

buffet[2] /ˈbʌfɪt/ *vt* (*pt/pp* **buffeted**) sferzare

buffoon /bəˈfuːn/ *n* buffone, -a *mf*

bug /bʌɡ/ *n* (*insect*) insetto *m*; *Comput* bug *m inv*; (*fam: device*) cimice *f* ● *vt* (*pt/pp* **bugged**)

fam installare le microspie in (*room*); mettere sotto controllo (*telephone*); (*fam: annoy*) scocciare

buggy /ˈbʌɡɪ/ *n* (*baby*) ~ passeggino *m*

bugle /ˈbjuːɡl/ *n* tromba *f*

build /bɪld/ *n* (*of person*) corporatura *f* ● *vt/i* (*pt/pp* **built**) costruire. **build on** *vt* aggiungere (*extra storey*); sviluppare (*previous work*). **build up** *vt* ~ **up one's strength** rimettersi in forza ● *vi* (*pressure, traffic:*) aumentare; (*excitement, tension:*) crescere

builder /ˈbɪldə(r)/ *n* (*company*) costruttore *m*; (*worker*) muratore *m*

building /ˈbɪldɪŋ/ *n* edificio *m*. ~ **site** *n* cantiere *m* [di costruzione]. ~ **society** *n* istituto *m* di credito immobiliare

'build-up *n* (*of gas etc*) accumulo *m*; *fig* battage *m inv* pubblicitario

built /bɪlt/ *see* **build**. **~-in** *a* (*unit*) a muro; (*fig: feature*) incorporato. **~-up area** *n Auto* centro *m* abitato

bulb /bʌlb/ *n* bulbo *m*; *Electr* lampadina *f*

bulge /bʌldʒ/ *n* rigonfiamento *m* ● *vi* esser gonfio (**with** di); (*stomach, wall:*) sporgere; (*eyes, with surprise:*) uscire dalle orbite. **~ing** *a* gonfio; (*eyes*) sporgente

bulk /bʌlk/ *n* volume *m*; (*greater part*) grosso *m*; **in** ~ in grande quantità; (*loose*) sfuso. **~y** *a* voluminoso

bull /bʊl/ *n* toro *m*

'bulldog *n* bulldog *m inv*

bulldozer /ˈbʊldəʊzə(r)/ *n* bulldozer *m inv*

bullet /ˈbʊlɪt/ *n* pallottola *f*

bulletin /ˈbʊlɪtɪn/ *n* bollettino *m*. ~ **board** *n Comput* bacheca *f* elettronica

'bullet-proof *a* antiproiettile *inv*; (*vehicle*) blindato

'bullfight *n* corrida *f*. **~er** *n* torero *m*

bullion /'bʊlɪən/ n gold ● oro m in lingotti

bullock /'bʊlək/ n manzo m

bull: ~**ring** n arena f. ~**'s-eye** n centro m del bersaglio; **score** a ~**'s-eye** fare centro

bully /'bʊlɪ/ n prepotente mf ● vt fare il/la prepotente con. ~**ing** n prepotenze fpl

bum[1] /bʌm/ n sl sedere m

bum[2] n Am fam vagabondo, -a mf ● **bum around** vi fam vagabondare

bumble-bee /'bʌmbl-/ n calabrone m

bump /bʌmp/ n botta f; (swelling) bozzo m, gonfiore m; (in road) protuberanza f ● vt sbattere. **bump into** vt sbattere contro; (meet) imbattersi in. **bump off** vt fam far fuori

bumper /'bʌmpə(r)/ n Auto paraurti m inv ● a abbondante

bumpkin /'bʌmpkɪn/ n **country** ~ zoticone, -a mf

bumptious /'bʌmpʃəs/ a presuntuoso

bumpy /'bʌmpɪ/ a (road) accidentato; (flight) turbolento

bun /bʌn/ n focaccia f (dolce); (hair) chignon m inv

bunch /bʌntʃ/ n (of flowers, keys) mazzo m; (of bananas) casco m; (of people) gruppo m; ~ **of grapes** grappolo m d'uva

bundle /'bʌndl/ n fascio m; (of money) mazzetta f; **a** ~ **of nerves** fam un fascio di nervi ● vt ~ [**up**] affastellare

bung /bʌŋ/ vt fam (throw) buttare. **bung up** vt (block) otturare

bungalow /'bʌŋgələʊ/ n bungalow m inv

bungle /'bʌŋgl/ vt fare un pasticcio di

bunion /'bʌnjən/ n Med callo m all'alluce

bunk /bʌŋk/ n cuccetta f. ~**-beds** npl letti mpl a castello

bunny /'bʌnɪ/ n fam coniglietto m

buoy /bɔɪ/ n boa f

buoyan|cy /'bɔɪənsɪ/ n galleggiabilità f. ~**t** a (boat) galleggiante; (water) che aiuta a galleggiare

burden /'bɜːdn/ n carico m ● vt caricare. ~**some** /-səm/ a gravoso

bureau /'bjʊərəʊ/ n (pl -**x** /-əʊz/ ~**s**) (desk) scrivania f; (office) ufficio m

bureaucracy /bjʊə'rɒkrəsɪ/ n burocrazia f

bureaucrat /'bjʊərəkræt/ n burocrate mf. ~**ic** /-'krætɪk/ a burocratico

burger /'bɜːgə(r)/ n hamburger m inv

burglar /'bɜːglə(r)/ n svaligiatore, -trice mf. ~ **alarm** n antifurto m inv

burglar|ize /'bɜːgləraɪz/ vt Am svaligiare. ~**y** n furto m con scasso

burgle /'bɜːgl/ vt svaligiare

Burgundy /'bɜːgəndɪ/ n Borgogna f

burial /'berɪəl/ n sepoltura f. ~ **ground** n cimitero m

burlesque /bɜː'lesk/ n parodia f

burly /'bɜːlɪ/ a (-**ier, -iest**) corpulento

Burm|a /'bɜːmə/ n Birmania f. ~**ese** /-'miːz/ a & n birmano, -a mf

burn /bɜːn/ n bruciatura f ● v (pt/pp **burnt** or **burned**) ● vt bruciare ● vi bruciare. **burn down** vt/i bruciare. **burn out** vi fig esaurirsi. ~**er** n (on stove) bruciatore m

burnish /'bɜːnɪʃ/ vt lucidare

burnt /bɜːnt/ see **burn**

burp /bɜːp/ n fam rutto m ● vi fam ruttare

burrow /'bʌrəʊ/ n tana f ● vt scavare (hole)

bursar /'bɜːsə(r)/ n economo, -a mf. ~**y** n borsa f di studio

burst /bɜːst/ n (of gunfire, energy, laughter) scoppio m; (of speed) scatto m ● v (pt/pp **burst**) ● vt far scoppiare ● vi scoppiare; ~ **into tears** scoppiare in lacrime; **she** ~ **into the room** ha fatto irruzione nella stanza. **burst out** vi ~ **out laughing/crying** scoppiare a ridere/piangere

bury /'berɪ/ vt (pt/pp -ied) seppellire; (hide) nascondere

bus /bʌs/ n autobus m inv, pullman m inv; (long distance) pullman m inv, corriera f

bush /bʊʃ/ n cespuglio m; (land) boscaglia f. **~y** a (-ier, -iest) folto

busily /'bɪzɪlɪ/ adv con grande impegno

business /'bɪznɪs/ n affare m; Comm affari mpl; (establishment) attività f di commercio; **on ~** per affari; **he has no ~** to non ha alcun diritto di; **mind one's own ~** farsi gli affari propri; **that's none of your ~** non sono affari tuoi. **~-like** a efficiente. **~man** n uomo m d'affari. **~woman** n donna f d'affari

busker /'bʌskə(r)/ n suonatore, -trice mf ambulante

bus station n stazione f degli autobus

bus-stop n fermata f d'autobus

bust¹ /bʌst/ n busto m; (chest) petto m

bust² a fam rotto; **go ~** fallire ● v (pt/pp busted or bust) fam ● vt far scoppiare ● vi scoppiare

bustle /'bʌsl/ n (activity) trambusto m ● **bustle about** vi affannarsi. **~ing** a animato

bust-up n fam lite f

busy /'bɪzɪ/ a (-ier, -iest) occupato; ⟨day, time⟩ intenso; ⟨street⟩ affollato; (with traffic) pieno di traffico; **be ~ doing** essere occupato a fare ● vt ● **oneself** darsi da fare

busybody n ficcanaso mf inv

but /bʌt/, atono /bət/ conj ma ● prep eccetto, tranne; **nobody ~ you** nessuno tranne te; **~ for** (without) se non fosse stato per; **the last ~ one** il penultimo; **the next ~ one** il secondo ● adv (only) soltanto; **there were ~ two** ce n'erano soltanto due

butcher /'bʊtʃə(r)/ n macellaio m; **~'s [shop]** macelleria f ● vt macellare; fig massacrare

butler /'bʌtlə(r)/ n maggiordomo m

butt /bʌt/ n (of gun) calcio m; (of cigarette) mozzicone m; (for water) barile m; (fig: target) bersaglio m ● vt dare una testata a; ⟨goat:⟩ dare una cornata a. **butt in** vi interrompere

butter /'bʌtə(r)/ n burro m ● vt imburrare. **butter up** vt fam arruffianarsi

butter: **~cup** n ranuncolo m. **~fingers** nsg fam **be a ~fingers** avere le mani di pasta frolla. **~fly** n farfalla f

buttocks /'bʌtəks/ npl natiche fpl

button /'bʌtn/ n bottone m ● vt **[up]** abbottonare ● vi abbottonarsi. **~hole** n occhiello m, asola f

buttress /'bʌtrɪs/ n contrafforte m

buxom /'bʌksəm/ a formósa

buy /baɪ/ n **good/bad** ~ buon/cattivo acquisto m ● vt (pt/pp bought) comprare; ~ **sb a drink** pagare da bere a qcno; **I'll ~ this one** (drink) questo, lo offro io. **~er** n compratore, -trice mf

buzz /bʌz/ n ronzio m; **give sb a ~** fam (on phone) dare un colpo di telefono a qcno; (excite) mettere in fermento qcno ● vi ronzare ● vt ~ **sb** chiamare qcno col cicalino. **buzz off** vi fam levarsi di torno

buzzer /'bʌzə(r)/ n cicalino m

by /baɪ/ prep (near, next to) vicino a; (at the latest) per; **by Mozart** di Mozart; **he was run over by a bus** è stato investito da un autobus; **by oneself** da solo; **by the sea** al mare; **by sea** via mare; **by car/bus** in macchina/autobus; **by day/night** di giorno/notte; **by the hour/metre** a ore/metri; **six metres by four** sei metri per quattro; **he won by six metres** ha vinto di sei metri; **I missed the train by a minute** ho perso il treno per un minuto; **I'll be home by six** sarò a casa per le sei; **by this time next week** a quest'ora tra una settimana; **he rushed by me** mi è

passato accanto di corsa ● *adv*
she'll be here by and by sarà qui
fra poco; **by and large** in complesso
bye[-bye] /baɪ'baɪ/ *int fam* ciao
by: ~**-election** *n* elezione *f* straordinaria indetta per coprire una carica rimasta vacante in Parlamento.
~**gone** *a* passato. ~**-law** *n* legge *f*
locale. ~**pass** *n* circonvallazione
f; *Med* by-pass *m inv* ● *vt* evitare.
~**-product** *n* sottoprodotto *m*.
~**stander** *n* spettatore, -trice *mf*.
~**word** *n* **be a** ~**word for** essere
sinonimo di

Cc

cab /kæb/ *n* taxi *m inv*; (*of lorry*,
train) cabina *f*
cabaret /'kæbəreɪ/ *n* cabaret *m inv*
cabbage /'kæbɪdʒ/ *n* cavolo *m*
cabin /'kæbɪn/ *n* (*of plane, ship*) cabina *f*; (*hut*) capanna *f*
cabinet /'kæbɪnɪt/ *n* armadietto
m; (**display**) vetrina *f*; **C~** *Pol*
consiglio *m* dei ministri. ~**-maker**
n ebanista *mf*
cable /'keɪbl/ *n* cavo *m*. ~ '**railway**
n funicolare *f*. ~ '**television** *n* televisione *f* via cavo
cache /kæʃ/ *n* nascondiglio *m*; ~
of arms deposito *m* segreto di
armi
cackle /'kækl/ *vi* ridacchiare
cactus /'kæktəs/ *n* (*pl* **-ti** /-taɪ/ *or*
-tuses) cactus *m inv*
caddie /'kædɪ/ *n* portabastoni *m*
inv
caddy /'kædɪ/ *n* [**tea-**]~ barattolo
m del tè
cadet /kə'det/ *n* cadetto *m*
cadge /kædʒ/ *vt/i fam* scroccare
Caesarean /sɪ'zeərɪən/ *n* parto *m*
cesareo
café /'kæfeɪ/ *n* caffè *m inv*

cafeteria /kæfə'tɪərɪə/ *n* tavola *f*
calda
caffeine /'kæfi:n/ *n* caffeina *f*
cage /keɪdʒ/ *n* gabbia *f*
cagey /'keɪdʒɪ/ *a fam* riservato
(**about** su)
cajole /kə'dʒəʊl/ *vt* persuadere
con le lusinghe
cake /keɪk/ *n* torta *f*; (*small*) pasticcino *m*. ~**d** *a* incrostato (**with**
di)
calamity /kə'læmətɪ/ *n* calamità *f*
inv
calcium /'kælsɪəm/ *n* calcio *m*
calculate /'kælkjʊleɪt/ *vt* calcolare. ~**ing** *a fig* calcolatore. ~**ion**
/-'leɪʃn/ *n* calcolo *m*. ~**or** *n* calcolatrice *f*
calendar /'kælɪndə(r)/ *n* calendario *m*
calf[1] /kɑ:f/ *n* (*pl* **calves**) vitello *m*
calf[2] *n* (*pl* **calves**) *Anat* polpaccio *m*
calibre /'kælɪbə(r)/ *n* calibro *m*
call /kɔ:l/ *n* grido *m*; *Teleph* telefonata *f*; (*visit*) visita *f*; **be on** ~
(*doctor:*) essere di guardia ● *vt*
chiamare; indire (*strike*); **be** ~**ed**
chiamarsi ● *vi* chiamare; **[in** *or*
round] passare. **call back** *vt/i* richiamare. **call for** *vt* (*ask for*) chiedere; (*require*) richiedere; (*fetch*)
passare a prendere. **call off** *vt* richiamare (*dog*); disdire (*meeting*);
revocare (*strike*). **call on** *vt* chiamare; (*appeal to*) fare un appello
a; (*visit*) visitare. **call out** *vt* chiamare ad alta voce (*names*) ● *vi*
chiamare ad alta voce. **call together** *vt* riunire. **call up** *vt Mil*
chiamare alle armi; *Teleph* chiamare
call: ~**-box** *n* cabina *f* telefonica.
~**er** *n* visitatore, -trice *mf*; *Teleph*
persona *f* che telefona. ~**ing** *n* vocazione *f*
callous /'kæləs/ *a* insensibile
'**call-up** *n Mil* chiamata *f* alle armi
calm /kɑ:m/ *a* calmo *m* ● *n* calma
f. **calm down** *vt* calmare ● *vi* calmarsi. ~**ly** *adv* con calma

calorie /'kælərɪ/ n caloria f
calves /kɑːvz/ npl see **calf¹ & ²**
camber /'kæmbə(r)/ n curvatura f
Cambodia /kæm'bəʊdɪə/ n Cambogia f. **~n** a & n cambogiano, -a mf
camcorder /'kæmkɔːdə(r)/ n videocamera f
came /keɪm/ see **come**
camel /'kæml/ n cammello m
camera /'kæmərə/ n macchina f fotografica; TV telecamera f. **~man** n operatore m [televisivo], cameraman m inv
camouflage /'kæməflɑːʒ/ n mimetizzazione f ● vt mimetizzare
camp /kæmp/ n campeggio f; Mil campo m ● vi campeggiare; Mil accamparsi
campaign /kæm'peɪn/ n campagna f ● vi fare una campagna
camp: **~-bed** n letto m da campo. **~er** n campeggiatore, -trice mf; Auto camper m inv. **~ing** n campeggio m. **~site** n campeggio m
campus /'kæmpəs/ n (pl **-puses**) Univ città f universitaria, campus m inv
can¹ /kæn/ n (for petrol) latta f; (tin) scatola f; **~ of beer** lattina f di birra ● vt mettere in scatola
can² /kæn/, atono /kən/ v aux (pres **can**; pt **could**) (be able to) potere; (know how to) sapere; **I cannot** or **can't go** non posso andare; **he could not** or **couldn't go** non poteva andare; **she can't swim** non sa nuotare; **I ~ smell something burning** sento odor di bruciato
Canada /'kænədə/ n Canada m. **~ian** /kə'neɪdɪən/ a & n canadese mf
canal /kə'næl/ n canale m
Canaries /kə'neərɪz/ npl Canarie fpl
canary /kə'neərɪ/ n canarino m
cancel /'kænsl/ v (pt/pp **cancelled**) ● vt disdire (meeting, newspaper); revocare (contract, order); annullare (reservation, appointment, stamp). **~lation**

/-ə'leɪʃn/ n (of meeting, contract) revoca f; (in hotel, restaurant, for flight) cancellazione f
cancer /'kænsə(r)/ n cancro m; **C~** Astr Cancro m. **~ous** /-rəs/ a canceroso
candelabra /kændə'lɑːbrə/ n candelabro m
candid /'kændɪd/ a franco
candidate /'kændɪdət/ n candidato, -a mf
candle /'kændl/ n candela f. **~stick** n portacandele m inv
candour /'kændə(r)/ n franchezza f
candy /'kændɪ/ n Am caramella f; **a [piece of] ~** una caramella. **~floss** /-flɒs/ n zucchero m filato
cane /keɪn/ n (stick) bastone m; Sch bacchetta f ● vt prendere a bacchettate (pupil)
canine /'keɪnaɪn/ a canino. **~ tooth** n canino m
canister /'kænɪstə(r)/ n barattolo m (di metallo)
cannabis /'kænəbɪs/ n cannabis f
canned /kænd/ a in scatola; **~ music** fam musica f registrata
cannibal /'kænɪbl/ n cannibale mf. **~ism** n cannibalismo m
cannon /'kænən/ n inv cannone m. **~-ball** n palla f di cannone
cannot /'kænɒt/ see **can²**
canny /'kænɪ/ a astuto
canoe /kə'nuː/ n canoa f ● vi andare in canoa
'can-opener n apriscatole m inv
canopy /'kænəpɪ/ n baldacchino f; (of parachute) calotta f
can't /kɑːnt/ = **cannot** see **can²**
cantankerous /kæn'tæŋkərəs/ a stizzoso
canteen /kæn'tiːn/ n mensa f; **~ of cutlery** servizio m di posate
canter /'kæntə(r)/ vi andare a piccolo galoppo
canvas /'kænvəs/ n tela f; (painting) dipinto m su tela
canvass /'kænvəs/ vi Pol fare propaganda elettorale. **~ing** n sollecitazione f di voti

canyon /'kænjən/ *n* canyon *m inv*

cap /kæp/ *n* berretto *m*; (*nurse's*) cuffia *f*; (*top, lid*) tappo *m* ● *vt* (*pt/pp* **capped**) (*fig: do better than*) superare

capability /keɪpə'bɪlətɪ/ *n* capacità *f*

capab|le /'keɪpəbl/ *a* capace; (*skilful*) abile; **be ~e of doing sth** essere capace di fare qcsa. **~y** *adv* con abilità

capacity /kə'pæsətɪ/ *n* capacità *f*; (*function*) qualità *f*; **in my ~ as** in qualità di

cape[1] /keɪp/ *n* (*cloak*) cappa *f*

cape[2] *n Geog* capo *m*

caper[1] /'keɪpə(r)/ *vi* saltellare ● *n fam* birichinata *f*

caper[2] *n Culin* cappero *m*

capital /'kæpɪtl/ *n* (*town*) capitale *f*; (*money*) capitale *m*; (*letter*) lettera *f* maiuscola. **~ city** *n* capitale *f*

capital|ism /'kæpɪtalɪzm/ *n* capitalismo *m*. **~ist** /-ɪst/ *a* & *n* capitalista *mf*. **~ize** /-aɪz/ *vi* **~ize on** *fig* trarre vantaggio da. **~ 'letter** *n* lettera *f* maiuscola. **~ 'punishment** *n* pena *f* capitale

capitulat|e /kə'pɪtjʊleɪt/ *vi* capitolare. **~ion** /-'leɪʃn/ *n* capitolazione *f*

capricious /kə'prɪʃəs/ *a* capriccioso

Capricorn /'kæprɪkɔːn/ *n Astr* Capricorno *m*

capsize /kæp'saɪz/ *vi* capovolgersi ● *vt* capovolgere

capsule /'kæpsjuːl/ *n* capsula *f*

captain /'kæptɪn/ *n* capitano *m* ● *vt* comandare (*team*)

caption /'kæpʃn/ *n* intestazione *f*; (*of illustration*) didascalia *f*

captivate /'kæptɪveɪt/ *vt* incantare

captiv|e /'kæptɪv/ *a* prigioniero; **hold/take ~e** tenere/fare prigioniero ● *n* prigioniero, -a *mf*. **~ity** /-'tɪvətɪ/ *n* prigionia *f*; (*animals*) cattività *f*

capture /'kæptʃə(r)/ *n* cattura *f* ● *vt* catturare; attirare (*attention*)

car /kɑː(r)/ *n* macchina *f*; **by ~** in macchina

carafe /kə'ræf/ *n* caraffa *f*

caramel /'kærəmel/ *n* (*sweet*) caramella *f* al mou; *Culin* caramello *m*

carat /'kærət/ *n* carato *m*

caravan /'kærəvæn/ *n* roulotte *f inv*; (*horse-drawn*) carovana *f*

carbohydrate /kɑːbə'haɪdreɪt/ *n* carboidrato *m*

carbon /'kɑːbən/ *n* carbonio *m*

carbon: **~ copy** *n* copia *f* in carta carbone; (*fig: person*) ritratto *m*. **~ di'oxide** *n* anidride *f* carbonica. **~ paper** *n* carta *f* carbone

carburettor /kɑːbju'retə(r)/ *n* carburatore *m*

carcass /'kɑːkəs/ *n* carcassa *f*

card /kɑːd/ *n* (*for birthday, Christmas etc*) biglietto *m* di auguri; (*playing* ~) carta *f* [da gioco]; (*membership* ~) tessera *f*; (*business* ~) biglietto *m* da visita; (*credit* ~) carta *f* di credito; *Comput* scheda *f*

'cardboard *n* cartone *m*. **~ 'box** *n* scatola *f* di cartone; (*large*) scatolone *m*

'card-game *n* gioco *m* di carte

cardiac /'kɑːdɪæk/ *a* cardiaco

cardigan /'kɑːdɪgən/ *n* cardigan *m inv*

cardinal /'kɑːdɪnl/ *a* cardinale; **~ number** numero *m* cardinale ● *n Relig* cardinale *m*

card 'index *n* schedario *m*

care /keə(r)/ *n* cura *f*; (*caution*) attenzione *f*; (*worry*) preoccupazione *f*; **~ of** (*on letter abbr* **c/o**) presso; **take ~** (*be cautious*) fare attenzione; **bye, take ~** ciao, stammi bene; **take ~ of** occuparsi di; **be taken into ~** essere preso in custodia da un ente assistenziale ● *vi* **~ about** interessarsi di; **~ for** (*feel affection for*) volere bene a; (*look after*) aver cura di; **I don't ~ for chocolate** non mi piace il cioccolato; **I don't ~** non me ne importa; **who ~s?** chi se ne frega?

career /kə'rɪə(r)/ *n* carriera *f*;

(*profession*) professione *f* ● *vi* andare a tutta velocità

care: **~free** a spensierato. **~ful** a attento; ⟨*driver*⟩ prudente. **~fully** *adv* con attenzione. **~less** a irresponsabile; (*in work*) trascurato; ⟨*work*⟩ fatto con poca cura; ⟨*driver*⟩ distratto. **~lessly** *adv* negligentemente. **~lessness** n trascuratezza *f*. **~r** n persona *f* che accudisce a un anziano o a un malato

caress /kəˈres/ n carezza *f* ● *vt* accarezzare

'caretaker n custode *mf*; (*in school*) bidello m

'car ferry n traghetto m (*per il trasporto di auto*)

cargo /ˈkɑːgəʊ/ n (*pl* **-es**) carico m

Caribbean /kærɪˈbiːən/ n **the ~** (*sea*) il Mar dei Caraibi ● a caraibico

caricature /ˈkærɪkətjʊə(r)/ n caricatura *f*

caring /ˈkeərɪŋ/ a ⟨*parent*⟩ premuroso; ⟨*attitude*⟩ altruista; **the ~ professions** le attività assistenziali

carnage /ˈkɑːnɪdʒ/ n carneficina *f*

carnal /ˈkɑːnl/ a carnale

carnation /kɑːˈneɪʃn/ n garofano m

carnival /ˈkɑːnɪvl/ n carnevale m

carnivorous /kɑːˈnɪvərəs/ a carnivoro

carol /ˈkærəl/ n [**Christmas**] **~** canzone *f* natalizia

carp[1] /kɑːp/ n *inv* carpa *f*

carp[2] *vi* **~ at** trovare da ridire su

'car park n parcheggio m

carpent|er /ˈkɑːpɪntə(r)/ n falegname m. **~ry** n falegnameria *f*

carpet /ˈkɑːpɪt/ n tappeto m; (*wall-to-wall*) moquette *f inv* ● *vt* mettere la moquette in ⟨*room*⟩

'car phone n telefono m in macchina

carriage /ˈkærɪdʒ/ n carrozza *f*; (*of goods*) trasporto m; (*cost*) spese *fpl* di trasporto; (*bearing*) portamento m; **~way** n strada *f* carrozzabile; **north-bound ~way** carreggiata *f* nord

carrier /ˈkærɪə(r)/ n (*company*) impresa *f* di trasporti; *Aeron* compagnia *f* di trasporto aereo; (*of disease*) portatore m. **~** [**bag**] n borsa *f* [per la spesa]

carrot /ˈkærət/ n carota *f*

carry /ˈkærɪ/ *v* (*pt/pp* **-ied**) ● *vt* portare; (*transport*) trasportare; **get carried away** *fam* lasciarsi prender la mano ● *vi* ⟨*sound*⟩ trasmettersi. **carry off** *vt* portare via; vincere ⟨*prize*⟩. **carry on** *vi* continuare; (*fam: make scene*) fare delle storie; **~ on with sth** continuare qcsa; **~ on with sb** *fam* intendersela con qcno ● *vt* mantenere ⟨*business*⟩. **carry out** *vt* portare fuori; eseguire ⟨*instructions, task*⟩; mettere in atto ⟨*threat*⟩; effettuare ⟨*experiment, survey*⟩

'carry-cot n porte-enfant m *inv*

cart /kɑːt/ n carretto m ● *vt* (*fam: carry*) portare

cartilage /ˈkɑːtɪlɪdʒ/ n *Anat* cartilagine *f*

carton /ˈkɑːtn/ n scatola *f* di cartone; (*for drink*) cartone m; (*of cream, yoghurt*) vasetto m; (*of cigarettes*) stecca *f*

cartoon /kɑːˈtuːn/ n vignetta *f*; (*strip*) vignette *fpl*; (*film*) cartone m animato; (*in art*) bozzetto m. **~ist** n vignettista *mf*; (*for films*) disegnatore, -trice *mf* di cartoni animati

cartridge /ˈkɑːtrɪdʒ/ n cartuccia *f*; (*for film*) bobina *f*; (*of record player*) testina *f*

carve /kɑːv/ *vt* scolpire; tagliare ⟨*meat*⟩

carving /ˈkɑːvɪŋ/ n scultura *f*. **~-knife** n trinciante m

'car wash n autolavaggio m *inv*

case[1] /keɪs/ n caso m; **in any ~** in ogni caso; **in that ~** in questo caso; **just in ~** per sicurezza; **in ~ he comes** nel caso in cui venisse

case[2] n (*container*) scatola *f*;

(*crate*) cassa *f*; (*for spectacles*) astuccio *m*; (*suitcase*) valigia *f*; (*for display*) vetrina *f*

cash /kæʃ/ *n* denaro *m* contante; (*fam: money*) contanti *mpl*; **pay [in]** ~ pagare in contanti; ~ **on delivery** pagamento alla consegna ●*vt* incassare (*cheque*). ~ **desk** *n* cassa *f*

cashier /kæˈʃɪə(r)/ *n* cassiere, -a *mf*

'cash register *n* registratore *m* di cassa

casino /kəˈsiːnəʊ/ *n* casinò *m inv*

casket /ˈkɑːskɪt/ *n* scrigno *m*; (*Am: coffin*) bara *f*

casserole /ˈkæsərəʊl/ *n* casseruola *f*; (*stew*) stufato *m*

cassette /kəˈset/ *n* cassetta *f*. ~ **recorder** *n* registratore *n* (a cassette)

cast /kɑːst/ *n* (*mould*) forma *f*; *Theat* cast *m inv*; [*plaster*] ~ *Med* ingessatura *f* ●*vt* (*pt/pp* **cast**) dare (*vote*); *Theat* assegnare le parti di (*play*); fondere (*metal*); (*throw*) gettare; ~ **an actor as** dare ad un attore il ruolo di; ~ **a glance at** lanciare uno sguardo a. **cast off** *vi Naut* sganciare gli ormeggi ●*vt* (*in knitting*) diminuire. **cast on** *vt* (*in knitting*) avviare

castaway /ˈkɑːstəweɪ/ *n* naufrago, -a *mf*

caste /kɑːst/ *n* casta *f*

caster /ˈkɑːstə(r)/ *n* (*wheel*) rotella *f*. ~ **sugar** *n* zucchero *m* raffinato

cast 'iron *n* ghisa *f*

cast-'iron *a* di ghisa; *fig* solido

castle /ˈkɑːsl/ *n* castello *m*; (*in chess*) torre *f*

'cast-offs *npl* abiti *mpl* smessi

castor /ˈkɑːstə(r)/ *n* (*wheel*) rotella *f*. ~ **oil** *n* olio *m* di ricino. ~ **sugar** *n* zucchero *m* raffinato

castrat|e /kæˈstreɪt/ *vt* castrare. ~**ion** /-eɪʃn/ *n* castrazione *f*

casual /ˈkæʒʊəl/ *a* (*chance*) casuale; (*remark*) senza importanza; (*glance*) di sfuggita; (*attitude, approach*) disinvolto; (*chat*) infor-

male; (*clothes*) casual *inv*; (*work*) saltuario; ~ **wear** abbigliamento *m* casual. ~**ly** *adv* (*dress*) casual; (*meet*) casualmente

casualty /ˈkæʒʊəltɪ/ *n* (*injured person*) ferito *m*; (*killed*) vittima *f*. ~ [**department**] *n* pronto soccorso *m*

cat /kæt/ *n* gatto *m*; *pej* arpia *f*

catalogue /ˈkætəlɒg/ *n* catalogo *m* ●*vt* catalogare

catalyst /ˈkætəlɪst/ *n* *Chem & fig* catalizzatore *m*

catalytic /kætəˈlɪtɪk/ *a* ~ **converter** *Auto* marmitta *f* catalitica

catapult /ˈkætəpʌlt/ *n* catapulta *f*; (*child's*) fionda *f* ●*vt* *fig* catapultare

cataract /ˈkætərækt/ *n* *Med* cataratta *f*

catarrh /kəˈtɑː(r)/ *n* catarro *m*

catastroph|e /kəˈtæstrəfɪ/ *n* catastrofe *f*. ~**ic** /kætəˈstrɒfɪk/ *a* catastrofico

catch /kætʃ/ *n* (*of fish*) pesca *f*; (*fastener*) fermaglio *m*; (*on door*) fermo *m*; (*on window*) gancio *m*; (*fam: snag*) tranello *m* ●*v* (*pt/pp* **caught**) ●*vt* acchiappare (*ball*); (*grab*) afferrare; prendere (*illness, fugitive, train*); ~ **a cold** prendersi un raffreddore; ~ **sight of** scorgere; **I caught him stealing** l'ho sorpreso mentre rubava; ~ **one's finger in the door** chiudersi il dito nella porta; ~ **sb's eye** *or* **attention** attirare l'attenzione di qcno ●*vi* (*fire*) prendere; (*get stuck*) impigliarsi. **catch on** *vi fam* (*understand*) afferrare; (*become popular*) diventare popolare. **catch up** *vt* raggiungere ●*vi* recuperare; (*runner:*) riguadagnare terreno; ~ **up with** raggiungere (*sb*); mettersi in pari con (*work*)

catching /ˈkætʃɪŋ/ *a* contagioso

catch: ~**-phrase** *n* tormentone *m*. ~**word** *n* slogan *m inv*

catchy /ˈkætʃɪ/ *a* (**-ier, -iest**) orecchiabile

categor|ical /ˌkætɪˈgɒrɪkl/ a categorico. **~y** /ˈkætɪgəri/ n categoria f

cater /ˈkeɪtə(r)/ vi **~ for** provvedere a (needs); fig venire incontro alle esigenze di. **~ing** n (trade) ristorazione f; (food) rinfresco m

caterpillar /ˈkætəpɪlə(r)/ n bruco m

cathedral /kəˈθiːdrl/ n cattedrale f

Catholic /ˈkæθəlɪk/ a & n cattolico, -a mf. **~ism** /kəˈθɒlɪsɪzm/ n cattolicesimo m

cat's eyes npl catarifrangente msg (inserito nell'asfalto)

cattle /ˈkætl/ npl bestiame msg

catty /ˈkætɪ/ a (-ier, -iest) dispettoso

catwalk /ˈkætwɔːk/ n passerella f

caught /kɔːt/ see catch

cauliflower /ˈkɒlɪ-/ n cavolfiore m

cause /kɔːz/ n causa f ● vt causare; **~ sb to do sth** far fare qcsa a qcno

'causeway n strada f sopraelevata

caustic /ˈkɔːstɪk/ a caustico

caution /ˈkɔːʃn/ n cautela f; (warning) ammonizione f ● vt mettere in guardia; Jur ammonire

cautious /ˈkɔːʃəs/ a cauto

cavalry /ˈkævəlrɪ/ n cavalleria f

cave /keɪv/ n caverna f ● **cave in** vi (roof:) crollare; (fig: give in) capitolare

cavern /ˈkævən/ n caverna f

caviare /ˈkævɪɑː(r)/ n caviale m

caving /ˈkeɪvɪŋ/ n speleologia f

cavity /ˈkævətɪ/ n cavità f inv; (in tooth) carie f inv

cavort /kəˈvɔːt/ vi saltellare

CD n CD m inv. **~ player** n lettore m [di] compact

CD-Rom /siːdiːˈrɒm/ n CD-Rom m inv. **~ drive** n lettore m [di] CD-Rom

cease /siːs/ n without **~** incessantemente ● vt/i cessare. **~-fire** n cessate il fuoco m inv. **~less** a incessante

cedar /ˈsiːdə(r)/ n cedro m

cede /siːd/ vt cedere

ceiling /ˈsiːlɪŋ/ n soffitto m; fig tetto m [massimo]

celebrat|e /ˈselɪbreɪt/ vt festeggiare (birthday, victory) ● vi far festa. **~ed** a celebre (for per). **~ion** /-ˈbreɪʃn/ n celebrazione f

celebrity /sɪˈlebrətɪ/ n celebrità f inv

celery /ˈselərɪ/ n sedano m

celiba|cy /ˈselɪbəsɪ/ n celibato m. **~te** a (man) celibe; (woman) nubile

cell /sel/ n cella f; Biol cellula f

cellar /ˈselə(r)/ n scantinato m; (for wine) cantina f

cellist /ˈtʃelɪst/ n violoncellista mf

cello /ˈtʃeləʊ/ n violoncello m

Cellophane® /ˈseləfeɪn/ n cellofan m inv

cellular phone /seljʊləˈfəʊn/ n [telefono m] cellulare m

celluloid /ˈseljʊlɔɪd/ n celluloide f

Celsius /ˈselsɪəs/ a Celsius

Celt /kelt/ n celta mf. **~ic** a celtico

cement /sɪˈment/ n cemento m; (adhesive) mastice m ● vt cementare; fig consolidare

cemetery /ˈsemətrɪ/ n cimitero m

censor /ˈsensə(r)/ n censore m ● vt censurare. **~ship** n censura f

censure /ˈsenʃə(r)/ vt biasimare

census /ˈsensəs/ n censimento m

cent /sent/ n (coin) centesimo m

centenary /senˈtiːnərɪ/ n, Am **centennial** /senˈtenɪəl/ n centenario m

center /ˈsentə(r)/ n Am = centre

centi|grade /ˈsentɪ-/ a centigrado. **~metre** n centimetro m. **~pede** /-piːd/ n centopiedi m inv

central /ˈsentrl/ a centrale. **~ heating** n riscaldamento m autonomo. **~ize** vt centralizzare. **~ly** adv al centro; **~ly heated** con riscaldamento autonomo. **~ reservation** n Auto banchina f spartitraffico

centre /ˈsentə(r)/ n centro m ● v (pt/pp centred) ● vt centrare ● vi **~ on** fig incentrarsi su. **~-'forward** n centravanti m inv

centrifugal | channel

centrifugal /sentrɪ'fjuːgl/ a ~ **force** forza f centrifuga

century /'sentʃərɪ/ n secolo m

ceramic /sɪ'ræmɪk/ a ceramico. **~s** n (art) ceramica fsg; (objects) ceramiche fpl

cereal /'sɪərɪəl/ n cereale m

cerebral /'serɪbrl/ a cerebrale

ceremon|ial /serɪ'məʊnɪəl/ a da cerimonia ● n cerimoniale m. **~ious** /-ɪəs/ a cerimonioso

ceremony /'serɪmənɪ/ n cerimonia f

certain /'sɜːtn/ a certo; **for ~** di sicuro; **make ~** accertarsi ; **he is ~ to win** è certo di vincere; **it's not ~ whether he'll come** non è sicuro che venga. **~ly** adv certamente; **~ly not!** no di certo! **~ty** /-tɪ/ n certezza f; **it's a ~ty** è una cosa certa

certificate /sə'tɪfɪkət/ n certificato m

certify /'sɜːtɪfaɪ/ vt (pt/pp -ied) certificare; (declare insane) dichiarare malato di mente

cessation /se'seɪʃn/ n cessazione f

cesspool /'ses-/ n pozzo m nero

cf abbr (compare) cf, cfr

chafe /tʃeɪf/ vt irritare

chain /tʃeɪn/ n catena f ● vt incatenare (prisoner); attaccare con la catena (dog) (to a). **chain up** vt legare alla catena (dog)

chain: **~ re'action** n reazione f a catena. **~-smoke** vi fumare una sigaretta dopo l'altra. **~-smoker** n fumatore, -trice mf accanito, -a. **~ store** n negozio m appartenente a una catena

chair /tʃeə(r)/ n sedia f; Univ cattedra f ● vt presiedere. **~-lift** n seggiovia f. **~man** n presidente m

chalet /'ʃæleɪ/ n chalet m inv; (in holiday camp) bungalow m inv

chalice /'tʃælɪs/ n Relig calice m

chalk /tʃɔːk/ n gesso m. **~y** a gessoso

challeng|e /'tʃælɪndʒ/ n sfida f; Mil intimazione f ● vt sfidare; Mil intimare il chi va là a; fig mettere in dubbio (statement). **~er** n sfi-

dante mf. **~ing** a (job) impegnativo

chamber /'tʃeɪmbə(r)/ n C~ **of Commerce** camera f di commercio

chamber: **~maid** n cameriera f [d'albergo]. **~music** n musica f da camera

chamois¹ /'ʃæmwɑː/ n inv (animal) camoscio m

chamois² /'ʃæmɪ/ n **[-leather]** [pelle f di] camoscio m

champagne /ʃæm'peɪn/ n champagne m inv

champion /'tʃæmpɪən/ n Sport campione m; (of cause) difensore, difenditrice mf ● vt (defend) difendere; (fight for) lottare per. **~ship** n Sport campionato m

chance /tʃɑːns/ n caso m; (possibility) possibilità f inv; (opportunity) occasione f; **by ~** per caso; **take a ~** provarci; **give sb a second ~** dare un'altra possibilità a qcno ● attrib fortuito ● vt **I'll ~ it** fam corro il rischio

chancellor /'tʃɑːnsələ(r)/ n cancelliere m; Univ rettore m; **C~ of the Exchequer** ≈ ministro m del tesoro

chancy /'tʃɑːnsɪ/ a rischioso

chandelier /ʃændə'lɪə(r)/ n lampadario m

change /tʃeɪndʒ/ n cambiamento m; (money) resto m; (small coins) spiccioli mpl; for a ~ tanto per cambiare; **a ~ of clothes** un cambio di vestiti; **the ~ [of life]** la menopausa ● vt cambiare; (substitute) scambiare (for con); **~ one's clothes** cambiarsi [i vestiti]; **~ trains** cambiare treno ● vi cambiare; (~ clothes) cambiarsi; **all ~!** stazione terminale!

changeable /'tʃeɪndʒəbl/ a mutevole; (weather) variabile

'changing-room n camerino m; (for sports) spogliatoio m

channel /'tʃænl/ n canale m; **the [English] C~** la Manica; **the C~ Islands** le Isole del Canale ● vt (*pt/pp* **channelled**) ~ **one's energies into sth** convogliare le proprie energie in qcsa

chant /tʃɑ:nt/ *n* cantilena *f*; (*of demonstrators*) slogan *m inv* di protesta ● *vt* cantare; (*demonstrators:*) gridare

chaos /'keɪɒs/ *n* caos *m*. ~**tic** /-'ɒtɪk/ *a* caotico

chap /tʃæp/ *n fam* tipo *m*

chapel /'tʃæpl/ *n* cappella *f*

chaperon /'ʃæpərəʊn/ *n* chaperon *f inv* ● *vt* fare da chaperon a (*sb*)

chaplain /'tʃæplɪn/ *n* cappellano *m*

chapped /tʃæpt/ *a* (*skin, lips*) screpolato

chapter /'tʃæptə(r)/ *n* capitolo *m*

char¹ /tʃɑ:(r)/ *n fam* donna *f* delle pulizie

char² *vt* (*pt/pp* **charred**) (*burn*) carbonizzare

character /'kærɪktə(r)/ *n* carattere *m*; (*in novel, play*) personaggio *m*; **quite a** ~ *fam* un tipo particolare

characteristic /kærɪktə'rɪstɪk/ *a* caratteristico ● *n* caratteristica *f*. ~**ally** *adv* tipicamente

characterize /'kærɪktəraɪz/ *vt* caratterizzare

charade /ʃə'rɑ:d/ *n* farsa *f*

charcoal /'tʃɑ:-/ *n* carbonella *f*

charge /tʃɑ:dʒ/ *n* (*cost*) prezzo *m*; *Electr, Mil* carica *f*; *Jur* accusa *f*; **free of** ~ gratuito; **be in** ~ essere responsabile (**of** di); **take** ~ assumersi la responsabilità; **take** ~ **of** occuparsi di ● *vt* far pagare (*fee*); far pagare a (*person*); *Electr, Mil* caricare; *Jur* accusare (**with** di); ~ **sb for sth** far pagare qcsa a qcno; ~ **it to my account** lo addebiti sul mio conto ● *vi* (*attack*) caricare

chariot /'tʃærɪət/ *n* cocchio *m*

charisma /kə'rɪzmə/ *n* carisma *m*. ~**tic** /kærɪz'mætɪk/ *a* carismatico

charitable /'tʃærɪtəbl/ *a* caritatevole; (*kind*) indulgente

charity /'tʃærətɪ/ *n* carità *f*; (*organization*) associazione *f* di bene-

ficenza; **concert given for** ~ concerto *m* di beneficenza; **live on** ~ vivere di elemosina

charm /tʃɑ:m/ *n* fascino *m*; (*object*) ciondolo *m* ● *vt* affascinare. ~**ing** *a* affascinante

chart /tʃɑ:t/ *n* carta *f* nautica; (*table*) tabella *f*

charter /'tʃɑ:tə(r)/ *n* ~ **[flight]** [volo *m*] charter *m inv* ● *vt* noleggiare. ~**ed accountant** *n* commercialista *mf*

charwoman /'tʃɑ:-/ *n* donna *f* delle pulizie

chase /tʃeɪs/ *n* inseguimento *m* ● *vt* inseguire. **chase away** *or* **off** *vt* cacciare via

chasm /'kæz(ə)m/ *n* abisso *m*

chassis /'ʃæsɪ/ *n* (*pl* **chassis** /-sɪz/) telaio *m*

chaste /tʃeɪst/ *a* casto

chastity /'tʃæstɪtɪ/ *n* castità *f*

chat /tʃæt/ *n* chiacchierata *f*; **have a** ~ with fare quattro chiacchiere con ● *vi* (*pt/pp* **chatted**) chiacchierare. ~ **show** *n* talk show *m inv*

chatter /'tʃætə(r)/ *n* chiacchiere *fpl* ● *vi* chiacchierare; (*teeth:*) battere. ~**box** *n fam* chiacchierone, -a *mf*

chatty /'tʃætɪ/ *a* (**-ier, -iest**) chiacchierone; (*style*) familiare

chauffeur /'ʃəʊfə(r)/ *n* autista *mf*

chauvin|ism /'ʃəʊvɪnɪzm/ *n* sciovinismo *m*. ~**ist** *n* sciovinista *mf*. **male** ~**ist** *n fam* maschilista *m*

cheap /tʃi:p/ *a* a buon mercato; (*rate*) economico; (*vulgar*) grossolano; (*of poor quality*) scadente ● *adv* a buon mercato. ~**ly** *adv* a buon mercato

cheat /tʃi:t/ *n* imbroglione, -a *mf*; (*at cards*) baro *m* ● *vt* imbrogliare; ~ **sb out of sth** sottrarre qcsa a qcno con l'inganno ● *vi* imbrogliare; (*at cards*) barare. **cheat on** *vt fam* tradire (*wife*)

check¹ /tʃek/ *a* (*pattern*) a quadri ● *n* disegno *m* a quadri

check² n verifica f; (of tickets) controllo m; (in chess) scacco m; (Am: bill) conto m; (Am: cheque) assegno m; (Am: tick) segnetto m; keep a ~ on controllare; keep in ~ tenere sotto controllo ● vt verificare; controllare (tickets); (restrain) contenere; (stop) bloccare ● vi controllare; ~ on sth controllare qcsa. check in vi registrarsi all'arrivo (in albergo); Aeron fare il check-in ● vt registrare all'arrivo (in albergo). check out vi (of hotel) saldare il conto ● vt (fam: investigate) controllare. check up vi accertarsi; ~ up on prendere informazioni su

checked /ed/ a quadri. ~ers n Am dama f

check: ~-in n (in airport: place) banco m accettazione, check-in m inv; ~-in time check-in m inv. ~mark n Am segnetto m. ~mate int scacco matto! ~-out n (in supermarket) cassa f. ~room n Am deposito m bagagli. ~-up n Med visita f di controllo, check-up m inv

cheek /tʃi:k/ n guancia f; (impudence) sfacciataggine f. ~y a sfacciato

cheep /tʃi:p/ vi pigolare

cheer /tʃɪə(r)/ n evviva m inv; three ~s tre urrà; ~s! salute!; (goodbye) arrivederci!; (thanks) grazie! ● vt/i acclamare. cheer up vt tirare su [di morale] ● vi tirarsi su [di morale]; ~ up! su con la vita!. ~ful a allegro. ~fulness n allegria f. ~ing n acclamazione f

cheerio /tʃɪərɪ'əʊ/ int fam arrivederci

'cheerless a triste, tetro

cheese /tʃi:z/ n formaggio m. ~cake n dolce m al formaggio

chef /ʃef/ n cuoco, -a mf, chef mf inv

chemical /'kemɪkl/ a chimico ● n prodotto m chimico

chemist /'kemɪst/ n (pharmacist) farmacista mf; (scientist) chimico,

-a mf; ~'s [shop] farmacia f. ~ry n chimica f

cheque /tʃek/ n assegno m. ~-book n libretto m degli assegni. ~ card n carta f assegni

cherish /'tʃerɪʃ/ vt curare teneramente; (love) avere caro; nutrire 〈hope〉

cherry /'tʃerɪ/ n ciliegia f; (tree) ciliegio m

cherub /'tʃerəb/ n cherubino m

chess /tʃes/ n scacchi mpl

chess: ~board n scacchiera f. ~-man n pezzo m degli scacchi. ~player n scacchista mf

chest /tʃest/ n petto m; (box) cassapanca f

chestnut /'tʃesnʌt/ n castagna f; (tree) castagno m

chest of 'drawers n cassettone m

chew /tʃu:/ vt masticare. ~ing-gum n gomma f da masticare

chic /ʃi:k/ a chic inv

chick /tʃɪk/ n pulcino m; (fam: girl) ragazza f

chicken /'tʃɪkɪn/ n pollo m ● attrib 〈soup, casserole〉 di pollo ● a fam fifone ● chicken out vi fam he ~ed out gli è venuta fifa. ~pox n varicella f

chicory /'tʃɪkərɪ/ n cicoria f

chief /tʃi:f/ a principale ● n capo m. ~ly adv principalmente

chilblain /'tʃɪlbleɪn/ n gelone m

child /tʃaɪld/ n (pl ~ren) bambino, -a mf; (son/daughter) figlio, -a mf

child: ~birth n parto m. ~hood n infanzia f. ~ish a infantile. ~ishness n puerilità f. ~less a senza figli. ~like a ingenuo. ~minder n baby-sitter mf inv

children /'tʃɪldrən/ see child

Chile /'tʃɪlɪ/ n Cile m. ~an a & n cileno, -a mf

chill /tʃɪl/ n freddo m; (illness) infreddatura f ● vt raffreddare

chilli /'tʃɪlɪ/ n (pl ~es) [pepper] peperoncino m

chilly /'tʃɪlɪ/ a freddo

chime /tʃaɪm/ vi suonare

chimney /'tʃɪmnɪ/ n camino m.

~-pot n comignolo m. **~-sweep** n spazzacamino m

chimpanzee /tʃɪmpæn'zi:/ n scimpanzé m inv

chin /tʃɪn/ n mento m

china /'tʃaɪnə/ n porcellana f

Chin|a n Cina f. **~-ese** /-'ni:z/ a & n cinese mf; (language) cinese m; **the ~ese** pl i cinesi

chink[1] /tʃɪŋk/ n (slit) fessura f

chink[2] n (noise) tintinnio m

chip /tʃɪp/ n (fragment) scheggia f; (in china, paintwork) scheggiatura f; Comput chip m inv; (in gambling) fiche f inv. **~s** pl Br Culin patatine fpl fritte; Am Culin patatine fpl ●vt (pt/pp chipped) (damage) scheggiare. **chip in** vi fam intromettersi; (with money) contribuire. **~ped** a (damaged) scheggiato

chiropod|ist /kɪ'rɒpədɪst/ n podiatra mf inv. **~y** n podiatria f

chirp /tʃɜ:p/ vi cinguettare; (cricket) fare cri cri. **~y** a fam pimpante

chisel /'tʃɪzl/ n scalpello m

chival|rous /'ʃɪvlrəs/ a cavalleresco. **~ry** n cavalleria f

chives /tʃaɪvz/ npl erba f cipollina

chlorine /'klɔ:ri:n/ n cloro m

chloroform /'klɒrəfɔ:m/ n cloroformio m

chock-a-block /tʃɒkə'blɒk/, **chock-full** /tʃɒk'ful/ a pieno zeppo

chocolate /'tʃɒkələt/ n cioccolato m; (drink) cioccolata f; **a ~** un cioccolatino

choice /tʃɔɪs/ n scelta f ●a scelto

choir /'kwaɪə(r)/ n coro m. **~boy** n corista m

choke /tʃəʊk/ n Auto aria f ●vt/i soffocare

cholera /'kɒlərə/ n colera m

cholesterol /kə'lestərɒl/ n colesterolo m

choose /tʃu:z/ vt/i (pt chose, pp chosen) scegliere; **as you ~** come vuoi

choos[e]y /'tʃu:zɪ/ a fam difficile

chop /tʃɒp/ n (blow) colpo m (d'ascia); Culin costata f ●vt (pt/pp chopped) tagliare. **chop down** vt abbattere (tree). **chop off** vt spaccare

chop|per /'tʃɒpə(r)/ n accetta f; fam elicottero m. **~py** n increspato

'chopsticks npl bastoncini mpl cinesi

choral /'kɔ:rəl/ a corale

chord /kɔ:d/ n Mus corda f

chore /tʃɔ:(r)/ n corvé f inv; [household] **~s** faccende fpl domestiche

choreograph|er /kɒrɪ'ɒgrəfə(r)/ n coreografo, -a mf. **~y** /-ɪ/ n coreografia f

chortle /'tʃɔ:tl/ vi ridacchiare

chorus /'kɔ:rəs/ n coro m; (of song) ritornello m

chose, chosen /tʃəʊz, 'tʃəʊzn/ see **choose**

Christ /kraɪst/ n Cristo m

christen /'krɪsn/ vt battezzare. **~ing** n battesimo m

Christian /'krɪstʃən/ a & n cristiano, -a mf. **~ity** /-stɪ'ænɪtɪ/ n cristianesimo m. **~ name** n nome m di battesimo

Christmas /'krɪsməs/ n Natale m ●attrib di Natale. **'~ card** n biglietto m d'auguri di Natale. **'Day** n il giorno di Natale. **~ Eve** n la vigilia di Natale. **'~ present** n regalo m di Natale. **~ 'pudding** dolce m natalizio a base di frutta candita e liquore. **'~ tree** n albero m di Natale

chrome /krəʊm/ n, **chromium** /'krəʊmɪəm/ n cromo m

chromosome /'krəʊməsəʊm/ n cromosoma m

chronic /'krɒnɪk/ a cronico

chronicle /'krɒnɪkl/ n cronaca f

chronological /krɒnə'lɒdʒɪkl/ a cronologico. **~ly** adv (ordered) in ordine cronologico

chrysanthemum /krɪ'sænθəməm/ n crisantemo m

chubby /'tʃʌbɪ/ a (-ier, -iest) paffuto

chuck /tʃʌk/ vt fam buttare. **chuck out** vt fam buttare via ‹object›; buttare fuori ‹person›

chuckle /'tʃʌkl/ vi ridacchiare

chug /tʃʌg/ vi (pt/pp **chugged**) **the train ~ged out of the station** il treno è uscito dalla stazione sbuffando

chum /tʃʌm/ n amico, -a mf. **~my** a fam **be ~my with** essere amico di

chunk /tʃʌŋk/ n grosso pezzo m

church /tʃɜːtʃ/ n chiesa f. **~yard** n cimitero m

churlish /'tʃɜːlɪʃ/ a sgarbato

churn /tʃɜːn/ vt churn out sfornare

chute /ʃuːt/ n scivolo m; (for rubbish) canale m di scarico

CID n abbr (**Criminal Investigation Department**) polizia f giudiziaria

cider /'saɪdə(r)/ n sidro m

cigar /sɪ'gɑː(r)/ n sigaro m

cigarette /sɪgə'ret/ n sigaretta f

cine-camera /'sɪnɪ-/ n cinepresa f

cinema /'sɪnɪmə/ n cinema m inv

cinnamon /'sɪnəmən/ n cannella f

circle /'sɜːkl/ n cerchio m; Theat galleria f; **in a ~** in cerchio ● vt girare intorno a; cerchiare (mistake) ● vi descrivere dei cerchi

circuit /'sɜːkɪt/ n circuito m; (lap) giro m; **~ board** n circuito m stampato. **~ous** /sə'kjuːɪtəs/ a **~ous route** percorso m lungo e indiretto

circular /'sɜːkjʊlə(r)/ a circolare ● n circolare f

circulat|e /'sɜːkjʊleɪt/ vt far circolare ● vi circolare. **~ion** /-'leɪʃn/ n circolazione f; (of newspaper) tiratura f

circumcis|e /'sɜːkəmsaɪz/ vt circoncidere. **~ion** /-'sɪʒn/ n circoncisione f

circumference /sə'kʌmfərəns/ n conferenza f

circumstance /'sɜːkəmstəns/ n

circostanza f; **~s** pl (financial) condizioni fpl finanziarie

circus /'sɜːkəs/ n circo m

CIS n abbr (**Commonwealth of Independent States**) CSI f

cistern /'sɪstən/ n (tank) cisterna f; (of WC) serbatoio m

cite /saɪt/ vt citare

citizen /'sɪtɪzn/ n cittadino, -a mf; (of town) abitante mf. **~ship** n cittadinanza f

citrus /'sɪtrəs/ n **~ [fruit]** agrume m

city /'sɪtɪ/ n città f inv; **the C~** la City (di Londra)

civic /'sɪvɪk/ a civico

civil /'sɪvl/ a civile

civilian /sɪ'vɪljən/ a civile; **in ~ clothes** in borghese ● n civile mf

civiliz|ation /sɪvɪlaɪ'zeɪʃn/ n civiltà f inv. **~e** /'sɪvɪlaɪz/ vt civilizzare

civil: ~ 'servant n impiegato, -a mf statale. **C~ 'Service** n pubblica amministrazione f

clad /klæd/ a vestito (in di)

claim /kleɪm/ n richiesta f; (right) diritto m; (assertion) dichiarazione f; **lay ~ to sth** rivendicare qcsa ● vt richiedere; reclamare (lost property); rivendicare (ownership); **~ that** sostenere che. **~ant** n richiedente mf

clairvoyant /kleə'vɔɪənt/ n chiaroveggente mf

clam /klæm/ n Culin vongola f ● **clam up** vi (pt/pp **clammed**) zittirsi

clamber /'klæmbə(r)/ vi arrampicarsi

clammy /'klæmɪ/ a (-ier, -iest) appiccicaticcio

clamour /'klæmə(r)/ n (protest) rimostranza f ● vi **~ for** chiedere a gran voce

clamp /klæmp/ n morsa f ● vt ammorsare; Auto mettere i ceppi bloccaruote a. **clamp down** vi fam essere duro; **~ down on** reprimere

clan /klæn/ n clan m inv

clandestine /klæn'destɪn/ a clandestino

clang /klæŋ/ n suono m metallico. **~er** n fam gaffe f inv

clank /klæŋk/ n rumore m metallico

clap /klæp/ n give sb a ~ applaudire qcno; ~ of thunder tuono m ●vt/i (pt/pp clapped) applaudire; ~ one's hands applaudire. **~ping** n applausi mpl

clari|fication /klærɪfɪ'keɪʃn/ n chiarimento m. **~fy** /'klærɪfaɪ/ vt/i (pt/pp -ied) chiarire

clarinet /klærɪ'net/ n clarinetto m

clarity /'klærətɪ/ n chiarezza f

clash /klæʃ/ n scontro m; (noise) fragore m ●vi scontrarsi; (colours:) stonare; (events:) coincidere

clasp /klɑːsp/ n chiusura f ● vt agganciare; (hold) stringere

class /klɑːs/ n classe f; (lesson) corso m ● vt classificare

classic /'klæsɪk/ a classico ● n classico m; **~s** pl Univ lettere fpl classiche. **~al** a classico

classi|fication /klæsɪfɪ'keɪʃn/ n classificazione f. **~fy** /'klæsɪfaɪ/ vt (pt/pp -ied) classificare

classroom n aula f

classy /'klɑːsɪ/ a (-ier, -iest) fam d'alta classe

clatter /'klætə(r)/ n fracasso m ● vi far fracasso

clause /klɔːz/ n clausola f; Gram proposizione f

claustrophob|ia /klɔːstrə'fəʊbɪə/ n claustrofobia f

claw /klɔː/ n artiglio m; (of crab, lobster, & Techn) tenaglia f ● vt (cat:) graffiare

clay /kleɪ/ n argilla f

clean /kliːn/ a pulito, lindo ● adv completamente ● vt pulire (shoes, windows); ~ one's teeth lavarsi i denti; have a coat ~ed portare un cappotto in lavanderia. **clean up** vt pulire ● vi far pulizia

cleaner /'kliːnə(r)/ n uomo m/donna f delle pulizie; (substance) de-

tersivo m; [dry] **~'s** lavanderia f, tintoria f

cleanliness /'klenlɪnɪs/ n pulizia f

cleanse /klenz/ vt pulire. **~r** n detergente m

clean-shaven a sbarbato

cleansing cream /klenz-/ n latte m detergente

clear /klɪə(r)/ a chiaro; (conscience) pulito; (road) libero; (profit, advantage, majority) netto; (sky) sereno; (water) limpido; (glass) trasparente; **make sth ~** mettere qcsa in chiaro; **have I made myself ~?** mi sono fatto capire?; **five ~ days** cinque giorni buoni ● adv stand ~ of allontanarsi da; **keep ~ of** tenersi alla larga da ● vt sgombrare (room, street); sparecchiare (table); (acquit) scagionare; (authorize) autorizzare; scavalcare senza toccare (fence, wall); guadagnare (sum of money); passare (Customs); ~ one's throat schiarirsi la gola ● vi (face, sky:) rasserenarsi; (fog:) dissiparsi. **clear away** vt metter via. **clear off** vi fam filar via. **clear out** vt sgombrare ● vi fam filar via. **clear up** vt (tidy) mettere a posto; chiarire (mystery) ● vi (weather:) schiarirsi

clearance /'klɪərəns/ n (space) spazio m libero; (authorization) autorizzazione f; (Customs) sdoganamento m. ~ **sale** n liquidazione f

clear|ing /'klɪərɪŋ/ n radura f. **~ly** adv chiaramente. ~ **way** n Auto strada f con divieto di sosta

cleavage /'kliːvɪdʒ/ n (woman's) décolleté m inv

cleft /kleft/ n fenditura f

clench /klentʃ/ vt serrare

clergy /'klɜːdʒɪ/ npl clero m. **~man** n ecclesiastico m

cleric /'klerɪk/ n ecclesiastico m. **~al** a impiegatizio; Relig clericale

clerk /klɑːk/, Am /klɜːk/ n impiegato, -a *mf*; (*Am: shop assistant*) commesso, -a *mf*

clever /'klevə(r)/ *a* intelligente; (*skilful*) abile

cliché /'kli:ʃeɪ/ *n* cliché *m inv*

click /klɪk/ *vi* scattare ● *n* Comput click *m*. **click on** *vt* Comput cliccare su

client /'klaɪənt/ *n* cliente *mf*

clientele /kli:ɒn'tel/ *n* clientela *f*

cliff /klɪf/ *n* scogliera *f*

climate /'klaɪmət/ *n* clima *f*. ~**ic** /-'mætɪk/ *a* climatico

climax /'klaɪmæks/ *n* punto *m* culminante

climb /klaɪm/ *n* salita *f* ● *vt* scalare (*mountain*); arrampicarsi su (*ladder, tree*) ● *vi* arrampicarsi; (*rise*) salire; (*road:*) salire. **climb down** *vi* scendere; (*from ladder, tree*) scendere; *fig* tornare sui propri passi

climber /'klaɪmə(r)/ *n* alpinista *mf*; (*plant*) rampicante *m*

clinch /klɪntʃ/ *vt fam* concludere (*deal*) ● *n* (*in boxing*) clinch *m inv*

cling /klɪŋ/ *vi* (*pt/pp* clung) aggrapparsi; (*stick*) aderire. ~ **film** *n* pellicola *f* trasparente

clinic /'klɪnɪk/ *n* ambulatorio *m*. ~**al** *a* clinico

clink /klɪŋk/ *n* tintinnio *m*; (*fam: prison*) galera *f* ● *vi* tintinnare

clip¹ /klɪp/ *n* fermaglio *m*; (*jewellery*) spilla *f* ● *vt* (*pt/pp* clipped) attaccare

clip² *n* (*extract*) taglio *m* ● *vt* obliterare (*ticket*). ~**board** *n* fermabloc *m inv*. ~**pers** *npl* (*for hair*) rasoio *m*; (*for hedge*) tosasiepi *m inv*; (*for nails*) tronchesina *f*. ~**ping** *n* (*from newspaper*) ritaglio *m*

clique /kli:k/ *n* cricca *f*

cloak /kləʊk/ *n* mantello *m*. ~**room** *n* guardaroba *m inv*; (*toilet*) bagno *m*

clock /klɒk/ *n* orologio *m*; (*fam: speedometer*) tachimetro *m* ● **clock in** *vi* attaccare. **clock out** *vi* staccare

clock: ~ **tower** *n* torre *f* dell'orologio. ~**wise** *a & adv* in senso orario. ~**work** *n* meccanismo *m*

clod /klɒd/ *n* zolla *f*

clog /klɒg/ *n* zoccolo *m* ● *vt* (*pt/pp* clogged) ~ [**up**] intasare (*drain*); inceppare (*mechanism*) ● *vi* (*drain:*) intasarsi

cloister /'klɔɪstə(r)/ *n* chiostro *m*

clone /kləʊn/ *n* clone *m*

close¹ /kləʊs/ *a* vicino; (*friend*) intimo; (*weather*) afoso; **have a** ~ **shave** *fam* scamparla bella; **be** ~ **to sb** essere unito a qcno ● *adv* vicino; ~ **by** vicino; **it's** ~ **on five o'clock** sono quasi le cinque

close² /kləʊz/ *n* fine *f* ● *vt* chiudere ● *vi* chiudersi; (*shop:*) chiudere. **close down** *vt* chiudere ● *vi* (*TV station:*) interrompere la trasmissione; (*factory:*) chiudere

closely /'kləʊslɪ/ *adv* da vicino; (*watch, listen*) attentamente

closet /'klɒzɪt/ *n Am* armadio *m*

close-up /'kləʊs-/ *n* primo piano *m*

closure /'kləʊʒə(r)/ *n* chiusura *f*

clot /klɒt/ *n* grumo *m*; (*fam: idiot*) tonto, -a *mf* ● *vi* (*pt/pp* clotted) (*blood:*) coagularsi

cloth /klɒθ/ *n* (*fabric*) tessuto *m*; (*duster etc*) straccio *m*

clothe /kləʊð/ *vt* vestire

clothes /kləʊðz/ *npl* vestiti *mpl*, abiti *mpl*. ~**brush** *n* spazzola *f* per abiti. ~**line** *n* corda *f* stendibiancheria

clothing /'kləʊðɪŋ/ *n* abbigliamento *m*

cloud /klaʊd/ *n* nuvola *f* ● **cloud over** *vi* rannuvolarsi. ~**burst** *n* acquazzone *m*

cloudy /'klaʊdɪ/ *a* (-**ier**, -**iest**) nuvoloso; (*liquid*) torbido

clout /klaʊt/ *n fam* colpo *m*; (*influence*) impatto *m* (**with** su) ● *vt fam* colpire

clove /kləʊv/ *n* chiodo *m* di garofano; ~ **of garlic** spicchio *m* d'aglio

clover /'kləʊvə(r)/ *n* trifoglio *m*

clown /klaʊn/ *n* pagliaccio *m* ● *vi* ~ [**about**] fare il pagliaccio

club /klʌb/ n club m inv; (weapon)
clava f; Sport mazza f; ~s pl
(Cards) fiori mpl ●vt (pt/pp
clubbed) ●vt bastonare. **club
together** vi unirsi

cluck /klʌk/ vi chiocciare

clue /klu:/ n indizio m; (in
crossword) definizione f; **I haven't
a** ~ fam non ne ho idea

clump /klʌmp/ n gruppo m

clumsiness /ˈklʌmzɪnɪs/ n goffag-
gine f

clumsy /ˈklʌmzɪ/ a (-ier, -iest)
maldestro; (tool) scomodo; (remark) senza tatto

clung /klʌŋ/ see **cling**

cluster /ˈklʌstə(r)/ n gruppo m
●vi raggrupparsi (**round** intorno
a)

clutch /klʌtʃ/ n stretta f; Auto frizione f; **be in sb's** ~s essere in
balia di qcno ●vt stringere; (grab)
afferrare ●vi ~ **at** afferrare

clutter /ˈklʌtə(r)/ n caos m ●vt ~
[up] ingombrare

c/o abbr (care of) c/o, presso

coach /kəʊtʃ/ n pullman m inv;
Rail vagone m; (horse-drawn) carrozza f; Sport allenatore, -trice mf
●vt fare esercitare; Sport allenare

coagulate /kəʊˈægjʊleɪt/ vi coagularsi

coal /kəʊl/ n carbone m

coalition /kəʊəˈlɪʃn/ n coalizione f

'coal-mine n miniera f di carbone

coarse /kɔːs/ a grossolano; (joke)
spinto

coast /kəʊst/ n costa f ●vi
(freewheel) scendere a ruota libera; Auto scendere in folle. ~**al** a
costiero. ~**er** n (mat) sottobic-
chiere m inv

coast: ~**guard** n guardia f costiera. ~**line** n litorale m

coat /kəʊt/ n cappotto m; (of
animal) manto m; (of paint) mano
f; ~ **of arms** stemma f ●vt coprire; (with paint) ricoprire.
~**hanger** n gruccia f. ~**hook** n
gancio m [appendiabiti]

coating /ˈkəʊtɪŋ/ n rivestimento
m; (of paint) stato m

coax /kəʊks/ vt convincere con le
moine

cob /kɒb/ n (of corn) pannocchia f

cobble /ˈkɒbl/ vt ~ **together** raf-
fazzonare. ~**r** n ciabattino m

'cobblestones npl ciottolato msg

cobweb /ˈkɒb-/ n ragnatela f

cocaine /kəˈkeɪn/ n cocaina f

cock /kɒk/ n gallo m; (any male
bird) maschio m ●vt sollevare il
grilletto a (gun); ~ **its ears**
(animal:) drizzare le orecchie

cockerel /ˈkɒkərəl/ n galletto m

cock-'eyed a fam storto; (absurd)
assurdo

cockle /ˈkɒkl/ n cardio m

cockney /ˈkɒknɪ/ n (dialect) dialetto m londinese; (person) abitante
mf dell'est di Londra

cock: ~**pit** n Aeron cabina f.
~**roach** /-rəʊtʃ/ n scarafaggio m.
~**tail** n cocktail m inv. ~**up** n sl
make a ~**up** fare un casino (**of**
con)

cocky /ˈkɒkɪ/ a (-ier, -iest) fam
presuntuoso

cocoa /ˈkəʊkəʊ/ n cacao m

coconut /ˈkəʊkənʌt/ n noce f di
cocco

cocoon /kəˈkuːn/ n bozzolo m

COD abbr (cash on delivery) pagamento m alla consegna

cod /kɒd/ n inv merluzzo m

code /kəʊd/ n codice m. ~**d** a codi-
ficato

coeducational /kəʊ-/ a misto

coerce /kəʊˈɜːs/ vt costringere.
~**ion** /-ˈɜːʃn/ n coercizione f

coexist vi coesistere. ~**ence** n
coesistenza f

coffee /ˈkɒfɪ/ n caffè m inv

coffee: ~**grinder** n macinacaffè
m inv. ~**pot** n caffettiera f.
~**table** n tavolino m

coffin /ˈkɒfɪn/ n bara f

cog /kɒg/ n Techn dente m (di ruota)

cogent /ˈkəʊdʒənt/ a convincente

cog-wheel n ruota f dentata

cohabit /kəʊ'hæbɪt/ vi Jur convivere

coherent /kəʊ'hɪərənt/ a coerente; (when speaking) logico

coil /kɔɪl/ n rotolo m; Electr bobina f; ~s pl spire fpl ● vt ~ [up] avvolgere

coin /kɔɪn/ n moneta f ● vt coniare (word)

coincide /kəʊɪn'saɪd/ vi coincidere

coinciden|ce /kəʊ'ɪnsɪdəns/ n coincidenza f. ~tal /-'dentl/ a casuale. ~tally adv casualmente

Coke® n Coca[-cola]®f

coke /kəʊk/ n [carbone m] coke m

cold /kəʊld/ a freddo; I'm ~ ho freddo ● n freddo m; Med raffreddore m

cold: ~-'blooded a spietato. ~-'hearted a insensibile. ~ly adv fig freddamente. ~ meat n salumi mpl. ~ness n freddezza f

coleslaw /'kəʊlslɔː/ n insalata f di cavolo crudo, cipolle e carote in maionese

colic /'kɒlɪk/ n colica f

collaborat|e /kə'læbəreɪt/ vi collaborare; ~e on sth collaborare in qcsa. ~ion /-'reɪʃn/ n collaborazione f; (with enemy) collaborazionismo m. ~or n collaboratore, -trice mf; (with enemy) collaborazionista mf

collaps|e /kə'læps/ n crollo m ● vi (person:) svenire; (roof, building:) crollare. ~ible a pieghevole

collar /'kɒlə(r)/ n colletto m; (for animal) collare m. ~-bone n clavicola f

colleague /'kɒliːg/ n collega mf

collect /kə'lekt/ vt andare a prendere (person); ritirare (parcel, tickets); riscuotere (taxes); raccogliere (rubbish); (as hobby) collezionare ● vi riunirsi ● adv call ~ Am telefonare a carico del destinatario. ~ed /-ɪd/ a controllato

collection /kə'lekʃn/ n collezione f; (in church) questua f; (of rubbish) raccolta f; (of post) levata f

collective /kə'lektɪv/ a collettivo

collector /kə'lektə(r)/ n (of stamps etc) collezionista mf

college /'kɒlɪdʒ/ n istituto m parauniversitario; C~ of... Scuola f di...

collide /kə'laɪd/ vi scontrarsi

colliery /'kɒlɪərɪ/ n miniera f di carbone

collision /kə'lɪʒn/ n scontro m

colloquial /kə'ləʊkwɪəl/ a colloquiale. ~ism n espressione f colloquiale

cologne /kə'ləʊn/ n colonia f

colon /'kəʊlən/ n due punti mpl; Anat colon m inv

colonel /'kɜːnl/ n colonnello m

colonial /kə'ləʊnɪəl/ a coloniale

colon|ize /'kɒlənaɪz/ vt colonizzare. ~y n colonia f

colossal /kə'lɒsl/ a colossale

colour /'kʌlə(r)/ n colore m; (complexion) colorito m; ~s pl (flag) bandiera fsg; off ~ fam giù di tono ● vt colorare; ~ [in] colorare ● vi (blush) arrossire

colour: ~ bar n discriminazione f razziale. ~-blind a daltonico. ~ed a colorato; (person) di colore ● n (person) persona f di colore. ~-fast a dai colori resistenti. ~ film n film m inv a colori. ~ful a pieno di colore. ~less a incolore. ~ television n televisione f a colori

colt /kəʊlt/ n puledro m

column /'kɒləm/ n colonna f. ~ist /-nist/ n giornalista mf che cura una rubrica

coma /'kəʊmə/ n coma m inv

comb /kəʊm/ n pettine m; (for wearing) pettinino m ● vt pettinare; (fig: search) setacciare; ~ one's hair pettinarsi i capelli

combat /'kɒmbæt/ n combattimento m ● vt (pt/pp combated) combattere

combination /kɒmbɪ'neɪʃn/ n combinazione f

combine[1] /kəm'baɪn/ vt unire; ~ **a job with being a mother** conciliare il lavoro con il ruolo di madre ● vi ⟨chemical elements:⟩ combinarsi

combine[2] /'kɒmbaɪn/ n Comm associazione f. ~ (**harvester**) n mietitrebbia f

combustion /kəm'bʌstʃn/ n combustione f

come /kʌm/ vi (pt **came**, pp **come**) venire; **where do you ~ from?** di dove vieni?; ~ **to** (reach) arrivare a; **that ~s to £10** fanno 10 sterline; ~ **into money** ricevere dei soldi; ~ **true** poav verificarsi/aprirsi; ~ **first** arrivare primo; fig venire prima di tutto; ~ **in two sizes** esistere in due misure; **the years to ~** gli anni a venire; **how ~?** fam come mai? **come about** vi succedere. **come across** vi ~ **across as being** fam dare l'impressione di essere ● vt (find) imbattersi in. **come along** vi venire; ⟨job, opportunity:⟩ presentarsi; ⟨progress⟩ andare bene. **come apart** vi smontarsi; (break) rompersi. **come away** vi venir via; ⟨button, fastener:⟩ staccarsi. **come back** vi ritornare. **come by** vi passare ● vt (obtain) avere. **come down** vi scendere; ~ **down to** (reach) arrivare a. **come in** vi entrare; (in race) arrivare; ⟨tide:⟩ salire. **come in for** ~ **in for criticism** essere criticato. **come off** vi staccarsi; ⟨take place⟩ esserci; (succeed) riuscire. **come on** vi (make progress) migliorare; ~ **on!** (hurry) dai!; (indicating disbelief) ma va là!. **come out** vi venir fuori; ⟨book, sun:⟩ uscire; ⟨stain:⟩ andar via. **come over** vi venire. **come round** vi venire; (after fainting) riaversi; ⟨change one's mind⟩ farsi convincere. **come to** vi (after fainting) riaversi. **come up** vi salire; ⟨sun:⟩ sorgere; ⟨plant:⟩ crescere; **something came up** (I was prevented) ho avuto un imprevisto. **come up with** vt tirar fuori

'**come-back** n ritorno m

comedian /kə'miːdɪən/ n comico m

'**come-down** n passo m indietro

comedy /'kɒmədɪ/ n commedia f

comet /'kɒmɪt/ n cometa f

come-uppance /kʌm'ʌpəns/ n **get one's ~** fam avere quel che si merita

comfort /'kʌmfət/ n benessere m; (consolation) conforto m ● vt confortare

comfortable /'kʌmfətəbl/ a comodo; **be ~e** ⟨person:⟩ stare comodo; (fig: in situation) essere a proprio agio; (financially) star bene. **~y** adv comodamente

'**comfort station** n Am bagno m pubblico

comfy /'kʌmfɪ/ a fam comodo

comic /'kɒmɪk/ a comico ● n comico, -a m/f; (periodical) fumetto m. **~al** a comico. **~ strip** n striscia f di fumetti

coming /'kʌmɪŋ/ n venuta f; **~s and goings** viavai m

comma /'kɒmə/ n virgola f

command /kə'mɑːnd/ n comando m; (order) ordine m; (mastery) padronanza f ● vt ordinare; comandare ⟨army⟩

commandeer /kɒmən'dɪə(r)/ vt requisire

commander /kə'mɑːndə(r)/ n comandante m. **~ing** a ⟨view⟩ imponente; (lead) dominante. **~ing officer** n comandante m. **~ment** n comandamento m

commemorate /kə'meməreɪt/ vt commemorare. **~ion** /-'reɪʃn/ n commemorazione f. **~ive** /-ətɪv/ a commemorativo

commence /kə'mens/ vt/i cominciare. **~ment** n inizio m

commend /kə'mend/ vt complimentarsi con (**on** per); (recommend) raccomandare (**to** a). **~able** /-əbl/ a lodevole

commensurate /kə'menʃərət/ a proporzionato (**with** a)

comment /'kɒment/ n commento m ●vi fare commenti (**on** su)

commentary /'kɒməntri/ n commento m; [**running**] ~ (**on** radio, TV) cronaca f diretta

commentate /'kɒmenteit/ vt ~e **on** TV, Radio fare la cronaca di. ~**or** n cronista mf

commerce /'kɒmɜːs/ n commercio m

commercial /kə'mɜːʃl/ a commerciale ●n TV pubblicità f inv. ~**ize** vt commercializzare

commiserate /kə'mɪzəreit/ vi esprimere il proprio rincrescimento (**with** a)

commission /kə'mɪʃn/ n commissione f; **receive one's** ~ Mil essere promosso ufficiale; **out of** ~ fuori uso ● vt commissionare

commissionaire /kəmɪʃə'neə(r)/ n portiere m

commissioner /kə'mɪʃənə(r)/ n commissario m

commit /kə'mɪt/ vt (pt/pp **committed**) commettere; (to prison, hospital) affidare (**to** a); impegnare (funds); ~ **oneself** impegnarsi. ~**ment** n impegno m; (involvement) compromissione f. ~**ted** a impegnato

committee /kə'mɪti/ n comitato m

commodity /kə'mɒdəti/ n prodotto m

common /'kɒmən/ a comune; (vulgar) volgare ●n prato m pubblico; **have in** ~ avere in comune; **House of C~s** Camera f dei Comuni. ~**er** n persona f non nobile

common: ~**law** n diritto m consuetudinario. ~**ly** adv comunemente. **C~ 'Market** n Mercato m Comune. ~**place** a banale. ~**room** n sala f dei professori o degli studenti. ~**'sense** n buon senso m

commotion /kə'məʊʃn/ n confusione f

communal /'kɒmjʊnl/ a comune

communicate /kə'mjuːnɪkeit/ vt/i comunicare

communication /kəmjuːnɪ'keiʃn/ n comunicazione f; (of disease) trasmissione f; **be in** ~ **with sb** essere in contatto con qcno; ~**s** pl (technology) telecomunicazioni fpl. ~ **cord** n fermata f d'emergenza

communicative /kə'mjuːnɪkətɪv/ a comunicativo

Communion /kə'mjuːnɪən/ n [Holy] ~ comunione f

communiqué /kə'mjuːnɪkeɪ/ n comunicato m stampa

Communism /'kɒmjʊnɪzm/ n comunismo m. ~**t** /-ɪst/ a & n comunista mf

community /kə'mjuːnəti/ n comunità f. ~ **centre** n centro m sociale

commute /kə'mjuːt/ vi fare il pendolare ● vt Jur commutare. ~**r** n pendolare mf

compact¹ /kəm'pækt/ a compatto

compact² /'kɒmpækt/ n portacipria m inv. ~ **disc** n compact disc m inv

companion /kəm'pænjən/ n compagno, -a mf. ~**ship** n compagnia f

company /'kʌmpəni/ n compagnia f; (guests) ospiti mpl. ~ **car** n macchina f della ditta

comparable /'kɒmpərəbl/ a paragonabile

comparative /kəm'pærətɪv/ a comparativo; (relative) relativo ●n Gram comparativo m. ~**ly** adv relativamente

compare /kəm'peə(r)/ vt paragonare (**with/to** a) ● vi essere paragonato

comparison /kəm'pærɪsn/ n paragone m

compartment /kəm'pɑːtmənt/ n compartimento m; Rail scompartimento m

compass /'kʌmpəs/ n bussola f. ~**es** npl, **pair of** ~**es** compasso msg

compassion /kəm'pæʃn/ n compassione f. **~ate** /-ʃənət/ a compassionevole

compatible /kəm'pætəbl/ a compatibile

compatriot /kəm'pætrɪət/ n compatriota mf

compel /kəm'pel/ vt (pt/pp **compelled**) costringere. **~ling** a (reason) inconfutabile

compensat|e /'kɒmpənseɪt/ vt risarcire ● vi **~ for** fig compensare di. **~ion** /-'seɪʃn/ n risarcimento m; (fig: comfort) consolazione f

compère /'kɒmpeə(r)/ n presentatore, -trice mf

compete /kəm'piːt/ vi competere; (take part) gareggiare

competen|ce /'kɒmpɪtəns/ n competenza f. **~t** a competente

competition /kɒmpə'tɪʃn/ n concorrenza f; (contest) gara f

competitive /kəm'petɪtɪv/ a competitivo; **~ prices** prezzi mpl concorrenziali

competitor /kəm'petɪtə(r)/ n concorrente mf

complacen|cy /kəm'pleɪsənsɪ/ n compiacimento m. **~t** a compiaciuto

complain /kəm'pleɪn/ vi lamentarsi (about di); (formally) reclamare; **~ of** Med accusare. **~t** n lamentela f; (formal) reclamo m; Med disturbo m

complement¹ /'kɒmplɪmənt/ n complemento m

complement² /'kɒmplɪment/ vt complementare; **~ each other** complementarsi a vicenda. **~ary** /-'mentərɪ/ a complementare

complete /kəm'pliːt/ a completo; (utter) finito ● vt completare; compilare (form). **~ly** adv completamente

completion /kəm'pliːʃn/ n fine f

complex /'kɒmpleks/ a complesso ● n complesso m

complexion /kəm'plekʃn/ n carnagione f

complexity /kəm'pleksətɪ/ n complessità f inv

compliance /kəm'plaɪəns/ n accettazione f; (with rules) osservanza f; **in ~ with** in osservanza a (law); conformemente a (request)

complicat|e /'kɒmplɪkeɪt/ vt complicare. **~ed** a complicato. **~ion** /-'keɪʃn/ n complicazione f

compliment /'kɒmplɪmənt/ n complimento m; **~s** pl omaggi mpl ● vt complimentare. **~ary** /-'mentərɪ/ a complimentoso; (given free) in omaggio

comply /kəm'plaɪ/ vi (pt/pp **-ied**) **~ with** conformarsi a

component /kəm'pəʊnənt/ a & n [part] componente m

compose /kəm'pəʊz/ vt comporre; **~ oneself** ricomporsi; **be ~d of** essere composto da. **~d** a (calm) composto. **~r** n compositore, -trice mf

composition /kɒmpə'zɪʃn/ n posizione f; (essay) tema m

compost /'kɒmpɒst/ n composta f

composure /kəm'pəʊʒə(r)/ n calma f

compound /'kɒmpaʊnd/ a composto. **~ fracture** n frattura esposta. **~ 'interest** n interesse m composto ● n Chem composto m; Gram parola f composta; (enclosure) recinto m

comprehen|d /kɒmprɪ'hend/ vt comprendere. **~sible** /-'hensəbl/ a comprensibile. **~sion** /-'henʃn/ n comprensione f

comprehensive /kɒmprɪ'hensɪv/ a & n comprensivo; **~ [school]** scuola f media in cui gli allievi hanno capacità d'apprendimento diverse. **~ insurance** n Auto polizza f casco

compress¹ /'kɒmpres/ n compressa f

compress² /kəm'pres/ vt comprimere; **~ed air** aria f compressa

comprise /kəm'praɪz/ vt comprendere; (form) costituire

compromise /'kɒmprəmaɪz/ n compromesso m ● vt compromettere ● vi fare un compromesso

compuls|ion /kəm'pʌlʃn/ n desiderio m irresistibile. **~ive** /-sɪv/ a Psych patologico. **~ive eating** voglia f ossessiva di mangiare. **~ory** /-sərɪ/ a obbligatorio

comput|er /kəm'pju:tə(r)/ n computer m inv. **~erize** vt computerizzare. **~ing** n informatica f

comrade /'kɒmreɪd/ n camerata m; Pol compagno, -a mf. **~ship** n cameratismo m

con[1] /kɒn/ see pro

con[2] n fam fregatura f ● vt (pt/pp conned) fam fregare

concave /'kɒnkeɪv/ a concavo

conceal /kən'si:l/ vt nascondere

concede /kən'si:d/ vt (admit) ammettere; (give up) rinunciare a; lasciar fare (goal)

conceit /kən'si:t/ n presunzione f. **~ed** a presuntuoso

conceivable /kən'si:vəbl/ a concepibile

conceive /kən'si:v/ vt Biol concepire ● vi aver figli. **conceive of** vt fig concepire

concentrat|e /'kɒnsəntreɪt/ vt concentrare ● vi concentrarsi. **~ion** /-'treɪʃn/ n concentrazione f. **~ion camp** n campo m di concentramento

concept /'kɒnsept/ n concetto m. **~ion** /kən'sepʃn/ n concezione f; (idea) idea f

concern /kən'sɜːn/ n preoccupazione f; Comm attività f inv ● vt (be about, affect) riguardare; (worry) preoccupare; **be ~ed about** essere preoccupato per; **~ oneself with** preoccuparsi di; **as far as I am ~ed** per quanto mi riguarda. **~ing** prep riguardo a

concert /'kɒnsət/ n concerto m. **~ed** /kən'sɜːtɪd/ a collettivo

concertina /kɒnsə'ti:nə/ n piccola fisarmonica f

'concertmaster n Am primo violino m

concerto /kən'tʃeətəʊ/ n concerto m

concession /kən'seʃn/ n conces-

sione f; (reduction) sconto m. **~ary** a (reduced) scontato

conciliation /kənsɪlɪ'eɪʃn/ n conciliazione f

concise /kən'saɪs/ a conciso

conclu|de /kən'klu:d/ vt concludere ● vi concludersi. **~ding** a finale

conclusion /kən'klu:ʒn/ n conclusione f; **in ~** per concludere

conclusive /kən'klu:sɪv/ a definitivo. **~ly** adv in modo definitivo

concoct /kən'kɒkt/ vt confezionare; fig inventare. **~ion** /-ɒkʃn/ n mistura f; (drink) intruglio m

concourse /'kɒŋkɔːs/ n atrio m

concrete /'kɒŋkriːt/ a concreto ● n calcestruzzo m

concur /kən'kɜː(r)/ vi (pt/pp concurred) essere d'accordo

concurrently /kən'kʌrəntlɪ/ adv contemporaneamente

concussion /kən'kʌʃn/ n commozione f cerebrale

condemn /kən'dem/ vt condannare; dichiarare inagibile (building). **~ation** /kɒndem'neɪʃn/ n condanna f

condensation /kɒnden'seɪʃn/ n condensazione f

condense /kən'dens/ vt condensare; Phys condensare ● vi condensarsi. **~d milk** n latte m condensato

condescend /kɒndɪ'send/ vi degnarsi. **~ing** a condiscendente

condition /kən'dɪʃn/ n condizione f; **on ~ that** a condizione che ● vt Psych condizionare. **~al** (acceptance) condizionato; Gram condizionale ● n Gram condizionale m. **~er** n balsamo m; (for fabrics) ammorbidente m

condolences /kən'dəʊlənsɪz/ npl condoglianze fpl

condom /'kɒndəm/ n preservativo m

condo[minium] /'kɒndə('mɪnɪəm)/ n Am condominio m

condone /kən'dəʊn/ vt passare sopra a

conducive /kənˈdjuːsɪv/ *a* be ~ to contribuire a

conduct¹ /ˈkɒndʌkt/ *n* condotta *f*

conduct² /kənˈdʌkt/ *vt* condurre; dirigere ⟨*orchestra*⟩. ~**or** *n* direttore *m* ⟨*d'orchestra*⟩; (*of bus*) bigliettaio *m*; *Phys* conduttore *m*. ~**ress** *n* bigliettaia *f*

cone /kəʊn/ *n* cono *m*; *Bot* pigna *f*; *Auto* birillo *m* ● **cone off** *vt* be ~**d off** *Auto* essere chiuso da birilli

confectioner /kənˈfekʃənə(r)/ *n* pasticciere, -a *mf*. ~**y** *n* pasticceria *f*

confederation /kənfedəˈreɪʃn/ *n* confederazione *f*

confer /kənˈfɜː(r)/ *v* (*pt/pp* conferred) ● *vt* conferire ⟨on a⟩ ● *vi* (*discuss*) conferire

conference /ˈkɒnfərəns/ *n* conferenza *f*

confess /kənˈfes/ *vt* confessare ● *vi* confessare; *Relig* confessarsi. ~**ion** /-eʃn/ *n* confessione *f*. ~**ional** /-eʃənəl/ *n* confessionale *m*. ~**or** *n* confessore *m*

confetti /kənˈfeti/ *n* coriandoli *mpl*

confide /kənˈfaɪd/ *vt* confidare. **confide in** *vt* ~ **in sb** fidarsi di qcno

confidence /ˈkɒnfɪdəns/ *n* (*trust*) fiducia *f*; (*self-assurance*) sicurezza *f* di sé; (*secret*) confidenza *f*; **in** ~ in confidenza. ~ **trick** *n* truffa *f*

confident /ˈkɒnfɪdənt/ *a* fiducioso; (*self-assured*) sicuro di sé. ~**ly** *adv* con aria fiduciosa

confidential /kɒnfɪˈdenʃl/ *a* confidenziale

confine /kənˈfaɪn/ *vt* rinchiudere; (*limit*) limitare; **be** ~**d to bed** essere confinato a letto. ~**d** *a* ⟨*space*⟩ limitato. ~**ment** *n* detenzione *f*; *Med* parto *m*

confines /ˈkɒnfaɪnz/ *npl* confini *mpl*

confirm /kənˈfɜːm/ *vt* confermare; *Relig* cresimare. ~**ation** /kɒnfəˈmeɪʃn/ *n* conferma *f*; *Relig*

cresima *f*. ~**ed** *a* incallito; ~**ed bachelor** scapolo *m* impenitente

confiscat|e /ˈkɒnfɪskeɪt/ *vt* confiscare. ~**ion** /-ˈkeɪʃn/ *n* confisca *f*

conflict¹ /ˈkɒnflɪkt/ *n* conflitto *m*

conflict² /kənˈflɪkt/ *vi* essere in contraddizione. ~**ing** *a* contraddittorio

conform /kənˈfɔːm/ *vi* ⟨*person*:⟩ conformarsi; ⟨*thing*:⟩ essere conforme (**to** a). ~**ist** *n* conformista *mf*

confound /kənˈfaʊnd/ *vt* confondere. ~**ed** *a* fam maledetto

confront /kənˈfrʌnt/ *vt* affrontare; **the problems** ~**ing us** i problemi che dobbiamo affrontare. ~**ation** /kɒnfrʌnˈteɪʃn/ *n* confronto *m*

confus|e /kənˈfjuːz/ *vt* confondere. ~**ing** *a* che confonde. ~**ion** /-juːʒn/ *n* confusione *f*

congeal /kənˈdʒiːl/ *vi* ⟨*blood*:⟩ coagularsi

congenial /kənˈdʒiːnɪəl/ *a* congeniale

congenital /kənˈdʒenɪtl/ *a* congenito

congest|ed /kənˈdʒestɪd/ *a* congestionato. ~**ion** /-estʃn/ *n* congestione *f*

congratulat|e /kənˈgrætjʊleɪt/ *vt* congratularsi (**on** per). ~**ions** /-ˈeɪʃnz/ *npl* congratulazioni *fpl*

congregat|e /ˈkɒŋgrɪgeɪt/ *vi* radunarsi. ~**ion** /-ˈgeɪʃn/ *n* *Relig* assemblea *f*

congress /ˈkɒŋgres/ *n* congresso *m*. ~**man** *n* *Am Pol* membro *m* del congresso

conical /ˈkɒnɪkl/ *a* conico

conifer /ˈkɒnɪfə(r)/ *n* conifera *f*

conjecture /kənˈdʒektʃə(r)/ *n* congettura *f*

conjugal /ˈkɒndʒʊgl/ *a* coniugale

conjugat|e /ˈkɒndʒʊgeɪt/ *vt* coniugare. ~**ion** /-ˈgeɪʃn/ *n* coniugazione *f*

conjunction /kənˈdʒʌŋkʃn/ *n* congiunzione *f*; **in** ~ **with** insieme a

conjunctivitis /kəndʒʌŋktɪˈvaɪtɪs/ *n* congiuntivite *f*

conjur|e /ˈkʌndʒə(r)/ vi ~ing **tricks** npl giochi mpl di prestigio. ~or n prestigiatore, -trice mf. **conjure up** vt evocare ‹image›; tirar fuori dal nulla ‹meal›

conk /kɒŋk/ vi ~ **out** fam ‹machine:› guastarsi; ‹person:› crollare

'con-man n fam truffatore m

connect /kəˈnekt/ vt collegare; **be ~ed with** avere legami con; ‹be related to› essere imparentato con; **be well ~ed** aver conoscenze influenti ● vi essere collegato **(with a)** ‹train:› fare coincidenza

connection /kəˈnekʃn/ n ‹between ideas› nesso m; ‹in travel› coincidenza f; Electr collegamento m; **in ~ with** con riferimento a. ~s pl ‹people› conoscenze fpl

connoisseur /kɒnəˈsɜː(r)/ n intenditore, -trice mf

conquer /ˈkɒŋkə(r)/ vt conquistare; fig superare ‹fear›. ~or n conquistatore m

conquest /ˈkɒŋkwest/ n conquista f

conscience /ˈkɒnʃəns/ n coscienza f

conscientious /kɒnʃɪˈenʃəs/ a coscienzioso. ~ **ob'jector** n obiettore m di coscienza

conscious /ˈkɒnʃəs/ a conscio; ‹decision› meditato; **[fully]** ~ cosciente; **be/become ~ of sth** rendersi conto di qcsa. ~ly adv consapevolmente. ~ness n consapevolezza f; Med conoscenza f

conscript[1] /ˈkɒnskrɪpt/ n coscritto m

conscript[2] /kənˈskrɪpt/ vt Mil chiamare alle armi. ~ion /-ˈɪpʃn/ n coscrizione f, leva f

consecrat|e /ˈkɒnsɪkreɪt/ vt consacrare. ~ion /-ˈkreɪʃn/ n consacrazione f

consecutive /kənˈsekjʊtɪv/ a consecutivo

consensus /kənˈsensəs/ n consenso m

consent /kənˈsent/ n consenso m ● vi acconsentire

consequen|ce /ˈkɒnsɪkwəns/ n conseguenza f; ‹importance› importanza f. ~t a conseguente. ~tly adv di conseguenza

conservation /kɒnsəˈveɪʃn/ n conservazione f. ~ist n fautore, -trice mf della tutela ambientale

conservative /kənˈsɜːvətɪv/ a conservativo; ‹estimate› ottimistico. **C~** Pol a conservatore ● n conservatore, -trice mf

conservatory /kənˈsɜːvətrɪ/ n spazio m chiuso da vetrate adiacente alla casa

conserve /kənˈsɜːv/ vt conservare

consider /kənˈsɪdə(r)/ vt considerare; ~ **doing sth** considerare la possibilità di fare qcsa. ~able /-əbl/ a considerevole. ~ably adv considerevolmente

consider|ate /kənˈsɪdərət/ a pieno di riguardo. ~ately adv con riguardo. ~ation /-ˈreɪʃn/ n considerazione f; ‹thoughtfulness› attenzione f; ‹respect› riguardo m; ‹payment› compenso m; **take sth into ~ation** prendere qcsa in considerazione. ~ing prep considerando

consign /kənˈsaɪn/ vt affidare. ~ment n consegna f

consist /kənˈsɪst/ vi ~ **of** consistere di

consisten|cy /kənˈsɪstənsɪ/ n coerenza f; ‹density› consistenza f. ~t a coerente; ‹loyalty› costante. ~tly adv coerentemente; ‹late, loyal› costantemente

consolation /kɒnsəˈleɪʃn/ n consolazione f. ~ **prize** n premio m di consolazione

console /kənˈsəʊl/ vt consolare

consolidate /kənˈsɒlɪdeɪt/ vt consolidare

consonant /ˈkɒnsənənt/ n consonante f

consort /kənˈsɔːt/ vi ~ **with** frequentare

consortium /kən'sɔːtɪəm/ n consorzio m

conspicuous /kən'spɪkjʊəs/ a facilmente distinguibile

conspiracy /kən'spɪrəsɪ/ n cospirazione f

conspire /kən'spaɪə(r)/ vi cospirare

constable /'kʌnstəbl/ n agente m [di polizia]

constant /'kɒnstənt/ a costante. **~ly** adv costantemente

constellation /kɒnstə'leɪʃn/ n costellazione f

consternation /kɒnstə'neɪʃn/ n costernazione f

constipat|ed /'kɒnstɪpeɪtɪd/ a stitico. **~ion** /-'peɪʃn/ n stitichezza f

constituency /kən'stɪtjʊənsɪ/ n area f elettorale di un deputato nel Regno Unito

constituent /kən'stɪtjʊənt/ n costituente m; Pol elettore, -trice mf

constitute /'kɒnstɪtjuːt/ vt costituire. **~ion** /-'tjuːʃn/ n costituzione f **~ional** /-'tjuːʃənl/ a costituzionale

constrain /kən'streɪn/ vt costringere. **~t** n costrizione f; (restriction) restrizione f; (strained manner) disagio m

construct /kən'strʌkt/ vt costruire. **~ion** /-ʌkʃn/ n costruzione f; **under ~ion** in costruzione. **~ive** /-ɪv/ a costruttivo

construe /kən'struː/ vt interpretare

consul /'kɒnsl/ n console m. **~ar** /'kɒnsjʊlə(r)/ a consolare. **~ate** /'kɒnsjʊlət/ n consolato m

consult /kən'sʌlt/ vt consultare. **~ant** n consulente mf; Med specialista mf. **~ation** /kɒnsl'teɪʃn/ n consultazione f; Med consulto m

consume /kən'sjuːm/ vt consumare. **~r** n consumatore, -trice m/f. **~r goods** npl beni mpl di consumo. **~r organization** n organizzazione f per la tutela dei consumatori

consumerism /kən'sjuːmərɪzm/ n consumismo m

consummate /'kɒnsəmeɪt/ vt consumare

consumption /kən'sʌmpʃn/ n consumo m

contact /'kɒntækt/ n contatto m; (person) conoscenza f ● vt mettersi in contatto con. **~ 'lenses** npl lenti fpl a contatto

contagious /kən'teɪdʒəs/ a contagioso

contain /kən'teɪn/ vt contenere; **~ oneself** controllarsi. **~er** n recipiente m; (for transport) container m inv

contaminat|e /kən'tæmɪneɪt/ vt contaminare. **~ion** /-'neɪʃn/ n contaminazione f

contemplat|e /'kɒntəmpleɪt/ vt contemplare; (consider) considerare; **~e doing sth** considerare di fare qcsa. **~ion** /-'pleɪʃn/ n contemplazione f

contemporary /kən'tempərərɪ/ a & n contemporaneo, -a mf

contempt /kən'tempt/ n disprezzo m; **beneath ~** più che vergognoso; **~ of court** oltraggio m alla Corte. **~ible** /-əbl/ a spregevole. **~uous** /-tjʊəs/ a sprezzante

contend /kən'tend/ vi **~ with** occuparsi di ● vt (assert) sostenere. **~er** n concorrente mf

content¹ /'kɒntent/ n contenuto m

content² /kən'tent/ a soddisfatto ● vt **~ oneself** accontentarsi (with di). **~ed** a soddisfatto. **~edly** adv con aria soddisfatta

contention /kən'tenʃn/ n (assertion) opinione f

contentment /kən'tentmənt/ n soddisfazione f

contents /'kɒntents/ npl contenuto m

contest¹ /'kɒntest/ n gara f

contest² /kən'test/ vt contestare (statement); impugnare (will); Pol (candidates:) contendersi; (one candidate:) aspirare a. **~ant** n concorrente mf

context /'kɒntekst/ n contesto m

continent /'kɒntɪnənt/ n continente m; **the C~** l'Europa f continentale

continental /kɒntɪ'nentl/ a continentale. **~ breakfast** n prima colazione f a base di pane, burro, marmellata, croissant, ecc. **~ quilt** n piumone m

contingency /kən'tɪndʒənsɪ/ n eventualità f inv

continual /kən'tɪnjʊəl/ a continuo

continuation /kəntɪnjʊ'eɪʃn/ n continuazione f

continue /kən'tɪnjuː/ vt continuare; **~ doing** or **to do sth** continuare a fare qcsa; **to be ~d** continued ● vi continuare. **~d** a continuo

continuity /kɒntɪ'njuːətɪ/ n continuità f

continuous /kən'tɪnjʊəs/ a continuo

contort /kən'tɔːt/ vt contorcere. **~ion** /-ɔːʃn/ n contorsione f. **~ionist** n contorsionista mf

contour /'kɒntʊə(r)/ n contorno m; (line) curva f di livello

contraband /'kɒntrəbænd/ n contrabbando m

contraception /kɒntrə'sepʃn/ n contraccezione f. **~tive** /-tɪv/ n contraccettivo m

contract¹ /'kɒntrækt/ n contratto m

contract² /kən'trækt/ vi (get smaller) contrarsi ● vt contrarre (illness). **~ion** /-ækʃn/ n contrazione f. **~or** n imprenditore, -trice mf

contradict /kɒntrə'dɪkt/ vt contraddire. **~ion** /-ɪkʃn/ n contraddizione f. **~ory** a contraddittorio

contra-flow /'kɒntrəfləʊ/ n utilizzazione f di una corsia nei due sensi di marcia durante lavori stradali

contralto /kən'træltəʊ/ n contralto m

contraption /kən'træpʃn/ n fam aggeggio m

contrary¹ /'kɒntrərɪ/ a contrario

● adv **~ to** contrariamente a ● n contrario m; **on the ~** al contrario

contrary² /kən'treərɪ/ a disobbediente

contrast¹ /'kɒntrɑːst/ n contrasto m

contrast² /kən'trɑːst/ vt confrontare ● vi contrastare. **~ing** a contrastante

contraven|e /kɒntrə'viːn/ vt trasgredire. **~tion** /-'venʃn/ n trasgressione f

contribut|e /kən'trɪbjuːt/ vt/i contribuire. **~ion** /kɒntrɪ'bjuːʃn/ n contribuzione f; (what is contributed) contributo m. **~or** n contributore, -trice mf

contrive /kən'traɪv/ vt escogitare; **~ to do sth** riuscire a fare qcsa

control /kən'trəʊl/ n controllo m; **~s** pl (of car, plane) comandi mpl; **get out of ~** sfuggire al controllo ● vt (pt/pp controlled) controllare; **~ oneself** controllarsi

controvers|ial /kɒntrə'vɜːʃl/ a controverso. **~y** /'kɒntrəvɜːsɪ/ n controversia f

conurbation /kɒnə'beɪʃn/ n conurbazione f

convalesce /kɒnvə'les/ vi essere in convalescenza

convalescent /kɒnvə'lesənt/ a convalescente. **~ home** n convalescenziario m

convector /kən'vektə(r)/ n **~ [heater]** convettore m

convene /kən'viːn/ vt convocare ● vi riunirsi

convenience /kən'viːnɪəns/ n convenienza f; **[public] ~** gabinetti mpl pubblici; **with all modern ~s** con tutti i comfort

convenient /kən'viːnɪənt/ a comodo; **be ~ for sb** andar bene per qcno; **if it is ~ [for you]** se ti va bene. **~ly** adv comodamente; **~ly located** in una posizione comoda

convent /'kɒnvənt/ n convento m

convention /kən'venʃn/ n conven-

zione f; (assembly) convegno m.
~al a convenzionale

converge /kən'vɜːdʒ/ vi convergere

conversant /kən'vɜːsənt/ a ~ with pratico di

conversation /kɒnvə'seɪʃn/ n conversazione f. ~al a di conversazione. ~alist n conversatore, -trice mf

converse[1] /kən'vɜːs/ vi conversare

converse[2] /'kɒnvɜːs/ n inverso m. ~ly adv viceversa

conversion /kən'vɜːʃn/ n conversione f

convert[1] /'kɒnvɜːt/ n convertito, -a mf

convert[2] /kən'vɜːt/ vt convertire (into in); sconsacrare (church). ~ible /-əbl/ a convertibile ● n Auto macchina f decappottabile

convex /'kɒnveks/ a convesso

convey /kən'veɪ/ vt portare; trasmettere (idea, message). ~or belt n nastro m trasportatore

convict[1] /'kɒnvɪkt/ n condannato, -a mf

convict[2] /kən'vɪkt/ vt giudicare colpevole. ~ion /-kʃn/ n condanna f; (belief) convinzione f; previous ~ precedente m penale

convinc|e /kən'vɪns/ vt convincere. ~ing a convincente

convivial /kən'vɪvɪəl/ a conviviale

convoluted /'kɒnvəluːtɪd/ a contorto

convoy /'kɒnvɔɪ/ n convoglio m

convuls|e /kən'vʌls/ vt sconvolgere; be ~ed with laughter contorcersi dalle risa. ~ion /-ʌlʃn/ n convulsione f

coo /kuː/ vi tubare

cook /kʊk/ n cuoco, -a mf ● vt cucinare; is it ~ed? è cotto?; ~ the books fam truccare i libri contabili ● vi (food:) cuocere; (person:) cucinare. ~book n libro m di cucina

cooker /'kʊkə(r)/ n cucina f; (apple) mela f da cuocere. ~y n cu-

cina f. ~y book n libro m di cucina

cookie /'kʊkɪ/ n Am biscotto m

cool /kuːl/ a fresco; (calm) calmo; (unfriendly) freddo ● n fresco m ● vt rinfrescare ● vi rinfrescarsi. ~-box n borsa f termica. ~ness n freddezza f

coop /kuːp/ n stia f ● vt ~ up rinchiudere

co-operat|e /kəʊ'ɒpəreɪt/ vi cooperare. ~ion /-'reɪʃn/ n cooperazione f

co-operative /kəʊ'ɒpərətɪv/ a cooperativo ● n cooperativa f

co-opt /kəʊ'ɒpt/ vt eleggere

co-ordinat|e /kəʊ'ɔːdmeɪt/ vt coordinare. ~ion /-'neɪʃn/ n coordinazione f

cop /kɒp/ n fam poliziotto m

cope /kəʊp/ vi fam farcela; can she ~ by herself? ce la fa da sola?; ~ with farcela con

copious /'kəʊpɪəs/ a abbondante

copper[1] /'kɒpə(r)/ n rame m; ~s pl monete fpl da uno o due pence ● attrib di rame

copper[2] n fam poliziotto m

coppice /'kɒpɪs/ n, **copse** /kɒps/ n boschetto m

copulat|e /'kɒpjʊleɪt/ vi accoppiarsi. ~ion /-'leɪʃn/ n copulazione f

copy /'kɒpɪ/ n copia f ● vt (pt/pp -ied) copiare

copy: ~right n diritti mpl d'autore. ~writer n copywriter mf inv

coral /'kɒrəl/ n corallo m

cord /kɔːd/ n corda f; (thinner) cordoncino m; (fabric) velluto m a coste; ~s pl pantaloni mpl di velluto a coste

cordial /'kɔːdɪəl/ a cordiale ● n analcolico m

cordon /'kɔːdn/ n cordone m (di persone) ● **cordon off** vt mettere un cordone (di persone) intorno a

corduroy /'kɔːdərɔɪ/ n velluto m a coste

core /kɔː(r)/ n (of apple, pear) torsolo *m*; (*fig: of organization*) cuore *m*; (*of problem, theory*) nocciolo *m*

cork /kɔːk/ *n* sughero *m*; (*for bottle*) turacciolo *m*. **~screw** *n* cavatappi *m inv*

corn[1] /kɔːn/ *n* grano *m*; (*Am: maize*) granturco *m*

corn[2] *n Med* callo *m*

cornea /ˈkɔːnɪə/ *n* cornea *f*

corned beef /kɔːnd'biːf/ *n* manzo *m* sotto sale

corner /ˈkɔːnə(r)/ *n* angolo *m*; (*football*) calcio *m* d'angolo, corner *m inv* ● *vt fig* bloccare; *Comm* accaparrarsi ⟨*market*⟩

cornet /ˈkɔːnɪt/ *n Mus* cornetta *f*; (*for ice-cream*) cono *m*

corn: ~flour *n*, *Am* **~starch** *n* farina *f* di granturco

corny /ˈkɔːnɪ/ *a* (**-ier, -iest**) ⟨*fam: joke, film*⟩ scontato; ⟨*person*⟩ banale; (*sentimental*) sdolcinato

coronary /ˈkɒrənərɪ/ *a* coronario ● *n* **~** [**thrombosis**] trombosi *f* coronarica

coronation /kɒrəˈneɪʃn/ *n* incoronazione *f*

coroner /ˈkɒrənə(r)/ *n* coroner *m inv* (*nel diritto britannico, ufficiale incaricato delle indagini su morti sospette*)

corporal[1] /ˈkɔːpərəl/ *n Mil* caporale *m*

corporal[2] *a* corporale; **~ punishment** punizione *f* corporale

corporate /ˈkɔːpərət/ *a* ⟨*decision, policy, image*⟩ aziendale; **~ life** la vita in un'azienda

corporation /kɔːpəˈreɪʃn/ *n* ente *m*; (*of town*) consiglio *m* comunale

corps /kɔː(r)/ *n* (*pl* **corps** /kɔːz/) corpo *m*

corpse /kɔːps/ *n* cadavere *m*

corpulent /ˈkɔːpjʊlənt/ *a* corpulento

corpuscle /ˈkɔːpʌsl/ *n* globulo *m*

correct /kəˈrekt/ *a* corretto; **be ~** ⟨*person*⟩ aver ragione; **~!** esatto! ● *vt* correggere. **~ion** /-ekʃn/ *n* correzione *f*. **~ly** *adv* correttamente

correlation /kɒrəˈleɪʃn/ *n* correlazione *f*

correspond /kɒrɪˈspɒnd/ *vi* corrispondere (**to** a); (*two things*:) corrispondere; (*write*) scriversi. **~ence** *n* corrispondenza *f*. **~ent** *n* corrispondente *mf*. **~ing** *a* corrispondente. **~ingly** *adv* in modo corrispondente

corridor /ˈkɒrɪdɔː(r)/ *n* corridoio *m*

corroborate /kəˈrɒbəreɪt/ *vt* corroborare

corro|de /kəˈrəʊd/ *vt* corrodere ● *vi* corrodersi. **~sion** /-ˈrəʊʒn/ *n* corrosione *f*

corrugated /ˈkɒrəgeɪtɪd/ *a* ondulato. **~ iron** *n* lamiera *f* ondulata

corrupt /kəˈrʌpt/ *a* corrotto ● *vt* corrompere. **~ion** /-ʌpʃn/ *n* corruzione *f*

corset /ˈkɔːsɪt/ *n* & **-s** *pl* busto *m*

Corsica /ˈkɔːsɪkə/ *n* Corsica *f*. **~n** *a* & *n* corso, -a *mf*

cortège /kɔːˈteɪʒ/ *n* [**funeral**] **~** corteo *m* funebre

cosh /kɒʃ/ *n* randello *m*

cosmetic /kɒzˈmetɪk/ *a* cosmetico ● *n* **~s** *pl* cosmetici *mpl*

cosmic /ˈkɒzmɪk/ *a* cosmico

cosmonaut /ˈkɒzmənɔːt/ *n* cosmonauta *mf*

cosmopolitan /kɒzməˈpɒlɪtən/ *a* cosmopolita

cosmos /ˈkɒzmɒs/ *n* cosmo *m*

cosset /ˈkɒsɪt/ *vt* coccolare

cost /kɒst/ *n* costo *m*; **~s** *pl Jur* spese *fpl* processuali; **at all ~s** a tutti i costi; **I learnt to my ~s** ho imparato a mie spese ● *vt* (*pt/pp* **cost**) costare; **it ~ me £20** mi è costato 20 sterline ● *vt* (*pt/pp* **costed**) **~** [**out**] stabilire il prezzo di

costly /ˈkɒstlɪ/ *a* (**-ier, -iest**) costoso

cost: ~ of living *n* costo *m* della vita. **~ price** *n* prezzo *m* di costo

costume /ˈkɒstjuːm/ *n* costume *m*. **~ jewellery** *n* bigiotteria *f*

cosy /ˈkəʊzɪ/ *a* (**-ier, -iest**) ⟨*pub*,

cot /kɒt/ *n* lettino *m*; (*Am:* camp*bed*) branda *f*

cottage /ˈkɒtɪdʒ/ *n* casetta *f*. ~ **'cheese** *n* fiocchi *mpl* di latte

cotton /ˈkɒtn/ *n* cotone *m* ● *attrib* di cotone ● **cotton on** *vi fam* capire

cotton 'wool *n* cotone *m* idrofilo

couch /kaʊtʃ/ *n* divano *m*. ~ **pota-to** *n* pantofolaio, -a *f*

couchette /kuːˈʃet/ *n* cuccetta *f*

cough /kɒf/ *n* tosse *f* ● *vi* tossire. **cough up** *vt/i* sputare; (*fam: pay*) sborsare

'cough mixture *n* sciroppo *m* per la tosse

could /kʊd, *atono* /kəd/ *v aux* (*see also* can²) ● **I have a glass of water?** potrei avere un bicchier d'acqua?; **I** ~**n't do it even if I wanted to** non potrei farlo nemmeno se lo volessi; **I** ~**n't care less** non potrebbe importarmene di meno; **he** ~**n't have done it without help** non avrebbe potuto farlo senza aiuto; **you** ~ **have phoned** avresti potuto telefonare

council /ˈkaʊnsl/ *n* consiglio *m*. ~ **house** *n* casa *f* popolare

councillor /ˈkaʊnsələ(r)/ *n* consigliere, -a *mf*

'council tax *n* imposta *f* locale sugli immobili

counsel /ˈkaʊnsl/ *n* consigli *mpl*; *Jur* avvocato *m* ● *vt* (*pt/pp* coun-selled) consigliare a (*person*). ~**lor** *n* consigliere, -a *mf*

count¹ /kaʊnt/ *n* (*nobleman*) conte *m*

count² *n* conto *m*; **keep** ~ tenere il conto ● *vt/i* contare. **count on** *vt* contare su

countdown /ˈkaʊntdaʊn/ *n* conto *m* alla rovescia

countenance /ˈkaʊntənəns/ *n* espressione *f* ● *vt* approvare

counter¹ /ˈkaʊntə(r)/ *n* banco *m*; (*in games*) gettone *m*

counter² *adv* ~ **to** contro, in con-

trasto a; **go** ~ **to** sth andare contro qcsa ● *vt/i* opporre (*measure, effect*); parare (*blow*)

counter'act *vt* neutralizzare

counter-attack *n* contrattacco *m*

counter-'espionage *n* contro-spionaggio *m*

counterfeit /-fɪt/ *a* contraffatto ● *n* contraffazione *f* ● *vt* contraf-fare

'counterfoil *n* matrice *f*

'counterpart *n* equivalente *mf*

counter-pro'ductive *a* contro-produttivo

'countersign *vt* controfirmare

countess /ˈkaʊntɪs/ *n* contessa *f*

countless /ˈkaʊntlɪs/ *a* innumere-vole

country /ˈkʌntrɪ/ *n* nazione *f*, pae-se *m*; (*native land*) patria *f*; (*countryside*) campagna *f*; **in the** ~ in campagna; **go to the** ~ anda-re in campagna; *Pol* indire le ele-zioni politiche. ~**man** *n* uomo *m* di campagna; (*fellow* ~man) com-patriota *m*. ~**side** *n* campagna *f*

county /ˈkaʊntɪ/ *n* contea *f* (*unità amministrativa britannica*)

coup /kuː/ *n Pol* colpo *m* di stato

couple /ˈkʌpl/ *n* coppia *f*; **a** ~ **of** un paio di

coupon /ˈkuːpɒn/ *n* tagliando *m*; (*for discount*) buono *m* sconto

courage /ˈkʌrɪdʒ/ *n* coraggio *m*. ~**ous** /kəˈreɪdʒəs/ *a* coraggioso

courgette /kʊəˈʒet/ *n* zucchino *m*

courier /ˈkʊrɪə(r)/ *n* corriere *m*; (*for tourists*) guida *f*

course /kɔːs/ *n Sch* corso *m*; *Naut* rotta *f*; *Culin* portata *f*; (*for golf*) campo *m*; ~ **of treatment** *Med* se-rie *f inv* di cure; **of** ~ naturalmen-te; **in the** ~ **of** durante; **in due** ~ a tempo debito

court /kɔːt/ *n* tribunale *m*; *Sport* campo *m*; **take sb to** ~ citare qcno in giudizio ● *vt* fare la corte a (*woman*); sfidare (*danger*); ~**ing couples** coppiette *fpl*

courteous /ˈkɜːtɪəs/ *a* cortese

courtesy /ˈkɜːtəsɪ/ *n* cortesia *f*

court: ~ **'martial** n (pl ~**s martial**) corte f marziale ● ~**-martial** vt (pt ~**martialled**) portare davanti alla corte marziale; ~**yard** n cortile m

cousin /'kʌzn/ n cugino, -a mf

cove /kəʊv/ n insenatura f

cover /'kʌvə(r)/ n copertura f; (of cushion, to protect sth) fodera f; (of book, magazine) copertina f; **take** ~ mettersi al riparo; **under separate** ~ a parte ● vt coprire; foderare (cushion); Journ fare un servizio su. **cover up** vt coprire; fig soffocare (scandal)

coverage /'kʌvərɪdʒ/ n Journ **it got a lot of** ~ i media gli hanno dedicato molto spazio

cover: ~ **charge** n coperto m. ~**ing** n copertura f; (for floor) rivestimento m; ~**ing letter** lettera f d'accompagnamento. ~**-up** n messa f a tacere

covet /'kʌvɪt/ vt bramare

cow /kaʊ/ n vacca f, mucca f

coward /'kaʊəd/ n vigliacco, -a mf. ~**ice** /-ɪs/ n vigliaccheria f. ~**ly** a da vigliacco

'**cowboy** n cowboy m inv; fam buffone m

cower /'kaʊə(r)/ vi acquattarsi

'**cowshed** n stalla f

cox /kɒks/ n, **coxswain** /'kɒksn/ n timoniere, -a mf

coy /kɔɪ/ a falsamente timido; (flirtatiously) civettuolo; **be** ~ **about sth** essere evasivo su qcsa

crab /kræb/ n granchio m

crack /kræk/ n (in wall) crepa f; (in china, glass, bone) incrinatura f; (noise) scoppio m; (fam: joke) battuta f; **have a** ~ (try) fare un tentativo ● a (fam: best) di prim'ordine ● vt incrinare (china, glass); schiacciare (nut); decifrare (code); fam risolvere (problem); ~ **a joke** fam fare una battuta ● vi (china, glass:) incrinarsi; (whip:) schioccare. **crack down** vi fam prendere seri provvedimenti.

crack down on vt fam prendere seri provvedimenti contro

cracked /krækt/ a (plaster) crepato; (skin) screpolato; (rib) incrinato; (fam: crazy) svitato

cracker /'krækə(r)/ n (biscuit) cracker m inv; (firework) petardo m; [Christmas] ~ tubo m di cartone colorato contenente una sorpresa

crackers /'krækəz/ a fam matto

crackle /'krækl/ vi crepitare

cradle /'kreɪdl/ n culla f

craft[1] /krɑːft/ n inv (boat) imbarcazione f

craft[2] n mestiere m; (technique) arte f. ~**sman** n artigiano m

crafty /'krɑːftɪ/ a (-ier, -iest) astuto

crag /kræg/ n rupe f. ~**gy** a scosceso; (face) dai lineamenti marcati

cram /kræm/ v (pt/pp **crammed**) ● vt stipare (into in) ● vi (for exams) sgobbare

cramp /kræmp/ n crampo m. ~**ed** a (room) stretto; (handwriting) appiccicato

crampon /'kræmpən/ n rampone m

cranberry /'krænbərɪ/ n Culin mirtillo m rosso

crane /kreɪn/ n (at docks, bird) gru f inv ● vt ~ **one's neck** allungare il collo

crank[1] /kræŋk/ n tipo, -a mf strampalato, -a

crank[2] n Techn manovella f. ~**shaft** n albero m a gomiti

cranky /'kræŋkɪ/ a strampalato; (Am: irritable) irritabile

cranny /'krænɪ/ n fessura f

crash /kræʃ/ n (noise) fragore m; Auto, Aeron incidente m; Comm crollo m ● vi schiantarsi (into contro); (plane:) precipitare ● vt schiantare (car)

crash: ~ **course** n corso m intensivo. ~**helmet** n casco m. ~**landing** n atterraggio m di fortuna

crate /kreɪt/ n (for packing) cassa f

crater /ˈkreɪtə(r)/ n cratere m

crav|e /kreɪv/ vt morire dalla voglia di. **~ing** n voglia f smodata

crawl /krɔːl/ n (swimming) stile m libero; **do the ~** nuotare a stile libero; **at a ~** a passo di lumaca ● vi andare carponi; **~lebst** vi andare carponi, **-lebst** slither di. **~er lane** n Auto corsia f riservata al traffico lento

crayon /ˈkreɪən/ n pastello m a cera; (pencil) matita f colorata

craze /kreɪz/ n mania f

crazy /ˈkreɪzɪ/ a (-ier, -iest) matto; **be ~ about** andar matto per

creak /kriːk/ n scricchiolio m ● vi scricchiolare

cream /kriːm/ n crema f; (fresh) panna f ● a (colour) [bianco] panna inv ● vt Culin sbattere. **~ 'cheese** n formaggio m cremoso. **~y** a cremoso

crease /kriːs/ n piega f ● vt stropicciare ● vi stropicciarsi. **~-resistant** a che non si stropiccia

creat|e /kriːˈeɪt/ vt creare. **~ion** /-ˈeɪʃn/ n creazione f. **~ive** /-tɪv/ a creativo. **~or** n creatore, -trice mf

creature /ˈkriːtʃə(r)/ n creatura f

crèche /kreʃ/ n asilo m nido

credentials /krɪˈdenʃlz/ npl credenziali fpl

credibility /kredəˈbɪlətɪ/ n credibilità f

credible /ˈkredəbl/ a credibile

credit /ˈkredɪt/ n credito m; (honour) merito m; **take the ~ for** prendersi il merito di ● vt (pt/pp **credited**) accreditare; **~ sb with sth** Comm accreditare qcsa a qcno; fig attribuire qcsa a qcno. **~able** /-əbl/ a lodevole

credit: ~ card n carta f di credito. **~or** n creditore, -trice mf

creed /kriːd/ n credo m inv

creek /kriːk/ n insenatura f; (Am: stream) torrente m

creep /kriːp/ vi (pt/pp **crept**) muoversi furtivamente ● n fam tipo m

viscido. **~er** n pianta f rampicante. **~y** a che fa venire i brividi

cremat|e /krɪˈmeɪt/ vt cremare. **~ion** /-ˈmeɪʃn/ n cremazione f

crematorium /kreməˈtɔːrɪəm/ n crematorio m

crêpe /kreɪp/ n (fabric) crespo m

crept /krept/ see **creep**

crescent /ˈkresənt/ n mezzaluna f

cress /kres/ n crescione m

crest /krest/ n cresta f; (coat of arms) cimiero m

Crete /kriːt/ n Creta f

crevasse /krɪˈvæs/ n crepaccio m

crevice /ˈkrevɪs/ n crepa f

crew /kruː/ n equipaggio m; (gang) équipe f inv. **~ cut** n capelli mpl a spazzola. **~ neck** n girocollo m

crib¹ /krɪb/ n (for baby) culla f

crib² /krɪb/ vt/i (pt/pp **cribbed**) fam copiare

crick /krɪk/ n **~ in the neck** torcicollo m

cricket¹ /ˈkrɪkɪt/ n (insect) grillo m

cricket² n cricket m. **~er** n giocatore m di cricket

crime /kraɪm/ n crimine m; (criminality) criminalità f

criminal /ˈkrɪmɪnl/ a criminale; (law, court) penale ● n criminale mf

crimson /ˈkrɪmzn/ a cremisi inv

cringe /krɪndʒ/ vi (cower) acquattarsi; (at bad joke etc) fare una smorfia

crinkle /ˈkrɪŋkl/ vt spiegazzare ● vi spiegazzarsi

cripple /ˈkrɪpl/ n storpio, -a mf ● vt storpiare; fig danneggiare. **~d** a (person) storpio; (ship) danneggiato

crisis /ˈkraɪsɪs/ n (pl **-ses** /-siːz/) crisi f inv

crisp /krɪsp/ a croccante; (air) frizzante; (style) incisivo. **~bread** n crostini mpl di pane. **~s** npl patatine fpl

criterion /kraɪˈtɪərɪən/ n (pl **-ria** /-rɪə/) criterio m

critic /ˈkrɪtɪk/ n critico, -a mf. **~al**

a critico. **~ally** *adv* in modo critico; **~ally ill** gravemente malato

criticism /'krɪtɪsɪzm/ *n* critica *f*; **he doesn't like ~** non ama le critiche

criticize /'krɪtɪsaɪz/ *vt* criticare

croak /krəʊk/ *vi* gracchiare; ⟨frog:⟩ gracidare

crochet /'krəʊʃeɪ/ *n* lavoro *m* all'uncinetto ● *vt* fare all'uncinetto. **~-hook** *n* uncinetto *m*

crock /krɒk/ *n fam* **old ~** ⟨person⟩ rudere *m*; ⟨car⟩ macinino *m*

crockery /'krɒkərɪ/ *n* terrecotte *fpl*

crocodile /'krɒkədaɪl/ *n* coccodrillo *m*. **~ tears** lacrime *fpl* di coccodrillo

crocus /'krəʊkəs/ *n* (*pl* **-es**) croco *m*

crony /'krəʊnɪ/ *n* compare *m*

crook /krʊk/ *n* ⟨*fam: criminal*⟩ truffatore, -trice *mf*

crooked /'krʊkɪd/ *a* storto; ⟨limb⟩ storpiato; ⟨*fam: dishonest*⟩ disonesto

crop /krɒp/ *n* raccolto *m*, *fig* quantità *f inv* ● *v* (*pt/pp* **cropped**) ● *vt* coltivare. **crop up** *vi fam* presentarsi

croquet /'krəʊkeɪ/ *n* croquet *m*

croquette /krəʊ'ket/ *n* crocchetta *f*

cross /krɒs/ *a* ⟨annoyed⟩ arrabbiato; **talk at ~ purposes** fraintendersi ● *n* croce *f*; Bot, Zool incrocio *m* ● *vt* sbarrare ⟨cheque⟩; incrociare ⟨road, animals⟩; **~ oneself** farsi il segno della croce; **~ one's arms** incrociare le braccia; **~ one's legs** accavallare le gambe; **keep one's fingers ~ed for sb** tenere le dita incrociate per qcno; **it ~ed my mind** mi è venuto in mente ● *vi* ⟨go across⟩ attraversare; ⟨lines:⟩ incrociarsi. **cross out** *vt* depennare

cross: **~bar** *n* (*of goal*) traversa *f*; (*on bicycle*) canna *f*. **~-country** *n* Sport corsa *f* campestre. **~-examine** *vt* sottoporre a contro-

interrogatorio. **~-exami'nation** *n* controinterrogatorio *m*. **~-'eyed** *a* strabico. **~-fire** *n* fuoco *m* incrociato. **~-ing** *n* (*for pedestrians*) passaggio *m* pedonale; (*sea journey*) traversata *f*. **~-'reference** *n* rimando *m*. **~-roads** *n* incrocio *m*. **~-'section** *n* sezione *f*; (*of community*) campione *m*. **~-wise** *adv* in diagonale. **~word** *n* **~word** [puzzle] parole *fpl* crociate

crotchet /'krɒtʃɪt/ *n Mus* seminima *f*

crotchety /'krɒtʃɪtɪ/ *a* irritabile

crouch /kraʊtʃ/ *vi* accovacciarsi

crow /krəʊ/ *n* corvo *m*; **as the ~ flies** in linea d'aria ● *vi* cantare. **~bar** *n* piede *m* di porco

crowd /kraʊd/ *n* folla *f* ● *vt* affollare ● *vi* affollarsi. **~ed** /'kraʊdɪd/ *a* affollato

crown /kraʊn/ *n* corona *f* ● *vt* incoronare; incapsulare ⟨tooth⟩

crucial /'kru:ʃl/ *a* cruciale

crucifix /'kru:sɪfɪks/ *n* crocifisso *m*

cruci|fixion /kru:sɪ'fɪkʃn/ *n* crocifissione *f*. **~fy** /'kru:sɪfaɪ/ *vt* (*pt/pp* **-ied**) crocifiggere

crude /kru:d/ *a* (*oil*) greggio; ⟨language⟩ crudo; ⟨person⟩ rozzo

cruel /krʊəl/ *a* (**crueller**, **cruellest**) crudele (**to** verso). **~ly** *adv* con crudeltà. **~ty** *n* crudeltà *f*

cruis|e /kru:z/ *n* crociera *f* ● *vi* fare una crociera; ⟨car:⟩ andare a velocità di crociera. **~er** *n Mil* incrociatore *m*; (*motor boat*) motoscafo *m*. **~ing speed** *n* velocità *m inv* di crociera

crumb /krʌm/ *n* briciola *f*

crumbl|e /'krʌmbl/ *vt* sbriciolare ● *vi* sbriciolarsi; ⟨building, society:⟩ sgretolarsi. **~ly** *a* friabile

crumple /'krʌmpl/ *vt* spiegazzare ● *vi* spiegazzarsi

crunch /krʌntʃ/ *n fam* **when it comes to the ~** quando si viene al dunque ● *vt* sgranocchiare ● *vi* ⟨snow:⟩ scricchiolare

crusade /kru:'seɪd/ *n* crociata *f*.
~r *n* crociato *m*

crush /krʌʃ/ *n* (*crowd*) calca *f*;
have a ~ on sb essersi preso una
cotta per qcno ● *vt* schiacciare;
sgualcire ⟨*clothes*⟩

crust /krʌst/ *n* crosta *f*

crutch /krʌtʃ/ *n* gruccia *f*; Anat
inforcatura *f*

crux /krʌks/ *n* fig punto *m* cruciale

cry /kraɪ/ *n* grido *m*; have a ~ farsi un pianto; a far ~ from fig tutta
un'altra cosa rispetto a ● *vi* (*pt/pp*
cried) (*weep*) piangere; (*call*) gridare

crypt /krɪpt/ *n* cripta *f*. ~ic *a*
criptico

crystal /'krɪstl/ *n* cristallo *m*;
(*glassware*) cristalli *mpl*. ~lize *vi*
(*become clear*) concretizzarsi

cub /kʌb/ *n* (*animal*) cucciolo *m*;
C~ [Scout] lupetto *m*

Cuba /'kju:bə/ *n* Cuba *f*

cubby-hole /'kʌbɪ-/ *n* (*compartment*) scomparto *m*; (*room*) ripo-
stiglio *m*

cub|e /kju:b/ *n* cubo *m*. ~ic *a* cubi-
co

cubicle /'kju:bɪkl/ *n* cabina *f*

cuckoo /'kʊku:/ *n* cuculo *m*. ~
clock *n* orologio *m* a cucù

cucumber /'kju:kʌmbə(r)/ *n* ce-
triolo *m*

cuddl|e /'kʌdl/ *vt* coccolare ● *vi*
~e up to starsene accoccolato insieme a ● *n* have a ~e ⟨*child*⟩: farsi coccolare; ⟨*lovers*⟩: abbracciarsi. ~y *a* tenerone; (*wanting
cuddles*) coccolone. ~y 'toy *n*
peluche *m inv*

cudgel /'kʌdʒl/ *n* randello *m*

cue¹ /kju:/ *n* segnale *m*; Theat battuta *f* d'entrata

cue² *n* (*in billiards*) stecca *f*. ~ ball
n pallino *m*

cuff /kʌf/ *n* polsino *m*; (*Am*: *turn-
up*) orlo *m*; (*blow*) scapaccione *m*;
off the ~ improvvisando ● *vt*
dare una pacca a. ~-link *n* gemello *m*

cul-de-sac /'kʌldəsæk/ *n* vicolo *m*
cieco

culinary /'kʌlɪnərɪ/ *a* culinario

cull /kʌl/ *vt* scegliere ⟨*flowers*⟩;
(*kill*) selezionare e uccidere

culminat|e /'kʌlmɪneɪt/ *vi* culmi-
nare. ~ion /-'neɪʃn/ *n* culmine *m*

culottes /kju:'lɒts/ *npl* gonna *fsg*
pantalone

culprit /'kʌlprɪt/ *n* colpevole *mf*

cult /kʌlt/ *n* culto *m*

cultivate /'kʌltɪveɪt/ *vt* coltivare;
fig coltivarsi ⟨*person*⟩

cultural /'kʌltʃərəl/ *a* culturale

culture /'kʌltʃə(r)/ *n* cultura *f*. ~d
a colto

cumbersome /'kʌmbəsəm/ *a* ingombrante

cumulative /'kju:mjʊlətɪv/ *a* cu-
mulativo

cunning /'kʌnɪŋ/ *a* astuto ● *n* astuzia *f*

cup /kʌp/ *n* tazza *f*; (*prize*, *of bra*)
coppa *f*

cupboard /'kʌbəd/ *n* armadio *m*.
~ love *fam* amore *m* interessato

Cup 'Final *n* finale *f* di coppa

Cupid /'kju:pɪd/ *n* Cupido *m*

curable /'kjʊərəbl/ *a* curabile

curate /'kjʊərət/ *n* curato *m*

curator /kjʊə'reɪtə(r)/ *n* direttore,
-trice *mf* (*di museo*)

curb /kɜ:b/ *vt* tenere a freno

curdle /'kɜ:dl/ *vi* coagularsi

cure /kjʊə(r)/ *n* cura *f* ● *vt* curare;
(*salt*) mettere sotto sale; (*smoke*)
affumicare

curfew /'kɜ:fju:/ *n* coprifuoco *m*

curio /'kjʊərɪəʊ/ *n* curiosità *f inv*

curiosity /kjʊərɪ'ɒsətɪ/ *n* curiosità
f

curious /'kjʊərɪəs/ *a* curioso. ~ly
adv (*strangely*) curiosamente

curl /kɜ:l/ *n* ricciolo *m* ● *vt* arricciare ● *vi* arricciarsi. **curl up** *vi*
raggomitolarsi

curler /'kɜ:lə(r)/ *n* bigodino *m*

curly /'kɜ:lɪ/ *a* (-ier, -iest) riccio

currant /'kʌrənt/ *n* (*dried*) uvetta *f*

currency /'kʌrənsɪ/ *n* valuta *f*; (*of*

word) ricorrenza *f*; **foreign ~** valuta *f* estera

current /'kʌrənt/ *a* corrente ● *n* corrente *f*. **~ affairs** *or* **events** *npl* attualità *fsg*. **~ly** *adv* attualmente

curriculum /kə'rɪkjʊləm/ *n* programma *m* di studi. **~ vitae** /'vi:taɪ/ *n* curriculum vitae *m inv*

curry /'kʌrɪ/ *n* curry *m inv*; (*meal*) piatto *m* cucinato nel curry ● *vt* (*pt/pp -ied*) **~ favour with sb** cercare d'ingraziarsi qcno

curse /kɜ:s/ *n* maledizione *f*; imprecazione *f* ● *vt* maledire ● *vi* imprecare

cursor /'kɜ:sə(r)/ *n* cursore *m*

cursory /'kɜ:sərɪ/ *a* sbrigativo

curt /kɜ:t/ *a* brusco

curtail /kɜ:'teɪl/ *vt* ridurre

curtain /'kɜ:tn/ *n* tenda *f*; *Theat* sipario *m*

curtsy /'kɜ:tsɪ/ *n* inchino *m* ● *vi* (*pt/pp -ied*) fare l'inchino

curve /kɜ:v/ *n* curva *f* ● *vi* curvare; **~ to the right/left** curvare a destra/sinistra. **~d** *a* curvo

cushion /'kʊʃn/ *n* cuscino *m* ● *vt* attutire; (*protect*) proteggere

cushy /'kʊʃɪ/ *a* (**-ier, -iest**) *fam* facile

custard /'kʌstəd/ *n* (*liquid*) crema *f* pasticcera

custodian /kʌ'stəʊdɪən/ *n* custode *mf*

custody /'kʌstədɪ/ *n* (*of child*) custodia *f*; (*imprisonment*) detenzione *f* preventiva

custom /'kʌstəm/ *n* usanza *f*; *Jur* consuetudine *f*; *Comm* clientela *f*. **~ary** *a* (*habitual*) abituale; **it's ~ to...** è consuetudine.... **~er** *n* cliente *mf*

customs /'kʌstəmz/ *npl* dogana *f*. **~ officer** *n* doganiere *m*

cut /kʌt/ *n* (*with knife etc, of clothes*) taglio *m*; (*reduction*) riduzione *f*; (*in public spending*) taglio *m* ● *vt/i* (*pt/pp* **cut**, *pres p* **cutting**) tagliare; (*reduce*) ridurre; **~ one's finger** tagliarsi il dito; **~ sb's hair**

tagliare i capelli a qcno ● *vi* (*with cards*) alzare. **cut back** *vt* tagliare ⟨*hair*⟩; potare ⟨*hedge*⟩; (*reduce*) ridurre. **cut down** *vt* abbattere ⟨*tree*⟩; (*reduce*) ridurre. **cut off** *vt* tagliar via; (*disconnect*) interrompere; *fig* isolare; **I was ~ off** *Teleph* la linea è caduta. **cut out** *vt* ritagliare; (*delete*) eliminare; **be ~ out for** *fam* essere tagliato per; **~ it out!** *fam* dacci un taglio!. **cut up** *vt* (*slice*) tagliare a pezzi

'cut-back *n* riduzione *f*; (*in government spending*) taglio *m*

cute /kju:t/ *a fam* (*in appearance*) carino; (*clever*) acuto

cuticle /'kju:tɪkl/ *n* cuticola *f*

cutlery /'kʌtlərɪ/ *n* posate *fpl*

cutlet /'kʌtlɪt/ *n* cotoletta *f*

'cut-price *a* a prezzo ridotto; ⟨*shop*⟩ che fa prezzi ridotti

'cut-throat *a* spietato

cutting /'kʌtɪŋ/ *a* ⟨*remark*⟩ tagliente ● *n* (*from newspaper*) ritaglio *m*; (*of plant*) talea *f*

CV *n abbr* **curriculum vitae**

cyanide /'saɪənaɪd/ *n* cianuro *m*

cybernetics /saɪbə'netɪks/ *n* cibernetica *f*

cycle /'saɪkl/ *n* ciclo *m*; (*bicycle*) bicicletta *f*, bici *f inv fam* ● *vi* andare in bicicletta. **~ing** *n* ciclismo *m*. **~ist** *n* ciclista *m*

cyclone /'saɪkləʊn/ *n* ciclone *m*

cylind|er /'sɪlɪndə(r)/ *n* cilindro *m*. **~rical** /-'lɪndrɪkl/ *a* cilindrico

cymbals /'sɪmblz/ *npl Mus* piatti *mpl*

cynic /'sɪnɪk/ *n* cinico, -a *mf*. **~al** *a* cinico. **~ism** /-sɪzm/ *n* cinismo *m*

cypress /'saɪprəs/ *n* cipresso *m*

Cypriot /'sɪprɪət/ *n* cipriota *mf*

Cyprus /'saɪprəs/ *n* Cipro *m*

cyst /sɪst/ *n* ciste *f*. **~itis** /-'staɪtɪs/ *n* cistite *f*

Czech /tʃek/ *a* ceco; **~ Republic** Repubblica *f* Ceca ● *n* ceco, -a *mf*

Czechoslovak /tʃekə'sləʊvæk/ *a* cecoslovacco. **~ia** /-'vækɪə/ *n* Cecoslovacchia *f*

Dd

dab /dæb/ n colpetto m; **a ~ of** un pochino di ● vt (pt/pp **dabbed**) toccare leggermente ⟨eyes⟩. **dab on** vt mettere un po' di ⟨paint etc⟩

dabble /dæbl/ vi **~ in sth** fig occuparsi di qcsa a tempo perso

dachshund /dækshʊnd/ n bassotto m

dad[dy] /dæd[ɪ]/ n fam papà m inv, babbo m

daddy-'long-legs n zanzarone m [dei boschi]; (Am: spider) ragno m

daffodil /dæfədɪl/ n giunchiglia f

daft /dɑ:ft/ a sciocco

dagger /dægə(r)/ n stiletto m

dahlia /deɪlɪə/ n dalia f

daily /deɪlɪ/ a giornaliero ● adv giornalmente ● n ⟨newspaper⟩ quotidiano m; (fam: cleaner) donna f delle pulizie

dainty /deɪntɪ/ a (-ier, -iest) grazioso; ⟨movement⟩ delicato

dairy /deərɪ/ n caseificio m; (shop) latteria f. **~ cow** n mucca f da latte. **~ products** npl latticini mpl

dais /deɪs/ n pedana f

daisy /deɪzɪ/ n margheritina f; (larger) margherita f

dale /deɪl/ n liter valle f

dam /dæm/ n diga f ● vt (pt/pp **dammed**) costruire una diga su

damage /dæmɪdʒ/ n danno m (**to** a); **~es** pl Jur risarcimento msg ● vt danneggiare; fig nuocere a. **~ing** a dannoso

dame /deɪm/ n liter dama f; Am sl donna f

damn /dæm/ a fam maledetto ● adv ⟨lucky, late⟩ maledettamente ● n **I don't care** or **give a ~** fam non me ne frega un accidente ● vt dannare. **~ation** /-'neɪʃn/ n dannazione f ● int fam accidenti!

damp /dæmp/ a umido ● n umidità f ● vt = **dampen**

damp|en /dæmpən/ vt inumidire; fig raffreddare ⟨enthusiasm⟩. **~ness** n umidità f

dance /dɑ:ns/ n ballo m ● vt/i ballare. **~hall** n sala f da ballo. **~ music** n musica f da ballo

dancer /dɑ:nsə(r)/ n ballerino, -a mf

dandelion /dændɪlaɪən/ n dente m di leone

dandruff /dændrʌf/ n forfora f

Dane /deɪn/ n danese mf; **Great ~** danese m

danger /deɪndʒə(r)/ n pericolo m; **in/out of ~** in/fuori pericolo. **~ous** /-rəs/ a pericoloso. **~ously** adv pericolosamente; **~ously ill** in pericolo di vita

dangle /dæŋgl/ vi penzolare ● vt far penzolare

Danish /deɪnɪʃ/ a & n danese m. **~ 'pastry** n dolce m a base di pasta sfoglia contenente pasta di mandorle, mele ecc

dank /dæŋk/ a umido e freddo

Danube /dænju:b/ n Danubio m

dare /deə(r)/ vt/i osare; ⟨challenge⟩ sfidare (**to** a); **~ [to] do sth** osare fare qcsa; **I ~ say!** molto probabilmente! ● n sfida f. **~devil** n spericolato, -a f

daring /deərɪŋ/ a audace ● n audacia f

dark /dɑ:k/ a buio; **~ blue/brown** blu/marrone scuro; **it's getting ~** sta cominciando a fare buio; **~ horse** fig (in race, contest) vincitore m imprevisto; (not much known about) misterioso m; **keep sth ~** fig tenere qcsa nascosto ● n **after ~** col buio; **in the ~** al buio; **keep sb in the ~** fig tenere qcno all'oscuro

dark|en /dɑ:kn/ vt oscurare ● vi oscurarsi. **~ness** n buio m

'dark-room n camera f oscura

darling /dɑ:lɪŋ/ a adorabile; **my ~ Joan** carissima Joan ● n tesoro m

darn /dɑ:n/ vt rammendare. **~ing-needle** n ago m da rammendo

dart /dɑːt/ n dardo m; (in sewing) pince f inv; **~s** sg (game) freccette fpl ● vi lanciarsi

dartboard /dɑːtbɔːd/ n bersaglio m [per freccette]

dash /dæʃ/ n Typ trattino m; (in Morse) linea f; **a ~** of milk un goccio di latte; **make a ~ for** lanciarsi verso ● vi **I must ~** devo scappare ● vt far svanire ⟨hopes⟩. **dash off** vi scappar via ● vt ⟨write quickly⟩ buttare giù. **dash out** vi uscire di corsa

'dashboard n cruscotto m

dashing /dæʃɪŋ/ a ⟨bold⟩ ardito; (in appearance) affascinante

data /deɪtə/ npl & sg dati mpl. **~base** n base [di] dati f, database m inv. **~comms** /kɒmz/ n telematica f. **~ processing** n elaborazione f [di] dati

date[1] /deɪt/ n (fruit) dattero m

date[2] n data f; (meeting) appuntamento m; **to ~** fino ad oggi; **out of ~** ⟨not fashionable⟩ fuori moda; ⟨expired⟩ scaduto; ⟨information⟩ non aggiornato; **make a ~ with sb** dare un appuntamento a qcno; **be up to ~** essere aggiornato ● vt/i datare; ⟨go out with⟩ uscire con. **date back to** vi risalire a

dated /deɪtɪd/ a fuori moda; ⟨language⟩ antiquato

'date-line n linea f [del cambiamento] di data

daub /dɔːb/ vt imbrattare ⟨walls⟩

daughter /dɔːtə(r)/ n figlia f. **~-in-law** n (pl **~s-in-law**) nuora f

daunt /dɔːnt/ vt scoraggiare; **nothing ~ed** per niente scoraggiato. **~less** a intrepido

dawdle /dɔːdl/ vi bighellonare; ⟨over work⟩ cincischiarsi

dawn /dɔːn/ n alba f; **at ~** all'alba ● vi albeggiare; **it ~ed on me** fig mi è apparso chiaro

day /deɪ/ n giorno m; (whole day) giornata f; (period) epoca f; **these ~s** oggigiorno; **in those ~s** a quei tempi; **it's had its ~** fam ha fatto il suo tempo

day: **~break** n **at ~break** allo spuntar del giorno. **~dream** n sogno m ad occhi aperti ● vi sognare ad occhi aperti. **~light** n luce f del giorno. **~re'turn** n (ticket) biglietto m di andata e ritorno con validità giornaliera. **~time** n giorno m; **in the ~time** di giorno

daze /deɪz/ n **in a ~** stordito; fig sbalordito. **~d** a stordito; fig sbalordito

dazzle /dæzl/ vt abbagliare

deacon /diːkn/ n diacono m

dead /ded/ a morto; (numb) intorpidito; **~ body** morto m; **~ centre** pieno centro m ● adv **~ tired** stanco morto; **~ slow/easy** lentissimo/facilissimo; **you're ~ right** hai perfettamente ragione; **stop ~** fermarsi di colpo; **be ~ on time** essere in perfetto orario ● n **the ~** pl i morti; **in the ~ of night** nel cuore della notte

deaden /dedn/ vt attutire ⟨sound⟩; calmare ⟨pain⟩

dead: **~ 'end** n vicolo m cieco. **'heat** n it was a **~ heat** è finita a pari merito. **~line** n scadenza f. **~lock** n reach **~lock** fig giungere a un punto morto

deadly /dedlɪ/ a (**-ier, -iest**) mortale; ⟨fam: dreary⟩ barboso; **~ sins** peccati mpl capitali

deadpan /dedpæn/ a impassibile; ⟨humour⟩ all'inglese

deaf /def/ a sordo; **~ and dumb** sordomuto. **~-aid** n apparecchio m acustico

deaf|en /defn/ vt assordare; ⟨permanently⟩ render sordo. **~ening** a assordante. **~ness** n sordità f

deal /diːl/ n (agreement) patto m; (in business) accordo m; **whose ~?** (in cards) a chi tocca dare le carte?; **a good** or **great ~** molto; **get a raw ~** fam ricevere un trattamento ingiusto ● vt (pt/pp **dealt** /delt/) (in cards) dare; **~ sb a blow** dare un colpo a qcno. **deal in** vt trattare in. **deal out** vt ⟨hand

out) distribuire. **deal with** *vt*
(handle) occuparsi di; trattare con
(company); *(be about)* trattare di;
that's been ~t with è stato risolto
deal|er /'di:lə(r)/ *n* commerciante
mf. *(in drugs)* spacciatore, -trice
mf. **~ings** *npl* **have ~ings with**
avere a che fare con
dean /di:n/ *n* decano *m; Univ* ≈
preside *mf* di facoltà
dear /dɪə(r)/ *a* caro; *(in letter)*
Caro; *(formal)* Gentile ● *n* caro, -a
mf ● *int* oh **~! Dio mio!**. **~ly** *adv*
(love) profondamente; *(pay)* pro-
fumatamente
dearth /dɜ:θ/ *n* penuria *f*
death /deθ/ *n* morte *f*. **~ certificate** *n* certificato *m* di morte. **~
duty** *n* tassa *f* di successione
deathly /'deθlɪ/ *a* **~ silence** silenzio *m* di tomba ● *adv* **~ pale** di un
pallore cadaverico
death: ~ penalty *n* pena *f* di morte. **~-trap** *n* trappola *f* mortale
debar /dɪ'bɑ:(r)/ *vt (pt/pp* **de-
barred)** escludere
debase /dɪ'beɪs/ *vt* degradare
debatable /dɪ'beɪtəbl/ *a* discutibile
debate /dɪ'beɪt/ *n* dibattito *m* ● *vt*
discutere; *(in formal debate)* dibattere ● *vi* **~ whether to...** considerare se...
debauchery /dɪ'bɔ:tʃərɪ/ *n* disso-
lutezza *f*
debility /dɪ'bɪlɪtɪ/ *n* debilitazione *f*
debit /'debɪt/ *n* debito *m* ● *vt*
(pt/pp **debited)** *Comm* addebitare
(sum)
debris /'debri:/ *n* macerie *fpl*
debt /det/ *n* debito *m*; **be in ~** avere dei debiti. **~or** *n* debitore, -trice
mf
début /'deɪbu:/ *n* debutto *m*
decade /dekeɪd/ *n* decennio *m*
decaden|ce /'dekədəns/ *n* deca-
denza *f*. **~t** *a* decadente
decaffeinated /di:'kæfɪneɪtɪd/ *a*
decaffeinato
decant /dɪ'kænt/ *vt* travasare. **~er**
n caraffa *f (di cristallo)*

decapitate /dɪ'kæpɪteɪt/ *vt* decapitare
decay /dɪ'keɪ/ *n (also fig)* decaden-
za *f*; *(rot)* decomposizione *f*; *(of
tooth)* carie *f inv* ● *vi* imputridire;
(rot) decomporsi; *(tooth:)* cariarsi
deceased /dɪ'si:st/ *a* defunto ● *n*
the ~d il defunto; la defunta
deceit /dɪ'si:t/ *n* inganno *m*. **~ful** *a*
falso
deceive /dɪ'si:v/ *vt* ingannare
December /dɪ'sembə(r)/ *n* dicembre *m*
decency /'di:sənsɪ/ *n* decenza *f*
decent /'di:sənt/ *a* decente;
(respectable) rispettabile; **very ~
of you** molto gentile da parte tua.
~ly *adv* decentemente; *(kindly)*
gentilmente
decentralize /di:'sentrəlaɪz/ *vt*
decentrare
decep|tion /dɪ'sepʃn/ *n* inganno
m. **~ive** /-tɪv/ *a* ingannevole.
~ively *adv* ingannevolmente; **it
looks ~ively easy** sembra facile,
ma non lo è
decibel /'desɪbel/ *n* decibel *m inv*
decide /dɪ'saɪd/ *vt* decidere ● *vi*
decidere **(on** di)
decided /dɪ'saɪdɪd/ *a* risoluto.
~ly *adv* risolutamente; *(without
doubt)* senza dubbio
deciduous /dɪ'sɪdjʊəs/ *a* a foglie
decidue
decimal /'desɪml/ *a* decimale ● *n*
numero *m* decimale. **~ 'point** *n*
virgola *f*
decimate /'desɪmeɪt/ *vt* decimare
decipher /dɪ'saɪfə(r)/ *vt* decifrare
decision /dɪ'sɪʒn/ *n* decisione *f*
decisive /dɪ'saɪsɪv/ *a* decisivo
deck¹ /dek/ *vt* abbigliare
deck² *n Naut* ponte *m*; **on ~** in coperta; **top ~** *(of bus)* piano *m* di sopra; **~ of cards** mazzo *m*. **~-chair**
n [sedia *f* a] sdraio *f inv*
declaration /deklə'reɪʃn/ *n* dichiarazione *f*
declare /dɪ'kleə(r)/ *vt* dichiarare;
anything to ~? niente da dichiarare?

declension /dɪˈklenʃn/ n declinazione f

decline /dɪˈklaɪn/ n declino m ● vt also Gram declinare ● vi (decrease) diminuire; ⟨health:⟩ deperire; ⟨say no⟩ rifiutare

decode /diːˈkəʊd/ vt decifrare; Comput decodificare

decompose /diːkəmˈpəʊz/ vi decomporsi

décor /ˈdeɪkɔː(r)/ n decorazione f; (including furniture) arredamento m

decorat|e /ˈdekəreɪt/ vt decorare; (paint) pitturare; (wallpaper) tappezzare. **~ion** /-ˈreɪʃn/ n decorazione f. **~ive** /-rətɪv/ a decorativo. **~or** n painter and **~or** imbianchino m

decorum /dɪˈkɔːrəm/ n decoro m

decoy¹ /ˈdiːkɔɪ/ n esca f

decoy² /dɪˈkɔɪ/ vt adescare

decrease¹ /ˈdiːkriːs/ n diminuzione f

decrease² /dɪˈkriːs/ vt/i diminuire

decree /dɪˈkriː/ n decreto m ● vt (pt/pp **decreed**) decretare

decrepit /dɪˈkrepɪt/ a decrepito

dedicat|e /ˈdedɪkeɪt/ vt dedicare. **~ed** a ⟨person⟩ scrupoloso. **~ion** /-ˈkeɪʃn/ n dedizione f; (in book) dedica f

deduce /dɪˈdjuːs/ vt dedurre (**from** da)

deduct /dɪˈdʌkt/ vt dedurre

deduction /dɪˈdʌkʃn/ n deduzione f

deed /diːd/ n azione f; Jur atto m di proprietà

deem /diːm/ vt ritenere

deep /diːp/ a profondo; **go off the ~ end** fam arrabbiarsi. **~ly** adv profondamente

deepen /ˈdiːpn/ vt approfondire; scavare più profondamente ⟨trench⟩ ● vi approfondirsi; ⟨fig: mystery:⟩ infittirsi

deep-'freeze n congelatore m

deer /dɪə(r)/ n inv cervo m

deface /dɪˈfeɪs/ vt sfigurare ⟨picture⟩; deturpare ⟨monument⟩

defamat|ion /defəˈmeɪʃn/ n diffamazione f. **~ory** /dɪˈfæmətərɪ/ a diffamatorio

default /dɪˈfɔːlt/ n ⟨Jur: nonpayment⟩ morosità f; (failure to appear) contumacia f; **win by ~** Sport vincere per abbandono dell'avversario; **in ~ of** per mancanza di ● a **~ drive** Comput lettore m di default ● vi (not pay) venir meno a un pagamento

defeat /dɪˈfiːt/ n sconfitta f ● vt sconfiggere; ⟨frustrate⟩ vanificare ⟨attempts⟩; **that ~s the object** questo fa fallire l'obiettivo

defect¹ /dɪˈfekt/ vi Pol fare defezione

defect² /ˈdiːfekt/ n difetto m. **~ive** /dɪˈfektɪv/ a difettoso

defence /dɪˈfens/ n difesa f. **~less** a indifeso

defend /dɪˈfend/ vt difendere; ⟨justify⟩ giustificare. **~ant** n Jur imputato, -a mf

defensive /dɪˈfensɪv/ a difensivo ● n difensiva f; **on the ~** sulla difensiva

defer /dɪˈfɜː(r)/ v (pt/pp **deferred**) ● vt (postpone) rinviare ● vi **~ to sb** rimettersi a qcno

deferen|ce /ˈdefərəns/ n deferenza f. **~tial** /-ˈrenʃl/ a deferente

defian|ce /dɪˈfaɪəns/ n sfida f; **in ~ce of** sfidando. **~t** a ⟨person⟩ ribelle; ⟨gesture, attitude⟩ di sfida. **~tly** adv con aria di sfida

deficien|cy /dɪˈfɪʃənsɪ/ n insufficienza f. **~t** a insufficiente; **be ~t** in mancare di

deficit /ˈdefɪsɪt/ n deficit m inv

defile /dɪˈfaɪl/ vt fig contaminare

define /dɪˈfaɪn/ vt definire

definite /ˈdefɪnɪt/ a definito; ⟨certain⟩ ⟨answer, yes⟩ definitivo; ⟨improvement, difference⟩ netto; **he was ~ about it** è stato chiaro in proposito. **~ly** adv sicuramente

definition /defɪˈnɪʃn/ n definizione f

definitive /dɪˈfɪnɪtɪv/ a definitivo

deflat|e /dɪˈfleɪt/ vt sgonfiare.
~ion /-eɪʃn/ n Comm deflazione f

deflect /dɪˈflekt/ vt deflettere

deform|ed /dɪˈfɔːmd/ a deforme.
~ity n deformità f inv

defraud /dɪˈfrɔːd/ vt defraudare

defrost /diːˈfrɒst/ vt sbrinare
⟨fridge⟩; scongelare ⟨food⟩

deft /deft/ a abile

defunct /dɪˈfʌŋkt/ a morto e sepol-
to; ⟨law⟩ caduto in disuso

defuse /diːˈfjuːz/ vt disinnescare;
calmare ⟨situation⟩

defy /dɪˈfaɪ/ vt (pt/pp **-ied**) ⟨chal-
lenge⟩ sfidare; resistere a ⟨at-
tempt⟩; ⟨not obey⟩ disobbedire a

degenerate¹ /dɪˈdʒenəreɪt/ vi degenerare; **~ into** fig degenerare
in

degenerate² /dɪˈdʒenərət/ a dege-
nerato

degrading /dɪˈgreɪdɪŋ/ a degradante

degree /dɪˈgriː/ n grado m; Univ
laurea f; **20 ~s** 20 gradi; **not to
be the same ~** non allo stesso
livello

dehydrate /diːˈhaɪdreɪt/ vt disi-
dratare. **~d** /-ɪd/ a disidratato

de-ice /diːˈaɪs/ vt togliere il ghiaccio da

deign /deɪn/ vi **~ to do sth** degnarsi di fare qcsa

deity /ˈdiːɪti/ n divinità f inv

dejected /dɪˈdʒektɪd/ a demoraliz-
zato

delay /dɪˈleɪ/ n ritardo m; **without
~** senza indugio ● vt ritardare; **be
~ed** ⟨person:⟩ essere trattenuto;
⟨train, aircraft:⟩ essere in ritardo
● vi indugiare

delegate¹ /ˈdelɪgət/ n delegato, -a
mf

delegat|e² /ˈdelɪgeɪt/ vt delegare.
~ion /-ˈgeɪʃn/ n delegazione f

delet|e /dɪˈliːt/ vt cancellare. **~ion**
/-iːʃn/ n cancellatura f

deliberate¹ /dɪˈlɪbərət/ a delibera-
to; ⟨slow⟩ posato. **~ly** adv
deliberatamente; ⟨slowly⟩ in modo
posato

deliberat|e² /dɪˈlɪbəreɪt/ vt/i deli-

berare. **~ion** /-ˈreɪʃn/ n delibera-
zione f

delicacy /ˈdelɪkəsɪ/ n delicatezza f;
⟨food⟩ prelibatezza f

delicate /ˈdelɪkət/ a delicato

delicatessen /delɪkəˈtesn/ n negozio m di specialità gastronomiche

delicious /dɪˈlɪʃəs/ a delizioso

delight /dɪˈlaɪt/ n piacere m ● vt
deliziare ● vi **~ in** dilettarsi con.
~ed a lieto. **~ful** a delizioso

delinquen|cy /dɪˈlɪŋkwənsɪ/ n de-
linquenza f. **~t** a delinquente ● n
delinquente mf

deli|rious /dɪˈlɪrɪəs/ a **be ~rious**
delirare; ⟨fig: very happy⟩ essere
pazzo di gioia. **~rium** /-rɪəm/ n de-
lirio m

deliver /dɪˈlɪvə(r)/ vt consegnare;
recapitare ⟨post, newspaper⟩; tenere ⟨speech⟩; dare ⟨message⟩; tirare ⟨blow⟩; ⟨set free⟩ liberare; **~ a
baby** far nascere un bambino.
~ance n liberazione f. **~y** n consegna f; ⟨of post⟩ distribuzione f;
Med parto m; **cash on ~y** pagamento m alla consegna

delude /dɪˈluːd/ vt ingannare; **~
oneself** illudersi

deluge /ˈdeljuːdʒ/ n diluvio m ⟨vt⟩
⟨fig: with requests etc⟩ inondare

delusion /dɪˈluːʒn/ n illusione f

de luxe /dəˈlʌks/ a di lusso

delve /delv/ vi **~ into** ⟨into pocket
etc⟩ frugare in; ⟨into notes, the
past⟩ fare ricerche in

demand /dɪˈmɑːnd/ n richiesta f;
Comm domanda f; **in ~** richiesto;
on ~ a richiesta ● vt esigere
⟨of/from da⟩. **~ing** a esigente

demarcation /diːmɑːˈkeɪʃn/ n
demarcazione f

demean /dɪˈmiːn/ vt **~ oneself** abbassarsi ⟨**to** a⟩

demeanour /dɪˈmiːnə(r)/ n comportamento m

demented /dɪˈmentɪd/ a demente

demise /dɪˈmaɪz/ n decesso m

demister /diːˈmɪstə(r)/ n Auto
sbrinatore m

demo /ˈdeməʊ/ n (pl **~s**) fam manifestazione *f*; **~ disk** *Comput* demodisk *m inv*

democracy /dɪ'mɒkrəsɪ/ *n* democrazia *f*

democrat /'deməkræt/ *n* democratico, -a *mf*. **~ic** /-'krætɪk/ *a* democratico

demolish /dɪ'mɒlɪʃ/ *vt* demolire. **~lition** /-lɪʃn/ *n* demolizione *f*

demon /'di:mən/ *n* demonio *m*

demonstrate /'demənstreɪt/ *vt* dimostrare; fare una dimostrazione sull'uso di ⟨appliance⟩ ● *vi Pol* manifestare. **~ion** /-'streɪʃn/ *n* dimostrazione *f*; *Pol* manifestazione *f*

demonstrative /dɪ'mɒnstrətɪv/ *a Gram* dimostrativo; **be ~** essere espansivo

demonstrator /'demənstreɪtə(r)/ *n Pol* manifestante *mf*; ⟨for product⟩ dimostratore, -trice *mf*

demoralize /dɪ'mɒrəlaɪz/ *vt* demoralizzare

demote /dɪ'məʊt/ *vt* retrocedere di grado; *Mil* degradare

demure /dɪ'mjʊə(r)/ *a* schivo

den /den/ *n* tana *f*; ⟨room⟩ rifugio *m*

denial /dɪ'naɪəl/ *n* smentita *f*

denim /'denɪm/ *n* [tessuto *m*] jeans *m*; **~s** *pl* [blue]jeans *mpl*

Denmark /'denmɑ:k/ *n* Danimarca *f*

denomination /dɪnɒmɪ'neɪʃn/ *n Relig* confessione *f*; ⟨money⟩ valore *f*

denounce /dɪ'naʊns/ *vt* denunciare

dense /dens/ *a* denso; ⟨crowd, forest⟩ fitto; ⟨stupid⟩ ottuso. **~ly** *adv* ⟨populated⟩ densamente; **~ly wooded** fittamente ricoperto di alberi. **~ity** *n* densità *f inv*; ⟨of forest⟩ fittezza *f*

dent /dent/ *n* ammacatura *f* ● *vt* ammacare; **~ed** *a* ammacato

dental /'dentl/ *a* dei denti; ⟨treatment⟩ dentistico; ⟨hygiene⟩ dentale. **~ surgeon** *n* odontoiatra *mf*, medico *m* dentista

dentist /'dentɪst/ *n* dentista *mf*. **~ry** *n* odontoiatria *f*

dentures /'dentʃəz/ *npl* dentiera *fsg*

denunciation /dɪnʌnsɪ'eɪʃn/ *n* denuncia *f*

deny /dɪ'naɪ/ *vt* (*pt/pp* **-ied**) negare; ⟨officially⟩ smentire; **~ sb sth** negare qcsa a qcno

deodorant /di:'əʊdərənt/ *n* deodorante *m*

depart /dɪ'pɑ:t/ *vi* ⟨plane, train:⟩ partire; ⟨liter: person⟩ andare via; ⟨deviate⟩ allontanarsi (**from** da)

department /dɪ'pɑ:tmənt/ *n* reparto *m*; *Pol* ministero *m*; ⟨of company⟩ sezione *f*, *Univ* dipartimento *m*. **~ store** *n* grande magazzino *m*

departure /dɪ'pɑ:tʃə(r)/ *n* partenza *f*; ⟨from rule⟩ allontanamento *m*; **new ~** svolta *f*

depend /dɪ'pend/ *vi* dipendere (**on** da); ⟨rely⟩ contare (**on** su); **it all ~s** dipende; **~ing on what he says** a seconda di quello che dice. **~able** /-əbl/ *a* fidato. **~ant** *n* persona *f* a carico. **~ence** *n* dipendenza *f*. **~ent** *a* dipendente (**on** da)

depict /dɪ'pɪkt/ *vt* ⟨in writing⟩ dipingere; ⟨with picture⟩ rappresentare

depilatory /dɪ'pɪlətərɪ/ *n* ⟨cream⟩ crema *f* depilatoria

deplete /dɪ'pli:t/ *vt* ridurre; **totally ~d** completamente esaurito

deplorable /dɪ'plɔ:rəbl/ *a* deplorevole. **~e** *vt* deplorare

deploy /dɪ'plɔɪ/ *vt Mil* spiegare ● *vi* schierarsi

deport /dɪ'pɔ:t/ *vt* deportare. **~ation** /di:pɔ:'teɪʃn/ *n* deportazione *f*

depose /dɪ'pəʊz/ *vt* deporre

deposit /dɪ'pɒzɪt/ *n* deposito *m*; ⟨against damage⟩ cauzione *f*; ⟨first instalment⟩ acconto *m* ● *vt* (*pt/pp* **deposited**) depositare. **~ account** *n* libretto *m* di risparmio;

(*without instant access*) conto *m* vincolato

depot /'depəʊ/ *n* deposito *m*; *Am Rail* stazione *f* ferroviaria

deprave /dɪ'preɪv/ *vt* depravare. **~ed** *a* depravato. **~ity** /-'prævətɪ/ *n* depravazione *f*

depreciat|e /dɪ'priːʃɪeɪt/ *vi* deprezzarsi. **~ion** /-'eɪʃn/ *n* deprezzamento *f*

depress /dɪ'pres/ *vt* deprimere; (*press down*) premere. **~ed** *a* depresso; **~ed area** zona *f* depressa. **~ing** *a* deprimente. **~ion** /-eʃn/ *n* depressione *f*

deprivation /deprɪ'veɪʃn/ *n* privazione *f*

deprive /dɪ'praɪv/ *vt* **~ sb of sth** privare qcno di qcsa. **~d** *a* (*area, childhood*) disagiato

depth /depθ/ *n* profondità *f inv*; **in ~** (*study, analyse*) in modo approfondito; **in the ~s** of winter in pieno inverno; **be out of one's ~** (*in water*) non toccare il fondo; *fig* sentirsi in alto mare

deputation /depjʊ'teɪʃn/ *n* deputazione *f*

deputize /'depjʊtaɪz/ *vi* **~ for** fare le veci di

deputy /'depjʊtɪ/ *n* vice *mf*; (*temporary*) sostituto, -a *mf* ● *attrib* **~ leader** ≈ vicesegretario, -a *mf*; **~ chairman** vicepresidente *mf*

derail /dɪ'reɪl/ *vt* **be ~ed** (*train*) essere deragliato. **~ment** *n* deragliamento *m*

deranged /dɪ'reɪndʒd/ *a* squilibrato

derelict /'derəlɪkt/ *a* abbandonato

deri|de /dɪ'raɪd/ *vt* deridere. **~sion** /-'rɪʒn/ *n* derisione *f*

derisory /dɪ'raɪsərɪ/ *a* (*laughter*) derisorio; (*offer*) irrisorio

derivation /derɪ'veɪʃn/ *n* derivazione *f*

derivative /dɪ'rɪvətɪv/ *a* derivato ● *n* derivato *m*

derive /dɪ'raɪv/ *vt* (*obtain*) derivare; **be ~d from** (*word*) derivare da

dermatologist /dɜːmə'tɒlədʒɪst/ *n* dermatologo, -a *mf*

derogatory /dɪ'rɒgətrɪ/ *a* (*comments*) peggiorativo

descend /dɪ'send/ *vi* scendere ● *vt* scendere da; **be ~ed from** discendere da. **~ant** *n* discendente *mf*

descent /dɪ'sent/ *n* discesa *f*; (*lineage*) origine *f*

describe /dɪ'skraɪb/ *vt* descrivere

descrip|tion /dɪ'skrɪpʃn/ *n* descrizione *f*; **they had no help of any ~tion** non hanno avuto proprio nessun aiuto. **~tive** /-tɪv/ *a* descrittivo; (*vivid*) vivido

desecrat|e /'desɪkreɪt/ *vt* profanare. **~ion** /-'kreɪʃn/ *n* profanazione *f*

desert[1] /'dezət/ *n* deserto *m* ● *a* deserto; **~ island** isola *f* deserta

desert[2] /dɪ'zɜːt/ *vt* abbandonare ● *vi* disertare. **~ed** *a* deserto. **~er** *n Mil* disertore *m*. **~ion** /-'zɜːʃn/ *n Mil* diserzione *f*; (*of family*) abbandono *m*

deserts /dɪ'zɜːts/ *npl* **get one's just ~** ottenere ciò che ci si merita

deserv|e /dɪ'zɜːv/ *vt* meritare. **~ing** *a* meritevole; **~ing cause** opera *f* meritoria

design /dɪ'zaɪn/ *n* progettazione *f*; (*fashion ~, appearance*) design *m inv*; (*pattern*) modello *m*; (*aim*) proposto *m* ● *vt* progettare; disegnare (*clothes, furniture, model*); **be ~ed for** essere fatto per

designat|e /'dezɪgneɪt/ *vt* designare. **~ion** /-'neɪʃn/ *n* designazione *f*

designer /dɪ'zaɪnə(r)/ *n* progettista *mf*; (*of clothes*) stilista *mf*; (*Theat: of set*) scenografo, -a *mf*

desirable /dɪ'zaɪərəbl/ *a* desiderabile

desire /dɪ'zaɪə(r)/ *n* desiderio *m* ● *vt* desiderare

desk /desk/ *n* scrivania *f*; (*in school*) banco *m*; (*in hotel*) reception *f inv*; (*cash ~*) cassa *f*. **~top** 'publishing *n* desktop publishing *m*, editoria *f* da tavolo

desolat|e /'desələt/ a desolato.
~ion /-'leɪʃn/ n desolazione f

despair /dɪ'speə(r)/ n disperazione
f; **in ~** disperato; (say) per disperazione ● *vi* **l ~ of that boy** quel
ragazzo mi fa disperare

desperat|e /'despərət/ a dispera-
to; **be ~e** (criminal:) essere disperato; **be ~e for** morire dalla voglia di. **~ely** adv disperatamente; **he said ~ely** ha detto, disperato. **~ion** /-'reɪʃn/ n disperazione f; **in ~ion** per disperazione

despicable /dɪ'spɪkəbl/ a disprez-
zevole

despise /dɪ'spaɪz/ vt disprezzare

despite /dɪ'spaɪt/ prep malgrado

despondent /dɪ'spɒndənt/ a abbattuto

despot /'despɒt/ n despota m

dessert /dɪ'zɜːt/ n dolce m. **~-
spoon** n cucchiaio m da dolce

destination /destɪ'neɪʃn/ n desti-
nazione f

destine /'destɪn/ vt destinare; **be
~d for sth** essere destinato a qcsa

destiny /'destɪnɪ/ n destino m

destitute /'destɪtjuːt/ a bisognoso

destroy /dɪ'strɔɪ/ vt distruggere.
~er n Naut cacciatorpediniere m

destruc|tion /dɪ'strʌkʃn/ n distru-
zione f. **~tive** /-tɪv/ a distruttivo;
(fig: criticism) negativo

detach /dɪ'tætʃ/ vt staccare. **~able**
/-əbl/ a separabile. **~ed** a fig di-
staccato; **~ed house** villetta f

detachment /dɪ'tætʃmənt/ n distacco m; Mil distaccamento m

detail /'diːteɪl/ n particolare m,
dettaglio m; **in ~** particolareg-
giatamente ● vt esporre con tutti i
particolari; Mil assegnare. **~ed** a
particolareggiato, dettagliato

detain /dɪ'teɪn/ vt (police:) trattenere; (delay) far ritardare. **~ee**
/diːteɪ'niː/ n detenuto, -a mf

detect /dɪ'tekt/ vt individuare;
(perceive) percepire. **~ion** /-ekʃn/
n scoperta f

detective /dɪ'tektɪv/ n investigato-

re, -trice mf. **~ story** n racconto m
poliziesco

detector /dɪ'tektə(r)/ n (for metal)
metal detector m inv

detention /dɪ'tenʃn/ n detenzione
f; Sch punizione f

deter /dɪ'tɜː(r)/ vt (pt/pp **deterred**)
impedire; **~ sb from doing sth**
impedire a qcno di fare qcsa

detergent /dɪ'tɜːdʒənt/ n detersivo m

deteriorat|e /dɪ'tɪərɪəreɪt/ vi dete-
riorarsi. **~ion** /-'reɪʃn/ n deterio-
ramento m

determination /dɪtɜːmɪ'neɪʃn/ n
determinazione f

determine /dɪ'tɜːmɪn/ vt (ascertain) determinare; **~ to** (resolve)
decidere di. **~d** a deciso

deterrent /dɪ'terənt/ n deterren-
te m

detest /dɪ'test/ vt detestare. **~able**
/-əbl/ a detestabile

detonat|e /'detəneɪt/ vt far detona-
re ● vi detonare. **~or** n detonatore m

detour /'diːtʊə(r)/ n deviazione f

detract /dɪ'trækt/ vi **~ from** sminuire (merit); rovinare (pleasure,
beauty)

detriment /'detrɪmənt/ n **to the ~
of** a danno di. **~al** /-'mentl/ a dannoso

deuce /djuːs/ n Tennis deuce m inv

devaluation /diːvæljʊ'eɪʃn/ n sva-
lutazione f

de'value vt svalutare (currency)

devastat|e /'devəsteɪt/ vt devasta-
re. **~ed** a fam sconvolto. **~ing** a
devastante; (news) sconvolgente.
~ion /-'steɪʃn/ n devastazione f

develop /dɪ'veləp/ vt sviluppare;
contrarre (illness); (add to value
of) valorizzare (area) ● vi svilupparsi; **~ into** divenire. **~er** n
[property] **~er** imprenditore,
-trice mf edile

de'veloping country n paese m
in via di sviluppo

development /dɪ'veləpmənt/ n

sviluppo *m*; *(of vaccine etc)* messa *f* a punto

deviant /'di:vɪənt/ *a* deviato

deviate /'di:vɪeɪt/ *vi* deviare. **~ion** /-'eɪʃn/ *n* deviazione *f*

device /dɪ'vaɪs/ *n* dispositivo *m*

devil /'devl/ *n* diavolo *m*

devious /'di:vɪəs/ *a (person)* subdolo; *(route)* tortuoso

devise /dɪ'vaɪz/ *vt* escogitare

devoid /dɪ'vɔɪd/ *a* **~ of** privo di

devolution /di:və'lu:ʃn/ *n (of power)* decentramento *m*

devote /dɪ'vəʊt/ *vt* dedicare. **~ed** *a (daughter etc)* affezionato; **be ~ed to sth** consacrarsi a qcsa. **~ee** /devə'ti:/ *n* appassionato, -a *mf*

devotion /dɪ'vəʊʃn/ *n* dedizione *f*; **~s** *pl Relig* devozione *fsg*

devour /dɪ'vaʊə(r)/ *vt* divorare

devout /dɪ'vaʊt/ *a* devoto

dew /dju:/ *n* rugiada *f*

dexterity /dek'sterətɪ/ *n* destrezza *f*

diabetes /daɪə'bi:ti:z/ *n* diabete *m*. **~ic** /-'betɪk/ *a* diabetico **●***n* diabetico, -a *mf*

diabolical /daɪə'bɒlɪkl/ *a* diabolico

diagnose /daɪəg'nəʊz/ *vt* diagnosticare

diagnosis /daɪəg'nəʊsɪs/ *n (pl -oses* /-si:z/*)* diagnosi *f inv*

diagonal /daɪ'æɡənl/ *a* diagonale **●***n* diagonale *f*

diagram /'daɪəɡræm/ *n* diagramma *m*

dial /'daɪəl/ *n (of clock, machine)* quadrante *m*; *Teleph* disco *m* combinatore **●***v (pt/pp* **dialled)** **●***vi Teleph* fare il numero; **~ direct** chiamare in teleselezione **●***vt* fare *(number)*

dialect /'daɪəlekt/ *n* dialetto *m*

dialling: ~ code *n* prefisso *m*. **~ tone** *n* segnale *m* di linea libera

dialogue /'daɪəlɒɡ/ *n* dialogo *m*

'dial tone *n Am Teleph* segnale *m* di linea libera

diameter /daɪ'æmɪtə(r)/ *n* diametro *m*

diametrically /daɪə'metrɪklɪ/ *adv* **~ opposed** diametralmente opposto

diamond /'daɪəmənd/ *n* diamante *m*, brillante *m*; *(shape)* losanga *f*; **~s** *pl (in cards)* quadri *mpl*

diaper /'daɪəpə(r)/ *n Am* pannolino *m*

diaphragm /'daɪəfræm/ *n* diaframma *m*

diarrhoea /daɪə'rɪə/ *n* diarrea *f*

diary /'daɪərɪ/ *n (for appointments)* agenda *f*; *(for writing in)* diario *m*

dice /daɪs/ *n inv* dadi *mpl* **●***vt Culin* tagliare a dadini

dicey /'daɪsɪ/ *a fam* rischioso

dictate /dɪk'teɪt/ *vt/i* dettare. **~ion** /-eɪʃn/ *n* dettato *m*

dictator /dɪk'teɪtə(r)/ *n* dittatore *m*. **~ial** /-tə'tɔ:rɪəl/ *a* dittatoriale. **~ship** *n* dittatura *f*

dictionary /'dɪkʃənrɪ/ *n* dizionario *m*

did /dɪd/ *see* **do**

didactic /dɪ'dæktɪk/ *a* didattico

diddle /'dɪdl/ *vt fam* gabbare

didn't /'dɪdnt/ **= did not**

die /daɪ/ *vi (pres p* **dying)** morire *(of* di); **be dying to do sth** *fam* morire dalla voglia di fare qcsa. **die down** *vi* calmarsi; *(fire, flames:)* spegnersi. **die out** *vi* estinguersi; *(custom:)* morire

diesel /'di:zl/ *n* diesel *m*

diet /'daɪət/ *n* regime *m* alimentare; *(restricted)* dieta *f*; **be on a ~** essere a dieta **●***vi* essere a dieta

differ /'dɪfə(r)/ *vi* differire; *(disagree)* non essere d'accordo

difference /'dɪfrəns/ *n* differenza *f*; *(disagreement)* divergenza *f*

different /'dɪfrənt/ *a* diverso, differente; *(various)* diversi; **be ~ from** essere diverso da

differential /dɪfə'renʃl/ *a* differenziale **●***n* differenziale *m*

differentiate /dɪfə'renʃɪeɪt/ *vt* distinguere **(between** fra); *(dis-*

criminate) discriminare (**between** fra); (*make differ*) differenziare

differently /'dɪfrəntlɪ/ *adv* in modo diverso; ~ **from** diversamente da

difficult /'dɪfɪkəlt/ *a* difficile. ~**y** *n* difficoltà *f inv*; **with** ~**y** con difficoltà

diffuse[1] /dɪ'fju:s/ *a* diffuso; (*wordy*) prolisso

diffuse[2] /dɪ'fju:z/ *vt* Phys diffondere

dig /dɪg/ *n* (*poke*) spinta *f*; (*remark*) frecciata *f*; Archaeol scavo *m*; ~s *pl fam* camera *fsg* ammobiliata ● *vt/i* (*pt/pp* **dug**, *pres p* **digging**) scavare (*hole*); vangare (*garden*); (*thrust*) conficcare; ~ **sb in the ribs** dare una gomitata a qcno. **dig out** *vt fig* tirar fuori. **dig up** *vt* scavare (*garden, street, object*); sradicare (*plant*); (*fig: find*) scovare

digest[1] /'daɪdʒest/ *n* compendio *m*

digest[2] /daɪ'dʒest/ *vt* digerire. ~**ible** *a* digeribile. ~**ion** /-estʃn/ *n* digestione *f*

digger /'dɪgə(r)/ *n* Techn scavatrice *f*

digit /'dɪdʒɪt/ *n* cifra *f*; (*finger*) dito *m*

digital /'dɪdʒɪtl/ *a* digitale; ~ **clock** orologio *m* digitale

dignified /'dɪgnɪfaɪd/ *a* dignitoso

dignitary /'dɪgnɪtərɪ/ *n* dignitario *m*

dignity /'dɪgnɪtɪ/ *n* dignità *f*

digress /daɪ'gres/ *vi* divagare. ~**ion** /-eʃn/ *n* digressione *f*

dike /daɪk/ *n* diga *f*

dilapidated /dɪ'læpɪdeɪtɪd/ *a* cadente

dilate /daɪ'leɪt/ *vi* dilatarsi

dilemma /dɪ'lemə/ *n* dilemma *m*

dilettante /dɪlɪ'tæntɪ/ *n* dilettante *mf*

dilly-dally /'dɪlɪdælɪ/ *vi* (*pt/pp -*led*) *fam* tentennare

dilute /daɪ'lu:t/ *vt* diluire

dim /dɪm/ *a* (**dimmer**, **dimmest**) debole (*light*); (*dark*) scuro; (*prospect, chance*) scarso; (*indis-*

tinct) impreciso; (*fam: stupid*) tonto ● *vt/i* (*pt/pp* **dimmed**) affievolire. ~**ly** *adv* (*see, remember*) indistintamente; (*shine*) debolmente

dime /daɪm/ *n* Am moneta *f* da dieci centesimi

dimension /daɪ'menʃn/ *n* dimensione *f*

diminish /dɪ'mɪnɪʃ/ *vt/i* diminuire

diminutive /dɪ'mɪnjʊtɪv/ *a* minuscolo ● *n* diminutivo *m*

dimple /'dɪmpl/ *n* fossetta *f*

din /dɪn/ *n* baccano *m*

dine /daɪn/ *vi* pranzare. ~**r** *n* (*Am: restaurant*) tavola *f* calda; **the last ~r in the restaurant** l'ultimo cliente nel ristorante

dinghy /'dɪŋgɪ/ *n* dinghy *m*; (*inflatable*) canotto *m* pneumatico

dingy /'dɪndʒɪ/ *a* (**-ier, -iest**) squallido e tetro

dining /'daɪnɪŋ/: ~**-car** *n* carrozza *f* ristorante. ~**-room** *n* sala *f* da pranzo. ~**-table** *n* tavolo *m* da pranzo

dinner /'dɪnə(r)/ *n* cena *f*; (*at midday*) pranzo *m*. ~**-jacket** *n* smoking *m inv*

dinosaur /'daɪnəsɔ:(r)/ *n* dinosauro *m*

dint /dɪnt/ *n* **by ~ of** a forza di

diocese /'daɪəsɪs/ *n* diocesi *f inv*

dip /dɪp/ *n* (*in ground*) inclinazione *f*; Culin salsina *f*; **go for a ~** andare a fare una nuotata ● *v* (*pt/pp* **dipped**) ● *vt* (*in liquid*) immergere; abbassare (*head, headlights*) ● *vi* (*land*:) formare un avvallamento. **dip into** *vt* scorrere (*book*)

diphtheria /dɪf'θɪərɪə/ *n* difterite *f*

diphthong /'dɪfθɒŋ/ *n* dittongo *m*

diploma /dɪ'pləʊmə/ *n* diploma *m*

diplomacy /dɪ'pləʊməsɪ/ *n* diplomazia *f*

diplomat /'dɪpləmæt/ *n* diplomatico, -a *mf*. ~**ic** /-'mætɪk/ *a* diplomatico. ~**ically** *adv* con diplomazia

'dip-stick *n* Auto astina *f* dell'olio

dire /'daɪə(r)/ *n* (*situation, consequences*) terribile

direct /dɪˈrekt/ a diretto ● adv direttamente ● vt ⟨aim⟩ rivolgere ⟨attention, criticism⟩; ⟨control⟩ dirigere; fare la regia di ⟨film, play⟩; ~ **sb** ⟨show the way⟩ indicare la strada a qcno; ~ **sb to do sth** ordinare a qcno di fare qcsa. ~ **'current** n corrente m continua

direction /dɪˈrekʃn/ n direzione f; ⟨of play, film⟩ regia f; ~**s** pl indicazioni fpl

directly /dɪˈrektlɪ/ adv direttamente; ⟨at once⟩ immediatamente ● conj [non] appena

director /dɪˈrektə(r)/ n Comm direttore, -trice mf; ⟨of play, film⟩ regista mf

directory /dɪˈrektərɪ/ n elenco m; Teleph elenco m [telefonico]; ⟨of streets⟩ stradario m

dirt /dɜːt/ n sporco m; ~ **cheap** fam a [un] prezzo stracciato

dirty /ˈdɜːtɪ/ a (-**ier**, -**iest**) sporco; ~ **trick** brutto scherzo m; ~ **word** parolaccia f ● vt (pt/pp -**ied**) sporcare

dis|**a**'**bility** /dɪs-/ n infermità f inv. ~**abled** /dɪˈseɪbld/ a invalido

disad'van|tage n svantaggio m; **at a** ~ **tage** in una posizione di svantaggio. ~**taged** a svantaggiato. ~**tageous** a svantaggioso

disa'gree vi non essere d'accordo; ~ **with** ⟨food:⟩ far male a

disa'greeable a sgradevole

disa'greement n disaccordo f; ⟨quarrel⟩ dissidio m

disal'low vt annullare ⟨goal⟩

disap'pear vi scomparire. ~**ance** n scomparsa f

disap'point vt deludere; **I'm** ~**ed** sono deluso. ~**ing** a deludente. ~**ment** n delusione f

disap'proval n disapprovazione f

disap'prov|**e** vi disapprovare; ~ **of sb/sth** disapprovare qcno/qcsa

dis'arm vt disarmare ● vi Mil disarmarsi. ~**ament** n disarmo m. ~**ing** a ⟨frankness etc⟩ disarmante

disar'ray n in ~ in disordine

disast|**er** /dɪˈzɑːstə(r)/ n disastro m. ~**rous** /-rəs/ a disastroso

dis'band vt smobilitare ● vi scogliersi; ⟨regiment:⟩ essere smobilitato

disbe'lief n incredulità f; **in** ~ con incredulità

disc /dɪsk/ n disco m; ⟨CD⟩ compact disc m inv

discard /dɪˈskɑːd/ vt scartare; ⟨throw away⟩ eliminare; scaricare ⟨boyfriend⟩

discern /dɪˈsɜːn/ vt discernere. ~**ible** a discernibile. ~**ing** a perspicace

'discharge¹ n Electr scarica f; ⟨dismissal⟩ licenziamento m; Mil congedo m; ⟨Med: of blood⟩ emissione f; ⟨of cargo⟩ scarico m

dis'charge² vt scaricare ⟨battery, cargo⟩; ⟨dismiss⟩ licenziare; Mil congedare; Jur assolvere ⟨accused⟩; dimettere ⟨patient⟩ ● vi Electr scaricarsi

disciple /dɪˈsaɪpl/ n discepolo m

disciplinary /ˈdɪsɪplɪnərɪ/ a disciplinare

discipline /ˈdɪsɪplɪn/ n disciplina f ● vt disciplinare; ⟨punish⟩ punire

'disc jockey n disc jockey m inv

dis'claim vt disconoscere. ~**er** n rifiuto m

dis'clos|**e** vt svelare. ~**ure** n rivelazione f

disco /ˈdɪskəʊ/ n discoteca f

dis'colour vt scolorire ● vi scolorirsi

dis'comfort n scomodità f; fig disagio m

disconcert /dɪskənˈsɜːt/ vt sconcertare

discon'nect vt disconnettere

disconsolate /dɪsˈkɒnsələt/ a sconsolato

discon'tent n scontentezza f. ~**ed** a scontento

discon'tinue vt cessare, smettere; Comm sospendere la produzione di; ~**d line** fine f serie

'discord n discordia f; Mus disso-

dissident /ˈdɪsɪdənt/ n dissidente mf

dis'similar a dissimile (**to** da)

dissociate /dɪˈsəʊʃɪeɪt/ vt dissociare; ~ **oneself from** dissociarsi da

dissolute /ˈdɪsəluːt/ a dissoluto

dissolution /dɪsəˈluːʃn/ n scioglimento m

dissolve /dɪˈzɒlv/ vt dissolvere ● vi dissolversi

dissuade /dɪˈsweɪd/ vt dissuadere

distance /ˈdɪstəns/ n distanza f; **it's a short ~ from here to the station** la stazione non è lontana da qui; **in the ~** in lontananza; **from a ~** da lontano

distant /ˈdɪstənt/ a distante; (relative) lontano

dis'taste n avversione f. ~**ful** a spiacevole

distil /dɪˈstɪl/ vt (pt/pp distilled) distillare. ~**lation** /-ˈleɪʃn/ n distillazione f. ~**lery** /-ərɪ/ n distilleria f

distinct /dɪˈstɪŋkt/ a chiaro; (different) distinto. ~**ion** /-ɪŋkʃn/ n distinzione f; Sch massimo m dei voti. ~**ive** /-tɪv/ a caratteristico. ~**ly** adv chiaramente

distinguish /dɪˈstɪŋgwɪʃ/ vt/i distinguere; ~ **oneself** distinguersi. ~**ed** a rinomato; (appearance) distinto; (career) brillante

distort /dɪˈstɔːt/ vt distorcere. ~**ion** /-ɔːʃn/ n distorsione f

distract /dɪˈstrækt/ vt distrarre. ~**ed** /-ɪd/ a assente; (fam: worried) preoccupato. ~**ing** a che distoglie. ~**ion** /-ækʃn/ n distrazione f; (despair) disperazione f; **drive sb to ~** portare qcno alla disperazione

distraught /dɪˈstrɔːt/ a sconvolto

distress /dɪˈstres/ n angoscia f; (pain) sofferenza f; (danger) difficoltà f ● vt sconvolgere; (sadden) affliggere. ~**ing** a penoso; (shocking) sconvolgente. ~ **signal** n segnale m di richiesta di soccorso

distribute /dɪˈstrɪbjuːt/ vt distribuire. ~**ion** /-ˈbjuːʃn/ n distribuzione f. ~**or** n distributore m

district /ˈdɪstrɪkt/ n regione f; Admin distretto m. ~ **nurse** n infermiera, -a mf che fa visite a domicilio

dis'trust n sfiducia f ● vt non fidarsi di. ~**ful** a diffidente

disturb /dɪˈstɜːb/ vt disturbare; (emotionally) turbare; spostare (papers). ~**ance** n disturbo m; ~**ances** (pl: rioting etc) disordini mpl. ~**ed** a turbato; (mentally) ~**ed** malato di mente. ~**ing** a inquietante

dis'used a non utilizzato

ditch /dɪtʃ/ n fosso m ● vt (fam: abandon) abbandonare (plan, car); piantare (lover)

dither /ˈdɪðə(r)/ vi titubare

divan /dɪˈvæn/ n divano m

dive /daɪv/ n tuffo m; Aeron picchiata f; (fam: place) bettola f ● vi tuffarsi; (when in water) immergersi; Aeron scendere in picchiata; (fam: rush) precipitarsi

diver /ˈdaɪvə(r)/ n (from board) tuffatore, -trice mf; (scuba) sommozzatore, -trice mf; (deep sea) palombaro m

diverge /daɪˈvɜːdʒ/ vi divergere. ~**gent** /-ənt/ a divergente

diverse /daɪˈvɜːs/ a vario

diversify /daɪˈvɜːsɪfaɪ/ vt/i (pt/pp -ied) diversificare

diversion /daɪˈvɜːʃn/ n deviazione f; (distraction) diversivo m

diversity /daɪˈvɜːsətɪ/ n varietà f

divert /daɪˈvɜːt/ vt deviare (traffic); distogliere (attention)

divest /daɪˈvest/ vt privare (**of** di)

divide /dɪˈvaɪd/ vt dividere (**by** per); **six ~d by two** sei diviso due ● vi dividersi

dividend /ˈdɪvɪdend/ n dividendo m; **pay ~s** fig ripagare

divine /dɪˈvaɪn/ a divino

diving /ˈdaɪvɪŋ/ n (from board) tuffi mpl; (scuba) immersione f. ~**-board** n trampolino m. ~ **mask**

n maschera *f* [subacquea]. **~-suit**
n muta *f*; (*deep sea*) scafandro *m*

divinity /dɪ'vɪnɪtɪ/ *n* divinità *f inv*;
(*subject*) teologia *f*; (*at school*) religione *f*

divisible /dɪ'vɪzɪbl/ *a* divisibile (**by** *per*)

division /dɪ'vɪʒn/ *n* divisione *f*; (*in sports league*) serie *f*

divorce /dɪ'vɔːs/ *n* divorzio *m* ● *vt* divorziare da. **~d** *a* divorziato; **get ~d** divorziare

divorcee /dɪvɔː'siː/ *n* divorziato, -a *mf*

divulge /dɪ'vʌldʒ/ *vt* rendere pubblico

DIY *n abbr* **do-it-yourself**

dizziness /'dɪzɪnɪs/ *n* giramenti *mpl* di testa

dizzy /'dɪzɪ/ *a* (**-ier**, **-iest**) vertiginoso; **I feel ~** mi gira la testa

do /duː/ *n* (*pl* **dos** *or* **do's**) *fam* festa *f* ● *v* (*3 sg pres tense* **does**; *pt* **did**; *pp* **done**) ● *vt* fare; (*fam: cheat*) fregare; **be done** *Culin* essere cotto; **well done** bravo; *Culin* ben cotto; **do the flowers** sistemare i fiori; **do the washing up** lavare i piatti; **do one's hair** farsi i capelli ● *vi* (*be suitable*) andare; (*be enough*) bastare; **this will do** questo va bene; **that will do!** basta così!; **do well/badly** cavarsela bene/male; **how is he doing?** come sta? ● *v aux* **do you speak Italian?** parli italiano?; **you don't like him, do you?** non ti piace, vero?; (*expressing astonishment*) non dirmi che ti piace!; **yes, I do** sì; (*emphatic*) invece sì; **no, I don't** no; **I don't smoke** non fumo; **don't you/doesn't he?** vero?; **so do I** anch'io; **do come in, John** entra, John; **how do you do?** piacere. **do away with** *vt* abolire (*rule*). **do for** *vt* **done for** *fam* rovinato. **do in** *vt* (*fam: kill*) uccidere; farsi male (*a back*); **done in** *fam* esausto. **do up** *vt* (*fasten*) abbottonare; (*renovate*) rimettere a nuovo; (*wrap*) avvolgere. **do with** *vt* **I**

could do with a spanner mi ci vorrebbe una chiave inglese. **do without** *vt* fare a meno di

docile /'dəʊsaɪl/ *a* docile

dock[1] /dɒk/ *n Jur* banco *m* degli imputati

dock[2] *n Naut* bacino *m* ● *vi* entrare in porto; (*spaceship*) congiungersi. **~er** *n* portuale *m*. **~s** *npl* porto *m*. **~yard** *n* cantiere *m* navale

doctor /'dɒktə(r)/ *n* dottore *m*, dottoressa *f* ● *vt* alterare (*drink*); castrare (*cat*). **~ate** /-ət/ *n* dottorato *m*

doctrine /'dɒktrɪn/ *n* dottrina *f*

document /'dɒkjʊmənt/ *n* documento *m*. **~ary** /-'mentərɪ/ *a* documentario ● *n* documentario *m*

doddery /'dɒdərɪ/ *a fam* barcollante

dodge /dɒdʒ/ *n fam* trucco *m* ● *vt* schivare (*blow*); evitare (*person*) ● *vi* scansarsi; **~ out of the way** scansarsi

dodgems /'dɒdʒəmz/ *npl* autoscontro *msg*

dodgy /'dɒdʒɪ/ *a* (**-ier**, **-iest**) (*fam: dubious*) sospetto

doe /dəʊ/ *n* femmina *f* (*di daino, renna, lepre*); (*rabbit*) coniglia *f*

does /dʌz/ *see* **do**

doesn't /'dʌznt/ = **does not**

dog /dɒg/ *n* cane *m* ● *vt* (*pt/pp* **dogged**) (*illness, bad luck:*) perseguitare

dog: **~-biscuit** *n* biscotto *m* per cani. **~-collar** *n* collare *m* (*per cani*); *Relig fam* collare *m* del prete. **~-eared** *a* con le orecchie

dogged /'dɒgɪd/ *a* ostinato

'dog house *n* **in the ~** *fam* in disgrazia

dogma /'dɒgmə/ *n* dogma *m*. **~tic** /-'mætɪk/ *a* dogmatico

'dogsbody *n fam* tirapiedi *mf inv*

doily /'dɔɪlɪ/ *n* centrino *m*

do-it-yourself /'duːɪtjə'self/ *n* fai da te *m*, bricolage *m*. **~ shop** *n* negozio *m* di bricolage

doldrums /'dɒldrəmz/ *npl* **be in**

the ~ essere giù di corda;
⟨business⟩ essere in fase di stasi

dole /dəʊl/ n sussidio m di disoccupazione; **be on the ~** essere disoccupato ● **dole out** vt distribuire

doleful /ˈdəʊlfl/ a triste

doll /dɒl/ n bambola f ● **doll oneself up** vt fam mettersi in ghingheri

dollar /ˈdɒlə(r)/ n dollaro m

dollop /ˈdɒləp/ n fam cucchiaiata f

dolphin /ˈdɒlfɪn/ n delfino m

dome /dəʊm/ n cupola f

domestic /dəˈmestɪk/ a domestico; Pol interno; Comm nazionale. **~ animal** n animale m domestico

domesticated /dəˈmestɪkeɪtɪd/ a ⟨animal⟩ addomesticato

domestic: ~ flight n volo m nazionale. **~ 'servant** n domestico, -a mf

dominant /ˈdɒmɪnənt/ a dominante

dominat|e /ˈdɒmɪneɪt/ vt/i dominare. **~ion** /-ˈneɪʃn/ n dominio m

domineering /dɒmɪˈnɪərɪŋ/ a autoritario

dominion /dəˈmɪnjən/ n Br Pol dominion m inv

domino /ˈdɒmɪnəʊ/ n (pl -es) tessera f del domino; **~es** sg ⟨game⟩ domino m

don[1] /dɒn/ vt (pt/pp donned) liter indossare

don[2] n docente mf universitario, -a

donat|e /dəʊˈneɪt/ vt donare. **~ion** /-eɪʃn/ n donazione f

done /dʌn/ see **do**

donkey /ˈdɒŋkɪ/ n asino m; **~'s years** fam secoli mpl. **~-work** n sgobbata f

donor /ˈdəʊnə(r)/ n donatore, -trice mf

don't /dəʊnt/ = **do not**

doodle /ˈduːdl/ vi scarabocchiare

doom /duːm/ n fato m; ⟨ruin⟩ rovina f ● vt **be ~ed** ⟨to failure⟩ essere destinato al fallimento; ⟨and ship⟩ destinato ad affondare

door /dɔː(r)/ n porta f; ⟨of car⟩ portiera f; **out of ~s** all'aperto

door: ~man n portiere m. **~mat** n zerbino m. **~step** n gradino m della porta. **~way** n vano m della porta

dope /dəʊp/ n fam ⟨drug⟩ droga f leggera; ⟨information⟩ indiscrezioni fpl; ⟨idiot⟩ idiota mf ● vt drogare; Sport dopare

dopey /ˈdəʊpɪ/ a fam addormentato

dormant /ˈdɔːmənt/ a latente; ⟨volcano⟩ inattivo

dormer /ˈdɔːmə(r)/ n ~ [window] abbaino m

dormitory /ˈdɔːmɪtərɪ/ n dormitorio m

dormouse /ˈdɔː-/ n ghiro m

dosage /ˈdəʊsɪdʒ/ n dosaggio m

dose /dəʊs/ n dose f

doss /dɒs/ vi sl accamparsi. **~er** n barbone, -a mf. **~-house** n dormitorio m pubblico

dot /dɒt/ n punto m; **at 8 o'clock on the ~** alle 8 in punto

dote /dəʊt/ vi ~ **on** stravedere per

dotted /ˈdɒtɪd/ a ~ **line** linea f punteggiata; **be ~ with** essere punteggiato di

dotty /ˈdɒtɪ/ a (-ier, -iest) fam tocco; ⟨idea⟩ folle

double /ˈdʌbl/ a doppio ● adv cost ~ costare il doppio; **see ~** vedere doppio; ~ **the amount** la quantità doppia ● n doppio m; ⟨person⟩ sosia m inv; **~s** pl Tennis doppio m; **at the ~** di corsa ● vt raddoppiare; ⟨fold⟩ piegare in due ● vi raddoppiare. **double back** vi ⟨go back⟩ fare dietro front. **double up** vi ⟨bend⟩ piegarsi in due ⟨with per⟩; ⟨share⟩ dividere una stanza

double: ~-bass n contrabbasso m. ~ **'bed** n letto m matrimoniale. **~-breasted** a a doppio petto. ~ **'chin** n doppio mento m. **~-'cross** vt ingannare. **~-'decker** n autobus m inv a due piani. ~ **'Dutch** n fam ostrogoto m. ~ **'glazing** n doppiovetro m. ~ **'room** n camera f doppia

doubly /ˈdʌblɪ/ adv doppiamente

doubt /daʊt/ n dubbio m ● vt dubitare di. **~ful** a dubbio; (having doubts) in dubbio. **~fully** adv con aria dubbiosa. **~less** adv indubbiamente

dough /dəʊ/ n pasta f; (for bread) impasto m; (fam: money) quattrini mpl. **~nut** n bombolone m, krapfen m inv

douse /daʊs/ vt spegnere

dove /dʌv/ n colomba f. **~tail** n Techn incastro m a coda di rondine

dowdy /'daʊdɪ/ a (-ier, -iest) trasandato

down[1] /daʊn/ n (feathers) piumino m

down[2] adv giù; go/come ~ scendere; ~ there laggiù; sales are ~ le vendite sono diminuite; £50 ~ 50 sterline d'acconto; ~ 10% ridotto del 10%; ~ with...! abbasso...! ● prep walk ~ the road camminare per strada; ~ the stairs giù per le scale; fall ~ the stairs cadere giù dalle scale; get that ~ you! fam butta giù!; be ~ the pub fam essere al pub ● vt bere tutto d'un fiato ⟨drink⟩

down: **~-and-'out** n spiantato, -a mf. **~cast** a abbattuto. **~fall** n caduta f; (of person) rovina f. **~'grade** vt (in seniority) degradare. **~-'hearted** a scoraggiato. **~'hill** adv in discesa; go **~hill** fig essere in declino. **~ payment** n deposito m. **~pour** n acquazzone m. **~right** a (absolute) totale; ⟨lie⟩ bell'e buono; ⟨idiot⟩ perfetto ● adv (completely) completamente. **~'stairs** adv al piano di sotto ● a /'-/ del piano di sotto. **~'stream** adv a valle. **~-to-'earth** a ⟨person⟩ con i piedi per terra. **~town** adv Am in centro. **~trodden** a oppresso. **~ward[s]** a verso il basso; ⟨slope⟩ in discesa ● adv verso il basso

dowry /'daʊrɪ/ n dote f

doze /dəʊz/ n sonnellino m ● vi

sonnecchiare. **doze off** vi assopirsi

dozen /'dʌzn/ n dozzina f; **~s of books** libri a dozzine

Dr abbr **doctor**

drab /dræb/ a spento

draft[1] /drɑːft/ n abbozzo m; Comm cambiale f; Am Mil leva f ● vt abbozzare; Am Mil arruolare

draft[2] n Am = **draught**

drag /dræg/ n fam scocciatura f; in ~ fam ⟨man⟩ travestito da donna ● vt (pt/pp **dragged**) trascinare; dragare ⟨river⟩. **drag on** vi ⟨time, meeting⟩ trascinarsi

dragon /'drægən/ n drago m. **~-fly** n libellula f

drain /dreɪn/ n tubo m di scarico; (grid) tombino m; **the ~s** pl le fognature; **be a ~ on sb's finances** prosciugare le finanze di qcno ● vt drenare ⟨land, wound⟩; scolare ⟨liquid, vegetables⟩; svuotare ⟨tank, glass, person⟩ ● vi ~ [away] andar via

drain|age /'dreɪnɪdʒ/ n (system) drenaggio m; (of land) scolo m. **~ing board** n scolapiatti m inv. **~pipe** n tubo m di scarico

drake /dreɪk/ n maschio m dell'anatra

drama /'drɑːmə/ n arte f drammatica; ⟨play⟩ opera f teatrale; (event) dramma m

dramatic /drə'mætɪk/ a drammatico

dramat|ist /'dræmətɪst/ n drammaturgo, -a mf. **~ize** vt adattare per il teatro; fig drammatizzare

drank /dræŋk/ see **drink**

drape /dreɪp/ n Am tenda f ● vt appoggiare (over su)

drastic /'dræstɪk/ a drastico; **~ally** adv drasticamente

draught /drɑːft/ n corrente f d'aria]; **~s** sg (game) [gioco m della] dama f sg

draught: **~ beer** n birra f alla spina. **~sman** n disegnatore, -trice mf

draughty /'drɔːftɪ/ a pieno di correnti d'aria; **it's ~** c'è corrente

draw /drɔː/ n (attraction) attrazione f; Sport pareggio m; (in lottery) sorteggio m ● vt (pt drew, pp drawn) ● vt tirare; (attract) attirare; disegnare (picture); tracciare (line); ritirare (money); **~ lots** tirare a sorte ● vi (tea:) essere in infusione; Sport pareggiare; **~ near** avvicinarsi. **draw back** vt tirare indietro; ritirare (hand); tirare (curtains) ● vi (recoil) tirarsi indietro. **draw in** vt ritrarre (claws etc) ● vi (train:) arrivare; (days:) accorciarsi. **draw out** vt (pull out) tirar fuori; ritirare (money) ● vi (train:) partire; (days:) allungarsi. **draw up** vt redigere (document); accostare (chair); **~ oneself up to one's full height** farsi grande ● vi (stop) fermarsi

draw: **~back** n inconveniente m. **~bridge** n ponte m levatoio

drawer /drɔː(r)/ n cassetto m

drawing /'drɔːɪŋ/ n disegno m

drawing: **~-board** n tavolo m da disegno; fig **go back to the ~-board** ricominciare da capo. **~-pin** n puntina f. **~-room** n salotto m

drawl /drɔːl/ n pronuncia f strascicata

drawn /drɔːn/ see **draw**

dread /dred/ n terrore m ● vt aver il terrore di

dreadful /'dredful/ a terribile. **~ly** adv terribilmente

dream /driːm/ n sogno m ● attrib di sogno ● vt/i (pt/pp **dreamt** /dremt/ or **dreamed**) sognare (about/of di)

dreary /'drɪərɪ/ a (-ier, -iest) tetro; (boring) monotono

dredge /dredʒ/ vt/i dragare

dregs /dregz/ npl feccia fsg

drench /drentʃ/ vt **get ~ed** inzupparsi; **~ed** zuppo

dress /dres/ n (woman's) vestito m; (clothing) abbigliamento m ● vt

vestire; (decorate) adornare; Culin condire; Med fasciare; **~ oneself, get ~ed** vestirsi ● vi vestirsi. **dress up** vi mettersi elegante; (in disguise) travestirsi (**as** da)

dress: **~ circle** n Theat prima galleria f. **~er** n (furniture) credenza f; (Am: dressing-table) toilette f inv

dressing /'dresɪŋ/ n Culin condimento m; Med fasciatura f

dressing: **~-gown** n vestaglia f. **~-room** n (in gym) spogliatoio m; Theat camerino m. **~-table** n toilette f inv

dress: **~maker** n sarta f. **~ rehearsal** n prova f generale

dressy /'dresɪ/ a (-ier, -iest) elegante

drew /druː/ see **draw**

dribble /'drɪbl/ vi gocciolare; (baby:) sbavare; Sport dribblare

dribs and drabs /drɪbzən'dræbz/ npl **in ~** alla spicciolata

dried /draɪd/ a (food) essiccato

drier /'draɪə(r)/ n asciugabiancheria m inv

drift /drɪft/ n movimento m lento; (of snow) cumulo m; (meaning) senso m ● vi (off course) andare alla deriva; (snow:) accumularsi; (fig: person:) procedere senza meta. **drift apart** vi (people:) allontanarsi l'uno dall'altro

drill /drɪl/ n trapano m; Mil esercitazione f ● vt trapanare; Mil fare esercitare ● vi Mil esercitarsi; **~ for oil** trivellare in cerca di petrolio

drily /'draɪlɪ/ adv seccamente

drink /drɪŋk/ n bevanda f; (alcoholic) bicchierino m; **have a ~** bere qualcosa; **a ~ of water** un po' d'acqua ● vt/i (pt **drank**, pp **drunk**) bere. **drink up** vt finire ● vi finire il bicchiere

drinkable /'drɪŋkəbl/ a potabile. **~er** n bevitore, -trice mf

'drinking-water n acqua f potabile

drip /drɪp/ n gocciolamento m; (drop) goccia f; Med flebo f inv;

⟨fam: person⟩ mollaccione, -a mf ●vi ⟨pt/pp dripped⟩ gocciolare. ~·'dry a che non si stira. ~ping n ⟨from meat⟩ grasso m d'arrosto ●a ~ping [wet] fradicio

drive /draɪv/ n ⟨in car⟩ giro m; ⟨entrance⟩ viale m; ⟨energy⟩ grinta f; Psych pulsione f; ⟨organized effort⟩ operazione f; Techn motore m; Comput lettore m ●v ⟨pt drove, pp driven⟩ ●vt portare ⟨person by car⟩; guidare ⟨car⟩; ⟨Sport: hit⟩ mandare; Techn far funzionare; ~ sb mad far diventare matto qcno ●vi guidare. drive at vt what are you driving at? dove vuoi arrivare? drive away vt portare via in macchina; ⟨chase⟩ cacciare ●vi andare via in macchina. drive in vt piantare ⟨nail⟩ ●vi arrivare [in macchina]. drive off vt portare via in macchina; ⟨chase⟩ cacciare ●vi andare via in macchina. drive on vi proseguire ⟨in macchina⟩. drive up vi arrivare ⟨in macchina⟩

drivel /'drɪvl/ n fam sciocchezze fpl

driven /'drɪvn/ see drive

driver /'draɪvə(r)/ n guidatore, -trice mf; ⟨of train⟩ conducente mf

driving /'draɪvɪŋ/ a ⟨rain⟩ violento; ⟨force⟩ motore ●n guida f

driving: ~ **lesson** n lezione f di guida. ~ **licence** n patente f di guida. ~ **school** n scuola f guida. ~ **test** n esame m di guida

drizzle /'drɪzl/ n pioggerella f ●vi piovigginare

drone /drəʊn/ n ⟨bee⟩ fuco m; ⟨sound⟩ ronzio m

droop /druːp/ vi abbassarsi; ⟨flowers:⟩ afflosciarsi

drop /drɒp/ n ⟨of liquid⟩ goccia f; ⟨fall⟩ caduta f; ⟨in price, temperature⟩ calo m ●v ⟨pt/pp dropped⟩ ●vt far cadere; sganciare ⟨bomb⟩; ⟨omit⟩ omettere; ⟨give up⟩ abbandonare ●vi cadere; ⟨price, temperature, wind:⟩ calare; ⟨ground:⟩ essere in pendenza. drop in vi passare. drop off vt depositare ⟨person⟩ ●vi cadere; ⟨fall asleep⟩

assopirsi. drop out vi cadere; ⟨of race, society⟩ ritirarsi; ~ out of school lasciare la scuola

'drop-out n persona f contro il sistema sociale

droppings /'drɒpɪŋz/ npl sterco m

drought /draʊt/ n siccità f

drove /drəʊv/ see drive

droves /drəʊvz/ npl in ~ in massa

drown /draʊn/ vi annegare ●vt annegare; coprire ⟨noise⟩; he was ~ed è annegato

drowsy /'draʊzɪ/ a sonnolento

drudgery /'drʌdʒərɪ/ n lavoro m pesante e noioso

drug /drʌg/ n droga f; Med farmaco m; take ~s drogarsi ●vt ⟨pt/pp drugged⟩ drogare

drug: ~ **addict** n tossicomane, -a mf. ~ **dealer** n spacciatore, -trice mf ⟨di droga⟩. ~**gist** n Am farmacista mf. ~**store** n Am negozio m di generi vari, inclusi medicinali, che funge anche da bar; ⟨dispensing⟩ farmacia f

drum /drʌm/ n tamburo m; ⟨for oil⟩ bidone m; ~s ⟨pl: in pop-group⟩ batteria f ●v ⟨pt/pp drummed⟩ ●vi suonare il tamburo; ⟨in pop-group⟩ suonare la batteria ●vt ~ sth into sb ripetere qcsa a qcno cento volte. ~**mer** n percussionista mf; ⟨in pop-group⟩ batterista mf. ~**stick** n bacchetta f; ⟨of chicken, turkey⟩ coscia f

drunk /drʌŋk/ see drink ●a ubriaco; get ~ ubriacarsi ●n ubriaco, -a mf

drunk|ard /'drʌŋkəd/ n ubriacone, -a mf. ~**en** a ubriaco; ~**en driving** guida f in stato di ebbrezza

dry /draɪ/ a ⟨drier, driest⟩ asciutto; ⟨climate, country⟩ secco ●v/i ⟨pt/pp dried⟩ asciugare; ~ one's eyes asciugarsi le lacrime. dry up vi seccarsi; ⟨fig: source:⟩ prosciugarsi; ⟨fam: be quiet⟩ stare zitto; ⟨do dishes⟩ asciugare i piatti

dry: ~**'clean** vt pulire a secco. ~-'cleaner's n ⟨shop⟩ tintoria f. ~**ness** n secchezza f

DTP n abbr (**desktop publishing**) desktop publishing m

dual /'dju:əl/ a doppio

dual: ~ **'carriageway** n strada f a due carreggiate. ~-'**purpose** a a doppio uso

dub /dʌb/ vt (pt/pp dubbed) doppiare ⟨film⟩; ⟨name⟩ soprannominare

dubious /'dju:bɪəs/ a dubbio; **be** ~ **about** avere dei dubbi riguardo

duchess /'dʌtʃɪs/ n duchessa f

duck /dʌk/ n anatra f ● vt (in water) immergere; ~ **one's head** abbassare la testa ● vi abbassarsi. ~**ling** n anatroccolo m

duct /dʌkt/ n condotto m; Anat dotto m

dud /dʌd/ fam a Mil disattivato; ⟨coin⟩ falso; ⟨cheque⟩ a vuoto ● n ⟨banknote⟩ banconota f falsa

due /dju:/ a dovuto; **be** ~ ⟨train⟩ essere previsto; **the baby is** ~ **next week** il bambino dovrebbe nascere la settimana prossima; ~ **to** ⟨owing to⟩ a causa di; **be** ~ **to** ⟨causally⟩ essere dovuto a; **I'm** ~ **to...** dovrei...; **in** ~ **course** a tempo debito ● adv ~ **north** direttamente a nord

duel /'dju:əl/ n duello m

dues /dju:z/ npl quota f [di iscrizione]

duet /dju:'et/ n duetto m

dug /dʌg/ see **dig**

duke /dju:k/ n duca m

dull /dʌl/ a (overcast, not bright) cupo; (not shiny) opaco; (sound) soffocato; (boring) monotono; (stupid) ottuso ● vt intorpidire ⟨mind⟩; attenuare ⟨pain⟩

duly /'dju:lɪ/ adv debitamente

dumb /dʌm/ a muto; (fam: stupid) ottuso. ~**founded** /dʌm'faʊndɪd/ a sbigottito

dummy /'dʌmɪ/ n (tailor's) manichino m; (for baby) succhiotto m; (model) riproduzione f

dump /dʌmp/ n (for refuse) scarico m; (fam: town) mortorio m; **be down in the** ~**s** fam essere depresso ● vt scaricare; (fam: put down) lasciare; (fam: get rid of) liberarsi di

dumpling /'dʌmplɪŋ/ n gnocco m

dunce /dʌns/ n zuccone, -a mf

dune /dju:n/ n duna f

dung /dʌŋ/ n sterco m

dungarees /dʌŋgə'ri:z/ npl tuta fsg

dungeon /'dʌndʒən/ n prigione f sotterranea

duo /'dju:əʊ/ n duo m inv; Mus duetto m

duplicate¹ /'dju:plɪkət/ a doppio ● n duplicato m; ⟨document⟩ copia f; **in** ~ in duplicato

duplicate² /'dju:plɪkeɪt/ vt fare un duplicato di; ⟨research⟩ essere una ripetizione di ⟨work⟩

durable /'djʊərəbl/ a resistente; durevole ⟨basis, institution⟩

duration /djʊə'reɪʃn/ n durata f

duress /djʊə'res/ n costrizione f; **under** ~ sotto minaccia

during /'djʊərɪŋ/ prep durante

dusk /dʌsk/ n crepuscolo m

dust /dʌst/ n polvere f ● vt spolverare; (sprinkle) cospargere ⟨cake⟩ (**with** di) ● vi spolverare

dust: ~**bin** n pattumiera f. ~**cart** n camion m della nettezza urbana. ~**er** n strofinaccio m. ~**jacket** n sopraccoperta f. ~**man** n spazzino m. ~**pan** n paletta f per la spazzatura

dusty /'dʌstɪ/ a (-ier, -iest) polveroso

Dutch /dʌtʃ/ a olandese; **go** ~ fam fare alla romana ● n (language) olandese m; **the** ~ pl gli olandesi. ~**man** n olandese m

dutiable /'dju:tɪəbl/ a soggetto a imposta

dutiful /'dju:tɪfl/ a rispettoso

duty /'dju:tɪ/ n dovere m; (task) compito m; (tax) dogana f; **be on** ~ essere di servizio. ~-**free** a esente da dogana

duvet /'du:veɪ/ n piumone m

dwarf /dwɔ:f/ n (pl -s or **dwarves**) nano, -a mf ● vt rimpicciolire

dwell /dwel/ *vi* (*pt/pp* **dwelt**) *liter* dimorare. **dwell on** *vt fig* soffermarsi su. **~ing** *n* abitazione *f*

dwindle /'dwindl/ *vi* diminuire

dye /daɪ/ *n* tintura *f* ● *vt* (*pres p* **dyeing**) tingere

dying /'daɪɪŋ/ *see* **die²**

dynamic /daɪ'næmɪk/ *a* dinamico

dynamite /'daɪnəmaɪt/ *n* dinamite *f*

dynamo /'daɪnəməʊ/ *n* dinamo *f* *inv*

dynasty /'dɪnəstɪ/ *n* dinastia *f*

dysentery /'dɪsəntrɪ/ *n* dissenteria *f*

dyslex|ia /dɪs'leksɪə/ *n* dislessia *f*. **~ic** *a* dislessico

....

Ee

....

each /iːtʃ/ *a* ogni ● *pron* ognuno; **£1 ~** una sterlina ciascuno; **they love/hate ~ other** si amano/odiano; **we lend ~ other money** ci prestiamo i soldi

eager /'iːgə(r)/ *a* ansioso (**to do it** di fare); ⟨*pupil*⟩ avido di sapere. **~ly** *adv* ⟨*wait*⟩ ansiosamente; ⟨*offer*⟩ premurosamente. **~ness** *n* premura *f*

eagle /'iːgl/ *n* aquila *f*

ear¹ /ɪə(r)/ *n* (*of corn*) spiga *f*

ear² *n* orecchio *m*. **~ache** *n* mal *m* d'orecchi. **~drum** *n* timpano *m*

earl /ɜːl/ *n* conte *m*

early /'ɜːlɪ/ *a* (**-ier, -iest**) (*before expected time*) in anticipo; ⟨*spring*⟩ prematuro; ⟨*reply*⟩ pronto; ⟨*works, writings*⟩ primo; **be here ~!** sii puntuale!; **you're ~!** sei in anticipo!; **~ morning walk** passeggiata *f* mattutina; **in the ~ morning** la mattina presto; **in the ~ spring** all'inizio della primavera; **~ retirement** prepensionamento *m* ● *adv* presto; (*ahead of time*) in anticipo; **~ in the morning** la mattina presto

earmark *vt* riservare (**for** a)

earn /ɜːn/ *vt* guadagnare; (*deserve*) meritare

earnest /'ɜːnɪst/ *a* serio ● *n* **in ~** sul serio. **~ly** *adv* con aria seria

earnings /'ɜːnɪŋz/ *npl* guadagni *mpl*; (*salary*) stipendio *m*

ear: ~phones *npl* cuffia *fsg*. **~-ring** *n* orecchino *m*. **~shot** *n* **within ~shot** a portata d'orecchio; **he is out of ~shot** non può sentire

earth /ɜːθ/ *n* terra *f* **where/what on ~?** dove/che diavolo? ● *vt Electr* mettere a terra

earthenware /'ɜːθn-/ *n* terraglia *f*

earthly /'ɜːθlɪ/ *a* terrestre; **be no ~ use** *fam* essere perfettamente inutile

earthquake *n* terremoto *m*

earthy /'ɜːθɪ/ *a* terroso; (*coarse*) grossolano

earwig /'ɪəwɪg/ *n* forbicina *f*

ease /iːz/ *n* at **~** a proprio agio; **at ~!** *Mil* riposo!; **ill at ~** a disagio; **with ~** con facilità ● *vt* calmare ⟨*pain*⟩; alleviare ⟨*tension, shortage*⟩; (*slow down*) rallentare; (*loosen*) allentare ● *vi* ⟨*pain, situation, wind*⟩ calmarsi

easel /'iːzl/ *n* cavalletto *m*

easily /'iːzɪlɪ/ *adv* con facilità; **the best** certamente il meglio

east /iːst/ *n* est *m*; **to the ~ of** a est di ● *a* dell'est ● *adv* verso est

Easter /'iːstə(r)/ *n* Pasqua *f*. **~ egg** *n* uovo *m* di Pasqua

east|erly /'iːstəlɪ/ *a* da levante. **~ern** *a* orientale. **~ward[s]** /-wəd[z]/ *adv* verso est

easy /'iːzɪ/ *a* (**-ier, -iest**) facile; **take it** *or* **things ~** prendersela con calma; **take it ~!** (*don't get excited*) calma!; **go ~ with** andarci piano con

easy: ~ chair *n* poltrona *f*. **~-going** *a* conciliante; **too ~-going** troppo accomodante

eat /iːt/ *vt/i* (*pt* **ate**, *pp* **eaten**) mangiare. **eat into** vt intaccare. **eat up** vt mangiare tutto ⟨food⟩; fig inghiottire ⟨profits⟩

eat|able /'i:təbl/ a mangiabile. **~er** n ⟨apple⟩ mela f da tavola; **be a big ~er** ⟨person:⟩ essere una buona forchetta

eau-de-Cologne /əʊdəkə'ləʊn/ n acqua f di Colonia

eaves /i:vz/ npl cornicione msg. **~drop** vi ⟨pt/pp **~dropped**⟩ origliare; **~drop on** ascoltare di nascosto

ebb /eb/ n ⟨tide⟩ riflusso m; **at a low ~** fig a terra ● vi rifluire; fig declinare

ebony /'ebənɪ/ n ebano m

EC n abbr (**European Community**) CE f

eccentric /ɪk'sentrɪk/ a & n eccentrico, -a mf

ecclesiastical /ɪkli:zɪ'æstɪkl/ a ecclesiastico

echo /'ekəʊ/ n (pl **-es**) eco f or m ● v ⟨pt/pp **echoed**, pres p **echoing**⟩ ● vt echeggiare; ripetere ⟨words⟩ ● vi risuonare (**with** di)

eclipse /ɪ'klɪps/ n Astr eclissi f inv ● vt fig eclissare

ecological /i:kə'lɒdʒɪkl/ a ecologico. **~y** /ɪ'kɒlədʒɪ/ n ecologia f

economic /i:kə'nɒmɪk/ a economico. **~al** a economico. **~ally** adv economicamente; ⟨thriftily⟩ in economia. **~s** n economia f

economist /ɪ'kɒnəmɪst/ n economista mf

economize /ɪ'kɒnəmaɪz/ vi economizzare (**on** su)

economy /ɪ'kɒnəmɪ/ n economia f

ecstasy /'ekstəsɪ/ n estasi f inv; ⟨drug⟩ ecstasy f

ecstatic /ɪk'stætɪk/ a estatico

ecu /'eɪkju:/ n ecu m inv

eczema /'eksɪmə/ n eczema m

edge /edʒ/ n bordo m; ⟨of knife⟩ filo m; ⟨of road⟩ ciglio m; **on ~** con i nervi tesi; **have the ~ on** fam avere un vantaggio su ● vt bordare. **edge forward** vi avanzare lentamente

edgeways /'edʒweɪz/ adv di fianco; **I couldn't get a word in ~** non ho potuto infilare neanche mezza parola nel discorso

edging /'edʒɪŋ/ n bordo m

edgy /'edʒɪ/ a nervoso

edible /'edɪbl/ a commestibile; **this pizza's not ~** questa pizza è immangiabile

edict /'i:dɪkt/ n editto m

edify /'edɪfaɪ/ vt ⟨pt/pp **-ied**⟩ edificare. **~ing** a edificante

edit /'edɪt/ vt ⟨pt/pp **edited**⟩ far la revisione di ⟨text⟩; curare l'edizione di ⟨anthology, dictionary⟩; dirigere ⟨newspaper⟩; montare ⟨film⟩; editare ⟨tape⟩; **~ed by** ⟨book⟩ a cura di

edition /ɪ'dɪʃn/ n edizione f

editor /'edɪtə(r)/ n ⟨of anthology, dictionary⟩ curatore, -trice mf; ⟨of newspaper⟩ redattore, -trice mf; ⟨of film⟩ responsabile mf del montaggio

editorial /edɪ'tɔ:rɪəl/ a redazionale ● n Journ editoriale m

educate /'edjʊkeɪt/ vt istruire; educare ⟨public, mind⟩; **be ~d at** Eton essere educato a Eton. **~d** a istruito

education /edjʊ'keɪʃn/ n istruzione f; ⟨culture⟩ cultura f, educazione f. **~al** a istruttivo; ⟨visit⟩ educativo; ⟨publishing⟩ didattico

eel /i:l/ n anguilla f

eerie /'ɪərɪ/ a (**-ier, -iest**) inquietante

effect /ɪ'fekt/ n effetto m; **in ~** in effetti; **take ~** ⟨law:⟩ entrare in vigore; ⟨medicine:⟩ fare effetto ● vt effettuare

effective /ɪ'fektɪv/ a efficace; ⟨striking⟩ che colpisce; ⟨actual⟩ di fatto; **~ from** in vigore a partire da. **~ly** adv efficacemente; ⟨actually⟩ di fatto. **~ness** n efficacia f

effeminate /ɪ'femɪnət/ a effeminato

effervescent /efə'vesnt/ a effervescente

efficiency /ɪˈfɪʃənsɪ/ n efficienza f; (of machine) rendimento m

efficient /ɪˈfɪʃənt/ a efficiente. **~ly** adv efficientemente

effort /ˈefət/ n sforzo m; **make an ~** sforzarsi. **~less** a facile. **~lessly** adv con facilità

effrontery /ɪˈfrʌntərɪ/ n sfrontatezza f

effusive /ɪˈfjuːsɪv/ a espansivo; (speech) caloroso

e.g. abbr (exempli gratia) per es.

egalitarian /ɪgælɪˈteərɪən/ a egalitario

egg[1] /eg/ vt **~ on** fam incitare

egg[2] n uovo m. **~-cup** n portauovo m inv. **~head** n fam intellettuale mf. **~shell** n guscio m d'uovo. **~timer** n clessidra f per misurare il tempo di cottura delle uova

ego /ˈiːgəʊ/ n ego m. **~centric** /-ˈsentrik/ a egocentrico. **~ism** n egoismo m. **~ist** n egoista m. **~tism** n egotismo m. **~tist** n egotista mf

Egypt /ˈiːdʒɪpt/ n Egitto m. **~ian** /ɪˈdʒɪpʃn/ a & n egiziano, -a mf

eiderdown /ˈaɪdə-/ n (quilt) piumino m

eight /eɪt/ a otto ●n otto m. **~teen** a diciotto. **~teenth** a diciottesimo

eighth /eɪtθ/ a ottavo ●n ottavo m

eightieth /ˈeɪtɪɪθ/ a ottantesimo

eighty /ˈeɪtɪ/ a ottanta

either /ˈaɪðə(r)/ a & pron - [of them] l'uno o l'altro; [of them like ~ [of them] non mi piace né l'uno né l'altro; on ~ side da tutte e due le parti ● adv I don't ~ nemmeno io; I don't like John or his brother ~ non mi piace John e nemmeno suo fratello ●conj ~ John or his brother will be there ci saranno o John o suo fratello; I don't like ~ John or his brother non mi piacciono né John né suo fratello; ~ you go to bed or [else]... o vai a letto o [altrimenti]...

eject /ɪˈdʒekt/ vt eiettare (pilot); espellere (tape, drunk)

eke /iːk/ vt **~ out** far bastare; (increase) arrotondare; **~ out a living** arrangiarsi

elaborate[1] /ɪˈlæbərət/ a elaborato

elaborate[2] /ɪˈlæbəreɪt/ vi entrare nei particolari (**on** di)

elapse /ɪˈlæps/ vi trascorrere

elastic /ɪˈlæstɪk/ a elastico ●n elastico m. **~ 'band** n elastico m

elasticity /ɪlæˈstɪsətɪ/ n elasticità f

elated /ɪˈleɪtɪd/ a esultante

elbow /ˈelbəʊ/ n gomito m

elder[1] /ˈeldə(r)/ n (tree) sambuco m

elder[2] a maggiore ● the ~ il/la maggiore. **~ly** a anziano. **~est** a maggiore ●n the ~est il/la maggiore

elect /ɪˈlekt/ a **the president ~** il futuro presidente ● vt eleggere; **~ to do sth** decidere di fare qcsa. **~ion** /-ekʃn/ n elezione f

elector /ɪˈlektə(r)/ n elettore, -trice mf. **~al** a elettorale; **~al roll** liste fpl elettorali. **~ate** /-rət/ n elettorato m

electric /ɪˈlektrɪk/ a elettrico

electrical /ɪˈlektrɪk(ə)l/ a elettrico; **~ engineering** n elettrotecnica f

electric: ~ 'blanket n termocoperta f. **~ 'fire** n stufa f elettrica

electrician /ɪlekˈtrɪʃn/ n elettricista m

electricity /ɪlekˈtrɪsətɪ/ n elettricità f

electrify /ɪˈlektrɪfaɪ/ vt (pt/pp -ied) elettrificare; fig elettrizzare. **~ing** a fig elettrizzante

electrocute /ɪˈlektrəkjuːt/ vt fulminare; (execute) giustiziare sulla sedia elettrica

electrode /ɪˈlektrəʊd/ n elettrodo m

electron /ɪˈlektrɒn/ n elettrone m

electronic /ɪlekˈtrɒnɪk/ a elettronico. **~ mail** n posta f elettronica. **~s** n elettronica f

elegance /ˈelɪgəns/ n eleganza f

elegant /ˈelɪgənt/ a elegante

elegy /ˈelɪdʒɪ/ n elegia f

element /'elɪmənt/ n elemento m. **~ary** /-'mentərɪ/ a elementare

elephant /'elɪfənt/ n elefante m

elevat|e /'elɪveɪt/ vt elevare. **~ion** /-'veɪʃn/ n elevazione f; (height) altitudine f; (angle) alzo m

elevator /'elɪveɪtə(r)/ n Am ascensore m

eleven /ɪ'levn/ a undici ● n undici m. **~th** a undicesimo; **at the ~th hour** fam all'ultimo momento

elf /elf/ n (pl **elves**) elfo m

elicit /ɪ'lɪsɪt/ vt ottenere

eligible /'elɪdʒəbl/ a eleggibile; **~ young man** buon partito; **be ~ for** aver diritto a

eliminate /ɪ'lɪmɪneɪt/ vt eliminare

élite /eɪ'liːt/ n fior fiore m

ellip|se /ɪ'lɪps/ n ellisse f. **~tical** a ellittico

elm /elm/ n olmo m

elocution /elə'kjuːʃn/ n elocuzione f

elope /ɪ'ləʊp/ vi fuggire [per sposarsi]

eloquen|ce /'eləkwəns/ n eloquenza f. **~t** a eloquente. **~tly** adv con eloquenza

else /els/ adv altro; **who ~?** e chi altro?; **he did of course, who ~?** l'ha fatto lui e chi, se no?; **nothing ~** nient'altro; **or ~** altrimenti; **someone ~** qualcun altro; **somewhere ~** da qualche altra parte; **anyone ~** chiunque altro; (as question) nessun'altro?; **anything ~** qualunque altra cosa; (as question) altro?. **~where** adv altrove

elucidate /ɪ'luːsɪdeɪt/ vt delucidare

elude /ɪ'luːd/ vt eludere; (avoid) evitare; **the name ~s me** il nome mi sfugge

elusive /ɪ'luːsɪv/ a elusivo

emaciated /ɪ'meɪsɪeɪtɪd/ a emaciato

e-mail /'iːmeɪl/ n posta f elettronica ● vt spedire via posta elettronica

emanate /'eməneɪt/ vi emanare

emancipat|ed /ɪ'mænsɪpeɪtɪd/ a

emancipato. **~ion** /-'peɪʃn/ n emancipazione f; (of slaves) liberazione f

embankment /ɪm'bæŋkmənt/ n argine m; Rail massicciata f

embargo /em'bɑːgəʊ/ n (pl **-es**) embargo m

embark /ɪm'bɑːk/ vi imbarcarsi; **~ on** intraprendere. **~ation** /embɑː-'keɪʃn/ n imbarco m

embarrass /em'bærəs/ vt imbarazzare. **~ed** a imbarazzato. **~ing** a imbarazzante. **~ment** n imbarazzo m

embassy /'embəsɪ/ n ambasciata f

embedded /ɪm'bedɪd/ a (in concrete) cementato; (traditions, feelings) radicato

embellish /ɪm'belɪʃ/ vt abbellire

embers /'embəz/ npl braci fpl

embezzle /ɪm'bezl/ vt appropriarsi indebitamente di. **~ment** n appropriazione f indebita

embitter /ɪm'bɪtə(r)/ vt amareggiare

emblem /'embləm/ n emblema m

embody /ɪm'bɒdɪ/ vt (pt/pp **-ied**) incorporare; **~ what is best in...** rappresentare quanto c'è di meglio di...

emboss /ɪm'bɒs/ vt sbalzare (metal); stampare in rilievo (paper). **~ed** a in rilievo

embrace /ɪm'breɪs/ n abbraccio m ● vt abbracciare ● vi abbracciarsi

embroider /ɪm'brɔɪdə(r)/ vt ricamare (design); fig abbellire. **~y** n ricamo m

embryo /'embrɪəʊ/ n embrione m

emerald /'emərəld/ n smeraldo m

emer|ge /ɪ'mɜːdʒ/ vi emergere; (come into being: nation) nascere; (sun, flowers) spuntare fuori. **~gence** /-əns/ n emergere m; (of new country) nascita f

emergency /ɪ'mɜːdʒənsɪ/ n emergenza f; **in an ~** in caso di emergenza. **~ exit** n uscita f di sicurezza

emery /'emərɪ/: **~ board** n limetta f [per le unghie]

emigrant /'emɪgrənt/ n emigrante mf

emigrat|e /'emɪgreɪt/ vi emigrare. **~ion** /-'greɪʃn/ n emigrazione f. **~ly** adv eminentemente

eminent /'emɪnənt/ a eminente. **~ly** adv eminentemente

emission /ɪ'mɪʃn/ n emissione f; (of fumes) esalazione f

emit /ɪ'mɪt/ vt (pt/pp **emitted**) emettere; esalare (fumes)

emotion /ɪ'məʊʃn/ n emozione f. **~al** a denso di emozione; (person, reaction) emotivo; **become ~al** avere una reazione emotiva

emotive /ɪ'məʊtɪv/ a emotivo

empathize /'empəθaɪz/ vi ~ **with sb** immedesimarsi nei problemi di qcno

emperor /'empərə(r)/ n imperatore m

emphasis /'emfəsɪs/ n enfasi f; **put the ~ on sth** accentuare qcsa

emphasize /'emfəsaɪz/ vt accentuare (word, syllable); sottolineare (need)

emphatic /ɪm'fætɪk/ a categorico

empire /'empaɪə(r)/ n impero m

empirical /ɪm'pɪrɪkl/ a empirico

employ /em'plɔɪ/ vt impiegare; fig usare (tact). **~ee** /emplɔɪ'i:/ n impiegato, -a mf. **~er** n datore m di lavoro. **~ment** n occupazione f; (work) lavoro m. **~ment agency** n ufficio m di collocamento

empower /ɪm'paʊə(r)/ vt autorizzare; (enable) mettere in grado

empress /'emprɪs/ n imperatrice f

empties /'emptɪz/ npl vuoti mpl

emptiness /'emptɪnɪs/ n vuoto m

empty /'emptɪ/ a vuoto; (promise, threat) vano ● v (pt/pp -ied) ● vt vuotare (con-tainer). ● vi vuotarsi

emulate /'emjʊleɪt/ vt emulare

emulsion /ɪ'mʌlʃn/ n emulsione f

enable /ɪ'neɪbl/ vt ~ **sb to** mettere qcno in grado di

enact /ɪ'nækt/ vt Theat rappresentare; decretare (law)

enamel /ɪ'næml/ n smalto m ● vt (pt/pp **enamelled**) smaltare

enchant /ɪn'tʃɑ:nt/ vt incantare.

~ing a incantevole. **~ment** n incanto m

encircle /ɪn'sɜ:kl/ vt circondare

enclave /'enkleɪv/ n enclave f inv; fig territorio m

enclos|e /ɪn'kləʊz/ vt circondare (land); (in letter) allegare (with a). **~ed** a (space) chiuso; (in letter) allegato. **~ure** /-ʒə(r)/ n (at zoo) recinto m; (in letter) allegato m

encompass /ɪn'kʌmpəs/ vt (include) comprendere

encore /'ɒŋkɔ:(r)/ n & int bis m inv

encounter /ɪn'kaʊntə(r)/ n incontro m; (battle) scontro m ● vt incontrare

encourag|e /ɪn'kʌrɪdʒ/ vt incoraggiare; promuovere (the arts, independence). **~ement** n incoraggiamento m; (of the arts) promozione f. **~ing** a incoraggiante; (smile) di incoraggiamento

encroach /ɪn'krəʊtʃ/ vi ~ **on** invadere (land, privacy); abusare di (time); interferire con (rights)

encumber /ɪn'kʌmbə(r)/ vt **~ed with** essere carico di (children, suitcases); ingombro di (furniture). **~rance** /-rəns/ n peso m

encyclop[a]ed|ia /ɪnsaɪklə'pi:dɪə/ n enciclopedia f. **~ic** a enciclopedico

end /end/ n fine f; (of box, table, piece of string) estremità f; (of town, room) parte f; (purpose) fine m; **in the ~** alla fine; **at the ~ of May** alla fine di maggio; **at the ~ of the street/garden** in fondo alla strada/al giardino; **on ~** (upright) in piedi; **for days on ~** per giorni e giorni; **for six days on ~** per sei giorni di fila; **put an ~ to sth** mettere fine a qcsa; **make ~s meet** fam sbarcare il lunario; **no ~ of** fam un sacco di ● vt/i finire. **end up** vi finire; **~ up doing sth** finire col fare qcsa

endanger /ɪn'deɪndʒə(r)/ vt rischiare (one's life); mettere a repentaglio (sb else, success of sth)

endear|ing /ɪn'dɪərɪŋ/ a accat-

tivante. **~ment** n term of **~ment**
vezzeggiativo m

endeavour /ɪn'devə(r)/ n tentativo
m ● vi sforzarsi (**to** di)

ending /'endɪŋ/ n fine f; Gram desinenza f

endive /'endɪv/ n indivia f

endless /'endlɪs/ a interminabile;
⟨patience⟩ infinito. **~ly** adv continuamente; ⟨patient⟩ infinitamente

endorse /ɪn'dɔːs/ vt girare
⟨cheque⟩; ⟨sports personality:⟩ fare
pubblicità a ⟨product⟩; approvare
⟨plan⟩. **~ment** n ⟨of cheque⟩ girata
f; ⟨of plan⟩ conferma f; ⟨on driving
licence⟩ registrazione f su patente
di un'infrazione

endow /ɪn'daʊ/ vt dotare

endur|able /ɪn'djʊərəbl/ a sopportabile. **~ance** /-rəns/ n resistenza
f; **it is beyond ~ance** è insopportabile

endur|e /ɪn'djʊə(r)/ vt sopportare
● vi durare. **~ing** a duraturo

'end user n utente m finale

enemy /'enəmɪ/ n nemico, -a mf
● attrib nemico

energetic /enə'dʒetɪk/ a energico

energy /'enədʒɪ/ n energia f

enforce /ɪn'fɔːs/ vt far rispettare
⟨law⟩. **~d** a forzato

engage /ɪn'geɪdʒ/ vt assumere
⟨staff⟩; Theat ingaggiare; Auto ingranare ⟨gear⟩ ● vi Techn ingranare; **~ in** impegnarsi in. **~d** a ⟨in
use, busy⟩ occupato; ⟨person⟩ impegnato; ⟨to be married⟩ fidanzato;
get ~d fidanzarsi (**to** con); **~d
tone** Teleph segnale m di occupato. **~ment** n fidanzamento m;
⟨appointment⟩ appuntamento m;
Mil combattimento m; **~ment
ring** anello m di fidanzamento

engaging /ɪn'geɪdʒɪŋ/ a attraente

engender /ɪn'dʒendə(r)/ vt fig generare

engine /'endʒɪn/ n motore m; Rail
locomotrice f. **~-driver** n macchinista m

engineer /endʒɪ'nɪə(r)/ n ingegne-

re m; ⟨service, installation⟩ tecnico
m; Naut, Am Rail macchinista m
● vt fig architettare. **~ing** n ingegneria f

England /'ɪŋglənd/ n Inghilterra f

English /'ɪŋglɪʃ/ a inglese; the **~
Channel** la Manica ● n ⟨language⟩
inglese m; the **~** pl gli inglesi.
~man n inglese m. **~woman** n inglese f

engrav|e /ɪn'greɪv/ vt incidere.
~ing n incisione f

engross /ɪn'grəʊs/ vt **~ed in** assorto in

engulf /ɪn'gʌlf/ vt ⟨fire, waves:⟩
inghiottire

enhance /ɪn'hɑːns/ vt accrescere
⟨beauty, reputation⟩; migliorare
⟨performance⟩

enigma /ɪ'nɪgmə/ n enigma m.
~tic /enɪg'mætɪk/ a enigmatico

enjoy /ɪn'dʒɔɪ/ vt godere di ⟨good
health⟩; **~** oneself divertirsi; **I ~
cooking/painting** mi piace cucinare/dipingere; **~ your meal**
buon appetito. **~able** /-əbl/ a piacevole. **~ment** n piacere m

enlarge /ɪn'lɑːdʒ/ vt ingrandire
● vi **~ upon** dilungarsi su. **~ment**
n ingrandimento m

enlighten /ɪn'laɪtn/ vt illuminare.
~ed a progressista. **~ment** n The
E~ment l'Illuminismo m

enlist /ɪn'lɪst/ vt Mil reclutare; **~**
sb's help farsi aiutare da qcno
● vi Mil arruolarsi

enliven /ɪn'laɪvn/ vt animare

enmity /'enmətɪ/ n inimicizia f

enormity /ɪ'nɔːmətɪ/ n enormità f

enormous /ɪ'nɔːməs/ a enorme.
~ly adv estremamente; ⟨grateful⟩
infinitamente

enough /ɪ'nʌf/ a & n abbastanza; **I
didn't bring ~ clothes** non ho portato abbastanza vestiti; **have you
had ~?** ⟨to eat/drink⟩ hai mangiato/bevuto abbastanza?; **I've had
~!** fam ne ho abbastanza!; **is that
~?** basta?; **that's ~!** basta così!;
£50 isn't ~ 50 sterline non
sono sufficienti ● adv abbastanza;

you're not working fast ~ non lavori abbastanza in fretta; **funnily** ~ stranamente

enquir|e /ɪnˈkwaɪə(r)/ vi domandare; **~e about** chiedere informazioni su. **~y** n domanda f; (investigation) inchiesta f

enrage /ɪnˈreɪdʒ/ vt fare arrabbiare

enrich /ɪnˈrɪtʃ/ vt arricchire; (improve) migliorare (vocabulary)

enrol /ɪnˈrəʊl/ vi (pt/pp -rolled) (for exam, in club) iscriversi (for, in a). **~ment** n iscrizione f

ensemble /ɒnˈsɒmbl/ n (clothing & Mus) complesso m

enslave /ɪnˈsleɪv/ vt render schiavo

ensu|e /ɪnˈsjuː/ vi seguire; **the ~ing discussion** la discussione che ne è seguita

ensure /ɪnˈʃʊə(r)/ vt assicurare; **~ that** (person:) assicurarsi che; (measure:) garantire che

entail /ɪnˈteɪl/ vt comportare; **what does it ~?** in che cosa consiste?

entangle /ɪnˈtæŋgl/ vt **get ~d in** rimanere impigliato in; fig rimanere coinvolto in

enter /ˈentə(r)/ vt entrare in; iscrivere (horse, runner in race); cominciare a (university); partecipare a (competition); Comput immettere (data); (write down) scrivere ● vi entrare; Theat entrare in scena; (register as competitor) iscriversi; (take part) partecipare (in a)

enterprise /ˈentəpraɪz/ n impresa f; (quality) iniziativa f. **~ing** a intraprendente

entertain /entəˈteɪn/ vt intrattenere; (invite) ricevere; nutrire (ideas, hopes); prendere in considerazione (possibility) ● vi intrattenersi; (have guests) ricevere. **~er** n artista mf. **~ing** a (person) di gradevole compagnia; (evening, film, play) divertente. **~ment** n (amusement) intrattenimento m

enthral /ɪnˈθrɔːl/ vt (pt/pp en-thralled) **be ~led** essere affascinato (by da)

enthusias|m /ɪnˈθjuːzɪæzm/ n entusiasmo m. **~t** n entusiasta mf. **~tic** /-ˈæstɪk/ a entusiastico

entice /ɪnˈtaɪs/ vt attirare. **~ment** n (incentive) incentivo m

entire /ɪnˈtaɪə(r)/ a intero. **~ly** adv del tutto; **I'm not ~ly satisfied** non sono completamente soddisfatto. **~ty** /-rəti/ n **in its ~ty** nell'insieme

entitled /ɪnˈtaɪtld/ a (book) intitolato; **be ~ to sth** aver diritto a qcsa

entitlement /ɪnˈtaɪtlmənt/ n diritto m

entity /ˈentəti/ n entità f

entrance¹ /ˈentrəns/ n entrata f; Theat entrata f in scena; (right to enter) ammissione f; **'no ~'** 'ingresso vietato'. **~ examination** n esame m di ammissione. **~ fee** n how much is the **~ fee?** quanto costa il biglietto di ingresso?

entrance² /ɪnˈtrɑːns/ vt estasiare

entrant /ˈentrənt/ n concorrente mf

entreat /ɪnˈtriːt/ vt supplicare

entrenched /ɪnˈtrentʃt/ a (ideas, views) radicato

entrust /ɪnˈtrʌst/ vt **~ sb with sth, ~ sth to sb** affidare qcsa a qcno

entry /ˈentrɪ/ n ingresso m; (way in) entrata f; (in directory etc) voce f; (in appointment diary) appuntamento m; **no ~** ingresso vietato; Auto accesso vietato. **~ form** n modulo m di ammissione. **~ visa** n visto m di ingresso

enumerate /ɪˈnjuːməreɪt/ vt enumerare

enunciate /ɪˈnʌnsɪeɪt/ vt enunciare

envelop /ɪnˈveləp/ vt (pt/pp enveloped) avviluppare

envelope /ˈenvələʊp/ n busta f

enviable /ˈenvɪəbl/ a invidiabile

envious /ˈenvɪəs/ a invidioso. **~ly** adv con invidia

environment /ɪn'vaɪrənmənt/ n ambiente m

environmental /ɪnvaɪrən'mentl/ a ambientale. **~ist** n ambientalista mf. **~ly** adv **~ly friendly** che rispetta l'ambiente

envisage /ɪn'vɪzɪdʒ/ vt prevedere

envoy /'envɔɪ/ n inviato, -a m

envy /'envɪ/ n invidia ● vt (pt/pp **-ied**) **~ sb sth** invidiare qcno per qcsa

enzyme /'enzaɪm/ n enzima m

epic /'epɪk/ a epico ● n epopea f

epidemic /epɪ'demɪk/ n epidemia f

epilep|sy /'epɪlepsɪ/ n epilessia f. **~tic** /-'leptɪk/ a & n epilettico, -a mf

epilogue /'epɪlɒg/ n epilogo m

episode /'epɪsəʊd/ n episodio m

epitaph /'epɪtɑːf/ n epitaffio m

epithet /'epɪθet/ n epiteto m

epitom|e /ɪ'pɪtəmɪ/ n epitome f. **~ize** vt essere il classico esempio di

epoch /'iːpɒk/ n epoca f

equal /'iːkwl/ a (parts, amounts) uguale; **of ~ height** della stessa altezza; **be ~ to the task** essere a l'altezza del compito ● n pari m inv ● vt (pt/pp **equalled**) (be same in quantity as) essere pari a; (rival) uguagliare; **5 plus 5 ~s 10** 5 più 5 [è] uguale a 10. **~ity** /ɪ'kwɒlətɪ/ n uguaglianza f

equalize /'iːkwəlaɪz/ vi Sport pareggiare. **~r** n Sport pareggio m

equally /'iːkwəlɪ/ adv (divide) in parti uguali; **~ intelligent** della stessa intelligenza; **~,...** allo stesso tempo...

equanimity /ekwə'nɪmətɪ/ n equanimità f

equat|e /ɪ'kweɪt/ vt **~e sth with sth** equiparare qcsa a qcsa. **~ion** /-eɪʒn/ n Math equazione f

equator /ɪ'kweɪtə(r)/ n equatore m

equestrian /ɪ'kwestrɪən/ a equestre

equilibrium /iːkwɪ'lɪbrɪəm/ n equilibrio m

equinox /'iːkwɪnɒks/ n equinozio m

equip /ɪ'kwɪp/ vt (pt/pp **equipped**) equipaggiare; attrezzare (kitchen, office). **~ment** n attrezzatura f

equitable /'ekwɪtəbl/ a giusto

equity /'ekwɪtɪ/ n (justness) equità f; Comm azioni fpl

equivalent /ɪ'kwɪvələnt/ a equivalente; **be ~ to** equivalere a ● n equivalente m

equivocal /ɪ'kwɪvəkl/ a equivoco

era /'ɪərə/ n età f; (geological) era f

eradicate /ɪ'rædɪkeɪt/ vt eradicare

erase /ɪ'reɪz/ vt cancellare. **~r** n gomma f [da cancellare]; (for blackboard) cancellino m

erect /ɪ'rekt/ a eretto ● vt erigere. **~ion** /-ekʃn/ n erezione f

ero|de /ɪ'rəʊd/ vt (water:) erodere; (acid:) corrodere. **~sion** /-əʊʒn/ n erosione f; (by acid) corrosione f

erotic /ɪ'rɒtɪk/ a erotico. **~ism** /-tɪsɪzm/ n erotismo m

err /ɜː(r)/ vi errare; (sin) peccare

errand /'erənd/ n commissione f

erratic /ɪ'rætɪk/ a irregolare; (person, moods) imprevedibile; (exchange rate) incostante

erroneous /ɪ'rəʊnɪəs/ a erroneo

error /'erə(r)/ n errore m; **in ~** per errore

erudit|e /'eruːdaɪt/ a erudito. **~ion** /-'dɪʃn/ n erudizione f

erupt /ɪ'rʌpt/ vi eruttare; (spots:) spuntare; (fig: in anger) dare in escandescenze. **~ion** /-ʌpʃn/ n eruzione f; fig scoppio m

escalat|e /'eskəleɪt/ vi intensificarsi ● vt intensificare. **~ion** /-'leɪʃn/ n escalation f inv. **~or** n scala f mobile

escapade /'eskəpeɪd/ n scappatella f

escape /ɪ'skeɪp/ n fuga f; (from prison) evasione f; **have a narrow ~** cavarsela per un pelo ● vi (prisoner:) evadere (from da); sfuggire (from sb alla sorveglianza di qcno); (animal:) scappare; (gas:) fuoriuscire ● vt **~ notice**

passare inosservato; **the name ~s me** mi sfugge il nome

escapism /ɪˈskeɪpɪzm/ *n* evasione *f* [dalla realtà]

escort¹ /ˈeskɔːt/ *n* (*of person*) accompagnatore, -trice *mf*; *Mil etc* scorta *f*

escort² /ɪˈskɔːt/ *vt* accompagnare; *Mil etc* scortare

Eskimo /ˈeskɪməʊ/ *n* esquimese *mf*

esoteric /esəˈterɪk/ *a* esoterico

especial /ɪˈspeʃl/ *a* speciale. **~ly** *adv* specialmente; (*kind*) particolarmente

espionage /ˈespɪənɑːʒ/ *n* spionaggio *m*

essay /ˈeseɪ/ *n* saggio *m*; *Sch* tema *f*

essence /ˈesns/ *n* essenza *f*; **in ~** in sostanza

essential /ɪˈsenʃl/ *a* essenziale ● *npl* **the ~s** l'essenziale *m*. **~ly** *adv* essenzialmente

establish /ɪˈstæblɪʃ/ *vt* stabilire (*contact, lead*); fondare (*firm*); (*prove*) accertare; **~ oneself as** affermarsi come. **~ment** *n* (*firm*) azienda *f*; **the E~ment** l'ordine *m* costituito

estate /ɪˈsteɪt/ *n* tenuta *f*; (*possessions*) patrimonio *m*; (*housing*) quartiere *m* residenziale. **~ agent** *n* agente *m* immobiliare. **~ car** *n* giardiniera *f*

esteem /ɪˈstiːm/ *n* stima *f* ● *vt* stimare; (*consider*) giudicare

estimate¹ /ˈestɪmət/ *n* valutazione *f*; *Comm* preventivo *m*; **at a rough ~** a occhio e croce

estimate² /ˈestɪmeɪt/ *vt* stimare. **~ion** /-ˈmeɪʃn/ *n* (*esteem*) stima *f*; **in my ~ion** (*judgement*) a mio giudizio

estuary /ˈestjʊərɪ/ *n* estuario *m*

etc /etˈsetərə/ *abbr* (*et cetera*) ecc

etching /ˈetʃɪŋ/ *n* acquaforte *f*

eternal /ɪˈtɜːnl/ *a* eterno

eternity /ɪˈtɜːnətɪ/ *n* eternità *f*

ethic /ˈeθɪk/ *n* etica *f*. **~al** *a* etico. **~s** *n* etica *f*

Ethiopia /iːθɪˈəʊpɪə/ *n* Etiopia *f*

ethnic /ˈeθnɪk/ *a* etnico

etiquette /ˈetɪket/ *n* etichetta *f*

EU *n abbr* (**European Union**) UE *f*

eucalyptus /juːkəˈlɪptəs/ *n* eucalipto *m*

eulogy /ˈjuːlədʒɪ/ *n* elogio *m*

euphemism /ˈjuːfəmɪzm/ *n* eufemismo *m*. **~tic** /-ˈmɪstɪk/ *a* eufemistico

euphoria /juːˈfɔːrɪə/ *n* euforia *f*

Euro- /ˈjʊərəʊ-/ *pref* **~cheque** *n* eurochèque *m inv*. **~dollar** *n* eurodollaro *m*

Europe /ˈjʊərəp/ *n* Europa *f*

European /jʊərəˈpɪən/ *a* europeo; **~ Community** Comunità *f* Europea; **~ Union** Unione *f* Europea ● *n* europeo, -a *mf*

evacuate /ɪˈvækjʊeɪt/ *vt* evacuare (*building, area*). **~ion** /-ˈeɪʃn/ *n* evacuazione *f*

evade /ɪˈveɪd/ *vt* evadere (*taxes*); evitare (*the enemy, authorities*); **the issue** evitare l'argomento

evaluate /ɪˈvæljʊeɪt/ *vt* valutare

evange|lical /iːvænˈdʒelɪkl/ *a* evangelico. **~list** /ɪˈvændʒəlɪst/ *n* evangelista *m*

evaporat|e /ɪˈvæpəreɪt/ *vi* evaporare; *fig* svanire. **~ion** /-ˈreɪʃn/ *n* evaporazione *f*

evasion /ɪˈveɪʒn/ *n* evasione *f*

evasive /ɪˈveɪsɪv/ *a* evasivo

eve /iːv/ *n liter* vigilia *f*

even /ˈiːvn/ *a* (*level*) piatto; (*same, equal*) uguale; (*regular*) regolare; (*number*) pari; **get ~ with** vendicarsi di; **now we're ~** adesso siamo pari ● *adv* anche, ancora; **~ if** anche se; **~ so** con tutto ciò; **not ~** nemmeno; **~ bigger/hotter** ancora più grande/caldo ● *vt* ~ **the score** *Sport* pareggiare. **even out** *vi* livellarsi. **even up** *vt* livellare

evening /ˈiːvnɪŋ/ *n* sera *f*; (*whole evening*) serata *f*; **this ~** stasera; **in the ~** la sera. **~ class** *n* corso *m* serale. **~ dress** *n* (*man's*) abito *m* scuro; (*woman's*) abito *m* da sera

evenly /ˈiːvnlɪ/ *adv* (*distributed*) uniformemente; (*breathe*) regolarmente; (*divided*) in uguali parti

event /ɪ'vent/ n avvenimento m; (*function*) manifestazione f; Sport gara f; **in the ~ of** nell'eventualità di; **in the ~** alla fine; **~ful** a movimentato

eventual /ɪ'ventjʊəl/ a the ~ **winner was....** alla fine il vincitore è stato.... **~ity** /-'ælətɪ/ n eventualità f. **~ly** adv alla fine; **~ly!** finalmente!

ever /'evə(r)/ adv mai; **I haven't ~....** non ho mai...; **for ~** per sempre; **hardly ~** quasi mai; **~ since** da quando; (*since that time*) da allora; **~ so** fam veramente

'evergreen n sempreverde m

ever'lasting a eterno

every /'evrɪ/ a ogni; **~ one** ciascuno; **~ other day** un giorno sì un giorno no

every: **~body** pron tutti pl. **~day** a quotidiano, di ogni giorno. **~one** pron tutti pl; **~one else** tutti gli altri. **~thing** pron tutto; **~thing else** tutto il resto. **~where** adv dappertutto; (*wherever*) dovunque

evict /ɪ'vɪkt/ vt sfrattare. **~ion** /-ɪkʃn/ n sfratto m

eviden|ce /'evɪdəns/ n evidenza f; Jur testimonianza f; **give ~ce** testimoniare. **~t** a evidente. **~tly** adv evidentemente

evil /'i:vl/ a cattivo ● n male m

evocative /ɪ'vɒkətɪv/ a evocativo; **be ~ of** evocare

evoke /ɪ'vəʊk/ vt evocare

evolution /i:və'lu:ʃn/ n evoluzione f

evolve /ɪ'vɒlv/ vt evolvere ● vi evolversi

ewe /ju:/ n pecora f

exacerbate /ɪg'zæsəbeɪt/ vt esacerbare (*situation*)

exact /ɪg'zækt/ a esatto ● vt esigere. **~ing** a esigente. **~itude** /-ɪtju:d/ n esattezza f. **~ly** adv esattamente; **not ~ly** non proprio. **~ness** n precisione f

exaggerat|e /ɪg'zædʒəreɪt/ vt/i esagerare. **~ion** /-'reɪʃn/ n esagerazione f

exam /ɪg'zæm/ n esame m

examination /ɪgzæmɪ'neɪʃn/ n esame m; (*of patient*) visita f

examine /ɪg'zæmɪn/ vt esaminare; visitare (*patient*). **~r** n Sch esaminatore, -trice mf

example /ɪg'zɑ:mpl/ n esempio m; **for ~** per esempio; **make an ~ of sb** punire qcno per dare un esempio; **be an ~ to sb** dare il buon esempio a qcno

exasperat|e /ɪg'zæspəreɪt/ vt esasperare. **~ion** /-'reɪʃn/ n esasperazione f

excavat|e /'ekskəveɪt/ vt scavare; Archaeol fare gli scavi di. **~ion** /-'veɪʃn/ n scavo m

exceed /ɪk'si:d/ vt eccedere. **~ingly** adv estremamente

excel /ɪk'sel/ v (pt/pp **excelled**) ● vi eccellere ● vt **~ oneself** superare se stessi

excellen|ce /'eksələns/ n eccellenza f. **E~cy** n (title) Eccellenza f. **~t** a eccellente

except /ɪk'sept/ prep eccetto, tranne; **~ for** eccetto, tranne; **~ that....** eccetto che... ● vt eccettuare. **~ing** prep eccetto, tranne

exception /ɪk'sepʃn/ n eccezione f; **take ~ to** fare obiezioni a. **~al** a eccezionale. **~ally** adv eccezionalmente

excerpt /'eksɜ:pt/ n estratto m

excess /ɪk'ses/ n eccesso m; **in ~ of** oltre. **~ baggage** n bagaglio m in eccedenza. **~ 'fare** n supplemento m

excessive /ɪk'sesɪv/ a eccessivo. **~ly** adv eccessivamente

exchange /ɪks'tʃeɪndʒ/ n scambio m; Teleph centrale f; Comm cambio m; (*stock*) ~ borsa f valori; **in ~** in cambio (**for** di) ● vt scambiare (**for** con); cambiare (*money*). **~ rate** n tasso m di cambio

exchequer /ɪks'tʃekə(r)/ n Pol tesoro m

excise[1] /'eksaɪz/ n dazio m; ~ **duty** dazio m

excise[2] /ek'saɪz/ vt recidere

excitable /ɪk'saɪtəbl/ a eccitabile

excit|e /ɪk'saɪt/ vt eccitare. ~**ed** a eccitato; **get** ~**ed** eccitarsi. ~**edly** adv tutto eccitato. ~**ement** n eccitazione f. ~**ing** a eccitante; (story, film) appassionante; (holiday) entusiasmante

exclaim /ɪk'skleɪm/ vt/i esclamare

exclamation /eksklə'meɪʃn/ n esclamazione f. ~ **mark** n, Am ~ **point** n punto m esclamativo

exclud|e /ɪk'sklu:d/ vt escludere. ~**ding** pron escluso. ~**sion** /-ʒn/ n esclusione f

exclusive /ɪk'sklu:sɪv/ a (rights, club) esclusivo; (interview) in esclusiva; ~ **of...** ...escluso. ~**ly** adv esclusivamente

excommunicate /ekskə'mju:nɪkeɪt/ vt comunicare

excrement /'ekskrɪmənt/ n escremento m

excruciating /ɪk'skru:ʃieɪtɪŋ/ a atroce (pain); (fam: very bad) spaventoso

excursion /ɪk'skɜ:ʃn/ n escursione f

excusable /ɪk'skju:zəbl/ a perdonabile

excuse[1] /ɪk'skju:s/ n scusa f

excuse[2] /ɪk'skju:z/ vt scusare; ~ **from** esonerare da; ~ **me!** (to get attention) scusi!; (to get past) permesso!, scusi!; (indignant) come ha detto?

ex-di'rectory a be ~ non figurare sull'elenco telefonico

execute /'eksɪkju:t/ vt eseguire; (put to death) giustiziare; attuare (plan)

execution /eksɪ'kju:ʃn/ n esecuzione f; (of plan) attuazione f. ~**er** n boia m or f

executive /ɪg'zekjutɪv/ a esecutivo ● n dirigente mf; Pol esecutivo m

executor /ɪg'zekjutə(r)/ n Jur esecutore, -trice mf

exemplary /ɪg'zemplərɪ/ a esemplare

exemplify /ɪg'zemplɪfaɪ/ vt (pt/pp -ied) esemplificare

exempt /ɪg'zempt/ a esente ● vt esentare (**from** da). ~**ion** /-empʃn/ n esenzione f

exercise /'eksəsaɪz/ n esercizio m; Mil esercitazione f; **physical** ~**s** ginnastica f; **take** ~ fare del moto ● vt esercitare (muscles, horse); portare a spasso (dog); mettere in pratica (skills) ● vi esercitarsi. ~ **book** n quaderno m

exert /ɪg'zɜ:t/ vt esercitare; ~ **oneself** sforzarsi. ~**ion** /-ɜ:ʃn/ n sforzo m

exhale /eks'heɪl/ vt/i esalare

exhaust /ɪg'zɔ:st/ n Auto scappamento m; (pipe) tubo m di scappamento; ~ **fumes** fumi mpl di scarico m ● vt esaurire. ~**ed** a esausto. ~**ing** a estenuante; (climate, person) sfibrante. ~**ion** /-ɔ:stʃn/ n esaurimento m. ~**ive** /-ɪv/ a fig esauriente

exhibit /ɪg'zɪbɪt/ n oggetto m esposto; Jur reperto m ● vt esporre; fig dimostrare

exhibition /eksɪ'bɪʃn/ n mostra f; (of strength, skill) dimostrazione f. ~**ist** n esibizionista m

exhibitor /ɪg'zɪbɪtə(r)/ n espositore, -trice mf

exhilarat|ed /ɪg'zɪləreɪtɪd/ a rallegrato. ~**ing** a stimolante; (mountain air) tonificante. ~**ion** /-'reɪʃn/ n allegria f

exhort /ɪg'zɔ:t/ vt esortare

exhume /ɪg'zju:m/ vt esumare

exile /'eksaɪl/ n esilio m; (person) esule m ● vt esiliare

exist /ɪg'zɪst/ vi esistere. ~**ence** /-əns/ n esistenza f; **in** ~ esistente; **be in** ~**ence** esistere. ~**ing** a attuale

exit /'eksɪt/ n uscita f; Theat uscita f di scena ● vi Theat uscire di scena; Comput uscire

exonerate /ɪg'zɒnəreɪt/ vt esonerare

exorbitant /ɪɡˈzɔːbɪtənt/ a esorbitante

exorcize /ˈeksɔːsaɪz/ vt esorcizzare

exotic /ɪɡˈzɒtɪk/ a esotico

expand /ɪkˈspænd/ vt espandere ● vi espandersi; Comm svilupparsi; ⟨metal⟩ dilatarsi; ~ on ⟨fig: explain better⟩ approfondire

expans|e /ɪkˈspæns/ n estensione f. ~ion /-ænʃn/ n espansione f; Comm sviluppo m; (of metal) dilatazione f. ~ive /-ɪv/ a espansivo

expatriate /eksˈpætrɪət/ n espatriato, -a mf

expect /ɪkˈspekt/ vt aspettare ⟨letter, baby⟩; ⟨suppose⟩ pensare; ⟨demand⟩ esigere; I ~ so penso di sì; be ~ing essere in stato interessante

expectan|cy /ɪkˈspektənsɪ/ n aspettativa f. ~t a in attesa; ~t mother donna f incinta. ~tly adv con impazienza

expectation /ekspekˈteɪʃn/ n aspettativa f, speranza f

expedient /ɪkˈspiːdɪənt/ a conveniente ● n espediente m

expedition /ekspɪˈdɪʃn/ n spedizione f. ~ary a Mil di spedizione

expel /ɪkˈspel/ vt (pt/pp expelled) espellere

expend /ɪkˈspend/ vt consumare. ~able /-əbl/ a sacrificabile

expenditure /ɪkˈspendɪtʃə(r)/ n spesa f

expense /ɪkˈspens/ n spesa f; business ~s pl spese fpl; at my ~ a mie spese; at the ~ of fig a spese di

expensive /ɪkˈspensɪv/ a caro, costoso. ~ly adv costosamente

experience /ɪkˈspɪərɪəns/ n esperienza f ● vt provare ⟨sensation⟩; avere ⟨problem⟩. ~d a esperto

experiment /ɪkˈsperɪmənt/ n esperimento ● /-ˈment/ vi sperimentare. ~al /-ˈmentl/ a sperimentale

expert /ˈekspɜːt/ a & n esperto, -a mf. ~ly adv abilmente

expertise /ekspɜːˈtiːz/ n competenza f

expire /ɪkˈspaɪə(r)/ vi scadere

expiry /ɪkˈspaɪərɪ/ n scadenza f. ~ date n data f di scadenza

explain /ɪkˈspleɪn/ vt spiegare

explana|tion /ekspləˈneɪʃn/ n spiegazione f. ~tory /ɪkˈsplænətərɪ/ a esplicativo

expletive /ɪkˈspliːtɪv/ n imprecazione f

explicit /ɪkˈsplɪsɪt/ a esplicito. ~ly adv esplicitamente

explode /ɪkˈspləʊd/ vi esplodere ● vt far esplodere

exploit[1] /ˈeksplɔɪt/ n impresa f

exploit[2] /ɪkˈsplɔɪt/ vt sfruttare. ~ation /eksplɔɪˈteɪʃn/ n sfruttamento m

explora|tion /eksplɔːˈreɪʃn/ n esplorazione f. ~tory /ɪkˈsplɔːrətərɪ/ a esplorativo

explore /ɪkˈsplɔː(r)/ vt esplorare; fig studiare ⟨implications⟩. ~r n esploratore, -trice mf

explos|ion /ɪkˈspləʊʒn/ n esplosione f. ~ive /-sɪv/ a & n esplosivo m

exponent /ɪkˈspəʊnənt/ n esponente mf

export /ˈekspɔːt/ n esportazione f ● vt /-ˈspɔːt/ esportare. ~er n esportatore, -trice mf

expose /ɪkˈspəʊz/ vt esporre; ⟨reveal⟩ svelare; smascherare ⟨traitor etc⟩. ~ure /-ʒə(r)/ n esposizione f; Med esposizione f prolungata al freddo/caldo; (of crimes) smascheramento m; 24 ~ures Phot 24 pose

expound /ɪkˈspaʊnd/ vt esporre

express /ɪkˈspres/ a espresso ● adv ⟨send⟩ per espresso ● n ⟨train⟩ espresso m ● vt esprimere; ~ oneself esprimersi. ~ion /-ʃn/ n espressione f. ~ive /-ɪv/ a espressivo. ~ly adv espressamente

expulsion /ɪkˈspʌlʃn/ n espulsione f

exquisite /ekˈskwɪzɪt/ a squisito

ex-'serviceman n ex-combattente m

extend /ɪkˈstend/ vt prolungare

(visit, road); prorogare *(visa, contract)*; ampliare *(building, knowledge)*; *(stretch out)* allungare; tendere *(hand)* ● *vi (garden, knowledge:)* estendersi

extension /ɪkˈstenʃn/ *n* prolungamento *m*; *(of visa, contract)* proroga *f*; *(of treaty)* ampliamento *m*; *(part of building)* annesso *m*; *(length of cable)* prolunga *f*; *Teleph* interno *m*; **~ 226** interno 226

extensive /ɪkˈstensɪv/ *a* ampio, vasto. **~ly** *adv* ampiamente

extent /ɪkˈstent/ *n (scope)* portata *f*; **to a certain ~** fino a un certo punto; **to such an ~ that...** fino al punto che...

extenuating /ɪkˈstenjʊeɪtɪŋ/ *a* **~ circumstances** attenuanti *fpl*

exterior /ɪkˈstɪərɪə(r)/ *a* & *n* esterno *m*

exterminat|e /ɪkˈstɜːmɪneɪt/ *vt* sterminare. **~ion** /-ˈneɪʃn/ *n* sterminio *m*

external /ɪkˈstɜːnl/ *a* esterno; **for ~ use only** *Med* per uso esterno. **~ly** *adv* esternamente

extinct /ɪkˈstɪŋkt/ *a* estinto. **~ion** /-ɪŋkʃn/ *n* estinzione *f*

extinguish /ɪkˈstɪŋgwɪʃ/ *vt* estinguere. **~er** *n* estintore *m*

extort /ɪkˈstɔːt/ *vt* estorcere. **~ion** /-ɔːʃn/ *n* estorsione *f*

extortionate /ɪkˈstɔːʃənət/ *a* esorbitante

extra /ˈekstrə/ *a* in più; *(train)* straordinario; **an ~ £10** 10 sterline extra, 10 sterline in più ● *adv* in più; *(especially)* più; **pay ~** pagare in più, pagare extra; **~ strong/busy** fortissimo/occupatissimo ● *n Theat* comparsa *f*; **~s** *pl* extra *mpl*

extract[1] /ˈekstrækt/ *n* estratto *m*

extract[2] /ɪkˈstrækt/ *vt* estrarre *(tooth, oil)*; strappare *(secret)*; ricavare *(truth)*. **~or** /-ter/ *n* *[fan]* **~** aspiratore *m*

extradit|e /ˈekstrədaɪt/ *Jur vt* estradare. **~ion** /-ˈdɪʃn/ *n* estradizione *f*

extramarital *a* extraconiugale

extraordinar|y /ɪkˈstrɔːdɪnərɪ/ *a* straordinario. **~ily** /-ɪlɪ/ *adv* straordinariamente

extravagan|ce /ɪkˈstrævəgəns/ *n (with money)* prodigalità *f*; *(of behaviour)* stravaganza *f*. **~t** *a* spendaccione; *(bizarre)* stravagante; *(claim)* esagerato

extrem|e /ɪkˈstriːm/ *a* estremo ● *n* estremo *m*; **in the ~** e al massimo. **~ely** *adv* estremamente. **~ist** *n* estremista *mf*

extremity /ɪkˈstremətɪ/ *n (end)* estremità *f inv*

extricate /ˈekstrɪkeɪt/ *vt* districare

extrovert /ˈekstrəvɜːt/ *n* estroverso, -a *mf*

exuberant /ɪgˈzjuːbərənt/ *a* esuberante

exude /ɪgˈzjuːd/ *vt also fig* trasudare

exult /ɪgˈzʌlt/ *vi* esultare

eye /aɪ/ *n* occhio *m*; *(of needle)* cruna *f*; **keep an ~ on** tener d'occhio; **see ~ to ~** vedere le stesse idee ● *vt (pt/pp* **eyed**, *pres p* **ey[e]ing)** guardare

eye: **~ball** *n* bulbo *m* oculare. **~brow** *n* sopracciglio *m* (*pl* sopracciglia *f*). **~lash** *n* ciglio *m* (*pl* ciglia *f*). **~lid** *n* palpebra *f*. **~-opener** *n* rivelazione *f*. **~-shadow** *n* ombretto *m*. **~sight** *n* vista *f*. **~sore** *n fam* pugno *m* nell'occhio. **~witness** *n* testimone *mf* oculare

Ff

fable /ˈfeɪbl/ *n* favola *f*

fabric /ˈfæbrɪk/ *n also fig* tessuto *m*

fabrication /fæbrɪˈkeɪʃn/ *n* invenzione *f*; *(manufacture)* fabbricazione *f*

fabulous /ˈfæbjʊləs/ *a fam* favoloso

façade /fə'sɑːd/ n (of building, person) facciata f

face /feɪs/ n faccia f, viso m; (grimace) smorfia f; (surface) faccia f; (of clock) quadrante m; **pull ~s** far boccacce; **in the ~ of** di fronte a; **on the ~ of it** in apparenza ●vt essere di fronta a; (confront) affrontare; **~ north** (house:) dare a nord; **~ the fact that** arrendersi al fatto che. **face up to** vt accettare (facts); affrontare (person)

face: ~flannel n ≈ guanto m di spugna. **~less** a anonimo. **~-lift** n plastica f facciale

facet /'fæsɪt/ n sfaccettatura f, fig aspetto m

facetious /fə'siːʃəs/ a spiritoso. **~ remarks** spiritosaggini mpl

'face value n (of money) valore m nominale; **take sb/sth at ~** fermarsi alle apparenze

facial /'feɪʃl/ a facciale ●n trattamento m di bellezza al viso

facile /'fæsaɪl/ a semplicistico

facilitate /fə'sɪlɪteɪt/ vt rendere possibile; (make easier) facilitare

facilit|y /fə'sɪlətɪ/ n facilità f; **~ies** pl (of area, in hotel etc) attrezzature fpl

facing /'feɪsɪŋ/ prep ~ **the sea** (house:) che dà sul mare; **the person ~ me** la persona di fronte a me

facsimile /fæk'sɪməlɪ/ n facsimile m

fact /fækt/ n fatto m; **in ~** infatti

faction /'fækʃn/ n fazione f

factor /'fæktə(r)/ n fattore m

factory /'fæktərɪ/ n fabbrica f

factual /'fæktʃʊəl/ a be ~ attenersi ai fatti. **~ly** adv (inaccurate) dal punto di vista dei fatti

faculty /'fækəltɪ/ n facoltà f inv

fad /fæd/ n capriccio m

fade /feɪd/ vi sbiadire; (sound, light:) affievolirsi; (flower:) appassire. **fade in** vt cominciare in dissolvenza (picture). **fade out** vt finire in dissolvenza (picture)

fag /fæg/ n (chore) fatica f; (fam: cigarette) sigaretta f; (Am sl: homosexual) frocio m. **~ end** n fam cicca f

fagged /fægd/ a ~ **out** fam stanco morto

Fahrenheit /'færənhaɪt/ a Fahrenheit

fail /feɪl/ n **without ~** senz'altro ●vi (attempt:) fallire; (eyesight, memory:) indebolirsi; (engine, machine:) guastarsi; (marriage:) andare a rotoli; (in exam) essere bocciato; **~ to do sth** non fare qcsa; **I tried but I ~ed** ho provato ma non ci sono riuscito ●vt non superare (exam); bocciare (candidate); (disappoint) deludere; **words ~ me** mi mancano le parole

failing /'feɪlɪŋ/ n difetto m ●prep ~ **that** altrimenti

failure /'feɪljə(r)/ n fallimento m; (mechanical) guasto m; (person) incapace mf

faint /feɪnt/ a leggero; (memory) vago; **feel ~** sentirsi mancare ●n svenimento m ●vi svenire

faint: ~-hearted a timido. **~ly** adv (slightly) leggermente. **~ness** n (physical) debolezza f

fair¹ /feə(r)/ n fiera f

fair² /feə(r)/ a (hair, person) biondo; (skin) chiaro; (weather) bello; (just) giusto; (quite good) discreto; Sch abbastanza bene; a ~ **amount** abbastanza ●adv play ~ fare un gioco pulito. **~ly** adv con giustizia; (rather) discretamente, abbastanza. **~ness** n giustizia f. ~ **play** n fair play m inv

fairy /'feərɪ/ n fata f; ~ **story**, **~-tale** n fiaba f

faith /feɪθ/ n fede f; (trust) fiducia f; **in good/bad ~** in buona/mala fede

faithful /'feɪθfl/ a fedele. **~ly** adv fedelmente; **yours ~ly** distinti saluti. **~ness** n fedeltà f

'faith-healer n guaritore, -trice mf

fake /feɪk/ a falso ● n falsificazione f; (person) impostore m ● vt falsificare; (pretend) fingere

falcon /ˈfɔːlkən/ n falcone m

fall /fɔːl/ n caduta f; (in prices) ribasso m; (Am: autumn) autunno m; have a ~ fare una caduta ● vi (pt fell, pp fallen) cadere; (night:) scendere; ~ in love innamorarsi. **fall about** vi (with laughter) morire dal ridere. **fall back on** vt ritornare su. **fall for** vt fam innamorarsi di (person); cascare in (sth, trick). **fall down** vi cadere; (building:) crollare. **fall in** vi caderci dentro; (collapse) crollare; Mil mettersi in riga; ~ in with concordare con (suggestion, plan). **fall off** vi cadere; (diminish) diminuire. **fall out** vi (quarrel) litigare; **his hair is ~ing out** perde i capelli. **fall over** vi cadere. **fall through** vi (plan:) andare a monte

fallacy /ˈfæləsɪ/ n errore m

fallible /ˈfæləbl/ a fallibile

'fall-out n pioggia f radioattiva

false /fɔːls/ a falso; ~ **bottom** doppio fondo m; ~ **start** Sport falsa partenza f. ~**hood** n menzogna f. ~**ness** n falsità f

false 'teeth npl dentiera f

falsify /ˈfɔːlsɪfaɪ/ vt (pt/pp -ied) falsificare

falter /ˈfɔːltə(r)/ vi vacillare; (making speech) esitare

fame /feɪm/ n fama f

familiar /fəˈmɪljə(r)/ a familiare; **be ~ with** (know) conoscere. ~**ity** /-lɪˈærɪtɪ/ n familiarità f. ~**ize** vt familiarizzare; ~**ize oneself with** familiarizzarsi con

family /ˈfæməlɪ/ n famiglia f

family: ~ **al'lowance** n assegni mpl familiari. ~ **'doctor** n medico m di famiglia. ~ **'life** n vita f familiare. ~ **'planning** n pianificazione f familiare. ~ **'tree** n albero m genealogico

famine /ˈfæmɪn/ n carestia f

famished /ˈfæmɪʃt/ a **be ~** fam avere una fame da lupo

famous /ˈfeɪməs/ a famoso

fan¹ /fæn/ n ventilatore m; (handheld) ventaglio m ● vt (pt/pp fanned) far vento a; ~ **oneself** sventagliarsi; fig ~ **the flames** soffiare sul fuoco. **fan out** vi spiegarsi a ventaglio

fan² n (admirer) ammiratore, -trice mf; Sport tifoso m; (of Verdi etc) appassionato, -a mf

fanatic /fəˈnætɪk/ n fanatico, -a mf. ~**al** a fanatico. ~**ism** /-sɪzm/ n fanatismo m

'fan belt n cinghia f per ventilatore

fanciful /ˈfænsɪfl/ a fantasioso

fancy /ˈfænsɪ/ n fantasia f; **I've taken a real ~ to him** mi è molto simpatico; **as the ~ takes you** come ti pare ● a (a) fantasia ● vt (pt/pp -ied) (believe) credere; (fam: want) aver voglia di; **he fancies you** fam gli piaci; ~ **that!** ma guarda un po'! ~ **'dress** n costume m (per maschera)

fanfare /ˈfænfeə(r)/ n fanfara f

fang /fæŋ/ n zanna f; (of snake) dente m

fan: ~ **heater** n termoventilatore m. ~**light** n lunetta f

fantas|ize /ˈfæntəsaɪz/ vi fantasticare. ~**tic** /-ˈtæstɪk/ a fantastico. ~**y** n fantasia f

far /fɑː(r)/ adv lontano; (much) molto; **by ~** di gran lunga; ~ **away** lontano; **as ~ as the church** fino alla chiesa; **how ~ is it from here?** quanto dista da qui?; **as ~ as I know** per quanto io sappia ● a (end, side) altro; **the F~ East** l'Estremo Oriente m

farc|e /fɑːs/ n farsa f. ~**ical** a ridicolo

fare /feə(r)/ n tariffa f; (food) vitto m. ~**-dodger** /-dɒdʒə(r)/ n passeggero, -a mf senza biglietto

farewell /feəˈwel/ int liter addio! ● n addio m

far-'fetched a improbabile

farm /fɑːm/ n fattoria f ● vi lavorare

l'agricoltore ● vt coltivare ⟨land⟩. ~er n agricoltore m

farm: **~house** n casa f colonica. **~ing** n agricoltura f. **~yard** n aia f

far: **~'reaching** a di larga portata. **~'sighted** a fig prudente; (Am: long-sighted) presbite

fart /fɑ:t/ fam n scoreggia f ● vi scoreggiare

farther /'fɑ:ðə(r)/ adv più lontano ● a **at the** ~ **end** of all'altra estremità di

fascinat|e /'fæsɪneɪt/ vt affascinare. **~ing** a affascinante. **~ion** /-'neɪʃn/ n fascino m

fascis|m /'fæʃɪzm/ n fascismo m. **~t** n fascista mf ● a fascista

fashion /'fæʃn/ n moda f; (manner) maniera f ● vt modellare. **~able** /-əbl/ a di moda; **be ~able** essere alla moda. **~ably** adv alla moda

fast[1] /fɑ:st/ a veloce; ⟨colour⟩ indelebile; **be ~** ⟨clock:⟩ andare avanti ● adv velocemente; (firmly) saldamente; **~er!** più in fretta!; **be ~ asleep** dormire profondamente

fast[2] n digiuno m ● vi digiunare

fasten /'fɑ:sn/ vt allacciare; chiudere ⟨window⟩; (stop flapping) mettere un fermo a ● vt allacciarsi. **~er**, **~ing** n chiusura f

fastidious /fə'stɪdɪəs/ a esigente

fat /fæt/ a (fatter, fattest) ⟨person, cheque⟩ grasso ● n grasso m

fatal /'feɪtl/ a mortale; ⟨error⟩ fatale. **~ism** /-təlɪzm/ n fatalismo m. **~ist** /-təlɪst/ n fatalista mf. **~ity** /fə'tælətɪ/ n morte f. **~ly** adv mortalmente

fate /feɪt/ n destino m. **~ful** a fatidico

'fat-head n fam zuccone, -a mf

father /'fɑ:ðə(r)/ n padre m; **F~ Christmas** Babbo m Natale ● vt generare ⟨child⟩

father: **~hood** n paternità f. **~-in-law** n (pl **~s-in-law**) suocero m. **~ly** a paterno

fathom /'fæð(ə)m/ n Naut braccio m ● vt ~ [**out**] comprendere

fatigue /fə'ti:g/ n fatica f.

fatten /'fætn/ vt ingrassare ⟨animal⟩. **~ing a cream is ~ing** la panna fa ingrassare

fatty /'fætɪ/ a grasso ● n fam ciccione, -a mf

fatuous /'fætjʊəs/ a fatuo

faucet /'fɔːsɪt/ n Am rubinetto m

fault /fɔːlt/ n difetto m; Geol faglia f; Tennis fallo m; **be at ~** avere torto; **find ~ with** trovare da ridire su; **it's your ~** è colpa tua ● vt criticare. **~less** a impeccabile

faulty /'fɔːltɪ/ a difettoso

fauna /'fɔːnə/ n fauna f

favour /'feɪvə(r)/ n favore m; **be in ~ of sth** essere a favore di qcsa; **do sb a ~** fare un piacere a qcno ● vt (prefer) preferire. **~able** /-əbl/ a favorevole

favourite /'feɪvərɪt/ a preferito ● n preferito, -a mf; Sport favorito, -a mf. **~ism** n favoritismo m

fawn /fɔːn/ a fulvo ● n (animal) cerbiatto m

fax /fæks/ n (document, machine) fax m inv; **by ~** per fax ● vt faxare. **~ machine** n fax m inv. **~modem** n modem-fax m inv, fax-modem m inv

fear /fɪə(r)/ n paura f; **no ~!** fam vai tranquillo! ● vt temere ● vi ~ **for sth** temere per qcsa

fearful /'fɪəfl/ a pauroso; (awful) terribile. **~less** a impavido. **~some** /-səm/ a spaventoso

feas|ibility /fi:zɪ'bɪlɪtɪ/ n praticabilità f. **~ible** /-ɪble/ a fattibile; (possible) probabile

feast /fi:st/ n festa f; (banquet) banchetto m ● vi banchettare. ~ **on** godersi

feat /fiːt/ n impresa f

feather /'feðə(r)/ n piuma f

feature /'fiːtʃə(r)/ n (quality) caratteristica f; Journ articolo m; **~s** (pl: of face) lineamenti mpl ● vt (film:) avere come protagonista ● vi (on a list etc) comparire. **~ film** n lungometraggio m

February /'februərɪ/ n febbraio m

fed /fed/ *see* **feed** ● *a* **be ~ up** *fam* essere stufo (**with** di)

federal /ˈfed(ə)rəl/ *a* federale

federation /fedəˈreɪʃn/ *n* federazione *f*

fee /fiː/ *n* tariffa *f*; (*lawyer's, doctor's*) onorario *m*; (*for membership, school*) quota *f*

feeble /ˈfiːbl/ *a* debole; (*excuse*) fiacco

feed /fiːd/ *n* mangiare *m*; (*for baby*) pappa *f* ● *v* (*pt/pp* **fed**) ● *vt* dar da mangiare a (*animal*); (*support*) nutrire; **~ sth into** sth inserire qcsa in qcsa ● *vi* mangiare

'feedback *n* controreazione *f*; (*of information*) reazione *f*, feedback *m*

feel /fiːl/ *v* (*pt/pp* **felt**) ● *vt* sentire; (*experience*) provare; (*think*) pensare; (*touch: searching*) tastare; (*touch: for texture*) toccare ● *vi* **~ soft/hard** essere duro/morbido al tatto; **~ hot/hungry** aver caldo/fame; **~ ill** sentirsi male; **I don't ~ like it** non ne ho voglia; **how do you ~ about it?** (*opinion*) che te ne pare?; **it doesn't ~ right** non mi sembra giusto. **~er** *n* (*of animal*) antenna *f*; **put out ~ers** *fig* tastare il terreno. **~ing** *n* sentimento *m*; (*awareness*) sensazione *f*

feet /fiːt/ *see* **foot**

feign /feɪn/ *vt* simulare

feline /ˈfiːlaɪn/ *a* felino

fell[1] /fel/ *vt* (*knock down*) abbattere

fell[2] *see* **fall**

fellow /ˈfeləʊ/ *n* (*of society*) socio *m*; (*fam: man*) tipo *m*

fellow: **~-'countryman** *n* compatriota *m*. **~men** *npl* prossimi *mpl*. **~ship** *n* cameratismo *m*; (*group*) associazione *f*; *Univ* incarico *m* di ricercatore, -trice

felony /ˈfelənɪ/ *n* delitto *m*

felt[1] /felt/ *see* **feel**

felt[2] /felt/ *n* feltro *m*. **~ [-tipped] 'pen** /[-tɪpt]/ *n* pennarello *m*

female /ˈfiːmeɪl/ *a* femminile; **the ~ antelope** l'antilope femmina ● *n* femmina *f*

feminine /ˈfemɪnɪn/ *a* femminile ● *n* *Gram* femminile *m*. **~inity** /-ˈnɪnətɪ/ *n* femminilità *f*. **~ist** *a* & *n* femminista *mf*

fenc|e /fens/ *n* recinto *m*; (*fam: person*) ricettatore *m* ● *vi* *Sport* tirar di scherma. **fence in** *vt* chiudere in un recinto. **~er** *n* schermidore *m*. **~ing** *n* steccato *m*; *Sport* scherma *f*

fend /fend/ *vi* **~ for oneself** badare a se stesso. **fend off** *vt* parare; difendersi da (*criticisms*)

fender /ˈfendə(r)/ *n* parafuoco *m* *inv*; (*Am: on car*) parafango *m*

fennel /ˈfenl/ *n* finocchio *m*

ferment[1] /ˈfɜːment/ *n* fermento *m*

ferment[2] /fəˈment/ *vi* fermentare ● *vt* far fermentare. **~ation** /fɜːmenˈteɪʃn/ *n* fermentazione *f*

fern /fɜːn/ *n* felce *f*

feroc|ious /fəˈrəʊʃəs/ *a* feroce. **~ity** /-ˈrɒsətɪ/ *n* ferocia *f*

ferret /ˈferɪt/ *n* furetto *m* ● **ferret out** *vt* scovare

ferry /ˈferɪ/ *n* traghetto *m* ● *vt* traghettare

fertil|e /ˈfɜːtaɪl/ *a* fertile. **~ity** /fɜːˈtɪlətɪ/ *n* fertilità *f*

fertilize /ˈfɜːtɪlaɪz/ *vt* fertilizzare (*land, ovum*). **~r** *n* fertilizzante *m*

fervent /ˈfɜːvənt/ *a* fervente

fervour /ˈfɜːvə(r)/ *n* fervore *m*

fester /ˈfestə(r)/ *vi* suppurare

festival /ˈfestɪvl/ *n* *Mus*, *Theat* festival *m*; *Relig* festa *f*

festive /ˈfestɪv/ *a* festivo; **~e season** periodo *m* delle feste natalizie. **~ities** /feˈstɪvətɪz/ *npl* festeggiamenti *mpl*

festoon /feˈstuːn/ *vt* **~ with** ornare di

fetch /fetʃ/ *vt* andare/venire a prendere; (*be sold for*) raggiungere [il prezzo di]

fetching /ˈfetʃɪŋ/ *a* attraente

fête /feɪt/ *n* festa *f* ● *vt* festeggiare

fetish /ˈfetɪʃ/ *n* feticcio *m*

fetter /ˈfetə(r)/ *vt* incatenare

fettle /ˈfetl/ n **in fine** ~ in buona forma

feud /fjuːd/ n faida f

feudal /ˈfjuːdl/ a feudale

fever /ˈfiːvə(r)/ n febbre f. ~**ish** a febbricitante; *fig* febbrile

few /fjuː/ a pochi; **every** ~ **days** ogni due o tre giorni; **a** ~ **people** alcuni; ~**er reservations** meno prenotazioni; **the** ~**est number** il numero più basso ● *pron* pochi; ~ **of us** pochi di noi; **a** ~ alcuni; **quite a** ~ parecchi; ~**er than last year** meno dell'anno scorso

fiancé /fɪˈɒnseɪ/ n fidanzato m. ~**e** n fidanzata f

fiasco /fɪˈæskəʊ/ n fiasco m

fib /fɪb/ n storia f; **tell a** ~ raccontare una storia

fibre /ˈfaɪbə(r)/ n fibra f. ~**glass** n fibra f di vetro

fickle /ˈfɪkl/ a incostante

fiction /ˈfɪkʃn/ n [**works of**] ~ narrativa f; (*fabrication*) finzione f. ~**al** a immaginario

fictitious /fɪkˈtɪʃəs/ a fittizio

fiddle /ˈfɪdl/ n *fam* violino m; (*cheating*) imbroglio m ● vi gingillarsi (**with** con) ● vt *fam* truccare (*accounts*)

fiddly /ˈfɪdlɪ/ a intricato

fidelity /fɪˈdelətɪ/ n fedeltà f

fidget /ˈfɪdʒɪt/ vi agitarsi. ~**y** a agitato

field /fiːld/ n campo m

field: ~ **events** npl atletica f leggera. ~**glasses** npl binocolo msg. **F~** 'Marshal n feldmaresciallo m. ~**work** n ricerche fpl sul terreno

fiend /fiːnd/ n demonio m

fierce /fɪəs/ a feroce. ~**ness** n ferocia f

fiery /ˈfaɪərɪ/ a (**-ier, -iest**) focoso

fifteen /fɪfˈtiːn/ a & n quindici m. ~**th** a quindicesimo

fifth /fɪfθ/ a quinto

fiftieth /ˈfɪftɪəθ/ a cinquantesimo

fifty /ˈfɪftɪ/ a cinquanta

fig /fɪg/ n fico m

fight /faɪt/ n lotta f; (*brawl*) zuffa f;

(*argument*) litigio m; (*boxing*) incontro m ● v (*pt/pp* **fought**) ● vt *also fig* combattere ● vi combattere; (*brawl*) azzuffarsi; (*argue*) litigare. ~**er** n combattente mf; *Aeron* caccia m inv. ~**ing** n combattimento m

figment /ˈfɪgmənt/ n **it's a** ~ **of your imagination** questo è tutta una tua invenzione

figurative /ˈfɪgjərətɪv/ a (*sense*) figurato; (*art*) figurativo

figure /ˈfɪgə(r)/ n (*digit*) cifra f; (*carving, sculpture, illustration, form*) figura f; (*body shape*) linea f; ~ **of speech** modo m di dire ● vi (*appear*) figurare ● vt (*Am: think*) pensare. **figure out** vt dedurre; capire (*person*)

figure: ~**head** n figura f simbolica. ~ **skating** n pattinaggio m artistico

filament /ˈfɪləmənt/ n filamento m

file[1] /faɪl/ n scheda f; (*set of documents*) incartamento m; (*folder*) cartellina f; *Comput* file m inv ● vt archiviare (*documents*)

file[2] n (*line*) fila f; **in single** ~ in fila

file[3] n *Techn* lima f ● vt limare

filing cabinet /ˈfaɪlɪŋkæbɪnət/ n schedario m, classificatore m

filings /ˈfaɪlɪŋz/ npl limatura fsg

fill /fɪl/ n **eat one's** ~ mangiare a sazietà ● vt riempire; otturare (*tooth*) ● vi riempirsi. **fill in** vt compilare (*form*). **fill out** vt compilare (*form*). **fill up** vi (*room, tank:*) riempirsi; *Auto* far il pieno ● vt riempire

fillet /ˈfɪlɪt/ n filetto m ● vt (*pt/pp* **filleted**) disossare

filling /ˈfɪlɪŋ/ n *Culin* ripieno m; (*of tooth*) piombatura f. ~ **station** n stazione f di rifornimento

filly /ˈfɪlɪ/ n puledra f

film /fɪlm/ n *Cinema* film m inv; *Phot* pellicola f; [**cling**] ~ pellicola f per alimenti ● vt/i filmare. ~ **star** n star f inv, divo, -a mf

filter /ˈfɪltə(r)/ n filtro m ● vt filtrare. **filter through** *vi* ⟨*news.*⟩ trapelare. ~ **tip** *n* filtro *m*; ⟨*cigarette*⟩ sigaretta *f* col filtro

filth /fɪlθ/ *n* sudiciume *m*. ~**y** *a* (*-ier*, *-iest*) sudicio; ⟨*word*⟩ sconcio

fin /fɪn/ *n* pinna *f*

final /ˈfaɪnl/ *a* finale; ⟨*conclusive*⟩ decisivo ● *n* Sport finale *f*; ~**s** *pl* Univ esami *mpl* finali

finale /fɪˈnɑːlɪ/ *n* finale *m*

finalist /ˈfaɪnəlɪst/ *n* finalista *mf*. ~**ity** /-ˈnælɪtɪ/ *n* finalità *f*

finalize /ˈfaɪnəlaɪz/ *vt* mettere a punto ⟨*text*⟩; definire ⟨*agreement*⟩. ~**ly** *adv* ⟨*at last*⟩ finalmente; ⟨*at the end*⟩ alla fine; ⟨*to conclude*⟩ per finire

finance /ˈfaɪnæns/ *n* finanza *f* ● *vt* finanziare

financial /faɪˈnænʃl/ *a* finanziario

finch /fɪntʃ/ *n* fringuello *m*

find /faɪnd/ *n* scoperta *f* ● *vt* (*pt/pp* **found**) trovare; ⟨*establish*⟩ scoprire; ~ **sb guilty** Jur dichiarare qcno colpevole. **find out** *vt* scoprire ● *vi* ⟨*enquire*⟩ informarsi

findings /ˈfaɪndɪŋz/ *npl* conclusioni *fpl*

fine¹ /faɪn/ *n* ⟨*penalty*⟩ multa *f* ● *vt* multare

fine² *a* bello; ⟨*slender*⟩ fine; **he's ~** ⟨*in health*⟩ sta bene. ~ **arts** *fpl* belle arti *fpl*. ● *adv* bene; **that's cutting it ~** non ci lascia molto tempo ● *int* [va] bene. ~**ly** *adv* ⟨*cut*⟩ finemente

finery /ˈfaɪnərɪ/ *n* splendore *m*

finesse /fɪˈnes/ *n* finezza *f*

finger /ˈfɪŋɡə(r)/ *n* dito *m* (*pl* dita *f*) ● *vt* tastare

finger: ~**mark** *n* ditata *f*. ~**nail** *n* unghia *f*. ~**print** *n* impronta *f* digitale. ~**tip** *n* punta *f* del dito; **have sth at one's ~tips** sapere qcsa a menadito; ⟨*close at hand*⟩ avere qcsa a portata di mano

finicky /ˈfɪnɪkɪ/ *a* ⟨*person*⟩ pignolo; ⟨*task*⟩ intricato

finish /ˈfɪnɪʃ/ *n* fine *f*; ⟨*finishing line*⟩ traguardo *m*; ⟨*of product*⟩ finitura *f*; **have a good ~** ⟨*runner:*⟩

avere un buon finale ● *vt* finire; ~ **reading** finire di leggere ● *vi* finire

finite /ˈfaɪnaɪt/ *a* limitato

Finland /ˈfɪnlənd/ *n* Finlandia *f*

Finn /fɪn/ *n* finlandese *mf*. ~**ish** *a* finlandese ● *n* ⟨*language*⟩ finnico *m*

fiord /fjɔːd/ *n* fiordo *m*

fir /fɜː(r)/ *n* abete *m*

fire /ˈfaɪə(r)/ *n* fuoco *m*; ⟨*forest, house*⟩ incendio *m*; **be on ~** bruciare; **catch ~** prendere fuoco; **set ~ to** dar fuoco a; **under ~** sotto il fuoco ● *vt* cuocere ⟨*pottery*⟩; sparare ⟨*shot*⟩; tirare ⟨*gun*⟩; ⟨*fam: dismiss*⟩ buttar fuori ● *vi* sparare (**at** a)

fire: ~ **alarm** *n* allarme *m* antincendio. ~**arm** *n* arma *f* da fuoco. ~ **brigade** *n* vigili *mpl* del fuoco. ~**engine** *n* autopompa *f*. ~**escape** *n* uscita *f* di sicurezza. ~ **extinguisher** *n* estintore *m*. ~**man** *n* pompiere *m*, vigile *m* del fuoco. ~**place** *n* caminetto *m*. ~**side** *n* **by** *or* **at the ~side** accanto al fuoco. ~ **station** *n* caserma *f* dei pompieri. ~**wood** *n* legna *f* ⟨*da ardere*⟩. ~**work** *n* fuoco *m* d'artificio; ~**works** *pl* ⟨*display*⟩ fuochi *mpl* d'artificio

'firing squad *n* plotone *m* d'esecuzione

firm¹ /fɜːm/ *n* ditta *f*, azienda *f*

firm² *a* fermo; ⟨*soil*⟩ compatto; ⟨*stable, properly fixed*⟩ solido; ⟨*resolute*⟩ risoluto. ~**ly** *adv* ⟨*hold*⟩ stretto; ⟨*say*⟩ con fermezza

first /fɜːst/ *a* & *n* primo, -a *mf*; **at ~** all'inizio; **who's ~?** chi è il primo?; **from the ~** [fin] dall'inizio ● *adv* ⟨*arrive, leave*⟩ per primo; ⟨*beforehand*⟩ prima; ⟨*in listing*⟩ prima di tutto, innanzitutto

first: ~ **'aid** *n* pronto soccorso *m*. ~**'aid kit** *n* cassetta *f* di pronto soccorso. ~**class** *a* di prim'ordine; Rail in prima classe ● *adv* ⟨*travel*⟩ in prima classe. ~ **'floor** *n* primo piano *m*; ⟨*Am: ground floor*⟩

pianterreno m. **~ly** adv in primo luogo. **~ name** n nome m di battesimo. **~rate** a ottimo

fish /fɪʃ/ n pesce m ● vt/i pescare. **fish out** vt tirar fuori

fish: **~bone** n lisca f. **~erman** n pescatore m. **~farm** n vivaio m. **~ finger** n bastoncino m di pesce

fishing /ˈfɪʃɪŋ/ n pesca f. **~ boat** n peschereccio m. **~-rod** n canna f da pesca

fish: **~monger** /-mʌŋgə(r)/ n pescivendolo m. **~-slice** n paletta f per fritti. **~y** a (fam: suspicious) sospetto

fission /ˈfɪʃn/ n Phys fissione f

fist /fɪst/ n pugno m

fit[1] /fɪt/ n (attack) attacco m; (of rage) accesso m; (of generosity) slancio m

fit[2] a (fitter, fittest) (suitable) adatto; (healthy) in buona salute; Sport in forma; **be ~ to do sth** essere in grado di fare qcsa; **~ to eat** buono da mangiare; **keep ~** tenersi in forma

fit[3] n (of clothes) taglio m; **it's a good ~** (coat etc.) ti/le sta bene ● v (pt/pp **fitted**) ● vi (be the right size) andare bene; **it won't ~** (no room) non ci sta ● vt (fix) applicare (**to** a); (install) installare; **it doesn't ~ me** (coat etc.) non mi va bene; **~ with** fornire di. **fit in** vi (person:) adattarsi; **it won't ~ in** (no room)₊ non ci sta ● vt (in schedule, vehicle) trovare un buco per

fit·ful /ˈfɪtfl/ a irregolare. **~fully** adv (sleep) a sprazzi. **~ments** npl (in house) impianti mpl fissi. **~ness** n (suitability) capacità f; [**physical**] **~ness** forma f, fitness m

fitted: **~ 'carpet** n moquette f inv. **~ 'cupboard** n armadio m a muro; (smaller) armadietto m a muro. **~'kitchen** n cucina f componibile. **~ 'sheet** n lenzuolo m con angoli

fitter /ˈfɪtə(r)/ n installatore, -trice mf

fitting /ˈfɪtɪŋ/ a appropriato ● n (of clothes) prova f; Techn montaggio m; **~s** pl accessori mpl. **~ room** n camerino m

five /faɪv/ a & n cinque m. **~r** n fam biglietto m da cinque sterline

fix /fɪks/ n (sl: drug) pera f; **be in a ~** fam essere nei guai ● vt fissare; (repair) aggiustare; preparare (meal). **fix up** vt fissare (meeting)

fixation /fɪkˈseɪʃn/ n fissazione f

fixed /fɪkst/ a fisso

fixture /ˈfɪkstʃə(r)/ n Sport incontro m; **~s and fittings** impianti mpl fissi

fizz /fɪz/ vi frizzare

fizzle /ˈfɪzl/ vi **~ out** finire in nulla

fizzy /ˈfɪzɪ/ a gassoso. **~ drink** n bibita f gassata

flabbergasted /ˈflæbəgɑːstɪd/ a **be ~** rimanere a bocca aperta

flabby /ˈflæbɪ/ a floscio

flag[1] /flæg/ n bandiera f ● **flag down** vt (pt/pp **flagged**) far segno di fermarsi a (taxi)

flag[2] vi (pt/pp **flagged**) cedere

flag-pole n asta f della bandiera

flagrant /ˈfleɪgrənt/ a flagrante

flagship n Naut nave f ammiraglia; fig fiore m all'occhiello

flagstone n pietra f da lastricare

flair /fleə(r)/ n (skill) talento m; (style) stile m

flake /fleɪk/ n fiocco m ● vi **~ [off]** cadere in fiocchi

flaky /ˈfleɪkɪ/ a a scaglie. **~ pastry** n pasta f sfoglia

flamboyant /flæmˈbɔɪənt/ a (personality) brillante; (tie) sgargiante

flame /fleɪm/ n fiamma f

flammable /ˈflæməbl/ a infiammabile

flan /flæn/ n (fruit) ~ crostata f

flank /flæŋk/ n fianco m ● vt fiancheggiare

flannel /ˈflæn(ə)l/ n flanella f; (for washing) ≈ guanto m di spugna;

(trousers) pantaloni *mpl* di flanella

flannelette /ˌflænəˈlet/ *n* flanella *f* di cotone

flap /flæp/ *n (of pocket, envelope)* risvolto *m*; *(of table)* ribalta *f*; **in a ~ fam** in grande agitazione ● *v (pt/pp* **flapped)** ● *vi* sbattere; *fam* agitarsi ● *vt* ~ **its wings** battere le ali

flare /fleə(r)/ *n* fiammata *f*; *(device)* razzo *m* ● **flare up** *vi (rash:)* venire fuori; *(fire:)* fare una fiammata; *(person, situation:)* esplodere. ~**d** *a (garment)* svasato

flash /flæʃ/ *n* lampo *m*; **in a ~ fam** in un attimo ● *vi* lampeggiare; ~ **past** passare come un bolide ● *vt* lanciare *(smile)*; ~ **one's headlights** lampeggiare; ~ **a torch** at puntare una torcia su

flash: ~**back** *n* scena *f* retrospettiva. ~**bulb** *n Phot* flash *m inv*. ~**er** *n Auto* lampeggiatore *m*. ~**light** *n Phot* flash *m inv*; *(Am: torch)* torcia *f* [elettrica]. ~**y** *a* vistoso

flask /flɑːsk/ *n* fiasco *m*; *(vacuum ~)* termos *m inv*

flat /flæt/ *a* (**flatter, flattest**) piatto; *(refusal)* reciso; *(beer)* sgasato; *(battery)* scarico; *(tyre)* a terra; **A ~** *Mus* la bemolle ● *n* appartamento *m*; *Mus* bemolle *m*; *(puncture)* gomma *f* a terra

flat: ~ **'feet** *npl* piedi *mpl* piatti. ~**fish** *n* pesce *m* piatto. ~**ly** *adv (refuse)* categoricamente. ~ **rate** *n* tariffa *f* unica

flatten /ˈflætn/ *vt* appiattire

flatter /ˈflætə(r)/ *vt* adulare. ~**ing** *a (comments)* lusinghiero; *(colour, dress)* che fa sembrare più bello. ~**y** *n* adulazione *f*

flat 'tyre *n* gomma *f* a terra

flaunt /flɔːnt/ *vt* ostentare

flautist /ˈflɔːtɪst/ *n* flautista *mf*

flavour /ˈfleɪvə(r)/ *n* sapore *m* ● *vt* condire; **chocolate ~ed** *a* al sapore di cioccolato. ~**ing** *n* condimento *m*

---

flaw /flɔː/ *n* difetto *m*. ~**less** *a* perfetto

flax /flæks/ *n* lino *m*. ~**en** *a (hair)* biondo platino

flea /fliː/ *n* pulce *m*. ~ **market** *n* mercato *m* delle pulci

fleck /flek/ *n* macchiolina *f*

fled /fled/ *see* **flee**

flee /fliː/ *vt/i (pt/pp* **fled)** fuggire *(from da)*

fleece /fliːs/ *n* pelliccia *f* ● *vt fam* spennare. ~**y** *a (lining)* felpato

fleet /fliːt/ *n* flotta *f*; *(of cars)* parco *m*

fleeting /ˈfliːtɪŋ/ *a* **catch a ~ glance of sth** intravedere qcsa; **for a ~ moment** per un attimo

flesh /fleʃ/ *n* carne *f*; **in the ~** in persona. ~**y** *a* carnoso

flew /fluː/ *see* **fly²**

flex¹ /fleks/ *vt* flettere *(muscle)*

flex² *n Electr* filo *m*

flexibility /fleksrˈbɪlətɪ/ *n* flessibilità *f*. ~**le** *a* flessibile

flexitime /ˈfleksɪ-/ *n* orario *m* flessibile

flick /flɪk/ *vt* dare un buffetto a; ~ **sth off sth** togliere qcsa da qcsa con un colpetto. **flick through** *vt* sfogliare

flicker /ˈflɪkə(r)/ *vi* tremolare

flier /ˈflaɪə(r)/ *n* = **flyer**

flight¹ /flaɪt/ *n (fleeing)* fuga *f*; **take ~** darsi alla fuga

flight² *n (flying)* volo *m*; ~ **of stairs** rampa *f*

flight: ~ **path** *n* traiettoria *f* di volo. ~ **recorder** *n* registratore *m* di volo

flighty /ˈflaɪtɪ/ *a* (**-ier, -iest**) frivolo

flimsy /ˈflɪmzɪ/ *a* (**-ier, -iest**) *(material)* leggero; *(shelves)* poco robusto; *(excuse)* debole

flinch /flɪntʃ/ *vi (wince)* sussultare; *(draw back)* ritirarsi; ~ **from a task** *fig* sottrarsi a un compito

fling /flɪŋ/ *n* **have a ~** *(fam: affair)* aver un'avventura ● *vt (pt/pp* **flung)** gettare

flint /flɪnt/ *n* pietra *f* focaia; *(for lighter)* pietrina *f*

flip /flɪp/ v (pt/pp **flipped**) ● vt dare un colpetto a; buttare in aria ‹coin› ● vi fam uscire dai gangheri; ‹go mad› impazzire. **flip through** vt sfogliare

flippant /ˈflɪpənt/ a irriverente

flipper /ˈflɪpə(r)/ n pinna f

flirt /flɜːt/ n civetta f ● vi flirtare

flirtat|ion /flɜːˈteɪʃn/ n flirt m inv. **~ious** /-ʃəs/ a civettuolo

flit /flɪt/ vi (pt/pp **flitted**) volteggiare

float /fləʊt/ n galleggiante m; (in procession) carro m; (money) riserva f di cassa ● vi galleggiare; Fin fluttuare

flock /flɒk/ n gregge m; (of birds) stormo m ● vi affollarsi

flog /flɒg/ vt (pt/pp **flogged**) bastonare; (fam: sell) vendere

flood /flʌd/ n alluvione f; (of river) straripamento m; (fig: of replies, letters, tears) diluvio m; **be in ~** ‹river:› essere straripato ● vt allagare ● vi ‹river:› straripare

floodlight /ˈflʌdlaɪt/ n riflettore m ● vt (pt/pp **floodlit**) illuminare con riflettori

floor /flɔː(r)/ n pavimento m; (storey) piano m; (for dancing) pista f ● vt (baffle) confondere; (knock down) stendere ‹person›

floor: ~ board n asse f del pavimento. **~-polish** n cera f per il pavimento. **~ show** n spettacolo m di varietà

flop /flɒp/ n fam (failure) tonfo m; Theat fiasco m ● vi (pt/pp **flopped**) (fam: fail) far fiasco. **flop down** vi accasciarsi

floppy /ˈflɒpɪ/ a floscio. **~ disk** n floppy disk m inv. **~ [disk] drive** n lettore di floppy m

flora /ˈflɔːrə/ n flora f

floral /ˈflɔːrəl/ a floreale

Florence /ˈflɒrəns/ n Firenze f

florid /ˈflɒrɪd/ a ‹complexion› florido; ‹style› troppo ricercato

florist /ˈflɒrɪst/ n fioraio, -a mf

flounce /flaʊns/ n balza f ● vi ~ **out** uscire con aria melodrammatica

flounder[1] /ˈflaʊndə(r)/ vi dibattersi; (speaker:) impappinarsi

flounder[2] n (fish) passera f di mare

flour /ˈflaʊə(r)/ n farina f

flourish /ˈflʌrɪʃ/ n gesto m drammatico; (scroll) ghirigoro m ● vi prosperare ● vt brandire

floury /ˈflaʊərɪ/ a farinoso

flout /flaʊt/ vt fregarsene di ‹rules›

flow /fləʊ/ n flusso m ● vi scorrere; (hang loosely) ricadere

flower /ˈflaʊə(r)/ n fiore m ● vi fiorire

flower: ~-bed n aiuola f. **~ed** a a fiori. **~pot** n vaso m [per i fiori]. **~y** a fiorito

flown /fləʊn/ see **fly**[2]

flu /fluː/ n influenza f

fluctuat|e /ˈflʌktjʊeɪt/ vi fluttuare. **~ion** /-ˈeɪʃn/ n fluttuazione f

fluent /ˈfluːənt/ a spedito; **speak ~ Italian** parlare correntemente l'italiano. **~ly** adv speditamente

fluff /flʌf/ n peluria f ● vt fam ‹chance›. **~y** a (-ier, -iest) vaporoso; (toy) di peluche

fluid /ˈfluːɪd/ n fluido m ● a fluido m

fluke /fluːk/ n colpo m di fortuna

flung /flʌŋ/ see **fling**

flunk /flʌŋk/ vt Am fam essere bocciato in

fluorescent /flʊəˈresnt/ a fluorescente

fluoride /ˈflʊəraɪd/ n fluoruro m

flurry /ˈflʌrɪ/ n (snow) raffica f; fig agitazione f

flush /flʌʃ/ n (blush) [vampata f di] rossore m ● vi arrossire ● vt lavare con un getto d'acqua; **~ the toilet** tirare l'acqua ● a a livello (with di); (fam: affluent) a soldi

flustered /ˈflʌstəd/ a in agitazione; **get ~** mettersi in agitazione

flute /fluːt/ n flauto m

flutter /ˈflʌtə(r)/ n battito m ● vi svolazzare

flux /flʌks/ n **in a state of ~** in uno stato di flusso

fly[1] /flaɪ/ n (pl **flies**) mosca f

fly[2] v (pt **flew**, pp **flown**) ● vi volare; ⟨go by plane⟩ andare in aereo; ⟨flag:⟩ sventolare; ⟨rush⟩ precipitarsi; ~ **open** spalancarsi ● vt pilotare ⟨plane⟩; trasportare [in aereo] ⟨troops, supplies⟩; volare con ⟨Alitalia etc⟩

fly³ n & **flies** pl ⟨on trousers⟩ patta f

flyer /'flaɪə(r)/ n aviatore m; ⟨leaflet⟩ volantino m

flying /'flaɪɪŋ/: ~ **'buttress** arco m rampante. ~ **'colours: with** ~ **colours** a pieni voti. ~ **'saucer** n disco m volante. ~ **'start** n **get off to a** ~ **start** fare un'ottima partenza. ~ **'visit** n visita f lampo

fly: ~ **leaf** n risguardo m. ~ **over** n cavalcavia m inv

foal /fəʊl/ n puledro m

foam /fəʊm/ n schiuma f; ⟨synthetic⟩ gommapiuma® f ● vi spumare; ~ **at the mouth** far la bava alla bocca. ~ **'rubber** n gommapiuma® f

fob /fɒb/ vt (pt/pp **fobbed**) ~ **sth off** affibbiare qcsa (**on sb** a qcno); ~ **sb off** liquidare qcno

focal /'fəʊkl/ a focale

focus /'fəʊkəs/ n fuoco m; **in** ~ a fuoco; **out of** ~ sfocato ● v (pt/pp **focused** or **focussed**) ● vt fig concentrare (**on** su) ● vi Phot mettere a fuoco; fig concentrarsi (**on** su)

fodder /'fɒdə(r)/ n foraggio m

foe /fəʊ/ n nemico, -a mf

foetus /'fiːtəs/ n (pl **-tuses**) feto m

fog /fɒg/ n nebbia f

fogey /'fəʊgɪ/ n **old** ~ persona f antiquata

foggy /'fɒgɪ/ a (**foggier, foggiest**) nebbioso; **it's** ~ c'è nebbia

'fog-horn n sirena f da nebbia

foil¹ /fɔɪl/ n lamina f di metallo

foil² vt ⟨thwart⟩ frustrare

foil³ n ⟨sword⟩ fioretto m

foist /fɔɪst/ vt appioppare (**on sb** a qcno)

fold¹ /fəʊld/ n ⟨for sheep⟩ ovile m

fold² n piega f ● vt piegare; ~ **one's arms** incrociare le braccia ● vi piegarsi; ⟨fail⟩ crollare. **fold up** vt

ripiegare ⟨chair⟩ ● vi essere pieghevole; ⟨fam: business:⟩ collassare

fold|er /'fəʊldə(r)/ n cartella f. ~**ing** a pieghevole

foliage /'fəʊlɪdʒ/ n fogliame m

folk /fəʊk/ n gente f; **my** ~s ⟨family⟩ i miei; **hello there** ~s ciao a tutti

folk: ~-**dance** n danza f popolare. ~**lore** n folclore m. ~-**song** n canto m popolare

follow /'fɒləʊ/ vt/i seguire; **it doesn't** ~ non è necessariamente così; ~ **suit** fig fare lo stesso; **as** ~s come segue. **follow up** vt fare seguito a ⟨letter⟩

follow|er /'fɒləʊə(r)/ n seguace mf. ~**ing** a seguente ● n seguito m; ⟨supporters⟩ seguaci mpl ● prep in seguito a

folly /'fɒlɪ/ n follia f

fond /fɒnd/ a affezionato; ⟨hope⟩ vivo; **be** ~ **of** essere appassionato di ⟨music⟩; **I'm** ~ **of...** ⟨food, person⟩ mi piace moltissimo...

fondle /'fɒndl/ vt coccolare

fondness /'fɒndnɪs/ n affetto m; ⟨for things⟩ amore m

font /fɒnt/ n fonte f battesimale; Typ carattere m di stampa

food /fuːd/ n cibo m; ⟨for animals, groceries⟩ mangiare m; **let's buy some** ~ compriamo qualcosa da mangiare

food: ~ **mixer** n frullatore m. ~ **poisoning** n intossicazione f alimentare.~ **processor** n tritatutto m inv elettrico

fool¹ /fuːl/ n sciocco, -a mf; **she's no** ~ non è una stupida; **make a** ~ **of oneself** rendersi ridicolo ● vt prendere in giro ● vi ~ **around** giocare; ⟨husband, wife:⟩ avere l'amante

fool² n Culin crema f

'fool|hardy a temerario. ~**ish** a stolto. ~**ishly** adv scioccamente. ~**ishness** n sciocchezza f. ~**proof** a facilissimo

foot /fʊt/ n (pl **feet**) piede m; ⟨of

animal) zampa *f*; (*measure*) piede *m* (= 30,48 cm); **on ~** a piedi; **on one's feet** in piedi; **put one's ~** in it *fam* fare una gaffe

foot: **~-and-'mouth disease** *n* afta *f* epizootica; **~ball** *n* calcio *m*; (*ball*) pallone *m*. **~baller** *n* giocatore *m* di calcio. **~ball pools** *npl* totocalcio *m*. **~brake** *n* freno *m* a pedale. **~bridge** *n* passerella *f*. **~hills** *npl* colline *fpl* pedemontane. **~hold** *n* punto *m* d'appoggio. **~ing** *n* lose one's **~ing** perdere l'appiglio; **on an equal ~ing** in condizioni di parità. **~man** *n* valletto *m*. **~note** *n* nota *f* a piè di pagina. **~path** *n* sentiero *m*. **~print** *n* orma *f*. **~step** *n* passo *m*; follow in sb's **~steps** *fig* seguire l'esempio di qcno. **~stool** *n* sgabello *m*. **~wear** *n* calzature *fpl*

for /fə(r)/, *accentato* /fɔː(r)/ *prep* per; **~ this reason** per questa ragione; **I have lived here ~ ten years** vivo qui da dieci anni; **~ supper** per cena; **~ all that** nonostante questo; **what ~?** a che pro?; **send ~ a doctor** chiamare un dottore; **fight ~ a cause** lottare per una causa; **go ~ a walk** andare a fare una passeggiata; **there's no need ~ you to go** non c'è bisogno che tu vada; **it's not ~ me to say** non sta a me dirlo; **now you're ~ it** ora sei nei pasticci ● *conj* poiché, perché

forage /'fɒrɪdʒ/ *n* foraggio *m* ● *vi* **~ for** cercare

forbade /fə'bæd/ *see* **forbid**

forbear|ance /fɔː'beərəns/ *n* pazienza *f*. **~ing** *a* tollerante

forbid /fə'bɪd/ *vt* (*pt* **forbade**, *pp* **forbidden**) proibire. **~ding** *a* (*prospect*) che spaventa; (*stern*) severo

force /fɔːs/ *n* forza *f*; **in ~** in vigore; (*in large numbers*) in massa; **come into ~** entrare in vigore; **the** [**armed**] **~s** *pl* le forze armate ● *vt* forzare; **~ sth on sb** ⟨*decision*⟩ imporre qcsa a qcno;

⟨*drink*⟩ costringere qcno a fare qcsa

forced /fɔːst/ *a* forzato

force: **~-'feed** *vt* (*pt*/*pp* **-fed**) nutrire a forza. **~ful** *a* energico. **~fully** *adv* ⟨*say, argue*⟩ con forza

forceps /'fɔːseps/ *npl* forcipe *m*

forcible /'fɔːsɪbl/ *a* forzato

ford /fɔːd/ *n* guado *m* ● *vt* guadare

fore /fɔː(r)/ *n* **to the ~** in vista; **come to the ~** salire alla ribalta

fore: **~arm** *n* avambraccio *m*. **~boding** /·'bəʊdɪŋ/ *n* presentimento *m*. **~cast** *n* previsione *f* ● *vt* (*pt*/*pp* **-cast**) prevedere. **~court** *n* cortile *m* anteriore. **~fathers** *npl* antenati *mpl*. **~finger** *n* [dito *m*] indice *m*. **~front** *n* **be in the ~front** essere all'avanguardia. **~gone** *a* **be a ~gone conclusion** essere una cosa scontata. **~ground** *n* primo piano *m*. **~head** /'fɒrɪd, 'fɔːhed/ *n* fronte *f*. **~hand** *n* Tennis diritto *m*

foreign /'fɒrən/ *a* straniero; ⟨*trade*⟩ estero; (*not belonging*) estraneo; **he is ~** è uno straniero. **~ currency** *n* valuta *f* estera. **~er** *n* straniero, -a *mf*. **~ language** *n* lingua *f* straniera

Foreign: **~ Office** *n* ministero *m* degli [affari] esteri. **~ 'Secretary** *n* ministro *m* degli esteri

fore: **~man** *n* caporeparto *m*. **~most** *a* principale ● *adv* **first and ~most** in primo luogo. **~name** *n* nome *m* di battesimo

forensic /fə'rensɪk/ *a* **~ medicine** medicina *f* legale

'forerunner *n* precursore *m*

fore'see (*pt* **-saw**, *pp* **-seen**) prevedere. **~able** /-əbl/ *a* **in the ~able future** in futuro per quanto si possa prevedere

'foresight *n* previdenza *f*

forest /'fɒrɪst/ *n* foresta *f*. **~er** *n* guardia *f* forestale

fore'stall *vt* prevenire

forestry /'fɒrɪstrɪ/ *n* silvicoltura *f*

'foretaste *n* pregustazione *f*

fore'tell *vt* (*pt*/*pp* **-told**) predire

forever /fə'revə(r)/ adv per sempre; **he's ~ complaining** si lamenta sempre

fore'warn vt avvertire

foreword /'fɔːwɜːd/ n prefazione f

forfeit /'fɔːfɪt/ n (in game) pegno m; Jur penalità f ● vt perdere

forgave /fə'geɪv/ see **forgive**

forge¹ /fɔːdʒ/ vi ~ **ahead** (runner:) lasciarsi indietro gli altri; fig farsi strada

forge² n fucina f ● vt fucinare; (counterfeit) contraffare. ~r n contraffattore m. ~ry n contraffazione f

forget /fə'get/ vt/i (pt -got, pp -gotten, pres p -getting) dimenticare; dimenticarsi di ⟨language, skill⟩. ~ful a smemorato. ~fulness n smemoratezza f. ~-me-not n non-ti-scordar-di-mé m inv. ~table /-əbl/ a ⟨day, film⟩ da dimenticare

forgive /fə'gɪv/ vt (pt -gave, pp -given) ~ **sb for sth** perdonare qcno per qcsa. ~**ness** n perdono m

forgo /fɔː'gəʊ/ vt (pt -went, pp -gone) rinunciare a

forgot(ten) /fə'gɒt(n)/ see **forget**

fork /fɔːk/ n forchetta f, (for digging) forca f, (in road) bivio m ● vi ⟨road:⟩ biforcarsi; ~ **right** prendere a destra. **fork out** vt fam sborsare

fork-lift 'truck n elevatore m

forlorn /fə'lɔːn/ a ⟨look⟩ perduto; ⟨place⟩ derelitto; ~ **hope** speranza f vana

form /fɔːm/ n forma f, (document) modulo m; Sch classe f ● vt formare; formulare ⟨opinion⟩ ● vi formarsi

formal /'fɔːml/ a formale. ~**ity** /-'mælətɪ/ n formalità f inv. ~**ly** adv in modo formale; (officially) ufficialmente

format /'fɔːmæt/ n formato m ● vt formattare ⟨disk, page⟩

formation /fɔː'meɪʃn/ n formazione f

formative /'fɔːmətɪv/ a ~ **years** anni mpl formativi

former /'fɔːmə(r)/ a precedente; ⟨PM, colleague⟩ ex; **the ~, the latter** il primo, l'ultimo. ~**ly** adv precedentemente; (in olden times) in altri tempi

formidable /'fɔːmɪdəbl/ a formidabile

formula /'fɔːmjʊlə/ n (pl -ae /-liː/ or -s) formula f

formulate /'fɔːmjʊleɪt/ vt formulare

forsake /fə'seɪk/ vt (pt -sook /-sʊk/, pp -saken) abbandonare

fort /fɔːt/ n Mil forte m

forth /fɔːθ/ adv **back and ~** avanti e indietro; **and so ~** e così via

forth: ~'**coming** a prossimo; (communicative) comunicativo; **no response was ~** non arrivava nessuna risposta. ~**right** a schietto. ~'**with** adv immediatamente

fortieth /'fɔːtɪθ/ a quarantesimo

fortification /fɔːtɪfɪ'keɪʃn/ n fortificazione f

fortify /'fɔːtɪfaɪ/ vt (pt/pp -ied) fortificare; fig rendere forte

fortnight /'fɔːt-/ Br n quindicina f. ~**ly** a bimensile ● adv ogni due settimane

fortress /'fɔːtrɪs/ n fortezza f

fortuitous /fɔː'tjuːɪtəs/ a fortuito

fortunate /'fɔːtʃənət/ a fortunato; **that's ~!** meno male!. ~**ly** adv fortunatamente

fortune /'fɔːtʃuːn/ n fortuna f. ~-**teller** n indovino, -a mf

forty /'fɔːtɪ/ a & n quaranta m

forum /'fɔːrəm/ n foro m

forward /'fɔːwəd/ adv avanti; (towards the front) in avanti ● a in avanti; (presumptuous) sfacciato ● n Sport attaccante m ● vt inoltrare ⟨letter⟩; spedire ⟨goods⟩. ~**s** adv avanti

fossil /'fɒsl/ n fossile m. ~**ized** a fossile; ⟨ideas⟩ fossilizzato

foster /'fɒstə(r)/ vt allevare ⟨child⟩.

~child *n* figlio, -a *mf* in affidamento. **~mother** *n* madre *f* affidataria

fought /fɔːt/ *see* **fight**

foul /faʊl/ *a* ⟨*smell, taste*⟩ cattivo; ⟨*air*⟩ viziato; ⟨*language*⟩ osceno; ⟨*mood, weather*⟩ orrendo; **~ play** *Jur* delitto *m* ● *n Sport* fallo *m* ● *vt* inquinare ⟨*water*⟩; *Sport* commettere un fallo contro; ⟨*nets, rope*⟩ impigliarsi in. **~-smelling** *a* puzzo

found¹ /faʊnd/ *see* **find**

found² *vt* fondare

foundation /faʊnˈdeɪʃn/ *n* ⟨*basis*⟩ fondamento *m*; ⟨*charitable*⟩ fondazione *f*; **~s** *pl* ⟨*of building*⟩ fondamenta *fpl*; **lay the ~-stone** porre la prima pietra

founder¹ /ˈfaʊndə(r)/ *n* fondatore, -trice *mf*

founder² *vi* ⟨*ship*⟩ affondare

foundry /ˈfaʊndrɪ/ *n* fonderia *f*

fountain /ˈfaʊntɪn/ *n* fontana *f*. **~-pen** *n* penna *f* stilografica

four /fɔː(r)/ *a* & *n* quattro *m*

four: **~-'poster** *n* letto *m* a baldacchino. **~some** /ˈfɔːsəm/ *n* quartetto *m*. **~teen** *a* & *n* quattordici *m*. **~'teenth** *a* quattordicesimo

fourth /fɔːθ/ *a* quarto

fowl /faʊl/ *n* pollame *m*

fox /fɒks/ *n* volpe *f* ● *vt* ⟨*puzzle*⟩ ingannare

foyer /ˈfɔɪeɪ/ *n* *Theat* ridotto *m*; ⟨*in hotel*⟩ salone *m* d'ingresso

fraction /ˈfrækʃn/ *n* frazione *f*

fracture /ˈfræktʃə(r)/ *n* frattura *f* ● *vt* fratturare ● *vi* fratturarsi

fragile /ˈfrædʒaɪl/ *a* fragile

fragment /ˈfrægmənt/ *n* frammento *m*. **~ary** *a* frammentario

fragran|ce /ˈfreɪɡrəns/ *n* fragranza *f*. **~t** *a* fragrante

frail /freɪl/ *a* gracile

frame /freɪm/ *n* ⟨*of picture, door, window*⟩ cornice *f*; ⟨*of spectacles*⟩ montatura *f*; *Anat* ossatura *f*; ⟨*structure, of bike*⟩ telaio *m*; **~ of mind** stato *m* d'animo ● *vt* incorniciare ⟨*picture*⟩; *fig* formulare;

⟨*sl: incriminate*⟩ montare. **~work** *n* struttura *f*

franc /fræŋk/ *n* franco *m*

France /frɑːns/ *n* Francia *f*

franchise /ˈfræntʃaɪz/ *n* *Pol* diritto *m* di voto; *Comm* franchigia *f*

frank¹ /fræŋk/ *vt* affrancare ⟨*letter*⟩

frank² *a* franco. **~ly** *adv* francamente

frankfurter /ˈfræŋkfɜːtə(r)/ *n* würstel *m inv*

frantic /ˈfræntɪk/ *a* frenetico; **be ~ with** worry essere agitatissimo. **~ally** *adv* freneticamente

fraternal /frəˈtɜːnl/ *a* fraterno

fraud /frɔːd/ *n* frode *f*; ⟨*person*⟩ impostore *m*. **~ulent** /-jʊlənt/ *a* fraudolento

fraught /frɔːt/ *a* **~ with** pieno di

fray¹ /freɪ/ *n* mischia *f*

fray² *vi* sfilacciarsi

frayed /freɪd/ *a* ⟨*cuffs*⟩ sfilacciato; ⟨*nerves*⟩ a pezzi

freak /friːk/ *n* fenomeno *m*; ⟨*person*⟩ scherzo *m* di natura; ⟨*fam: weird person*⟩ tipo *m* strambo ● *a* anormale. **~ish** *a* strambo

freckle /ˈfrekl/ *n* lentiggine *f*. **~d** *a* lentigginoso

free /friː/ *a* ⟨*freer, freest*⟩ libero; ⟨*ticket, copy*⟩ gratuito; ⟨*lavish*⟩ generoso; **~ of charge** gratuito; **set ~** liberare ● *vt* ⟨*pt/pp* **freed**⟩ liberare

free: **~dom** *n* libertà *f*. **~hand** *adv* a mano libera. **~hold** *n* proprietà *f* ⟨*fondiaria*⟩ assoluta. **~'kick** *n* calcio *m* di punizione. **~lance** *a* & *adv* indipendente. **~ly** *adv* liberamente; ⟨*generously*⟩ generosamente; **I ~ly admit that...** devo ammettere che... **F~mason** *n* massone *m*. **~range** *a* **~range egg** uovo *m* di gallina ruspante. **~sample** *n* campione *m* gratuito. **~style** *n* stile *m* libero. **~way** *n Am* autostrada *f*. **~'wheel** *vi* ⟨*car:*⟩ ⟨*in neutral*⟩ andare in folle; ⟨*with engine switched off*⟩ andare a motore spento; ⟨bicycle:⟩ andare a ruota libera

freez|e /friːz/ vt (pt **froze**, pp **frozen**) gelare; bloccare ⟨wages⟩ ● vi ⟨water:⟩ gelare; **it's ~ing** si gela; **my hands are ~ing** ho le mani congelate

freez|er /'friːzə(r)/ n freezer m inv, congelatore m. **~ing** a gelido ● n **below ~ing** sotto zero

freight /freɪt/ n carico m. **~er** n nave f da carico. **~ train** n Am treno m merci

French /frentʃ/ a francese ● n ⟨language⟩ francese m; **the ~** pl i francesi mpl

French: **~ beans** npl fagiolini mpl ⟨verdi⟩. **~ 'bread** n filone m ⟨di pane⟩. **~ 'fries** npl patate fpl fritte. **~man** n francese m. **~'window** n porta-finestra f. **~woman** n francese f

frenzied /'frenzɪd/ a frenetico

frenzy /'frenzɪ/ n frenesia f

frequency /'friːkwənsɪ/ n frequenza f

frequent[1] /'friːkwənt/ a frequente. **~ly** adv frequentemente

frequent[2] /frɪ'kwent/ vt frequentare

fresco /'freskəʊ/ n affresco m

fresh /freʃ/ a fresco; ⟨new⟩ nuovo; ⟨Am: cheeky⟩ sfacciato. **~ly** adv di recente

freshen /'freʃn/ vi ⟨wind:⟩ rinfrescare. **freshen up** vt dare una rinfrescata a ● vi rinfrescarsi

freshness /'freʃnɪs/ n freschezza f

'freshwater a di acqua dolce

fret /fret/ vi (pt/pp **fretted**) inquietarsi. **~ful** a irritabile

'fretsaw n seghetto m da traforo

friar /'fraɪə(r)/ n frate m

friction /'frɪkʃn/ n frizione f

Friday /'fraɪdeɪ/ n venerdì m inv

fridge /frɪdʒ/ n frigo m

fried /fraɪd/ see **fry** ● a fritto; **~ egg** uovo m fritto

friend /frend/ n amico, -a mf. **~ly** a (**-ier**, **-iest**) ⟨relations, meeting, match⟩ amichevole; ⟨neighbour-

hood, smile⟩ piacevole; ⟨software⟩ di facile uso; **be ~ly with** essere amico di. **~ship** n amicizia f

frieze /friːz/ n fregio m

fright /fraɪt/ n paura f; **take ~** spaventarsi

frighten /'fraɪtn/ vt spaventare. **~ed** a spaventato; **be ~ed** aver paura ⟨of di⟩. **~ing** a spaventoso

frightful /'fraɪtfl/ a terribile

frigid /'frɪdʒɪd/ a frigido. **~ity** /-'dʒɪdətɪ/ n freddezza f, Psych frigidità f

frill /frɪl/ n volant m inv. **~y** a ⟨dress⟩ con tanti volant

fringe /frɪndʒ/ n frangia f; ⟨of hair⟩ frangetta f; ⟨fig: edge⟩ margine m. **~ benefits** npl benefici mpl supplementari

frisk /frɪsk/ vt ⟨search⟩ perquisire

frisky /'frɪskɪ/ a (**-ier**, **-iest**) vispo

fritter /'frɪtə(r)/ n frittella f. **● fritter away** vt sprecare

frivol|ity /frɪ'vɒlɪtɪ/ n frivolezza f. **~ous** /'frɪvələs/ a frivolo

frizzy /'frɪzɪ/ a crespo

fro /frəʊ/ see **to**

frock /frɒk/ n abito m

frog /frɒg/ n rana f. **~man** n uomo m rana

frolic /'frɒlɪk/ vi (pt/pp **frolicked**) ⟨lambs:⟩ sgambettare; ⟨people:⟩ folleggiare

from /frɒm/ prep da; **~ Monday** da lunedì; **~ that day** da quel giorno; **he's ~ London** è di Londra; **this is a letter ~ my brother** questa è una lettera di mio fratello; **documents ~ the 16th century** documenti del XVI secolo; **~ made** fatto con; **she felt ill ~ fatigue** si sentiva male dalla stanchezza; **~ now on** d'ora in poi

front /frʌnt/ n parte f anteriore; ⟨fig: organization etc⟩ facciata f; ⟨of garment⟩ davanti m; ⟨sea~⟩ lungomare m; Mil, Pol, Meteorol fronte m; **in ~** davanti a; **in or at the ~** davanti; **to the ~** avanti ● a davanti; ⟨page, row, wheel⟩ anteriore

frontal /'frʌntl/ a frontale

front: ~ '**door** n porta f d'entrata. ~ '**garden** n giardino m d'avanti

frontier /'frʌntɪə(r)/ n frontiera f

front-wheel 'drive n trazione f anteriore

frost /frɒst/ n gelo m; (hoar~) brina f. ~**bite** n congelamento m. ~**bitten** a congelato

frost|ed /'frɒstɪd/ a ~**ed glass** vetro m smerigliato. ~**ily** adv gelidamente. ~**ing** n Am Culin glassa f. ~**y** a also fig gelido

froth /frɒθ/ n schiuma f • vi far schiuma. ~**y** a schiumoso

frown /fraʊn/ n cipiglio m • vi aggrottare le sopraciglia. **frown on** vt disapprovare

froze /frəʊz/ see **freeze**

frozen /'frəʊzn/ see **freeze** • a (corpse, hand) congelato; (wastes) gelido; Culin surgelato; **I'm** ~ sono gelato. ~ **food** n surgelati mpl

frugal /'fru:gl/ a frugale

fruit /fru:t/ n frutto m; (collectively) frutta f; **eat more** ~ mangia più frutta. ~ **cake** n dolce m con frutta candita

fruit|erer /'fru:tərə(r)/ n fruttivendolo, -a mf. ~**ful** a fig fruttuoso

fruition /fru:'ɪʃn/ n **come to** ~ dare i frutti

fruit: ~ **juice** n succo m di frutta. ~**less** a infruttuoso. ~ **machine** n macchinetta f mangiasoldi. ~ '**salad** n macedonia f [di frutta]

frumpy /'frʌmpɪ/ a scialbo

frustrat|e /frʌ'streɪt/ vt frustrare; rovinare (plans). ~**ing** a frustrante. ~**ion** /-'eɪʃn/ n frustrazione f

fry[1] vt/i (pt/pp **fried**) friggere

fry[2] /fraɪ/ n inv **small** ~ fig pesce m piccolo

frying pan n padella f

fuck /fʌk/ vulg vt/i scopare • int cazzo. ~**ing** a del cazzo

fuddy-duddy /'fʌdɪdʌdɪ/ n fam matusa mf inv

fudge /fʌdʒ/ n caramella f a base di zucchero, burro e latte

fuel /'fju:əl/ n carburante m; fig nutrimento m • vt fig alimentare

fugitive /'fju:dʒɪtɪv/ n fuggiasco, -a mf

fugue /fju:g/ n Mus fuga f

fulfil /fʊl'fɪl/ vt (pt/pp -**filled**) soddisfare (conditions, need); realizzare (dream, desire); ~ **oneself** realizzarsi. ~**ling** a soddisfacente. ~**ment** n **sense of** ~**ment** senso m di appagamento

full /fʊl/ a pieno (**of** di); (detailed) esauriente; (bus, hotel) completo; (skirt) ampio; **at** ~ **speed** a tutta velocità; **in** ~ **swing** in pieno fervore • n **in** ~ per intero

full: ~ '**moon** n luna f piena. ~**-scale** a (model) in scala reale; (alert) di massima gravità. ~ '**stop** n punto m. ~**-time** a & adv a tempo pieno

fully /'fʊlɪ/ adv completamente; (in detail) dettagliatamente; ~ **booked** (hotel, restaurant) tutto prenotato

fumble /'fʌmbl/ vi ~ **in** rovistare in; ~ **with** armeggiare con; ~ **for one's keys** rovistare alla ricerca delle chiavi

fume /fju:m/ vi (be angry) essere furioso

fumes /fju:mz/ npl fumi mpl; (from car) gas mpl di scarico

fumigate /'fju:mɪgeɪt/ vt suffumicare

fun /fʌn/ n divertimento m; **for** ~ per ridere; **make** ~ **of** prendere in giro; **have** ~ divertirsi

function /'fʌŋkʃn/ n funzione f; (event) cerimonia f • vi funzionare; ~ **as** (serve as) funzionare da. ~**al** a funzionale

fund /fʌnd/ n fondo m; fig pozzo m; ~**s** pl fondi mpl • vt finanziare

fundamental /fʌndə'mentl/ a fondamentale

funeral /'fju:nərəl/ n funerale m

funeral: ~ **directors** n impresa f di pompe funebri. ~ **home** Am, ~ **parlour** n camera f ardente.

march n marcia f funebre. **~
service** n rito m funebre
'funfair n luna park m inv
fungus /'fʌŋgəs/ n (pl -gi /-gaɪ/)
fungo m
funicular /fjuː'nɪkjʊlə(r)/ n funicolare f
funnel /'fʌnl/ n imbuto m; (on ship)
ciminiera f
funnily /'fʌnɪlɪ/ adv comicamente;
(oddly) stranamente; **~ enough**
strano a dirsi
funny /'fʌnɪ/ a (-ier, -iest) buffo;
(odd) strano. **~ business** n affare
m losco
fur /fɜː(r)/ n pelo m; (for clothing)
pelliccia f; (in kettle) deposito m.
~ 'coat n pelliccia f
furious /'fjʊərɪəs/ a furioso
furnace /'fɜːnɪs/ n fornace f
furnish /'fɜːnɪʃ/ vt ammobiliare
(flat); fornire (supplies). **~ed** a
~ed room stanza f ammobiliata.
~ings npl mobili mpl
furniture /'fɜːnɪtʃə(r)/ n mobili
mpl
furred /fɜːd/ a (tongue) impastato
furrow /'fʌrəʊ/ n solco m
furry /'fɜːrɪ/ a (animal) peloso;
(toy) di peluche
further /'fɜːðə(r)/ a (additional) ulteriore; **at the ~ end** all'altra
estremità; **until ~ notice** fino a
nuovo avviso ● adv più lontano;
~... inoltre,...; **~ off** più lontano
● vt promuovere
further-: ~ edu'cation n ≈ formazione f parauniversitaria. **~'more**
adv per di più
furthest /'fɜːðɪst/ a più lontano
● adv più lontano
furtive /'fɜːtɪv/ a furtivo
fury /'fjʊərɪ/ n furore m
fuse¹ /fjuːz/ n (of bomb) detonatore
m; (cord) miccia f
fuse² n Electr fusibile m ● vt fondere; Electr far saltare ● vi fondersi;
Electr saltare; **the lights have ~d**
sono saltate le luci. **~-box** n scatola
f dei fusibili

fuselage /'fjuːzəlɑːʒ/ n Aeron fusoliera f
fusion /'fjuːʒn/ n fusione f
fuss /fʌs/ n storie fpl; **make a ~**
fare storie; **make a ~ of** colmare
di attenzioni ● vi fare storie
fussy /'fʌsɪ/ a (-ier, -iest) (person)
difficile da accontentare; (clothes
etc) pieno di fronzoli
fusty /'fʌstɪ/ a che odora di stantio;
(smell) di stantio
futile /'fjuːtaɪl/ a inutile. **~ity**
/-'tɪlətɪ/ n futilità f
future /'fjuːtʃə(r)/ a & n futuro; **in
~** in futuro. **~ perfect** futuro m
anteriore
futuristic /fjuːtʃə'rɪstɪk/ a futuristico
fuzz /fʌz/ n **the ~** (sl: police) la pula
fuzzy /'fʌzɪ/ a (-ier, -iest) (hair)
crespo; (photo) sfuocato

Gg

gab /gæb/ n fam **have the gift of
the ~** avere la parlantina
gabble /'gæb(ə)l/ vi parlare troppo
in fretta
gad /gæd/ vi (pt/pp gadded) **~
about** andarsene in giro
gadget /'gædʒɪt/ n aggeggio m
Gaelic /'geɪlɪk/ a & n gaelico m
gaffe /gæf/ n gaffe f inv
gag /gæg/ n bavaglio m; (joke) battuta f ● vt (pt/pp gagged) imbavagliare
gaily /'geɪlɪ/ adv allegramente
gain /geɪn/ n guadagno m; (increase) aumento m ● vt acquisire;
~ weight aumentare di peso; **~
access** accedere ● vi (clock:) andare avanti. **~ful** a **~ful employment** lavoro m remunerativo
gait /geɪt/ n andatura f
gala /'gɑːlə/ n gala f; (swimming ~)
manifestazione f di nuoto ● attrib
di gala

galaxy /'gæləksɪ/ n galassia f

gale /geɪl/ n bufera f

gall /gɔːl/ n (impudence) impudenza f

gallant /'gælənt/ a coraggioso; (chivalrous) galante. **~ry** n coraggio m

'gall-bladder n cistifellea f

gallery /'gælərɪ/ n galleria f

galley /'gælɪ/ n (ship's kitchen) cambusa f; **~** [**proof**] bozza f in colonna

gallivant /'gælɪvænt/ vi fam andare in giro

gallon /'gælən/ n gallone m (= 4,5 l; Am = 3,7 l)

gallop /'gæləp/ n galoppo m ● vi galoppare

gallows /'gæləʊz/ n forca f

'gallstone n calcolo m biliare

galore /gə'lɔː(r)/ adv a bizzeffe

galvanize /'gælvənaɪz/ vt Techn galvanizzare; fig stimolare (**into** a)

gambit /'gæmbɪt/ n prima mossa f

gamble /'gæmbl/ n (risk) azzardo m ● vi giocare; (on Stock Exchange) speculare; **~e on** (rely) contare su. **~er** n giocatore, -trice mf [d'azzardo]. **~ing** n gioco m [d'azzardo]

game /geɪm/ n gioco m; (match) partita f; (animals, birds) selvaggina f; **~s** Sch ≈ ginnastica f ● a (brave) coraggioso; **are you ~?** ti va?; **be ~ for** essere pronto per. **~keeper** n guardacaccia m inv

gammon /'gæmən/ n coscia f di maiale

gamut /'gæmət/ n fig gamma f

gander /'gændə(r)/ n oca f maschio

gang /gæŋ/ n banda f; (of workmen) squadra f ● **gang up** vi far comunella (**on** contro)

gangling /'gæŋglɪŋ/ a spilungone

gangrene /'gæŋgriːn/ n cancrena f

gangster /'gæŋstə(r)/ n gangster m inv

gangway /'gæŋweɪ/ n passaggio m; Naut, Aeron passerella f

gaol /dʒeɪl/ n carcere m ● vt incarcerare. **~er** n carceriere m

gap /gæp/ n spazio m; (in ages, between teeth) scarto m; (in memory) vuoto m; (in story) punto m oscuro

gape /geɪp/ vi stare a bocca aperta; (be wide open) spalancarsi; **~e at** guardare a bocca aperta. **~ing** a aperto

garage /'gærɑːʒ/ n garage m inv; (for repairs) meccanico m; (for petrol) stazione f di servizio

garbage /'gɑːbɪdʒ/ n immondizia f; (nonsense) idiozie fpl. **~ can** n Am bidone m dell'immondizia

garbled /'gɑːbld/ a confuso

garden /'gɑːdn/ n giardino m; [**public**] **~s** pl giardini mpl pubblici ● vi fare giardinaggio. **~ centre** n negozio m di piante e articoli da giardinaggio. **~er** n giardiniere, -a mf. **~ing** n giardinaggio m

gargle /'gɑːgl/ n gargarismo m ● vi fare gargarismi

gargoyle /'gɑːgɔɪl/ n gargouille f inv

garish /'geərɪʃ/ a sgargiante

garland /'gɑːlənd/ n ghirlanda f

garlic /'gɑːlɪk/ n aglio m. **~ bread** n pane m condito con aglio

garment /'gɑːmənt/ n indumento m

garnish /'gɑːnɪʃ/ n guarnizione f ● vt guarnire

garrison /'gærɪsn/ n guarnigione f

garter /'gɑːtə(r)/ n giarrettiera f; (Am: for man's socks) reggicalze m inv da uomo

gas /gæs/ n gas m inv; (Am fam: petrol) benzina f ● v (pt/pp **gassed**) ● vt asfissiare ● vi fam blaterare. **~ cooker** n cucina f a gas. **~ 'fire** n stufa f a gas

gash /gæʃ/ n taglio m ● vt tagliare

gasket /'gæskɪt/ n Techn guarnizione f

gas: ~ mask n maschera f antigas. **~-meter** n contatore m del gas

gasoline /'gæsəliːn/ n Am benzina f

gasp /gɑːsp/ vi avere il fiato mozzato

'**gas station** n Am distributore m di benzina

gastric /'gæstrɪk/ a gastrico. ~ '**flu** n influenza f gastro-intestinale. ~ '**ulcer** n ulcera f gastrica

gastronomy /gæˈstrɒnəmɪ/ n gastronomia f

gate /geɪt/ n cancello m; (at airport) uscita f

gâteau /'gætəʊ/ n torta f

gate: ~**crash** vt entrare senza invito a. ~**crasher** n intruso, -a mf. ~**way** n ingresso m

gather /'gæðə(r)/ vt raccogliere; (conclude) dedurre; (in sewing) arricciare; ~ **speed** acquistare velocità; ~ **together** radunare ⟨people, belongings⟩; (obtain gradually) acquistare ● vi ⟨people:⟩ radunarsi. ~**ing** n family ~**ing** ritrovo m di famiglia

gaudy /'gɔːdɪ/ a (**-ier**, **-iest**) pacchiano

gauge /geɪdʒ/ n calibro m; Rail scartamento m; (device) indicatore m ● vt misurare; fig stimare

gaunt /gɔːnt/ a (thin) smunto

gauze /gɔːz/ n garza f

gave /geɪv/ see **give**

gawky /'gɔːkɪ/ a (**-ier**, **-iest**) sgraziato

gawp /gɔːp/ vi ~ [**at**] fam guardare con aria di ebete

gay /geɪ/ a gaio; (homosexual) omosessuale; ⟨bar, club⟩ gay

gaze /geɪz/ n sguardo m fisso ● vi guardare; ~ **at** fissare

GB abbr (**Great Britain**) GB

gear /gɪə(r)/ n equipaggiamento m; Techn ingranaggio m; Auto marcia f; **in** ~ con la marcia innestata; **change** ~ cambiare marcia ● vt finalizzare (**to** a)

gear: ~**box** n Auto scatola f del cambio. ~**lever**, Am ~**shift** n leva f del cambio

geese /giːs/ see **goose**

geezer /'giːzə(r)/ n sl tipo m

gel /dʒel/ n gel m inv

gelatine /'dʒelətɪn/ n gelatina f

gelignite /'dʒelɪgnaɪt/ n gelatina f esplosiva f

gem /dʒem/ n gemma f

Gemini /'dʒemɪnaɪ/ n Astr Gemelli mpl

gender /'dʒendə(r)/ n Gram genere m

gene /dʒiːn/ n gene m

genealogy /dʒiːnɪˈælədʒɪ/ n genealogia f

general /'dʒenrəl/ a generale ● n generale m; **in** ~ in generale. ~ **e'lection** n elezioni fpl politiche

generalization /dʒenrəlaɪˈzeɪʃn/ n generalizzazione f. ~**e** /'dʒenrəlaɪz/ vi generalizzare

generally /'dʒenrəlɪ/ adv generalmente

general prac'titioner n medico m generico

generate /'dʒenəreɪt/ vt generare

generation /dʒenəˈreɪʃn/ n generazione f

generator /'dʒenəreɪtə(r)/ n generatore m

generic /dʒɪˈnerɪk/ a ~ **term** termine m generico

generosity /dʒenəˈrɒsɪtɪ/ n generosità f

generous /'dʒenərəs/ a generoso. ~**ly** adv generosamente

genetic /dʒɪˈnetɪk/ a genetico. ~ **engineering** n ingegneria f genetica. ~**s** n genetica f

Geneva /dʒɪˈniːvə/ n Ginevra f

genial /'dʒiːnɪəl/ a gioviale

genitals /'dʒenɪtlz/ npl genitali mpl

genitive /'dʒenɪtɪv/ a & n ~ [**case**] genitivo m

genius /'dʒiːnɪəs/ n (pl **-uses**) genio m

genocide /'dʒenəsaɪd/ n genocidio m

genre /'ʒɑ̃rə/ n genere m [letterario]

gent /dʒent/ n fam signore m; **the** ~**s** sg il bagno per uomini

genteel /dʒenˈtiːl/ a raffinato

gentle /'dʒentl/ a delicato; ⟨breeze, tap, slope⟩ leggero
gentleman /'dʒentlmən/ n signore m; (well-mannered) gentiluomo m
gent|leness /'dʒentlnɪs/ n delicatezza f. **~ly** adv delicatamente
genuine /'dʒenjʊm/ a genuino. **~ly** adv ⟨sorry⟩ sinceramente
geograph|ical /dʒɪə'græfɪkl/ a geografico. **~y** /dʒɪ'ɒgrəfi/ n geografia f
geological /dʒɪə'lɒdʒɪkl/ a geologico
geolog|ist /dʒɪ'ɒlədʒɪst/ n geologo, -a mf. **~y** n geologia f
geometr|ic[al] /dʒɪə'metrɪk(l)/ a geometrico. **~y** /dʒɪ'ɒmɪtrɪ/ n geometria f
geranium /dʒə'reɪnɪəm/ n geranio m
geriatric /dʒerɪ'ætrɪk/ a geriatrico; **~ ward** n reparto m geriatria. **~s** n geriatria f
germ /dʒɜːm/ n germe m; **~s** pl microbi mpl
German /'dʒɜːmən/ n & a tedesco, -a mf; (language) tedesco m
Germanic /dʒə'mænɪk/ a germanico
German: **~ 'measles** n rosolia f. **~ 'shepherd** n pastore m tedesco
Germany /'dʒɜːmənɪ/ n Germania f
germinate /'dʒɜːmɪneɪt/ vi germogliare
gesticulate /dʒe'stɪkjʊleɪt/ vi gesticolare
gesture /'dʒestʃə(r)/ n gesto m
get /get/ v ⟨pt/pp got, pp Am also gotten, pres p getting⟩ ● vt (receive) ricevere; (obtain) ottenere; trovare ⟨job⟩; (buy, catch, fetch) prendere; (transport, deliver to airport etc) portare; (reach on telephone) trovare; (fam: understand) comprendere; prepare ⟨meal⟩; **~ sb to do sth** far fare qcsa a qcno **~** (fetch) andare a prendere; **I'm ~ting hungry** mi sta venendo fame; **~**

dressed/married vestirsi/sposarsi; **~ sth ready** preparare qcsa.; **~ nowhere** non concludere nulla; **this is ~ting us nowhere** questo non ci è di nessun aiuto; **~ to** (reach) arrivare a. **get at** vi (criticize) criticare; **I see what you're ~ting at** ho capito cosa vuoi dire; **what are you ~ting at?** dove vuoi andare a parare?. **get away** vi (leave) andarsene; (escape) scappare. **get back** vi tornare ● vt (recover) riavere; **~ one's own back** rifarsi. **get by** vi passare; (manage) cavarsela. **get down** vi scendere; **~ down to work** mettersi al lavoro ● vt (depress) buttare giù. **get in** vi entrare ● vt mettere dentro ⟨washing⟩; far venire ⟨plumber⟩. **get off** vi scendere; (from work) andarsene; Jur essere assolto; **~ off the bus/one's bike** scendere dal pullman/dalla bici ● vt (remove) togliere. **get on** vi salire; (be on good terms) andare d'accordo; (make progress) andare avanti; (in life) riuscire; **~ on the bus/one's bike** salire sul pullman/sulla bici; **how are you ~ting on?** come va?. **get out** vi uscire; (of car) scendere; **~ out!** fuori!; **~ out of** (avoid doing) evitare ● vt togliere ⟨cork, stain⟩. **get over** vi andare di là ● vt fig riprendersi da ⟨illness⟩. **get round** vt aggirare ⟨rule⟩; rigirare ⟨person⟩ ● vi **I never ~ round to it** non mi sono mai deciso a farlo. **get through** vi (on telephone) prendere la linea. **get up** vi alzarsi; (climb) salire; **~ up a hill** salire su una collina **get: ~away** n fuga f. **~-up** n tenuta f
geyser /'giːzə(r)/ n scaldabagno m; Geol geyser m inv
ghastly /'gɑːstlɪ/ a (-ier, -iest) terribile; **feel ~** sentirsi da cani
gherkin /'gɜːkɪn/ n cetriolino m
ghetto /'getəʊ/ n ghetto m

ghost /gəʊst/ n fantasma m. ~ly a spettrale

ghoulish /'gu:lɪʃ/ a macabro

giant /'dʒaɪənt/ n gigante m ● a gigante

gibberish /'dʒɪbərɪʃ/ n stupidaggini fpl

gibe /dʒaɪb/ n malignità fm

giblets /'dʒɪblɪts/ npl frattaglie fpl

giddiness /'gɪdɪnɪs/ n vertigini fpl

giddy /'gɪdɪ/ a (-ler, -iest) vertiginoso; **feel ~** avere le vertigini

gift /gɪft/ n dono m; (to charity) donazione f. **~ed** /-ɪd/ a dotato. **~-wrap** vt impacchettare in carta da regalo

gig /gɪg/ n Mus fam concerto m

gigantic /dʒaɪ'gæntɪk/ a gigantesco

giggle /'gɪgl/ n risatina f ● vi ridacchiare

gild /gɪld/ vt dorare

gills /gɪlz/ npl branchia fsg

gilt /gɪlt/ a dorato ● n doratura f. **~-edged stock** n investimento m sicuro

gimmick /'gɪmɪk/ n trovata f

gin /dʒɪn/ n gin m inv

ginger /'dʒɪndʒə(r)/ a rosso fuoco inv; (cat) rosso ● n zenzero m. **~ ale** n, **~ beer** n bibita f allo zenzero. **~bread** n panpepato m

gingerly /'dʒɪndʒəlɪ/ adv con precauzione

gipsy /'dʒɪpsɪ/ n = **gypsy**

giraffe /dʒɪ'rɑ:f/ n giraffa f

girder /'gɜ:də(r)/ n Techn trave f

girl /gɜ:l/ n ragazza f; (female child) femmina f. **~friend** n amica f; (of boy) ragazza f. **~ish** a da ragazza

giro /'dʒaɪərəʊ/ n bancogiro m; (cheque) sussidio m di disoccupazione

girth /gɜ:θ/ n circonferenza f

gist /dʒɪst/ n **the ~** la sostanza

give /gɪv/ n elasticità f ● v (pt gave, pp given) ● vt dare; (as present) regalare (**to** a); fare (lecture, present, shriek); donare (blood); **~ birth** partorire ● vi (to charity) fare delle donazioni;

(yield) cedere. **give away** vt dar via; (betray) tradire; (distribute) assegnare; **~ away the bride** portare la sposa all'altare. **give back** vt restituire. **give in** vt consegnare ● vi (yield) arrendersi. **give off** vt emanare. **give over** vi **~ over!** piantala!. **give up** vt rinunciare a; **~ oneself up** arrendersi ● vi rinunciare. **give way** vi cedere; Auto dare la precedenza; (collapse) crollare

given /'gɪvn/ see **give** ● a **~ name** nome m di battesimo

glacier /'glæsɪə(r)/ n ghiacciaio m

glad /glæd/ a contento (**of** di). **~den** /'glædn/ vt rallegrare

glade /gleɪd/ n radura f

gladly /'glædlɪ/ adv volentieri

glamor|ize /'glæməraɪz/ vt rendere affascinante. **~ous** a affascinante

glamour /'glæmə(r)/ n fascino m

glance /glɑ:ns/ n sguardo m ● vi **~ at** dare un'occhiata a. **glance up** vi alzare gli occhi

gland /glænd/ n glandola f

glandular /'glændjʊlə(r)/ a ghiandolare. **~ fever** n Med mononucleosi f

glare /gleə(r)/ n bagliore m; (look) occhiataccia f ● vi **~ at** dare un'occhiataccia a

glaring /'gleərɪŋ/ a sfolgorante; (mistake) madornale

glass /glɑ:s/ n vetro m; (for drinking) bicchiere m; **~es** (pl: spectacles) occhiali mpl. **~y** a vitreo

glaze /gleɪz/ n smalto m ● vt mettere i vetri a (door, window); smaltare (pottery); Culin spennellare. **~d** a (eyes) vitreo

glazier /'gleɪzɪə(r)/ n vetraio m

gleam /gli:m/ n luccichio m ● vi luccicare

glean /gli:n/ vt racimolare (information)

glee /gli:/ n gioia f. **~ful** a gioioso

glen /glen/ n vallone m

glib /glɪb/ a pej insincero

glid|e /glaɪd/ vi scorrere; (*through the air*) planare. **~er** n aliante m

glimmer /'glɪmə(r)/ n barlume m ● vi emettere un barlume

glimpse /glɪmps/ n occhiata f; **catch a ~ of** intravedere ● vt intravedere

glint /glɪnt/ n luccichio m ● vi luccicare

glisten /'glɪsn/ vi luccicare

glitter /'glɪtə(r)/ vi brillare

gloat /gləʊt/ vi gongolare (**over** su)

global /'gləʊbl/ a mondiale

globe /gləʊb/ n globo m; (*map*) mappamondo m

gloom /gluːm/ n oscurità f; (*sadness*) tristezza f. **~ily** adv (*sadly*) con aria cupa

gloomy /'gluːmɪ/ a (**-ier, -iest**) cupo

glorify /'glɔːrɪfaɪ/ vt (*pt/pp* **-ied**) glorificare; **a ~ied waitress** niente più che una cameriera

glorious /'glɔːrɪəs/ a splendido; (*deed, hero*) glorioso

glory /'glɔːrɪ/ n gloria f; (*splendour*) splendore m; (*cause for pride*) vanto m ● vi (*pt/pp* **-ied**) **~ in** vantarsi di

gloss /glɒs/ n lucentezza f. **~ paint** n vernice f lucida ● **gloss over** vt sorvolare su

glossary /'glɒsərɪ/ n glossario m

glossy /'glɒsɪ/ a (**-ier, -iest**) lucido; **~ [magazine]** rivista f femminile

glove /glʌv/ n guanto m. **~ compartment** n Auto cruscotto m

glow /gləʊ/ n splendore m; (*in cheeks*) rossore m; (*of candle*) luce f soffusa ● vi risplendere; (*candle:*) brillare; (*person:*) avvampare. **~ing** a ardente; (*account*) entusiastico

'glow-worm n lucciola f

glucose /'gluːkəʊs/ n glucosio m

glue /gluː/ n colla f ● vt (*pres p* gluing) incollare

glum /glʌm/ a (**glummer, glummest**) tetro

glut /glʌt/ n eccesso m

glutton /'glʌtən/ n ghiottone, -a

mf. **~ous** /-əs/ a ghiotto. **~y** n ghiottoneria f

gnarled /nɑːld/ a nodoso

gnash /næʃ/ vt **~ one's teeth** digrignare i denti

gnat /næt/ n moscerino m

gnaw /nɔː/ vt rosicchiare

gnome /nəʊm/ n gnomo m

go /gəʊ/ n (*pl* **goes**) energia f; (*attempt*) tentativo m; **on the go** in movimento; **at one go** in una sola volta; **it's your go** tocca a te; **make a go of it** riuscire ● vi (*pt* **went**, *pp* **gone**) andare; (*leave*) andare via; (*vanish*) sparire; (*become*) diventare; (*be sold*) vendersi; **go and see** andare a vedere; **go swimming/shopping** andare a nuotare/fare spese; **where's the time gone?** come ha fatto il tempo a volare così?; **it's all gone** è finito; **be going to do** stare per fare; **I'm not going to** non ne ho nessuna intenzione; **to go** ‹Am: hamburgers etc› da asporto; **a coffee to go** un caffè da portar via. **go about** vi andare in giro. **go away** vi andarsene. **go back** vi ritornare. **go by** vi passare. **go down** vi scendere; (*sun:*) tramontare; (*ship:*) affondare; (*swelling:*) diminuire. **go for** vt andare a prendere; andare a cercare (*doctor*); (*choose*) optare per; (*fam: attack*) aggredire; **he's not the kind I go for** non è il genere che mi attira. **go in** vi entrare. **go in for** vt partecipare a (*competition*); darsi a (*tennis*). **go off** vi andarsene; (*alarm:*) scattare; (*gun, bomb:*) esplodere; (*food, milk:*) andare a male; **go off well** riuscire. **go on** vi andare avanti; **what's going on?** cosa succede? **go on at** vt fam scocciare. **go out** vi uscire; (*light, fire:*) spegnersi. **go over** vi andare ● vt (*check*) controllare. **go round** vi andare in giro; (*visit*) andare; (*turn*) girare; **is there enough to go round?** ce n'è abbastanza per tutti? **go through** vi (*bill, proposal:*) passare ● vt (*suffer*) subire; (check) controllare; (read) leggere. **go under** vi passare sotto; (ship, swimmer:) andare sott'acqua; (fail) fallire. **go up** vi salire; (Theat: curtain:) aprirsi. **go with** vt accompagnare. **go without** vt fare a meno di (supper, sleep) ● vi fare senza

goad /gəʊd/ vt spingere (**into** a); (taunt) spronare

'go-ahead a (person, company) intraprendente ● n okay m

goal /gəʊl/ n porta f; (point scored) gol m inv; (in life) obiettivo m; **score a ~** segnare. **~ie** fam, **~keeper** n portiere m. **~post** n palo m

goat /gəʊt/ n capra f

gobble /'gɒbl/ vt ~ [**down, up**] tranguiare

'go-between n intermediario, -a mf

God, god /gɒd/ n Dio m, dio m

god: **~child** n figlioccio, -a mf. **~daughter** n figlioccia f. **~dess** n dea f. **~father** n padrino m. **~fearing** a timorato di Dio. **~forsaken** a dimenticato da Dio. **~mother** n madrina f. **~parents** npl padrino m e madrina f. **~send** n manna f. **~son** n figlioccio m

go-getter /'gəʊgetə(r)/ n ambizioso, -a mf

goggle /'gɒgl/ vi fam ~ **at** fissare con gli occhi sgranati. **~s** npl occhiali mpl; (of swimmer) occhialini mpl [da piscina]; (of worker) occhiali mpl

going /'gəʊɪŋ/ a (price, rate) corrente; **~ concern** azienda f florida ● n it's hard ~ è una faticaccia; while the ~ is good finché si può. **~s-'on** npl avvenimenti mpl

gold /gəʊld/ n oro m ● a d'oro

golden /'gəʊldn/ a dorato. **~ 'handshake** n buonuscita f (al termine di un rapporto di lavoro). **~ mean** n giusto mezzo m. **~ 'wedding** n nozze fpl d'oro

gold: **~fish** n inv pesce m rosso. **~-mine** n miniera f d'oro.

~-plated a placcato d'oro. **~smith** n orefice m

golf /gɒlf/ n golf m

golf: **~-club** n circolo m di golf; (implement) mazza f da golf. **~-course** n campo m di golf. **~er** n giocatore, -trice mf di golf

gondola /'gɒndələ/ n gondola f. **~lier** /-'lɪə(r)/ n gondoliere m

gone /gɒn/ see **go**

gong /gɒŋ/ n gong m inv

good /gʊd/ a (**better, best**) buono; (child, footballer, singer) bravo; (holiday, film) bello; **at** bravo in; **a ~ deal of** molta rabbia; **as ~ as** (almost) quasi; **~ morning, ~ afternoon** buon giorno; **~ evening** buona sera; **~ night** buonanotte; **have a ~ time** divertirsi ● n bene m; **for ~** per sempre; **do ~** far del bene; **do sb ~** far bene a qcno; **it's no ~** è inutile; **be up to no ~** combinare qualcosa

goodbye /gʊd'baɪ/ int arrivederci

good: **~-for-nothing** n buono, -a mf a nulla. **G~ 'Friday** n Venerdì m Santo

good: **~-'looking** a bello. **~-'natured** a **be ~-natured** avere un buon carattere

goodness /'gʊdnɪs/ n bontà f; **my ~!** santo cielo!; **thank ~!** grazie al cielo!

goods /gʊdz/ npl prodotti mpl. **~ train** n treno m merci

good'will n buona volontà f; Comm avviamento m

goody /'gʊdɪ/ n (fam: person) buono m. **~-goody** n santarellino, -a mf

gooey /'gu:ɪ/ a fam appiccicaticcio; fig sdolcinato

goof /gu:f/ vi fam cannare

goose /gu:s/ n (pl **geese**) oca f

gooseberry /'gʊzbərɪ/ n uva f spina

goose /gu:s/: **~-flesh** n, **~-pimples** npl pelle fsg d'oca

gore¹ /gɔ:(r)/ n sangue m

gore² vt incornare

gorge /gɔːdʒ/ n Geog gola f ● vt ~ **oneself** ingozzarsi

gorgeous /ˈgɔːdʒəs/ a stupendo

gorilla /gəˈrɪlə/ n gorilla m inv

gormless /ˈgɔːmlɪs/ a fam stupido

gorse /gɔːs/ n ginestrone m

gory /ˈgɔːrɪ/ a (**-ier, -iest**) cruento

gosh /gɒʃ/ int fam caspita

gospel /ˈgɒspl/ n vangelo m. ~ **truth** n sacrosanta verità f

gossip /ˈgɒsɪp/ n pettegolezzi mpl; (person) pettegolo, -a mf ● vi pettegolare. **~y** a pettegolo

got /gɒt/ see **get**; **have** ~ avere; **have** ~ **to do sth** dover fare qcsa

Gothic /ˈgɒθɪk/ a gotico

gotten /ˈgɒtn/ Am see **get**

gouge /gaʊdʒ/ vt ~ **out** cavare

gourmet /ˈgʊəmeɪ/ n buongustaio, -a mf

gout /gaʊt/ n gotta f

govern /ˈgʌv(ə)n/ vt/i governare; (determine) determinare

government /ˈgʌvnmənt/ n governo m. **~al** /-ˈmentl/ a governativo

governor /ˈgʌvənə(r)/ n governatore m; (of school) membro m de consiglio di istituto; (of prison) direttore, -trice mf; (fam: boss) capo m

gown /gaʊn/ n vestito m; Univ, Jur toga f

GP n abbr **general practitioner**

grab /græb/ vt (pt/pp **grabbed**) ~ **[hold of]** afferrare

grace /greɪs/ n grazia f; (before meal) benedicite m inv; **with good** ~ volentieri; **three days'** ~ tre giorni di proroga. **~ful** a aggraziato. **~fully** adv con grazia

gracious /ˈgreɪʃəs/ a cortese; (elegant) lussuoso

grade /greɪd/ n livello m; Comm qualità f; Sch voto m; (Am Sch: class) classe f; Am = **gradient** ● vt Comm classificare; Sch dare il voto a. **~ crossing** n Am passaggio m a livello

gradient /ˈgreɪdɪənt/ n pendenza f

gradual /ˈgrædʒʊəl/ a graduale. **~ly** adv gradualmente

graduate[1] /ˈgrædʒʊət/ n laureato, -a mf

graduate[2] /ˈgrædʒʊeɪt/ vi Univ laurearsi

graduation /grædʒʊˈeɪʃn/ n laurea f

graffiti /grəˈfiːtɪ/ npl graffiti mpl

graft /grɑːft/ n (Bot, Med) innesto m; (Med: organ) trapianto m; (fam: hard work) duro lavoro m; (fam: corruption) corruzione f ● vt innestare; trapiantare (organ)

grain /greɪn/ n (of sand, salt) granello m; (of rice) chicco m; (cereals) cereali mpl; (in wood) venatura f; **it goes against the** ~ fig è contro la mia/sua natura

gram /græm/ n grammo m

grammar /ˈgræmə(r)/ n grammatica f. ~ **school** n ≈ liceo m

grammatical /grəˈmætɪkl/ a grammaticale

granary /ˈgrænərɪ/ n granaio m

grand /grænd/ a grandioso; fam eccellente

grandad /ˈgrændæd/ n fam nonno m

grandchild n nipote mf

granddaughter n nipote f

grandeur /ˈgrændʒə(r)/ n grandiosità f

grandfather n nonno m. ~ **clock** n pendolo m (che poggia a terra)

grandiose /ˈgrændɪəʊs/ a grandioso

grand: **~mother** n nonna f. **~parents** npl nonni mpl. **~piano** n pianoforte m a coda. **~son** n nipote m. **~stand** n tribuna f

granite /ˈgrænɪt/ n granito m

granny /ˈgrænɪ/ n fam nonna f

grant /grɑːnt/ n (money) sussidio m; Univ borsa f di studio ● vt accordare; (admit) ammettere; **take sth for ~ed** dare per scontato qcsa

granulated /ˈgrænjʊleɪtɪd/ a ~ **sugar** zucchero m semolato

granule /ˈgrænjuː/ n granello m

grape /greɪp/ n acino m; ~s pl uva f sg

grapefruit /'greɪp-/ n inv pompelmo m

graph /grɑːf/ n grafico m

graphic /'græfɪk/ a grafico; (vivid) vivido. ~s n grafica f

'graph paper n carta f millimetrata

grapple /'græpl/ vi ~ with also fig essere alle prese con

grasp /grɑːsp/ n stretta f; (understanding) comprensione f ● vt afferrare. ~ing a avido

grass /grɑːs/ n erba f; at the ~ roots alla base. ~hopper n cavalletta f. ~land n prateria f

grassy /'grɑːsɪ/ a erboso

grate[1] /greɪt/ n grata f

grate[2] vt Culin grattugiare ● vi stridere

grateful /'greɪtfl/ a grato. ~ly adv con gratitudine

grater /'greɪtə(r)/ n Culin grattugia f

gratify /'grætɪfaɪ/ vt (pt/pp -ied) appagare. ~ied a appagato. ~ying a appagante

grating /'greɪtɪŋ/ n grata f

gratis /'grɑːtɪs/ adv gratis

gratitude /'grætɪtjuːd/ n gratitudine f

gratuitous /grə'tjuːɪtəs/ a gratuito

gratuity /grə'tjuːɪtɪ/ n gratifica f

grave[1] /greɪv/ a grave

grave[2] n tomba f

gravel /'grævl/ n ghiaia f

grave: ~stone n lapide f. ~yard n cimitero m

gravitate /'græviteɪt/ vi gravitare

gravity /'grævɪtɪ/ n gravità f

gravy /'greɪvɪ/ n sugo m della carne

gray /greɪ/ a Am = grey

graze[1] /greɪz/ vi (animal:) pascolare

graze[2] n escoriazione f ● vt (touch lightly) sfiorare; (scrape) escoriare; sbucciarsi (knee)

grease /griːs/ n grasso m ● vt un-

gere. ~-proof 'paper n carta f oleata

greasy /'griːsɪ/ a (-ier, -iest) untuoso; (hair, skin) grasso

great /greɪt/ a grande; (fam: marvellous) eccezionale

great: ~-'aunt n prozia f. G~ 'Britain n Gran Bretagna f. ~-'grandchildren npl pronipoti mpl. ~-'grandfather n bisnonno m. ~-'grandmother n bisnonna f

great|ly /'greɪtlɪ/ adv enormemente. ~ness n grandezza f

great-'uncle n prozio m

Greece /griːs/ n Grecia f

greed /griːd/ n avidità f; (for food) ingordigia f

greed|ily /'griːdɪlɪ/ adv avidamente; (eat) con ingordigia. ~y a (-ier, -iest) avido; (for food) ingordo

Greek /griːk/ a & n greco, -a mf; (language) greco m

green /griːn/ a verde; (fig: inexperienced) immaturo ● n verde m; ~s pl verdura f; the G~s pl Pol i verdi. ~ belt n zona f verde intorno a una città. ~ card n Auto carta f verde

greenery /'griːnərɪ/ n verde m

green fingers npl have ~ ~ avere il police verde

greenfly n afide m

green: ~grocer n fruttivendolo, -a mf. ~house n serra f. ~house effect n effetto m serra. ~ light n fam verde m

greet /griːt/ vt salutare; (welcome) accogliere. ~ing n saluto m; (welcome) accoglienza f. ~ings card n biglietto m d'auguri

gregarious /grɪ'geərɪəs/ a gregario; (person) socievole

grenade /grɪ'neɪd/ n granata f

grew /gruː/ see **grow**

grey /greɪ/ a grigio; (hair) bianco ● n grigio m. ~hound n levriero m

grid /grɪd/ n griglia f; (on map) reticolato m; Electr rete f

grief /griːf/ n dolore m; **come to ~**
⟨plans:⟩ naufragare

grievance /ˈgriːvəns/ n lamentela f

grieve /griːv/ vt addolorare ● vi essere addolorato

grill /grɪl/ n graticola f; ⟨for
grilling⟩ griglia f; **mixed ~**
griglia f mista ● vt/i cuocere
alla griglia; ⟨interrogate⟩ sottoporre al terzo grado

grille /grɪl/ n grata f

grim /grɪm/ a (**grimmer, grim-
mest**) arcigno; ⟨determination⟩
accanito

grimace /grɪˈmeɪs/ n smorfia f ● vi
fare una smorfia

grime /graɪm/ n sudiciume m

grimy /ˈgraɪmɪ/ a (**-ier, -iest**) sudicio

grin /grɪn/ n sorriso m ● vi (pt/pp
grinned) fare un gran sorriso

grind /graɪnd/ n (fam: hard work)
sfacchinata f ● vt (pt/pp **ground**)
macinare; affilare ⟨knife⟩; (Am:
mince) tritare; **~ one's teeth** digrignare i denti

grip /grɪp/ n presa f; fig controllo
m; ⟨bag⟩ borsone m; **get a ~ of
oneself** controllarsi ● vt (pt/pp
gripped) afferrare; ⟨tyres:⟩ far
presa su; tenere avvinto ⟨atten-
tion⟩

gripe /graɪp/ vi (fam: grumble) lagnarsi

gripping /ˈgrɪpɪŋ/ a avvincente

grisly /ˈgrɪzlɪ/ a (**-ier, -iest**) raccapricciante

gristle /ˈgrɪsl/ n cartilagine f

grit /grɪt/ n graniglia f; ⟨for roads⟩
sabbia f; ⟨courage⟩ coraggio m ● vt
(pt/pp **gritted**) spargere sabbia su
⟨road⟩; **~ one's teeth** serrare i
denti

grizzle /ˈgrɪzl/ vi piagnucolare

groan /grəʊn/ n gemito m ● vi ge-
mere

grocer /ˈgrəʊsə(r)/ n droghiere, -a
mf; **~'s** [**shop**] drogheria f. **~ies**
npl generi mpl alimentari

groggy /ˈgrɒgɪ/ a (**-ier, -iest**)
stordito; ⟨unsteady⟩ barcollante

groin /grɔɪn/ n Anat inguine m

groom /gruːm/ n sposo m; ⟨for
horse⟩ stalliere m ● vt strigliare
⟨horse⟩; fig preparare; **well-~ed**
ben curato

groove /gruːv/ n scanalatura f

grope /grəʊp/ vi brancolare; **~ for**
cercare a tastoni

gross /grəʊs/ a obeso; ⟨coarse⟩ volgare; ⟨glaring⟩ grossolano; ⟨sal-
ary, weight⟩ lordo ● n inv grossa f.
~ly adv (very) enormemente

grotesque /grəʊˈtesk/ a grottesco

grotto /ˈgrɒtəʊ/ n (pl **-es**) grotta f

grotty /ˈgrɒtɪ/ a (**-ier, -iest**) ⟨fam:
flat, street⟩ squallido

ground[1] /graʊnd/ see **grind**

ground[2] n terra f; Sport terreno m;
⟨reason⟩ ragione f; **~s** pl ⟨park⟩
giardini mpl; ⟨of coffee⟩ fondi mpl
● vi ⟨ship:⟩ arenarsi ● vt bloccare
a terra ⟨aircraft⟩; Am Electr mettere a terra

ground: **~ floor** n pianterreno m.
~ing n base f. **~less** a infondato.
~sheet n telone m impermeabile.
~work n lavoro m di preparazione

group /gruːp/ n gruppo m ● vt rag-
gruppare ● vi raggrupparsi

grouse[1] /graʊs/ n inv gallo m
cedrone

grouse[2] vi fam brontolare

grovel /ˈgrɒvl/ vi (pt/pp **grovelled**)
strisciare. **~ling** a leccapiedi inv

grow /grəʊ/ v (pt **grew**, pp **grown**)
● vi crescere; ⟨become⟩ diventare;
⟨unemployment, fear:⟩ aumentare;
⟨town⟩ ingrandirsi ● vt coltivare;
~ one's hair farsi crescere i capelli. **grow up** vi crescere; ⟨town:⟩
svilupparsi

growl /graʊl/ n grugnito m ● vi
ringhiare

grown /grəʊn/ see **grow** ● a adulto.
~-up a & n adulto, -a mf

growth /grəʊθ/ n crescita f;
⟨increase⟩ aumento m; Med tumore m

grub /grʌb/ n larva f; (fam: food) mangiare m

grubby /ˈgrʌbɪ/ a (-ier, -iest) sporco

grudg|e /grʌdʒ/ n rancore m; **bear sb a ~e** portare rancore a qcno ●vt dare a malincuore. **~ing** a reluttante. **~ingly** adv a malincuore

gruelling /ˈgruːəlɪŋ/ a estenuante

gruesome /ˈgruːsəm/ a macabro

gruff /grʌf/ a burbero

grumble /ˈgrʌmbl/ vi brontolare (**at** contro)

grumpy /ˈgrʌmpɪ/ a (-ier, -iest) scorbutico

grunt /grʌnt/ n grugnito m ●vi fare un grugnito

guarant|ee /gærənˈtiː/ n garanzia f ●vt garantire. **~or** n garante mf

guard /gɑːd/ n guardia f; (security) guardiano m; (on train) capotreno m; Techn schermo m protettivo; **be on ~** essere di guardia ●vt sorvegliare; (protect) proteggere. **guard against** vt guardarsi da. **~-dog** n cane m da guardia

guarded /ˈgɑːdɪd/ a guardingo

guardian /ˈgɑːdɪən/ n (of minor) tutore, -trice mf

guerrilla /gəˈrɪlə/ n guerrigliero, -a mf. **~ warfare** n guerriglia f

guess /ges/ n supposizione f ●vt indovinare ●vi indovinare; (Am: suppose) supporre. **~work** n supposizione f

guest /gest/ n ospite mf; (in hotel) cliente mf. **~-house** n pensione f

guffaw /gʌˈfɔː/ n sghignazzata f ●vi sghignazzare

guidance /ˈgaɪdəns/ n guida f; (advice) consigli mpl

guide /gaɪd/ n guida f; [**Girl**] **G~** giovane esploratrice f ●vt guidare. **~book** n guida f turistica

guided /ˈgaɪdɪd/ a **~ missile** missile m teleguidato; **~ tour** giro m guidato

guide: **~-dog** n cane m per ciechi. **~lines** npl direttive fpl

guild /gɪld/ n corporazione f

guile /gaɪl/ n astuzia f

guillotine /ˈgɪlətiːn/ n ghigliottina f; (for paper) taglierina f

guilt /gɪlt/ n colpa f. **~ily** adv con aria colpevole

guilty /ˈgɪltɪ/ a (-ier, -iest) colpevole; **have a ~ conscience** avere la coscienza sporca

guinea-pig /ˈgɪnɪ-/ n porcellino m d'India; (fig: used for experiments) cavia f

guise /gaɪz/ n **in the ~ of** sotto le spoglie di

guitar /gɪˈtɑː(r)/ n chitarra f. **~ist** n chitarrista mf

gulf /gʌlf/ n Geog golfo m; fig abisso m

gull /gʌl/ n gabbiano m

gullet /ˈgʌlɪt/ n esofago m; (throat) gola f

gullible /ˈgʌlɪbl/ a credulone

gully /ˈgʌlɪ/ n burrone m; (drain) canale m di scolo

gulp /gʌlp/ n azione f di deglutire; (of food) boccone m; (of liquid) sorso m ●vt deglutire. **gulp down** vt tranguigiare (food); scolarsi (liquid)

gum¹ /gʌm/ n Anat gengiva f

gum² n gomma f; (chewing-gum) gomma f da masticare, chewing-gum m inv ●vt (pt/pp **gummed**) ingommare (**to** a)

gummed /gʌmd/ see **gum²** ●a (label) adesivo

gumption /ˈgʌmpʃn/ n fam buon senso m

gun /gʌn/ n pistola f; (rifle) fucile m; (cannon) cannone m ●**gun down** vt (pt/pp **gunned**) freddare

gun: **~fire** n spari mpl; (of cannon) colpi mpl (di cannone). **~man** uomo m armato

gun: **~powder** n polvere f da sparo. **~shot** n colpo m [di pistola]

gurgle /ˈgɜːgl/ vi gorgogliare; (baby:) fare degli urletti

gush /gʌʃ/ vi sgorgare; (enthuse) parlare con troppo entusiasmo (**over** di). **gush out** vi sgorgare

~ing a eccessivamente entusiastico

gust /gʌst/ n (of wind) raffica f

gusto /'gʌstəʊ/ n with ~ con trasporto

gusty /'gʌstɪ/ a ventoso

gut /gʌt/ n intestino m; **~s** pl pancia f; (fam: courage) fegato m ● vt (pt/pp gutted) Culin svuotare delle interiora; **~ted by fire** sventrato da un incendio

gutter /'gʌtə(r)/ n canale m di scolo; (on roof) grondaia f; fig bassifondi mpl

guttural /'gʌtərəl/ a gutturale

guy /gaɪ/ n fam tipo m, tizio m

guzzle /'gʌzl/ vt ingozzarsi con ⟨food⟩; **he's ~d the lot** si è sbafato tutto

gym /dʒɪm/ n fam palestra f; ⟨gymnastics⟩ ginnastica f

gymnasium /dʒɪm'neɪzɪəm/ n palestra f

gymnast /'dʒɪmnæst/ n ginnasta mf. **~ics** /-'næstɪks/ n ginnastica f

gym: **~ shoes** npl scarpe fpl da ginnastica. **~-slip** n Sch grembiule m (da bambina)

gynaecolog|ist /gaɪnɪˈkɒlədʒɪst/ n ginecologo, a mf. **~y** n ginecologia f

gypsy /'dʒɪpsɪ/ n zingaro, a m

gyrate /dʒaɪ'reɪt/ vi roteare

Hh

haberdashery /hæbəˈdæʃərɪ/ n merceria f; Am negozio m d'abbigliamento da uomo

habit /'hæbɪt/ n abitudine f; (Relig: costume) tonaca f; **be in the ~ of doing sth** avere l'abitudine di fare qcsa

habitable /'hæbɪtəbl/ a abitabile

habitat /'hæbɪtæt/ n habitat m inv

habitation /hæbɪˈteɪʃn/ n **unfit for human ~** inagibile

habitual /həˈbɪtjʊəl/ a abituale; ⟨smoker, liar⟩ inveterato. **~ly** adv regolarmente

hack[1] /hæk/ n (writer) scribacchino, a mf

hack[2] vt tagliare; **~ to pieces** tagliare a pezzi

hackneyed /'hæknɪd/ a trito [e ritrito]

'hacksaw n seghetto m

had /hæd/ see **have**

haddock /'hædək/ n inv eglefino m

haemorrhage /'hemərɪdʒ/ n emorragia f

haemorrhoids /'hemərɔɪdz/ npl emorroidi fpl

hag /hæg/ n **old ~** vecchia befana f

haggard /'hægəd/ a sfatto

haggle /'hægl/ vi contrattare (**over** per)

hail[1] /heɪl/ vt salutare; far segno a ⟨taxi⟩ ● vi **~ from** provenire da

hail[2] n grandine f ● vi grandinare. **~stone** n chicco m di grandine. **~storm** n grandinata f

hair /heə(r)/ n capelli mpl; (on body, of animal) pelo m

hair: **~brush** n spazzola f per capelli. **~cut** n taglio m di capelli; **have a ~cut** farsi tagliare i capelli. **~-do** n fam pettinatura f. **~dresser** n parrucchiere, a mf. **~dryer** n fon m inv; (with hood) casco m [asciugacapelli]. **~-grip** n molletta f. **~pin** n forcina f. **~pin 'bend** n tornante m, curva f a gomito. **~-raising** a terrificante. **~style** n acconciatura f

hairy /'heərɪ/ a (-ier, -iest) peloso; (fam: frightening) spaventoso

hale /heɪl/ a **~ and hearty** in piena forma

half /hɑːf/ n (pl **halves**) metà f inv; **cut in ~** tagliare a metà; **one and a ~** uno e mezzo; **~ a dozen** mezza dozzina; **~ an hour** mezz'ora ● a mezzo; **[at] ~ price** [a] metà prezzo ● adv a metà; **~ past two** le due e mezza

half: **~ board** n mezza pensione f. **~-'hearted** a esitante. **~-'hourly**

& *adv* ogni mezz'ora. ~ '**mast** *n* at ~ **mast** a mezz'asta. ~ **measures** *npl* mezze misure *fpl*. ~**open** *a* socchiuso; **say** ~ to salutare a metà trimestre. ~**'term** *n* vacanza *f* di metà trimestre. ~**'time** *n* Sport intervallo *m*. ~ '**way** *a* the ~**way mark/stage** il livello principale ● *adv* a metà strada; **get** ~**way** *fig* arrivare a metà. ~**wit** *n* idiota *mf*

hall /hɔ:l/ *n* (*entrance*) ingresso *m*; (*room*) sala *f*; (*mansion*) residenza *f* di campagna; ~ **of residence** *Univ* casa *f* dello studente

'**hallmark** *n* marchio *m* di garanzia; *fig* marchio *m*

hallo /hə'ləʊ/ *int* ciao!; (*on telephone*) pronto!; **say** ~ to salutare

Hallowe'en /hæləʊ'i:n/ *n* vigilia *f* d'Ognissanti *e* notte delle streghe, *celebrata soprattutto dai bambini*

hallucination /həluːsɪ'neɪʃn/ *n* allucinazione *f*

halo /'heɪləʊ/ *n* (*pl* -**es**) aureola *f*; *Astr* alone *m*

halt /hɔ:lt/ *n* alt *m* *inv*; **come to a** ~ fermarsi; (*traffic*) bloccarsi ● *vi* fermarsi; ~**!** alt! ● *vt* fermare. ~**ing** *a* esitante

halve /hɑ:v/ *vt* dividere a metà; (*reduce*) dimezzare

ham /hæm/ *n* prosciutto *m*; *Theat* attore, -trice *mf* da strapazzo

hamburger /'hæmbɜ:gə(r)/ *n* hamburger *m* *inv*

hamlet /'hæmlɪt/ *n* paesino *m*

hammer /'hæmə(r)/ *n* martello *m* ● *vt* martellare ● *vi* ~ **at/on** picchiare a

hammock /'hæmək/ *n* amaca *f*

hamper[1] /'hæmpə(r)/ *n* cesto *m*; (*gift*) cestino *m*

hamper[2] *vt* ostacolare

hamster /'hæmstə(r)/ *n* criceto *m*

hand /hænd/ *n* mano *f*; (*of clock*) lancetta *f*; (*writing*) scrittura *f*; (*worker*) manovale *m*; **at** ~, **to** ~ a portata di mano; **on the one** ~ da un lato; **on the other** ~ d'altra parte; **out of** ~ incontrollabile; (*summarily*) su due piedi; **give sb a** ~ dare una mano a qcno ● *vt*

porgere. **hand down** *vt* tramandare. **hand in** *vt* consegnare. **hand out** *vt* distribuire. **hand over** *vt* passare; (*to police*) consegnare

hand: ~**bag** *n* borsa *f* (*da signora*). ~**book** *n* manuale *m*. ~**brake** *n* freno *m* a mano. ~**cuffs** *npl* manette *fpl*. ~**ful** *n* manciata *f*; **be [quite] a** ~**ful** *fam* essere difficile da tenere a freno

handicap /'hændɪkæp/ *n* handicap *m* *inv*. ~**ped** *a* **mentally/physically** ~**ped** mentalmente/fisicamente handicappato

handicraft /'hændɪkrɑːft/ *n* artigianato *m*. ~**work** *n* opera *f*

handkerchief /'hæŋkətʃɪf/ *n* (*pl* ~**s** & -**chieves**) fazzoletto *m*

handle /'hændl/ *n* manico *m*; (*of door*) maniglia *f*; **fly off the** ~ *fam* perdere le staffe ● *vt* maneggiare; occuparsi di (*problem, customer*); prendere (*difficult person*); trattare (*subject*). ~**bars** *npl* manubrio *m*

hand: ~**luggage** *n* bagaglio *m* a mano. ~**made** *a* fatto a mano. ~**out** *n* (*at lecture*) foglio *m* informativo; (*fam: money*) elemosina *f*. ~**rail** *n* corrimano *m*. ~**shake** *n* stretta *f* di mano

handsome /'hænsəm/ *a* bello; (*fig: generous*) generoso

hand: ~**stand** *n* verticale *f*. ~**writing** *n* calligrafia *f*. ~**written** *a* scritto a mano

handy /'hændɪ/ *a* (-**ier**, -**iest**) utile; (*person*) abile; **have/keep** ~ avere/tenere a portata di mano. ~**man** *n* tuttofare *m* *inv*

hang /hæŋ/ *vt* (*pt/pp* hung) appendere (*picture*); (*pt/pp* hanged) impiccare (*criminal*); ~ **oneself** impiccarsi ● *vi* (*pt/pp* hung) pendere; (*hair:*) scendere ● *n* **get the** ~ **of it** *fam* afferrare. **hang about** *vi* gironzolare. **hang on** *vi* tenersi stretto; (*fam: wait*) aspettare; *Teleph* restare in linea. **hang on to** *vt* tenersi stretto a; (*keep*) tenere. **hang out** *vi* spuntare; **where**

does he usually ~ out? *fam* dove bazzica di solito? ● *vt* stendere ⟨*washing*⟩. **hang up** *vt* appendere; *Teleph* riattaccare ● *vi* essere appeso; *Teleph* riattaccare

hangar /'hæŋə(r)/ *n* *Aeron* hangar *m inv*

hanger /'hæŋə(r)/ *n* gruccia *f*. **~-on** *n* leccapiedi *mf*

hang: **~-glider** *n* deltaplano *m*. **~-gliding** *n* deltaplano *m*. **~man** *n* boia *m*. **~over** *n* postumi *mpl* della sbornia. **~-up** *n* *fam* complesso *m*

hanker /'hæŋkə(r)/ *vi* ~ **after sth** smaniare per qcsa

hanky /'hæŋkɪ/ *n* fazzoletto *m*

hanky-panky /hæŋkɪ'pæŋkɪ/ *n* *fam* qualcosa *m* di losco

haphazard /hæp'hæzəd/ *a* casaccio

happen /'hæpn/ *vi* capitare, succedere; **as it ~s** per caso; **I ~ed to meet him** mi è capitato di incontrarlo; **what has ~ed to him?** cosa gli è capitato?; ⟨*become of*⟩ che fine ha fatto? **~ing** *n* avvenimento *m*

happily /'hæpɪlɪ/ *adv* felicemente; ⟨*fortunately*⟩ fortunatamente. **~ness** *n* felicità *f*

happy /'hæpɪ/ *a* (**-ier, -iest**) contento, felice. **~-go-'lucky** *a* spensierato

harass /'hærəs/ *vt* perseguitare. **~ed** *a* stressato. **~ment** *n* persecuzione *f*; **sexual ~ment** molestie *fpl* sessuali

harbour /'hɑːbə(r)/ *n* porto *m* ● *vt* dare asilo a; nutrire ⟨*grudge*⟩

hard /hɑːd/ *a* duro; ⟨*question, problem*⟩ difficile; **~ of hearing** duro d'orecchi; **be ~ on sb** ⟨*person*⟩ essere duro con qcno ● *adv* ⟨*work*⟩ duramente; ⟨*pull, hit, rain, snow*⟩ forte; **~ hit by unemployment** duramente colpito dalla disoccupazione; **take sth ~** non accettare qcsa; **think ~!** pensaci bene!; **try ~** mettercela tutta; **try ~er** metterci più impe-

gno; **~ done by** *fam* trattato ingiustamente

hard: **~back** *n* edizione *f* rilegata. **~-boiled** *a* ⟨*egg*⟩ sodo. **~ copy** *n* copia *f* stampata. **~ disk** *n* hard disk *m inv*, disco *m* rigido

harden /'hɑːdn/ *vi* indurirsi

hard: **~-'headed** *a* ⟨*businessman*⟩ dal sangue freddo. **~-'hearted** *a* dal cuore duro. **~ line** *n* linea *f* dura; **~ lines!** ció sfortuna!. **~line** *a* duro. **~liner** *n* fautore, -trice *mf* della linea dura. **~ luck** *n* sfortuna *f*

hardly /'hɑːdlɪ/ *adv* appena; **~ ever** quasi mai. **~ness** *n* durezza *f*. **~ship** *n* avversità *f inv*

hard: **~ 'shoulder** *n* *Auto* corsia *f* d'emergenza. **~ up** *a fam* a corto di soldi; **~ up for sth** a corto di qcsa. **~ware** *n* ferramenta *fpl*; *Comput* hardware *m inv*. **~-'wearing** *a* resistente. **~-'working** *a* **be ~working** essere un gran lavoratore

hardy /'hɑːdɪ/ *a* (**-ier, -iest**) dal fisico resistente; ⟨*plant*⟩ che sopporta il gelo

hare /heə(r)/ *n* lepre *f*. **~-brained** *a fam* ⟨*scheme*⟩ da scervellati

hark /hɑːk/ *vi* **~ back to** *fig* ritornare su

harm /hɑːm/ *n* male *m*; ⟨*damage*⟩ danni *mpl*; **out of ~'s way** in un posto sicuro; **it won't do any ~** non farà certo male ● *vt* far male a; ⟨*damage*⟩ danneggiare. **~ful** *a* dannoso. **~less** *a* innocuo

harmonica /hɑː'mɒnɪkə/ *n* armonica *f* [a bocca]

harmonious /hɑː'məʊnɪəs/ *a* armonioso. **~ly** *adv* in armonia

harmonize /'hɑːmənaɪz/ *vi* *fig* armonizzare. **~y** *n* armonia *f*

harness /'hɑːnɪs/ *n* finimenti *mpl*; ⟨*of parachute*⟩ imbracatura *f* ● *vt* bardare ⟨*horse*⟩; sfruttare ⟨*resources*⟩

harp /hɑːp/ *n* arpa *f* ● **harp on** *vi* *fam* insistere (**about** su). **~ist** *n* arpista *mf*

harpoon /ˈhɑːˈpuːn/ n arpione m

harpsichord /ˈhɑːpsɪkɔːd/ n clavicembalo m

harrowing /ˈhærəʊŋ/ a straziante

harsh /hɑːʃ/ a duro; (light) abbagliante. **~ness** n durezza f

harvest /ˈhɑːvɪst/ n raccolta f; (of grapes) vendemmia f; (crop) raccolto m ● vt raccogliere

has /hæz/ see have

hash /hæʃ/ n make a ~ of fam fare un casino con

hashish /ˈhæʃɪʃ/ n hascish m

hassle /ˈhæsl/ n fam rottura f ● vt rompere le scatole a

haste /heɪst/ n fretta f

hasty /ˈheɪstɪ/ a (-ier, -iest) frettoloso; (decision) affrettato. **~ily** adv frettolosamente

hat /hæt/ n cappello m

hatch¹ /hætʃ/ n (for food) sportello m passavivande m; Naut boccaporto m

hatch² vi ~[out] rompere il guscio; (egg:) schiudersi ● vt covare; tramare (plot)

'hatchback n tre/cinque porte m inv; (door) porta f del bagagliaio

hatchet /ˈhætʃɪt/ n ascia f

hate /heɪt/ n odio m ● vt odiare. **~ful** a odioso

hatred /ˈheɪtrɪd/ n odio m

haughty /ˈhɔːtɪ/ a (-ier, -iest) altezzoso. **~ily** adv altezzosamente

haul /hɔːl/ n (fish) pescata f; (loot) bottino m; (pull) tirata f ● vt tirare; trasportare (goods) ● vi ~ on tirare. **~age** /-ɪdʒ/ n trasporto m. **~ier** /-ɪə(r)/ n autotrasportatore m

haunt /hɔːnt/ n ritrovo m ● vt frequentare; (linger in the mind) perseguitare; **this house is ~ed** questa casa è abitata dai fantasmi

have /hæv/ vt (3 sg pres tense has; pt/pp had) avere; fare (breakfast, bath, walk etc); **~ a drink** bere qualcosa; **~ lunch/dinner** pranzare/cenare; **~ a rest** riposarsi; **I had my hair cut** mi sono tagliata i capelli; **we had the house painted** abbiamo fatto tinteggiare la casa; **I had it made** l'ho fatto fare; **~ to do sth** dover fare qcsa; **~ him telephone me tomorrow** digli di telefonarmi domani; **he has** or **he's got two houses** ha due case; **you've got the money, ~n't you?** hai i soldi, no? ● v aux avere; (with verbs of motion & some others) essere; **I ~ seen him** l'ho visto; **he has never been there** non ci è mai stato. **have on** vt (be wearing) portare; (dupe) prendere in giro; **I've got something on tonight** ho un impegno stasera. **have out** vt ~ **it out with sb** chiarire le cose con qcno ● npl the ~s **and the ~-nots** i ricchi e i poveri

haven /ˈheɪvn/ n fig rifugio m

haversack /ˈhævə-/ n zaino m

havoc /ˈhævək/ n strage f; **play ~ with** fig scombussolare

haw /hɔː/ see **hum**

hawk /hɔːk/ n falco m

hay /heɪ/ n fieno m. ~ **fever** n raffreddore m da fieno. **~stack** n pagliaio m

'haywire a fam **go ~** dare i numeri; (plans:) andare all'aria

hazard /ˈhæzəd/ n (risk) rischio m ● vt rischiare; ~ **a guess** azzardare un'ipotesi. **~ous** /-əs/ a rischioso. **~ [warning] lights** npl Auto luci fpl d'emergenza

haze /heɪz/ n foschia f

hazel /ˈheɪz(ə)l/ n nocciolo m; (colour) [color m] nocciola m. **~nut** n nocciola f

hazy /ˈheɪzɪ/ a (-ier, -iest) nebbioso; (fig: person) confuso; (memories) vago

he /hiː/ pron lui; **he's tired** è stanco; **I'm going but he's not** io vengo, ma lui no

head /hed/ n testa f; (of firm) capo m; (of primary school) direttore, -trice mf; (of secondary school) preside mf; (on beer) schiuma f; **be off one's ~** essere fuori di testa; **have a good ~ for business** avere il senso degli affari; **have a good ~**

for heights non soffrire di vertigini; **10 pounds a** ~ sterline a testa; **20 ~ of cattle** 20 capi di bestiame; ~ **first** a capofitto; ~ **over heels in love** innamorato pazzo; **~s or tails?** testa o croce? ● *vt* essere a capo di; essere in testa a ⟨*list*⟩; colpire di testa ⟨*ball*⟩ ● *vi* ~ **for** dirigersi verso.

head: **~ache** n mal *m* di testa. **~dress** n acconciatura *f*. ~er /'heda(r)/ n rinvio *m* di testa; ⟨*dive*⟩ tuffo *m* di testa. **~hunter** n cacciatore, -trice *mf* di teste. **~ing** n (*in list etc*) titolo *m*. **~lamp** n Auto fanale *m*. **~land** n promontorio *m*. **~light** n Auto fanale *m*. **~line** n titolo *m*. **~long** a & adv a capofitto. **~master** n (*of primary school*) direttore *m*; (*of secondary school*) preside *m*. **~mistress** n (*of primary school*) direttrice *f*; (*of secondary school*) preside *f*. **~ office** n sede *f* centrale. **~-on** a ⟨*collision*⟩ frontale ● adv frontalmente. **~phones** npl cuffie *fpl*. **~quarters** npl sede *fsg*; Mil quartier *m* generale *msg*. **~-rest** n poggiatesta *m inv*. **~room** n sottotetto *m*; (*of bridge*) altezza *f* libera di passaggio. **~scarf** n foulard *m inv*, fazzoletto *m*. **~strong** a testardo. ~ **'waiter** n capocameriere *m*. **~way** n progresso *m*. **~wind** n vento *m* di prua

heady /'hedɪ/ a che dà alla testa
heal /hiːl/ vt/i guarire
health /helθ/ n salute *f*

health: ~ **farm** n centro *m* di rimessa in forma. **~ foods** npl alimenti *mpl* macrobiotici. **~-food shop** n negozio *m* di macrobiotica. **~ insurance** n assicurazione *f* contro malattie
health|y /'helθɪ/ a (**-ier, -iest**) sano. **~ily** adv in modo sano
heap /hiːp/ n mucchio *m*; **~s of** *fam* un sacco di ● vt ~ [**up**] ammucchiare; **~ed teaspoon** un cucchiaino abbondante

hear /hɪə(r)/ vt/i (*pt/pp* **heard**) sentire; ~, ~! bravo! ~ **from** *vi* aver notizie di. **hear of** *vi* sentir parlare di; **he would not ~ of it** non ne ha voluto sentir parlare
hearing /'hɪərɪŋ/ n udito *m*; Jur udienza *f*. **~-aid** n apparecchio *m* acustico
'hearsay n **from ~** per sentito dire
hearse /hɜːs/ n carro *m* funebre
heart /hɑːt/ n cuore *m*; **~s** pl (*in cards*) cuori *mpl*; **by ~** a memoria
heart: **~ache** n pena *f*. ~ **attack** n infarto *m*. **~beat** n battito *m* cardiaco. **~-break** n afflizione *f*. **~-breaking** a straziante. **~-broken** a **be ~-broken** avere il cuore spezzato. **~-burn** n mal *m* di stomaco. **~en** vt rincuorare. **~felt** a di cuore
hearth /hɑːθ/ n focolare *m*.
heart|ily /'hɑːtɪlɪ/ adv di cuore; ⟨*eat*⟩ con appetito; **be ~ily sick of sth** non poterne più di qcsa. **~less** a spietato. **~-searching** n esame *m* di coscienza. **~-to-** ~ n conversazione a cuore aperto ● a a cuore aperto. **~y** a caloroso; ⟨*meal*⟩ copioso; ⟨*person*⟩ gioviale
heat /hiːt/ n calore *m*; Sport prova *f* eliminatoria ● vt scaldare ● vi scaldarsi. **~ed** a ⟨*swimming pool*⟩ riscaldato; ⟨*discussion*⟩ animato. **~er** n (*for room*) stufa *f*; (*for water*) boiler *m inv*; Auto riscaldamento *m*
heath /hiːθ/ n brughiera *f*
heathen /'hiːðn/ a & n pagano, -a *mf*
heather /'heðə(r)/ n erica *f*
heating /'hiːtɪŋ/ n riscaldamento *m*
heat: **~-stroke** n colpo *m* di sole. ~ **wave** n ondata *f* di calore
heave /hiːv/ vt tirare; (*lift*) tirare su; (*fam: throw*) gettare; emettere ⟨*sigh*⟩ ● vi tirare
heaven /'hev(ə)n/ n paradiso *m*; ~ **help you if...** Dio ti scampi se...; **H~s!** santo cielo!. **~ly** a celeste; *fam* delizioso

heavy/'hevɪ/ a (-ier, -iest) pesante; ⟨traffic⟩ intenso; ⟨rain, cold⟩ forte; **be a ~y smoker/drinker** essere un gran fumatore/bevitore. **~ily** adv pesantemente; ⟨smoke, drink etc⟩ molto. **~yweight** n peso m massimo

Hebrew /'hi:bru:/ a ebreo

heckle /'hekl/ vt interrompere di continuo. **~r** n disturbatore, -trice mf

hectic /'hektɪk/ a frenetico

hedge /hedʒ/ n siepe f ● vi fig essere evasivo. **~hog** n riccio m

heed /hi:d/ n pay **~ to** prestare ascolto a ● vt prestare ascolto a. **~less** a noncurante

heel[1] /hi:l/ n tallone m; ⟨of shoe⟩ tacco m; **take to one's ~s** fam darsela a gambe

heel[2] vi **~ over** Naut inclinarsi

hefty /'heftɪ/ a (-ier, -iest) massiccio

heifer /'hefə(r)/ n giovenca f

height /haɪt/ n altezza f; ⟨of plane⟩ altitudine f; ⟨of season, fame⟩ culmine m. **~en** vt fig accrescere

heir /eə(r)/ n erede mf. **~ess** n ereditiera f. **~loom** n cimelio m di famiglia

held /held/ see **hold**[2]

helicopter /'helɪkɒptə(r)/ n elicottero m

hell /hel/ n inferno m; **go to ~!** sl va' al diavolo! ● int porca miseria!

hello /hə'ləʊ/ int & n = **hallo**

helm /helm/ n timone m; **at the ~** fig al timone

helmet /'helmɪt/ n casco m

help /help/ n aiuto m; ⟨employee⟩ aiuto m domestico; **that's no ~** non è d'aiuto ● vt aiutare; **~ oneself to sth** servirsi di qcsa; **~ yourself** ⟨at table⟩ servìti pure; **I could not ~ laughing** non ho potuto trattenermi dal ridere; **it cannot be ~ed** non c'è niente da fare; **I can't ~ it** non ci posso far niente ● vi aiutare

helper /'helpə(r)/ n aiutante mf. **~ful** a ⟨person⟩ di aiuto; ⟨advice⟩

utile. **~ing** n porzione f. **~less** a ⟨unable to manage⟩ incapace; ⟨powerless⟩ impotente

helter-skelter /heltə'skeltə(r)/ adv in fretta e furia ● n scivolo m a spirale nei luna park

hem /hem/ n orlo m ● vt ⟨pt/pp hemmed⟩ orlare. **hem in** vt intrappolare

hemisphere /'hemɪ-/ n emisfero m

hemp /hemp/ n canapa f

hen /hen/ n gallina f; ⟨any female bird⟩ femmina f

hence /hens/ adv ⟨for this reason⟩ quindi. **~forth** adv d'ora innanzi

henchman /'hentʃmən/ n pej tirapiedi m

hen: **~-party** n fam festa f di addio al celibato per sole donne. **~pecked** a tiranneggiato dalla moglie

her /hɜ:(r)/ poss a il suo m, la sua f, i suoi mpl, le sue fpl; **~ mother/father** sua madre/suo padre ● pers pron ⟨direct object⟩ la; ⟨indirect object⟩ le; ⟨after prep⟩ lei; **I know ~** la conosco; **give ~ the money** dalle i soldi; **give it to ~** daglielo; **I came with ~** sono venuto con lei; **it's ~** è lei; **I've seen ~**, **but not him** ho visto lei, ma non lui

herald /'herəld/ vt annunciare

herb /hɜ:b/ n erba f

herbal /'hɜ:b(ə)l/ a alle erbe; **~ tea** tisana f

herbs /hɜ:bz/ npl ⟨for cooking⟩ aromi mpl ⟨da cucina⟩; ⟨medicinal⟩ erbe fpl

herd /hɜ:d/ n gregge m ● vt ⟨tend⟩ sorvegliare; ⟨drive⟩ far muovere; fig ammassare

here /hɪə(r)/ adv qui, qua; **in ~** qui dentro; **come/bring ~** vieni/porta qui; **~ is...**, **~ are...** ecco...; **~ you are!** ecco qua!. **~'after** adv in futuro. **~'by** adv con la presente

hereditary /hə'redɪtərɪ/ a ereditario. **~y** n eredità f

heresy /'herəsɪ/ n eresia f. **~tic** n eretico, -a mf

here'with *adv Comm* con la presente

heritage /'herɪtɪdʒ/ *n* eredità *f*

hermetic /hɜ:'metɪk/ *a* ermetico. **~ally** *adv* ermeticamente

hermit /'hɜ:mɪt/ *n* eremita *mf*

hernia /'hɜ:nɪə/ *n* ernia *f*

hero /'hɪərəʊ/ *n* (*pl* **-es**) eroe *m*

heroic /hɪ'rəʊɪk/ *a* eroico

heroin /'herəʊɪn/ *n* eroina *f* (*droga*)

heroine /'herəʊɪn/ *n* eroina *f*. **~ism** *n* eroismo *m*

heron /'herən/ *n* airone *m*

herring /'herɪŋ/ *n* aringa *f*

hers /hɜ:z/ *poss pron* il suo *m*, la sua *f*, i suoi *mpl*, le sue *fpl*; **a friend of ~** un suo amico; **friends of ~** dei suoi amici; **that is ~** quello è suo; (*as opposed to mine*) quello è il suo

her'self *pers pron* (*reflexive*) si; (*emphatic*) lei stessa; (*after prep*) sé, se stessa; **she poured ~ a drink** si è versata da bere; **she told me so ~** me lo ha detto lei stessa; **she's proud of ~** è fiera di sé; **by ~** da sola

hesitant /'hezɪtənt/ *a* esitante. **~ly** *adv* con esitazione

hesitate /'hezɪteɪt/ *vi* esitare. **~ion** /-'teɪʃn/ *n* esitazione *f*

het /het/ *a* **~ up** *fam* agitato

hetero'sexual /hetərəʊ-/ *a* eterosessuale

hexagon /'heksəgən/ *n* esagono *m*. **~al** /hek'sægənl/ *a* esagonale

hey /heɪ/ *int* ehi

heyday /'heɪ-/ *n* tempi *mpl* d'oro

hi /haɪ/ *int* ciao!

hiatus /haɪ'eɪtəs/ *n* (*pl* **-tuses**) iato *m*

hibernate /'haɪbəneɪt/ *vi* andare in letargo. **~ion** /-'neɪʃn/ *n* letargo *m*

hiccup /'hɪkʌp/ *n* singhiozzo *m*; (*fam: hitch*) intoppo *m* • *vi* fare un singhiozzo

hid /hɪd/, **hidden** /'hɪdn/ *see* **hide²**

hide¹ /haɪd/ *n* (*leather*) pelle *f* (*di animale*)

hide² *vt* (*pt* **hid**, *pp* **hidden**)

nascondere • *vi* nascondersi. **~-and-'seek** *n* play **~-and-seek** giocare a nascondino

hideous /'hɪdɪəs/ *a* orribile

'hide-out *n* nascondiglio *m*

hiding¹ /'haɪdɪŋ/ *n* (*fam: beating*) bastonata *f*; (*defeat*) batosta *f*

hiding² *n* **go into ~** sparire dalla circolazione

hierarchy /'haɪərɑ:kɪ/ *n* gerarchia *f*

hieroglyphics /haɪərə'glɪfɪks/ *npl* geroglifici *mpl*

hi-fi /'haɪfaɪ/ *n fam* stereo *m*, hi-fi *m inv* • *a fam* ad alta fedeltà

higgledy-piggledy /hɪgldɪ'pɪgldɪ/ *adv* alla rinfusa

high /haɪ/ *a* alto; (*meat*) che comincia ad andare a male; (*wind*) forte; (*on drugs*) fatto; **it's ~ time we did something about it** è ora di fare qualcosa in proposito • *adv* in alto; **~ and low** in lungo e in largo • *n* massimo *m*; (*temperature*) massima *f*; **be on a ~** *fam* essere fatto

high: ~brow *a & n* intellettuale *mf*. **~ chair** *n* seggiolone *m*. **~er education** *n* formazione *f* universitaria. **~-'handed** *a* dispotico. **~-'heeled** *a* coi tacchi alti. **~ heels** *npl* tacchi *mpl* alti. **~ jump** *n* salto *m* in alto

highlight /'haɪlaɪt/ *n fig* momento *m* clou; **~s** *pl* (*in hair*) mèche *fpl* • *vt* (*emphasize*) evidenziare. **~er** *n* (*marker*) evidenziatore *m*

highly /'haɪlɪ/ *adv* molto; **speak ~ of** lodare; **think ~ of** avere un'alta opinione di. **~'strung** *a* nervoso

Highness /'haɪnɪs/ *n* altezza *f*; **Your ~** Sua Altezza

high: ~-rise *a* (*building*) molto alto • *n* edificio *m* molto alto. **~ school** *n* ≈ scuola *f* superiore. **~ season** *n* alta stagione *f*. **~ street** *n* strada *f* principale. **~ tea** *n* pasto *m* pomeridiano servito insieme al tè. **~'tide** *n* alta marea *f*. **~way code** *n* codice *m* stradale

hijack /'haɪdʒæk/ *vt* dirottare • *n*

dirottamento m. **~er** n dirottatore, ·trice mf

hike /haɪk/ n escursione f a piedi ● vi fare un'escursione a piedi. **~r** n escursionista mf

hilarious /hɪ'leərɪəs/ a esilarante

hill /hɪl/ n collina f; (mound) collinetta f; (slope) altura f

hill: **~side** n pendio m. **~y** a collinoso

hilt /hɪlt/ n impugnatura f; **to the ~** (fam: support) fino in fondo; (mortgaged) fino al collo

him /hɪm/ pers pron (direct object) lo; (indirect object) gli; (with prep) lui; **I know ~** lo conosco; **give ~ the money** dagli i soldi; **give it to ~** daglielo; **I spoke to ~** gli ho parlato; **it's ~** è lui; **she loves ~** lo ama; **she loves ~, not you** ama lui, non te. **~'self** pers pron (reflexive) si; (emphatic) lui stesso; (after prep) sé, se stesso; **he poured ~ a drink** si è versato da bere; **he told me to ~self** me lo ha detto lui stesso; **he's proud of ~self** è fiero di sé; **by ~self** da solo

hind /haɪnd/ a posteriore

hinder /'hɪndə(r)/ vt intralciare. **~rance** /·rəns/ n intralcio m

hindsight /'haɪnd·/ n **with ~** con il senno del poi

Hindu /'hɪnduː/ n indù mf inv ● a indù. **~ism** n induismo m

hinge /hɪndʒ/ n cardine m ● vi **~ on** fig dipendere da

hint /hɪnt/ n (clue) accenno m; (advice) suggerimento m; (indirect suggestion) allusione f; (trace) tocco m ● vt **~ that...** far capire che... ● vi **~ at** alludere a

hip /hɪp/ n fianco m

hippie /'hɪpɪ/ n hippy mf inv

hippo /'hɪpəʊ/ n ippopotamo m

hip 'pocket n tasca f posteriore

hippopotamus /hɪpə'pɒtəməs/ n (pl -muses or -mi /-maɪ/) ippopotamo m

hire /haɪə(r)/ vt affittare; assumere ⟨person⟩; **~** [**out**] affittare ● n noleggio m; **'for ~'** 'affittasi'. **~**

car n macchina f a noleggio. **~ purchase** n acquisto m rateale

his /hɪz/ poss a il suo m, la sua f, i suoi mpl, le sue fpl; **~ mother/father** sua madre/suo padre ● poss pron il suo m, la sua f, i suoi mpl, le sue fpl; **a friend of ~** un suo amico; **friends of ~** dei suoi amici; **that is ~** questo è suo; (as opposed to mine) questo è il suo

hiss /hɪs/ n sibilo m; (of disapproval) fischio m ● vt fischiare ● vi sibilare; (in disapproval) fischiare

historian /hɪ'stɔːrɪən/ n storico, ·a mf

historic /hɪ'stɒrɪk/ a storico. **~al** a storico. **~ally** adv storicamente

history /'hɪstərɪ/ n storia f; **make ~** passare alla storia

hit /hɪt/ n (blow) colpo m; (fam: success) successo m; **score a direct ~** ⟨missile:⟩ colpire in pieno ● vt/i (pt/pp **hit**, pres p **hitting**) colpire; **~ one's head on the table** battere la testa contro il tavolo; **the car ~ the wall** la macchina ha sbattuto contro il muro; **~ the roof** fam perdere le staffe. **hit off** vt **~ it off** andare d'accordo. **hit on** vt fig trovare

hitch /hɪtʃ/ n intoppo m; **technical ~** problema m tecnico ● vt attaccare; **~ a lift** chiedere un passaggio. **hitch up** vt tirarsi su ⟨trousers⟩. **~hike** vi fare l'autostop. **~hiker** n autostoppista mf

hit-or-'miss a on a very **~ basis** all'improvvisata

hither /'hɪðə(r)/ adv **~ and thither** di qua e di là. **~to** adv finora

hive /haɪv/ n alveare m; **~ of industry** fucina f di lavoro ● **hive off** vt Comm separare

hoard /hɔːd/ n provvista f; (of money) gruzzolo m ● vt accumulare

hoarding /'hɔːdɪŋ/ n palizzata f; (with advertisements) tabellone m per manifesti pubblicitari

hoarse /hɔːs/ *a* rauco. **~ly** *adv* con voce rauca. **~ness** *n* raucedine *f*

hoax /həʊks/ *n* scherzo *m*; (*false alarm*) falso allarme *m*. **~er** *n* burlone. *-a mf*

hob /hɒb/ *n* piano *m* di cottura

hobble /ˈhɒbl/ *vi* zoppicare

hobby /ˈhɒbɪ/ *n* hobby *m inv*. **~-horse** *n* fig fissazione *f*

hockey /ˈhɒkɪ/ *n* hockey *m*

hoe /həʊ/ *n* zappa *f*

hog /hɒg/ *n* maiale *m* ● *vt* (*pt/pp* **hogged**) *fam* monopolizzare

hoist /hɔɪst/ *n* montacarichi *m inv*; (*fam: push*) spinta *f* in su ● *vt* sollevare; innalzare (*flag*); levare (*anchor*)

hold¹ /həʊld/ *n* Naut, Aeron stiva *f*

hold² *n* presa *f*; (*fig: influence*) ascendente *m*; **get ~ of** trovare; procurarsi (*information*) ● *v* (*pt/pp* **held**) ● *vt* tenere; (*container:*) contenere; essere titolare di (*licence, passport*); trattenere (*breath, suspect*); mantenere vivo (*interest*); (*civil servant etc.:*) occupare (*position*); (*retain*) mantenere; **~ sb's hand** tenere qcno per mano; **~ one's tongue** tenere la bocca chiusa; **~ sb responsible** considerare qcno responsabile; **~ that** (*believe*) ritenere ● *vi* tenere; (*weather, luck:*) durare; (*offer:*) essere valido; Teleph restare in linea; **I don't ~ with the idea that** fam non sono d'accordo sul fatto che. **hold back** *vt* rallentare ● *vi* esitare. **hold down** *vt* tenere a bada (*sb*). **hold on** *vi* (*wait*) aspettare; (*keep*) tenersi. **hold on to** *vt* aggrapparsi a; (*keep*) tenersi. **hold out** *vt* porgere (*hand*); fig offrire (*possibility*) ● *vi* (*resist*) resistere. **hold up** *vt* tenere su; (*delay*) rallentare; (*rob*) assalire; **~ one's head up** fig tenere la testa alta

'hold: ~all *n* borsone *m*. **~er** *n* titolare *mf*; (*of record*) detentore, -trice *mf*; (*container*) astuccio *m*. **~ing** *n* (*land*) terreno *m* in affitto;

Comm azioni *fpl*. **~up** *n* ritardo *m*; (*attack*) rapina *f* a mano armata

hole /həʊl/ *n* buco *m*

holiday /ˈhɒlɪdeɪ/ *n* vacanza *f*; (*public*) giorno *m* festivo; (*day off*) giorno *m* di ferie; **go on ~** andare in vacanza ● *vi* andare in vacanza. **~-maker** *n* vacanziere *m*

holiness /ˈhəʊlɪnɪs/ *n* santità *f*; **Your H~** Sua Santità

Holland /ˈhɒlənd/ *n* Olanda *f*

hollow /ˈhɒləʊ/ *a* cavo; (*promise*) a vuoto; (*voice*) assente; (*cheeks*) infossato ● *n* cavità *f inv*; (*in ground*) affossamento *m*

holly /ˈhɒlɪ/ *n* agrifoglio *m*

holocaust /ˈhɒləkɔːst/ *n* olocausto *m*

hologram /ˈhɒləgræm/ *n* ologramma *m*

holster /ˈhəʊlstə(r)/ *n* fondina *f*

holy /ˈhəʊlɪ/ *a* (**-ier, -est**) santo; (*water*) benedetto. **H~ Ghost** or **Spirit** *n* Spirito *m* Santo. **H~ Scriptures** *npl* sacre scritture *fpl*. **H~ Week** *n* settimana *f* santa

homage /ˈhɒmɪdʒ/ *n* omaggio *m*; **pay ~** to rendere omaggio a

home /həʊm/ *n* casa *f*; (*for children*) istituto *m*; (*for old people*) casa *f* di riposo; (*native land*) patria *f* ● *adv* **at ~** a casa; (*football*) in casa; **feel at ~** sentirsi a casa propria; **come/go ~** venire/andare a casa; **drive a nail ~** piantare un chiodo a fondo ● *a* domestico; (*movie, video*) casalingo; (*team*) ospitante; Pol nazionale

home: ~ ad'dress *n* indirizzo *m* di casa. **~ com'puter** *n* computer *m inv* da casa. **H~ Counties** *npl* contee *fpl* intorno a Londra. **~ game** *n* gioco *m* in casa. **~ help** *n* aiuto *m* domestico (*per persone non autosufficienti*). **~land** *n* patria *f*. **~less** *a* senza tetto

homely /ˈhəʊmlɪ/ *a* (**-ier, -iest**) semplice; (*atmosphere*) familiare; (*Am: ugly*) bruttino

home: ~-'made *a* fatto in casa. **H~ Office** *n Br* ministero *m* degli interni. **H~ 'Secretary** *n Br* ministro *m* degli interni. **~sick** *a* **be ~sick** avere nostalgia (**for** di). **~sickness** *n* nostalgia *f* di casa. ~'**town** *n* città *f* *inv* natia. **~ward** *a* di ritorno ● *adv* verso casa. **~work** *n Sch* compiti *mpl*

homicide /'hɒmɪsaɪd/ *n* (*crime*) omicidio *m*

homoeopath|ic /həʊmɪə'pæθɪk/ *a* omeopatico. **~y** /-'ɒpəθɪ/ *n* omeopatia *f*

homogeneous /hɒmə'dʒiːnɪəs/ *a* omogeneo

homo'sexual *a & n* omosessuale *mf*

honest /'ɒnɪst/ *a* onesto; (*frank*) sincero. **~ly** *adv* onestamente; (*frankly*) sinceramente; **~ly!** ma insomma! **~y** *n* onestà *f*; (*frankness*) sincerità *f*

honey /'hʌnɪ/ *n* miele *m*; (*fam: darling*) tesoro *m*

honey: **~comb** *n* favo *m*. **~moon** *n* luna *f* di miele. **~suckle** *n* caprifoglio *m*

honk /hɒŋk/ *vi* Aut clacsonare

honorary /'ɒnərərɪ/ *a* onorario

honour /'ɒnə(r)/ *n* onore *m* ● *vt* onorare. **~able** /-əbl/ *a* onorevole. **~ably** *adv* con onore. **~s degree** *n* ≈ diploma *m* di laurea

hood /hʊd/ *n* cappuccio *m*; (*of pram*) tettuccio *m*; (*over cooker*) cappa *f*; *Am Auto* cofano *m*

hoodlum /'huːdləm/ *n* teppista *m*

'hoodwink *vt fam* infinocchiare

hoof /huːf/ *n* (*pl* **~s** *or* **hooves**) zoccolo *m*

hook /hʊk/ *n* gancio *m*; (*for fishing*) amo *m*; **off the ~** *Teleph* staccato; *fig* fuori pericolo ● *vt* agganciare ● *vi* agganciarsi

hook|ed /hʊkt/ *a* (*nose*) adunco. **~ed on** (*fam: drugs*) dedito a; **be ~ed on skiing** essere un fanatico dello sci. **~er** *n Am sl* battona *f*

hookey /'hʊkɪ/ *n* **play ~** *Am fam* marinare la scuola

hooligan /'huːlɪgən/ *n* teppista *mf*. **~ism** *n* teppismo *m*

hoop /huːp/ *n* cerchio *m*

hooray /hʊ'reɪ/ *int & n* = **hurrah**

hoot /huːt/ *n* colpo *m* di clacson; (*of siren*) ululato *m*; (*of owl*) grido *m* ● *vi* gridare; (*car:*) clacsonare; (*siren:*) ululare; (*jeer*) fischiare. **~er** *n* (*of factory*) sirena *f*; *Auto* clacson *m inv*

hoover® /'huːvə(r)/ *n* aspirapolvere *m inv* ● *vt* passare l'aspirapolvere su (*carpet*); passare l'aspirapolvere in (*room*)

hop /hɒp/ *n* saltello *m* ● *vi* (*pt/pp* **hopped**) saltellare; **~ it!** *fam* tela! **hop in** *vi fam* saltar su

hope /həʊp/ *n* speranza *f* ● *vi* sperare (**for** in); **I ~ so**/**not** spero di sì/no ● *vt* **~ that** sperare che

hope|ful /'həʊpfl/ *a* pieno di speranza; (*promising*) promettente; **be ~ful that** avere buone speranze che. **~fully** *adv* con speranza; (*it is hoped*) se tutto va bene. **~less** *a* senza speranza; (*useless*) impossibile; (*incompetent*) incapace. **~lessly** *adv* disperatamente; (*inefficient, lost*) completamente. **~lessness** *n* disperazione *f*

horde /hɔːd/ *n* orda *f*

horizon /hə'raɪzn/ *n* orizzonte *m*

horizontal /hɒrɪ'zɒntl/ *a* orizzontale

hormone /'hɔːməʊn/ *n* ormone *m*

horn /hɔːn/ *n* corno *m*; *Auto* clacson *m inv*

horny /'hɔːnɪ/ *a* calloso; *fam* arrapato

horoscope /'hɒrəskəʊp/ *n* oroscopo *m*

horribl|e /'hɒrɪbl/ *a* orribile. **~y** *adv* spaventosamente

horrid /'hɒrɪd/ *a* orrendo

horrific /hə'rɪfɪk/ *a* raccapricciante; (*fam: accident, prices, story*) terrificante

horrify /'hɒrɪfaɪ/ *vt* (*pt/pp* **-led**) far inorridire; **I was horrified** ero sconvolto. **~ing** *a* terrificante

horror /'hɒrə(r)/ n orrore m. ~ **film** n film m dell'orrore

hors-d'œuvre /ɔː'dɜːvr/ n antipasto m

horse /hɔːs/ n cavallo m

horse: ~**back** n on ~**back** a cavallo. ~**man** n cavaliere m. ~**play** n gioco m pesante. ~**power** n cavallo m [vapore]. ~**racing** n corse fpl di cavalli. ~**shoe** n ferro m di cavallo

horti'cultural /hɔːtɪ-/ a di orticoltura

'**horticulture** n orticoltura f

hose /həʊz/ n (pipe) manichetta f ● **hose down** vt lavare con la manichetta

hospice /'hɒspɪs/ n (for the terminally ill) ospedale m per i malati in fase terminale

hospitabl|e /hɒ'spɪtəbl/ a ospitale. ~**y** adv con ospitalità

hospital /'hɒspɪtl/ n ospedale m

hospitality /hɒspɪ'tælətɪ/ n ospitalità f

host¹ /həʊst/ n a ~ **of** una moltitudine di

host² n ospite m

host³ n Relig ostia f

hostage /'hɒstɪdʒ/ n ostaggio m; **hold sb** ~ tenere qcno in ostaggio

hostel /'hɒstl/ n ostello m

hostess /'həʊstɪs/ n padrona f di casa; Aeron hostess f inv

hostile /'hɒstaɪl/ a ostile

hostilit|y /hɒ'stɪlətɪ/ n ostilità f; ~**ies** pl ostilità fpl

hot /hɒt/ a (hotter, hottest) caldo; (spicy) piccante; **I am** or **feel** ~ ho caldo; **it is** ~ fa caldo

'**hotbed** n fig focolaio m

hotchpotch /'hɒtʃpɒtʃ/ n miscuglio m

'**hot-dog** n hot dog m inv

hotel /həʊ'tel/ n albergo m. ~**ier** /-ɪə(r)/ n albergatore, -trice mf

hot: ~**head** n persona f impetuosa. ~**house** n serra f. ~**ly** adv fig accanitamente. ~**plate** n piastra f riscaldante ● **tap** n rubinetto m dell'acqua calda. ~'**tempered** a ira-

scibile. ~'**water bottle** n borsa f dell'acqua calda

hound /haʊnd/ n cane m da caccia ● vt fig perseguire

hour /'aʊə(r)/ n ora f. ~**ly** a di ogni ora; (pay, rate) a ora ● adv ogni ora

house¹ /haʊs/ n casa f; Pol camera f; Theat sala f; **at my** ~ a casa mia, da me

house² /haʊz/ vt alloggiare (person)

house /haʊs/: ~**boat** n casa f galleggiante. ~**breaking** n furto m con scasso. ~**hold** n casa f, famiglia f. ~**holder** n capo m di famiglia. ~**keeper** n governante f di casa. ~**keeping** n governo m della casa; (money) soldi mpl per le spese di casa. ~**plant** n pianta f da appartamento. ~**trained** a che non sporca in casa. ~**warming** [**party**] n festa f di inaugurazione della nuova casa. ~**wife** n casalinga f. ~**work** n lavoro m domestico

housing /'haʊzɪŋ/ n alloggio m. ~ **estate** n zona f residenziale

hovel /'hɒvl/ n tugurio m

hover /'hɒvə(r)/ vi librarsi; (linger) indugiare. ~**craft** n hovercraft m inv

how /haʊ/ adv come; ~ **are you?** come stai?; ~ **about a coffee/going on holiday?** che ne diresti di un caffè/di andare in vacanza?; ~ **do you do?** molto lieto!; ~ **old are you?** quanti anni hai?; ~ **long** quanto tempo; ~ **many** quanti; ~ **much** quanto; ~ **often** ogni quanto; **and** ~! eccome!; ~ **odd!** che strano!

how'ever adv (nevertheless) comunque; ~ **small** per quanto piccolo

howl /haʊl/ n ululato m ● vi ululare; (cry, with laughter) singhiozzare. ~**er** n fam strafalcione m

HP n abbr hire purchase; n abbr (horse power) C.V.

hub /hʌb/ n mozzo m; fig centro m

hubbub /'hʌbʌb/ n baccano m

'**hub-cap** n coprimozzo m

huddle /'hʌdl/ *vi* ~ **together** rannicchiarsi

hue[1] /hju:/ *n* colore *m*

hue[2] *n* ~ **and cry** clamore *m*

huff /hʌf/ *n* **be in/go into a** ~ fare il broncio

hug /hʌg/ *n* abbraccio *m* ● *vt* (*pt/pp* **hugged**) abbracciare; (*keep close to*) tenersi vicino a

huge /hju:dʒ/ *a* enorme

hulking /'hʌlkɪŋ/ *a* fam grosso

hull /hʌl/ *n* Naut scafo *m*

hullo /hə'ləʊ/ *int* = **hallo**

hum /hʌm/ *n* ronzio *m* ● *v* (*pt/pp* **hummed**) *vt* canticchiare ● *vi* (*motor:*) ronzare; *fig* fervere (*di attività*); ~ **and haw** esitare

human /'hju:mən/ *a* umano ● *n* essere *m* umano. ~ **'being** *n* essere *m* umano

humane /hju:'meɪn/ *a* umano

humanitarian /hju:mænɪ'teərɪən/ *a* & *n* umanitario, -a *mf*

humanit|y /hju:'mænətɪ/ *n* umanità *f*; **~ies** *pl* Univ dottrine *fpl* umanistiche

humb|le /'hʌmbl/ *a* umile ● *vt* umiliare

'humdrum *a* noioso

humid /'hju:mɪd/ *a* umido. **~ifier** /-'mɪdɪfaɪə(r)/ *n* umidificatore *m*. **~ity** /-'mɪdətɪ/ *n* umidità *f*

humiliat|e /hju:'mɪlɪeɪt/ *vt* umiliare. **~ion** /-'eɪʃn/ *n* umiliazione *f*

humility /hju:'mɪlətɪ/ *n* umiltà *f*

humorous /'hju:mərəs/ *a* umoristico. **~ly** *adv* con spirito

humour /'hju:mə(r)/ *n* umorismo *m*; (*mood*) umore *m*; **have a sense of** ~ avere il senso dell'umorismo ● *vt* compiacere

hump /hʌmp/ *n* protuberanza *f*; (*of camel, hunchback*) gobba *f*

hunch /hʌntʃ/ *n* (*idea*) intuizione *f*

'hunch|back *n* gobbo, -a *mf*. **~ed** *a* **~ed up** incurvato

hundred /'hʌndrəd/ *a* **one/a** ~ cento ● *n* cento *m*; **~s of** centinaia di. **~th** *a* centesimo ● *n* centesimo *m*. **~weight** *n* cinquanta chili *m*

hung /hʌŋ/ *see* **hang**

Hungarian /hʌŋ'geərɪən/ *n* & *a* ungherese *mf*; (*language*) ungherese *m*

Hungary /'hʌŋgərɪ/ *n* Ungheria *f*

hunger /'hʌŋgə(r)/ *n* fame *f*. **~-strike** *n* sciopero *m* della fame *m*

hungr|y /'hʌŋgrɪ/ *a* (**-ier, -iest**) affamato; **be ~y** aver fame. **~ily** *adv* con appetito

hunk /hʌŋk/ *n* [grosso] pezzo *m*

hunt /hʌnt/ *n* caccia *f* ● *vt* andare a caccia di (*animal*); dare la caccia a (*criminal*) ● *vi* andare a caccia; ~ **for** cercare. **~er** *n* cacciatore *m*. **~ing** *n* caccia *f*

hurdle /'hɜ:dl/ *n* Sport & *fig* ostacolo *m*. **~r** *n* ostacolista *mf*

hurl /hɜ:l/ *vt* scagliare

hurrah /hʊ'rɑ:/, **hurray** /hʊ'reɪ/ *int* urrà! ● *n* urrà *m*

hurricane /'hʌrɪkən/ *n* uragano *m*

hurried /'hʌrɪd/ *a* affrettato; (*job*) fatto in fretta. **~ly** *adv* in fretta

hurry /'hʌrɪ/ *n* fretta *f*; **be in a** ~ aver fretta ● *vi* (*pt/pp* **-ied**) affrettarsi. **hurry up** *vi* sbrigarsi ● *vt* fare sbrigare (*person*); accelerare (*things*)

hurt /hɜ:t/ *v* (*pt/pp* **hurt**) ● *vt* far male a; (*offend*) ferire ● *vi* far male; **my leg** ~**s** mi fa male la gamba. **~ful** *a* *fig* offensivo

hurtle /'hɜ:tl/ *vi* ~ **along** andare a tutta velocità

husband /'hʌzbənd/ *n* marito *m*

hush /hʌʃ/ *n* silenzio *m* ● **hush up** *vt* mettere a tacere. **~ed** *a* (*voice*) sommesso. **~-'hush** *a* fam segretissimo

husky /'hʌskɪ/ *a* (**-ier, -iest**) (*voice*) rauco

hustle /'hʌsl/ *vt* affrettare ● *n* attività *f* incessante; ~ **and bustle** trambusto *m*

hut /hʌt/ *n* capanna *f*

hybrid /'haɪbrɪd/ *a* ibrido ● *n* ibrido *m*

hydrant /'haɪdrənt/ *n* [**fire**] ~ idrante *m*

hydraulic /haɪ'drɔ:lɪk/ *a* idraulico

hydroe'lectric /haɪdrəʊ-/ a idroelettrico

hydrofoil /'haɪdrə-/ n aliscafo m

hydrogen /'haɪdrədʒən/ n idrogeno m

hyena /haɪ'i:nə/ n iena f

hygien|e /'haɪdʒi:n/ n igiene m. **~ic** /haɪ'dʒi:nɪk/ a igienico

hymn /hɪm/ n inno m. **~-book** n libro m dei canti

hypermarket /'haɪpəmɑ:kɪt/ n ipermercato m

hyphen /'haɪfn/ n lineetta f. **~ate** vt unire con lineetta

hypno|sis /hɪp'nəʊsɪs/ n ipnosi f. **~tic** /-'nɒtɪk/ a ipnotico

hypno|tism /'hɪpnətɪzm/ n ipnotismo m. **~tist** /-tɪst/ n ipnotizzatore, -trice mf. **~tize** vt ipnotizzare

hypochondriac /haɪpə'kɒndrɪæk/ a ipocondriaco ● n ipocondriaco, -a mf

hypocrisy /hɪ'pɒkrəsɪ/ n ipocrisia f

hypocrit|e /'hɪpəkrɪt/ n ipocrita mf. **~ical** /-'krɪtɪkl/ a ipocrita

hypodermic /haɪpə'dɜ:mɪk/ a & n **~ [syringe]** siringa f ipodermica

hypothe|sis /haɪ'pɒθəsɪs/ n ipotesi f inv. **~tical** /-ə'θetɪkl/ a ipotetico. **~tically** adv in teoria; ⟨speak⟩ per ipotesi

hyster|ia /hɪ'stɪərɪə/ n isterismo m. **~ical** /-'sterɪkl/ a isterico. **~ically** adv istericamente; **~ically funny** da morir dal ridere. **~ics** /hɪ'sterɪks/ npl attacco m isterico

....

Ii

I /aɪ/ pron io; **I'm tired** sono stanco; **he's going, but I'm not** lui va, ma io no

ice /aɪs/ n ghiaccio m ● vt glassare ⟨cake⟩. **ice over/up** vi ghiacciarsi

ice: **~ age** n era f glaciale. **~-axe** n piccozza f per il ghiaccio. **~berg** /-bɜ:g/ n iceberg m inv. **~box** n Am frigorifero m. **~'cream** n gelato m. **~'cream parlour** n gelateria f. **~-cube** n cubetto m di ghiaccio. **~ hockey** n hockey m su ghiaccio

Iceland /'aɪslənd/ n Islanda f. **~er** n islandese mf; **~ic** /-'lændɪk/ a & n islandese m

ice: **~'lolly** n ghiacciolo m. **~ rink** n pista f di pattinaggio. **~ skater** n pattinatore, -trice mf sul ghiaccio. **~ skating** n pattinaggio m su ghiaccio

icicle /'aɪsɪkl/ n ghiacciolo m

icing /'aɪsɪŋ/ n glassa f. **~ sugar** n zucchero m a velo

icon /'aɪkɒn/ n icona f

ic|y /'aɪsɪ/ a (-ier, -iest) ghiacciato; fig gelido. **~ily** adv gelidamente

idea /aɪ'dɪə/ n idea f; **I've no ~!** non ne ho idea!

ideal /aɪ'dɪəl/ a ideale ● n ideale m. **~ism** n idealismo m. **~ist** n idealista mf. **~istic** /-'lɪstɪk/ a idealistico. **~ize** vt idealizzare. **~ly** adv idealmente

identical /aɪ'dentɪkl/ a identico

identi|fication /aɪdentɪfɪ'keɪʃn/ n identificazione f; (proof of identity) documento m di riconoscimento. **~fy** /aɪ'dentɪfaɪ/ vt (pt/pp -ied) identificare

identikit® /aɪ'dentɪkɪt/ n identikit m inv

identity /aɪ'dentətɪ/ n identità f inv. **~ card** n carta f d'identità

ideolog|ical /aɪdɪə'lɒdʒɪkl/ a ideologico. **~y** /aɪdɪ'ɒlədʒɪ/ n ideologia f

idiom /'ɪdɪəm/ n idioma f. **~atic** /-'mætɪk/ a idiomatico

idiosyncrasy /ɪdɪə'sɪŋkrəsɪ/ n idiosincrasia f

idiot /'ɪdɪət/ n idiota mf. **~ic** /-'ɒtɪk/ a idiota

idl|e /'aɪd(ə)l/ a (lazy) pigro, ozioso; (empty) vano; ⟨machine⟩ fermo ● vi oziare; ⟨engine:⟩ girare a vuoto. **~eness** n ozio m. **~y** adv oziosamente

idol /'aɪdl/ n idolo m. **~ize** /'aɪdəlaɪz/ vt idolatrare

idyllic /ɪ'dɪlɪk/ a idillico

i.e. abbr (**id est**) cioè

if /ɪf/ conj se; **as if** come se

ignite /ɪg'naɪt/ vt dar fuoco a ● vi prender fuoco

ignition /ɪg'nɪʃn/ n Auto accensione f. **~ key** n chiave f d'accensione

ignoramus /ɪgnə'reɪməs/ n ignorante mf

ignoran|ce /'ɪgnərəns/ n ignoranza f. **~t** a (lacking knowledge) ignaro; (rude) ignorante

ignore /ɪg'nɔː(r)/ vt ignorare

ill /ɪl/ a ammalato; **feel ~ at ease** sentirsi a disagio ● adv male ● n male m. **~-advised** a avventato. **~-bred** a maleducato

illegal /ɪ'liːgl/ a illegale

illegible /ɪ'ledʒɪbl/ a illeggibile

illegitima|cy /ɪlɪ'dʒɪtɪməsɪ/ n illegittimità f. **~te** /-mət/ a illegittimo

illicit /ɪ'lɪsɪt/ a illecito

illitera|cy /ɪ'lɪtərəsɪ/ n analfabetismo m. **~te** /-rət/ a & n analfabeta mf

illness /'ɪlnɪs/ n malattia f

illogical /ɪ'lɒdʒɪkl/ a illogico

ill-treat /ɪl'triːt/ vt maltrattare. **~ment** n maltrattamento m

illuminat|e /ɪ'luːmɪneɪt/ vt illuminare. **~ing** a chiarificatore. **~ion** /-'neɪʃn/ n illuminazione f

illusion /ɪ'luːʒn/ n illusione f; **be under the ~ that** avere l'illusione che

illusory /ɪ'luːsərɪ/ a illusorio

illustrat|e /'ɪləstreɪt/ vt illustrare. **~ion** /-'streɪʃn/ n illustrazione f. **~or** n illustratore, -trice mf

illustrious /ɪ'lʌstrɪəs/ a illustre

ill 'will n malanimo m

image /'ɪmɪdʒ/ n immagine f; (exact likeness) ritratto m

imagin|able /ɪ'mædʒɪnəbl/ a immaginabile. **~ary** /-ərɪ/ a immaginario

imaginat|ion /ɪmædʒɪ'neɪʃn/ n immaginazione f, fantasia f; **it's your ~ion** è solo una tua idea. **~ive** /ɪ'mædʒɪnətɪv/ a fantasioso. **~ively** adv con fantasia or immaginazione

imagine /ɪ'mædʒɪn/ vt immaginare; (wrongly) inventare

im'balance n squilibrio m

imbecile /'ɪmbəsiːl/ n imbecille mf

imbibe /ɪm'baɪb/ vt ingerire

imbue /ɪm'bjuː/ vt **~d with** impregnato di

imitat|e /'ɪmɪteɪt/ vt imitare. **~ion** /-'teɪʃn/ n imitazione f. **~or** n imitatore, -trice mf

immaculate /ɪ'mækjʊlət/ a immacolato. **~ly** adv immacolatamente

imma'terial a (unimportant) irrilevante

imma'ture a immaturo

immediate /ɪ'miːdɪət/ a immediato; (relative) stretto; **in the ~ vicinity** nelle immediate vicinanze. **~ly** adv immediatamente; **~ly next to** subito accanto a ● conj [non] appena

immemorial /ɪmɪ'mɔːrɪəl/ a **from time ~** da tempo immemorabile

immense /ɪ'mens/ a immenso

immers|e /ɪ'mɜːs/ vt immergere; **be ~ed in** fig essere immerso in. **~ion** /-ʒn/ n immersione f. **~ion heater** n scaldabagno m elettrico

immigrant /'ɪmɪgrənt/ n immigrante mf

immigrat|e /'ɪmɪgreɪt/ vi immigrare. **~ion** /-'greɪʃn/ n immigrazione f

imminent /'ɪmɪnənt/ a imminente

immobil|e /ɪ'məʊbaɪl/ a immobile. **~ize** /-bɪlaɪz/ vt immobilizzare

immoderate /ɪ'mɒdərət/ a smodato

immodest /ɪ'mɒdɪst/ a immodesto

immoral /ɪ'mɒrəl/ a immorale. **~ity** /ɪmə'rælɪtɪ/ n immoralità f

immortal /ɪ'mɔːtl/ a immortale. **~ity** /-'tælətɪ/ n immortalità f. **~ize** vt immortalare

immovable /ɪˈmuːvəbl/ *a fig* irremovibile

immune /ɪˈmjuːn/ *a* immune (**to/from** da). **~ system** *n* sistema m immunitario

immunity /ɪˈmjuːnəti/ *n* immunità *f*

immuniz|e /ˈɪmjʊnaɪz/ *vt* immunizzare

imp /ɪmp/ *n* diavoletto m

impact /ˈɪmpækt/ *n* impatto m

impair /ɪmˈpeə(r)/ *vt* danneggiare

impale /ɪmˈpeɪl/ *vt* impalare

impart /ɪmˈpɑːt/ *vt* impartire

im'partial *a* imparziale. **~ality** *n* imparzialità *f*

im'passable *a* impraticabile

impasse /æmˈpɑːs/ *n fig* impasse *f inv*

impassioned /ɪmˈpæʃnd/ *a* appassionato

im'passive *a* impassibile

im'patien|ce *n* impazienza *f*. **~t** *a* impaziente. **~tly** *adv* impazientemente

impeccab|le /ɪmˈpekəbl/ *a* impeccabile. **~y** *adv* in modo impeccabile

impede /ɪmˈpiːd/ *vt* impedire

impediment /ɪmˈpedɪmənt/ *n* impedimento m; (*in speech*) difetto m

impel /ɪmˈpel/ *vt* (*pt/pp* **impelled**) costringere; **feel ~led to** sentire l'obbligo di

impending /ɪmˈpendɪŋ/ *a* imminente

impenetrable /ɪmˈpenɪtrəbl/ *a* impenetrabile

imperative /ɪmˈperətɪv/ *a* imperativo ●*n Gram* imperativo m

imper'ceptible *a* impercettibile

im'perfect *a* imperfetto; (*faulty*) difettoso ●*n Gram* imperfetto m. **~ion** /-ˈfekʃn/ *n* imperfezione *f*

imperial /ɪmˈpɪərɪəl/ *a* imperiale. **~ism** *n* imperialismo m. **~ist** *n* imperialista *m*

imperious /ɪmˈpɪərɪəs/ *a* imperioso

im'personal *a* impersonale

impersonat|e /ɪmˈpɜːsəneɪt/ *vt* impersonare. **~or** *n* imitatore, -trice *mf*

impertinen|ce /ɪmˈpɜːtɪnəns/ *n* impertinenza *f*. **~t** *a* impertinente

imperturbable /ɪmpəˈtɜːbəbl/ *a* imperturbabile

impervious /ɪmˈpɜːvɪəs/ *a* **~ to** *fig* indifferente a

impetuous /ɪmˈpetjʊəs/ *a* impetuoso. **~ly** *adv* impetuosamente

impetus /ˈɪmpɪtəs/ *n* impeto m

implacable /ɪmˈplækəbl/ *a* implacabile

im'plant[1] *vt* trapiantare; *fig* inculcare

'implant[2] *n* trapianto m

implement[1] /ˈɪmplɪmənt/ *n* attrezzo m

implement[2] /ˈɪmplɪmənt/ *vt* mettere in atto

implicat|e /ˈɪmplɪkeɪt/ *vt* implicare. **~ion** /-ˈkeɪʃn/ *n* implicazione *f*; **by ~ion** implicitamente

implicit /ɪmˈplɪsɪt/ *a* implicito; (*absolute*) assoluto

implore /ɪmˈplɔː(r)/ *vt* implorare

imply /ɪmˈplaɪ/ *vt* (*pt/pp* **-ied**) implicare; **what are you ~ing?** che cosa vorresti insinuare?

impo'lite *a* sgarbato

import[1] /ˈɪmpɔːt/ *n Comm* importazione *f*

import[2] /ɪmˈpɔːt/ *vt* importare

importan|ce /ɪmˈpɔːtəns/ *n* importanza *f*. **~t** *a* importante

importer /ɪmˈpɔːtə(r)/ *n* importatore, -trice *mf*

impos|e /ɪmˈpəʊz/ *vt* imporre (**on** a) ●*vi* imporsi; **~e on** abusare di. **~ing** *a* imponente. **~ition** /ɪmpəˈzɪʃn/ *n* imposizione *f*

impossi'bility *n* impossibilità *f*

im'possible *a* impossibile

impostor /ɪmˈpɒstə(r)/ *n* impostore, -trice *mf*

impoten|ce /ˈɪmpətəns/ *n* impotenza *f*. **~t** *a* impotente

impound /ɪmˈpaʊnd/ *vt* confiscare

impoverished /ɪmˈpɒvərɪʃt/ *a* impoverito

im'practicable *a* impraticabile

im'practical *a* non pratico

impre'cise *a* impreciso

impregnable /ɪmˈpregnəbl/ *a* imprendibile

impregnate /ˈɪmpregneɪt/ *vt* impregnare (**with** di); *Biol* fecondare

im'press *vt* imprimere; *fig* colpire (*positivamente*); ~ **sth** [up]**on sb** fare capire qcsa a qcno

impression /ɪmˈpreʃn/ *n* impressione *f*; (*imitation*) imitazione *f*. ~**able** *a* (*child*, *mind*) influenzabile. ~**ism** *n* impressionismo *m*. ~**ist** *n* imitatore, -trice *mf*; (*artist*) impressionista *mf*

impressive /ɪmˈpresɪv/ *a* imponente

'imprint[1] *n* impressione *f*

im'print[2] *vt* imprimere; ~**ed on my mind** impresso nella mia memoria

im'prison *vt* incarcerare. ~**ment** *n* reclusione *f*

im'probable *a* improbabile

impromptu /ɪmˈprɒmptjuː/ *a* improvvisato

im'proper *a* (*use*) improprio; (*behaviour*) scorretto. ~**ly** *adv* scorrettamente

impro'priety *n* scorrettezza *f*

improve /ɪmˈpruːv/ *vt*/*i* migliorare. **improve** [up]**on** *vt* perfezionare. ~**ment** /-mənt/ *n* miglioramento *m*

improvise /ˈɪmprəvaɪz/ *vt*/*i* improvvisare

im'prudent *a* imprudente

impudence /ˈɪmpjʊdəns/ *n* sfrontatezza *f*. ~**t** *a* sfrontato

impulse /ˈɪmpʌls/ *n* impulso *m*; **on** [**an**] ~**e** impulsivamente. ~**ive** /-ˈpʌlsɪv/ *a* impulsivo

impunity /ɪmˈpjuːnəti/ *n* **with** ~ impunemente

im'pure *a* impuro. ~**ity** *n* impurità *f inv*; ~**ities** *pl* impurità *fpl*

impute /ɪmˈpjuːt/ *vt* imputare (**to** a)

in /ɪn/ *prep* in; (*with names of towns*) a; **in the garden** in giardi-

no; **in the street** in *or* per strada; **in bed/hospital** a letto/all'ospedale; **in the world** nel mondo; **in the rain** sotto la pioggia; **in the sun** al sole; **in this heat** con questo caldo; **in summer/winter** in estate/inverno; **in 1995** nel 1995; **in the evening** la sera; **he's arriving in two hours time** arriva fra due ore; **deaf in one ear** sordo da un orecchio; **in the army** nell'esercito; **in English/Italian** in inglese/italiano; **in ink/pencil** a penna/matita; **in red** (*dressed*, *circled*) di rosso; **the man in the raincoat** l'uomo con l'impermeabile; **in a soft/loud voice** a voce bassa/alta; **one in ten people** una persona su dieci; **in doing this, he...** nel far questo,...; **in itself** in sé; **in that** in quanto ● *adv* (*at home*) a casa; (*indoors*) dentro; **he's not in yet** non è ancora arrivato; **in there/here** lì/qui dentro; **ten in all** dieci in tutto; **day in, day out** giorno dopo giorno; **have it in for sb** *fam* avercela con qcno; **send him in** fallo entrare; **come in** entrare; **bring in the washing** portare dentro i panni ● *a* (*fam*: *in fashion*) di moda ● *n* **the ins and outs** i dettagli

ina'bility *n* incapacità *f*

inac'cessible *a* inaccessibile

in'accuracy *n* inesattezza *f*. ~**te** *a* inesatto

in'active *a* inattivo. ~'**tivity** *n* inattività *f*

in'adequate *a* inadeguato. ~**ly** *adv* inadeguatamente

inad'missible *a* inammissibile

inadvertently /ɪnədˈvɜːtəntli/ *adv* inavvertitamente

inad'visable *a* sconsigliabile

inane /ɪˈneɪn/ *a* stupido

in'animate *a* esanime

in'applicable *a* inapplicabile

inap'propriate *a* inadatto

inar'ticulate *a* inarticolato

in'attentive *a* disattento

in'audible *a* impercettibile

inaugural /ɪˈnɔ:gjʊrəl/ a inaugurale

inaugurat|e /ɪˈnɔ:gjʊreɪt/ vt inaugurare. **~ion** /-ˈreɪʃn/ n inaugurazione f

inau'spicious a infausto

inborn /ˈmbɔ:n/ a innato

inbred /m'bred/ a congenito

incalculable /ɪnˈkælkjʊləbl/ a incalcolabile

in'capable a incapace

incapacitate /ɪnkəˈpæsɪteɪt/ vt rendere incapace

incarnat|e /ɪnˈkɑ:nət/ a **the devil ~e** il diavolo in carne e ossa

incendiary /ɪnˈsendɪərɪ/ a incendiario

incense¹ /ˈɪnsens/ n incenso m

incense² /ɪnˈsens/ vt esasperare

incentive /ɪnˈsentɪv/ n incentivo m

incessant /ɪnˈsesənt/ a incessante

incest /ˈɪnsest/ n incesto m

inch /ɪntʃ/ n pollice m (= 2.54 cm) ● vi ~ **forward** avanzare gradatamente

inciden|ce /ˈɪnsɪdəns/ n incidenza f. **~t** n incidente m

incidental /ɪnsɪˈdentl/ a incidentale; ~ **expenses** spese fpl accessorie. **~ly** adv incidentalmente; (by the way) a proposito

incinerat|e /ɪnˈsɪnəreɪt/ vt incenerire. **~or** n inceneritore m

incision /ɪnˈsɪʒn/ n incisione f

incisive /ɪnˈsaɪsɪv/ a incisivo

incisor /ɪnˈsaɪzə(r)/ n incisivo m

incite /ɪnˈsaɪt/ vt incitare. **~ment** n incitamento m

inclination /ɪnklɪˈneɪʃn/ n inclinazione f

incline¹ /ɪnˈklaɪn/ vt inclinare; **be ~d to** do sth essere propenso a fare qcsa

incline² /ˈɪnklaɪn/ n pendio m

includ|e /ɪnˈklu:d/ vt includere. **~ding** prep incluso. **~sion** /-u:ʒn/ n inclusione f

inclusive /ɪnˈklu:sɪv/ a incluso; ~ **of** comprendente; **be ~ of** comprendere ● adv incluso

incognito /ɪnkɒgˈni:təʊ/ adv incognito

inco'herent a incoerente; (because drunk etc) incomprensibile

income /ˈɪnkʌm/ n reddito m. ~ **tax** n imposta f sul reddito

'incoming a in arrivo. ~ **tide** n marea f montante

in'comparable a incomparabile

incompati'bility n incompatibilità f

incom'patible a incompatibile

incompeten|ce n incompetenza f. **~t** a incompetente

incom'plete a incompleto

incompre'hensible a incomprensibile

incon'ceivable a inconcepibile

incon'clusive a inconcludente

incongruous /ɪnˈkɒŋgrʊəs/ a contrastante

inconsequential /ɪnkɒnsɪˈkwen-ʃl/ a senza importanza

incon'siderate a trascurabile

incon'sistency n incoerenza f

incon'sistent a incoerente; **be ~ with** non essere coerente con. **~ly** adv in modo incoerente

inconsolable /ɪnkənˈsəʊləbl/ a inconsolabile

incon'spicuous a non appariscente. **~ly** adv modestamente

incontinen|ce /ɪnˈkɒntɪnəns/ n incontinenza f. **~t** a incontinente

incon'venien|ce n scomodità f; (drawback) inconveniente m; **put sb to ~ce** dare disturbo a qcno. **~t** a scomodo; (time, place) inopportuno. **~tly** adv in modo inopportuno

incorporate /ɪnˈkɔ:pəreɪt/ vt incorporare; (contain) comprendere

incor'rect a incorretto. **~ly** adv scorrettamente

incorrigible /ɪnˈkɒrɪdʒəbl/ a incorreggibile

incorruptible /ɪnkəˈrʌptəbl/ a incorruttibile

increase¹ /ˈɪnkri:s/ n aumento m; **on the ~** in aumento

increas|e² /ɪnˈkri:s/ vt/i aumentare. **~ing** a ⟨impatience etc⟩ crescente; ⟨numbers⟩ in aumento. **~ingly** adv sempre più

in'**credible** a incredibile

in**credulous** /ɪn'kredjʊləs/ a incredulo

increment /'ɪnkrɪmənt/ n incremento m

in**criminate** /ɪn'krɪmɪneɪt/ vt Jur incriminare

incubat|e /'ɪŋkjʊbeɪt/ vt incubare. **~ion** /-'beɪʃn/ n incubazione f. **~ion period** n Med periodo m di incubazione. **~or** n ⟨for baby⟩ incubatrice f

in**cumbent** /ɪn'kʌmbənt/ a **be ~ on sb** incombere a qcno

in**cur** /ɪn'kɜː(r)/ vt ⟨pt/pp incurred⟩ incorrere; contrarre ⟨debts⟩

in'**curable** a incurabile

in**cursion** /ɪn'kɜːʃn/ n incursione f

in**debted** /ɪn'detɪd/ a obbligato ⟨to verso⟩

in'**decent** a indecente

inde'**cision** n indecisione f

inde'**cisive** a indeciso. **~ness** n indecisione f

in**deed** /ɪn'diːd/ adv ⟨in fact⟩ difatti; **yes ~!** sì, certamente!; **I am/do** veramente!; **very much ~** moltissimo; **thank you very much ~** grazie infinite; **~?** davvero?

inde**fatigable** /ɪndɪ'fætɪgəbl/ a instancabile

inde'**finable** a indefinibile

in'**definite** a indefinito. **~ly** adv indefinitamente; ⟨postpone⟩ a tempo indeterminato

in**delible** /ɪn'delɪbl/ a indelebile

in**demnity** /ɪn'demnɪtɪ/ n indennità f inv

indent¹ /'ɪndent/ n Typ rientranza f dal margine

indent² /ɪn'dent/ vt Typ fare rientrare dal margine. **~ation** /-'teɪʃn/ n ⟨notch⟩ intaccatura f

inde'**penden|ce** n indipendenza f. **~t** a indipendente. **~tly** adv indipendentemente

inde**scribable** /ɪndɪ'skraɪbəbl/ a indescrivibile

inde**structible** /ɪndɪ'strʌktəbl/ a indistruttibile

inde**terminate** /ɪndɪ'tɜːmɪnət/ a indeterminato

index /'ɪndeks/ n indice m

index: ~ card n scheda f. **~ finger** n dito m indice. **~·linked** a ⟨pension⟩ legato al costo della vita

indic**at|e** /'ɪndɪkeɪt/ vt indicare; ⟨register⟩ segnare ● vi Auto mettere la freccia. **~ion** /-'keɪʃn/ n indicazione f

in**dicative** /ɪn'dɪkətɪv/ a **be ~ of** essere indicativo di ● n Gram indicativo m

indicator /'ɪndɪkeɪtə(r)/ n Auto freccia f

in**dict** /ɪn'daɪt/ vt accusare. **~ment** n accusa f

in**differen|ce** n indifferenza f. **~t** a indifferente; ⟨not good⟩ mediocre

in**digenous** /ɪn'dɪdʒɪnəs/ a indigeno

indi'**gest|ible** a indigesto. **~ion** n indigestione f

indig**na|nt** /ɪn'dɪgnənt/ a indignato. **~ntly** adv con indignazione. **~tion** /-'neɪʃn/ n indignazione f

in**dignity** n umiliazione f

indi'**rect** a indiretto. **~ly** adv indirettamente

indi'**screet** a indiscreto

indis**cretion** n indiscrezione f

indis**criminate** /ɪndɪ'skrɪmɪnət/ a indiscriminato. **~ly** adv senza distinzione

indi'**spensable** a indispensabile

indis**posed** /ɪndɪ'spəʊzd/ a indisposto

indis**putable** /ɪndɪ'spjuːtəbl/ a indisputabile

indi'**stinct** a indistinto

indistinguishable /ɪndɪˈstɪŋgwɪʃəbl/ *a* indistinguibile

individual /ɪndɪˈvɪdjʊəl/ *a* individuale ● *n* individuo *m.* **~ity** /-ˈæljtɪ/ *n* individualità *f*

indi'visible *a* indivisibile

indoctrinate /ɪnˈdɒktrɪneɪt/ *vt* indottrinare

indomitable /ɪnˈdɒmɪtəbl/ *a* indomito

indoor /ˈɪndɔː(r)/ *a* interno; ⟨shoes⟩ per casa; ⟨plant⟩ da appartamento; ⟨swimming pool etc⟩ coperto. **~s** /-ˈdɔːz/ *adv* dentro

induce /ɪnˈdjuːs/ *vt* indurre (**to** a); ⟨produce⟩ causare. **~ment** *n* (*incentive*) incentivo *m*

indulge /ɪnˈdʌldʒ/ *vt* soddisfare; viziare ⟨child⟩ ● *vi* ~ **in** concedersi. **~nce** /-əns/ *n* lusso *m*; ⟨leniency⟩ indulgenza *f.* **~nt** *a* indulgente

industrial /ɪnˈdʌstrɪəl/ *a* industriale; **take ~ action** scioperare. **~ist** *n* industriale *mf.* **~ized** *a* industrializzato

industrious /ɪnˈdʌstrɪəs/ *a* industrioso. **~y** /ˈɪndəstrɪ/ *n* industria *f*; ⟨zeal⟩ operosità *f*

inebriated /ɪˈniːbrɪeɪtɪd/ *a* ebbro

in'edible *a* immangiabile

ineffective *a* inefficace

ineffectual /ɪnɪˈfektʃʊəl/ *a* inutile; ⟨person⟩ inconcludente

inefficien|cy *n* inefficienza *f.* **~t** *a* inefficiente

in'eligible *a* inadatto

inept /ɪˈnept/ *a* inetto

ine'quality *n* ineguaglianza *f*

inert /ɪˈnɜːt/ *a* inerte. **~ia** /ɪˈnɜːʃə/ *n* inerzia *f*

inescapable /ɪnɪˈskeɪpəbl/ *a* inevitabile

inestimable /ɪnˈestɪməbl/ *a* inestimabile

inevitabl|e /ɪnˈevɪtəbl/ *a* inevitabile. **~y** *adv* inevitabilmente

ine'xact *a* inesatto

inex'cusable *a* imperdonabile

inexhaustible /ɪnɪgˈzɔːstəbl/ *a* inesauribile

inexorable /ɪnˈeksərəbl/ *a* inesorabile

inex'pensive *a* poco costoso

inex'perience *n* inesperienza *f.* **~d** *a* inesperto

inexplicable /ɪnɪkˈsplɪkəbl/ *a* inesplicabile

in'fallible *a* infallibile

infam|ous /ˈɪnfəməs/ *a* infame; ⟨person⟩ famigerato. **~y** *n* infamia *f*

infan|cy /ˈɪnfənsɪ/ *n* infanzia *f*; **in its ~cy** *fig* agli inizi. **~t** *n* bambino, -a *mf* piccolo, -a. **~tile** *a* infantile

infantry /ˈɪnfəntrɪ/ *n* fanteria *f*

infatuat|ed /ɪnˈfætʃʊeɪtɪd/ *a* infatuato (**with** di). **~ion** *n* infatuazione *f*

infect /ɪnˈfekt/ *vt* infettare; **become ~ed** ⟨wound⟩ infettarsi. **~ion** /-ˈfekʃn/ *n* infezione *f.* **~ious** /-ˈfekʃəs/ *a* infettivo

infer /ɪnˈfɜː(r)/ *vt* (*pt/pp* **inferred**) dedurre (**from** da); ⟨imply⟩ implicare. **~ence** /ˈɪnfərəns/ *n* deduzione *f*

inferior /ɪnˈfɪərɪə(r)/ *a* inferiore; ⟨goods⟩ scadente; (*in rank*) subalterno ● *n* inferiore *mf*; (*in rank*) subalterno, -a *mf*

inferiority /ɪnfɪərɪˈɒrətɪ/ *n* inferiorità *f.* **~ complex** *n* complesso *m* di inferiorità

infern|al /ɪnˈfɜːnl/ *a* infernale. **~o** *n* inferno *m*

in'fertil|e *a* sterile. **~'tility** *n* sterilità *f*

infest /ɪnˈfest/ *vt* **be ~ed with** essere infestato di

infi'delity *n* infedeltà *f*

infighting /ˈɪnfaɪtɪŋ/ *n* *fig* lotta *f* per il potere

infiltrate /ˈɪnfɪltreɪt/ *vt* infiltrare; *Pol* infiltrarsi in

infinite /ˈɪnfɪnət/ *a* infinito

infinitive /ɪnˈfɪnətɪv/ *n* *Gram* infinito *m*

infinity /ɪnˈfɪnətɪ/ *n* infinità *f*

infirm /ɪnˈfɜːm/ *a* debole. **~ary** *n* infermeria *f.* **~ity** *n* debolezza *f* inflame /ɪnˈfleɪm/ vt infiammare. ~d a infiammato; become ~d infiammarsi

inflammable a infiammabile

inflammation /ɪnfləˈmeɪʃn/ n infiammazione f

inflammatory /ɪnˈflæmətrɪ/ a incendiario

inflatable /ɪnˈfleɪtəbl/ a gonfiabile

inflat|e /ɪnˈfleɪt/ vt gonfiare. ~ion /-eɪʃn/ n inflazione f. ~ionary /-eɪʃənərɪ/ a inflazionario

inflexible a inflessibile

inflexion /ɪnˈflekʃn/ n inflessione f

inflict /ɪnˈflɪkt/ vt infliggere (on a)

influen|ce /ˈɪnflʊəns/ n influenza f ● vt influenzare. ~tial /-ˈenʃl/ a influente

influenza /ɪnflʊˈenzə/ n influenza f

influx /ˈɪnflʌks/ n affluenza f

inform /ɪnˈfɔːm/ vt informare; keep sb ~ed tenere qcno al corrente ● vi ~ against denunziare

infor|mal /ɪnˈfɔːml/ a informale; (agreement) ufficioso. ~mally adv in modo informale. ~'mality n informalità f inv

informant /ɪnˈfɔːmənt/ n informatore, -trice mf

informat|ion /ɪnfəˈmeɪʃn/ n informazioni fpl; a piece of ~ion un'informazione. ~ion highway n autostrada f telematica. ~ion technology n informatica f. ~ive /ɪnˈfɔːmətɪv/ a informativo; (film, book) istruttivo

informer /ɪnˈfɔːmə(r)/ n informatore, -trice mf; Pol delatore, -trice mf

infra-'red /ɪnfrə-/ a infrarosso

infrastructure /ˈɪnfrəstrʌktʃə(r)/ n infrastruttura f

infringe /ɪnˈfrɪndʒ/ vt ~ on usurpare. ~ment n violazione f

infuriat|e /ɪnˈfjʊərɪeɪt/ vt infuriare. ~ing a esasperante

infusion /ɪnˈfjuːʒn/ n (drink) infusione f; (of capital, new blood) afflusso m

ingenious /ɪnˈdʒiːnɪəs/ a ingegnoso

ingenuity /ɪndʒɪˈnjuːətɪ/ n ingegnosità f

ingenuous /ɪnˈdʒenjʊəs/ a ingenuo

ingot /ˈɪŋɡət/ n lingotto m

ingrained /ɪnˈɡreɪnd/ a (in person) radicato; (dirt) incrostato

ingratiate /ɪnˈɡreɪʃɪeɪt/ vt ~ oneself with sb ingraziarsi qcno

in'gratitude n ingratitudine f

ingredient /ɪnˈɡriːdɪənt/ n ingrediente m

ingrowing /ˈɪnɡrəʊɪŋ/ a (nail) incarnito

inhabit /ɪnˈhæbɪt/ vt abitare. ~ant n abitante mf

inhale /ɪnˈheɪl/ vt aspirare; Med inalare ● vi inspirare; (when smoking) aspirare. ~r n (device) inalatore m

inherent /ɪnˈhɪərənt/ a inerente

inherit /ɪnˈherɪt/ vt ereditare. ~ance /-əns/ n eredità f inv

inhibit /ɪnˈhɪbɪt/ vt inibire. ~ed a inibito. ~ion /-ˈbɪʃn/ n inibizione f

inho'spitable a inospitale

in'human a disumano

initial /ɪˈnɪʃl/ a iniziale ● n iniziale f ● vt (pt/pp initialled) siglare. ~ly adv all'inizio

initiat|e /ɪˈnɪʃɪeɪt/ vt iniziare. ~ion /-ˈeɪʃn/ n iniziazione f

initiative /ɪˈnɪʃətɪv/ n iniziativa f

inject /ɪnˈdʒekt/ vt iniettare. ~ion /-ekʃn/ n iniezione f

injur|e /ˈɪndʒə(r)/ vt ferire; (wrong) nuocere. ~y n ferita f; (wrong) torto m

in'justice n ingiustizia f; do sb an ~ giudicare qcno in modo sbagliato

ink /ɪŋk/ n inchiostro m

inkling /ˈɪŋklɪŋ/ n sentore m

inlaid /ɪnˈleɪd/ a intarsiato

inland /ˈɪnlənd/ a interno ● adv all'interno. I~ Revenue n fisco m

in-laws /ˈɪnlɔːz/ npl fam parenti mpl acquisiti

inlay /ˈɪnleɪ/ n intarsio m

inlet /ˈɪnlet/ n insenatura f; Techn entrata f

inmate /'ɪnmeɪt/ n (of hospital) degente mf; (of prison) carcerato, -a mf

inn /ɪn/ n locanda f

innate /ɪ'neɪt/ a innato

inner /'ɪnə(r)/ a interno. ~**most** a il più profondo. ~ **tube** camera f d'aria

'innkeeper n locandiere, -a mf

innocen|ce /'ɪnəsəns/ n innocenza f. ~**t** a innocente

innocuous /ɪ'nɒkjʊəs/ a innocuo

innovat|e /'ɪnəveɪt/ vi innovare. ~**ion** /-'veɪʃn/ n innovazione f. ~**ive** /'ɪnəvətɪv/ a innovativo. ~**or** /'ɪnəveɪtə(r)/ n innovatore, -trice mf

innuendo /ɪnjʊ'endəʊ/ n (pl -es) insinuazione f

innumerable /ɪ'njuːmərəbl/ a innumerevole

inoculat|e /ɪ'nɒkjʊleɪt/ vt vaccinare. ~**ion** /-'leɪʃn/ n vaccinazione f

inof'fensive a inoffensivo

in'operable a inoperabile

in'opportune a inopportuno

inordinate /ɪ'nɔːdɪnət/ a smodato

inor'ganic a inorganico

'in-patient n degente mf

input /'ɪnpʊt/ n input m inv, ingresso m

inquest /'ɪnkwest/ n inchiesta f

inquir|e /ɪn'kwaɪə(r)/ vi informarsi (**about** su). ~**e into** far indagini su ● vt domandare. ~**y** n domanda f; (investigation) inchiesta f

inquisitive /ɪn'kwɪzətɪv/ a curioso

inroad /'ɪnrəʊd/ n make ~**s into** intaccare (savings); cominciare a risolvere (problem)

in'sane a pazzo; fig insensato

in'sanitary a malsano

in'sanity n pazzia f

insatiable /ɪn'seɪʃəbl/ a insaziabile

inscri|be /ɪn'skraɪb/ vt iscrivere. ~**ption** /-'skrɪpʃn/ n iscrizione f

inscrutable /ɪn'skruːtəbl/ a impenetrabile

insect /'ɪnsekt/ n insetto m.

~**icide** /-'sektɪsaɪd/ n insetticida m

inse'cur|e a malsicuro; ⟨fig: person⟩ insicuro. ~**ity** n mancanza f di sicurezza

insemination /ɪnsemɪ'neɪʃn/ n inseminazione f

in'sensitive a insensibile

in'separable a inseparabile

insert¹ /'ɪnsɜːt/ n inserto m

insert² /ɪn'sɜːt/ vt inserire. ~**ion** /-ʃn/ n inserzione f

inside /ɪn'saɪd/ n interno m. ~**s** npl fam pancia f ● attrib Aut ~ **lane** n corsia f interna ● adv dentro; ~ **out** a rovescio; (thoroughly) a fondo ● prep dentro; (of time) entro

insidious /ɪn'sɪdɪəs/ a insidioso

insight /'ɪnsaɪt/ n intuito m (**into** per); an ~ **into** un quadro di

insignia /ɪn'sɪgnɪə/ npl insegne fpl

insig'nificant a insignificante

insin'cer|e a poco sincero. ~**ity** /-'serɪtɪ/ n mancanza f di sincerità

insinuat|e /ɪn'sɪnjʊeɪt/ vt insinuare. ~**ion** /-'eɪʃn/ n insinuazione f

insipid /ɪn'sɪpɪd/ a insipido

insist /ɪn'sɪst/ vi insistere (**on** per) ● vt ~ **that** insistere che. ~**ence** f n insistenza f. ~**ent** a insistente

insole n soletta f

insolen|ce /'ɪnsələns/ n insolenza f. ~**t** a insolente

in'soluble a insolubile

in'solven|cy n insolvenza f. ~**t** a insolvente

insomnia /ɪn'sɒmnɪə/ n insonnia f

inspect /ɪn'spekt/ vt ispezionare; controllare (ticket). ~**ion** /-ekʃn/ n ispezione f; (of ticket) controllo m. ~**or** n ispettore, -trice mf; (of tickets) controllore m

inspiration /ɪnspə'reɪʃn/ n ispirazione f

inspire /ɪn'spaɪə(r)/ vt ispirare

insta'bility n instabilità f

install /ɪn'stɔːl/ vt installare. ~**ation** /-stə'leɪʃn/ n installazione f

instalment /ɪn'stɔːlmənt/ n Comm

rata *f*; (*of serial*) puntata *f*; (*of publication*) fascicolo *m*

instance /'ɪnstəns/ *n* (*case*) caso *m*; (*example*) esempio *m*; **in the first ~** in primo luogo; **for ~** per esempio

instant /'ɪnstənt/ *a* immediato; *Culin* espresso ●*n* istante *m*. **~aneous** /-'teɪnɪəs/ *a* istantaneo

instant 'coffee *n* caffè *m inv* solubile

instantly /'ɪnstəntlɪ/ *adv* immediatamente

instead /ɪn'sted/ *adv* invece; **~ of doing** anziché fare; **~ of me** al mio posto; **~ of going** invece di andare

instep /'ɪnstep/ *n* collo *m* del piede

instigat|e /'ɪnstɪgeɪt/ *vt* istigare. **~ion** /-'geɪʃn/ *n* istigazione *f*; **at his ~ion** dietro suo suggerimento. **~or** *n* istigatore, -trice *mf*

instil /ɪn'stɪl/ *vt* (*pt/pp* **instilled**) inculcare (**into** in)

instinct /'ɪnstɪŋkt/ *n* istinto *m*. **~ive** /ɪn'stɪŋktɪv/ *a* istintivo

institut|e /'ɪnstɪtjuːt/ *n* istituto *m* ●*vt* istituire (*scheme*); iniziare (*search*); intentare (*legal action*). **~ion** /-'tjuːʃn/ *n* istituzione *f*; (*home for elderly*) istituto *m* per anziani; (*for mentally ill*) istituto *m* per malati di mente

instruct /ɪn'strʌkt/ *vt* istruire; (*order*) ordinare. **~ion** /-ʌkʃn/ *n* istruzione *f*; **~s** (*orders*) ordini *mpl*. **~ive** /-ɪv/ *a* istruttivo. **~or** *n* istruttore, -trice *mf*

instrument /'ɪnstrʊmənt/ *n* strumento *m*. **~al** /-'mentl/ *a* strumentale; **be ~al in** contribuire a. **~alist** *n* strumentista *mf*

insu'bordin|ate *a* insubordinato. **~nation** /-'neɪʃn/ *n* insubordinazione *f*

in'sufferable *a* insopportabile

insuf'ficient *a* insufficiente

insular /'ɪnsjʊlə(r)/ *a fig* gretto

insulat|e /'ɪnsjʊlert/ *vt* isolare. **~ing tape** *n* nastro *m* isolante. **~ion** /-'leɪʃn/ *n* isolamento *m*

insulin /'ɪnsjʊlɪn/ *n* insulina *f*

insult[1] /'ɪnsʌlt/ *n* insulto *m*

insult[2] /ɪn'sʌlt/ *vt* insultare

insuperable /ɪn'suːpərəbl/ *a* insuperabile

insur|ance /ɪn'ʃʊərəns/ *n* assicurazione *f*. **~e** *vt* assicurare

insurrection /ɪnsə'rekʃn/ *n* insurrezione *f*

intact /ɪn'tækt/ *a* intatto

'intake *n* immissione *f*; (*of food*) consumo *m*

in'tangible *a* intangibile

integral /'ɪntɪɡrəl/ *a* integrale

integrat|e /'ɪntɪgreɪt/ *vt* integrare ●*vi* integrarsi. **~ion** /-'greɪʃn/ *n* integrazione *f*

integrity /ɪn'tegrətɪ/ *n* integrità *f*

intellect /'ɪntəlekt/ *n* intelletto *m*. **~ual** /-'lektjʊəl/ *a & n* intellettuale *mf*

intelligen|ce /ɪn'telɪdʒəns/ *n* intelligenza *f*; *Mil* informazioni *fpl*. **~t** *a* intelligente

intelligentsia /ɪntelɪ'dʒentsɪə/ *n* intellighenzia *f*

intelligible /ɪn'telɪdʒəbl/ *a* intelligibile

intend /ɪn'tend/ *vt* destinare; (*have in mind*) aver intenzione di; **be ~ed for** essere destinato a. **~ed** *a* (*effect*) voluto ●*n* **my ~ed** *fam* il mio/la mia fidanzato, -a

intense /ɪn'tens/ *a* intenso; (*person*) dai sentimenti intensi. **~ly** *adv* intensamente; (*very*) estremamente

intensi|fication /ɪntensɪfɪ'keɪʃn/ *n* intensificazione *f*. **~fy** /-'tensɪfaɪ/ *v* (*pt/pp* **-ied**) ●*vt* intensificare ●*vi* intensificarsi

intensity /ɪn'tensətɪ/ *n* intensità *f*

intensive /ɪn'tensɪv/ *a* intensivo. **~ care** (*for people in coma*) rianimazione *f*; **~ care [unit]** terapia *f* intensiva

intent /ɪn'tent/ *a* intento; **~ on** (*absorbed in*) preso da; **be ~ on doing sth** essere intento a fare qcsa ●*n* intenzione *f*; **to all ~s**

and purposes a tutti gli effetti. **~ly** adv attentamente

intention /ɪnˈtenʃn/ n intenzione f. **~al** a intenzionale. **~ally** adv intenzionalmente

inter•acti|on n cooperazione f. **~ve** a interattivo

intercede /ɪntəˈsiːd/ vi intercedere **(on behalf of** a favore di)

intercept /ɪntəˈsept/ vt intercettare

'**interchange** n scambio m; Auto raccordo m [autostradale]

inter•changeable a a interscambiabile

intercom /ˈɪntəkɒm/ n citofono m

'**intercourse** n (sexual) rapporti mpl [sessuali]

interest /ˈɪntrəst/ n interesse m; **have an ~** in Comm essere cointeressato in; **be of ~** essere interessante; **~ rate** n tasso m di interesse ● vt interessare. **~ed** a interessato. **~ing** a interessante

interface /ˈɪntəfeɪs/ n interfaccia f ● vt interfacciare ● vi interfacciarsi

interfere /ɪntəˈfɪə(r)/ vi interferire; **~ with** interferire con. **~nce** /-əns/ n interferenza f

interim /ˈɪntərɪm/ a temporaneo; **~ payment** acconto m ● n **in the ~** nel frattempo

interior /ɪnˈtɪərɪə(r)/ a interiore ● n interno m. **~ designer** n arredatore, -trice mf

interject /ɪntəˈdʒekt/ vt intervenire. **~ion** /-ekʃn/ n Gram interiezione f; (remark) intervento m

interloper /ˈɪntələʊpə(r)/ n intruso, -a mf

interlude /ˈɪntəluːd/ n intervallo m

inter•marry vi sposarsi tra parenti; (different groups:) contrarre matrimoni misti

intermediary /ɪntəˈmiːdɪərɪ/ n intermediario, -a mf

intermediate /ɪntəˈmiːdɪət/ a intermedio

interminable /ɪnˈtɜːmɪnəbl/ a interminabile

intermission /ɪntəˈmɪʃn/ n intervallo m

intermittent /ɪntəˈmɪtənt/ a intermittente

intern /ɪnˈtɜːn/ vt internare

internal /ɪnˈtɜːnl/ a interno. **~ly** adv internamente; (deal with) all'interno

inter•national a internazionale ● n (game) incontro m internazionale; (player) competitore, -trice mf in gare internazionali. **~ly** adv internazionalmente

Internet /ˈɪntənet/ n Internet m

internist /ɪnˈtɜːnɪst/ n Am internista mf

internment /ɪnˈtɜːnmənt/ n internamento m

'**interplay** n azione f reciproca

interpret /ɪnˈtɜːprɪt/ vt interpretare ● vi fare l'interprete. **~ation** /-ˈteɪʃn/ n interpretazione f. **~er** n interprete mf

interre•lated a (facts) in correlazione

interrogat|e /ɪnˈterəgeɪt/ vt interrogare. **~ion** /-ˈgeɪʃn/ n interrogazione f; (by police) interrogatorio m

interrogative /ɪntəˈrɒgətɪv/ a & n **~ [pronoun]** interrogativo m

interrupt /ɪntəˈrʌpt/ vt/i interrompere. **~ion** /-ʌpʃn/ n interruzione f

intersect /ɪntəˈsekt/ vi intersecarsi ● vt intersecare. **~ion** /-ekʃn/ n intersezione f; (of street) incrocio m

interspersed /ɪntəˈspɜːst/ a **~ with** inframmezzato di

inter•twine vi attorcigliarsi

interval /ˈɪntəvl/ n intervallo m; **bright ~s** pl schiarite fpl

interven|e /ɪntəˈviːn/ vi intervenire. **~tion** /-ˈvenʃn/ n intervento m

interview /ˈɪntəvjuː/ n Journ intervista f; (for job) colloquio m [di lavoro] ● vt intervistare. **~er** n intervistatore, -trice mf

intestin|**e** /ɪnˈtestɪn/ n intestino m. **~al** a intestinale

intimacy /ˈɪntɪməsɪ/ n intimità f

intimate[1] /ˈɪntɪmət/ a intimo. **~ly** adv intimamente

intimate[2] /ˈɪntɪmeɪt/ vt far capire; (imply) suggerire

intimidat|**e** /ɪnˈtɪmɪdeɪt/ vt intimidire. **~ion** /-ˈdeɪʃn/ n intimidazione f

into /ˈɪntə/, /ˈɪntʊ/ prep dentro, in; **go ~ the house** andare dentro [casa] o in casa; **be ~** (fam: like) essere appassionato di; **I'm not ~ that** questo non mi piace; **7 ~ 21 goes 3** il 7 nel 21 ci sta 3 volte; **translate ~ French** tradurre in francese; **get ~ trouble** mettersi nei guai

in'tolerable a intollerabile

in'toleran|**ce** n intolleranza f. **~t** a intollerante

intonation /ɪntəˈneɪʃn/ n intonazione f

intoxicat|**ed** /ɪnˈtɒksɪkeɪtɪd/ a inebriato. **~ion** /-ˈkeɪʃn/ n ebbrezza f

intractable /ɪnˈtræktəbl/ a intrattabile; (problem) insolubile

intransigent /ɪnˈtrænzɪdʒənt/ a intransigente

in'transitive a intransitivo

intravenous /ɪntrəˈviːnəs/ a endovenoso. **~ly** adv per via endovenosa

intrepid /ɪnˈtrepɪd/ a intrepido

intricate /ˈɪntrɪkət/ a complesso

intrigu|**e** /ɪnˈtriːɡ/ n intrigo m ● vt intrigare ● vi tramare. **~ing** a intrigante

intrinsic /ɪnˈtrɪnsɪk/ a intrinseco

introduce /ɪntrəˈdjuːs/ vt presentare; (bring in, insert) introdurre

introduct|**ion** /ɪntrəˈdʌkʃn/ n introduzione f; (to person) presentazione f; (to book) prefazione f. **~ory** /-tərɪ/ a introduttivo

introspective /ɪntrəˈspektɪv/ a introspettivo

introvert /ˈɪntrəvɜːt/ n introverso, -a mf

intru|**de** /ɪnˈtruːd/ vi intrometter-

si. **~der** n intruso, -a mf. **~sion** /-uːʒn/ n intrusione f

intuit|**ion** /ɪntjʊˈɪʃn/ n intuito m. **~ive** /-ˈtjuːɪtɪv/ a intuitivo

inundate /ˈɪnʌndeɪt/ vt fig inondare (with di)

invade /ɪnˈveɪd/ vt invadere. **~r** n invasore m

invalid[1] /ˈɪnvəlɪd/ n invalido, -a mf

invalid[2] /ɪnˈvælɪd/ a non valido. **~ate** vt invalidare

in'valuable a prezioso; (priceless) inestimabile

in'variab|**le** a invariabile. **~y** adv invariabilmente

invasion /ɪnˈveɪʒn/ n invasione f

invective /ɪnˈvektɪv/ n invettiva f

invent /ɪnˈvent/ vt inventare. **~ion** /-enʃn/ n invenzione f. **~ive** /-tɪv/ a inventivo. **~or** n inventore, -trice mf

inventory /ˈɪnvəntrɪ/ n inventario m

inverse /ɪnˈvɜːs/ a inverso ● n inverso m

invert /ɪnˈvɜːt/ vt invertire; **in ~ed commas** tra virgolette

invest /ɪnˈvest/ vt investire ● vi fare investimenti; **~ in** (fam: buy) comprarsi

investigat|**e** /ɪnˈvestɪgeɪt/ vt investigare. **~ion** /-ˈɡeɪʃn/ n investigazione f

invest|**ment** /ɪnˈvestmənt/ n investimento m. **~or** n investitore, -trice mf

inveterate /ɪnˈvetərət/ a inveterato

invidious /ɪnˈvɪdɪəs/ a ingiusto; (position) antipatico

invigilat|**e** /ɪnˈvɪdʒɪleɪt/ vi Sch sorvegliare lo svolgimento di un esame. **~or** n persona f che sorveglia lo svolgimento di un esame

invigorate /ɪnˈvɪɡəreɪt/ vt rinvigorire

invigorating /ɪnˈvɪɡəreɪtɪŋ/ a tonificante

invincible /ɪnˈvɪnsəbl/ a invincibile

inviolable /m'vaɪələbl/ a inviolabile

in'visible a invisibile

invitation /mvɪ'teɪʃn/ n invito m

invit|e /m'vaɪt/ vt invitare; (attract) attirare. **~ing** a invitante

invoice /'mvɔɪs/ n fattura f ● vt ~ sb emettere una fattura a qcno

invoke /m'vəʊk/ vt invocare

in'voluntary a involontario

involve /m'vɒlv/ vt comportare; (affect, include) coinvolgere; (entail) implicare; **get ~d with sb** (romantically) legarsi sentimentalmente a qcno. **~d** a complesso. **~ment** n coinvolgimento m

in'vulnerable a invulnerabile; (position) inattaccabile

inward /'mwəd/ a interno; (thoughts etc) interiore; **~ invest-ment** Comm investimento m straniero. **~ly** adv interiormente. **~[s]** adv verso l'interno

iodine /'aɪədi:n/ n iodio m

iota /aɪ'əʊtə/ n briciolo m

IOU n abbr (I owe you) pagherò m inv

IQ n abbr (intelligence quotient) Q.I.

IRA n abbr (Irish Republican Army) I.R.A. f

Iran /ɪ'rɑ:n/ n Iran m. **~ian** /ɪ'reɪnɪən/ a & n iraniano, -a m

Iraq /ɪ'rɑ:k/ n Iraq m. **~i** /ɪ'rɑ:kɪ/ a & n iracheno, -a mf

irascible /ɪ'ræsəbl/ a irascibile

irate /aɪ'reɪt/ a adirato

Ireland /'aɪələnd/ n Irlanda f

iris /'aɪrɪs/ n Anat iride f; Bot iris f inv

Irish /'aɪrɪʃ/ a irlandese ● **n the ~** pl gli irlandesi mpl. **~man** n irlandese m. **~woman** n irlandese f

iron /'aɪən/ n ferro m; **~ Curtain** n cortina f di ferro ● n ferro m; (appliance) ferro m [da stiro] ● vt/i stirare. **iron out** vt eliminare stirando; fig appianare

ironic[al] /aɪ'rɒnɪk[l]/ a ironico

ironing /'aɪənɪŋ/ n stirare m;

(articles) roba f da stirare; **do the ~** stirare. **~-board** n asse f da stiro

'ironmonger /-mʌŋgə(r)/ n **~'s** [shop] negozio m di ferramenta

irony /'aɪrənɪ/ n ironia f

irradiate /ɪ'reɪdɪeɪt/ vt irradiare

irrational /ɪ'ræʃənl/ a irrazionale

irreconcilable /ɪ'rekənsaɪləbl/ a irreconciliabile

irrefutable /ɪ'rfju:təbl/ a irrefutabile

irregular /ɪ'regjʊlə(r)/ a irregolare. **~ity** /-'lærɪtɪ/ n irregolarità f inv

irrelevant /ɪ'reləvənt/ a non pertinente

irreparab|le /ɪ'repərəbl/ a irreparabile. **~y** adv irreparabilmente

irreplaceable /ɪrɪ'pleɪsəbl/ a insostituibile

irrepressible /ɪrɪ'presəbl/ a irrefrenabile; (person) incontenibile

irresistible /ɪrɪ'zɪstəbl/ a irresistibile

irresolute /ɪ'rezəlu:t/ a irresoluto

irrespective /ɪrɪ'spektɪv/ a **~ of** senza riguardo per

irresponsible /ɪrɪ'spɒnsɪbl/ a irresponsabile

irreverent /ɪ'revərənt/ a irreverente

irreversible /ɪrɪ'vɜ:səbl/ a irreversibile

irrevocab|le /ɪ'revəkəbl/ a irrevocabile. **~y** adv irrevocabilmente

irrigat|e /'ɪrɪgeɪt/ vt irrigare. **~ion** /-'geɪʃn/ n irrigazione f

irritability /ɪrɪtə'bɪlətɪ/ n irritabilità f

irritable /'ɪrɪtəbl/ a irritabile

irritant /'ɪrɪtənt/ n sostanza f irritante

irritat|e /'ɪrɪteɪt/ vt irritare. **~ing** a irritante. **~ion** /-'teɪʃn/ n irritazione f

is /ɪz/ see **be**

Islam /'ɪzlɑ:m/ n Islam m. **~ic** /-'læmɪk/ a islamico

island /'aɪlənd/ n isola f; (in road)

isola f spartitraffico. **~er** n isolano, -a mf

isle /aɪl/ n isola f

isolat|e /'aɪsəleɪt/ vt isolare. **~ed** a isolato. **~ion** /-'leɪʃn/ n isolamento m

Israel /'ɪzreɪl/ n Israele m. **~i** /ɪz'reɪlɪ/ a & n israeliano, -a mf

issue /'ɪʃuː/ n (outcome) risultato m; (of magazine) numero m; (of stamps etc) emissione f; (off-spring) figli mpl; (matter, question) questione f; **at ~** in questione; **take ~ with sb** prendere posizione contro qcno ● vt distribuire (supplies); rilasciare (passport); emettere (stamps, order); pubblicare (book); **be ~d with sth** ricevere qcsa ● vi **~ from** uscire da

isthmus /'ɪsməs/ n (pl **-muses**) istmo m

it /ɪt/ pron (direct object) lo m, la f; (indirect object) gli m, le f; **it's broken** è rotto/rotta; **will it be enough?** basterà?; **it's hot** fa caldo; **it's raining** piove; **it's me** sono io; **who is it?** chi è?; **it's two o'clock** sono le due; **I doubt it** ne dubito; **take it with you** prendilo con te; **give it a wipe** dagli una pulita

Italian /ɪ'tæljən/ a & n italiano, -a mf; (language) italiano m

italic /ɪ'tælɪk/ a italico. **~s** npl corsivo msg

Italy /'ɪtəlɪ/ n Italia f

itch /ɪtʃ/ n prurito m ● vi avere prurito, prudere; **be ~ing to** fam avere una voglia matta di. **~y** a che prude; **my foot is ~y** ho prurito al piede

item /'aɪtəm/ n articolo m; (on agenda, programme) punto m; (on invoice) voce f; **~ [of news]** notizia f. **~ize** vt dettagliare (bill)

itinerant /aɪ'tɪnərənt/ a itinerante

itinerary /aɪ'tɪnərərɪ/ n itinerario m

its /ɪts/ poss pron suo m, sua f, suoi mpl, sue fpl; **~ mother/cage** sua madre/la sua gabbia

it's = **it is**, **it has**

itself /ɪt'self/ pron (reflexive) si; (emphatic) essa stessa; **the baby looked at ~ in the mirror** il bambino si è guardato nello specchio; **by ~** da solo; **the machine in ~ is simple** la macchina di per sé è semplice

ITV n abbr (**Independent Television**) stazione f televisiva privata britannica

ivory /'aɪvərɪ/ n avorio m

ivy /'aɪvɪ/ n edera f

Jj

jab /dʒæb/ n colpo m secco; (fam: injection) puntura f ● vt (pt/pp **jabbed**) punzecchiare

jabber /'dʒæbə(r)/ vi borbottare

jack /dʒæk/ n Auto cric m inv; (in cards) fante m, jack m inv ● **jack up** vt Auto sollevare [con il cric]

jackdaw /'dʒækdɔː/ n taccola f

jacket /'dʒækɪt/ n giacca f; (of book) sopraccoperta f. **~ po'tato** n patata f cotta al forno con la buccia

'jackpot n premio m (di una lotteria); **win the ~** vincere alla lotteria; **hit the ~** fig fare un colpo grosso

jade /dʒeɪd/ n giada f ● attrib di giada

jaded /'dʒeɪdɪd/ a spossato

jagged /'dʒægɪd/ a dentellato

jail /dʒeɪl/ = **gaol**

jalopy /dʒə'lɒpɪ/ n fam vecchia carretta f

jam¹ /dʒæm/ n marmellata f

jam² /dʒæm/ n Auto ingorgo m; (fam: difficulty) guaio m ● v (pt/pp **jammed**) ● vt (cram) pigiare; disturbare (broadcast); inceppare (mechanism, drawer etc); **be ~med** (roads:) essere congestio-

nato ● *vi* 〈mechanism:〉 inceppar-
si; 〈window, drawer:〉 incastrarsi
Jamaica /dʒə'meɪkə/ *n* Giamaica *f*.
~n *a* & *n* giamaicano, -a *mf*
jam-·packed *a fam* pieno zeppo
jangle /'dʒæŋgl/ *vt* far squillare
● *vi* squillare
janitor /'dʒænɪtə(r)/ *n* 〈caretaker〉
custode *m*; 〈in school〉 bidello, -a
mf
January /'dʒænjʊərɪ/ *n* gennaio *m*
Japan /dʒə'pæn/ *n* Giappone *m*.
~ese /dʒæpə'niːz/ *a* & *n* giapponese *mf*; 〈language〉 giapponese *m*
jar[1] /dʒɑː(r)/ *n* 〈glass〉 barattolo *m*
jar[2] *vi* 〈pt/pp **jarred**〉 〈sound:〉 stridere
jargon /'dʒɑːgən/ *n* gergo *m*
jaundice /'dʒɔːndɪs/ *n* itterizia *f*.
~d *a fig* inacidito
jaunt /dʒɔːnt/ *n* gita *f*
jaunty /'dʒɔːntɪ/ *a* (**-ier, -iest**) sba-
razzino
javelin /'dʒævlɪn/ *n* giavellotto *m*
jaw /dʒɔː/ *n* mascella *f*; 〈bone〉 man-
dibola *f*
jay-·walker /'dʒeɪwɔːkə(r)/ *n* pedone *m* distratto
jazz /dʒæz/ *n* jazz *m* ● **jazz up** *vt*
ravvivare. **~y** *a* vistoso
jealous /'dʒeləs/ *a* geloso. **~y** *n* ge-
losia *f*
jeans /dʒiːnz/ *npl* 〈blue〉 jeans *mpl*
jeep /dʒiːp/ *n* jeep *f inv*
jeer /dʒɪə(r)/ *n* scherno *m* ● *vi*
schernire; ~ at prendersi gioco di
● *vt* 〈boo〉 fischiare
jell /dʒel/ *vi* concretarsi
jelly /'dʒelɪ/ *n* gelatina *f*. **~fish** *n*
medusa *f*
jeopar·dize /'dʒepədaɪz/ *vt* mettere in pericolo. **~dy** /-dɪ/ *n* **in ~dy**
in pericolo
jerk /dʒɜːk/ *n* scatto *m*, scossa *f* ● *vt*
scattare ● *vi* sobbalzare; 〈limb,
muscle:〉 muoversi a scatti. **~ily**
adv a scatti. **~y** *a* traballante
jersey /'dʒɜːzɪ/ *n* maglia *f*; Sport
maglietta *f*; 〈fabric〉 jersey *m*
jest /dʒest/ *n* scherzo *m*; **in ~** per
scherzo ● *vi* scherzare

Jesus /'dʒiːzəs/ *n* Gesù *m*
jet[1] /dʒet/ *n* 〈stone〉 giaietto *m*
jet[2] *n* 〈of water〉 getto *m*; 〈nozzle〉
becco *m*; 〈plane〉 aviogetto *m*, jet
m inv
jet: ~·black *a* nero ebano. **~·lag** *n*
scombussolamento da fuso ora-
rio. **~·pro·pelled** *a* a reazione
jettison /'dʒetɪsn/ *vt* gettare a
mare; *fig* abbandonare
jetty /'dʒetɪ/ *n* molo *m*
Jew /dʒuː/ *n* ebreo *m*
jewel /'dʒuːəl/ *n* gioiello *m*. **~ler** *n*
gioielliere *m*; **~ler's [shop]** gioiel-
leria *f*. **~lery** *n* gioielli *mpl*
Jew·ess /'dʒuːɪs/ *n* ebrea *f*. **~ish** *a*
ebreo
jiffy /'dʒɪfɪ/ *n fam* **in a ~** in un batter d'occhio
jigsaw /'dʒɪgsɔː/ *n* ~ **[puzzle]**
puzzle *m inv*
jilt /dʒɪlt/ *vt* piantare
jingle /'dʒɪŋgl/ *n* 〈rhyme〉 canzon-
cina *f* pubblicitaria ● *vi* tintinnare
jinx /dʒɪŋks/ *n* 〈person〉 iettatore,
-trice *mf*; **it's got a ~ on it** è iella-
to
jitter·s /'dʒɪtəz/ *npl fam* **have the
~s** aver una gran fifa. **~y** *a fam* in
preda alla fifa
job /dʒɒb/ *n* lavoro *m*; **this is going
to be quite a ~** *fam* 〈questa〉 non
sarà un'impresa facile; **it's a good
~ that...** meno male che.... **~
centre** *n* ufficio *m* statale di collocamento. **~less** *a* senza lavoro
jockey /'dʒɒkɪ/ *n* fantino *m*
jocular /'dʒɒkjʊlə(r)/ *a* scherzoso
jog /dʒɒg/ *n* colpetto *m*; **at a ~** in
un balzo; Sport **go for a ~** andare
a fare jogging ● *vt* 〈pt/pp **jogged**〉
● *vt* 〈hit〉 urtare; **~ sb's memory**
farlo ritornare in mente a qcno
● *vi* Sport fare jogging. **~ging** *n*
jogging *m*
john /dʒɒn/ *n* 〈Am fam: toilet〉 gabinetto *m*
join /dʒɔɪn/ *n* giuntura *f* ● *vt* raggiungere, unire; raggiungere
〈person〉; 〈become member of〉

iscriversi a; entrare in (*firm*) ● *vi* (*roads*:) congiungersi. **join in** *vi* partecipare. **join up** *vi Mil* arruolarsi ● *vt* unire

joiner /ˈdʒɔɪnə(r)/ *n* falegname *m*

joint /dʒɔɪnt/ *a* comune ● *n* articolazione *f*; (*in wood, brickwork*) giuntura *f*; *Culin* arrosto *m*; (*fam: bar*) bettola *f*; (*sl:drug*) spinello *m*. **~ly** *adv* unitamente

joist /dʒɔɪst/ *n* travetto *m*

jok|e /dʒəʊk/ *n* (*trick*) scherzo *m*; (*funny story*) barzelletta *f* ● *vi* scherzare. **~er** *n* burlone, -a *mf*; (*in cards*) jolly *m inv*. **~ing** *n* **~ing apart** scherzi a parte. **~ingly** *adv* per scherzo

jolly /ˈdʒɒlɪ/ *a* (**-ier, -iest**) allegro ● *adv fam* molto

jolt /dʒəʊlt/ *n* scossa *f*, sobbalzo *m* ● *vt* far sobbalzare ● *vi* sobbalzare

Jordan /ˈdʒɔːdn/ *n* Giordania *f*; (*river*) Giordano *m*. **~ian** /-ˈdeɪnɪən/ *a* & *n* giordano, -a *mf*

jostle /ˈdʒɒsl/ *vt* spingere

jot /dʒɒt/ *n* nulla *f* ● **jot down** *vt* (*pt/pp* jotted) annotare. **~ter** *n* taccuino *m*

journal /ˈdʒɜːnl/ *n* giornale *m*; (*diary*) diario *m*. **~ese** /-əˈliːz/ *n* gergo *m* giornalistico. **~ism** *n* giornalismo *m*. **~ist** *n* giornalista *mf*

journey /ˈdʒɜːnɪ/ *n* viaggio *m*

jovial /ˈdʒəʊvɪəl/ *a* gioviale

joy /dʒɔɪ/ *n* gioia *f*. **~ful** *a* gioioso. **~ride** *n fam* giro *m* con una macchina rubata. **~stick** *n* Comput joystick *m inv*

jubil|ant /ˈdʒuːbɪlənt/ *a* giubilante. **~ation** /-ˈleɪʃn/ *n* giubilo *m*

jubilee /ˈdʒuːbɪliː/ *n* giubileo *m*

judder /ˈdʒʌdə(r)/ *vi* vibrare violentemente

judge /dʒʌdʒ/ *n* giudice *m* ● *vt* giudicare; (*estimate*) valutare; (*consider*) ritenere ● *vi* giudicare (**by** da). **~ment** *n* giudizio *m*; *Jur* sentenza *f*

judic|ial /dʒuːˈdɪʃl/ *a* giudiziario. **~iary** /-ʃərɪ/ *n* magistratura *f*. **~ious** /-ʃəs/ *a* giudizioso

judo /ˈdʒuːdəʊ/ *n* judo *m*

jug /dʒʌg/ *n* brocca *f*; (*small*) bricco *m*

juggernaut /ˈdʒʌgənɔːt/ *n fam* grosso autotreno *m*

juggle /ˈdʒʌgl/ *vi* fare giochi di destrezza. **~r** *n* giocoliere, -a *mf*

juice /dʒuːs/ *n* succo *m*

juicy /ˈdʒuːsɪ/ *a* (**-ier, -iest**) succoso; (*fam: story*) piccante

juke-box /ˈdʒuːk-/ *n* juke-box *m inv*

July /dʒuˈlaɪ/ *n* luglio *m*

jumble /ˈdʒʌmbl/ *n* accozzaglia *f* ● *vt* **~ [up]** mischiare. **~ sale** *n* vendita *f* di beneficenza

jumbo /ˈdʒʌmbəʊ/ *n* **~ [jet]** jumbo jet *m inv*

jump /dʒʌmp/ *n* salto *m*; (*in prices*) balzo *m*; (*in horse racing*) ostacolo *m* ● *vi* saltare; (*with fright*) sussultare; (*prices*:) salire rapidamente; **~ to conclusions** *vt* saltare alle conclusioni ● *vt* saltare; **the gun** *fig* precipitarsi; **~ the queue** non rispettare la fila. **jump at** *vt fig* accettare con entusiasmo (*offer*). **jump up** *vi* rizzarsi in piedi

jumper /ˈdʒʌmpə(r)/ *n* (*sweater*) golf *m inv*

jumpy /ˈdʒʌmpɪ/ *a* nervoso

junction /ˈdʒʌŋkʃn/ *n* (*of roads*) incrocio *m*; (*of motorway*) uscita *f*; Rail nodo *m* ferroviario

juncture /ˈdʒʌŋktʃə(r)/ *n* **at this ~** a questo punto

June /dʒuːn/ *n* giugno *m*

jungle /ˈdʒʌŋgl/ *n* giungla *f*

junior /ˈdʒuːnɪə(r)/ *a* giovane; (*in rank*) subalterno; *Sport* junior *inv* ● *n* the **~s** *Sch* i più giovani. **~ school** *n* scuola *f* elementare

junk /dʒʌŋk/ *n* cianfrusaglie *fpl*. **~ food** *n fam* cibo *m* poco sano, porcherie *fpl*. **~ mail** posta *f* spazzatura

junkie /ˈdʒʌŋkɪ/ *n sl* tossico, -a *mf*

junk-shop *n* negozio *m* di rigattiere

jurisdiction /dʒʊərɪsˈdɪkʃn/ *n* giurisdizione *f*

juror /'dʒʊərə(r)/ n giurato, -a mf

jury /'dʒʊərɪ/ n giuria f

just /dʒʌst/ a giusto ● adv (barely) appena; (simply) solo; (exactly) esattamente; **~ as tall** altrettanto alto; **~ as I was leaving** proprio quando stavo andando via; **I've ~ seen her** l'ho appena vista; **it's ~ as well** meno male; **~ at that moment** proprio in quel momento; **~ listen!** ascolta!; **I'm ~ going** sto andando proprio ora

justice /'dʒʌstɪs/ n giustizia f; **do ~ to** rendere giustizia a; **J~ of the Peace** giudice m conciliatore

justifiable /'dʒʌstɪfaɪəbl/ a giustificabile

justi|fication /dʒʌstɪfɪ'keɪʃn/ n giustificazione f. **~fy** /'dʒʌstɪfaɪ/ vt (pt/pp -ied) giustificare

justly /'dʒʌstlɪ/ adv giustamente

jut /dʒʌt/ vi (pt/pp jutted) **~ out** sporgere

juvenile /'dʒuːvənaɪl/ a giovanile; (childish) infantile; (for the young) per i giovani ● n giovane mf. **~ delinquency** n delinquenza f giovanile

juxtapose /dʒʌkstə'pəʊz/ vt giustapporre

••••••••••••••••••••••••••••••••••

Kk

••••••••••••••••••••••••••••••••••

kangaroo /kæŋgə'ruː/ n canguro m

karate /kə'rɑːtɪ/ n karate m

kebab /kɪ'bæb/ n Culin spiedino m di carne

keel /kiːl/ n chiglia f ● **keel over** vi capovolgersi

keen /kiːn/ a (intense) acuto; (interest) vivo; (eager) entusiastico; (competition) feroce; (wind, knife) tagliente; **~ on** entusiasta di; **she's ~ on him** le piace molto; **be ~ to do sth** avere voglia di fare qcsa. **~ness** n entusiasmo m

keep /kiːp/ n (maintenance) mantenimento m; (of castle) maschio m; **for ~s** per sempre ● v (pt/pp **kept**) ● vt tenere; (not throw away) conservare; (detain) trattenere; mantenere (family, promise); avere (shop); allevare (animals); rispettare (law, rules); **~ sth hot** tenere qcsa in caldo; **~ sb from doing sth** impedire a qcno di fare qcsa; **~ sb waiting** far aspettare qcno; **~ sth to oneself** tenere qcsa per sé; **~ sth from sb** tenere nascosto qcsa a qcno ● vi (remain) rimanere; (food:) conservarsi; **~ calm** rimanere calmo; **~ left/right** tenere la destra/la sinistra; **~ [on] doing sth** continuare a fare qcsa. **keep back** vt trattenere (person); **~ sth back from sb** tenere nascosto qcsa a qcno ● vi tenersi indietro. **keep in with** vt mantenersi in buoni rapporti con. **keep on** vi fam assillare (**at sb** qcno). **keep up** vi stare al passo ● vt (continue) continuare

keeper /'kiːpə(r)/ n custode mf. **~-fit** n ginnastica f. **~ing** n custodia f; **be in ~ing with** essere in armonia con. **~sake** n ricordo m

keg /keg/ n barilotto m

kennel /'kenl/ n canile m; **~s** pl (boarding) canile m; (breeding) allevamento m di cani

Kenya /'kenjə/ n Kenia m. **~n** a e n keniota mf

kept /kept/ see keep

kerb /kɜːb/ n bordo m del marciapiede

kernel /'kɜːnl/ n nocciolo m

kerosene /'kerəsiːn/ n Am cherosene m

ketchup /'ketʃʌp/ n ketchup m

kettle /'ket(ə)l/ n bollitore m; **put the ~ on** mettere l'acqua a bollire

key /kiː/ n also Mus chiave f; (of piano, typewriter) tasto m ● vt ~ **[in]** digitare (character); **could you ~ this?** puoi battere questo?

key: **~board** n Comput, Mus tastiera f. **~boarder** n tastierista mf.

~ed-up *a* (*anxious*) estremamente agitato; (*ready to act*) psicologicamente preparato. **~hole** *n* buco *m* della serratura. **~ring** *n* portachiavi *m inv*

khaki /'kɑːkɪ/ *a* cachi *inv* ● *n* cachi *m*

kick /kɪk/ *n* calcio *m*; (*fam: thrill*) piacere *m*; **for ~s** *fam* per spasso ● *vt* dar calci a; **~ the bucket** *fam* crepare ● *vi* ⟨*animal:*⟩ scalciare; ⟨*person:*⟩ dare calci. **kick off** *vi* *Sport* dare il calcio d'inizio; *fam* iniziare. **kick up** *vt* **~ up a row** fare una scenata

'**kickback** *n* (*fam: percentage*) tangente *f*

'**kick-off** *n* *Sport* calcio *m* d'inizio

kid /kɪd/ *n* capretto *m*; (*fam: child*) ragazzino, -a *mf* ● *v* (*pt/pp* **kidded**) ● *vt* *fam* prendere in giro ● *vi* *fam* scherzare

kidnap /'kɪdnæp/ *vt* (*pt/pp* **-napped**) rapire, sequestrare. **~per** *n* sequestratore, -trice *mf*, rapitore, -trice *mf*. **~ping** *n* rapimento *m*, sequestro *m* [di persona]

kidney /'kɪdnɪ/ *n* rene *m*; *Culin* rognone *m*. **~ machine** *n* rene *m* artificiale

kill /kɪl/ *vt* uccidere; *fig* metter fine a; ammazzare ⟨*time*⟩. **~er** *n* assassino, -a *mf*. **~ing** *n* uccisione *f*; (*murder*) omicidio *m*; **make a ~ing** *fig* fare un colpo grosso

'**killjoy** *n* guastafeste *mf inv*

kiln /kɪln/ *n* fornace *f*

kilo /'kiːləʊ/ *n* chilo *m*

kilo /'kɪlə/: **~byte** *n* kilobyte *m inv*. **~gram** *n* chilogrammo *m*. **~metre** /kɪ'lɒmɪtə(r)/ *n* chilometro *m*. **~watt** *n* chilowatt *m inv*

kilt /kɪlt/ *n* kilt *m inv* (*gonnellino degli scozzesi*)

kin /kɪn/ *n* congiunti *mpl*; **next of ~** parente *m* stretto; parenti *mpl* stretti

kind¹ /kaɪnd/ *n* genere *m*, specie *f*; (*brand, type*) tipo *m*; **~ of** *fam* al-

quanto; **two of a ~** due della stessa specie

kind² *a* gentile, buono; **~ to animals** amante degli animali; **~ regards** cordiali saluti

kindergarten /'kɪndəgɑːtn/ *n* asilo *m* infantile

kindle /'kɪndl/ *vt* accendere

kind|ly /'kaɪndlɪ/ *a* (**-ier**, **-iest**) benevolo ● *adv* gentilmente; (*if you please*) per favore. **~ness** *n* gentilezza *f*

kindred /'kɪndrɪd/ *a* **she's a ~ spirit** è la mia/sua/tua anima gemella

kinetic /kɪ'netɪk/ *a* cinetico

king /kɪŋ/ *n* re *m inv*. **~dom** *n* regno *m*

king: **~fisher** *n* martin *m inv* pescatore. **~-sized** *a* (*cigarette*) king-size *inv*; lungo; ⟨*bed*⟩ matrimoniale grande

kink /kɪŋk/ *n* nodo *m*. **~y** *a* *fam* bizzarro

kiosk /'kiːɒsk/ *n* chiosco *m*; *Teleph* cabina *f* telefonica

kip /kɪp/ *n* *fam* pisolino *m*; **have a ~** schiacciare un pisolino ● *vi* (*pt/pp* **kipped**) *fam* dormire

kipper /'kɪpə(r)/ *n* aringa *f* affumicata

kiss /kɪs/ *n* bacio *m*; **~ of life** respirazione *f* bocca a bocca ● *vt* baciare ● *vi* baciarsi

kit /kɪt/ *n* equipaggiamento *m*, kit *m inv*; (*tools*) attrezzi *mpl*; (*construction* ~) pezzi *mpl* da montare, kit *m inv* ● **kit out** *vt* (*pt/pp* **kitted**) equipaggiare. **~bag** *n* sacco *m* a spalla

kitchen /'kɪtʃɪn/ *n* cucina *f* ● *attrib* di cucina. **~ette** /kɪtʃɪ'net/ *n* cucinino *m*

kitchen: **~ 'garden** *n* orto *m*. **~ roll** *or* **towel** Scottex® *m inv*. **~'sink** *n* lavello *m*

kite /kaɪt/ *n* aquilone *m*

kitten /'kɪtn/ *n* gattino *m*

kitty /'kɪtɪ/ *n* (*money*) cassa *f* comune

kleptomaniac /kleptə'memɪæk/ n cleptomane mf

knack /næk/ n tecnica f; **have the ~ for doing sth** avere la capacità di fare qcsa

knead /ni:d/ vt impastare

knee /ni:/ n ginocchio m. **~cap** n rotula f

kneel /ni:l/ vi (pt/pp **knelt**) ~ **[down]** inginocchiarsi; **be ~ing** essere inginocchiato

knelt /nelt/ see **kneel**

knew /nju:/ see **know**

knickers /'nɪkəz/ npl mutandine fpl

knick-knacks /'nɪknæks/ npl ninnoli mpl

knife /naɪf/ n (pl **knives**) coltello m ● vt fam accoltellare

knight /naɪt/ n cavaliere m; (in chess) cavallo m ● vt nominare cavaliere

knit /nɪt/ vt/i (pt/pp **knitted**) lavorare a maglia; ~ **one, purl one** un diritto, un rovescio. **~ting** n lavorare m a maglia; (work) lavoro m a maglia. **~ting-needle** n ferro m da calza. **~wear** n maglieria f

knives /naɪvz/ see **knife**

knob /nɒb/ n pomello m; (of stick) pomo m; (of butter) noce f. **~bly** a nodoso; (bony) spigoloso

knock /nɒk/ n colpo m; **there was a ~ at the door** hanno bussato alla porta ● vt bussare a 〈door〉; (fam: criticize) denigrare; ~ **a hole in sth** fare un buco in qcsa; ~ **one's head** battere la testa (**on** contro) ● vi (at door) bussare. **knock about** vt malmenare ● vi fam girovagare. **knock down** vt far cadere; (with car) investire con un pugno; (in car) investire; (demolish) abbattere; (fam: reduce) ribassare 〈price〉. **knock off** vt (fam: steal) fregare; (fam: complete quickly) fare alla bell'e meglio ● vi (fam: cease work) staccare. **knock out** vt eliminare; (make unconscious) mettere K.O.; (fam: anaesthetize) addormenta-

re. **knock over** vt rovesciare; (in car) investire

knock: **~-down** a **~-down price** prezzo m stracciato. **~er** n battente m. **~-kneed** /-'ni:d/ a con gambe storte. **~-out** n (in boxing) knock-out m inv

knot /nɒt/ n nodo m ● vt (pt/pp **knotted**) annodare

knotty /'nɒtɪ/ a (**-ier, -iest**) fam spinoso

know /nəʊ/ v (pt **knew**, pp **known**) ● vt sapere; conoscere 〈person, place〉; (recognize) riconoscere; **get to ~ sb** conoscere qcno; ~ **how to swim** sapere nuotare ● vi sapere; **did you ~ about this?** lo sapevi? ● n **in the ~** fam al corrente

know: **~-all** n fam sapientone, -a mf. **~-how** n abilità f. **~ing** a d'intesa. **~ingly** adv (intentionally) consapevolmente; 〈smile etc〉 con un'aria d'intesa

knowledge /'nɒlɪdʒ/ n conoscenza f. **~able** /-əbl/ a ben informato

known /nəʊn/ see **know** ● a noto

knuckle /'nʌkl/ n nocca f ● **knuckle down** vi darci sotto (**to** con). **knuckle under** vi sottomettersi

Koran /kə'rɑːn/ n Corano m

Korea /kə'rɪə/ n Corea f. **~n** a & n coreano, -a mf

kosher /'kəʊʃə(r)/ a kasher inv

kowtow /kaʊ'taʊ/ vi piegarsi

kudos /'kju:dɒs/ n fam gloria f

LI

lab /læb/ n fam laboratorio m

label /'leɪbl/ n etichetta f ● vt (pt/pp **labelled**) mettere un'etichetta a; fig etichettare 〈person〉

laboratory /lə'bɒrətrɪ/ n laboratorio m

laborious /lə'bɔːrɪəs/ a laborioso

labour /'leɪbə(r)/ n lavoro m; (workers) manodopera f; Med doglie fpl; **be in ~** avere le doglie; **L~** Pol partito m laburista ●attrib Pol laburista ●vi lavorare ●vt ~ **the point** fig ribadire il concetto. **~er** n manovale m

labour-saving a che fa risparmiare lavoro e fatica

labyrinth /'læbərɪnθ/ n labirinto m

lace /leɪs/ n pizzo m; (of shoe) laccio m ●attrib di pizzo ●vt allacciare (shoes); correggere (drink)

lacerate /'læsəreɪt/ vt lacerare

lack /læk/ n mancanza f ●vt mancare di; **I ~ the time** mi manca il tempo ●vi **be ~ing** mancare; **be ~ing in sth** mancare di qcsa

lackadaisical /lækə'deɪzɪkl/ a senza entusiasmo

laconic /lə'kɒnɪk/ a laconico

lacquer /'lækə(r)/ n lacca f

lad /læd/ n ragazzo m

ladder /'lædə(r)/ n scala f; (in tights) sfilatura f

laden /'leɪdn/ a carico (with di)

ladle /'leɪdl/ n mestolo m ●vt ~ [out] versare (col mestolo)

lady /'leɪdɪ/ n signora f; (title) Lady; **ladies** [room] bagno m per donne

lady: **~bird** n, Am **~bug** n coccinella f. **~like** a signorile

lag¹ /læg/ vi (pt/pp **lagged**) ~ **behind** restare indietro

lag² vt (pt/pp **lagged**) isolare (pipes)

lager /'lɑːgə(r)/ n birra f chiara

lagoon /lə'guːn/ n laguna f

laid /leɪd/ see **lay³**

lain /leɪn/ see **lie²**

lair /leə(r)/ n tana f

lake /leɪk/ n lago m

lamb /læm/ n agnello m

lame /leɪm/ a zoppo; fig (argument) zoppicante; (excuse) traballante

lament /lə'ment/ n lamento m ●vt lamentare ●vi lamentarsi

lamentable /'læməntəbl/ a deplorevole

laminated /'læmɪneɪtɪd/ a laminato m

lamp /læmp/ n lampada f; (in street) lampione m. **~post** n lampione m. **~shade** n paralume m

lance /lɑːns/ n fiocina f ●vt Med incidere. **~-'corporal** n appuntato m

land /lænd/ n terreno m; (country) paese m; (as opposed to sea) terra f; **plot of ~** pezzo m di terreno ●vt Naut sbarcare; (fam: obtain) assicurarsi; **be ~ed with sth** fam ritrovarsi fra capo e collo qcsa ●vi Aeron atterrare; (fall) cadere. **land up** fam finire

landing /'lændɪŋ/ n Naut sbarco m; Aeron atterraggio m; (top of stairs) pianerottolo m. **~-stage** n pontile m da sbarco. **~ strip** n pista f d'atterraggio di fortuna

land: **~lady** n proprietaria f; (of flat) padrona f di casa. **~-locked** a privo di sbocco sul mare. **~lord** n proprietario m; (of flat) padrone m di casa. **~mark** n punto m di riferimento; fig pietra f miliare. **~owner** n proprietario, -a mf terriero. **~scape** /-skeip/ n paesaggio m. **~slide** n frana f; Pol valanga f di voti

lane /leɪn/ n sentiero m; Auto, Sport corsia f

language /'læŋgwɪdʒ/ n lingua f; (speech, style) linguaggio m. **~ laboratory** n laboratorio m linguistico

languid /'læŋgwɪd/ a languido

languish /'læŋgwɪʃ/ vi languire

lank /læŋk/ a (hair) diritto

lanky /'læŋkɪ/ a (-ier, -iest) allampanato

lantern /'læntən/ n lanterna f

lap¹ /læp/ n grembo m

lap² n (of journey) tappa f; Sport giro m ●vi (pt/pp **lapped**) (water:) ~ **against** lambire ●vi Sport doppiare

lap³ vt (pt/pp **lapped**) ~ **up** bere avidamente; bersi completamente

⟨lies⟩; credere ciecamente a ⟨praise⟩

lapel /lə'pel/ n bavero m

lapse /læps/ n sbaglio m; (moral) sbandamento m [morale]; (of time) intervallo m ● vi (expire) scadere; (morally) scivolare; ~ **into** cadere in

laptop /'læptɒp/ n ~ [**computer**] computer m inv portabile, laptop m inv

larceny /'lɑ:sənɪ/ n furto m

lard /lɑ:d/ n strutto m

larder /'lɑ:də(r)/ n dispensa f

large /lɑ:dʒ/ a grande; ⟨number, amount⟩ grande, grosso; **by and** ~ in complesso; **at** ~ in libertà; (in general) ampiamente; ~**ly** adv ampiamente; ~**ly because of** in gran parte a causa di

lark[1] /lɑ:k/ n ⟨bird⟩ allodola f

lark[2] n ⟨joke⟩ burla f ● **lark about** vi giocherellare

larva /'lɑ:və/ n (pl -**vae** /-vi:/) larva f

laryngitis /lærɪn'dʒaɪtɪs/ n laringite f

larynx /'lærɪŋks/ n laringe f

lascivious /lə'sɪvɪəs/ a lascivo

laser /'leɪzə(r)/ n laser m inv. ~ [**printer**] n stampante f laser

lash /læʃ/ n frustata f; ⟨eyelash⟩ ciglio m ● vt ⟨whip⟩ frustare; ⟨tie⟩ legare fermamente. **lash out** vi attaccare; (spend) sperperare (**on** in)

lashings /'læʃɪŋz/ npl ~ **of** fam una marea di

lass /læs/ n ragazzina f

lasso /læ'su:/ n lazo m

last[1] /lɑ:st/ a (final) ultimo; (recent) scorso; ~ **year** l'anno scorso; ~ **night** ieri sera; **at** ~ alla fine; **at** ~! finalmente!; **that's the** ~ **straw** fam questa è l'ultima goccia ● n ultimo, -a mf; **the** ~ **but one** il penultimo ● adv per ultimo; (last time) l'ultima volta ● vi durare. ~**ing** a durevole. ~**ly** adv infine

late /leɪt/ a (delayed) in ritardo; (at a late hour) tardo; (deceased) de-

funto; **it's** ~ (at night) è tardi; **in** ~ **November** alla fine di Novembre ● adv tardi; **stay up** ~ stare alzati fino a tardi. ~**comer** n ritardatario, -a mf; (to political party etc) nuovo, -a arrivato, -a mf. ~**ly** adv recentemente. ~**ness** n ora f tarda; (delay) ritardo m

latent /'leɪtnt/ a latente

later /'leɪtə(r)/ a ⟨train⟩ che parte più tardi; ⟨edition⟩ più recente ● adv più tardi; ~ **on** più tardi, dopo

lateral /'lætərəl/ a laterale

latest /'leɪtɪst/ a ultimo; (most recent) più recente; **the** ~ [**news**] le ultime notizie ● n **six o'clock at the** ~ alle sei al più tardi

lathe /leɪð/ n tornio m

lather /'lɑ:ðə(r)/ n schiuma f ● vt insaponare ● vi far schiuma

Latin /'lætɪn/ a latino ● n latino m. ~ **America** n America f Latina. ~ **American** a & n latino-americano, -a mf

latitude /'lætɪtju:d/ n Geog latitudine f; fig libertà f d'azione

latter /'lætə(r)/ a ultimo ● **the** ~ quest'ultimo. ~**ly** adv ultimamente

lattice /'lætɪs/ n traliccio m

Latvia /'lætvɪə/ n Lettonia f. ~**n** a & n lettone mf

laudable /'lɔ:dəbl/ a lodevole

laugh /lɑ:f/ n risata f ● vi ridere (**at/about** di); ~ **at sb** (mock) prendere in giro qcno. ~**able** /-əbl/ a ridicolo. ~**ing-stock** n zimbello m

laughter /'lɑ:ftə(r)/ n risata f

launch[1] /lɔ:ntʃ/ n ⟨boat⟩ varo m

launch[2] n lancio m; (of ship) varo m ● vt lanciare ⟨rocket, product⟩; varare ⟨ship⟩; sferrare ⟨attack⟩

launder /'lɔ:ndə(r)/ vt lavare e stirare; ~ **money** fig riciclare denaro sporco. ~**ette** /-'dret/ n lavanderia f automatica

laundry /'lɔ:ndrɪ/ n lavanderia f; (clothes) bucato m

laurel /'lʊrəl/ *n* lauro *m*; *fig* rest on one's ~s dormire sugli allori

lava /'lɑːvə/ *n* lava *f*

lavatory /'lævətrɪ/ *n* gabinetto *m*

lavender /'lævəndə(r)/ *n* lavanda *f*

lavish /'lævɪʃ/ *a* copioso; *(wasteful)* prodigo; **on a ~ scale** su vasta scala ● *vt* ~ **sth on sb** ricoprire qcno di qcsa. **~ly** *adv* copiosamente

law /lɔː/ *n* legge *f*; **study** ~ studiare giurisprudenza, studiare legge; ~ **and order** ordine *m* pubblico

law: **~-abiding** *a* che rispetta la legge. **~court** *n* tribunale *m*. **~less** *a* senza legge. **~ school** *n* facoltà *f* di giurisprudenza

lawn /lɔːn/ *n* prato *m* [all'inglese]. **~-mower** *n* tosaerba *m inv*

'law suit *n* causa *f*

lawyer /'lɔːjə(r)/ *n* avvocato *m*

lax /læks/ *a* negligente; *(morals etc)* lassista

laxative /'læksətɪv/ *n* lassativo *m*

laxity /'læksətɪ/ *n* lassismo *m*

lay¹ /leɪ/ *a* laico; *fig* profano

lay² *see* **lie²**

lay³ *vt* *(pt/pp* laid*)* porre, mettere; apparecchiare *(table)* ● *vi (hen.)* fare le uova. **lay down** *vt* posare; stabilire *(rules, conditions)*. **lay off** *vt* licenziare *(workers)* ● *vi (fam: stop)* ~ **off!** smettila! **lay out** *vt (display, set forth)* esporre; *(plan)* pianificare *(garden)*; *(spend)* sborsare; *Typ* impaginare

lay: **~about** *n* fannullone, -a *mf*. **~-by** *n* corsia *f* di sosta

layer /'leɪə(r)/ *n* strato *m*

lay: **~man** *n* profano *m*. **~out** *n* disposizione *f*; *Typ* impaginazione *f*, layout *m inv*

laze /leɪz/ *vi* ~ **[about]** oziare

laziness /'leɪzɪnɪs/ *n* pigrizia *f*

lazy /'leɪzɪ/ *a* (**-ier**, **-iest**) pigro. **~-bones** *n* poltrone, -a *mf*

lb *abbr* (**pound**) libbra

lead¹ /led/ *n* piombo *m*; *(of pencil)* mina *f*

lead² /liːd/ *n* guida *f*; *(leash)*

giunzaglio *m*; *(flex)* filo *m*; *(clue)* indizio *m*; *Theat* parte *f* principale; *(distance ahead)* distanza *f* *(over* su*)*; **in the** ~ in testa ● *v (pt/pp* led*)* ● *vt* condurre; *(expedition, party etc)*; *(induce)* indurre; ~ **the way** mettersi in testa ● *vi (be in front)* condurre; *(in race, competition)* essere in testa; *(at cards)* giocare *(per* primo*)*

lead away *vt* portar via. **lead to** *vt* portare a. **lead up to** *vt* preludere; **what's this ~ing up to?** dove porta questo?

leaded /'ledɪd/ *a* con piombo

leader /'liːdə(r)/ *n* capo *m*; *(of orchestra)* primo violino *m*; *(in newspaper)* articolo *m* di fondo. **~ship** *n* direzione *f*, leadership *f inv*; **show ~ship** mostrare capacità di comando

lead-'free *a* senza piombo

leading /'liːdɪŋ/ *a* principale; ~ **lady/man** attrice *f*/attore *m* principale; ~ **question** domanda *f* tendenziosa

leaf /liːf/ *n* *(pl* **leaves**) foglia *f*; *(of table)* asse *f* ● **leaf through** *vt* sfogliare. **~let** *n* dépliant *m inv*; *(advertising)* dépliant *m inv* pubblicitario; *(political)* manifestino *m*

league /liːg/ *n* lega *f*; *Sport* campionato *m*; **be in ~ with** essere in combutta con

leak /liːk/ *n* *(hole)* fessura *f*; *Naut* falla *f*; *(of gas & fig)* fuga *f* ● *vi* colare; *(ship:)* fare acqua; *(liquid, gas:)* fuoriuscire ● *vt* ~ **sth to sb** *fig* far trapelare qcsa a qcno. **~y** *a* che perde; *Naut* che fa acqua

lean¹ /liːn/ *a* magro

lean² *v (pt/pp* leaned *or* leant /lent/*)* ● *vt* appoggiare **(against/on** contro/su*)* ● *vi* appoggiarsi **(against/on** contro/su*)*; *(not be straight)* pendere; **be ~ing against** essere appoggiato contro; ~ **on sb** *(depend on)* appoggiarsi a qcno; *(fam: exert pressure on)* stare alle calcagne di qcno. **lean**

back *vi* sporgersi indietro. **lean forward** *vi* piegarsi in avanti. **lean out** *vi* sporgersi. **lean over** *vi* piegarsi

leaning /'li:nɪŋ/ *a* pendente; **the L~ Tower of Pisa** la torre di Pisa, la torre pendente ● *n* tendenza *f*

leap /li:p/ *n* salto *m* ● *vi* (*pt/pp* **leapt** /lept/ *or* **leaped**) saltare; **he leapt at it** *fam* l'ha preso al volo. **~-frog** *n* cavallina *f*. **~ year** *n* anno *m* bisestile

learn /lɜ:n/ *vt* (*pt/pp* **learnt** *or* **learned**) imparare; **~ to swim** imparare a nuotare; **I have ~ed that...** (*heard*) sono venuto a sapere che... ● *vi* imparare

learn|ed /'lɜ:nɪd/ *a* colto. **~er** *n* also *Auto* principiante *mf*. **~ing** *n* cultura *f*

lease /li:s/ *n* contratto *m* d'affitto; (*rental*) affitto *m* ● *vt* affittare

leash /li:ʃ/ *n* guinzaglio *m*

least /li:st/ *a* più piccolo; (*amount*) minore; **you've got ~ luggage** hai meno bagagli di tutti ● *n* the **~** il meno; **at ~** almeno; **not in the ~** niente affatto ● *adv* meno; **the ~ expensive wine** il vino meno caro

leather /'leðə(r)/ *n* pelle *f*; (*of soles*) cuoio *m* ● *attrib* di pelle/cuoio. **~y** *a* (*meat, skin*) duro

leave /li:v/ *n* (*holiday*) congedo *m*; *Mil* licenza *f*; **on ~** in congedo/licenza ● *v* (*pt/pp* **left**) ● *vt* lasciare; uscire da (*house, office*); (*forget*) dimenticare; **there is nothing left** non è rimasto niente ● *vi* andare via; (*train, bus:*) partire. **leave behind** *vt* lasciare; (*forget*) dimenticare. **leave out** *vt* omettere; (*not put away*) lasciare fuori

leaves /li:vz/ *see* **leaf**

Leban|on /'lebənən/ *n* Libano *m* **~ese** /-'ni:z/ *a & n* libanese *mf*

lecherous /'letʃərəs/ *a* lascivo

lectern /'lektɜ:n/ *n* leggio *m*

lecture /'lektʃə(r)/ *n* conferenza *f*; *Univ* lezione *f*; (*reproof*) ramanzi-

na *f* ● *vi* fare una conferenza (**on** su); *Univ* insegnare (**on sth** qcsa) ● *vt* **~ sb** rimproverare qcno. **~r** *n* conferenziere, -a *mf*; *Univ* docente *mf* universitario, -a

led /led/ *see* **lead**[2]

ledge /ledʒ/ *n* cornice *f*; (*of window*) davanzale *m*

ledger /'ledʒə(r)/ *n* libro *m* mastro

leech /li:tʃ/ *n* sanguisuga *f*

leek /li:k/ *n* porro *m*

leer /lɪə(r)/ *n* sguardo *m* libidinoso ● *vi* **~ [at]** guardare in modo libidinoso

leeway /'li:weɪ/ *n* fig libertà *f* di azione

left[1] /left/ *see* **leave**

left[2] *a* a sinistra ● *adv* a sinistra ● *n* also *Pol* sinistra *f*; **on the ~** a sinistra;

left: **~-'handed** *a* mancino. **~-'luggage [office]** *n* deposito *m* bagagli. **~overs** *npl* rimasugli *mpl*. **~-'wing** *a Pol* di sinistra

leg /leg/ *n* gamba *f*; (*of animal*) zampa *f*; (*of journey*) tappa *f*; *Culin* (*of chicken*) coscia *f*; (*of lamb*) cosciotto *m*

legacy /'legəsɪ/ *n* lascito *m*

legal /'li:gl/ *a* legale; **take ~ action** intentare un'azione legale. **~ly** *adv* legalmente

legality /lɪ'gælətɪ/ *n* legalità *f*

legalize /'li:gəlaɪz/ *vt* legalizzare

legend /'ledʒənd/ *n* leggenda *f*. **~ary** *a* leggendario

legible /'ledʒəbl/ *a* leggibile. **~ly** *adv* in modo leggibile

legislat|e /'ledʒɪsleɪt/ *vi* legiferare. **~ion** /-'leɪʃn/ *n* legislazione *f*

legislat|ive /'ledʒɪslətɪv/ *a* legislativo. **~ure** /-leɪtʃə(r)/ *n* legislatura *f*

legitimate /lɪ'dʒɪtɪmət/ *a* legittimo; (*excuse*) valido

leisure /'leʒə(r)/ *n* tempo *m* libero; **at your ~** con comodo. **~ly** *a* senza fretta

lemon /'lemən/ *n* limone *m*. **~ade** /-'neɪd/ *n* limonata *f*

lend /lend/ *vt* (*pt/pp* **lent**) prestare;

~ **a hand** *fig* dare una mano. **~ing library** *n* biblioteca *f* per il prestito

length /leŋθ/ *n* lunghezza *f*; *(piece)* pezzo *m*; *(of wallpaper)* parte *f*; *(of visit)* durata *f*; **at** ~ a lungo; *(at last)* alla fine

length|en /'leŋθən/ *vt* allungare ● *vi* allungarsi. **~ways** *adv* per lungo

lengthy /'leŋθɪ/ *a* (**-ier, -iest**) lungo

lenien|ce /'li:nɪəns/ *n* indulgenza *f*. **~t** *a* indulgente

lens /lenz/ *n* lente *f*; *Phot* obiettivo *m*; *(of eye)* cristallino *m*

Lent *n* Quaresima *f*

lent /lent/ *see* **lend**

lentil /'lentl/ *n* *Bot* lenticchia *f*

Leo /'li:əʊ/ *n* *Astr* Leone *m*

leopard /'lepəd/ *n* leopardo *m*

leotard /'li:əta:d/ *n* body *m* *inv*

leprosy /'leprəsɪ/ *n* lebbra *f*

lesbian /'lezbɪən/ *a* lesbico ● *n* lesbica *f*

less /les/ *a* meno di; ~ **and** ~ sempre meno ● *adv* & *prep* meno ● *n* meno *m*

lessen /'lesn/ *vt/i* diminuire

lesser /'lesə(r)/ *a* minore

lesson /'lesn/ *n* lezione *f*

lest /lest/ *conj liter* per timore che

let /let/ *vt* (*pt/pp* **let**, *pres p* **letting**) lasciare, permettere; *(rent)* affittare; ~ **alone** *(not to mention)* tanto meno; '**to** ~' 'affittasi'; ~ **us go** andiamo; ~ **sb do sth** lasciare fare qcsa a qcno, permettere a qcno di fare qcsa; ~ **me know** fammi sapere; **just** ~ **him try!** che ci provi solamente!; ~ **oneself in for sth** *fam* impelagarsi in qcsa. **let down** *vt* sciogliersi *(hair)*; abbassare *(blinds)*; *(lengthen)* allungare; *(disappoint)* deludere; **don't** ~ **me down** conto su di te. **let in** *vt* far entrare. **let off** *vt* far partire; *(not punish)* perdonare; ~ **sb off doing sth** abbonare qcsa a qcno. **let out** *vt* far uscire; *(make larger)* allargare; emettere

(scream, groan). **let through** *vt* far passare. **let up** *vi* *fam* diminuire

let-down *n* delusione *f*

lethal /'li:θl/ *a* letale

letharg|ic /lɪ'θɑ:dʒɪk/ *a* apatico. **~y** /'leθədʒɪ/ *n* apatia *f*

letter /'letə(r)/ *n* lettera *f*. **~-box** *n* buca *f* per le lettere. **~-head** *n* carta *f* intestata. **~ing** *n* caratteri *mpl*

lettuce /'letɪs/ *n* lattuga *f*

let-up *n* *fam* pausa *f*

leukaemia /lu:'ki:mɪə/ *n* leucemia *f*

level /'levl/ *a* piano; *(in height, competition)* allo stesso livello; *(spoonful)* raso; **draw** ~ **with sb** affiancare qcno ● *n* livello *m*; **on the** ~ *fam* giusto ● *vt* *(pt/pp* **levelled)** livellare; *(aim)* puntare *(at* su)

level: ~ **'crossing** *n* passaggio *m* a livello. **~-headed** *a* posato

lever /'li:və(r)/ *n* leva *f* ● **lever up** *vt* sollevare *(con una leva).* **~age** /-rɪdʒ/ *n* azione *f* di una leva; *fig* influenza *f*

levy /'levɪ/ *vt* *(pt/pp* **levied)** imporre *(tax)*

lewd /lju:d/ *a* osceno

liability /laɪə'bɪlətɪ/ *n* responsabilità *f*; *(fam: burden)* peso *m*; **~ies** *pl* debiti *mpl*

liable /'laɪəbl/ *a* responsabile *(for* di); **be** ~ **to** *(rain, break etc)* rischiare di; *(tend to)* tendere a

liaise /lɪ'eɪz/ *vi* *fam* essere in contatto

liaison /lɪ'eɪzɒn/ *n* contatti *mpl*; *Mil* collegamento *m*; *(affair)* relazione *f*

liar /'laɪə(r)/ *n* bugiardo, -a *mf*

libel /'laɪbl/ *n* diffamazione *f* ● *vt* *(pt/pp* **libelled)** diffamare. **~lous** *a* diffamatorio

liberal /'lɪb(ə)rəl/ *a* *(tolerant)* di larghe vedute; *(generous)* generoso. **L~** *a* *Pol* liberale ● *n* liberale *mf*

liberat|e /'lɪbəreɪt/ *vt* liberare. **~ed** *a (woman)* emancipata. **~ion** /-'reɪʃn/ *n* liberazione *f*; *(of* *women*) emancipazione *f*. ~**or** *n* liberatore, -trice *mf*

liberty /'lɪbətɪ/ *n* libertà *f*; **take the** ~ **of doing sth** prendersi la libertà di fare qcsa; **be at** ~ **to do sth** essere libero di fare qcsa

Libra /'liːbrə/ *n Astr* Bilancia *f*

librarian /laɪˈbreərɪən/ *n* bibliotecario, -a *mf*

library /'laɪbrərɪ/ *n* biblioteca *f*

Libya /'lɪbɪə/ *n* Libia *f*. ~**n** *a* & *n* libico, -a *mf*

lice /laɪs/ *see* **louse**

licence /'laɪsns/ *n* licenza *f*; (*for TV*) canone *m* televisivo; (*for driving*) patente *f*; (*freedom*) sregolatezza *f*. ~**-plate** *n* targa *f*

license /'laɪsns/ *vt* autorizzare; **be** ~**d** (*car*:) avere il bollo; (*restaurant*:) essere autorizzato alla vendita di alcolici

licentious /laɪˈsenʃəs/ *a* licenzioso

lick /lɪk/ *n* leccata *f*; **a** ~ **of paint** una passata leggera di pittura ● *vt* leccare; (*fam: defeat*) battere; **leccarsi** (*lips*)

lid /lɪd/ *n* coperchio *m*; (*of eye*) palpebra *f*

lie[1] /laɪ/ *n* bugia *f*; **tell a** ~ **mentire** ● *vi* (*pt/pp* **lied**, *pres p* **lying**) mentire

lie[2] *vi* (*pt* **lay**, *pp* **lain**, *pres p* **lying**) (*person*:) sdraiarsi; (*object*:) stare; (*remain*) rimanere; **leave sth lying about** *or* **around** lasciare qcsa in giro. **lie down** *vi* sdraiarsi

'**lie:** ~**-down** *n* **have a** ~**-down** fare un riposino. ~**-in** *n fam* **have a** ~**-in** restare a letto fino a tardi

lieu /ljuː/ *n* **in** ~ **of** in luogo di

lieutenant /lefˈtenənt/ *n* tenente *m*

life /laɪf/ *n* (*pl* **lives**) vita *f*

life: ~**belt** *n* salvagente *m*. ~**boat** *n* lancia *f* di salvataggio; (*on ship*) scialuppa *f* di salvataggio. ~**buoy** *n* salvagente *m*. ~**guard** *n* bagnino *m*. ~**insurance** *n* assicurazione *f* sulla vita. ~**jacket** *n* giubbotto *m* di salvataggio. ~**less** *a* inanimato. ~**like** *a* realistico. ~**long** *a*

di tutta la vita. ~**size[d]** *a* in grandezza naturale. ~**time** *n* vita *f*; **the chance of a** ~**time** un'occasione unica

lift /lɪft/ *n* ascensore *m*; *Auto* passaggio *m* ● *vt* sollevare; revocare (*restrictions*); (*fam: steal*) rubare ● *vi* (*fog*:) alzarsi. **lift up** *vt* sollevare

'**lift-off** *n* decollo *m* (*di razzo*)

ligament /'lɪgəmənt/ *n Anat* legamento *m*

light[1] /laɪt/ *a* (*not dark*) luminoso; ~ **green** verde chiaro ● *n* luce *f*; (*lamp*) lampada *f*; **in the** ~ **of** *fig* alla luce di; **have you got a** ~? ha da accendere?; **come to** ~ essere rivelato ● *vt* (*pt/pp* **lit** *or* **lighted**) accendere; (*illuminate*) illuminare. **light up** *vi* (*face*:) illuminarsi

light[2] *a* (*not heavy*) leggero ● *adv* **travel** ~ viaggiare con poco bagaglio

'**light-bulb** *n* lampadina *f*

lighten[1] /'laɪtn/ *vt* illuminare

lighten[2] *vt* alleggerire (*load*)

lighter /'laɪtə(r)/ *n* accendino *m*

light: ~**-fingered** *a* svelto di mano. ~**-headed** *a* sventato. ~**-hearted** *a* spensierato. ~**house** *n* faro *m*. ~**ing** *n* illuminazione *f*. ~**ly** *adv* leggermente; (*accuse*) con leggerezza; (*without concern*) senza dare importanza alla cosa; **get off** ~**ly** cavarsela a buon mercato. ~**ness** *n* leggerezza *f*

lightning /'laɪtnɪŋ/ *n* lampo *m*, fulmine *m*. ~**-conductor** *n* parafulmine *m*

light: ~**weight** *a* leggero ● *n* (*in boxing*) peso *m* leggero. ~ **year** *n* anno *m* luce

like[1] /laɪk/ *a* simile ● *prep* come; ~ **this/that** così; **what's he** ~? com'è? ● *conj* (*fam: as*) come; (*Am: as if*) come se

like[2] *vt* piacere, gradire; **I should/ would** ~ vorrei, gradirei; **I** ~ **him** mi piace; **I** ~ **this car** mi piace questa macchina; **I** ~ **dancing** mi piace ballare; **I** ~ **that!** *fam* questa

mi è piaciuto ● *n* ~s **and dislikes**
pl gusti *mpl*
like|able /'laɪkəbl/ *a* simpatico.
~**lihood** /-lɪhʊd/ *n* probabilità *f*.
~**ly** *a* (-ier, -iest) probabile ● *adv*
probabilmente; **not** ~**ly!** *fam* neanche per sogno!

like-'minded *a* con gusti affini
liken /'laɪkən/ *vt* paragonare (**to**).
like|ness /'laɪknɪs/ *n* somiglianza
f. '~**wise** *adv* lo stesso

liking /'laɪkɪŋ/ *n* gusto *m*; **is it to
your** ~? è di suo gusto?; **take a** ~
to sb prendere qcno in simpatia
lilac /'laɪlək/ *n* lillà *m* ● *a* color lillà

lily /'lɪlɪ/ *n* giglio *m*. ~ **of the valley**
n mughetto *m*

limb /lɪm/ *n* arto *m*
limber /'lɪmbə(r)/ *vi* ~ **up** sciogliersi i muscoli

lime[1] /laɪm/ *n* (*fruit*) cedro *m*;
(*tree*) tiglio *m*

lime[2] *n* calce *f*. '~**light** *n* **be in the**
~**light** essere molto in vista.
'~**stone** *n* calcare *m*

limit /'lɪmɪt/ *n* limite *m*; **that's
the** ~! *fam* questo è troppo! ● *vt*
limitare (**to** a). ~**ation** /-'teɪʃn/ *n*
limite *m*. ~**ed** *a* ristretto; ~**ed
company** società *f* anonima
limousine /'lɪməziːn/ *n* limousine
f inv

limp[1] /lɪmp/ *n* andatura *f* zoppicante; **have a** ~ zoppicare ● *vi* zoppicare

limp[2] *a* floscio
line[1] /laɪn/ *n* linea *f*; (*length of rope,
cord*) filo *m*; (*of writing*) riga *f*; (*of
poem*) verso *m*; (*row*) fila *f*;
(*wrinkle*) ruga *f*; (*of business*) settore *m*; (*Am: queue*) coda *f*; **in** ~
with in conformità con ● *vt* segnare; fiancheggiare (*street*). **line up**
vi allinearsi ● *vt* allineare

line[2] *vt* foderare (*garment*)
linear /'lɪnɪə(r)/ *a* lineare
lined[1] /laɪnd/ *a* (*face*) rugoso;
(*paper*) a righe
lined[2] *a* (*garment*) foderato

linen /'lɪnɪn/ *n* lino *m*; (*articles*)
biancheria *f* ● *attrib* di lino
liner /'laɪnə(r)/ *n* nave *f* di linea
linesman /'laɪnzmən/ *n Sport* guardalinee *m*
inv
linger /'lɪŋɡə(r)/ *vi* indugiare
lingerie /'lɒʒərɪ/ *n* biancheria *f* intima (*da donna*)
linguist /'lɪŋɡwɪst/ *n* linguista *m*
linguistic /lɪŋ'ɡwɪstɪk/ *a* linguistico. ~**s** *n* linguistica *fsg*
lining /'laɪnɪŋ/ *n* (*of garment*) fodera *f*; (*of brakes*) guarnizione *f*
link /lɪŋk/ *n* (*of chain*) anello *m*; *fig*
legame *m* ● *vt* collegare. **link up** *vi*
unirsi (**with** a); *TV* collegarsi
lino /'laɪnəʊ/ *n*, **linoleum**
/lɪ'nəʊlɪəm/ *n* linoleum *m*
lint /lɪnt/ *n* garza *f*
lion /'laɪən/ *n* leone *m*. ~**ess** *n*
leonessa *f*
lip /lɪp/ *n* labbro *m* (*pl* labbra *f*);
(*edge*) bordo *m*
lip: ~-**read** *vi* leggere le labbra;
~-**reading** *n* lettura *f* delle labbra.
~-**service** *n* **pay** ~-**service to** approvare soltanto a parole. ~**salve**
n burro *m* [di] cacao. ~**stick** *n* rossetto *m*
liqueur /lɪ'kjʊə(r)/ *n* liquore *m*
liquid /'lɪkwɪd/ *n* liquido *m* ● *a* liquido
liquidat|e /'lɪkwɪdeɪt/ *vt* liquidare.
~**ion** /-'deɪʃn/ *n* liquidazione *f*;
Comm **go into** ~**ion** andare in liquidazione
liquidize /'lɪkwɪdaɪz/ *vt* rendere liquido. ~**r** *n Culin* frullatore *m*
liquor /'lɪkə(r)/ *n* bevanda *f* alcoolica
liquorice /'lɪkərɪs/ *n* liquirizia *f*
liquor store *n Am* negozio *m* di alcolici
lisp /lɪsp/ *n* pronuncia *f* con la lisca
● *vi* parlare con la lisca
list[1] /lɪst/ *n* lista *f* ● *vt* elencare
list[2] *vi* (*ship:*) inclinarsi
listen /'lɪsn/ *vi* ascoltare; ~ **to**
ascoltare. ~**er** *n* ascoltatore, -trice
mf

listings /'lɪstɪŋz/ npl TV programma m

listless /'lɪstlɪs/ a svogliato

lit /lɪt/ see **light**[1]

literacy /'lɪtərəsɪ/ n alfabetizzazione f

literal /'lɪtərəl/ a letterale. **~ly** adv letteralmente

literary /'lɪtərərɪ/ a letterario

literate /'lɪtərət/ a be ~ saper leggere e scrivere

literature /'lɪtrətʃə(r)/ n letteratura f

Lithuania /lɪθjʊ'eɪnɪə/ n Lituania f. **~n** a & n lituano, -a mf

litigation /lɪtɪ'geɪʃn/ n causa f [giudiziaria]

litre /'liːtə(r)/ n litro m

litter /'lɪtə(r)/ n immondizie fpl; Zool figliata f ● vt be ~ed with essere ingombrato di. **~bin** n bidone m della spazzatura

little /'lɪtl/ a piccolo; (not much) poco ● adv & n poco m; **a ~** un po'; **a ~ water** un po' d'acqua; **a ~ better** un po' meglio; **by ~ by ~** a poco a poco

liturgy /'lɪtədʒɪ/ n liturgia f

live[1] /laɪv/ a vivo; (ammunition) carico; **~ broadcast** trasmissione f in diretta; **be ~** Electr essere sotto tensione; **~ wire** n fig persona f dinamica ● adv (broadcast) in diretta

live[2] /lɪv/ vi vivere; (reside) abitare; **~ with** convivere con. **live down** vt far dimenticare. **live off** vt vivere alle spalle di. **live on** vt vivere di ● vi sopravvivere. **live up** ~ **it up** far la bella vita. **live up to** vt essere all'altezza di

livelihood /'laɪvlɪhʊd/ n mezzi mpl di sostentamento. **~ness** n vivacità f

lively /'laɪvlɪ/ a (-ier, -iest) vivace

liven /'laɪvn/ vt ~ **up** vivacizzare ● vi vivacizzarsi

liver /'lɪvə(r)/ n fegato m

lives /laɪvz/ see **life**

livestock /'laɪv-/ n bestiame m

livid /'lɪvɪd/ a fam livido

living /'lɪvɪŋ/ a vivo ● n earn one's ~ guadagnarsi da vivere; **the ~** pl i vivi. **~-room** n soggiorno m

lizard /'lɪzəd/ n lucertola f

load /ləʊd/ n carico m; **~s of** fam un sacco di ● vt caricare. **~ed** a carico; (fam: rich) ricchissimo

loaf[1] /ləʊf/ n (pl **loaves**) pagnotta f

loaf[2] vi oziare

loan /ləʊn/ n prestito m; **on ~** in prestito ● vt prestare

loath /ləʊθ/ a **be ~ to do sth** essere restio a fare qcsa

loathe /ləʊð/ vt detestare. **~ing** n disgusto m. **~some** a disgustoso

loaves /ləʊvz/ see **loaf**

lobby /'lɒbɪ/ n atrio m; Pol gruppo m di pressione, lobby f inv

lobster /'lɒbstə(r)/ n aragosta f

local /'ləʊkl/ a locale; **I'm not ~** non sono del paese ● n abitante mf del luogo; (fam: public house) pub m locale. **~ authority** n autorità f locale. **~ call** n Teleph telefonata f urbana. **~ government** n autorità f inv locale

locality /ləʊ'kælətɪ/ n zona f

localized /'ləʊkəlaɪzd/ a localizzato

locally /'ləʊkəlɪ/ adv localmente; (live, work) nei paraggi

local network n Comput rete f locale

locate /ləʊ'keɪt/ vt situare; trovare (person); **be ~ed** essere situato. **~ion** /-ʃn/ n posizione f; **filmed on ~ion** girato in esterni

lock[1] /lɒk/ n (hair) ciocca f

lock[2] n (on door) serratura f; (on canal) chiusa f ● vt chiudere a chiave; bloccare (wheels) ● vi chiudersi. **lock in** vt chiudere dentro. **lock out** vt chiudere fuori. **lock up** vt (in prison) mettere dentro ● vi chiudere

locker /'lɒkə(r)/ n armadietto m

locket /'lɒkɪt/ n medaglione m

lock: **~-out** n serrata f. **~smith** n fabbro m

locomotive /ləʊkə'məʊtɪv/ n locomotiva f

locum /ˈləʊkəm/ n sostituto, -a mf

locust /ˈləʊkəst/ n locusta f

lodge /lɒdʒ/ n ⟨porter's⟩ portineria f; ⟨masonic⟩ loggia f ● vt presentare ⟨claim, complaint⟩; ⟨with bank, solicitor⟩ depositare; **be ~d** essersi conficcato ● vi essere a pensione ⟨with da⟩; ⟨become fixed⟩ conficcarsi. **~r** n inquilino, -a mf

lodgings /ˈlɒdʒɪŋz/ npl camere fpl in affitto

loft /lɒft/ n soffitta f

lofty /ˈlɒftɪ/ a (-ier, -iest) alto; ⟨haughty⟩ altezzoso

log /lɒg/ n ceppo m; Auto libretto m di circolazione; Naut giornale m di bordo ● vt (pt logged) registrare. **log on to** vt Comput connettersi a

logarithm /ˈlɒgərɪðm/ n logaritmo m

log-book n Naut giornale m di bordo; Auto libretto m di circolazione

loggerheads /ˈlɒgə-/ npl **be at ~** fam essere in totale disaccordo

logic /ˈlɒdʒɪk/ n logica f. **~al** a logico. **~ally** adv logicamente

logistics /ləˈdʒɪstɪks/ npl logistica f

logo /ˈləʊgəʊ/ n logo m inv

loin /lɔɪn/ n Culin lombata f

loiter /ˈlɔɪtə(r)/ vi gironzolare

loll|ipop /ˈlɒlɪpɒp/ n lecca-lecca m inv. **~y** n lecca-lecca m; ⟨fam: money⟩ quattrini mpl

London /ˈlʌndən/ n Londra f ● attrib londinese, di Londra. **~er** n londinese mf

lone /ləʊn/ a solitario. **~liness** n solitudine f

lonely /ˈləʊnlɪ/ a (-ier, -iest) solitario; ⟨person⟩ solo

lone|r /ˈləʊnə(r)/ n persona f solitaria. **~some** a solo

long¹ /lɒŋ/ a lungo; **a ~ time** molto tempo; **a ~ way** distante; **in the ~ run** a lungo andare; ⟨in the end⟩ alla fin fine ● adv a lungo, lungamente; **how ~ is?** quanto è lungo?; ⟨in time⟩ quanto dura?; **all**

day ~ tutto il giorno; **not ~ ago** non molto tempo fa; **before ~** fra breve; **he's no ~er here** non è più qui; **as** or **so ~as** finché; ⟨provided that⟩ purché; **so ~!** fam ciao!; **will you be ~?** [ti] ci vuole molto?

long² vi **~ for** desiderare ardentemente

long-'distance a a grande distanza; Sport di fondo; ⟨call⟩ interurbano

'longhand n **in ~** in scrittura ordinaria

longing /ˈlɒŋɪŋ/ a desideroso ● n brama f. **~ly** adv con desiderio

longitude /ˈlɒŋgɪtjuːd/ n Geog longitudine f

long: **~ jump** n salto m in lungo. **~-life 'milk** n latte m a lunga conservazione. **~-lived** /-lɪvd/ a longevo. **~-range** a Mil, Aeron a lunga portata; ⟨forecast⟩ a lungo termine. **~-sighted** a presbite. **~-sleeved** a a maniche lunghe. **~-suffering** a infinitamente paziente. **~-term** a a lunga scadenza. **~ wave** n onde fpl lunghe. **~-winded** /-ˈwɪndɪd/ a prolisso

loo /luː/ n fam gabinetto m

look /lʊk/ n occhiata f; ⟨appearance⟩ aspetto m; **[good] ~s** pl bellezza f; **have a ~ at** dare un'occhiata a ● vi guardare; ⟨seem⟩ sembrare; **~ here!** mi ascolti bene!; **~ at** guardare; **~ for** cercare; **~ like** ⟨resemble⟩ assomigliare a. **look after** vt badare a. **look down** vi guardare in basso; **look down on sb** fig guardare dall'alto in basso qcno. **look forward to** vt essere impaziente di. **look in on** vt passare da. **look into** vt ⟨examine⟩ esaminare. **look on to** vt ⟨room:⟩ dare su. **look out** vi guardare fuori; ⟨take care⟩ fare attenzione; **~ out for** cercare; **~ out!** attento! **look round** vi girarsi; ⟨in shop, town etc⟩ dare un'occhiata. **look through** vt dare un'occhiata a ⟨script, notes⟩. **look up** vi guardare in alto; ~ **up to sb** *fig* rispettare qcno ● *vt* cercare [nel dizionario] ⟨*word*⟩; ⟨*visit*⟩ andare a trovare

'look-out *n* guardia *f*, ⟨*prospect*⟩ prospettiva *f*; **be on the ~ for** tenere gli occhi aperti per

loom /lu:m/ *vi* apparire; *fig* profilarsi

loony /'lu:nɪ/ *a & n fam* matto, -a *mf*. ~ **bin** *n* manicomio *m*

loop /lu:p/ *n* cappio *m*; ⟨*on garment*⟩ passante *m*. ~**hole** *n* (*in the law*) scappatoia *f*

loose /lu:s/ *a* libero; ⟨*knot*⟩ allentato; ⟨*page*⟩ staccato; ⟨*clothes*⟩ largo; ⟨*morals*⟩ dissoluto; ⟨*inexact*⟩ vago; **be at a ~ end** non sapere cosa fare; **come ~** ⟨*knot*⟩ sciogliersi; **set ~** liberare. ~ **change** *n* spiccioli *mpl*. ~**ly** *adv* scorrevolmente; ⟨*defined*⟩ vagamente

loosen /'lu:sn/ *vt* sciogliere

loot /lu:t/ *n* bottino *m* ● *vt/i* depredare. ~**er** *n* predatore, -trice *mf*. ~**ing** *n* saccheggio *m*

lop /lɒp/ ~ **off** *vt* (*pt/pp* lopped) potare

lop'sided *a* sbilenco

lord /lɔ:d/ *n* signore *m*; ⟨*title*⟩ Lord *m*; **House of L~s** Camera *f* dei Lords; **the L~'s Prayer** il Padrenostro; **good L~!** Dio mio!

lore /lɔ:(r)/ *n* tradizioni *fpl*

lorry /'lɒrɪ/ *n* camion *m inv*; ~ **driver** camionista *mf*

lose /lu:z/ *v* (*pt/pp* lost) ● *vt* perdere ● *vi* perdere; ⟨*clock*⟩ essere indietro; **get lost** perdersi; **get lost!** *fam* va a quel paese! ~**r** *n* perdente *mf*

loss /lɒs/ *n* perdita *f*; *Comm* ~**es** perdite *fpl*; **be at a ~** essere perplesso; **be at a ~ for words** non trovare le parole

lost /lɒst/ *see* **lose** ● *a* perduto. ~ **'property office** *n* ufficio *m* oggetti smarriti

lot¹ /lɒt/ *n* (*at auction*) lotto *m*; **draw ~s** tirare a sorte

lot² *n* **the** ~ il tutto; **a ~ of, ~s of**

molti; **the ~ of you** tutti voi; **it has changed a ~** è cambiato molto

lotion /'ləʊʃn/ *n* lozione *f*

lottery /'lɒtərɪ/ *n* lotteria *f*. ~ **ticket** *n* biglietto *m* della lotteria

loud /laʊd/ *a* alto; ⟨*colours*⟩ sgargiante ● *adv* forte; **out** ~ ad alta voce. ~ **'hailer** *n* megafono *m*. ~**ly** *adv* forte. ~ **'speaker** *n* altoparlante *m*

lounge /laʊndʒ/ *n* salotto *m*; (*in hotel*) salone *m* ● *vi* poltrire. ~ **suit** *n* vestito *m* da uomo, completo *m* da uomo

louse /laʊs/ *n* (*pl* **lice**) pidocchio *m*

lousy /'laʊzɪ/ *a* (**-ier, -iest**) *fam* schifoso

lout /laʊt/ *n* zoticone *m*. ~**ish** *a* rozzo

lovable /'lʌvəbl/ *a* adorabile

love /lʌv/ *n* amore *m*; *Tennis* zero *m*; **in** ~ innamorato (**with** di) ● *vt* amare ⟨*person, country*⟩; **I** ~ **watching tennis** mi piace molto guardare il tennis. ~**-affair** *n* relazione *f* [sentimentale]. ~ **letter** *n* lettera *f* d'amore

lovely /'lʌvlɪ/ *a* (**-ier, -iest**) bello; (*in looks*) bello, attraente; (*in character*) piacevole; ⟨*meal*⟩ delizioso; **have a ~ time** divertirsi molto

lover /'lʌvə(r)/ *n* amante *mf*

love: ~ **song** *n* canzone *f* d'amore. ~ **story** *n* storia *f* d'amore

loving /'lʌvɪŋ/ *a* affettuoso

low /ləʊ/ *a* basso; (*depressed*) giù *inv* ● *adv* basso; **feel** ~ sentirsi giù ● *n* minimo *m*; *Meteorol* depressione *f*; **at an all-time** ~ ⟨*prices etc*⟩ al livello minimo

low: ~**brow** *a* di scarsa cultura. ~**-cut** *a* ⟨*dress*⟩ scollato

lower /'ləʊə(r)/ *a & adv see* **low** ● *vt* abbassare; ~ **oneself** abbassarsi

low: ~**-'fat** *a* magro. ~**-'grade** *a* di qualità inferiore. ~**-key** *fig* moderato. ~**lands** /-ləndz/ *npl* pianure *fpl*. ~**'tide** *n* bassa marea *f*

loyal /'lɔɪəl/ *a* leale. ~**ty** *n* lealtà *f*

lozenge /'lɒzɪndʒ/ n losanga f; (tablet) pastiglia f

LP n abbr **long-playing record**

Ltd abbr (**Limited**) s.r.l.

lubricant /'lu:brɪkənt/ n lubrificante m

lubricat|e /'lu:brɪkeɪt/ vt lubrificare. **~ion** /-'keɪʃn/ n lubrificazione f

lucid /'lu:sɪd/ a ‹explanation› chiaro; ‹sane› lucido. **~ity** /-'sɪdətɪ/ n lucidità f; (of explanation) chiarezza f

luck /lʌk/ n fortuna f; **bad ~** sfortuna f; **good ~!** buona fortuna! **~ily** adv fortunatamente

lucky /'lʌkɪ/ a (**-ier, -iest**) fortunato; **be ~** essere fortunato; ‹thing:› portare fortuna. **~ 'charm** n portafortuna m inv

lucrative /'lu:krətɪv/ a lucrativo

ludicrous /'lu:dɪkrəs/ a ridicolo. **~ly** adv ‹expensive, complex› eccessivamente

lug /lʌg/ vt (pt/pp **lugged**) fam trascinare

luggage /'lʌgɪdʒ/ n bagaglio m; **~-rack** n portabagagli m inv. **~ trolley** n carrello m portabagagli. **~-van** n bagagliaio m

lukewarm /'lu:k-/ a tiepido; fig poco entusiasta

lull /lʌl/ n pausa f ● vt **~ to sleep** cullare

lullaby /'lʌləbaɪ/ n ninna nanna f

lumbago /lʌm'beɪgəʊ/ n lombaggine f

lumber /'lʌmbə(r)/ n cianfrusaglie fpl; (Am: timber) legname m ● vt fam **~ sb with sth** affibbiare qcsa a qcno. **~ jack** n tagliaboschi m inv

luminous /'lu:mɪnəs/ a luminoso

lump¹ /lʌmp/ n (of sugar) zolletta f; (swelling) gonfiore m; (in breast) nodulo m; (in sauce) grumo m ● vt **~ together** ammucchiare

lump² vt **~ it** fam **you'll just have**

to ~ it che ti piaccia o no è così

lump sum n somma f globale

lumpy /'lʌmpɪ/ a (**-ier, -iest**) grumoso

lunacy /'lu:nəsɪ/ n follia f

lunar /'lu:nə(r)/ a lunare

lunatic /'lu:nətɪk/ n pazzo, -a mf

lunch /lʌntʃ/ n pranzo m ● vi pranzare

luncheon /'lʌntʃn/ n (formal) pranzo m. **~ meat** n carne f in scatola. **~ voucher** n buono m pasto

lunch: ~-hour n intervallo m per il pranzo. **~-time** n ora f di pranzo

lung /lʌŋ/ n polmone m. **~ cancer** n cancro m al polmone

lunge /lʌndʒ/ vi lanciarsi (at su)

lurch¹ /lɜ:tʃ/ n **leave in the ~** fam lasciare nei guai

lurch² vi barcollare

lure /lʊə(r)/ n esca f; fig lusinga f ● vt adescare

lurid /'lʊərɪd/ a (gaudy) sgargiante; (sensational) sensazionalistico

lurk /lɜ:k/ vi appostarsi

luscious /'lʌʃəs/ a saporito; fig sexy inv

lush /lʌʃ/ a lussureggiante

lust /lʌst/ n lussuria f ● vi **~ after** desiderare [fortemente]. **~ful** a lussurioso

lusty /'lʌstɪ/ a (**-ier, -iest**) vigoroso

lute /lu:t/ n liuto m

luxuriant /lʌg'ʒʊərɪənt/ a lussureggiante

luxurious /lʌg'ʒʊərɪəs/ a lussuoso

luxury /'lʌkʃərɪ/ n lusso m ● attrib di lusso

lying /'laɪɪŋ/ see **lie¹** & **²** ● n mentire m

lymph gland /'lɪmf/ n linfoghiandola f

lynch /lɪntʃ/ vt linciare

lynx /lɪŋks/ n lince f

lyric /'lɪrɪk/ a lirico. **~al** a lirico; (fam: enthusiastic) entusiasta. **~s** npl parole fpl

Mm

mac /mæk/ n fam impermeabile m

macabre /mə'ka:br/ a macabro

macaroni /mækə'rəʊnɪ/ n maccheroni mpl

mace[1] /meɪs/ n (staff) mazza f

mace[2] n (spice) macis m o f

machinations /mækɪ'neɪʃnz/ npl macchinazioni fpl

machine /mə'ʃi:n/ n macchina f ● vt (sew) cucire a macchina; Techn lavorare a macchina. **~-gun** n mitragliatrice f

machinery /mə'ʃi:nərɪ/ n macchinario m

machinist /mə'ʃi:nɪst/ n macchinista mf; (on sewing machine) lavorante mf adetto alla macchina da cucire

machismo /mə'tʃɪzməʊ/ n machismo m

macho /'mætʃəʊ/ a macho inv

mackerel /'mækrəl/ n inv sgombro m

mackintosh /'mækɪntɒʃ/ n impermeabile m

mad /mæd/ a (madder, maddest) pazzo, matto; (fam: angry) furioso (at sth); **like ~** fam come un pazzo; **be ~ about sb/sth** (fam: keen on) andare matto per qcno/qcsa

madam /'mædəm/ n signora f

madden /'mædən/ vt (make angry) far diventare matto

made /meɪd/ see **make**; **~ to measure** [fatto] su misura

Madeira cake /mə'dɪərə/ n dolce m di pan di Spagna

mad|ly /'mædlɪ/ adv fam follemente; **~ly in love** innamorato follemente. **~man** n pazzo m. **~ness** n pazzia f

madonna /mə'dɒnə/ n madonna f

magazine /mægə'zi:n/ n rivista f; Mil, Phot magazzino m

maggot /'mægət/ n verme m

Magi /'meɪdʒaɪ/ npl the **~** i Re mpl Magi

magic /'mædʒɪk/ n magia f; (tricks) giochi mpl di prestigio ● a magico; (trick) di prestigio. **~al** a magico

magician /mə'dʒɪʃn/ n mago, -a mf; (entertainer) prestigiatore, -trice mf

magistrate /'mædʒɪstreɪt/ n magistrato m

magnanim|ity /mægnə'nɪmətɪ/ n magnanimità f. **~ous** /-'nænɪməs/ a magnanimo

magnet /'mægnɪt/ n magnete m, calamita f. **~ic** /-'netɪk/ a magnetico. **~ism** n magnetismo m

magnification /mægnɪfɪ'keɪʃn/ n ingrandimento m

magnificen|ce /mæg'nɪfɪsəns/ n magnificenza f. **~t** a magnifico

magnify /'mægnɪfaɪ/ vt (pt/pp -ied) ingrandire; (exaggerate) ingigantire. **~ing glass** n lente f d'ingrandimento

magnitude /'mægnɪtju:d/ n grandezza f; (importance) importanza f

magpie /'mægpaɪ/ n gazza f

mahogany /mə'hɒgənɪ/ n mogano m ● a attrib di mogano

maid /meɪd/ n cameriera f; old **~** pej zitella f

maiden /'meɪdn/ n (liter) fanciulla f ● a (speech, voyage) inaugurale. **~ 'aunt** n zia f zitella. **~ name** n nome m da ragazza

mail /meɪl/ n posta f ● vt impostare

mail: **~-bag** n sacco m postale. **~-box** n Am cassetta f delle lettere; (e-mail) casella f di posta elettronica. **~ing list** n elenco m d'indirizzi per un mailing. **~man** n Am postino m. **~ order** n vendita f per corrispondenza. **~-order firm** n ditta f di vendita per corrispondenza

mailshot /'meɪlʃɒt/ n mailing m inv

maim /meɪm/ vt menomare

main[1] /meɪn/ n (water, gas, electricity) conduttura f principale

main[2] a principale; **the ~ thing is**

to... la cosa essenziale è di... ● *n* **in the** ~ in complesso

main /meɪn/ a principale. ● *n* **in the** ~ in complesso

main /meɪn/ ·lənd/ *n* continente *m*. **~ly** *adv* principalmente. **~stay** *n fig* pilastro *m*. **~ street** *n* via *f* principale

maintain /meɪnˈteɪn/ *vt* mantenere; (*keep in repair*) curare la manutenzione di; (*claim*) sostenere

maintenance /ˈmeɪntənəns/ *n* mantenimento *m*; (*care*) manutenzione *f*; (*allowance*) alimenti *mpl*

maisonette /meɪzəˈnet/ *n* appartamento *m* a due piani

majestic /məˈdʒestɪk/ *a* maestoso

majesty /ˈmædʒəstɪ/ *n* maestà *f*; **His/Her M~** Sua Maestà

major /ˈmeɪdʒə(r)/ *a* maggiore; **~ road** strada *f* con diritto di precedenza ● *n Mil, Mus* maggiore *m* ● *vi Am* ~ **in** specializzarsi in

Majorca /məˈjɔːkə/ *n* Maiorca *f*

majority /məˈdʒɒrətɪ/ *n* maggioranza *f*; **be in the** ~ avere la maggioranza

make /meɪk/ *n* (*brand*) marca *f* ● *v* (*pt/pp* **made**) ● *vt* fare; (*earn*) guadagnare; rendere (*happy, clear*) prendere (*decision*); ~ **sb laugh** far ridere qcno; ~ **sb do sth** far fare qcsa a qcno; ~ **it** (*to party, top of hill etc*) farcela; **what time do you** ~ **it?** che ore fai? ● *vi* ~ **as if to** fare per. **make do** *vi* arrangiarsi. **make for** *vt* dirigersi verso. **make off** *vi* fuggire. **make out** *vt* (*distinguish*) distinguere; (*write out*) rilasciare (*cheque*); compilare (*list*); (*claim*) far credere. **make over** *vt* cedere. **make up** *vt* (*constitute*) comporre; (*complete*) completare; (*invent*) inventare; (*apply cosmetics to*) truccare; fare (*parcel*); ~ **up one's mind** decidersi; ~ **it up** (*after quarrel*) riconciliarsi ● *vi* (*after quarrel*) fare la pace; ~ **up for** compensare; ~ **up for lost time** recuperare il tempo perso

'make-believe *n* finzione *f*

maker /ˈmeɪkə(r)/ *n* fabbricante *mf*; **M~** Creatore *m*

make: ~ **shift** *a* di fortuna ● *n* espediente *m*. **~up** *n* trucco *m*; (*character*) natura *f*

making /ˈmeɪkɪŋ/ *n* **have the** ~**s of** aver la stoffa di

maladjusted /mælæˈdʒʌstɪd/ *a* disadattato

malaise /məˈleɪz/ *n fig* malessere *m*

malaria /məˈleərɪə/ *n* malaria *f*

Malaysia /məˈleɪzɪə/ *n* Malesia *f*

male /meɪl/ *a* maschile ● *n*. maschio *m*. ~ **nurse** *n* infermiere *m*

malevolence /məˈlevələns/ *n* malevolenza *f*. ~**t** *a* malevolo

malfunction /mælˈfʌŋkʃn/ *n* funzionamento *m* imperfetto ● *vi* funzionare male

malice /ˈmælɪs/ *n* malignità *f*; **bear sb** ~ voler del male a qcno

malicious /məˈlɪʃəs/ *a* maligno

malign /məˈlaɪn/ *vt* malignare su

malignancy /məˈlɪgnənsɪ/ *n* malignità *f*. ~**t** *a* maligno

malinger /məˈlɪŋgə(r)/ *vi* fingersi malato. ~**er** *n* scansafatiche *mf inv*

malleable /ˈmælɪəbl/ *a* malleabile

mallet /ˈmælɪt/ *n* martello *m* di legno

malnutrition /mæl-/ *n* malnutrizione *f*

malpractice *n* negligenza *f*

malt /mɔːlt/ *n* malto *m*

Malta /ˈmɔːltə/ *n* Malta *f*. **~ese** /-iːz/ *a* & *n* maltese *mf*

maltreat /mæl-/ *vt* maltrattare. **~ment** *n* maltrattamento *m*

mammal /ˈmæml/ *n* mammifero *m*

mammoth /ˈmæməθ/ *a* mastodontico ● *n* mammut *m inv*

man /mæn/ *n* (*pl* **men**) uomo *m*; (*chess, draughts*) pedina *f* ● *vt* (*pt/pp* **manned**) equipaggiare; essere di servizio a (*counter, telephones*)

manage /ˈmænɪdʒ/ *vt* dirigere; gestire (*shop, affairs*); (*cope with*) farcela; ~ **to do sth** riuscire a fare qcsa ● *vi* riuscire; *(cope)* farcela (on con). ~**able** /-əbl/ *a (hair)* docile; *(size)* maneggevole. ~**ment** /-mənt/ *n* gestione *f*; **the ~ment** la direzione

manager /'mænɪdʒə(r)/ *n* direttore *m*; *(of shop, bar)* gestore *m*; *Sport* manager *m inv.* ~**ess** /-'res/ *n* direttrice *f*. ~**ial** /-'dʒɪərɪəl/ *a* ~**ial staff** personale *m* direttivo

managing /'mænɪdʒɪŋ/ ~ **director** direttore, -trice *mf* generale

mandarin /'mændərɪn/ *n* ~ **[orange]** mandarino *m*

mandate /'mændeɪt/ *n* mandato *m*. ~**ory** /-dətrɪ/ *a* obbligatorio

mane /meɪn/ *n* criniera *f*

mangle /'mæŋgl/ *vt (damage)* maciullare

mango /'mæŋgəʊ/ *n (pl* **-es)** mango *m*

mangy /'meɪndʒɪ/ *a (dog)* rognoso

man: ~**handle** *vt* malmenare. ~**hole** *n* botola *f*. ~**hole cover** *n* tombino *m*. ~**hood** *n* età *f* adulta; *(quality)* virilità *f*. ~**hour** *n* ora *f* lavorativa. ~**-hunt** *n* caccia *f* all'uomo

man|ia /'meɪnɪə/ *n* mania *f*. ~**iac** /-ɪæk/ *n* maniaco, -a *mf*

manicure /'mænɪkjʊə(r)/ *n* manicure *f* ● *vt* fare la manicure a

manifest /'mænɪfest/ *a* manifesto ● *vt* ~ **itself** manifestarsi. ~**ly** *adv* palesemente

manifesto /mænɪ'festəʊ/ *n* manifesto *m*

manifold /'mænɪfəʊld/ *a* molteplice

manipulat|e /mə'nɪpjʊleɪt/ *vt* manipolare. ~**ion** /-'leɪʃn/ *n* manipolazione *f*

man'kind *n* genere *m* umano

manly /'mænlɪ/ *a* virile

'man-made *a* artificiale. ~ **fibre** *n* fibra *f* sintetica

manner /'mænə(r)/ *n* maniera *f*; **in this ~** in questo modo; **have no ~s** avere dei pessimi modi;

good/bad ~s buone/cattive maniere *fpl*. ~**ism** *n* affettazione *f*

manœuvre /mə'nuːvə(r)/ *n* manovra *f* ● *vt* fare manovra con *(vehicle)*; manovrare *(person)*

manor /'mænə(r)/ *n* maniero *m*

'manpower *n* manodopera *f*

mansion /'mænʃn/ *n* palazzo *m*

'manslaughter *n* omicidio *m* colposo

mantelpiece /'mæntl-/ *n* mensola *f* di caminetto

manual /'mænjʊəl/ *a* manuale ● *n* manuale *m*

manufacture /mænjʊ'fæktʃə(r)/ *vt* fabbricare ● *n* manifattura *f*. ~**r** *n* fabbricante *m*

manure /mə'njʊə(r)/ *n* concime *m*

manuscript /'mænjʊskrɪpt/ *n* manoscritto *m*

many /'menɪ/ *a* & *pron* molti; **there are as ~ boys as girls** ci sono tanti ragazzi quante ragazze; **as ~ as 500** ben 500; **as ~ as that** così tanti; **as ~** altrettanti; **very ~, a good/great ~** moltissimi; **~ a time** molte volte

map /mæp/ *n* carta *f* geografica; *(of town)* mappa *f* ● **map out** *vt (pt/pp* **mapped)** *fig* programmare

maple /'meɪpl/ *n* acero *m*

mar /mɑː(r)/ *vt (pt/pp* **marred)** rovinare

marathon /'mærəθən/ *n* maratona *f*

marble /'mɑːbl/ *n* marmo *m*; *(for game)* pallina *f* ● *attrib* di marmo

March *n* marzo *m*

march *n* marcia *f*; *(protest)* dimostrazione *f* ● *vi* marciare ● *vt* far marciare; **~ sb off** scortare qcno fuori

mare /meə(r)/ *n* giumenta *f*

margarine /mɑːdʒə'riːn/ *n* margarina *f*

margin /'mɑːdʒɪn/ *n* margine *m*. ~**al** *a* marginale. ~**ally** *adv* marginalmente

marigold /'mærɪgəʊld/ *n* calendula *f*

marijuana /mærʊ'wɑːnə/ n marijuana f

marina /mə'riːnə/ n porticciolo m

marinade /mærɪ'neɪd/ n marinata f ● vt marinare

marine /mə'riːn/ a marino ●n (sailor) soldato m di fanteria marina

marionette /mærɪə'net/ n marionetta f

marital /'mærɪtl/ a coniugale. ~ **status** stato m civile

maritime /'mærɪtaɪm/ a marittimo

mark¹ /mɑːk/ n (currency) marco m

mark² n (stain) macchia f; (sign, indication) segno m; Sch voto m ● vt segnare; (stain) macchiare; Sch correggere; Sport marcare; ~ **time** Mil segnare il passo; fig non far progressi; ~ **my words** ricordati quello che dico. **mark out** vt delimitare; fig designare

marked /mɑːkt/ a marcato. ~**ly** /-kɪdlɪ/ adv notevolmente

marker /'mɑːkə(r)/ n (for highlighting) evidenziatore m; Sport marcatore m; (of exam) esaminatore, -trice mf

market /'mɑːkɪt/ n mercato m ● vt vendere al mercato; (launch) commercializzare; **on the** ~ sul mercato. ~**ing** n marketing m. ~ **re'search** n ricerca f di mercato

marksman /'mɑːksmən/ n tiratore m scelto

marmalade /'mɑːməleɪd/ n marmellata f d'arance

maroon /mə'ruːn/ a marrone rossastro

marooned /mə'ruːnd/ a abbandonato

marquee /mɑː'kiː/ n tendone m

marquis /'mɑːkwɪs/ n marchese m

marriage /'mærɪdʒ/ n matrimonio m

married /'mærɪd/ a sposato; (life) coniugale

marrow /'mærəʊ/ n Anat midollo m; (vegetable) zucca f

marr|y /'mærɪ/ vt (pt/pp **married**) sposare; **get** ~**ied** sposarsi ● vi sposarsi

marsh /mɑːʃ/ n palude f

marshal /'mɑːʃl/ n (steward) cerimoniere m ● vt (pt/pp **marshalled**) fig organizzare (arguments)

marshy /'mɑːʃɪ/ a paludoso

marsupial /mɑː'suːpɪəl/ n marsupiale m

martial /'mɑːʃl/ a marziale

martyr /'mɑːtə(r)/ n martire mf ● vt martoriare. ~**dom** /-dəm/ n martirio m. ~**ed** a fam da martire

marvel /'mɑːvl/ n meraviglia f ● vi (pt/pp **marvelled**) meravigliarsi (at di). ~**lous** /-vələs/ a meraviglioso

Marxis|m /'mɑːksɪzm/ n marxismo m. ~**t** a & n marxista mf

marzipan /'mɑːzɪpæn/ n marzapane m

mascara /mæ'skɑːrə/ n mascara m inv

mascot /'mæskət/ n mascotte f inv

masculin|e /'mæskjʊlɪn/ a maschile ●n Gram maschile m. ~**ity** /-'lɪnɪtɪ/ n mascolinità f

mash /mæʃ/ vt impastare. ~**ed potatoes** npl purè m inv di patate

mask /mɑːsk/ n maschera f ● vt mascherare

masochis|m /'mæsəkɪzm/ n masochismo m. ~**t** /-ɪst/ n masochista mf

Mason n massone m. ~**ic** /mə'sɒnɪk/ a massonico

mason /'meɪsn/ n muratore m

masonry /'meɪsnrɪ/ n massoneria f

masquerade /mæskə'reɪd/ n fig mascherata f ● vi ~ **as** (pose) farsi passare per

mass¹ /mæs/ n Relig messa f

mass² n massa f; ~**es of** fam un sacco di ● vi ammassarsi

massacre /'mæsəkə(r)/ n massacro m ● vt massacrare

massage /'mæsɑːʒ/ n massaggio

m ● *vt* massaggiare; *fig* manipolare ⟨*statistics*⟩

masseu|**r** /mæˈsɜː(r)/ *n* massaggiatore *m*. **~se** /-ˈsɜːz/ *n* massaggiatrice *f*

massive /ˈmæsɪv/ *a* enorme

mass: ~ **media** *npl* mezzi *mpl* di comunicazione di massa, mass media *mpl*. **~-pro'duce** *vt* produrre in serie. **~-pro'duction** *n* produzione *f* in serie

mast /mɑːst/ *n* Naut albero *m*; ⟨*for radio*⟩ antenna *f*

master /ˈmɑːstə(r)/ *n* maestro *m*, padrone *m*; ⟨*teacher*⟩ professore *m*; ⟨*of ship*⟩ capitano *m*; **M~** ⟨*boy*⟩ signorino *m*

master: **~-key** *n* passe-partout *m inv*. **~-ly** *a* magistrale. **~-mind** *n* cervello *m* ● *vt* ideare e dirigere. **~piece** *n* capolavoro *m*. **~-stroke** *n* colpo *m* da maestro. **~y** *n* ⟨*of subject*⟩ padronanza *f*

masturbat|**e** /ˈmæstəbeɪt/ *vi* masturbarsi. **~ion** /-ˈbeɪʃn/ *n* masturbazione *f*

mat /mæt/ *n* stuoia *f*; ⟨*on table*⟩ sottopiatto *m*

match[1] /mætʃ/ *n* Sport partita *f*; ⟨*equal*⟩ uguale *mf*; ⟨*marriage*⟩ matrimonio *m*; ⟨*person to marry*⟩ partito *m*; **be a good ~** ⟨*colours:*⟩ intonarsi bene; **be no ~ for** non essere dello stesso livello di ● *vt* ⟨*equal*⟩ uguagliare; ⟨*be like*⟩ andare bene con ● *vi* intonarsi

match[2] *n* fiammifero *m*. **~box** *n* scatola *f* di fiammiferi

matching /ˈmætʃɪŋ/ *a* intonato

mate[1] /meɪt/ *n* compagno, -a *mf*; ⟨*assistant*⟩ aiuto *m*; Naut secondo *m*; ⟨*fam: friend*⟩ amico, -a *mf* ● *vi* accoppiarsi ● *vt* accoppiare

mate[2] *n* ⟨*in chess*⟩ scacco *m* matto

material /məˈtɪərɪəl/ *n* materiale *m*; ⟨*fabric*⟩ stoffa *f*; **raw ~s** materie *fpl* prime ● *a* materiale

material|**ism** /məˈtɪərɪəlɪzm/ *n* materialismo *m*. **~istic** /-ˈlɪstɪk/ *a* materialistico. **~ize** /-aɪz/ *vi* materializzarsi

maternal /məˈtɜːnl/ *a* materno

maternity /məˈtɜːnəti/ *n* maternità *f*. ~ **clothes** *npl* abiti *mpl* premaman. ~ **ward** *n* maternità *f inv*

matey /ˈmeɪti/ *a* fam amichevole

mathematic|**al** /mæθəˈmætɪkl/ *a* matematico. **~ian** /-məˈtɪʃn/ *n* matematico, -a *mf*

mathematics /mæθəˈmætɪks/ *n* matematica *fsg*

maths /mæθs/ *n* fam matematica *fsg*

matinée /ˈmætɪneɪ/ *n* Theat matinée *m*

mating /ˈmeɪtɪŋ/ *n* accoppiamento *m*; ~ **season** stagione *f* degli amori

matriculat|**e** /məˈtrɪkjuleɪt/ *vi* immatricolarsi. **~ion** /-ˈleɪʃn/ *n* immatricolazione *f*

matrix /ˈmeɪtrɪks/ *n* ⟨*pl* **matrices** /-siːz/⟩ *n* matrice *f*

matted /ˈmætɪd/ *a* ~ **hair** capelli *mpl* tutti appiccicati tra loro

matter /ˈmætə(r)/ *n* ⟨*affair*⟩ faccenda *f*; ⟨*question*⟩ questione *f*; ⟨*pus*⟩ pus *m*; ⟨*phys: substance*⟩ materia *f*; **as a ~ of fact** a dire la verità; **what is the ~?** che cosa c'è? ● *vi* importare; ~ **to sb** essere importante per qcno; **it doesn't ~** non importa. **~-of-fact** *a* pratico

mattress /ˈmætrɪs/ *n* materasso *m*

matur|**e** /məˈtʃʊə(r)/ *a* maturo; Comm in scadenza ● *vi* maturare ● *vt* far maturare. **~ity** *n* maturità *f*; Fin maturazione *f*

maul /mɔːl/ *vt* malmenare

Maundy /ˈmɔːndi/ *n* ~ **Thursday** giovedì *m* santo

mauve /məʊv/ *a* malva

maxim /ˈmæksɪm/ *n* massima *f*

maximum /ˈmæksɪməm/ *a* massimo; **ten minutes** ~ dieci minuti al massimo ● *n* ⟨*pl* **-ima**⟩ massimo *m*

May /meɪ/ *n* maggio *m*

may /meɪ/ *v aux* ⟨*solo al presente*⟩ potere; ~ **I come in?** posso entrare?; **if I ~ say so** se mi posso permettere; ~ **you both be very**

happy siate felici; **I ~ as well** stay potrei anche rimanere; **it ~ be true** potrebbe esser vero; **she ~ be old, but...** sarà anche vecchia, ma...

maybe /'meibi/ adv forse, può darsi

'May Day n il primo maggio

mayonnaise /meiə'neiz/ n maionese f

mayor /'meə(r)/ n sindaco m. **~ess** n sindaco m. **(wife of mayor)** moglie f del sindaco

maze /meiz/ n labirinto m

me /mi:/ pron (object) mi; (with preposition) me; **she called me** mi ha chiamato; **she called me, not you** ha chiamato me, non te; **give me the money** dammi i soldi; **give it to me** dammelo; **he gave it to me** me lo ha dato; **it's ~** sono io

meadow /'medəʊ/ n prato m

meagre /'mi:gə(r)/ a scarso

meal¹ /mi:l/ n pasto m

meal² n (grain) farina f

mealy-mouthed /mi:lɪ'maʊðd/ a ambiguo

mean¹ /mi:n/ a avaro; (unkind) meschino

mean² /mi:n/ a medio ●n (average) media f. **Greenwich ~ time** ora f media di Greenwich

mean³ /mi:n/ vt (pt/pp meant) voler dire; (signify) significare; (intend) intendere; **I ~ it** lo dico seriamente; **~ well** avere buone intenzioni; **be meant for** (present:) essere destinato a; (remark:) essere riferito a

meander /mi'ændə(r)/ vi vagare

meaning /'mi:nɪŋ/ n significato m. **~ful** a significativo. **~less** a senza senso

means /mi:nz/ n mezzo m; **~ of transport** mezzo m di trasporto; **by ~ of** per mezzo di; **by all ~!** certamente!; **by no ~** niente affatto ●npl (resources) mezzi mpl

meant /ment/ see **mean³**

meantime n **in the ~** nel frattempo ●adv intanto

'meanwhile adv intanto

measles /'mi:zlz/ n morbillo m

measly /'mi:zlɪ/ a fam misero

measurable /'meʒərəbl/ a misurabile

measure /'meʒə(r)/ n misura f ● vt/i misurare. **measure up to** vt fig essere all'altezza di. **~d** a misurato. **~ment** /-mənt/ n misura f

meat /mi:t/ n carne f. **~ ball** n Culin polpetta f di carne. **~ loaf** n polpettone m

mechanic /mɪ'kænɪk/ n meccanico m. **~ical** a meccanico; **~ical engineering** ingegneria f meccanica. **~ically** adv meccanicamente. **~ics** n meccanica f ●npl meccanismo msg

mechanism /'mekənɪzm/ n meccanismo m. **~ize** vt meccanizzare

medal /'medl/ n medaglia f

medallion /mɪ'dælɪən/ n medaglione m

medallist /'medəlɪst/ n vincitore, -trice mf di una medaglia

meddle /'medl/ vi immischiarsi (in di); (tinker) armeggiare (with con)

media /'mi:dɪə/ see **medium** ●npl **the ~** i mass media

median /'mi:dɪən/ a **~ strip** Am banchina f spartitraffico

mediate /'mi:dɪeɪt/ vi fare da mediatore. **~ion** /-'eɪʃn/ n mediazione f. **~or** n mediatore, -trice mf

medical /'medɪkl/ a medico ● n visita f medica. **~ insurance** n assicurazione f sanitaria. **~ student** n studente, -essa mf di medicina

medicated /'medɪkeɪtɪd/ a medicato. **~ion** /-'keɪʃn/ n (drugs) medicinali mpl

medicinal /mɪ'dɪsɪnl/ a medicinale

medicine /'medsən/ n medicina f

medieval /medɪ'i:vl/ a medievale

mediocre /mi:dɪ'əʊkə(r)/ a mediocre. **~ity** /-'ɒkrətɪ/ n mediocrità f

meditate /'medɪteɪt/ vi meditare (on su). **~ion** /-'teɪʃn/ n meditazione f

Mediterranean /medɪtəˈreɪnɪən/ n the ~ [Sea] il [mare m] Mediterraneo m ● a mediterraneo

medium /ˈmiːdɪəm/ a medio; Culin di media cottura ● n (pl **media**) mezzo m; (pl **-s**) (person) medium mf inv

medium: ~**-sized** a di taglia media. ~ **wave** n onde fpl medie

medley /ˈmedlɪ/ n miscuglio m; Mus miscellanea f

meek /miːk/ a mite, mansueto. ~**ly** adv docilmente

meet /miːt/ v (pt/pp **met**) ● vt incontrare; (at station, airport) andare incontro a; (for first time) far la conoscenza di; pagare (bill); soddisfare (requirements) ● vi incontrarsi; (committee:) riunirsi; ~ **with** incontrare (problem); incontrarsi con (person) ● n raduno m [sportivo]

meeting /ˈmiːtɪŋ/ n riunione f, meeting m inv; (large) assemblea f; (by chance) incontro m

megabyte /ˈmegəbaɪt/ n megabyte m

megalomania /megələˈmeɪnɪə/ n megalomania f

megaphone /ˈmegəfəʊn/ n megafono m

melancholy /ˈmelənkəlɪ/ a malinconico ● n malinconia f

mellow /ˈmeləʊ/ a (wine) generoso; (sound, colour) caldo; (person) dolce ● vi (person:) addolcirsi

melodic /mɪˈlɒdɪk/ a melodico

melodrama /ˈmelə-/ n melodramma m. ~**tic** /-drəˈmætɪk/ a melodrammatico

melody /ˈmelədɪ/ n melodia f

melon /ˈmelən/ n melone m

melt /melt/ vt sciogliere ● vi sciogliersi. **melt down** vt fondere. ~**ing-pot** n fig crogiuolo m

member /ˈmembə(r)/ n membro m; ~ **countries** paesi mpl membri; **M~ of Parliament** eurodeputato, -a mf; **M~ of the European Parliament** eurodeputato, -a mf.

~ship n iscrizione f; (members) soci mpl

membrane /ˈmembreɪn/ n membrana f

memo /ˈmeməʊ/ n promemoria m inv

memoirs /ˈmemwɑːz/ n ricordi mpl

memorable /ˈmemərəbl/ a memorabile

memorandum /meməˈrændəm/ n promemoria m inv

memorial /mɪˈmɔːrɪəl/ n monumento m. ~ **service** n funzione f commemorativa

memorize /ˈmeməraɪz/ vt memorizzare

memory /ˈmemərɪ/ n also Comput memoria f; (thing remembered) ricordo m; **from** ~ a memoria; **in** ~ **of** in ricordo di

men /men/ see **man**

menace /ˈmenəs/ n minaccia f; (nuisance) piaga f ● vt minacciare. ~**ing** a minaccioso

mend /mend/ vt riparare; (darn) rammendare ● n on the ~ in via di guarigione

'menfolk n uomini mpl

menial /ˈmiːnɪəl/ a umile

meningitis /menɪnˈdʒaɪtɪs/ n meningite f

menopause /ˈmenə-/ n menopausa f

menstruat|e /ˈmenstruet/ vi mestruare. ~**ion** /-ˈeɪʃn/ n mestruazione f

mental /ˈmentl/ a mentale; (fam: mad) pazzo. ~ **a'rithmetic** n calcolo m mentale. ~ **illness** n malattia f mentale

mental|ity /menˈtælətɪ/ n mentalità f inv. ~**ly** adv mentalmente; ~**ly ill** malato di mente

mention /ˈmenʃn/ n menzione f ● vt menzionare; **don't** ~ **it** non c'è di che

menu /ˈmenjuː/ n menu m inv

MEP n abbr Member of the European Parliament

mercenary /'mɜːsɪnərɪ/ a mercenario ● n mercenario m

merchandise /'mɜːtʃəndaɪz/ n merce f

merchant /'mɜːtʃənt/ n commerciante mf. ~ **bank** n banca f d'affari. ~ '**navy** n marina f mercantile

merci|ful /'mɜːsɪfl/ a misericordioso. ~**fully** adv fam grazie a Dio. ~**less** a spietato

mercury /'mɜːkjʊrɪ/ n mercurio m

mercy /'mɜːsɪ/ n misericordia f; **be at sb's** ~ essere alla mercè di qcno, essere in balia di qcno

mere /mɪə(r)/ a solo. ~**ly** adv solamente

merest /'mɪərɪst/ a minimo

merge /mɜːdʒ/ vi fondersi

merger /'mɜːdʒə(r)/ n fusione f

meringue /mə'ræŋ/ n meringa f

merit /'merɪt/ n merito m; (advantage) pregio f inv ● vt meritare

mermaid /'mɜːmeɪd/ n sirena f

merri|ly /'merɪlɪ/ adv allegramente. ~**ment** /-mənt/ n baldoria f

merry /'merɪ/ a (-**ier**, -**iest**) allegro; ~ **Christmas!** Buon Natale!

merry: ~-**go-round** n giostra f. ~-**making** n festa f

mesh /meʃ/ n maglia f

mesmerize /'mezməraɪz/ vt ipnotizzare. ~**d** a fig ipnotizzato

mess /mes/ n disordine m, casino m fam; (trouble) guaio m; (something spilt) sporco m; Mil mensa f; **make a** ~ **of** (botch) fare un pasticcio di ● **mess about** vi perder tempo; ~ **about with** armeggiare con ● vt prendere in giro (person).

mess up vt mettere in disordine, incasinare fam; (botch) mandare all'aria

message /'mesɪdʒ/ n messaggio m

messenger /'mesɪndʒə(r)/ n messaggero m

Messiah /mɪ'saɪə/ n Messia m

Messrs /'mesəz/ npl (on letter) ~ **Smith** Spett. ditta Smith

messy /'mesɪ/ a (-**ier**, -**iest**) disordinato; (in dress) sciatto

met /met/ see **meet**

metal /'metl/ n metallo m ● a di metallo. ~**lic** /mɪ'tælɪk/ a metallico

metamorphosis /metə'mɔːfəsɪs/ n (pl -**phoses** /-siːz/) metamorfosi f inv

metaphor /'metəfə(r)/ n metafora f. ~**ical** /-'fɒrɪkl/ a metaforico

meteor /'miːtɪə(r)/ n meteora f. ~**ic** /-'ɒrɪk/ a fig fulmineo

meteorological /miːtɪərə'lɒdʒɪkl/ a meteorologico

meteorolog|ist /miːtɪə'rɒlədʒɪst/ n meteorologo, -a mf. ~**y** n meteorologia f

meter[1] /'miːtə(r)/ n contatore m

meter[2] n Am = **metre**

method /'meθəd/ n metodo m

methodical /mɪ'θɒdɪkl/ a metodico. ~**ly** adv metodicamente

Methodist /'meθədɪst/ n metodista mf

meths /meθs/ n fam alcol m denaturato

methylated /'meθɪleɪtɪd/ a ~ **spirit[s]** alcol m denaturato

meticulous /mɪ'tɪkjʊləs/ a meticoloso. ~**ly** adv meticolosamente

metre /'miːtə(r)/ n metro m

metric /'metrɪk/ a metrico

metropolis /mɪ'trɒpəlɪs/ n metropoli f inv

metropolitan /metrə'pɒlɪtən/ a metropolitano

mew /mjuː/ n miao m ● vi miagolare

Mexican /'meksɪkən/ a & n messicano, -a mf. '**Mexico** n Messico m

miaow /mɪ'aʊ/ n miao m ● vi miagolare

mice /maɪs/ see **mouse**

mickey /'mɪkɪ/ n **take the** ~ **out of** prendere in giro

microbe /'maɪkrəʊb/ n microbo m

micro /'maɪkrəʊ/: ~**chip** n chip m. ~**computer** n microcomputer m. ~**film** n microfilm m. ~**phone** microfono m. ~**processor** n microprocesso-

re *m*. **~scope** *n* microscopio *m*.
~scopic /-'skɒpɪk/ *a* microscopico. **~wave** *n* microonda *f*; *(oven)*
forno *m* a microonde

mid /mɪd/ *a* ~ **May** metà maggio; **in
~ air** a mezz'aria

midday /mɪd'deɪ/ *n* mezzogiorno
m

middle /mɪdl/ *a* di centro; **the M~
Ages** il medioevo; **the ~
class[es]** la classe media; **the ~
East** il Medio Oriente ● *n* mezzo
m; **in the ~ of** *(room, floor etc)* in
mezzo a; **in the ~ of the night** nel
pieno della notte, a notte piena

middle: **~-aged** *a* di mezza età.
~-class *a* borghese. **~man** *n*
Comm intermediario *m*

middling /mɪdlɪŋ/ *a* discreto

midge /mɪdʒ/ *n* moscerino *m*

midget /mɪdʒɪt/ *n* nano, -a *mf*

Midlands /mɪdləndz/ *npl* **the ~**
l'Inghilterra *fsg* centrale

'midnight *n* mezzanotte *f*

midriff /mɪdrɪf/ *n* diaframma *m*

midst /mɪdst/ *n* **in the ~ of** in mezzo a; **in our ~** fra di noi, in mezzo
a noi

mid: **~summer** *n* mezza estate *f*
~way *adv* a metà strada. **~wife** *n*
ostetrica *f*. **~wifery** /-wɪfrɪ/ *n* ostetricia *f*. **~'winter** *n* pieno inverno *m*

might[1] /maɪt/ *v aux* **I ~** potrei; **will
you come? – I ~** vieni? – può dar-
si; **it ~ be true** potrebbe essere
vero; **I ~ as well stay** potrei anche restare; **you ~ have drowned**
avresti potuto affogare; **you ~
have said so!** avresti potuto dirlo!

might[2] *n* potere *m*

mighty /maɪtɪ/ *a* (**-ier, -iest**) potente ● *adv* fam molto

migraine /miːgreɪn/ *n* emicrania *f*

migrant /maɪgrənt/ *a* migratore
● *n* *(bird)* migratore, -trice *mf*;
(person: for work) emigrante *m*

migrat|e /maɪgreɪt/ *vi* migrare.
~ion /-greɪʃn/ *n* migrazione *f*

mike /maɪk/ *n* fam microfono *m*

Milan /mɪˈlæn/ *n* Milano *f*

mild /maɪld/ *a* *(weather)* mite;
(person) dolce; *(flavour)* delicato;
(illness) leggero

mildew /mɪldjuː/ *n* muffa *f*

mild|ly /maɪldlɪ/ *adv* moderata-
mente; *(say)* dolcemente; **to put it
~ly** a dir poco, senza esagerazio-
ne. **~ness** *n (of person, words)* dolcezza *f*; *(of weather)* mitezza *f*

mile /maɪl/ *n* miglio *m* (= *1,6 km*);
~s nicer *fam* molto più bello

mile|age /-ɪdʒ/ *n* chilometraggio
m. **~stone** *n* pietra *f* miliare

militant /mɪlɪtənt/ *a* & *n* militante
mf

military /mɪlɪtrɪ/ *a* militare. **~
service** *n* servizio *m* militare

militate /mɪlɪteɪt/ *vi* **~ against** opporsi a

militia /mɪˈlɪʃə/ *n* milizia *f*

milk /mɪlk/ *n* latte *m* ● *vt* mungere

milk: **~man** *n* lattaio *m*. **~ shake** *n*
frappé *m inv*

milky /mɪlkɪ/ *a* (**-ier, -iest**) latteo;
(tea etc) con molto latte. **M~ Way**
n *Astr* Via *f* Lattea

mill /mɪl/ *n* mulino *m*; *(factory)*
fabbrica *f*; *(for coffee etc)* macinino
m ● *vt* macinare *(grain)*. **mill
about, mill around** *vi* brulicare

millennium /mɪˈlenɪəm/ *n* millen-
nio *m*

miller /mɪlə(r)/ *n* mugnaio *m*

milligram /mɪlɪ-/ *n* milligrammo
m. **~metre** *n* millimetro *m*

million /mɪljən/ *n* milione *m*; **a ~
pounds** un milione di sterline.
~aire /-'neə(r)/ *n* miliardario, -a
mf

'millstone *n fig* peso *m*

mime /maɪm/ *n* mimo *m* ● *vt*
mimare

mimic /mɪmɪk/ *n* imitatore, -trice
mf ● *vt* (*pt/pp* **mimicked**) imitare.
~ry *n* mimetismo *m*

mimosa /mɪˈməʊzə/ *n* mimosa *f*

mince /mɪns/ *n* carne *f* tritata ● *vt*
Culin tritare; **not ~ one's words**
parlare senza mezzi termini

mince: **~meat** *n* miscuglio *m* di
frutta secca; **make ~meat of** *fig*

demolire. **~'pie** n pasticcino m a base di frutta secca

mincer /'mɪnsə(r)/ n tritacarne m inv

mind /maɪnd/ n mente f; (sanity) ragione f; **to my** ~ a mio parere; **give sb a piece of one's** ~ dire chiaro e tondo a qcno quello che si pensa; **make up one's** ~ decidersi; **have sth in** ~ avere qcsa in mente; **bear sth in** ~ tenere presente qcsa; **have something on one's** ~ essere preoccupato; **have a good** ~ **to** avere una grande voglia di; **I have changed my** ~ ho cambiato idea; **in two** ~s indeciso; **are you out of your** ~? sei diventato matto? ● vt (look after) occuparsi di; **I don't** ~ **the noise** il rumore non mi dà fastidio; **I don't** ~ **what we do** non mi importa quello che facciamo; ~ **the step!** attenzione al gradino! ● vi **I don't** ~ non mi importa; **never** ~! non importa!; **do you** ~ **if...**? ti dispiace se...? **mind out** vi ~ **out!** [fai] attenzione!

minder /'maɪndə(r)/ n (Br: bodyguard) gorilla m; (for child) baby-sitter mf inv

mindful ~**ful of** attento a. **~less** a noncurante

mine¹ /maɪn/ poss pron il mio m, la mia f; i miei mpl, le mie fpl; **a friend of** ~ un mio amico; **friends of** ~ dei miei amici; **that is** ~ questo è mio; (as opposed to yours) questo è il mio

mine² n miniera f; (explosive) mina f ● vt estrarre; Mil minare. ~ **detector** n rivelatore m di mine. ~**field** n campo m minato

miner /'maɪnə(r)/ n minatore m

mineral /'mɪnərəl/ n minerale m ● a minerale. ~ **water** n acqua f minerale

minesweeper /'maɪn-/ n dragamine m inv

mingle /'mɪŋgl/ vi ~ **with** mescolarsi a

mini /'mɪnɪ/ n (skirt) mini f

miniature /'mɪnɪtʃə(r)/ a in miniatura ● n miniatura f

mini|bus /'mɪnɪ-/ n minibus m, pulmino m. ~**cab** n taxi m inv

minim /'mɪnɪm/ n Mus minima f

minim|al /'mɪnɪməl/ a minimo. ~**ize** vt minimizzare. ~**um** n (pl -ima) minimo m ● a minimo; **ten minutes** ~**um** minimo dieci minuti

mining /'maɪnɪŋ/ n estrazione f ● a estrattivo

miniskirt /'mɪnɪ-/ n minigonna f

minist|er /'mɪnɪstə(r)/ n ministro m; Relig pastore m. ~**erial** /-'stɪərɪəl/ a ministeriale

ministry /'mɪnɪstrɪ/ n Pol ministero m; **the** ~ Relig il ministero sacerdotale

mink /mɪŋk/ n visone m

minor /'maɪnə(r)/ a minore ● n minorenne mf

minority /maɪ'nɒrətɪ/ n minoranza f; (age) minore età f

minor road n strada f secondaria

mint¹ /mɪnt/ n fam patrimonio m ● a in ~ **condition** in condizione perfetta

mint² n (herb) menta f

minus /'maɪnəs/ prep meno; (fam: without) senza ● n ~ [**sign**] meno m

minute¹ /'mɪnɪt/ n minuto m; **in a** ~ (shortly) in un minuto; ~**s** pl (of meeting) verbale msg

minute² /maɪ'njuːt/ a minuto; (precise) minuzioso

mirac|le /'mɪrəkl/ n miracolo m. ~**ulous** /-'rækjʊləs/ a miracoloso

mirage /'mɪrɑːʒ/ n miraggio m

mirror /'mɪrə(r)/ n specchio m ● vt rispecchiare

mirth /mɜːθ/ n ilarità f

misad'venture /mɪs-/ n disavventura f

misanthropist /mɪ'zænθrəpɪst/ n misantropo, -a mf

misapprehension n malinteso m; **be under a** ~ avere frainteso

misbe'have vi comportarsi male

mis'calcu|late vt/i calcolare male. **~ 'lation** n calcolo m sbagliato

'miscarriage n aborto m spontaneo; **~ of justice** errore m giudiziario. **mis'carry** vi abortire

miscellaneous /mɪsə'leɪnɪəs/ a assortito

mischief /'mɪstʃɪf/ n malefatta f; (harm) danno m

mischievous /'mɪstʃɪvəs/ a (naughty) birichino; (malicious) dannoso

miscon'ception n concetto m erroneo

mis'conduct n cattiva condotta f

misde'meanour n reato m

miser /'maɪzə(r)/ n avaro m

miserabl|e /'mɪzrəbl/ a (unhappy) infelice; (wretched) miserabile; (fig: weather) deprimente. **~y** adv (live, fail) miseramente; (say) tristemente

miserly /'maɪzəlɪ/ a avaro; (amount) ridicolo

misery /'mɪzərɪ/ n miseria f; (fam: person) piagnone, -a mf

mis'fire vi (gun:) far cilecca; (plan etc:) non riuscire

'misfit n disadattato, -a mf

mis'fortune n sfortuna f

mis'givings npl dubbi mpl

mis'guided a fuorviato

mishap /'mɪshæp/ n disavventura f

misin'terpret vt fraintendere

mis'judge vt giudicar male; (estimate wrongly) valutare male

mis'lay vt (pt/pp **-laid**) smarrire

mis'lead vt (pt/pp **-led**) fuorviare. **~ing** a fuorviante

mis'manage vt amministrare male. **~ment** n cattiva amministrazione f

misnomer /mɪs'nəʊmə(r)/ n termine m improprio

'misprint n errore m di stampa

mis'quote vt citare erroneamente

misrepre'sent vt rappresentare male

Miss n (pl **-es**) signorina f

miss /mɪs/ n colpo m mancato ● vt

(fail to hit or find) mancare; perdere (train, bus, class); (feel the loss of) sentire la mancanza di; **I ~ed that part** (failed to notice) mi è sfuggita quella parte ● vi **but he ~ed** (failed to hit) ma l'ha mancato. **miss out** vt saltare, omettere

misshapen /mɪs'ʃeɪpən/ a malformato

missile /'mɪsaɪl/ n missile m

missing /'mɪsɪŋ/ a mancante; (person) scomparso; Mil disperso; **be ~** essere introvabile

mission /'mɪʃn/ n missione f

missionary /'mɪʃənrɪ/ n missionario, -a mf

mis'spell vt (pt/pp **-spelled**, **-spelt**) sbagliare l'ortografia di

mist /mɪst/ n (fog) foschia f ● **mist up** vi appannarsi, annebbiarsi

mistake /mɪ'steɪk/ n sbaglio m; **by ~** per sbaglio ● vt (pt **mistook**, pp **mistaken**) sbagliare (road, house); fraintendere (meaning, words); **~ for** prendere per

mistaken /mɪ'steɪkən/ a sbagliato; **be ~** sbagliarsi; **~ identity** errore m di persona. **~ly** adv erroneamente

mistletoe /'mɪsltəʊ/ n vischio m

mistress /'mɪstrɪs/ n padrona f; (teacher) maestra f; (lover) amante f

mis'trust n sfiducia f ● vt non aver fiducia in

misty /'mɪstɪ/ a (**-ier, -iest**) nebbioso

misunder'stand vt (pt/pp **-stood**) fraintendere. **~ing** n malinteso m

misuse[1] /mɪs'ju:z/ vt usare male

misuse[2] /mɪs'ju:s/ n cattivo uso m

mite /maɪt/ n (child) piccino, -a mf

mitigat|e /'mɪtɪɡeɪt/ vt attenuare. **~ing** a attenuante

mitten /'mɪtn/ n manopola f, muffola m

mix /mɪks/ n (combination) mescolanza f; Culin miscuglio m; (readymade) preparato m ● vt mischiare ● vi mischiarsi; (person:) inserirsi; **~ with** (associate with) frequentare. **mix up** *vt* mescolare ⟨papers⟩; ⟨confuse, mistake for⟩ confondere

mixed /mɪkst/ *a* misto; **~ up** ⟨person⟩ confuso

mixer /'mɪksə(r)/ *n* Culin frullatore *m*, mixer *m inv*; **he's a good ~** è un tipo socievole

mixture /'mɪkstʃə(r)/ *n* mescolanza *f*; ⟨medicine⟩ sciroppo *m*; Culin miscela *f*

'mix-up *n* ⟨confusion⟩ confusione *f*; ⟨mistake⟩ pasticcio *m*

moan /məʊn/ *n* lamento *m* ● *vi* lamentarsi; ⟨complain⟩ lagnarsi

moat /məʊt/ *n* fossato *m*

mob /mɒb/ *n* folla *f*; ⟨rabble⟩ gentaglia *f*; ⟨fam: gang⟩ banda *f* ● *vt* ⟨pt/pp **mobbed**⟩ assalire

mobile /'məʊbaɪl/ *a* mobile ● *n* composizione *f* mobile. **~ 'home** *n* casa *f* roulotte. **~ [phone]** *n* [telefono *m*] cellulare *m*

mobility /mə'bɪlətɪ/ *n* mobilità *f*

mock /mɒk/ *a* finto ● *vt* canzonare. **~ery** *n* derisione *f*

'mock-up *n* modello *m* in scala

mode /məʊd/ *n* modo *m*; Comput modalità *f*

model /'mɒdl/ *n* modello *m*; ⟨fashion⟩ ~ indossatore, -trice *mf*, modello, -a *mf* ● *a* ⟨yacht, plane⟩ in miniatura; ⟨pupil, husband⟩ esemplare, modello ● *v* ⟨pt/pp **modelled**⟩ ● *vt* indossare ⟨clothes⟩ ● *vi* fare l'indossatore, -trice *mf*; ⟨for artist⟩ posare

modem /'məʊdem/ *n* modem *m inv*

moderate¹ /'mɒdəreɪt/ *vt* moderare ● *vi* moderarsi

moderate² /'mɒdərət/ *a* moderato ● *n* Pol moderato, -a *mf*. **~ly** *adv* ⟨drink, speak etc⟩ moderatamente; ⟨good, bad etc⟩ relativamente

moderation /mɒdə'reɪʃn/ *n* moderazione *f*; **in ~** con moderazione

modern /'mɒdn/ *a* moderno. **~ize** *vt* modernizzare

modest /'mɒdɪst/ *a* modesto. **~y** *n* modestia *f*

modicum /'mɒdɪkəm/ *n* **a ~ of** un po' di

modification /mɒdɪfɪ'keɪʃn/ *n* modificazione *f*. **~y** /'mɒdɪfaɪ/ *vt* ⟨pt/pp **-fied**⟩ modificare

module /'mɒdjuːl/ *n* modulo *m*

moist /mɔɪst/ *a* umido

moisten /'mɔɪsn/ *vt* inumidire

moisture /'mɔɪstʃə(r)/ *n* umidità *f*. **~izer** *n* [crema *f*] idratante *m*

molar /'məʊlə(r)/ *n* molare *m*

molasses /mə'læsɪz/ *n Am* melassa *f*

mole¹ /məʊl/ *n* ⟨on face etc⟩ neo *m*

mole² /məʊl/ *n Zool* talpa *f*

molecule /'mɒlɪkjuːl/ *n* molecola *f*

molest /mə'lest/ *vt* molestare

mollycoddle /'mɒlɪkɒdl/ *vt* tenere nella bambagia

molten /'məʊltən/ *a* fuso

mom /mɒm/ *n Am fam* mamma *f*

moment /'məʊmənt/ *n* momento *m*; **at the ~** in questo momento. **~arily** *adv* momentaneamente. **~ary** *a* momentaneo

momentous /mə'mentəs/ *a* molto importante

momentum /mə'mentəm/ *n* impeto *m*

monarch /'mɒnək/ *n* monarca *m*. **~y** *n* monarchia *f*

monastery /'mɒnəstrɪ/ *n* monastero *m*. **~ic** /mə'næstɪk/ *a* monastico

Monday /'mʌndeɪ/ *n* lunedì *m inv*

monetary /'mʌnɪtrɪ/ *a* monetario

money /'mʌnɪ/ *n* denaro *m*

money: ~-box *n* salvadanaio *m*. **~-lender** *n* usuraio *m*

mongrel /'mʌŋgrəl/ *n* bastardo *m*

monitor /'mɒnɪtə(r)/ *n* Techn monitor *m inv* ● *vt* controllare

monk /mʌŋk/ *n* monaco *m*

monkey /'mʌŋkɪ/ *n* scimmia *f*. **~-nut** *n* nocciolina *f* americana. **~-wrench** *n* chiave *f* inglese a rullino

mono /'mɒnəʊ/ *n* mono *m*

monogram /'mɒnəgræm/ *n* monogramma *m*

monologue /'mɒnəlɒg/ n monologo m

monopol|ize /mə'nɒpəlaɪz/ vt monopolizzare. **~y** n monopolio m

monosyllabic /mɒnəsɪ'læbɪk/ a monosillabico

monotone /'mɒnətəʊn/ n **speak in a ~** parlare con tono monotono

monoton|ous /mə'nɒtənəs/ a monotono. **~y** n monotonia f

monsoon /mɒn'suːn/ n monsone m

monster /'mɒnstə(r)/ n mostro m

monstrosity /mɒn'strɒsətɪ/ n mostruosità f

monstrous /'mɒnstrəs/ a mostruoso

month /mʌnθ/ n mese m. **~ly** a mensile ● adv mensilmente ● n (periodical) mensile m

monument /'mɒnjʊmənt/ n monumento m. **~al** /-'mentl/ a fig monumentale

moo /muː/ n muggito m ● vi (pt/pp mooed) muggire

mooch /muːtʃ/ vi **~ about** fam gironzolare (**the house** per casa)

mood /muːd/ n umore m; **be in a good/bad ~** essere di buon/cattivo umore; **be in the ~ for** essere in vena di

moody /'muːdɪ/ a (-ier, -iest) (variable) lunatico; (bad-tempered) di malumore

moon /muːn/ n luna f; **over the ~** fam al settimo cielo

moon: ~light n chiaro m di luna ● vi fam lavorare in nero. **~lit** a illuminato dalla luna

moor[1] /mʊə(r)/ n brughiera f

moor[2] vt Naut ormeggiare

moose /muːs/ n (pl **moose**) alce m

moot /muːt/ a **it's a ~ point** è un punto controverso

mop /mɒp/ n straccio m (per i pavimenti); **~ of hair** zazzera f ● vt (pt/pp mopped) lavare con lo straccio. **mop up** vt (dry) asciugare con lo straccio; (clean) pulire con lo straccio

mope /məʊp/ vi essere depresso

moped /'məʊped/ n ciclomotore m

moral /'mɒrəl/ a morale ● n morale f. **~ly** adv moralmente. **~s** pl moralità f

morale /mə'rɑːl/ n morale m

morality /mə'rælətɪ/ n moralità f

morbid /'mɔːbɪd/ a morboso

more /mɔː(r)/ a più; **a few ~ books** un po' di più di libri; **some ~ tea?** ancora un po' di tè?; **there's no ~ bread** non c'è più pane; **there are no ~ apples** non ci sono più mele; **one ~ word and...** ancora una parola e... ● pron di più; **would you like some ~?** ne vuoi ancora?; **no ~, thank you** non ne voglio più, grazie ● adv più; **~ interesting** più interessante; **~ [and ~] quickly** [sempre] più veloce; **~ than** più di; **I don't love him any ~** no lo amo più; **once ~** ancora una volta; **~ or less** più o meno; **the ~ I see him, the ~ I like him** più lo vedo, più mi piace

moreover /mɔː'rəʊvə(r)/ adv inoltre

morgue /mɔːg/ n obitorio m

moribund /'mɒrɪbʌnd/ a moribondo

morning /'mɔːnɪŋ/ n mattino m, mattina f; **in the ~** del mattino; (tomorrow) domani mattina

Morocc|o /mə'rɒkəʊ/ n Marocco m ● a **~an** a & n marocchino, -a ● a -a mf

moron /'mɔːrɒn/ n fam deficiente mf

morose /mə'rəʊs/ a scontroso

morphine /'mɔːfiːn/ n morfina f

Morse /mɔːs/ n **~ [code]** (codice m) Morse m

morsel /'mɔːsl/ n (food) boccone m

mortal /'mɔːtl/ a & n mortale mf. **~ity** /mɔː'tælətɪ/ n mortalità f. **~ly** adv ⟨wounded, offended⟩ a morte; ⟨afraid⟩ da morire

mortar /'mɔːtə(r)/ n mortaio m

mortgage /'mɔːgɪdʒ/ n mutuo m; (on property) ipoteca f ● vt ipotecare

mortuary /'mɔːtjʊərɪ/ n camera f mortuaria

mosaic /məʊ'zeɪɪk/ n mosaico m

Moscow /'mɒskəʊ/ n Mosca f

Moslem /'mɒzlɪm/ a & n musulmano, -a f

mosque /mɒsk/ n moschea f

mosquito /mɒs'kiːtəʊ/ n (pl -es) zanzara f

moss /mɒs/ n muschio m. **~y** a muschioso

most /məʊst/ a (majority) la maggior parte di; **for the ~ part** per lo più ●adv più, maggiormente; (very) estremamente, molto; **the ~ interesting day** la giornata più interessante; **a ~ interesting day** una giornata estremamente interessante; **the ~ beautiful woman in the world** la donna più bella del mondo; **~ unlikely** veramente improbabile ●pron **~ of them** la maggior parte di loro; **at** [**the**] **~** al massimo; **make the ~ of** sfruttare al massimo; **~ of the time** la maggior parte del tempo. **~ly** adv per lo più

MOT n revisione f obbligatoria di autoveicoli

motel /məʊ'tel/ n motel m inv

moth /mɒθ/ n falena f. [**clothes-**] **~** tarma f

moth: ~ball n pallina f di naftalina. **~-eaten** a tarmato

mother /'mʌðə(r)/ n madre f; **M~'s Day** la festa della mamma ●vt fare da madre a

mother: ~board n Comput scheda f madre. **~hood** n maternità f. **~-in-law** n (pl **~s-in-law**) suocera f. **~ly** a materno. **~-of-pearl** n madreperla f. **~-to-be** n futura mamma f. **~ tongue** n madrelingua f

mothproof /'mɒθ-/ a antitarmico

motif /məʊ'tiːf/ n motivo m

motion /'məʊʃn/ n moto m; (proposal) mozione f; (gesture) gesto m ●vt/i **~ [to] sb to come in** fare segno a qcno di entrare. **~less** a immobile. **~lessly** adv senza alcun movimento

motivate /'məʊtɪveɪt/ vt motivare. **~ion** /-'veɪʃn/ n motivazione f

motive /'məʊtɪv/ n motivo m

motley /'mɒtlɪ/ a disparato

motor /'məʊtə(r)/ n motore m; (car) macchina f ●a a motore; Anat motore ●vi andare in macchina

Motorail /'məʊtəreɪl/ n treno m per trasporto auto

motor: ~ bike n fam moto f inv. **~ boat** n motoscafo m. **~cade** /-keɪd/ n Am corteo m di auto. **~ car** n automobile f. **~ cycle** n motocicletta f. **~-cyclist** n motociclista mf. **~ing** n automobilismo m. **~ist** n automobilista mf. **~ racing** n corse fpl automobilistiche. **~ vehicle** n autoveicolo m. **~way** n autostrada f

mottled /'mɒtld/ a chiazzato

motto /'mɒtəʊ/ n (pl -es) motto m

mould[1] /məʊld/ n (fungus) muffa f

mould[2] n stampo m ●vt foggiare; fig formare. **~ing** n Archit cornice f

mouldy /'məʊldɪ/ a ammuffito; (fam: worthless) ridicolo

moult /məʊlt/ vi ⟨bird:⟩ fare la muta; ⟨animal:⟩ perdere il pelo

mound /maʊnd/ n mucchio m; (hill) collinetta f

mount /maʊnt/ n (horse) cavalcatura f; (of jewel, photo, picture) montatura f ●vt montare a ⟨horse⟩; salire su ⟨bicycle⟩; incastonare ⟨jewel⟩; incorniciare ⟨photo, picture⟩ ●vi aumentare.
mount up vi aumentare

mountain /'maʊntɪn/ n montagna f. **~ bike** n mountain bike f inv

mountaineer /maʊntɪ'nɪə(r)/ n alpinista mf. **~ing** n alpinismo m

mountainous /'maʊntɪnəs/ a montagnoso

mourn /mɔːn/ vt lamentare ●vi **~ for** piangere la morte di. **~er** n persona f che partecipa a un funerale. **~ful** a triste. **~ing** n **in ~ing** in lutto

mouse /maʊs/ n (pl mice) topo m;

Comput mouse *m inv*. **~trap** *n* trappola *f* [per topi]

mousse /muːs/ *n Culin* mousse *f inv*

moustache /məˈstɑːʃ/ *n* baffi *mpl*

mousy /ˈmaʊsɪ/ *a* (*colour*) grigio topo

mouth[1] /maʊð/ *vt* ~ sth dire qcsa silenziosamente muovendo solamente le labbra

mouth[2] /maʊθ/ *n* bocca *f*; (*of river*) foce *f*

mouth: **~ful** *n* boccone *m*. **~organ** *n* armonica *f* [a bocca]. **~piece** *n* imboccatura *f*; (*fig: person*) portavoce *m inv*. **~wash** *n* acqua *f* dentifricia. **~watering** *a* che fa venire l'acquolina in bocca

movable /ˈmuːvəbl/ *a* mobile

move /muːv/ *n* mossa *f*; (*moving house*) trasloco *m*; **on the ~** in movimento; **get a ~ on** darsi una mossa ● *vt* muovere; (*emotionally*) commuovere; spostare 〈*car, furniture*〉; (*transfer*) trasferire; (*propose*) proporre; ● *house* traslocare ● *vi* muoversi; (*move house*) traslocare. **move along** *vi* andare avanti ● *vt* muovere in avanti. **move away** *vi* allontanarsi; (*move house*) trasferirsi ● *vt* allontanare. **move forward** *vi* avanzare ● *vt* spostare avanti. **move in** *vi* (*to a house*) trasferirsi. **move off** *vi* (*vehicle*): muoversi. **move out** *vi* (*of house*) andare via. **move over** *vi* spostarsi ● *vt* spostare. **move up** *vi* muoversi; (*advance, increase*) avanzare

movement /ˈmuːvmənt/ *n* movimento *m*

movie /ˈmuːvɪ/ *n* film *m inv*; **go to the ~s** andare al cinema

moving /ˈmuːvɪŋ/ *a* mobile; (*touching*) commovente

mow /məʊ/ *vt* (*pt* mowed, *pp* mown *or* mowed) tagliare 〈*lawn*〉. **mow down** *vt* (*destroy*) sterminare

mower /ˈməʊə(r)/ *n* tosaerbe *m inv*

MP *n abbr* Member of Parliament

Mr /ˈmɪstə(r)/ *n* (*pl* **Messrs**) Signor *m*

Mrs /ˈmɪsɪz/ *n* Signora *f*

Ms /mɪz/ *n* Signora *f* (*modo m formale di rivolgersi ad una donna quando non si vuole connotarla come sposata o nubile*)

much /mʌtʃ/ *a, adv & pron* molto; **~ as** per quanto; **I love you just as ~ as before**/**him** ti amo quanto prima/lui; **as ~ as £5 million** ben cinque milioni di sterline; **as ~ as that** così tanto; **very ~** tantissimo, moltissimo; **the same** quasi uguale

muck /mʌk/ *n* (*dirt*) sporcizia *f*; (*farming*) letame *m*; (*fam: filth*) porcheria *f*. **muck about** *vi fam* perder tempo; **~ about with** trafficare con. **muck up** *vt fam* rovinare; (*make dirty*) sporcare

mucky /ˈmʌkɪ/ *a* (**-ier, -iest**) sudicio

mucus /ˈmjuːkəs/ *n* muco *m*

mud /mʌd/ *n* fango *m*

muddle /ˈmʌdl/ *n* disordine *m*; (*mix-up*) confusione *f* ● *vt* ~ **[up]** confondere 〈*dates*〉

muddy /ˈmʌdɪ/ *a* (**-ier, -iest**) 〈*path*〉 fangoso; 〈*shoes*〉 infangato

'mudguard *n* parafango *m*

muesli /ˈmuːzlɪ/ *n* muesli *m inv*

muffle /ˈmʌfl/ *vt* smorzare 〈*sound*〉. **muffle** **[up]** *vt* (*for warmth*) imbacuccare

muffler /ˈmʌflə(r)/ *n* sciarpa *f*; *Am Auto* marmitta *f*

mug[1] /mʌg/ *n* tazza *f*; (*for beer*) boccale *m*; (*fam: face*) muso *m*; (*fam: simpleton*) pollo *m*

mug[2] *vt* (*pt*/*pp* mugged) aggredire e derubare. **~ger** *n* assalitore, -trice *mf*. **~ging** *n* aggressione *f* per furto

muggy /ˈmʌgɪ/ *a* (**-ier, -iest**) afoso

mule /mjuːl/ *n* mulo *m*

mull /mʌl/ *vt* ~ **over** rimuginare su

mulled /mʌld/ *a* ~ **wine** vin brulé *m inv*

multi /ˈmʌltɪ/: **~coloured** *a* variopinto. **~lingual** /-'lɪŋgwəl/ a multilingue *inv*. **~'media** n multimedia *mpl* ● a multimediale. **~national** a multinazionale ● n multinazionale f

multiple /'mʌltɪpl/ a multiplo

multiplication /mʌltɪplɪ'keɪʃn/ n moltiplicazione f

multiply /'mʌltɪplaɪ/ v t (pt/pp -ied) ● vt moltiplicare (by per) ● vi moltiplicarsi

multi'storey a ~ car park parcheggio m a più piani

mum[1] /mʌm/ a keep ~ fam non aprire bocca

mum[2] n fam mamma f

mumble /'mʌmbl/ vt/i borbottare

mummy[1] /'mʌmɪ/ n fam mamma f

mummy[2] n Archaeol mummia f

mumps /mʌmps/ n orecchioni mpl

munch /mʌntʃ/ vt/i sgranocchiare

mundane /mʌn'deɪn/ a (everyday) banale

municipal /mjʊ'nɪsɪpl/ a municipale

mural /'mjʊərəl/ n dipinto m murale

murder /'mɜ:də(r)/ n assassinio m ● vt assassinare; (fam: ruin) massacrare. **~er** n assassino, -a mf. **~ous** /-rəs/ a omicida

murky /'mɜ:kɪ/ a (-ier, -iest) oscuro

murmur /'mɜ:mə(r)/ n mormorio m ● vt/i mormorare

muscle /'mʌsl/ n muscolo m ● muscle in vi sl intromettersi (on in)

muscular /'mʌskjʊlə(r)/ a muscolare; (strong) muscoloso

muse /mju:z/ vi meditare (on su)

museum /mju:'zɪəm/ n museo m

mushroom /'mʌʃrʊm/ n fungo m ● vi fig spuntare come funghi

music /'mju:zɪk/ n musica f; (written) spartito m.

musical /'mju:zɪkl/ a musicale; ⟨person⟩ dotato di senso musicale ● n commedia f musicale. **~ box** n

carillon m inv. **~ instrument** n strumento m musicale

music: ~ box n carillon m inv. **~ centre** impianto m stereo; '**~-hall** n teatro m di varietà

musician /mju:'zɪʃn/ n musicista mf

Muslim /'mʊzlɪm/ a & n musulmano, -a mf

mussel /'mʌsl/ n cozza f

must /mʌst/ v aux (solo al presente) dovere; **you ~ not be late** non devi essere in ritardo; **she ~ have finished by now** (probability) deve aver finito ormai ● n a ~ fam una cosa da non perdere

mustard /'mʌstəd/ n senape f

musty /'mʌstɪ/ a (-ier, -iest) stantio

mutation /mju:'teɪʃn/ n Biol mutazione f

mute /mju:t/ a muto

muted /'mju:tɪd/ a smorzato

mutilat|e /'mju:tɪleɪt/ vt mutilare. **~ion** /-'leɪʃn/ n mutilazione f

mutin|ous /'mju:tɪnəs/ a ammutinato. **~y** n ammutinamento m ● vi (pt/pp -ied) ammutinarsi

mutter /'mʌtə(r)/ vt/i borbottare

mutton /'mʌtn/ n carne f di montone

mutual /'mju:tjʊəl/ a reciproco; (fam: common) comune. **~ly** adv reciprocamente

muzzle /'mʌzl/ n (of animal) muso m; (of firearm) bocca f; (for dog) museruola f ● vt fig mettere il bavaglio a

my /maɪ/ a il mio m, la mia f, i miei mpl, le mie fpl; **my mother/father** mia madre/mio padre

myself /maɪ'self/ pron (reflexive) mi; (emphatic) me stesso; (after prep) me; **I've seen it ~** l'ho visto io stesso; **by ~** da solo; **I thought to ~** ho pensato tra me e me; **I'm proud of ~** sono fiero di me

mysterious /mɪ'stɪərɪəs/ a misterioso. **~ly** adv misteriosamente

mystery /'mɪstərɪ/ n mistero m; **~ [story]** racconto m del mistero

mysti|c[al] /'mɪstɪk[l]/ *a* mistico. **~cism** /-sɪzm/ *n* misticismo *m*

mystified /'mɪstɪfaɪd/ *a* disorientato

mystify /'mɪstɪfaɪ/ *vt* (*pt/pp* **-ied**) disorientare

mystique /mɪ'sti:k/ *n* mistica *f*

myth /mɪθ/ *n* mito *m*. **~ical** *a* mitico

mythology /mɪ'θɒlədʒɪ/ *n* mitologia *f*

nome di. **~less** *a* senza nome. **~ly** *adv* cioè

name: **~-plate** *n* targhetta *f*. **~sake** *n* omonimo, -a *mf*

nanny /'nænɪ/ *n* bambinaia *f*. **~-goat** *n* capra *f*

nap /næp/ *n* pisolino *m*; **have a ~** fare un pisolino ● *vi* (*pt/pp* **napped**) **catch sb ~ping** cogliere qcno alla sprovvista

nape /neɪp/ *n* ~ [**of the neck**] nuca *f*

napkin /'næpkɪn/ *n* tovagliolo *m*

Naples /'neɪplz/ *n* Napoli *f*

nappy /'næpɪ/ *n* pannolino *m*

narcotic /nɑː'kɒtɪk/ *a* & *n* narcotico *m*

Nn

nab /næb/ *vt* (*pt/pp* **nabbed**) *fam* beccare

naff /næf/ *a Br fam* banale

nag¹ /næg/ *n* (*horse*) ronzino *m*

nag² (*pp/pp* **nagged**) *vt* assillare ● *vi* essere insistente ● *n* (*person*) brontolone, -a *mf*. **~ging** *a* (*pain*) persistente

nail /neɪl/ *n* chiodo *m*; (*of finger, toe*) unghia *f* ● **nail down** *vt* inchiodare; **~ sb down to a time/price** far fissare a qcno un'ora/un prezzo

nail: **~-brush** *n* spazzolino *m* da unghie. **~-file** *n* limetta *f* da unghie. **~ polish** *n* smalto *m* [per unghie]. **~ scissors** *npl* forbicine *fpl* da unghie. **~ varnish** *n* smalto *m* [per unghie]

naïve /naɪ'i:v/ *a* ingenuo. **~ty** /-ətɪ/ *n* ingenuità *f*

naked /'neɪkɪd/ *a* nudo; **with the ~ eye** a occhio nudo

name /neɪm/ *n* nome *m*; **what's your ~?** come ti chiami?; **my ~ is Matthew** mi chiamo Matthew; **I know her by ~** la conosco di nome; **by the ~ of Bates** di nome Bates; **call sb ~s** *fam* insultare qcno ● *vt* (*to position*) nominare; chiamare (*baby*); (*identify*) citare; **be ~d after** essere chiamato col

narrat|e /nə'reɪt/ *vt* narrare. **~ion** /-eɪʃn/ *n* narrazione *f*

narrative /'nærətɪv/ *a* narrativo ● *n* narrazione *f*

narrator /nə'reɪtə(r)/ *n* narratore, -trice *mf*

narrow /'nærəʊ/ *a* stretto; (*fig: views*) ristretto; (*margin, majority*) scarso ● *vi* restringersi. **~ly** *adv* **~ly escape death** evitare la morte per un pelo. **~-'minded** *a* di idee ristrette

nasal /'neɪzl/ *a* nasale

nastily /'nɑːstɪlɪ/ *adv* (*spitefully*) con cattiveria

nasty /'nɑːstɪ/ *a* (**-ier, -iest**) (*smell, person, remark*) cattivo; (*injury, situation, weather*) brutto; **turn ~** (*person:*) diventare cattivo

nation /'neɪʃn/ *n* nazione *f*

national /'næʃənl/ *a* nazionale ● *n* cittadino, -a *mf*

national: **~ 'anthem** *n* inno *m* nazionale. **N~ 'Health Service** *n Br* servizio *m* sanitario. **N~ In'surance** *n* ≈ Previdenza *f* sociale

nationalism /'næʃənəlɪzm/ *n* nazionalismo *m*

nationality /næʃə'nælətɪ/ *n* nazionalità *f inv*

national|ization /næʃənəlaɪ'zeɪʃn/ *n* nazionalizzazione. **~ize** /'næʃənəlaɪz/ *vt* nazionalizzare.

~ly /'næʃənəlɪ/ *adv* a livello nazionale

'nation-wide *a* su scala nazionale

native /'neɪtɪv/ *a* nativo; *(innate)* innato ●*n* nativo, -a *mf*; *(local inhabitant)* abitante *mf* del posto; *(outside Europe)* indigeno, -a *mf*; **she's a ~ of Venice** è originaria di Venezia

native: **~ 'land** *n* paese *m* nativo. **~ 'language** *n* lingua *f* madre

Nativity /nə'trvətɪ/ *n* **the ~** la Natività *f*. **~ play** *n* rappresentazione *f* sulla nascita di Gesù

natter /'nætə(r)/ *vi fam* chiacchierare

natural /'nætʃrəl/ *a* naturale

natural: **~ 'gas** *n* metano *m*. **~ 'history** *n* storia *f* naturale

naturalist /'nætʃ(ə)rəlɪst/ *n* naturalista *mf*

natural|ization /nætʃ(ə)rəlaɪ'zeɪʃn/ *n* naturalizzazione *f*. **~ize** /'nætʃ(ə)rəlaɪz/ *vt* naturalizzare

naturally /'nætʃ(ə)rəlɪ/ *adv* (*of course*) naturalmente; *(by nature)* per natura

nature /'neɪtʃə(r)/ *n* natura *f*; **by ~** per natura. **~ reserve** *n* riserva *f* naturale

naughtily /'nɔ:tɪlɪ/ *adv* male

naughty /'nɔ:tɪ/ *a* (-**ier**, -**iest**) monello; *(slightly indecent)* spinto

nausea /'nɔ:zɪə/ *n* nausea *f*

nause|ate /'nɔ:zɪeɪt/ *vt* nauseare. **~ating** *a* nauseante. **~ous** /-ɪəs/ *a* **I feel ~ous** ho la nausea

nautical /'nɔ:tɪkl/ *a* nautico. **~ mile** *n* miglio *m* marino

naval /'neɪvl/ *a* navale

nave /neɪv/ *n* navata *f* centrale

navel /'neɪvl/ *n* ombelico *m*

navigable /'nævɪgəbl/ *a* navigabile

navigat|e /'nævɪgeɪt/ *vi* navigare; *Auto* fare da navigatore ●*vt* navigare su *(river)*. **~ion** /-'geɪʃn/ *n* navigazione *f*. **~or** *n* navigatore *m*

navy /'neɪvɪ/ *n* marina *f* ● **~ [blue]** *a* blu marine *inv* ●*n* blu *m inv* marine

Neapolitan /nɪə'pɒlɪtən/ *a & n* napoletano, -a *mf*

near /nɪə(r)/ *a* vicino; *(future)* prossimo; **the ~est bank** la banca più vicina ●*adv* vicino; **draw ~** avvicinarsi; **~ at hand** a portata di mano ●*prep* vicino a; **he was ~ to tears** aveva le lacrime agli occhi ●*vt* avvicinarsi a

near: **~'by** *a & adv* vicino. **~ly** *adv* quasi; **it's not ~ly enough** non è per niente sufficiente. **~ness** *n* vicinanza *f*. **~ side** *a Auto (wheel)* (*left*) sinistro; (*right*) destro. **~-sighted** *a Am* miope

neat /ni:t/ *a* (*tidy*) ordinato; (*clever*) efficace; (*undiluted*) liscio. **~ly** *adv* ordinatamente; (*cleverly*) efficacemente. **~ness** *n* (*tidiness*) ordine *m*

necessarily /nesə'serɪlɪ/ *adv* necessariamente

necessary /'nesəsərɪ/ *a* necessario

necessit|ate /nɪ'sesɪteɪt/ *vt* rendere necessario. **~y** *n* necessità *f inv*

neck /nek/ *n* collo *m*; (*of dress*) colletto *m*; **~ and ~** testa a testa

necklace /'neklɪs/ *n* collana *f*

neck: **~line** *n* scollatura *f*. **~tie** *n* cravatta *f*

neé /neɪ/ *a* **~ Brett** nata Brett

need /ni:d/ *n* bisogno *m*; **be in ~ of** avere bisogno di; **if ~ be** se ce ne fosse bisogno; **there is a ~ for** c'è bisogno di; **there is no ~ for that** non ce n'è bisogno; **there is no ~ for you to go** non c'è bisogno che tu vada ●*vt* aver bisogno di; **I ~ to know** devo saperlo; **it ~s to be done** bisogna farlo ●*v aux* you **~ not go** non c'è bisogno che tu vada; **~ I come?** devo [proprio] venire?

needle /'ni:dl/ *n* ago *m*; (*for knitting*) uncinetto *m*; (*of record player*) puntina *f* ●*vt* (*fam: annoy*) punzecchiare

needless /'ni:dlɪs/ *a* inutile

'needlework *n* cucito *m*

needy /'niːdɪ/ a (-ier, -iest) bisognoso

negation /nɪ'geɪʃn/ n negazione f

negative /'negətɪv/ a negativo ● n negazione f; *Phot* negativo m; **in the** ~ *Gram* alla forma negativa

neglect /nɪ'glekt/ n trascuratezza f; **state of** ~ stato m di abbandono ● vt trascurare; **he** ~**ed to write** non si è curato di scrivere. **~ed** a trascurato. **~ful** a negligente; **be** ~**ful of** trascurare

négligée /'neglɪʒeɪ/ n négligé m inv

negligence /'neglɪdʒəns/ n negligenza f. **~t** a negligente

negligible /'neglɪdʒəbl/ a trascurabile

negotiable /nɪ'gəʊʃəbl/ a ‹road› transitabile; *Comm* negoziabile; **not** ~ ‹cheque› non trasferibile

negotiat|e /nɪ'gəʊʃɪeɪt/ vt negoziare; *Auto* prendere ‹bend› ● vi negoziare. **~ion** /-'eɪʃn/ n negoziato m. **~or** n negoziatore, -trice mf

Negro /'niːgrəʊ/ a & n (pl **-es**) negro, -a mf

neigh /neɪ/ vi nitrire

neighbour /'neɪbə(r)/ n vicino, -a mf. **~hood** n vicinato m; **in the** ~**hood of** nei dintorni di; *fig* circa. **~ing** a vicino. **~ly** a amichevole

neither /'naɪðə(r)/ a & pron nessuno dei due, né l'uno né l'altro ● adv ~**...** nor né... né ● conj nemmeno, neanche; ~ **do/did I** nemmeno io

neon /'niːɒn/ n neon m. ~ **light** n luce f al neon

nephew /'nevjuː/ n nipote m

nerve /nɜːv/ n nervo m; (fam: courage) coraggio m; (fam: impudence) faccia f tosta; **lose one's** ~ perdersi d'animo. **~-racking** a logorante

nervous /'nɜːvəs/ a nervoso; **he makes me** ~ mi mette in agitazione; **be a** ~ **wreck** avere i nervi a pezzi. ~ **breakdown** n esaurimento m nervoso. **~ly** adv nervo-

samente. **~ness** n nervosismo m; (before important event) tensione f

nervy /'nɜːvɪ/ a (-ier, -iest) nervoso; (Am: impudent) sfacciato

nest /nest/ n nido m ● vi fare il nido. **~-egg** n gruzzolo m

nestle /'nesl/ vi accoccolarsi

net[1] /net/ n rete f ● vt (pt/pp **netted**) (catch) prendere (con la rete)

net[2] a netto ● vt (pt/pp **netted**) incassare un utile netto di

'netball n sport m inv femminile, simile a pallacanestro

Netherlands /'neðələndz/ npl **the** ~ i Paesi mpl Bassi

netting /'netɪŋ/ n [wire] ~ reticolato m

nettle /'netl/ n ortica f

'network n rete f

neuralgia /njʊə'rældʒə/ n nevralgia f

neurolog|ist /njʊə'rɒlədʒɪst/ n neurologo, -a mf

neur|osis /njʊə'rəʊsɪs/ n (pl **-oses** /-siːz/) nevrosi f inv. **~otic** /-'rɒtɪk/ a nevrotico

neuter /'njuːtə(r)/ a *Gram* neutro ● n *Gram* neutro m ● vt sterilizzare

neutral /'njuːtrəl/ a neutro; (country, person) neutrale ● n **in** ~ *Auto* in folle. **~ity** /-'trælɪtɪ/ n neutralità f. **~ize** vt neutralizzare

never /'nevə(r)/ adv [non...] mai; (fam: expressing disbelief) ma va; ~ **again** mai più; **well I** ~! chi l'avrebbe detto!. **~-ending** a interminabile

nevertheless /nevəðə'les/ adv tuttavia

new /njuː/ a nuovo

new: **~born** n neonato. **~comer** n nuovo, -a arrivato, -a mf. **~fangled** /-'fæŋgld/ a pej modernizzante. **~-laid** a fresco

newly adv (recently) di recente; **~-built** costruito di recente. **~-weds** npl sposini mpl

new: ~ **'moon** n luna f nuova. **~ness** n novità f

news /njuːz/ n notizie fpl; TV telegiornale m; Radio giornale m radio; **piece of** ~ notizia f

news: ~**agent** n giornalaio, -a mf. ~ **bulletin** n notiziario m. ~**caster** n giornalista mf televisivo, -a/radiofonico, -a. ~**flash** n notizia f flash. ~**letter** n bollettino m d'informazione. ~**paper** n giornale m; (material) carta f di giornale. ~**reader** n giornalista mf televisivo, -a/radiofonico, -a

new: ~ **year** n (next year) anno m nuovo; N~ **Year's Day** n Capodanno m. N~ **Year's Eve** n vigilia f di Capodanno. N~ **Zealand** /'ziːlənd/ n Nuova Zelanda f. N~ **Zealander** n neozelandese mf

next /nekst/ a prossimo. (adjoining) vicino; **who's** ~? a chi tocca?; ~ **door** accanto; ~ **to nothing** quasi niente; **the** ~ **day** il giorno dopo; ~ **week** la settimana prossima; **the week after** ~ fra due settimane ● adv dopo; **when will you see him** ~? quando lo rivedi la prossima volta?; ~ **to** accanto a ● n seguente mf; ~ **of kin** parente m prossimo

NHS n abbr **National Health Service**

nib n pennino m

nibble /'nɪbl/ vt/i mordicchiare

nice /naɪs/ a (day, weather, holiday) bello; (person) gentile, simpatico; (food) buono; **it was** ~ **meeting you** è stato un piacere conoscerla. ~**ly** adv gentilmente; (well) bene. ~**ties** /'naɪsətɪz/ npl finezze fpl

niche /niːʃ/ n nicchia f

nick /nɪk/ n tacca f; (on chin etc) taglietto m; (fam: prison) galera f; (fam: police station) centrale f [di polizia]; **in the** ~ **of time** fam appena in tempo ● vt intaccare; (fam: steal) fregare; (fam: arrest) beccare; ~ **one's chin** farsi un taglietto nel mento

nickel /'nɪkl/ n nichel m; Am moneta f da cinque centesimi

nickname n soprannome m ● vt soprannominare

nicotine /'nɪkətiːn/ n nicotina f

niece /niːs/ n nipote f

Nigeria /naɪ'dʒɪərɪə/ n Nigeria f. ~**n** a & n nigeriano, -a mf

niggling /'nɪglɪŋ/ a (detail) insignificante; (pain) fastidioso; (doubt) persistente

night /naɪt/ n notte f; (evening) sera f; **at** ~ la notte, di notte; (in the evening) la sera, di sera; **Monday** ~ lunedì notte/sera ● a di notte

night: ~**cap** n papalina f; (drink) bicchierino m bevuto prima di andare a letto. ~**club** n locale m notturno, night[-club] m inv. ~**dress** n camicia f da notte. ~**fall** n crepuscolo m. ~**gown**, fam ~**ie** /'naɪtɪ/ n camicia f da notte

nightingale /'naɪtɪŋgeɪl/ n usignolo m

night: ~**life** n vita f notturna. ~**ly** a di notte, di sera ● adv ogni notte, ogni sera. ~**mare** n incubo f. ~**school** scuola f serale. ~**time** n at ~**time** di notte, la notte. ~-'**watchman** n guardiano m notturno

nil /nɪl/ n nulla m; Sport zero m

nimble /'nɪmbl/ a agile. ~**y** adv agilmente

nine /naɪn/ a nove inv ● n nove m. ~**teen** a diciannove inv ● n diciannove. ~'**teenth** a & n diciannovesimo, -a mf

ninetieth /'naɪntɪɪθ/ a & n novantesimo, -a mf

ninety /'naɪntɪ/ a novanta inv ● n novanta m

ninth /naɪnθ/ a & n nono, -a mf

nip /nɪp/ n pizzicotto m; (bite) morso m ● vt pizzicare; (bite) mordere; ~ **in the bud** fig stroncare sul nascere ● vi (fam: run) fare un salto

nipple /'nɪpl/ n capezzolo m; (Am: on bottle) tettarella f

nippy /'nɪpɪ/ a (-ier, -iest) fam (cold) pungente; (quick) svelto

nitrogen /'naɪtrədʒən/ n azoto m

nitwit /'nɪtwɪt/ n fam imbecille mf

no /nəʊ/ adv no ● n (pl **noes**) no m inv ● a nessuno; **I have no time** non ho tempo; **in no time** in un baleno; **'no parking'** 'sosta vietata'; **'no smoking'** 'vietato fumare'; **no one** = **nobody**

nobility /nə'bɪlətɪ/ n nobiltà f

noble /'nəʊbl/ a nobile. **~man** n nobile m

nobody /'nəʊbədɪ/ pron nessuno; **he knows ~** non conosce nessuno ● n **he's a ~** non è nessuno

nocturnal /nɒk'tɜːnl/ a notturno

nod /nɒd/ n cenno m del capo ● vi (pt/pp **nodded**) fare un cenno col capo; (in agreement) fare di sì col capo ● vt **~ one's head** fare di sì col capo. **nod off** vi assopirsi

nodule /'nɒdjuːl/ n nodulo m

noise /nɔɪz/ n rumore m; (loud) rumore m, chiasso m. **~less** a silenzioso. **~lessly** adv silenziosamente

noisy /'nɔɪzɪ/ a (-ier, -iest) rumoroso

nomad /'nəʊmæd/ n nomade mf. **~ic** /-'mædɪk/ a nomade

nominal /'nɒmɪnl/ a nominale

nominate /'nɒmɪneɪt/ vt proporre come candidato; (appoint) designare. **~ion** /-'neɪʃn/ n nomina f; (person nominated) candidato, -a mf

nominative /'nɒmɪnətɪv/ a & n Gram **~** [**case**] nominativo m

nominee /nɒmɪ'niː/ n persona f nominata

nonchalant /'nɒnʃələnt/ a disinvolto

non-com'missioned /nɒn-/ a **~ officer** sottufficiale m

non-com'mittal /nɒn-/ a che non si sbilancia

nondescript /'nɒndɪskrɪpt/ a qualunque

none /nʌn/ pron (person) nessuno; (thing) niente; **~ of us** nessuno di

noi; **~ of this** niente di questo; **there's ~ left** non ce n'è più ● adv **she's ~ too pleased** non è per niente soddisfatta; **I'm ~ the wiser** non ne so più di prima

nonentity /nɒ'nentətɪ/ n nullità f

non-event n delusione f

non-ex'istent a inesistente

non-'fiction n saggistica f

non-'iron a che non si stira

nonplussed /nɒn'plʌst/ a perplesso

nonsens|e /'nɒnsəns/ n sciocchezze fpl. **~ical** /-'sensɪkl/ a assurdo

non-'smoker n non fumatore, -trice mf; (compartment) scompartimento m non fumatori

non-'stick a antiaderente

non-'stop a **~ 'flight** volo m diretto ● adv senza sosta; (fly) senza scalo

non-'violent a non violento

noodles /'nuːdlz/ npl taglierini mpl

nook /nʊk/ n cantuccio m

noon /nuːn/ n mezzogiorno m; **at ~** a mezzogiorno

noose /nuːs/ n nodo m scorsoio

nor /nɔː(r)/ adv & conj né; **~ do I** neppure io

Nordic /'nɔːdɪk/ a nordico

norm /nɔːm/ n norma f

normal /'nɔːml/ a normale. **~ity** /-'mælətɪ/ n normalità f. **~ly** adv (usually) normalmente

north /nɔːθ/ n nord m; **to the ~ of** a nord di ● a del nord, settentrionale ● adv a nord

north: N~ America n America f del Nord. **~bound** a Auto in direzione nord. **~-east** a di nord-est, nordorientale ● n nord-est m ● adv a nord-est; (travel) verso nord-est

norther|ly /'nɔːðəlɪ/ a ‹direction› nord; ‹wind› del nord. **~n** a del nord, settentrionale. **N~n Ireland** n Irlanda f del Nord

north: N~ 'Pole n polo m nord. **N~ 'Sea** n Mare m del Nord. **~ward[s]** /-wəd[z]/ adv verso nord. **~-west** a di nord-ovest, nordoccidentale ● n nord-ovest m ● adv a nord-ovest; ⟨travel⟩ verso nord-ovest

Nor|way /'nɔːweɪ/ n Norvegia f. **~wegian** /-'wiːdʒn/ a & n norvegese mf

nose /nəʊz/ n naso m

nose: ~bleed n emorragia f nasale. **~-dive** n Aeron picchiata f

nostalg|ia /nɒ'stældʒɪə/ n nostalgia f. **~ic** a nostalgico

nostril /'nɒstrəl/ n narice f

nosy /'nəʊzɪ/ a (-ier, -iest) fam ficcanaso inv

not /nɒt/ adv non; **he is ~ Italian** non è italiano; **I hope ~** spero di no; **~ all of us have been invited** non siamo stati tutti invitati; **if ~** se no; **~ at all** niente affatto; **~ a bit** per niente; **~ even** neanche; **~ yet** non ancora; **~ only... but also...** non solo... ma anche...

notabl|e /'nəʊtəbl/ a (remarkable) notevole. **~y** adv (in particular) in particolare

notary /'nəʊtərɪ/ n notaio m; **~ public** notaio m

notch /nɒtʃ/ n tacca f ● **notch up** vt (score) segnare

note /nəʊt/ n nota f; (short letter, banknote) biglietto m; (memo, written comment etc) appunto m; **of ~** ⟨person⟩ di spicco; ⟨comments, event⟩ degno di nota; **make a ~ of** prendere nota di; **take ~ of** (notice) prendere nota di ● vt (notice) notare; (write) annotare. **note down** vt annotare

'notebook n taccuino m; Comput notebook m inv

noted /'nəʊtɪd/ a noto, celebre (**for** per)

note: ~paper n carta f da lettere. **~worthy** a degno di nota

nothing /'nʌθɪŋ/ pron niente, nulla ● adv niente affatto. **for ~** (free, in vain) per niente; (with no reason) senza motivo; **~ but** niente altro che; **~ much** poco o nulla; **~ interesting** niente di interes-

sante; **it's ~ to do with you** non ti riguarda

notice /'nəʊtɪs/ n (on board) avviso m; (review) recensione f; (termination of employment) licenziamento m; (advance) ~ preavviso m; **two months** ~ due mesi di preavviso; **at short** ~ con breve preavviso; **until further** ~ fino nuovo avviso; **give [in one's]** ~ ⟨employee:⟩ dare le dimissioni; **give an employee** ~ dare il preavviso a un impiegato; **take no** ~ **of** non fare caso a; **take no** ~! non farci caso! ● vt notare. **~able** /-əbl/ a evidente. **~ably** adv sensibilmente. **~-board** n bacheca f

noti|fication /nəʊtɪfɪ'keɪʃn/ n notifica f. **~fy** /'nəʊtɪfaɪ/ vt (pt/pp -ied) notificare

notion /'nəʊʃn/ n idea f, nozione f; **~s** pl (Am: haberdashery) merceria f

notoriety /nəʊtə'raɪətɪ/ n notorietà f

notorious /nəʊ'tɔːrɪəs/ a famigerato; **be ~ for** essere tristemente famoso per

notwith'standing prep malgrado ● adv ciononostante

nougat /'nuːgɑː/ n torrone m

nought /nɔːt/ n zero m

noun /naʊn/ n nome m, sostantivo m

nourish /'nʌrɪʃ/ vt nutrire. **~ing** a nutriente. **~ment** n nutrimento m

novel /'nɒvl/ a insolito ● n romanzo m. **~ist** n romanziere, -a mf. **~ty** n novità f; **~ties** pl (objects) oggettini mpl

November /nəʊ'vembə(r)/ n novembre m

novice /'nɒvɪs/ n novizio, -a mf

now /naʊ/ adv ora, adesso; **by ~** ormai; **just ~** proprio ora; **right ~** subito; **~ and again, ~ and then** ogni tanto; **~, ~!** su! ● conj **~ [that]** ora che, adesso che

'nowadays adv oggigiorno

nowhere /'nəʊ-/ adv in nessun posto, da nessuna parte

noxious /'nɒkʃəs/ a nocivo

nozzle /'nɒzl/ n bocchetta f

nuance /'nju:ɒs/ n sfumatura f

nuclear /'nju:klɪə(r)/ a nucleare

nucleus /'nju:klɪəs/ n (pl **-lei** /-lɪaɪ/) nucleo m

nude /nju:d/ a nudo ● n nudo m; **in the ~** nudo

nudge /nʌdʒ/ n colpetto m di gomito ● vt dare un colpetto col gomito a

nudism /'nju:dɪzm/ n nudismo m

nud|ist /'nju:dɪst/ n nudista mf. **~ity** n nudità f

nugget /'nʌgɪt/ n pepita f

nuisance /'nju:sns/ n seccatura f; (person) piaga f; **what a ~!** che seccatura!

null /nʌl/ a **~ and void** nullo

numb /nʌm/ a intorpidito; **~ with cold** intirizzito dal freddo

number /'nʌmbə(r)/ n numero m; **a ~ of people** un certo numero di persone ● vt numerare; (include) annoverare. **~-plate** n targa f

numeral /'nju:mərəl/ n numero m, cifra f

numerate /'nju:mərət/ a **be ~** saper fare i calcoli

numerical /nju:'merɪkl/ a numerico; **in ~ order** in ordine numerico

numerous /'nju:mərəs/ a numeroso

nun /nʌn/ n suora f

nurse /nɜ:s/ n infermiere, -a mf; **children's ~** bambinaia f ● vt curare

nursery /'nɜ:səri/ n stanza f dei bambini; (for plants) vivaio m; [day] ~ asilo m. **~ rhyme** n filastrocca f. **~ school** n scuola f materna

nursing /'nɜ:sɪŋ/ n professione f d'infermiere. **~ home** n casa f di cura per anziani

nurture /'nɜ:tʃə(r)/ vt allevare; fig coltivare

nut /nʌt/ n Techn dado m; (fam: head) zucca f; **~s** npl frutta f secca; **be ~s** fam essere svitato.

~crackers npl schiaccianoci m inv. **~meg** n noce f moscata

nutrition /nju:'trɪʃn/ n nutrizione f. **~ious** /-ʃəs/ a nutriente

'nutshell n **in a ~** fig in parole povere

nuzzle /'nʌzl/ vt (horse, dog:) strofinare il muso contro

nylon /'naɪlɒn/ n nailon m; **~s** pl calze fpl di nailon ● attrib di nailon

Oo

O /əʊ/ n Teleph zero m

oaf /əʊf/ n (pl **oafs**) zoticone, -a mf

oak /əʊk/ n quercia f ● attrib di quercia

OAP n abbr (**old-age pensioner**) pensionato, -a mf

oar /ɔ:(r)/ n remo m. **~sman** n vogatore m

oasis /əʊ'eɪsɪs/ n (pl **oases** /-si:z/) oasi f inv

oath /əʊθ/ n giuramento m; (swearword) bestemmia f

oatmeal /'əʊt-/ n farina f d'avena

oats /əʊts/ npl avena fsg; Culin [rolled] ~ fiocchi mpl di avena

obedien|ce /ə'bi:dɪəns/ n ubbidienza f. **~t** a ubbidiente

obese /ə'bi:s/ a obeso. **~ity** n obesità f

obey /ə'beɪ/ vt ubbidire a; osservare (instructions, rules) ● vi ubbidire

obituary /ə'bɪtjʊərɪ/ n necrologio m

object¹ /'ɒbdʒɪkt/ n oggetto m; Gram complemento m oggetto; **money is no ~** i soldi non sono un problema

object² /əb'dʒekt/ vi (be against) opporsi (**to** a); **~ that...** obiettare che...

objection /əb'dʒekʃn/ n obiezione f. **have no ~** non avere niente in contrario. **~able** /-əbl/ a discutibile; ⟨person⟩ sgradevole

objective /əb'dʒektɪv/ a oggettivo ●n obiettivo m. **~ely** adv obiettivamente. **~ity** /-'tɪvɪtɪ/ n oggettività f

obligation /ɒblɪ'geɪʃn/ n obbligo m; **be under an ~** avere un obbligo; **without ~** senza impegno

obligatory /ə'blɪgətrɪ/ a obbligatorio

oblig|e /ə'blaɪdʒ/ vt ⟨compel⟩ obbligare; **much ~ed** grazie mille. **~ing** a disponibile

oblique /ə'bli:k/ a obliquo; fig indiretto ●n ~ **[stroke]** barra f

obliterate /ə'blɪtəreɪt/ vt obliterare

oblivion /ə'blɪvɪən/ n oblio m

oblivious /ə'blɪvɪəs/ a be ~ essere dimentico (**of,** to di)

oblong /'ɒblɒŋ/ a oblungo ●n rettangolo m

obnoxious /əb'nɒkʃəs/ a detestabile

oboe /'əʊbəʊ/ n oboe m inv

obscen|e /əb'si:n/ a osceno; ⟨profits, wealth⟩ vergognoso. **~ity** /-'senɪtɪ/ n oscenità f inv

obscur|e /əb'skjʊə(r)/ a oscuro ●vt oscurare; ⟨confuse⟩ mettere in ombra. **~ity** n oscurità f

obsequious /əb'si:kwɪəs/ a ossequioso

observa|nce /əb'zɜ:vəns/ n ⟨of custom⟩ osservanza f. **~nt** a attento. **~tion** /ɒbzə'veɪʃn/ n osservazione f

observatory /əb'zɜ:vətrɪ/ n osservatorio m

observe /əb'zɜ:v/ vt osservare; ⟨notice⟩ notare; ⟨keep, celebrate⟩ celebrare. **~r** n osservatore, -trice mf

obsess /əb'ses/ vt **be ~ed by** essere fissato con. **~ion** /-eʃn/ n fissazione f. **~ive** /-ɪv/ a ossessivo

obsolete /'ɒbsəli:t/ a obsoleto; ⟨word⟩ desueto

obstacle /'ɒbstəkl/ n ostacolo m

obstetrician /ɒbstə'trɪʃn/ n

ostetrico, -a mf. **obstetrics** /əb'stetrɪks/ n ostetricia f

obstina|cy /'ɒbstɪnəsɪ/ n ostinazione f. **~te** /-nət/ a ostinato

obstreperous /əb'strepərəs/ a turbolento

obstruct /əb'strʌkt/ vt ostruire; ⟨hinder⟩ ostacolare. **~ion** /-ʌkʃn/ n ostruzione f; ⟨obstacle⟩ ostacolo m. **~ive** /-ɪv/ a be ~ive ⟨person:⟩ creare dei problemi

obtain /əb'teɪn/ vt ottenere. **~able** /-əbl/ a ottenibile

obtrusive /əb'tru:sɪv/ a ⟨object⟩ stonato

obtuse /əb'tju:s/ a ottuso

obvious /'ɒbvɪəs/ a ovvio. **~ly** adv ovviamente

occasion /ə'keɪʒn/ n occasione f; ⟨event⟩ evento m; **on ~** talvolta; **on the ~ of** in occasione di

occasional /ə'keɪʒənl/ a saltuario; **he has the ~ glass of wine** ogni tanto beve un bicchiere di vino. **~ly** adv ogni tanto

occult /ɒ'kʌlt/ a occulto

occupant /'ɒkjʊpənt/ n occupante mf; ⟨of vehicle⟩ persona f a bordo

occupation /ɒkjʊ'peɪʃn/ n occupazione f; ⟨job⟩ professione f **~al** a professionale

occupier /'ɒkjʊpaɪə(r)/ n residente mf

occupy /'ɒkjʊpaɪ/ vt ⟨pt/pp occupied⟩ occupare; ⟨keep busy⟩ tenere occupato

occur /ə'kɜ:(r)/ vi ⟨pt/pp occurred⟩ accadere; ⟨exist⟩ trovarsi; **it ~red to me that** mi è venuto in mente che. **~rence** /ə'kʌrəns/ n ⟨event⟩ fatto m

ocean /'əʊʃn/ n oceano m

o'clock /ə'klɒk/ adv **it's 7 ~** sono le sette; **at 7 ~** alle sette;

octave /'ɒktɪv/ n Mus ottava f

October /ɒk'təʊbə(r)/ n ottobre m

octopus /'ɒktəpəs/ n ⟨pl -puses⟩ polpo m

odd /ɒd/ a ⟨number⟩ dispari; ⟨not of set⟩ scompagnato; ⟨strange⟩ strano; **forty ~** quaranta e rotti;

~ **jobs** lavoretti *mpl*; **the ~ one out** l'eccezione; **at ~ moments** a tempo perso; **have the ~ glass of wine** avere un bicchiere di vino ogni tanto

odd|ity /ˈɒdɪtɪ/ *n* stranezza *f*. **~ly** *adv* stranamente; **~ly** anche stranamente. **~ment** *n* (*of fabric*) scampolo *m*

odds /ɒdz/ *npl* (*chances*) probabilità *fpl*; **at ~** in disaccordo; **~ and ends** cianfrusaglie *fpl*; **it makes no ~** non fa alcuna differenza

ode /əʊd/ *n* ode *f*

odour /ˈəʊdə(r)/ *n* odore *m*. **~less** *a* inodore

of /ɒv/, /əv/ *prep* di; **a cup of tea/coffee** una tazza di tè/caffè; **the hem of my skirt** l'orlo della mia gonna; **the summer of 1989** l'estate del 1989; **the two of us** noi due; **made of** di; **that's very kind of you** è molto gentile da parte tua; **a friend of mine** un mio amico; **a child of three** un bambino di tre anni; **the fourth of January** il quattro gennaio; **within a year of their divorce** a circa un anno dal loro divorzio; **half of it** la metà; **the whole of the room** tutta la stanza

off /ɒf/ *prep* da; (*distant from*) lontano da; **take £10 ~ the price** ridurre il prezzo di 10 sterline; **the coast** presso la costa; **a street ~ the main road** una traversa della via principale; (*near*) una strada vicino alla via principale; **get ~ the ladder** scendere dalla scala; **get off the bus** uscire dall'autobus; **leave the lid ~ the saucepan** lasciare la pentola senza il coperchio ● *adv* ⟨*button, handle*⟩ staccato; ⟨*light, machine*⟩ spento; ⟨*brake*⟩ tolto; ⟨*tap*⟩ chiuso; **'off'** (*on appliance*) 'off'; **2 kilometres ~** a due chilometri di distanza; **a long way ~** molto distante; (*time*) lontano; **~ and on** di tanto in tanto; **with his hat/ coat ~** senza il cappello/cappot-

to; **with the light ~** a luce spenta; **20% ~** 20% di sconto; **be ~** (*leave*) andar via; *Sport* essere partito; (*food:*) essere andato a male; (*all gone*) essere finito; ⟨*wedding, engagement:*⟩ essere cancellato; **I'm ~ alcohol** ho smesso di bere; **be ~ one's food** non avere appetito; **she's ~ today** (*on holiday*) è in ferie oggi; (*ill*) è malata oggi; **I'm ~ home** vado a casa; **you'd be better ~ doing..** faresti meglio a fare...; **have a day ~** avere un giorno di vacanza; **drive/sail ~** andare via

offal /ˈɒfl/ *n Culin* frattaglie *fpl*

off-beat *a* insolito

off-chance *n* possibilità *f* remota

off-'colour *a* (*not well*) giù di forma; ⟨*joke, story*⟩ sporco

offence /əˈfens/ *n* (*illegal act*) reato *m*; **give ~** offendere; **take ~** offendersi (**at** per)

offend /əˈfend/ *vt* offendere. **~er** *n Jur* colpevole *mf*

offensive /əˈfensɪv/ *a* offensivo ● *n* offensiva *f*.

offer /ˈɒfə(r)/ *n* offerta *f* ● *vt* offrire; opporre ⟨*resistance*⟩; **~ sb sth** offrire qcsa a qcno; **~ to do sth** offrirsi di fare qcsa. **~ing** *n* offerta *f*

off'hand *a* (*casual*) spiccio ● *adv* su due piedi

office /ˈɒfɪs/ *n* ufficio *m*; (*post, job*) carica *f*. **~ hours** *pl* orario *m* d'ufficio

officer /ˈɒfɪsə(r)/ *n* ufficiale *m*; (*police*) agente *m* [di polizia]

official /əˈfɪʃl/ *a* ufficiale ● *n* funzionario, -a *mf*; *Sport* dirigente *m*. **~ly** *adv* ufficialmente

officiate /əˈfɪʃɪeɪt/ *vi* officiare

'offing *n* **in the ~** in vista

off-licence *n* negozio *m* per la vendita di alcolici

off-load *vt* scaricare

off-putting *a fam* scoraggiante

offset *vt* (*pt/pp* **-set**, *pres p* **-setting**) controbilanciare

'offshoot *n* ramo *m*; *fig* diramazione *f*

'**offshore** a ⟨wind⟩ di terra; ⟨company, investment⟩ offshore. ~ **rig** n piattaforma f petrolifera, offshore m inv

off:side a Sport (in fuori gioco; ⟨wheel etc⟩ (left) sinistro; (right) destro

'**offspring** n prole m

off:stage adv dietro le quinte

off-'white a bianco sporco

often /'ɒfn/ adv spesso; **how ~** ogni quanto; **every so ~** una volta ogni tanto

ogle /'əʊgl/ vt mangiarsi con gli occhi

oh /əʊ/ int oh!; **~ dear** oh Dio!

oil /ɔɪl/ n olio m; ⟨petroleum⟩ petrolio m; ⟨for heating⟩ nafta f ● vt oliare

oil: **~field** n giacimento m di petrolio. **~-painting** n pittura f a olio. **~ refinery** n raffineria f di petrolio. **~ rig** piattaforma f per trivellazione subacquea. **~skins** npl vestiti mpl di tela cerata. **~-slick** n chiazza f di petrolio. **~-tanker** n petroliera f. **~ well** n pozzo m petrolifero

oily /'ɔɪlɪ/ a (-ier, -iest) unto; fig untuoso

ointment /'ɔɪntmənt/ n pomata f

OK /əʊ'keɪ/ int va bene, o.k. ● a **if that's OK with you** se ti va bene; **she's OK** (well) sta bene; **is the milk still OK?** il latte è ancora buono? ● adv (well) bene ● vt (anche **okay**) (pt/pp **okayed**) dare l'o.k.

old /əʊld/ a vecchio; ⟨girlfriend⟩ ex; **how ~ is she?** quanti anni ha?; **she is ten years ~** ha dieci anni

old: **~ 'age** n vecchiaia f. **~-age 'pensioner** n pensionato, -a mf. **~ boy** n Sch ex-allievo m. **~-'fashioned** a antiquato. **~ girl** n ex-allieva f. **~ 'maid** n zitella f

olive /'ɒlɪv/ n ⟨fruit, colour⟩ oliva f; ⟨tree⟩ olivo m ● a d'oliva; ⟨colour⟩ olivastro. **~ branch** n fig ramoscello m d'olivo. **~ 'oil** n olio m di oliva

Olympic /ə'lɪmpɪk/ a olimpico; **~s, ~ Games** Olimpiadi fpl inv

omelette /'ɒmlɪt/ n omelette f inv

omen /'əʊmən/ n presagio m

ominous /'ɒmɪnəs/ a sinistro

omission /ə'mɪʃn/ n omissione f

omit /ə'mɪt/ vt (pt/pp **omitted**) omettere; **~ to do sth** tralasciare di fare qcsa

omnipotent /ɒm'nɪpətənt/ a onnipotente

on /ɒn/ prep su; ⟨on horizontal surface⟩ su, sopra; **on Monday** lunedì; **on Mondays** di lunedì; **on the first of May** il primo di maggio; **on arriving** all'arrivo; **on one's finger** (cut) nel dito; ⟨ring⟩ al dito; **on foot** a piedi; **on the right/left** a destra/sinistra; **on the Rhine/Thames** sul Reno/Tamigi; **on the radio/television** alla radio/televisione; **on the bus/train** in autobus/treno; **go on the bus/train** andare in autobus/treno; **get on the bus/train** salire sull'autobus/sul treno; **on me** (with me) con me; **it's on me** fam tocca a me ● adv (further on) dopo; (switched on) acceso; ⟨brake⟩ inserito; (in operation) in funzione; **'on'** (on machine) 'on'; **he had his hat/coat on** portava il cappello/cappotto; **without his hat/coat on** senza cappello/cappotto; **with/ without the lid on** con/senza coperchio; **be on** ⟨film, programme, event:⟩ esserci; **it's not on** fam non è giusto; **be on at** fam tormentare (to per); **on and on** senza sosta; **on and off** a intervalli; **and so on** e così via; **go on** continuare; **drive on** spostarsi (con la macchina); **stick on** attaccare; **sew on** cucire

once /wʌns/ adv una volta; (formerly) un tempo; **~ upon a time there was** c'era una volta; **at ~** subito; (at the same time) contemporaneamente; **~ and for all** una volta per tutte ● conj [non] appena. **~over** n fam give sb/sth the **~over** (look, check) dare un'occhiata veloce a qcno/qcsa

'**oncoming** a che si avvicina dalla direzione opposta

one /wʌn/ a uno, una; not ~ **person** nemmeno una persona ●n uno m ●pron uno; (impersonal) si; ~ **another** l'un l'altro; ~ **by** ~ [a] uno a uno; ~ **never knows** non si sa mai

one: **~-eyed** a con un occhio solo. **~-off** a unico. **~-parent 'family** n famiglia f con un solo genitore. **~self** pron (reflexive) si; (emphatic) sé, se stesso; **by ~self** da solo; **be proud of ~self** essere fieri di sé. **~-sided** a unilaterale. **~-way** a (street) a senso unico; (ticket) di sola andata

onion /'ʌnjən/ n cipolla f

'**onlooker** n spettatore, -trice mf

only /'əʊnlɪ/ a solo; ~ **child** figlio, -a mf unico, -a ●adv & conj solo, solamente; ~ **just** appena

on/'off switch n pulsante m di accensione

'**onset** n (beginning) inizio m

'**onslaught** /'ɒnslɔːt/ n attacco m

onus /'əʊnəs/ n **the ~ is on me** spetta a me la responsabilità (**to** di)

onward[s] /'ɒnwəd[z]/ adv in avanti; **from then ~** da allora [in poi]

ooze /uːz/ vi fluire

opal /'əʊpl/ n opale f

opaque /əʊ'peɪk/ a opaco

open /'əʊpən/ a aperto; (free to all) pubblico; (job) vacante; **in the ~ air** all'aperto; **in the ~** all'aperto; fig alla luce del sole ●vt aprire ●vi aprirsi; (shop:) aprire; (flower:) sbocciare. **open up** vt aprire ●vi aprirsi

open: **~-air 'swimming pool** n piscina f all'aperto. **~ day** n giorno m di apertura al pubblico

'**opener** /ˈəʊpnə(r)/ n (for tins) apriscatole m inv; (for bottles) apribottiglie m inv

opening /'əʊpənɪŋ/ n apertura f; (beginning) inizio m; (job) posto m libero; **~ hours** npl orario m d'apertura

openly /'əʊpənlɪ/ adv apertamente

open: **~-'minded** a aperto; (broadminded) di vedute larghe. **~-plan** a a pianta aperta. ~ '**sandwich** n tartina f. ~ **secret** segreto m di Pulcinella. ~ **ticket** biglietto m aperto. **O~ University** corsi mpl universitari per corrispondenza

opera /'ɒpərə/ n opera f

operable /'ɒpərəbl/ a operabile

opera: **~-glasses** npl binocolo msg da teatro. **~-house** n teatro m lirico. **~-singer** n cantante mf lirico, -a

operate /'ɒpəreɪt/ vt far funzionare (machine, lift); azionare (lever, brake); mandare avanti (business) ●vi Techn funzionare; (be in action) essere in funzione; Mil, fig operare; ~ **on** Med operare

operatic /ɒpə'rætɪk/ a lirico, operistico

operation /ɒpə'reɪʃn/ n operazione f; Tech funzionamento m; **in ~** Techn in funzione; **come into ~** fig entrare in funzione; (law:) entrare in vigore; **have an ~** Med subire un'operazione. **~al** a operativo; (law etc) in vigore

operative /'ɒpərətɪv/ a operativo

operator /'ɒpəreɪtə(r)/ n (user) operatore, -trice mf; Teleph centralinista mf

operetta /ɒpə'retə/ n operetta f

opinion /ə'pɪnjən/ n opinione f; **in my ~** secondo me. **~ated** a dogmatico

opponent /ə'pəʊnənt/ n avversario, -a mf

opportune /'ɒpətjuːn/ a opportuno. **~ist** ~'tjuːnɪst/ n opportunista mf. **~istic** a opportunistico

opportunity /ɒpə'tjuːnətɪ/ n opportunità f inv

oppose /ə'pəʊz/ vt opporsi a; **be ~ed to sth** essere contrario a qcsa; **as ~ed to** al contrario di.

~ing *a* avversario; *(opposite)* opposto

opposite /'ɒpəzɪt/ *a* opposto; *(house)* di fronte; *~* **number** *fig* controparte *f;* **the ~ sex** l'altro sesso ● *n* contrario *m* ● *adv* di fronte ● *prep* di fronte a

opposition /ɒpə'zɪʃn/ *n* opposizione *f*

oppress /ə'pres/ *vt* opprimere. **~ion** /- eʃn/ *n* oppressione *f.* **~ive** /-ɪv/ *a* oppressivo; *(heat)* opprimente. **~or** *n* oppressore *m*

opt /ɒpt/ *vi* **~ for** optare per; **~ out** dissociarsi *(of* da)

optical /'ɒptɪkl/ *a* ottico; **~ illusion** *a* ottica

optician /ɒp'tɪʃn/ *n* ottico, -a *mf*

optimism /'ɒptɪmɪzm/ *n* ottimismo *m.* **~t** /-mɪst/ *n* ottimista *mf.* **~tic** /-'mɪstɪk/ *a* ottimistico

optimum /'ɒptɪməm/ *a* ottimale ● *n* (*pl* **-ima**) optimum *m*

option /'ɒpʃn/ *n* scelta *f;* *Comm* opzione *f.* **~al** *a* facoltativo; **~al extras** *pl* optional *m inv*

opulen|ce /'ɒpjʊləns/ *n* opulenza *f.* **~t** *a* opulento

or /ɔː(r)/ *conj* o, oppure; *(after negative)* né; **or [else]** se no; **in a year or two** fra un anno o due

oracle /'ɒrəkl/ *n* oracolo *m*

oral /'ɔːrəl/ *a* orale ● *n fam* esame *m* orale. **~ly** *adv* oralmente

orange /'ɒrɪndʒ/ *n* arancia *f;* *(colour)* arancione *m* ● *a* arancione. **~ade** /-'dʒeɪd/ *n* aranciata *f.* **~ juice** *n* succo *m* d'arancia

orator /'ɒrətə(r)/ *n* oratore, -trice *mf*

oratorio /ɒrə'tɔːrɪəʊ/ *n* oratorio *m*

oratory /'ɒrətərɪ/ *n* oratorio *m*

orbit /'ɔːbɪt/ *n* orbita *f* ● *vt* orbitare. **~al** *a* **~al road** tangenziale *f*

orchard /'ɔːtʃəd/ *n* frutteto *m*

orchestra /'ɔːkɪstrə/ *n* orchestra *f.* **~tral** /-'kestrəl/ *a* orchestrale. **~trate** *vt* orchestrare

orchid /'ɔːkɪd/ *n* orchidea *f*

ordain /ɔː'deɪn/ *vt* decretare; *Relig* ordinare

ordeal /ɔː'diːl/ *n fig* terribile esperienza *f*

order /'ɔːdə(r)/ *n* ordine *m;* *Comm* ordinazione *f;* **out of ~** *(machine)* fuori servizio; **in ~ that** affinché; **in ~ to** per ● *vt* ordinare

orderly /'ɔːdəlɪ/ *a* ordinato ● *n Mil* attendente *m;* *Med* inserviente *m*

ordinary /'ɔːdɪnərɪ/ *a* ordinario

ordination /ɔːdɪ'neɪʃn/ *n Relig* ordinazione *f*

ore /ɔː(r)/ *n* minerale *m* grezzo

organ /'ɔːgən/ *n Anat, Mus* organo *m*

organic /ɔː'gænɪk/ *a* organico; *(without chemicals)* biologico. **~ally** *adv* organicamente; **~ally grown** coltivato biologicamente

organism /'ɔːgənɪzm/ *n* organismo *m*

organist /'ɔːgənɪst/ *n* organista *mf*

organization /ɔːgənaɪ'zeɪʃn/ *n* organizzazione *f*

organize /'ɔːgənaɪz/ *vt* organizzare. **~r** *n* organizzatore, -trice *mf*

orgasm /'ɔːgæzm/ *n* orgasmo *m*

orgy /'ɔːdʒɪ/ *n* orgia *f*

Orient /'ɔːrɪənt/ *n* Oriente *m.* **o~al** /-'entl/ *a* orientale ● *n* orientale *mf*

orient|ate /'ɔːrɪənteɪt/ *vt* **~ate oneself** orientarsi. **~ation** /-'teɪʃn/ *n* orientamento *m*

origin /'ɒrɪdʒɪn/ *n* origine *f*

original /ə'rɪdʒɪn(ə)l/ *a* originario; *(not copied, new)* originale ● *n* originale *m;* **in the ~** in versione originale. **~ity** /-'nælətɪ/ *n* originalità *f.* **~ly** *adv* originariamente

originate /ə'rɪdʒɪneɪt/ *vi* **~e in** avere origine in. **~or** *n* ideatore, -trice *mf*

ornament /'ɔːnəmənt/ *n* ornamento *m;* *(on mantelpiece etc)* soprammobile *m.* **~al** /-'mentl/ *a* ornamentale. **~ation** /-'teɪʃn/ *n* decorazione *f*

ornate /ɔː'neɪt/ *a* ornato

orphan /'ɔːfn/ *n* orfano, -a *mf* ● *vt*

rendere orfano; **be ~ed** rimanere
orfano. **~age** /-ɪdʒ/ *n* orfanotrofio *m*

orthodox /'ɔ:θədɒks/ *a* ortodosso

orthopaedic /ɔ:θə'pi:dɪk/ *a* ortopedico

oscillate /'ɒsɪleɪt/ *vi* oscillare

ostensibl|e /ɒ'stensəbl/ *a* apparente. **~y** *adv* apparentemente

ostentat|ion /ɒsten'teɪʃn/ *n* ostentazione *f*. **~ious** /-ʃəs/ *a* ostentato

osteopath /'ɒstɪəpæθ/ *n* osteopata *mf*

ostracize /'ɒstrəsaɪz/ *vt* bandire

ostrich /'ɒstrɪtʃ/ *n* struzzo *m*

other /'ʌðə(r)/ *a, pron* & *n* altro,
-a *mf*. **the ~ [one]** l'altro, -a *mf*;
the ~ two gli altri due; **two ~s**
altri due; **~ people** gli altri; **any
~ questions?** altre domande?;
every ~ day (*alternate days*) a
giorni alterni; **the ~ day** l'altro
giorno; **the ~ evening** l'altra
sera; **someone/something or ~**
qualcuno/qualcosa ● *adv* **~ than
him** tranne lui; **somehow or ~** in
qualche modo; **somewhere or ~**
da qualche parte

'otherwise *adv* altrimenti; (*differently*) diversamente

otter /'ɒtə(r)/ *n* lontra *f*

ouch /aʊtʃ/ *int* ahi!

ought /ɔ:t/ *v aux* **I/we ~ to stay**
dovrei/dovremmo rimanere; **he ~
not to have done it** non avrebbe
dovuto farlo; **that ~ to be enough**
questo dovrebbe bastare

ounce /aʊns/ *n* oncia *f* (= 28, 35 g)

our /'aʊə(r)/ *a* il nostro *m*, la nostra
f, i nostri *mpl*, le nostre *fpl*; **~
mother/father** nostra madre/nostro padre

ours /'aʊəz/ *poss pron* il nostro *m*,
la nostra *f*, i nostri *mpl*, le nostre
fpl; **a friend of ~** un nostro amico;
friends of ~ dei nostri amici; **that
is ~** quello è nostro; (*as opposed to
yours*) quello è il nostro

ourselves /aʊə'selvz/ *pron* (*reflexive*) ci; (*emphatic*) noi, noi stessi;
we poured ~ a drink ci siamo

versati da bere; **we heard it ~**
l'abbiamo sentito noi stessi; **we
are proud of ~** siamo fieri di noi;
by ~ da soli

out /aʊt/ *adv* fuori; (*not alight*)
spento; **be ~** (*flower:*) essere sbocciato; (*workers:*) essere in sciopero; (*calculation:*) essere sbagliato;
Sport essere fuori; (*unconscious*)
aver perso i sensi; (*fig: not feasible*) fuori questione; **the sun is ~** è
uscito il sole; **~ and about** in piedi; **get ~!** *fam* fuori!; **you should
get ~ more** dovresti uscire più
spesso; **~ with it!** *fam* sputa il rospo!; ● *prep* **~ of** fuori da; **~ of
date** non aggiornato; (*passport*) scaduto; **~ of order** guasto;
~ of print/stock esaurito; **be ~ of
bed** (*the room* fuori dal letto/dalla
stanza; **~ of breath** senza fiato;
~ of danger fuori pericolo; **~ of
work** disoccupato; **nine ~ of ten**
nove su dieci; **be ~ of sugar/bread**
rimanere senza zucchero/pane; **go
~ of the room** uscire dalla stanza

out'bid *vt* (*pt/pp* **-bid**, *pres p*
-bidding) **~ sb** rilanciare l'offerta
di qcno

outboard *a* **~ motor** motore *m*

'outbreak *n* (*of war*) scoppio *m*; (*of
disease*) insorgenza *f*

'outbuilding *n* costruzione *f* annessa

'outburst *n* esplosione *f*

'outcome *n* risultato *m*

'outcry *n* protesta *f*

out'dated *a* sorpassato

out'do *vt* (*pt* **-did**, *pp* **-done**) superare

'outdoor *a* (*life, sports*) all'aperto;
~ clothes *pl* vestiti per uscire;
swimming pool piscina *f* scoperta

out'doors *adv* all'aria aperta; **go
~** uscire [all'aria aperta]

'outer *a* esterno

'outfit *n* equipaggiamento *m*;
(*clothes*) completo *m*; (*fam: organization*) organizzazione *f*. **~ter
n** **men's ~ter's** negozio *m* di
abbigliamento maschile

'outgoing a (president) uscente; (mail) in partenza; (sociable) estroverso. **~s** npl uscite fpl

out'grow vi (pt **-grew**, pp **-grown**) diventare troppo grande per

'outhouse n costruzione f annessa

outing /'autɪŋ/ n gita f

outlandish /aut'lændɪʃ/ a stravagante

'outlaw n fuorilegge mf inv ● vt dichiarare illegale

'outlay n spesa f

'outlet n sbocco m; fig sfogo m; Comm punto m [di] vendita

'outline n contorno m; (summary) sommario m ● vt tracciare il contorno di; (describe) descrivere

out'live vt sopravvivere a

'outlook n vista f; (future prospect) prospettiva f; (attitude) visione f

'outlying a **~ areas** fpl zone fpl periferiche

out'number vt superare in numero

'out-patient n paziente mf esterno, -a; **~s' department** ambulatorio m

'output n produzione f

'outrage n oltraggio m ● vt oltraggiare. **~ous** /-'reɪdʒəs/ a oltraggioso; (price) scandaloso

'outright¹ a completo; (refusal) netto

out'right² adv completamente; (at once) immediatamente; (frankly) francamente

'outset n inizio m; **from the ~** fin dall'inizio

'outside¹ a esterno ● n esterno m; **from the ~** dall'esterno; **at the ~** al massimo

out'side² adv all'esterno, fuori; (out of doors) fuori ● prep fuori da; (in front of) davanti a

out'sider n estraneo, -a mf

'outskirts npl sobborghi mpl

out'spoken a schietto

out'standing a eccezionale; (landmark) prominente; (not settled) in sospeso

out'stretched a allungato

out'strip vt (pt/pp **-stripped**) superare

out'vote vt mettere in minoranza

'outward /-wəd/ a esterno; (journey) di andata ● adv verso l'esterno. **~ly** adv esternamente. **~s** adv verso l'esterno

out'weigh vt aver maggior peso di

out'wit vt (pt/pp **-witted**) battere in astuzia

oval /'əʊvl/ a ovale ● n ovale m

ovary /'əʊvərɪ/ n Anat ovaia f

ovation /əʊ'veɪʃn/ n ovazione f

oven /'ʌvn/ n forno m. **~-ready** a pronto da mettere in forno

over /'əʊvə(r)/ prep sopra; (across) al di là di; (during) durante; (more than) più di; **~ the phone** al telefono; **~ the page** alla pagina seguente; **all ~ Italy** in tutta [l']Italia; (travel) per l'Italia ● adv (all over) dappertutto; **it's all ~** è tutto finito; **I ache all ~** ho male dappertutto; **come/bring ~** venire/portare; **turn ~** girare

over- pref (too) troppo

overall¹ /'əʊvərɔːl/ n grembiule m; **~s** pl tuta fsg [da lavoro]

overall² /əʊvər'ɔːl/ a complessivo; (general) generale ● adv complessivamente

over'balance vi perdere l'equilibrio

over'bearing a prepotente

over'board adv Naut in mare

'overcast a coperto

over'charge vt **~ sb** far pagare più del dovuto a qcno ● vi far pagare più del dovuto

'overcoat n cappotto m

over'come vt (pt **-came**, pp **-come**) vincere; **be ~ by** essere sopraffatto da

over'crowded a sovraffollato

over'do vt (pt **-did**, pp **-done**) esagerare; (cook too long) stracuocere; ~ **it** (*fam: do too much*) strafare

'overdose n overdose f inv

'overdraft n scoperto m; **have an** ~ avere il conto scoperto

over'draw vt (*pt* -**drew**, *pp* -**drawn**) ~ **one's account** andare allo scoperto; **be ~n by** (*account:*) essere [allo] scoperto di

over'due a in ritardo

over'estimate vt sopravvalutare

'overflow¹ n (*water*) acqua f che deborda; (*people*) pubblico m in eccesso; (*outlet*) scarico m

over'flow² vi debordare

over'grown a (*garden*) coperto di erbacce

'overhaul¹ n revisione f

over'haul² vt Techn revisionare

over'head¹ adv in alto

'overhead² a aereo; (*railway*) sopraelevato; (*lights*) da soffitto. **~s** npl spese fpl generali

over'hear vt (*pt/pp* -**heard**) sentire per caso (*conversation*)

over'heat vi Auto surriscaldarsi ● vt surriscaldare

over'joyed a felicissimo

'overland a & adv via terra; ~ **route** via f terrestre

over'lap v (*pt/pp* -**lapped**) ● vi sovrapporsi ● vt sovrapporre

over'leaf adv sul retro

over'load vt sovraccaricare

over'look vt dominare; (*fail to see, ignore*) lasciarsi sfuggire

overly /'əʊvəlɪ/ adv eccessivamente

over'night¹ adv per la notte; **stay** ~ fermarsi a dormire

'overnight² a notturno; ~ **bag** piccola borsa f da viaggio; ~ **stay** sosta f per la notte

'overpass n cavalcavia m inv

over'pay vt (*pt/pp* -**paid**) strapagare

over'populated a sovrappopolato

over'power vt sopraffare. **~ing** a insostenibile

over'priced a troppo caro

overpro'duce vt produrre in eccesso

over'rate vt sopravvalutare. **~d** a sopravvalutato

over'reach vt ~ **oneself** puntare troppo in alto

overre'act vi avere una reazione eccessiva. **~ion** n reazione f eccessiva

over'rid|e vt (*pt* -**rode**, *pp* -**ridden**) passare sopra a. **~ing** a prevalente

over'rule vt annullare (*decision*)

over'run vt (*pt* -**ran**, *pp* -**run**, *pres p* -**running**) invadere; oltrepassare (*time*); **be ~ with** essere invaso da

over'seas¹ adv oltremare

'overseas² a d'oltremare

over'see vt (*pt* -**saw**, *pp* -**seen**) sorvegliare

over'shadow vt adombrare

over'shoot vt (*pt/pp* -**shot**) oltrepassare

'oversight n disattenzione f; **an** ~ una svista

over'sleep vi (*pt/pp* -**slept**) svegliarsi troppo tardi

over'step vt (*pt/pp* -**stepped**) ~ **the mark** oltrepassare ogni limite

overt /əʊ'vɜːt/ a palese

over'tak|e vt/i (*pt* -**took**, *pp* -**taken**) sorpassare. **~ing** n sorpasso m; **no ~ing** divieto di sorpasso

over'tax vt fig abusare di

'overthrow¹ n Pol rovesciamento m

over'throw² vt (*pt* -**threw**, *pp* -**thrown**) Pol rovesciare

'overtime n lavoro m straordinario ● adv **work** ~ fare lo straordinario

over'tired a sovraffaticato

'overtone n fig sfumatura f

'overture /'əʊvətjʊə(r)/ n Mus preludio m; **~s** pl fig approccio msg

over'turn vt ribaltare ● vi ribaltarsi

'overweight a sovrappeso

over'whelm /-'welm/ vt sommergere (**with** di); (*with emotion*) con-

fondere. **~ing** a travolgente;
〈victory, majority〉 schiacciante
over'work n lavoro m eccessivo
● vt far lavorare eccessivamente
● vi lavorare eccessivamente
owe /əʊ/ vt also fig dovere (**[to]** sb
a qcno); **~ sb sth** dovere qcsa a
qcno. **~ing** a be **~ing** 〈money〉: essere da pagare ● prep **~ing to** a
causa di
owl /aʊl/ n gufo m
own[1] /əʊn/ a proprio ● pron a car
of my ~ una macchina per conto
mio; **on one's ~** da solo; **hold
one's ~ with** tener testa a; **get
one's ~ back** fam prendersi una
rivincita
own[2] vt possedere; 〈confess〉 ammettere; **I don't ~ it** non mi appartiene. **own up** vi confessare (**to
sth** qcsa)
owner /'əʊnə(r)/ n proprietario, -a
mf. **~ship** n proprietà f
ox /ɒks/ n (pl **oxen**) bue m (pl buoi)
oxide /'ɒksaɪd/ n ossido m
oxygen /'ɒksɪdʒən/ n ossigeno m;
~ mask maschera f a ossigeno
oyster /'ɔɪstə(r)/ n ostrica f
ozone /'əʊzəʊn/ n ozono m.
~-'friendly a che non danneggia
l'ozono. **~ layer** n fascia f d'ozono

Pp

PA abbr (**per annum**) all'anno
pace /peɪs/ n passo m; 〈speed〉 ritmo m; **keep ~ with** camminare di
pari passo con ● vi **~ up and
down** camminare avanti e indietro. **~-maker** n Med pacemaker m;
〈runner〉 battistrada m
Pacific /pə'sɪfɪk/ a & n the **~
[Ocean]** l'oceano m Pacifico, il Pacifico
pacifier /'pæsɪfaɪə(r)/ n Am ciuccio m, succhiotto m
pacifist /'pæsɪfɪst/ n pacifista mf

pacify /'pæsɪfaɪ/ vt (pt/pp **-ied**)
placare 〈person〉; pacificare
〈country〉
pack /pæk/ n 〈of cards〉 mazzo m;
〈of hounds〉 muta f; 〈of wolves,
thieves〉 branco m; 〈of cigarettes
etc〉 pacchetto m. **a ~ of lies** un
mucchio di bugie ● vt impacchettare 〈article〉; fare 〈suitcase〉; mettere in valigia 〈swimsuit etc〉;
〈press down〉 comprimere; **~ed
[out]** 〈crowded〉 pieno zeppo ● vi
fare i bagagli; **send sb ~ing** fam
mandare qcno a stendere. **pack
up** vt impacchettare ● vi fam
〈machine〉 piantare in asso
package /'pækɪdʒ/ n pacco m ● vt
impacchettare. **~ deal** offerta f
tutto compreso. **~ holiday** n vacanza f organizzata. **~ tour** viaggio m organizzato
packaging /'pækɪdʒɪŋ/ n confezione f
packed 'lunch n pranzo m al sacco
packet /'pækɪt/ n pacchetto m;
cost a ~ fam costare un sacco
packing /'pækɪŋ/ n imballaggio m
pact /pækt/ n patto m
pad[1] /pæd/ n imbottitura f; 〈for
writing〉 bloc-notes m, taccuino m;
〈fam: home〉 [piccolo] appartamento m ● vt (pt/pp **padded**) imbottire. **pad out** vt gonfiare
pad[2] vi (pt/pp **padded**) camminare
con passo felpato
padded /'pædɪd/ a **~ bra** reggiseno m imbottito
padding /'pædɪŋ/ n imbottitura f;
〈in written work〉 fronzoli mpl
paddle[1] /'pædl/ n pagaia f ● vi
〈row〉 spingere remando
paddle[2] vi 〈wade〉 sguazzare
paddock /'pædək/ n recinto m
padlock /'pædlɒk/ n lucchetto m
● vt chiudere con lucchetto
paediatrician /pi:dɪə'trɪʃn/ n pediatra mf
paediatrics /pi:dɪ'ætrɪks/ n pediatria f
page[1] /peɪdʒ/ n pagina f

page² n (boy) paggetto m; (in hotel) fattorino m ● vt far chiamare ‹person›

pageant /'pædʒənt/ n parata f. **~ry** n cerimoniale m

pager /'peidʒə(r)/ n cercapersone m inv

paid /peid/ see **pay** ● a ~ employment lavoro m remunerato; put ~ to mettere un termine a

pail /peil/ n secchio m

pain /pein/ n dolore m; be in ~ soffrire; take ~s darsi un gran d'affare; ~ in the neck fam spina f nel fianco

pain: ~ful a doloroso; (laborious) penoso.. ~killer n calmante m. ~less a indolore

painstaking /'peinzteikiŋ/ a minuzioso

paint /peint/ n pittura f; ~s colori mpl ● vt/i pitturare; ‹artist:› dipingere. **~brush** n pennello m. **~er** n pittore, -trice mf; (decorator) imbianchino m. **~ing** n pittura f; (picture) dipinto m. **~work** n pittura f

pair /peə(r)/ n paio m; (of people) coppia f; ~ of trousers paio m di pantaloni; ~ of scissors paio m di forbici

pajamas /pə'dʒɑːməz/ npl Am pigiama msg

Pakistan /pɑːkɪ'stɑːn/ n Pakistan m. **~i** a pakistano ● n pakistano, -a mf

pal /pæl/ n fam amico, -a mf

palace /'pælɪs/ n palazzo m

palatable /'pælətəbl/ a gradevole (al gusto)

palate /'pælət/ n palato m

palatial /pə'leɪʃl/ a sontuoso

palaver /pə'lɑːvə(r)/ n (fam: fuss) storie fpl

pale /peil/ a pallido

Palestin|e /'pælɪstaɪn/ n Palestina f. **~ian** /pælɪ'stɪnɪən/ a palestinese ● n palestinese mf

palette /'pælɪt/ n tavolozza f

pall|id /'pælɪd/ a pallido. **~or** n pallore m

palm /pɑːm/ n palmo m; (tree) palma f; P~ 'Sunday n Domenica f delle Palme ● **palm off** vt ~ sth off on sb rifilare qcsa a qcno

palpable /'pælpəbl/ a palpabile; (perceptible) tangibile

palpitat|e /'pælpɪteɪt/ vi palpitare. **~ions** /-'teɪʃnz/ npl palpitazioni fpl

paltry /'pɔːltrɪ/ a (-ier, -iest) insignificante

pamper /'pæmpə(r)/ vt viziare

pamphlet /'pæmflɪt/ n opuscolo m

pan /pæn/ n tegame m, pentola f; (for frying) padella f; (of scales) piatto m ● vt (pt/pp **panned**) (fam: criticize) stroncare

panache /pə'næʃ/ n stile m

pancake n crêpe f inv, frittella f

pancreas /'pæŋkrɪəs/ n pancreas m inv

panda /'pændə/ n panda m inv. ~ car n macchina f della polizia

pandemonium /pændɪ'məunɪəm/ n pandemonio m

pander /'pændə(r)/ vi ~ to sb compiacere qcno

pane /pein/ n ~ [of glass] vetro m

panel /'pænl/ n pannello m; (group of people) giuria f; ~ of experts gruppo m di esperti. **~ling** n pannelli mpl

pang /pæŋ/ n ~s of hunger morsi mpl della fame; ~s of conscience rimorsi mpl di coscienza

panic /'pænɪk/ n panico m ● vi (pt/pp **panicked**) lasciarsi prendere dal panico. **~-stricken** a in preda al panico

panoram|a /pænə'rɑːmə/ n panorama m. **~ic** /-'ræmɪk/ a panoramico

pansy /'pænzɪ/ n viola f del pensiero; (fam: effeminate man) finocchio m

pant /pænt/ vi ansimare

panther /'pænθə(r)/ n pantera f

panties /'pæntɪz/ npl mutandine fpl

pantomime /'pæntəmaɪm/ n pantomima f

pantry /'pæntrɪ/ n dispensa f

pants /pænts/ npl (underwear) mutande fpl; (woman's) mutandine fpl; (trousers) pantaloni mpl

'pantyhose n Am collant m inv

papal /'peɪpl/ a papale

paper /'peɪpə(r)/ n carta f; (wallpaper) carta f da parati; (newspaper) giornale m; (exam) esame m; (treatise) saggio m; **~s** pl (documents) documenti mpl; (for identification) documento m [d'identità]; **on ~** in teoria; **put down on ~** mettere per iscritto ● attrib di carta ● vt tappezzare

paper: **~back** n edizione f economica. **~clip** n graffetta f. **~knife** n tagliacarte m inv. **~weight** n fermacarte m inv. **~work** n lavoro m d'ufficio

par /pɑː(r)/ n (in golf) par m inv; **on a ~ with** alla pari con; **feel below ~** essere un po' giù di tono

parable /'pærəbl/ n parabola f

parachut|e /'pærəʃuːt/ n paracadute m ● vi lanciarsi col paracadute. **~ist** n paracadutista m

parade /pə'reɪd/ n (military) parata f militare ● vi sfilare ● vt (show off) far sfoggio di

paradise /'pærədaɪs/ n paradiso m

paradox /'pærədɒks/ n paradosso m. **~ical** /-'dɒksɪkl/ a paradossale. **~ically** adv paradossalmente

paraffin /'pærəfɪn/ n paraffina f

paragon /'pærəgən/ n **~ of virtue** modello m di virtù

paragraph /'pærəgrɑːf/ n paragrafo m

parallel /'pærəlel/ a & adv parallelo. **~ bars** npl parallele fpl. **~ port** n Comput porta f parallela n Geog, fig parallelo m; (line) parallela f ● vt essere paragonabile a

paralyse /'pærəlaɪz/ vt also fig paralizzare

paralysis /pə'ræləsɪs/ n (pl -ses) /-siːz/ paralisi f inv

parameter /pə'ræmɪtə(r)/ n parametro m

paramount /'pærəmaʊnt/ a supremo; **be ~** essere essenziale

paranoia /pærə'nɔɪə/ n paranoia f

paranoid /'pærənɔɪd/ a paranoico

paraphernalia /pærəfə'neɪljə/ n armamentario m

paraphrase /'pærəfreɪz/ n parafrasi f ● vt parafrasare

paraplegic /pærə'pliːdʒɪk/ a paraplegico n paraplegico, -a mf

parasite /'pærəsaɪt/ n parassita mf

parasol /'pærəsɒl/ n parasole m

paratrooper /'pærətruːpə(r)/ n paracadutista m

parcel /'pɑːsl/ n pacco m

parch /pɑːtʃ/ vt disseccare; **be ~ed** (person:) morire dalla sete

pardon /'pɑːdn/ n perdono m; Jur grazia f; **~?** prego?; **I beg your ~** fml chiedo scusa?; **I do beg your ~** (sorry) chiedo scusa! ● vt perdonare; Jur graziare

pare /peə(r)/ vt (peel) pelare

parent /'peərənt/ n genitore, -trice mf; **~s** pl genitori mpl. **~al** /pə'rentl/ a dei genitori

parenthesis /pə'renθəsɪs/ n (pl -ses /-siːz/) parentesi f inv

Paris /'pærɪs/ n Parigi f

parish /'pærɪʃ/ n parrocchia f. **~ioner** /pə'rɪʃənə(r)/ n parrocchiano, -a mf

Parisian /pə'rɪzɪən/ a & n parigino, -a mf

parity /'pærətɪ/ n parità f

park /pɑːk/ n parco m ● vt/i Auto posteggiare, parcheggiare; **~ oneself** fam installarsi

parka /'pɑːkə/ n parka m inv

parking /'pɑːkɪŋ/ n parcheggio m, posteggio m; **'no ~'** 'divieto di sosta'. **~lot** n Am posteggio m, parcheggio m. **~meter** n parchimetro m. **~ space** n posteggio m, parcheggio m

parliament /'pɑːləmənt/ n parlamento m. **~ary** /-'mentərɪ/ a parlamentare

parlour /'pɑːlə(r)/ n salotto m

parochial /pə'rəʊkɪəl/ a parrocchiale; *fig* ristretto

parody /'pærədɪ/ n parodia f ●vt (*pt/pp* -**ied**) parodiare

parole /pə'rəʊl/ n **on** ~ in libertà condizionale● vt mettere in libertà condizionale

parquet /'pɑːkeɪ/ n ~ **floor** parquet m

parrot /'pærət/ n pappagallo m

parry /'pærɪ/ vt (*pt/pp* -**ied**) parare (*blow*); (*in fencing*) eludere

parsimonious /pɑːsɪ'məʊnɪəs/ a parsimonioso

parsley /'pɑːslɪ/ n prezzemolo m

parsnip /'pɑːsnɪp/ n pastinaca f

parson /'pɑːsn/ n pastore m

part /pɑːt/ n parte f; (*of machine*) pezzo m; **for my** ~ per quanto mi riguarda; **on the** ~ **of** da parte di; **take sb's** ~ prendere le parti di qcno; **take** ~ **in** prendere parte a ● adv in parte ● vt ~ **one's hair** farsi la riga ● vi (*people:*) separarsi; ~ **with** separarsi da

part-ex'change n take in ~ prendere indietro

partial /'pɑːʃl/ a parziale; **be** ~ **to** aver un debole per. ~**ly** adv parzialmente

participant /pɑː'tɪsɪpənt/ n partecipante mf. ~**ate** /-pet/ vi partecipare (**in** a). ~**ation** /-'peɪʃn/ n partecipazione f

participle /'pɑːtɪsɪpl/ n participio m; **present/past** ~ participio m presente/passato

particle /'pɑːtɪkl/ n Phys, Gram particella f

particular /pə'tɪkjʊlə(r)/ a particolare; (*precise*) meticoloso; pej noioso; **in** ~ in particolare. ~**ly** adv particolarmente. ~**s** npl particolari mpl

parting /'pɑːtɪŋ/ n separazione f; (*in hair*) scriminatura f ● attrib di commiato

partisan /pɑːtɪ'zæn/ n partigiano, -a mf

partition /pɑː'tɪʃn/ n (*wall*) parete f divisoria; Pol divisione f ● vt dividere (*in parti*). **partition off** vt separare

partly /'pɑːtlɪ/ adv in parte

partner /'pɑːtnə(r)/ n Comm socio, -a mf; (*sport, in relationship*) compagno, -a mf. ~**ship** n Comm società f

partridge /'pɑːtrɪdʒ/ n pernice f

part-'time a & adv part time; **be** or **work** ~ lavorare part time

party /'pɑːtɪ/ n ricevimento m, festa f; (*group*) gruppo m; Pol partito m; Jur parte f (*in causa*); **be** ~ **to** essere parte attiva in

'party line[1] n Teleph duplex m inv

'party line[2] n Pol linea f del partito

pass /pɑːs/ n lasciapassare m inv; (*in mountains*) passo m; Sport passaggio m; Sch (*mark*) [voto m] sufficiente m; **make a** ~ **at** fam fare delle avances a ● vt passare; (*overtake*) sorpassare; (*approve*) far passare; fare (*remark*); Jur pronunciare (*sentence*). ~ **the time** passare il tempo ● vi passare; (*in exam*) essere promosso. **pass away** vi mancare. **pass down** vt passare; fig trasmettere. **pass out** vi fam svenire. **pass round** vt far passare. **pass through** vt attraversare. **pass up** vt passare; (*fam: miss*) lasciarsi scappare

passable /'pɑːsəbl/ a (*road*) praticabile; (*satisfactory*) passabile

passage /'pæsɪdʒ/ n passaggio m; (*corridor*) corridoio m; (*voyage*) traversata f

passenger /'pæsɪndʒə(r)/ n passeggero, -a mf. ~ **seat** n posto m accanto al guidatore

passer-by /pɑːsə'baɪ/ n (*pl* ~**s-by**) passante mf

'passing place n piazzola f di sosta per consentire il transito dei veicoli nei due sensi

passion /'pæʃn/ n passione f. ~**ate** /-ət/ a appassionato

passive /'pæsɪv/ a passivo ● n passivo m. ~**ness** n passività f '**pass-mark** n Sch [voto m] sufficiente m

Passover /'pɑːsəʊvə(r)/ n Pasqua f ebraica

pass: ~**port** n passaporto m. ~**word** n parola f d'ordine

past /pɑːst/ a passato; (former) ex; **in the ~** few days nei giorni scorsi; **that's all ~** tutto questo è passato; **the ~ week** la settimana scorsa ● n passato m ● prep oltre; **at ten ~ two** alle due e dieci ● adv oltre; **go/come ~** passare

pasta /'pæstə/ n pasta[sciutta] f

paste /peɪst/ n pasta f, (dough) impasto m; (adhesive) colla f ● vt incollare

pastel /'pæstl/ n pastello m ● attrib pastello

pasteurize /'pɑːstʃəraɪz/ vt pastorizzare

pastille /'pæstɪl/ n pastiglia f

pastime /'pɑːstaɪm/ n passatempo m

pastoral /'pɑːstərəl/ a pastorale

pastrami /pæ'strɑːmɪ/ n carne f di manzo affumicata

pastry /'peɪstrɪ/ n pasta f; ~**ies** pl pasticcini mpl

pasture /'pɑːstʃə(r)/ n pascolo m

pasty[1] /'pæstɪ/ n ≈ pasticcio m

pasty[2] /'peɪstɪ/ a smorto

pat /pæt/ n buffetto m; (of butter) pezzetto m ● adv **have sth off ~** conoscere qcsa a menadito ● vt (pt/pp **patted**) dare un buffetto a; ~ **sb on the back** fig congratularsi con qcno

patch /pætʃ/ n toppa f; (spot) chiazza f; (period) periodo m; **not a ~ on** fam molto inferiore a ● vt mettere una toppa su. **patch up** vt riparare alla bell'e meglio; appianare (quarrel)

patchy /'pætʃɪ/ a incostante

pâté /'pæteɪ/ n pâté m inv

patent /'peɪtnt/ a palese ● n brevetto m ● vt brevettare. ~ **leather shoes** npl scarpe fpl di vernice. ~**ly** adv in modo palese

paternal /pə'tɜːnl/ a paterno. ~**ity** n paternità f inv

path /pɑːθ/ n (pl ~**s** /pɑːðz/) sentiero m; (orbit) traiettoria m; fig strada f

pathetic /pə'θetɪk/ a patetico; (fam: very bad) penoso

pathological /pæθə'lɒdʒɪkl/ a patologico. ~**ist** /pə'θɒlədʒɪst/ n patologo, -a mf. ~**y** patologia f

pathos /'peɪθɒs/ n pathos m

patience /'peɪʃns/ n pazienza f; (game) solitario m

patient /'peɪʃnt/ a paziente ● n paziente mf. ~**ly** adv pazientemente

patio /'pætɪəʊ/ n terrazza f

patriot /'pætrɪət/ n patriota mf. ~**ic** /-'ɒtɪk/ a patriottico. ~**ism** n patriottismo m

patrol /pə'trəʊl/ n pattuglia f ● vt/i pattugliare. ~ **car** n autopattuglia f

patron /'peɪtrən/ n patrono m; (of charity) benefattore, -trice mf; (of the arts) mecenate mf; (customer) cliente mf

patronize /'pætrənaɪz/ vt frequentare abitualmente; fig trattare con condiscendenza. ~**ing** a condiscendente. ~**ingly** adv con condiscendenza

patter[1] /'pætə(r)/ n picchiettio m ● vi picchiettare

patter[2] n (of salesman) chiacchiere fpl

pattern /'pætn/ n disegno m (stampato); (for knitting, sewing) modello m

paunch /pɔːntʃ/ n pancia f

pause /pɔːz/ n pausa f ● vi fare una pausa

pave /peɪv/ vt pavimentare; ~ **the way** preparare la strada (**for** a). ~**ment** n marciapiede m

pavilion /pə'vɪljən/ n padiglione m

paw /pɔː/ n zampa f ● vt fam mettere le zampe addosso a

pawn[1] /pɔːn/ n (in chess) pedone m; fig pedina f

pawn[2] vt impegnare ● n **in ~** in pegno. ~**broker** n prestatore, -trice *mf* su pegno. **~shop** *n* monte *m* di pietà

pay /peɪ/ *n* paga *f*; **in the ~** of al soldo di ● *v* (*pt/pp* **paid**) ● *vt* pagare; prestare ⟨*attention*⟩; fare ⟨*compliment, visit*⟩; **~ cash** pagare in contanti ● *vi* pagare; (*be profitable*) rendere; **it doesn't ~ to...** *fig* è fatica sprecata...; **~ for sth** pagare per qcsa. **pay back** *vt* ripagare. **pay in** *vt* versare. **pay off** *vt* saldare ⟨*debt*⟩ ● *vi fig* dare dei frutti. **pay up** *vi* pagare

payable /'peɪəbl/ *a* pagabile; **make ~ to** intestare a

payee /peɪ'iː/ *n* beneficiario *m* (*di una somma*)

payment /'peɪmənt/ *n* pagamento *m*

pay: **~ packet** *n* busta *f* paga. **~ phone** *n* telefono *m* pubblico

PC *n abbr* (**personal computer**) PC *m inv*

pea /piː/ *n* pisello *m*

peace /piːs/ *n* pace *f*; **~ of mind** tranquillità *f*

peace|able /'piːsəbl/ *a* pacifico. **~ful** *a* calmo, sereno. **~fully** *adv* in pace. **~maker** *n* mediatore, -trice *mf*

peach /piːtʃ/ *n* pesca *f*; ⟨*tree*⟩ pesco *m*

peacock /'piːkɒk/ *n* pavone *m*

peak /piːk/ *n* picco *m*; *fig* culmine *m*. **~ed 'cap** *n* berretto *m* a punta. **~ hours** *npl* ore *fpl* di punta

peaky /'piːkɪ/ *a* malaticcio

peal /piːl/ *n* (*of bells*) scampanio *m*; **~s of laughter** fragore *m* di risate

peanut *n* nocciolina *f* [americana]; **~s** *fam* miseria *f*

pear /peə(r)/ *n* pera *f*; ⟨*tree*⟩ pero *m*

pearl /pɜːl/ *n* perla *f*

peasant /'peznt/ *n* contadino, -a *mf*

pebble /'pebl/ *n* ciottolo *m*

peck /pek/ *n* beccata *f*; (*kiss*) bacetto *m* ● *vt* beccare; (*kiss*) dare un bacetto a. **~ing order** *n* gerarchia *f*. **peck at** *vt* beccare

peckish /'pekɪʃ/ *a* **be ~** *fam* avere un languorino [allo stomaco]

---

peculiar /pɪ'kjuːlɪə(r)/ *a* strano; (*special*) particolare; **~** tipico di. **~ity** /-'ærətɪ/ *n* stranezza *f*; (*feature*) particolarità *f inv*

pedal /'pedl/ *n* pedale *m* ● *vi* pedalare. **~ bin** *n* pattumiera *f* a pedale

pedantic /pɪ'dæntɪk/ *a* pedante

pedestal /'pedɪstl/ *n* piedistallo *m*

pedestrian /pɪ'destrɪən/ *n* pedone *m* ● *a fig* scadente. **~ 'crossing** *n* passaggio *m* pedonale. **~ 'precinct** *n* zona *f* pedonale

pedicure /'pedɪkjʊə(r)/ *n* pedicure *f inv*

pedigree /'pedɪgriː/ *n* pedigree *m inv*; (*of person*) lignaggio *m* ● *attrib* ⟨*animal*⟩ di razza, con pedigree

pee /piː/ *vi* (*pt/pp* **peed**) *fam* fare [la] pipì

peek /piːk/ *vi fam* sbirciare

peel /piːl/ *n* buccia *f* ● *vt* sbucciare ● *vi* ⟨*nose etc:*⟩ spellarsi; ⟨*paint:*⟩ staccarsi

peep /piːp/ *n* sbirciata *f* ● *vi* sbirciare

peer[1] /pɪə(r)/ *vi* **~ at** scrutare

peer[2] *n* nobile *m*; **his ~s** *pl* (*in rank*) i suoi pari *mpl*; (*in age*) i suoi coetanei *mpl*. **~age** *n* nobiltà *f*

peeved /piːvd/ *a fam* irritato

peg /peg/ *n* (*hook*) piolo *m*; (*for tent*) picchetto *m*; (*for clothes*) molletta *f*; **off the ~** *fam* prêt-à-porter

pejorative /pɪ'dʒɒrətɪv/ *a* peggiorativo

pelican /'pelɪkən/ *n* pellicano *m*

pellet /'pelɪt/ *n* pallottola *f*

pelt /pelt/ *vt* bombardare ● *vi* (*fam: run fast*) catapultarsi; **~ [down]** ⟨*rain:*⟩ venir giù a fiotti

pelvis /'pelvɪs/ *n Anat* bacino *m*

pen[1] /pen/ *n* (*for animals*) recinto *m*

pen[2] *n* penna *f*; (*ball-point*) penna *f* a sfera

penal /'piːnl/ *a* penale. **~ize** *vt* penalizzare

penalty /'penltɪ/ *n* sanzione *f*;

penance /'penəns/ n penitenza f
pence /pens/ see **penny**
pencil /'pensl/ n matita f.
~**-sharpener** n temperamatite m
inv
pendant /'pendənt/ n ciondolo m
pending /'pendɪŋ/ a in sospeso
● prep in attesa di
pendulum /'pendjʊləm/ n pendolo m
penetrate /'penɪtreɪt/ vt/i penetrare. ~**ing** a acuto; (sound, stare) penetrante. ~**ion** /-'treɪʃn/ n penetrazione f
'**penfriend** n amico, -a m f di penna
penguin /'peŋgwɪn/ n pinguino m
penicillin /penɪ'sɪlɪn/ n penicillina f
peninsula /pɪ'nɪnsjʊlə/ n penisola f
penis /'piːnɪs/ n pene m
peniten|ce /'penɪtəns/ n penitenza f. ~**t** a penitente ● n penitente mf
penitentiary /penɪ'tenʃərɪ/ n Am penitenziario m
pen: ~**knife** n temperino m.
~**-name** n pseudonimo m
pennant /'penənt/ n bandiera f
penniless /'penɪlɪs/ a senza un soldo
penny /'penɪ/ n (pl **pence**; single coins **pennies**) penny m; Am centesimo m; **spend a** ~ fam andare in bagno
pension /'penʃn/ n pensione f. ~**er** n pensionato, -a mf
pensive /'pensɪv/ a pensoso
Pentecost /'pentɪkɒst/ n Pentecoste f
pent-up /pent'ʌp/ a represso
penultimate /pɪ'nʌltɪmət/ a penultimo
people /'piːpl/ npl persone fpl, gente fsg; (citizens) popolo msg; **a lot of** ~ una marea di gente; **the** ~ la gente; **English** ~ gli inglesi; **the** ~ **say** si dice; **for four** ~ per quattro ● vt popolare

pepper /'pepə(r)/ n pepe m; (vegetable) peperone m ● vt (season) pepare
pepper: ~**corn** n grano m di pepe.
~ **mill** n macinapepe m inv. ~**mint** n menta f peperita; (sweet) caramella f alla menta. ~**pot** n pepiera f
per /pɜː(r)/ prep per; ~ **annum** all'anno; ~ **cent** percento
perceive /pə'siːv/ vt percepire; (interpret) interpretare
percentage /pə'sentɪdʒ/ n percentuale f
perceptible /pə'septəbl/ a percettibile; (difference) sensibile
percepti|on /pə'sepʃn/ n percezione f. ~**ive** /-tɪv/ a perspicace
perch /pɜːtʃ/ n pertica f ● vi (bird:) appollaiarsi
percolator /'pɜːkəleɪtə(r)/ n caffettiera f a filtro
percussion /pə'kʌʃn/ n percussione f. ~ **instrument** n strumento m a percussione
peremptory /pə'remptərɪ/ a perentorio
perennial /pə'renɪəl/ a perenne
● n pianta f perenne
perfect[1] /'pɜːfɪkt/ a perfetto ● n Gram passato m prossimo
perfect[2] /-'fekt/ vt perfezionare. ~**ion** /-ekʃn/ n perfezione f; **to** ~**ion** alla perfezione. ~**ionist** n perfezionista mf
perfectly /'pɜːfɪktlɪ/ adv perfettamente
perforat|e /'pɜːfəreɪt/ vt perforare. ~**ed** a perforato; (ulcer) perforante. ~**ion** n perforazione f
perform /pə'fɔːm/ vt compiere, fare; eseguire (operation, sonata); recitare (role); mettere in scena (play) ● vi Theat recitare; Techn funzionare. ~**ance** n esecuzione f; (at theatre, cinema) rappresentazione f; Techn rendimento m. ~**er** n artista mf
perfume /'pɜːfjuːm/ n profumo m
perfunctory /pə'fʌŋktərɪ/ a superficiale
perhaps /pə'hæps/ adv forse

peril /'perɪl/ n pericolo m. **~ous** /-əs/ a pericoloso

perimeter /pə'rɪmɪtə(r)/ n perimetro m

period /'pɪərɪəd/ n periodo m; (menstruation) mestruazioni fpl; Sch ora f di lezione; (full stop) punto m fermo ● attrib (costume) d'epoca; (furniture) in stile. **~ic** /-'ɒdɪk/ a periodico. **~ical** /-'ɒdɪk/ n periodico m, rivista f

peripher|al /pə'rɪfərəl/ a periferico. **~y** n periferia f

periscope /'perɪskəʊp/ n periscopio m

perish /'perɪʃ/ vi (rot) deteriorarsi; (die) perire. **~able** /-əbl/ a deteriorabile

perjur|e /'pɜːdʒə(r)/ vt **~e oneself** spergiurare. **~y** n spergiuro m

perk /pɜːk/ n fam vantaggio m

perk up vt tirare su ● vi tirarsi su

perky /'pɜːkɪ/ a allegro

perm /pɜːm/ n permanente f ● vt **~ sb's hair** fare la permanente a qno

permanent /'pɜːmənənt/ a permanente; (job, address) stabile. **~ly** adv stabilmente

permeate /'pɜːmɪeɪt/ vt impregnare

permissible /pə'mɪsəbl/ a ammissibile

permission /pə'mɪʃn/ n permesso m

permissive /pə'mɪsɪv/ a permissivo

permit¹ /pə'mɪt/ vt (pt/pp -mitted) permettere; **~ sb to do sth** permettere a qcno di fare qcsa

permit² /'pɜːmɪt/ n autorizzazione f

perpendicular /pɜːpən'dɪkjʊlə(r)/ a perpendicolare ● n perpendicolare f

perpetual /pə'petjʊəl/ a perenne. **~ly** adv perennemente

perpetuate /pə'petjʊeɪt/ vt perpetuare

perplex /pə'pleks/ vt lasciare perplesso. **~ed** a perplesso. **~ity** n perplessità f inv

persecut|e /'pɜːsɪkjuːt/ vt perseguitare. **~ion** /-'kjuːʃn/ n persecuzione f

perseverance /pɜːsɪ'vɪərəns/ n perseveranza f

persever|e /pɜːsɪ'vɪə(r)/ vi perseverare. **~ing** a assiduo

Persian /'pɜːʃn/ a persiano

persist /pə'sɪst/ vi persistere; **~ in doing sth** persistere nel fare qcsa. **~ence** n persistenza f. **~ent** a persistente. **~ently** adv persistentemente

person /'pɜːsn/ n persona f; **in ~** di persona

personal /'pɜːsənl/ a personale. **'hygiene** n igiene f personale. **~ly** adv personalmente. **~ organizer** n Comput agenda f elettronica

personality /pɜːsə'nælətɪ/ n personalità f inv; (on TV) personaggio m

personnel /pɜːsə'nel/ n personale m

perspective /pə'spektɪv/ n prospettiva f

persp|iration /pɜːspɪ'reɪʃn/ n sudore m. **~ire** /-'spaɪə(r)/ vi sudare

persua|de /pə'sweɪd/ vt persuadere. **~sion** /-eɪʒn/ n persuasione f; (belief) convinzione f

persuasive /pə'sweɪsɪv/ a persuasivo. **~ly** adv in modo persuasivo

pertinent /'pɜːtɪnənt/ a pertinente (to a)

perturb /pə'tɜːb/ vt perturbare

peruse /pə'ruːz/ vt leggere

perva|de /pə'veɪd/ vt pervadere. **~sive** /-sɪv/ a pervasivo

pervers|e /pə'vɜːs/ a irragionevole. **~ion** /-ʃn/ n perversione f

pervert /'pɜːvɜːt/ n pervertito, -a mf

perverted /pə'vɜːtɪd/ a perverso

pessimis|m /'pesɪmɪzm/ n pessimismo m. **~t** /-mɪst/ n pessimista mf. **~tic** /-'mɪstɪk/ a pessimistico. **~tically** adv in modo pessimistico

pest /pest/ n piaga f; (fam: person) peste f

pester /'pestə(r)/ vt molestare

pesticide /'pestɪsaɪd/ n pesticida m

pet /pet/ n animale m domestico; (favourite) cocco, -a mf ● a prediletto ● v (pt/pp **petted**) ● vt coccolare ● vi (couple:) praticare il petting

petal /'petl/ n petalo m

peter /'piːtə(r)/ vi ~ **out** finire

petite /pə'tiːt/ a minuto

petition /pə'tɪʃn/ n petizione f

pet 'name n vezzeggiativo m

petrify /'petrɪfaɪ/ vt (pt/pp -**ied**) pietrificare. **~ied** a (frightened) pietrificato

petrol /'petrəl/ n benzina f

petroleum /pɪ'trəʊliəm/ n petrolio m

petrol: **~-pump** n pompa f di benzina. **~ station** n stazione f di servizio. **~ tank** n serbatoio m della benzina

'pet shop n negozio m di animali [domestici]

petticoat /'petɪkəʊt/ n sottoveste f

petty /'petɪ/ a (-**ier**, -**iest**) insignificante; (mean) meschino. **'cash** n cassa f per piccole spese

petulant /'petjʊlənt/ a petulante

pew /pjuː/ n banco m (di chiesa)

pewter /'pjuːtə(r)/ n peltro m

phallic /'fælɪk/ a fallico

phantom /'fæntəm/ n fantasma m

pharmaceutical /fɑːmə'sjuːtɪkl/ a farmaceutico

pharmacist /'fɑːməsɪst/ n farmacista mf. **~y** n farmacia f

phase /feɪz/ n fase f ● vt phase in/out introdurre/eliminare gradualmente

Ph.D. n abbr (Doctor of Philosophy) ≈ dottorato m di ricerca

pheasant /'feznt/ n fagiano m

phenomenal /fɪ'nɒmɪnl/ a fenomenale; (incredible) incredibile. **~ally** adv incredibilmente. **~on** n (pl -na) fenomeno m

philanderer /fɪ'lændərə(r)/ n donnaiolo m

philanthropic /fɪlən'θrɒpɪk/ a filantropico. **~ist** /fɪ'lænθrəpɪst/ n filantropo, -a f

philately /fɪ'lætəlɪ/ n filatelia f. **~ist** n filatelico, -a mf

philharmonic /fɪlhɑː'mɒnɪk/ n (orchestra) orchestra f filarmonica ● a filarmonico

Philippines /'fɪlɪpiːnz/ npl Filippine fpl

philistine /'fɪlɪstaɪn/ n filisteo, -a mf

philosopher /fɪ'lɒsəfə(r)/ n filosofo, -a mf. **~ical** /fɪlə'sɒfɪkl/ a filosofico. **~ically** adv con filosofia. **~y** n filosofia f

phlegm /flem/ n Med flemma f

phlegmatic /fleg'mætɪk/ a flemmatico

phobia /'fəʊbɪə/ n fobia f

phone /fəʊn/ n telefono m; **be on the** ~ avere il telefono; (be phoning) essere al telefono ● vt telefonare a ● vi telefonare. **phone back** vt/i richiamare. **~ book** n guida f del telefono. **~ box** n cabina f telefonica. **~ card** n scheda f telefonica. **~ call** telefonata f. **~-in** n trasmissione f con chiamate in diretta. **~ number** n numero m telefonico

phonetic /fə'netɪk/ a fonetico. **~s** n fonetica f

phoney /'fəʊnɪ/ a (-**ier**, -**iest**) fasullo

phosphorus /'fɒsfərəs/ n fosforo m

photo /'fəʊtəʊ/ n foto f; **~ album** album m inv di fotografie. **~copier** n fotocopiatrice f. **~copy** n fotocopia f ● vt fotocopiare

photogenic /fəʊtəʊ'dʒenɪk/ a fotogenico

photograph /'fəʊtəgrɑːf/ n fotografia f ● vt fotografare

photographer /fə'tɒgrəfə(r)/ n fotografo, -a mf. **~ic** /fəʊtə'græfɪk/ a fotografico. **~y** n fotografia f

phrase /freɪz/ n espressione f ● vt

esprimere. **~-book** *n* libro *m* di fraseologia

physical /'fɪzɪkl/ *a* fisico. **~ edu'cation** *n* educazione *f* fisica. **~ly** *adv* fisicamente

physician /fɪ'zɪʃn/ *n* medico *m*

physic|ist /'fɪzɪsɪst/ *n* fisico, -a *mf*. **~s** *n* fisica *f*

physiology /fɪzɪ'ɒlədʒɪ/ *n* fisiologia *f*

physio'therap|ist /fɪzɪəʊ-/ *n* fisioterapista *mf*. **~y** *n* fisioterapia *f*

physique /fɪ'zi:k/ *n* fisico *m*

pianist /'pɪənɪst/ *n* pianista *mf*

piano /pɪ'ænəʊ/ *n* piano *m*

pick[1] /pɪk/ *n* (*tool*) piccone *m*

pick[2] *n* scelta *f*; **take your ~** prendi quello che vuoi ● *vt* (*select*) scegliere; cogliere (*flowers*); scassinare (*lock*); borseggiare (*pockets*); **~ and choose** fare il difficile; **~ one's nose** mettersi le dita nel naso; **~ a quarrel** attaccar briga; **~ holes in** *fam* criticare; **~ at one's food** spilluzzicare. **pick on** *vt* (*fam: nag*) assillare; **he always ~s on me** ce l'ha con me. **pick out** *vt* (*identify*) individuare. **pick up** *vt* sollevare; (*off the ground, information*) raccogliere; prendere in braccio (*baby*); (*learn*) imparare; prendersi (*illness*); (*buy*) comprare; captare (*signal*); (*collect*) andare/venire a prendere; prendere (*passengers, habit*); (*police:*) arrestare (*criminals*); *fam* rimorchiare (*girl*); **~ oneself up** riprendersi ● *vi* (*improve*) recuperare; (*weather:*) rimettersi

pickaxe *n* piccone *m*

picket /'pɪkɪt/ *n* picchettista *mf* ● *vt* picchettare. **~ line** *n* picchetto *m*

pickle /'pɪkl/ *n* **~s** *pl* sottaceti *mpl*; **in a ~** *fig* nei pasticci ● *vt* mettere sottaceto

pick: ~pocket *n* borsaiolo *m*. **~-up** *n* (*truck*) furgone *m*; (*on record-player*) pickup *m inv*

picnic /'pɪknɪk/ *n* picnic *m* ● *vi* (*pt/pp* **-nicked**) fare un picnic

picture /'pɪktʃə(r)/ *n* (*painting*) quadro *m*; (*photo*) fotografia *f*; (*drawing*) disegno *m*; (*film*) film *m inv*; **put sb in the ~** *fig* mettere qcno al corrente; **the ~s** il cinema ● *vt* (*imagine*) immaginare

picturesque /pɪktʃə'resk/ *a* pittoresco

pie /paɪ/ *n* torta *f*

piece /piːs/ *n* pezzo *m*; (*in game*) pedina *f*; **a ~ of bread/paper** un pezzo di pane/carta; **a ~ of news/ advice** una notizia/un consiglio; **take to ~s** smontare. **~meal** *adv* un po' alla volta. **~-work** *n* lavoro *m* a cottimo ● **piece together** *vt* montare; *fig* ricostruire

pier /pɪə(r)/ *n* molo *m*; (*pillar*) pilastro *m*

pierc|e /pɪəs/ *vt* perforare; **~e a hole in sth** fare un buco in qcsa. **~ing** *a* penetrante

pig /pɪg/ *n* maiale *m*

pigeon /'pɪdʒɪn/ *n* piccione *m*. **~-hole** *n* casella *f*

piggy /'pɪgɪ/ **~back** *n* **give sb a ~back** portare qcno sulle spalle. **~ bank** *n* salvadanaio *m*

pig'headed *a fam* cocciuto

pig: ~skin *n* pelle *f* di cinghiale. **~tail** *n* (*plait*) treccina *f*

pile *n* (*heap*) pila *f* ● *vt* **~ sth on to sth** appilare qcsa su qcsa. **pile up** *vt* accatastare ● *vi* ammucchiarsi

piles /paɪlz/ *npl* emorroidi *fpl*

'pile-up *n* tamponamento a catena

pilfering /'pɪlfərɪŋ/ *n* piccoli furti *mpl*

pilgrim /'pɪlgrɪm/ *n* pellegrino, -a *mf*. **~age** /-ɪdʒ/ *n* pellegrinaggio *m*

pill /pɪl/ *n* pillola *f*

pillage /'pɪlɪdʒ/ *vt* saccheggiare

pillar /'pɪlə(r)/ *n* pilastro *m*. **~-box** *n* buca *f* delle lettere

pillion /'pɪljən/ *n* sellino *m* posteriore; **ride ~** viaggiare dietro

pillory /'pɪlərɪ/ *vt* (*pt/pp* **-led**) *fig* mettere alla berlina

pillow /ˈpɪləʊ/ n guanciale m. **~case** n federa f

pilot /ˈpaɪlət/ n pilota mf ● vt pilotare. **~-light** n fiamma f di sicurezza

pimp /pɪmp/ n protettore m

pimple /ˈpɪmpl/ n foruncolo m

pin /pɪn/ n spillo m; Electr spinotto m; Med chiodo m; **I have ~s and needles in my leg** mi formicola una gamba ● vt (pt/pp pinned) (sewing) fissare con gli spilli; (hold down) immobilizzare; **~ sb down to a date** ottenere un appuntamento da qcno; **~ sth on sb** fam addossare la colpa di qcsa. **pin up** vt appuntare; (on wall) affiggere

pinafore /ˈpɪnəfɔː(r)/ n grembiule m. **~ dress** n scamiciato m

pincers /ˈpɪnsəz/ npl tenaglie fpl

pinch /pɪntʃ/ n pizzicotto m; (of salt) presa f; **at a ~** fam in caso di bisogno ● vt pizzicare; (fam: steal) fregare ● vi (shoe:) stringere

pincushion n puntaspilli m inv

pine¹ /paɪn/ n (tree) pino m

pine² vi **she is pining for you** le manchi molto. **pine away** vi deperire

pineapple /ˈpaɪn-/ n ananas m inv

ping /pɪŋ/ n rumore m metallico

ping-pong n ping-pong m

pink /pɪŋk/ a rosa m

pinnacle /ˈpɪnəkl/ n guglia f

PIN number n codice m segreto

pin: **~point** vt definire con precisione. **~stripe** a gessato

pint /paɪnt/ n pinta f (= 0,571, Am: 0,47 l); **a ~** fam una birra media

pin-up n ragazza f da copertina, pin-up f inv

pioneer /paɪəˈnɪə(r)/ n pioniere, -a mf ● vt essere un pioniere di

pious /ˈpaɪəs/ a pio

pip /pɪp/ n (seed) seme m

pipe /paɪp/ n tubo m; (for smoking) pipa f; **the ~s** Mus la cornamusa ● vt far arrivare con tubature

(water, gas etc). **pipe down** vi fam abbassare la voce

pipe: **~-cleaner** n scovolino m. **~-dream** n illusione f. **~line** n conduttura f; **in the ~line** fam in cantiere

piper /ˈpaɪpə(r)/ n suonatore m di cornamusa

piping /ˈpaɪpɪŋ/ a **~ hot** bollente

pirate /ˈpaɪrət/ n pirata m

Pisces /ˈpaɪsiːz/ n Astr Pesci mpl

piss /pɪs/ vi sl pisciare

pistol /ˈpɪstl/ n pistola f

piston /ˈpɪstn/ n Techn pistone m

pit /pɪt/ n fossa f; (mine) miniera f; (for orchestra) orchestra f ● vt (pt/pp pitted) fig opporre (against a)

pitch¹ /pɪtʃ/ n (tone) tono m; (level) altezza f; (in sport) campo m; (fig: degree) grado m ● vt montare (tent). **pitch in** vi fam mettersi sotto

pitch² n **~-'black** a nero come la pece. **~-'dark** a buio pesto

pitchfork n forca f

piteous /ˈpɪtɪəs/ a pietoso

pitfall n fig trabocchetto m

pith /pɪθ/ n (of lemon, orange) interno m della buccia

pithy /ˈpɪθɪ/ a (-ier, -iest) fig conciso

pitiful /ˈpɪtɪfl/ a pietoso. **~less** a spietato

pittance /ˈpɪtns/ n miseria f

pity /ˈpɪtɪ/ n pietà f; **[what a] ~!** che peccato!; **take ~ on** avere compassione di ● vt aver pietà di

pivot /ˈpɪvət/ n perno m; fig fulcro m ● vi imperniarsi (on su)

pizza /ˈpiːtsə/ n pizza f

placard /ˈplækɑːd/ n cartellone m

placate /pləˈkeɪt/ vt placare

place /pleɪs/ n posto m; (fam: house) casa f; (in book) segno m; **feel out of ~** sentirsi fuori posto; **take ~** aver luogo; **all over the ~** dappertutto ● vt collocare; (remember) identificare; **~ an order** fare un'ordinazione; **be ~d** (in *race*) piazzarsi. **~-mat** *n* sotto-piatto *m*

placid /'plæsɪd/ *a* placido

plagiar|ism /'pleɪdʒərɪzm/ *n* plagio *m*. **~ize** *vt* plagiare

plague /pleɪg/ *n* peste *f*

plaice /pleɪs/ *n inv* platessa *f*

plain /pleɪn/ *a* chiaro; (*simple*) semplice; (*not pretty*) scialbo; (*not patterned*) normale; (*chocolate*) fondente; **in ~ clothes** in borghese ● *adv* (*simply*) semplicemente ● *n* pianura *f*. **~ly** *adv* francamente; (*simply*) semplicemente; (*obviously*) chiaramente

plaintiff /'pleɪntɪf/ *n Jur* parte *f* lesa

plaintive /'pleɪntɪv/ *a* lamentoso

plait /plæt/ *n* treccia *f* ● *vt* intrecciare

plan /plæn/ *n* progetto *m*, piano *m* ● *vt* (*pt/pp* **planned**) progettare; (*intend*) prevedere

plane[1] /pleɪn/ *n* (*tree*) platano *m*

plane[2] *n* aeroplano *m*

plane[3] *n* (*tool*) pialla *f* ● *vt* piallare

planet /'plænɪt/ *n* pianeta *m*

plank /plæŋk/ *n* asse *f*

planning /'plænɪŋ/ *n* pianificazione *f*. **~ permission** *n* licenza *f* edilizia

plant /plɑːnt/ *n* pianta *f*; (*machinery*) impianto *m*; (*factory*) stabilimento *m* ● *vt* piantare. **~ation** /plænˈteɪʃn/ *n* piantagione *f*

plaque /plɑːk/ *n* placca *f*

plasma /'plæzmə/ *n* plasma *m*

plaster /'plɑːstə(r)/ *n* intonaco *m*; *Med* gesso *m*; (*sticking ~*) cerotto *m*; **~ of Paris** gesso *m* ● *vt* intonacare (*wall*); (*cover*) ricoprire. **~ed** *a sl* sbronzo. **~er** *n* intonacatore *m*

plastic /'plæstɪk/ *n* plastica *f* ● *a* plastico

Plasticine® /'plæstɪsiːn/ *n* plastilina® *f*

plastic: ~ 'surgeon *n* chirurgo *m* plastico. **~ surgery** *n* chirurgia *f* plastica

plate /pleɪt/ *n* piatto *m*; (*flat sheet*)

placca *f*; (*gold and silverware*) argenteria *f*; (*in book*) tavola *f* [fuori testo] ● *vt* (*cover with metal*) placcare

plateau /'plætəʊ/ *n* (*pl* **~x** /-əʊz/) altopiano *m*

platform /'plætfɔːm/ *n* (*stage*) palco *m*; *Rail* marciapiede *m*; *Pol* piattaforma *f*; **~ 5** binario 5

platinum /'plætɪnəm/ *n* platino *m* ● *attrib* di platino

platitude /'plætɪtjuːd/ *n* luogo *m* comune

platonic /pləˈtɒnɪk/ *a* platonico

platoon /pləˈtuːn/ *n Mil* plotone *m*

platter /'plætə(r)/ *n* piatto *m* da portata

plausible /'plɔːzəbl/ *a* plausibile

play /pleɪ/ *n* gioco *m*; *Theat*, *TV* rappresentazione *f*; *Radio* sceneggiato *m* radiofonico; **~ on words** gioco *m* di parole ● *vt* giocare a; (*act*) recitare; suonare (*instrument*); giocare (*card*) ● *vi* giocare; *Mus* suonare; **~ safe** non prendere rischi. **play down** *vt* minimizzare. **play up** *vi fam* fare i capricci

play: ~boy *n* playboy *m inv*. **~er** *n* giocatore, -trice *mf*. **~ful** *a* scherzoso. **~ground** *n Sch* cortile *m* (*per la ricreazione*). **~group** *n* asilo *m*

playing: ~-card *n* carta *f* da gioco. **~-field** *n* campo *m* da gioco

play: ~mate *n* compagno, -a *mf* di gioco. **~-pen** *n* box *m inv*. **~thing** *n* giocattolo *m*. **~wright** /-raɪt/ *n* drammaturgo, -a *mf*

plc *n abbr* (**public limited company**) s.r.l.

plea /pliː/ *n* richiesta *f*; **make a ~ for** fare un appello a

plead /pliːd/ *vi* fare appello (**for** a); **~ guilty** dichiararsi colpevole; **~ with sb** implorare qcno

pleasant /'plez(ə)nt/ *a* piacevole. **~ly** *adv* piacevolmente; (*say*, *smile*) cordialmente

pleas|e /pliːz/ *adv* per favore; **~ do** prego ● *vt* far contento; **~e oneself** fare il proprio comodo;

~e yourself! come vuoi!; *pej* fai come ti pare!. **~ed** a lieto; **~ed with/about** contento di. **~ing** a gradevole

pleasurable /'pleʒərəbl/ a gradevole

pleasure /'pleʒə(r)/ n piacere m; **with ~** con piacere, volentieri

pleat /pli:t/ n piega f ● vt pieghettare. **~ed 'skirt** n gonna f a pieghe

pledge /pledʒ/ n pegno m; (*promise*) promessa f ● vt impegnarsi a; (*pawn*) impegnare

plentiful /'plentɪfl/ a abbondante

plenty /'plentɪ/ n abbondanza f; **~ of money** soldi *mpl*; **~ of people** molta gente; **I've got ~** ne ho in abbondanza

pliable /'plaɪəbl/ a flessibile

pliers /'plaɪəz/ npl pinze fpl

plight /plaɪt/ n condizione f

plimsolls /'plɪmsəlz/ npl scarpe fpl da ginnastica

plinth /plɪnθ/ n plinto m

plod /plɒd/ vi (*pt/pp* **plodded**) trascinarsi; (*work hard*) sgobbare

plonk /plɒŋk/ n fam vino m mediocre

plot /plɒt/ n complotto m; (*of novel*) trama f; **~ of land** appezzamento m [di terreno] ● vt/i complottare

plough /plaʊ/ n aratro m ● vt/i arare. **~man's [lunch]** piatto m di formaggi e sottaceti, servito con pane. **plough back** vt Comm reinvestire

ploy /plɔɪ/ n fam manovra f

pluck /plʌk/ n fegato m ● vt strappare; depilare (*eyebrows*); spennare (*bird*); cogliere (*flower*). **pluck up** vt **~ up courage** farsi coraggio

plucky /'plʌkɪ/ a (**-ier, -iest**) coraggioso

plug /plʌg/ n tappo m; Electr spina f; Auto candela f; (*fam: advertisement*) pubblicità f inv ● vt (*pt/pp* **plugged**) tappare; (*fam: advertise*) pubblicizzare con insistenza. **plug in** vt Electr inserire la spina di

plum /plʌm/ n prugna f; (*tree*) prugno m

plumage /'plu:mɪdʒ/ n piumaggio m

plumb /plʌm/ a verticale ● adv esattamente ● **plumb in** vt collegare

plumb|er /'plʌmə(r)/ n idraulico m. **~ing** n impianto m idraulico

'plumb-line n filo m a piombo

plume /plu:m/ n piuma f

plummet /'plʌmɪt/ vi precipitare

plump /plʌmp/ a paffuto ● **plump for** vt scegliere

plunge /plʌndʒ/ n tuffo m; **take the ~** fam buttarsi ● vt tuffare; fig sprofondare ● vi tuffarsi

plunging /'plʌndʒɪŋ/ a (*neckline*) profondo

plu'perfect /plu:-/ n trapassato m prossimo

plural /'plʊərəl/ a plurale ● n plurale m

plus /plʌs/ prep più ● a in più; **500 ~** più di 500 ● n più m; (*advantage*) extra m inv

plush /plʌʃ[ɪ]/ a lussuoso

plutonium /plʊ'təʊnɪəm/ n plutonio m

ply /plaɪ/ vt (*pt/pp* **plied**) **~ sb with drink** continuare a offrire da bere a qcno. **~wood** n compensato m

PM n abbr Prime Minister

p.m. abbr (**post meridiem**) del pomeriggio

pneumatic /nju:'mætɪk/ a pneumatico. **~ 'drill** n martello m pneumatico

pneumonia /nju:'məʊnɪə/ n polmonite f

P.O. abbr Post Office

poach /pəʊtʃ/ vt Culin bollire; cacciare di frodo (*deer*); pescare di frodo (*salmon*). **~ed egg** uovo m in camicia. **~er** n bracconiere m

pocket /'pɒkɪt/ n tasca f; **be out of ~** rimetterci ● vt intascare. **~book** n taccuino m; (*wallet*) portafoglio m. **~money** n denaro m per le piccole spese

pod /pɒd/ n baccello m

podgy /'pɒdʒɪ/ a (-ier, -iest) grassoccio

poem /'pəʊɪm/ n poesia f

poet /'pəʊɪt/ n poeta m. **~ic** /-'etɪk/ a poetico

poetry /'pəʊɪtrɪ/ n poesia f

poignant /'pɔɪnjənt/ a emozionante

point /pɔɪnt/ n punto m; (sharp end) punta f; (meaning, purpose) senso m; Electr presa f [di corrente]; **~s** pl Rail scambio m; **~ of view** punto m di vista; **good/bad ~s** aspetti mpl positivi/negativi; **what is the ~?** a che scopo?; **the ~ is** il fatto è; **I don't see the ~** non vedo il senso; **up to a ~** fino a un certo punto; **be on the ~ of doing sth** essere sul punto di fare qcsa ● vt puntare (at verso) ● vi (with finger) puntare il dito; **~ at/to** (person:) mostrare col dito; (indicator:) indicare. **point out** vt far notare (fact); **~ sth out to sb** far notare qcsa a qcno

point-blank a a bruciapelo

point|ed /'pɔɪntɪd/ a appuntito; (question) diretto. **~ers** npl (advice) consigli mpl. **~less** a inutile

poise /pɔɪz/ n padronanza f. **~d** a in equilibrio; **~d to** sul punto di

poison /'pɔɪzn/ n veleno m ● vt avvelenare. **~ous** a velenoso

poke /pəʊk/ n [piccola] spinta f ● vt spingere; (fire) attizzare; (put) ficcare; **~ fun at** prendere in giro. **poke about** vi frugare

poker[1] /'pəʊkə(r)/ n attizzatoio m

poker[2] n (Cards) poker m

poky /'pəʊkɪ/ a (-ier, -iest) angusto

Poland /'pəʊlənd/ n Polonia f

polar /'pəʊlə(r)/ a polare. **~ bear** n orso m bianco. **~ize** vt polarizzare

Pole /pəʊl/ n polacco, -a mf

pole[1] n palo m

pole[2] n (Geog, Electr) polo m

'pole-star n stella f polare

'pole-vault n salto m con l'asta

police /pə'liːs/ npl polizia f ● vt pattugliare (area)

police: ~man n poliziotto m. **~ state** n stato m militarista. **~ station** n commissariato m. **~woman** n donna f poliziotto

policy[1] /'pɒlɪsɪ/ n politica f

policy[2] n (insurance) polizza f

polio /'pəʊlɪəʊ/ n polio f

Polish /'pəʊlɪʃ/ a polacco ● n (language) polacco m

polish /'pɒlɪʃ/ n (shine) lucentezza f; (substance) lucido m; (for nails) smalto m; fig raffinatezza f ● vt lucidare; fig smussare. **polish off** vt fam finire in fretta; spazzolare (food)

polished /'pɒlɪʃt/ a (manner) raffinato; (performance) senza sbavature

polite /pə'laɪt/ a cortese. **~ly** adv cortesemente. **~ness** n cortesia f

politic /'pɒlɪtɪk/ a prudente

politic|al /pə'lɪtɪkl/ a politico. **~ally** adv dal punto di vista politico. **~ian** /-'tɪʃn/ n politico m

politics /'pɒlɪtɪks/ n politica f

poll /pəʊl/ n votazione f; (election) elezioni fpl; (opinion ~) sondaggio m d'opinione; **go to the ~s** andare alle urne ● vt ottenere (votes)

pollen /'pɒlən/ n polline m

polling /'pəʊlɪŋ/ **~-booth** n cabina f elettorale. **~-station** n seggio m elettorale

'poll tax n imposta f locale sulle persone fisiche

pollutant /pə'luːtənt/ n sostanza f inquinante

pollut|e /pə'luːt/ vt inquinare. **~ion** /-uːʃn/ n inquinamento m

polo /'pəʊləʊ/ n polo m. **~-neck** n collo m alto. **~-shirt** n dolcevita f

polyester /pɒlɪ'estə(r)/ n poliestere m

polystyrene® /pɒlɪ'staɪriːn/ n polistirolo m

polytechnic /pɒlɪ'teknɪk/ n politecnico m

polythene /'pɒlɪθiːn/ n politene m. **~ bag** n sacchetto m di plastica

polyun'saturated a polinsaturo

pomegranate /ˈpɒmɪgrænɪt/ n melagrana f

pomp /pɒmp/ n pompa f

pompon /ˈpɒmpɒn/ n pompon m

pompous /ˈpɒmpəs/ a pomposo

pond /pɒnd/ n stagno m

ponder /ˈpɒndə(r)/ vt/i ponderare

pong /pɒŋ/ n fam puzzo m

pontiff /ˈpɒntɪf/ n pontefice m

pony /ˈpəʊnɪ/ n pony m. **~-tail** n coda f di cavallo. **~-trekking** n escursioni fpl col pony

poodle /ˈpuːdl/ n barboncino m

pool¹ /puːl/ n (of water, blood) pozza f, [swimming] ~ piscina f

pool² n (common fund) cassa f comune; (in cards) piatto m; (game) biliardo m a buca. **~s** npl ≈ totocalcio msg ● vt mettere insieme

poor /pʊə(r)/ a povero; (not good) scadente; **in ~ health** in cattiva salute ● npl **the ~** i poveri. **~ly** a **be ~ly** non stare bene ● adv male

pop¹ /pɒp/ n botto m; (drink) bibita f gasata ● v (pt/pp **popped**) ● vt (fam: put) mettere; (burst) far scoppiare ● vi (burst) scoppiare. **pop in/out** vi fam fare un salto/un salto fuori

pop² n fam musica f pop ● attrib pop

'popcorn n popcorn m inv

pope /pəʊp/ n papa m

poplar /ˈpɒplə(r)/ n pioppo m

poppy /ˈpɒpɪ/ n papavero m

popular /ˈpɒpjʊlə(r)/ a popolare; (belief) diffuso. **~ity** /-ˈlærəti/ n popolarità f inv

populat|e /ˈpɒpjʊleɪt/ vt popolare. **~ion** /-ˈleɪʃn/ n popolazione f

porcelain /ˈpɔːsəlɪn/ n porcellana f

porch /pɔːtʃ/ n portico m; Am veranda f

porcupine /ˈpɔːkjʊpaɪn/ n porcospino m

pore¹ /pɔː(r)/ n poro m

pore² vi **~ over** immergersi in

pork /pɔːk/ n carne f di maiale

porn /pɔːn/ n fam porno m. **~o** a fam porno inv

pornograph|ic /pɔːnəˈgræfɪk/ a pornografico. **~y** /-ˈnɒgrəfɪ/ n pornografia f

porous /ˈpɔːrəs/ a poroso

porpoise /ˈpɔːpəs/ n focena f

porridge /ˈpɒrɪdʒ/ n farinata f di fiocchi d'avena

port¹ /pɔːt/ n porto m

port² n (Naut: side) babordo m

port³ n (wine) porto m

portable /ˈpɔːtəbl/ a portatile

porter /ˈpɔːtə(r)/ n portiere m; (for luggage) facchino m

portfolio /pɔːtˈfəʊlɪəʊ/ n cartella f; Comm portafoglio m

'porthole n oblò m inv

portion /ˈpɔːʃn/ n parte f; (of food) porzione f

portly /ˈpɔːtlɪ/ a (-ier, -iest) corpulento

portrait /ˈpɔːtrɪt/ n ritratto m

portray /pɔːˈtreɪ/ vt ritrarre; (represent) descrivere; (actor:) impersonare. **~al** n ritratto m

Portugal /ˈpɔːtjʊgl/ n Portogallo m. **~uese** /-ˈgiːz/ a portoghese ● n portoghese mf

pose /pəʊz/ n posa f ● vt porre (problem, question) ● vi (for painter) posare; **~ as** atteggiarsi a

posh /pɒʃ/ a fam lussuoso; (people) danaroso

position /pəˈzɪʃn/ n posizione f; (job) posto m; (status) ceto m [sociale] ● vt posizionare

positive /ˈpɒzɪtɪv/ a positivo; (certain) sicuro; (progress) concreto ● n positivo m. **~ly** adv positivamente; (decidedly) decisamente

possess /pəˈzes/ vt possedere. **~ion** /pəˈzeʃn/ n possesso m; **~ions** pl beni mpl

possess|ive /pəˈzesɪv/ a possessivo. **~iveness** n carattere m possessivo. **~or** n possessore, -ditrice mf

possibility /pɒsəˈbɪlətɪ/ n possibilità f inv

possib|le /ˈpɒsɪbl/ a possibile. **~ly** adv possibilmente; **I couldn't ~ly accept** non mi è possibile accettare; **he can't ~ly be right** non è possibile che abbia ragione; **could you ~ly...?** potrebbe per favore...?

post¹ /pəʊst/ n (pole) palo m ● vt affiggere ⟨notice⟩

post² n (place of duty) posto m ● vt appostare; (transfer) assegnare

post³ n (mail) posta f; **by ~** per posta ● vt spedire; (put in letter-box) imbucare; (as opposed to fax) mandare per posta; **keep sb ~ed** tenere qcno al corrente

post- pref dopo

postage /'pəʊstɪdʒ/ n affrancatura f. **~ stamp** n francobollo m

postal /'pəʊstl/ a postale. **~ order** n vaglia m postale

post: **~-box** n cassetta f delle lettere. **~card** n cartolina f. **~code** n codice m postale. **~-'date** vt postdatare

poster /'pəʊstə(r)/ n poster m inv; (advertising, election) cartellone m

posterior /pɒ'stɪərɪə/ n fam posteriore m

posterity /pɒ'sterətɪ/ n posterità f

posthumous /'pɒstjʊməs/ a postumo. **~ly** adv dopo la morte

post: **~man** n postino m. **~mark** n timbro m postale

post-mortem /-'mɔːtəm/ n autopsia f

'post office n ufficio m postale

postpone /pəʊst'pəʊn/ vt rimandare. **~ment** n rinvio m

posture /'pɒstʃə(r)/ n posizione f

post-'war a del dopoguerra

pot /pɒt/ n vaso m; (for tea) teiera f; (for coffee) caffettiera f; (for cooking) pentola f; **~s of money** fam un sacco di soldi; **go to ~** fam andare in malora

potassium /pə'tæsɪəm/ n potassio m

potato /pə'teɪtəʊ/ n (pl -es) patata f

poten|t /'pəʊtənt/ a potente. **~tate** n potentato m

potential /pə'tenʃl/ a potenziale ● n potenziale m. **~ly** adv potenzialmente

pot: **~-hole** n cavità f inv; (in road) buca f. **~-holer** n speleologo, -a mf. **~-luck** n take **~-luck** affidarsi alla sorte. **~ 'plant** n pianta f da appartamento. **~-shot** n take a **~-shot at** sparare a casaccio a

potted /'pɒtɪd/ a conservato; (shortened) condensato. **~ 'plant** n pianta f da appartamento

potter¹ /'pɒtə(r)/ vi **~ [about]** gingillarsi

potter² n vasaio, -a mf. **~y** n lavorazione f della ceramica; (articles) ceramiche fpl; (place) laboratorio m di ceramiche

potty /'pɒtɪ/ a (-ier, -iest) fam matto ● n vasino m

pouch /paʊtʃ/ n marsupio m

pouffe /puːf/ n pouf m inv

poultry /'pəʊltrɪ/ n pollame m

pounce /paʊns/ vi balzare; **~ on** saltare su

pound¹ /paʊnd/ n libbra f (= 0,454 kg); (money) sterlina f

pound² vt battere ● vi ⟨heart:⟩ battere forte; (run heavily) correre pesantemente

pour /pɔː(r)/ vt versare ● vi riversarsi; (with rain) piovere a dirotto. **pour out** vi riversarsi fuori ● vt versare ⟨drink⟩; sfogare ⟨troubles⟩

pout /paʊt/ vi fare il broncio ● n broncio m

poverty /'pɒvətɪ/ n povertà f

powder /'paʊdə(r)/ n polvere f; (cosmetic) cipria f ● vt polverizzare; (face) incipriare. **~y** a polveroso

power /'paʊə(r)/ n potere m; Electr corrente f [elettrica]; Math potenza f. **~ cut** n interruzione f di corrente. **~ed** a **~ed by electricity** dotato di corrente [elettrica]. **~ful** a potente. **~less** a impotente. **~-station** n centrale f elettrica

PR n abbr public relations

practicable /'præktɪkəbl/ a praticabile

practical /'præktɪkl/ a pratico. **~**

'**joke** n burla f. **~ly** adv praticamente

practice /'præktɪs/ n pratica f; (custom) usanza f; (habit) abitudine f; (exercise) esercizio m; Sport allenamento m; **in ~** (in reality) in pratica; **out of ~** fuori esercizio; **put into ~** mettere in pratica

practise /'præktɪs/ vt fare pratica in; (carry out) mettere in pratica; esercitare (profession) ● vi esercitarsi; (doctor:) praticare. **~d** a esperto

pragmatic /'prægmætɪk/ a pragmatico

praise /preɪz/ n lode f ● vt lodare. **~worthy** a lodevole

pram /præm/ n carrozzella f

prance /prɑːns/ vi saltellare

prank /præŋk/ n tiro m

prattle /'prætl/ vi parlottare

prawn /prɔːn/ n gambero m. **~ 'cocktail** n cocktail m inv di gamberetti

pray /preɪ/ vi pregare. **~er** /preə(r)/ n preghiera f

preach /priːtʃ/ vt/i predicare. **~er** n predicatore, -trice mf

preamble /priː'æmbl/ n preambolo m

pre-ar'range /priː-/ vt predisporre

precarious /prɪ'keərɪəs/ a precario. **~ly** adv in modo precario

precaution /prɪ'kɔːʃn/ n precauzione f; **as a ~** per precauzione. **~ary** a preventivo

precede /prɪ'siːd/ vt precedere

preceden|ce /'presɪdəns/ n precedenza f. **~t** n precedente m

preceding /prɪ'siːdɪŋ/ a precedente

precinct /'priːsɪŋkt/ n (traffic-free) zona f pedonale; (Am: district) circoscrizione f

precious /'preʃəs/ a prezioso; (style) ricercato ● adv fam **~ little** ben poco

precipice /'presɪpɪs/ n precipizio m

precipitate /prɪ'sɪpɪteɪt/ vt precipitare

précis /'preɪsiː/ n (pl précis /-siːz/) sunto m

precise /prɪ'saɪs/ a preciso. **~ely** adv precisamente. **~ion** /-'sɪʒn/ n precisione f

precursor /priː'kɜːsə(r)/ n precursore m

predator /'predətə(r)/ n predatore, -trice mf. **~y** a rapace

predecessor /'priːdɪsesə(r)/ n predecessore m

predicament /prɪ'dɪkəmənt/ n situazione f difficile

predicat|e /'predɪkət/ n Gram predicato m. **~ive** /prɪ'dɪkətɪv/ a predicativo

predict /prɪ'dɪkt/ vt predire. **~able** /-əbl/ a prevedibile. **~ion** /-'dɪkʃn/ n previsione f

pre'domin|ant /prɪ-/ a predominante. **~ate** vi predominare

pre-'eminent /priː-/ a preminente

preen /priːn/ vt lisciarsi; **~ oneself** fig farsi bello

pre|fab /'priːfæb/ n fam casa f prefabbricata. **~'fabricated** a prefabbricato

preface /'prefɪs/ n prefazione f

prefect /'priːfekt/ n Sch studente, -tessa mf della scuola superiore con responsabilità disciplinari, ecc

prefer /prɪ'fɜː(r)/ vt (pt/pp preferred) preferire

prefera|ble /'prefərəbl/ a preferibile (to a). **~bly** adv preferibilmente

preferen|ce /'prefərəns/ n preferenza f. **~tial** /-'renʃl/ a preferenziale

prefix /'priːfɪks/ n prefisso m

pregnan|cy /'pregnənsɪ/ n gravidanza f. **~t** a incinta

prehi'storic /priː-/ a preistorico

prejudice /'predʒʊdɪs/ n pregiudizio m ● vt influenzare (against contro); (harm) danneggiare. **~d** a prevenuto

preliminary /prɪ'lɪmɪnərɪ/ a preliminare

prelude /'prelju:d/ n preludio m

pre-'marital *a* prematrimoniale

premature /'prematjʊə(r)/ *a* prematuro

pre'meditated /pri:-/ *a* premeditato

premier /'premɪə(r)/ *a* primario ● *n Pol* primo ministro *m*, premier *m inv*

première /'premɪeə(r)/ *n* prima *f*

premises /'premɪsɪz/ *npl* locali *mpl*; **on the ~** sul posto

premium /'pri:mɪəm/ *n* premio *m*; **be at a ~** essere una cosa rara

premonition /premə'nɪʃn/ *n* presentimento *m*

preoccupied /pri:'ɒkjʊpaɪd/ *a* preoccupato

prep /prep/ *n Sch* compiti *mpl*

preparation /prepə'reɪʃn/ *n* preparazione *f*. **~s** preparativi *mpl*

preparatory /prɪ'pærətrɪ/ *a* preparatorio ● *adv* **~ to** per

prepare /prɪ'peə(r)/ *vt* preparare ● *vi* prepararsi (**for** per); **~d to** disposto a

pre'pay /pri:-/ *vt* (*pt/pp* **-paid**) pagare in anticipo

preposition /prepə'zɪʃn/ *n* preposizione *f*

prepossessing /pri:pə'zesɪŋ/ *a* attraente

preposterous /prɪ'pɒstərəs/ *a* assurdo

prerequisite /pri:'rekwɪzɪt/ *n* condizione *f* sine qua non

prescribe /prɪ'skraɪb/ *vt* prescrivere

prescription /prɪ'skrɪpʃn/ *n Med* ricetta *f*

presence /'prezns/ *n* presenza *f*; **~ of mind** presenza *f* di spirito

present[1] /'preznt/ *a* presente ● *n* presente *m*; **at ~** attualmente

present[2] *n* (*gift*) regalo *m*; **give sb sth as a ~** regalare qcsa a qcno

present[3] /prɪ'zent/ *vt* presentare; **~ sb with an award** consegnare un premio a qcno. **~able** /-əbl/ *a* **be ~able** essere presentabile

presentation /prezn'teɪʃn/ *n* presentazione *f*

presently /'prezntlɪ/ *adv* fra poco; (*Am: now*) attualmente

preservation /prezə'veɪʃn/ *n* conservazione *f*

preservative /prɪ'zɜ:vətɪv/ *n* conservante *m*

preserve /prɪ'zɜ:v/ *vt* preservare; (*maintain, Culin*) conservare ● *n* (*in hunting & fig*) riserva *f*; (*jam*) marmellata *f*

preside /prɪ'zaɪd/ *vi* presiedere (**over** a)

presidency /'prezɪdənsɪ/ *n* presidenza *f*

president /'prezɪdənt/ *n* presidente *m*. **~ial** /-'denʃl/ *a* presidenziale

press /pres/ *n* (*machine*) pressa *f*; (*newspapers*) stampa *f* ● *vt* premere; pressare (*flower*); (*iron*) stirare; (*squeeze*) stringere ● *vi* (*urge*) incalzare. **press for** *vi* fare pressione per; **be ~ed for** essere a corto di. **press on** *vi* andare avanti

press: **~ conference** *n* conferenza *f* stampa. **~ cutting** *n* ritaglio *m* di giornale. **~ing** *a* urgente. **~-stud** *n* [bottone *m*] automatico *m*. **~-up** *n* flessione *f*

pressure /'preʃə(r)/ *n* pressione *f* ● *vt* = **pressurize**. **~-cooker** *n* pentola *f* a pressione. **~ group** *n* gruppo *m* di pressione

pressurize /'preʃəraɪz/ *vt* far pressione su. **~d** *a* pressurizzato

prestige /pre'sti:ʒ/ *n* prestigio *m*. **~ious** /-'stɪdʒəs/ *a* prestigioso

presumably /prɪ'zju:məblɪ/ *adv* presumibilmente

presume /prɪ'zju:m/ *vt* presumere; **~ to do sth** permettersi di fare qcsa

presumpt|ion /prɪ'zʌmpʃn/ *n* presunzione *f*; (*boldness*) impertinenza *f*. **~uous** /-'zʌmptjʊəs/ *a* impertinente

presup'pose /pri:-/ *vt* presupporre

pretence /prɪ'tens/ *n* finzione *f*; (*pretext*) pretesto *m*; **it's all ~** è tutta una scena

pretend /prɪ'tend/ *vt* fingere; (*claim*) pretendere ● *vi* fare finta

pretentious /prɪ'tenʃəs/ *a* pretenzioso

pretext /'pri:tekst/ *n* pretesto *m*

pretty /'prɪtɪ/ *a* (**-ier, -iest**) carino ● *adv* (*fam: fairly*) abbastanza

prevail /prɪ'veɪl/ *vi* prevalere; ~ **on sb to do sth** convincere qcno a fare qcsa. **~ing** *a* prevalente

prevalen|ce /'prevələns/ *n* diffusione *f*. **~t** *a* diffuso

prevent /prɪ'vent/ *vt* impedire; ~ **sb [from] doing sth** impedire a qcno di fare qcsa. **~ion** /-enʃn/ *n* prevenzione *f*. **~ive** /-ɪv/ *a* preventivo

preview /'pri:vju:/ *n* anteprima *f*

previous /'pri:vɪəs/ *a* precedente. **~ly** *adv* precedentemente

pre-'war /pri:-/ *a* anteguerra

prey /preɪ/ *n* preda *f*; **bird of** ~ uccello *m* rapace ● *vi* ~ **on** far preda di; ~ **on sb's mind** attanagliare qcno

price /praɪs/ *n* prezzo *m* ● *vt* Comm fissare il prezzo di. **~less** *a* inestimabile; (*fam: amusing*) spassosissimo. **~y** *a fam* caro

prick /prɪk/ *n* puntura *f* ● *vt* pungere. **prick up** *vt* ~ **up one's ears** rizzare le orecchie

prickl|e /'prɪkl/ *n* spina *f*; (*sensation*) formicolio *m*. **~y** *a* pungente; (*person*) irritabile

pride /praɪd/ *n* orgoglio *m* ● *vt* ~ **oneself on** vantarsi di

priest /pri:st/ *n* prete *m*

prim /prɪm/ *a* (**primmer, primmest**) perbenino

primarily /'praɪmərɪlɪ/ *adv* in primo luogo

primary /'praɪmərɪ/ *a* primario; (*chief*) principale. ~ **school** *n* scuola *f* elementare

prime[1] /praɪm/ *a* principale, primo; (*first-rate*) eccellente ● *n* **be in one's** ~ essere nel fiore degli anni

prime[2] *vt* preparare (*surface, person*)

Prime Minister *n* Primo *m* Ministro

primeval /praɪ'mi:vl/ *a* primitivo

primitive /'prɪmɪtɪv/ *a* primitivo

primrose /'prɪmrəʊz/ *n* primula *f*

prince /prɪns/ *n* principe *m*

princess /prɪn'ses/ *n* principessa *f*

principal /'prɪnsəpl/ *a* principale ● *n* Sch preside *m*

principality /prɪnsɪ'pælətɪ/ *n* principato *m*

principally /'prɪnsəplɪ/ *adv* principalmente

principle /'prɪnsəpl/ *n* principio *m*; **in** ~ in teoria; **on** ~ per principio

print /prɪnt/ *n* (*mark, trace*) impronta *f*; Phot copia *f*; (*picture*) stampa *f*; **in** ~ (*printed out*) stampato; (*book*) in commercio; **out of** ~ esaurito ● *vt* stampare; (*write in capitals*) scrivere in stampatello. **~ed matter** *n* stampe *fpl*

print|er /'prɪntə(r)/ *n* stampante *f*; Typ tipografo, -a *mf*. **~er port** *n* Comput porta *f* per la stampante. **~ing** *n* tipografia *f*

'printout *n* Comput stampa *f*

prior /'praɪə(r)/ *a* precedente. ~ **to** *prep* prima di

priority /praɪ'ɒrətɪ/ *n* precedenza *f*; (*matter*) priorità *f inv*

prise /praɪz/ *vt* ~ **open/up** forzare

prison /'prɪz(ə)n/ *n* prigione *f*. **~er** *n* prigioniero, -a *mf*

privacy /'prɪvəsɪ/ *n* privacy *f inv*

private /'praɪvət/ *a* privato; (*car, secretary, letter*) personale ● *n* Mil soldato *m* semplice; **in** ~ in privato: **~ly** *adv* (*funded, educated etc*) privatamente; (*in secret*) in segreto; (*confidentially*) in privato; (*inwardly*) interiormente

privation /praɪ'veɪʃn/ *n* privazione *f*; **~s** *npl* stenti *mpl*

privatize /'praɪvətaɪz/ *vt* privatizzare

privilege /'prɪvəlɪdʒ/ *n* privilegio *m*. **~d** *a* privilegiato

privy /'prɪvɪ/ *a* **be** ~ **to** essere al corrente di

prize /praɪz/ n premio m ● a (idiot etc) perfetto ● vt apprezzare. **~-giving** n premiazione f. **~-winner** n vincitore, -trice mf. **~-winning** a vincente

pro /prəʊ/ n (fam: professional) professionista mf; the **~s and cons** il pro e il contro

probability /prɒbə'bɪlətɪ/ n probabilità f inv

probabl|e /'prɒbəbl/ a probabile. **~y** adv probabilmente

probation /prə'beɪʃn/ n prova f; Jur libertà f vigilata. **~ary** a in prova; **~ary period** periodo m di prova

probe /prəʊb/ n sonda f; (fig: investigation) indagine f ● vt sondare; (investigate) esaminare a fondo

problem /'prɒbləm/ n problema m ● a difficile. **~atic** a /-'mætɪk/ a problematico

procedure /prə'si:dʒə(r)/ n procedimento m

proceed /prə'si:d/ vi procedere ● vt → **to do sth** proseguire facendo qcsa

proceedings /prə'si:dɪŋz/ npl (report) atti mpl; Jur azione fsg legale

proceeds /'prəʊsi:dz/ npl ricavato msg

process /'prəʊses/ n processo m; (procedure) procedimento m; in the **~** nel far ciò ● vt trattare; Admin occuparsi di; Phot sviluppare

procession /prə'seʃn/ n processione f

proclaim /prə'kleɪm/ vt proclamare

procure /prə'kjʊə(r)/ vt ottenere

prod /prɒd/ n colpetto m ● vt (pt/pp prodded) punzecchiare; fig incitare

prodigal /'prɒdɪgl/ a prodigo

prodigious /prə'dɪdʒəs/ a prodigioso

prodigy /'prɒdɪdʒɪ/ n [infant] **~** bambino m prodigio

produce¹ /'prɒdju:s/ n prodotti mpl; **~ of Italy** prodotto in Italia

produce² /prə'dju:s/ vt produrre; (bring out) tirar fuori; (cause) causare; (fam: give birth to) fare. **~r** n produttore m

product /'prɒdʌkt/ n prodotto m. **~ion** /prə'dʌkʃn/ n produzione f; Theat spettacolo m

productiv|e /prə'dʌktɪv/ a produttivo. **~ity** /-'tɪvətɪ/ n produttività f

profane /prə'feɪn/ a profano; (blasphemous) blasfemo. **~ity** /-'fænətɪ/ n (oath) bestemmia f

profession /prə'feʃn/ n professione f. **~al** a professionale; (not amateur) professionista; (piece of work) da professionista; (man) di professione ● n professionista mf. **~ally** adv professionalmente

professor /prə'fesə(r)/ n professore m [universitario]

proficien|cy /prə'fɪʃnsɪ/ n competenza f. **~t** a be **~t** in essere competente in

profile /'prəʊfaɪl/ n profilo m

profit /'profɪt/ n profitto m ● vi **~ from** trarre profitto da. **~able** /-əbl/ a proficuo. **~ably** adv in modo proficuo

profound /prə'faʊnd/ a profondo. **~ly** adv profondamente

profuse /prə'fju:s/ a **~e apologies/flowers** una profusione di scuse/fiori. **~ion** /-ju:ʒn/ n profusione f; **in ~ion** in abbondanza

progeny /'prɒdʒənɪ/ n progenie f inv

prognosis /prɒg'nəʊsɪs/ n (pl -oses) prognosi f inv

program /'prəʊgræm/ n programma m ● vt (pt/pp programmed) programmare

programme /'prəʊgræm/ n Br programma m. **~r** n Comput programmatore, -trice mf

progress¹ /'prəʊgres/ n progresso m; **in ~** in corso; **make ~** fig fare progressi

progress² /prə'gres/ vi progredire; fig fare progressi

progressive /prə'gresɪv/ a progressivo; (reforming) progressista. **~ly** adv progressivamente

prohibit /prə'hɪbɪt/ vt proibire. **~ive** -IV/ a proibitivo

project¹ /'prɒdʒekt/ n progetto m; Sch ricerca f

project² /prə'dʒekt/ vt proiettare (film, image) ● vi (jut out) sporgere

projectile /prə'dʒektaɪl/ n proiettile m

projector /prə'dʒektə(r)/ n proiettore m

prolific /prə'lɪfɪk/ a prolifico

prologue /'prəʊlɒg/ n prologo m

prolong /prə'lɒŋ/ vt prolungare

promenade /prɒmə'nɑːd/ n lungomare m inv

prominent /'prɒmɪnənt/ a prominente; (conspicuous) di rilievo

promiscuity /prɒmɪ'skjuːətɪ/ n promiscuità f. **~ous** /prə'mɪskjʊəs/ a promiscuo

promis|e /'prɒmɪs/ n promessa f ● vt promettere; **~e sb that** promettere a qcno che; **I ~ed to** l'ho promesso. **~ing** a promettente

promot|e /prə'məʊt/ vt promuovere; **be ~ed** Sport essere promosso. **~ion** -əʊʃn/ n promozione f

prompt /prɒmpt/ a immediato; (punctual) puntuale ● adv in punto ● vt incitare (to a); Theat suggerire a ● vi suggerire. **~er** n suggeritore, -trice mf. **~ly** adv puntualmente

Proms /prɒmz/ npl rassegna f di concerti estivi di musica classica presso l'Albert Hall a Londra

prone /prəʊn/ a **be ~ to do sth** essere incline a fare qcsa

prong /prɒŋ/ n dente m (di forchetta)

pronoun /'prəʊnaʊn/ n pronome m

pronounce /prə'naʊns/ vt pronunciare; (declare) dichiarare. **~d** a (noticeable) pronunciato

pronunciation /prənʌnsɪ'eɪʃn/ n pronuncia f

proof /pruːf/ n prova f; Typ bozza f, prova f ● a **~ against** a prova di

prop¹ /prɒp/ n puntello m ● vt (pt/pp **propped**) **~ open** tenere aperto; **~ against** (lean) appoggiare a. **prop up** vt sostenere

prop² n Theat, fam accessorio m di scena

propaganda /prɒpə'gændə/ n propaganda f

propel /prə'pel/ vt (pt/pp **propelled**) spingere. **~ler** n elica f

proper /'prɒpə(r)/ a corretto; (suitable) adatto; (fam: real) vero [e proprio]. **~ly** adv correttamente. **~ 'name**, **~ 'noun** n nome m proprio

property /'prɒpətɪ/ n proprietà f inv. **~ developer** n agente m immobiliare. **~ market** n mercato m immobiliare

prophecy /'prɒfəsɪ/ n profezia f

prophesy /'prɒfɪsaɪ/ vt (pt/pp -ied) profetizzare

prophet /'prɒfɪt/ n profeta m. **~ic** /prə'fetɪk/ a profetico

proportion /prə'pɔːʃn/ n proporzione f; (share) parte f; **~s** pl (dimensions) proporzioni fpl. **~al** a proporzionale. **~ally** adv in proporzione

proposal /prə'pəʊzl/ n proposta f; (of marriage) proposta f di matrimonio

propose /prə'pəʊz/ vt proporre; (intend) proporsi ● vi fare una proposta di matrimonio

proposition /prɒpə'zɪʃn/ n proposta f; (fam: task) impresa f

proprietor /prə'praɪətə(r)/ n proprietario, -a mf

prosaic /prə'zeɪɪk/ a prosaico

prose /prəʊz/ n prosa f

prosecut|e /'prɒsɪkjuːt/ vt intentare azione contro. **~ion** -'kjuːʃn/ n azione f giudiziaria; **the ~ion** l'accusa f. **~or** n [**Public**] **P~or** il Pubblico Ministero m

prospect¹ /'prɒspekt/ n (expectation) prospettiva f

prospect[2] /prə'spekt/ *vi* ~ **for** cercare

prospective /prə'spektiv/ *a* (*future*) futuro; (*possible*) potenziale. ~**or** *n* cercatore *m*

prospectus /prə'spektəs/ *n* prospetto *m*

prosper /'prɒspə(r)/ *vi* prosperare; (*person:*) stare bene finanziariamente. ~**ity** /-'sperəti/ *n* prosperità *f*

prosperous /'prɒspərəs/ *a* prospero

prostitute /'prɒstitju:t/ *n* prostituta *f*. ~**ion** /-'tju:ʃn/ *n* prostituzione *f*

prostrate /'prɒstreɪt/ *a* prostrato; ~ **with grief** *fig* prostrato dal dolore

protagonist /prəʊ'tægənɪst/ *n* protagonista *mf*

protect /prə'tekt/ *vt* proteggere (**from** da). ~**ion** /-ekʃn/ *n* protezione *f*. ~**ive** /-ɪv/ *a* protettivo. ~**or** *n* protettore, -trice *mf*

protégé /'prɒtiʒeɪ/ *n* protetto *m*

protein /'prəʊti:n/ *n* proteina *f*

protest[1] /'prəʊtest/ *n* protesta *f*

protest[2] /prə'test/ *vt/i* protestare

Protestant /'prɒtɪstənt/ *a* protestante ● *n* protestante *mf*

protester /prə'testə(r)/ *n* contestatore, -trice *mf*

protocol /'prəʊtəkɒl/ *n* protocollo *m*

prototype /'prəʊtə-/ *n* prototipo *m*

protract /prə'trækt/ *vt* protrarre

protrude /prə'tru:d/ *vi* sporgere

proud /praʊd/ *a* fiero (**of**). ~**ly** *adv* fieramente

prove /pru:v/ *vt* provare ● *vi* ~ **to be a lie** rivelarsi una bugia. ~**n** *a* dimostrato

proverb /'prɒvɜ:b/ *n* proverbio *m*. ~**ial** /prə'vɜ:bɪəl/ *a* proverbiale

provide /prə'vaɪd/ *vt* fornire; ~ **sb with sth** fornire qcsa a qcno ● *vi* ~ **for** (*law:*) prevedere

provided /prə'vaɪdɪd/ *conj* ~ [**that**] purché

providen|ce /'prɒvɪdəns/ *n* prov-

videnza *f*. ~**tial** /-'denʃl/ *a* provvidenziale

providing /prə'vaɪdɪŋ/ *conj* = **provided**

provinc|e /'prɒvɪns/ *n* provincia *f*; *fig* campo *m*. ~**ial** /prə'vɪnʃl/ *a* provinciale

provision /prə'vɪʒn/ *n* (*of food, water*) approvvigionamento *m* (**of** di); (*of law*) disposizione *f*; ~**s** *pl* provviste *fpl*. ~**al** *a* provvisorio

proviso /prə'vaɪzəʊ/ *n* condizione *f*

provoca|tion /prɒvə'keɪʃn/ *n* provocazione *f*. ~**tive** /prə'vɒkətɪv/ *a* provocatorio; (*sexually*) provocante. ~**tively** *adv* in modo provocatorio

provoke /prə'vəʊk/ *vt* provocare

prow /praʊ/ *n* prua *f*

prowess /'praʊɪs/ *n* abilità *f inv*

prowl /praʊl/ *vi* aggirarsi ● *n* **on the** ~ in cerca di preda. ~**er** *n* tipo *m* sospetto

proximity /prɒk'sɪmətɪ/ *n* prossimità *f*

proxy /'prɒksɪ/ *n* procura *f*; (*person*) persona *f* che agisce per procura

prude /pru:d/ *n* **be a** ~ essere eccessivamente pudico

pruden|ce /'pru:dəns/ *n* prudenza *f*. ~**t** *a* prudente; (*wise*) oculatezza *f*

prudish /'pru:dɪʃ/ *a* eccessivamente pudico

prune[1] /pru:n/ *n* prugna *f* secca

prune[2] /pru:n/ *vt* potare

pry /praɪ/ *vi* (*pt/pp* **pried**) ficcare il naso

psalm /sɑ:m/ *n* salmo *m*

pseudonym /'sju:dənɪm/ *n* pseudonimo *m*

psychiatric /saɪkɪ'ætrɪk/ *a* psichiatrico

psychiatr|ist /saɪ'kaɪətrɪst/ *n* psichiatra *mf*. ~**y** *n* psichiatria *f*

psychic /'saɪkɪk/ *a* psichico; **I'm not** ~ non sono un indovino

psycho'analyse /saɪkəʊ-/ *vt* psicanalizzare. ~**a'nalysis** *n* psicnalisi f. ~'analyst n psicanalista mf

psychological /saɪkə'lɒdʒɪkl/ a psicologico

psychologist /saɪ'kɒlədʒɪst/ n psicologo, -a mf. ~y n psicologia f

psychopath /'saɪkəpæθ/ n psicopatico, -a mf

P.T.O. abbr (please turn over) vedi retro

pub /pʌb/ n fam pub m inv

puberty /'pju:bətɪ/ n pubertà f

public /'pʌblɪk/ a pubblico ●n the ~ il pubblico; **in** ~ in pubblico. ~**ly** adv pubblicamente

publican /'pʌblɪkən/ n gestore, -trice mf/proprietario, -a mf di un pub

publication /pʌblɪ'keɪʃn/ n pubblicazione f

public: ~ **con'venience** n gabinetti mpl pubblici. ~ **'holiday** n festa f nazionale. ~ **'house** n pub m

publicity /pʌb'lɪsətɪ/ n pubblicità f

publicize /'pʌblɪsaɪz/ vt pubblicizzare

public: ~ **'library** n biblioteca f pubblica. ~ **relations** pubbliche relazioni fpl. ~ **'school** n scuola f privata; Am scuola f pubblica. ~-**'spirited** a be ~-**spirited** essere dotato di senso civico. ~ **'transport** n mezzi mpl pubblici

publish /'pʌblɪʃ/ vt pubblicare. ~**er** n editore m; (firm) editore m, casa f editrice. ~**ing** n editoria f

pudding /'pʊdɪŋ/ n dolce m cotto al vapore; (course) dolce m

puddle /'pʌdl/ n pozzanghera f

pudgy /'pʌdʒɪ/ a (-ier, -iest) grassoccio

puff /pʌf/ n (of wind) soffio m; (of smoke) tirata f; (for powder) piumino m ● vt sbuffare. **puff at** vt tirare boccate da (pipe). **puff out** vt lasciare senza fiato (person); spegnere (candle). ~**ed** a (out of breath) senza fiato. ~ **pastry** n pasta f sfoglia

puffy /'pʌfɪ/ a gonfio

pull /pʊl/ n trazione f; (fig: attrac-

tion) attrazione f; (fam: influence) influenza f ● vt tirare; estrarre (tooth); stirarsi (muscle); ~ **faces** far boccacce; ~ **oneself together** cercare di controllarsi; ~ **one's weight** mettercela tutta; ~ **sb's leg** fam prendere in giro qcno. **pull down** vt (demolish) demolire. **pull in** vi Auto accostare. **pull off** vt togliere; fam azzeccare. **pull out** vt tirar fuori ● vi Auto spostarsi; (of competition) ritirarsi. **pull through** vi (recover) farcela. **pull up** vt sradicare (plant); (reprimand) rimproverare ● vi Auto fermarsi

pulley /'pʊlɪ/ n Techn puleggia f

pullover /'pʊləʊvə(r)/ n pullover m

pulp /pʌlp/ n poltiglia f; (of fruit) polpa f; (for paper) pasta f

pulpit /'pʊlpɪt/ n pulpito m

pulsate /pʌl'seɪt/ vi pulsare

pulse /pʌls/ n polso m

pulses /'pʌlsɪz/ npl legumi mpl secchi

pulverize /'pʌlvəraɪz/ vt polverizzare

pumice /'pʌmɪs/ n pomice f

pummel /'pʌml/ vt (pt/pp pummelled) prendere a pugni

pump /pʌmp/ n pompa f ● vt pompare; fam cercare di estrorcere da. **pump up** vt (inflate) gonfiare

pumpkin /'pʌmpkɪn/ n zucca f

pun /pʌn/ n gioco m di parole

punch[1] /pʌntʃ/ n pugno m; (device) pinza f per forare ● vt dare un pugno a; forare (ticket); perforare (hole)

punch[2] n (drink) ponce m inv

punch: ~ **line** n battuta f finale. ~-**up** n rissa f

punctual /'pʌŋktjʊəl/ a puntuale. ~**ity** /-'ælətɪ/ n puntualità f. ~**ly** adv puntualmente

punctuate /'pʌŋktjʊeɪt/ vt punteggiare. ~**ion** /-'eɪʃn/ n punteggiatura f. ~**ion mark** n segno m di interpunzione

puncture /'pʌŋktʃə(r)/ n foro m; (tyre) foratura f ● vt forare

pungent /ˈpʌndʒənt/ *a* acre
punish /ˈpʌnɪʃ/ *vt* punire. **~able**
/-əbl/ *a* punibile. **~ment** *n* punizione *f*
punitive /ˈpjuːnɪtɪv/ *a* punitivo
punk /pʌŋk/ *n* punk *m inv*
punnet /ˈpʌnɪt/ *n* cestello *m* (*per frutta*)
punt /pʌnt/ *n* (*boat*) barchino *m*
punter /ˈpʌntə(r)/ *n* (*gambler*) scommettitore, **-trice** *mf*; (*client*) consumatore, **-trice** *mf*
puny /ˈpjuːnɪ/ *a* (**-ier, -iest**) striminzito
pup /pʌp/ *n* = **puppy**
pupil /ˈpjuːpl/ *n* alunno, **-a** *mf*; (*of eye*) pupilla *f*
puppet /ˈpʌpɪt/ *n* marionetta *f*; (*glove ~, fig*) burattino *m*
puppy /ˈpʌpɪ/ *n* cucciolo *m*
purchase /ˈpɜːtʃəs/ *n* acquisto *m*; (*leverage*) presa *f* ● *vt* acquistare. **~r** *n* acquirente *mf*
pure /pjʊə(r)/ *a* puro. **~ly** *adv* puramente
purée /ˈpjʊəreɪ/ *n* purè *m*
purgatory /ˈpɜːgətrɪ/ *n* purgatorio *m*
purge /pɜːdʒ/ *Pol n* epurazione *f* ● *vt* epurare
puri|fication /pjʊərɪfɪˈkeɪʃn/ *n* purificazione *f*. **~fy** /ˈpjʊərɪfaɪ/ *vt* (*pt/pp* **-ied**) purificare
puritan /ˈpjʊərɪtən/ *n* puritano, **-a** *mf*. **~ical** *a* puritano
purity /ˈpjʊərɪtɪ/ *n* purità *f*
purple /ˈpɜːpl/ *a* viola
purpose /ˈpɜːpəs/ *n* scopo *m*; (*determination*) fermezza *f*; **on ~** apposta. **~-built** *a* costruito ad hoc. **~ful** *a* deciso. **~fully** *adv* con decisione. **~ly** *adv* apposta
purr /pɜː(r)/ *vi* (*cat:*) fare le fusa
purse /pɜːs/ *n* borsellino *m*; (*Am: handbag*) borsa *f* ● *vt* increspare (*lips*)
pursue /pəˈsjuː/ *vt* inseguire; *fig* proseguire. **~r** /-ə(r)/ *n* inseguitore, **-trice** *mf*
pursuit /pəˈsjuːt/ *n* inseguimento *m*; (*fig: of happiness*) ricerca *f*;

(*pastime*) attività *f inv*; **in ~** all'inseguimento
pus /pʌs/ *n* pus *m*
push /pʊʃ/ *n* spinta *f*; (*fig: effort*) sforzo *m*; (*drive*) iniziativa *f*; **at a ~** in caso di bisogno; **get the ~** *fam* essere licenziato ● *vt* spingere; premere (*button*); (*pressurize*) far pressione su; **be ~ed for time** *fam* non avere tempo ● *vi* spingere. **push aside** *vt* scostare. **push back** *vt* respingere. **push off** *vt* togliere ● *vi* (*fam: leave*) levarsi dai piedi. **push on** *vi* (*continue*) continuare. **push up** *vt* alzare (*price*)
push: **~-button** *n* pulsante *m*. **~-chair** *n* passeggino *m*. **~-over** *n* *fam* bazzecola *f*. **~-up** *n* flessione *f*
pushy /ˈpʊʃɪ/ *a* *fam* troppo intraprendente
puss /pʊs/ *n*, **pussy** /ˈpʊsɪ/ *n* micio *m*
put /pʊt/ *vt* (*pt/pp* put, *pres p* putting) mettere; ~ **the cost of sth at** valutare il costo di qcsa ● *vi* ~ **to sea** salpare. **put aside** *vt* mettere da parte. **put away** *vt* mettere via. **put back** *vt* rimettere; mettere indietro (*clock*). **put by** *vt* mettere da parte. **put down** *vt* mettere giù; (*suppress*) reprimere; (*kill*) sopprimere; (*write*) annotare; ~ **one's foot down** *fam* essere fermo; *Auto* dare un'accelerata; (*attribute*) attribuire. **put forward** *vt* avanzare; mettere avanti (*clock*). **put in** *vt* (*insert*) introdurre; (*submit*) presentare ● *vi* ~ **in for** fare domanda di. **put off** *vt* spegnere (*light*); (*postpone*) rimandare; ~ **sb off** tenere a bada qcno; (*deter*) smontare qcno; (*disconcert*) distrarre qcno; ~ **sb off sth** (*disgust*) disgustare qcno di qcsa. **put on** *vt* mettersi (*clothes*); mettere (*brake*); *Culin* mettere su; accendere (*light*); mettere in scena (*play*); mettere in scena (*accent*); ~ **on weight** mettere su qualche chilo. **put out** *vt* spegnere (*fire, light*); tendere

⟨hand⟩; (inconvenience) creare degli inconvenienti a. **put through** vt far passare; Teleph **I'll ~ you through to him** glielo passo. **put up** vt alzare; erigere ⟨building⟩; montare ⟨tent⟩; aprire ⟨umbrella⟩; affiggere ⟨notice⟩; aumentare ⟨price⟩; ospitare ⟨guest⟩; **~ sb up to sth** mettere qcno in testa a qcno ● vi (at hotel) stare; **~ up with** sopportare ● a **stay ~!** rimani lì!

putty /'pʌtɪ/ n mastice m

put-up /'pʊtʌp/ a **~ job** truffa f

puzzl|e /'pʌzl/ n enigma m; (jigsaw) puzzle m inv ● vt lasciare perplesso ● vi **~e over** scervellarsi su. **~ing** a inspiegabile

pygmy /'pɪgmɪ/ n pigmeo, -a mf

pyjamas /pə'dʒɑːməz/ npl pigiama msg

pylon /'paɪlən/ n pilone m

pyramid /'pɪrəmɪd/ n piramide f

python /'paɪθn/ n pitone m

Qq

quack¹ /kwæk/ n qua qua m inv ● vi fare qua qua

quack² n (doctor) ciarlatano m

quad /kwɒd/ n (fam: court) = **quadrangle. ~s** pl = **quadruplets**

quadrangle /'kwɒdræŋgl/ n quadrangolo m; (court) cortile m quadrangolare

quadruped /'kwɒdroped/ n quadrupede m

quadruple /'kwɒdrʊpl/ a quadruplo ● vt quadruplicare ● vi quadruplicarsi. **~ts** -plɪts/ npl quattro gemelli mpl

quagmire /'kwɒgmaɪə(r)/ n pantano m

quaint /kweɪnt/ a pittoresco; (odd) bizzarro

quake /kweɪk/ n fam terremoto m ● vi tremare

qualification /kwɒlɪfɪ'keɪʃn/ n qualifica f. **~ied** -faɪd/ a qualificato; (limited) con riserva

qualify /'kwɒlɪfaɪ/ v (pt/pp -ied) ● vt (course:) dare la qualifica a (as di); (entitle) dare diritto a; (limit) precisare ● vi ottenere la qualifica; Sport qualificarsi

quality /'kwɒlɪtɪ/ n qualità f inv; **in ~** in qualità m

qualm /kwɑːm/ n scrupolo m

quandary /'kwɒndərɪ/ n dilemma m

quantity /'kwɒntɪtɪ/ n quantità f inv; **in ~** in grande quantità

quarantine /'kwɒrəntiːn/ n quarantena f

quarrel /'kwɒrəl/ n lite f ● vi (pt/pp **quarrelled**) litigare. **~some** a litigioso

quarry¹ /'kwɒrɪ/ n (prey) preda f

quarry² n cava f

quart /kwɔːt/ n 1.14 litro

quarter /'kwɔːtə(r)/ n quarto m; (of year) trimestre m; Am 25 centesimi mpl; **~s** pl Mil quartiere msg; **at [a] ~ to six** alle sei meno un quarto ● vt dividere in quattro. **~-'final** n quarto m di finale

quarterly /'kwɔːtəlɪ/ a trimestrale ● adv trimestralmente

quartet /kwɔː'tet/ n quartetto m

quartz /kwɔːts/ n quarzo m. **~ watch** n orologio m al quarzo

quash /kwɒʃ/ vt annullare; soffocare ⟨rebellion⟩

quaver /'kweɪvə(r)/ vi tremolare

quay /kiː/ n banchina f

queasy /'kwiːzɪ/ a **I feel ~** ho la nausea

queen /kwiːn/ n regina f. **~ mother** n regina f madre

queer /kwɪə(r)/ a strano; (dubious) sospetto; (fam: homosexual) finocchio ● n fam finocchio m

quell /kwel/ vt reprimere

quench /kwentʃ/ vt **~ one's thirst** dissetarsi

query /'kwɪərɪ/ n domanda f, (question mark) punto m interrogativo ● vt (pt/pp -ied) interrogare; (doubt) mettere in dubbio

quest /kwest/ n ricerca f (**for** di)

question /'kwestʃn/ n domanda f; (for discussion) questione f; **out of the** ~ fuori discussione; **without** ~ senza dubbio; **in** ~ in questione ● vt interrogare; (doubt) mettere in dubbio. **~able** /-əbl/ a discutibile. **~ mark** n punto m interrogativo

questionnaire /kwestʃə'neə(r)/ n questionario m

queue /kju:/ n coda f, fila f ● vi ~ [up] mettersi in coda (**for** per)

quick /kwɪk/ a veloce; **be** ~ sbrigati!; **have a** ~ **meal** fare uno spuntino ● adv in fretta ● n **be cut to the** ~ fig essere punto sul vivo. **~ly** adv in fretta. **~-tempered** a collerico

quid /kwɪd/ n inv fam sterlina f

quiet /'kwaɪət/ a (calm) tranquillo; (silent) silenzioso; (voice, music) basso; **keep** ~ **about** fam non raccontare a nessuno ● n quiete f; **on the** ~ di nascosto. **~ly** adv (peacefully) tranquillamente; (say) a bassa voce

quiet|en /'kwaɪətn/ vt calmare. **quieten down** vi calmarsi. **~ness** n quiete f

quilt /kwɪlt/ n piumino m. **~ed** a trapuntato

quins /kwɪnz/ npl fam = **quintuplets**

quintet /kwɪn'tet/ n quintetto m

quintuplets /'kwɪntjuplɪts/ npl cinque gemelli mpl

quirk /kwɜ:k/ n stranezza f

quit /kwɪt/ v (pt/pp **quitted**, **quit**) ● vt lasciare; (give up) smettere (**doing** di fare) ● vi (fam: resign) andarsene; Comput uscir [e]; **tell sb notice to** ~ (landlord:) dare a qcno il preavviso di sfratto

quite /kwaɪt/ adv (fairly) abbastanza; (completely) completamente; (really) veramente; ~ [**so**]! proprio così!; ~ **a few** parecchi

quits /kwɪts/ a pari

quiver /'kwɪvə(r)/ vi tremare

quiz /kwɪz/ n (game) quiz m inv ● vt (pt/pp **quizzed**) interrogare

quota /'kwəʊtə/ n quota f

quotation /kwəʊ'teɪʃn/ n citazione f; (price) preventivo m; (of shares) quota f. ~ **marks** npl virgolette fpl

quote /kwəʊt/ n fam = **quotation**; **in** ~s tra virgolette ● vt citare; quotare (price)

Rr

rabbi /'ræbaɪ/ n rabbino m; (title) rabbi

rabbit /'ræbɪt/ n coniglio m

rabble /'ræbl/ n **the** ~ la plebaglia

rabies /'reɪbi:z/ n rabbia f

race¹ /reɪs/ n (people) razza f

race² n corsa f ● vi correre ● vt gareggiare con; fare correre (horse)

race: ~**course** n ippodromo m. ~**horse** n cavallo m da corsa. ~**track** n pista f

racial /'reɪʃl/ a razziale. ~**ism** n razzismo m

racing /'reɪsɪŋ/ n corse fpl; (horse-) corse fpl dei cavalli. ~ **car** n macchina f da corsa. ~ **driver** n corridore m automobilistico

racis|m /'reɪsɪzm/ n razzismo m. ~**t** /-ɪst/ a razzista ● n razzista mf

rack¹ /ræk/ n (for plates) rastrelliera f; (for luggage) portabagagli m inv; (for plates) scolapiatti m inv ● vt ~ **one's brains** scervellarsi

rack² n **go to** ~ **and ruin** andare in rovina

racket¹ /'rækɪt/ n Sport racchetta f

racket² n (din) chiasso m; (swindle) truffa f; (crime) racket m inv, giro m

radar /'reɪdɑ:(r)/ n radar m inv

radian|ce /'reɪdɪəns/ n radiosità f inv. ~**t** a raggiante

radiat|e /'reɪdɪeɪt/ vt irradiare ● vi

⟨*heat:*⟩ irradiarsi. **~ion** /-'eɪʃn/ n radiazione f

radiator /'reɪdɪeɪtə(r)/ n radiatore m

radical /'rædɪkl/ a radicale ●n radicale mf. **~ly** adv radicalmente

radio /'reɪdɪəʊ/ n radio f inv

radio'active a radioattivo. **~'tivity** n radioattività f

radiograph|er /reɪdɪ'ɒgrəfə(r)/ n radiologo, -a mf. **~y** n radiografia f

radio'therapy n radioterapia f

radish /'rædɪʃ/ n ravanello m

radius /'reɪdɪəs/ n (pl **-dii** /-dɪaɪ/) raggio m

raffle /'ræfl/ n lotteria f

raft /rɑ:ft/ n zattera f

rafter /'rɑ:ftə(r)/ n trave f

rag /ræg/ n straccio m; ⟨*pej: newspaper*⟩ giornalaccio m; **in ~s** straccioso

rage /reɪdʒ/ n rabbia f; **all the ~** fam all'ultima moda ●vi infuriarsi; ⟨*storm:*⟩ infuriare; ⟨*epidemic:*⟩ imperversare

ragged /'rægɪd/ a logoro; ⟨*edge*⟩ frastagliato

raid /reɪd/ n ⟨*by thieves*⟩ rapina f; Mil incursione f, raid m inv; ⟨*police*⟩ irruzione f ●vt Mil fare un'incursione in; ⟨*police, burglars:*⟩ fare irruzione in. **~er** n ⟨*of bank*⟩ rapinatore, -trice mf

rail /reɪl/ n ringhiera f; ⟨*hand~*⟩ ringhiera f; Naut parapetto m; **by ~** per ferrovia

'railroad n Am = **railway**

'railway n ferrovia f. **~man** n ferroviere m. **~ station** n stazione f ferroviaria

rain /reɪn/ n pioggia f ●vi piovere

rain: **~bow** n arcobaleno m. **~coat** n impermeabile m. **~fall** n precipitazione f [atmosferica]

rainy /'reɪnɪ/ a (**-ier, -iest**) piovoso

raise /reɪz/ n Am aumento m ●vt alzare; levarsi ⟨*hat*⟩; allevare ⟨*children, animals*⟩; sollevare ⟨*question*⟩; ottenere ⟨*money*⟩

raisin /'reɪzn/ n uva f passa

rake /reɪk/ n rastrello m ●vt rastrellare. **rake up** vt raccogliere col rastrello; fam rivangare

rally /'rælɪ/ n raduno m; Auto rally m inv; Tennis scambio m ●vt ⟨*pt/pp* **-ied**⟩ radunare ●vi radunarsi; ⟨*recover strength*⟩ riprendersi

RAM /ræm/ n [memoria f] RAM f

ram /ræm/ n montone m; Astr Ariete m ●vt ⟨*pt/pp* **rammed**⟩ cozzare contro

rambl|e /'ræmbl/ n escursione f ●vi gironzolare; ⟨*in speech*⟩ divagare. **~er** n escursionista mf; ⟨*rose*⟩ rosa f rampicante. **~ing** a ⟨*in speech*⟩ sconnesso; ⟨*club*⟩ escursionistico

ramp /ræmp/ n rampa f; Aeron scaletta f mobile ⟨*di aerei*⟩

rampage /'ræmpeɪdʒ/ n be/go on the **~** scatenarsi ●vi **~ through the streets** scatenarsi per le strade

rampant /'ræmpənt/ a dilagante

rampart /'ræmpɑ:t/ n bastione f

ramshackle /'ræmʃækl/ a sgangherato

ran /ræn/ see **run**

ranch /rɑ:ntʃ/ n ranch m

rancid /'rænsɪd/ a rancido

rancour /'ræŋkə(r)/ n rancore m

random /'rændəm/ a casuale; **~ sample** campione m a caso ●n **at ~** a casaccio

randy /'rændɪ/ a (**-ier, -iest**) fam eccitato

rang /ræŋ/ see **ring²**

range /reɪndʒ/ n serie f; Comm, Mus gamma f; ⟨*of mountains*⟩ catena f; ⟨*distance*⟩ raggio m; ⟨*for shooting*⟩ portata f; ⟨*stove*⟩ cucina f economica; **at a ~ of** a una distanza di ●vi estendersi; **~ from... to...** andare da... a.... **~r** n guardia f forestale

rank /ræŋk/ n ⟨*row*⟩ riga f; Mil grado m; ⟨*social position*⟩ rango m; **the ~ and file** la base f; **the ~s** Mil i soldati mpl semplici ●vt (*place*) annoverare (**among** tra) ● *vi* (*be placed*) collocarsi

rankle /'ræŋkl/ *vi fig* bruciare

ransack /'rænsæk/ *vt* rovistare; (*pillage*) saccheggiare

ransom /'rænsəm/ *n* riscatto *m*; **hold sb to ~** tenere qcno in ostaggio (*per il riscatto*)

rant /rænt/ *vi* ~ [**and rave**] inveire; **what's he ~ing on about?** cosa sta blaterando?

rap /ræp/ *n* colpo *m* [secco]; *Mus* rap *m* ● *v* (*pt/pp* **rapped**) ● *vt* dare colpetti a ● *vi* ~ **at** bussare a

rape /reɪp/ *n* (*sexual*) stupro *m* ● *vt* violentare, stuprare

rapid /'ræpɪd/ *a* rapido. **~ity** /rə'pɪdəti/ *n* rapidità *f*. **~ly** *adv* rapidamente

rapids /'ræpɪdz/ *npl* rapida *fsg*

rapist /'reɪpɪst/ *n* violentatore *m*

rapport /ræ'pɔ:(r)/ *n* rapporto *m* di intesa

rapture /'ræptʃə(r)/ *n* estasi *f*. **~ous** /-rəs/ *a* entusiastico

rare[1] /reə(r)/ *a* raro. **~ly** *adv* raramente

rare[2] *a Culin* al sangue

rarefied /'reərɪfaɪd/ *a* rarefatto

rarity /'reərəti/ *n* rarità *f inv*

rascal /'rɑːskl/ *n* mascalzone *m*

rash[1] /ræʃ/ *n Med* eruzione *f*

rash[2] *a* avventato. **~ly** *adv* avventatamente

rasher /'ræʃə(r)/ *n* fetta *f* di pancetta

rasp /rɑːsp/ *n* (*noise*) stridio *m*. **~ing** *a* stridente

raspberry /'rɑːzbərı/ *n* lampone *m*

rat /ræt/ *n* topo *m*; (*fam: person*) carogna *f*; **smell a ~** *fam* sentire puzzo di bruciato

rate /reɪt/ *n* (*speed*) velocità *f*; (*of payment*) tariffa *f*; (*of exchange*) tasso *m*; **~s** *pl* (*taxes*) imposte *fpl* comunali sui beni immobili; **at any ~** in ogni caso; **at this ~** di questo passo ● *vt* stimare; **~ among** annoverare tra ● *vi* **~ as** essere considerato

rather /'rɑːðə(r)/ *adv* piuttosto; **~!** eccome!; **~ too...** un po' troppo...

ratification /rætɪfɪ'keɪʃn/ *n* ratifica *f*. **~fy** /'rætɪfaɪ/ *vt* (*pt/pp* **-ied**) ratificare

rating /'reɪtɪŋ/ *n* **~s** *pl* Radio, TV indice *m* d'ascolto, audience *f inv*

ratio /'reɪʃɪəʊ/ *n* rapporto *m*

ration /'ræʃn/ *n* razione *f* ● *vt* razionare

rational /'ræʃənl/ *a* razionale. **~ize** *vi/t* razionalizzare

'rat race *n fam* corsa *f* al successo

rattle /'rætl/ *n* tintinnio *m*; (*toy*) sonaglio *m* ● *vi* tintinnare ● *vt* (*shake*) scuotere; *fam* innervosire. **rattle off** *vt fam* sciorinare

'rattlesnake *n* serpente *m* a sonagli

raucous /'rɔːkəs/ *a* rauco

rave /reɪv/ *vi* vaneggiare; **~ about** andare in estasi per

raven /'reɪvn/ *n* corvo *m* imperiale

ravenous /'rævənəs/ *a* (*person*) affamato

ravine /rə'viːn/ *n* gola *f*

raving /'reɪvɪŋ/ *a* **~ mad** *fam* matto da legare

ravishing /'rævɪʃɪŋ/ *a* incantevole

raw /rɔː/ *a* crudo; (*not processed*) grezzo; (*weather*) gelido; (*inexperienced*) inesperto; **get a ~ deal** *fam* farsi fregare. **~ ma'terials** *npl* materie *fpl* prime

ray /reɪ/ *n* raggio *m*; **~ of hope** barlume *m* di speranza

raze /reɪz/ *vt* **~ to the ground** radere al suolo

razor /'reɪzə(r)/ *n* rasoio *m*. **~ blade** *n* lametta *f* da barba

re /riː/ *prep* con riferimento a

reach /riːtʃ/ *n* portata *f*; **within ~** a portata di mano; **out of ~ of** fuori dalla portata di; **within easy ~** facilmente raggiungibile ● *vt* arrivare a (*place, decision*); (*contact*) contattare; (*pass*) passare; **I can't ~** it non ci arrivo ● *vi* arrivare (**to** a); **~ for** allungare la mano per prendere

re'act /rɪ-/ *vi* reagire

re'action /rɪ-/ *n* reazione *f.* ~ary *a* reazionario, -a *mf*

reactor /rɪ'æktə(r)/ *n* reattore *m*

read /ri:d/ *vt* (*pt/pp* read /red/) leggere; *Univ* studiare ● *vi* leggere; ⟨*instrument:*⟩ indicare. read out *vt* leggere ad alta voce

readable /'ri:dəbl/ *a* piacevole a leggersi; (*legible*) leggibile

reader /'ri:də(r)/ *n* lettore, -trice *mf*; (*book*) antologia *f*

readily /'redɪlɪ/ *adv* volentieri; (*easily*) facilmente. ~ness *n* disponibilità *f inv*; ~ness pronto

reading /'ri:dɪŋ/ *n* lettura *f*

rea'djust /ri:ə-/ *vt* regolare di nuovo ● *vi* riabituarsi (to a)

ready /'redɪ/ *a* (-ier, -iest) pronto; (*quick*) veloce; get ~ prepararsi

ready: ~-'made *a* confezionato. ~ 'money *n* contanti *mpl.* ~-to-'wear *a* prêt-à-porter

real /ri:l/ *a* vero; (*increase*) reale ● *adv Am fam* veramente. ~ esta-te *n* beni *mpl* immobili

realism /'rɪəlɪzm/ *n* realismo *m.* ~t /-lɪst/ *n* realista *mf.* ~tic /-'lɪstɪk/ *a* realistico

reality /rɪ'ælətɪ/ *n* realtà *f inv*

realization /rɪəlaɪ'zeɪʃn/ *n* realizzazione *f*

realize /'rɪəlaɪz/ *vt* realizzare

really /'rɪəlɪ/ *adv* davvero

realm /relm/ *n* regno *m*

realtor /'rɪəltə(r)/ *n Am* agente *mf* immobiliare

reap /ri:p/ *vt* mietere

reap'pear /ri:-/ *vi* riapparire

rear¹ /rɪə/ *n* posteriore *m*; *Auto* di dietro; ~ end *fam* didietro *m* ● *in* the ~ (of *building*) il retro *m*; (of *bus, plane*) la parte *f* posteriore; from the ~ = da dietro

rear² *vt* allevare ● *vi* ~ [up] ⟨*horse:*⟩ impennarsi

'rear-light *n* luce *f* posteriore

re'arm /ri:-/ *vt* riarmare ● *vi* riarmarsi

rear'range /ri:-/ *vt* cambiare la disposizione di

rear-view 'mirror *n Auto* specchietto *m* retrovisore

reason /'ri:zn/ *n* ragione *f*; within ~ nei limiti del ragionevole ● *vi* ragionare; ~ with cercare di far ragionare. ~able /-əbl/ *a* ragionevole. ~ably /-əblɪ/ *adv* (in *reasonable way, fairly*) ragionevolmente

reas'sur|ance /ri:-/ *n* rassicurazione *f.* ~e *vt* rassicurare; ~e sb of sth rassicurare qcno su qcsa. ~ing *a* rassicurante

rebate /'ri:beɪt/ *n* rimborso *m*; (*discount*) deduzione *f*

rebel¹ /'rebl/ *n* ribelle *mf*

rebel² /rɪ'bel/ *vi* (*pt/pp* rebelled) ribellarsi. ~lion /-jən/ *n* ribellione *f.* ~lious /-jəs/ *a* ribelle

re'bound¹ /rɪ-/ *vi* rimbalzare; *fig* ricadere

'rebound² /rɪ-/ *n* rimbalzo *m*

rebuff /rɪ'bʌf/ *n* rifiuto *m*

re'build /ri:-/ *vt* (*pt/pp* -built) ricostruire

rebuke /rɪ'bju:k/ *vt* rimproverare

rebuttal /rɪ'bʌtl/ *n* rifiuto *m*

re'call /rɪ-/ *n* richiamo *m*; beyond ~ irrevocabile ● *vt* richiamare; riconvocare (*diplomat, parliament*); (*remember*) rievocare

recap /'ri:kæp/ *vt/i fam* = recapitulate ● *n* ricapitolazione *f*

recapitulate /ri:kə'pɪtjʊleɪt/ *vt/i* ricapitolare

re'capture /ri:-/ *vt* riconquistare; ricatturare ⟨*person, animal*⟩

recede /rɪ'si:d/ *vi* allontanarsi. ~ing *a* ⟨*forehead, chin*⟩ sfuggente; have ~ing hair essere stempiato

receipt /rɪ'si:t/ *n* ricevuta *f*; (*receiving*) ricezione *f*; ~s *pl Comm* entrate *fpl*

receive /rɪ'si:v/ *vt* ricevere. ~r *n Teleph* ricevitore *m*; *Radio, TV* apparecchio *m* ricevente; (of *stolen goods*) ricettatore, -trice *mf*

recent /'ri:snt/ *a* recente. ~ly *adv* recentemente

receptacle /rɪ'septəkl/ *n* recipiente *m*

reception /rɪ'sepʃn/ n ricevimento m; (welcome) accoglienza f; Radio ricezione f; **~ [desk]** (in hotel) reception f inv. **~ist** n persona f alla reception

receptive /rɪ'septɪv/ a ricettivo

recess /rɪ'ses/ n rientranza f; (holiday) vacanza f; Am Sch intervallo m

recession /rɪ'seʃn/ n recessione f

re'charge /riː-/ vt ricaricare

recipe /'resəpɪ/ n ricetta f

recipient /rɪ'sɪpɪənt/ n (of letter) destinatario, -a mf; (of money) beneficiario, -a mf

reciprocal /rɪ'sɪprəkl/ a reciproco. **~cate** /-kett/ vt ricambiare

recital /rɪ'saɪtl/ n recital m inv

recite /rɪ'saɪt/ vt recitare; (list) elencare

reckless /'reklɪs/ a (action, decision) sconsiderato; **be a ~ driver** guidare in modo spericolato. **~ly** adv in modo sconsiderato. **~ness** n sconsideratezza f

reckon /'rekən/ vt calcolare; (consider) pensare. **reckon on/ with** vt fare i conti con

re'claim /riː-/ vt reclamare; bonificare (land)

recline /rɪ'klaɪn/ vi sdraiarsi. **~ing** a (seat) reclinabile

recluse /rɪ'kluːs/ n recluso, -a mf

recognition /rekəg'nɪʃn/ n riconoscimento m; **beyond ~** irriconoscibile

recognize /'rekəgnaɪz/ vt riconoscere

re'coil /rɪ-/ vi (in fear) indietreggiare

recollect /rekə'lekt/ vt ricordare. **~ion** /-ekʃn/ n ricordo m

recommend /rekə'mend/ vt raccomandare. **~ation** /-'deɪʃn/ n raccomandazione f

recompense /'rekəmpens/ n ricompensa f

reconcile /'rekənsaɪl/ vt riconciliare; conciliare (facts); **~cile oneself to** rassegnarsi a.

~ciliation /-sɪlɪ'eɪʃn/ n riconciliazione f

recondition /riː-/ vt ripristinare. **~ed engine** n motore m che ha subito riparazioni

reconnaissance /rɪ'kɒnɪsns/ n Mil ricognizione f

reconnoitre /rekə'nɔɪtə(r)/ vi (pres p -tring) fare una ricognizione

recon'sider /riː-/ vt riconsiderare

recon'struct /riː-/ vt ricostruire. **~ion** n ricostruzione f

record[1] /rɪ'kɔːd/ vt registrare; (make a note of) annotare

record[2] /'rekɔːd/ n (file) documentazione f; Mus disco m; Sport record m inv; **~s** pl (files) schedario msg; **keep a ~ of** tener nota di; **off the ~** in via ufficiosa; **have a [criminal] ~** avere la fedina penale sporca

recorder /rɪ'kɔːdə(r)/ n Mus flauto m dolce

recording /rɪ'kɔːdɪŋ/ n registrazione f

'record-player n giradischi m inv

re'count /rɪ'kaʊnt/ vt raccontare

're-count[1] /riː-/ vt ricontare

're-count[2] /'riː-/ n Pol nuovo conteggio m

recoup /rɪ'kuːp/ vt rifarsi di (losses)

recourse /rɪ'kɔːs/ n **have ~ to** ricorrere a

re'cover /riː-/ vt rifoderare

recover /rɪ'kʌvə(r)/ vt/i recuperare. **~y** n recupero m; (of health) guarigione f

recreation /rekrɪ'eɪʃn/ n ricreazione f. **~al** a ricreativo

recrimination /rɪkrɪmɪ'neɪʃn/ n recriminazione f

recruit /rɪ'kruːt/ n Mil recluta f; **new ~** (member) nuovo, -a adepto, a mf; (worker) neoassunto, -a mf ● vt assumere (staff). **~ment** n assunzione f

rectangle /'rektæŋgl/ n rettangolo m. **~ular** /-'tæŋgjʊlə(r)/ a rettangolare

rectify /'rektɪfaɪ/ *vt* (*pt/pp* -**ied**) rettificare

recuperate /rɪ'kuːpəreɪt/ *vi* ristabilirsi

recur /rɪ'kɜː(r)/ *vi* (*pt/pp* **recurred**) ricorrere; ⟨*illness:*⟩ ripresentarsi

recurren|ce /rɪ'kʌrəns/ *n* ricorrenza *f*; (*of illness*) ricomparsa *f*. ~**t** *a* ricorrente

recycle /riː'saɪkl/ *vt* riciclare

red /red/ *a* (**redder, reddest**) rosso ●*n* rosso *m*; **in the ~** (*account*) scoperto. **R~ Cross** *n* Croce *f* rossa

redd|en /'redn/ *vt* arrossare ●*vi* arrossire. ~**ish** *a* rossastro

re'decorate /riː-/ *vt* (*paint*) ridipingere; (*wallpaper*) ritappezzare

redeem /rɪ'diːm/ *vt* ~**ing quality** unico aspetto *m* positivo

redemption /rɪ'dempʃn/ *n* riscatto *m*

rede'ploy /riː-/ *vt* ridistribuire

red: ~**-haired** *a* con i capelli rossi. ~**-'handed** *a* **catch sb** ~**-handed** cogliere qcno con le mani nel sacco. ~ **'herring** *n* diversione *f*. ~**-hot** *a* rovente

red: ~ **'light** *n* *Auto* semaforo *m* rosso

re'double /riː-/ *vt* raddoppiare

redress /rɪ'dres/ *n* riparazione *f* ●*vt* ristabilire ⟨*balance*⟩

red 'tape *n fam* burocrazia *f*

reduc|e /rɪ'djuːs/ *vt* ridurre; *Culin* far consumare. ~**tion** /-'dʌkʃn/ *n* riduzione *f*

redundan|cy /rɪ'dʌndənsɪ/ *n* licenziamento *m*; (*payment*) cassa *f* integrazione. ~**t** *a* superfluo; **make** ~**t** licenziare; **be made** ~**t** essere licenziato

reed /riːd/ *n Bot* canna *f*

reef /riːf/ *n* scogliera *f*

reek /riːk/ *vi* puzzare (**of** di)

reel /riːl/ *n* bobina *f* ●*vi* (*stagger*) vacillare. **reel off** *vt fig* snocciolare

refectory /rɪ'fektərɪ/ *n* refettorio *m*; *Univ* mensa *f* universitaria

refer /rɪ'fɜː(r)/ *v* (*pt/pp* **referred**) ●*vt* rinviare ⟨*matter*⟩ (**to** a); indirizzare ⟨*person*⟩ ●*vi* ~ **to** fare allusione a; (*consult*) rivolgersi a ⟨*book*⟩

referee /refə'riː/ *n* arbitro *m*; (*for job*) garante *mf* ●*vt/i* (*pt/pp* **refereed**) arbitrare

reference /'refərəns/ *n* riferimento *m*; (*in book*) nota *f* bibliografica; (*for job*) referenza *f*; *Comm* '**your** ~' 'riferimento'; **with** ~ **to** con riferimento a; **make** [**a**] ~ **to** fare riferimento a. ~ **book** *n* libro *m* di consultazione. ~ **number** *n* numero *m* di riferimento

referendum /refə'rendəm/ *n* referendum *m inv*

re'fill[1] /riː-/ *vt* riempire di nuovo; ricaricare ⟨*pen, lighter*⟩

refill[2] /'riː-/ *n* (*for pen*) ricambio *m*

refine /rɪ'faɪn/ *vt* raffinare. ~**d** *a* raffinato. ~**ment** *n* raffinatezza *f*; *Techn* raffinazione *f*. ~**ry** /-ərɪ/ *n* raffineria *f*

reflect /rɪ'flekt/ *vt* riflettere; **be** ~**ed in** essere riflesso in ●*vi* (*think*) riflettere (**on** su); ~ **badly on sb** *fig* mettere in cattiva luce qcno. ~**ion** /-ekʃn/ *n* riflessione *f*; (*image*) riflesso *m*; **on** ~**ion** dopo riflessione. ~**ive** /-ɪv/ *a* riflessivo. ~**or** *n* riflettore *m*

reflex /'riːfleks/ *n* riflesso *m* ● *attrib* di riflesso

reflexive /rɪ'fleksɪv/ *a* riflessivo

reform /rɪ'fɔːm/ *n* riforma *f* ●*vt* riformare ●*vi* correggersi. **R~ation** /refə'meɪʃn/ *n Relig* riforma *f*. ~**er** *n* riformatore, -trice *mf*

refrain[1] /rɪ'freɪn/ *n* ritornello *m*

refrain[2] /rɪ'freɪn/ *vi* astenersi (**from** da)

refresh /rɪ'freʃ/ *vt* rinfrescare. ~**ing** *a* rinfrescante. ~**ments** *npl* rinfreschi *mpl*

refrigerat|e /rɪ'frɪdʒəreɪt/ *vt* conservare in frigo. ~**or** *n* frigorifero *m*

re'fuel /ri:-/ v (pt/pp -fuelled) ● vt rifornire (di carburante) ● vi fare rifornimento

refuge /'refju:dʒ/ n rifugio m; **take ~** rifugiarsi

refugee /refju'dʒi:/ n rifugiato, -a mf

'refund¹ /'ri:-/ n rimborso m

re'fund² /ri:-/ vt rimborsare

refurbish /ri:'fɜ:bɪʃ/ vt rimettere a nuovo

refusal /rɪ'fju:zl/ n rifiuto m

refuse¹ /rɪ'fju:z/ vt/i rifiutare; **~ to do sth** rifiutare di fare qcsa

refuse² /'refju:s/ n rifiuti mpl. **~ collection** n raccolta f dei rifiuti

refute /rɪ'fju:t/ vt confutare

re'gain /rɪ-/ vt riconquistare

regal /'ri:gl/ a regale

regalia /rɪ'geɪlɪə/ npl insegne fpl reali

regard /rɪ'gɑ:d/ n ⟨heed⟩ riguardo m; ⟨respect⟩ considerazione f; **~s** pl saluti mpl; **send/give my ~s to your brother** salutami tuo fratello ● vt ⟨consider⟩ considerare ⟨as come⟩; **as ~s** riguardo a. **~ing** prep riguardo a. **~less** adv lo stesso; **~ of** senza badare a

regatta /rɪ'gætə/ n regata f

regenerate /rɪ'dʒenəreɪt/ vt rigenerare ● vi rigenerarsi

regime /reɪ'ʒi:m/ n regime m

regiment /'redʒɪmənt/ n reggimento m. **~al** /-'mentl/ a reggimentale. **~ation** /-mən'teɪʃn/ n irreggimentazione f

region /'ri:dʒən/ n regione f; **in the ~ of** fig approssimativamente. **~al** a regionale

register /'redʒɪstə(r)/ n registro m ● vt registrare; mandare per raccomandata ⟨letter⟩; assicurare ⟨luggage⟩; immatricolare ⟨vehicle⟩; mostrare ⟨feeling⟩ ● vi ⟨instrument:⟩ funzionare; ⟨student:⟩ iscriversi (**for** a); **~ with** iscriversi nella lista di ⟨doctor⟩

registrar /redʒɪ'strɑ:(r)/ n ufficiale m di stato civile

registration /redʒɪ'streɪʃn/ n ⟨of vehicle⟩ immatricolazione f; ⟨of letter⟩ raccomandazione f; ⟨of luggage⟩ assicurazione f; ⟨for course⟩ iscrizione f. **~ number** n Auto targa f

registry office /'redʒɪstrɪ-/ n anagrafe f

regret /rɪ'gret/ n rammarico m ● vt (pp regretted) rimpiangere; **I ~ that** mi rincresce che. **~fully** adv con rammarico

regrettab|le /rɪ'gretəbl/ a spiacevole. **~ly** adv spiacevolmente; ⟨before adjective⟩ deplorevolmente

regular /'regjʊlə(r)/ a regolare; ⟨usual⟩ abituale ● n cliente mf abituale. **~ity** /-'lærətɪ/ n regolarità f. **~ly** adv regolarmente

regulat|e /'regjʊleɪt/ vt regolare. **~ion** /-'leɪʃn/ n ⟨rule⟩ regolamento m

rehabilitat|e /ri:hə'bɪlɪteɪt/ vt riabilitare. **~ion** /-'teɪʃn/ n riabilitazione f

rehears|al /rɪ'hɜ:sl/ n Theat prova f. **~e** vt/i provare

reign /reɪn/ n regno m ● vi regnare

reimburse /ri:ɪm'bɜ:s/ vt **~ sb for sth** rimborsare qcsa a qcno

rein /reɪn/ n redine f

reincarnation /ri:ɪnkɑ:'neɪʃn/ n reincarnazione f

reinforce /ri:ɪn'fɔ:s/ vt rinforzare. **~d 'concrete** n cemento m armato. **~ment** n rinforzo m

reinstate /ri:ɪn'steɪt/ vt reintegrare

reiterate /ri:'ɪtəreɪt/ vt reiterare

reject /rɪ'dʒekt/ vt rifiutare. **~ion** /-ekʃn/ n rifiuto m; Med rigetto m

rejoic|e /rɪ'dʒɔɪs/ vi liter rallegrarsi. **~ing** n gioia f

rejuvenate /rɪ'dʒu:vəneɪt/ vt ringiovanire

relapse /rɪ'læps/ n ricaduta f ● vi ricadere

relate /rɪ'leɪt/ vt ⟨tell⟩ riportare; ⟨connect⟩ collegare ● vi **~ to** riferirsi a; identificarsi con ⟨person⟩. **~d** a imparentato (**to** a); ⟨ideas etc⟩ affine

relation /rɪˈleɪʃn/ *n* rapporto *m*; (*person*) parente *mf*. **~ship** *n* rapporto *m* (*blood tie*) parentela *f*; (*affair*) relazione *f*

relative /ˈrelətɪv/ *n* parente *mf* ● *a* relativo. **~ly** *adv* relativamente

relax /rɪˈlæks/ *vt* rilassare; allentare ⟨*pace, grip*⟩ ● *vi* rilassarsi. **~ation** /rɪˌlækˈseɪʃn/ *n* rilassamento *m*, relax *m inv*; (*recreation*) svago *m*. **~ing** *a* rilassante

relay[1] /ˈriːleɪ/ *vt* (*pt/pp* **-laid**) ritrasmettere; *Radio, TV* trasmettere

relay[2] /ˈriːleɪ/ *n Electr* relais *m inv*; **work in ~s** fare i turni. **~ [race]** *n* [corsa *f* a] staffetta *f*

release /rɪˈliːs/ *n* rilascio *m*; (*of film*) distribuzione *f* ● *vt* liberare; lasciare ⟨*hand*⟩; togliere ⟨*brake*⟩; distribuire ⟨*film*⟩; rilasciare (*information etc*)

relegate /ˈrelɪɡeɪt/ *vt* relegare; **be ~d** *Sport* essere retrocesso

relent /rɪˈlent/ *vi* cedere. **~less** *a* inflessibile; (*unceasing*) incessante. **~lessly** *adv* incessantemente

relevan|ce /ˈrelǝvǝns/ *n* pertinenza *f*. **~t** *a* pertinente (**to** a)

reliab|ility /rɪlaɪǝˈbɪlǝtɪ/ *n* affidabilità *f*. **~le** /-ˈlaɪǝbl/ *a* affidabile a. **~ly** *adv* in modo affidabile; **be ~ly informed** sapere da fonte certa

relian|ce /rɪˈlaɪǝns/ *n* fiducia *f* (**on** in). **~t** *a* fiducioso (**on** a)

relic /ˈrelɪk/ *n Relig* reliquia *f*; **~s** *npl* resti *mpl*

relief /rɪˈliːf/ *n* sollievo *m*; (*assistance*) soccorso *m*; (*distraction*) diversivo *m*; (*replacement*) cambio *m*; (*in art*) rilievo *m*; **in ~** in rilievo. **~ map** *n* carta *f* in rilievo. **~ train** *n* treno *m* supplementare

relieve /rɪˈliːv/ *vt* alleviare; (*take over from*) dare il cambio a; **~ of** liberare da ⟨*burden*⟩

religion /rɪˈlɪdʒǝn/ *n* religione *f*

religious /rɪˈlɪdʒǝs/ *a* religioso. **~ly** *adv* (*conscientiously*) scrupolosamente

relinquish /rɪˈlɪŋkwɪʃ/ *vt* abbandonare; **~ sth to sb** rinunciare a qcsa in favore di qcno

relish /ˈrelɪʃ/ *n* gusto *m*; *Culin* salsa *f* ● *vt fig* apprezzare

relo'cate /riː-/ *vt* trasferire

reluctan|ce /rɪˈlʌktǝns/ *n* riluttanza *f*. **~t** *a* riluttante. **~tly** *adv* a malincuore

rely /rɪˈlaɪ/ *vi* (*pt/pp* **-ied**) **~ on** dipendere da; (*trust*) contare su

remain /rɪˈmeɪn/ *vi* restare. **~der** *n* resto *m*. **~ing** *a* restante. **~s** *npl* resti *mpl*; (*dead body*) spoglie *fpl*

remand /rɪˈmɑːnd/ **~ on** in custodia cautelare ● *vt* **~ in custody** rinviare con detenzione provvisoria

remark /rɪˈmɑːk/ *n* osservazione *f* ● *vt* osservare. **~able** /-ǝbl/ *a* notevole. **~ably** *adv* notevolmente

remarry /riː-/ *vi* risposarsi

remedial /rɪˈmiːdɪǝl/ *a* correttivo; *Med* curativo

remedy /ˈremǝdɪ/ *n* rimedio *m* (**for** contro) ● *vt* (*pt/pp* **-ied**) rimediare a

remember /rɪˈmembǝ(r)/ *vt* ricordare, ricordarsi; **~ to do sth** ricordarsi di fare qcsa; **~ me to him** salutamelo *vi* ricordarsi

remind /rɪˈmaɪnd/ *vt* **~ sb of sth** ricordare qcsa a qcno. **~er** *n* ricordo *m*; (*memo*) promemoria *m*; (*letter*) lettera *f* di sollecito

reminisce /remɪˈnɪs/ *vi* rievocare il passato. **~nces** /-ǝnsɪz/ *npl* reminiscenze *fpl*. **~nt** *a* **be ~nt of** richiamare alla memoria

remiss /rɪˈmɪs/ *a* negligente

remission /rɪˈmɪʃn/ *n* remissione *f*; (*of sentence*) condono *m*

remit /rɪˈmɪt/ *vt* (*pt/pp* **remitted**) rimettere ⟨*money*⟩. **~tance** *n* rimessa *f*

remnant /ˈremnǝnt/ *n* resto *m*; (*of material*) scampolo *m*; (*trace*) traccia *f*

remonstrate /ˈremǝnstreɪt/ *vi* fare rimostranze; **~ with sb** fare rimostranze a qcno

remorse /rɪˈmɔːs/ n rimorso m. **~ful** a pieno di rimorso. **~less** a spietato. **~lessly** adv senza pietà

remote /rɪˈməʊt/ a remoto; (slight) minimo. **~ access** n Comput accesso m remoto. **~ conˈtrol** n telecomando m. **~-conˈtrolled** a telecomandato. **~ly** adv lontanamente; **be not ~ly...** non essere lontanamente...

reˈmovable /rɪ-/ a rimovibile

removal /rɪˈmuːvl/ n rimozione f; (from house) trasloco m. **~ van** n camion m inv da trasloco

remove /rɪˈmuːv/ vt togliere; togliersi (clothes); eliminare (stain, doubts)

remuneratˈion /rɪmjuːnəˈreɪʃn/ n rimunerazione f. **~ive** /-ˈmjuːnərətɪv/ a rimunerativo

render /ˈrendə(r)/ vt rendere (service)

rendering /ˈrend(ə)rɪŋ/ n Mus interpretazione f

renegade /ˈrenɪgeɪd/ n rinnegato, -a m f

renew /rɪˈnjuː/ vt rinnovare (contract). **~al** n rinnovo m

renounce /rɪˈnaʊns/ vt rinunciare a

renovatˈe /ˈrenəveɪt/ vt rinnovare. **~ion** /-ˈveɪʃn/ n rinnovo m

renown /rɪˈnaʊn/ n fama f. **~ed** a rinomato

rent /rent/ n affitto m ● vt affittare; **~ [out]** dare in affitto. **~al** n affitto m

renunciation /rɪnʌnsɪˈeɪʃn/ n rinuncia f

reˈopen /riː-/ vt/i riaprire

reˈorganize /riː-/ vt riorganizzare

rep /rep/ n Comm fam rappresentante mf; Theat ≈ teatro m stabile

repair /rɪˈpeə(r)/ n riparazione f, **in good/bad ~** in cattive/buone condizioni ● vt riparare

repatriatˈe /riːˈpætrɪeɪt/ vt rimpatriare. **~ion** /-ˈeɪʃn/ n rimpatrio m

reˈpay /riː-/ vt (pt/pp -paid) ripagare. **~ment** n rimborso m

repeal /rɪˈpiːl/ n abrogazione f ● vt abrogare

repeat /rɪˈpiːt/ n TV replica f ● vt/i ripetere; **~ oneself** ripetersi. **~ed** a ripetuto. **~edly** adv ripetutamente

repel /rɪˈpel/ vt (pt/pp repelled) respingere; fig ripugnare. **~lent** a ripulsivo

repent /rɪˈpent/ vi pentirsi. **~ance** n pentimento m. **~ant** a pentito

repercussions /riːpəˈkʌʃnz/ npl ripercussioni fpl

repertoire /ˈrepətwɑː(r)/ n repertorio m

repetitˈion /repɪˈtɪʃn/ n ripetizione f. **~ive** /rɪˈpetɪtɪv/ a ripetitivo

reˈplace /rɪ-/ vt (put back) rimettere a posto; (take the place of) sostituire; **~ sth with sth** sostituire qcsa con qcsa. **~ment** n sostituzione f; (person) sostituto, -a m f. **~ment part** n pezzo m di ricambio

reˈplay /ˈriː-/ n Sport partita f ripetuta; [action] **~** replay m inv

replenish /rɪˈplenɪʃ/ vt rifornire (stocks); (refill) riempire di nuovo

replica /ˈreplɪkə/ n copia f

reply /rɪˈplaɪ/ n risposta f (to a) ● vt/i (pt/pp replied) rispondere

report /rɪˈpɔːt/ n rapporto m; TV, Radio servizio m; Journ cronaca f; Sch pagella f; (rumour) diceria f ● vt riportare; **~ sb to the police** denunciare qcno alla polizia ● vi riportare; (present oneself) presentarsi (to a). **~edly** adv secondo quanto si dice. **~er** n cronista mf, reporter mf inv

repose /rɪˈpəʊz/ n riposo m

reposˈsess /riː-/ vt riprendere possesso di

reprehensible /reprɪˈhensəbl/ a riprovevole

represent /reprɪˈzent/ vt rappresentare

representative /reprɪˈzentətɪv/ a rappresentativo ● n rappresentante mf

repress /rɪˈpres/ vt reprimere.

~ion /-ɪʃn/ n repressione f. ~ive /-ɪv/ a repressivo

reprieve /rɪ'priːv/ n commutazione f della pena capitale; (postponement) sospensione f della pena capitale; fig tregua f ● vt sospendere la sentenza a; fig risparmiare

reprimand /'reprɪmɑːnd/ n rimprovero m ● vt rimproverare

'reprint¹ /riː-/ n ristampa f

re'print² /riː-/ vt ristampare

reprisal /rɪ'praɪzl/ n rappresaglia f; in ~ for per rappresaglia contro

reproach /rɪ'prəʊtʃ/ n ammonimento m ● vt ammonire. ~ful a di provevole. ~fully adv con aria di rimprovero

repro'duc|e /riː-/ vt riprodurre ● vi riprodursi. ~tion /-'dʌkʃn/ n riproduzione f. ~tive /-'dʌktɪv/ a riproduttivo

reprove /rɪ'pruːv/ vt rimproverare

reptile /'reptaɪl/ n rettile m

republic /rɪ'pʌblɪk/ n repubblica f. ~an a repubblicano ● n repubblicano, -a mf

repudiate /rɪ'pjuːdɪeɪt/ vt ripudiare; respingere (view, suggestion)

repugnan|ce /rɪ'pʌgnəns/ n ripugnanza f. ~t a ripugnante

repuls|ion /rɪ'pʌlʃn/ n repulsione f. ~ive /-ɪv/ a ripugnante

reputable /'repjʊtəbl/ a affidabile

reputation /repjʊ'teɪʃn/ n reputazione f

repute /rɪ'pjuːt/ n reputazione f. ~d /-ɪd/ a presunto; he is ~d to be si presume che sia. ~dly adv presumibilmente

request /rɪ'kwest/ n richiesta f ● vt richiedere. ~ stop n fermata f a richiesta

require /rɪ'kwaɪə(r)/ vt (need) necessitare di; (demand) esigere. ~d a richiesto; I am ~d to do si esige che io faccia. ~ment n esigenza f; (condition) requisito m

requisite /'rekwɪzɪt/ a necessario ● n toilet/travel ~s pl articoli mpl da toilette/viaggio

re'sale /riː-/ n rivendita f

rescue /'reskjuː/ n salvataggio m ● vt salvare. ~r n salvatore, -trice mf

research /rɪ'sɜːtʃ/ n ricerca f ● vt fare ricerche su; Journ fare un'inchiesta su ● vi ~ into fare ricerche su. ~er n ricercatore, -trice mf

resemb|lance /rɪ'zembləns/ n rassomiglianza f. ~le /-bl/ vt rassomigliare a

resent /rɪ'zent/ vt risentirsi per. ~ful a pieno di risentimento. ~fully adv con risentimento. ~ment n risentimento m

reservation /rezə'veɪʃn/ n (booking) prenotazione f; (doubt, enclosure) riserva f

reserve /rɪ'zɜːv/ n riserva f; (shyness) riserbo m ● vt riservare; riservarsi (right). ~d a riservato

reservoir /'rezəvwɑː(r)/ n bacino m idrico

re'shape /riː-/ vt ristrutturare

re'shuffle /riː-/ n Pol rimpasto m ● vt Pol rimpastare

reside /rɪ'zaɪd/ vi risiedere

residence /'rezɪdəns/ n residenza f; (stay) soggiorno m. ~ permit n permesso m di soggiorno

resident /'rezɪdənt/ a residente ● n residente mf. ~ial /-'denʃl/ a residenziale

residue /'rezɪdjuː/ n residuo m

resign /rɪ'zaɪn/ vt dimettersi da; ~ oneself to rassegnarsi a ● vi dare le dimissioni. ~ation /rezɪg'neɪʃn/ n rassegnazione f; (from job) dimissioni fpl. ~ed a rassegnato

resilien|ce /rɪ'zɪlɪənt/ n elasticità f; fig con buone capacità di ripresa

resin /'rezɪn/ n resina f

resist /rɪ'zɪst/ vt resistere a ● vi resistere. ~ance n resistenza f. ~ant a resistente

resolute /'rezəluːt/ a risoluto. ~ely adv con risolutezza. ~ion /-'luːʃn/ n risolutezza f

resolve /rɪ'zɒlv/ vt ~ **to do** decidere di fare

resonan|ce /'rezənəns/ n risonanza f. ~**t** a risonante

resort /rɪ'zɔːt/ n (place) luogo m di villeggiatura; **as a last** ~ come ultima risorsa ● vi ~ **to** ricorrere a

resound /rɪ'zaʊnd/ vi risonare (**with** di). ~**ing** (success) risonante

resource /rɪ'sɔːs/ n ~**s** pl risorse fpl. ~**ful** a pieno di risorse; (solution) ingegnoso. ~**fulness** n ingegnosità f inv

respect /rɪ'spekt/ n rispetto m; (aspect) aspetto m; **with** ~ **to** per quanto riguarda ● vt rispettare

respectab|ility /rɪspektə'bɪlətɪ/ n rispettabilità f inv

respect|able /rɪ'spektəbl/ a rispettabile. ~**ably** adv rispettabilmente. ~**ful** a rispettoso

respective /rɪ'spektɪv/ a rispettivo. ~**ly** adv rispettivamente

respiration /respɪ'reɪʃn/ n respirazione f

respite /'respaɪt/ n respiro m

respond /rɪ'spɒnd/ vi rispondere; (react) reagire (**to** a); (patient:) rispondere (**to** a)

response /rɪ'spɒns/ n risposta f. (reaction) reazione f

responsibility /rɪspɒnsɪ'bɪlətɪ/ n responsabilità f inv

responsib|le /rɪ'spɒnsəbl/ a responsabile; (job) impegnativo

responsive /rɪ'spɒnsɪv/ a **be** ~ (audience etc:) reagire; (brakes:) essere sensibile

rest¹ /rest/ n riposo m; Mus pausa f; **have a** ~ riposarsi ● vt riposare; (lean) appoggiare (**on** su); (place) appoggiare ● vi riposarsi; (elbows:) appoggiarsi; (hopes:) riposare

rest² n **the** ~ il resto m; (people) gli altri mpl ● vi **it** ~**s with you** sta a te

restaurant /'restərɒnt/ n ristorante m. ~ **car** n vagone m ristorante

restful /'restfl/ a riposante

restive /'restɪv/ a irrequieto

restless /'restlɪs/ a nervoso

restoration /restə'reɪʃn/ n (of building) restauro m

restore /rɪ'stɔː(r)/ vt ristabilire; restaurare (building); (give back) restituire

restrain /rɪ'streɪn/ vt trattenere; ~ **oneself** controllarsi. ~**ed** a controllato. ~**t** n restrizione f; (moderation) ritegno m

restrict /rɪ'strɪkt/ vt limitare; ~ **to** limitarsi a. ~**ion** /-kʃn/ n limitazione m; (restraint) restrizione f. ~**ive** /-ɪv/ a limitativo

'rest room n Am toilette f inv

result /rɪ'zʌlt/ n risultato m; **as a** ~ a causa (**of** di) ● vi ~ **from** risultare da; ~ **in** portare a

resume /rɪ'zjuːm/ vt/i riprendere

résumé /'rezjʊmeɪ/ n riassunto m; Am curriculum vitae m inv

resumption /rɪ'zʌmpʃn/ n ripresa f

resurgence /rɪ'sɜːdʒəns/ n rinascita f

resurrect /rezə'rekt/ vt fig risuscitare. ~**ion** /-ekʃn/ n **the R~ion** Relig la Risurrezione

resuscitat|e /rɪ'sʌsɪteɪt/ vt rianimare. ~**ion** /-'teɪʃn/ n rianimazione f

retail /'riːteɪl/ n vendita f al minuto o al dettaglio ● a & adv al minuto ● vt vendere al minuto ● vi ~ **at** essere venduto al pubblico al prezzo di. ~**er** n dettagliante mf

retain /rɪ'teɪn/ vt conservare; (hold back) trattenere

retaliat|e /rɪ'tælɪeɪt/ vi vendicarsi. ~**ion** /-'eɪʃn/ n rappresaglia f; **in** ~**ion for** per rappresaglia contro

retarded /rɪ'tɑːdɪd/ a ritardato

retentive /rɪ'tentɪv/ a (memory) buono

rethink /riː'θɪŋk/ vt (pt/pp rethought) ripensare

reticen|ce /'retɪsəns/ n reticenza f. ~**t** a reticente

retina /'retɪnə/ n retina f

retinue /'retɪnjuː/ n seguito m

retire /rɪˈtaɪə(r)/ vi andare in pensione; (withdraw) ritirarsi ● vt mandare in pensione (employee). ~d a in pensione. ~ment n pensione f; **since my ~ment** da quando sono andato in pensione

retiring /rɪˈtaɪərɪŋ/ a riservato

retort /rɪˈtɔːt/ n replica f ● vt ribattere

re'touch /riː-/ vt Phot ritoccare

re'trace /rɪ-/ vt ripercorrere; ~ **one's steps** ritornare sui propri passi

retract /rɪˈtrækt/ vt ritirare; ritrattare (statement, evidence) ● vi ritrarsi

re'train /riː-/ vt riqualificare ● vi riqualificarsi

retreat /rɪˈtriːt/ n ritirata f; (place) ritiro m ● vi ritirarsi; Mil battere in ritirata

re'trial /riː-/ n nuovo processo m

retribution /retrɪˈbjuːʃn/ n castigo m

retrieval /rɪˈtriːvəl/ n recupero m

retrieve /rɪˈtriːv/ vt recuperare

retrograde /ˈretrəɡreɪd/ a retrogrado

retrospect /ˈretrəspekt/ n **in ~** guardando indietro. ~**ive** /-ˈspektɪv/ a retrospettivo; (legislation) retroattivo ● n retrospettiva f

return /rɪˈtɜːn/ n ritorno m; (giving back) restituzione f; Comm profitto m; (ticket) biglietto m di andata e ritorno; **by ~** [of post] a stretto giro di posta; **in ~** in cambio (for di); **many happy ~s!** cento di questi giorni! ● vi ritornare ● vt (give back) restituire; ricambiare (affection, invitation); (put back) rimettere; (send back) mandare indietro; (elect) eleggere

return: ~ flight n volo m di andata e ritorno. **~ match** n rivincita f. **~ ticket** n biglietto m di andata e ritorno

reunion /riːˈjuːnjən/ n riunione f

reunite /riːjʊˈnaɪt/ vt riunire

re'us|able /riː-/ a riutilizzabile. ~**e** vt riutilizzare

rev /rev/ n Auto, fam giro m (di motore) ● v (pt/pp revved) ● vt ~ **[up]** far andare su di giri ● vi andare su di giri

reveal /rɪˈviːl/ vt rivelare; (dress:) scoprire. ~**ing** a rivelatore; (dress) osé

revel /ˈrevl/ vi (pt/pp revelled) ~ **in sth** godere di qcsa

revelation /revəˈleɪʃn/ n rivelazione f

revelry /ˈrevlrɪ/ n baldoria f

revenge /rɪˈvendʒ/ n vendetta f; Sport rivincita f; **take ~** vendicarsi ● vt vendicare

revenue /ˈrevənjuː/ n reddito m

reverberate /rɪˈvɜːbəreɪt/ vi riverberare

revere /rɪˈvɪə(r)/ vt riverire. ~**nce** /ˈrevərəns/ n riverenza f

Reverend /ˈrevərənd/ a reverendo

reverent /ˈrevərənt/ a riverente

reverse /rɪˈvɜːs/ a opposto; **in ~ order** in ordine inverso ● n contrario m; (back) rovescio m; Auto marcia m indietro ● vt invertire; ~ **the car into the garage** entrare in garage a marcia indietro; ~ **the charges** Teleph fare una telefonata a carico ● vi Auto fare marcia indietro

revert /rɪˈvɜːt/ vi ~ **to** tornare a

review /rɪˈvjuː/ n (survey) rassegna f; (re-examination) riconsiderazione f; Mil rivista f; (of book, play) recensione f ● vt riesaminare (situation); Mil passare in rivista; recensire (book, play). ~**er** n critico, -a mf

revile /rɪˈvaɪl/ vt ingiuriare

revis|e /rɪˈvaɪz/ vt rivedere; (for exam) ripassare. ~**ion** /-ˈvɪʒn/ n revisione f; (for exam) ripasso m

revival /rɪˈvaɪvl/ n ritorno m; (of patient) recupero m; (from coma) risveglio m

revive /rɪˈvaɪv/ vt resuscitare; rianimare (person) ● vi riprendersi; (person:) rianimarsi

revoke /rɪ'vəʊk/ vt revocare

revolt /rɪ'vəʊlt/ n rivolta f ● vi ribellarsi ● vt rivoltare. **~ing** a rivoltante

revolution /revə'luːʃn/ n rivoluzione f; Auto **~s** per minute giri mpl al minuto. **~ary** /-ərɪ/ a & n rivoluzionario, -a mf. **~ize** vt rivoluzionare

revolve /rɪ'vɒlv/ vi ruotare; **~ around** girare intorno

revolv|er /rɪ'vɒlvə(r)/ n rivoltella f, revolver m inv. **~ing** a ruotante

revue /rɪ'vjuː/ n rivista f

revulsion /rɪ'vʌlʃn/ n ripulsione f

reward /rɪ'wɔːd/ n ricompensa f ● vt ricompensare. **~ing** a gratificante

re'write /riː-/ vt (pt rewrote, pp rewritten) riscrivere

rhapsody /'ræpsədɪ/ n rapsodia f

rhetoric /'retərɪk/ n retorica f. **~al** /rɪ'tɒrɪkl/ a retorico

rheuma|tic /ruː'mætɪk/ a reumatico. **~tism** /'ruːmətɪzm/ n reumatismo m

Rhine /raɪn/ n Reno m

rhinoceros /raɪ'nɒsərəs/ n rinoceronte m

rhubarb /'ruːbɑːb/ n rabarbaro m

rhyme /raɪm/ n rima f; (poem) filastrocca f ● vi rimare

rhythm /'rɪðm/ n ritmo m. **~ic[al]** a ritmico. **~ically** adv con ritmo

rib /rɪb/ n costola f

ribald /'rɪbld/ a spinto

ribbon /'rɪbən/ n nastro m; **in ~s** a brandelli

rice /raɪs/ n riso m

rich /rɪtʃ/ a ricco; (food) pesante ● n the **~ pl** i ricchi mpl; **~es** pl ricchezze fpl. **~ly** adv riccamente; (deserve) largamente

rickety /'rɪkɪtɪ/ a malfermo

ricochet /'rɪkəʃeɪ/ vi rimbalzare ● n rimbalzo m

rid /rɪd/ vt (pt/pp rid, pres p ridding) sbarazzare (of di); **get ~ of** sbarazzarsi di

riddance /'rɪdns/ n **good ~!** che liberazione!

ridden /'rɪdn/ see ride

riddle /'rɪdl/ n enigma m

riddled /'rɪdld/ a **~ with** crivellato di

ride /raɪd/ n (on horse) cavalcata f; (in vehicle) giro m; (journey) viaggio m; **take sb for a ~ fam** prendere qcno in giro ● v (pt rode, pp ridden) ● vt montare (horse); andare su (bicycle) ● vi andare a cavallo; (jockey, showjumper:) cavalcare; (cyclist:) andare in bicicletta; (in vehicle) viaggiare. **~r** n cavallerizzo, -a mf; (in race) fantino m; (on bicycle) ciclista mf; (in document) postilla f

ridge /rɪdʒ/ n spigolo m; (on roof) punta f; (of mountain) cresta f

ridicule /'rɪdɪkjuːl/ n ridicolo m ● vt mettere in ridicolo

ridiculous /rɪ'dɪkjʊləs/ a ridicolo

riding /'raɪdɪŋ/ n equitazione f ● attrib d'equitazione

rife /raɪf/ a **be ~** essere diffuso; **~ with** pieno di

riff-raff /'rɪfræf/ n marmaglia f

rifle /'raɪfl/ n fucile m; **~-range** tiro m al bersaglio ● vt **~ [through]** mettere a soqquadro

rift /rɪft/ n fessura f; fig frattura f

rig¹ /rɪg/ n equipaggiamento m; (at sea) piattaforma f per trivellazioni subacquee ● rig out vt (pt/pp rigged) equipaggiare. **rig up** vt allestire

rig² vt (pt/pp rigged) manovrare (election)

right /raɪt/ a giusto; (not left) destro; **be ~** (person:) aver ragione; (clock:) essere giusto; **put ~** mettere all'ora (clock); correggere (person); rimediare a (situation); **that's ~!** proprio cosi! ● adv (correctly) bene; (not left) a destra; (directly) proprio; (completely) completamente; **~ away** immediatamente ● n giusto m; (not left) destra f; (what is due) diritto m; **on/to the ~** a destra; **be in the ~** essere nel giusto; **know ~ from wrong** distinguere il bene dal male; **by ~s** secondo giustizia; **the R~** *Pol* la destra *f* ● *vt* raddrizzare; **~ a wrong** *fig* riparare a un torto. **~ angle** *n* angolo *m* retto

rightful /ˈraɪtfl/ *a* legittimo

right: **~-'handed** *a* che usa la mano destra. **~-hand 'man** *n fig* braccio *m* destro

rightly /ˈraɪtlɪ/ *adv* giustamente

right: **~ of way** *n* diritto *m* di transito; (*path*) passaggio *m*. **~'wing** *a Pol* di destra. ● *n Sport* ala *f* destra

rigid /ˈrɪdʒɪd/ *a* rigido. **~ity** /-ˈdʒɪdətɪ/ *n* rigidità *f inv*

rigmarole /ˈrɪgmərəʊl/ *n* trafila *f*; (*story*) tiritera *f*

rigorous /ˈrɪgərəs/ *a* rigoroso

rile /raɪl/ *vt fam* irritare

rim /rɪm/ *n* bordo *m*; (*of wheel*) cerchione *m*

rind /raɪnd/ *n* (*on fruit*) scorza *f*; (*on cheese*) crosta *f*; (*on bacon*) cotenna *f*

ring[1] /rɪŋ/ *n* (*circle*) cerchio *m*; (*on finger*) anello *m*; (*boxing*) ring *m inv*; (*for circus*) pista *f*; **stand in a ~** essere in cerchio

ring[2] *n* suono *m*; **give sb a ~** *Teleph* dare un colpo di telefono a qcno ● *v* (*pt* rang, *pp* rung) ● *vt* suonare; **~ [up]** *Teleph* telefonare a ● *vi* suonare; *Teleph* **~ [up]** telefonare. **ring back** *vt/i Teleph* richiamare. **ring off** *vi Teleph* riattaccare

ring: **~leader** *n* capobanda *m*. **~ road** *n* circonvallazione *f*

rink /rɪŋk/ *n* pista *f* di pattinaggio

rinse /rɪns/ *n* risciacquo *m*; (*hair colour*) cachet *m inv* ● *vt* sciacquare

riot /ˈraɪət/ *n* rissa *f*; (*of colour*) accozzaglia *f*; **~s** *pl* disordini *mpl*; **run ~** impazzare ● *vi* creare disordini. **~er** *n* dimostrante *mf*. **~ous** /-əs/ *a* sfrenato

rip /rɪp/ *n* strappo *m* ● *vt* (*pt/pp* ripped) strappare; **~ open** aprire con uno strappo. **rip off** *vt fam* fregare

ripe /raɪp/ *a* maturo; (*cheese*) stagionato

ripen /ˈraɪpn/ *vi* maturare; (*cheese:*) stagionarsi ● *vt* far maturare; stagionare (*cheese*)

ripeness /ˈraɪpnɪs/ *n* maturità *f inv*

'rip-off *n fam* frode *f*

ripple /ˈrɪpl/ *n* increspatura *f*; (*sound*) mormorio *m* ●

rise /raɪz/ *n* (*of sun*) levata *f*; (*fig:* to fame, power) ascesa *f*; (*increase*) aumento *m*; **give ~ to** dare adito a ● *vi* (*pt* rose, *pp* risen) alzarsi; (*sun:*) sorgere; (*dough:*) lievitare; (*prices, water level:*) aumentare; (*to power, position*) arrivare (**to** a). **~r** *n* **early ~r** persona *f* mattiniera

rising /ˈraɪzɪŋ/ *a* (*sun*) levante; **~ generation** nuova generazione *f* ● *n* (*revolt*) sollevazione *f*

risk /rɪsk/ *n* rischio *m*; **at one's own ~** a proprio rischio e pericolo ● *vt* rischiare

risky /ˈrɪskɪ/ *a* (-ier, -iest) rischioso

risqué /ˈrɪskeɪ/ *a* spinto

rite /raɪt/ *n* rito *m*; **last ~s** estrema unzione *f*

ritual /ˈrɪtjʊəl/ *a* rituale ● *n* rituale *m*

rival /ˈraɪvl/ *a* rivale ● *n* rivale *mf*; **~s** *pl Comm* concorrenti *mpl* ● *vt* (*pt/pp* rivalled) rivaleggiare con. **~ry** *n* rivalità *f inv*; *Comm* concorrenza *f*

river /ˈrɪvə(r)/ *n* fiume *m*. **~-bed** *n* letto *m* del fiume

rivet /ˈrɪvɪt/ *n* rivetto *m* ● *vt* rivettare; **~ed by** *fig* inchiodato da

Riviera /rɪvɪˈeərə/ *n* **the Italian ~** la riviera ligure

road /rəʊd/ *n* strada *f*, via *f*; **be on the ~** viaggiare

road: **~-block** *n* blocco *m* stradale. **~-hog** *n fam* pirata *m* della strada. **~-map** *n* carta *f* stradale. **~ safety** *n* sicurezza *f* sulle strade. **~ sense** *n* prudenza *f* (per stra-

da). **~side** *n* bordo *m* della strada. **~sign** cartello *m* stradale. **~way** *n* carreggiata *f,* corsia *f.* **~works** *npl* lavori *mpl* stradali. **~worthy** *a* sicuro

roam /rəʊm/ *vi* girovagare

roar /rɔː(r)/ *n* ruggito *m;* **~s of laughter** scroscio *msg* di risa ● *vi* ruggire; ⟨*lorry, thunder:*⟩ rombare; **~ with laughter** ridere fragorosamente. **~ing** *a* **do a ~ing trade** *fam* fare affari d'oro

roast /rəʊst/ *a* arrosto; **~ pork** arrosto *m* di maiale ● *n* arrosto *m* ● *vt* arrostire ⟨*meat*⟩ ● *vi* arrostirsi

rob /rɒb/ *vt (pt/pp* **robbed***)* derubare (**of** di); svaligiare ⟨*bank*⟩. **~ber** *n* rapinatore *m.* **~bery** *n* rapina *f*

robe /rəʊb/ *n* tunica *f;* (*Am: bathrobe*) accappatoio *m*

robin /'rɒbɪn/ *n* pettirosso *m*

robot /'rəʊbɒt/ *n* robot *m inv*

robust /rəʊ'bʌst/ *a* robusto

rock[1] /rɒk/ *n* roccia *f;* (*in sea*) scoglio *m;* (*sweet*) zucchero *m* candito. **on the ~s** ⟨*ship:*⟩ incagliato; ⟨*marriage:*⟩ finito; ⟨*drink:*⟩ con ghiaccio

rock[2] *vt* cullare ⟨*baby:*⟩; (*shake*) far traballare; (*shock*) scuotere ● *vi* dondolarsi

rock[3] *n Mus* rock *m inv*

rock-'bottom *a* bassissimo ● *n* livello *m* più basso

rockery /'rɒkərɪ/ *n* giardino *m* roccioso

rocket /'rɒkɪt/ *n* razzo *m* ● *vi* salire alle stelle

rocking /'rɒkɪŋ/: **~-chair** *n* sedia *f* a dondolo. **~-horse** *n* cavallo *m* a dondolo

rocky /'rɒkɪ/ *a* (**-ier, -iest**) roccioso; *fig* traballante

rod /rɒd/ *n* bacchetta *f;* (*for fishing*) canna *f*

rode /rəʊd/ *see* **ride**

rodent /'rəʊdnt/ *n* roditore *m*

roe /rəʊ/ *n* (*pl* **roe** o **roes**) **~[-deer]** capriolo *m*

rogue /rəʊg/ *n* farabutto *m*

role /rəʊl/ *n* ruolo *m*

roll /rəʊl/ *n* rotolo *m;* (*bread*) panino *m;* (*list*) lista *f;* (*of ship, drum*) rullio *m* ● *vi* rotolare; **be ~ing in money** *fam* nuotare nell'oro ● *vt* spianare ⟨*lawn, pastry*⟩. **roll over** *vi* rigirarsi. **roll up** *vt* arrotolare; rimboccarsi ⟨*sleeves*⟩ ● *vi* *fam* arrivare

'roll-call *n* appello *m*

roller /'rəʊlə(r)/ *n* rullo *m;* (*for hair*) bigodino *m.* **~ blind** *n* tapparella *f.* **~-coaster** *n* montagne *fpl* russe. **~-skate** *n* pattino *m* a rotelle

'rolling-pin *n* mattarello *m*

Roman /'rəʊmən/ *a* romano ● *n* romano, -a *mf.* **~ Catholic** *a* cattolico ● *n* cattolico, -a *mf*

romance /rəʊ'mæns/ *n* (*love-affair*) storia *f* d'amore; (*book*) romanzo *m* rosa

Romania /rəʊ'meɪnɪə/ *n* Romania *f.* **~n** *a* rumeno ● *n* rumeno, -a *mf*

romantic /rəʊ'mæntɪk/ *a* romantico. **~ally** *adv* romanticamente. **~ism** /-tɪsɪzm/ *n* romanticismo *m*

Rome /rəʊm/ *n* Roma *f*

romp /rɒmp/ *n* gioco *m* rumoroso ● *vi* giocare rumorosamente. **~ers** *npl* pagliaccetto *msg*

roof /ruːf/ *n* tetto *m;* (*of mouth*) palato *m* ● *vt* mettere un tetto su. **~-rack** *n* portabagagli *m inv.* **~-top** *n* tetto *m*

rook /rʊk/ *n* corvo *m;* (*in chess*) torre *f*

room /ruːm/ *n* stanza *f;* (*bedroom*) camera *f;* (*for functions*) sala *f;* (*space*) spazio *m.* **~y** *a* spazioso; ⟨*clothes*⟩ ampio

roost /ruːst/ *vi* appollaiarsi

root[1] /ruːt/ *n* radice *f.* **take ~** mettere radici ● *vt* **root out** *vt fig* scovare

root[2] *vi* **~ about** grufolare; **~ for sb** *Am* fare il tifo per qcno

rope /rəʊp/ *n* corda *f;* **know the ~s** *fam* conoscere i trucchi del mestiere ● **rope in** *vt fam* coinvolgere

rosary /'rəʊzərɪ/ n rosario m
rose[1] /rəʊz/ n rosa f; (of watering-can) bocchetta f
rose[2] see **rise**
rosé /'rəʊzeɪ/ n [vino m] rosé m inv
rosemary /'rəʊzmərɪ/ n rosmarino m
rosette /rəʊ'zet/ n coccarda f
roster /'rɒstə(r)/ n tabella f dei turni
rostrum /'rɒstrəm/ n podio m
rosy /'rəʊzɪ/ a (-ier, -iest) roseo
rot /rɒt/ n marciume m; (fam: nonsense) sciocchezze fpl ● vi (pt/pp rotted) marcire
rota /'rəʊtə/ n tabella f dei turni
rotary /'rəʊtərɪ/ a rotante
rotate /rəʊ'teɪt/ vt far ruotare; avvicendare (crops) ● vi ruotare. **~ion** /-eɪʃn/ n rotazione f; **in ~ion** a turno
rote /rəʊt/ n **by ~** meccanicamente
rotten /'rɒtn/ a marcio; fam schifoso; (person) penoso
rotund /rəʊ'tʌnd/ a paffuto
rough /rʌf/ a (not smooth) ruvido; (ground) accidentato; (behaviour) rozzo; (sport) violento; (area) malfamato; (crossing, time) brutto; (estimate) approssimativo ● adv (play) grossolanamente; **sleep ~** dormire sotto i ponti ● vt ~ **it** vivere senza comfort. **rough out** vt abbozzare
roughage /'rʌfɪdʒ/ n fibre fpl
rough 'draft n abbozzo m
rough|ly /'rʌflɪ/ adv rozzamente; (more or less) pressappoco. **~ness** n ruvidità f; (of behaviour) rozzezza f
rough paper n carta f da brutta
roulette /ru:'let/ n roulette f inv
round /raʊnd/ a rotondo ● n tondo m; (slice) fetta f; (of visits, drinks) giro m; (of competition) partita f; (boxing) ripresa f, round m inv; **do one's ~s** (doctor:) fare il giro delle visite ● prep intorno a; **open ~ the clock** aperto ventiquatt'ore ● adv **all** ~ tutt'intorno; **ask sb ~** invitare qcno; **go/come ~ to** (a

friend etc) andare da; **turn/look ~** girarsi; ~ **about** (approximately) intorno a ● vt arrotondare; girare (corner). **round down** vt arrotondare (per difetto). **round off** vt (end) terminare. **round on** vt aggredire. **round up** vt radunare; arrotondare (prices)
roundabout /'raʊndəbaʊt/ a indiretto ● n giostra f; (for traffic) rotonda f
round: ~ 'trip n viaggio m di andata e ritorno
rous|e /raʊz/ vt svegliare; risvegliare (suspicion, interest). **~ing** a di incoraggiamento
route /ru:t/ n itinerario m; Naut, Aeron rotta f; (of bus) percorso m
routine /ru:'ti:n/ a di routine ● n routine f inv; Theat numero m
rov|e /rəʊv/ vi girovagare. **~ing** a (reporter, ambassador) itinerante
row[1] /rəʊ/ n (line) fila f; **three years in a ~** tre anni di fila
row[2] vi (in boat) remare
row[3] /raʊ/ n fam (quarrel) litigata f; (noise) baccano m ● vi fam litigare
rowdy /'raʊdɪ/ a (-ier, -iest) chiassoso
rowing boat /'rəʊɪŋ-/ n barca f a remi
royal /'rɔɪəl/ a reale
royal|ty /'rɔɪəltɪ/ n appartenenza f alla famiglia reale; (persons) i membri mpl della famiglia reale. **~ies** npl (payments) diritti mpl d'autore
rpm abbr **revolutions per minute**
rub /rʌb/ n **give sth a ~** dare una sfregata a qcsa ● vt (pt/pp rubbed) sfregare. **rub in** vt **don't ~ it in** fam non rigirare il coltello nella piaga. **rub off** vt mandar via sfregando (stain); (from blackboard) cancellare ● vi andar via; ~ **off on** essere trasmesso a. **rub out** vt cancellare
rubber /'rʌbə(r)/ n gomma f; (eraser) gomma f [da cancellare].

~ band n elastico m. **~y** a gommoso

rubbish /'rʌbɪʃ/ n immondizie fpl; ⟨fam: nonsense⟩ idiozie fpl; ⟨fam: junk⟩ robaccia f ● vt fam fare a pezzi. **~ bin** n pattumiera f. **~ dump** n discarica f; ⟨official⟩ discarica f comunale

rubble /'rʌbl/ n macerie fpl

ruby /'ru:bɪ/ n rubino m ● attrib di rubini; ⟨lips⟩ scarlatta

rucksack /'rʌksæk/ n zaino m

rudder /'rʌdə(r)/ n timone m

ruddy /'rʌdɪ/ a (**-ier, -iest**) rubicondo; fam maledetto

rude /ru:d/ a scortese; ⟨improper⟩ spinto. **~ly** adv scortesemente. **~ness** n scortesia f

rudiment /'ru:dɪmənt/ n **~s** pl rudimenti mpl. **~ary** /-'mentərɪ/ a rudimentale

rueful /'ru:fl/ a rassegnato

ruffian /'rʌfɪən/ n farabutto m

ruffle /'rʌfl/ n gala f ● vt scompigliare ⟨hair⟩

rug /rʌg/ n tappeto m; ⟨blanket⟩ coperta f

rugby /'rʌgbɪ/ n — [**football**] rugby m

rugged /'rʌgɪd/ a ⟨coastline⟩ roccioso

ruin /'ru:ɪn/ n rovina f; **in ~s** in rovina ● vt rovinare. **~ous** /-əs/ a estremamente costoso

rule /ru:l/ n regola f; ⟨control⟩ ordinamento m; ⟨for measuring⟩ metro m; **~s** regolamento msg; **as a ~** generalmente ● vt governare; dominare ⟨colony, behaviour⟩; **~ that** stabilire che ● vi governare.
rule out vt escludere

ruled /ru:ld/ a ⟨paper⟩ a righe

ruler /'ru:lə(r)/ n capo m di Stato; ⟨sovereign⟩ sovrano, -a mf; ⟨measure⟩ righello m, regolo m

ruling /'ru:lɪŋ/ a ⟨class⟩ dirigente; ⟨party⟩ di governo ● n decisione f

rum /rʌm/ n rum m inv

rumble /'rʌmbl/ n rombo m; ⟨of stomach⟩ brontolio m ● vi rombare; ⟨stomach:⟩ brontolare

rummage /'rʌmɪdʒ/ vi rovistare (**in/through** in)

rummy /'rʌmɪ/ n ramino m

rumour /'ru:mə(r)/ n diceria f ● vt **it is ~ed that** si dice che

rump /rʌmp/ n natiche fpl. **~ steak** n bistecca f di girello

rumpus /'rʌmpəs/ n fam baccano m

run /rʌn/ n ⟨on foot⟩ corsa f; ⟨distance to be covered⟩ tragitto m; ⟨outing⟩ giro m; Theat rappresentazioni fpl; ⟨in skiing⟩ pista f; ⟨Am: ladder⟩ smagliatura f ⟨in calze⟩; **at a ~** di corsa; **of bad luck** periodo m sfortunato; **on the ~** in fuga; **have the ~** of avere a disposizione ● v (pt **ran**, pp **run**, pres p **running**) ● vi correre; ⟨river:⟩ scorrere; ⟨nose, makeup:⟩ colare; ⟨bus:⟩ fare servizio; ⟨play:⟩ essere in cartellone; ⟨colours:⟩ sbiadire; ⟨in election⟩ presentarsi [come candidato] ● vt ⟨manage⟩ dirigere; tenere ⟨house⟩; ⟨drive⟩ dare un passaggio a; correre ⟨risk⟩; Comput lanciare; Journ pubblicare ⟨article⟩; ⟨pass⟩ far scorrere ⟨eyes, hand⟩; **~ a bath** far scorrere l'acqua per il bagno. **run across** vi ⟨meet, find⟩ imbattersi in. **run away** vi scappare [via]. **run down** vi scaricarsi; ⟨clock:⟩ scaricarsi; ⟨stocks:⟩ esaurirsi ● vt Auto investire; ⟨reduce⟩ esaurire; ⟨fam: criticize⟩ denigrare. **run in** vi entrare di corsa. **run into** vt ⟨meet⟩ imbattersi in; ⟨knock against⟩ urtare. **run off** vi andare via di corsa ● vt stampare ⟨copies⟩. **run out** vi uscire di corsa; ⟨supplies, money:⟩ esaurirsi; **~ out of** rimanere senza. **run over** vi ⟨overflow⟩ traboccare ● vt Auto investire. **run through** vi scorrere. **run up** vi salire di corsa; ⟨towards⟩ arrivare di corsa ● vt accumulare ⟨debts, bill⟩; ⟨sew⟩ cucire

'runaway n fuggitivo, -a mf

run-'down *a* ⟨*area*⟩ in abbandono; ⟨*person*⟩ esaurito ● *n* analisi *f*

rung¹ /rʌŋ/ *n* (*of ladder*) piolo *m*

rung² *see* **ring²**

runner /'rʌnə(r)/ *n* podista *mf*; (*in race*) corridore, -trice *mf*; (*on sledge*) pattino *m*. **~ bean** *n* fagiolino *m*. **~-up** *n* secondo, -a *mf* classificato, -a

running /'rʌnɪŋ/ *a* in corsa; ⟨*water*⟩ corrente; **four times ~** quattro volte di seguito ● *n* corsa *f*; (*management*) direzione *f*; **be in the ~** essere in lizza. **~ 'commentary** *n* cronaca *f*

runny /'rʌnɪ/ *a* semiliquido; **~ nose** naso che cola

run: **~-of-the-'mill** *a* ordinario. **~-up** *n* Sport rincorsa *f*; **the ~-up to** il periodo precedente. **~way** *n* pista *f*

rupture /'rʌptʃə(r)/ *n* rottura *f*; Med ernia *f* ● *vt* rompere; **~ oneself** farsi venire l'ernia ● *vi* rompersi

rural /'rʊərəl/ *a* rurale

ruse /ru:z/ *n* astuzia *f*

rush¹ /rʌʃ/ *n* Bot giunco *m*

rush² *n* fretta *f*; **in a ~** di fretta ● *vi* precipitarsi ● *vt* far premura a; **~ sb to hospital** trasportare qcno di corsa all'ospedale. **~-hour** *n* ora *f* di punta

rusk /rʌsk/ *n* biscotto *m*

Russia /'rʌʃə/ *n* Russia *f*. **~n** *a* & *n* russo, -a *mf*; (*language*) russo *m*

rust /rʌst/ *n* ruggine *f* ● *vi* arrugginirsi

rustic /'rʌstɪk/ *a* rustico

rustle /'rʌsl/ *vi* frusciare ● *vt* far frusciare; Am rubare ⟨*cattle*⟩. **rustle up** *vt fam* rimediare

'rustproof *a* a prova di ruggine

rusty /'rʌstɪ/ *a* (-ier, -iest) arrugginito

rut /rʌt/ *n* solco *m*; **in a ~** *fam* nella routine

ruthless /'ru:θlɪs/ *a* spietato. **~ness** *n* spietatezza *f*

rye /raɪ/ *n* segale *f*

Ss

sabbath /'sæbəθ/ *n* domenica *f*; (*Jewish*) sabato *m*

sabbatical /sə'bætɪkl/ *n* Univ anno *m* sabbatico

sabot|age /'sæbətɑːʒ/ *n* sabotaggio *m* ● *vt* sabotare. **~eur** /-'tɜː(r)/ *n* sabotatore, -trice *mf*

saccharin /'sækərɪn/ *n* saccarina *f*

sachet /'sæʃeɪ/ *n* bustina *f*; (*scented*) sacchetto *m* profumato

sack¹ /sæk/ *vt* (*plunder*) saccheggiare

sack² *n* sacco *m*; **get the ~** *fam* essere licenziato ● *vt fam* licenziare. **~ing** *n* tela *f* per sacchi; (*fam: dismissal*) licenziamento *m*

sacrament /'sækrəmənt/ *n* sacramento *m*

sacred /'seɪkrɪd/ *a* sacro

sacrifice /'sækrɪfaɪs/ *n* sacrificio *m* ● *vt* sacrificare

sacrilege /'sækrɪlɪdʒ/ *n* sacrilegio *m*

sad /sæd/ *a* (**sadder, saddest**) triste. **~den** *vt* rattristare

saddle /'sædl/ *n* sella *f* ● *vt* sellare; **I've been ~d with...** *fig* mi hanno affibbiato...

sadis|m /'seɪdɪzm/ *n* sadismo *m*. **~t** /-dɪst/ *n* sadico, -a *mf*. **~tic** /sə'dɪstɪk/ *a* sadico

sad|ly /'sædlɪ/ *adv* tristemente; (*unfortunately*) sfortunatamente. **~ness** *n* tristezza *f*

safe /seɪf/ *a* sicuro; (*out of danger*) salvo; (*object*) al sicuro; **~ and sound** sano e salvo ● *n* cassaforte *f*. **~guard** *n* protezione *f* ● *vt* proteggere. **~ly** *adv* in modo sicuro; (*arrive*) senza incidenti; (*assume*) con certezza

safety /'seɪftɪ/ *n* sicurezza *f*. **~-belt** *n* cintura *f* di sicurezza. **~-deposit box** *n* cassetta *f* di sicurezza. **~-pin** *n* spilla *f* di sicurezza *o da balia.* **~-valve** *n* valvola *f* di sicurezza

sag /sæg/ *vi* (*pt/pp* **sagged**) abbassarsi

saga /'sɑ:gə/ *n* saga *f*

sage /seɪdʒ/ *n* (*herb*) salvia *f*

Sagittarius /sædʒɪ'teərɪəs/ *n* Sagittario *m*

said /sed/ *see* **say**

sail /seɪl/ *n* vela *f*; (*trip*) giro *m* in barca a vela ● *vi* navigare; *Sport* praticare la vela; (*leave*) salpare ● *vt* pilotare

'sailboard *n* tavola *f* del windsurf. **~ing** *n* windsurf *m inv*

sailing /'seɪlɪŋ/ *n* vela *f*. **~-boat** *n* barca *f* a vela. **~-ship** *n* veliero *m*

sailor /'seɪlə(r)/ *n* marinaio *m*

saint /seɪnt/ *n* santo, -a *mf*. **~ly** *a* da santo

sake /seɪk/ *n* **for the ~ of** (*person*) per il bene di; (*peace*) per amor di; **for the ~ of it** per il gusto di farlo

salad /'sæləd/ *n* insalata *f*. **~ bowl** *n* insalatiera *f*. **~ cream** *n* salsa *f* per condire l'insalata. **~-dressing** *n* condimento *m* per insalata

salary /'sælərɪ/ *n* stipendio *m*

sale /seɪl/ *n* vendita *f*; (*at reduced prices*) svendita *f*; **for/on ~** in vendita

sales|man /'seɪlzmən/ *n* venditore *m*; (*traveller*) rappresentante *m*. **~woman** *n* venditrice *f*

salient /'seɪlɪənt/ *a* saliente

saliva /sə'laɪvə/ *n* saliva *f*

sallow /'sæləʊ/ *a* giallastro

salmon /'sæmən/ *n* salmone *m*

saloon /sə'lu:n/ *n* *Auto* berlina *f*; (*Am: bar*) bar *m*

salt /sɔ:lt/ *n* sale *m* ● *a* salato; (*fish, meat*) sotto sale ● *vt* salare; (*cure*) mettere sotto sale. **~-cellar** *n* saliera *f*. **~ 'water** *n* acqua *f* di mare. **~y** *a* salato

salutary /'sæljʊtərɪ/ *a* salutare

salute /sə'lu:t/ *Mil* *n* saluto *m* ● *vt* salutare ● *vi* fare il saluto

salvage /'sælvɪdʒ/ *n* *Naut* recupero *m* ● *vt* recuperare

salvation /sæl'veɪʃn/ *n* salvezza *f*. **S~ 'Army** *n* Esercito *m* della Salvezza

salvo /'sælvəʊ/ *n* salva *f*

same /seɪm/ *a* stesso (**as** di) ● *pron* **the ~** lo stesso; **be all the ~** essere tutti uguali ● *adv* **the ~** nello stesso modo; **all the ~** (*however*) lo stesso; **the ~ to you** altrettanto

sample /'sɑ:mpl/ *n* campione *m*. ● *vt* testare

sanatorium /sænə'tɔ:rɪəm/ *n* casa *f* di cura

sanctimonious /sæŋktɪ'məʊnɪəs/ *a* moraleggiante

sanction /'sæŋkʃn/ *n* (*approval*) autorizzazione *f*; (*penalty*) sanzione *f* ● *vt* autorizzare

sanctity /'sæŋktətɪ/ *n* santità *f inv*

sanctuary /'sæŋktjʊərɪ/ *n Relig* santuario *m*; (*refuge*) asilo *m*; (*for wildlife*) riserva *f*

sand /sænd/ *n* sabbia *f* ● *vt* **~ [down]** cartegiare

sandal /'sændl/ *n* sandalo *m*

sand: **~bank** *n* banco *m* di sabbia. **~paper** *n* carta *f* vetrata ● *vt* cartavetrare. **~-pit** *n* recinto *m* contenente sabbia dove giocano i bambini

sandwich /'sænwɪdʒ/ *n* tramezzino *m* ● *vt* **~ed between** schiacciato tra

sandy /'sændɪ/ *a* (**-ier, -iest**) (*beach, soil*) sabbioso; (*hair*) biondiccio

sane /seɪn/ *a* (*not mad*) sano di mente; (*sensible*) sensato

sang /sæŋ/ *see* **sing**

sanitary /'sænɪtərɪ/ *a* igienico; (*system*) sanitario. **~ napkin** *n* *Am*, **~ towel** *n* assorbente *m* igienico

sanitation /sænɪ'teɪʃn/ *n* impianti *mpl* igienici

sanity /'sænətɪ/ *n* santità *f inv* di mente; (*common sense*) buon senso *m*

sank /sæŋk/ *see* **sink**

sapphire /'sæfaɪə(r)/ *n* zaffiro *m* ● *a* blu zaffiro

sarcas|m /'sɑːkæzm/ n sarcasmo m. **~tic** /-'kæstɪk/ a sarcastico

sardine /sɑː'diːn/ n sardina f

Sardinia /sɑː'dɪnɪə/ n Sardegna f. **~n** a & n sardo, -a mf

sardonic /sɑː'dɒnɪk/ a sardonico

sash /sæʃ/ n fascia f; (for dress) fusciacca f

sat /sæt/ see **sit**

satanic /sə'tænɪk/ a satanico

satchel /'sætʃl/ n cartella f

satellite /'sætəlaɪt/ n satellite m. **~ dish** n antenna f parabolica. **television** n televisione f via satellite

satin /'sætɪn/ n raso m ●attrib di raso

satire /'sætaɪə(r)/ n satira f

satirical /sə'tɪrɪkl/ a satirico

satir|ist /'sætɪrɪst/ n scrittore, -trice mf satirico, -a; (comedian) comico, -a mf satirico, -a. **~ize** vt satireggiare

satisfaction /sætɪs'fækʃn/ n soddisfazione f; **be to sb's ~** soddisfare qcno

satisfactor|y /sætɪs'fæktərɪ/ a soddisfacente. **~ily** adv in modo soddisfacente

satisf|y /'sætɪsfaɪ/ vt (pt/pp -fied) soddisfare; (convince) convincere; **be ~ied** essere soddisfatto. **~ying** a soddisfacente

saturat|e /'sætʃəreɪt/ vt inzuppare (with di); Chem, fig saturare (with di). **~ed** a saturo

Saturday /'sætədeɪ/ n sabato m

sauce /sɔːs/ n salsa f; (cheek) impertinenza f. **~pan** n pentola f

saucer /'sɔːsə(r)/ n piattino m

saucy /'sɔːsɪ/ a (-ier, -iest) impertinente

Saudi Arabia /saʊdɪə'reɪbɪə/ n Arabia f Saudita

sauna /'sɔːnə/ n sauna f

saunter /'sɔːntə(r)/ vi andare a spasso

sausage /'sɒsɪdʒ/ n salsiccia f; (dried) salame m

savage /'sævɪdʒ/ a feroce; (tribe, custom) selvaggio ●n selvaggio,

-a mf ●vt fare a pezzi. **~ry** n ferocia f

save /seɪv/ n Sport parata f ●vt salvare (from da); (keep, collect) tenere; risparmiare (time, money); (avoid) evitare; Sport parare (goal); Comput salvare, memorizzare ●vi ~ **[up]** risparmiare ●prep salvo

saver /'seɪvə(r)/ n risparmiatore, -trice mf

savings /'seɪvɪŋz/ npl (money) risparmi mpl. **~ account** n libretto m di risparmio. **~ bank** n cassa f di risparmio

saviour /'seɪvjə(r)/ n salvatore m

savour /'seɪvə(r)/ n sapore m ●vt assaporare. **~y** a salato; fig rispettabile

saw¹ /sɔː/ see **see¹**

saw² n sega f ●vt/i (pt sawed, pp sawn or sawed) segare. **~dust** n segatura f

saxophone /'sæksəfəʊn/ n sassofono m

say /seɪ/ n **I have one's ~** dire la propria; **have a ~** avere voce in capitolo ●vt/i (pt/pp said) dire; **that is to ~** cioè; **that goes without ~ing** questo è ovvio; **when all is said and done** alla fine dei conti. **~ing** n proverbio m

scab /skæb/ n crosta f; pej crumiro m

scaffold /'skæfəld/ n patibolo m. **~ing** n impalcatura f

scald /skɔːld/ vt scottare; (milk) scaldare ●n scottatura f

scale¹ /skeɪl/ n (of fish) scaglia f

scale² n scala f; **on a grand ~** su vasta scale ●vt (climb) scalare. **scale down** vt diminuire

scales /skeɪlz/ npl (for weighing) bilancia fsg

scallop /'skɒləp/ n (shellfish) pettine m

scalp /skælp/ n cuoio m capelluto

scalpel /'skælpl/ n bisturi m inv

scam /skæm/ n fam fregatura f

scamper /'skæmpə(r)/ vi ~ **away** sgattaiolare via

scampi /'skæmpɪ/ *npl* scampi *mpl*

scan /skæn/ *n Med* scanning *m inv*, scansioscintigrafia *f* ● *vt* (*pt/pp* **scanned**) scrutare; (*quickly*) dare una scorsa a; *Med* fare uno scanning di

scandal /'skændl/ *n* scandalo *m*; (*gossip*) pettegolezzi *mpl*. **~ize** /-d(ə)laɪz/ *vt* scandalizzare. **~ous** /-əs/ *a* scandaloso

Scandinavia /skændɪ'neɪvɪə/ *n* Scandinavia *f*. **~n** *a* & *n* scandinavo, -a *mf*

scanner /'skænə(r)/ *n Comput* scanner *m inv*

scant /skænt/ *a* scarso

scant|y /'skæntɪ/ *a* (**-ier, -iest**) scarso; (*clothing*) succinto. **~ily** *adv* scarsamente; (*clothed*) succintamente

scapegoat /'skeɪp-/ *n* capro *m* espiatorio

scar /skɑː(r)/ *n* cicatrice *f* ● *vt* (*pt/pp* **scarred**) lasciare una cicatrice a

scarc|e /skeəs/ *a* scarso; *fig* raro; **make oneself ~e** *fam* svignarsela. **~ely** *adv* appena; **~ely anything** quasi niente. **~ity** *n* scarsezza *f*

scare /skeə(r)/ *n* spavento *m*; (*panic*) panico *m* ● *vt* spaventare; **be ~d** aver paura (**of** di)

'scarecrow *n* spaventapasseri *m inv*

scarf /skɑːf/ *n* (*pl* **scarves**) sciarpa *f*; (*square*) foulard *m inv*

scarlet /'skɑːlət/ *a* scarlatto. **~ 'fever** *n* scarlattina *f*

scary /'skeərɪ/ *a* be **~** far paura

scathing /'skeɪðɪŋ/ *a* mordace

scatter /'skætə(r)/ *vt* spargere; (*disperse*) disperdere ● *vi* disperdersi. **~-brained** *a fam* scervellato. **~ed** *a* sparso

scatty /'skætɪ/ *a* (**-ier, -iest**) *fam* svitato

scavenge /'skævɪndʒ/ *vi* frugare nella spazzatura. **~r** *n* persona *f* che fruga nella spazzatura

scenario /sɪ'nɑːrɪəʊ/ *n* scenario *m*

scene /siːn/ *n* scena *f*; (*quarrel*) scenata *f*; **behind the ~s** dietro le quinte

scenery /'siːnərɪ/ *n* scenario *m*

scenic /'siːnɪk/ *a* panoramico

scent /sent/ *n* odore *m*; (*trail*) scia *f*; (*perfume*) profumo *m*. **~ed** *a* profumato (**with** di)

sceptic|al /'skeptɪkl/ *a* scettico. **~ism** /-tɪsɪzm/ *n* scetticismo *m*

schedule /'ʃedjuːl/ *n* piano *m*, programma *m*; (*of work*) programma *m*; (*timetable*) orario *m*; **behind ~** indietro; **on ~** nei tempi previsti; **according ~** secondo i tempi previsti ● *vt* prevedere. **~d flight** *n* volo *m* di linea

scheme /skiːm/ *n* (*plan*) piano *m*; (*plot*) macchinazione *f* ● *vi pej* macchinare

schizophren|ia /skɪtsə'friːnɪə/ *n* schizofrenia *f*. **~ic** /-'frenɪk/ *a* schizofrenico

scholar /'skɒlə(r)/ *n* studioso, -a *mf*. **~ly** *a* erudito. **~ship** *n* erudizione *f*; (*grant*) borsa *f* di studio

school /skuːl/ *n* scuola *f*; (*in university*) facoltà *f*; (*of fish*) branco *m*

school: ~boy *n* scolaro *m*. **~girl** *n* scolara *f*. **~ing** *n* istruzione *f*. **~teacher** *n* insegnante *mf*

sciatica /saɪ'ætɪkə/ *n* sciatica *f*

scien|ce /'saɪəns/ *n* scienza *f*; **~ce fiction** fantascienza *f*. **~tific** /-'tɪfɪk/ *a* scientifico. **~tist** *n* scienziato, -a *mf*

scintillating /'sɪntɪleɪtɪŋ/ *a* brillante

scissors /'sɪzəz/ *npl* forbici *fpl*

scoff[1] /skɒf/ *vi* **~ at** schernire

scoff[2] *vt fam* divorare

scold /skəʊld/ *vt* sgridare. **~ing** *n* sgridata *f*

scone /skɒn/ *n* pasticcino *m* da tè

scoop /skuːp/ *n* paletta *f*; *Journ* scoop *m inv* ● **scoop out** *vt* svuotare. **scoop up** *vt* tirar su

scoot /skuːt/ *vi fam* filare. **~er** *n* motoretta *f*

scope /skəʊp/ n portata f;
(*opportunity*) opportunità f inv

scorch /skɔːtʃ/ vt bruciare. **~er** n
fam giornata f torrida. **~ing** a
caldissimo

score /skɔː(r)/ n punteggio m;
(*individual*) punteggio m; Mus
partitura f; (*for film, play*) musica
f; **a ~** [of] (*twenty*) una ventina
[di]; **keep** [the] **~** tenere il punteggio; **on that ~** a questo proposito ● vt segnare (*goal*); (*cut*) incidere ● vi far punti; (*in football etc*)
segnare; (*keep score*) tenere il punteggio. **~r** n segnapunti m inv; (*of
goals*) giocatore, -trice mf che segna

scorn /skɔːn/ n disprezzo m ● vt
disprezzare. **~ful** a sprezzante

Scorpio /ˈskɔːpɪəʊ/ n Scorpione m

scorpion /ˈskɔːpɪən/ n scorpione m

Scot /skɒt/ n scozzese mf

Scotch /skɒtʃ/ a scozzese ● n
(*whisky*) whisky m [scozzese]

scotch vt far cessare

scot-'free a **get off ~** cavarsela
impunemente

Scot|land /ˈskɒtlənd/ n Scozia f.
~s, ~tish a scozzese

scoundrel /ˈskaʊndrəl/ n mascalzone m

scour[1] /ˈskaʊə(r)/ vt (*search*) perlustrare

scour[2] vt (*clean*) strofinare

scourge /skɜːdʒ/ n flagello m

Scout n [Boy] **~** [boy]scout m inv

scout /skaʊt/ n Mil esploratore m
● vi **~ for** andare in cerca di

scowl /skaʊl/ n sguardo m torvo
● vi guardare [di] storto

Scrabble® /ˈskræbl/ n Scara-
beo® m

scraggy /ˈskrægɪ/ a (**-ier, -iest**) pej
scarno

scram /skræm/ vi fam levarsi dai
piedi

scramble /ˈskræmbl/ n (*climb*) ar-
rampicata f ● vi (*clamber*) arrampicarsi; **~ for** azzuffarsi per ● vt

Teleph creare delle interferenze
in; (*eggs*) strapazzare

scrap[1] /skræp/ n (*fam: fight*) litigio m

scrap[2] n pezzetto m; (*metal*)
ferraglia f; **~s** pl (*of food*) avanzi
mpl ● vt (*pt/pp scrapped*) buttare
via

'scrap-book n album m inv

scrape /skreɪp/ vt raschiare;
(*damage*) graffiare. **scrape
through** vi passare per un pelo.
scrape together vt racimolare

scraper /ˈskreɪpə(r)/ n raschiet-
to m

scrappy /ˈskræpɪ/ a frammentario

'scrap-yard n deposito m di
ferraglia; (*for cars*) cimitero m
delle macchine

scratch /skrætʃ/ n graffio m; (*to
relieve itch*) grattata f; **start from
~** partire da zero; **up to ~** (*work*)
all'altezza ● vt graffiare; (*to
relieve itch*) grattare ● vi grattarsi

scrawl /skrɔːl/ n scarabocchio m
● vt/i scarabocchiare

scrawny /ˈskrɔːnɪ/ a (**-ier, -iest**)
pej magro

scream /skriːm/ n strillo m ● vt/i
strillare

screech /skriːtʃ/ n stridore m ● vi
stridere ● vt strillare

screen /skriːn/ n paravento m;
Cinema, TV schermo m ● vt proteggere; (*conceal*) riparare; proiettare (*film*); (*candidates*) passare al
setaccio; Med sottoporre a visita
medica. **~ing** n Med visita f medica; (*of film*) proiezione f. **~-play** n
sceneggiatura f

screw /skruː/ n vite f ● vt avvitare.
screw up vt (*crumple*) ac-
cartocciare; strizzare (*eyes*); storcere (*face*); (*sl: bungle*) mandare
all'aria

'screwdriver n cacciavite m

screwy /ˈskruːɪ/ a (**-ier, -iest**) fam
svitato

scribble /ˈskrɪbl/ n scarabocchio
m ● vt/i scarabocchiare

script /skrɪpt/ n scrittura f (a mano); (of film) sceneggiatura f

'script-writer n sceneggiatore, -trice mf

scroll /skrəʊl/ n rotolo m (di pergamena); (decoration) voluta f

scrounge /skraʊndʒ/ vt/i scroccare. **~r** n scroccone, -a mf

scrub[1] /skrʌb/ n (land) boscaglia f

scrub[2] vt/i (pt/pp **scrubbed**) strofinare; (fam: cancel) cancellare (plan)

scruff /skrʌf/ n **by the ~ of the neck** per la collottola

scruffy /skrʌfɪ/ a (-ier, -iest) trasandato

scrum /skrʌm/ n (in rugby) mischia f

scruple /skruːpl/ n scrupolo m

scrupulous /skruːpjʊləs/ a scrupoloso

scrutin|ize /skruːtɪnaɪz/ vt scrutinare. **~y** n (look) esame m minuzioso

scuffle /skʌfl/ n tafferuglio m

sculpt /skʌlpt/ vt/i scolpire. **~or** /ˈskʌlptə(r)/ n scultore m. **~ure** /-tʃə(r)/ n scultura f

scum /skʌm/ n schiuma f; (people) feccia f

scurrilous /skʌrɪləs/ a scurrile

scurry /skʌrɪ/ vi (pt/pp **-ied**) affrettare il passo

scuttle /skʌtl/ vi (hurry) **~ away** correre via

sea /siː/ n mare m; **at ~** in mare; fig confuso; **by ~** via mare. **~board** n costiera f. **~food** n frutti mpl di mare. **~gull** n gabbiano m

seal[1] /siːl/ n Zool foca f

seal[2] n sigillo m; Techn chiusura f ermetica ● vt sigillare; Techn chiudere ermeticamente. **seal off** vt bloccare (area)

'sea-level n livello m del mare

seam /siːm/ n cucitura f; (of coal) strato m

'seaman n marinaio m

seamless /siːmlɪs/ a senza cucitura

seamy /siːmɪ/ a sordido; (area) malfamato

seance /seɪɑːns/ n seduta f spiritica

sea: **~plane** n idrovolante m. **~port** n porto m di mare

search /sɜːtʃ/ n ricerca f; (official) perquisizione f; **in ~ of** alla ricerca di ● vt frugare (for alla ricerca di); perlustrare (area); (officially) perquisire ● vi **~ for** cercare. **~ing** a penetrante

search: **~light** n riflettore m. **~party** n squadra f di ricerca

sea: **~sick** a be/get **~** avere il mal di mare. **~side** n at/to the **~side** al mare. **~side resort** n stazione f balneare. **~side town** città f di mare

season /siːzn/ n stagione f ● vt (flavour) condire. **~able** /-əbl/ a, **~al** a stagionale. **~ing** n condimento m

'season ticket n abbonamento m

seat /siːt/ n (chair) sedia f; (in car) sedile m; (place to sit) posto m [a sedere]; (bottom) didietro m; (of government) sede f; **take a ~** sedersi ● vt mettere a sedere; (have seats for) aver posti [a sedere] per; **remain ~ed** mantenere il proprio posto. **~-belt** n cintura f di sicurezza

sea: **~weed** n alga f marina. **~worthy** a in stato di navigare

secateurs /sekətɜːz/ npl cesoie fpl

seclu|de /sɪˈkluːd/ a appartato. **~sion** /-ʒn/ n isolamento m

second[1] /sɪˈkɒnd/ vt (transfer) distaccare

second[2] /sekənd/ a secondo; **on ~ thoughts** ripensandoci meglio ● n secondo m; **~s** pl (goods) merce fsg di seconda scelta; **have ~s** (at meal) fare il bis; **John the S~** Giovanni Secondo ● adv (in race) al secondo posto ● vt assistere; appoggiare (proposal)

secondary /sekəndrɪ/ a secondario. **~ school** n scuola f media (inferiore e superiore)

second: ~-**best** a secondo dopo il migliore; **be** ~-**best** pej essere un ripiego. ~ '**class** adv ⟨travel, send⟩ in seconda classe. ~-**class** a di seconda classe

'**second hand** n ⟨on clock⟩ lancetta f dei secondi

second-'hand a & adv di seconda mano

secondly /'sekəndlɪ/ adv in secondo luogo

second-'rate a di second'ordine

secrecy /'si:krəsɪ/ n segretezza f; **in** ~ in segreto

secret /'si:krɪt/ a segreto ● n segreto m

secretarial /sekrə'teərɪəl/ a ⟨work, staff⟩ di segreteria

secretary /'sekrətərɪ/ n segretario, -a mf

secret|e /sɪ'kri:t/ vt secernere ⟨poison⟩. ~**ion** /-i:ʃn/ n secrezione f

secretive /'si:krətɪv/ a riservato. ~**ness** n riserbo m

secretly /'si:krɪtlɪ/ adv segretamente

sect /sekt/ n setta f. ~**arian** a settario

section /'sekʃn/ n sezione f

sector /'sektə(r)/ n settore m

secular /'sekjʊlə(r)/ a secolare; ⟨education⟩ laico

secure /sɪ'kjʊə(r)/ a sicuro ● vt proteggere; chiudere bene ⟨door⟩; rendere stabile ⟨ladder⟩; ⟨obtain⟩ assicurarsi. ~**ly** adv saldamente

securit|y /sɪ'kjʊərətɪ/ n sicurezza f; ⟨for loan⟩ garanzia f. ~**ies** npl titoli mpl

sedate[1] /sɪ'deɪt/ a posato

sedate[2] vt somministrare sedativi a

sedation /sɪ'deɪʃn/ n somministrazione f di sedativi; **be under** ~ essere sotto l'effetto di sedativi

sedative /'sedətɪv/ a sedativo ● n sedativo m

sedentary /'sedəntərɪ/ a sedentario

sediment /'sedɪmənt/ n sedimento m

seduce /sɪ'dju:s/ vt sedurre

seduct|ion /sɪ'dʌkʃn/ n seduzione f. ~**ive** /-tɪv/ a seducente

see[1] /si:/ v ⟨pt saw, pp seen⟩ ● vt vedere; ⟨understand⟩ capire; ⟨escort⟩ accompagnare; **go and** ~ andare a vedere; ⟨visit⟩ andare a trovare; ~ **you!** ci vediamo!; ~ **you later!** a più tardi!; ~**ing that** visto che ● vi vedere; ⟨understand⟩ capire; ~ **that** ⟨make sure⟩ assicurarsi che; ~ **about** occuparsi di. **see off** vt veder partire; ⟨chase away⟩ mandar via. **see through** vi vedere attraverso; fig non farsi ingannare da ● vt portare a buon fine. **see to** vi occuparsi di

see[2] n sede f

seed /si:d/ n seme m; Tennis testa f di serie; **go to** ~ fare seme; fig lasciarsi andare. ~**ed player** n Tennis testa f di serie. ~**ling** n pianticella f

seedy /'si:dɪ/ a (-**ier**, -**iest**) squallido

seek /si:k/ vt ⟨pt/pp **sought**⟩ cercare

seem /si:m/ vi sembrare. ~**ingly** adv apparentemente

seen /si:n/ see **see**[1]

seep /si:p/ vi filtrare

see-saw /'si:sɔ:/ n altalena f

seethe /si:ð/ vi ~ **with anger** ribollire di rabbia

'**see-through** a trasparente

segment /'segmənt/ n segmento m; ⟨of orange⟩ spicchio m

segregat|e /'segrɪgeɪt/ vt segregare. ~**ion** /-'geɪʃn/ n segregazione f

seize /si:z/ vt afferrare; Jur confiscare. **seize up** vi Techn bloccarsi

seizure /'si:ʒə(r)/ n Jur confisca f; Med colpo m [apoplettico]

seldom /'seldəm/ adv raramente

select /sɪ'lekt/ a scelto; ⟨exclusive⟩ esclusivo ● vt scegliere; selezionare ⟨team⟩. ~**ion** /-ekʃn/ n selezione f. ~**ive** /-ɪv/ a selettivo. ~**or** n Sport selezionatore, -trice mf

self /self/ n io m

self: **~-ad'dressed** *a* con il proprio indirizzo. **~-ad'hesive** *a* autoadesivo. **~-as'surance** *n* sicurezza *f* di sé. **~-as'sured** *a* sicuro di sé. **~-'catering** *a* in appartamento attrezzato di cucina. **~-'centred** *a* egocentrico. **~-con'fidence** *n* fiducia *f* in se stesso. **~-'confident** *a* sicuro di sé. **~-'conscious** *a* impacciato. **~-con'tained** *a* (flat) con ingresso indipendente. **~-con'trol** *n* autocontrollo *m*. **~-de'fence** *n* autodifesa *f*, *Jur* legittima difesa *f*. **~-de'nial** *n* abnegazione *f*. **~-determi'nation** *n* autodeterminazione *f*. **~-em'ployed** *a* che lavora in proprio. **~-es'teem** *n* stima *f* di sé. **~-'evident** *a* ovvio. **~-'governing** *a* autonomo. **~-'help** *n* iniziativa *f* personale. **~-in'dulgent** *a* indulgente con se stesso. **~-'interest** *n* interesse *m* personale

self:ish /'selfɪʃ/ *a* egoista. **~ishness** *n* egoismo *m*. **~less** *a* disinteressato

self: **~-'made** *a* che si è fatto da sé. **~-pity** *n* autocommiserazione *f*. **~-'portrait** *n* autoritratto *m*. **~-pos'sessed** *a* padrone di sé. **~-preser'vation** *n* istinto *m* di conservazione. **~-re'spect** *n* amor *m* proprio. **~-'righteous** *a* presuntuoso. **~-'sacrifice** *n* abnegazione *f*. **~-'satisfied** *a* compiaciuto di sé. **~-'service** *n* self-service *m inv* ● *attrib* self-service. **~-suf'ficient** *a* autosufficiente. **~-'willed** *a* ostinato

sell /sel/ *v* (*pt/pp* **sold**) ● *vt* vendere; **be sold out** essere esaurito ● *vi* vendersi. **sell off** *vt* liquidare

seller /'selə(r)/ *n* venditore, -trice *mf*

Sellotape® /'seləʊ-/ *n* nastro *m* adesivo, scotch® *m*

'sell-out *n* (*fam: betrayal*) tradimento *m*; **be a** ~ (*concert:*) fare il tutto esaurito

selves /selvz/ *pl of* **self**

semblance /'sembləns/ *n* parvenza *f*

semen /'si:mən/ *n Anat* liquido *m* seminale

semester /sɪ'mestə(r)/ *n Am* semestre *m*

semi /'semɪ/: **~breve** /'sembrɪv/ *n* semibreve *f*. **~circle** /'semɪs3:k(ə)l/ *n* semicerchio *m*. **~'circular** *a* semicircolare. **~'colon** *n* punto e virgola *m*. **~-de'tached** *a* gemella ● *n* casa *f* gemella. **~'final** *n* semifinale *f*

seminar /'semɪnɑ:(r)/ *n* seminario *m*. **~y** /-nərɪ/ *n* seminario *m*

semolina /semə'li:nə/ *n* semolino *m*

senate /'senət/ *n* senato *m*. **~or** *n* senatore *m*

send /send/ *vt/i* (*pt/pp* **sent**) mandare; **~ for** mandare a chiamare (*person*); far venire (*thing*). **~er** *n* mittente *mf*. **~-off** *n* commiato *m*

senil|e /'si:naɪl/ *a* arteriosclerotico; *Med* senile. **~ity** /sɪ'nɪlətɪ/ *n* senilismo *m*

senior /'si:nɪə(r)/ *a* più vecchio; (*in rank*) superiore ● *n* (*in rank*) superiore *mf*; (*in sport*) senior *mf*; **she's two years my ~** è più vecchia di me di due anni. **~ citizen** *n* anziano, -a *mf*

seniority /si:nɪ'ɒrətɪ/ *n* anzianità *f inv* di servizio

sensation /sen'seɪʃn/ *n* sensazione *f*. **~al** *a* sensazionale. **~ally** *adv* in modo sensazionale

sense /sens/ *n* senso *m*; (*common* ~) buon senso *m*; **in a** ~ in un certo senso; **make** ~ aver senso ● *vt* sentire. **~less** *a* insensato; (*unconscious*) privo di sensi

sensib|le /'sensəbl/ *a* sensato; (*suitable*) appropriato. **~y** *adv* in modo appropriato

sensiti|ve /'sensɪtɪv/ *a* sensibile; (*touchy*) suscettibile. **~ely** *adv* con sensibilità. **~ity** /-'tɪvɪtɪ/ *n* sensibilità *f inv*

sensory /'sensərɪ/ *a* sensoriale

sensual /ˈsensjʊəl/ a sensuale.
~ity /-ˈælətɪ/ n sensualità f inv

sensuous /ˈsensjʊəs/ a voluttuoso

sent /sent/ see **send**

sentence /ˈsentəns/ n frase f; Jur
sentenza f; (punishment) condanna f ●vt ~ **to** condannare a

sentiment /ˈsentɪmənt/ n sentimento m; (opinion) opinione f;
(sentimentality) sentimentalismo
m. ~**al** /-ˈmentl/ a sentimentale; pej sentimentalista. ~**ality**
/-ˈtælətɪ/ n sentimentalità f inv

sentry /ˈsentrɪ/ n sentinella f

separable /ˈsepərəbl/ a separabile

separate[1] /ˈsepərət/ a separato.
~**ly** adv separatamente

separate[2] /ˈsepəreɪt/ vt separare
●vi separarsi. ~**ion** /-ˈreɪʃn/ n separazione f

September /sepˈtembə(r)/ n settembre m

septic /ˈseptɪk/ a settico; **go ~ in-**
fettarsi. **~ tank** n fossa f biologica

sequel /ˈsiːkwəl/ n seguito m

sequence /ˈsiːkwəns/ n sequenza f

sequin /ˈsiːkwɪn/ n lustrino m,
paillette f inv

serenade /serəˈneɪd/ n serenata f
●vt fare una serenata a

seren|**e** /sɪˈriːn/ a sereno. ~**ity**
/-ˈrenətɪ/ n serenità f inv

sergeant /ˈsɑːdʒənt/ n sergente m

serial /ˈsɪərɪəl/ n racconto m a puntate; TV sceneggiato m a puntate;
Radio commedia f radiofonica.
~**ize** vt pubblicare a puntate; Radio, TV trasmettere a puntate. ~
killer n serial killer mf inv. ~
number n numero m di serie. ~
port n Comput porta f seriale

series /ˈsɪəriːz/ n serie f inv

serious /ˈsɪərɪəs/ a serio; (illness,
error) grave. ~**ly** adv seriamente;
(ill) gravemente; **take ~ly** prendere sul serio. ~**ness** n serietà f
inv; (of situation) gravità f inv

sermon /ˈsɜːmən/ n predica f

serpent /ˈsɜːpənt/ n serpente m

serrated /seˈreɪtɪd/ a dentellato

serum /ˈsɪərəm/ n siero m

servant /ˈsɜːvənt/ n domestico, -a
mf

serve /sɜːv/ n Tennis servizio m
●vt servire; scontare (sentence);
~ **its purpose** servire al proprio
scopo; **it ~s you right!** ben ti sta!;
~**s two per due persone** ●vi prestare servizio; Tennis servire; ~
as servire da

server /ˈsɜːvə(r)/ n Comput server
m inv

service /ˈsɜːvɪs/ n servizio m;
Relig funzione f; (maintenance) revisione f; ~**s** pl forze fpl armate;
(on motorway) area f di servizio;
in the ~s sotto le armi; **of ~ to**
utile a; **out of ~** (machine:)
guasto ●vt Techn revisionare. ~**able** /-əbl/ a utilizzabile; (hardwearing) resistente; (practical)
pratico

service: ~ area n area f di servizio. **~ charge** n servizio m. ~**man**
n militare m. **~ provider** n
fornitore, -trice mf di servizi. ~
station n stazione f di servizio

serviette /sɜːvɪˈet/ n tovagliolo m

servile /ˈsɜːvaɪl/ a servile

session /ˈseʃn/ n seduta f; Jur sessione f; Univ anno m accademico

set /set/ n serie f, set m inv; (of
crockery, cutlery) servizio m; TV,
Radio apparecchio m; Math insieme m; Theat scenario m; Cinema,
Tennis set m inv; (of people) circolo m; (of hair) messa f in piega ●a
(ready) pronto; (rigid) fisso;
(book) in programma; **be ~ on
doing sth** essere risoluto a fare
qcsa; **be ~ in one's ways** essere
abitudinario ●v /pt/pp **set**, pres p
setting ●vt mettere, porre; mettere (alarm clock); assegnare
(task, homework); fissare (date,
limit); chiedere (questions); montare (gem); assestare (bone); apparecchiare (table); ~ **fire to** dare
fuoco a; ~ **free** liberare ●vi (sun:)
tramontare; (jelly, concrete:) solidificare; ~ **about doing sth** mettersi a fare qcsa. **set back** vt met-

tere indietro; (hold up) ritardare; (fam: cost) costare a. **set off** vi partire ●vt avviare; mettere (alarm;) fare esplodere (bomb).

set out vi partire; ~ **out to do sth** proporsi di fare qcsa ●vt disporre; (state) esporre. **set to** vi mettersi all'opera. **set up** vt fondare (company); istituire (committee)

'set-back n passo m indietro

set 'meal n menù m inv fisso

settee /se'ti:/ n divano m

setting /'setɪŋ/ n scenario m; (position) posizione f; (of sun) tramonto m; (of jewel) montatura f

settle /'setl/ vt (decide) definire; risolvere (argument;) fissare (date); calmare (nerves;) saldare (bill) ●vi (to live) stabilirsi; (snow, dust, bird:) posarsi; (subside) assestarsi; (sediment:) depositarsi. **settle down** vi sistemarsi; (stop making noise) calmarsi. **settle for** vt accontentarsi di. **settle up** vi regolare i conti

settlement /'setlmənt/ n (agreement) accordo m; (of bill) saldo m; (colony) insediamento m

settler /'setlə(r)/ n colonizzatore, -trice mf

'set-to n fam zuffa f; (verbal) batti-becco m

'set-up n situazione f

seven /'sevn/ a sette. ~**teen** a diciassette. ~**teenth** a diciassettesimo

seventh /'sevnθ/ a settimo

seventieth /'sevntɪɪθ/ a settantesimo

seventy /'sevntɪ/ a settanta

sever /'sevə(r)/ vt troncare (relations)

several /'sevrəl/ a & pron parecchi

severe /sɪ'vɪə(r)/ a severo; (pain) violento; (illness) grave; (winter) rigido. ~**ly** adv severamente; (ill) gravemente. ~**ity** /-'verətɪ/ n severità f inv; (of pain) violenza f; (of illness) gravità f; (of winter) rigore m

sew /səʊ/ vt/i (pt sewed, pp sewn

or sewed) cucire. **sew up** vt ricucire

sewage /'su:ɪdʒ/ n acque fpl di scolo

sewer /'su:ə(r)/ n fogna f

sewing /'səʊɪŋ/ n cucito m; (work) lavoro m di cucito. ~ **machine** n macchina f da cucire

sewn /səʊn/ see **sew**

sex /seks/ n sesso m; **have** ~ avere rapporti sessuali. ~**ist** a sessista. ~ **offender** n colpevole mf di delitti a sfondo sessuale

sexual /'seksjʊəl/ a sessuale. ~ **'intercourse** n rapporti mpl sessuali. ~**ity** /-'æləti/ n sessualità f inv. ~**ly** adv sessualmente

sexy /'seksɪ/ a (-ier, -iest) sexy

shabby /'ʃæbɪ/ a (-ier, -iest) scialbo; (treatment) meschino. ~**iness** n trasandatezza f; (of treatment) meschinità f inv

shack /ʃæk/ n catapecchia f ●**shack up with** vt fam vivere con

shade /ʃeɪd/ n ombra f; (of colour) sfumatura f; (for lamp) paralume m; (Am: for window) tapparella f; **a** ~ **better** un tantino meglio ●vt riparare dalla luce; (draw lines on) ombreggiare. ~**s** npl fam occhiali mpl da sole

shadow /'ʃædəʊ/ n ombra f; **S**~ **Cabinet** governo m ombra ●vt (follow) pedinare. ~**y** a ombroso

shady /'ʃeɪdɪ/ a (-ier, -iest) ombroso; (fam: disreputable) losco

shaft /ʃɑ:ft/ n Techn albero m; (of light) raggio m; (of lift, mine) pozzo m; ~**s** pl (of cart) stanghe fpl

shaggy /'ʃægɪ/ a (-ier, -iest) irsuto; (animal) dal pelo arruffato

shake /ʃeɪk/ n scrollata f ●v (pt shook, pp shaken) ●vt scuotere; agitare (bottle); far tremare (building); ~ **hands with** stringere la mano a ●vi tremare. **shake off** vt scrollarsi di dosso. ~**up** n Pol rimpasto m; Comm ristrutturazione f

shaky /'ʃeɪkɪ/ a (-ier, -iest) tre-

mante; ⟨table etc⟩ traballante;
(unreliable) vacillante

shall /ʃæl/ v aux **I ~** go andrò; **we
~ see** vedremo; **what ~ I do?**
cosa faccio?; **I'll come too, ~ I?**
vengo anch'io, no?; **thou shalt not
kill** liter non uccidere

shallow /ʃæləʊ/ a basso, poco profondo; ⟨dish⟩ poco profondo; fig
superficiale

sham /ʃæm/ a falso ● n finzione f;
(person) spaccone, -a mf ● vt
(pt/pp **shammed**) simulare

shambles /ʃæmblz/ n baraonda
fsg

shame /ʃeɪm/ n vergogna f; **it's a
~ that** è un peccato che; **what a
~!** che peccato! **~-faced** a vergognoso

shame|ful /ʃeɪmfl/ a vergognoso.
~less a spudorato

shampoo /ʃæmpu:/ n shampoo m
inv ● vt fare uno shampoo a

shandy /ʃændɪ/ n bevanda f a base
di birra e gassosa

shan't /ʃɑ:nt/ = **shall not**

shanty town /ʃæntɪtaʊn/ n
bidonville f inv, baraccopoli f inv

shape /ʃeɪp/ n forma f; (figure)
ombra f; **take ~** prendere forma;
get back in ~ ritornare in forma
● vt dare forma a (into di) ● vi ~
[up] mettere la testa a posto; **~ up
nicely** mettersi bene. **~less** a in-
forme

shapely /ʃeɪplɪ/ a (-ier, -iest) ben
fatto

share /ʃeə(r)/ n porzione f; Comm
azione f ● vt dividere; condividere
⟨views⟩ ● vi dividere. **~holder** n
azionista mf

shark /ʃɑ:k/ n squalo m, pescecane
m; fig truffatore, -trice mf

sharp /ʃɑ:p/ a ⟨knife etc⟩ tagliente;
⟨pencil⟩ appuntito; ⟨drop⟩ a picco;
⟨reprimand⟩ severo; ⟨outline⟩
marcato; ⟨alert⟩ acuto; ⟨unscrupu-
lous⟩ senza scrupoli; **~ pain** fitta f
● adv in punto; Mus fuori tono;
look ~! sbrigati! ● n Mus diesis m

inv. **~en** vt affilare ⟨knife⟩; appun-
tire ⟨pencil⟩

shatter /ʃætə(r)/ vt frantumare;
fig mandare in frantumi; **~ed**
(fam: exhausted) a pezzi ● vi fran-
tumarsi

shav|e /ʃeɪv/ n rasatura f; **have a
~e** farsi la barba ● vt radere ● vi
radersi. **~er** n rasoio m elettrico.
~ing-brush n pennello m da barba; **~ing foam** n schiuma f da barba; **~ing soap** n sapone m da barba

shawl /ʃɔ:l/ n scialle m

she /ʃi:/ pron lei

sheaf /ʃi:f/ n (pl **sheaves**) fascio m

shear /ʃɪə(r)/ vt (pt **sheared**, pp
shorn or **sheared**) tosare

shears /ʃɪəz/ npl (for hedge) cesoie
fpl

sheath /ʃi:θ/ n (pl **~s** /ʃi:ðz/) guai-
na f

shed[1] /ʃed/ n baracca f; (for cattle)
stalla f

shed[2] vt (pt/pp **shed**, pres p **shedding**) perdere; versare ⟨blood,
tears⟩; **~ light on** far luce su

sheen /ʃi:n/ n lucentezza f

sheep /ʃi:p/ n inv pecora f. **~-dog**
n cane m da pastore

sheepish /ʃi:pɪʃ/ a imbarazzato.
~ly adv con aria imbarazzata

sheepskin n [pelle f di] montone m

sheer /ʃɪə(r)/ a puro; (steep) a picco; (transparent) trasparente
● adv a picco

sheet /ʃi:t/ n lenzuolo m; (of paper)
foglio m; (of glass, metal) lastra f

shelf /ʃelf/ n (pl **shelves**) ripiano
m; (set of shelves) scaffale m

shell /ʃel/ n conchiglia f; (of egg,
snail, tortoise) guscio m; (of crab)
corazza f; (of unfinished building)
ossatura f; Mil granata f ● vt sgusciare ⟨peas⟩; Mil bombardare.
shell out vi fam sborsare

shellfish n inv mollusco m; Culin
frutti mpl di mare

shelter /ʃeltə(r)/ n rifugio m; (air
raid ~) rifugio m antiaereo ● vt ri-

parare (*from* da); *fig* mettere al riparo; (*give lodging to*) dare asilo a ● *vi* rifugiarsi. **~ed** *a* (*spot*) riparato; (*life*) ritirato

shelve /ʃelv/ *vt* accantonare (*project*)

shelves /ʃelvz/ *see* **shelf**

shelving /ʃelvɪŋ/ *n* (*shelves*) ripiani *mpl*

shepherd /ʃepəd/ *n* pastore *m* ● *vt* guidare. **~'s pie** *n* pasticcio *m* di carne tritata e patate

sherry /ʃerɪ/ *n* sherry *m*

shield /ʃiːld/ *n* scudo *m*; (*for eyes*) maschera *f*; *Techn* schermo *m* ● *vt* proteggere (*from* da)

shift /ʃɪft/ *n* cambiamento *m*; (*in position*) spostamento *m*; (*at work*) turno *m* ● *vt* spostare; (*take away*) togliere; riversare (*blame*) ● *vi* spostarsi; (*wind:*) cambiare; (*fam: move quickly*) darsi una mossa

'shift work *n* turni *mpl*

shifty /ʃɪftɪ/ *a* (**-ier, -iest**) *pej* losco; (*eyes*) sfuggente

shilly-shally /ʃɪlɪʃælɪ/ *vi* titubare

shimmer /ʃɪmə(r)/ *n* luccichio *m* ● *vi* luccicare

shin /ʃɪn/ *n* stinco *m*

shine /ʃaɪn/ *n* lucentezza *f*; **give sth a ~** dare una lucidata a qcsa ● *v* (*pt/pp* **shone**) ● *vi* splendere; (*reflect light*) brillare; (*hair, shoes:*) essere lucido ● *vt* **~ a light on** puntare una luce su

shingle /ʃɪŋgl/ *n* (*pebbles*) ghiaia *f*

shingles /ʃɪŋglz/ *n* *Med* fuochi *mpl* di Sant'Antonio

shiny /ʃaɪnɪ/ *a* (**-ier, -iest**) lucido

ship /ʃɪp/ *n* nave *f* ● *vt* (*pt/pp* **shipped**) spedire; (*by sea*) spedire via mare

ship: **~ment** *n* spedizione *f*; (*consignment*) carico *m*. **~per** *n* spedizioniere *m*. **~ping** *n* trasporto *m*; (*traffic*) imbarcazioni *fpl*. **~shape** *a & adv* in perfetto ordine. **~wreck** *n* naufragio *m*. **~wrecked** *a* naufragato. **~yard** *n* cantiere *m* navale

shirk /ʃɜːk/ *vt* scansare. **~er** *n* scansafatiche *mf* *inv*

shirt /ʃɜːt/ *n* camicia *f*. **in ~-sleeves** in maniche di camicia

shit /ʃɪt/ *vulg* *n* & *int* merda *f* ● *vi* (*pt/pp* **shit**) cagare

shiver /ʃɪvə(r)/ *n* brivido *m* ● *vi* rabbrividire

shoal /ʃəʊl/ *n* (*of fish*) banco *m*

shock /ʃɒk/ *n* (*impact*) urto *m*; *Electr* scossa *f* [elettrica]; *fig* colpo *m*, shock *m inv*; *Med* shock *m inv*; **get a ~** *Electr* prendere la scossa ● *vt* scioccare. **~ing** *a* scioccante; (*fam: weather, handwriting etc*) tremendo

shod /ʃɒd/ *see* **shoe**

shoddy /ʃɒdɪ/ *a* (**-ier, -iest**) scadente

shoe /ʃuː/ *n* scarpa *f*; (*of horse*) ferro *m* ● *vt* (*pt/pp* **shod**, *pres* *p* **shoeing**) ferrare (*horse*)

shoe: **~horn** *n* calzante *m*. **~-lace** *n* laccio *m* da scarpa. **~maker** *n* calzolaio *m*. **~shop** *n* calzoleria *f*. **~string** *n* **on a ~-string** *fam* con una miseria

shone /ʃɒn/ *see* **shine**

shoo /ʃuː/ *vt* **~ away** cacciar via ● *int* sciò

shook /ʃʊk/ *see* **shake**

shoot /ʃuːt/ *n* *Bot* germoglio *m*; (*hunt*) battuta *f* di caccia ● *v* (*pt/pp* **shot**) ● *vt* sparare; girare (*film*) ● *vi* (*hunt*) andare a caccia.
shoot down *vt* abbattere. **shoot out** *vi* (*rush*) precipitarsi fuori. **shoot up** *vi* (*grow*) crescere in fretta; (*prices:*) salire di colpo

'shooting-range *n* poligono *m* di tiro

shop /ʃɒp/ *n* negozio *m*; (*workshop*) officina *f*; **talk ~** *fam* parlare di lavoro ● *vi* (*pt/pp* **shopped**) far compere; **go ~ping** andare a fare compere. **shop around** *vi* confrontare i prezzi

shop: **~ assistant** *n* commesso, -a *mf*. **~keeper** *n* negoziante *mf*. **~-lifter** *n* taccheggiatore, -trice *mf.* **~-lifting** *n* taccheggio *m;* **~per** *n* compratore, -trice *mf*

shopping /ˈʃɒpɪŋ/ *n* compere *fpl;* (*articles*) acquisti *mpl;* **do the ~** fare la spesa. **~ bag** *n* borsa *f* per la spesa. **~ centre** *n* centro *m* commerciale. **~ trolley** *n* carrello *m*

shop: **~-steward** *n* rappresentante *mf* sindacale. **~-window** *n* vetrina *f*

shore /ʃɔː(r)/ *n* riva *f*

shorn /ʃɔːn/ *see* **shear**

short /ʃɔːt/ *a* corto; (*not lasting*) breve; (*person*) basso; (*curt*) brusco; **a ~ time ago** poco tempo fa; **be ~ of** essere a corto di; **be in ~ supply** essere scarso; *fig* essere raro; **Mick is ~ for Michael** Mick è il diminutivo di Michael ● *adv* bruscamente; **in ~** in breve; **~ of doing** a meno di fare; **go ~** essere privato (**of** di); **stop ~ of doing sth** non arrivare fino a fare qcsa; **cut ~** interrompere (*meeting, holiday*); **to cut a long story ~** per farla breve

shortage /ˈʃɔːtɪdʒ/ *n* scarsità *f inv*

short: **~bread** *n* biscotto *m* di pasta frolla. **~ circuit** *n* corto *m* circuito. **~coming** *n* difetto *m.* **'cut** *n* scorciatoia *f*

shorten /ˈʃɔːtn/ *vt* abbreviare; accorciare (*garment*)

short: **~hand** *n* stenografia *f.* **~-handed** *a* a corto di personale. **~hand 'typist** *n* stenodattilografo, -a *mf.* **~ list** *n* lista *f* dei candidati selezionati per un lavoro. **~-lived** /-lɪvd/ *a* di breve durata

shortly /ˈʃɔːtlɪ/ *adv* presto; **~ly before/after** poco prima/dopo. **~ness** *n* brevità *f inv;* (*of person*) bassa statura *f*

short-range *a* di breve portata

shorts /ʃɔːts/ *npl* calzoncini *mpl* corti

short: **~-sighted** *a* miope. **~-sleeved** *a* a maniche corte. **~-staffed** *a* a corto di personale. **~ story** *n* racconto *m,* novella *f.*

~-tempered *a* irascibile. **~-term** *a* a breve termine. **~ wave** *n* onde *fpl* corte

shot /ʃɒt/ *see* **shoot** ● *n* colpo *m;* (*person*) tiratore *m; Phot* foto *f;* (*injection*) puntura *f;* (*attempt*) prova *f;* **like a ~** *fam* come un razzo. **~gun** *n* fucile *m* da caccia

should /ʃʊd/ *v aux* **I ~ go** dovrei andare; **I ~ have seen him** avrei dovuto vederlo; **I ~ like** mi piacerebbe; **this ~ be enough** questo dovrebbe bastare; **if he ~ come** se dovesse venire

shoulder /ˈʃəʊldə(r)/ *n* spalla *f* ● *vt* mettersi in spalla; *fig* accollarsi. **~-bag** *n* borsa *f* a tracolla. **~-blade** *n* scapola *f.* **~-strap** *n* spallina *f;* (*of bag*) tracolla *f*

shout /ʃaʊt/ *n* grido *m* ● *vt/i* gridare. **shout at** *vi* alzar la voce con. **shout down** *vt* azzittire gridando

shouting /ˈʃaʊtɪŋ/ *n* grida *fpl*

shove /ʃʌv/ *n* spintone *m* ● *vt* spingere; (*fam: put*) ficcare ● *vi* spingere. **shove off** *vi fam* togliersi di torno

shovel /ˈʃʌvl/ *n* pala *f* ● *vt* (*pt/pp* **shovelled**) spalare

show /ʃəʊ/ *n* (*display*) manifestazione *f;* (*exhibition*) mostra *f;* (*ostentation*) ostentazione *f; Theat, TV* spettacolo *m;* (*programme*) programma *m;* **on ~** esposto ● *v* (*pt* **showed**, *pp* **shown**) ● *vt* mostrare; (*put on display*) esporre; proiettare (*film*) ● *vi* (*film:*) essere proiettato; **your slip is ~ing** ti si vede la sottoveste. **show in** *vt* fare accomodare. **show off** *vi fam* mettersi in mostra ● *vt* mettere in mostra. **show up** *vi* risaltare; (*fam: arrive*) farsi vedere ● *vt* (*fam: embarrass*) far fare una brutta figura a

'show-down *n* regolamento *m* dei conti

shower /ˈʃaʊə(r)/ *n* doccia *f;* (*of rain*) acquazzone *m;* **have a ~** fare la doccia ● *vt* ~ **with** coprire di ● *vi* fare la doccia. **~proof** *a* impermeabile. **~y** *a* da acquazzoni

'**show-jumping** *n* concorso *m* ippico

shown /ʃəʊn/ *see* **show**

'**show-off** *n* esibizionista *mf*

showy /'ʃəʊɪ/ *a* appariscente

shrank /ʃræŋk/ *see* **shrink**

shred /ʃred/ *n* brandello *m*; *fig* briciolo *m* ● *vt* (*pt/pp* **shredded**) fare a brandelli; *Culin* tagliuzzare. **~der** *n* distruttore *m* di documenti

shrewd /ʃruːd/ *a* accorto. **~ness** *n* accortezza *f*

shriek /ʃriːk/ *n* strillo *m* ● *vt/i* strillare

shrift /ʃrɪft/ *n* **give sb short ~** liquidare qcno rapidamente

shrill /ʃrɪl/ *a* penetrante

shrimp /ʃrɪmp/ *n* gamberetto *m*

shrine /ʃraɪn/ *n* (*place*) santuario *m*

shrink /ʃrɪŋk/ *vi* (*pt* **shrank**, *pp* **shrunk**) restringersi; (*draw back*) ritrarsi (**from** da)

shrivel /'ʃrɪvl/ *vi* (*pt/pp* **shrivelled**) raggrinzare

shroud /ʃraʊd/ *n* sudario *m*; *fig* manto *m*

Shrove /ʃrəʊv/ *n* ~ '**Tuesday** martedì *m* grasso

shrub /ʃrʌb/ *n* arbusto *m*

shrug /ʃrʌɡ/ *n* scrollata *f* di spalle ● *vt/i* (*pt/pp* **shrugged**) ~ [**one's shoulders**] scrollare le spalle

shrunk /ʃrʌŋk/ *see* **shrink**. **~en** *a* rimpicciolito

shudder /'ʃʌdə(r)/ *n* fremito *m* ● *vi* fremere

shuffle /'ʃʌfl/ *vi* strascicare i piedi ● *vt* mescolare (*cards*)

shun /ʃʌn/ *vt* (*pt/pp* **shunned**) rifuggire

shunt /ʃʌnt/ *vt* smistare

shush /ʃʊʃ/ *int* zitto!

shut /ʃʌt/ *v* (*pt/pp* **shut**, *pres p* **shutting**) ● *vt* chiudere ● *vi* chiudersi; (*shop:*) chiudere. **shut down** *vt/i* chiudere. **shut up** *vt*

chiudere; *fam* far tacere ● *vi fam* stare zitto; ~ **up!** stai zitto!

'**shut-down** *n* chiusura *f*

shutter /'ʃʌtə(r)/ *n* serranda *f*; *Phot* otturatore *m*

shuttle /'ʃʌtl/ *n* navetta *f* ● *vi* far la spola

shuttle: **~cock** *n* volano *m*. **~ service** *n* servizio *m* pendolare

shy /ʃaɪ/ *a* (*timid*) timido. **~ness** *n* timidezza *f*

Siamese /saɪə'miːz/ *a* siamese

sibling /'sɪblɪŋz/ *n* (*brother*) fratello *m*; (*sister*) sorella *f*; **~s** fratelli *mpl*

Sicily /'sɪsɪlɪ/ *n* Sicilia *f*. **~ian** *a & n* siciliano, -a *mf*

sick /sɪk/ *a* ammalato; (*humour*) macabro; **be** ~ (*vomit*) vomitare; **be** ~ **of sth** *fam* essere stufo di qcosa; **feel** ~ aver la nausea

sicken /'sɪkn/ *vt* disgustare ● *vi* **be ~ing for something** covare qualche malanno. **~ing** *a* disgustoso

sickly /'sɪklɪ/ *a* (**-ier, -iest**) malaticcio. **~ness** *n* malattia *f*; (*vomiting*) vomitevole. **~ness benefit** *n* indennità *f* di malattia

side /saɪd/ *n* lato *m*; (*of person, mountain*) fianco *m*; (*of road*) bordo *m*; **on the** ~ (*as sideline*) come attività secondaria; **~ by** ~ fianco a fianco; **take** ~**s** immischiarsi; **take sb's** ~ prendere le parti di qcno; **be on the safe** ~ andare sul sicuro ● *attrib* laterale ● *vi* ~ **with** parteggiare per

side: **~board** *n* credenza *f*. **~burns** *npl* basette *fpl*. **~effect** *n* effetto *m* collaterale. **~lights** *npl* luci *fpl* di posizione. **~line** *n* attività *f inv* complementare. **~-show** *n* attrazione *f*. **~-step** *vt* schivare. **~-track** *vt* sviare. **~walk** *n Am* marciapiede *m*. **~ways** *adv* obliquamente

siding /'saɪdɪŋ/ *n* binario *m* di raccordo

sidle /'saɪdl/ *vi* camminare furtivamente (**up to** verso)

siege /siːdʒ/ *n* assedio *m*

sieve /sɪv/ n setaccio m ●vt setacciare

sift /sɪft/ vt setacciare; ~ [through] fig passare al setaccio

sigh /saɪ/ n sospiro m ●vi sospirare

sight /saɪt/ n vista f; (on gun) mirino m; **the ~s** pl le cose da vedere; **at first ~** a prima vista; **be within/out of ~** essere/non essere in vista; **lose ~ of** perdere di vista; **know by ~** conoscere di vista; **have bad ~** vederci male ●vt avvistare

'**sightseeing** n **go ~** andare a visitare posti

sign /saɪn/ n segno m; (notice) insegna f ● vt/i firmare. **sign on** vi (as unemployed) presentarsi all'ufficio di collocamento; Mil arruolarsi

signal /ˈsɪɡnl/ n segnale m ● v (pt/pp **signalled**) ● vt segnalare ● vi fare segnali; ~ **to sb** far segno a qcno (**to** do). ~-**box** n cabina f di segnalazione

signature /ˈsɪɡnɪtʃə(r)/ n firma f. ~ **tune** n sigla f [musicale]

signet-ring /ˈsɪɡnɪt-/ n anello m con sigillo

significan|ce /sɪɡˈnɪfɪkəns/ n significato m. ~**t** a significativo

signify /ˈsɪɡnɪfaɪ/ vt (pt/pp -**ied**) indicare

sign-language n linguaggio m dei segni

signpost /ˈsaɪn-/ n segnalazione f stradale

silence /ˈsaɪləns/ n silenzio m ● vt far tacere. ~ n (on gun) silenziatore m; Auto marmitta f

silent /ˈsaɪlənt/ a silenzioso; (film) muto; **remain ~** rimanere in silenzio. ~**ly** adv silenziosamente

silhouette /sɪluˈet/ n sagoma f, silhouette f inv ● vt **be ~d** profilarsi

silicon /ˈsɪlɪkən/ n silicio m. ~ **chip** n piastrina f di silicio

silk /sɪlk/ n seta f ● attrib di seta. ~**worm** n baco m da seta

silky /ˈsɪlkɪ/ a (-**ier**, -**iest**) come la seta

sill /sɪl/ n davanzale m

silly /ˈsɪlɪ/ a (-**ier**, -**iest**) sciocco

silo /ˈsaɪləʊ/ n silo m

silt /sɪlt/ n melma f

silver /ˈsɪlvə(r)/ a d'argento; (paper) argentato ● n argento m; (silverware) argenteria f

silver: ~-**plated** a placcato d'argento. ~**ware** n argenteria f. ~**wedding** n nozze fpl d'argento

similar /ˈsɪmɪlə(r)/ a simile. ~**ity** /-'lærətɪ/ n somiglianza f. ~**ly** adv in modo simile

simile /ˈsɪmɪlɪ/ n similitudine f

simmer /ˈsɪmə(r)/ vi bollire lentamente ● vt far bollire lentamente. **simmer down** vi calmarsi

simple /ˈsɪmpl/ a semplice; (person) sempliciotto. ~-'**minded** a semplicistico

simplicity /sɪmˈplɪsətɪ/ n semplicità f inv

simpli|fication /sɪmplɪfɪˈkeɪʃn/ n semplificazione f. ~**fy** /ˈsɪmplɪfaɪ/ vt (pt/pp -**ied**) semplificare

simply /ˈsɪmplɪ/ adv semplicemente

simulat|e /ˈsɪmjuleɪt/ vt simulare. ~**ion** /-'leɪʃn/ n simulazione f

simultaneous /sɪmlˈteɪnɪəs/ a simultaneo

sin /sɪn/ n peccato m ● vi (pt/pp **sinned**) peccare

since /sɪns/ prep da ● adv da allora ● conj da quando; (because) siccome

sincere /sɪnˈsɪə(r)/ a sincero. ~**ly** adv sinceramente; **Yours** ~**ly** distinti saluti

sincerity /sɪnˈserətɪ/ n sincerità f inv

sinful /ˈsɪnfl/ a peccaminoso

sing /sɪŋ/ vt/i (pt **sang**, pp **sung**) cantare

singe /sɪndʒ/ vt (pres p **singeing**) bruciacchiare

singer /ˈsɪŋə(r)/ n cantante mf

single /ˈsɪŋɡl/ a solo; (not double) semplice; (unmarried) celibe;

⟨woman⟩ nubile; ⟨room⟩ singolo; ⟨bed⟩ a una piazza ● n ⟨ticket⟩ biglietto m di sola andata; ⟨record⟩ singolo m; ~s pl Tennis singolo m ● **single out** vt scegliere; ⟨distinguish⟩ distinguere

single: ~-**breasted** a a un petto. ~-**handed** a & adv da solo. ~-**minded** a risoluto. ● '**parent** n 'parent m genitore m che alleva il figlio da solo

singly /'sɪŋglɪ/ adv singolarmente

singular /'sɪŋgjolə(r)/ a Gram singolare ● n singolare m. ~**ly** adv singolarmente

sinister /'sɪnɪstə(r)/ a sinistro

sink /sɪŋk/ n lavandino m ● v ⟨pt **sank**, pp **sunk**⟩ ● vi affondare ● vt affondare ⟨ship⟩; scavare ⟨shaft⟩; investire ⟨money⟩. **sink in** vi rientrare; **it took a while to ~ in** ⟨fam: be understood⟩ c'è voluto un po' a capirlo

sinner /'sɪnə(r)/ n peccatore, -trice mf

sinus /'saɪnəs/ n seno m paranasale. ~**itis** n sinusite f

sip /sɪp/ n sorso m ● vt ⟨pt/pp **sipped**⟩ sorseggiare

siphon /'saɪfn/ n ⟨bottle⟩ sifone m ● **siphon off** vt travasare ⟨con sifone⟩

sir /sɜ:(r)/ n signore m; **S~** ⟨title⟩ Sir m; **Dear S~s** Spettabile ditta

siren /'saɪrən/ n sirena f

sissy /'sɪsɪ/ n femminuccia f

sister /'sɪstə(r)/ n sorella f; ⟨nurse⟩ [infermiera f] caposala f. ~-**in-law** n ⟨pl ~**s-in-law**⟩ cognata f. ~**ly** a da sorella

sit /sɪt/ v ⟨pt/pp **sat**, pres p **sitting**⟩ ● vi essere seduto; ⟨sit down⟩ sedersi; ⟨committee:⟩ riunirsi ● vt sostenere ⟨exam⟩. **sit back** vi fig starsene con le mani in mano. **sit down** vi mettersi a sedere. **sit up** vi mettersi seduto; ⟨not slouch⟩ star seduto diritto; ⟨stay up⟩ stare alzato

site /saɪt/ n posto m; Archaeol sito

m; ⟨building ~⟩ cantiere m ● vt collocare

sit-in /'sɪtɪn/ n occupazione f ⟨di fabbrica, ecc.⟩

sitting /'sɪtɪŋ/ n seduta f; ⟨for meals⟩ turno m. ~-**room** n salotto m

situat|e /'sɪtjʊeɪt/ vt situare. ~**ed** a situato. ~**ion** /-'eɪʃn/ n situazione f; ⟨location⟩ posizione f; ⟨job⟩ posto m

six /sɪks/ a sei. ~**teen** a sedici. ~**teenth** a sedicesimo

sixth /sɪksθ/ a sesto

sixtieth /'sɪkstɪɪθ/ a sessantesimo

sixty /'sɪkstɪ/ a sessanta

size /saɪz/ n dimensioni fpl; ⟨of clothes⟩ taglia f, misura f; ⟨of shoes⟩ numero m; **what ~ is the room?** che dimensioni ha la stanza? ● **size up** vt fam valutare

sizeable /'saɪzəbl/ a piuttosto grande

sizzle /'sɪzl/ vi sfrigolare

skate¹ /skeɪt/ n inv ⟨fish⟩ razza f

skate² /skeɪt/ n pattino m ● vi pattinare

skateboard /'skeɪtbɔːd/ n skateboard m inv

skater /'skeɪtə(r)/ n pattinatore, -trice mf

skating /'skeɪtɪŋ/ n pattinaggio m. ~-**rink** n pista f di pattinaggio

skeleton /'skelɪtn/ n scheletro m. ~ **key** n passe-partout m inv. ~ **staff** n personale m ridotto

sketch /sketʃ/ n schizzo m; Theat sketch m inv ● vt fare uno schizzo di

sketch|y /'sketʃɪ/ a (-**ier**, -**iest**) abbozzato. ~**ily** adv in modo abbozzato

skewer /'skjʊə(r)/ n spiedo m

ski /skiː/ n sci m inv ● vi ⟨pt/pp **skied**, pres p **skiing**⟩ sciare; **go ~ing** andare a sciare

skid /skɪd/ n slittata f ● vi ⟨pt/pp **skidded**⟩ slittare

skier /'skiːə(r)/ n sciatore, -trice mf

skiing /'skiːɪŋ/ n sci m

skilful /'skɪlfl/ a abile

'**ski-lift** n impianto m di risalita

skill /skɪl/ n abilità f inv. **~ed** a dotato; ⟨worker⟩ specializzato

skim /skɪm/ vt ⟨pt/pp **skimmed**⟩ schiumare; scremare ⟨milk⟩. **skim off** vt togliere. **skim through** vt scorrere

skimp /skɪmp/ vi **~ on** lesinare su

skimpy /ˈskɪmpɪ/ a (-ier, -iest) succinto

skin /skɪn/ n pelle f; ⟨on fruit⟩ buccia f ● vt ⟨pt/pp **skinned**⟩ spellare

skin-: **~-deep** a superficiale. **~-diving** n nuoto m subacqueo

skinflint /ˈskɪnflɪnt/ n miserabile mf

skinny /ˈskɪnɪ/ a (-ier, -iest) molto magro

skip[1] /skɪp/ n ⟨container⟩ benna f

skip[2] n salto m ● v ⟨pt/pp **skipped**⟩ ● vi saltellare; ⟨with rope⟩ saltare la corda ● vt omettere

skipper /ˈskɪpə(r)/ n skipper m inv

skipping-rope /ˈskɪpɪŋrəʊp/ n corda f per saltare

skirmish /ˈskɜːmɪʃ/ n scaramuccia f

skirt /skɜːt/ n gonna f ● vt costeggiare

skit /skɪt/ n bozzetto m comico

skittle /ˈskɪtl/ n birillo m

skive /skaɪv/ vi fam fare lo scansafatiche

skulk /skʌlk/ vi aggirarsi furtivamente

skull /skʌl/ n cranio m

skunk /skʌŋk/ n moffetta f

sky /skaɪ/ n cielo m. **~light** n lucernario m. **~scraper** n grattacielo m

slab /slæb/ n lastra f; ⟨slice⟩ fetta f; ⟨of chocolate⟩ tavoletta f

slack /slæk/ a lento; ⟨person⟩ fiacco ● vi fare lo scansafatiche. **slack off** vi rilassarsi

slacken /ˈslækn/ vi allentare; **~** [off] ⟨trade:⟩ rallentare; ⟨speed, rain:⟩ diminuire ● vt allentare; diminuire ⟨speed⟩

slacks /slæks/ npl pantaloni mpl sportivi

slag /slæg/ n scorie fpl ● **slag off**

vt ⟨pt/pp **slagged**⟩ Br fam criticare

slain /sleɪn/ see **slay**

slam /slæm/ v ⟨pt/pp **slammed**⟩ ● vt sbattere; ⟨fam: criticize⟩ stroncare ● vi sbattere

slander /ˈslɑːndə(r)/ n diffamazione f ● vt diffamare. **~ous** /-rəs/ a diffamatorio

slang /slæŋ/ n gergo m. **~y** a gergale

slant /slɑːnt/ n pendenza f; ⟨point of view⟩ angolazione f; **on the ~** in pendenza ● vt pendere; fig distorcere ⟨report⟩ ● vi pendere

slap /slæp/ n schiaffo m ● vt ⟨pt/pp **slapped**⟩ schiaffeggiare; ⟨put⟩ schiaffare ● adv in pieno

slap: **~-dash** a fam frettoloso. **~-up** a fam di prim'ordine

slash /slæʃ/ n taglio m ● vt tagliare; ridurre drasticamente ⟨prices⟩

slat /slæt/ n stecca f

slate /sleɪt/ n ardesia f ● vt fam fare a pezzi

slaughter /ˈslɔːtə(r)/ n macello m; ⟨of people⟩ massacro m ● vt macellare; massacrare ⟨people⟩. **~house** n macello m

Slav /slɑːv/ a slavo ● n slavo, -a mf

slave /sleɪv/ n schiavo, -a mf ● vi **[away]** lavorare come un negro. **~-driver** n schiavista mf

slav|ery /ˈsleɪvərɪ/ n schiavitù f inv. **~ish** a servile

Slavonic /slaˈvɒnɪk/ a slavo

slay /sleɪ/ vt ⟨pt slew, pp slain⟩ ammazzare

sleazy /ˈsliːzɪ/ a (-ier, -iest) sordido

sledge /sledʒ/ n slitta f. **~-hammer** n martello m

sleek /sliːk/ a liscio, lucente; ⟨well-fed⟩ pasciuto

sleep /sliːp/ n sonno m; **go to ~** addormentarsi; **put to ~** far addormentare ● v ⟨pt/pp **slept**⟩ ● vi dormire ● vt **~s six** ha sei posti letto. **~er** n Rail treno m con vagoni letto; ⟨compartment⟩ vagone m letto; **be a light/heavy ~er** avere il sonno leggero/pesante

sleeping: **~-bag** n sacco m a pelo. **~-car** n vagone m letto. **~-pill** n sonnifero m

sleep: **~less** a insonne. **~lessness** n insonnia f. **~-walker** n sonnambulo, -a mf. **~-walking** n sonnambulismo m

sleepy /'sliːpi/ a (-ier, -iest) assonnato; **be ~** aver sonno

sleet /sliːt/ n nevischio m ● vi **it is ~ing** nevischia

sleeve /sliːv/ n manica f; (for record) copertina f. **~less** a senza maniche

sleigh /sleɪ/ n slitta f

sleight /slaɪt/ n **~ of hand** gioco m di prestigio

slender /'slendə(r)/ a snello; (fingers, stem) affusolato; fig scarso; (chance) magro

slept /slept/ see **sleep**

sleuth /sluːθ/ n investigatore m, detective m inv

slew[1] /sluː/ vi girare

slew[2] see **slay**

slice /slaɪs/ n fetta f ● vt affettare; **~d bread** pane m a cassetta

slick /slɪk/ a liscio; (cunning) astuto ● n (of oil) chiazza f di petrolio

slid|e /slaɪd/ n scivolata f; (in playground) scivolo m; (for hair) fermaglio m (per capelli); Phot diapositiva f ● v (pt/pp **slid**) ● vi scivolare ● vt far scivolare. **~-rule** n regolo m calcolatore. **~ing** a scorrevole; (door, seat) scorrevole; **~ing scale** scala f mobile

slight /slaɪt/ a leggero; (importance) poco; (slender) esile. **~est** a minimo; **not in the ~est** niente affatto ● vt offendere ● n offesa f. **~ly** adv leggermente

slim /slɪm/ a (slimmer, slimmest) snello; fig scarso; (chance) magro ● vi dimagrire

slim|e /slaɪm/ n melma f. **~y** a melmoso; fig viscido

sling /slɪŋ/ n Med benda f al collo ● vt (pt/pp **slung**) fam lanciare

slip /slɪp/ n scivolata f; (mistake) lieve errore m; (petticoat) sottoveste f; (for pillow) federa f; (paper) scontrino m; **give sb the ~** fam sbarazzarsi di qcno; **~ of the tongue** lapsus m inv ● v (pt/pp **slipped**) ● vi scivolare; (go quickly) sgattaiolare; (decline) retrocedere ● vt **he ~ped it into his pocket** se l'è infilato in tasca; **~ sb's mind** sfuggire di mente a qcno. **slip away** vi sgusciar via; (time:) sfuggire. **slip into** vi infilarsi (clothes). **slip up** vi fam sbagliare

slipped 'disc n Med ernia f del disco

slipper /'slɪpə(r)/ n pantofola f

slippery /'slɪpərɪ/ a scivoloso

slip-road n bretella f

slipshod /'slɪpʃɒd/ a trascurato

'slip-up n fam sbaglio m

slit /slɪt/ n spacco m; (tear) strappo m; (hole) fessura f ● vt (pt/pp **slit**) tagliare

slither /'slɪðə(r)/ vi scivolare

sliver /'slɪvə(r)/ n scheggia f

slobber /'slɒbə(r)/ vi sbavare

slog /slɒg/ n [hard] **~** sgobbata f ● vi (pt/pp **slogged**) (work) sgobbare

slogan /'sləʊgən/ n slogan m inv

slop /slɒp/ v (pt/pp **slopped**) ● vt versare. **slop over** vi versarsi

slop|e /sləʊp/ n pendenza f; (ski ~) pista f ● vi essere inclinato, inclinarsi. **~ing** a in pendenza

sloppy /'slɒpɪ/ a (-ier, -iest) (work) trascurato; (worker) negligente; (in dress) sciatto; (sentimental) sdolcinato

slosh /slɒʃ/ vi fam (person, feet:) sguazzare; (water:) scrosciare ● vt (fam: hit) colpire

sloshed /slɒʃt/ a fam sbronzo

slot /slɒt/ n fessura f; (time-~) spazio m ● v (pt/pp **slotted**) ● vt infilare. **slot in** vi incastrarsi

'slot-machine n distributore m automatico; (for gambling) slot-machine f inv

slouch /slaʊtʃ/ vi (in chair) stare scomposto

sloven|ly /'slʌvnlı/ a sciatto. **~iness** n sciatteria f

slow /sləʊ/ a lento; **be** ~ ⟨clock:⟩ essere indietro; **in** ~ **motion** al rallentatore ● adv lentamente ● **slow down/up** vt/i rallentare

slow: **~coach** n fam tartaruga f. **~ly** adv lentamente. **~ness** n lentezza f

sludge /slʌdʒ/ n fanghiglia f

slug /slʌg/ n lumacone m; (bullet) pallottoia f

sluggish /'slʌgıʃ/ a lento

sluice /sluːs/ n chiusa f

slum /slʌm/ n (house) tugurio m; **~s** pl bassifondi mpl

slumber /'slʌmbə(r)/ vi dormire

slump /slʌmp/ n crollo m; (economic) depressione f ● vi crollare

slung /slʌŋ/ see **sling**

slur /slɜː(r)/ n (discredit) calunnia f ● vt (pt/pp slurred) biascicare

slurp /slɜːp/ vt/i bere rumorosamente

slush /slʌʃ/ n pantano m nevoso; fig sdolcinatezza f. **~ fund** n fondi mpl neri

slushy /'slʌʃı/ a fangoso; (sentimental) sdolcinato

slut /slʌt/ n sgualdrina f

sly /slaı/ a (-er, -est) scaltro ● n **on the ~** di nascosto

smack¹ /smæk/ n (on face) schiaffo m; (on bottom) sculaccione m ● vt (on face) schiaffeggiare; (on bottom) sculacciare; **~ one's lips** far schioccare le labbra ● adv fam in pieno

smack² vi **~ of** fig sapere di

small /smɔːl/ a piccolo; **be out/ work/etc until the ~ hours** fare le ore piccole ● adv **chop up** ~ fare a pezzettini ● n **the ~ of the back** le reni fpl

small: **~ ads** npl annunci mpl [commerciali]. **~ 'change** n spiccioli mpl. **~-holding** n piccola te-

nuta f. **~pox** n vaiolo m. **~ talk** n chiacchiere fpl

smarmy /'smɑːmı/ a (-ier, -iest) fam untuoso

smart /smɑːt/ a elegante; (clever) intelligente; (brisk) svelto; **be** ~ (fam: cheeky) fare il furbo ● vi (hurt) bruciare

smarten /'smɑːtn/ vt **~oneself up** farsi bello

smash /smæʃ/ n fragore m, (collision) scontro m; Tennis schiacciata f ● vt spaccare; Tennis schiacciare ● vi spaccarsi; (crash) schiantarsi (into contro). **~ [hit]** n successo m. **~ing** a fam fantastico

smattering /'smætərıŋ/ n infarinatura f

smear /smıə(r)/ n macchia f; Med striscio m ● vt imbrattare; (coat) spalmare (with di); fig calunniare

smell /smel/ n odore m, (sense) odorato m ● v (pt/pp smelt o smelled) ● vt odorare; (sniff) annusare ● vi odorare (of di)

smelly /'smelı/ a (-ier, -iest) puzzolente

smelt¹ /smelt/ see **smell**

smelt² vt fondere

smile /smaıl/ n sorriso m ● vi sorridere; **~ at** sorridere a ⟨sb⟩; sorridere di ⟨sth⟩

smirk /smɜːk/ n sorriso m compiaciuto

smithereens /smıðə'riːnz/ npl **to/in** ~ in mille pezzi

smitten /'smıtn/ a ~ **with** tutto preso da

smock /smɒk/ n grembiule m

smog /smɒg/ n smog m inv

smoke /sməʊk/ n fumo m ● vt/i fumare. **~less** a senza fumo; ⟨fuel⟩ che non fa fumo

smoker /'sməʊkə(r)/ n fumatore, -trice mf; Rail vagone m fumatori

'smoke-screen n cortina f di fumo

smoking /'sməʊkıŋ/ n fumo m; **'no ~'** 'vietato fumare'

smoky /'sməʊkı/ a (-ier, -iest) fumoso; ⟨taste⟩ di fumo

smooth /smu:ð/ a liscio; ⟨movement⟩ scorrevole; ⟨sea⟩ calmo; ⟨manners⟩ mellifluo ● vt lisciare. **smooth out** vt lisciare. **~ly** adv in modo scorrevole

smother /'smʌðə(r)/ vt soffocare

smoulder /'smǝʊldə(r)/ vi fumare; ⟨with rage⟩ consumarsi

smudge /smʌdʒ/ n macchia f ● vt/i imbrattare

smug /smʌg/ a (**smugger**, **smuggest**) compiaciuto. **~ly** adv con aria compiaciuta

smuggl|e /'smʌgl/ vt contrabbandare. **~er** n contrabbandiere, a, mf. **~ing** n contrabbando m

smut /smʌt/ n macchia f di fuliggine; fig sconcezza f

smutty /'smʌtɪ/ a (-**ier**, -**iest**) fuligginoso; fig sconcio

snack /snæk/ n spuntino m. **~-bar** n snack bar m inv

snag /snæg/ n ⟨problem⟩ intoppo m

snail /sneɪl/ n lumaca f; at a **~'s pace** a passo di lumaca

snake /sneɪk/ n serpente m

snap /snæp/ n colpo m secco; ⟨photo⟩ istantanea f ● attrib ⟨decision⟩ istantaneo ● v (pt/pp **snapped**) ● vi ⟨break⟩ spezzarsi; **~ at** ⟨dog:⟩ cercare di azzannare; ⟨person:⟩ parlare seccamente ● vt ⟨break⟩ spezzare; ⟨say⟩ dire seccamente; Phot fare un'istantanea di. **snap up** vt afferrare

snappy /'snæpɪ/ a (-**ier**, -**iest**) scorbutico; ⟨smart⟩ elegante; **make it ~!** sbrigati!

snapshot n istantanea f

snare /sneǝ(r)/ n trappola f

snarl /snɑ:l/ n ringhio m ● vi ringhiare

snatch /snætʃ/ n strappo m; ⟨fragment⟩ brano m; ⟨theft⟩ scippo m; **make a ~ at** cercare di afferrare qcsa ● vt strappare [di mano] ⟨from⟩ a); ⟨steal⟩ scippare; rapire ⟨child⟩

sneak /sni:k/ n fam spia m/f ● vi ⟨fam: tell tales⟩ fare la spia ● vt ⟨take⟩ rubare; **~ a look at** dare una sbirciata a. **sneak in/out** vi sgattaiolare dentro/fuori

sneakers /'sni:kǝz/ npl Am scarpe fpl da ginnastica

sneaking /'sni:kɪŋ/ a furtivo; ⟨suspicion⟩ vago

sneaky /'sni:kɪ/ a sornione

sneer /snɪə(r)/ n ghigno m ● vi sogghignare; ⟨mock⟩ ridere di

sneeze /sni:z/ n starnuto m ● vi starnutire

snide /snaɪd/ a fam insinuante

sniff /snɪf/ n ⟨of dog⟩ annusata f ● vi tirare su col naso ● vt odorare ⟨flower⟩; sniffare ⟨glue, cocaine⟩; ⟨dog:⟩ annusare

snigger /'snɪgə(r)/ n risatina f soffocata ● vi ridacchiare

snip /snɪp/ n taglio m; ⟨fam: bargain⟩ affare m ● vt/i (pt/pp **snipped**) **~ [at]** tagliare

snipe /snaɪp/ vi **~ at** tirare su; fig sparare a zero su. **~r** n cecchino m

snippet /'snɪpɪt/ n a **~ of information/news** una breve notizia/informazione

snivel /'snɪvl/ vi (pt/pp **snivelled**) piagnucolare. **~ling** a piagnucoloso

snob /snɒb/ n snob mf. **~bery** n snobismo m. **~bish** a da snob

snooker /'snu:kǝ(r)/ n snooker m

snoop /snu:p/ n spia f ● vi fam curiosare

snooty /'snu:tɪ/ a fam sdegnoso

snooze /snu:z/ n sonnellino m ● vi fare un sonnellino

snore /snɔ:(r)/ vi russare

snorkel /'snɔ:kl/ n respiratore m

snort /snɔ:t/ n sbuffo m ● vi sbuffare

snout /snaʊt/ n grugno m

snow /snǝʊ/ n neve f ● vi nevicare; **~ed under with** fig sommerso da

snow: **~ball** n palla f di neve ● vi fare a palle di neve. **~drift** n cumulo m di neve. **~drop** n bucaneve m. **~fall** n nevicata f. **~flake** n fiocco m di neve. **~man** n pupazzo m di neve. **~-plough** n spazzaneve _m._ **~storm** _n_ tormenta _f._ **~y** _a_ nevoso

snub /snʌb/ _n_ sgarbo _m_ ● _vt_ (_pt/pp_ **snubbed**) snobbare

'snub-nosed _a_ dal naso all'insù

snuff /snʌf/ _n_ tabacco _m_ da fiuto

snug /snʌg/ _a_ (**snugger**, **snuggest**) comodo; (_tight_) aderente

snuggle /'snʌgl/ _vi_ rannicchiarsi (**up to** accanto a)

so /səʊ/ _adv_ così; **so far** finora; **so am I** anch'io; **so I see** così pare; **that is so** è così; **so much** così tanto; **so much the better** tanto meglio; **so it is** proprio così; **if so** se è così; **so as to** in modo da; **so long!** fam a presto! ● _pron_ **I hope/think/am afraid so** spero/penso/temo di sì; **I told you so** te l'ho detto; **because I say so** perché lo dico io; **I did so!** è vero!; **so saying/doing,...** così dicendo/facendo,...; **or so** circa; **very much so** sì, molto; **and so forth** or **on e** così via ● _conj_ (_therefore_) perciò; (_in order that_) così; **so there ecco!**; **so what!** e allora?; **so where have you been?** allora, dove sei stato?

soak /səʊk/ _vt_ mettere a bagno ● _vi_ stare a bagno; **~ into** (_liquid:_) penetrare. **soak up** _vt_ assorbire

soaking /'səʊkɪŋ/ _n_ ammollo _m_ ● _a_ & _adv_ **~ [wet]** _fam_ inzuppato

so-and-so /'səʊənsəʊ/ _n_ Tal dei Tali _mf_; (_euphemism_) specie _f_ di imbecille

soap /səʊp/ _n_ sapone _m_. **~ opera** _n_ telenovela _f_, soap opera _f inv_. **~ powder** _n_ detersivo _m_ in polvere

soapy /'səʊpɪ/ _a_ (**-ier**, **-iest**) insaponato

soar /sɔː(r)/ _vi_ elevarsi; (_prices:_) salire alle stelle

sob /sɒb/ _n_ singhiozzo _m_ ● _vi_ (_pt/pp_ **sobbed**) singhiozzare

sober /'səʊbə(r)/ _a_ sobrio; (_serious_) serio ● **sober up** _vi_ ritornare sobrio

'so-called _a_ cosiddetto

soccer /'sɒkə(r)/ _n_ calcio _m_

sociable /'səʊʃəbl/ _a_ socievole

social /'səʊʃl/ _a_ sociale; (_sociable_) socievole

socialis|m /'səʊʃəlɪzm/ _n_ socialismo _m._ **~t** /-ɪst/ _a_ socialista ● _n_ socialista _mf_

socialize /'səʊʃəlaɪz/ _vi_ socializzare

socially /'səʊʃəlɪ/ _adv_ socialmente; **know sb ~** frequentare qcno

social: ~ se'curity _n_ previdenza _f_ sociale. **~ work** _n_ assistenza _f_ sociale. **~ worker** _n_ assistente _mf_ sociale

society /sə'saɪətɪ/ _n_ società _f inv_

sociolog|ist /səʊsɪ'ɒlədʒɪst/ _n_ sociologo, -a _mf._ **~y** _n_ sociologia _f_

sock¹ /sɒk/ _n_ calzino _m_; (_kneelength_) calza _f_

sock² /sɒk/ _n fam_ pugno _m_ ● _vt fam_ dare un pugno a

socket /'sɒkɪt/ _n_ (_wall plug_) presa _f_ [di corrente]; (_for bulb_) portalampada _m inv_

soda /'səʊdə/ _n_ soda _f_; _Am_ gazzosa _f._ **~ water** _n_ seltz _m inv_

sodden /'sɒdn/ _a_ inzuppato

sodium /'səʊdɪəm/ _n_ sodio _m_

sofa /'səʊfə/ _n_ divano _m._ **~ bed** _n_ divano _m_ letto

soft /sɒft/ _a_ morbido, soffice; (_voice_) sommesso; (_light, colour_) tenue; (_not strict_) indulgente; (_fam: silly_) stupido; **have a ~ spot for sb** avere un debole per qcno. **~ drink** _n_ bibita _f_ analcolica

soften /'sɒfn/ _vt_ ammorbidire; _fig_ attenuare ● _vi_ ammorbidirsi

softly /'sɒftlɪ/ _adv_ (_say_) sottovoce; (_treat_) con indulgenza; (_play music_) in sottofondo

soft: ~ toy _n_ pupazzo _m_ di peluche. **~ware** _n_ software _m_

soggy /'sɒgɪ/ _a_ (**-ier**, **-iest**) zuppo

soil¹ /sɔɪl/ _n_ suolo _m_

soil² _vt_ sporcare

solar /'səʊlə(r)/ _a_ solare

sold /səʊld/ _see_ **sell**

solder /'səʊldə(r)/ _n_ lega _f_ da saldatura ● _vt_ saldare

soldier /'səʊldʒə(r)/ n soldato m
● **soldier on** vi perseverare

sole¹ /səʊl/ n (of foot) pianta f; (of shoe) suola f

sole² n (fish) sogliola f

sole³ a unico, solo. **~ly** adv unicamente

solemn /'sɒləm/ a solenne. **~ity** /sə'lemnəti/ n solennità f inv

solicit /sə'lɪsɪt/ vt sollecitare • vi (prostitute:) adescare

solicitor /sə'lɪsɪtə(r)/ n avvocato m

solid /'sɒlɪd/ a solido; (oak, gold) massiccio ● n (figure) solido m; **~s** pl (food) cibi mpl solidi

solidarity /sɒlɪ'dærətɪ/ n solidarietà f inv

solidify /sə'lɪdɪfaɪ/ vi (pt/pp -ied) solidificarsi

soliloquy /sə'lɪləkwɪ/ n soliloquio m

solitaire /sɒlɪ'teə/ n solitario m

solitary /'sɒlɪtərɪ/ a solitario; (sole) solo. **~ con'finement** n cella f di isolamento

solitude /'sɒlɪtjuːd/ n solitudine f

solo /'səʊləʊ/ n Mus assolo m ● a (flight) in solitario ● adv in solitario. **~ist** n solista mf

solstice /'sɒlstɪs/ n solstizio m

soluble /'sɒljəbl/ a solubile

solution /sə'luːʃn/ n soluzione f

solve /sɒlv/ vt risolvere

solvent /'sɒlvənt/ a solvente ● n solvente m

sombre /'sɒmbə(r)/ a tetro; (clothes) scuro

some /sʌm/ a (a certain amount of) del; (a certain number of) qualche, alcuni; **~ day** un giorno o l'altro; **I need ~ money/books** ho bisogno di soldi/libri; **do ~ shopping** fare qualche acquisto ● pron (a certain amount) un po'; (a certain number) alcuni; **I want ~** ne voglio

some: **~body** /-bɒdɪ/ pron & n qualcuno m. **~how** adv in qualche modo; **~how or other** in un modo o nell'altro. **~one** pron & n = **somebody**

somersault /'sʌməsɔːlt/ n capriola f; **turn a ~** fare una capriola

'something pron qualche cosa, qualcosa; **~ different** qualcosa di diverso; **~ like** un po' come; (approximately) qualcosa come; **see ~ of sb** vedere qcno un po'

some: **~time** adv un giorno o l'altro; **~time last summer** durante l'estate scorsa. **~times** adv qualche volta. **~what** adv piuttosto. **~where** adv da qualche parte ● pron **~where to eat** un posto in cui mangiare

son /sʌn/ n figlio m

sonata /sə'nɑːtə/ n sonata f

song /sɒŋ/ n canzone f

sonic /'sɒnɪk/ a sonico. **~ 'boom** n bang m inv sonico

'son-in-law n (pl **~s-in-law**) genero m

sonnet /'sɒnɪt/ n sonetto m

soon /suːn/ adv presto; (in a short time) tra poco; **as ~ as** [non] appena; **as ~ as possible** il più presto possibile; **~er or later** prima o poi; **the ~er the better** prima è, meglio è; **no ~er had I arrived than...** ero appena arrivato quando...; **I would ~er go** preferirei andare; **~ after** subito dopo

soot /sʊt/ n fuliggine f

soothe /suːð/ vt calmare

sooty /'sʊtɪ/ a fuligginoso

sophisticated /sə'fɪstɪkeɪtɪd/ a sofisticato

soporific /sɒpə'rɪfɪk/ a soporifero

sopping /'sɒpɪŋ/ a & adv be **~** [wet] essere bagnato fradicio

soppy /'sɒpɪ/ a (-ier, -iest) fam svenevole

soprano /sə'prɑːnəʊ/ n soprano m

sordid /'sɔːdɪd/ a sordido m

sore /sɔː(r)/ a dolorante; (Am: vexed) arrabbiato; **it's ~** fa male; **have a ~ throat** avere mal di gola ● n piaga f. **~ly** adv (tempted) seriamente

sorrow /'sɒrəʊ/ n tristezza f. **~ful** a triste

sorry /'sɒrɪ/ a (-ier, -iest) (sad)

spiacente; (*wretched*) pietoso;
you'll be ~! te ne pentirai!; **I am ~** mi dispiace; **be** *or* **feel ~ for** provare compassione per; **~!** scusa!; (*more polite*) scusi!

sort /sɔːt/ *n* specie *f*, (*fam*: *person*) tipo *m*; **it's a ~ of fish** è un tipo di pesce; **be out of ~s** (*fam*: *unwell*) stare poco bene ● *vt* classificare. **sort out** *vt* selezionare (*papers*); *fig* risolvere (*problem*); occuparsi di (*person*)

'so ɔ *a & adv* così così

sought /sɔːt/ *see* **seek**

soul /səʊl/ *n* anima *f*

sound¹ /saʊnd/ *a* sano; (*sensible*) saggio; (*secure*) solido; (*thrashing*) clamoroso ● *adv* **~ asleep** profondamente addormentato

sound² *n* suono *m*; (*noise*) rumore *m*; **I don't like the ~ of it** *fam* non mi suona bene ● *vi* suonare; (*seem*) aver l'aria ● *vt* (*pronounce*) pronunciare; *Med* ascoltare (*chest*). **~ barrier** *n* muro *m* del suono. **~ card** *n Comput* scheda *f* sonora. **~less** *a* silenzioso. **sound out** *vt fig* sondare

soundly /'saʊndlɪ/ *adv* (*sleep*) profondamente; (*defeat*) clamorosamente

'sound: ~proof *a* impenetrabile al suono. **~track** *n* colonna *f* sonora

soup /suːp/ *n* minestra *f*. **~ed-up** *a fam* (*engine*) truccato

soup: ~-plate *n* piatto *m* fondo. **~-spoon** *n* cucchiaio *m* da minestra

sour /'saʊə(r)/ *a* agro; (*not fresh & fig*) acido

source /sɔːs/ *n* fonte *f*

south /saʊθ/ *n* sud *m*; **to the ~ of** a sud di ● *a* del sud, meridionale ● *adv* verso il sud

south: S~ 'Africa *n* Sudafrica *m*. **S~ A'merica** *n* America *f* del Sud. **S~ American** *a & n* sud-americano, -a *mf*. **~-'east** *n* sud-est *m*

southerly /'sʌðəlɪ/ *a* del sud

southern /'sʌðən/ *a* del sud, meri-

dionale; **~ Italy** il Mezzogiorno *m*. **~er** *n* meridionale *mf*

South 'Pole *n* polo *m* Sud

'southward[s] /-wəd[z]/ *adv* verso sud

souvenir /suːvə'nɪə(r)/ *n* ricordo *m*, souvenir *m* *inv*

sovereign /'sɒvrɪn/ *a* sovrano ● *n* sovrano, -a *mf*. **~ty** *n* sovranità *f* *inv*

Soviet /'səʊvɪət/ *a* sovietico; **~ Un:on** Unione *f* Sovietica

sow¹ /saʊ/ *n* scrofa *f*

sow² /səʊ/ *vt* (*pt* **sowed**, *pp* **sown** *or* **sowed**) seminare

soya /'sɔɪə/ *n* **~ bean** soia *f*

spa /spɑː/ *n* stazione *f* termale

space /speɪs/ *n* spazio *m* ● *a* (*research etc*) spaziale ● *vt* ~ [**out**] distanziare

space: ~craft *n* navetta *f* spaziale. **~ship** *n* astronave *f*

spacious /'speɪʃəs/ *a* spazioso

spade /speɪd/ *n* vanga *f*; (*for child*) paletta *f*; **~s** *pl* (*in cards*) picche *fpl*. **~work** *n* lavoro *m* preparatorio

Spain /speɪn/ *n* Spagna *f*

span¹ /spæn/ *n* spanna *f*; (*of arch*) luce *f*; (*of time*) arco *m*; (*of wings*) apertura *f* ● *vt* (*pt/pp* **spanned**) estendersi su

span² *see* **spick**

Span:iard /'spænjəd/ *n* spagnolo, -a *mf*. **~ish** *a* spagnolo ● *n* (*language*) spagnolo *m*; **the ~ish** *pl* gli spagnoli

spank /spæŋk/ *vt* sculacciare. **~ing** *n* sculacciata *f*

spanner /'spænə(r)/ *n* chiave *f* inglese

spar /spɑː(r)/ *vi* (*pt/pp* **sparred**) (*boxing*) allenarsi; (*argue*) litigare

spare /speə(r)/ *a* (*surplus*) in più; (*additional*) di riserva ● *n* (*part*) ricambio *m* ● *vt* risparmiare; (*do without*) fare a meno di; **can you ~ five minutes?** avresti cinque minuti?; **to ~** (*surplus*) in eccedenza. **~ part** *n* pezzo *m* di ricam-

563   sparing | spell

bio. ~ time n tempo m libero. ~ 'wheel n ruota f di scorta

sparing /'speərɪŋ/ a parco (**with** di). ~**ly** adv con parsimonia

spark /spɑːk/ n scintilla f. ~**ing-plug** n Auto candela f

spark|le /'spɑːkl/ n scintillio m ● vi scintillare. ~**ing** a frizzante; ⟨wine⟩ spumante

sparrow /'spærəʊ/ n passero m

sparse /spɑːs/ a rado. ~**ly** adv scarsamente; ~**ly populated** a bassa densità di popolazione

spartan /'spɑːtn/ a spartano

spasm /'spæzm/ n spasmo m. ~**odic** /-'mɒdɪk/ a spasmodico

spastic /'spæstɪk/ a spastico ● n spastico, -a mf

spat /spæt/ see **spit**[1]

spate /speɪt/ n (series) successione f; **be in full** ~ essere in piena

spatial /'speɪʃl/ a spaziale

spatter /'spætə(r)/ vt schizzare

spatula /'spætjʊlə/ n spatola f

spawn /spɔːn/ n uova fpl (di pesci, rane, ecc.) ● vi deporre le uova ● vt fig generare

spay /speɪ/ vt sterilizzare

speak /spiːk/ v (pt **spoke**, pp **spoken**) ● vi parlare (**to** a); ~**ing!** Teleph sono io! ● vt dire; ~**one's mind** dire quello che si pensa. **speak for** vi parlare a nome di. **speak up** vi parlare più forte; ~ **up for oneself** parlare a favore di

speaker /'spiːkə(r)/ n parlante mf; (in public) oratore, -trice mf; (of stereo) cassa f

spear /spɪə(r)/ n lancia f

spec /spek/ n **on** ~ fam senza certezza

special /'speʃl/ a speciale. ~**ist** n specialista mf. ~**ity** /-ʃɪ'ælətɪ/ n specialità f inv

special|ize /'speʃəlaɪz/ vi specializzarsi. ~**ly** adv specialmente; (particularly) particolarmente

species /'spiːʃiːz/ n specie f inv

specific /spə'sɪfɪk/ a specifico. ~**ally** adv in modo specifico

specifications /spesɪfɪ'keɪʃnz/ npl descrizione f

specify /'spesɪfaɪ/ vt (pt/pp -**ied**) specificare

specimen /'spesɪmən/ n campione m

speck /spek/ n macchiolina f; (particle) granello m

speckled /'spekld/ a picchiettato

specs /speks/ npl fam occhiali mpl

spectacle /'spektəkl/ n (show) spettacolo m. ~**s** npl occhiali mpl

spectacular /spek'tækjʊlə(r)/ a spettacolare

spectator /spek'teɪtə(r)/ n spettatore, -trice mf

spectre /'spektə(r)/ n spettro m

spectrum /'spektrəm/ n (pl -**tra**) spettro m; fig gamma f

specula|te /'spekjʊleɪt/ vi speculare. ~**ion** /-'leɪʃn/ n speculazione f. ~**ive** /-ɪv/ a speculativo. ~**or** n speculatore, -trice mf

sped /sped/ see **speed**

speech /spiːtʃ/ n linguaggio m; (address) discorso m. ~**less** a senza parole

speed /spiːd/ n velocità f inv; (gear) marcia f; **at** ~ a tutta velocità ● vi (pt/pp **sped**) andare veloce; (pt/pp **speeded**) (go too fast) andare a velocità eccessiva. **speed up** (pt/pp **speeded up**) vt/i accelerare

speed: ~**boat** n motoscafo m. ~**ing** n eccesso m di velocità. ~**limit** n limite m di velocità

speedometer /spiː'dɒmɪtə(r)/ n tachimetro m

speed|y /'spiːdɪ/ a (-**ier**, -**iest**) rapido. ~**ily** adv rapidamente

spell[1] /spel/ n (turn) turno m; (of weather) periodo m

spell[2] /spel/ v (pt/pp **spelled**, **spelt**) ● vt how do you ~...? come si scrive...?; **could you** ~ **that for me?** me lo può compitare?; ~ **disaster** essere disastroso ● vi **he can't** ~ fa molti errori d'ortografia

spell[3] /spel/ n (magic) incantesimo m. ~**bound** a affascinato

spelling /'spelɪŋ/ n ortografia f

spelt /spelt/ see **spell²**

spend /spend/ vt/i (pt/pp **spent**) spendere; passare (time)

spent /spent/ see **spend**

sperm /spɜːm/ n spermatozoo m; (semen) sperma m

spew /spju:/ vt/i vomitare

sphere /sfɪə(r)/ n sfera f. **~ical** /'sferɪkl/ a sferico

spice /spaɪs/ n spezia f; fig pepe m

spick /spɪk/ a **~ and span** lindo

spicy /'spaɪsɪ/ a piccante

spider /'spaɪdə(r)/ n ragno m

spike /spaɪk/ n punta f; Bot, Zool spina f; (on shoe) chiodo m. **~y** a (plant) pungente

spill /spɪl/ v (pt/pp **spilt** or **spilled**) ● vt versare (blood) ● vi rovesciarsi

spin /spɪn/ v (pt/pp **spun**, pres p **spinning**) ● vt far girare; filare (wool); centrifugare (washing) ● vi girare; (washing machine:) centrifugare ● n rotazione f; (short drive) giretto m. **spin out** vt far durare

spinach /'spɪnɪdʒ/ n spinaci mpl

spinal /'spaɪnl/ a spinale. **~ 'cord** n midollo m spinale

spindle /'spɪndl/ n fuso m. **~y** a affusolato

spin-'drier n centrifuga f

spine /spaɪn/ n spina f dorsale; (of book) dorso m; Bot, Zool spina f. **~less** a fig smidollato

spinning /'spɪnɪŋ/ n filatura f. **~-wheel** n filatoio m

'spin-off n ricaduta f

spiral /'spaɪrəl/ a a spirale ● n spirale f ● vi (pt/pp **spiralled**) formare una spirale. **~ 'staircase** n scala f a chiocciola

spire /'spaɪə(r)/ n guglia f

spirit /'spɪrɪt/ n spirito m; (courage) ardore m; **~s** pl (alcohol) liquori mpl; **in good ~s** di buon umore; **in low ~s** abbattuto

spirited /'spɪrɪtɪd/ a vivace; (courageous) pieno d'ardore

spirit: **~-level** n livella f a bolla

d'aria. **~ stove** n fornellino m [da campeggio]

spiritual /'spɪrɪtjʊəl/ a spirituale ● n spiritual m. **~ism** /-tɪzm/ n spiritismo m. **~ist** /-ɪst/ n spiritista mf

spit¹ /spɪt/ n (for roasting) spiedo m

spit² n sputo m ● vt/i (pt/pp **spat**, pres p **spitting**) sputare; (cat:) soffiare; (fat:) sfrigolare; **it's ~ting [with rain]** piovviggina; **the ~ting image of** il ritratto spiccicato di

spite /spaɪt/ n dispetto m; **in ~ of** malgrado ● vt far dispetto a. **~ful** a indispettito

spittle /'spɪtl/ n saliva f

splash /splæʃ/ n schizzo m; (of colour) macchia f; (fam: drop) goccio m ● vt schizzare; **~ sb with** sth schizzare qcno di qcsa ● vi schizzare. **splash about** vi schizzarsi. **splash down** vi (spacecraft:) ammarare

spleen /spli:n/ n Anat milza f

splendid /'splendɪd/ a splendido

splendour /'splendə(r)/ n splendore m

splint /splɪnt/ n Med stecca f

splinter /'splɪntə(r)/ n scheggia f ● vi scheggiarsi

split /splɪt/ n fessura f; (quarrel) rottura f; (division) scissione f; (tear) strappo m ● v (pt/pp **split**, pres p **splitting**) ● vt spaccare; (share, divide) dividere; (tear) strappare ● vi spaccarsi; (tear) strapparsi; (divide) dividersi; **~ on sb fam** denunciare qcno ● a **~ second** una frazione f di secondo. **split up** vt dividersi ● vi (couple:) separarsi

splutter /'splʌtə(r)/ vi farfugliare

spoil /spɔɪl/ **~s** pl bottino msg ● v (pt/pp **spoilt** or **spoiled**) ● vt rovinare; viziare (person) ● vi andare a male. **~sport** n guastafeste mf inv

spoke¹ /spəʊk/ n raggio m

spoke², **spoken** /'spəʊkn/ see **speak**

'spokesman n portavoce m inv

sponge /spʌndʒ/ n spugna f ● vt pulire (con la spugna) ● vi ~ **on** scroccare da. **~-cake** n pan m di Spagna

spong|er /'spʌndʒə(r)/ n scroccone, -a mf. **~y** a spugnoso

sponsor /'spɒnsə(r)/ n garante m; Radio, TV sponsor m inv; (godparent) padrino m, madrina f; (for membership) socio, -a mf garante ● vt sponsorizzare. **~ship** n sponsorizzazione f

spontaneous /spɒn'teɪnɪəs/ a spontaneo

spoof /spuːf/ n fam parodia f

spooky /'spuːkɪ/ a (-ier, -iest) fam sinistro

spool /spuːl/ n bobina f

spoon /spuːn/ n cucchiaio m ● vt mettere col cucchiaio. **~-feed** vt (pt/pp -fed) fig imboccare. **~ful** n cucchiaiata f

sporadic /spə'rædɪk/ a sporadico

sport /spɔːt/ n sport m inv ● vt sfoggiare. **~ing** a sportivo; **~ing chance** possibilità f inv

sports: **~car** n automobile f sportiva. **~ coat** n, **~ jacket** n giacca f sportiva. **~man** n sportivo m. **~woman** n sportiva f

sporty /'spɔːtɪ/ a (-ier, -iest) sportivo

spot /spɒt/ n macchia f; (pimple) brufolo m; (place) posto m; (in pattern) pois m inv; (of rain) goccia f; (of water) goccio m; ~ **pl** (rash) sfogo msg; a **~ of** fam un po' di; a **~ of bother** qualche problema; **on the ~** sul luogo; (immediately) immediatamente; **in a [tight] ~** fam in difficoltà ● vt (pt/pp **spotted**) macchiare; (fam: notice) individuare

spot: ~ **'check** n (without warning) controllo m a sorpresa; **do a ~ check on sth** dare una controllata a qcsa. **~less** a immacolato. **~light** n riflettore m

spotted /'spɒtɪd/ a (material) a pois

spotty /'spɒtɪ/ a (-ier, -iest) (pimply) brufoloso

spouse /spaʊz/ n consorte mf

spout /spaʊt/ n becco m ● vi zampillare (**from** da)

sprain /spreɪn/ n slogatura f ● vt slogare

sprang /spræŋ/ see **spring²**

sprawl /sprɔːl/ vi (in chair) stravaccarsi; (city etc:) estendersi; **go ~ing** (fall) cadere disteso

spray /spreɪ/ n spruzzo m; (preparation) spray m inv; (container) spruzzatore m inv ● vt spruzzare. **~-gun** n pistola f a spruzzo

spread /spred/ n estensione f; (of disease) diffusione f; (paste) crema f; (fam: feast) banchetto m ● v (pt/pp **spread**) ● vt spargere; spalmare (butter, jam); stendere (cloth, arms); diffondere (news, disease); dilazionare (payments); ~ **sth with** spalmare qcsa di ● vi spargersi; (butter:) spalmarsi; (disease:) diffondersi. **~sheet** n Comput foglio m elettronico. **spread out** vt sparpagliare ● vi sparpagliarsi

spree /spriː/ n fam **go on a ~** far baldoria; **go on a shopping ~** fare spese folli

sprig /sprɪg/ n rametto m

sprightly /'spraɪtlɪ/ a (-ier, -iest) vivace

spring¹ /sprɪŋ/ n primavera f ● attrib primaverile

spring² /sprɪŋ/ n (jump) balzo m; (water) sorgente f; (device) molla f; (elasticity) elasticità f inv ● v (pt **sprang**, pp **sprung**) ● vi balzare; (arise) provenire (**from** da). **he just sprang it on me** me l'ha detto a cose fatte compiuto. **spring up** vi balzare; fig spuntare

spring: **~board** n trampolino m. **~-'cleaning** n pulizie fpl di Pasqua. **~time** n primavera f

sprinkl|e /'sprɪŋkl/ vt (scatter) spruzzare (liquid); spargere (flour, cocoa); ~ **sth with** spruzzare qcsa di (liquid); cospargere qcsa di *(flour, cocoa)*. **~er** *n* sprinkler *m inv*; *(for lawn)* irrigatore *m*. **~ing** *n* (*of liquid*) spruzzatina *f*; (*of pepper, salt*) pizzico *m*; (*of flour, sugar*) spolveratina *f*; (*of knowledge*) infarinatura *f*; (*of people*) pugno *m*

sprint /sprint/ *n* sprint *m inv* ●*vi* fare uno sprint; *Sport* sprintare. **~er** sprinter *mf inv*

sprout /spraʊt/ *n* germoglio *m*; [**Brussels**] **~s** *pl* cavolini *mpl* di Bruxelles ●*vi* germogliare

spruce /spruːs/ *a* elegante ●*n* abete *m*

sprung /sprʌŋ/ *see* **spring**² ●*a* molleggiato

spud /spʌd/ *n fam* patata *f*

spun /spʌn/ *see* **spin**

spur /spɜː(r)/ *n* sperone *m*; (*stimulus*) stimolo *m*; (*road*) svincolo *m*; **on the ~ of the moment** su due piedi ●*vt* (*pt/pp* **spurred**) **~** [**on**] *fig* spronare [a]

spurious /'spjʊərɪəs/ *a* falso

spurn /spɜːn/ *vt* sdegnare

spurt /spɜːt/ *n* getto *m*; *Sport* scatto *m*; **put on a ~** fare uno spurt ●*vi* sprizzare; (*increase speed*) scattare

spy /spaɪ/ *n* spia *f* ●*v* (*pt/pp* **spied**) ●*vi* spiare ●*vt* (*fam*: *see*) spiare. **spy on** *vi* spiare

spying /'spaɪɪŋ/ *n* spionaggio *m*

squabble /'skwɒbl/ *n* bisticcio *m* ●*vi* bisticciare

squad /skwɒd/ *n* squadra *f*; *Sport* squadra

squadron /'skwɒdrən/ *n* *Mil* squadrone *m*; *Aeron, Naut* squadriglia *f*

squalid /'skwɒlɪd/ *a* squallido

squalor /'skwɒlə(r)/ *n* squallore *m*

squander /'skwɒndə(r)/ *vt* sprecare

square /skweə(r)/ *a* quadrato; (*meal*) sostanzioso; (*fam*: *old-fashioned*) vecchio stampo; **all ~** *fam pari* ●*n* quadrato *m*; (*in city*) piazza *f*; (*on chessboard*) riquadro *m* ●*vt* (*settle*) far quadrare; *Math*

elevare al quadrato ●*vi* (*agree*) armonizzare

squash /skwɒʃ/ *n* (*drink*) spremuta *f*; (*sport*) squash *m*; (*vegetable*) zucca *f* ●*vt* schiacciare; soffocare (*rebellion*)

squat /skwɒt/ *a* tarchiato ●*n fam* edificio *m* occupato abusivamente ●*vi* (*pt/pp* **squatted**) accovacciarsi; **~ in** occupare abusivamente. **~ter** *n* occupante *mf* abusivo, -a

squawk /skwɔːk/ *n* gracchio *m* ●*vi* gracchiare

squeak /skwiːk/ *n* squittio *m*; (*of hinge, brakes*) scricchiolio *m* ●*vi* squittire; (*hinge, brakes*:) scricchiolare

squeal /skwiːl/ *n* strillo *m*; (*of brakes*) cigolio *m* ●*vi* strillare; *sl* spifferare

squeamish /'skwiːmɪʃ/ *a* dallo stomaco delicato

squeeze /skwiːz/ *n* stretta *f*; (*crush*) pigia pigia *m inv* ●*vt* premere; (*to get juice*) spremere; stringere (*hand*); (*force*) spingere a forza; (*fam*: *extort*) estorcere (*out of* da). **squeeze in/out** *vi* sgusciare dentro/fuori. **squeeze up** *vi* stringersi

squelch /skweltʃ/ *vi* sguazzare

squid /skwɪd/ *n* calamaro *m*

squiggle /'skwɪɡl/ *n* scarabocchio *m*

squint /skwɪnt/ *n* strabismo *m* ●*vi* essere strabico

squire /skwaɪə(r)/ *n* signorotto *m* di campagna

squirm /skwɜːm/ *vi* contorcersi; (*feel embarrassed*) sentirsi imbarazzato

squirrel /'skwɪrəl/ *n* scoiattolo *m*

squirt /skwɜːt/ *n* spruzzo *m*; (*fam*: *person*) presuntuoso *m* ●*vt/i* spruzzare

St *abbr* (**Saint**) S; *abbr* **Street**

stab /stæb/ *n* pugnalata *f*, coltellata *f*; (*sensation*) fitta *f*; (*fam*: *attempt*) tentativo *m* ●*vt* (*pt/pp* **stabbed**) pugnalare, accoltellare

stability /stə'bɪlətɪ/ n stabilità f inv

stabilize /'steɪbɪlaɪz/ vt stabilizzare ● vi stabilizzarsi

stable¹ /'steɪbl/ a stabile

stable² n stalla f; (establishment) scuderia f

stack /stæk/ n catasta f; (of chimney) comignolo m; (chimney) ciminiera f; (fam: large quantity) montagna f ● vt accatastare

stadium /'steɪdɪəm/ n stadio m

staff /stɑːf/ n (stick) bastone m; (employees) personale m; (teachers) corpo m insegnante; Mil Stato m Maggiore ● vt fornire di personale. **~-room** n Sch sala f insegnanti

stag /stæg/ n cervo m

stage /steɪdʒ/ n palcoscenico m; (profession) teatro m; (in journey) tappa f; (in process) stadio m; **go on the ~** darsi al teatro; **by** or **in ~s** a tappe ● vt mettere in scena; (arrange) organizzare

stage: ~ **door** n ingresso m degli artisti. ~ **fright** n panico m da scena. ~ **manager** n direttore, -trice mf di scena

stagger /'stægə(r)/ vi barcollare ● vt sbalordire; scaglionare (holidays etc); **I was ~ed** sono rimasto sbalordito ● n vacillamento m. **~ing** a sbalorditivo

stagnant /'stægnənt/ a stagnante

stagnate /stæg'neɪt/ vi fig [ristagnare. **~ion** /-'neɪʃn/ n fig inattività f

'stag party n addio m al celibato

staid /steɪd/ a posato

stain /steɪn/ n macchia f; (for wood) mordente m ● vt macchiare; (wood) dare il mordente a; **~ed glass** vetro m colorato; **~ed-glass window** vetrata f colorata. **~less** a senza macchia; (steel) inossidabile. **~ remover** n smacchiatore m

stair /steə(r)/ n gradino m; **~s** pl scale fpl. **~case** n scale fpl

stake /steɪk/ n palo m; (wager) po-

sta f; Comm partecipazione f; **at ~** in gioco ● vt puntellare; (wager) scommettere

stale /steɪl/ a stantio; (air) viziato; (uninteresting) trito [e ritrito]. **~mate** n (in chess) stallo m; (deadlock) situazione f di stallo

stalk¹ /stɔːk/ n gambo m

stalk² vt inseguire ● vi camminare impettito

stall /stɔːl/ n box m inv; **~s** pl Theat platea f; (in market) bancarella f ● vi (engine:) spegnersi; fig temporeggiare ● vt far spegnere (engine); tenere a bada (person)

stallion /'stæljən/ n stallone m

stalwart /'stɔːlwət/ a fedele

stamina /'stæmɪnə/ n [capacità f inv di] resistenza f

stammer /'stæmə(r)/ n balbettio m ● vt/i balbettare

stamp /stæmp/ n (postage ~) francobollo m; (instrument) timbro m; fig impronta f ● vt affrancare (letter); timbrare (bill); battere (feet). **stamp out** vt spegnere; fig soffocare

stampede /stæm'piːd/ n fuga f precipitosa; fam fuggi-fuggi m ● vi fuggire precipitosamente

stance /stɑːns/ n posizione f

stand /stænd/ n (for bikes) rastrelliera f; (at exhibition) stand m inv; (in market) bancarella f; (in stadium) gradinata f inv; fig posizione f ● vi (pt/pp **stood**) ● vi stare in piedi; (rise) alzarsi [in piedi]; (be) trovarsi; (be candidate) essere candidato (**for** a); (stay valid) rimanere valido; **~ still** non muoversi; **I don't know where I ~** non so qual'è la mia posizione; **~ firm** fig tener duro; **~ together** essere solidali; **~ to lose/gain** rischiare di perdere/vincere; **~ to reason** essere logico ● vt (withstand) resistere a; (endure) sopportare; (place) mettere; **~ a chance** avere una possibilità; **~ one's ground** tener duro; **~ the test of time** superare la prova del tempo; **~ sb a**

beer offrire una birra a qcno.
stand by *vi* stare a guardare; *(be ready)* essere pronto ● *vt (support)* appoggiare. **stand down** *vi (retire)* ritirarsi. **stand for** *vt (mean)* significare; *(tolerate)* tollerare. **stand in for** *vt* sostituire. **stand out** *vi* spiccare. **stand up** *vi* alzarsi [in piedi]. **stand up for** *vt* prendere le difese di; **~ up for oneself** farsi valere. **stand up to** *vt* affrontare.

standard /'stændəd/ *a* standard; **be ~ practice** essere pratica corrente ● *n* standard *m inv*; *Techn* norma *f*; *(level)* livello *m*; *(quality)* qualità *f inv*; *(flag)* stendardo *m*; **~s** *pl (morals)* valori *mpl*; **~ of living** tenore *m* di vita. **~ize** *vt* standardizzare

'standard lamp *n* lampada *f* a stelo

'stand-by *n* riserva *f*; **on ~** *(at airport)* in lista d'attesa

'stand-in *n* controfigura *f*

standing /'stændɪŋ/ *a (erect)* in piedi; *(permanent)* permanente ● *n* posizione *f*; *(duration)* durata *f*. **~ 'order** *n* addebitamento *m* diretto. **~-room** *n* posti *mpl* in piedi

stand: **~-offish** /stænd'ɒfɪʃ/ *a* scostante. **~point** *n* punto *m* di vista. **~still** *n* come to a **~still** fermarsi; **at a ~still** in un periodo di stasi

stank /stæŋk/ *see* **stink**

staple¹ /'steɪpl/ *n (product)* prodotto *m* principale

staple² *n* graffa *f* ● *vt* pinzare. **~r** *n* pinzatrice *f*, cucitrice *f*

star /stɑ:(r)/ *n* stella *f*, *(asterisk)* asterisco *m*; *Theat, Cinema, Sport* divo. **~a** *m/f*, stella *f* ● *vi (pt/pp* **starred)** essere l'interprete principale

starboard /'stɑ:bəd/ *n* tribordo *m*

starch /stɑ:tʃ/ *n* amido *m* ● *vt* inamidare. **~y** *a* ricco di amido; *fig* compito

stare /steə(r)/ *n* sguardo *m* fisso ● *vi* **it's rude to ~** è da maleducati

fissare la gente; **~ at** fissare; **~ into space** guardare nel vuoto

'starfish *n* stella *f* di mare

stark /stɑ:k/ *a* austero; *(contrast)* forte ● *adv* completamente; **~ naked** completamente nudo

starling /'stɑ:lɪŋ/ *n* storno *m*

starlit *a* stellato

starry /'stɑ:rɪ/ *a* stellato

start /stɑ:t/ *n* inizio *m*; *(departure)* partenza *f*; *(jump)* sobbalzo *m*; **from the ~** *(fin)* dall'inizio; **for a ~** tanto per cominciare; **give sb a ~** *Sport* dare un vantaggio a qcno ● *vi* [in]cominciare; *(set out)* avviarsi; *(engine, car.)* partire; *(jump)* trasalire; **to ~ with,...** tanto per cominciare,... ● *vt* [in]cominciare; *(cause)* dare inizio a; *(found)* mettere su; *(engine, car.)* mettere in moto; mettere in giro *(rumour)*. **~er** *n* *Culin* primo *m* [piatto *m*]; *(in race: giving signal)* starter *m inv*; *(participant)* concorrente *mf*; *Auto* motorino *m* d'avviamento. **~ing-point** *n* punto *m* di partenza

startle /'stɑ:tl/ *vt* far trasalire; *(news:)* sconvolgere

starvation /stɑ:'veɪʃn/ *n* fame *f*

starve /stɑ:v/ *vi* morire di fame ● *vt* far morire di fame

stash /stæʃ/ *vt* fam **~ [away]** nascondere

state /steɪt/ *n* stato *m*; *(grand style)* pompa *f*; **~ of play** punteggio *m*; **be in a ~** *(person:)* essere agitato; **lie in ~** essere esposto ● *attrib* di Stato; *Sch* pubblico; *(with ceremony)* di gala ● *vt* dichiarare; *(specify)* precisare. **~less** *a* apolide

stately /'steɪtlɪ/ *a* (**-ier, -iest**) maestoso. **~ 'home** *n* dimora *f* signorile

statement /'steɪtmənt/ *n* dichiarazione *f*; *Jur* deposizione *f*; *(in banking)* estratto *m* conto; *(account)* rapporto *m*

'statesman *n* statista *mf*

static /'stætɪk/ *a* statico

station /'steɪʃn/ n stazione f; (police) commissariato m ● vt appostare (guards); **be ~ed in** Germany essere di stanza in Germania. **~ary** /-ərɪ/ a immobile

stationer /'steɪʃənə(r)/ n **~'s [shop]** cartoleria f. **~y** n cartoleria f

'station-wagon n Am familiare f

statistic|al /stə'tɪstɪkl/ a statistico. **~s** n & pl statistica f

statue /'stætjuː/ n statua f

stature /'stætʃə(r)/ n statura f

status /'steɪtəs/ n condizione f; (high rank) alto rango m. **~ symbol** n status symbol m inv

statut|e /'stætjuːt/ n statuto m. **~ory** a statutario

staunch /stɔːntʃ/ a fedele. **~ly** adv fedelmente

stave /steɪv/ vt **~ off** tenere lontano

stay /steɪ/ n soggiorno m ● vi restare, rimanere; (reside) alloggiare; **~ the night** passare la notte; **~ put** non muoverti ● vt **~ the course** resistere fino alla fine. **stay away** vi stare lontano. **stay behind** vi non andare con gli altri. **stay in** vi (at home) stare in casa; Sch restare a scuola dopo le lezioni. **stay up** vi stare su; (person:) stare alzato

stead /sted/ n **in his ~** in sua vece; **stand sb in good ~** tornare utile a qcno. **~fast** a fedele; (refusal) fermo

steadily /'stedɪlɪ/ adv (continually) continuamente

steady /'stedɪ/ a (-ier, -iest) saldo, fermo; (breathing) regolare; (job, boyfriend) fisso; (dependable) serio

steak /steɪk/ n (for stew) spezzatino m; (for grilling, frying) bistecca f

steal /stiːl/ v (pt **stole**, pp **stolen**) ● vt rubare (from da). **steal in/out** vi entrare/uscire furtivamente

stealth /stelθ/ n **by ~** di nascosto. **~y** a furtivo

steam /stiːm/ n vapore m; **under one's own ~** fam da solo ● vt Culin cucinare a vapore ● vi fumare. **steam up** vi appannarsi

'steam-engine n locomotiva f

steamer /'stiːmə(r)/ n piroscafo m; (saucepan) pentola f a vapore

'steamroller n rullo m compressore

steamy /'stiːmɪ/ a appannato

steel /stiːl/ n acciaio m ● vt **~ oneself** temprarsi

steep¹ /stiːp/ vt (soak) lasciare a bagno

steep² a ripido; (fam: price) esorbitante. **~ly** adv ripidamente

steeple /'stiːpl/ n campanile m. **~chase** n corsa f ippica a ostacoli

steer /stɪə(r)/ vt/i guidare; **~ clear of** stare alla larga da. **~ing** n Auto sterzo m. **~ing-wheel** n volante m

stem¹ /stem/ n stelo m; (of glass) gambo m; (of word) radice f ● vi (pt/pp **stemmed**) **~ from** derivare da

stem² vt (pt/pp **stemmed**) contenere

stench /stentʃ/ n fetore m

step /step/ n passo m; (stair) gradino m; **~s** pl (ladder) scala f portatile; **in ~** al passo; **be out of ~** non stare al passo; **~ by ~** un passo alla volta ● vi (pt/pp **stepped**) **~ into** entrare in; **~ out** uscire da; **~ out of line** sgarrare. **step down** vi fig dimettersi. **step forward** vi farsi avanti. **step in** vi fig intervenire. **step up** vt (increase) aumentare

step: **~brother** n fratellastro m. **~child** n figliastro, -a mf. **~daughter** n figliastra f. **~father** n patrigno m. **~ladder** n scala f portatile. **~mother** n matrigna f

'stepping-stone n pietra f per guadare; fig trampolino m

step: **~sister** n sorellastra f. **~son** n figliastro m

stereo /'sterɪəʊ/ n stereo m; **in ~**

in stereofonia. **~phonic** /-'fɒnɪk/ a stereofonico

stereotype /'sterɪətaɪp/ n stereotipo m. **~d** a stereotipato

steril|e /'steraɪl/ a sterile. **~ity** /stə'rɪlɪtɪ/ n sterilità f inv

steriliz|ation /sterəlaɪ'zeɪʃn/ n sterilizzazione f. **~e** /'ster-/ vt sterilizzare

sterling /'stɜːlɪŋ/ a fig apprezzabile; **~ silver** argento m pregiato ● n sterlina f

stern[1] /stɜːn/ a severo

stern[2] n (of boat) poppa f

stethoscope /'steθəskəʊp/ n stetoscopio m

stew /stjuː/ n stufato m; **in a ~** fam agitato ● vt/i cuocere in umido; **~ed fruit** frutta f cotta

steward /'stjuːəd/ n (at meeting) organizzatore, -trice mf; (on ship, aicraft) steward m inv. **~ess** n hostess f inv

stick[1] /stɪk/ n bastone m; (of celery, rhubarb) gambo m; Sport mazza f

stick[2] v (pt/pp **stuck**) ● vt (stab) [con]ficcare; (glue) attaccare; (fam: put) mettere; (fam: endure) sopportare ● vi (adhere) attaccarsi (**to** a); (jam) bloccarsi; **~ to** attenersi a ⟨facts⟩; mantenere ⟨story⟩; perseverare in ⟨task⟩; **~ at it** fam tener duro; **~ at nothing** fam non fermarsi di fronte a niente; **be stuck** ⟨vehicle, person:⟩ essere bloccato; ⟨drawer:⟩ essere incastrato; **be stuck with sth** fam farsi incastrare con qcsa. **stick out** vi ⟨project⟩ sporgere; (fam: catch the eye) risaltare ● vt fam fare ⟨tongue⟩. **stick up for** vt fam difendere

sticker /'stɪkə(r)/ n autoadesivo m

'sticking plaster n cerotto m

stick-in-the-mud n retrogrado m

stickler /'stɪklə(r)/ n **be a ~ for** tenere molto a

sticky /'stɪkɪ/ a (-ier, -iest) appiccicoso; ⟨adhesive⟩ adesivo; (fig: difficult) difficile

stiff /stɪf/ a rigido; ⟨brush, task⟩

duro; ⟨person⟩ controllato; ⟨drink⟩ forte; ⟨penalty⟩ severo; ⟨price⟩ alto; **bored ~** fam annoiato a morte; **~ neck** torcicollo m. **~en** vt irrigidire ● vi irrigidirsi. **~ness** n rigidità f inv

stifl|e /'staɪfl/ vt soffocare. **~ing** a soffocante

stigma /'stɪgmə/ n marchio m

stiletto /strˈletəʊ/ n stiletto m; **~ heels** tacchi mpl a spillo; **~s** ⟨shoes⟩ scarpe fpl coi tacchi a spillo

still[1] /stɪl/ n distilleria f

still[2] a fermo; ⟨drink⟩ non gasato; **keep/stand ~** stare fermo ● n quiete f; ⟨photo⟩ posa f ● adv ancora; ⟨nevertheless⟩ nondimeno, comunque; **I'm ~ not sure** non sono ancora sicuro

'stillborn a nato morto

still 'life n natura f morta

stilted /'stɪltɪd/ a artificioso

stilts /stɪlts/ npl trampoli mpl

stimulant /'stɪmjʊlənt/ n eccitante m

stimulat|e /'stɪmjʊleɪt/ vt stimolare. **~ion** /-'leɪʃn/ n stimolo m

stimulus /'stɪmjʊləs/ n (pl -li /-laɪ/) stimolo m

sting /stɪŋ/ n puntura f; ⟨from nettle, jellyfish⟩ sostanza f irritante; ⟨organ⟩ pungiglione m ● v (pt/pp **stung**) ● vt pungere; ⟨jellyfish:⟩ pizzicare ● vi pungere. **~ing nettle** n ortica f

stingy /'stɪndʒɪ/ a (-ier, -iest) tirchio

stink /stɪŋk/ n puzza f ● vi (pt **stank**, pp **stunk**) puzzare

stint /stɪnt/ n lavoro m; **do one's ~** fare la propria parte ● vt **~ on** lesinare su

stipulat|e /'stɪpjʊleɪt/ vt porre come condizione. **~ion** /-'leɪʃn/ n condizione f

stir /stɜː(r)/ n mescolata f; (commotion) trambusto m ● v (pt/pp **stirred**) ● vt muovere; (mix) mescolare ● vi muoversi

stirrup /'stɪrəp/ n staffa f

stitch /stɪtʃ/ n punto m; (in knitting) maglia f; (pain) fitta f; **have sb in ~es** fam far ridere qcno a crepapelle ● vt cucire

stock /stɒk/ n (for use or selling) scorta f, stock m inv; (livestock) bestiame m; (lineage) stirpe f; Fin titoli mpl; Culin brodo m; **in ~** disponibile; **out of ~** esaurito; **take ~ fig** fare il punto ● a solito ● vt (shop:) vendere; approvvigionare (shelves). **stock up** vi far scorta (**with** di)

stock: ~broker n agente m di cambio. **~ cube** n dado m [da brodo]. **S~ Exchange** n Borsa f Valori

stocking /'stɒkɪŋ/ n calza f

stockist /'stɒkɪst/ n rivenditore m

stock: ~market n mercato m azionario. **~pile** vt fare scorta di ● n riserva f. **~·still** a immobile. **~-taking** n Comm inventario m

stocky /'stɒkɪ/ a (-ier, -iest) tarchiato

stodgy /'stɒdʒɪ/ a indigesto

stoic /'stəʊɪk/ n stoico, -a mf. **~al** a stoico. **~ism** /-sɪzm/ stoicismo m

stoke /stəʊk/ vt alimentare

stole¹ /stəʊl/ n stola f

stole², **stolen** /'stəʊln/ see **steal**

stolid /'stɒlɪd/ a apatico

stomach /'stʌmək/ n pancia f; Anat stomaco m ● vt fam reggere. **~-ache** n mal m di pancia

stone /stəʊn/ n pietra f; (in fruit) nocciolo m; Med calcolo m; (weight) 6,348 kg ● a di pietra; (wall, Age) della pietra ● vt snocciolare (fruit). **~-cold** a gelido. **~·deaf** a fam sordo come una campana

stony /'stəʊnɪ/ a pietroso; (glare) glaciale

stood /stʊd/ see **stand**

stool /stu:l/ n sgabello m

stoop /stu:p/ n curvatura f ● vi stare curvo; (bend down) chinarsi; fig abbassarsi

stop /stɒp/ n (break) sosta f; (for bus, train) fermata f; Gram punto m; **come to a ~** fermarsi; **put a ~**

to sth mettere fine a qcsa ● v (pt/pp **stopped**) ● vt fermare; arrestare (machine); (prevent) impedire; **~ sb doing sth** impedire a qcno di fare qcsa; **~ doing sth** smettere di fare qcsa; **~ that!** smettila! ● int fermo!. **stop off** vi fare una sosta. **stop up** vt otturare (sink); tappare (hole). **stop with** vi (fam: stay with) fermarsi da

stop: ~gap n palliativo m; (person) tappabuchi m inv. **~-over** n sosta f, Aeron scalo m

stoppage /'stɒpɪdʒ/ n ostruzione f; (strike) interruzione f; (deduction) trattenute fpl

stopper /'stɒpə(r)/ n tappo m

stop: ~-press n ultimissime fpl. **~-watch** n cronometro m

storage /'stɔ:rɪdʒ/ n deposito m; (in warehouse) immagazzinaggio m; Comput memoria f

store /stɔ:(r)/ n (stock) riserva f; (shop) grande magazzino m; (depot) deposito m; **in ~** in deposito; **what the future has in ~ for me** cosa mi riserva il futuro; **set great ~ by** tenere in gran conto ● vt tenere; (in warehouse, Comput) immagazzinare. **~-room** n magazzino m

storey /'stɔ:rɪ/ n piano m

stork /stɔ:k/ n cicogna f

storm /stɔ:m/ n temporale m; (with thunder) tempesta f ● vt prendere d'assalto. **~y** a tempestoso

story /'stɔ:rɪ/ n storia f; (in newspaper) articolo m

stout /staʊt/ a (shoes) resistente; (fat) robusto; (defence) strenuo

stove /stəʊv/ n stufa f; (for cooking) cucina f [economica]

stow /stəʊ/ vt metter via. **~away** n passeggero, -a mf clandestino, -a

straddle /'strædl/ vt stare a cavalcioni su; (standing) essere a cavallo su

straggle /'strægl/ vi crescere disordinatamente; (dawdle) rimanere indietro. **~n** persona *f* che rimane indietro. **~y** *a* in disordine

straight /streɪt/ *a* diritto, dritto; ⟨*answer, question, person*⟩ diretto; ⟨*tidy*⟩ in ordine; ⟨*drink, hair*⟩ liscio ● *adv* diritto, dritto; ⟨*directly*⟩ direttamente; **~ away** immediatamente; **~ on** *or* **ahead** diritto; **~ out** *fig* apertamente; **go ~** *fam* rigare diritto; **put sth ~** mettere qcsa in ordine; **sit/stand up ~** stare diritto

straighten /ˈstreɪtn/ *vt* raddrizzare ● *vi* raddrizzarsi; **~ [up]** ⟨*person.*⟩ mettersi diritto. **straighten out** *vt fig* chiarire ⟨*situation*⟩

straight·forward *a* franco; ⟨*simple*⟩ semplice

strain¹ /streɪn/ *n* ⟨*streak*⟩ vena *f*; *Bot* varietà *f inv*; ⟨*of virus*⟩ forma *f*

strain² *n* tensione *f*; ⟨*injury*⟩ stiramento *m*; **~s** *pl* ⟨*of music*⟩ note *fpl* ● *vt* tirare; sforzare ⟨*eyes, voice*⟩; stirarsi ⟨*muscle*⟩; *Culin* scolare ● *vi* sforzarsi. **~ed** *a* ⟨*relations*⟩ teso. **~er** *n* colino *m*

strait /streɪt/ *n* stretto *m*; **in dire ~s** in serie difficoltà. **~·jacket** *n* camicia *f* di forza. **~·laced** *a* puritano

strand¹ /strænd/ *n* ⟨*of thread*⟩ gugliata *f*; ⟨*of beads*⟩ filo *m*; ⟨*of hair*⟩ capello *m*

strand² *vt* **be ~ed** rimanere bloccato

strange /streɪndʒ/ *a* strano; ⟨*not known*⟩ sconosciuto; ⟨*unaccustomed*⟩ estraneo. **~ly** *adv* stranamente; **~ly enough** curiosamente. **~r** *n* estraneo, -a *mf*

strangle /ˈstræŋgl/ *vt* strangolare; *fig* reprimere

strangulation /stræŋgjʊˈleɪʃn/ *n* strangolamento *m*

strap /stræp/ *n* cinghia *f* ⟨*to grasp in vehicle*⟩ maniglia *f*; ⟨*of watch*⟩ cinturino *m*; ⟨*shoulder* ~⟩ bretella *f*, spallina *f* ● *vt* ⟨*pt/pp* **strapped**⟩ legare; **~ in** *or* **down** assicurare

strapping /ˈstræpɪŋ/ *a* robusto

strata /ˈstrɑːtə/ *npl see* **stratum**

stratagem /ˈstrætədʒəm/ *n* stratagemma *m*

strategic /strəˈtiːdʒɪk/ *a* strategico

strategy /ˈstrætədʒɪ/ *n* strategia *f*

stratum /ˈstrɑːtəm/ *n* ⟨*pl* **strata**⟩ strato *m*

straw /strɔː/ *n* paglia *f*; ⟨*single piece*⟩ fuscello *m*; ⟨*for drinking*⟩ cannuccia *f*; **the last ~** l'ultima goccia

strawberry /ˈstrɔːbərɪ/ *n* fragola *f*

stray /streɪ/ *a* ⟨*animal*⟩ randagio ● *n* randagio *m* ● *vi* andarsene per conto proprio; ⟨*deviate*⟩ deviare ⟨*from* da⟩

streak /striːk/ *n* striatura *f*; ⟨*fig: trait*⟩ vena *f* ● *vi* sfrecciare. **~y** *a* striato; ⟨*bacon*⟩ grasso

stream /striːm/ *n* ruscello *m*; ⟨*current*⟩ corrente *f*; ⟨*of blood, people*⟩ flusso *m*; *Sch* classe *f* ● *vi* scorrere. **stream in/out** *vi* entrare/uscire a fiotti

streamer /ˈstriːmə(r)/ *n* ⟨*paper*⟩ stella *f* filante; ⟨*flag*⟩ pennone *m*

'streamline *vt* rendere aerodinamico; ⟨*simplify*⟩ snellire. **~d** *a* aerodinamico

street /striːt/ *n* strada *f*. **~·car** *n* *Am* tram *m inv*. **~·lamp** *n* lampione *m*

strength /streŋθ/ *n* forza *f*; ⟨*of wall, bridge etc*⟩ solidità *f inv*; **~s** punti *mpl* forti; **on the ~ of** grazie a. **~en** *vt* rinforzare

strenuous /ˈstrenjʊəs/ *a* faticoso; ⟨*attempt, denial*⟩ energico

stress /stres/ *n* ⟨*emphasis*⟩ insistenza *f*; *Gram* accento *m* tonico; ⟨*mental*⟩ stress *m inv*; *Mech* spinta *f* ● *vt* ⟨*emphasize*⟩ insistere su; *Gram* mettere l'accento [tonico] su. **~ed** *a* ⟨*mentally*⟩ stressato. **~ful** *a* stressante

stretch /stretʃ/ *n* stiramento *m*; ⟨*period*⟩ periodo *m* di tempo; ⟨*of road*⟩ estensione *f*; ⟨*elasticity*⟩ elasticità *f inv*; **at a ~** di fila; **have a ~** stirarsi ● *vt* tirare; allargare ⟨*shoes, arms etc*⟩; ⟨*person:*⟩ allungare ● *vi* (*become wider*) allargarsi; (*extend*) estendersi; ⟨*person:*⟩ stirarsi. ~**er** *n* barella *f*

strew /stru:/ *vt* (*pp* **strewn** or **strewed**) sparpagliare

stricken /'strɪkn/ *a* prostrato; ~ **with** affetto da (*illness*)

strict /strɪkt/ *a* severo; (*precise*) preciso. ~**ly** *adv* severamente; ~**ly speaking** in senso stretto

stride /straɪd/ *n* [lungo] passo *m*; **take sth in one's** ~ accettare qcsa con facilità ● *vi* (*pt* **strode**, *pp* **stridden**) andare a gran passi

strident /'straɪdənt/ *a* stridente; ⟨*colour*⟩ vistoso

strife /straɪf/ *n* conflitto *m*

strike /straɪk/ *n* sciopero *m*; *Mil* attacco *m*; **on** ~ in sciopero ● *v* (*pt/pp* **struck**) ● *vt* colpire; accendere (*match*); trovare (*oil, gold*); (*delete*) depennare; (*occur to*) venire in mente a; *Mil* attaccare ● *vi* ⟨*lightning:*⟩ cadere; ⟨*clock:*⟩ suonare; *Mil* attaccare; ⟨*workers:*⟩ scioperare; ~ **lucky** azzeccarla. **strike off**, **strike out** *vt* eliminare. **strike up** *vt* fare ⟨*friendship*⟩; attaccare ⟨*conversation*⟩. ~**-breaker** *n* persona *f* che non aderisce a uno sciopero

striker /'straɪkə(r)/ *n* scioperante *mf*

striking /'straɪkɪŋ/ *a* impressionante; (*attractive*) affascinante

string /strɪŋ/ *n* spago *m*; (*of musical instrument, racket*) corda *f*; (*of pearls*) filo *m*; (*of lies*) serie *f*; **the** ~**s** *Mus* gli archi; **pull** ~**s** *fam* usare le proprie conoscenze ● *vt* (*pt/pp* **strung**) (*thread*) infilare (*beads*). ~**ed** *a* (*instrument*) a corda

stringent /'strɪndʒnt/ *a* rigido

strip /strɪp/ *n* striscia *f* ● *v* (*pt/pp* **stripped**) ● *vt* spogliare; togliere le lenzuola da (*bed*); scrostare (*wood, furniture*); smontare (*machine*); (*deprive*) privare (**of** di) ● *vi* (*undress*) spogliarsi. ~ **car-**

toon *n* striscia *f*. ~ **club** *n* locale *m* di strip-tease

stripe /straɪp/ *n* striscia *f*; *Mil* gallone *m*. ~**d** *a* a strisce

'striplight *n* tubo *m* al neon

stripper /'strɪpə(r)/ *n* spogliarellista *mf*; (*solvent*) sverniciatore *m*

strip-'tease *n* spogliarello *m*, strip-tease *m inv*

strive /straɪv/ *vi* (*pt* **strove**, *pp* **striven**) sforzarsi (**to** di); ~ **for** sforzarsi di ottenere

strode /strəʊd/ *see* **stride**

stroke[1] /strəʊk/ *n* colpo *m*; (*of pen*) tratto *m*; (*in swimming*) bracciata *f*; *Med* ictus *m inv*; ~ **of luck** colpo *m* di fortuna; **put sb off his** ~ far perdere il filo a qcno

stroke[2] *vt* accarezzare

stroll /strəʊl/ *n* passeggiata *f* ● *vi* passeggiare. ~**er** *n* (*Am:* pushchair) passeggino *m*

strong /strɒŋ/ *a* (**-er** /-gə(r)/, **-est** /-gɪst/) forte; (*argument*) valido

strong: ~**-box** *n* cassaforte *f*. ~**hold** *n* roccaforte *f*. ~**ly** *adv* fortemente. ~**-minded** *a* risoluto. ~**-room** *n* camera *f* blindata

stroppy /'strɒpɪ/ *a* scorbutico

strove /strəʊv/ *see* **strive**

struck /strʌk/ *see* **strike**

structural /'strʌktʃərəl/ *a* strutturale. ~**ly** *adv* strutturalmente

structure /'strʌktʃə(r)/ *n* struttura *f*

struggle /'strʌgl/ *n* lotta *f*; **with a** ~ lottare con ● *vi* lottare; ~ **for breath** respirare con fatica; ~ **to do sth** fare fatica a fare qcsa; ~ **to one's feet** alzarsi con fatica

strum /strʌm/ *vt/i* (*pt/pp* **strummed**) strimpellare

strung /strʌŋ/ *see* **string**

strut[1] /strʌt/ *n* (*component*) puntello *m*

strut[2] *vi* (*pt/pp* **strutted**) camminare impettito

stub /stʌb/ *n* mozzicone *m*; (*counterfoil*) matrice *f* ● *vt* (*pt/pp* **stubbed**) ~ **one's toe** sbattere il

stubble | subjugate

dito del piede (**on** contro). **stub out** vt spegnere ⟨cigarette⟩

stubb|le /'stʌbl/ n barba f ispida. **~ly** a ispido

stubborn /'stʌbən/ a testardo; ⟨refusal⟩ ostinato

stubby /'stʌbɪ/ a (-ier, -iest) tozzo

stucco /'stʌkəʊ/ n stucco m

stuck /stʌk/ see **stick²**. **~-'up** a fam snob

stud¹ /stʌd/ n (on boot) tacchetto m; (on jacket) borchia f; (for ear) orecchino m [a bottone]

stud² n (of horses) scuderia f

student /'stju:dənt/ n studente m, studentessa f; (school child) scolaro, -a m f. **~ nurse** n studente, studentessa infermiere, -a

studied /'stʌdɪd/ a intenzionale; ⟨politeness⟩ studiato

studio /'stju:dɪəʊ/ n studio m

studious /'stju:dɪəs/ a studioso; ⟨attention⟩ studiato

study /'stʌdɪ/ n studio m ●vt/i (pt/pp **studied**) studiare

stuff /stʌf/ n materiale m; (fam: things) roba f ●vt riempire; (with padding) imbottire; Culin farcire; **~ sth into a drawer/one's pocket** ficcare qcsa alla rinfusa in un cassetto/in tasca. **~ing** n (padding) imbottitura f; Culin ripieno m

stuffy /'stʌfɪ/ a (-ier, -iest) che sa di chiuso; (old-fashioned) antiquato

stumbl|e /'stʌmbl/ vi inciampare; **~e across** or **on** imbattersi in. **~ing-block** n ostacolo m

stump /stʌmp/ n ceppo m; (of limb) moncone m. **~ed** a fam perplesso ●**stump up** vt/i fam sganciare

stun /stʌn/ vt (pt/pp **stunned**) stordire; ⟨astonish⟩ sbalordire

stung /stʌŋ/ see **sting**

stunk /stʌŋk/ see **stink**

stunning /'stʌnɪŋ/ a fam favoloso; ⟨blow, victory⟩ sbalorditivo

stunt¹ /stʌnt/ n fam trovata f pubblicitaria

stunt² vt arrestare lo sviluppo di. **~ed** a stentato

stupendous /stju:'pendəs/ a stupendo. **~ly** adv stupendamente

stupid /'stju:pɪd/ a stupido. **~ity** /-'pɪdɪtɪ/ n stupidità f. **~ly** adv stupidamente

stupor /'stju:pə(r)/ n torpore m

sturdy /'stɜ:dɪ/ a (-ier, -iest) robusto; ⟨furniture⟩ solido

stutter /'stʌtə(r)/ n balbuzie f ●vt/i balbettare

sty, stye /staɪ/ n (pl **styes**) Med orzaiolo m

style /staɪl/ n stile m; ⟨fashion⟩ moda f; (sort) tipo m; ⟨hair~⟩ pettinatura f. **in ~** in grande stile

stylish /'staɪlɪʃ/ a elegante. **~ly** adv con eleganza

stylist /'staɪlɪst/ n stilista mf; ⟨hair-~⟩ parrucchiere, -a mf. **~ic** /-'lɪstɪk/ a stilistico

stylized /'staɪlaɪzd/ a stilizzato

stylus /'staɪləs/ n (on record player) puntina f

suave /swɑ:v/ a dai modi garbati

sub'conscious /sʌb-/ a subcosciente ●n subcosciente m. **~ly** adv in modo inconscio

subcon'tract vt subappaltare (to a). **~or** n subappaltatore m

'subdivi|de vt suddividere. **~sion** n suddivisione f

subdue /səb'dju:/ vt sottomettere; (make quieter) attenuare. **~d** a ⟨light⟩ attenuato; ⟨person, voice⟩ pacato

subhuman /sʌb'hju:mən/ a disumano

subject¹ /'sʌbdʒɪkt/ a **~ to** soggetto a; (depending on) subordinato a; **~ to availability** nei limiti della disponibilità ●n soggetto m; (of ruler) suddito, -a mf; Sch materia f

subject² /səb'dʒekt/ vt (to attack, abuse) sottoporre; assoggettare ⟨country⟩

subjective /səb'dʒektɪv/ a soggettivo. **~ly** adv soggettivamente

subjugate /'sʌbdʒʊgeɪt/ vt soggiogare

subjunctive /səbˈdʒʌŋktɪv/ a & n congiuntivo m

sub'let vt (pt/pp -let) subaffittare

sublime /səˈblaɪm/ a sublime. **~ly** adv sublimemente

subliminal /səˈblɪmɪnl/ a subliminale

sub-ma'chine-gun n mitraglietta f

subma'rine n sommergibile m

submerge /səbˈmɜːdʒ/ vt immergere; **be ~d** essere sommerso ● vi immergersi

submiss|ion /səbˈmɪʃn/ n sottomissione f. **~ive** /-sɪv/ a sottomesso

submit /səbˈmɪt/ v (pt/pp -mitted, pres p -mitting) ● vt sottoporre ● vi sottomettersi

subordinate /səˈbɔːdɪnət/ vt subordinare (**to** a)

subscribe /səbˈskraɪb/ vi contribuire; **~ to** abbonarsi a ⟨newspaper⟩; sottoscrivere ⟨fund⟩; fig aderire a. **~r** n abbonato, -a mf

subscription /səbˈskrɪpʃn/ n ⟨to club⟩ sottoscrizione f; ⟨to newspaper⟩ abbonamento m

subsequent /ˈsʌbsɪkwənt/ a susseguente. **~ly** adv in seguito

subservient /səbˈsɜːvɪənt/ a subordinato; ⟨servile⟩ servile. **~ly** adv servilmente

subside /səbˈsaɪd/ vi sprofondare; ⟨ground:⟩ avvallarsi; ⟨storm:⟩ placarsi

subsidiary /səbˈsɪdɪərɪ/ a secondario ● n **[company]** filiale f

subsid|ize /ˈsʌbsɪdaɪz/ vt sovvenzionare. **~y** n sovvenzione f

subsist /səbˈsɪst/ vi vivere (**on** di). **~ence** n sussistenza f

substance /ˈsʌbstəns/ n sostanza f

sub'standard a di qualità inferiore

substantial /səbˈstænʃl/ a solido; ⟨meal⟩ sostanzioso; ⟨considerable⟩ notevole. **~ly** adv notevolmente; ⟨essentially⟩ sostanzialmente

substantiate /səbˈstænʃɪeɪt/ vt comprovare

substitut|e /ˈsʌbstɪtjuːt/ n sostituto m ● vt **~e A for B** sostituire B con A ● vi **~e for sb** sostituire qcno. **~ion** /-ˈtjuːʃn/ n sostituzione f

subterranean /sʌbtəˈreɪnɪən/ a sotterraneo

'subtitle n sottotitolo m

subt|le /ˈsʌtl/ a sottile; ⟨taste, perfume⟩ delicato. **~tlety** n sottigliezza f. **~tly** adv sottilmente

subtract /səbˈtrækt/ vt sottrarre. **~ion** /-ækʃn/ n sottrazione f

suburb /ˈsʌbɜːb/ n sobborgo m; **in the ~s** in periferia. **~an** /səˈbɜːbən/ a suburbano. **~ia** /səˈbɜːbɪə/ n i sobborghi mpl

subversive /səbˈvɜːsɪv/ a sovversivo

'subway n sottopassaggio m; ⟨Am: railway⟩ metropolitana f

succeed /səkˈsiːd/ vi riuscire; ⟨follow⟩ succedere a; **~ in doing** riuscire a fare ● vt succedere a ⟨king⟩. **~ing** a successivo

success /səkˈses/ n successo m; **be a ~** ⟨in life⟩ aver successo. **~ful** a riuscito; ⟨businessman, artist etc⟩ di successo. **~fully** adv con successo

succession /səkˈseʃn/ n successione f; **in ~** di seguito

successive /səkˈsesɪv/ a successivo. **~ly** adv successivamente

successor /səkˈsesə(r)/ n successore m

succinct /səkˈsɪŋkt/ a succinto

succulent /ˈsʌkjʊlənt/ a succulento

succumb /səˈkʌm/ vi soccombere (**to** a)

such /sʌtʃ/ a tale; **~ a** tale; **~ a book** un libro di questo genere; **~ a thing** una cosa di questo genere; **~ a long time ago** talmente tanto tempo fa; **there is no ~ thing** non esiste una cosa così; **there is no ~ person** non esiste una persona così ● pron **as ~** come tale; **~ as** chi; **and ~** e simili; **~ as it is** così com'è. **~like** *pron fam* di tal genere

suck /sʌk/ *vt* succhiare. **suck up** *vt* assorbire. **suck up to** *vt fam* fare il lecchino con

sucker /'sʌkə(r)/ *n Bot* pollone *m*; *(fam: person)* credulone, -a *mf*

suction /'sʌkʃn/ *n* aspirazione *f*

sudden /'sʌdn/ *a* improvviso ● *n* **all of a ~** all'improvviso. **~ly** *adv* improvvisamente

sue /su:/ *vt (pres p suing)* fare causa a *(for per)* ● *vi* fare causa

suede /sweɪd/ *n* pelle *f* scamosciata

suet /'su:ɪt/ *n* grasso *m* di rognone

suffer /'sʌfə(r)/ *vi* soffrire *(from* per) ● *vt* soffrire; subire *(loss etc)*; *(tolerate)* subire. **~ing** *n* sofferenza *f*

suffice /sə'faɪs/ *vi* bastare

sufficient /sə'fɪʃənt/ *a* sufficiente. **~ly** *adv* sufficientemente

suffix /'sʌfɪks/ *n* suffisso *m*

suffocat|e /'sʌfəkeɪt/ *vt/i* soffocare. **~ion** /-'keɪʃn/ *n* soffocamento *m*

sugar /'ʃʊgə(r)/ *n* zucchero *m* ● *vt* zuccherare. **~ basin**, **~bowl** *n* zuccheriera *f*. **~y** *a* zuccheroso; *fig* sdolcinato

suggest /sə'dʒest/ *vt* suggerire; *(indicate, insinuate)* fare pensare a. **~ion** /-estʃən/ *n* suggerimento *m*; *(trace)* traccia *f*. **~ive** /-ɪv/ *a* allusivo. **~ively** *adv* in modo allusivo

suicidal /su:ɪ'saɪdl/ *a* suicida

suicide /'su:ɪsaɪd/ *n* suicidio *m*; *(person)* suicida *mf*; **commit ~** suicidarsi

suit /su:t/ *n* vestito *m*; *(woman's)* tailleur *m inv*; *(in cards)* seme *m*; *Jur* causa *f*; **follow ~** *fig* fare lo stesso ● *vt* andar bene a; *(adapt)* adattare *(to* a); *(be convenient for)* andare bene per; **be ~ed to** *or* for essere adatto a; **~ yourself!** fa' come vuoi!

suitab|le /'su:təbl/ *a* adatto. **~y** *adv* convenientemente

'suitcase *n* valigia *f*

suite /swi:t/ *n* suite *f inv*; *(of furniture)* divano *m* e poltrone *fpl* assortiti

sulk /sʌlk/ *vi* fare il broncio. **~y** *a* imbronciato

sullen /'sʌlən/ *a* svogliato

sulphur /'sʌlfə(r)/ *n* zolfo *m*. **~ic** /-'fjuːrɪk/ **~ic acid** *n* acido *m* solforico

sultana /sʌl'tɑːnə/ *n* uva *f* sultanina

sultry /'sʌltrɪ/ *a* (**-ier**, **-iest**) *(weather)* afoso; *fig* sensuale

sum /sʌm/ *n* somma *f*; *Sch* addizione *f* ● **sum up** *(pt/pp summed)* *vi* riassumere ● *vt* valutare

summar|ize /'sʌmərʌɪz/ *vt* riassumere. **~y** *n* sommario *m* ● *a* sommario; *(dismissal)* sbrigativo

summer /'sʌmə(r)/ *n* estate *f*. **~house** *n* padiglione *m*. **~time** *n* *(season)* estate *f*

summery /'sʌmərɪ/ *a* estivo

summit /'sʌmɪt/ *n* cima *f*. **~ conference** *n* vertice *m*

summon /'sʌmən/ *vt* convocare; *Jur* citare. **summon up** *vt* raccogliere *(strength)*; rievocare *(memory)*

summons /'sʌmənz/ *n Jur* citazione *f* ● *vt* citare in giudizio

sump /sʌmp/ *n Auto* coppa *f* dell'olio

sumptuous /'sʌmptjʊəs/ *a* sontuoso. **~ly** *adv* sontuosamente

sun /sʌn/ *n* sole *m* ● *vt* *(pt/pp sunned)* **~ oneself** prendere il sole

sun: **~bathe** *vi* prendere il sole. **~bed** *n* lettino *m* solare. **~burn** *n* scottatura *f* *(solare)*. **~burnt** *a* scottato *(dal sole)*

sundae /'sʌndeɪ/ *n* gelato *m* guarnito

Sunday /'sʌndeɪ/ *n* domenica *f*

'sundial *n* meridiana *f*

sundry /'sʌndrɪ/ *a* svariati; **all and ~** tutti quanti

'sunflower *n* girasole *m*

sung /sʌŋ/ *see* **sing**

'**sun-glasses** *npl* occhiali *mpl* da sole

sunk /sʌŋk/ *see* **sink**

sunken /'sʌŋkn/ *a* incavato

'**sunlight** *n* [luce *f* del] sole *m*

sunny /'sʌnɪ/ *a* (**-ier, -iest**) assolato

sun: ~**rise** *n* alba *f*. ~**roof** *n* Auto tettuccio *m* apribile. ~**set** *n* tramonto *m*. ~**shade** *n* parasole *m*. ~**shine** *n* [luce *f* del] sole *m*. ~**stroke** *n* insolazione *f*. ~**tan** *n* abbronzatura *f*. ~**tanned** *a* abbronzato. ~**tan oil** *n* olio *m* solare

super /'su:pə(r)/ *a fam* fantastico

superb /su'pɜ:b/ *a* splendido

supercilious /su:pə'sɪlɪəs/ *a* altezzoso

superficial /su:pə'fɪʃl/ *a* superficiale. ~**ly** *adv* superficialmente

superfluous /su'pɜ:fluəs/ *a* superfluo

super'human *a* sovrumano

superintendent /su:pərɪn'tendənt/ *n* (*of police*) commissario *m* di polizia

superior /su:'pɪərɪə(r)/ *a* superiore ● *n* superiore, -a *mf*. ~**ity** /-'ɒrətɪ/ *n* superiorità *f*

superlative /su:'pɜ:lətɪv/ *a* eccellente ● *n* superlativo *m*

'**superman** *n* superuomo *m*

'**supermarket** *n* supermercato *m*

'**supermodel** *n* top model *f inv*

super'natural *a* soprannaturale

'**superpower** *n* superpotenza *f*

supersede /su:pə'si:d/ *vt* rimpiazzare

super'sonic *a* supersonico

superstition /su:pə'stɪʃn/ *n* superstizione *f*. ~**ous** /-'stɪʃəs/ *a* superstizioso

supervis|e /'su:pəvaɪz/ *vt* supervisionare. ~**ion** /-'vɪʒn/ *n* supervisione *f*. ~**or** *n* supervisore *m*

supper /'sʌpə(r)/ *n* cena *f*

supple /'sʌpl/ *a* slogato

supplement /'sʌplɪmənt/ *n* supplemento *m* ● *vt* integrare. ~**ary** /-'mentərɪ/ *a* supplementare

supplier /sə'plaɪə(r)/ *n* fornitore, -trice *mf*

supply /sə'plaɪ/ *n* fornitura *f*; (*in economics*) offerta *f*; **supplies** *pl* Mil approvvigionamenti *mpl* ● *vt* (*pt/pp* **-ied**) fornire; ~ **sb with sth** fornire qcsa a qcno

support /sə'pɔ:t/ *n* sostegno *m*; (*base*) supporto *m*; (*keep*) sostentamento *m* ● *vt* sostenere; mantenere (*family*); (*give money to*) mantenere finanziariamente; Sport fare il tifo per. ~**er** *n* sostenitore, -trice *mf*; Sport tifoso, -a *mf*. ~**ive** /-ɪv/ *a* incoraggiante

suppose /sə'pəuz/ *vt* (*presume*) supporre; (*imagine*) pensare; **be** ~**d to** do dover fare; **not be** ~**d to** *fam* non avere il permesso di; **I** ~ **so** suppongo di sì. ~**dly** /-ɪdlɪ/ *adv* presumibilmente

suppress /sə'pres/ *vt* sopprimere. ~**ion** /-eʃn/ *n* soppressione *f*

supremacy /su:'preməsɪ/ *n* supremazia *f*

supreme /su:'pri:m/ *a* supremo

surcharge /'sɜ:tʃɑ:dʒ/ *n* supplemento *m*

sure /ʃuə(r)/ *a* sicuro, certo; **make** ~ accertarsi; **be** ~ **to do it** mi raccomando di farlo ● *adv* Am *fam* certamente; ~ **enough** infatti. ~**ly** *adv* certamente; (*Am: gladly*) volentieri

surety /'ʃuərətɪ/ *n* garanzia *f*; **stand** ~ **for** garantire

surf /sɜ:f/ *n* schiuma *f* ● *vt* Comput ~ **the Net** surfare in Internet

surface /'sɜ:fɪs/ *n* superficie *f*; **on the** ~ *fig* in apparenza ● *vi* (*emerge*) emergere. ~ **mail** *n* **by** ~ **mail** per posta ordinaria

'**surfboard** *n* tavola *f* da surf

surfing /'sɜ:fɪŋ/ *n* surf *m inv*

surge /sɜ:dʒ/ *n* (*of sea*) ondata *f*; (*of interest*) aumento *m*; (*in demand*) impennata *f*; (*of anger, pity*) impeto *m* ● *vi* riversarsi; ~ **forward** buttarsi in avanti

surgeon /'sɜ:dʒən/ *n* chirurgo *m*

surgery /'sɜ:dʒərɪ/ *n* chirurgia *f*; (*place, consulting room*) ambulatorio *m*; (*hours*) ore *fpl* di visita;

have ~ subire un'intervento [chirurgico]

surgical /'sɜːdʒɪkl/ a chirurgico

surly /'sɜːlɪ/ a (-ier, -iest) scontroso

surmise /səˈmaɪz/ vt supporre

surmount /səˈmaʊnt/ vt sormontare

surname /'sɜːneɪm/ n cognome m

surpass /səˈpɑːs/ vt superare

surplus /'sɜːpləs/ a d'avanzo ● n sovrappiù m

surpris|e /səˈpraɪz/ n sorpresa f ● vt sorprendere; **be ~ed** essere sorpreso (**at** da). **~ing** a sorprendente. **~ingly** adv sorprendentemente

surrender /səˈrendə(r)/ n resa f ● vi arrendersi ● vt cedere

surreptitious /sʌrəpˈtɪʃəs/ a ● adv di nascosto

surrogate /'sʌrəgət/ n surrogato m. ~ **'mother** n madre f surrogata

surround /səˈraʊnd/ vt circondare. **~ing** a circostante. **~ings** npl dintorni mpl

surveillance /səˈveɪləns/ n sorveglianza f

survey¹ /'sɜːveɪ/ n sguardo m; (poll) sondaggio m; (investigation) indagine f; (of land) rilevamento m; (of house) perizia f

survey² /səˈveɪ/ vt esaminare; fare un rilevamento di ⟨land⟩; fare una perizia di ⟨building⟩. **~or** n perito m; (of land) topografo, -a mf

survival /səˈvaɪvl/ n sopravvivenza f; (relic) resto m

surviv|e /səˈvaɪv/ vt sopravvivere a ● vi sopravvivere. **~or** n superstite mf; **be a ~or** fam riuscire sempre a cavarsela

susceptible /səˈseptəbl/ a influenzabile; **~ to** sensibile a

suspect¹ /səˈspekt/ vt sospettare; (assume) supporre

suspect² /'sʌspekt/ a & n sospetto, -a mf

suspend /səˈspend/ vt appendere; (stop, from duty) sospendere. **~er belt** n reggicalze m inv. **~ders** npl

giarrettiere fpl; (Am: braces) bretelle mpl

suspense /səˈspens/ n tensione f; (in book etc) suspense f

suspension /səˈspenʃn/ n Auto sospensione f. ~ **bridge** n ponte m sospeso

suspici|on /səˈspɪʃn/ n sospetto m; (trace) pizzico m; **under ~on** sospettato. **~ous** /-ɪʃəs/ a sospettoso; (arousing suspicion) sospetto. **~ously** adv sospettosamente; (arousing suspicion) in modo sospetto

sustain /səˈsteɪn/ vt sostenere; mantenere ⟨life⟩; subire ⟨injury⟩

sustenance /'sʌstɪnəns/ n nutrimento m

swab /swɒb/ n Med tampone m

swagger /'swægə(r)/ vi pavoneggiarsi

swallow¹ /'swɒləʊ/ vt/i inghiottire. **swallow up** vt divorare; ⟨earth, crowd:⟩ inghiottire

swallow² n (bird) rondine f

swam /swæm/ see **swim**

swamp /swɒmp/ n palude f ● vt fig sommergere. **~y** a paludoso

swan /swɒn/ n cigno m

swap /swɒp/ n fam scambio m ● vt (pt/pp **swapped**) fam scambiare (**for** con) ● vi fare cambio

swarm /swɔːm/ n sciame m ● vi sciamare; **be ~ing with** brulicare di

swarthy /'swɔːðɪ/ a (-ier, -iest) di carnagione scura

swastika /'swɒstɪkə/ n svastica f

swat /swɒt/ vt (pt/pp **swatted**) schiacciare

sway /sweɪ/ n fig influenza f ● vi oscillare; ⟨person:⟩ ondeggiare ● vt (influence) influenzare

swear /sweə(r)/ v (pt **swore**, pp **sworn**) ● vt giurare ● vi giurare; (curse) dire parolacce; **~ at sb** imprecare contro qcno; **~ by** fam credere ciecamente in. **~-word** n parolaccia f

sweat /swet/ n sudore m ● vi sudare

sweater /'swetə(r)/ n golf m inv

sweaty /'swetɪ/ a sudato

swede /swiːd/ n rapa f svedese

Swede n svedese mf. **~en** n Svezia f. **~ish** a svedese

sweep /swiːp/ n scopata f, spazzata f; (curve) curva f; (movement) movimento m ampio; **make a clean ~** fig fare piazza pulita ● v (pt/pp **swept**) ● vt scopare, spazzare; (wind:) spazzare ● vi (go swiftly) andare rapidamente; (wind:) soffiare. **sweep away** vt fig spazzare via. **sweep up** vt spazzare

sweeping /'swiːpɪŋ/ a (gesture) ampio; (statement) generico; (changes) radicale

sweet /swiːt/ a dolce; **have a ~ tooth** essere goloso ● n caramella f; (dessert) dolce m. **~ corn** n mais m

sweeten /'swiːtn/ vt addolcire. **~er** n dolcificante m

sweet: ~heart n innamorato, -a mf; **hi, ~heart** ciao, tesoro. **~ness** n dolcezza f. **~ pea** n pisello m odoroso. **~shop** n negozio m di dolciumi

swell /swel/ ● v (pt **swelled**, pp **swollen** or **swelled**) ● vi gonfiarsi; (increase) aumentare ● vt gonfiare; (increase) far salire. **~ing** n gonfiore m

swelter /'sweltə(r)/ vi soffocare [dal caldo]

swept /swept/ see **sweep**

swerve /swɜːv/ vi deviare bruscamente

swift /swɪft/ a rapido. **~ly** adv rapidamente

swig /swɪɡ/ n fam sorso m ● vt (pt/pp **swigged**) fam scolarsi

swill /swɪl/ n (for pigs) brodaglia f ● vt **~** [**out**] risciacquare

swim /swɪm/ n have a **~** fare una nuotata ● v (pt **swam**, pp **swum**) ● vi nuotare; (room:) girare; **my head is ~ming** mi gira la testa ● vt percorrere a nuoto. **~mer** n nuotatore, -trice mf

swimming /'swɪmɪŋ/ n nuoto m.

~-baths npl piscina fsg. **~ costume** n costume m da bagno. **~-pool** n piscina f. **~ trunks** npl calzoncini mpl da bagno

'swim-suit n costume m da bagno

swindle /'swɪndl/ n truffa f ● vt truffare. **~r** n truffatore, -trice mf

swine /swaɪn/ n fam porco m

swing /swɪŋ/ n oscillazione f; (shift) cambiamento m; (seat) altalena f; Mus swing m; **in full ~** in piena attività ● v (pt/pp **swung**) ● vi oscillare; (on swing, sway) dondolare; (dangle) penzolare; (turn) girare ● vt oscillare; far deviare (vote). **~-'door** n porta f a vento

swingeing /'swɪndʒɪŋ/ a (increase) drastico

swipe /swaɪp/ n fam botta f ● vt fam colpire; (steal) rubare; far passare nella macchinetta (credit card)

swirl /swɜːl/ n (of smoke, dust) turbine m ● vi (water:) fare mulinello

swish /swɪʃ/ a fam chic ● vi schioccare

Swiss /swɪs/ a & n svizzero, -a mf; **the ~** pl gli svizzeri. **~ 'roll** n rotolo m di pan di Spagna ripieno di marmellata

switch /swɪtʃ/ n interruttore m; (change) mutamento m ● vt cambiare; (exchange) scambiare ● vi cambiare; **to ~** to passare a. **switch off** vt spegnere. **switch on** vt accendere

switch: ~back n montagne fpl russe. **~board** n centralino m

Switzerland /'swɪtsələnd/ n Svizzera f

swivel /'swɪvl/ v (pt/pp **swivelled**) ● vt girare ● vi girarsi

swollen /'swəʊlən/ see **swell** ● a gonfio. **~-'headed** a presuntuoso

swoop /swuːp/ n (by police) incursione f ● vi **~** [**down**] (bird:) piombare; fig fare un'incursione

sword /sɔːd/ n spada f

swore /swɔː(r)/ *see* **swear**

sworn /swɔːn/ *see* **swear**

swot /swɒt/ *n fam* sgobbone, -a *mf* ● *vt* (*pt/pp* **swotted**) *fam* sgobbare

swum /swʌm/ *see* **swim**

swung /swʌŋ/ *see* **swing**

syllable /'sɪləbl/ *n* sillaba *f*

syllabus /'sɪləbəs/ *n* programma *m* [dei corsi]

symbol /'sɪmbl/ *n* simbolo *m* (**of** di). **~ic** /·'bɒlɪk/ *a* simbolico. **~ism** /·-ɪzm/ *n* simbolismo *m*. **~ize** *vt* simboleggiare

symmetr|ical /sɪ'metrɪkl/ *a* simmetrico. **~y** /'sɪmətrɪ/ *n* simmetria *f*

sympathetic /sɪmpə'θetɪk/ *a* (*understanding*) comprensivo; (*showing pity*) compassionevole. **~ally** *adv* con comprensione/compassione

sympathize /'sɪmpəθaɪz/ *vi* capire; (*in grief*) solidarizzare; **~ with sb** capire qcno/solidarizzare con qcno. **~r** *n Pol* simpatizzante *mf*

sympathy /'sɪmpəθɪ/ *n* comprensione *f*; (*pity*) compassione *f*; (*condolences*) condoglianze *fpl*; **in ~ with** (*strike*) per solidarietà con

symphony /'sɪmfənɪ/ *n* sinfonia *f*

symptom /'sɪmptəm/ *n* sintomo *m*. **~atic** /·'mætɪk/ *a* sintomatico (**of** di)

synagogue /'sɪnəgɒg/ *n* sinagoga *f*

synchronize /'sɪŋkrənaɪz/ *vt* sincronizzare

syndicate /'sɪndɪkət/ *n* gruppo *m*

syndrome /'sɪndrəʊm/ *n* sindrome *f*

synonym /'sɪnənɪm/ *n* sinonimo *m*. **~ous** /·'ɪnɪməs/ *a* sinonimo

synopsis /sɪ'nɒpsɪs/ *n* (*pl* -**opses** /-sɪːz/) (*of opera, ballet*) trama *f*; (*of book*) riassunto *m*

syntax /'sɪntæks/ *n* sintassi *f inv*

synthesize /'sɪnθəsaɪz/ *vt* sintetizzare. **~r** *n Mus* sintetizzatore *m*

synthetic /sɪn'θetɪk/ *a* sintetico ● *n* fibra *f* sintetica

Syria /'sɪrɪə/ *n* Siria *f*. **~n** *a* & *n* siriano, -a *mf*

syringe /sɪ'rɪndʒ/ *n* siringa *f*

syrup /'sɪrəp/ *n* sciroppo *m*; *Br* tipo *m* di melassa

system /'sɪstəm/ *n* sistema *m*. **~atic** /·'mætɪk/ *a* sistematico

....

Tt

....

tab /tæb/ *n* linguetta *f*; (*with name*) etichetta *f*; **keep ~s on** *fam* sorvegliare; **pick up the ~** *fam* pagare il conto

tabby /'tæbɪ/ *n* gatto *m* tigrato

table /'teɪbl/ *n* tavolo *m*; (*list*) tavola *f*; **at** [**the**] **~** a tavola; **~ of contents** tavola *f* delle materie ● *vt* proporre. **~-cloth** *n* tovaglia *f*. **~-spoon** *n* cucchiaio *m* da tavola. **~spoon[ful]** *n* cucchiaiata *f*

tablet /'tæblɪt/ *n* pastiglia *f*; (*slab*) lastra *f*; **~ of soap** saponetta *f*

table tennis *n* tennis *m* da tavolo; (*everyday level*) ping pong *m*

tabloid /'tæblɔɪd/ *n* [giornale *m* formato] tabloid *m inv*; *pej* giornale *m* scandalistico

taboo /tə'buː/ *a* tabù *inv* ● *n* tabù *m inv*

tacit /'tæsɪt/ *a* tacito

taciturn /'tæsɪtɜːn/ *a* taciturno

tack /tæk/ *n* (*nail*) chiodino *m*; (*stitch*) imbastitura *f*; *Naut* virata *f*; *fig* linea *f* di condotta ● *vt* inchiodare; (*sew*) imbastire ● *vi* *Naut* virare

tackle /'tækl/ *n* (*equipment*) attrezzatura *f*; (*football etc*) contrasto *m*, tackle *m inv* ● *vt* affrontare

tacky /'tækɪ/ *a* (*paint*) non ancora asciutto; (*glue*) appiccicoso; *fig* pacchiano

tact /tækt/ *n* tatto *m*. **~ful** *a* pieno di tatto; (*remark*) delicato. **~fully** *adv* con tatto

tactic|al /'tæktɪkl/ *a* tattico. **~s** *npl* tattica *fsg*

tactless /'tæktlɪs/ *a* privo di tatto.

~ly *adv* senza tatto. **~ness** *n* mancanza *f* di tatto; *(of remark)* indelicatezza *f*

tadpole /'tædpəυl/ *n* girino *m*

tag¹ /tæg/ *n* *(label)* etichetta *f* ● *vt* *(pt/pp* **tagged)** attaccare l'etichetta a. **tag along** *vi* seguire passo passo

tag² *n* *(game)* acchiapparello *m*

tail /teɪl/ *n* coda *f;* **~s** *pl* *(tailcoat)* frac *m inv* ● *vt* *(fam: follow)* pedinare. **tail off** *vi* diminuire

tail: **~back** *n* coda *f.* **~end** *n* parte *f* finale; *(of train)* coda *f.* **~ light** *n* fanalino *m* di coda

tailor /'teɪlə(r)/ *n* sarto *m.* **~-made** *a* fatto su misura

'tail wind *n* vento *m* di coda

taint /teɪnt/ *vt* contaminare

take /teɪk/ *n* Cinema ripresa *f* ● *v* *(pt* **took,** *pp* **taken)** ● *vt* prendere; *(to a place)* portare *(person, object);* *(contain)* contenere *(passengers* etc*);* *(endure)* sopportare; *(require)* occorrere; *(teach)* insegnare; *(study)* studiare *(subject);* fare *(exam, holiday, photograph, walk, bath);* sentire *(pulse);* misurare *(sb's temperature);* **~ sb prisoner** fare prigioniero qcno; **be ~n ill** ammalarsi; **~ sth calmly** prendere con calma qcsa ● *vi* *(plant:)* attecchire. **take after** *vt* assomigliare a. **take away** *vt* *(with one)* portare via; *(remove)* togliere; *(subtract)* sottrarre; **'to ~ away'** 'da asporto'. **take back** *vt* riprendere; ritirare *(statement);* *(return)* riportare [indietro]. **take down** *vt* portare giù; *(remove)* tirare giù; *(write down)* prendere nota di. **take in** *vt* *(bring indoors)* portare dentro; *(to one's home)* ospitare; *(understand)* capire; *(deceive)* ingannare; riprendere *(garment);* *(include)* includere. **take off** *vt* togliersi *(clothes);* *(deduct)* togliere; *(mimic)* imitare; **~ time off** prendere delle vacanze; **~ oneself off** andarsene ● *vi* Aeron decollare. **take on** *vt* farsi

carico di; assumere *(employee);* *(as opponent)* prendersela con. **take out** *vt* portare fuori; togliere *(word, stain);* *(withdraw)* ritirare *(money, books);* **~ out a subscription to sth** abbonarsi a qcsa; **~ it out on sb** *fam* prendersela con qcno. **take over** *vt* assumere il controllo di *(firm)* ● *vi* **~ over from sb** sostituire qcno; *(permanently)* succedere a qcno. **take to** *vt* *(as a habit)* darsi a; **I took to her** *(liked)* mi è piaciuta. **take up** *vt* portare su; accettare *(offer);* intraprendere *(profession);* dedicarsi a *(hobby);* prendere *(time);* occupare *(space);* tirare su *(floorboards);* accorciare *(dress);* **~ sth up with sb** discutere qcsa con qcno ● *vi* **~ up with sb** legarsi a qcno

take: **~-away** *n* *(meal)* piatto *m* da asporto; *(restaurant)* ristorante *m* che prepara piatti da asporto. **~-off** *n* Aeron decollo *m.* **~-over** *n* rilevamento *m.* **~-over bid** offerta *f* di assorbimento

takings /'teɪkɪŋz/ *npl* incassi *mpl*

talcum /'tælkəm/ *n* **~ [powder]** talco *m*

tale /teɪl/ *n* storia *f;* *pej* fandonia *f*

talent /'tælənt/ *n* talento *m.* **~ed** *a* [ricco] di talento

talk /tɔ:k/ *n* conversazione *f;* *(lecture)* conferenza *f;* *(gossip)* chiacchere *fpl;* **make small ~** *vi* parlare del più e del meno ● *vi* parlare ● *vt* parlare di *(politics* etc*);* **~ sb into sth** convincere qcno di qcsa. **talk over** *vt* discutere

talkative /'tɔ:kətɪv/ *a* loquace

'talking-to *n* sgridata *f*

talk show *n* talk show *m inv*

tall /tɔ:l/ *a* alto. **~boy** *n* cassettone *m.* **~ order** *n* impresa *f* difficile. **~ 'story** *n* frottola *f*

tally /'tælɪ/ *n* conteggio *m;* **keep a ~ of** tenere il conto di ● *vi* coincidere

tambourine /tæmbə'ri:n/ *n* tamburello *m*

tame /teɪm/ a ‹animal› domestico; ‹dull› insulso ● vt domare. **~ly** adv docilmente. **~r** n domatore, ·trice mf

tamper /ˈtæmpə(r)/ vi **~ with** manomettere

tampon /ˈtæmpɒn/ n tampone m

tan /tæn/ a marrone rossiccio ● n marrone m rossiccio; ‹from sun› abbronzatura f ● v (pt/pp **tanned**). ● vt conciare ‹hide› ● vi abbronzarsi

tang /tæŋ/ n sapore m forte; ‹smell› odore m penetrante

tangent /ˈtændʒənt/ n tangente f

tangible /ˈtændʒɪbl/ a tangibile

tangle /ˈtæŋgl/ n groviglio m; ‹in hair› nodo m ● vt **~[up]** aggrovigliare ● vi aggrovigliarsi

tango /ˈtæŋgəʊ/ n tango m inv

tank /tæŋk/ n contenitore m; ‹for petrol› serbatoio m; ‹fish ~› acquario m; Mil carro m armato

tankard /ˈtæŋkəd/ n boccale m

tanker /ˈtæŋkə(r)/ n nave f cisterna; ‹lorry› autobotte f

tanned /tænd/ a abbronzato

tantaliz|e /ˈtæntəlaɪz/ vt tormentare. **~ing** a allettante; ‹smell› stuzzicante

tantamount /ˈtæntəmaʊnt/ a **~ to** equivalente a

tantrum /ˈtæntrəm/ n scoppio m d'ira

tap /tæp/ n rubinetto m; ‹knock› colpo m; **on ~** a disposizione ● v (pt/pp **tapped**) ● vt dare un colpetto a; sfruttare ‹resources›; mettere sotto controllo ‹telephone› ● vi picchiettare. **~-dance** n tip tap m ● vi ballare il tip tap

tape /teɪp/ n nastro m; ‹recording› cassetta f ● vt legare con nastro; ‹record› registrare

'**tape: ~ backup drive** n Comput unità f di backup a nastro. **~-deck** n piastra f. **~-measure** n metro m [a nastro]

taper /ˈteɪpə(r)/ n candela f sottile ● **taper off** vi assottigliarsi

'**tape: ~ recorder** n registratore m. **~ recording** n registrazione f

tapestry /ˈtæpɪstrɪ/ n arazzo m

'**tap water** n acqua f del rubinetto

tar /tɑː(r)/ n catrame m ● vt (pt/pp **tarred**) incatramare

tardy /ˈtɑːdɪ/ a (-ier, -iest) tardivo

target /ˈtɑːgɪt/ n bersaglio m; fig obiettivo m

tariff /ˈtærɪf/ n ‹price› tariffa f; ‹duty› dazio m

Tarmac® /ˈtɑːmæk/ n macadam m al catrame. **tarmac** n Aeron pista f di decollo

tarnish /ˈtɑːnɪʃ/ vi ossidarsi ● vt ossidare; fig macchiare

tarpaulin /tɑːˈpɔːlɪn/ n telone m impermeabile

tart¹ /tɑːt/ a aspro; fig acido

tart² n crostata f; ‹individual› crostatina f; ‹sl: prostitute› donnaccia f ● **tart up** vt fam **~ oneself up** agghindarsi

tartan /ˈtɑːtn/ n tessuto m scozzese, tartan m inv ● attrib di tessuto scozzese

tartar /ˈtɑːtə(r)/ n ‹on teeth› tartaro m

tartar 'sauce /tɑːtə-/ n salsa f tartara

task /tɑːsk/ n compito m; **take sb to ~** riprendere qcno. **~ force** n Pol commissione f; Mil task-force f inv

tassel /ˈtæsl/ n nappa f

taste /teɪst/ n gusto m; ‹sample› saggio m; **get a ~ of sth** fig assaporare il gusto di qcsa ● vt sentire il sapore di; ‹sample› assaggiare ● vi sapere (**of** di); **it ~s lovely** è ottimo. **~ful** a ‹di [buon] gusto. **~fully** adv con gusto. **~less** a senza gusto. **~lessly** adv con cattivo gusto

tasty /ˈteɪstɪ/ a (-ier, -iest) saporito

tat /tæt/ see **tit²**

tatter|ed /ˈtætəd/ a cencioso ‹pages› stracciato. **~s** npl **in ~s** a brandelli

tattoo¹ /təˈtuː/ n tatuaggio m ● vt tatuare

tattoo² *n Mil* parata *f* militare

tatty /'tætɪ/ *a* (-ier, -iest) ⟨clothes, person⟩ trasandato; ⟨book⟩ malandato

taught /tɔːt/ *see* **teach**

taunt /tɔːnt/ *n* scherno *m* ● *vt* schernire

Taurus /'tɔːrəs/ *n* Toro *m*

taut /tɔːt/ *a* teso

tawdry /'tɔːdrɪ/ *a* (-ier, -iest) pacchiano

tax /tæks/ *n* tassa *f*; (on income) imposte *fpl*; **before** ~ ⟨price⟩ tasse escluse; ⟨salary⟩ lordo ● *vt* tassare; *fig* mettere alla prova; ~ **with** accusare di. **~able** /-əbl/ *a* tassabile. **~ation** /-'seɪʃn/ *n* tasse *fpl*. **~ evasion** *n* evasione *f* fiscale. **~-free** *a* esentasse. **~ haven** *n* paradiso *m* fiscale

taxi /'tæksɪ/ *n* taxi *m inv* ● *vi* (pt/pp **taxied**, pres p **taxiing**) ⟨aircraft:⟩ rullare. **~ driver** *n* tassista *mf*. **~ rank** *n* posteggio *m* per taxi

'taxpayer *n* contribuente *mf*

tea /tiː/ *n* tè *m inv*. **~-bag** *n* bustina *f* di tè. **~-break** *n* intervallo *m* per il tè

teach /tiːtʃ/ *vt/i* (pt/pp **taught**) insegnare; ~ **sb sth** insegnare qcsa a qcno. **~er** *n* insegnante *mf*; ⟨primary⟩ maestro, -a *mf*. **~ing** *n* insegnamento *m*

tea: **~cloth** *n* (for drying) asciugapiatti *m inv*. **~cup** *n* tazza *f* da tè

teak /tiːk/ *n* tek *m*

'tea-leaves *npl* tè *m inv* sfuso; (when infused) fondi *mpl* di tè

team /tiːm/ *n* squadra *f*; *fig* équipe *f inv* ● **team up** *vi* unirsi

'team-work *n* lavoro *m* di squadra; *fig* lavoro *m* d'équipe

'teapot *n* teiera *f*

tear¹ /teə(r)/ *n* strappo *m* ● *v* (pt **tore**, pp **torn**) ● *vt* strappare ● *vi* strapparsi; ⟨material:⟩ strapparsi; (run) precipitarsi. **tear apart** *vt* (fig: criticize) fare a pezzi; (separate) dividere. **tear away** *vt* ~

oneself away andare via; ~ **oneself away from** staccarsi da ⟨television⟩. **tear open** *vt* aprire strappando. **tear up** *vt* strappare; rompere ⟨agreement⟩

tear² /tɪə(r)/ *n* lacrima *f*. **~ful** *a* ⟨person⟩ in lacrime; ⟨farewell⟩ lacrimevole. **~fully** *adv* in lacrime. **~gas** *n* gas *m* lacrimogeno

tease /tiːz/ *vt* prendere in giro ⟨person⟩; tormentare ⟨animal⟩

tea: **~-set** *n* servizio *m* da tè. **~ shop** *n* sala *f* da tè. **~spoon** *n* cucchiaino *m* [da tè]. **~spoon[ful]** *n* cucchiaino *m*

teat /tiːt/ *n* capezzolo *m*; (on bottle) tettarella *f*

'tea-towel *n* strofinaccio *m* [per i piatti]

technical /'teknɪkl/ *a* tecnico. **~ity** /-'kælətɪ/ *n* tecnicismo *m*; *Jur* cavillo *m* giuridico. **~ly** *adv* tecnicamente; (strictly) strettamente

technician /tek'nɪʃn/ *n* tecnica *f* -a *mf*

technique /tek'niːk/ *n* tecnica *f*

technological /teknə'lɒdʒɪkl/ *a* tecnologico

technology /tek'nɒlədʒɪ/ *n* tecnologia *f*

teddy /'tedɪ/ *n* ~ [**bear**] orsacchiotto *m*

tedious /'tiːdɪəs/ *a* noioso

tedium /'tiːdɪəm/ *n* tedio *m*

tee /tiː/ *n* (in golf) tee *m inv*

teem /tiːm/ *vi* (rain) piovere a dirotto; **be ~ing with** (full of) pullulare di

teenage /'tiːneɪdʒ/ *a* per ragazzi; **~ boy/girl** adolescente *mf*. **~r** *n* adolescente *mf*

teens /tiːnz/ *npl* **the ~** l'adolescenza *fsg*; **be in one's ~** essere adolescente

teeny /'tiːnɪ/ *a* (-ier, -iest) piccolissimo

teeter /'tiːtə(r)/ *vi* barcollare

teeth /tiːθ/ *see* **tooth**

teethe /tiːð/ *vi* mettere i [primi] denti. **~ing troubles** *npl fig* difficoltà *fpl* iniziali

teetotal /tiːˈtəʊtl/ a astemio. **~ler** n astemio, -a mf

telecommunications /telɪkəmjuːnɪˈkeɪʃnz/ npl telecomunicazioni fpl

telegram /ˈtelɪgræm/ n telegramma m

telegraph /ˈtelɪgrɑːf/ n telegrafo m. **~ic** /-ˈgræfɪk/ a telegrafico. **~ pole** n palo m del telegrafo

telepathy /tɪˈlepəθɪ/ n telepatia f

telephone /ˈtelɪfəʊn/ n telefono m; **be on the ~** avere il telefono; (be telephoning) essere al telefono ● vt telefonare a ● vi telefonare

telephone: **~ book** n elenco m telefonico. **~ booth** n, **~ box** n cabina f telefonica. **~ directory** n elenco m telefonico. **~ number** n numero m di telefono

telephonist /tɪˈlefənɪst/ n telefonista mf

'telephoto /ˈtelɪ-/ a **~ lens** teleobiettivo m

telescop|e /ˈtelɪskəʊp/ n telescopio m. **~ic** /-ˈskʊpɪk/ a telescopico

televise /ˈtelɪvaɪz/ vt trasmettere per televisione

television /ˈtelɪvɪʒn/ n televisione f; **watch ~** guardare la televisione. **~ set** n televisore m

telex /ˈteleks/ n telex m inv

tell /tel/ vt (pt/pp **told**) dire; raccontare (story); (distinguish) distinguere (**from** da); **~ sb sth** dire qcsa a qcno; **~ the time** dire l'ora; **I couldn't ~ why...** non sapevo perché... ● vi (produce an effect) avere effetto; **time will ~** il tempo ce lo dirà; **his age is beginning to ~** l'età comincia a farsi sentire [per lui]; **you mustn't ~** non devi dire niente. **tell off** vt sgridare

teller /ˈtelə(r)/ n (in bank) cassiere, -a mf

telling /ˈtelɪŋ/ a significativo; (argument) efficace

telly /ˈtelɪ/ n fam tv f inv

temerity /tɪˈmerətɪ/ n audacia f

temp /temp/ n fam impiegato, -a mf temporaneo, -a

temper /ˈtempə(r)/ n (disposition) carattere m; (mood) umore m; (anger) collera f; **lose one's ~** arrabbiarsi; **be in a ~** essere arrabbiato; **keep one's ~** mantenere la calma

temperament /ˈtemprəmənt/ n temperamento m. **~al** /-ˈmentl/ a (moody) capriccioso

temperate /ˈtempərət/ a (climate) temperato

temperature /ˈtemprətʃə(r)/ n temperatura f; **have a ~** avere la febbre

tempest /ˈtempɪst/ n tempesta f. **~uous** /-ˈpestjʊəs/ a tempestoso

temple[1] /ˈtempl/ n tempio m

temple[2] n Anat tempia f

tempo /ˈtempəʊ/ n ritmo m; Mus tempo m

temporar|y /ˈtempərərɪ/ a temporaneo; (measure, building) provvisorio. **~ily** adv temporaneamente; (introduced, erected) provvisoriamente

tempt /tempt/ vt tentare; sfidare (fate); **~ sb to** indurre qcno a; **be ~ed** essere tentato (**to** di); **I am ~ed by the offer** l'offerta mi tenta. **~ation** /-ˈteɪʃn/ n tentazione f. **~ing** a allettante; (food, drink) invitante

ten /ten/ a dieci

tenable /ˈtenəbl/ a fig sostenibile

tenaci|ous /tɪˈneɪʃəs/ a tenace. **~ty** /-ˈnæsətɪ/ n tenacia f

tenant /ˈtenənt/ n inquilino, -a mf; Comm locatario, -a mf

tend[1] /tend/ vt (look after) prendersi cura di

tend[2] vi **~ to do sth** tendere a far qcsa

tendency /ˈtendənsɪ/ n tendenza f

tender[1] /ˈtendə(r)/ n Comm offerta f; **be legal ~** avere corso legale ● vt offrire; presentare (resignation)

tender[2] a tenero; (painful) dolorante. **~ly** adv teneramente. **~ness** n tenerezza f; (painfulness) dolore m

tendon /'tendən/ *n* tendine *m*

tenement /'tenəmənt/ *n* casamento *m*

tenner /'tenə(r)/ *n fam* biglietto *m* da dieci sterline

tennis /'tenɪs/ *n* tennis *m*. **~-court** *n* campo *m* da tennis. **~ player** *n* tennista *mf*

tenor /'tenə(r)/ *n* tenore *m*

tense[1] /tens/ *n Gram* tempo *m*

tense[2] *a* teso ● *vt* tendere *(muscle)*. **tense up** *vi* tendersi

tension /'tenʃn/ *n* tensione *f*

tent /tent/ *n* tenda *f*

tentacle /'tentəkl/ *n* tentacolo *m*

tentative /'tentətɪv/ *a* provvisorio; *(smile, gesture)* esitante. **~ly** *adv* timidamente; *(accept)* provvisoriamente

tenterhooks /'tentəhʊks/ *npl* **be on ~** essere sulle spine

tenth /tenθ/ *a* decimo ● *n* decimo, -a *mf*

tenuous /'tenjʊəs/ *a fig* debole

tepid /'tepɪd/ *a* tiepido

term /tɜ:m/ *n* periodo *m*; *Sch Univ* trimestre *m*; *(expression)* termine *m*; **~s** *pl (conditions)* condizioni *fpl*; **~ of office** carica *f*; **in the short/long ~** a breve/lungo termine; **be on good/bad ~s** essere in buoni/cattivi rapporti; **come to ~s with** accettare *(past, fact)*; **easy ~s** facilità *f* di pagamento

terminal /'tɜ:mɪn(ə)l/ *a* finale; *Med* terminale ● *n Aeron* terminal *m inv*; *Rail* stazione *f* di testa; *(of bus)* capolinea *m*; *(on battery)* morsetto *m*; *Comput* terminale *m*. **~ly** *adv* **be ~ly ill** essere in fase terminale

terminat|e /'tɜ:mɪneɪt/ *vt* terminare; rescindere *(contract)*; interrompere *(pregnancy)* ● *vi* terminare; **~e in** finire in. **~ion** /-'neɪʃn/ *n* termine *m*; *Med* interruzione *f* di gravidanza

terminology /tɜ:mɪ'nɒlədʒɪ/ *n* terminologia *f*

terminus /'tɜ:mɪnəs/ *n* (*pl* **-ni** /-naɪ/) *(for bus)* capolinea *m*; *(for train)* stazione *f* di testa

terrace /'terəs/ *n* terrazza *f*; *(houses)* fila *f* di case a schiera; **the ~s** *Sport* le gradinate. **~d house** *n* casa *f* a schiera

terrain /te'reɪn/ *n* terreno *m*

terrible /'terəbl/ *a* terribile. **~y** *adv* terribilmente

terrier /'terɪə(r)/ *n* terrier *m inv*

terrific /tə'rɪfɪk/ *a fam (excellent)* fantastico; *(huge)* enorme. **~ally** *adv fam* terribilmente

terrify /'terɪfaɪ/ *vt (pt/pp* **-ied)** atterrire; **be ~fied** essere terrorizzato. **~fying** *a* terrificante

territorial /terɪ'tɔ:rɪəl/ *a* territoriale

territory /'terɪtərɪ/ *n* territorio *m*

terror /'terə(r)/ *n* terrore *m*. **~ism** /-ɪzm/ *n* terrorismo *m*. **~ist** /-ɪst/ *n* terrorista *mf*. **~ize** *vt* terrorizzare

terse /tɜ:s/ *a* conciso

test /test/ *n* esame *m*; *(in laboratory)* esperimento *m*; *(of friendship, machine)* prova *m*; *(of intelligence, aptitude)* test *m inv*; **put to the ~** mettere alla prova ● *vt* esaminare; provare *(machine)*

testament /'testəmənt/ *n* testamento *m*; **Old/New T~** Antico/Nuovo Testamento *m*

testicle /'testɪkl/ *n* testicolo *m*

testify /'testɪfaɪ/ *vt/i (pt/pp* **-ied)** testimoniare

testimonial /testɪ'məʊnɪəl/ *n* lettera *f* di referenze

testimony /'testɪmənɪ/ *n* testimonianza *f*

'test: **~ match** *n* partita *f* internazionale. **~-tube** *n* provetta *f*. **~-tube 'baby** *n fam* bambino, -a *mf* in provetta

tetanus /'tetənəs/ *n* tetano *m*

tether /'teðə(r)/ *n* **be at the end of one's ~** non poterne più

text /tekst/ *n* testo *m*. **~book** *n* manuale *m*

textile /'tekstaɪl/ *a* tessile ● *n* stoffa *f*

texture /'tekstʃə(r)/ *n (of skin)* grana f; (of food) consistenza f; **of a smooth ~** (to the touch) soffice al tatto

Thai /taɪ/ a & n tailandese mf. **~land** n Tailandia f

Thames /temz/ n Tamigi m

than /ðən/, accentato /ðæn/ conj che; (with numbers, names) di; older ~ me più vecchio di me

thank /θæŋk/ vt ringraziare; ~ **you** [very much] grazie [mille]. **~ful** a grato. **~fully** adv con gratitudine; (happily) fortunatamente. **~less** a ingrato

thanks /θæŋks/ npl ringraziamenti mpl; **~!** fam grazie!; ~ **to** grazie a

that /ðæt/ a & pron (pl those) quel, quei pl; (before s + consonant, gn, ps and z) quello, quegli pl; (before vowel) quell' pl, quegli mpl, quelle fpl; ~ **one** quello; **I don't like those** quelli non mi piacciono; ~ **is** cioè; **is ~ you?** sei tu?; **who is ~?** chi è?; **what did you do after ~?** cosa hai fatto dopo?; **like ~** in questo modo, così; **a man like ~** un uomo così; ~ **is why** ecco perché; **~'s it!** (you've understood) ecco!; (I've finished) ecco fatto!; (I've had enough) basta così!; (there's nothing more) tutto qui!; **~'s ~** (with job) ecco fatto!; (with relationship) è tutto finito!; **and ~'s ~** il punto e basta! **all ~ I know** tutto quello che so **all ~ I wasn't ~** good non era poi così buono ● rel pron che; **the man ~ I spoke to** l'uomo con cui ho parlato; **the day ~ I saw him** il giorno in cui l'ho visto; **all ~ I know** tutto quello che so ● conj che; **I think ~...** penso che...

thatch /θætʃ/ n tetto m di paglia. **~ed** a coperto di paglia

thaw /θɔː/ n disgelo m ● vt fare scongelare (food) ● vi (food): scongelarsi; **it's ~ing** sta sgelando

the /ðə/, di fronte a una vocale /ðiː/ def art il, la f; i mpl, le fpl; (before s + consonant, gn, ps and z) lo, gli

mpl; (before vowel) l' mf, gli mpl, le fpl; **at ~ cinema/station** al cinema/alla stazion; **from ~ cinema/ station** dal cinema/dalla stazione ● adv ~ **more ~ better** più ce n'è meglio è; (with reference to pl) più ce ne sono, meglio è; **all ~ better** tanto meglio

theatre /'θɪətə(r)/ n teatro m; Med sala f operatoria

theatrical /θɪ'ætrɪkl/ a teatrale; (showy) melodrammatico

theft /θeft/ n furto m

their /ðeə(r)/ a il loro m, la loro f, i loro mpl, le loro fpl; ~ **mother/ father** la loro madre/il loro padre

theirs /ðeəz/ poss pron il loro m, la loro f, i loro mpl, le loro fpl; **a friend of ~** un loro amico; **friends of ~** dei loro amici; **those are ~** quelli sono loro; (as opposed to ours) quelli sono i loro

them /ðem/ pron (direct object) li m, le f; (indirect object) gli, loro fml; (after prep: with people) loro; (after preposition: with things) essi; **we haven't seen ~** non li/le abbiamo visti/viste; **give ~ the money** dai loro or dagli i soldi; **give it to ~** daglielo; **I've spoken to ~** ho parlato con loro; **it's ~** sono loro

theme /θiːm/ n tema m. ~ **song** n motivo m conduttore

them'selves pron (reflexive) si; (emphatic) se stessi; **they poured ~ a drink** si sono versati da bere; **they said so ~** lo hanno detto loro stessi; **they kept it to ~** se lo sono tenuti per sé; **by ~** da soli

then /ðen/ adv allora; (next) poi; **by ~** (in the past) ormai; (in the future) per allora; **since ~** sin da allora; **before ~** prima di allora; **from ~ on** da allora in poi; **now and ~** ogni tanto; **there and ~** all'istante ● a di allora

theologian /θɪə'ləʊdʒɪən/ n teologo, -a mf. **~y** /-'ɒlədʒɪ/ n teologia f

theorem /'θɪərəm/ n teorema m

theoretical /θɪə'retɪkl/ a teorico

theory /'θɪərɪ/ n teoria f; **in ~** in teoria

therapeutic /θerə'pju:tɪk/ a terapeutico

therap|ist /'θerəpɪst/ n terapista mf. **~y** n terapia f

there /ðeə(r)/ adv là, lì; **down/up ~** laggiù/lassù; **~ is/are** c'è/ci sono; **~ he/she is** eccolo/eccola ● int ~, **~!** dai, su!

there: **~abouts** adv [or] **~abouts** (roughly) all'incirca. **~'after** adv dopo di che. **~by** adv in tal modo. **~fore** /-fɔ:(r)/ adv perciò

thermal /'θɜ:m(ə)l/ a termale; **~ 'underwear** n biancheria f che mantiene la temperatura corporea

thermometer /θə'mɒmɪtə(r)/ n termometro m

Thermos® /'θɜ:məs/ n **~ [flask]** termos m inv

thermostat /'θɜ:məstæt/ n termostato m

thesaurus /θɪ'sɔ:rəs/ n dizionario m dei sinonimi

these /ði:z/ see **this**

thesis /'θi:sɪs/ n (pl **-ses** /-si:z/) tesi f inv

they /ðeɪ/ pron loro; **~ are tired** sono stanchi; **we're going, but ~ are not** noi andiamo, ma loro no; **~ say** (generalizing) si dice; **~ are building a new road** stanno costruendo una nuova strada

thick /θɪk/ a spesso; (forest) fitto; (liquid) denso; (hair) folto; (fam: stupid) ottuso; (fam: close) molto unito; **be 5 mm ~** essere 5 mm di spessore ● adv densamente ● **in the ~ of** nel mezzo di. ● **en** vt ispessire (sauce) ● vi ispessirsi; (fog:) infittirsi. **~ly** adv densamente; (cut) a fette spesse. **~ness** n spessore m

thick: **~set** a tozzo. **~-'skinned** a fam insensibile

thief /θi:f/ n (pl **thieves**) ladro, -a mf

thieving /'θi:vɪŋ/ a ladro ● n furti mpl

thigh /θaɪ/ n coscia f

thimble /'θɪmbl/ n ditale m

thin /θɪn/ a (**thinner, thinnest**) sottile; (shoes, sweater) leggero; (liquid) liquido; (person) magro; (fig: excuse, plot) inconsistente ● adv **~ly** ● v (pt/pp **thinned**) ● vt diluire (liquid) ● vi diradarsi. **thin out** vi diradarsi. **~ly** adv (populated) scarsamente; (disguised) leggermente; (cut) a fette sottili

thing /θɪŋ/ n cosa f; **~s** pl (belongings) roba fsg; **for one ~** in primo luogo; **the right ~** la cosa giusta; **just the ~!** proprio quel che ci vuole!; **how are ~s?** come vanno le cose?; **the latest ~** fam l'ultima cosa; **the best ~ would be** la cosa migliore sarebbe; **poor ~!** poveretto!

think /θɪŋk/ vt/i (pt/pp **thought**) pensare; (believe) credere; **I ~ so** credo di sì; **what do you ~?** (what is your opinion?) cosa ne pensi?; **~ of/about** pensare a; **what do you ~ of it?** cosa ne pensi di questo?. **think over** vt riflettere su. **think up** vt escogitare

third /θɜ:d/ a & n terzo, -a mf. **~ly** adv terzo. **~-rate** a scadente

thirst /θɜ:st/ n sete f. **~ily** adv con sete. **~y** a assetato; **be ~y** aver sete

thirteen /θɜ:'ti:n/ a tredici. **~th** a tredicesimo

thirtieth /'θɜ:tɪɪθ/ a trentesimo

thirty /'θɜ:tɪ/ a trenta

this /ðɪs/ a (pl **these**) questo; **~ man/woman** quest'uomo/questa donna; **these men/women** questi uomini/queste donne; **~ one** questo; **~ morning/evening** stamattina/stasera ● pron (pl **these**) questo; **we talked about ~ and that** abbiamo parlato del più e del meno; **like ~** così; **~ is Peter** questo è Peter; Teleph sono Peter; **who is ~?** chi è?; Teleph chi parla? ● adv così; **~ big** così grande

thistle /'θɪsl/ n cardo m

thorn /θɔ:n/ n spina f. **~y** a spinoso

thorough /ˈθʌrə/ *a* completo;
⟨knowledge⟩ profondo; ⟨clean,
search, training⟩ a fondo; ⟨person⟩
scrupoloso

thorough: ~**bred** *n* purosangue *m
inv*. ~**fare** *n* via *f* principale; **'no
~fare'** 'strada non transitabile '

thoroughly /ˈθʌrəlɪ/ *adv* ⟨clean,
search, know sth⟩ a fondo; ⟨extremely⟩ estremamente. ~**ness** *n*
completezza *f*

those /ðəʊz/ *see* **that**

though /ðəʊ/ *conj* sebbene; **as** ~
come **se** ● *adv* fam tuttavia

thought /θɔːt/ *see* **think** ● *n* pensiero *m*; (idea) idea *f*. ~**ful** *a*
pensieroso; (considerate) premuroso. ~**fully** *adv* pensierosamente;
(considerately) premurosamente.
~**less** *a* (inconsiderate) sconsiderato. ~**lessly** *adv* con noncuranza

thousand /ˈθaʊznd/ *a* **one** ~ mille *m inv* ● *n* mille *m inv*; ~**s of** migliaia *fpl* di. ~**th** *a* millesimo ● *n* millesimo, -a *mf*

thrash /θræʃ/ *vt* picchiare; (defeat)
sconfiggere. **thrash out** *vt* mettere
a punto

thread /θred/ *n* filo *m*; (of screw) filetto *m* ● *vt* infilare ⟨beads⟩; ~
one's way through farsi strada
fra. ~**bare** *a* logoro

threat /θret/ *n* minaccia *f*

threaten /ˈθretn/ *vt* minacciare (**to
do** di fare) ● *vi* fig incalzare. ~**ing**
a minaccioso; ⟨sky, atmosphere⟩ sinistro

three /θriː/ *a* tre. ~**fold** *a* & *adv* triplo. ~**some** /-səm/ *n* trio *m*

thresh /θreʃ/ *vt* trebbiare

threshold /ˈθreʃəʊld/ *n* soglia *f*

threw /θruː/ *see* **throw**

thrift /θrɪft/ *n* economia *f*. ~**y** *a*
parsimonioso

thrill /θrɪl/ *n* emozione *f*; (of fear)
brivido *m* ● *vt* entusiasmare; **be
~ed with** essere entusiasta di.
~**er** *n* ⟨book⟩ [romanzo *m*] giallo
m; ⟨film⟩ [film *m*] giallo *m*. ~**ing** *a*
eccitante

thrive /θraɪv/ *vi* (*pt* **thrived** *or*

throve, *pp* **thrived** *or* **thriven**
/ˈθrɪvn/) ⟨business:⟩ prosperare;
⟨child, plant:⟩ crescere bene; **I ~
on pressure** mi piace essere sotto
tensione

throat /θrəʊt/ *n* gola *f*; **sore** ~ mal
m di gola

throb /θrɒb/ *n* pulsazione *f*; (of
heart) battito *m* ● *vi* (*pt/pp*
throbbed) (vibrate) pulsare;
⟨heart:⟩ battere

throes /θrəʊz/ *npl* **in the** ~ **of** fig
alle prese con

thrombosis /θrɒmˈbəʊsɪs/ *n* trombosi *f*

throne /θrəʊn/ *n* trono *m*

throng /θrɒŋ/ *n* calca *f*

throttle /ˈθrɒtl/ *n* (on motorbike)
manopola *f* di accelerazione ● *vt*
strozzare

through /θruː/ *prep* attraverso;
(during) durante; (by means of)
tramite; (thanks to) grazie a;
Saturday ~ Tuesday *Am* da sabato a martedì incluso ● *adv* attraverso; ~ **and** ~ fino in fondo; **wet**
~ completamente bagnato; **read
sth** ~ dare una lettura a qcsa; **let**
~ lasciar passare ⟨sb⟩ ● *a* ⟨train⟩
diretto; **be** ~ (finished) aver finito; *Teleph* avere la comunicazione

throughout /θruːˈaʊt/ *prep* per tutto ● *adv* completamente; (time)
per tutto il tempo

throw /θrəʊ/ *n* tiro *m* ● *vt* (*pt*
threw, *pp* **thrown**) lanciare;
(throw away) gettare; azionare
⟨switch⟩; disarcionare ⟨rider⟩;
(fam: disconcert) disorientare;
fam dare ⟨party⟩. **throw away** *vt*
gettare via. **throw out** *vt* gettare
via; rigettare ⟨plan⟩; buttare fuori
⟨person⟩. **throw up** *vt* alzare ● *vi*
(vomit) vomitare

'throw-away *a* ⟨remark⟩ buttato lì;
⟨paper cup⟩ usa e getta *inv*

thrush /θrʌʃ/ *n* tordo *m*

thrust /θrʌst/ *n* spinta *f* ● *vt* (*pt/pp*
thrust) (push) spingere; (insert)
conficcare; ~ **[up]on** imporre a

thud /θʌd/ *n* tonfo *m*

thug /θʌg/ n deliquente m

thumb /θʌm/ n pollice m; **as a rule of** ~ come regola generale; **under sb's** ~ succube di qcno ● vt ~ **a lift** fare l'autostop. **~-index** n indice m a rubrica. **~tack** n Am puntina f da disegno

thump /θʌmp/ n colpo m; (noise) tonfo m ● vt battere su ⟨table, door⟩; battere ⟨fist⟩; colpire ⟨person⟩ ● vi battere ⟨on su⟩; ⟨heart:⟩ battere forte. **thump about** vi camminare pesantemente

thunder /'θʌndə(r)/ n tuono m; (loud noise) rimbombo m ● vi tuonare; (make loud noise) rimbombare. **~clap** n rombo m di tuono. **~storm** n temporale m. **~y** a temporalesco

Thursday /'θɜːzdeɪ/ n giovedì m inv

thus /ðʌs/ adv così

thwart /θwɔːt/ vt ostacolare

thyme /taɪm/ n timo m

Tiber /'taɪbə(r)/ n Tevere m

tick /tɪk/ n (sound) ticchettio m; (mark) segno m; (fam: instant) attimo m ● vi ticchettare. **tick off** vt spuntare; fam sgridare. **tick over** vi ⟨engine:⟩ andare al minimo

ticket /'tɪkɪt/ n biglietto m; (for item deposited, library) tagliando m; (label) cartellino m; (fine) multa f. **~collector** n controllore m. **~-office** n biglietteria f

tick|le /'tɪkl/ n solletico m ● vt fare il solletico a; (amuse) divertire ● vi fare prurito. **~lish** /'tɪklɪʃ/ a che soffre il solletico

tidal /'taɪdl/ a ⟨river, harbour⟩ di marea. ~ **wave** n onda f di marea

tiddly-winks /'tɪdlɪwɪŋks/ n gioco m della pulci

tide /taɪd/ n marea f; (of events) corso m; **the** ~ **is in/out** c'è alta/bassa marea ● **tide over** vt ~ **sb over** aiutare qcno a andare avanti

tidily /'taɪdɪlɪ/ adv in modo ordinato

tidiness /'taɪdɪnɪs/ n ordine m

tidy /'taɪdɪ/ a (-ier, -iest) ordinato; (fam: amount) bello ● vt (pt/pp -ied) ~ [up] ordinare; ~ **oneself up** mettersi in ordine

tie /taɪ/ n cravatta f; (cord) legaccio m; (fig: bond) legame m; (restriction) impedimento m; Sport pareggio m ● v (pres p tying) ● vt legare; fare ⟨knot⟩; **be** ~**d** ⟨in competition⟩ essere in parità ● vi pareggiare. **tie in with** vi corrispondere a. **tie up** vt legare; vincolare ⟨capital⟩; **be** ~**d up** ⟨busy⟩ essere occupato

tier /tɪə(r)/ n fila f; (of cake) piano m; (in stadium) gradinata f

tiff /tɪf/ n battibecco m

tiger /'taɪgə(r)/ n tigre f

tight /taɪt/ a stretto; (taut) teso; (fam: drunk) sbronzo; (fam: mean) spilorcio; ~ **corner** fam brutta situazione f ● adv strettamente; ⟨hold⟩ forte; ⟨closed⟩ bene

tighten /'taɪtn/ vt stringere; avvitare ⟨screw⟩; intensificare ⟨control⟩ ● vi stringersi

tight: ~**-'fisted** a tirchio. ~**-fitting** a aderente. ~**ly** adv strettamente; ⟨hold⟩ forte; ⟨closed⟩ bene. ~**rope** n fune f (da funamboli)

tights /taɪts/ npl collant m inv

tile /taɪl/ n mattonella f; (on roof) tegola f ● vt rivestire di mattonelle ⟨wall⟩

till[1] /tɪl/ prep & conj = until

till[2] n cassa f

tiller /'tɪlə(r)/ n barra f del timone

tilt /tɪlt/ n inclinazione f; **at full** ~ a tutta velocità ● vt inclinare ● vi inclinarsi

timber /'tɪmbə(r)/ n legname m

time /taɪm/ n tempo m; (occasion) volta f; (by clock) ora f; **two** ~**s four** due volte quattro; **at any** ~ in qualsiasi momento; **this** ~ questa volta; **at** ~**s, from** ~ **to** ~ ogni tanto; ~ **and again** cento volte; **two at a** ~ due alla volta; **on** ~ in orario; **in** ~ in tempo; (eventually) col tempo; **in no** ~ **at all** velocemente; **in a year's** ~ fra un anno;

behind ~ in ritardo; **behind the ~s** antiquato; **for the ~ being** per il momento; **what is the ~?** che ora è?; **by the ~ we arrive** quando arriviamo; **did you have a nice ~?** ti sei divertito?; **have a good ~!** divertiti! ● *vt* scegliere il momento per; cronometrare ‹race›; **be well ~d** essere ben calcolato **time:** ~ **bomb** *n* bomba *f* a orologeria. ~**-lag** *n* intervallo *m* di tempo. ~**less** *a* eterno. ~**ly** *a* opportuno. ~**-switch** *n* interruttore *m* a tempo. ~**-table** *n* orario *m*

timid /ˈtɪmɪd/ *a* ‹shy› timido; ‹fearful› timoroso

timing /ˈtaɪmɪŋ/ *n* Sport, Techn cronometraggio *m*; **the ~ of the election** il momento scelto per le elezioni

tin /tɪn/ *n* stagno *m*; ‹container› barattolo *m* ● *vt* ‹pt/pp **tinned**› inscatolare. ~ **foil** *n* [carta *f*] stagnola *f*

tinge /tɪndʒ/ *n* sfumatura *f* ● *vt* ~**d with** *fig* misto a

tingle /ˈtɪŋɡl/ *vi* pizzicare

tinker /ˈtɪŋkə(r)/ *vi* armeggiare

tinkle /ˈtɪŋkl/ *n* tintinnio *m*; ‹fam: phone call› colpo *m* di telefono ● *vi* tintinnare

tinned /tɪnd/ *a* in scatola

'**tin opener** *n* apriscatole *m* inv

tinsel /ˈtɪnsl/ *n* filo *m* d'argento

tint /tɪnt/ *n* tinta *f* ● *vt* tingersi ‹hair›

tiny /ˈtaɪnɪ/ *a* (-**ier**, -**iest**) minuscolo

tip[1] /tɪp/ *n* punta *f*

tip[2] *n* ‹money› mancia *f*; ‹advice› consiglio *m*; ‹for rubbish› discarica *f* ● *vt* ‹pt/pp **tipped**› ● *vt* ‹tilt› inclinare; ‹overturn› capovolgere; ‹pour› versare; ‹reward› dare una mancia a ● *vi* inclinarsi; ‹overturn› capovolgersi. **tip off** *vt* ~ **sb off** ‹inform› fare una soffiata a qcno. **tip out** *vt* rovesciare. **tip over** *vt* capovolgere ● *vi* capovolgersi

'**tip-off** *n* soffiata *f*

tipped /tɪpt/ *a* ‹cigarette› col filtro

tipsy /ˈtɪpsɪ/ *a* *fam* brillo

tiptoe /ˈtɪptəʊ/ *n* **on ~** in punta di piedi

tiptop /tɪpˈtɒp/ *a* *fam* in condizioni perfette

tire /ˈtaɪə(r)/ *vt* stancare ● *vi* stancarsi; ~**d** *a* stanco; ~**d of** stanco di; ~**d out** stanco morto. ~**less** *a* instancabile. ~**some** /-səm/ *a* fastidioso

tiring /ˈtaɪərɪŋ/ *a* stancante

tissue /ˈtɪʃuː/ *n* tessuto *m*; ‹handkerchief› fazzolettino *m* di carta. ~**-paper** *n* carta *f* velina

tit[1] /tɪt/ *n* ‹bird› cincia *f*

tit[2] *n* ~ **for tat** pan per focaccia

title /ˈtaɪtl/ *n* titolo *m*. ~**-deed** *n* atto *m* di proprietà. ~**-role** *n* ruolo *m* principale

tittle-tattle /ˈtɪtltætl/ *n* pettegolezzi *mpl*

to /tuː/, *atono* /tə/ *prep* a; ‹to countries› in; ‹towards› verso; ‹up to, until› fino a; **I'm going to John's/the butcher's** vado da John/dal macellaio; **come/go to sb** venire/andare da qcno; **to Italy/Switzerland** in Italia/Svizzera; **I've never been to Rome** non sono mai stato a Roma; **go to the market** andare al mercato; **to the toilet/my room** in bagno/camera mia; **to an exhibition** a una mostra; **to university** all'università; **twenty/quarter to eight** le otto meno venti/un quarto; **5 to 6 kilos** da 5 a 6 chili; **to the end** alla fine; **to this day** fino a oggi; **to the best of my recollection** per quanto mi possa ricordare; **give/say sth to sb** dare/dire qcsa a qcno; **give it to me** dammelo; **there's nothing to it** è una cosa da niente ● *verbal constructions* **to go** andare; **learn to swim** imparare a nuotare; **I want to/have to go** voglio/devo andare; **it's easy to forget** è facile da dimenticare; **too ill/tired to go** troppo malato/stanco per andare; **you have to do**; **I**

don't want to non voglio; **live to be 90** vivere fino a 90 anni; **he was the last to arrive** è stato l'ultimo ad arrivare; **to be honest,...** per essere sincero,... ● *adv* **pull to** chiudere; **to and fro** avanti e indietro

toad /təʊd/ *n* rospo *m.* **~stool** *n* fungo *m* velenoso

toast /təʊst/ *n* pane *m* tostato; (*drink*) brindisi *m* ● *vt* tostare (*bread*); (*drink a ~ to*) brindare a. **~er** *n* tostapane *m inv*

tobacco /tə'bækəʊ/ *n* tabacco *m.* **~nist's** [**shop**] *n* tabaccheria *f*

toboggan /tə'bɒgən/ *n* toboga *m* ● *vi* andare in toboga

today /tə'deɪ/ *n* & *adv* oggi *m*; **a week ~** una settimana a oggi; **~'s paper** il giornale di oggi

toddler /'tɒdlə(r)/ *n* bambino, -a *m/f* ai primi passi

to-do /tə'du:/ *n fam* baccano *m*

toe /təʊ/ *n* dito *m* del piede; (*of footwear*) punta *f*; **big ~** alluce *m* ● *vt* **~ the line** rigar diritto. **~nail** *n* unghia *f* del piede

toffee /'tɒfɪ/ *n* caramella *f* al mou

together /tə'geðə(r)/ *adv* insieme; (*at the same time*) allo stesso tempo; **~ with** insieme a

toilet /'tɔɪlɪt/ *n* (*lavatory*) gabinetto *m.* **~ paper** *n* carta *f* igienica

toiletries /'tɔɪlɪtrɪz/ *npl* articoli *mpl* da toilette

toilet: **~ roll** *n* rotolo *m* di carta igienica. **~ water** *n* acqua *f* di colonia

token /'təʊkən/ *n* segno *m*; (*counter*) gettone *m*; (*voucher*) buono *m* ● *attrib* simbolico

told /təʊld/ *see* **tell** ● *a* **all ~** in tutto

tolerab|le /'tɒl(ə)rəbl/ *a* tollerabile; (*not bad*) discreto. **~y** *adv* discretamente

toleran|ce /'tɒl(ə)r(ə)ns/ *n* tolleranza *f.* **~t** *a* tollerante. **~tly** *adv* con tolleranza

tolerate /'tɒləreɪt/ *vt* tollerare

toll¹ /təʊl/ *n* pedaggio *m*; **death ~** numero *m* di morti

toll² *vi* suonare a morto

tom /tɒm/ *n* (*cat*) gatto *m* maschio

tomato /tə'mɑːtəʊ/ *n* (*pl* **-es**) pomodoro *m.* **~ ketchup** *n* ketchup *m.* **~ purée** *n* concentrato *m* di pomodoro

tomb /tuːm/ *n* tomba *f*

tomboy /'tɒmbɔɪ/ *n* maschiaccio *m*

tombstone *n* pietra *f* tombale

tom-cat *n* gatto *m* maschio

tomfoolery /tɒm'fuːlərɪ/ *n* stupidaggini *fpl*

tomorrow /tə'mɒrəʊ/ *a* & *adv* domani *m*; **~ morning** domani mattina; **the day after ~** dopodomani; **see you ~!** a domani!

ton /tʌn/ *n* tonnellata *f* (*= 1,016 kg.*); **~s of** *fam* un sacco di

tone /təʊn/ *n* tono *m*; (*colour*) tonalità *f inv* ● *n* **tone down** *vt* attenuare. **tone up** *vt* tonificare (*muscles*)

toner /'təʊnə(r)/ *n* toner *m*

tongs /tɒŋz/ *npl* pinze *fpl*

tongue /tʌŋ/ *n* lingua *f*; **~ in cheek** (*fam: say*) ironicamente. **~-twister** *n* scioglilingua *m inv*

tonic /'tɒnɪk/ *n* tonico *m*; (*for hair*) lozione *f* per i capelli; *fig* toccasana *m inv*; **~** [**water**] acqua *f* tonica

tonight /tə'naɪt/ *adv* stanotte; (*evening*) stasera ● *n* questa notte *f*; (*evening*) questa sera *f*

tonne /tʌn/ *n* tonnellata *f* metrica

tonsil /'tɒnsl/ *n* *Anat* tonsilla *f.* **~litis** /-sə'laɪtɪs/ *n* tonsillite *f*

too /tuː/ *adv* troppo; (*also*) anche; **~ many** troppi; **~ much** troppo; **~ little** troppo poco

took /tʊk/ *see* **take**

tool /tuːl/ *n* attrezzo *m*

toot /tuːt/ *n* suono *m* di clacson ● *vi* *Auto* clacsonare

tooth /tuːθ/ *n* (*pl* **teeth**) dente *m*

tooth: **~ache** *n* mal *m* di denti. **~brush** *n* spazzolino *m* da denti. **~less** *a* sdentato. **~paste** *n* dentifricio *m.* **~pick** *n* stuzzicadenti *m inv*

top[1] /tɒp/ n (toy) trottola f

top[2] n cima f; Sch primo, -a mf; (upper part or half) parte f superiore; (of page, list, street) inizio m; (upper surface) superficie f; (lid) coperchio m; (of bottle) tappo m; (garment) maglia f; (blouse) camicia f; Auto marcia f più alta; **at the ~** fig al vertice; **at the ~ of one's voice** a squarciagola; **on ~ / on ~ of** sopra; **on ~ of that** (besides) per di più; **from ~ to bottom** da cima a fondo ● a in alto; (official, floor of building) superiore; (pupil, musician etc) migliore; (speed) massimo ● vt (pt/pp **topped**) essere in testa a (list); (exceed) sorpassare; **~ped with ice-cream** ricoperto di gelato. **top up** vt riempire

top: **~ 'floor** n ultimo piano m. **~ hat** n cilindro m. **~-heavy** a con la parte superiore sovraccarica

topic /'tɒpɪk/ n soggetto m; (of conversation) argomento m. **~al** a d'attualità

top: **~less** a & adv topless. **~most** a più alto

topple /'tɒpl/ vt rovesciare ● vi rovesciarsi. **topple off** vi cadere

top-'secret a segretissimo, top secret inv

topsy-turvy /tɒpsɪ'tɜːvɪ/ a & adv sottosopra

torch /tɔːtʃ/ n torcia f [elettrica]; (flaming) fiaccola f

tore /tɔː(r)/ see **tear**[1]

torment[1] /'tɔːment/ n tormento m

torment[2] /tɔː'ment/ vt tormentare

torn /tɔːn/ see **tear**[1] ● a bucato

tornado /tɔː'neɪdəʊ/ n (pl -es) tornado m inv

torpedo /tɔː'piːdəʊ/ n (pl -es) siluro m ● vt silurare

torrent /'tɒrənt/ n torrente m. **~ial** /tə'renʃl/ a (rain) torrenziale

torso /'tɔːsəʊ/ n torso m; (in art) busto m

tortoise /'tɔːtəs/ n tartaruga f

tortuous /'tɔːtʃʊəs/ a tortuoso

torture /'tɔːtʃə(r)/ n tortura f ● vt torturare

Tory /'tɔːrɪ/ a & n fam conservatore, -trice mf

toss /tɒs/ vt gettare; (into the air) lanciare in aria; (shake) scrollare; (horse) disarcionare; mescolare (salad); rivoltare facendo saltare in aria (pancake); **~ a coin** fare testa o croce ● vi **~ and turn** (in bed) rigirarsi; **let's ~ for it** facciamo testa o croce

tot[1] /tɒt/ n bimbetto, -a mf; (fam: of liquor) goccio m

tot[2] vt (pt/pp **totted**) **~ up** fam fare la somma di

total /'təʊtl/ a totale ● n totale m ● vt (pt/pp **totalled**) ammontare a; (add up) sommare

totalitarian /təʊtælɪ'teərɪən/ a totalitario

totally /'təʊtəlɪ/ adv totalmente

totter /'tɒtə(r)/ vi barcollare; (government:) vacillare

touch /tʌtʃ/ n tocco m; (sense) tatto m; (contact) contatto m; (trace) traccia f; (of irony, humour) tocco m; **get/be in ~** mettersi/essere in contatto ● vt toccare; (lightly) sfiorare; (equal) eguagliare; (fig: move) commuovere ● vi toccarsi. **touch down** vi Aeron atterrare. **touch on** vt fig accennare a. **touch up** vt ritoccare (painting)

touching /'tʌtʃɪŋ/ a commovente. **~y** a permaloso; (subject) delicato

tough /tʌf/ a duro; (severe, harsh) severo; (durable) resistente; (resilient) forte

toughen /'tʌfn/ vt rinforzare. **toughen up** vt rendere più forte (person)

tour /tʊə(r)/ n giro m; (of building, town) visita f; Theat, Sport tournée f inv; (of duty) servizio m ● vt visitare ● vi fare un giro turistico; Theat essere in tournée

tourism /'tʊərɪzm/ n turismo m. **~t** /-rɪst/ n turista mf ● attrib turistico. **~t office** n ufficio m turistico

tournament /'tʊənəmənt/ n torneo m

'**tour operator** n tour operator mf inv, operatore, -trice mf turistico

tousle /'taʊzl/ vt spettinare

tout /taʊt/ n (ticket ~) bagarino m; (horse-racing) informatore m ● vi ~ **for** sollecitare

tow /təʊ/ n rimorchio m; '**on ~**' 'a rimorchio; **in ~** fam al seguito ● vt rimorchiare. **tow away** vt portare via col carro attrezzi

toward[s] /tə'wɔːd(z)/ prep verso (with respect to) nei riguardi di

towel /'taʊəl/ n asciugamano m. **~ing** n spugna f

tower /'taʊə(r)/ n torre f ● vi ~ **above** dominare. **~ block** n palazzone m. **~ing** a torreggiante; (rage) violento

town /taʊn/ n città f inv. **~ 'hall** n municipio m

tow: **~-path** n strada f alzaia. **~-rope** n cavo m da rimorchio

toxic /'tɒksɪk/ a tossico

toxin /'tɒksɪn/ n tossina f

toy /tɔɪ/ n giocattolo m. **~shop** n negozio m di giocattoli. **toy with** vt giocherellare con

trace /treɪs/ n traccia f ● vt seguire le tracce di; (find) rintracciare; (draw) tracciare; (with tracing-paper) ricalcare

track /træk/ n traccia f; (path, Sport) pista f; Rail binario m; **keep ~ of** tenere d'occhio ● vt seguire le tracce di. **track down** vt scovare

'**track:** **~ball** n Comput trackball f inv. **~suit** n tuta f da ginnastica

tractor /'træktə(r)/ n trattore m

trade /treɪd/ n commercio m; (line of business) settore m; (craft) mestiere m; **by ~** di mestiere ● vt commerciare; **~ sth for sth** scambiare qcsa per qcsa ● vi commerciare. **trade in** vt (give in part exchange) dare in pagamento parziale

'**trade mark** n marchio m di fabbrica

trader /'treɪdə(r)/ n commerciante mf

trade: **~sman** n (joiner etc) operaio m. **~ 'union** n sindacato m. **~ 'unionist** n sindacalista mf

trading /'treɪdɪŋ/ n commercio m. **~ estate** n zona f industriale

tradition /trə'dɪʃn/ n tradizione f. **~al** a tradizionale. **~ally** adv tradizionalmente

traffic /'træfɪk/ n traffico m ● vi (pt/pp trafficked) trafficare

traffic: **~ circle** n Am isola f rotatoria. **~ jam** n ingorgo m. **~ lights** npl semaforo msg. **~ warden** n vigile m [urbano]; (woman) vigilessa f

tragedy /'trædʒədɪ/ n tragedia f

tragic /'trædʒɪk/ a tragico. **~ally** adv tragicamente

trail /treɪl/ n traccia f; (path) sentiero m ● vi strisciare; (plant:) arrampicarsi; ~ [**behind**] rimanere indietro; (in competition) essere in svantaggio ● vt trascinare

trailer /'treɪlə(r)/ n Auto rimorchio m; (Am: caravan) roulotte f inv; (film) presentazione f (di un film)

train /treɪn/ n treno m; ~ **of thought** filo m dei pensieri ● vt formare professionalmente; Sport allenare; (aim) puntare; educare (child); addestrare (animal, soldier) ● vi fare il tirocinio; Sport allenarsi. **~ed** a (animal) addestrato (**to do** a fare)

trainee /treɪ'niː/ n apprendista mf

train|er /'treɪnə(r)/ n Sport allenatore, -trice mf; (in circus) domatore, -trice mf; (of dog, racehorse) addestratore, -trice mf; **~ers** pl scarpe fpl da ginnastica. **~ing** n tirocinio m; Sport allenamento m; (of animal, soldier) addestramento m

traipse /treɪps/ vi ~ **around** fam andare in giro

trait /treɪt/ n caratteristica f

traitor /'treɪtə(r)/ n traditore, -trice mf

tram /træm/ n tram m inv. **~-lines** npl rotaie fpl del tram

tramp /træmp/ n (hike) camminata f; (vagrant) barbone, -a mf; (of feet) calpestio m ● vi camminare con passo pesante; (hike) percorrere a piedi

trample /'træmpl/ vt/i ~ [on] calpestare

trampoline /'træmpəli:n/ n trampolino m

trance /trɑːns/ n trance f inv

tranquil /'træŋkwɪl/ a tranquillo. **~lity** /-'kwɪlətɪ/ n tranquillità f

tranquillizer /'træŋkwɪlaɪzə(r)/ n tranquillante m

transact /træn'zækt/ vt trattare. **~ion** /-ækʃn/ n transazione f

transatlantic /trænzət'læntɪk/ a transatlantico

transcend /træn'send/ vt trascendere

transfer¹ /'trænsfɜː(r)/ n trasferimento m; Sport cessione f; (design) decalcomania f

transfer² /træns'fɜː(r)/ v (pt/pp transferred) ● vt trasferire; Sport cedere ● vi trasferirsi; (when travelling) cambiare. **~able** /-əbl/ a trasferibile

transform /træns'fɔːm/ vt trasformare. **~ation** /-fə'meɪʃn/ n trasformazione f. **~er** n trasformatore m

transfusion /træns'fjuːʒn/ n trasfusione f

transient /'trænzɪənt/ a passeggero

transistor /træn'zɪstə(r)/ n transistor m inv; (radio) radiolina f a transistor

transit /'trænzɪt/ n transito m; in ~ (goods) in transito

transition /træn'zɪʃn/ n transizione f. **~al** a di transizione

transitive /'trænzɪtɪv/ a transitivo

transitory /'trænzɪtərɪ/ a transitorio

translate /trænz'leɪt/ vt tradurre. **~ion** /-'leɪʃn/ n traduzione f. **~or** n traduttore, -trice mf

transmission /trænz'mɪʃn/ n trasmissione f

transmit /trænz'mɪt/ vt (pt/pp transmitted) trasmettere. **~ter** n trasmettitore m

transparen|cy /træn'spærənsɪ/ n Phot diapositiva f. **~t** a trasparente

transpire /træn'spaɪə(r)/ vi emergere; (fam: happen) accadere

transplant¹ /'trænspla:nt/ n trapianto m

transplant² /træns'pla:nt/ vt trapiantare

transport¹ /'trænspɔːt/ n trasporto m

transport² /træn'spɔːt/ vt trasportare. **~ation** /-'teɪʃn/ n trasporto m

transvestite /trænz'vestaɪt/ n travestito, -a mf

trap /træp/ n trappola f; (fam: mouth) boccaccia f ● vt (pt/pp trapped) intrappolare; schiacciare (finger in door). **~'door** n botola f

trapeze /trə'piːz/ n trapezio m

trash /træʃ/ n robaccia f; (rubbish) spazzatura f; (nonsense) schiocchezze fpl. **~can** n Am secchio m della spazzatura. **~y** a scadente

trauma /'trɔːmə/ n trauma m. **~tic** /-'mætɪk/ a traumatico. **~tize** /-taɪz/ traumatizzare

travel /'trævl/ n viaggi mpl ● v (pt/pp travelled) ● vi viaggiare; (to work) andare ● vt percorrere (distance). **~ agency** n agenzia f di viaggi. **~ agent** n agente m di viaggio

traveller /'trævələ(r)/ n viaggiatore, -trice mf; Comm commesso m viaggiatore; **~s** pl (gypsies) zingari mpl. **~'s cheque** n traveller's cheque m inv

trawler /'trɔːlə(r)/ n peschereccio m

tray /treɪ/ n vassoio m; (for baking) teglia f; (for documents) vaschetta f sparticarta; (of printer, photocopier) vassoio m

treacher|ous /'tretʃərəs/ a traditore; ⟨weather, currents⟩ pericoloso. **~y** n tradimento m

treacle /'triːkl/ n melassa f

tread /tred/ n andatura f; ⟨step⟩ gradino m; ⟨of tyre⟩ battistrada m inv ● v (pt **trod**, pp **trodden**) vi ⟨walk⟩ camminare. **tread on** vt calpestare ⟨grass⟩; pestare ⟨foot⟩

treason /'triːzn/ n tradimento m

treasure /'treʒə(r)/ n tesoro m ● vt tenere in gran conto. **~r** n tesoriere, -a mf

treasury /'treʒərɪ/ n **the T~** il Ministero del Tesoro

treat /triːt/ n piacere m; ⟨present⟩ regalo m; **give sb a ~** fare una sorpresa a qcno ● vt trattare; Med curare; **~ sb to sth** offrire qcsa a qcno

treatise /'triːtɪz/ n trattato m

treatment /'triːtmənt/ n trattamento m; Med cura f

treaty /'triːtɪ/ n trattato m

treble /'trebl/ a triplo ● n Mus ⟨voice⟩ voce f bianca ● vt triplicare ● vi triplicarsi. **~ clef** n chiave f di violino

tree /triː/ n albero m

trek /trek/ n scarpinata f; ⟨as holiday⟩ trekking m inv ● vi (pt/pp **trekked**) farsi una scarpinata; ⟨on holiday⟩ fare trekking

tremble /'trembl/ vi tremare

tremendous /trɪ'mendəs/ a ⟨huge⟩ enorme; ⟨fam: excellent⟩ formidabile. **~ly** adv ⟨very⟩ straordinariamente; ⟨a lot⟩ enormemente

tremor /'tremə(r)/ n tremito m; [earth] **~** scossa f [sismica]

trench /trentʃ/ n fosso m; Mil trincea f. **~ coat** n trench m inv

trend /trend/ n tendenza f; ⟨fashion⟩ moda f. **~y** a (-ier, -iest) fam di o alla moda

trepidation /trepɪ'deɪʃn/ n trepidazione f

trespass /'trespəs/ vi **~ on** introdursi abusivamente in; fig abusare di. **~er** n intruso, -a mf

trial /'traɪəl/ n Jur processo m;

⟨test, ordeal⟩ prova f; **on ~** in prova; Jur in giudizio; **by ~ and error** per tentativi

triangle /'traɪæŋgl/ n triangolo m. **~ular** /-'æŋgjʊlə(r)/ a triangolare

tribe /traɪb/ n tribù f inv

tribulation /trɪbjʊ'leɪʃn/ n tribolazione f

tribunal /traɪ'bjuːnl/ n tribunale m

tributary /'trɪbjʊtərɪ/ n affluente m

tribute /'trɪbjuːt/ n tributo m; **pay ~** rendere omaggio

trice /traɪs/ n **in a ~** in un attimo

trick /trɪk/ n trucco m; ⟨joke⟩ scherzo m; ⟨in cards⟩ presa f; **do the ~** fam funzionare; **play a ~ on** fare uno scherzo a ● vt imbrogliare

trickle /'trɪkl/ vi colare

trick|ster /'trɪkstə(r)/ n imbroglione, -a mf. **~y** a (-ier, -iest) a ⟨operation⟩ complesso; ⟨situation⟩ delicato

tricycle /'traɪsɪkl/ n triciclo m

tried /traɪd/ see **try**

trifle /'traɪfl/ n inezia f; Culin zuppa f inglese. **~ing** a insignificante

trigger /'trɪgə(r)/ n grilletto m ● vt **~ [off]** scatenare

trigonometry /trɪgə'nɒmɪtrɪ/ n trigonometria f

trim /trɪm/ a (**trimmer**, **trimmest**) curato; ⟨figure⟩ snello ● n ⟨of hair, hedge⟩ spuntata f; ⟨decoration⟩ rifinitura f; **in good ~** in buono stato; ⟨person⟩ in forma ● vt (pt/pp **trimmed**) spuntare ⟨hair etc⟩; ⟨decorate⟩ ornare; Naut orientare. **~ming** n bordo m; **~mings** pl ⟨decorations⟩ guarnizioni fpl; **with all the ~mings** Culin guarnito

trinket /'trɪŋkɪt/ n ninnolo m

trio /'triːəʊ/ n trio m

trip /trɪp/ n ⟨excursion⟩ gita f; ⟨journey⟩ viaggio m; ⟨stumble⟩ passo m falso ● v (pt/pp **tripped**) ● vt far inciampare ● vi inciampare ⟨on/over in⟩. **trip up** vt far inciampare

tripe /traɪp/ n trippa f; ⟨sl: nonsense⟩ fesserie fpl

triple /'trɪpl/ a triplo ● vt triplicare ● vi triplicarsi

triplets /'trɪplɪts/ npl tre gemelli mpl

triplicate /'trɪplɪkət/ n in ~ in triplice copia

tripod /'traɪpɒd/ n treppiede m

tripper /'trɪpə(r)/ n gitante mf

trite /traɪt/ a banale

triumph /'traɪʌmf/ n trionfo m ● vi trionfare (**over** su). **~ant** /-'ʌmf(ə)nt/ a trionfante. **~antly** adv (exclaim) con tono trionfante

trivial /'trɪvɪəl/ a insignificante. **~ity** /-'ælətɪ/ n banalità f inv

trod, trodden /trɒd, 'trɒdn/ see **tread**

trolley /'trɒlɪ/ n carrello m; (Am: tram) tram m inv. **~ bus** n filobus m inv

trombone /trɒm'bəʊn/ n trombone m

troop /tru:p/ n gruppo m; **~s** pl truppe fpl ● vi ~ **in/out** entrare/uscire in gruppo

trophy /'trəʊfɪ/ n trofeo m

tropic /'trɒpɪk/ n tropico m; **~s** pl tropici mpl. **~al** a tropicale

trot /trɒt/ n trotto m ● vi (pt/pp **trotted**) trottare

trouble /'trʌbl/ n guaio m; (difficulties) problemi mpl; (inconvenience, Med) disturbo m; (conflict) conflitto m; **be in** ~ essere nei guai; (swimmer, climber:) essere in difficoltà; **get into** ~ finire nei guai; **get sb into** ~ mettere qcno nei guai; **take the** ~ **to do sth** darsi la pena di far qcsa ● vt (worry) preoccupare; (inconvenience) disturbare; (conscience, old wound:) tormentare ● vi don't ~! non ti disturbare!. **~-maker** n be a ~-**maker** seminare zizzania. **~some** /-səm/ a fastidioso

trough /trɒf/ n trogolo m; (atmospheric) depressione f

trounce /traʊns/ vt (in competition) schiacciare

troupe /tru:p/ n troupe f inv

trousers /'traʊzəz/ npl pantaloni mpl

trout /traʊt/ n inv trota f

trowel /'traʊəl/ n (for gardening) paletta f; (for builder) cazzuola f

truant /'tru:ənt/ n **play** ~ marinare la scuola

truce /tru:s/ n tregua f

truck /trʌk/ n (lorry) camion m inv

trudge /trʌdʒ/ n camminata f faticosa ● vi arrancare

true /tru:/ a vero; **come** ~ avverarsi

truffle /'trʌfl/ n tartufo m

truism /'tru:ɪzm/ n truismo m

truly /'tru:lɪ/ adv veramente; **Yours** ~ distinti saluti

trump /trʌmp/ n (in cards) atout m inv

trumpet /'trʌmpɪt/ n tromba f. **~er** n trombettista mf

truncheon /'trʌntʃn/ n manganello m

trunk /trʌŋk/ n (of tree, body) tronco m; (of elephant) proboscide f; (for travelling, storage) baule m; (Am: of car) bagagliaio m; **~s** pl calzoncini mpl da bagno

truss /trʌs/ n Med cinto m erniario

trust /trʌst/ n fiducia f; (group of companies) trust m inv; (organization) associazione f. in ~ sulla parola ● vt fidarsi di; (hope) augurarsi ● vi ~ in credere in; ~ to affidarsi a. **~ed** a fidato

trustee /trʌs'ti:/ n amministratore, -trice mf fiduciario, -a

'trustful /'trʌstfl/ a fiducioso. **~ing** a fiducioso. **~worthy** a fidato

truth /tru:θ/ n (pl -**s** /tru:ðz/) verità f inv. **~ful** a veritiero. **~fully** adv sinceramente

try /traɪ/ n tentativo m, prova f; (in rugby) meta f ● v (pt/pp **tried**) vt provare; (be a strain on) mettere a dura prova; Jur processare (person); discutere (case); ~ **to do sth** provare a fare qcsa ● vi provare. **try on** vt provarsi (garment).

try out vt provare

trying /'traɪŋ/ a duro; ⟨person⟩ irritante

T-shirt /'tiː-/ n maglietta f

tub /tʌb/ n tinozza f, ⟨carton⟩ vaschetta f; ⟨bath⟩ vasca f da bagno

tuba /'tjuːbə/ n Mus tuba f

tubby /'tʌbɪ/ a (-ier, -iest) tozzo

tube /tjuːb/ n tubo m; ⟨of toothpaste⟩ tubetto m; Rail metro f

tuber /'tjuːbə(r)/ n tubero m

tuberculosis /tjʊːbɜːkjʊ'ləʊsɪs/ n tubercolosi f

tubular /'tjuːbjʊlə(r)/ a tubolare

tuck /tʌk/ n piega f ● vt ⟨put⟩ infilare. **tuck in** vt rimboccare; ~ **sb in** rimboccare le coperte a qcno ● vi ⟨fam: eat⟩ mangiare con appetito. **tuck up** vt rimboccarsi ⟨sleeves⟩; ⟨in bed⟩ rimboccare le coperte a

Tuesday /'tjuːzdeɪ/ n martedì m inv

tuft /tʌft/ n ciuffo m

tug /tʌg/ n strattone m; Naut rimorchiatore m ● v ⟨pt/pp tugged⟩ ● vt tirare ● vi dare uno strattone. ~ **of war** n tiro m alla fune

tuition /tjuː'ɪʃn/ n lezioni fpl

tulip /'tjuːlɪp/ n tulipano m

tumble /'tʌmbl/ n ruzzolone m ● vi ruzzolare. ~**down** a cadente. ~**-drier** n asciugabiancheria f

tumbler /'tʌmblə(r)/ n bicchiere m ⟨senza stelo⟩

tummy /'tʌmɪ/ n fam pancia f

tumour /'tjuːmə(r)/ n tumore m

tumult /'tjuːmʌlt/ n tumulto m. ~**uous** /-'mʌltjʊəs/ a tumultuoso

tuna /'tjuːnə/ n tonno m

tune /tjuːn/ n motivo m; **out of/in** ~ ⟨instrument⟩ scordato/accordato; ⟨person⟩ stonato/intonato; **to the** ~ **of** fam per la modesta somma di ● vt accordare ⟨instrument⟩; sintonizzare ⟨radio, TV⟩; mettere a punto ⟨engine⟩. **tune in** vt sintonizzare ● vi sintonizzarsi ⟨to su⟩. **tune up** vi ⟨orchestra:⟩ accordare gli strumenti

tuneful /'tjuːnfl/ a melodioso

tuner /'tjuːnə(r)/ n accordatore,

-trice mf; Radio, TV sintonizzatore m

tunic /'tjuːnɪk/ n tunica f; Mil giacca f; Sch ≈ grembiule m

Tunisia /tjuː'nɪzɪə/ n Tunisia f. ~**n** a & n tunisino, -a mf

tunnel /'tʌnl/ n tunnel m inv ● vi ⟨pt/pp tunnelled⟩ scavare un tunnel

turban /'tɜːbən/ n turbante m

turbine /'tɜːbaɪn/ n turbina f

turbulen|ce /'tɜːbjʊləns/ n turbolenza f. ~**t** a turbolento

turf /tɜːf/ n erba f; ⟨segment⟩ zolla f erbosa ● **turf out** vt fam buttar fuori

Turin /tjʊ'rɪn/ n Torino f

Turk /tɜːk/ n turco, -a mf

turkey /'tɜːkɪ/ n tacchino m

Turk|ey /'tɜːkɪ/ n Turchia f. ~**ish** a turco

turmoil /'tɜːmɔɪl/ n tumulto m

turn /tɜːn/ n ⟨rotation, short walk⟩ giro m; ⟨in road⟩ svolta f, curva f; ⟨development⟩ svolta f; Theat numero m; ⟨fam: attack⟩ crisi f inv; **a** ~ **for the better/worse** un miglioramento/peggioramento; **do sb a good** ~ rendere un servizio a qcno; **take** ~**s** fare a turno; **in** ~ **a** turno; **out of** ~ ⟨speak⟩ a sproposito; **it's your** ~ tocca a te ● vt girare; voltare ⟨back, eyes⟩; dirigere ⟨gun, attention⟩ ● vi girare; ⟨person:⟩ girarsi; ⟨leaves:⟩ ingiallire; ⟨become⟩ diventare; ~ **right/left** girare a destra/sinistra; ~ **sour** inacidirsi; ~ **to sb** girarsi verso qcno; fig rivolgersi a qcno. **turn against** vi diventare ostile a ● vt mettere contro. **turn away** vt mandare via ⟨people⟩; girare dall'altra parte ⟨head⟩ ● vi girarsi dall'altra parte. **turn down** vt piegare ⟨collar⟩; abbassare ⟨heat, gas, sound⟩; respingere ⟨person, proposal⟩. **turn in** vt ripiegare in dentro ⟨edges⟩; consegnare ⟨lost object⟩ ● vi ⟨fam: go to bed⟩ andare a letto; ~ **into the drive** entrare nel viale. **turn off** vt spegnere; chiudere ⟨tap, water⟩ ● vi ⟨car:⟩

girare. **turn on** *vt* accendere; aprire ⟨*tap, water*⟩; ⟨*fam: attract*⟩ eccitare ⟨*attack*⟩ attaccare. **turn out** *vt* ⟨*expel*⟩ mandar via; spegnere ⟨*light, gas*⟩; ⟨*produce*⟩ produrre; ⟨*empty*⟩ svuotare ⟨*room, cupboard*⟩ ⟨*vi* (*transpire*) risultare; **~ out well/badly** ⟨*cake, dress*⟩ riuscire bene/male; ⟨*situation:*⟩ andare bene/male. **turn over** *vt* girare ⟨*vt* girar; **please ~ over** vedi retro. **turn round** *vi* girarsi; ⟨*car:*⟩ girare. **turn up** *vt* tirare su ⟨*collar*⟩; alzare ⟨*heat, gas, sound, radio*⟩ ⟨*vi* farsi vedere

turning /'tɜːnɪŋ/ *n* svolta *f*. **~-point** *n* svolta *f* decisiva

turnip /'tɜːnɪp/ *n* rapa *f*

turn: ~-out *n* ⟨*of people*⟩ affluenza *f*. **~over** *n Comm* giro *m* d'affari; ⟨*of staff*⟩ ricambio *m*. **~pike** *n Am* autostrada *f*. **~stile** *n* cancelletto *m* girevole. **~table** *n* piattaforma *f* girevole; ⟨*on record-player*⟩ piatto *m* ⟨*di giradischi*⟩. **~-up** *n* ⟨*of trousers*⟩ risvolto *m*

turpentine /'tɜːpəntaɪn/ *n* trementina *f*

turquoise /'tɜːkwɔɪz/ *a* ⟨*colour*⟩ turchese ⟨*n* turchese *m*

turret /'tʌrɪt/ *n* torretta *f*

turtle /'tɜːtl/ *n* tartaruga *f* acquatica

tusk /tʌsk/ *n* zanna *f*

tussle /'tʌsl/ *n* zuffa *f* ⟨*vi* azzuffarsi

tutor /'tjuːtə(r)/ *n* insegnante *mf* privato, -a; ⟨*Univ* insegnante *mf* universitario, -a che segue individualmente un ristretto numero di studenti. **~ial** /-'tɔːrɪəl/ *n* discussione *f* col tutor

tuxedo /tʌk'siːdəʊ/ *n Am* smoking *m inv*

TV *n abbr* ⟨*television*⟩ tv *f inv*, tivù *f inv*

twaddle /'twɒdl/ *n* scemenze *fpl*

twang /twæŋ/ *n* ⟨*in voice*⟩ suono *m* nasale ⟨*vi* far vibrare

tweed /twiːd/ *n* tweed *m inv*

tweezers /'twiːzəz/ *npl* pinzette *fpl*

twelfth /twelfθ/ *a* dodicesimo

twelve /twelv/ *a* dodici

twentieth /'twentɪθ/ *a* ventesimo

twenty /'twentɪ/ *a* venti

twerp /twɜːp/ *n fam* stupido, -a *mf*

twice /twaɪs/ *adv* due volte

twiddle /'twɪdl/ *vt* giocherellare con; **~ one's thumbs** *fig* girarsi i pollici

twig¹ /twɪg/ *n* ramoscello *m*

twig² /twɪg/ *vt/i* ⟨*pt/pp* **twigged**⟩ *fam* intuire

twilight /'twaɪ-/ *n* crepuscolo *m*

twin /twɪn/ *n* gemello, -a *mf* ⟨*attrib* gemello. **~ beds** *npl* letti *mpl* gemelli

twine /twaɪn/ *n* spago *m* ⟨*vi* intrecciarsi; ⟨*plant:*⟩ attorcigliarsi ⟨*vt* intrecciare

twinge /twɪndʒ/ *n* fitta *f*; **~ of conscience** rimorso *m* di coscienza

twinkle /'twɪŋkl/ *n* scintillio *m* ⟨*vi* scintillare

twin 'town *n* città *f inv* gemellata

twirl /twɜːl/ *vt* far roteare ⟨*vi* volteggiare ⟨*n* piroetta *f*

twist /twɪst/ *n* torsione *f*; ⟨*curve*⟩ curva *f*; ⟨*in rope*⟩ attorcigliata *f*; ⟨*in book, plot*⟩ colpo *m* di scena ⟨*vt* attorcigliare ⟨*rope*⟩; torcere ⟨*metal*⟩; girare ⟨*knob, cap*⟩; ⟨*distort*⟩ distorcere; **~ one's ankle** storcersi la caviglia ⟨*vi* attorcigliarsi; ⟨*road:*⟩ essere pieno di curve

twit /twɪt/ *n fam* cretino, -a *mf*

twitch /twɪtʃ/ *n* tic *m inv*; ⟨*jerk*⟩ strattone *m* ⟨*vi* contrarsi

twitter /'twɪtə(r)/ *n* cinguettio *m* ⟨*vi* cinguettare; ⟨*person:*⟩ cianciare

two /tuː/ *a* due

two: ~-faced *a* falso. **~-piece** *a* ⟨*swimsuit*⟩ due pezzi *m inv*; ⟨*suit*⟩ completo *m*. **~some** /-səm/ *n* coppia *f*. **~-way** *a* ⟨*traffic*⟩ a doppio senso di marcia

tycoon /taɪ'kuːn/ *n* magnate *m*

tying /'taɪŋ/ *see* **tie**

type /taɪp/ *n* tipo *m*; ⟨*printing*⟩ carattere m [tipografico] ● vt scrivere a macchina ● vi scrivere a macchina. **~writer** n macchina f da scrivere. **~written** a dattiloscritto

typhoid /'taɪfɔɪd/ n febbre f tifoidea

typical /'tɪpɪkl/ a tipico. **~ly** adv tipicamente; (as usual) come al solito

typify /'tɪpɪfaɪ/ vt (pt/pp **-ied**) essere tipico di

typing /'taɪpɪŋ/ n dattilografia f

typist /'taɪpɪst/ n dattilografo, -a mf

typography /taɪ'pɒgrəfɪ/ n tipografia f

tyrannical /tɪ'rænɪkl/ a tirannico

tyranny /'tɪrənɪ/ n tirannia f

tyrant /'taɪrənt/ n tiranno, -a mf

tyre /'taɪə(r)/ n gomma f, pneumatico m

Uu

ubiquitous /ju:'bɪkwɪtəs/ a onnipresente

udder /'ʌdə(r)/ n mammella f (di vacca, capra etc)

ugl|iness /'ʌglɪnɪs/ n bruttezza f. **~y** a (**-ier, -iest**) brutto

UK n abbr **United Kingdom**

ulcer /'ʌlsə(r)/ n ulcera f

ulterior /ʌl'tɪərɪə(r)/ a **~ motive** secondo fine m

ultimate /'ʌltɪmət/ a definitivo; (final) finale; (fundamental) fondamentale. **~ly** adv alla fine

ultimatum /ʌltɪ'meɪtəm/ n ultimatum m inv

ultrasound /'ʌltrə-/ n Med ecografia f

ultra'violet a ultravioletto

umbilical /ʌm'bɪlɪkl/ a **~ cord** cordone m ombelicale

umbrella /ʌm'brelə/ n ombrello m

umpire /'ʌmpaɪə(r)/ n arbitro m ● vt/i arbitrare

umpteen /ʌmp'ti:n/ a fam innumerevole. **~th** a fam ennesimo; **for the ~th time** per l'ennesima volta

UN n abbr (**United Nations**) ONU f

un'able /ʌn-/ a **be ~ to do sth** non potere fare qcsa; (not know how) non sapere fare qcsa

una'bridged a integrale

unac'companied a non accompagnato; (luggage) incustodito

unac'countable a inspiegabile. **~y** adv inspiegabilmente

unac'customed a insolito; **be ~ to** non essere abituato a

una'dulterated a (water) puro; (wine) non sofisticato; fig assoluto

un'aided a senza aiuto

unanimity /ju:nə'nɪmətɪ/ n unanimità f

unanimous /ju:'nænɪməs/ a unanime. **~ly** adv all'unanimità

un'armed a disarmato; **~ combat** n lotta f senza armi

unas'suming a senza pretese

unat'tached a staccato; (person) senza legami

unat'tended a incustodito

un'authorized a non autorizzato

una'voidable a inevitabile

una'ware a **be ~ of sth** non rendersi conto di qcsa. **~s** /-eəz/ adv **catch sb ~s** prendere qcno alla sprovvista

un'balanced a non equilibrato; (mentally) squilibrato

un'bearabl|e a insopportabile. **~y** adv insopportabilmente

unbeat|able /ʌn'bi:təbl/ a imbattibile. **~en** a imbattuto

unbe'known /ʌnbɪ'nəʊn/ a fam **~ to me** a mia insaputa

unbe'lievable a incredibile

un'bend vi (pt/pp **-bent**) (relax) distendersi

un'biased a obiettivo

un'block vt sbloccare

un'bolt vt togliere il chiavistello di

un'breakable a infrangibile

un'bridled /ʌn'braɪdld/ a sfrenato

un'burden vt **~ oneself** fig sfogarsi (to con)

un'button *vt* sbottonare

uncalled-for /ˌʌnˈkɔːldfɔː(r)/ *a* a fuori luogo

un'canny *a* sorprendente; ⟨silence, feeling⟩ inquietante

un'ceasing *a* incessante

uncere'monious *a* (abrupt) brusco. ~ly *adv* senza tante cerimonie

un'certain *a* incerto; ⟨weather⟩ instabile; **in no ~ terms** senza mezzi termini. ~ty *n* incertezza *f*

un'changed *a* invariato

un'charitable *a* duro

uncle /'ʌŋkl/ *n* zio *m*

un'comfortable *a* scomodo; imbarazzante ⟨silence, situation⟩; **feel ~e** *fig* sentirsi a disagio. ~y *adv* ⟨sit⟩ scomodamente; ⟨causing alarm etc⟩ spaventosamente

un'common *a* insolito

un'compromising *a* intransigente

uncon'ditional *a* incondizionato. ~ly *adv* incondizionatamente

un'conscious *a* privo di sensi; (unaware) inconsapevole; **be ~ of sth** non rendersi conto di qcsa. ~ly *adv* inconsapevolmente

uncon'ventional *a* poco convenzionale

unco'operative *a* poco cooperativo

un'cork *vt* sturare

uncouth /ʌnˈkuːθ/ *a* zotico

un'cover *vt* scoprire; portare alla luce ⟨buried object⟩

unde'cided *a* indeciso; (not settled) incerto

undeniable /ˌʌndɪˈnaɪəbl/ *a* innegabile. ~y *adv* innegabilmente

under /'ʌndə(r)/ *prep* sotto; (less than) al di sotto di; ~ **there** li sotto; ~ **repair/construction** in riparazione/costruzione; ~ **way** *fig* in corso ● *adv* (~ water) sott'acqua; (unconscious) sotto anestesia

'undercarriage *n* Aeron carrello *m*

'underclothes *npl* biancheria *fsg* intima

'under'cover *a* clandestino

'undercurrent *n* corrente *f* sottomarina; *fig* sottofondo *m*

under'cut *vt* (pt/pp -cut) Comm vendere a minor prezzo di

'underdog *n* perdente *m*

under'done *a* ⟨meat⟩ al sangue

under'estimate *vt* sottovalutare

under'fed *a* denutrito

under'foot *adv* sotto i piedi; **trample** ~ calpestare

under'go *vt* (pt -went, pp -gone) subire ⟨operation, treatment⟩; ~ **repair** essere in riparazione

'undergraduate *n* studente, -tessa *mf* universitario, -a

under'ground¹ *adv* sottoterra

'underground² *a* sotterraneo; (secret) clandestino ● *n* (railway) metropolitana *f*. ~ **car park** *n* parcheggio *m* sotterraneo

'undergrowth *n* sottobosco *m*

'underhand *a* subdolo

'underlay *n* strato *m* di gomma o feltro posto sotto la moquette

under'lie *vt* (pt -lay, pp -lain, pres p -lying) *fig* essere alla base di

under'line *vt* sottolineare

underling /'ʌndəlɪŋ/ *n* pej subalterno, -a *mf*

under'lying *a* *fig* fondamentale

under'mine *vt* *fig* minare

underneath /ˌʌndə'niːθ/ *prep* sotto; ~ **it** sotto ● *adv* sotto

under'paid *a* mal pagato

'underpants *npl* mutande *fpl*

'underpass *n* sottopassaggio *m*

under'privileged *a* non abbiente

under'rate *vt* sottovalutare

'underseal *n* Auto antiruggine *m* *inv*

'undershirt *n* Am maglia *f* della pelle

under'staffed *a* a corto di personale

under'stand *vt* (pt/pp -stood) capire; **I ~ that...** (have heard) mi risulta che... ● *vi* capire. ~**able** /-əbl/ *a* comprensibile. ~**ably** /-əblɪ/ *adv* comprensibilmente

under'standing *a* comprensivo

●*n* comprensione *f*; *(agreement)* accordo *m*; **on the ~ that** a condizione che

'understatement *n* understatement *m inv*

'understudy *n* Theat sostituto, -a *mf*

under'take *vt* (*pt* -**took**, *pp* -**taken**) intraprendere; **~ to do sth** impegnarsi a fare qcsa

'undertaker *n* impresario *m* di pompe funebri; **[firm of] ~s** *n* impresa *f* di pompe funebri

under'taking *n* impresa *f*; *(promise)* promessa *f*

'undertone *n fig* sottofondo *m*; **in an ~** sottovoce

under'value *vt* sottovalutare

'underwater *a* subacqueo

under'water² *adv* sott'acqua

'underwear *n* biancheria *f* intima

under'weight *a* sotto peso

'underworld *n* *(criminals)* malavita *f*

'underwriter *n* assicuratore *m*

unde'sirable *a* indesiderato; *(person)* poco raccomandabile

undies /'ʌndɪz/ *npl fam* biancheria *fsg* intima (*da donna*)

un'dignified *a* non dignitoso

un'do *vt* (*pt* -**did**, *pp* -**done**) disfare; slacciare *(dress, shoes)*; sbottonare *(shirt)*; *fig*, *Comput* annullare

un'done *a (shirt, button)* sbottonato; *(shoes, dress)* slacciato; *(not accomplished)* non fatto; **leave ~** *(job)* tralasciare

un'doubted *a* indubbio; **~ly** *adv* senza dubbio

un'dress *vt* spogliare; **get ~ed** spogliarsi ●*vi* spogliarsi

un'due *a* eccessivo

undulating /'ʌndjʊleɪtɪŋ/ *a* ondulato; *(country)* collinoso

un'duly *adv* eccessivamente

un'dying *a* eterno

un'earth *vt* dissotterrare; *fig* scovare; scoprire *(secret)*. **~ly** *a* soprannaturale; **at an ~ly hour** *fam* a un'ora impossibile

un'ease *n* disagio *m*. **~y** *a* a disagio; *(person)* inquieto; *(feeling)* inquietante; *(truce)* precario

un'eatable *a* immangiabile

uneco'nomic *a* poco remunerativo

uneco'nomical *a* poco economico

unem'ployed *a* disoccupato ●*npl* **the ~** i disoccupati

unem'ployment *n* disoccupazione *f*. **~ benefit** *n* sussidio *m* di disoccupazione

un'ending *a* senza fine

un'equal *a* disuguale; *(struggle)* impari; **be ~ to a task** non essere all'altezza di un compito

unequivocal /ʌnɪ'kwɪvəkl/ *a* inequivocabile; *(person)* esplicito

unerring /ʌn'ɜːrɪŋ/ *a* infallibile

un'ethical *a* immorale

un'even *a* irregolare; *(distribution)* ineguale; *(number)* dispari

unex'pected *a* inaspettato. **~ly** *adv* inaspettatamente

un'failing *a* infallibile

un'fair *a* ingiusto. **~ly** *adv* ingiustamente. **~ness** *n* ingiustizia *f*

un'faithful *a* infedele

unfa'miliar *a* sconosciuto; **be ~ with** non conoscere

un'fasten *vt* slacciare; *(detach)* staccare

un'favourable *a* sfavorevole; *(impression)* negativo

un'feeling *a* insensibile

un'finished *a* da finire; *(business)* in sospeso

un'fit *a* inadatto; *(morally)* indegno; Sport fuori forma; **~ for work** non in grado di lavorare

unflinching /ʌn'flɪntʃɪŋ/ *a* risoluto

un'fold *vt* spiegare; *(spread out)* aprire; *fig* rivelare ●*vi* *(view:)* spiegarsi

unfore'seen *a* imprevisto

unfor'gettable /ʌnfə'getəbl/ *a* indimenticabile

unfor'givable /ʌnfə'gɪvəbl/ *a* imperdonabile

un'fortunate *a* sfortunato; *(regret-*

table) spiacevole; *(remark, choice)* infelice. **~ly** *adv* purtroppo

un'founded *a* infondato

un'furl /ʌn'fɜːl/ *vt* spiegare

un'furnished *a* non ammobiliato

un'gainly /ʌn'geɪnlɪ/ *a* sgraziato

un'godly /ʌn'gʊdlɪ/ *a* empio; **~ hour** *fam* ora *f* impossibile

un'grateful *a* ingrato. **~ly** *adv* senza riconoscenza

un'happi|ly *adv* infelicemente; *(unfortunately)* purtroppo. **~ness** *n* infelicità *f*

un'happy *a* infelice; *(not content)* insoddisfatto **(with** di)

un'harmed *a* incolume

un'healthy *a* poco sano; *(insanitary)* malsano

un'hook *vt* sganciare

un'hurt *a* illeso

unhy'gienic *a* non igienico

unifi'cation /juːnɪfɪ'keɪʃn/ *n* unificazione *f*

uniform /'juːnɪfɔːm/ *a* uniforme ● *n* uniforme *f*. **~ly** *adv* uniformemente

unify /'juːnɪfaɪ/ *vt* *(pt/pp* **-ied)** unificare

uni'lateral /juːnɪ-/ *a* unilaterale

unim'aginable *a* inimmaginabile

unim'portant *a* irrilevante

unin'habited *a* disabitato

unin'tentional *a* involontario. **~ly** *adv* involontariamente

union /'juːnɪən/ *n* unione *f*; *(trade* ~) sindacato *m*. **U~ Jack** *n* bandiera *f* del Regno Unito

unique /juː'niːk/ *a* unico. **~ly** *adv* unicamente

unison /'juːnɪsn/ *n* **in ~** all'unisono

unit /'juːnɪt/ *n* unità *f inv*; *(department)* reparto *m*; *(of furniture)* elemento *m*

unite /juː'naɪt/ *vt* unire ● *vi* unirsi

united /juː'naɪtɪd/ *a* unito. **U~ 'Kingdom** *n* Regno *m* Unito. **U~ 'Nations** *n* [Organizzazione *f* delle] Nazioni Unite *fpl*. **U~ States [of America]** *n* Stati *mpl* Uniti [d'America]

unity /'juːnɪtɪ/ *n* unità *f*; *(agreement)* accordo *m*

universal /juːnɪ'vɜːsl/ *a* universale. **~ly** *adv* universalmente

universe /'juːnɪvɜːs/ *n* universo *m*

university /juːnɪ'vɜːsətɪ/ *n* università *f* ● *attrib* universitario

un'just *a* ingiusto

unkempt /ʌn'kempt/ *a* trasandato; *(hair)* arruffato

un'kind *a* scortese. **~ly** *adv* in modo scortese. **~ness** *n* mancanza *f* di gentilezza

un'known *a* sconosciuto

un'lawful *a* illecito, illegale

un'leaded /ʌn'ledɪd/ *a* senza piombo

un'leash *vt fig* scatenare

unless /ən'les/ *conj* a meno che; **~ I am mistaken** se non mi sbaglio

un'like *a* *(not the same)* diversi ● *prep* diverso da; **that's ~ him** non è da lui; **~ me, he...** diversamente da me, lui...

un'likely *a* improbabile

un'limited *a* illimitato

un'load *vt* scaricare

un'lock *vt* aprire *(con chiave)*

un'lucky *a* sfortunato; **it's ~ to...** porta sfortuna...

un'manned *a* senza equipaggio

un'married *a* non sposato. **~ 'mother** *n* ragazza *f* madre

un'mask *vt fig* smascherare

unmistakab|le /ʌnmɪ'steɪkəbl/ *a* inconfondibile. **~y** *adv* chiaramente

un'mitigated *a* assoluto

un'natural *a* innaturale; *pej* anormale. **~ly** *adv* in modo innaturale; *pej* in modo anormale

un'necessar|y *a* inutile. **~ily** *adv* inutilmente

un'noticed *a* inosservato

unob'tainable *a* *(product)* introvabile; *(phone number)* non ottenibile

unob'trusive *a* discreto. **~ly** *adv* in modo discreto

unof'ficial *a* non ufficiale. **~ly** *adv* ufficiosamente

un'pack vi disfare le valigie ● vt svuotare (parcel); spacchettare (books); **~ one's case** disfare la valigia

un'paid a da pagare; (work) non retribuito

un'palatable a sgradevole

un'paralleled a senza pari

un'pick vt disfare

un'pleasant a sgradevole; (person) maleducato. **~ly** adv sgradevolmente; (behave) maleducatamente. **~ness** n (bad feeling) tensioni fpl

un'plug vt (pt/pp -**plugged**) staccare

un'popular a impopolare

un'precedented a senza precedenti

unpre'dictable a imprevedibile

unpre'meditated a involontario

unpre'pared a impreparato

unpre'tentious a senza pretese

un'principled a senza principi; (behaviour) scorretto

unpro'fessional a non professionale; **it's ~** è una mancanza di professionalità

un'profitable a non redditizio

un'qualified a non qualificato; (fig: absolute) assoluto

un'questionable a incontestabile

un'quote vi chiudere le virgolette

unravel /ʌn'rævl/ vt (pt/pp -**ravelled**) districare; (in knitting) disfare

un'real a irreale; fam inverosimile

un'reasonable a irragionevole

unre'lated a (fact) senza rapporto (to con); (person) non imparentato (to con)

unre'liable a inattendibile; (person) inaffidabile, che non dà affidamento

unre'quited /ʌnrɪ'kwaɪtɪd/ a non corrisposto

unre'servedly /ʌnrɪ'zɜːvɪdlɪ/ adv senza riserve; (frankly) francamente

un'rest n fermenti mpl

un'rivalled a ineguagliato

un'roll vt srotolare ● vi srotolarsi

unruly /ʌn'ruːlɪ/ a indisciplinato

un'safe a pericoloso

un'said a inespresso

un'salted a non salato

unsatis'factory a poco soddisfacente

un'savoury a equivoco

unscathed /ʌn'skeɪðd/ a illeso

un'screw vt svitare

un'scrupulous a senza scrupoli

un'seemly a indecoroso

un'selfish a disinteressato

un'settled a in agitazione; (weather) variabile; (bill) non saldato

unshakeable /ʌn'ʃeɪkəbl/ a categorico

unshaven /ʌn'ʃeɪvn/ a non rasato

unsightly /ʌn'saɪtlɪ/ a brutto

un'skilled a non specializzato. **~ worker** n manovale m

un'sociable a scontroso

unso'phisticated a semplice

un'sound a (building, reasoning) poco solido; (advice) poco sensato; **of ~ mind** malato di mente

unspeakable /ʌn'spiːkəbl/ a indicibile

un'stable a instabile; (mentally) squilibrato

un'steady a malsicuro

un'stuck a **come ~** staccarsi; (fam: project) andare a monte

unsuc'cessful a fallimentare; **be ~** (in attempt) non aver successo. **~ly** adv senza successo

un'suitable a (inappropriate) inadatto; (inconvenient) inopportuno

unsus'pecting a fiducioso

unthinkable /ʌn'θɪŋkəbl/ a impensabile

un'tidiness n disordine m

un'tidy a disordinato

un'tie vt slegare

until /ən'tɪl/ prep fino a; **not ~** non prima di; **~ the evening** fino alla sera; **~ his arrival** fino al suo arrivo ● conj finché, fino a quando; **not ~ you've seen it** non prima che tu l'abbia visto

untimely /ʌn'taɪmlɪ/ a inopportuno; ⟨premature⟩ prematuro

un'tiring a instancabile

un'told a ⟨wealth⟩ incalcolabile; ⟨suffering⟩ indescrivibile; ⟨story⟩ inedito

unto'ward a if nothing ~ happens se non capita un imprevisto

un'true a falso; that's ~ non è vero

unused[1] /ʌn'juːzd/ a non [ancora] usato

unused[2] /ʌn'juːst/ a be ~ to non essere abituato a

un'usual a insolito. ~ly adv insolitamente

un'veil vt scoprire

un'wanted a indesiderato

un'warranted a ingiustificato

un'welcome a sgradito

un'well a indisposto

unwieldy /ʌn'wiːldɪ/ a ingombrante

un'willing a riluttante. ~ly adv malvolentieri

un'wind v ⟨pt/pp unwound⟩ ● vt svolgere, srotolare ● vi svolgersi, srotolarsi; ⟨fam: relax⟩ rilassarsi

un'wise a imprudente

un'witting /ʌn'wɪtɪŋ/ a involontario; ⟨victim⟩ inconsapevole. ~ly adv involontariamente

un'worthy a non degno

un'wrap vt ⟨pt/pp -wrapped⟩ scartare ⟨present, parcel⟩

un'written a tacito

up /ʌp/ adv su; ⟨not in bed⟩ alzato; ⟨road⟩ smantellato; ⟨theatre curtain, blinds⟩ alzato; ⟨shelves, tent⟩ montato; ⟨notice⟩ affisso; ⟨building⟩ costruito; prices are up i prezzi sono aumentati; be up for sale essere in vendita; up here/there quassù/lassù; time's up tempo scaduto; what's up? fam cosa è successo?; up to ⟨as far as⟩ fino a; be up to essere all'altezza di ⟨task⟩; what's he up to? fam cosa sta facendo?; ⟨plotting⟩ cosa sta combinando?; I'm up to page 100 sono arrivato a pagina 100; feel up to it sentirsela; be one up on sb fam essere in vantaggio su qcno; go up salire; lift up alzare; up against fig alle prese con ● prep su; the cat ran/is up the tree il gatto è salito di corsa/è sull'albero; further up this road più avanti su questa strada; row up the river risalire il fiume; go up the stairs salire su per le scale; be up the pub fam essere al pub; be up on or in sth essere bene informato su qcsa ● n ups and downs npl alti mpl e bassi

'upbringing n educazione f

up'date[1] vt aggiornare

'update[2] n aggiornamento m

up'grade vt promuovere ⟨person⟩; modernizzare ⟨equipment⟩

upgradeable /ʌp'greɪdəbl/ a Comput upgradabile

upheaval /ʌp'hiːvl/ n scompiglio m

up'hill a in salita; fig arduo ● adv in salita

up'hold vt ⟨pt/pp upheld⟩ sostenere ⟨principle⟩; confermare ⟨verdict⟩

upholster /ʌp'həʊlstə(r)/ vt tappezzare. ~er n tappezziere, -a mf. ~y n tappezzeria f

'upkeep n mantenimento m

up-'market a di qualità

upon /ə'pɒn/ prep su; ~ arriving home una volta arrivato a casa

upper /'ʌpə(r)/ a superiore ● n ⟨of shoe⟩ tomaia f

upper: ~ circle n seconda galleria f. ~ class n alta borghesia f. ~ hand n have the ~ hand avere il sopravvento. ~most a più alto; that's ~most in my mind è la mia preoccupazione principale

upright a dritto; ⟨piano⟩ verticale; ⟨honest⟩ retto ● n montante m

uprising n rivolta f

uproar n tumulto m; be in an ~ essere in trambusto

up'root vt sradicare

up'set[1] vt ⟨pt/pp upset, pres p upsetting⟩ rovesciare; sconvolgere ⟨plan⟩; ⟨distress⟩ turbare; **get ~ about sth** prendersela per qcsa; **be very ~** essere sconvolto; **have an ~ stomach** avere l'intestino disturbato

upset² n scombussolamento m

upshot n risultato m

upside 'down adv sottosopra; **turn ~ ~** capovolgere

up'stairs¹ adv [al piano] di sopra

up'stairs² a del piano superiore

upstart n arrivato, -a mf

up'stream adv controcorrente

upsurge n (in sales) aumento m improvviso; ⟨of enthusiasm, crime⟩ ondata f

'uptake n **be slow on the ~** essere lento nel capire; **be quick on the ~** capire le cose al volo

up'tight a teso

up-to-'date a moderno; ⟨news⟩ ultimo; ⟨records⟩ aggiornato

upturn n ripresa f

upward /'ʌpwəd/ a verso l'alto, in su; **~ slope** salita f ● adv ~[**s**] verso l'alto; **~s of** oltre

uranium /jʊ'reɪnɪəm/ n uranio m

urban /'ɜːbən/ a urbano

urge /ɜːdʒ/ n forte desiderio m ● vt esortare (**to** a). **urge on** vt spronare

urgen|cy /'ɜːdʒənsɪ/ n urgenza f. **~t** a urgente

urinate /'jʊərɪneɪt/ vi urinare

urine /'jʊərɪn/ n urina f

urn /ɜːn/ n urna f; ⟨for tea⟩ contenitore m munito di cannella che si trova nei self-service, mense, ecc

us /ʌs/ pron ci; ⟨after prep⟩ noi; **they know us** ci conoscono; **give us the money** dateci i soldi; **give it to us** datecelo; **they showed it to us** ce l'hanno fatto vedere; **they meant us, not you** intendevano noi, non voi; **it's us** siamo noi; **she hates us** ci odia

US[A] n[pl] abbr (**United States** [**of America**]) U.S.A. mpl

usable /'juːzəbl/ a usabile

usage /'juːsɪdʒ/ n uso m

use¹ /juːs/ n uso m; **be of ~** essere utile; **be of no ~** essere inutile; **make ~ of** usare; ⟨exploit⟩ sfruttare; **it is no ~** è inutile; **what's the ~?** a che scopo?

use² /juːz/ vt usare. **use up** vt consumare

used¹ /juːzd/ a usato

used² /juːst/ pt **be ~ to sth** essere abituato a qcsa; **get ~ to** abituarsi a; **he ~ to live here** viveva qui

useful /'juːsfl/ a utile. **~ness** n utilità f

useless /'juːslɪs/ a inutile; ⟨fam: person⟩ incapace

user /'juːzə(r)/ n utente mf. **~-friendly** a facile da usare

usher /'ʌʃə(r)/ n Theat maschera f; Jur usciere m; ⟨at wedding⟩ persona f che accompagna gli invitati a un matrimonio ai loro posti in chiesa ● **usher in** vt fare entrare

usherette /ʌʃə'ret/ n maschera f

usual /'juːʒəl/ a usuale; **as ~** come al solito. **~ly** adv di solito

usurp /jʊ'zɜːp/ vt usurpare

utensil /juː'tensl/ n utensile m

uterus /'juːtərəs/ n utero m

utilitarian /jʊtɪlɪ'teərɪən/ a funzionale

utility /juː'tɪlətɪ/ n servizio m. **~ room** n stanza f in casa privata per il lavaggio, la stiratura dei panni, ecc

utilize /'juːtɪlaɪz/ vt utilizzare

utmost /'ʌtməʊst/ a estremo ● n **one's ~** tutto il possibile

utter¹ /'ʌtə(r)/ a totale. **~ly** adv completamente

utter² vt emettere ⟨sigh, sound⟩; proferire ⟨word⟩. **~ance** /-əns/ n dichiarazione f

U-turn /'juː-/ n Auto inversione f a U; fig marcia f in dietro

Vv

vacan|cy /'veɪk(ə)nsɪ/ n (job) posto m vacante; (room) stanza f disponibile. ~**t** a libero; (position) vacante; (look) assente

vacate /və'keɪt/ vt lasciare libero

vacation /və'keɪʃn/ n Univ & Am vacanza f

vaccinat|e /'væksɪneɪt/ vt vaccinare. ~**ion** /-'neɪʃn/ n vaccinazione f

vaccine /'væksiːn/ n vaccino m

vacuum /'vækjʊəm/ n vuoto m ● vt passare l'aspirapolvere in/ su. ~ **cleaner** n aspirapolvere m inv. ~ **flask** n thermos® m inv. ~**-packed** a confezionato sottovuoto

vagabond /'vægəbɒnd/ n vagabondo, -a mf

vagina /və'dʒaɪnə/ n Anat vagina f

vagrant /'veɪgrənt/ n vagabondo, -a mf

vague /veɪg/ a vago; (outline) impreciso; (absent-minded) distratto; **I'm still** ~ **about it** non ho ancora le idee chiare in proposito. ~**ly** adv vagamente

vain /veɪn/ a vanitoso; (hope, attempt) vano; **in** ~ invano. ~**ly** adv vanamente

valentine /'væləntaɪn/ n (card) biglietto m di San Valentino

valiant /'vælɪənt/ a valoroso

valid /'vælɪd/ a valido. ~**ate** vt (confirm) convalidare. ~**ity** /və'lɪdətɪ/ n validità f

valley /'vælɪ/ n valle f

valour /'vælə(r)/ n valore m

valuable /'væljʊəbl/ a di valore; fig prezioso. ~**s** npl oggetti mpl di valore

valuation /væljʊ'eɪʃn/ n valutazione f

value /'væljuː/ n valore m; (usefulness) utilità f ● vt valutare; (cher-

ish) apprezzare. ~ **'added tax** n imposta f sul valore aggiunto

valve /vælv/ n valvola f

vampire /'væmpaɪə(r)/ n vampiro m

van /væn/ n furgone m

vandal /'vændl/ n vandalo, -a mf. ~**ism** /-ɪzm/ n vandalismo m. ~**ize** vt vandalizzare

vanilla /və'nɪlə/ n vaniglia f

vanish /'vænɪʃ/ vi svanire

vanity /'vænətɪ/ n vanità f. ~ **bag** or **case** n beauty-case m inv

vantage-point /'vɑːntɪdʒ-/ n punto m d'osservazione; fig punto m di vista

vapour /'veɪpə(r)/ n vapore m

variable /'veərɪəbl/ a variabile; (adjustable) regolabile

variance /'veərɪəns/ n **be at** ~ essere in disaccordo

variant /'veərɪənt/ n variante f

variation /veərɪ'eɪʃn/ n variazione f

varicose /'værɪkəʊs/ a ~ **veins** vene fpl varicose

varied /'veərɪd/ a vario; (diet) diversificato; (life) movimentato

variety /və'raɪətɪ/ n varietà f inv

various /'veərɪəs/ a vario

varnish /'vɑːnɪʃ/ n vernice f; (for nails) smalto m ● vt verniciare; ~ **one's nails** mettersi lo smalto

vary /'veərɪ/ vt/i (pt/pp **-ied**) variare. ~**ing** a variabile; (different) diverso

vase /vɑːz/ n vaso m

vast /vɑːst/ a vasto; (difference, amusement) enorme. ~**ly** adv (superior) di gran lunga; (different, amused) enormemente

VAT /viːeɪˈtiː, væt/ n abbr (value added tax) I.V.A. f

vat /væt/ n tino m

vault¹ /vɔːlt/ n (roof) volta f; (in bank) caveau m inv; (tomb) cripta f

vault² n salto m ● vt/i ~ [over] saltare

VDU n abbr (visual display unit) VDU m

veal /viːl/ *n* carne *f* di vitello ● *attrib* di vitello

veer /vɪə(r)/ *vi* cambiare direzione; *Naut, Auto* virare

vegetable /'vedʒtəbl/ *n* (*food*) verdura *f*; (*when growing*) ortaggio *m* ● *attrib* (*oil, fat*) vegetale

vegetarian /vedʒɪ'teərɪən/ *a* & *n* vegetariano, -a *mf*

vegetate /'vedʒɪteɪt/ *vi* vegetare. **~ion** /-'teɪʃn/ *n* vegetazione *f*

vehemen|ce /'viːəmans/ *n* veemenza *f*. **~t** *a* veemente. **~tly** *adv* con veemenza

vehicle /'viːɪkl/ *n* veicolo *m*; (*fig: medium*) mezzo *m*

veil /veɪl/ *n* velo *m* ● *vt* velare

vein /veɪn/ *n* vena *f*; (*mood*) umore *m*; (*manner*) tenore *m*. **~ed** *a* venato

Velcro® /'velkrəʊ/ *n* **~ fastening** chiusura *f* con velcro®

velocity /vɪ'lɒsətɪ/ *n* velocità *f*

velvet /'velvɪt/ *n* velluto *m*. **~y** *a* vellutato

vendetta /ven'detə/ *n* vendetta *f*

vending-machine /'vendɪŋ-/ *n* distributore *m* automatico

veneer /və'nɪə(r)/ *n* impiallacciatura *f*; *fig* vernice *f*. **~ed** *a* impiallacciato

venereal /vɪ'nɪərɪəl/ *a* **~ disease** malattia *f* venerea

Venetian /vɪ'niːʃn/ *a* & *n* veneziano, -a *mf*. **v~ blind** *n* persiana *f* alla veneziana

vengeance /'vendʒəns/ *n* vendetta *f*; **with a ~** *fam* a più non posso

Venice /'venɪs/ *n* Venezia *f*

venison /'venɪsn/ *n* Culin carne *f* di cervo

venom /'venəm/ *n* veleno *m*. **~ous** /-əs/ *a* velenoso

vent¹ /vent/ *n* presa *f* d'aria; **give ~ to** *fig* dar libero sfogo a ● *vt fig* sfogare ⟨*anger*⟩

vent² *n* (*in jacket*) spacco *m*

ventilat|e /'ventɪleɪt/ *vt* ventilare. **~ion** /-'leɪʃn/ *n* ventilazione *f*; (*installation*) sistema *m* di ventilazione. **~or** *n* ventilatore *m*

ventriloquist /ven'trɪləkwɪst/ *n* ventriloquo, -a *mf*

venture /'ventʃə(r)/ *n* impresa *f* ● *vt* azzardare ● *vi* avventurarsi

venue /'venjuː/ *n* luogo *m* (di convegno, concerto, ecc.)

veranda /və'rændə/ *n* veranda *f*

verb /vɜːb/ *n* verbo *m*. **~al** *a* verbale

verbatim /vɜː'beɪtɪm/ *a* letterale ● *adv* parola per parola

verbose /vɜː'bəʊs/ *a* prolisso

verdict /'vɜːdɪkt/ *n* verdetto *m*; (*opinion*) parere *m*

verge /vɜːdʒ/ *n* orlo *m*; **be on the ~ of doing sth** essere sul punto di fare qcsa ● **verge on** *vt fig* rasentare

verger /'vɜːdʒə(r)/ *n* sagrestano *m*

verify /'verɪfaɪ/ *vt* (*pt/pp* -ied) verificare; (*confirm*) confermare

vermin /'vɜːmɪn/ *n* animali *mpl* nocivi

vermouth /'vɜːməθ/ *n* vermut *m*

vernacular /və'nækjʊlə(r)/ *n* vernacolo *m*

versatil|e /'vɜːsətaɪl/ *a* versatile. **~ity** /-'tɪlətɪ/ *n* versatilità *f*

verse /vɜːs/ *n* verso *m*; (*of Bible*) versetto *m*; (*poetry*) versi *mpl*

versed /vɜːst/ *a* **~ in** versato in

version /'vɜːʃn/ *n* versione *f*

versus /'vɜːsəs/ *prep* contro

vertebra /'vɜːtɪbrə/ *n* (*pl* -brae /-brɪ:/) *Anat* vertebra *f*

vertical /'vɜːtɪkl/ *a* & *n* verticale *m*

vertigo /'vɜːtɪgəʊ/ *n* *Med* vertigine *f*

verve /vɜːv/ *n* verve *f*

very /'verɪ/ *adv* molto; **~ much** molto; **~ little** pochissimo; **~ many** moltissimi; **~ few** pochissimi; **~ probably** molto probabilmente; **~ well** benissimo; **at the ~ most** tutt'al più; **at the ~ latest** al più tardi ● *a* the **~ first** il primissimo; **the ~ thing** proprio ciò che ci vuole; **at the ~ end/beginning** proprio alla fine/all'inizio; **that ~ day** proprio quel giorno; **the ~**

thought la sola idea; **only a ~ little** solo un pochino

vessel /'vesl/ n nave f

vest /vest/ n maglia f della pelle; (Am: waistcoat) gilè m inv. **~ed interest** n interesse m personale

vestige /'vestɪdʒ/ n (of past) vestigio m

vestment /'vestmənt/ n Relig paramento m

vestry /'vestrɪ/ n sagrestia f

vet /vet/ n veterinario, -a mf ● vt (pt/pp **vetted**) controllare minuziosamente

veteran /'vetərən/ n veterano, -a mf

veterinary /'vetərɪnərɪ/ a veterinario. **~ surgeon** n medico m veterinario

veto /'viːtəʊ/ n (pl **-es**) veto m ● vt proibire

vex /veks/ vt irritare. **~ation** /-'seɪʃn/ n irritazione f. **~ed** a irritato; **~ed question** questione f controversa

VHF n abbr (**very high frequency**) VHF

via /'vaɪə/ prep via; (by means of) attraverso

viable /'vaɪəbl/ a ⟨life form, relationship, company⟩ in grado di sopravvivere; ⟨proposition⟩ attuabile

viaduct /'vaɪədʌkt/ n viadotto m

vibrat|e /vaɪ'breɪt/ vi vibrare. **~ion** /-'breɪʃn/ n vibrazione f

vicar /'vɪkə(r)/ n parroco m (protestante). **~age** /-rɪdʒ/ n casa f parrocchiale

vicarious /vɪ'keərɪəs/ a indiretto

vice[1] /vaɪs/ n vizio m

vice[2] n Techn morsa f

vice 'chairman n vicepresidente mf

vice 'president n vicepresidente mf

vice versa /vaɪs'vɜːsə/ adv viceversa

vicinity /vɪ'sɪnətɪ/ n vicinanza f; **in the ~ of** nelle vicinanze di

vicious /'vɪʃəs/ a cattivo; ⟨attack⟩

brutale; ⟨animal⟩ pericoloso. **~ 'circle** n circolo m vizioso. **~ly** adv ⟨attack⟩ brutalmente

victim /'vɪktɪm/ n vittima f. **~ize** vt fare delle rappresaglie contro

victor /'vɪktə(r)/ n vincitore m

victor|ious /vɪk'tɔːrɪəs/ a vittorioso. **~y** /'vɪktərɪ/ n vittoria f

video /'vɪdɪəʊ/ n video m; (cassette) videocassetta f; (recorder) videoregistratore m ● attrib video ● vt registrare

video: ~ card n Comput scheda f video. **~ cas'sette** n videocassetta f. **~conference** n videoconferenza f. **~ game** n videogioco m. **~ recorder** n videoregistratore m. **~-tape** n videocassetta f

vie /vaɪ/ vi (pres p **vying**) rivaleggiare

view /vjuː/ n vista f; (photographed, painted) veduta f; (opinion) visione f; **look at the ~** guardare il panorama; **in my ~** secondo me; **in ~ of** in considerazione di; **on ~** esposto; **with a ~ to** con l'intenzione di ● vt visitare ⟨house⟩; (consider) considerare ● vi TV guardare. **~er** n TV telespettatore, -trice mf; Phot visore m

view: ~finder n Phot mirino m. **~point** n punto m di vista

vigil /'vɪdʒɪl/ n veglia f

vigilan|ce /'vɪdʒɪləns/ n vigilanza f. **~t** a vigile

vigorous /'vɪgərəs/ a vigoroso

vigour /'vɪgə(r)/ n vigore m

vile /vaɪl/ a disgustoso; ⟨weather⟩ orribile; ⟨temper, mood⟩ pessimo

villa /'vɪlə/ n (for holidays) casa f di villeggiatura

village /'vɪlɪdʒ/ n paese m. **~r** n paesano, -a mf

villain /'vɪlən/ n furfante m; (in story) cattivo m

vindicate /'vɪndɪkeɪt/ vt (from guilt) discolpare; **you are ~d** ti sei dimostrato nel giusto

vindictive /vɪn'dɪktɪv/ a vendicativo

vine /vaɪn/ n vite f

vinegar /'vɪnɪgə(r)/ n aceto m

vineyard /'vɪnjɑːd/ n vigneto m

vintage /'vɪntɪdʒ/ ⟨a wine⟩ d'annata ● n ⟨year⟩ annata f

viola /vɪ'əʊlə/ n Mus viola f

violat|e /'vaɪəleɪt/ vt violare. **~ion** /-'leɪʃn/ n violazione f

violen|ce /'vaɪələns/ n violenza f. **~t** a violento

violet /'vaɪələt/ a violetto ● n ⟨flower⟩ violetta f; ⟨colour⟩ violetto m

violin /vaɪə'lɪn/ n violino m. **~ist** n violinista mf

VIP n abbr (**very important person**) vip mf

virgin /'vɜːdʒɪn/ a vergine ● n vergine f. **~ity** /-'dʒɪnətɪ/ n verginità f

Virgo /'vɜːgəʊ/ n Vergine f

viril|e /'vɪraɪl/ a virile. **~ity** /-'rɪlətɪ/ n virilità f

virtual /'vɜːtjʊəl/ a effettivo. **~ reality** n realtà f virtuale. **~ly** adv praticamente

virtue /'vɜːtjuː/ n virtù f; ⟨advantage⟩ vantaggio m; **by** or **in ~ of** a causa di

virtuoso /vɜːtʊ'əʊzəʊ/ n (pl -si /-zi:/) virtuoso m

virtuous /'vɜːtjʊəs/ a virtuoso

virulent /'vɪrʊlənt/ a virulento

virus /'vaɪərəs/ n virus m inv

visa /'viːzə/ n visto m

vis-à-vis /viːzɑː'viː/ prep rispetto a

viscount /'vaɪkaʊnt/ n visconte m

viscous /'vɪskəs/ a vischioso

visibility /vɪzə'bɪlətɪ/ n visibilità f

visib|le /'vɪzəbl/ a visibile. **~y** adv visibilmente

vision /'vɪʒn/ n visione f; ⟨sight⟩ vista f

visit /'vɪzɪt/ n visita f ● vt andare a trovare ⟨person⟩; andare da ⟨doctor etc⟩; visitare ⟨town, building⟩. **~ing hours** npl orario m delle visite. **~or** n ospite mf; ⟨of town, museum⟩ visitatore, -trice mf; ⟨in hotel⟩ cliente mf

visor /'vaɪzə(r)/ n visiera f; Auto parasole m

vista /'vɪstə/ n ⟨view⟩ panorama m

visual /'vɪzjʊəl/ a visivo. **~ aids** npl supporto m visivo. **~ display unit** n visualizzatore m. **~ly** adv visualmente; **~ly handicapped** non vedente

visualize /'vɪzjʊəlaɪz/ vt visualizzare

vital /'vaɪtl/ a vitale. **~ity** /vaɪ'tælətɪ/ n vitalità f. **~ly** /'vaɪtəlɪ/ adv estremamente

vitamin /'vɪtəmɪn/ n vitamina f

vivacious /vɪ'veɪʃəs/ a vivace. **~ty** /-'væsətɪ/ n vivacità f

vivid /'vɪvɪd/ a vivido. **~ly** adv in modo vivido

vocabulary /və'kæbjʊlərɪ/ n vocabolario m; ⟨list⟩ glossario m

vocal /'vəʊkl/ a vocale; ⟨vociferous⟩ eloquente. **~ cords** npl corde fpl vocali

vocalist /'vəʊkəlɪst/ n cantante mf

vocation /və'keɪʃn/ n vocazione f. **~al** a di orientamento professionale

vociferous /və'sɪfərəs/ a vociante

vodka /'vɒdkə/ n vodka f inv

vogue /vəʊg/ n moda f; **in ~** in voga

voice /vɔɪs/ n voce f ● vt esprimere. **~mail** n posta f elettronica vocale

void /vɔɪd/ a ⟨not valid⟩ nullo; **~ of** privo di ● n vuoto m

volatile /'vɒlətaɪl/ a volatile; ⟨person⟩ volubile

volcanic /vɒl'kænɪk/ a vulcanico

volcano /vɒl'keɪnəʊ/ n vulcano m

volition /və'lɪʃn/ n of his own **~** di sua spontanea volontà

volley /'vɒlɪ/ n ⟨of gunfire⟩ raffica f; Tennis volée f inv

volt /vəʊlt/ n volt m inv. **~age** /-ɪdʒ/ n Electr voltaggio m

volubl|e /'vɒljʊbl/ a loquace

volume /'vɒljuːm/ n volume m; ⟨of work, traffic⟩ quantità f inv. **~ control** n volume m

voluntary /'vɒləntərɪ/ a volontario. **~y work** n volontariato m. **~ily** adv volontariamente

volunteer /vɒlən'tɪə(r)/ n volontario, -a mf ● vt offrire volontariamente ⟨information⟩ ● vi offrirsi volontario; Mil arruolarsi come volontario

voluptuous /və'lʌptjʊəs/ a voluttuoso

vomit /'vɒmɪt/ n vomito ● vt/i vomitare

voracious /və'reɪʃəs/ a vorace

vot|e /vəʊt/ n voto m; ⟨ballot⟩ votazione f; ⟨right⟩ diritto m di voto; **take a ~e on** votare su ● vi votare ● vt **~e sb president** eleggere qcno presidente. **~er** n elettore, -trice mf. **~ing** n votazione f

vouch /vaʊtʃ/ vi **~ for** garantire per. **~er** n buono m

vow /vaʊ/ n voto m ● vt giurare

vowel /'vaʊəl/ n vocale f

voyage /'vɔɪdʒ/ n viaggio m [marittimo]; ⟨in space⟩ viaggio m [nello spazio]

vulgar /'vʌlgə(r)/ a volgare. **~ity** /-'gærətɪ/ n volgarità f inv

vulnerable /'vʌlnərəbl/ a vulnerabile

vulture /'vʌltʃə(r)/ n avvoltoio m

vying /'vaɪɪŋ/ see **vie**

Ww

wad /wɒd/ n batuffolo m; ⟨bundle⟩ rotolo m. **~ding** n ovatta f

waddle /'wɒdl/ vi camminare ondeggiando

wade /weɪd/ vi guadare; **~ through** fam procedere faticosamente in ⟨book⟩

wafer /'weɪfə(r)/ n cialda f, wafer m inv; Relig ostia f

waffle¹ /'wɒfl/ vi fam blaterare

waffle² n Culin cialda f

waft /wɒft/ vt trasportare ● vi diffondersi

wag /wæg/ v ⟨pt/pp wagged⟩ ● vt agitare ● vi agitarsi

wage¹ /weɪdʒ/ vt dichiarare ⟨war⟩; lanciare ⟨campaign⟩

wage² n, & **~s** pl salario msg. **~-packet** n busta f paga

waggle /'wægl/ vt dimenare ● vi dimenarsi

wagon /'wægən/ n carro m; Rail vagone m merci

wail /weɪl/ n piagnucolio m; ⟨of wind⟩ lamento m; ⟨of baby⟩ vagito m ● vi piagnucolare; ⟨wind:⟩ lamentarsi; ⟨baby:⟩ vagire

waist /weɪst/ n vita f. **~coat** /'weɪskəʊt/ n gilè m inv; ⟨of man's suit⟩ panciotto m. **~line** n vita f

wait /weɪt/ n attesa f; **lie in ~ for** appostarsi per sorprendere ● vi aspettare; **~ for** aspettare ● vt **~ one's turn** aspettare il proprio turno. **wait on** vt servire

waiter /'weɪtə(r)/ n cameriere m

waiting: ~-list n lista f d'attesa. **~-room** n sala f d'aspetto

waitress /'weɪtrɪs/ n cameriera f

waive /weɪv/ vt rinunciare a ⟨claim⟩; non tener conto di ⟨rule⟩

wake¹ /weɪk/ n veglia f funebre ● v ⟨pt **woke**, pp **woken**⟩ **~ [up]** ● vt svegliare ● vi svegliarsi

wake² n Naut scia f; **in the ~ of** fig nella scia di

waken /'weɪkn/ vt svegliare ● vi svegliarsi

Wales /weɪlz/ n Galles m

walk /wɔːk/ n passeggiata f; ⟨gait⟩ andatura f; ⟨path⟩ sentiero m; **go for a ~** andare a fare una passeggiata ● vi camminare; ⟨as opposed to drive etc⟩ andare a piedi; ⟨ramble⟩ passeggiare ● vt portare a spasso ⟨dog⟩; percorrere ⟨streets⟩. **walk out** vi ⟨husband, employee:⟩ andarsene; ⟨workers:⟩ scioperare. **walk out on** vt lasciare

walker /'wɔːkə(r)/ n camminatore, -trice mf; ⟨rambler⟩ escursionista mf

walking /'wɔːkɪŋ/ n camminare m; ⟨rambling⟩ fare delle escursioni. **~-stick** n bastone m da passeggio

'**Walkman®** n Walkman® m inv

walk: ~**-out** n sciopero m. ~**-over** n fig vittoria f facile

wall /wɔːl/ n muro m; **go to the** ~ fam andare a rotoli; **drive sb up the** ~ fam far diventare matto qcno ● **wall up** vt murare

wallet /'wɒlɪt/ n portafoglio m

wallop /'wɒləp/ n fam colpo m ● vt (pt/pp walloped) fam colpire

wallow /'wɒləʊ/ vi sguazzare; (in self-pity, grief) crogiolarsi

'**wallpaper** n tappezzeria f ● vt tappezzare

walnut /'wɔːlnʌt/ n noce f

waltz /wɔːlts/ n valzer m inv ● vi ballare il valzer

wan /wɒn/ a esangue

wand /wɒnd/ n (magic ~) bacchetta f [magica]

wander /'wɒndə(r)/ vi girovagare; (fig: digress) divagare. **wander about** vi andare a spasso

wane /weɪn/ n be on the ~ essere in fase calante ● vi calare

wangle /'wæŋgl/ vt fam rimediare (invitation, holiday)

want /wɒnt/ n (hardship) bisogno m; (lack) mancanza f; (need) aver bisogno di; ~ **[to have]** sth volere qcsa; ~ **to do** sth voler fare qcsa; **we** ~ **to stay** vogliamo rimanere; **I** ~ **you to go** voglio che tu vada; **it** ~**s painting** ha bisogno d'essere dipinto; **you** ~ **to learn to swim** bisogna che impari a nuotare ● vi be ~ for mancare di. ~**ed** a ricercato. ~**ing** a be ~**ing** mancare; **be** ~**ing in** mancare di

wanton /'wɒntən/ a (cruelty, neglect) gratuito; (morally) debosciato

war /wɔː(r)/ n guerra f; fig lotta f (on contro); **at** ~ in guerra

ward /wɔːd/ n (in hospital) reparto m; (child) minore m sotto tutela ● **ward off** vt evitare; parare (blow)

warden /'wɔːdn/ n guardiano, -a mf

warder /'wɔːdə(r)/ n guardia f carceraria

wardrobe /'wɔːdrəʊb/ n guardaroba m

warehouse /'weəhaʊs/ n magazzino m

war: ~**fare** n guerra f. ~**head** n testata f. ~**like** a bellicoso

warm /wɔːm/ a caldo; (welcome) caloroso; **be** ~ (person:) aver caldo; **it is** ~ (weather) fa caldo ● vt scaldare. **warm up** vt scaldare ● vi scaldarsi; fig animarsi. ~**-hearted** a espansivo. ~**ly** adv (greet) calorosamente; (dress) in modo pesante

warmth /wɔːmθ/ n calore m

warn /wɔːn/ vt avvertire. ~**ing** n avvertimento m; (advance notice) preavviso m

warp /wɔːp/ vt deformare; fig distorcere ● vi deformarsi

'**war-path** n on the ~ sul sentiero di guerra

warped /wɔːpt/ a fig contorto; (sexuality) deviato; (view) distorto

warrant /'wɒrənt/ n (for arrest, search) mandato m ● vt (justify) giustificare; (guarantee) garantire

warranty /'wɒrəntɪ/ n garanzia f

warring /'wɔːrɪŋ/ a in guerra

warrior /'wɒrɪə(r)/ n guerriero, -a mf

'**warship** n nave f da guerra

wart /wɔːt/ n porro m

'**wartime** n tempo m di guerra

wary /'weərɪ/ a (-ier, -iest) (careful) cauto; (suspicious) diffidente. ~**ily** adv cautamente

was /wɒz/ see **be**

wash /wɒʃ/ n lavata f; (clothes) bucato m; (in washing machine) lavaggio m; **have a** ~ darsi una lavata ● vt lavare; (sea:) bagnare; ~ **one's hands** lavarsi le mani ● vi lavarsi. **wash out** vt sciacquare (soap); sciacquarsi (mouth). **wash up** vt lavare ● vi lavare i piatti; Am lavarsi

washable /ˈwɒʃəbl/ a lavabile
wash: ~**-basin** n lavandino m. ~ **cloth** n Am ≈ guanto m da bagno
washed 'out a (faded) scolorito; (tired) spossato
washer /ˈwɒʃə(r)/ n Techn guarnizione f; (machine) lavatrice f
washing /ˈwɒʃɪŋ/ n bucato m. ~**-machine** n lavatrice f. ~**-powder** n detersivo m. ~**:'up** n do the ~**-up** lavare i piatti. ~**-'up liquid** n detersivo m per i piatti
wash: ~**-out** n disastro m. ~**-room** n bagno m
wasp /wɒsp/ n vespa f
wastage /ˈweɪstɪdʒ/ n perdita f
waste /weɪst/ n spreco m; (rubbish) rifiuto m; ~ **of time** perdita f di tempo ● a (product) di scarto; (land) desolato; **lay** ~ devastare ● vt sprecare. **waste away** vi deperire
waste: ~**-di'sposal unit** n eliminatore m di rifiuti. ~**ful** a dispendioso. ~ **'paper** n carta f straccia. ~**:'paper basket** n cestino m per la carta (straccia)
watch /wɒtʃ/ n guardia f; (period of duty) turno m di guardia; (timepiece) orologio m; **be on the** ~ stare all'erta ● vt guardare (film, match, television); (be careful, look after) stare attento a ● vi guardare. **watch out** vi (be careful) stare attento (**for** a). **watch out for** vt (look for) fare attenzione all'arrivo di (person)
watch: ~**-dog** n cane m da guardia. ~**ful** a attento. ~**-maker** n orologiaio, -a mf. ~**-man** n guardiano m. ~**-strap** n cinturino m dell'orologio. ~**-word** n motto m
water /ˈwɔːtə(r)/ n acqua f ● vt annaffiare (garden, plant); (dilute) annacquare ● vi (eyes:) lacrimare; **my mouth was** ~**ing** avevo l'acquolina in bocca. **water down** vt diluire; fig attenuare
water: ~**-colour** n acquerello m. ~**cress** n crescione m. ~**-fall** n cascata f

'watering-can n annaffiatoio m
water: ~**-lily** n ninfea f. ~ **logged** a inzuppato. ~**-main** n conduttura f dell'acqua. ~ **polo** n pallanuoto f. ~**-power** n energia f idraulica. ~**proof** a impermeabile. ~**-shed** n spartiacque m inv; fig svolta f. ~**-skiing** n sci m nautico. ~**tight** stagno; fig irrefutabile. ~**way** n canale m navigabile
watery /ˈwɔːtərɪ/ a acquoso; (eyes) lacrimoso
watt /wɒt/ n watt m inv
wave /weɪv/ n onda f; (gesture) cenno m; fig ondata f ● vt agitare; ~ **one's hand** agitare la mano ● vi far segno; (flag:) sventolare. ~**length** n lunghezza f d'onda
waver /ˈweɪvə(r)/ vi vacillare; (hesitate) esitare
wavy /ˈweɪvɪ/ a ondulato
wax[1] /wæks/ vi (moon:) crescere; (fig: become) diventare
wax[2] n cera f; (in ear) cerume m ● vt dare la cera a. ~**works** n museo m delle cere
way /weɪ/ n percorso m; (direction) direzione f; (manner, method) modo m; ~**s** pl (customs) abitudini fpl; **be in the** ~ essere in mezzo; **on the** ~ **to Rome** andando a Roma; **I'll do it on the** ~ lo faccio mentre vado; **it's on my** ~ è sul mio percorso; **a long** ~ **off** lontano; **this** ~ da questa parte; (like this) così; **by the** ~ a proposito; **by** ~ **of** come; (via) via; **either** ~ (whatever we do) in un modo o nell'altro; **in some** ~**s** sotto certi aspetti; **in a** ~ in un certo senso; **in a bad** ~ (person) molto grave; **out of the** ~ fuori mano; **under** ~ in corso; **lead the** ~ far strada; fig aprire la strada; **make** ~ far posto (**for** a); **give** ~ Auto dare la precedenza; **go out of one's** ~ fig scomodarsi (**to** per); **get one's (own)** ~ averla vinta ● adv ~**s behind** molto indietro. ~ **in** n entrata f
way'lay vt (pt/pp -**laid**) aspettare al varco (person)

way 'out n uscita f; fig via f d'uscita

way-'out a fam eccentrico

wayward /'weɪwəd/ a capriccioso

WC n abbr WC; **the WC** il gabinetto m

we /wiː/ pron noi; **we're the last** siamo gli ultimi; **they're going, but we're not** loro vanno, ma noi no

weak /wiːk/ a debole; (liquid) leggero. **~en** vt indebolire ● vi indebolirsi. **~ling** n smidollato, -a mf. **~ness** n debolezza f; (liking) debole m

wealth /welθ/ n ricchezza f; fig gran quantità f. **~y** a (-ier, -iest) ricco

wean /wiːn/ vt svezzare

weapon /'wepən/ n arma f

wear /weə(r)/ n (clothing) abbigliamento m; **for everyday** ~ per portare tutti i giorni; ~ **[and tear]** usura f ● v (pt **wore**, pp **worn**) ● vt portare; (damage) consumare; ● vi **a hole in sth** logorare qcsa fino a fare un buco; **what shall I** ~? cosa mi metto? ● vi consumarsi; (last) durare. **wear off** vi scomparire; (effect:) finire. **wear out** vt consumare [fino in fondo]; (exhaust) estenuare ● vi estenuarsi

wearable /'weərəbl/ a portabile

weary /'wɪərɪ/ a (-ier, -iest) sfinito ● v (pt/pp **wearied**) ● vt sfinire ● vi **~y of** stancarsi di. **~ily** adv stancamente

weasel /'wiːzl/ n donnola f

weather /'weðə(r)/ n tempo m; **in this** ~ con questo tempo; **under the** ~ fam giù di corda ● vt sopravvivere a (storm)

weather: **~-beaten** a (face) segnato dalle intemperie. **~cock** n gallo m segnavento. **~ forecast** n previsioni fpl del tempo

weave[1] /wiːv/ vi (pt/pp **weaved**) (move) zigzagare

weave[2] n tessuto m ● vt (pt **wove**, pp **woven**) tessere; intrecciare (flowers etc); intrecciare le fila

di (story etc). **~r** n tessitore, -trice mf

web /web/ n rete f; (of spider) ragnatela f. **~bed feet** npl piedi mpl palmati. **W~ page** n Comput pagina f web. **W~ site** n Comput sito m web

wed /wed/ vt (pt/pp **wedded**) sposare ● vi sposarsi. **~ding** n matrimonio m

wedding: **~ cake** n torta f nuziale. **~ day** n giorno m del matrimonio. **~ dress** n vestito m da sposa. **~ring** n fede f

wedge /wedʒ/ n zeppa f; (for splitting wood) cuneo m; (of cheese) fetta f ● vt (fix) fissare

wedlock /'wedlɒk/ n **born out of** ~ nato fuori dal matrimonio

Wednesday /'wenzdeɪ/ n mercoledì m inv

wee[1] /wiː/ a fam piccolo

wee[2] vi fam fare la pipì

weed /wiːd/ n erbaccia f; (fam: person) mollusco m ● vt estirpare le erbacce da. **weed out** vt fig eliminare

'weed-killer n erbicida m

weedy /'wiːdɪ/ a fam mingherlino

week /wiːk/ n settimana f. **~day** n giorno m feriale. **~end** n fine settimana m

weekly /'wiːklɪ/ a (-ier, -iest) settimanale ● n settimanale m ● adv settimanalmente

weep /wiːp/ vi (pt/pp **wept**) piangere

weigh /weɪ/ vt/i pesare; ~ **anchor** levare l'ancora. **weigh down** vt fig piegare. **weigh up** vt fig soppesare; valutare (person)

weight /weɪt/ n peso m; **put on/lose** ~ ingrassare/dimagrire. **~ing** n (allowance) indennità f inv

weight: **~lessness** n assenza f di gravità. **~lifting** n sollevamento m pesi

weighty /'weɪtɪ/ a (-ier, -iest) pesante; (important) di un certo peso

weir /wɪə(r)/ n chiusa f

weird /wɪəd/ a misterioso; (bizarre) bizzarro

welcome /'welkəm/ a benvenuto; **you're ~!** prego!; **you're ~ to have it/to use** prendilo/vieni pure ● n accoglienza f ● vt accogliere; (appreciate) gradire

weld /weld/ vt saldare. **~er** n saldatore m

welfare /'welfeə(r)/ n benessere m; (aid) assistenza f. **W~ State** n Stato m assistenziale

well[1] /wel/ n pozzo m; (of staircase) tromba f

well[2] adv (better, best) bene; as ~ anche; as ~ as (in addition) oltre a; ~ **done!** bravo!; **very ~** benissimo ● a **he is not ~** non sta bene; **get ~ soon!** guarisci presto! ● int beh!; ~ **I never!** ma va!

well: ~-behaved a educato. **~-being** n benessere m. **~-bred** a beneducato. **~-heeled** a fam danaroso

wellingtons /'welɪŋtənz/ npl stivali mpl di gomma

well: ~-known a famoso. **~-meaning** a con buone intenzioni. **~-meant** a con le migliori intenzioni. **~-off** a benestante. **~-read** a colto. **~-to-do** a ricco

Welsh /welʃ/ a & n gallese; **the ~** pl i gallesi. **~man** n gallese m. **~ rabbit** n toast m inv al formaggio

went /went/ see **go**

wept /wept/ see **weep**

were /wɜ:(r)/ see **be**

west /west/ n ovest m; **to the ~ of** a ovest di; **the W~** l'Occidente m ● a occidentale ● adv verso occidente; **go ~** fam andare in malora. **~erly** a verso ovest; occidentale (wind). **~ern** a occidentale ● n western m inv

West: ~ 'Germany n Germania f Occidentale. **~ 'Indian** a & n antillese mf. **~ 'Indies** /'ɪndɪz/ npl Antille fpl

'westward[s] /-wəd[z]/ adv verso ovest

wet /wet/ a (wetter, wettest) bagnato; fresco (paint); (rainy) piovoso; (fam: person) smidollato;

get ~ bagnarsi ● vt (pt/pp wet, wetted) bagnare. **~ 'blanket** n guastafeste mf inv

whack /wæk/ n fam colpo m ● vt fam dare un colpo a. **~ed** a fam stanco morto. **~ing** a (fam: huge) enorme

whale /weɪl/ n balena f; **have a ~ of a time** fam divertirsi un sacco

wham /wæm/ int bum

wharf /wɔ:f/ n banchina f

what /wɒt/ pron che, [che] cosa; ~ **for?** perché?; ~ **is that for?** a che cosa serve?; ~ **is it?** (what do you want) cosa c'è?; ~ **is it like?** com'è?; ~ **is your name?** come ti chiami?; ~ **is the weather like?** com'è il tempo?; ~ **is the film about?** di cosa parla il film?; ~ **is he talking about?** di cosa sta parlando?; **he asked me ~ she had said** mi ha chiesto cosa ha detto; ~ **about going to the cinema?** e se andassimo al cinema?; ~ **about the children?** (what will they do) e i bambini?; ~ **if it rains?** e se piove? ● a quale, che; **take ~ books you want** prendi tutti i libri che vuoi; ~ **kind of a** che tipo di; **at ~ time?** a che ora? ● adv che; ~ **a lovely day!** che bella giornata! ● int ~! [che] cosa!; ~? [che] cosa?

what'ever /wɒt'evə(r)/ a & pron qualunque cosa; ~ **is it?** cos'è?; **he does** qualsiasi cosa faccia; ~ **happens** qualunque cosa succeda; **nothing ~** proprio niente

whatso'ever a & pron = **whatever**

wheat /wi:t/ n grano m, frumento m

wheedle /'wi:d(ə)l/ vt ~ **sth out of sb** ottenere qcsa da qualcuno con le lusinghe

wheel /wi:l/ n ruota f; (steering ~) volante m; **at the ~** al volante ● vt (push) spingere ● vi (circle) ruotare; ~ [round] ruotare

wheel: ~barrow n carriola f. **~chair** n sedia f a rotelle. **~clamp** n ceppo m bloccaruote

wheeze /wi:z/ vi ansimare

when /wen/ *adv & conj* quando; **the day ~** il giorno in cui; **~ swimming/reading** nuotando/leggendo

when'ever *adv & conj* in qualsiasi momento; (*every time that*) ogni volta che; **~ did it happen?** quando è successo?

where /weə(r)/ *adv & conj* dove; **the street ~ I live** la via in cui abito; **~ do you come from?** da dove vieni?

whereabouts¹ /weərə'bauts/ *adv* dove

'whereabouts² *n* nobody knows his **~** nessuno sa dove si trova

where'as *conj* dal momento che; (*in contrast*) mentre

where'by *adv* attraverso il quale

whereu'pon *adv* dopo di che

wher'ever *adv & conj* dovunque; **~ is he?** dov'è mai?; **~ possible** dovunque sia possibile

whet /wet/ *vt* (*pt/pp* whetted) aguzzare (*appetite*)

whether /'weðə(r)/ *conj* se; **~ you like it or not** che ti piaccia o no

which /wɪtʃ/ *a & pron* quale; **~ one?** quale?; **~ one of you?** chi di voi?; **~ way?** (*direction*) in che direzione? ● *rel pron* (*object*) che; **he does frequently** cosa che fa spesso; **after ~** dopo di che; **on/in ~** su/in cui

which'ever *a & pron* qualunque; **~ it is** qualunque sia; **~ one of you** chiunque tra voi

whiff /wɪf/ *n* zaffata *f*; **have a ~ of** sth odorare qcsa

while /waɪl/ *n* a long **~** un bel po'; **a little ~** un po' ● *conj* mentre; (*as long as*) finché; (*although*) sebbene ● **while away** *vt* passare (*time*)

whilst /waɪlst/ *conj* see **while**

whim /wɪm/ *n* capriccio *m*

whimper /'wɪmpə(r)/ *vi* piagnucolare; (*dog:*) mugolare

whimsical /'wɪmzɪkl/ *a* capriccioso; (*story*) fantasioso

whine /waɪn/ *n* lamento *m*; (*of dog*)

guaito *m* ● *vi* lamentarsi; (*dog:*) guaire

whip /wɪp/ *n* frusta *f*; (*pol: person*) parlamentare *mf* incaricato, -a di assicurarsi della presenza dei membri del suo partito alle votazioni ● *vt* (*pt/pp* whipped) frustare; Culin sbattere; (*snatch*) afferrare; (*fam: steal*) fregare. **whip up** *vt* (*incite*) stimolare; *fam* improvvisare (*meal*). **~ped 'cream** *n* panna *f* montata

whirl /wɜ:l/ *n* (*movement*) rotazione *f*; **my mind's in a ~** ho le idee confuse ● *vi* girare rapidamente ● *vt* far girare rapidamente. **~ pool** *n* vortice *m*. **~ wind** *n* turbine *m*

whirr /wɜ:(r)/ *vi* ronzare

whisk /wɪsk/ *n* Culin frullino *m* ● *vt* Culin frullare. **whisk away** *vt* portare via

whisker /'wɪskə(r)/ *n* **~s** (*of cat*) baffi *mpl*; (*on man's cheek*) basette *fpl*; **by a ~** per un pelo

whisky /'wɪskɪ/ *n* whisky *m inv*

whisper /'wɪspə(r)/ *n* sussurro *m*; (*rumour*) diceria *f* ● *vt/i* sussurrare

whistle /'wɪsl/ *n* fischio *m*; (*instrument*) fischietto *m* ● *vt* fischiettare ● *vi* fischiettare; (*referee*) fischiare

white /waɪt/ *a* bianco; **go ~** (*pale*) sbiancare ● *n* bianco *m*; (*of egg*) albume *m*; (*person*) bianco, -a *mf*

white: ~ 'coffee *n* caffè *m inv* macchiato. **~·'collar worker** *n* colletto *m* bianco

'Whitehall *n* strada *f* di Londra, sede degli uffici del governo britannico; *fig* amministrazione *f* britannica

white 'lie *n* bugia *f* pietosa

whiten /'waɪtn/ *vt* imbiancare ● *vi* sbiancare

whiteness /'waɪtnɪs/ *n* bianchezza *f*

'whitewash *n* intonaco *m*; *fig* copertura *f* ● *vt* dare una mano d'intonaco a; *fig* coprire

Whitsun /'wɪtsn/ n Pentecoste f

whittle /'wɪtl/ vt ~ **down** ridurre

whiz[z] /wɪz/ vi (pt/pp **whizzed**) sibilare. **~-kid** n fam giovane m prodigio

who /huː/ inter pron chi ● rel pron che; **the children,** ~ **were all tired,...** i bambini, che erano tutti stanchi,...

who'ever pron chiunque; ~ **he is** chiunque sia; ~ **can that be?** chi può mai essere?

whole /həʊl/ a tutto; (not broken) intatto; **the** ~ **truth** tutta la verità; **the** ~ **world** il mondo intero; **the** ~ **lot** (everything) tutto; (of) tutti; **the** ~ **lot of you** tutti voi ● n tutto m; as a ~ nell'insieme; on **the** ~ tutto considerato; **the** ~ **of Italy** tutta l'Italia

whole: **~food** n cibo m macrobiotico. **~-'hearted** a di tutto cuore. **~meal** a integrale

'wholesale a & adv all'ingrosso; fig in massa. **~r** n grossista m

wholesome /'həʊlsəm/ a sano

wholly /'həʊlɪ/ adv completamente

whom /huːm/ rel pron che; **the man** ~ **I saw** l'uomo che ho visto; **to/with** ~ a/con cui ● inter pron chi; **to** ~ **did you speak?** con chi hai parlato?

whooping cough /'huːpɪŋ/ n pertosse f

whopping /'wɒpɪŋ/ a fam enorme

whore /hɔː(r)/ n puttana f vulg

whose /huːz/ rel pron di cui; **people** ~ **name begins with D** le persone i cui nomi cominciano con la D ● inter pron di chi; ~ **is that?** di chi è quello? ● a ~ **car did you use?** di chi è la macchina che hai usato?

why /waɪ/ adv (inter) perché; **the reason** ~ la ragione per cui; **that's** ~ per questo ● int diamine

wick /wɪk/ n stoppino m

wicked /'wɪkɪd/ a cattivo; (mischievous) malizioso

wicker /'wɪkə(r)/ n vimini mpl ● attrib di vimini

wide /waɪd/ a largo; (experience, knowledge) vasto; (difference) profondo; (far from target) lontano; **10 cm** ~ largo 10 cm; **how** ~ **is it?** quanto è largo? ● adv (off target) lontano dal bersaglio; ~ **awake** del tutto sveglio; ~ **open** spalancato; **far and** ~ in lungo e in largo. **~ly** adv largamente; (known, accepted) generalmente; (different) profondamente

widen /'waɪdn/ vt allargare ● vi allargarsi

'widespread a diffuso

widow /'wɪdəʊ/ n vedova f. **~ed** a vedovo. **~er** n vedovo m

width /wɪdθ/ n larghezza f; (of material) altezza f

wield /wiːld/ vt maneggiare; esercitare (power)

wife /waɪf/ n (pl **wives**) moglie f

wig /wɪɡ/ n parrucca f

wiggle /'wɪɡl/ vi dimenarsi ● vt dimenare

wild /waɪld/ a selvaggio; (animal, flower) selvatico; (furious) furibondo; (applause) fragoroso; (idea) folle; (with joy) pazzo; (guess) azzardato; **be** ~ **about** (keen on) andare pazzo per ● adv **run** ~ crescere senza controllo ● n in the ~ allo stato naturale; **the** ~s pl le zone fpl sperdute

wilderness /'wɪldənɪs/ n deserto m; (fig: garden) giungla f

'wildfire n spread like ~ allargarsi a macchia d'olio

wild: **~'goose chase** n ricerca f inutile. **~life** n animali mpl selvatici

wilful /'wɪlfl/ a intenzionale; (person, refusal) ostinato. **~ly** adv intenzionalmente; (refuse) ostinatamente

will[1] /wɪl/ v aux **he** ~ **arrive tomorrow** arriverà domani; **I won't tell him** non glielo dirò; **you** ~ **be back soon, won't you?** tornerai presto, no?; ~ **he be there, won't he?** sarà là, no?; **she** ~ **be there by now** sarà là ormai; ~

you go? (*do you intend to go*) pensi di andare?; **~ you go to the baker's and buy...?** puoi andare dal panettiere a comprare...?; **~ you be quiet!** vuoi stare calmo!; **~ you have some wine?** vuoi del vino?; **the engine won't start** la macchina non parte

will[2] *n* volontà *f inv*; (*document*) testamento *m*

willing /'wɪlɪŋ/ *a* disposto; (*eager*) volonteroso; **~ly** *adv* volentieri. **~ness** *n* buona volontà *f*

willow /'wɪləʊ/ *n* salice *m*

'will-power *n* forza *f* di volontà

willy-'nilly *adv* (*at random*) a casaccio; (*wanting to or not*) volente o nolente

wilt /wɪlt/ *vi* appassire

wily /'waɪlɪ/ *a* (**-ier, -iest**) astuto

wimp /wɪmp/ *n* rammollito, -a *mf*

win /wɪn/ *n* vittoria *f*; **have a ~** riportare una vittoria ● *v* (*pt/pp* won; *pres p* winning) ● *vt* vincere; conquistare 〈*fame*〉 ● *vi* vincere. **win over** *vt* convincere

wince /wɪns/ *vi* contrarre il viso

winch /wɪntʃ/ *n* argano *m*

wind[1] /wɪnd/ *n* vento *m*; (*breath*) fiato *m*; (*fam: flatulence*) aria *f*; **get/have the ~ up** *fam* aver fifa; **get ~ of** aver sentore di; **in the ~** nell'aria ● *vt* ~sb lasciare qcno senza fiato

wind[2] /waɪnd/ *v* (*pt/pp* wound) ● *vt* (*wrap*) avvolgere; (*move by turning*) far girare; (*clock*) caricare ● *vi* 〈*road*〉 serpeggiare. **wind up** *vt* caricare 〈*clock*〉; concludere 〈*proceedings*〉; *fam* prendere in giro 〈*sb*〉

wind /wɪnd/ *n*: **~fall** *n fig* fortuna *f* inaspettata

winding /'waɪndɪŋ/ *a* tortuoso

wind: **~ instrument** *n* strumento *m* a fiato. **~mill** *n* mulino *m* a vento

window /'wɪndəʊ/ *n* finestra *f*; (*of car*) finestrino *m*; (*of shop*) vetrina *f*

window: **~-box** *n* cassetta *f* per

i fiori. **~-cleaner** *n* (*person*) lavavetri *mf inv*. **~-dresser** *n* vetrinista *mf*. **~-dressing** *n* vetrinistica *f*, *fig* fumo *m* negli occhi. **~-pane** *n* vetro *m*. **~-shopping** *n*: **go ~-shopping** andare in giro a vedere le vetrine. **~-sill** *n* davanzale *m*

'windscreen *n*, *Am* **'windshield** *n* parabrezza *m inv*. **~ washer** *n* getto *m* d'acqua. **~-wiper** *n* tergicristallo *m*

wind: **~ surfing** *n* windsurf *m inv*. **~swept** *a* esposto al vento; 〈*person*〉 scompigliato

windy /'wɪndɪ/ *a* (**-ier, -iest**) ventoso

wine /waɪn/ *n* vino *m*

wine: **~-bar** *n* ≈ enoteca *f*. **~glass** *n* bicchiere *m* da vino. **~-list** *n* carta *f* dei vini

winery /'waɪnərɪ/ *n Am* vigneto *m*

'wine-tasting *n* degustazione *f* di vini

wing /wɪŋ/ *n* ala *f*; *Auto* parafango *m*; **~s** *pl Theat* quinte *fpl*. **~er** *n Sport* ala *f*

wink /wɪŋk/ *n* strizzata *f* d'occhio; **not sleep a ~** non chiudere occhio ● *vi* strizzare l'occhio; 〈*light:*〉 lampeggiare

winner /'wɪnə(r)/ *n* vincitore, -trice *mf*

winning /'wɪnɪŋ/ *a* vincente; 〈*smile*〉 accattivante. **~-post** *n* linea *f* d'arrivo. **~s** *npl* vincite *fpl*

winter /'wɪntə(r)/ *n* inverno *m*. **~ry** *a* invernale

wipe /waɪp/ *n* passata *f*; (*to dry*) asciugata *f* ● *vt* strofinare; (*dry*) asciugare. **wipe off** *vt* asciugare; (*erase*) cancellare. **wipe out** *vt* annientare; eliminare 〈*debt*〉; estinguere 〈*debt*〉. **wipe up** *vt* asciugare 〈*dishes*〉

wire /'waɪə(r)/ *n* fil *m* di ferro; (*electrical*) filo *m* elettrico

wireless /'waɪəlɪs/ *n* radio *f*

wire 'netting *n* rete *f* metallica

wiring /'waɪərɪŋ/ *n* impianto *m* elettrico

wiry /'waɪərɪ/ a (-ier, -iest) ⟨person⟩ dal fisico asciutto; ⟨hair⟩ ispido

wisdom /'wɪzdəm/ n saggezza f; ⟨of action⟩ sensatezza f. ~ **tooth** n dente m del giudizio

wise /waɪz/ a saggio; ⟨prudent⟩ sensato. ~**ly** adv saggiamente; ⟨act⟩ sensatamente

wish /wɪʃ/ n desiderio m; **make a** ~ esprimere un desiderio; **with best** ~**es** con i migliori auguri ● vt desiderare; ~ **sb well** fare tanti auguri a qcno; **I** ~ **you every success** ti auguro buona fortuna; **I** ~ **you could stay** vorrei che tu potessi rimanere ● vi ~ **for sth** desiderare qcsa. ~**ful** a ~**ful thinking** illusione f

wishy-washy /'wɪʃɪwɒʃɪ/ a ⟨colour⟩ spento; ⟨personality⟩ insignificante

wisp /wɪsp/ n ⟨of hair⟩ ciocca f; ⟨of smoke⟩ filo m; ⟨of grass⟩ ciuffo m

wistful /'wɪstfl/ a malinconico

wit /wɪt/ n spirito m; ⟨person⟩ persona f di spirito; **be at one's** ~**s' end** non saper che pesci pigliare

witch /wɪtʃ/ n strega f. ~**craft** n magia f. ~**-hunt** n caccia f alle streghe

with /wɪð/ prep con; ⟨fear, cold, jealousy etc⟩ di; **I'm not** ~ **you** non ti seguo; **can I leave it** ~ **you?** ⟨task⟩ puoi occupartene tu?; ~ **no regrets/money** senza rimpianti/soldi; **be** ~ **it** fam essere al passo coi tempi; ⟨alert⟩ essere concentrato

with'draw v (pt -drew, pp -drawn) ● vt ritirare; prelevare ⟨money⟩ ● vi ritirarsi. ~**al** n ritiro m; ⟨of money⟩ prelevamento m; ⟨from drugs⟩ crisi f inv di astinenza; Psych chiusura f in se stessi. ~**al symptoms** npl sintomi mpl della crisi di astinenza

with'drawn see **withdraw** ● a ⟨person⟩ chiuso in se stesso

wither /'wɪðə(r)/ vi ⟨flower:⟩ appassire

with'hold vt (pt/pp -held) rifiutare ⟨consent⟩ (**from** a); nascondere ⟨information⟩ (**from** a); trattenere ⟨smile⟩

with'in prep in; ⟨before the end of⟩ entro; ~ **the law** legale ● adv all'interno

with'out prep senza; ~ **stopping** senza fermarsi

with'stand vt (pt/pp -stood) resistere a

witness /'wɪtnɪs/ n testimone mf ● vt autenticare ⟨signature⟩; essere testimone di ⟨accident⟩. ~**-box** n, Am ~**-stand** n banco m dei testimoni

witticism /'wɪtɪsɪzm/ n spiritosaggine f

wittingly /'wɪtɪŋlɪ/ adv consapevolmente

witty /'wɪtɪ/ a (-ier, -iest) spiritoso

wives /waɪvz/ see **wife**

wizard /'wɪzəd/ n mago m. ~**ry** n stregoneria f

wobb|le /'wɒbl/ vi traballare. ~**ly** a traballante

wodge /wɒdʒ/ n fam mucchio m

woe /wəʊ/ n afflizione f

woke, woken /wəʊk, 'wəʊkn/ see **wake[1]**

wolf /wʊlf/ n (pl wolves /wʊlvz/) lupo m; ⟨fam: womanizer⟩ donnaiolo m ● vt ~ [**down**] divorare. ~ **whistle** n fischio m ● vi ~**whistle at sb** fischiare dietro a qcno

woman /'wʊmən/ n (pl women) donna f. ~**izer** n donnaiolo m. ~**ly** a femmineo

womb /wu:m/ n utero m

women /'wɪmɪn/ see **woman**. **W~'s Libber** /'lɪbə(r)/ n femminista f. **W~'s Liberation** n movimento m femminista

won /wʌn/ see **win**

wonder /'wʌndə(r)/ n meraviglia f; ⟨surprise⟩ stupore m; **no** ~! non c'è da stupirsi!; **it's a** ~ **that...** è incredibile che... ● vi restare in ammirazione; ⟨be surprised⟩ essere sorpreso; **I** ~ **is** è quello che mi chiedo; **I** ~ **whether she is ill** mi chiedo se è malata?. **~ful** *a* meraviglioso. **~fully** *adv* meravigliosamente

won't /wəunt/ = **will not**

woo /wu:/ *vt* corteggiare; *fig* cercare di accattivarsi ⟨*voters*⟩

wood /wud/ *n* legno *m*; ⟨*for burning*⟩ legna *f*; ⟨*forest*⟩ bosco *m*; **out of the ~** *fig* fuori pericolo; **touch ~!** tocca ferro!

wood: **~ed** /-ɪd/ *a* boscoso. **~en** *a* di legno; *fig* legnoso. **~ wind** *n* strumenti *mpl* a fiato. **~work** *n* ⟨*wooden parts*⟩ parti *fpl* in legno; ⟨*craft*⟩ falegnameria *f*. **~worm** *n* tarlo *m*. **~y** *a* legnoso; ⟨*hill*⟩ boscoso

wool /wul/ *n* lana *f* ● *attrib* di lana. **~len** *a* di lana. **~lens** *npl* capi *mpl* di lana

woolly /'wulɪ/ *a* (**-ier, -iest**) ⟨*sweater*⟩ di lana; *fig* confuso

word /wɜ:d/ *n* parola *f*; ⟨*news*⟩ notizia *f*; **by ~ of mouth** a viva voce; **have a ~ with** dire due parole a; **have ~s** bisticciare; **in other ~s** in altre parole. **~ing** *n* parole *fpl*. **~ processor** *n* programma *m* di videoscrittura, word processor *m inv*

wore /wɔ:(r)/ *see* **wear**

work /wɜ:k/ *n* lavoro *m*; ⟨*of art*⟩ opera *f*. **~s** *pl* ⟨*factory*⟩ fabbrica *fsg*; ⟨*mechanism*⟩ meccanismo *msg*; **at ~** al lavoro; **out of ~** disoccupato ● *vi* lavorare; ⟨*machine, ruse*⟩ funzionare; ⟨*study*⟩ studiare ● *vt* far funzionare ⟨*machine*⟩; far lavorare ⟨*employee*⟩; far studiare ⟨*student*⟩. **work off** *vt* sfogare ⟨*anger*⟩; lavorare per estinguere ⟨*debt*⟩; fare sport per smaltire ⟨*weight*⟩. **work out** *vt* elaborare ⟨*plan*⟩; risolvere ⟨*problem*⟩; calcolare ⟨*bill*⟩; **I ~ed out how he did it** ho capito come l'ha fatto ● *vi* evolvere. **work up** *vt* **I've ~ed up an appetite** mi è venuto appetito; **don't get ~ed up** ⟨*anxious*⟩ non farti prendere dal panico; ⟨*angry*⟩ non arrabbiarti

workable /'wɜ:kəbl/ *a* ⟨*feasible*⟩ fattibile

workaholic /wɜ:kə'hɒlɪk/ *n* stacanovista *mf*

worker /'wɜ:kə(r)/ *n* lavoratore, -trice *mf*; ⟨*manual*⟩ operaio, -a *mf*

working /'wɜ:kɪŋ/ *a* ⟨*clothes etc*⟩ da lavoro; ⟨*day*⟩ feriale; **in ~ order** funzionante. **~ class** *n* classe *f* operaia. **~-class** *a* operaio

work: **~man** *n* operaio *m*. **~manship** *n* lavorazione *f*. **~-out** *n* allenamento *m*. **~shop** *n* officina *f*; ⟨*discussion*⟩ dibattito *m*

world /wɜ:ld/ *n* mondo *m*; **a ~ of difference** una differenza abissale; **out of this ~** favoloso; **think the ~ of sb** andare matto per qcno. **~ly** *a* materiale; ⟨*person*⟩ materialista. **~-wide** *a* mondiale ● *adv* mondialmente

worm /wɜ:m/ *n* verme *m* ● *vt* **~ one's way into sb's confidence** conquistarsi la fiducia di qcno in modo subdolo. **~-eaten** *a* tarlato

worn /wɔ:n/ *see* **wear** ● *a* sciupato. **~-out** *a* consumato; ⟨*person*⟩ sfinito

worried /'wʌrɪd/ *a* preoccupato

worry /'wʌrɪ/ *n* preoccupazione *f* ● *v* (*pt/pp* worried) ● *vt* preoccupare; ⟨*bother*⟩ disturbare ● *vi* preoccuparsi. **~ing** *a* preoccupante

worse /wɜ:s/ *a* & *adv* peggio ● *n* peggio *m*

worsen /'wɜ:sn/ *vt/i* peggiorare

worship /'wɜ:ʃɪp/ *n* culto *m*; ⟨*service*⟩ funzione *f*; **Your/His W~** ⟨*to judge*⟩ signor giudice/il giudice ● *v* (*pt/pp* -shipped) ● *vt* venerare ● *vi* andare a messa

worst /wɜ:st/ *a* peggiore ● *adv* peggio [di tutti] ● *n* the **~** il peggio; **get the ~ of it** avere la peggio; **if the ~ comes to the ~** nella peggiore delle ipotesi

worth /wɜ:θ/ *n* valore *m*; **£10 ~ of petrol** 10 sterline di benzina ● *a* **be ~** valere; **be it** *fig* valerne la pena; **it's ~ trying** vale la pena di provare; **it's ~ my while** mi conviene. **~less** a senza valore. **~while** a che vale la pena; ‹cause› lodevole

worthy /'wɜːðɪ/ a degno; ‹cause, motive› lodevole

would /wʊd/ v aux **I ~ do it** lo farei; **~ you go?** andresti?; **you mind if I opened the window?** ti dispiace se apro la finestra?; **he ~ come if he could** verrebbe se potesse; **he said he ~n't** ha detto di no; **~ you like a drink?** vuoi qualcosa da bere?; **what ~ you like to drink?** cosa prendi da bere?; **you ~n't, ~ you?** non lo faresti, vero?

wound² /wuːnd/ n ferita f ● vt ferire

wound² /waʊnd/ see **wind²**

wove, woven /wəʊv, 'wəʊvn/ see **weave²**

wrangle /'ræŋgl/ n litigio m ● vi litigare

wrap /ræp/ n ‹shawl› scialle m ● vt (pt/pp **wrapped**) ~ [**up**] avvolgere; ‹present› incartare; **be ~ped up in** fig essere completamente preso da ● vi ~ **up warmly** coprirsi bene. ~**per** n (for sweet) carta f [di caramella]. ~**ping** n materiale m da imballaggio. ~**ping paper** n carta f da pacchi; (for gift) carta f da regalo

wrath /rɒθ/ n ira f

wreak /riːk/ vt ~ **havoc with sth** scombussolare qcsa

wreath /riːθ/ n (pl ~**s** /-ðz/) corona f

wreck /rek/ n ‹of ship› relitto m; ‹of car› carcassa f; ‹person› rottame m ● vt far naufragare; demolire ‹car›. ~**age** /-ɪdʒ/ n rottami mpl; fig brandelli mpl

wrench /rentʃ/ n ‹injury› slogatura f; ‹tool› chiave f inglese; ‹pull› strattone m ● vt ‹pull› strappare; slogarsi ‹wrist, ankle etc›

wrest /rest/ vt strappare (**from** a)

wrestl|e /'resl/ vi lottare corpo a corpo; fig lottare. ~**er** n lottatore, -trice mf. ~**ing** n lotta f libera; ‹all-in› catch m

wretch /retʃ/ n disgraziato, -a mf. ~**ed** /-ɪd/ a odioso; ‹weather› orribile; **feel ~ed** a) (unhappy) essere triste; (ill) sentirsi malissimo

wriggle /'rɪgl/ n contorsione f ● vi contorcersi; ‹move forward› strisciare; ~ **out of sth** fam sottrarsi a qcsa

wring /rɪŋ/ vt (pt/pp **wrung**) torcere ‹sb's neck›; strizzare ‹clothes›; ~ **one's hands** torcersi le mani; ~**ing wet** inzuppato

wrinkle /'rɪŋkl/ n grinza f; ‹on skin› ruga f ● vt/i raggrinzire. ~**d** a ‹skin, face› rugoso; ‹clothes› raggrinzito

wrist /rɪst/ n polso m. ~-**watch** n orologio m da polso

writ /rɪt/ n Jur mandato m

write /raɪt/ vt/i (pt **wrote**, pp **written**, pres p **writing**) scrivere.

write down vt annotare. **write off** vt cancellare ‹debt›; distruggere ‹car›

write-off n ‹car› rottame m

writer /'raɪtə(r)/ n autore, -trice mf; **she's a ~** è una scrittrice

write-up n ‹review› recensione f

writhe /raɪð/ vi contorcersi

writing /'raɪtɪŋ/ n (occupation) scrivere m; (words) scritte fpl; (handwriting) scrittura f; **in ~** per iscritto. ~-**paper** n carta f da lettera

written /'rɪtn/ see **write**

wrong /rɒŋ/ a sbagliato; **be ~** ‹person› sbagliare; **what's ~?** cosa c'è che non va? ● adv ‹spelt› in modo sbagliato; **go ~** ‹person› sbagliare; ‹machine› funzionare male; ‹plan› andar male ● n ingiustizia f; **in the ~** dalla parte del torto; **know right from ~** distinguere il bene dal male ● vt fare torto a. ~**ful** a ingiusto. ~**ly** adv in modo sbagliato; ‹accuse, imagine› a torto; (informed) male

wrote /rəʊt/ see **write**

wrought iron /rɔːt-/ n ferro m battuto ● attrib di ferro battuto

wrung /rʌŋ/ see **wring**

wry /raɪ/ a (-er, -est) ⟨humour, smile⟩ beffardo

Xx

Xmas /'krɪsməs/ n fam Natale m

X-ray n (picture) radiografia f; **have an ~** farsi fare una radiografia ● vt passare ai raggi X

Yy

yacht /jɒt/ n yacht m inv; (for racing) barca f a vela. **~ing** n vela f

Yank n fam americano, -a mf

yank /jæŋk/ vt fam tirare

yap /jæp/ vi (pt/pp yapped) ⟨dog:⟩ guaire

yard[1] /jɑːd/ n cortile m; (for storage) deposito m

yard[2] n iarda f (= 91,44 cm). **~stick** n fig pietra f di paragone

yarn /jɑːn/ n filo m; (fam: tale) storia f

yawn /jɔːn/ vi sbadiglio m ● vi sbadigliare. **~ing** a **~ing gap** sbadiglio m

year /jɪə(r)/ n anno m; (of wine) annata f; **for ~s** fam da secoli. **~book** n annuario m. **~ly** a annuale ● adv annualmente

yearn /jɜːn/ vi struggersi. **~ing** n desiderio m struggente

yeast /jiːst/ n lievito m

yell /jel/ n urlo m ● vi urlare

yellow /'jeləʊ/ a & n giallo m

yelp /jelp/ n (of dog) guaito m ● vi ⟨dog:⟩ guaire

yen /jen/ n forte desiderio m (for di)

yes /jes/ adv sì ● n sì m inv

yesterday /'jestədeɪ/ a & adv ieri

m inv; **~'s paper** il giornale di ieri; **the day before ~** l'altroieri

yet /jet/ adv ancora; **as ~** fino ad ora; **not ~** non ancora; **the best ~** il migliore finora ● conj eppure

yew /juː/ n tasso m (albero)

yield /jiːld/ n produzione f; ⟨profit⟩ reddito m ● vt ⟨profit⟩ fruttare ⟨profit⟩ ● vi cedere; Am Auto dare la precedenza

yodel /'jəʊdl/ vi (pt/pp yodelled) cantare jodel

yoga /'jəʊgə/ n yoga m

yoghurt /'jɒgət/ n yogurt m inv

yoke /jəʊk/ n giogo m; (of garment) carré m inv

yokel /'jəʊkl/ n zotico, -a mf

yolk /jəʊk/ n tuorlo m

you /juː/ pron (subject) tu, voi pl; (formal) lei, voi pl; (direct/indirect object) ti, vi pl; (formal: direct object) la; (formal: indirect object) le; (after prep) te, voi pl; (formal: after prep) lei; **~ are very kind** (formal) è molto gentile; (pl & formal pl) siete molto gentili; **~ can stay, but he has to go** (sg) tu puoi rimanere, ma lui deve andarsene; (pl) voi potete rimanere, ma lui deve andarsene; **all of ~** tutti voi; **I'll give ~ the money** (sg) ti darò i soldi; (pl) vi darò i soldi; **I'll give it to ~** (sg) te/(pl) ve lo darò; **it was ~!** (sg) eri tu!; (pl) eravate voi!; **~ have to be careful** (one) si deve fare attenzione

young /jʌŋ/ a giovane ● npl (animals) piccoli mpl; **the ~** (people) i giovani mpl. **~ lady** n signorina f. **~ man** n giovanotto. **~ster** n ragazzo, -a mf; (child) bambino, -a mf

your /jɔː(r)/ a il tuo m, la tua f, i tuoi mpl, le tue fpl; (formal) il suo m, la sua f, i suoi mpl, le sue fpl; (pl & formal pl) il vostro m, la vostra f, i vostri mpl, le vostre fpl; **~ mother/father** tua madre/tuo padre; (*formal*) sua madre/suo padre; (*pl & formal pl*) vostra madre/vostro padre

yours /jɔːz/ *poss pron* il tuo *m*, la tua *f*, i tuoi *mpl*, le tue *fpl*; (*formal*) il suo *mpl*, la sua *f*, i suoi *mpl*, le sue *fpl*; **a friend of** ~ un tuo/suo/vostro amico; **friends of** ~ dei tuoi/vostri/suoi amici; **that is** ~ quello è tuo/vostro/suo; (*as opposed to mine*) quello è il tuo/il vostro/il suo

your'self *pron* (*reflexive*) ti; (*formal*) si; (*emphatic*) te stesso; (*formal*) sé, te stesso; **do pour ~ a drink** versati da bere; (*formal*) si versi da bere; **you said so** ~ lo hai detto tu stesso; (*formal*) lo ha detto lei stesso; **you can be proud of** ~ puoi essere fiero di te/di sé; **by** ~ da solo

your'selves *pron* (*reflexive*) vi; (*emphatic*) voi stessi; **do pour ~ a drink** versatevi da bere; **you said so** ~ lo avete detto voi stessi; **you can be proud of** ~ potete essere fieri di voi; **by** ~ da soli

youth /juːθ/ *n* (*pl* **youths** /-ðz/) gioventù *f inv*; (*boy*) giovanetto *m*; **the ~** (*young people*) i giovani *mpl*. **~ful** *a* giovanile. **~ hostel** *n* ostello *m* [della gioventù]

Yugoslav /ˈjuːgəslɑːv/ *a & n* jugoslavo, -a *mf*

Yugoslavia /-ˈslɑːvɪə/ *n* Jugoslavia *f*

....

Zz

....

zany /ˈzeɪnɪ/ *a* (**-ier, -iest**) demenziale

zeal /ziːl/ *n* zelo *m*

zealous /ˈzeləs/ *a* zelante. **~ly** *adv* con zelo

zebra /ˈzebrə/ *n* zebra *f*. **~'**

crossing *n* passaggio *m* pedonale, zebre *fpl*

zero /ˈzɪərəʊ/ *n* zero *m*

zest /zest/ *n* gusto *m*

zigzag /ˈzɪgzæg/ *n* zigzag *m inv* ● *vi* (*pt/pp* **-zagged**) zigzagare

zilch /zɪltʃ/ *n fam* zero *m* assoluto

zinc /zɪŋk/ *n* zinco *m*

zip /zɪp/ *n* ~ **[fastener]** cerniera *f* [lampo] ● *vt* (*pt/pp* **zipped**) ~ **[up]** chiudere con la cerniera [lampo]

'Zip code *n Am* codice *m* postale

zipper /ˈzɪpə(r)/ *n Am* cerniera *f* [lampo]

zodiac /ˈzəʊdɪæk/ *n* zodiaco *m*

zombie /ˈzɒmbɪ/ *n fam* zombi *mf inv*

zone /zəʊn/ *n* zona *f*

zoo /zuː/ *n* zoo *m inv*

zoolog|ist /zəʊˈɒlədʒɪst/ *n* zoologo, -a *mf*. **~y** zoologia *f*

zoom /zuːm/ *vi* sfrecciare. **~ lens** *n* zoom *m inv*

ITALIAN VERB TABLES

REGULAR VERBS:

1. in **-are** (*eg* compr|are)

   Present ~o, ~i, ~a, ~iamo, ~ate, ~ano
   Imperfect ~avo, ~avi, ~ava, ~avamo, ~avate, ~avano
   Past historic ~ai, ~asti, ~ò, ~ammo, ~aste, ~arono
   Future ~erò, ~erai, ~erà, ~eremo, ~erete, ~eranno
   Present subjunctive ~i, ~i, ~i, ~iamo, ~iate, ~ino
   Past subjunctive ~assi, ~assi, ~asse, ~assimo, ~aste, ~assero
   Present participle ~ando
   Past participle ~ato
   Imperative ~a (*fml* ~i), ~iamo, ~ate
   Conditional ~erei, ~eresti, ~erebbe, ~eremmo, ~ereste, ~erebbero

2. in **-ere** (*eg* vend|ere)

   Pres ~o, ~i, ~e, ~iamo, ~ete, ~ono
   Impf ~evo, ~evi, ~eva, ~evamo, ~evate, ~evano
   Past hist ~ei *or* ~etti, ~esti, ~è *or* ~ette, ~emmo, ~este, ~erono *or* ~ettero
   Fut ~erò, ~erai, ~erà, ~eremo, ~erete, ~eranno
   Pres sub ~a, ~a, ~a, ~iamo, ~iate, ~ano
   Past sub ~essi, ~essi, ~esse, ~essimo, ~este, ~essero
   Pres part ~endo
   Past part ~uto
   Imp ~i (*fml* ~a), ~iamo, ~ete
   Cond ~erei, ~eresti, ~erebbe, ~eremmo, ~ereste, ~erebbero

3. in **-ire** (*eg* dorm|ire)

   Pres ~o, ~i, ~e, ~iamo, ~ite, ~ono
   Impf ~ivo, ~ivi, ~iva, ~ivamo, ~ivate, ~ivano
   Past hist ~ii, ~isti, ~ì, ~immo, ~iste, ~irono
   Fut ~irò, ~irai, ~irà, ~iremo, ~irete, ~iranno
   Pres sub ~a, ~a, ~a, ~iamo, ~iate, ~ano
   Past sub ~issi, ~issi, ~isse, ~issimo, ~iste, ~issero
   Pres part ~endo
   Past part ~ito
   Imp ~i (*fml* ~a), ~iamo, ~ite
   Cond ~irei, ~iresti, ~irebbe, ~iremmo, ~ireste, ~irebbero

Notes

- Many verbs in the third conjugation take *isc* between the stem and the ending in the first, second, and third person singular and in the third person plural of the present, the present subjunctive, and the imperative: fin|ire **Pres** ~isco, ~isci, ~isce, ~iscono. **Pres sub** ~isca, ~iscano **Imp** ~isci.

- The three forms of the imperative are the same as the corresponding forms of the present for the second and third conjugation. In the first conjugation the forms are also the same except for the second person singular: present *compri*, imperative *compra*. The negative form of the second person singular is formed by putting *non* before the infinitive for all conjugations: *non comprare*. In polite forms the third person of the present subjunctive is used instead for all conjugations: *compri*.

Irregular verbs:

Certain forms of all irregular verbs are regular (except for *essere*). These are: the second person plural of the present, the past subjunctive, and the present participle. All forms not listed below are regular and can be derived from the parts given. Only those irregular verbs considered to be the most useful are shown in the tables.

accadere	*as* **cadere**
accendere	• **Past hist** accesi, accendesti • **Past part** acceso
affliggere	• **Past hist** afflissi, affliggesti • **Past part** afflitto
ammettere	*as* **mettere**
andare	• **Pres** vado, vai, va, andiamo, andate, vanno • **Fut** andrò *etc* • **Pres sub** vada, vadano • **Imp** va', vada, vadano
apparire	• **Pres** appaio *or* apparisco, appari *or* apparisci, appare *or* apparisce, appaiono *or* appariscono • **Past hist** apparvi *or* apparsi, apparisti, apparve *or* appari *or* apparse, apparvero *or* apparirono *or* apparsero • **Pres sub** appaia *or* apparisca
aprire	• **Pres** apro • **Past hist** aprii, apristi • **Pres sub** apra • **Past part** aperto

avere　● **Pres** ho, hai, ha, abbiamo, hanno ● **Past hist** ebbi, avesti, ebbe, avemmo, aveste, ebbero ● **Fut** avrò *etc* ● **Pres sub** abbia *etc* ● **Imp** abbi, abbia, abbiate, abbiano

bere　● **Pres** bevo *etc* ● **Impf** bevevo *etc* ● **Past hist** bevvi *or* bevetti, bevesti ● **Fut** berrò *etc* ● **Pres sub** beva *etc* ● **Past sub** bevessi *etc* ● **Pres part** bevendo ● **Cond** berrei *etc*

cadere　● **Past hist** caddi, cadesti ● **Fut** cadrò *etc*

chiedere　● **Past hist** chiesi, chiedesti ● **Pres sub** chieda *etc* ● **Past part** chiesto *etc*

chiudere　● **Past hist** chiusi, chiudesti ● **Past part** chiuso

cogliere　● **Pres** colgo, colgono ● **Past hist** colsi, cogliesti ● **Pres sub** colga ● **Past part** colto

correre　● **Past hist** corsi, corresti ● **Past part** corso

crescere　● **Past hist** crebbi ● **Past part** cresciuto

cuocere　● **Pres** cuocio, cuociamo, cuociono ● **Past hist** cossi, cocesti ● **Past part** cotto

dare　● **Pres** do, dai, da, diamo, danno ● **Past hist** diedi *or* detti, desti ● **Fut** darò *etc* ● **Pres sub** dia *etc* ● **Past sub** dessi *etc* ● **Imp** da' *(fml* dia*)*

dire　● **Pres** dico, dici, dice, diciamo, dicono ● **Impf** dicevo *etc* ● **Past hist** dissi, dicesti ● **Fut** dirò *etc* ● **Pres sub** dica, diciamo, diciate, dicano ● **Past sub** dicessi *etc* ● **Pres part** dicendo ● **Past part** detto ● **Imp** di' *(fml* dica*)*

dovere　● **Pres** devo *or* debbo, devi, deve, dobbiamo, devono *or* debbono ● **Fut** dovrò *etc* ● **Pres sub** deva *or* debba, dobbiamo, dobbiate, devano *or* debbano ● **Cond** dovrei *etc*

essere　● **Pres** sono, sei, è, siamo, siete, sono ● **Impf** ero, eri, era, eravamo, eravate, erano ● **Past hist** fui, fosti, fu, fummo, foste, furono ● **Fut** sarò *etc* ● **Pres sub** sia *etc* ● **Past sub** fossi, fossi, fosse, fossimo, foste, fossero ● **Past part** stato ● **Imp** sii *(fml* sia*)*, siate ● **Cond** sarei *etc*

fare　● **Pres** faccio, fai, fa, facciamo, fanno ● **Impf** facevo *etc* ● **Past hist** feci, facesti ● **Fut** farò *etc* ● **Pres sub** faccia *etc* ● **Past sub** facessi *etc* ● **Pres part** facendo ● **Past part** fatto ● **Imp** fa' *(fml* faccia*)* ● **Cond** farei *etc*

fingere • **Past hist** finsi, fingesti, finsero • **Past part** finto

giungere • **Past hist** giunsi, giungesti, giunsero • **Past part** giunto

leggere • **Past hist** lessi, leggesti • **Past part** letto

mettere • **Past hist** misi, mettesti • **Past part** messo

morire • **Pres** muoio, muori, muore, muoiono • **Fut** morirò or morrò etc • **Pres sub** muoia • **Past part** morto

muovere • **Past hist** mossi, movesti • **Past part** mosso

nascere • **Past hist** nacqui, nascesti • **Past part** nato

offrire • **Past hist** offersi or offrii, offristi • **Pres sub** offra • **Past part** offerto

parere • **Pres** paio, pari, pare, pariamo, paiono • **Past hist** parvi or parsi, paresti • **Fut** parrò etc • **Pres sub** paia, paiamo or pariamo, pariate, paiano • **Past part** parso

piacere • **Pres** piaccio, piaci, piace, piacciamo, piacciono • **Past hist** piacqui, piacesti, piacque, piacemmo, piaceste, piacquero • **Pres sub** piaccia etc • **Past part** piaciuto

porre • **Pres** pongo, poni, pone, poniamo, ponete, pongono • **Impf** ponevo etc • **Past hist** posi, ponesti • **Fut** porrò etc • **Pres sub** ponga, poniamo, poniate, pongano • **Past sub** ponessi etc

potere • **Pres** posso, puoi, può, possiamo, possono • **Fut** potrò etc • **Pres sub** possa, possiamo, possiate, possano • **Cond** potrei etc

prendere • **Past hist** presi, prendesti • **Past part** preso

ridere • **Past hist** risi, ridesti • **Past part** riso

rimanere • **Pres** rimango, rimani, rimane, rimaniamo, rimangono • **Past hist** rimasi, rimanesti • **Fut** rimarrò etc • **Pres sub** rimanga • **Past part** rimasto • **Cond** rimarrei

salire • **Pres** salgo, sali, sale, saliamo, salgono • **Pres sub** salga, saliate, salgano

sapere • **Pres** so, sai, sa, sappiamo, sanno • **Past hist** seppi, sapesti • **Fut** saprò etc • **Pres sub** sappia etc • **Imp** sappi (fml sappia), sappiate • **Cond** saprei etc

scegliere • **Pres** scelgo, scegli, sceglie, scegliamo, scelgono •
Past hist scelsi, scegliesti *etc* • **Past part** scelto

scrivere • **Past hist** scrissi, scrivesti *etc* • **Past part** scritto

sedere • **Pres** siedo *or* seggo, siedi, siede, siedono • **Pres sub**
sieda *or* segga

spegnere • **Pres** spengo, spengono • **Past hist** spensi, spegnesti •
Past part spento

stare • **Pres** sto, stai, sta, stiamo, stanno • **Past hist** stetti,
stesti • **Fut** starò *etc* • **Pres sub** stia *etc* • **Past sub**
stessi *etc* • **Past part** stato • **Imp** sta' *(fml* stia*)*

tacere • **Pres** taccio, tacciono • **Past hist** tacqui, tacque,
tacquero • **Pres sub** taccia

tendere • **Past hist** tesi • **Past part** teso

tenere • **Pres** tengo, tieni, tiene, tengono • **Past hist** tenni,
tenesti • **Fut** terrò *etc* • **Pres sub** tenga

togliere • **Pres** tolgo, tolgono • **Past hist** tolsi, tolse, tolsero •
Pres sub tolga, tolgano • **Past part** tolto • *Imp fml* tolga

trarre • **Pres** traggo, trai, trae, traiamo, traete, traggono • **Past
hist** trassi, traesti • **Fut** trarrò *etc* • **Pres sub** tragga •
Past sub traessi *etc* • **Past part** tratto

uscire • **Pres** esco, esci, esce, escono • **Pres sub** esca • **Imp**
esci *(fml* esca*)*

valere • **Pres** valgo, valgono • **Past hist** valsi, valesti • **Fut**
varrò *etc* • **Pres sub** valga, valgano • **Past part** valso •
Cond varrei *etc*

vedere • **Past hist** vidi, vedesti • **Fut** vedrò *etc* • **Past part**
visto *or* veduto • **Cond** vedrei *etc*

venire • **Pres** vengo, vieni, viene, vengono • **Past hist** venni,
venisti • **Fut** verrò *etc*

vivere • **Past hist** vissi, vivesti • **Fut** vivrò *etc* • **Past part**
vissuto • **Cond** vivrei *etc*

volere • **Pres** voglio, vuoi, vuole, vogliamo, volete, vogliono •
Past hist volli, volesti • **Fut** vorrò *etc* • **Pres sub** voglia
etc • **Imp** vogliate • **Cond** vorrei *etc*

ENGLISH IRREGULAR VERBS

Infinitive	Past Tense	Past Participle	Infinitive	Past Tense	Past Participle
Infinito	*Passato*	*Participio passato*	*Infinito*	*Passato*	*Participio passato*
arise	arose	arisen	**cling**	clung	clung
awake	awoke	awoken	**come**	came	come
be	was	been	**cost**	cost,	cost,
bear	bore	borne		costed (*vt*)	costed
beat	beat	beaten	**creep**	crept	crept
become	became	become	**cut**	cut	cut
begin	began	begun	**deal**	dealt	dealt ·
behold	beheld	beheld	**dig**	dug	dug
bend	bent	bent	**do**	did	done
beseech	beseeched	beseeched	**draw**	drew	drawn
	besought	besought	**dream**	dreamt,	dreamt,
bet	bet,	bet,		dreamed	dreamed
	betted	betted	**drink**	drank	drunk
bid	bade,	bidden,	**drive**	drove	driven
	bid	bid	**dwell**	dwelt	dwelt
bind	bound	bound	**eat**	ate	eaten
bite	bit	bitten	**fall**	fell	fallen
bleed	bled	bled	**feed**	fed	fed
blow	blew	blown	**feel**	felt	felt
break	broke	broken	**fight**	fought	fought
breed	bred	bred	**find**	found	found
bring	brought	brought	**flee**	fled	fled
build	built	built	**fling**	flung	flung
burn	burnt,	burnt,	**fly**	flew	flown
	burned	burned	**forbid**	forbade	forbidden
burst	burst	burst	**forget**	forgot	forgotten
bust	busted,	busted,	**forgive**	forgave	forgiven
	bust	bust	**forsake**	forsook	forsaken
buy	bought	bought	**freeze**	froze	frozen
cast	cast	cast	**get**	got	got,
catch	caught	caught			gotten *Am*
choose	chose	chosen	**give**	gave	given